Chap 32 - only start
Chap 33 - nowt
Chap 6 - v little
Chap 9 -
Chap 29 - Na
Chap 10 -

# Sensory Biology of Aquatic Animals

Conference participants. From left to right: *Front, row 1:* W. S. Wilcox, W. Saidel, W. Plassman, B. Fritsch, H. Zakon, M. Braford, P. Borroni, B. U. Budelmann, C. Platt, H. Bleckmann, C. Hawrwyshyn, J. Sivak. *Middle, row 2:* T. Waterman, B. Sokolowski, W. Heiligenberg, M. Swain, S. Coombs, E. Denton, R. G. Northcutt, W. Tavolga, J. Atema, R. Fay, A. Popper, J. Patton, P. Gorner, P. Moller, J. Caprio, J. Crawford, J. Blaxter, R. Voigt, J. Webb, R. Gleeson, C. Derby, I. Assip. *Back, row 3:* P. Rogers, M. Cox, M. Hagedorn, J. Wyneken, M. Salmon, C. McCormick, J. Song, M. Kreithen, M. Wullimann, J. Case, J. Gray, M. Powers, J. Levine, D. Hoekstra, T. Finger, P. Hamilton, D. Woodward, A. Kalmijn, T. Cronin, T. Bullock, M. Laverack, H. Munz, J. Douglass, S. Holderman. *Last, row 4:* C. Hopkins, W. Stachnik, T. Ream, P. Gilbert, J. Kendall, B. Ache, A. Elepfandt, R. Barlow, R. Brill, R. Fernald, R. Eaton, B. Zahuranec, W. Carr, J. Janssen, J. Lythgoe, J. R. Strickler, E. Hartwig, K. Wiese.

Jelle Atema
Richard R. Fay
Arthur N. Popper
William N. Tavolga
Editors

# Sensory Biology of Aquatic Animals

Springer-Verlag
New York Berlin Heidelberg
London Paris Tokyo

Jelle Atema, Boston University Marine Program, Marine Biological Laboratory, Woods Hole, Massachusetts 02543, USA

Richard R. Fay, Parmly Hearing Institute, Loyola University, Chicago, Illinois 60626, USA

Arthur N. Popper, Department of Zoology, University of Maryland, College Park, MD 20742, USA

William N. Tavolga, Mote Marine Laboratory, Sarasota, Florida 33577, USA

The cover illustration is a reproduction of Figure 13.3, p. 343 of this volume

Library of Congress Cataloging-in-Publication Data
Sensory biology of aquatic animals.
Papers based on presentations given at an International Conference on the Sensory Biology of Aquatic Animals held, June 24–28, 1985, at the Mote Marine Laboratory in Sarasota, Fla.
Bibliography: p.
Includes indexes.
1. Aquatic animals—Physiology—Congresses.
2. Senses and sensation—Congresses. I. Atema, Jelle.
II. International Conference on the Sensory Biology of Aquatic Animals (1985 : Sarasota, Fla.)
QL120.S46 1987 591.92 87–9632



Typeset by Arcata Graphics/Kingsport, Kingsport, Tennessee.
Printed and bound by Arcata Graphics/Halliday, West Hanover, Massachusetts.
Printed in the United States of America.

9 8 7 6 5 4 3 2 1

ISBN 0–387–96373–1 Springer-Verlag New York Berlin Heidelberg
ISBN 3–540–96373–1 Springer-Verlag Berlin Heidelberg New York

This volume is dedicated to

*Sven Dijkgraaf*
*Karl von Frisch*
*C. Judson Herrick*
*Erich von Holst*
*George H. Parker*
*Jacob von Uexkull*
*J. Z. Young*

Each of these investigators delved deeply and broadly into the sensory biology of aquatic animals. Their contributions to the study of sensory systems of aquatic animals have directly and indirectly influenced all of the work reported in this volume.

# Preface

This volume constitutes a series of invited chapters based on presentations given at an International Conference on the Sensory Biology of Aquatic Animals held June 24–28, 1985 at the Mote Marine Laboratory in Sarasota, Florida.

The immediate purpose of the conference was to spark an exchange of ideas, concepts, and techniques among investigators concerned with the different sensory modalities employed by a wide variety of animal species in extracting information from the aquatic environment. By necessity, most investigators of sensory biology are specialists in one sensory system: different stimulus modalities require different methods of stimulus control and, generally, different animal models. Yet, it is clear that all sensory systems have principles in common, such as stimulus filtering by peripheral structures, tuning of receptor cells, signal-to-noise ratios, adaption and disadaptation, and effective dynamic range. Other features, such as hormonal and efferent neural control, circadian reorganization, and receptor recycling are known in some and not in other senses. The conference afforded an increased awareness of new discoveries in other sensory systems that has effectively inspired a fresh look by the various participants at their own area of specialization to see whether or not similar principles apply. This inspiration was found not only in theoretical issues, but equally in techniques and methods of approach.

The myopy of sensory specialization was broken in one unexpected way by showing limitations of individual sense organs and their integration within each organism. For instance, studying vision, one generally chooses a visual animal as a model. Subsequent intense focus on that animal's visual performance easily leads to ignoring the fact that in the real world this animal uses other senses as well. Thus, the demands the animal places on vision are tempered in the context of other sensory information. The behavioral demands on visual detection range and acuity for instance may not be as severe as one might think, if it were known that acoustic or chemical senses serve to locate and identify the source of interest for that animal at greater distances, whereas vision is used mainly at shorter range. This multisensory conference thus allowed the different senses to show not only their individual strengths, but also their weaknesses and limitations, and their inherent interrelatedness as they appear in the aquatic environment. It is our hope that these issues will not be lost in book form.

The long-range purpose of the conference was to create a book that could serve as a conceptual framework for further investigations of the aquatic "Umwelt" and the intricate, often unexpected, methods and processes used by animals to utilize the information contained in the acquatic stimulus world. This book and its organization reflect the central premise that sensory systems are constrained by the stimulus environment (Section I), by the behavioral requirements of the organism (Section II), and by phylogeny (various chapters in Section III). In addition, there is the tacit assumption of economy in sensory design.

A major and unique effort was made to *stimulate the development of conceptual models of the aquatic stimulus environment per se*. This, we believe, is of paramount importance since only a detailed and comprehensive understanding of the natural distribution of stimuli and their noise backgrounds will allow the design of the proper experiments which, in turn, will lead to an understanding of sensory function in a natural context. This focus primarily serves to probe specific aquatic problems, but it also bears on terrestrial sensory problems by providing the aquatic contrast. The aquatic environment is a different medium than the terrestrial environment in nearly all aspects of sensory biology. Some differences are profound, such as the possibility of electroreception in the conductive aquatic medium. Other differences are less fundamental, such as the propagation speed of sound and the properties of water as a filter for light and a carrier for chemical stimuli. The sensory constraints imposed by the stimulus environment may emerge in clearest form when comparing aquatic and terrestrial solutions to similar sensory problems. Although a number of chapters refer explicitly to terrestrial animals and environments for contrast, a systematic comparison falls beyond the scope of this volume.

Intimately related to the direct constraints and opportunities presented by the sensory "Umwelt" are *indirect sensory constraints imposed by animal behavior,* the motoric "Umwelt." An animal's behavioral requirements invite, to various degrees, the development of sensory capabilities. The reverse may be true as well: extended sensory horizons may allow the expansion of motoric virtuosity. Thus, to understand the physiology of receptor cells and organs it is instructive to know the animal's behavioral capabilities. In the aquatic environment these capabilities are influenced to a large degree by the physical density and viscosity of water, regardless of whether the animal moves through the medium or moves water through or along its body.

An understanding of the constraints of behavior is a rich source of inspiration in the study of sensory function. *Ultimately all sensory processing stands in the service of making behavioral decisions*. These are often decisions of immediate life or death, and always decisions affecting competitive fitness. The central question concerns what an animal must know about its environment in order to survive and compete successfully. Behavioral ecology can show the knowledge that an animal must have had about its environment in order to have performed a certain behavior, or the information required to have followed a certain behavioral strategy. The acquisition of this knowledge is at least partially a matter of sensory information. Memory and expectation are also essential ingredients of this knowledge, and both memory and expectation require constant updating by sensory information.

In contrast to the acquisition of information, we recognize *stimulus acquisition behavior,* i.e., those behavioral actions and strategies that increase the animal's probabilities of interfacing with the proper sensory stimuli. These actions include search patterns, choice of location, current generation, acoustic and visual scanning, flicking and sniffing, probing, and any other behavior that appears designed specifically or primarily to serve the function of encountering further sensory cues that, in turn, may provide information about the source of the stimulus. Stimulus acquisition can then be seen as part of the sensory process of information acquisition leading to knowledge.

The opportunities and constraints afforded by *phylogeny* may emerge in the *comparison of solutions to similar sensory problems found by different species.* Most chapters in this volume dealing with comparative issues do so within animal groups related at the level of orders or classes. However, it is also instructive to compare such widely divergent groups as vertebrates and invertebrates in order to see common anatomical, physiological and behavioral solutions to various problems of signal extraction. For instance, "sniffing" is an efficient method for spatial and temporal sampling of the chemical environment for the enclosed olfactory organ of both aquatic and terrestrial vertebrates. To accomplish similar sampling with their non-enclosed olfactory organ the decapod crustacea "flick" an antennae with densely packed tufts of chemosensory hairs. The discovery of sniffing in certain fish and flicking in lobsters now forces us to consider the general importance of spatio-temporal sampling of the odor environment. Examples from vertebrates and invertebrates are presented in each section of the book, thus inviting the reader to implicit comparison.

In addition to stimulus environment, behavior, and phylogeny, the *constraint of economy* is implicit throughout the book. It is a pervasive concept in much of today's biology where it is recognized that there are costs and benefits to each process and that the multiplicity of processes that make up each organism forces a compromise between the cost–benefit ratios that may be theoretically optimal for each process by itself. Such compromises exist at molecular, cellular and behavioral levels.

Each of the sensory subsections includes a chapter on *central processing,* since the CNS is an integral part of any sensory system. The function of the CNS cannot be dissociated from the physiological properties of receptor cells. However, this book does not attempt to focus specifically on the overall sensory strategies of the CNS such as signal extraction, coding, and processing. This was not the goal of the symposium, primarily because it involves principles of sensory neurobiology that are not specifically aquatic.

Further, not all sensory systems are covered in this book. Only the best studied sensory modalities have a *base of information that allows conceptual development.* Even among the modalities that are included in this book, the base of information varies greatly. The result is that, for example, magnetic, thermal and hydrostatic pressure reception are not dealt with despite the fact that they may be of great importance for aquatic animals.

In examining the book, it is apparent that there are many chapters that could fit into more than one section. The organization that we have chosen for the book begins by first describing the physical and chemical stimulus nature of the environment, followed by a section on information and stimulus acquisition behavior and ends with several sections on receptor systems. Since many chapters could appear in different sections, we attempt to point out different possible organizations at the beginning of each section in order to show how chapters from various sections are tied together.

The organization of the presentations at the conference allowed substantial time for discussion of individual papers and groups of papers. These discussions were recorded but since they were extensive we have, instead of publishing them verbatim, provided them to the authors of individual chapters. The ideas and comments made during the discussion were then, as appropriate, added to chapters and are reflected in the papers in the volume.

In order to make this book more useful as a reference we provide two tables of contents and three indices. The short table of contents lists only the chapter titles, while the long table of contents lists also the subheadings within each chapter. This should facilitate the search process. In addition to the usual author and subject indices we include an animal index. This index serves the specific purpose of providing the reader with an opportunity to evaluate the multisensory capabilities of certain well-studied species. For instance, it will now be easier to find that catfish have been studied not only for their taste and smell capabilities but also for electric sensitivity. However, this index also shows how few and selective the animal species are that have been studied in detail for one, let alone for *more* than one sensory modality.

We hope that this book will be useful in identifying the strengths and weaknesses in our knowledge of sensory systems in the aquatic environment. In addition we hope that this book may inspire efforts to increase our understanding of a stimulus world that is foreign to us but so efficiently "read" and utilized by the aquatic participants of life on earth. Their methods of sensing should inspire not only our views of the world, but also our techniques of measuring the world.

Summer, 1987

J.A.
R.R.F.
A.N.P.
W.N.T.

# Acknowledgments

We would like to express our gratitude to the Office of Naval Research (contract N-00014-84-G-0107), the National Science Foundation, Sensory Physiology and Perception Program (BNS 84-20124), and The Mote Marine Laboratory for providing funds to support the conference. Without their generous support the conference would certainly not have been possible.

In addition to the financial support, the organizers received considerable assistance and moral support from a number of very dedicated individuals. Drs. Eric Hartwig and Bernard J. Zahuranec of the Oceanic Biology Program of the Office of Naval Research supported the conference with their enthusiastic interest. Mr. William Mote, Chairman of the Board of the Mote Marine Laboratory and Dr. Robert Dunn, Director of the Laboratory, provided unfailing support in every way imaginable from the initial planning to the closing banquet. We received superb cooperation from everyone at the Mote Marine Laboratory. In particular, we would like to express our gratitude to Ilene Assip, Susan Holderman, Peter Hull, Jeff Patton, and Howard St. Clair for untiring efforts in making every aspect of the meeting a success. We are also grateful to Mr. Hack Swain of Hack Swain Productions for recording the sessions and Ms. Laurie Fraser for doing the initial transcripts. The editors would like to thank Helen A. Popper for her invaluable editorial assistance.

Finally, on behalf of all of the participants, we would like to thank Chef Alain Mons of the French Affair Delicatessen in Sarasota who provided multisensory stimulation that made the lunches and coffee breaks significant parts of the overall study of sensory systems.

# Contents

## Part III. C. 3. Equilibrium



## Part III. D. Electroreception



## Part IV. Adaptation and Sensory Systems

# Detailed Chapter Contents

## Part III. C. Mechanoreception
## Part III. C. 1. Hydrodynamic Reception

## Part III. C. 2. Hearing

## Part III. C. 3. Equilibrium

## Part III. D. Electroreception

## Part IV. Adaptation and Sensory Systems

# Contributors

The following is a list of all the contributors to this volume. Complete affiliations are given as footnotes on the opening page of each chapter, and this page number is given after the name.

Barry W. Ache 387
Jelle Atema 29, 365
Robert B. Barlow Jr. 419
John H.S. Blaxter 203
Horst Bleckmann 619
Mark R. Bradford Jr. 733
Bernd-Ulrich Budelmann 757
Theodore H. Bullock 269
John Caprio 313
William E.S. Carr 3
Sheryl Coombs 553
Mardi Cox 131, 687
Thomas W. Cronin 403
Eric J. Denton 595
Charles D. Derby 365
Richard R. Fay 711
Russell D. Fernald 435
Thomas E. Finger 339
John A.B. Gray 595
Walter Heiligenberg 851
Carl D. Hopkins 234
John Janssen 553
Ad. J. Kalmijn 83, 151
Alan C. Kamil 189
Leonard C. Kass 419
Michael S. Laverack 287
John N. Lythgoe 57
Catherine A. McCormick 733
R. Glenn Northcutt 515, 869
Christopher Platt 783
Arthur N. Popper 687
Maureen K. Powers 419
Peter H. Rogers 131, 687
William M. Saidel 487, 687
Jacob G. Sivak 467
Jacqueline C. Webb 553
Konrad Wiese 665
R. Stimson Wilcox 643
Mario F. Wullimann 515
Harold Zakon 813

# ONR and Sensory Biology

In 1946, the Office of Naval Research (ONR) was established as the first government agency funding basic research. ONR, therefore, has the longest history of supporting innovative basic research programs, particularly in areas of long-term interest to the Navy. One of these long-term areas of interest has been sensory biology. The rationale behind the initiation of this research by the Oceanic Biology Program at ONR was to gain an understanding of the sensory capabilities of marine organisms. In pursuing this research, scientists have found that biosensory systems are more powerful and faster than engineered data processing systems and often surpass engineered systems in sensitivity.

At present, the working hypothesis behind ONR's support of aquatic biosensors is that the exact functional nature of a particular species' sensor capability is due to the environment in which it lives, its social organization, and its means of making a living (i.e., trophic level). For example, solitary, open ocean sharks are top level predators living in an environment of widely separated food items. Sharks have evolved sophisticated and powerful biosensory systems to detect, localize, pursue, and feed on their dispersed prey. Prey organisms, on the other hand, require biosensory systems that permit them to recognize signals rapidly in a very noisy environment and to respond rapidly.

The analogy to Naval systems and operations in the ocean is readily apparent. The Navy must continually use and develop new sensor systems to extract rapidly as much information as possible from all of the signals present in the environment. The basic research being done today on acoustic, optical, chemical, magnetic, electrical, and hydrodynamic biological sensors in the marine environment is expected to provide the Navy of the future with a new suite of sensors. These engineered sensor systems, based on biological principles and processes, will have more powerful and faster data processing capabilities, be more compact and sensititive, and extract new and more useful information than presently designed sensor systems.

This conference has provided a clearer understanding of biosensory modalities and from it, I believe, has emerged a new direction and focus for biosensory research. Heretofore, biosensory conferences have tended to focus on single topics and sensory modalities. The group at this conference represents almost all sensory modalities and the disciplines required to study these modalities. This diversity of the conference has fortified the growing perception that a system analysis is the only realistic technical approach for studying sensory modalities as functionally performing systems. Although almost all of today's signal detection systems have their origin in the human and animal senses, their actual design is based solely on physical engineering principles. Future sensory systems will need more sensitivity; increased precision; new approaches to signal processing in order to operate in an increasingly noisy environment; more efficient data processing capabilities; and the ability to monitor and optimize its own performance capabilities. Two approaches could rapidly advance the field providing these future sensors. The first is the discovery of new physical principles permitting better engineered systems. The second approach is to base detection systems on different principles and engineer new sensor systems on those principles. As the above desired needs of future sensor systems are contained within the abilities of biosensory systems, I suggest that the use of biological principles is the direction to be pursued.

Research focused on an understanding of the functioning of biosensory systems is required before these systems can be applied toward designing new sensors. A systems analysis approach will provide functional models of biosensors without necessarily requiring a complete understanding of all processes occurring within each part of the system. Below, I present a systems analysis conceptual approach for a better understanding of biosensor functional performance. With each component, two processes impacting it are given. Although this is not an inclusive list, it does serve to show the types of processes operating.

Signal (generation/propagation) → Conditioning (external/internal) →
→ Detection (transduction/coding) → Processing (multiple receptor
integration/relevant signals identified) → Feedback (receptor/behavior)

If one defines sensory detection as: "the sequence beginning with the occurrence of the signal in the environment and ending with the biologically meaningful response of the organism," research on biosensory systems will require input from a broad interdisciplinary base. No one scientist can cover all the aspects within even one sensory modality. Therefore, a program studying the functioning of biosensory systems must be coordinated to ensure that all required topics are covered and that each is relevant to the functional performance of the biosensory system as a whole. This conference is the beginning of such a coordinated scientific effort leading to the new principles upon which to base future sensory systems.

Eric O. Hartwig
Office of Naval Research

# Part I

# Physical and Chemical Characteristics of Signals in the Aquatic Environment

Animals are immersed in their physical and chemical environment, and their senses are in continuous interaction with environmental perturbations. A fundamental understanding of receptor function can begin only when the relevant stimuli and their noise backgrounds are known. It is most illuminating to consider separately the environment of the receptor cells, the receptor organs, and the whole animal. Because they operate in different environments, receptor cell physiology and animal behavior have their separate constraints and opportunities to acquire and filter stimuli from the great variety of signals that are typical of any environment.

This first part consists of six chapters that together describe much of what is known regarding the stimulus characteristics of the aquatic environment. Since stimuli are defined by responses to their environment, all chapters treat the physical nature of stimuli in relation to their known or potential biological uses.

In Chapter 1, Carr describes the great variety of organic compounds that are known to act as stimuli for aquatic animals and their receptors, and includes a comparison between receptors of the internal and external chemical environment. The physical processes that distribute chemical stimuli in the aquatic world are discussed in Chapter 2 by Atema. This chapter provides information on the poorly understood spatial and temporal patterns of chemical stimuli.

Lythgoe (Chapter 3) treats both spectral and quantal characteristics of the photic environment ranging from bright surface waters to the deep ocean where only few photons travel. His chapter demonstrates the relatively advanced state of knowledge about the coupling between photoreceptor characteristics and their stimulus world.

Kalmijn (Chapter 4) and Rogers and Cox (Chapter 5) focus on two aspects of mechanical disturbances acting as biological stimuli in the aquatic environment. Kalmijn focuses on the hydrodynamic flow fields that are created around moving objects. Very close to the source (''inner'' near field) the shear fields have properties that provide the differential stimulus required to activate the lateral line receptors. Further away from the stimulus (''outer'' near field) the flow field properties are better suited to stimulate the inner ear of fishes. Rogers and Cox concentrate on the progressive wave aspects of acoustic sources that dominate at still greater distances from the source (far field), and that stimulate the inner ear, often via accessory organs.

In Chapter 6, Kalmijn describes the physics and biological use of electric fields surrounding living organisms and those induced by movements through the earth's magnetic field. The formal similarity between animal-generated electric and mechanical flow fields is pointed out in Kalmijn's Chapter 4, and provides a link between the mechanoreceptive and electroreceptive aspects of the lateral line system.

# Chapter 1

# The Molecular Nature of Chemical Stimuli in the Aquatic Environment

WILLIAM E. S. CARR *

Many aquatic organisms possess well-developed chemosensory systems that are adapted to monitoring changes in the chemical composition of the aqueous environment in which they live. For aquatic organisms, specific chemicals in the environment may evoke highly predictable changes in behavior. Facets of behavior known to be affected by external chemicals include those associated with feeding, avoiding predators, recognizing a suitable habitat, reproducing, migrating, and interacting with conspecific organisms. General reviews of this material are provided by Mackie and Grant (1974), Daloze, Braekman, and Tursch (1980), and Atema (1985).

This chapter identifies the types of molecules that are known to serve as external chemical signals in the aquatic environment. The sources of these chemical signals are given, and a description is provided of how they are transmitted to the respondent organism. Finally, it will be shown that many of the substances serving as chemical signals in the external aquatic environment have parallel roles in the "internal aquatic environment," where they function as neuroactive agents, such as transmitters and modulators. This chapter does not use a phyletic approach, but instead cites studies of any type of aquatic organism providing a particularly good example of the topic. For reviews of chemoreception of specific groups of aquatic organisms, see the following: bacteria (Boyd and Simon 1982; Ordal 1985); coelenterates (Lenhoff and Lindstedt 1974; Lenhoff 1981); nematodes (Zuckerman and Jansson 1984); gastropods (Croll 1983); crustaceans (Ache 1982); fish (Hara 1975; Atema 1980; and Carr 1982).

Material on homing by migratory species and on nonsexual social behavior is omitted, because too little is yet known about the identity of the substances mediating these important facets of behavior. Recent reviews of these topics are as follows: homing (Stabell 1984); social behavior (Liley 1982).

* C. V. Whitney Laboratory and Department of Zoology, University of Florida, St. Augustine, FL 32086

# The Identity of Chemical Stimulants of Behavior

## Stimulants of Feeding Behavior

Potent stimulants of feeding behavior for aquatic animals are frequently obtained by preparing aqueous extracts of organisms eaten by the test species. In several studies of carnivorous and omnivorous fish and crustaceans, feeding stimulants in an excitatory extract were identified and shown to account for a major portion of the stimulatory capacity of the extract itself (see Table 1.1). Three important features of the feeding stimulants, are revealed in the table.

1. Common metabolites of low-molecular weight are the major stimulants in the extracts for the test species shown in the table. These common metabolites include amino acids, quaternary ammonium compounds, nucleosides and nucleotides, and organic acids (see Fig. 1.1). In crustaceans, molluscs, and echinoderms, there are occasional instances in which substances of high-molecular weight, such as proteins, are the principal stimulants of feeding behavior in an extract or a body fluid (Gurin and Carr 1971; Heeb 1973; Zimmer-Faust et al. 1984). In most cases, however it is clearly the low-molecular components of tissues that are the principal feeding stimulants. A similar conclusion was reached by Lindstedt (1971) for carnivores and omnivores in an earlier review of chemically stimulated feeding behavior in nine phyla of both aquatic and terrestrial organisms.

2. Different species may respond to different substances, even in the same extract. In Table 1.1, note that a similar squid extract was used by one group of investigators in studies with turbot, rainbow trout, and plaice. In this one extract, the major stimulants of feeding behavior were different for each of the fish species. The nucleoside inosine was the major stimulant for the turbot, whereas a mixture of amino acids was required by the trout. The plaice required an even more complex mixture of substances, including amino acids, quaternary ammonium compounds, nucleosides, nucleotides, and lactate.

3. The stimulation of feeding behavior by most extracts is due to a mixture of substances, rather than a single dominant one. Ten of the examples cited in the table required mixtures of substances to yield the synthetic stimulants with activities most similar to the natural extracts. Evidence for synergistic interactions among the components of mixtures has been reported (Carr and Derby 1986a).

The common metabolites cited in Table 1.1 serve as appropriate signals for initiating aspects of feeding behavior in many aquatic organisms. As a group, these substances are thoroughly water soluble and occur ubiquitously in tissues at concentrations that are many orders of magnitude higher than those reported in the aquatic environment. The concentrations at which these substances occur in a free, unbound form in the fresh tissues of representative marine organisms range as follows: amino acids (30–189 mM), nucleotides and related substances (0.9–9 mM), quaternary ammonium compounds (12–111 mM), and organic acids (3–34 mM). (See Table 1.2.) These high levels in tissues contrast dramatically with the low ambient levels of ca. $10^{-5}$ mM ($10^{-8}$ M) that have been reported for amino acids (Mopper and Lindroth 1982) and nucleotides (Ammerman and

Table 1.1. Studies of Chemically-Stimulated Feeding Behavior in Fish and Crustaceans in Which a Substantial Part of the Activity of an Extract Was Accounted for with a single Substance or a Mixture of Substances

| Animal Studied[a] | Substances Required to Mimic Activity of Extract | Source of Extract | Behavioral Response |
|---|---|---|---|
| Fish | Single Substances | | |
| 1. Dover sole | Betaine or Betaine + glycine | Mussel | Ingesting pellet |
| 2. Pigfish | Betaine | Oyster | Biting |
| 3. Turbot | Inosine | Squid | Ingesting pellet |
| | Mixture of Substances | | |
| 4. Rainbow trout | Amino acids | Squid | Ingesting pellet |
| 5. Japanese eel | Amino acids | Clam | Attraction |
| 6. European eel | Amino acids | Squid | Ingesting ration |
| 7. Red sea bream | Amino acids | Worm | Ingesting pellet |
| 8. Pigfish | Amino acids and betaine | Crab, shrimp | Biting |
| 9. Pinfish | Amino acids and betaine | Shrimp[b] | Biting |
| 10. Puffer | Amino acids and betaine | Clam | Ingesting pellet |
| 11. Plaice | Amino acids, quaternary ammonium compounds, nucleotides, nucleosides, and lactate | Squid | Ingesting pellet |
| Crustacea | | | |
| 12. Lobster | Amino acids, quaternary ammonium compounds, nucleotides, nucleosides, and lactate | Squid | Attraction |
| 13. Shrimp | Amino acids, quaternary ammonium compounds, nucleotides, nucleosides, and lactate | Crab, shrimp | Grasping delivery device |

*Source:* Carr and Derby (1986b).

[a] References for animal number cited: 1. Mackie and Adron (1980); Mackie and Mitchell (1982). 2. Carr, Blumenthal, and Netherton (1977). 3. Mackie and Adron (1978). 4. Mackie (1982). 5. Hashimoto et al. (1968). 6. Mackie and Mitchell (1983). 7. Fuke, Konosu, and Ina (1981). 8. Carr (1976); Carr, Blumenthal and Netherton (1977). 9. Carr and Chaney (1976). 10. Ohsugi, Hidaka, and Ikeda (1978); Hidaka (1982). 11. Mackie (1982). 12. Mackie (1973). 13. Carr, Netherton, and Milstead (1984), Carr and Derby (1986a).

[b] For the pinfish, a synthetic mixture of amino acids and betaine accounted for about 20% of extract's activity.

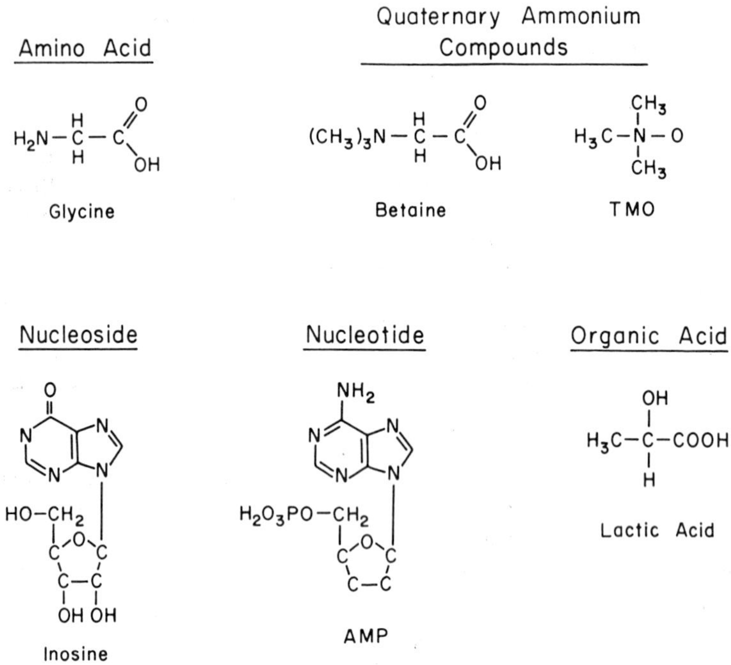


Figure 1.1 Types of common metabolites serving as feeding stimulants in aquatic animals.

Table 1.2. Concentrations of Common Metabolites in Tissues of Marine Organisms

| | Concentration (mM) in Tissue[a] | | | |
|---|---|---|---|---|
| Substance | Crab | Shrimp | Oyster | Mullet |
| Amino acids | | | | |
| Alanine | 12.80 | 21.20 | 7.32 | 2.64 |
| β-Alanine | | | 1.72 | 0.02 |
| α-Aminobutyric acid | | | | 0.03 |
| Arginine | 13.70 | 8.82 | 0.62 | 0.28 |
| Asparagine | 1.17 | 1.00 | 0.23 | 0.34 |
| Aspartic acid | 0.22 | 0.83 | 0.98 | 0.34 |
| Cysteine | 0.72 | | 0.13 | |
| Glutamic acid | 1.70 | 1.65 | 1.01 | 0.42 |
| Glutamine | 17.70 | 6.30 | 0.77 | 0.51 |
| Glycine | 68.00 | 93.70 | 4.79 | 3.05 |
| Histidine | 0.83 | 0.32 | 0.10 | 6.72 |
| Hydroxyproline | | | 0.25 | 0.06 |
| Isoleucine | 0.64 | 1.06 | 0.04 | 0.14 |
| Leucine | 1.47 | 1.90 | 0.08 | 0.26 |

Table 1.2. *Continued*

| Substance | Concentration (mM) in Tissue[a] | | | |
|---|---|---|---|---|
| | Crab | Shrimp | Oyster | Mullet |
| Lysine | 1.36 | 0.49 | 0.33 | 1.18 |
| Methionine | 2.35 | 1.04 | 0.02 | 0.07 |
| 3-Methylhistidine | | | | 0.12 |
| Ornithine | | | 0.19 | 0.27 |
| Phenylalanine | 0.56 | 0.54 | 0.01 | 0.08 |
| Phosphoserine | | | | 0.09 |
| Proline | 33.40 | 11.40 | 2.45 | 0.47 |
| Serine | 1.59 | 1.34 | 0.67 | 0.49 |
| Taurine | 11.30 | 33.00 | 33.10 | 11.40 |
| Threonine | 2.42 | 0.64 | 0.11 | 0.48 |
| Tryptophan | 0.79 | | | |
| Tyrosine | 0.61 | 0.92 | 0.05 | 0.04 |
| Valine | 1.91 | 2.30 | 0.08 | 0.28 |
| Total | 175.24 | 188.45 | 55.05 | 29.78 |
| Nucleotides, nucleosides, and related | | | | |
| Adenosine 5′-monophosphate | 0.15 | 4.88 | 0.59 | 0.02 |
| Adenosine 5′-diphosphate | 1.09 | 1.43 | 0.05 | 0.01 |
| Adenosine 5′-triphosphate | 5.27 | 0.40 | | |
| Guanosine 5′-monophosphate | | | 0.08 | 0.06 |
| Inosine 5′-monophosphate | 0.95 | 2.00 | 0.11 | 4.13 |
| Xanthosine 5′-monophosphate | | | | 0.17 |
| Hypoxanthine | 0.20 | 0.23 | | 0.07 |
| Inosine | 0.41 | 0.06 | 0.07 | 1.00 |
| Total | 8.07 | 9.00 | 0.90 | 5.46 |
| Quaternary ammonium compounds | | | | |
| Betaine | 22.20 | 50.40 | 20.30 | 6.20 |
| Homarine | 2.81 | 6.90 | 1.99 | |
| Trimethylamine oxide | 20.30 | 54.00 | | 5.90 |
| Total | 45.31 | 111.30 | 22.29 | 12.10 |
| Organic acids | | | | |
| L-Lactic acid | 26.10 | 27.50 | 0.55 | 34.30 |
| D-Lactic acid | | | 0.09 | |
| Succinic acid | | | 2.14 | |
| Grand total | 254.72 | 336.25 | 81.03 | 81.64 |

*Source:* Modified from Carr and Derby (1986a).

[a] Concentrations in tissues were obtained by correcting for the dilution of extracts shown in Carr and Derby (1986a). A blank space indicates the substance was not detected in the extract.

Azam 1985) in seawater. (Recent measurements of quaternary ammonium compounds and organic acids in seawater are not known.) Figure 1.2 shows that nucleotides (AMP, ADP and IMP), amino acids (glycine and taurine), and synthetic mixtures containing both of these types of substances are chemoattractants to a shrimp over a range of concentrations between the tissue levels and the ambient (noise) levels reported for nucleotides and amino acids in seawater.

The high tissue concentrations of many of the low molecular weight substances found to stimulate feeding relate to the role they play in intracellular osmotic regulation (Pierce 1982). These substances escape from organisms as a result of leakage, excretion, tissue damage, and decomposition (Poulet, Martin-Jezequel, and Head 1984). The rapid uptake by bacteria (McGrath and Sullivan 1981; Ferguson and Sunda 1984) and many higher organisms (Manahan, Wright, and

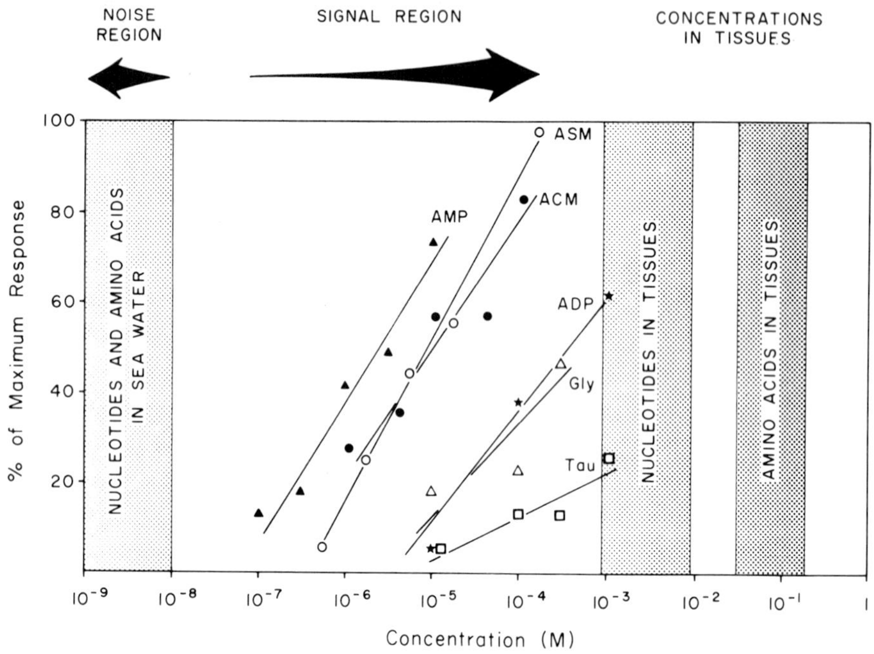


Figure 1.2. Chemical stimulants evoke behavioral responses over a range of stimulant concentrations that are less than those in food organisms, yet greater than ambient concentration (noise level) in aquatic environment. Dose response curves for behavioral chemoattractants of the shrimp *Palaemonetes pugio* are shown. Behavioral responses to stimulants in food odors occur within a range of concentrations from ca. $10^{-8}$ to $10^{-3}$ M. These concentrations are greater than noise level in seawater, yet less than the concentration in tissues of food organisms (see text). AMP = adenosine 5′-monophosphate; ADP = adenosine 5′-diphosphate; gly = glycine; tau = taurine; ACM = artificial crab mixture; ASM = artificial shrimp mixture. ACM and ASM contain the substances shown in Table 1.2 for crab and shrimp. Data from Carr and Derby (1986a).

Stephens 1983) serves as a damper that removes these substances from the water and assures that ambient concentrations (noise levels) remain low.

## Deterrents of Feeding Behavior

All aquatic organisms contain various concentrations of the common primary metabolites noted earlier to be feeding stimulants. Despite the presence of feeding stimulants in their tissues, however, many aquatic plants and animals are markedly ''unattractive'' to most predators because they contain repugnant chemicals that serve as feeding deterrents.

Chemical feeding deterrents compose a means of defense against predation that is especially evident in soft-bodied, slow-moving (or attached) organisms lacking obvious physical or mechanical defensive devices. In fact, the evolution of some of the ''defenseless'' soft-bodied forms may have been intimately coupled to the aquisition of chemical defenses against predation (Faulkner and Ghiselin 1983).

Chemical feeding deterrents are best known in the terrestrial environment, where many plants produce distasteful or noxious substances that negate or limit predation by insects (Whittaker and Feeny 1971). Similarly, aquatic plants have evolved chemical defenses against an array of predators. The analogy with the terrestrial environment goes at least two steps further to include the types of substances that deter feeding (Daloze, Brackman, and Tursch 1980), and the fact that many of the deterrents present in aquatic animals are sequestered from the plants in their diet (Norris and Fenical 1982).

Most feeding deterrents in both terrestrial and aquatic organisms are classified as ''secondary'' substances (secondary metabolites), so named because, unlike primary metabolites, they make no obvious contribution to the basic metabolic processes of the containing organism. Instead, the secondary substances are thought to be produced because of their apparent effects on other organisms (Whittaker 1970). Some of the types of substances identified as feeding deterrents in aquatic organisms are shown in Figure 1.3 by a representative terpene (A), alkaloid (B), halogenated compound (C), polyphenol (D), polyketide (E), and steroid glycoside (F). Other types of feeding deterrents include certain prostaglandins present in gorgonians (Gerhart 1984); a heavy metal, vanadium, present in high concentrations in some tunicates (Stoecker 1980); and inorganic acids that are secreted by several invertebrates (Thompson 1983).

Feeding deterrents are not usually released externally into the aquatic environment, but instead remain stationary in or on the organism employing them (Daloze, Braekman, and Tursch 1980). The signal is transmitted when a potential predator contacts or attempts to ingest the organism. In these situations noise is not a constraint to communication. To further facilitate the transmission of otherwise stationary signals, deterrents are frequently present in the body wall, as in the sea hare (Kinnel et al. 1979) and certain tunicates (Stoecker 1980), or are secreted by specialized glands onto an external surface, as in nudibranchs (Hochlowski et al. 1982; Cimino et al. 1985), sea cucumbers (Scheuer 1977), and a sole (Clark and George 1979).

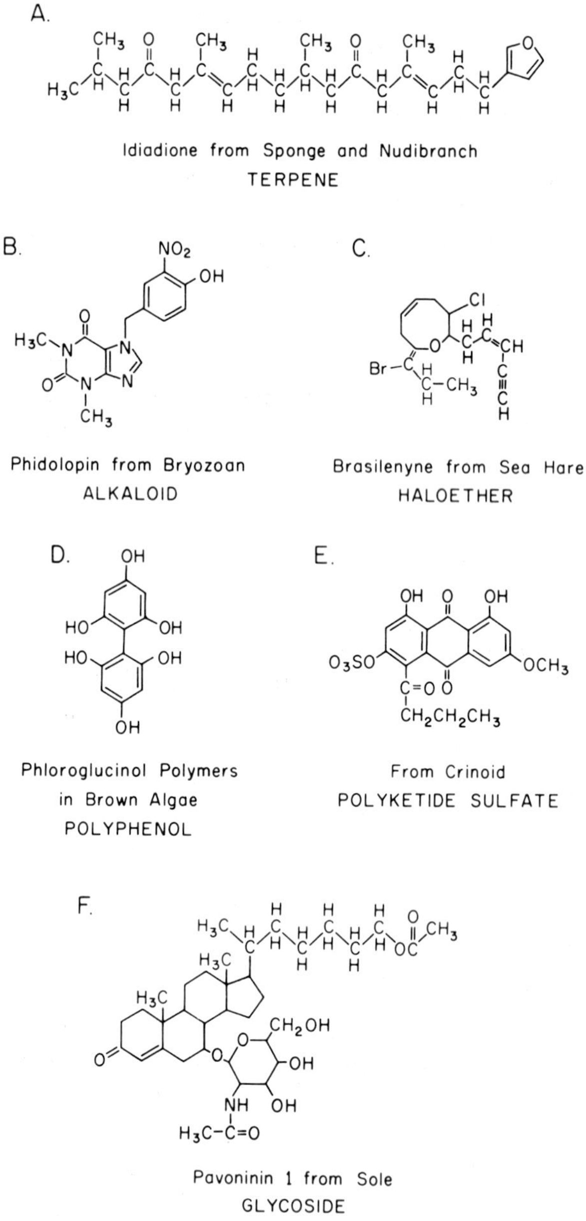


Figure 1.3. Examples of feeding deterrents in aquatic organisms. References: (A.) Walker, Thompson, and Faulkner (1980); (B.) Ayer, et al. (1984); (C.) Kinnel, et al. (1979); (D.) Geiselman and McConnell (1981); (E.) Rideout, Smith, and Sutherland (1979); (F.) Tachibana, Sakaitanai, and Nakanishi (1984).

## Substances Evoking Predator Avoidance and Alarm Responses

There are basically two sources of chemical signals available to prey species to warn them of the proximity of predators. The predator itself provides one source of a chemical signal, whereas the prey being physically damaged by the predator provides a second source. Chemical signals originating from both these sources have been identified. The term *alarm substance* ("schreckstoff") (von Frisch 1941) is applied in those cases in which the chemical signal originates from physically damaged prey. However, as will be shown later in this section, the "source" of the signal can depend on the recent feeding history of the predator.

Like the feeding deterrents discussed in the preceding section, the chemical signals evoking both predator avoidance and alarm responses deter predation. Consequently, it is not surprising that certain of the feeding deterrents may also serve in a second context as signals warning the prey of the proximity of a predator. Such is the case with several species of predatory starfish having substances that evoke dramatic avoidance responses in several molluscan prey species. The chemical signals in the starfish have been identified as steroid glycosides (saponins) present in the tube feet and elsewhere in the body of the predator (see Fig. 1.4) (Mackie 1970; Mackie, Singh, and Owen 1977). Although avoidance responses can sometimes be induced at a distance in molluscan prey, the strongest responses follow direct contact of the prey with the starfish. Interestingly, the eggs and larvae of at least one predatory starfish, *Acanthaster,* also contain large amounts of the same types of saponins found in the adults (Fig. 1.4B). These saponins play a role as feeding deterrents that make the eggs and larvae less susceptible to predation by planktivores (Lucas et al. 1979). Hence, the same signals that in the adult, forewarn potential prey about the proximity of this predator also serve the starfish by making its early developmental stages, as well as the adult itself, less susceptible to predation. This is clearly a case in which context is important, and one chemical signal may evoke two completely different forms of behavior.

As shown in examples given below, alarm substances from physically damaged prey are known to be either deposited in slime trails or released directly into the water. These substances stimulate others of the same species to disperse, bury themselves, or engage in other defensive maneuvers to avoid a predator. Alarm responses induced by chemical signals from injured conspecifics are known to come from aquatic animals as diverse as flatworms, sea anemones, snails, nudibranchs, fish, and amphibian larvae (Atema and Stenzler 1977).

As the nudibranch *Navanax* moves across the bottom, it leaves an insoluble slime trail of a mucopolysaccharide, which is normally followed by others of the species in search of food and mates. When an individual *Navanax* is molested in a fashion similar to that of an attack by a predator, it responds by secreting into its slime trail a mixture of at least three substances called *navenones* (Fig. 1.5A) (Sleeper, Paul, and Fenical 1980). When another trail-following *Navanax* encounters a trail containing navenones, an alarm response is evoked consisting of a contraction of the body, an abrupt turn of more than 90°, and retreat. The navenones are secreted from a gland located in the mantle cavity near the anus. Although slime trails can persist for several days, the navenones themselves do

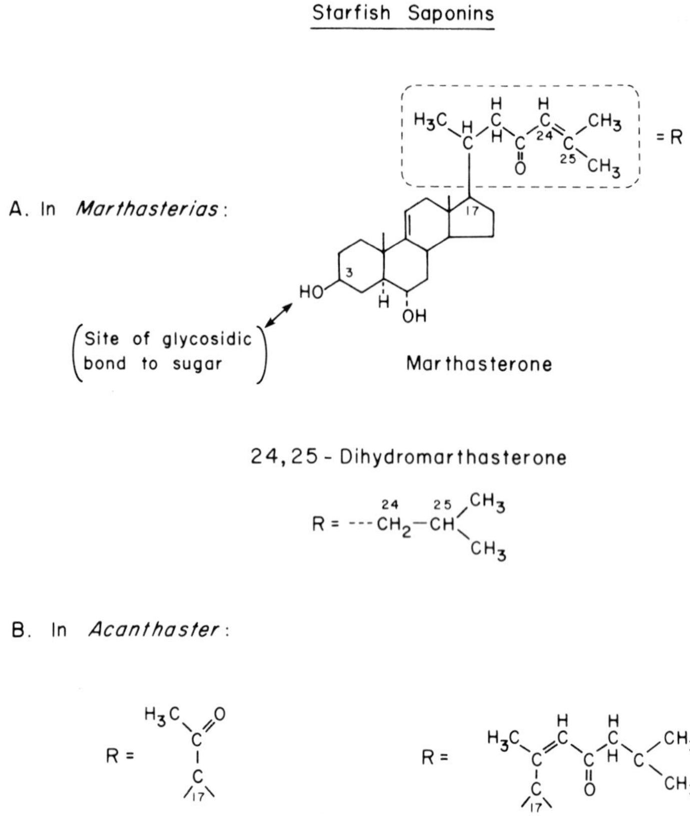


Figure 1.4. The structure of starfish saponins that evoke avoidance responses in prey organisms. Structures described by Mackie and Grant (1974).

not persist, because they are photochemically reactive. This reactivity results in their inactivation within a few hours and minimizes the potential for a high level of noise from signals contained in old trails (Sleeper, Paul, and Fenical 1980).

A second example of an alarm substance is provided by the betaine, anthopleurine, which is released directly into the water by the injured colonial sea anemone *Anthopleura* (Fig. 1.5B) (Howe and Sheikh 1975). Anthopleurine induces a withdrawal response in nearby individuals, thereby decreasing their susceptibility to predation. This chemical signal is quite species specific and potent, with a median effective concentration of about $3 \times 10^{-10}$M (Howe 1976). This substance is not known to occur in any other organism and hence is effectively nonexistent in the environment of *Anthopleura* except when others are being physically damaged (also see below). Since tests of structural analogs indicate that the alarm response is highly specific to anthopleurine, this substance provides an ideal signal for short-range communication in such a noise-free environment. Moreover, the turbulence of the habitat of *Anthopleura* ensures that old signals do not linger.

The above story of the alarm substance of the anemone, *Anthopleura,* has an

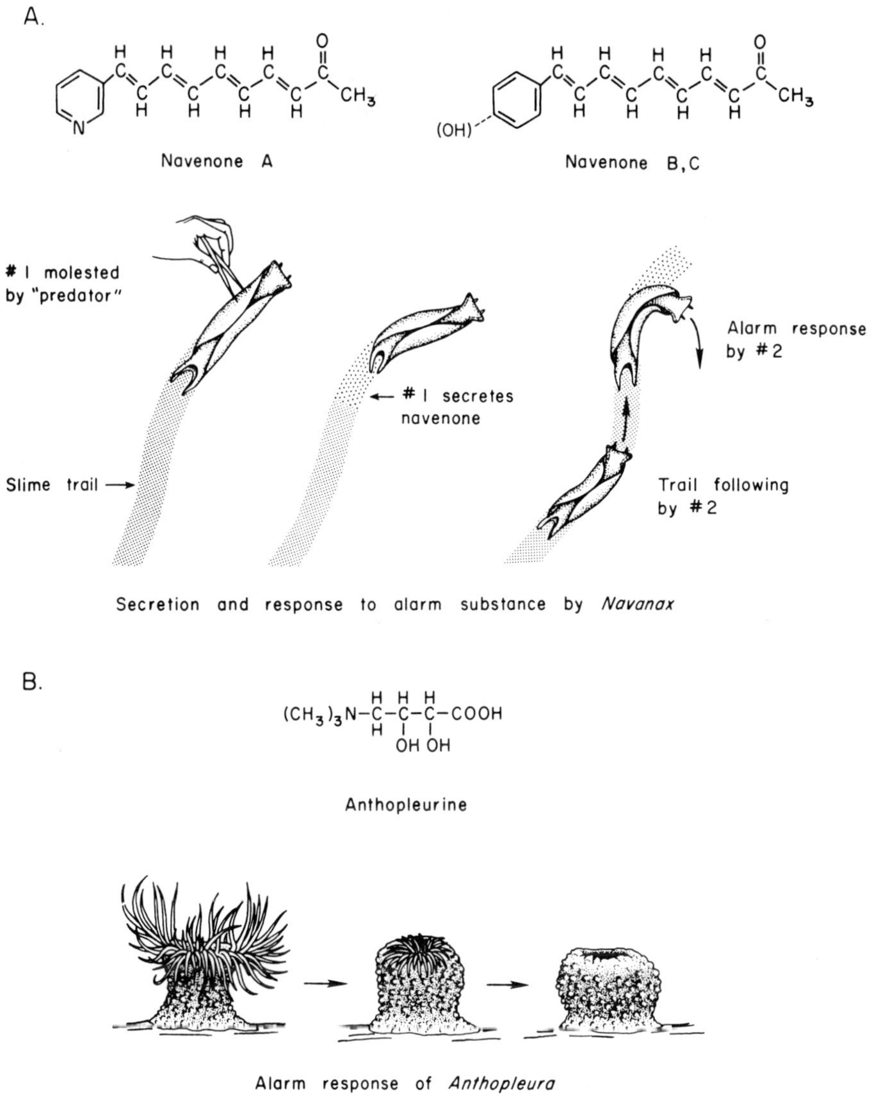


Figure 1.5. Alarm substances released by damaged or disturbed prey. (A.) Nudibranch *Navanax inermis* secretes navenones into slime trail when disturbed by predator. Navenones will evoke an alarm response in another *Navanax* encountering slime trail (Sleeper, Paul, and Fenical 1980). (B.) Anthopleurine released into the water by injured anemones *Anthopleura elegantissima* induces withdrawal response in others nearby, thereby decreasing their susceptibility to predation (Howe and Sheikh 1975).

interesting twist that relates to a transfer of the source of the signal. Major predators of *Anthopleura* are nudibranchs that, as a consequence of consuming the anemones, also accumulate anthopleurine from the tissues of their prey (Howe and Harris 1978). For several days after consuming the anemones, anthopleurine escapes from a predatory nudibranch in amounts sufficient to evoke alarm responses from *Anthopleura* that are approached by this predator. As a consequence, the predator now assumes the role of a signal amplifier that transmits information about the proximity of the predator even before additional members of the prey species are injured.

As a fitting close to this section, note that, after many years of study, the first alarm substance from a fish has recently been identified as the novel purine, hypoxanthine-3(N)-oxide (Pfeiffer et al. 1985).

## Sex Attractants

For dioecious organisms in which fertilization occurs externally in the aquatic environment, it is clearly advantageous to have systems of communication for ensuring the coordinated release of gametes and increasing the probability of their contact in the water. The communication of both types of information is affected by means of unique chemical signals in several species of brown algae (Phaeophyceae) (Muller, Gassmann, and Luning 1979; Boland et al. 1983). Furthermore, some species use apparent competitive ploys by releasing the chemical signal of a second species.

Several species of brown algae have motile female gametes that settle on a surface and broadcast their presence by releasing chemical signals that act over short distances (ca. 1 mm) to trigger the release and/or attraction of male gametes. The chemical signals are composed of one or a mixture of unsaturated cyclic

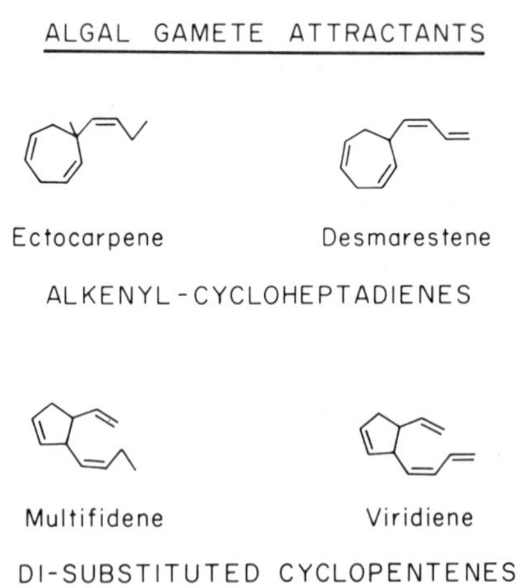


Figure 1.6. Unsaturated cyclic hydrocarbons used as gametic attractants by certain brown algae (Boland et al. 1984).

hydrocarbons that, in several species, are alkenyl-cycloheptadienes and substituted cyclopentenes (Fig. 1.6) (Boland et al. 1982; Boland et al. 1983; Boland et al. 1984). These substances have the following unique properties that contribute to their effectiveness as signals for precise, short-range communication in an aquatic environment. First, these substances are extremely potent and elicit responses at concentrations in the range of $10^{-9}$ to $10^{-10}$ M. The nonpolar nature of these small molecules makes them insoluble in water and also highly volatile. Subsequently, their concentration in the water is kept minimal owing to both insolubility and volatility, with the latter quickly eliminating the molecules from the water of the shallow photic zone inhabited by the algae. Finally, reactive double bonds make these molecules unstable and susceptible to rapid inactivation by both photo-oxygenation and metal-catalyzed hydration (Boland et al. 1982). Collectively, the above properties maintain a high signal-to-noise ratio that is essential for communication.

As noted above, the physicochemical properties of the algae gamete attractants are well adapted to the maintenance of low noise levels. However, it is known that at least some species "inject noise" into the system, apparently as a means of gaining a selective advantage. Female gametes of some species of brown algae release a bouquet of several cyclic hydrocarbons, of which only one is the specific attractant for the gametes of their own species. For example, female gametes of *Cutleria multifida* release not only multifidene, the attractant for male gametes of this species, but also ectocarpene, the attractant for male gametes of a second species, (see Fig. 1.6) (Boland et al. 1983). Since both species often occupy the same habitat, the mixture of attractants released by *Cutleria* not only enhances the reproductive success of that species, but also negatively affects the success of the competitor.

Progress in the identification of sex attractants (sex pheromones) for aquatic animals has lagged far behind the major advances realized with aquatic algae. Nevertheless, a brief summary of some findings with crustaceans and fishes will be provided. For comprehensive reviews, see Dunham (1978) for crustaceans and Colombo, et al. (1982) and Liley and Stacey (1983) for fishes.

Beginning with the observations of Ryan (1966) and Atema and Engstrom (1971), evidence has been accumulating that sexually mature females of certain crustaceans release a sex pheromone that evokes premating behavior in males of the same species (Fig. 1.7). An early report indicated that the steroid molting hormone, crustecdysone, was the sex pheromone for three species of crabs (Kittredge, Terry, and Takahashi 1971). Furthermore, a hypothesis was presented whereby this steroid hormone could provide the evolutionary basis for all crustacean sex pheromones (Kittredge and Takahashi 1972). Subsequent studies however, with other crustaceans, the American lobster (Atema and Gagosian 1973; Gagosian and Atema 1973) and the blue crab (Gleeson, Adams, and Smith 1984), have revealed that crustecdysone and related steroid metabolites are inactive in evoking premating behavior in these species. Hence, the question of the identity of crustacean sex pheromones remains unresolved. In fishes, however, evidence is accumulating that water-soluble glucuronides of steroid sex hormones are released from the ovaries of certain female fishes and that these modified steroids

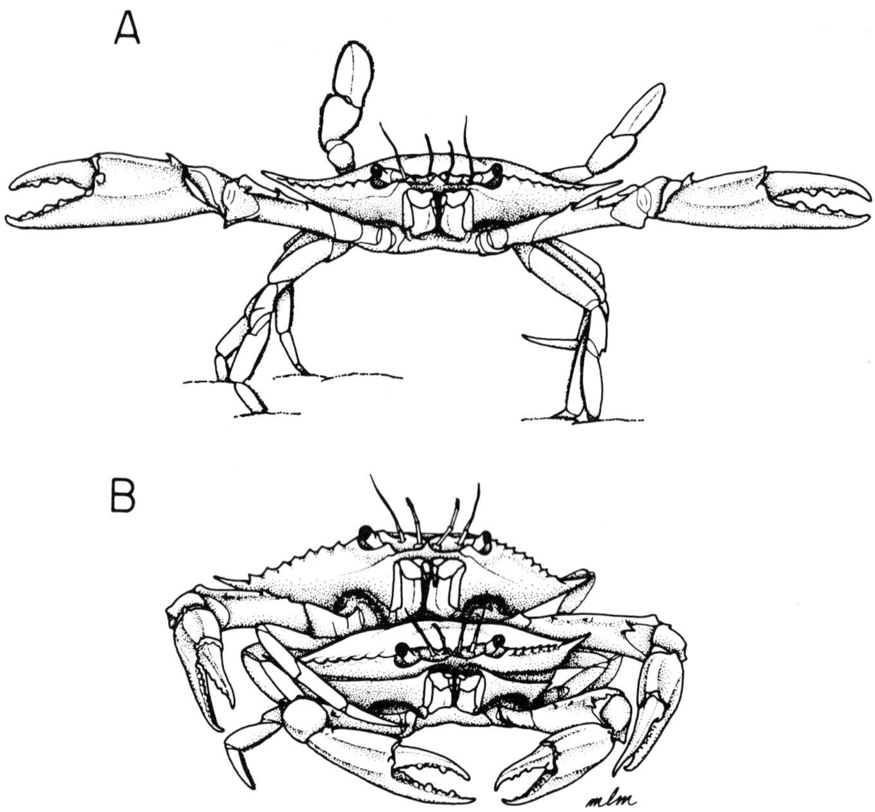


Figure 1.7. Behavior stimulated by sex pheromone of blue crab *Callinectes sapidus*. (A.) Male displaying premating behavior in the presence of sex pheromone. Pheromone-induced behavior includes walking on tips of dactyls, extending and displaying chelae, fanning swimming appendages (uropods), and attempting to pull any contacted crab into "cradle-carry" position illustrated in (B). Males carry (and defend) mature females in cradle-carry position (B) until molting, copulation, and hardening of new exoskeleton are completed (Gleeson 1980).

serve as sex pheromones, which are detected by the olfactory system and trigger courtship behavior in males (Colombo et al. 1982; van den Hurk and Lambert 1983).

The evidence presented above about the functioning of derivatives of sex hormones as sex pheromones in certain fishes deserves further comment. Certainly, sex hormones and their derivatives could function as timely messengers for communicating information about reproductive readiness. Yet messengers of this sort raise a question about species specificity, since many animals make many of the same hormones and derivatives. We learn from insects however, that substances made and used by many species can make important contributions to the sex pheromone of a particular species when the pheromone is composed of a mixture (blend) of substances (Weaver 1983). The specificity of a particular blend resides largely in the defined proportions in which the components occur. Moreover, the pheromonal blend released by a particular species frequently functions in concert with other chemical signals (i.e., odorous components from vegetation in the "proper" habitat) and with signals perceived by other modalities. Consequently, it is oftentimes a complement of sensory inputs that triggers courtship or mating behavior or both of them. The lessons learned from insects may prove to be applicable to an understanding of the use of hormonal derivatives as sex pheromones in aquatic organisms.

## Substances Influencing Habitat Selection by Larvae

Many aquatic invertebrates are bottom-dwelling or sessile as adults, yet have a free-swimming, planktonic larval stage, which disperses the population. Many of these larvae use external chemical signals to identify and select surfaces for settlement and metamorphosis. This selective process ensures that significant numbers of larvae metamorphose in a habitat suitable for the next stage of the life cycle. Invertebrates with larva using chemical signals for selection of habitat include various coelenterates, bryozoans, annelids, molluscs, echinoderms, arthropods (primarily barnacles), and tunicates (Morse 1984). A complete review of factors influencing larval settlement is provided by Crisp (1974).

The use of chemical signals by larvae has been thoroughly studied in the red abalone *Haliotis* (phylum Mollusca) (Morse et al. 1979; Morse 1984). As an adult, this large gastropod feeds on giant kelp, whereas the preferred foods of the post-larval juvenile stage are certain species of crustose red algae. The larvae of abalone placed in dishes of seawater containing the preferred red algae show a rapid settling and subsequent metamorphosis into the early juvenile form. Neither settlement nor metamorphosis is induced by other types of algae, including the giant kelp that is the principal food of the adult. The chemical signals inducing settlement by the larvae are not released into the water, but instead require direct physical contact with red algae, which will become the principal food for the next stage of the life cycle. The inducing substance is apparently a GABA-mimetic peptide associated with the phycobiliproteins that are "uniquely available" on the surface of the preferred algae (Morse and Morse 1984).

Proteins or peptides on other kinds of surfaces also serve as chemical signals that stimulate settlement and metamorphosis of other types of larvae. Specific proteins on the shells of adult barnacles (phylum Arthropoda) serve as signals that induce settlement of larval barnacles on the adults, thereby providing the gregariousness that is a prerequisite for a sessile species that must copulate to reproduce (Larman, Gabbott, and East 1982; Larman 1984). Likewise, a similar type of chemical signal is a basis for gregariousness in a species of sand dollar, *Dendraster* (phylum Echinodermata), in which a peptide released into the sand by the adults induces settlement and metamorphosis by the larvae of the same species (Burke 1984). Finally, settlement of the larvae of the polychaete *Janua* (phylum Annelida) on surfaces coated by bacterial films may be mediated by extracellular glycoproteins produced by the bacteria (Kirchman et al. 1982).

As noted above, the substances identified as natural stimulants of larval settlement have been found to be proteins or peptides. These chemical signals for larvae are adsorbed on or present in a surface that is appropriate for the next stage of the life cycle. In nature, the response of the larvae to the signal requires physical contact with the surface. Crisp (1974) noted that recognition of proteins or peptides on surfaces by a larval organism seemed analogous to an antigen-antibody reaction. A somewhat similar mechanism involving a lectin-glycoprotein reaction between the larva and the surface was proposed by Kirchman et al. (1982). If the above types of reactions indeed occur between a protein (lectin?) on the receptor of the larva and a complementary protein or peptide on the target surface, all that is required for this reaction to culminate in settlement is for the larva to recognize that stickiness, adhesion, or another form of molecular aggregation occurs on contact with a particular surface (Crisp 1974; Larman 1984). Such a precise means of communication between the organism and habitat may be appropriate to the magnitude of the decision that is made when an organism terminates its swimming stage and makes an irreversible commitment to a benthic habitat (Morse 1984).

Finally, it is noteworthy that the advantageous use of chemical signals for the recruitment of larvae may also have negative consequences. Recent evidence has been presented that certain "surface-marking" peptides produced by barnacles provide two quite different types of messages. When bound to a surface, the peptides stimulate settlement and metamorphosis by barnacle larvae. When small amounts of these same peptides pass into solution in the water, however, they now serve as attractants for a barnacle predator, the oyster drill (Rittshof and Bonaventura 1986). Presumably, evolution could permit the persistence of a signal produced by barnacles that both stimulates settlement by larvae and attracts a predator if the advantages of gregariousness (leading to increased reproduction) far exceed the disadvantages of increased predation. Several examples of comparable interspecific exploitations of intraspecific chemical signals are known to occur in insects (Whittaker and Feeny 1971).

## Similarities Between External Chemical Signals and Internal Neuroactive Agents

Many similarities exist between the monitoring of chemical signals in the external aquatic environment by chemoreceptors and the monitoring of neuroactive substances within an organism's internal environment by the highly specific receptors present in synapses, neuromuscular junctions, and elsewhere. Both external and internal receptors are adapted to monitor the chemical composition of their aquatic milieu. In fact, beginning with Haldane (1954), biologists have discussed the notion that internal receptors may trace their origins to external chemoreceptors of primitive aquatic organisms (Kittredge et al. 1974; Lenhoff and Heagy 1977; Boyd 1979; Carr and Thompson 1983). If this notion were correct, it might help to explain why the internal neuroactive, or regulatory, substances listed in Table 1.3 also function as external chemical signals in various aquatic organisms.

In regard to the functioning of substances as both external and internal chemical signals, some particularly interesting examples are provided by the three adenosine nucleotides cited in Table 1.3. The nucleotide cyclic AMP (cAMP) functions externally as a potent chemoattractant in cellular slime molds (Mato et al. 1978) and internally as a ubiquitous second messenger frequently engaged in the transfer of information across cell membranes (Greengard 1981). Regarding the other two nucleotides, 5'-AMP (AMP) is a potent external chemoattractant for a marine shrimp, *Palaemonetes pugio,* (Carr and Thompson 1983), whereas ATP induces gorging on fluids by a blood-sucking insect, *Rhodnius prolixus* (Friend and Smith 1982). Internally, in the vertebrates, both AMP and ATP play major roles in the brain, heart, and other organs as activators, inhibitors, or modulators of an array of internal activities (Olsson et al. 1979; Phillis and Wu 1981; Stone 1981). In fact, distinct types of receptors named *P1*- and *P2-types,* have been proposed for AMP (and adenosine) and ATP, respectively (Burnstock 1978). When one analyzes the particular roles of cAMP, AMP, and ATP as external chemical signals affecting behavior in diverse organisms, one finds that each substance evokes a response related either to food recognition or to feeding state. The

Table 1.3. Internal Neuroactive Agents for Which External Chemoreceptors Also Exist

| Substance Having Internal Neuroactive Role | Organism with External Chemoreceptors for the Substance[a] | Behavior Stimulated by Activation of External Receptor |
|---|---|---|
| 1. AMP | Shrimp, spiny lobster, mosquito larvae | Attraction for shrimp, mosquito larvae |
| 2. Cyclic AMP | Slime mold | Attraction following exhaustion of food supply |
| 3. ATP | Mosquito (insect)[b] | Gorging on liquid |
| 4. Acetylcholine | *Tetrahymena* (ciliated protozoan) | Reduction in forward swimming velocity |
| 5. γ-Aminobutyric acid (GABA) | Abalone larvae (mollusc) | Settlement and metamorphosis |
| 6. Glutamate | American lobster | Prey detection |
| 7. Glycine | Snail | Extending proboscis |
| 8. Taurine | Spiny lobster | Attraction |
| 9. α, β-ecdysone[c] | Spiny lobster | Not known |
| 10. Prostaglandin A2 | Reef fish | Rejection of food |

[a] Studies implicating the cited organism as having external chemoreceptors for the indicated substance. Substance: 1. Carr and Thompson (1983); Derby, Carr, and Ache (1984); Barber, Ellgard, and Herskowitz (1982). 2. Mato et al. (1978); Van Haastert et al. (1982). 3. Friend and Smith (1977) (1982). 4. Brown and Kerkut (1981). 5. Morse et al. (1979). 6. Derby and Atema (1982). 7. Carr (1967). 8. Fuzessery, Carr, and Ache (1978); Johnson and Ache (1978). 9. Spencer and Case (1984). 10. Gerhart (1984).

[b] Mosquito has an aquatic larval stage.

[c] These substances are steroid molting hormones of arthropods.

same is true of the substances glutamate, glycine, taurine, and prostaglandin A2, which are also shown in Table 1.3 as having dual roles as both external and internal chemical signals. The external role of each substance is again related to the recognition (or rejection) of food (Carr 1967; Fuzessery, Carr, and Ache 1978; Derby and Atema 1982; Gerhart 1984). One can speculate that the evolution of receptor types for all seven of the substances cited above may have occurred initially on outer membrane surfaces, where they monitored chemical signals related to food recognition in the external aquatic environment. With the evolution of internal cell layers, some of these same receptors, together with their associated capacity to affect membrane excitability, could have been conserved on the surface of cells in the interior of the organism. These cells and receptors now monitored the chemical signals released into the internal environment by other cells, rather than the signals released into the external environment by other organisms (see discussion by Lentz 1968, Chapter 5). It is noteworthy that, in all higher organisms, the nervous system, with its vast array of receptor types for transmitters, modulators, and hormones, traces its embryological origin to an outer cell layer, the ectoderm. Hence, before gastrulation, the cells destined to differentiate into the internal nervous system constitute a cell layer having outer membrane surfaces in intimate contact with an external aquatic environment.

Interestingly, the membrane surface of an amphibian oocyte possesses receptor sites for neurotransmitters, namely, acetylcholine and the catecholamines (Kusano, Miledi, and Stinnakre 1982). Likewise, both mice and human oocyte membranes respond to acetylcholine and another transmitter, GABA (Dolci, Eusebi, and Siracusa 1985). Since receptors on the surface of unfertilized oocytes detect chemical signals in their external aquatic environment, these findings lend considerable credence to the notion cited earlier by Haldane (1954) and others that receptors for neuroactive agents functioning in an organism's internal environment lead back to external chemoreceptors.

The following quotation provides a fitting close to this section:

> Can speculation about the evolutionary origin of the nervous system have any value? That we may never know exactly by what route the behaviour machinery of the first Metazoa was derived is only too probable, and the past history of such speculations does not encourage one to suppose that today will be enduring. But so long as we are careful not to confuse that which is speculative with that which is true, they can have a real if limited value. They encourage us from time to time to consider the capabilities of simple systems in the light of contemporary knowledge. (Pantin 1956, cited in de Ceccatty 1974)

## Summary

This chapter describes molecules known to serve as external chemical signals influencing the behavior of aquatic organisms. The natural sources of the chemicals are given, together with a description of how they are transmitted to the respondent organism.

Chemical signals evoking aspects of *feeding behavior* are simple common metabolites that are ubiquitous components of tissues. Chemicals affecting feeding behavior include amino acids, quaternary ammonium compounds, nucleosides, and nucleotides. These substances are usually more effective in mixtures than alone. As a group, feeding stimulants are very soluble substances that escape from organisms as a result of leakage, excretion, tissue damage, and decomposition.

Many potential prey organisms are not eaten by most predators because they contain distasteful or noxious substances, called *feeding deterrents,* that negate the excitatory effects of feeding stimulants. Most feeding deterrents in aquatic organisms are "secondary metabolites" similar to those serving a similar function in terrestrial organisms. These substances are not usually released into the water, but are present in or on the prey organism; the signal is transmitted by contact with the predator. Substances stimulating *predator avoidance* and *alarm responses* warn prey organisms of the proximity of predators and hence provide another type of defense against predation. Chemical signals evoking avoidance or alarm responses may be released by a predator or, alternatively, by prey species being attacked. These signals may be variously transmitted by contact with the predator, by a slime trail, or by release into the water. Like feeding deterrents, substances evoking avoidance responses are frequently secondary metabolites. Indeed, the same starfish saponins that provide a signal to starfish prey about the proximity of this predator also function as feeding deterrents, which make the eggs and larvae of the starfish itself less susceptible to predation.

The female gametes of several species of algae use chemical signals to trigger the release or attraction of male gametes or both. In several algal species these *sex attractants* are known to consist of one or a mixture of unsaturated cyclic hydrocarbons. These attractants serve as effective signals for precise, short-range communication because of their high potency, specificity, and short time in the water. Although it is well known that sex pheromones are also employed by some aquatic animals, the chemical structures of these signal molecules remain unknown.

Chemical signals used in the *selection of a surface* for the settlement and metamorphosis of larvae of some benthic invertebrates are known to be specific proteins or peptides that are adsorbed on or present in surfaces appropriate for the next stage of the life cycle. The larval response requires physical contact with the surface.

Finally, examples are provided of substances that serve as chemical signals in both the external and "internal" aquatic environments of various organisms. The hypothesis that the receptors for neurotransmitters and other internal neuroactive agents may have evolved from external chemoreceptors of primitive aquatic organisms is discussed.

*Acknowledgments.* I am grateful to Ms Marsha Lynn Milstead for preparing the illustrations. This work was supported in part by NSF Grant No. BNS-84–11693 and by the C.V. Whitney Marine Laboratory.

## References

Ache, B.W. 1982. "Chemoreception and Thermoreception." In *The Biology of Crustacea,* vol 3, edited by H.L. Atwood and D.C. Sandeman. New York: Academic Press.

Ammerman, J.W., and F. Azam. 1985. Bacterial 5′-nucleotidase in aquatic ecosystems: a novel mechanism of phosphorous regeneration. *Science* 227:1338–1340.

Atema, J. 1980. Chemical senses, chemical signals and feeding behavior in fishes. In *Fish Behavior and Its Use in the Capture and Culture of Fishes*. Manilla: Int. Center Living Aquatic Resources Mgt.

Atema, J. 1985. Chemoreception in the sea: adaptations of chemoreceptors and behavior to aquatic stimulus conditions. *Soc. Exp. Biol. Sympos.* 39:387–423.

Atema, J., and D.G. Engstrom. 1971. Sex pheromone in the lobster, *Homarus americanus*. *Nature* 232:261–263.

Atema, J., and R.B. Gagosian. 1973. Behavioral responses of male lobsters to ecdysones. *Mar. Behav. Physiol.* 2:15–20.

Atema, J., and D. Stenzler. 1977. Alarm substance of the marine mud snail, *Nassarius obsoletus:* biological characterization and possible evolution. *J. Chem. Ecol.* 3:173–187.

Ayer, S.W., R.J. Andersen, H. Cun-heng, and J. Clardy. 1984. Phidolopin, a new purine derivative from the bryozoan *Phidolopora pacifica*. *J. Org. Chem.* 49:3869–3870.

Barber, J.T., E.G. Ellgard, and K. Herskowitz. 1982. The attraction of larvae of *Culex pipiens quinquefasciatus* Say. to ribonucleic acids and nucleotides. *J. Insect Physiol.* 28:585–588.

Boland, W., R. Terlinden, L. Jaenicke, and D.G. Muller. 1982. Binding-mechanism and sensitivity in gamete chemotaxis of the Phaeophyta *Cutleria multifida*. *Eur. J. Biochem.* 126:173–179.

Boland, W., F.-J. Marner, L. Jaenicke, D.G. Muller, and E. Folster. 1983. Comparative receptor study in gamete chemotaxis of the seaweeds *Ectocarpus siliculosus* and *Cutleria multifida*. An approach to interspecific communication of algal gametes. *Eur. J. Biochem.* 134:97–103.

Boland, W., L. Jaenicke, D.G. Muller, and A. Peters. 1984. Differentiation of algal chemoreceptors. A comparative receptor study with two seaweed pairs: *Cutleria multifida/Syringoderma phinneyi* and *Desmarestia aculeata/Ectocarpus siliculosus* (Phaeophyceae). *Eur. J. Biochem.* 144:169–176.

Boyd, A., and M. Simon. 1982. Bacterial chemotaxis. *Annu. Rev. Physiol.* 44:501–507.

Boyd, C.A.R. 1979. Chemical neurotransmission: an hypothesis concerning the evolution of neurotransmitter substances. *J. Theor. Biol.* 76:415–417.

Brown, I.D., and G.A. Kerkut. 1981. A study of the chemokinetic effects of various pharmacological agents upon *Tetrahymena vorax*. *Comp. Biochem. Physiol.* 69C:275–280.

Burke, R.D. 1984. Pheromonal control of metamorphosis in the Pacific sand dollar, *Dendraster excentricus*. *Science* 225:442–443.

Burnstock, G. 1978. "A basis for distinguishing two types of purinergic receptors." In *Cell Membrane Receptors for Drugs and Hormones: A Multidisciplinary Approach*, edited by R.W. Straub and L. Bolis. New York: Raven Press.

Carr, W.E.S. 1967. Chemoreception in the mud snail *Nassarius obsoletus*. II. Identification of stimulatory substances. *Biol. Bull.* 133:106–127.

Carr, W.E.S. 1976. Chemoreception and feeding behavior in the pigfish, *Orthopristis chrysopterus:* characterization and identification of stimulatory substances in a shrimp extract. *Comp. Biochem. Physiol.* 155A:153–157.

Carr, W.E.S. 1982. "Chemical stimulation of feeding behavior." In *Chemoreception in Fishes*, edited by T.J. Hara. Amsterdam: Elsevier Scientific Publishing Co.

Carr, W.E.S., and T.B. Chaney. 1976. Chemical stimulation of feeding behavior in the pinfish, *Lagodon rhomboides:* characterization and identification of stimulatory substances extracted from shrimp. *Comp. Biochem. Physiol.* 54A:437–441.

Carr, W.E.S., and C.D. Derby. 1986a. Behavioral chemoattractants for the shrimp, *Palaemonetes pugio:* identification of active components in food extracts and evidence of synergistic mixture interactions. *Chem. Sens.* 11:49–64.

Carr, W.E.S., and C.D. Derby. 1986b. Chemically stimulated feeding behavior in marine animals: the importance of chemical mixtures and the involvement of mixture interactions. *J. Chem. Ecol.* 12:987–1009.

Carr, W.E.S., and H.W. Thompson. 1983. Adenosine 5′-monophosphate, an internal regulatory agent, is a potent chemoattractant for a marine shrimp. *J. Comp. Physiol.* 153:47–53.

Carr, W.E.S., K.M. Blumenthal, and J.C. Netherton III. 1977. Chemoreception in the pigfish, *Orthopristis chrysopterus:* the contribution of amino acids and betaine to stimulation of feeding behavior by extracts. *Comp. Biochem. Physiol.* 58A:69–73.

Carr, W.E.S., J.C. Netherton III, and M.L. Milstead. 1984. Chemoattractants of the shrimp, *Palaemonetes pugio:* variability in responsiveness and the stimulatory capacity of mixtures containing amino acids, quaternary ammonium compounds, purines and other substances. *Comp. Biochem. Physiol.* 77A:469–474.

Cimino, G., S. DeRosa, S. DeStefano, R. Morrone, and G. Sodano. 1985. The chemical defense of nudibranch molluscs. *Tetrahedron* 41:1093–1100.

Clark, E., and A. George. 1979. Toxic soles, *Pardachirus marmoratus* from the Red Sea and *P. pavoninus* from Japan, with notes on other species. *Environ. Biol. Fishes* 4:103–123.

Colombo, L., P.C. Belvedere, A. Marconato, and F. Bentivegna. 1982. "Pheromones in Teleost Fish." In *Proceedings of the International Symposium on Reproductive Physiology of Fish, 2–6 August 1982,* edited by C.J.J. Richter and H.J.T. Goos. Wageningen, The Netherlands: Centre for Agricultural Publishing and Documentation.

Crisp, D.J. 1974. "Factors Influencing the Settlement of Marine Invertebrate Larvae." In *Chemoreception in Marine Organisms,* edited by P.T. Grant and A.M. Mackie. London: Academic Press.

Croll, R.G. 1983. Gastropod chemoreception. *Biol. Rev.* 58:293–319.

Daloze, D., J.C. Braekman, and B. Tursch. 1980. "Chemical Communication in the Marine Environment." In *Animals and Environmental Fitness: Physiological and Biochemical Aspects of Adaptation and Ecology,* vol 1, edited by R. Gilles. Oxford: Pergamon Press.

de Ceccatty, M.P. 1974. The origin of the integrative systems: a change of view derived from research on coelenterates and sponges. *Perspect. Biol. Med.* Spring:379–390.

Derby, C.D., and J. Atema. 1982. Chemosensitivity of walking legs of the lobster *Homarus americanus:* neurophysiological response spectrum and thresholds. *J. Exp. Biol.* 98:303–315.

Derby, C.D., W.E.S. Carr, and B.W. Ache. 1984. Purinergic olfactory receptors of crustaceans are similar to internal purinergic receptors of vertebrates. *J. Comp. Physiol.* 155:341–349.

Dolci, S., F. Eusebi, and G. Siracusa. 1985. γ-Amino butyric-N-acid sensitivity of mouse and human oocytes. *Dev. Biol.* 109:242–246.

Dunham, P.H. 1978. Sex pheromones in crustacea. *Biol. Rev.* 53:555–583.

Faulkner, D.J., and M.T. Ghiselin. 1983. Chemical defense and evolutionary ecology of dorid nudibranchs and some other opisthobranch gastropods. *Mar. Ecol. Prog. Ser.* 13:295–301.

Fenical, W. 1975. Halogenation in the Rhodophyta. *J. Phycol.* 11:245–259.

Ferguson, R.L., and W.G. Sunda. 1984. Utilization of amino acids by planktonic marine bacteria: importance of clean technique and low substrate additions. *Limnol. Oceanogr.* 29:258–274.

Friend, M.G., and J.J.B. Smith. 1977. Factors affecting feeding by bloodsucking insects. *Annu. Rev. Entomol.* 22:309–331.

Friend, W.G., and J.J.B. Smith. 1982. ATP analogs and other phosphate compounds as gorging stimulants for *Rhodnius prolixus. J. Insect Physiol.* 28:371–376.

Fuke, S., S. Konosu, and K. Ina. 1981. Identification of feeding stimulants for red sea bream in the extract of marine worm *Perinereis brevicirrus. Bull. Jpn. Soc. Sci. Fish.* 47:1631–1635.

Fuzessery, Z.M., W.E.S. Carr, and B.W. Ache. 1978. Antennular chemosensitivity in the spiny lobster, *Panulirus argus:* studies of taurine sensitive receptors. *Biol. Bull.* 154:226–240.

Gagosian, R.B., and J. Atema. 1973. Behavioral responses of male lobsters to ecdysone metabolites. *Mar. Behav. Physiol.* 2:115–120.

Geiselman, J.A., and O.J. McConnell. 1981. Polyphenols in brown algae *Fucus vesiculosus* and *Ascophyllum nodosum:* chemical defenses against the marine herbivorous snail, *Littorina littorea. J. Chem. Ecol.* 7:1115–1133.

Gerhart, D.J. 1984. Prostaglandin A2: an agent of chemical defense in the Caribbean gorgonian *Plexaura homomalla. Mar. Ecol. Prog. Ser.* 19:181–187.

Gleeson, R.A. 1980. Pheromone communication in the reproductive behavior of the blue crab, *Callinectes sapidus. Mar. Behav. Physiol.* 7:119–134.

Gleeson, R.A., M.A. Adams, and A.B. Smith III. 1984. Characterization of a sex pheromone in the blue crab, *Callinectes sapidus:* crustecdysone studies. *J. Chem. Ecol.* 10:913–921.

Greengard, P.C. 1981. *Intracellular signals in the brain. The Harvey Lectures, Series 75.* New York: Academic Press.

Gurin, S, and W.E.S. Carr. 1971. Chemoreception in *Nassarius obsoletus:* the role of specific stimulatory proteins. *Science* 174:293–295.

Haldane, J.B.S. 1954. La signalisation animale. *Annee Biol.* 58:89–98.

Hara, T.J. 1975. Olfaction in fish. *Prog. Neurobiol.* 5:271–335.

Hashimoto, Y., S. Konosu, N. Fusetani, and T. Nose. 1968. Attractants for eels in the extracts of short-necked clams. I. Survey of constituents eliciting feeding behavior by the omission tests. *Bull. Jpn. Soc. Sci. Fish.* 34:78–83.

Heeb, M.A. 1973. Large molecules and chemical control of feeding behavior in the starfish *Asterias forbesi. Helgol. wiss. Meeresunters.* 24:425–435.

Hidaka, I. 1982. "Taste Receptor Stimulation and Feeding Behavior in the Puffer." In *Chemoreception in Fish,* edited by T.J. Hara. Amsterdam: Elsevier Scientific Publishing Co.

Hochlowski, J.E., R.P. Walker, C. Ireland, and D.J. Faulkner. 1982. Metabolites of four nudibranchs of the genus *Hypselodoris. J. Org. Chem.* 47:88–91.

Howe, N.R. 1976. Behavior of sea anemones evoked by the alarm pheromone anthopleurine. *J. Comp. Physiol.* 107:67–76.

Howe, N.R., and L.G. Harris. 1978. Transfer of the sea anemone pheromone, anthopleurine, by the nudibranch *Aeolidia papillosa.* J. Chem. Ecol. 4:551–561.

Howe, N.R., and Y.M. Sheikh. 1975. Anthopleurine: a sea anemone alarm pheromone. *Science* 189:386–388.

Johnson, B.R., and B.W. Ache. 1978. Antennular chemosensitivity in the spiny lobster, *Panulirus argus:* amino acids as feeding stimuli. Mar. Behav. Physiol. 5:145–157.

Kinnel, R.B., R.K. Dieter, J. Meinwald, D. Van Engen, J. Clardy, T. Eisner, M.O. Stallard, and W. Fenical. 1979. Brasilenyne and cis-dihydrorhodophytin: antifeedant medium-ring haloethers from a sea hare (*Aplysia brasiliana*). *Proc. Nat. Acad. Sci. USA* 76:3576–3579.

Kirchman, D., H. Graham, D. Reish, and R. Mitchell. 1982. Lectins may mediate in the settlement and metamorphosis of *Janua* (*Dexiospira*) *brasiliensis* Grube (Polychaeta: Spirorbidae [1982]). *Mar. Biol. Lett.* 3:131–142.

Kittredge, J.S., and F.T. Takahashi. 1972. The evolution of sex pheromone communication in the Arthropoda. *J. Theor. Biol.* 35:467–471.

Kittredge, J.S., M. Terry, and F.T. Takahashi. 1971. Sex pheromone activity of the molting hormone, crustecdysone, on male crabs. *Fish. Bull.* 69:337–343.

Kittredge, J.S., F.T. Takahashi, J. Lindsey, and R. Lasker. 1974. Chemical signals in the sea: marine allelochemics and evolution. *Fish. Bull.* 72:1–11.

Kusano, K., R. Miledi, and J. Stinnakre. 1982. Cholinergic and catecholaminergic receptors in the *Xenopus* oocyte membrane. *J. Physiol.* (*Lond.*) 328:143–170.

Larman, V.N. 1984. Protein extracts from some marine animals which promote barnacle settlement: possible relationship between a protein component of arthropod cuticle and actin. *Comp. Biochem. Physiol.* 77B:73–81.

Larman, V.N., P.A. Gabbott, and J. East. 1982. Physico-chemical properties of the settlement factor proteins from the barnacle *Balanus balanoides*. *Comp. Biochem. Physiol.* 72B:329–338.

Lenhoff, H.M. 1981. "Biology and Physical Chemistry of Feeding Response of Hydra." In *Biochemistry of Taste and Olfaction,* edited by R.H. Cagan and M.R. Kare. New York: Academic Press.

Lenhoff, H.M., and W. Heagy. 1977. Aquatic invertebrates: model systems for the study of receptor activation and evolution of receptor proteins. *Annu. Rev. Pharmacol. Toxicol.* 17:243–258.

Lenhoff, H.M., and K.J. Lindstedt. 1974. "Chemoreception in Aquatic Invertebrates with Special Emphasis on the Feeding Behavior of Coelenterates." In *Chemoreception in Marine Organisms,* edited by P.T. Grant and A.M. Mackie. New York: Academic Press.

Lentz, T.L. 1968. *Primitive Nervous Systems*. New Haven: Yale University Press.

Liley, N.R. 1982. Chemical communication in fish. *Can. J. Fish. Aquat. Sci.* 39:22–35.

Liley, N.R., and N.E. Stacey. 1983. "Hormones, Pheromones, and Reproductive Behavior in Fish." In *Fish Physiology,* vol. 9, edited by W.S. Hoar, D.J. Randall, and E.M. Donaldson. New York: Academic Press.

Lindstedt, K.J. 1971. Chemical control of feeding behavior. *Comp. Biochem. Physiol.* 39A:553–581.

Lucas, J.S., R.J. Hart, M.E. Howden, and R. Salathe. 1979. Saponins in eggs and larvae of *Acanthaster planci* (L.) (Asteroidea) as chemical defences against planktivorous fish. *J. Exp. Mar. Biol. Ecol.* 40:155–165.

Mackie, A.M. 1970. Avoidance reactions of marine invertebrates to either steroid glycosides of starfish or synthetic surface-active agents. *J. Exp. Mar. Biol. Ecol.* 5:63–69.

Mackie, A.M. 1973. The chemical basis of food detection in the lobster *Homarus gammarus*. *Mar. Biol.* 21:103–108.

Mackie, A.M. 1982. "Identification of the Gustatory Feeding Stimulants." In *Chemoreception in Fishes,* edited by T.J. Hara. Amsterdam: Elsevier Scientific Publishing Co.

Mackie, A.M., and J.W. Adron. 1978. Identification of inosine and inosine-5′-monophosphate as the gustatory feeding stimulants for the turbot, *Scophthalmus maximus*. *Comp. Biochem. Physiol.* 60A:79–83.

Mackie, A.M., and J.W. Adron. 1980. Chemical nature of feeding stimulants for the juvenile Dover sole *Solea solea* (L.). *J. Fish Biol.* 16:701–708.

Mackie, A.M., and P.T. Grant. 1974. "Interspecies and Intraspecies Communication by Marine Invertebrates." In *Chemoreception in Marine Organisms*, edited by P.T. Grant and A.M. Mackie. London: Academic Press.

Mackie, A.M., and A.I. Mitchell. 1982. Further studies on the chemical control of feeding behavior in the Dover sole, *Solea solea. Comp. Biochem. Physiol.* 73A:89–93.

Mackie, A.M., and A.I. Mitchell. 1983. Studies on the chemical nature of feeding stimulants for the juvenile European eel, *Anquilla anquilla* (L.). *J. Fish Biol.* 22:425–430.

Mackie, A.M., H.T. Singh, and J.M. Owen. 1977. Studies on the distribution, biosynthesis and function of steroidal saponins in echinoderms. *Comp. Biochem. Physiol.* 56B:9–14.

Manahan, D.T., S.H. Wright, and G.C. Stephens. 1983. Simultaneous determination of net uptake of 16 amino acids by a marine bivalve. *Am. J. Physiol.* 244:R832–R838.

Mato, J.M., B. Jastorff, M. Morr, and T.M. Konijn. 1978. A model for cyclic AMP-chemoreceptor interaction in *Dictyostelium discoideum. Biochem. Biophys. Acta* 544:309–314.

McGrath, S.M., and C.W. Sullivan. 1981. Community metabolism of adenylates by microheterotrophs from the Los Angeles harbor and Southern California coastal waters. *Mar. Biol.* 62:217–226.

Mopper, K., and P. Lindroth. 1982. Diel and depth variations in dissolved free amino acids and ammonium in the Baltic Sea determined by shipboard HPLC analysis. *Limnol. Oceanogr.* 27:336–347.

Morse, A.N.C., and D.E. Morse. 1984. Recruitment and metamorphosis of *Haliotis* larvae induced by chemicals uniquely available at the surfaces of crustose red algae. *J. Exp. Mar. Biol. Ecol.* 75:191–215.

Morse, D.E. 1984. "Biochemical Control of Larval Recruitment and Fouling." In Marine Biodeterioration: An Interdisciplinary Study, edited by J.D. Costlow and R.C.E. Tipper. Annapolis: Naval Institute Press.

Morse, D.E., N. Hooker, H. Duncan, and L. Jensen. 1979. γ-Aminobutyric acid, a neurotransmitter, induces planktonic abalone larvae to settle and begin metamorphosis. *Science* 204:407–410.

Muller, D.G., G. Gassmann, and K. Luning. 1979. Isolation of a spermatozoid-releasing substance from female gametophytes of *Laminaria digitata. Nature* 279:430–431.

Norris, J.N., and W. Fenical. 1982. "Chemical Defense in Tropical Marine Algae." In *The Atlantic Barrier Reef Ecosystem at Carrie Bow Bay, Belize. I. Structure and Communities*, edited by K. Rutzler and I.G. Macintyre. *Smithson. Contrib. Mar. Sci.* 12, Washington: Smithsonian Institution Press.

Ohsugi, T., I. Hidaka, and M. Ikeda. 1978. Taste receptor stimulation and feeding behavior in the puffer, *Fugu pardalis*. II. Effects produced by mixtures of constituents of clam extracts. *Chem. Sens. Flavour* 3:355–368.

Olsson, R.A., E.M. Khouri, J.L. Bedynek, Jr., and J. McLean. 1979. Coronary vasoactivity of adenosine in the conscious dog. *Circ. Res.* 45:468–478.

Ordal, G.W. 1985. Bacterial chemotaxis: biochemistry of behavior in a single cell. *CRC Crit. Rev. Microbiol.* 12:95–130.

Pfeiffer, W., G. Riegelbauer, G. Meier, and B. Schleiver. 1985. Effect of hypoxanthine-3(N)-oxide and hypoxanthine-1(N)-oxide on central nervous excitation of the black tetra *Gymnocorymbus ternetzi* (Characidae, Ostariophysi, Pisces) indicated by dorsal light response. *J. Chem. Ecol.* 11:507–523.

Phillis, J.W., and P.H. Wu. 1981. The role of adenosine and its nucleotides in central synaptic transmission. *Prog. Neurobiol.* 16:187–239.

Pierce, S.K. 1982. Invertebrate cell volume control mechanisms: a coordinated use of intracellular amino acids and inorganic ions as osmotic solute. *Biol. Bull.* 163:405–419.

Poulet, S.A., V. Martin-Jezequel, and R.N. Head. 1984. Distribution of dissolved free amino acids in the Ushant front region. *Mar. Ecol. Prog. Ser.* 18:49–55.

Rideout, J.A., N.B. Smith, and M.D. Sutherland. 1979. Chemical defense of crinoids by polyketide sulphates. *Experientia* 35:1273–1274.

Rittshof, D., and J. Bonaventura. 1986. Macromolecular cues in marine systems. *J. Chem. Ecol.* 12:1013–1023.

Ryan, E.P. 1966. Pheromone: evidence in a decapod crustacean. *Science* 151:340–341.

Scheuer, P.J. 1977. Chemical communication of marine invertebrates. *Bioscience* 27:664–668.

Sleeper, H.L., V.J. Paul, and W. Fenical. 1980. Alarm pheromones from the marine opisthobranch *Navanax inermis. J. Chem. Ecol.* 6:57–70.

Spencer, M., and J.F. Case. 1984. Exogenous ecdysteroids elicit low threshold sensory responses in spiny lobsters. *J. Exp. Zool.* 229:163–166.

Stabell, O.B. 1984. Homing and olfaction in salmonids: a critical review with special reference to the Atlantic salmon. *Biol. Rev.* 59:333–388.

Stoecker, D. 1980. Chemical defenses of ascidians against predators. *Ecology* 61:1327–1334.

Stone, T.W. 1981. Physiological roles for adenosine and adenosine 5′-triphosphate in the nervous system. *Neuroscience* 6:523–555.

Tachibana, K., M. Sakaitanai, and K. Nakanishi. 1984. Pavoninins: shark-repelling ichthyotoxins from the defense secretions of the Pacific sole. *Science* 226:703–705.

Thompson, T.E. 1983. Detection of epithelial acid secretions in marine mollusks: review of techniques, and new analytical methods. *Comp. Biochem. Physiol.* 74A:615–621.

van den Hurk, R., and J.G.D. Lambert. 1983. Ovarian steroid glucuronides function as sex pheromones for male zebra fish, *Brachydanio rerio*. Can. J. Zool. 61:2381–2387.

Van Haastert, P.J.M., B. Jastorff, J.E. Pinas, and T.M. Konijn. 1982. Analogs of cyclic AMP as chemoattractants and inhibitors of *Dictyostelium* chemotaxis. *J. Bacteriol.* 149:99–105.

von Frisch, K. 1941. Über einen Schreckstoff der Fischhaut und seine biologische Bedeutung. *Z. Vgl. Physiol.* 29:46–145.

Walker, R.P., J.E. Thompson, and D.J. Faulkner. 1980. Sesterterpenes from *Spongia idia. J. Org. Chem.* 45:4976–4979.

Weaver, N. 1983. "Pheromones and Behavior." In *Endocrinology of Insects*, edited by R.G.H. Downer and H. Laufer. New York: Alan R. Liss, Inc.

Whittaker, R.H. 1970. "The Biochemical Ecology of Higher Plants." In *Chemical Ecology*, edited by E. Sondheimer and J.B. Simeone. New York: Academic Press.

Whittaker, R.H., and P.P. Feeny 1971. Allelochemics: chemical interactions between species. *Science* 171:757–770.

Zimmer-Faust, R.K., W.C. Michel, J.E. Tyre, and J.F. Case. 1984. Chemical induction of feeding in California spiny lobster, *Panulirus interruptus* Randall: responses to molecular weight fractions of abalone. *J. Chem. Ecol.* 10:957–971.

Zuckerman, B.M., and H.B. Jansson. 1984. Nematode chemotaxis and possible mechanisms of host/prey recognition. *Annu. Rev. Phytopathol.* 22:95–113.

# Chapter 2

# Distribution of Chemical Stimuli

Jelle Atema*

## Introduction

### General Problem in Chemoreception

In the study of chemoreception, one encounters two principal difficulties that are unique to chemical stimuli. Stimulus qualities (i.e., the spectrum of all stimulatory chemical compounds) are not organized in one linear dimension such as the frequency spectra of mechanical and electromagnetic stimuli; and stimulus quantities (i.e., the distribution of the stimulus in space and time) cannot be expressed easily in mathematical formulations of stimulus dispersal comparable to the wave equation for radiating light and sound sources. Instead, the chemical spectrum of qualities consists of all reactive chemical compounds in the environment organized in what could be perceived as an unknown "*n*-dimensional hyperspace," and the distribution of chemical stimuli from their source of release through the environment depends on molecular diffusion and on fluid dynamics of the carrier medium. At the spatiotemporal scale of micro- to millimeters and seconds, diffusion dominates the distribution process; at various larger spatiotemporal scales, chemical stimuli are dispersed by fluid motion. This chapter will focus on the distribution of chemical stimuli and its consequences for chemoreceptive processes, such as receptor physiology and stimulus acquisition behavior.

Diffusion processes can be described statistically (Fick's law), leading to accurate estimates of dispersal (Carslaw and Jaeger 1959; Berg 1983; Futrelle 1984). The spatiotemporal scale of molecular processes in diffusion is so small and fast that statistical averaging is biologically allowable (Berg 1983; Berg and Purcell 1977). Fluid dynamic processes can also be described as time and space averages (Taylor 1921; Sutton 1953), but here such averaging is in conflict with the biologically relevant space-time scales (Elkinton and Cardé 1984): whenever the physical dispersal process takes place at about the same scale as sampling by chemoreceptor

*Boston University Marine Program, Marine Biological Laboratory, Woods Hole, MA 02543.

organs or organisms, instantaneous dispersal models are needed to predict responses. Motion of and within the fluid medium that contains the chemical stimulus is a complex process of mixing in which the energy injected into a part of the fluid mass (by some external agent, such as wind, convection, or animal behavior) is dissipated down a cascade of ever smaller eddies, until the smallest are wiped out by viscosity and dissipated in heat. Somewhere in the cascade, one can find the time-space scale appropriate for the chemoreceptive process being investigated: from the bacterial micrometer scale to the salmon's kilometer scale.

## Historic Perspective

In chemoreception the emphasis of research has been first on identifying chemical compounds that are stimulatory. Without such knowledge one cannot proceed. To date, many stimulatory compounds have been identified, ranging from simple and ubiquitous organic compounds (e.g., ammonium, glycine) to specially synthesized communication signals (e.g., pheromones) and proteins. For a review of aquatic stimuli see Carr (Chapter 1). Based on knowledge of the chemicals involved, subsequent investigations have addressed the cellular processes of stimulus recognition, including responses to mixtures (aquatic reviews in Caprio, Chapter 12 and Derby and Atema, Chapter 14) and coding of stimulus qualities in the CNS (review in Ache, Chapter 15). Characteristically, these studies include dose–response functions, which give a static representation of the response to different stimulus intensities. However, the far more important dynamic response properties have been investigated only recently (Voigt and Atema 1984). Besides the technical difficulties of stimulus delivery in precisely dosed temporal (and spatial) profiles, such studies are hampered by a nearly complete lack of knowledge of the natural dynamics of stimulus distributions and their critical features for chemoreception. Yet it is intuitively clear that chemoreceptor cells and organs are continuously faced with such dynamic stimulus properties as temporal changes in the composition of a signal mixture that must be measured against a fluctuating background of chemical noise.

Logically then, to understand chemoreception, one must describe in detail the spatiotemporal stimulus distributions that are relevant to the chemoreceptive process under investigation. Salient dynamic features of that distribution can then be abstracted and reproduced in the laboratory to measure biological responses of receptor cells and animal behavior. Once some of the quantitative relationships between stimulus patterns, receptor cell responses, and animal behavior have been established, selected stimulus-behavior correlations can be studied in the field. Focused studies of temporal and spatial stimulus characteristics matched to receptor and behavior characteristics will allow a more fundamental understanding of the intrinsic capabilities and limitations of chemoreception than is now available.

Studies of stimulus dispersal related to chemoreceptive responses are done more easily in the aquatic medium than in air, for a number of reasons. Most importantly, since water is so much denser than air, the same fluid dynamic processes take place on smaller space scales in water, thus allowing easier modeling (e.g., Vogel 1983). Furthermore, aquatic tracers of dispersal, such as dyes, are easier to measure than aerial tracers, such as ions. Finally, water-soluble organic stimulus compounds are relatively easy to measure and quantify at biologically relevant concentrations; this includes measurements of ambient chemical noise. Most of the principles derived from aquatic studies should be applicable to chemoreceptive processes where air is the stimulus carrier medium.

## Spatial and Temporal Frequency Filtering

Chemical stimuli[1] originate from a source, and they are removed by various physical, chemical, and biological processes. In between they become distributed by diffusion, laminar flow, and turbulence of various scales. If we were to fix this dynamic process at any moment, we would see the ''instantaneous'' chemical stimulus distribution. The spatial distribution pattern would appear as variously sized patches and plumes, clouds and currents, trails and trajectories. These terms refer to different geometric shapes resulting from distribution processes. I shall refer to them collectively as *patches,* since that is the established term in behavioral ecology and in oceanography and limnology. Patches of all shapes and sizes exist simultaneously, superimposed upon one another.

The patch size of interest to an organism determines how the larger and smaller scales will be viewed. Patches much smaller than the sampling volume used by the organism or its receptor organs will be imperceptible specks of noise (i.e., the chemical background). Much larger patches will be imperceptibly vast spaces (i.e., beyond the sensory and motoric activity sphere of the organism). For each organism the spatiotemporal scale of biological interest is bounded on the lower end (highest spatial frequencies) by the sampling volume and the adaptation/disadaptation time constants (flicker fusion) of its most phasic chemoreceptors and on the upper end (lowest spatial frequencies) by the adaptation rate (sensitivity to intensity ramps) of its most tonic receptors and the searching strategies and integration characteristics of the organism.

I assume that natural stimulus distribution patterns have been and are a driving force in ''shaping,'' over evolutionary time, the physiological properties of receptor cells and organs, as well as animal searching behavior. Great varieties of chemoreceptor organs exist (Laverack, Chapter 11) and—though phylogeny clearly is a major constraint—receptor organ morphology may well reflect to a yet unknown degree the spatial and temporal stimulus distributions that are characteristic for the microenvironments from which they must extract information useful for behavior (Atema 1985). The same can be said for physiological properties of receptor cells (e.g., tuning and adaptation) and for stimulus acquisition behavior (e.g., sniffing, search speed, and search patterns). This leads to the hypothesis

---

[1] In a true sense, chemical stimuli are not synonymous with stimulatory chemical compounds. To stimulate, a compound or a mixture must not only have the proper qualities to interact with a receptor cell, but it must also present itself to the receptor at a delivery rate to which the cell responds. In other words, the definition of a stimulus must include the temporal stimulus intensity profile. This is not commonly done, and I shall ignore this temporal aspect until a brief discussion of temporal frequency filtering in the section on turbulence and chemoreception.

that the *temporal* frequency filtering of the receptors is matched to those *spatial* frequencies of the stimulus that are important for the organism's competitive fitness. This hypothesis needs to be tested in chemoreception. Spectacular examples of stimulus-receptor-behavior frequency matching are known in vision, mechanoreception, and electroreception. Unfortunately, in chemoreception, little is known about the spatiotemporal characteristics of either the stimulus or the receptor, let alone their interactions and trade-offs. At the level of the whole organism, there is the additional possibility of spatial sampling of a stimulus distribution, either simultaneously with multiple dispersed receptors or by exploring a stimulus-containing space over a period of time. The latter has received attention in behavioral ecology where examples of gradient search and of foraging behavior among nutrient patches can be found. It is not unreasonable to expect that understanding the *temporal* sampling "techniques" used by various organisms to make estimates of *spatial* distributions may help us in developing practical measuring techniques for estimates of fluid dispersal and plume size and longevity.

In this chapter I discuss some of the physical and biological processes that distribute chemical stimuli and that lead to the spatial stimulus patterns that animals and their receptor organs have to deal with. I will then give some specific examples of stimulus distributions and biological sampling at different spatial and temporal scales. As much as possible, I chose examples in which a significant biological data base exists. It is convenient that sensory filtering results in sampling a limited physical range of stimulus distributions, so that in each example we can largely ignore the physical processes that take place simultaneously at scales several orders of magnitude larger and smaller. In addition, these physical processes are statistically uncoupled (Gregg and Cox 1972). By analogy when discussing the mechanics of human walking one may safely ignore the rotation of the earth and the movement of internal organs.

## Aquatic Stimulus Dispersal

It will be helpful to consider briefly the physical distribution processes in water and to give rough estimates of their size and time scales (Table 2.1). The scale of greatest potential importance for chemoreception can then be assessed from estimates of physiological and behavioral time scales of chemical sampling.

### Physical Transport Processes

Fluid motion in any body of water generally takes place at many different scales simultaneously. Depending on one's perspective and sampling methods, one can observe different ranges by ignoring larger or smaller scales or both. Any large-scale motion carries many small-scale motions within. At the largest scale are oceanic circulation patterns, mainly driven by wind, temperature, tides, and Coriolis forces. This flow fills ocean basins and is measured in space scales of $10^6$ m and in time scales of years. Temperature and density instabilities and solid

Table 2.1. Approximate Orders of Magnitude of Spatial and Temporal Frequencies of Stimuli and Biological Responses

| | Physical Oceanography (Scale) | Physical Stimulus Frequencies: Spatial (m) (Patch Size) | Physical Stimulus Frequencies: Temporal (s) (Patch Longevity) | Biological Response Frequencies: Temporal (s) (Sample Interval) | |
|---|---|---|---|---|---|
| Viscous domain | | $<10^{-6}$ | $<10^{-1}$ | $10^{-2}$–$10^{2}$ | Microorganisms, receptor physiology |
| | Micropatch | $10^{-6}$–$10^{-3}$ | $10^{-1}$–$10^{5}$ | | |
| | Microstructure | $10^{-3}$–10 | 1–$10^{5}$ (s–h) | 1–$10^{3}$ | Food search (macroscopic) |
| Inertial subrange | Fine Structure | 10–$10^{4}$ | $10^{5}$–$10^{6}$ (day–week) | Up to $10^{6}$ | Migration |
| | Meso (Eddy) | $10^{4}$–$10^{5}$ | $10^{6}$–$10^{7}$ (months) | | |
| | Large (Gyre) | $10^{5}$–$10^{6}$ | $10^{7}$–$10^{8}$ (years) | | |

obstacles create "wobble" in the large-scale flow, leading to eddy formation and turbulence. The largest known eddies are some $10^5$ m in horizontal diameter (typical North Atlantic warm core rings are 100–200 km wide), and their physical, chemical, and biological characteristics remain recognizable for periods of several months to over a year. By then their circulation has entrained much surrounding water, and its kinetic energy is transferred to smaller eddies with correspondingly shorter lifetimes. These phenomena constitute the oceanographic mesoscale. In turn, these secondary eddies generate smaller daughters, each time transferring mainly kinetic energy to a larger and larger population of smaller and smaller eddies. Eventually, at some small scale the size of which is dependent on the initial energy injected into the cascade, viscous forces become significant in the transfer of energy. This is where the so-called inertial subrange ends. Energy transfer in this range is theoretically characterized by exponential decay. Important for our purpose is the realization that, in the inertial subrange, mixing takes place only at the shearing edges of eddies and that eddy cores maintain their identity for time periods commensurate with their size (Table 2.1).

It is difficult to give an order-of-magnitude estimate for the eddy size at the lower end of the inertial subrange, because viscosity comes into play gradually and because each cascade starts with different amounts of energy. The more energy is injected into the system as a whole, the more inertial forces will dominate, leading in turn to finer-scaled turbulence. In addition, the concept of eddies at the small scale must be taken rather loosely: it "means no more than those parts of the motion that are coherent only over short distances" (Tritton 1977, p. 252). However, because of its potential importance to chemoreception, I shall

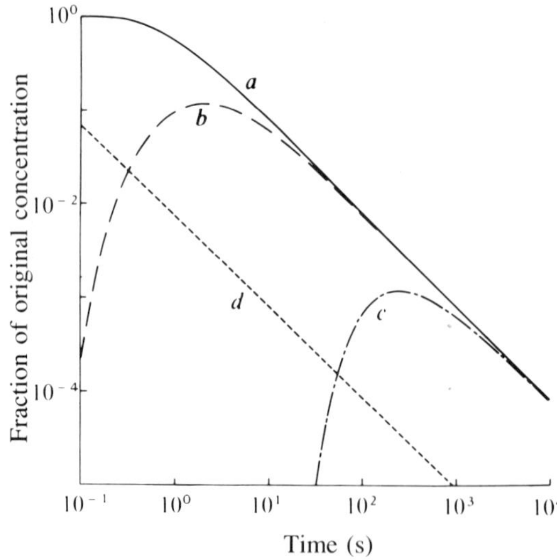


Figure 2.1. Concentration decay of a plume of metabolites as a function of time after a plume is released. The effect of a 100 μm wide plume, such as might be created by swimming copepod is given by *a* (at center of plume), *b* (100 μm radially from plume center), and *c* (1,000 μm from plume center). The effect of a 10 μm wide plume, such as might be created by swimming flagellate is given by *d*.

Decay curves are calculated from

$$\frac{C_t}{C_o} = \frac{1}{2}\left[erf\left(\frac{a-x}{2(D_t)^{1/2}}\right) + erf\left(\frac{a+x}{2(D_t)^{1/2}}\right)\right] \times erf\left(\frac{a}{2(D_t)^{1/2}}\right)$$

where $C_t/C_o$ = concentration at time $t$ after release as fraction of initial concentration $C_o$, $x$ = distance from center along center line, $a$ = half-width of initial plume, $D_t$ = $10^{-5}$ cm$^2$ s$^{-1}$, and *erf* is error function (see Carslaw and Jaeger 1959, p. 56). (From Jackson 1980; see original article for further details.)

try to set a practical limit on the effects of turbulence. Below $10^{-3}$ m, turbulence is essentially absent in most natural waters (Okubo 1980, p. 155; Lehman and Scavia 1982a). Only in extremely energetic environments such as waterfalls and surf zones will there be sufficient kinetic energy left to be significant below 1 mm. Furthermore, high-speed cinematography records of copepod feeding currents show that the flow extends about 8 mm and is essentially laminar (Strickler 1982). At this length scale the flow just begins to show signs of inertial instability (Strickler 1985). In practice this means that the feeding current is dominated by viscous forces, as a calculation of the Reynolds number at the flow maximum (Re = 0.75) confirms. Thus, in natural conditions, water will still be rather coherent even at $10^{-2}$ m. Viscosity will lose its grip above $10^{-1}$ m. Thus, the dynamic structure of $10^{-3}$–10 m, a very important range for many macroscopic animals, is determined by a mixture of viscous and inertial forces. The time scales connected with this range may be seconds to hours.

At the smaller ($<10^{-3}$ m) spatial scales, molecular diffusion operates very efficiently. Assuming the commonly used value for the diffusion constant ($D$ = $10^{-5}$ cm$^2$ s$^{-1}$) of small molecules in water at 20°C, Jackson (1980) calculated that the center of a chemical patch $10^{-4}$ m diameter would have diffused away to 0.1% of its original concentration within $10^3$ s. The same calculation for a $10^{-5}$ m diameter patch shows a drop to 0.1% in 10 s (Fig. 2.1). These very small patches are ephemeral phenomena, flickering in and out of existence in a matter of seconds. Locally increased viscosity, inversely proportional to molecular diffusion, and adsorption may be able to slow diffusion rates from suspended particles (Goldman 1984).

## Physiological and Behavioral Time Scales of Chemical Sampling

As seen above, physical distribution processes in the aquatic environment range in principle from molecular ($10^{-9}$ m) to oceanic scales ($10^6$ m). This large range can be broken down into a number of *physical* subranges, each one dominated by different physical transport processes. However, to determine the most important spatial stimulus dimensions, we might focus on the *biological* subranges that relate to chemoreception and animal behavior (Table 2.1). Fortunately, size scales and time scales of stimulus distributions are closely linked through the physical transport processes that lead to dissipation of inhomogeneities (i.e., large-scale bodies of water lose their identity slowly, small bodies lose it quickly). Thus, in order to establish the *spatial* frequencies of stimulus distributions that may be important for chemoreception, it will be of great help to consider the *temporal* frequencies over which chemoreceptors and animals sample.

I will use 10 ms as the theoretical lower time boundary of biologically relevant stimulus pulses. For lack of flicker fusion data in chemoreception this number is based on the following rationale: Berg and Purcell (1977) calculated that in this time an ideal measuring device could sample a spherical volume of 1-μm radius containing $10^{-6}$ M concentration of stimulus with 99% accuracy. On the commonly used logarithmic scale for stimulus intensity, stimulus concentrations of $10^{-6}$ M are in the middle of the known aquatic receptor range for important amino acid stimuli ($10^{-9}$–$10^{-3}$ M [Ache 1982]) and near the background concentration of some ($NH_4$ [Mopper and Lindroth 1982]). Actual measuring devices such as chemotactic bacteria approximate the ideal device by having many receptor sites evenly dispersed over their body surfaces (Berg and Purcell 1977). Receptor organs of macroscopic animals may still approximate this sampling ideal by creating a laminar flow over a large receptor membrane surface. In such organs, accuracy per unit sampling time would increase linearly with (the radius of) the sampling volume, but this gain may be offset by less than ideal odor access. Finally, Kelling and Halpern (1983) used well-measured salt pulses of 50-ms duration to which humans responded. Local diffusion near the tongue papillae may have caused suprathreshold concentrations to persist longer. Likewise, the rapid downstroke of a lobster antennule flick (i.e., one odor sniff) is on the order of 100 ms (personal observation).

Overall, 10 ms seems a reasonable estimate for a lower time limit on the temporal resolution of chemical stimulus pulses (i.e., the chemoreceptive flicker fusion frequency). This implies that stimulus concentration changes of shorter duration would not be resolved. Since, for equal accuracy, sampling time is inversely proportional to stimulus concentration, the same calculations (Berg and Purcell 1977) imply that stimulus concentrations less than $10^{-6}$ M would take a longer time to sample equally accurately, whereas higher concentrations can be sampled faster and/or more accurately.

For the higher physiological boundary I shall use 100 s. This is based on pyridine receptors in crayfish that, when stimulated with 10-s pulses, gave responses lasting up to 60 s (Hatt and Bauer 1982). However, the common observation in various laboratories including our own is that severe receptor cell adaptation takes place within a few seconds, probably starting within 1 s. Information similar to that obtained from lobster smell and taste receptors comes from crayfish amino acid taste receptors, in which 10-s stimulus pulses caused strong responses lasting about 1 s and adapting completely in a few seconds (Bauer, Dudel, and Hatt 1981). The temporal filtering properties of pyridine and amino acid receptor cell populations are clearly very different. In general, most physiologically relevant sampling times are probably not longer than 10–100 s. Bacteria (*Escherichia coli*) can detect gradients in 0.1–1 s, dependent on gradient steepness (Berg and Purcell 1977, p. 213). The center of this physiological range (i.e., about 1 s) is probably the temporal frequency filter window for many chemoreceptor cells. However, one must keep in mind that the critical experiments have not been done. The spatial stimulus frequency scales that correspond to this temporal receptor window are estimated in Table 2.1. Behavioral responses of microorganisms are considered within the range of receptor cell properties.

It is more difficult to establish behavioral sampling windows. Perhaps search persistence will be a useful guide. If we assume that an animal remains in the "search mode" after one stimulus encounter for 1–$10^3$ s, this implies that its behavior is geared to encountering the next stimulus "patch" within about 15 minutes at the higher end of the scale. This also implies that, if the next stimulus is not received within that time period, the animal loses interest and gives up its search. One might use search persistence ("give-up" time) to predict average stimulus patchiness for the animal. Such information can come from the foraging behavior literature (Kamil, Chapter 7). It is obvious that search persistence is influenced by many other factors, such as physiological state, social environment, and experience, which may obscure the correlation between interpatch distance and give-up time used in stimulus acquisition behavior. It is for the two reasons (i.e., variability between and within species) that I chose a time window for stimulus acquisition in search behavior of three orders of magnitude: 1–$10^3$ s. To set an upper limit I speculate that during migration, hormonal changes may cause the search for widely dispersed home odor patches to last for days ($10^6$ s) (Table 2.1). Such long-term processes are probably made up of a series of short-term mechanisms.

## Diffusion-Dominated Stimulus Dispersal

In this section I shall present examples of submillimeter scale stimulus distributions and their use by various organisms. Because these small-scale distributions are determined by molecular diffusion undisturbed by turbulence, the gradients are radially symmetrical and linear. That makes these distributions highly predictable for use in orientation. Unfortunately, these gradients have not been measured. We must infer their existence from diffusion theory and—perhaps, as I will argue—adaptative behavior and chemoreception of small organisms.

## "Micropatches"

Far from being the uniform mass of clear water we see when swimming in the ocean, every drop of it is in fact a teeming microcosm of chemical sources and sinks. In coastal euphotic parts of the ocean an average 1 ml of seawater may contain some $10^6$ bacteria, $10^3$ each of heterotrophic and phototrophic microflagellates, $10^3$ phytoplankton cells, and $10^{3-4}$ decomposing organic particles (Davis and Sieburth 1982; Azam 1984; Goldman 1984). In addition, there can be local blooms of invertebrate larvae with densities as high as three per milliliter (Jørgensen 1981). Many of the living organisms in this community are highly motile. Their metabolism makes each an active source and sink of organic matter. With efficient transport processes they concentrate organics from the surrounding water. Simultaneously, each organism passively leaks and actively excretes waste products. Thus, each one is surrounded by a cloud of changed chemistry, lower in nutrients and higher in waste products. Motile organisms are followed by a wake (Fig. 2.2) of various metabolites, among those organic compounds of low molecular weight (below 500 Daltons), such as ammonium, phosphates, amino acids, and monosaccharides. However, at this size scale, molecular diffusion keeps the clouds and wakes of these organic compounds very small (Figs. 2.1 and 2.2).

Whenever an organism dies, a source of proteins, polysaccharides, and other large organic compounds is created. Various authors estimate that the bulk of dissolved organic carbon (DOC) is in the range of 3,000–5,000 Daltons (Degens 1970) and that <10% of DOC is smaller than 500 Daltons (Billen et al. 1980) and thus directly available for bacterial uptake. Large molecules diffuse more slowly, allowing for longer-lasting gradients, which could aid chemotaxis. Various materials form microaggregates, which together with fecal pellets of zooplankton resemble a "marine snow" of decaying material and attached microbes. This snow acts not only as a source of organic release, but also as a sink: biological, chemical, and physical processes transport material to and from such centers of activity. When the water is turbid, inorganic clay particles provide active exchange surfaces. Even greater organic density is reached in benthic communities. In oligotrophic offshore regions the density of living and dead exchange surfaces is much lower. In all then, each drop of water contains hundreds of thousands of active exchange surfaces, creating an extremely heterogeneous environment of micropatches that we can imagine to flicker in and out of existence in time scales of $10^{-1}$–$10^5$ s (Table 2.1, Fig. 2.1). However, each patch contains a predictable diffusion gradient that can be used for chemical orientation.

Of importance to chemoreception are the precise space and time scales that define these micropatches. Unfortunately, they have never been measured. The

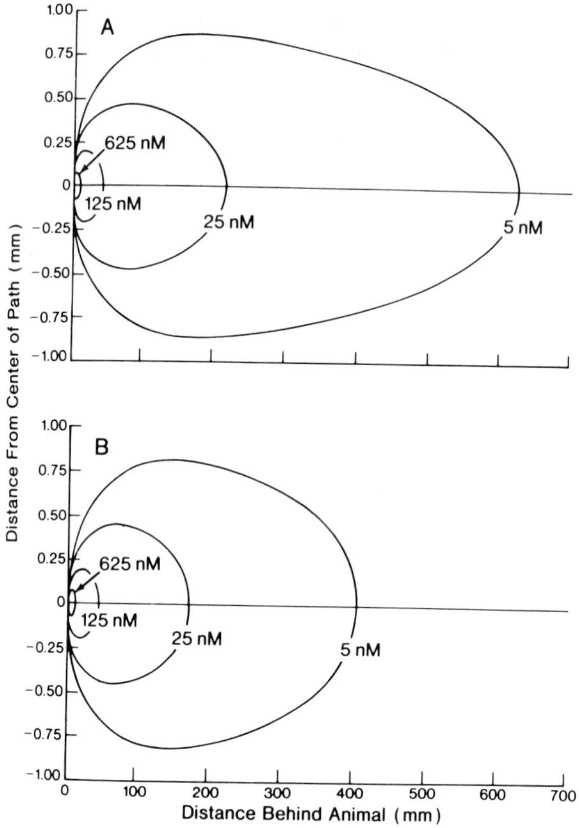


Figure 2.2 (A). Contour diagram for the steady-state plume of metabolites (numbers based on radiolabeled phosphorus release) behind swimming copepod *Daphnia pulex*, with body size of 2 mm. The source of release is swimming in the $-x$ direction with velocity u = 1.5 mm $s^{-1}$ and constantly releasing phosphorus at a rate $Q = 0.0618$ p mol $s^{-1}$.

Molecular diffusion plume is calculated from

$$\frac{\partial C}{\partial t} = -u\frac{\partial C}{\partial x} + D\left[\frac{1}{r}\frac{\partial}{\partial r}\left(r\frac{\partial C}{\partial r} + \frac{\partial^2 C}{\partial x^2}\right)\right]$$

with diffusion coefficient $D = 10^{-5}$ $cm^2$ $s^{-1}$, $t$ = time, and $C$ = concentration, and $C = (Q/4\ \pi\ Ds)\ e^{-u(s-x)/2D}$, where $s^2 = x^2 + r^2$ and $r$ is measured radially from the $x$-axis. (B) Same, but now including simultaneous phosphorus uptake by algal cells resulting in additional term to the equation above:

$$-\left[\frac{V_{max}\ C}{K_m + C}\right]N$$

where $N$ = number of algal cells $10^8$ $l^{-1}$, $V_{max}$ = maximum uptake rate, and $K_m$ = half saturation constant of membrane transport process. (From Lehman and Scavia 1982b; see original article for further details.)

most ephemeral would be the small constant release sources such as small molecules leaking from free-living bacteria. Diffusion will limit these bacterial wakes to micrometer scales. Nothing is known of the use of stimulus patches generated by single, free swimming bacteria. The size and longevity of these patches may be too small to be useful to any biological entity. However, bacteria themselves may exploit the larger scale patches generated by algal cells, ciliates, etc. Algal cells, in turn, may exploit the wakes of animals such as copepods. Also, bacterial clumps, microaggregates, and marine snow may attain sizes sufficient for production of detectable patches. Chemoreception—the ability to sense and respond to a local concentration change of a particular chemical compound—can be critical for competition in a patchy nutrient environment. It will be a challenge in the next few years to gain information on the dimension and composition of these very small chemical gradients. Meanwhile, we might estimate patch sizes from known behavior of microorganisms.

## Oceanic Bacteria and Nutrient Uptake

Currently there is a controversy in marine ecology in which one side argues that marine bacteria and algal cells exploit small patches of higher nutrient concentration to allow them to grow as efficiently as they do (McCarthy and Goldman 1979; Lehman and Scavia 1982a,b; Goldman 1984; Azam and Fuhrman 1984). The other side maintains that nutrient patches such as the ones produced by algal cells (for bacteria) and copepods (for algal cells) are too short-lived to be of any consequence (Jackson 1980; Williams and Muir 1981, Currie 1984). However, it is generally agreed that bacteria and algal cells grow well even in environments that, when sampled on a rather large scale such as the homogenized liters typical of oceanography, appear to some investigators too nutrient poor to sustain such growth. However, there is evidence from phosphorus uptake in freshwater phytoplankton that micropatches exist *and* are utilized (Lehman and Scavia 1982a,b, 1984). Theoretically, the organisms that can exploit patches would be at a competitive advantage in concentrating nutrients. For this, chemoreception and motility would be needed.

The average size of patches that may be important for bacteria can perhaps be estimated roughly from the known properties of bacterial motility and "chemotaxis." One assumes that different bacteria are well adapted to their specific environments. For instance, bacteria are small and offshore marine bacteria, in oligotrophic environments, tend to be smaller (0.2–0.6 μm in diameter) than coastal forms, which live in richer waters. The resulting increase in surface to volume ratio enhances uptake efficiency. In addition, they have high affinity binding and transmembrane transport systems to concentrate nutrients at low ambient concentrations: $K_t = 3 \times 10^{-8}$ M and $V_{max} = 10^{-12}$ mol ml$^{-1}$ g$^{-1}$ for amino acid uptake in marine bacteria. (Sepers 1977). Many marine bacteria are motile, an adaptation that may enhance their rate of encounter with ephemeral micropatches of higher nutrient concentration. Even purely random motility would help: it would increase their apparent size (area of exploration during the lifetime of a micropatch). Random walk models (Futrelle 1982; Berg 1983) show that small areas tend to be "explored" extensively at either molecular scales or bacterial scales. Marine bacteria move with very high velocities of up to 100 body lengths per second; typical values are 20–40 μm $s^{-1}$ (Azam and Ammerman 1984). In 30 s a roughly spherical patch of 1 nl = $(100\ \mu m)^3$ can be explored (Berg 1983, p. 80).

Finally, chemosensory control over motility would enhance uptake efficiency in two different situations. First, when an organism finds itself in an ephemeral patch of higher nutrient concentration, its chemoreceptivity would enable it to "stop" there for a moment to allow the membrane transport mechanisms to soak up the local nutrients. Without chemoreceptivity the organism would not know it was in an enriched spot and would continue its random motion pattern. The second situation is one of retarded diffusion and sustained chemical gradients. If gradients persist long enough, the organism can bias its random walk and move up the concentration gradient, guided by chemoreceptive mechanisms. This is the case of "bacterial chemotaxis" (Adler 1969; Berg and Purcell 1977). Not only the well-studied enteric bacteria, such as *E. coli,* but also various marine bacteria exhibit chemotactic behavior (Bell and Mitchell 1972). In their "run-and-tumble" behavior the average duration of a straight run is about 1 s. Behavior of *E. coli* in known chemical gradients shows that temporal comparisons of stimulus concentrations can be made in 1 s. If marine bacteria are similar in this respect, then they could measure a gradient over 20–40 μm. A 20–100 μm distance may reflect the order of magnitude of a typical marine patch size of interest to these organisms.

## Algal Gametes

On an order of magnitude larger in diameter than the marine bacteria, but still within the realm of predominantly diffusion-driven stimulus distributions, are a number of highly motile eukaryote microorganisms, such as small unicellular algae and heterotrophic flagellates. They are all in the size range of 1–10 μm. These organisms must be seen as surrounded by a small sphere of their metabolites. Their general problem is not only to obtain nutrients through molecular diffusion, but also to get rid of their metabolites so as to not suffocate in their own waste. At this small scale, diffusion is still efficient: if released in a pulse, the concentration of metabolites in the center of a plume left behind a motile 10-μm organism will drop to 1% of its initial concentration in 1 s (Fig. 2.1). More importantly, in the absence of turbulence at this size scale, the stimulus diffusion gradient is stable and thus predictable: it depends entirely on the output rate and quality of the source. For the purpose of chemical communication the source should be constantly releasing a unique (set of) compound(s). The opposite strategy should be followed if localization (by a predator) is to be avoided.

The so-called phycosphere (Bell and Mitchell 1972) is most clearly demonstrated in the pheromone release of female gametes of brown algae (Boland et al. 1982) and green algae (Maier and Müller 1986). The female gamete produces a pheromone that attracts male gametes at distances of up to 1 mm (Fig. 2.3B).

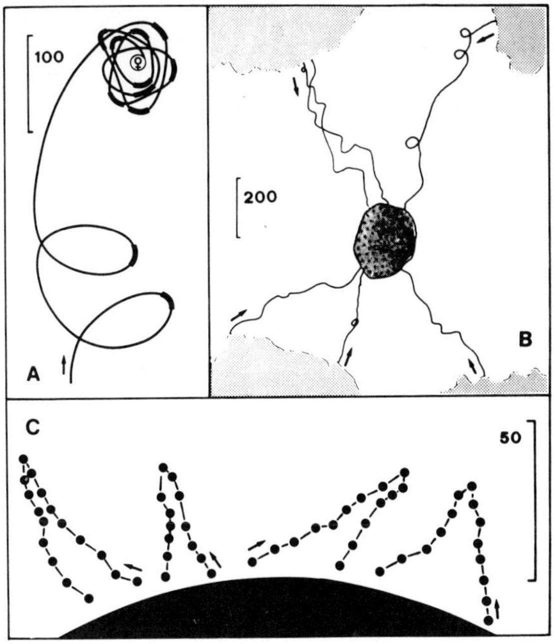


Figure 2.3. Tracks of male gametes approaching female gametes or artificial attractant sources. (A) Male gamete of *Ectocarpus siliculosus* in contact with coverslip (*below*) and in vicinity of female gamete (*above*). Both contact with solid object and detection of female attractant pheromone cause hind flagellum to bend, resulting in sharp turns (*broad sections of the track*). (B) Tracks of several freshly released spermatozoids from four male plants of *Laminaria digitata* (*corners*) approaching spherosil particle containing attractant pheromone lamoxirene (*center*). (C) Tracks of *Fucus spiralis* spermatozoids in the vicinity of a drop containing $10^{-5}$ M fucoserratene, female sex attractant. Time intervals between steps, 0.1 s. Scale units in micrometers. Derived from video recordings. (From Maier and Müller 1986.)

A very high signal-to-noise ratio is maintained by the lability of the pheromone, which is a volatile (these algae live near the water surface), nonpolar hydrocarbon that is easily photooxidated. Stimulus dispersal in volumes of 1 mm radius is dominated by diffusion even in the highly mixed surface layers of the ocean. This makes the localizing task easier for male gametes, since gradients are thought to be essentially radial. In laboratory assays the detection concentration threshold can be as low as 6.5 pM for *Cutleria* androgametes responding to multifidene $C_{11}H_{16}$. Thus, the pheromone concentration at the edge of the activity sphere is in the pM range. A concentration of 6.5 pM translates to an average of four molecules per nl. (A sphere with a radius of 62 μm contains 1 nl of water.) The androgametes are about 7 μm long with swimming speeds up to 350 μm $s^{-1}$. Contact with a surface or the pheromone slows them down immediately; in addition, at higher stimulus concentrations they swim in increasingly narrower curves (Fig. 2.3A). Slower and tightly curved swimming in response to increasing signal strength automatically leads to localization of the source of pheromone release. In some instances, gametes moving away from the source seem to respond, not to a gradient, but to a specific concentration of pheromones as they swim away from the source: at a critical distance they make a 180-degree turn that leads them to approach the female gamete (Fig. 2.3C). This constitutes an alternative search strategy that is efficient in a stable gradient.

The gamete examples illustrate that stimulus distribution at this size scale can be stable and predictable and that simple behavioral responses can be efficient. The demands of sexual communication and interspecific competition often result in specialized conditions of signaling and responding. Such exaggerations can serve as a model to understanding chemotactic responses of other motile organisms in the micrometer size range.

## Viscous Flow Fields

At size scales larger than 1–10 mm, diffusion is no longer an efficient transport process. Animals use currents to move chemical compounds for a variety of purposes, including chemoreception. The feeding current of calanoid copepods is a well-studied example (Strickler 1985). Other examples of viscous flow fields occur in and around chemoreceptor organs of larger animals, such as fish and large crustacea.

### Copepod Feeding Current

*Eucalanus crassus,* a calanoid copepod of about 3 mm body length, generates a current about 8 mm long with velocities of up to 10 mm $s^{-1}$ through the food capture area of 0.5 $mm^2$ (Strickler 1982). The Reynolds number of this current is 0.75 (Andrews 1983), and hence viscosity dominates the flow field, which is a double shear field shaped roughly like an hourglass. Algal cells that become entrained in this current can be captured. In addition, algal cells that would normally pass outside the capture area can be rerouted by altering the flow field. Rerouting demonstrates that algae can be perceived at some distance away. This distance can be as much as 1.25 mm, corresponding to 430 ms before capture (Strickler 1982). Based on these data, Andrews (1983) modeled the hypothesis that the flow field would deform the cloud of exudates surrounding the algal cell in a predictable fashion. The result (Fig. 2.4) shows that chemical stimuli identifying the alga would arrive well ahead of the algal cell, particularly in the area between 0 and 300 μm from the animal's body axis. This area is monitored by long antennae that—from morphological evidence—contain chemo- and mechanoreceptors. Both chemical and mechanical signals (Legier-Visser et al. 1986) could give advance warning of incoming food. Chemical signals in particular would provide good algal identity information that could be used in decisions to modify the flow field and draw only desirable algae into the capture area.

For a copepod, the spatial and temporal distribution of chemical stimuli is highly predictable. Since viscosity governs the flow field, the copepod (and not

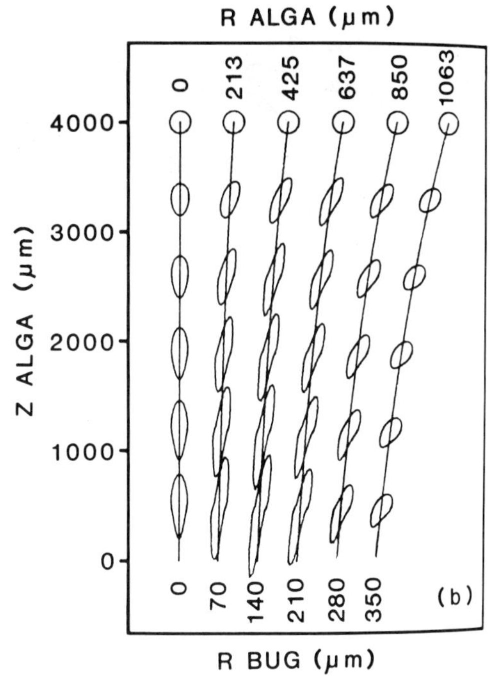


Figure 2.4. Model of "phycospheres" entrained in and predictably deformed by viscous feeding current of copepods. The plane of drawing is formed by copepod's longitudinal axis at $R = 0$ and one of its antennae at $Z = 0$. Flow direction from top to bottom. Six phycospheres (i.e., odor clouds surrounding single algal cells at their centers) are drawn at initial distance Z = 4,000 μm from copepod antenna. The antenna is extended at $Z = 0$ across the bottom of the figure. Streamlines connect algal centers. As the algal cells are "sucked" into the funnel-shaped feeding current, the odor clouds become deformed by the differential velocities of the shear field. The earliest warning of the impending arrival of an algal cell is at R bug = 140 μm, where suprathreshold odor concentration (i.e., the edge of the phycosphere) arrives about 500 μm ahead of the cell located in the cloud center. At $Z = 0$, current velocity may be 10 mm $s^{-1}$, giving the copepod 50 ms advance warning. *Z* alga = "anterior" distance between alga and antenna in flow direction. *R* alga = "lateral" distance between alga and copepod longitudinal axis along lines parallel to antenna. *R* bug = distance along antenna, located at $Z = O$. (Note gradually changing *R*-axes!) (From Andrews 1983.)

unpredictable turbulence) controls its shape. Larger flow fields would be more subject to disturbance by turbulence. In turbulent parts of the aquatic environment, smaller flow fields may be preferable. As Strickler (1985) points out, the animal and its flow field form one ecological unit, similar to a spider and its web, and under selection pressure the ability to control that field against local turbulence and to adapt to local food resources may well lead to niche separation and speciation. Chemoreception is well suited to identifying (preferred) food items (and other chemical sources) and to providing the information required for learning, but without a predictable flow field the animal could not reroute efficiently.

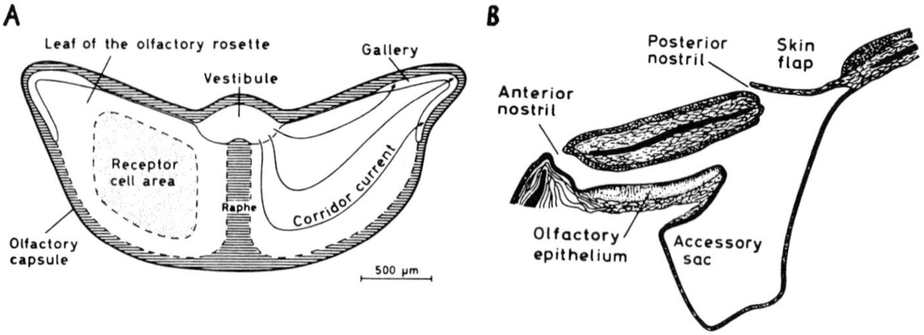


Figure 2.5. Olfactory organs of isosmate and cyclosmate fish. (A) Transverse section through the olfactory organ of an isosmate, the eel *Anguilla anguilla,* showing the vestibule, corridors, and the gallery which are compartments through which the water passes in sequential order, driven by cilia (from Døving and Thommesen 1977). (B) Sagittal section of the olfactory organ of a cyclosmate, a cyprinodont carp, with an accessory sac that pumps water (pump driven by jaw musculature) and a skin flap serving as unidirectional valve. (From Zeiske 1974.)

## Chemoreceptor Organs

Aquatic animals ranging in body size from $10^{-2}$ to 1 m have olfactory organs that range from $10^{-3}$ to $10^{-2}$ m. For instance, a 1 cm long mudsnail has an osphradium several millimeters long and a 1 m long eel has an olfactory organ about 15 mm long. This seems to indicate that there are size constraints on the design of a nose perhaps comparable to the size constraints on the vertebrate labyrinth. Olfactory organs show two main types of stimulus access: pulsed in sniffs or flicks, and continuous in laminar flow. Døving et al. (1977) referred to fish with the two types of noses as *cyclosmates* and *isosmates,* respectively (Fig. 2.5). The nose of isosmates pumps chemical stimuli in a steady, ciliary-driven flow over the olfactory epithelium. This flow has some similarities to the feeding current of copepods: it is a viscous, steady shear field under the animal's control. Surprisingly similar to the copepod feeding current, flow velocities of up to 15 mm $s^{-1}$ were measured by laser optics at the 0.06 $mm^2$ entry into the nose of a 60-mm fish (Kux, Zeiske, and Melinkat 1978). On entry this flow is divided into the "manifold" of the 1.5 mm long olfactory rosette. Between the lamellae of the rosette I estimate, from gross morphological measurements, the flow over the olfactory epithelium (maximum velocity of the shear field) to be 1–0.1 × the entry flow. The Reynolds number for fish nose flow may thus be similar to that of the copepod-feeding current. The similarity may point to functional constraints: in viscous flow, temporal patterns and intensity gradients are preserved, reflecting spatial stimulus patterns moving across the animal. This may allow it to infer spatial distributions of stimuli from temporal response patterns of receptor cells. Such spatial patterns may indicate the age of an odor plume and hence distance from the source, an important piece of information (see also the section on turbulence and chemoreception below). Laminar flow through the nose seems to be a common solution for slow-moving animals such as mud snails, catfish, eels, and dogfish.

In contrast, sniffing and flicking organs, such as the noses of tuna and lobster, sample a specified sniff volume. It is not known if, within this volume, spatial stimulus coherence is maintained. Sniffing in fast-moving fish may be advantageous in a turbulent environment. Phylogenetic constraints may be the determining factor in the flicking design of crustacean noses: the organ is not enclosed. But also here turbulent odor dispersal may be involved: lobsters deprived of one of their flicking antennules lost their ability to orient efficiently in a turbulently dispersed odor plume (Devine and Atema 1982). This will be discussed below in the section on orientation in a patchy odor field. In addition, lobsters can use a fan organ composed of the three exopodites of the maxillipeds to draw water towards themselves from an anterior hemisphere of radius $r = 5$ cm (Atema 1985). This slow flow is sampled by the 5 cm long antennules. Under calm conditions, not uncommon for this slow-moving animal, this flow appears nearly laminar, providing the lobster with predictable flow conditions in its olfactory sample area. Many other crustacea generate a variety of currents of yet unknown chemoreceptive function.

In general, physiological properties of olfactory receptor cells may be tuned not only to behaviorally important stimulus qualities, but also to prevalent and behaviorally important temporal frequencies of that stimulus. The olfactory CNS may show further enhancement of temporal processing, as is well known in other sense modalities. In contrast to olfaction, taste organs and other chemoreceptors appear to play behavioral roles that require only identification of stimulus quality, not its distribution in the environment. An exception is found in the external taste system of some catfishes, but here spatial, not temporal, analysis of spatial stimulus distributions is known to occur as reflected in body projections in the CNS. Similar spatial analysis is done by taste inside the mouths of fishes (Finger, Chapter 13).

## Turbulent Dispersal

A major problem for chemoreception appears with turbulent dispersal of odor. As a rule of thumb, aquatic systems are likely to be in turbulent motion at size scales larger than about $10^{-2}$ m. Although turbulence contains predictable patterns, for instance, when spatial or temporal averages are taken, the measuring devices and the behavioral constraints of animals may not always allow averaging. Different species have found different solutions to the problem of orientation in a turbulent odor field.

### Turbulence and Chemoreception

Turbulent flow is chaotic and is characterized by rapid, irregular fluctuations of fluid velocity in both space and time. Fluid patches follow highly irregular, distorted paths. Patches originating from the same source follow different paths, since the pattern of irregularities is changing all the time. These variations of the flow in time arise spontaneously even when all imposed conditions are held constant. Turbulence can therefore not be treated in a deterministic framework. Perhaps the same lack of deterministic predictability that forces us to describe turbulence in statistical terms has resulted in animal physiology and behavior patterns that utilize statistical features of turbulent flow fields. In detail, still very little is known about animal responses to turbulent flow.

One obvious feature of turbulent odor dispersal that animals may use is the onset rate of odor pulses. When moving through a turbulent odor cloud a point detector such as a nose will measure a spectrum of odor concentration changes representing its passing through patches; "typical" aquatic patches may pass by and last on the order of seconds (Atema 1985). The more recently the odor was released, the sharper the patch boundaries will be. Thus, an analysis of the rate of concentration change for each patch could provide information about nearness to the source. On the other hand, absolute stimulus concentration—actually, the signal-to-noise ratio of odor pulse peak height and local odor background—could give information on the strength of the source. Different size scales of ambient turbulence—which could be measured independently by mechanoreceptors—would modify odor distribution patterns. Over rough bottoms, for instance, eddies develop of the size scale of the bottom features. In general, each environment has a typical range of turbulent energy spectra. In order to use odor pulse shape, chemoreceptors with specific temporal response times are necessary. In general, the adaptation and disadaptation time courses of a receptor determine its frequency filter. As an organ, a nose is tuned to the frequency spectrum covered by the different temporal frequency filters of its receptor cells. Some odor concentration spectra in turbulent plumes have been measured in air (Murlis and Jones 1981) and in water (Atema 1985).

The simplest way to establish some rules of turbulence is to apply time averaging and to ignore local eddies at small spatiotemporal scales. The same approach is used in describing molecular diffusion in which the random walk behavior of individual molecules is ignored. In the case of a constantly emitting tracer source in various wind conditions and terrain roughness, Sutton (1953) assumed that tracer dispersal along any axis perpendicular to the mean wind direction would show a Gaussian distribution. The Sutton dispersion model was the basis for Bossert and Wilson's (1963) subsequent theoretical work on pheromone dispersal. A good review and critique of time-averaged dispersion models is found in Elkinton and Cardé (1984). Some animals such as the tsetse fly (see below) appear to average over some time (Bursell 1984). However, field tests with gypsy moths show that these animals do not behave according to the assumptions of time-averaging models: the moths are tracking constantly changing local conditions (Cardé 1984; David et al. 1982). In the case of the moth, instantaneous odor plume models are better predictors of behavior, as will be described in more detail below.

## Gypsy Moth and Tsetse Fly

Two examples from the terrestrial environment will serve to point out some solutions to the problem of orientation in a turbulent odor field. The male gypsy

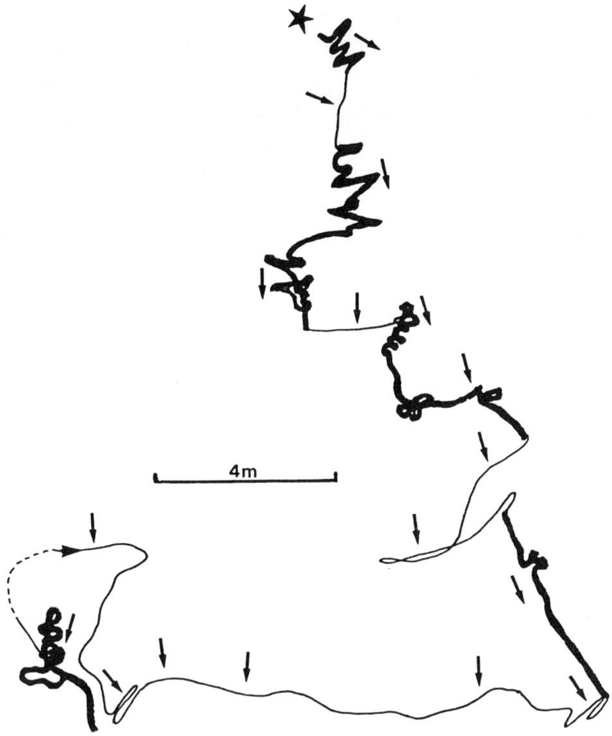


Figure 2.6. Orientation in turbulent odor field. Flight path of male gypsy moth toward a source of female attractant pheromone continuously released at the *star*. Average wind direction is from top to bottom of the figure, but the turbulence causes local wind direction to vary from place to place and from time to time. *Arrows* indicate local wind direction (visualized with soap bubbles) at the time of moth passage. The track is drawn thick when the moth is in a patch of odor. (Courtesy of C. T. David.)

moth responds in flight to the presence of female sex pheromone by orienting up the local wind, using visual information of ground drift to measure wind direction. Turbulence-induced changes in wind direction are measured and followed accurately, heading into the wind at all times as long as pheromone is detected. When contact with the pheromone is lost (i.e., the moth has flown out of a pheromone patch), the animal begins to fly across the local wind, casting about in ever larger zigzags. As soon as he enters another pheromone patch, the moth changes back to upwind flight. For a slow-flying animal this is an efficient search strategy, which leads in the majority of cases to successful localization of the source (David et al. 1982) (Fig. 2.6).

In contrast, the tsetse fly employs a strategy of determining average wind direction when it encounters a host odor (Bursell 1984). Perching allows the animal to use mechanoreceptors to measure average wind direction. After a certain integration time, this fast-flying insect takes off in the direction of the average wind, ignoring local wind shifts. After flying some distance it alights on vegetation again. If it overshot its goal and has lost contact with the odor, it will perch, measure wind direction, and then fly straight down the average wind direction. Usually after a few overshooting trials, the fly locates its host by flying up the average wind when odor patches are present and—if still in search behavior—down the average wind when patches are absent.

These examples illustrate contrasting methods of extracting information on source location from turbulent odor plumes. Perhaps its great flying speed allows the fly to use the average wind method, which is not practical for the slow-flying gypsy moth. It should be possible in the future to fill in aquatic examples.

Animals that live on or near the bottom can measure current direction and thus use either of the two insect methods described above. Which one is more efficient may depend on their swimming or walking speed. However, far away from the bottom it is more difficult to measure passive drift and current direction. Theoretically it can be measured from magnetically induced electric fields (elasmobranchs, catfish [Kalmijn, Chapter 6]), from magnetic fields (tuna [Walker et al. 1984]), and from lift generated by forward velocity across hydrofoils such as the pectoral fins of sharks and tuna, as suggested to me by Robert Olson (personal communication). None of these current meters appears well suited to quick determination of large-scale average current direction. Thus, despite great swimming speeds, oceanic fish may use local current search in gypsy moth fashion or ignore current altogether and perform area-restricted random walk. Any search will be restricted and thus facilitated by vertical stratification of the water column due to limited vertical mixing. This important feature of the aquatic environment might be a source of information in all cases of open ocean orientation, because it constrains odor dispersal (see the section below on oceanic fine structure and pancake theory of fish orientation).

## Orientation in Patchy Odor Field

Turbulent odor dispersal in water does appear to contain directional cues. In laboratory experiments lobsters were able to determine nearly perfectly the source direction of a very slowly moving, patchy odor plume. They lost this ability when one of the two flicking antennules was removed, causing their initial left-right direction choice to change from near-perfect to random (Devine and Atema 1982). Both flicking behavior and the presence of a dense tuft of chemoreceptor sensilla turned out to be critical: even when the major receptor sensilla of the flicking (lateral) antennule were shaved off, the lobsters still retained some orientation ability (66% correct). When one of the two nonflicking (medial) antennules was removed, orientation remained unaffected. The dense tuft of sensilla seems to be a morphological design to reduce odor access until the antennule is flicked: sensillum spacing is such that viscosity prevents flow between the sensilla. During a high-velocity flick, water flows between the sensilla and odor can be sampled from the local environment. Thus, the lobster smells in discrete temporal (and spatial) samples as shown physiologically by Schmitt and Ache (1979). It is not known how the lobster uses this discrete sampling method to infer the spatial distribution of odor in a turbulent plume.

Olfactory sampling by cyclosmate fishes, which pump pulses of water over their receptor epithelium, resembles the lobster's flicking method. Tuna show sharp, sudden jaw-snapping movements when entering odor plumes (Atema, Holland, and Ikehara 1980); jaw movement activates the pump that draws water over the nose of cyclosmate fishes (Fig. 2.5) (Zeiske 1974; Melinkat 1982). It is tempting to speculate that such animals use this "sniffing" to extract spatial information from a turbulent odor cloud, analogous to the lobster.

## Oceanic Fine Structure and Pancake Theory of Fish Orientation

It is well known that oceanic basins and lakes are vertically stratified in density layers resembling stacks of pancakes: the horizontal dimensions of a layer are some 100 times their vertical dimensions (1–10 m). These water masses vary in density, caused largely by temperature and salinity differences. Horizontal layers of this scale exist for days to weeks before losing their identiy in mixing processes. The layers have different current directions and velocities and thus "slide" over one another separated by thin shear layers of $10^{-2}$–$10^{-1}$ m scales. Depending on their origin and history, the pancake layers often also contain different populations of biological species and organic material (Fig. 2.7). Water masses of different origin, such as low-salinity, river input, or cold upwelling water, can be traced for long distances and times, even when broken up into patches by large-scale turbulence. Columbia River water can be traced in the northeast Pacific Ocean for hundreds of miles. The general location of traceable patches varies seasonally with large and mesoscale circulation patterns (Duxbury, Morse, and McGary 1966).

It is not difficult to see that patterns of vertical structure in aquatic basins could allow efficient sampling: a vertical dive of $10^2$ m through a stratified ocean could permit a discriminating nose to sample features of $10^5$ m in horizontal scale. Westerberg (1984) proposed just such a scheme for salmon homing. In tracking experiments with sonically tagged salmon, he and others (Døving, Westerberg, and Johnsen 1985) observed regular dives through several fine- and microstructure layers. The fish seemed to select certain depths. Such controlled diving behavior was absent in fish in which olfaction was eliminated experimentally. Similar vertical dives have been observed in other large fish (Carey and Robinson 1981). In theory, at least, fish can sample several layers in one vertical dive, choose the one that smells right, and swim horizontally in a selected "pancake" until it ends, then sample again by vertical dives, etc. This behavior would have certain features in common with the odor patch search strategy of gypsy moths. However, gypsy moths can easily measure wind direction by visual tracking of their ground speed. This is not possible for oceanic fish. However, an additional feature for fish is the possibility of detecting current shear in the thin layers separating the pancakes (Fig. 2.7). Detection of shear would allow the animal to measure the edge of a pancake and the relative motion of the two pancakes. Together with a coarse sun and/or magnetic compass this may be sufficient information for long-distance orientation at the mesoscale. Westerberg's "pancake

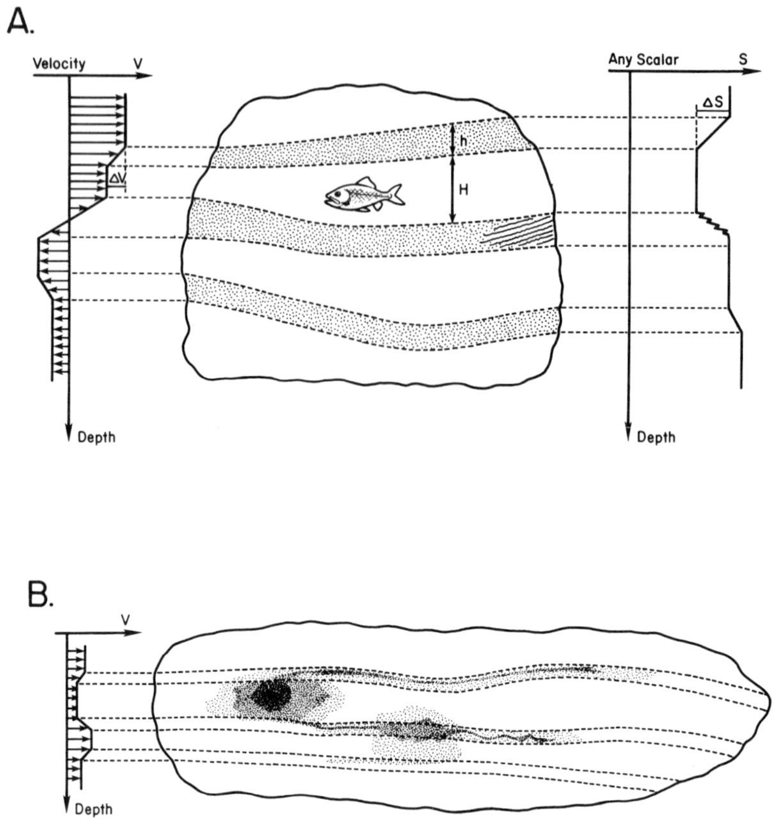


Figure 2.7 (A) Schematic representation of the hydrographic nearfield of mid-water fish. The horizontal scale is compressed; the thickness of the shear layers ($h$) is exaggerated. In a shear layer, the velocity profile changes ($\Delta V$). The water between shear layers forms a "pancake" (with thickness $H$) of relatively homogeneous velocity ($V$), as well as density, temperature, and chemical composition, expressed as "any scalar" ($S$). Shear layers separate layers of different density and composition. (B) Schematic section through an odor cloud spreading from a source located in homogeneous fine-structure layer. The horizontal scale is compressed. (From Westerberg 1984.)

theory of salmon orientation" is an attractive, but so far unproven, hypothesis. Its general principles could well apply at different size scales wherever vertical stratification exists at scales relevant to the animal in question.

Many animals are faced with the problem of orientation in a turbulent odor field. How and even if they use directional information is not known. For instance, can hyperiid shrimp extract directional information from the turbulent 1-m wake of their salp host to locate them (Diebel et al. 1985)? Can female lobsters use the turbulent 1–10 m odor plumes generated by "advertising" males to locate their shelters during courtship behavior (Atema 1986)? Can tuna determine directional clues in the turbulent $10^2$–$10^3$-m clouds of metabolites left behind traveling schools of their anchovy prey? Measurements of odor dispersal in aquatic environments will be needed at several different spatiotemporal scales. Together with observations on animal behavior and measurements of temporal filter characteristics of chemoreceptors, this will vastly improve our understanding of the chemical stimulus world.

## Information Currents

Thus far I have discussed the detection of chemical stimulus distributions. In this section I shall mention briefly the active dispersal of odor for communication purposes. Various larger crustacea produce substantial currents for locomotion, metabolic exchange, and chemical information. Information currents have been described in the hermit crab *Pagurus* (Brock 1926) and lobster *Homarus americanus* (Atema 1985). Although little studied, fish information currents probably exist, for instance, in fanning in a nest. In general, current is needed to broadcast a chemical stimulus at scales larger than those governed by molecular diffusion (e.g., gametes).

### Lobster Information Currents

Lobsters have three major current-generating organs. Dye visualization shows that all three currents are turbulent. The scaphognathite produces the gill current, which jets out of two nozzles and projects in mature animals up to seven body lengths forward. The Reynolds number at the nozzle is about 1,500. The gill is in part an excretory organ (Burger 1957), and thus the gill current contains waste products that can be detected (at least in principle) by other organisms. In addition, urine with socially important chemical stimuli (Atema and Cowan 1986) can be released into this current from two small nozzles just above the gill current exits. The gill current is injected into an external current regime that continues the dispersal of odor.

As mentioned earlier, the exopodites of the three maxillipeds form a fan organ that can deflect the gill current and direct it backwards over the animal. Simultaneously water is drawn into this exopodite current from an anterior hemisphere with radius approximately the length of the antennules. While the gill current must proceed continuously to avoid suffocation, the exopodite current is under the animal's control. Together they form a chemical information system by which body odors can be broadcast and environmental chemistry sampled by the antennules.

The third current is generated by the pleopods under the abdomen. It is the most powerful one of the three. In settling-stage lobsters this current serves to propel the animal forward. In older lobsters it gradually loses this function and serves, among other things, to move water through the shelter. In mature males this current may have a special function in advertising mating status. Pleopod fanning at the shelter entrance is most pronounced during courtship cohabitation and is correlated with female attraction to the male shelter (Atema 1986). This large odor cloud contains information on successful courtship in progress. The gill current and pleopod current are sketched in Fig. 2.8.

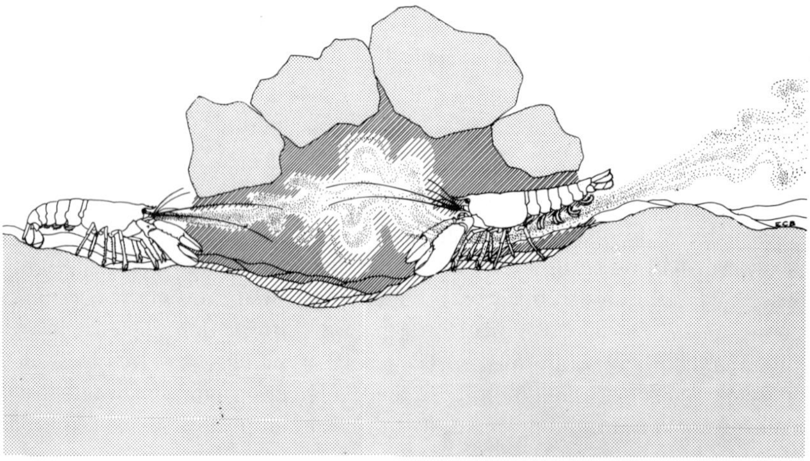

Figure 2.8. Chemical information currents. To distribute chemical signals at larger space scales (>1 mm) animals use currents. Here, in lobster courtship, three different currents are used. The female at left blows her gill current (in which she can inject pheromones) into male shelter. The male at right draws water through his shelter and fans it out into the environment, thus advertising his mating status. Ocean currents further distribute this signal. (From Atema 1986)

For lobster courtship the precise spatial distribution of the signal contained in the gill current is probably unimportant. In courtship, there is time to gather information. What matters here is accurate determination of signal quality, to be certain that this is the right mating partner. However, lobster parasites that must locate the source of release of lobster odor may need to gather information on plume shape to orient efficiently.

## Conclusion and Summary

Physical descriptions of odor dispersal show that predictable patterns exist in molecular diffusion gradients and in viscous flow fields (at length scales below $10^{-2}$ m). When flow becomes turbulent (generally at length scales somewhere above $10^{-3}$–$10^{-1}$ m), odor distribution patterns are no longer predictable except when averaged over time. Body size determines if an organism lives in a viscous, and thus predictable, environment or in a turbulent and rather unpredictable environment. Very small organisms, such as bacteria, flagellates, and gametes, can determine linear and radial spatial odor gradients from temporal sampling. Intermediate-sized organisms such as copepods can generate and control a viscous flow field and sample the predictable odor distribution patterns entrained in the flow. Larger organisms live in turbulence-dominated, and thus patchy, odor environments, where additional sensory information is required to improve search efficiency. This may be visual, mechanoreceptive, or electromagnetic information.

Depending, among other constraints, on its locomotor capabilities, such a larger animal can employ time averaging or, more commonly, develop other behavioral rules that improve stimulus acquisition behavior under the specific conditions that prevail for the environment of that animal through its evolutionary history. Flicking and sniffing may enhance spatiotemporal resolution in turbulent odor fields.

Olfactory chemoreceptor organs of larger animals operate at the scale of millimeter-sized animals. This allows such organs to utilize or even to control a viscous flow field. Theoretically this would result in temporal stimulus patterns that reflect accurately the spatial stimulus distribution immediately surrounding the animal. In general, temporal sampling by receptors could be used to estimate spatial stimulus distributions.

The distribution of chemical stimuli in the environment must be a powerful constraint on the evolution of chemoreceptor cells, chemoreceptor organs, and stimulus acquisition behavior. Very little is known about the critical features of odor distribution patterns and their relation to chemoreceptor physiology and animal behavior. Consequently, there is much to be gained from the study of the physics of the stimulus matched to the biology of the receptors.

*Acknowledgments.* I thank Adrianus Kalmijn for twenty-five years of scientific inspiration and enthusiasm. I thank Farooq Azam, Bruce Bryant, Cheryl Ann Butman, Joel Goldman, John Hobbie, Peter Johnsen, Donal Manahan, and Rudi Strickler for discussions, insights, and critical review of the manuscript. This chapter was written during a period of financial support by the Whitehall Foundation and the United States National Science Foundation, BNS-8512585.

## References

Ache, B.W. (1982) Chemoreception and thermoreception, in The Biology of Crustacea, vol. 3, Neurobiology: Structure and Function, Atwood, H.L., and Sandeman, D.C. (eds.), Academic Press, New York, pp. 369–393.

Adler, J. (1969) Chemoreceptors in bacteria, Science, 166:1588–1597.

Ammerman, J.W., and Azam, F. (1981) Dissolved cyclic adenosine monophosphate (cAMP) in the sea and uptake of cAMP by marine bacteria, Mar. Ecol. Prog. Ser., 5:85–89.

Andrews, J.C. (1983) Deformation of the active space in the low Reynolds number feeding current of calanoid copepods, Can. J. Fish. Aquat. Sci., 40:1293–1302.

Atema, J. (1985) Chemoreception in the sea: adaptations of chemoreceptors and behavior to aquatic stimulus conditions, Soc. Exp. Biol. Symp., 39:387–423.

Atema, J. (1986) Review of sexual selection and chemical communication in the lobster, *Homarus americanus,* Can. J. Fish. Aquat. Sci., 43:2283–2290.

Atema, J., and Cowan, D.F. (1986) Sex-identifying urine and molt signals in lobster (*Homarus americanus*), J. Chem. Ecol., 12:2065–2080.

Atema, J., Holland, K., and Ikehara, W. (1980) Olfactory responses of yellowfin tuna (*Thunnus albacares*) to prey odors: chemical search image, J. Chem. Ecol., 6:457–465.

Azam, F. (1984) Marine microbiology: recent advances in concepts and methodology, in International Training Course: Chemoreception in Aquatic Organisms, Shanghai, People's Republic of China, pp. 240–248.

Azam, F., and Ammerman, J.W. (1984) Cycling of organic matter by bacterioplankton in pelagic marine ecosystems; microenvironmental considerations, in Flows of Energy and Materials in Marine Ecosystems, Fashion, M.J.R. (ed.), Plenum Press, New York, pp. 345–360.

Azam, F., and Fuhrman, J.A. (1984) Measurement of bacterioplankton growth in the sea and its regulation by environmental conditions, in Heterotrophic Activity in the Sea, Hobbie, J.E., and Williams, P.J.L. (eds.), Plenum Press, New York, pp. 179–196.

Bauer, U., Dudel, J., and Hatt, H. (1981) Characteristics of single chemoreceptive units sensitive to amino acids and related substances in the crayfish leg, J. Comp. Physiol., 144:67–74.

Bell, W., and Mitchell, R. (1972) Chemotactic and growth responses of marine bacteria to algal extracellular products, Biol. Bull., 143:265–277.

Berg, H.C. (1983) Random Walks in Biology, Princeton University Press, Princeton, N.J.

Berg, H.C., and Purcell, E.M. (1977) Physics of chemoreception, Biophys. J., 20:193–219.

Billen, G., Joiris, C., Wynant, J., and Gillain, G. (1980) Concentration and microbiological utilization of small organic molecules in the Scheldt Estuary, the Belgian coastal zone of the North Sea and the English Channel, Estuarine Coastal Mar. Sci., 11:279–294.

Boland, W., Terlinden, R., Jaenicke, L., and Müller, D.G. (1982) Binding mechanism and sensitivity in gamete chemotaxis of the phaeophyte *Cutleria multifida*, Eur. J. Biochem., 125:173–179.

Bossert, W.H. and Wilson, E.O. (1963) The analysis of olfactory communication among animals. J. Theoret. Biol., 5:443–469.

Brock, F. (1926) Das Verhalten des Einsiedlerkrebses *Pagurus arrosor* Herbst während der Suche und Aufnahme der Nahrung, Z. Morphol. Oekol. Tiere, 6:415–552.

Burger, J.W. (1957) The general form of excretion in the lobster *Homarus*, Biol. Bull., 113:207–223.

Bursell, E. (1984) Observations on the orientation of tsetse flies (*Glossina pallidipes*) to wind-borne odours, Physiol. Entomol., 9:133–137.

Cardé, R.T. (1984) Chemo-orientation in flying insects, in Chemical Ecology of Insects, Bell, W.J. and Cardé, R.J. (eds.) Elsevier, Amsterdam, pp. 109–134.

Carey, F.G., and Robinson, B.H. (1981) Daily patterns in the activities of swordfish, *Xiphias gladius*, observed by acoustic telemetry, Fish. Bull., 79:277–292.

Carslaw, H.S., and Jaeger, J.C. (1959) Conduction of Heat in Solids, Oxford University Press, Oxford, U.K.

Currie, D. (1984) Microscale nutrient patches: do they matter to the phytoplankton? Limnol. Oceanogr., 29:211–214.

David, C.T., Kennedy, J.S., Ludlow, A.R., Perry, J.N., and Wall, C. (1982) A reappraisal of insect flight towards a distant point source of wind-borne odor, J. Chem. Ecol., 8:1207–1215.

Davis, P.G., and Sieburth, J.M. (1982) Differentiation of the phototrophic and heterotrophic nanoplankton populations in marine waters by epifluorescence microscopy, Ann. Inst. Oceanogr. (Paris), 58:249–260.

Degens, E.T. (1970) Molecular nature of nitrogenous compounds in sea water and recent marine sediments, in Organic Matter in Natural Waters, Occasional Publ. #1, Hoot, D. (ed.), Inst. Mar. Sci., Alaska, pp. 77–106.

Devine, D., and Atema, J. (1982) Function of chemoreceptor organs in spatial orientation of the lobster, *Homarus americanus:* differences and overlap, Biol. Bull., 163:144–153.

Diebel, C.E., Stolzenbach, K., Madin, L.P. and Atema, J. (1985) Multisensory trail detection in oceanic amphipods? Biol. Bull., 169:533–534.

Døving, K.B., Dubois-Dauphin, M., Holley, A., and Jourdan, F. (1977) Functional anatomy in the olfactory organ of fish and the ciliary mechanism of water transport, Acta Zool. (Stockh.), 58:245–255.

Døving, K.B., and Thommesen, G. (1977) Some properties of the fish olfactory system, in Olfaction and Taste VI, LeMagnen, J., and MacLeod, P. (eds.), Information Retrieval, London, pp. 175–183.

Døving, K.B., Westerberg, H., and Johnsen, P.B. (1985) Role of olfaction in the behavioral and neuronal responses of Atlantic salmon, *Salmo salar,* to hydrographic stratification, Can. J. Fish. Aquat. Sci., 42:1658–1667.

Duxbury, A.C., Morse, B.A., and McGary, N. (1966) The Columbia River effluent and its distribution at sea, 1961–63, U. of Washington, Dept. Oceanogr. Tech. Rept. No. 156 (ref. M66–31), Seattle.

Elkinton, J.S., and Cardé, R.T. (1984) Odor dispersion, in Chemical Ecology of Insects, Bell, W.J., and Cardé, R.T. (eds.), Sinauer Associates, Sunderland, Mass., pp. 73–91.

Futrelle, R.P. (1982) Dictyostelium chemotactic response to spatial and temporal gradients: theories of the limits of chemotactic sensitivity and of pseudochemotaxis, J. Cell. Biochem., 18:197–212.

Futrelle, R.P. (1984) How molecules get to their detectors: the physics of diffusion of insect pheromones, TINS, April: 116–120.

Goldman, J.C. (1984) Conceptual role for microaggregates in pelagic waters, Bull. Mar. Sci., 35:462–476.

Gregg, M.C. (1980) Microstructure patches in the thermocline, J. Phys. Oceanogr., 10:915–943.

Gregg, M.C., and Cox, C.S. (1972) The vertical microstructure of temperature and salinity, Deep-Sea Res. Oceanogr. Abstr., 19:355–376.

Hatt, H., and Bauer, U. (1982) Electrophysiological properties of pyridine receptors in the crayfish walking leg, J. Comp. Physiol., 148:221–224.

Jackson, G.A. (1980) Phytoplankton growth and zooplankton grazing in oligotrophic oceans, Nature, 284:439–441.

Jørgensen, C.B. (1981) Mortality, growth, and grazing impact of a cohort of bivalve larvae, *Mytilus edulis* L., Ophelia, 20:185–192.

Kelling, S.T., and Halpern, B.P. (1983) Taste flashes: reaction times, intensity, and quality, Science, 219:412–414.

Kux, J., Zeiske, E., and Melinkat, R. (1978) Measurement of flow velocity at the nares of the olfactory organ of fishes with a laser velocimeter, Laser & Electro-Optik, 1:21.

Legier-Visser, M.F., Mitchell, J.G., Okubo, A., and Fuhrman, J.A. (1986) Mechanoreception in calanoid copepods: a mechanism for prey detection, Mar. Biol., 90:529–535.

Lehman, J.T., and Scavia, D. (1982a) Microscale patchiness of nutrients in plankton communities, Science, 216:729–730.

Lehman, J.T., and Scavia, D. (1982b) Microscale nutrient patches produced by zooplankton, Proc. Nat. Acad. Sci. USA, 79:5001–5005.

Lehman, J.T., and Scavia, D. (1984) Measuring the ecological significance of microscale nutrient patches, Limnol. Oceanogr., 29:214–216.

Maier, I., and Müller, D.G. (1986) Sexual pheromones in algae, Biol. Bull., 170:145–175.

McCarthy, J.J., and Goldman, J.C. (1979) Nitrogenous nutrition of marine phytoplankton in nutrient-depleted waters, Science, 203:670–672.

Melinkat, R. (1982) Funktionsmorphologische Untersuchungen zur Ventilationsmechanik des Geruchsorgans bei ditremen Ährenfischartigen (Pisces, Atheriniformes), thesis, U. of Hamburg, Hamburg, W. Germany.

Mopper, K., and Lindroth, P. (1982) Diel and depth variations in dissolved free amino acids and ammonium in the Baltic Sea determined by shipboard HPLC analysis, Limnol. Oceanogr., 27:336–347.

Murlis, J., and Jones, C.D. (1981) Fine-scale structure of odour plumes in relation to insect orientation to distant pheromone and other attractant sources, Physiol. Entomol., 6:71–86.

Okubo, A. (1980) Diffusion and Ecological Problems: Mathematical Models, Springer-Verlag, New York.

Schmitt, B.C., and Ache, B.W. (1979) Olfaction: response enhancement by flicking in a decapod crustacean, Science, 205:204–206.

Sepers, A.B.J. (1977) The utilization of dissolved organic compounds in aquatic environments. Hydrobiologia 52:39–54.

Strickler, J.R. (1982) Calanoid copepods, feeding currents and the role of gravity, Science, 218:158–160.

Strickler, J.R. (1985) Feeding currents in calanoid copepods: two new hypotheses, Soc. Exp. Biol. Symp., 39:459–485.

Sutton, O.G. (1953) Micrometeorology, McGraw-Hill, New York.

Taylor, G. I. (1921) Diffusion by continuous movements, Proc. Lond. Math. Soc. Ser. 2, 10:196–212.

Tritton, D.J. (1977) Physical Fluid Dynamics, Van Nostrand Reinhold, U.K.

Vogel, S. (1981) Life in Moving Fluids: The Physical Biology of Flow, W. Grant Press, Boston.

Vogel, S. (1983) How much air passes through a silkmoth's antenna? J. Insect Physiol., 29:597–602.

Voigt, R., and Atema, J. (1984) Chemoreceptor adaptation: responses of glutamate cells to repeated stimulation, Biol. Bull., 167:534.

Walker, M.M., Kirschvink, J.L., Chang, S.B.R., and Dizon, A.E. (1984) A candidate magnetic sense organ in the yellowfin tuna, *Thunnus albacares,* Science, 224:751–753.

Westerberg, H. (1984) The orientation of fish and the vertical stratification at fine- and micro-structure scales, in Mechanisms of Migration in Fishes, McCleave, J.D., Arnold, G.P., Dodson, J.J., and Neill, W.H. (eds.), Plenum Press New York pp. 179–203.

Williams, P.J.L., and Muir, L.R. (1981) Diffusion as a constraint on the biological importance of microzones in the sea, in Ecohydrodynamics, Nihoul, J.C.J. (ed.), Elsevier, Amsterdam, pp. 209–218.

Zeiske, E. (1974) Morphologische und morphometrische Untersuchungen am Geruchsorgan oviparer Zahnkarpfen (Pisces), Z. Morphol, Tiere, 77:19–50.

# Chapter 3

# Light and Vision in the Aquatic Environment

John N. Lythgoe*

The conditions for vision underwater are more exacting than on land. As the depth increases, the daylight gets progressively less bright, and that which remains comes mostly from above and is restricted to a narrow and variable band of the spectrum. Deeper than about 1,000 m in even the clearest water, there is not enough daylight for vision, and animals must produce their own light. At all depths, light is more strongly scattered than is usual on land, and it is principally for this reason that it would be unusual indeed to be able to see farther than 100 m through the water. Both air and water scatter, refract, and absorb light; but the effects are quantitively much greater in water, and it is often possible to recognize visual adaptations to the particular conditions for vision underwater, and several of these will be considered in this chapter. However, an animal is concerned with the real problems of finding food, finding mates, and avoiding getting eaten, and several optical problems may be involved at any one time. If we are to understand how an animal is adapted to the real world around it, an attempt must be made to take a more integrated view of the problems it encounters in its visual environment.

This chapter begins by describing some of the main optical characteristics of water that affect the visual performance and adaptations of aquatic animals. These include the transfer of useful information from the visual scene to the observer and point to the importance of contrast, as well as light intensity, to an image-forming system like the eye. In the dim light that is the norm underwater, visual performance is limited because there are very few photons available to form an image, and by looking at visual tasks in terms of the quantal nature of light, it is possible to understand some of the inherent visual limitations and the compromises that have to be made in the perceptions of contrast, movement, detail, and color by aquatic animals. ''Noise'' is one of the most important factors that limit the performance of any sensory system, and this is abundantly true for underwater vision. In vision, noise is introduced both by the light-scattering

*Department of Zoology, Bristol University, Bristol, BS8 IUG, U.K.

nature of the aquatic medium and by the physiological mechanisms of the eye itself. In recent years, it has become possible to measure the photosensitive pigments within individual photoreceptors, and this has allowed us to look at the relationship between the spectral quality of the visual scene and the spectral sensitivity of the photoreceptors that receive light from it. There is little doubt that the visual pigments that an animal possesses are related to the color of the water in which it lives, but it is likely that the enhancement of sensitivity is not the only important factor. Color vision itself is more difficult to study, because it requires a high degree of neural processing of the raw data provided by the photoreceptors. Many aquatic animals that live in shallow or moderate depths do possess color vision, however, and for many, it is likely to be an important aspect of their visual sense.

## Light in the Sea

### Spectral Transmission of Water

The deep oceans far from land owe their color to the optical properties of the water itself, and a deep blue signifies only that there is little phytoplankton to contribute the green of chlorophyll and few yellow products of organic decay. Nutrient-poor inshore Mediterranean and Caribbean water is also blue, and this can be true as well—though rarely—of oligotrophic freshwater lakes. Freshwaters are more usually green, yellow-green, or brown, depending on the amount of phytoplankton and dissolved products of vegetable decay that they contain.

Because each body of natural water owes its color to a particular combination of chlorophyll-containing plankton, the yellow products of vegetable decay, and the blue of the pure water itself, it is possible to classify ocean water according to its color. This was first attempted, and with success, by Jerlov in 1951, and he refined it in 1976. The bio-optical basis for Jerlov's classification was worked out by Baker and Smith (1982), using the diffuse spectral attenuation of light (K) of pure water, phytoplankton assayed as chlorophyll, and dissolved organic matter (Fig. 3.1). The variation in water color can be very large; whereas the clearest water is a deep blue, transmitting most light around 475 nm, landlocked freshwater, which is replenished by water that has filtered through peat beds or forest detritus, may be brown or almost red and transmits most strongly in the near infrared (Muntz and Mouat 1984).

### The Scattering of Light

The most important factor that limits underwater visibility in shallow water is the scattering, rather than absorption, of light. The effect is like looking through fog. Non-image-forming light scattered by suspended particles between the visual

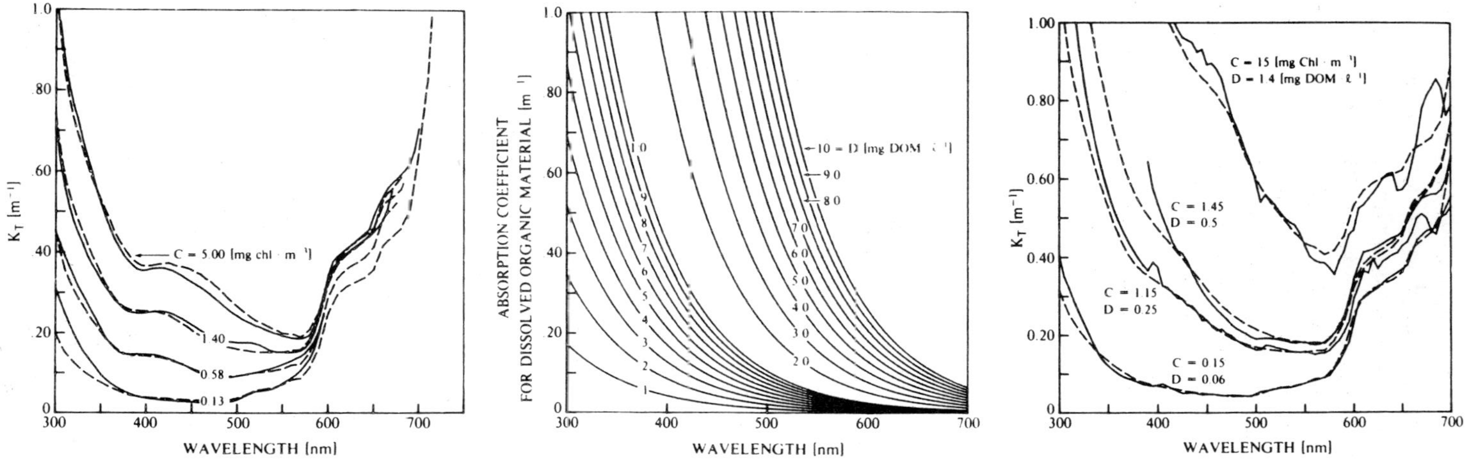


Figure 3.1. Optical classification of ocean water (Baker and Smith 1982). (A) Spectral diffuse attenuation ($K_T$) of ocean water that contains varying amounts of chlorophyll-containing plankton, but in which there is negligible amount of dissolved organic matter. At lowest chlorophyll concentrations, spectral attenuation is close to that of pure water. Solid lines represent actual field measurements; interrupted ones, calculated values of $K_T$. (B) Spectral attenuation coefficient of various concentrations of dissolved organic matter ("yellow substances," or *gelbstoffe*). (C) Spectral attenuation coefficient ($K_T$) of various ocean waters containing various concentrations of chlorophyll and dissolved organic matter.

scene and the eye creates a veil of brightness that reduces contrasts and hence reduces visibility. If there is little absorption of light and much scatter, as in fog, light comes almost equally from all directions. When absorption is more significant than scatter, light comes only from the direction of the source and hardly at all from other directions. There are anecdotes of divers' being unable to tell which direction is upward in turbid inshore waters, because it looks equally bright in all directions. In situ measurements of radiance show that, below about 200 m in the clear oceans, 98% of the daylight comes from above and only 2% from below (Tyler and Smith 1970; Smith 1974; Jerlov 1976). The difference in intensity between upwelling and downwelling light has a marked influence on the construction and position of the eyes of mesopelagic animals, with an even more marked effect on the distribution of photophores (Herring 1977).

## Surface Refractive Effects

From just under the surface to a depth of perhaps 50 m in the clearest water, the distribution (Fig. 3.2) and quality of light are strongly influenced by its refraction at the water surface, as described by Snell's equations (Jerlov 1976). On the rare day when the surface is completely flat, an aquatic animal looking up towards it can see the scene above the water, but it is all contained within the circular Snell's window, which has a radius of 48.5 degrees. An object seen directly overhead, perhaps a flying bird, will suffer little distortion, but objects near the horizon will be compressed and outlined by color fringes resulting from the refractive dispersion of light at the air–water interface. At angles greater than 48.5 degrees, there is total internal reflection from the underside of the water surface, meaning that an image of the bottom or the light reflected from deep water below is reflected in the surface. Looking out horizontally, the animal will see the unobstructed horizontal background space light; looking downwards, it will see directly into deeper water or the seabed, if sufficiently near. When the water is calm, the edge of Snell's window is sharply defined, and there is a strong contrast between the bright scene from above the surface and the much darker reflections of the bottom and deep water. It is against this sharp discontinuity of brightness that transparent animals are most apt to be visible, since they generally have a slightly different refractive index from that of water and will appear as slightly darker or brighter than their background (Lythgoe 1979). Indeed, there is some anatomical evidence that plankton-feeding fishes have an area of enhanced acuity on the part of the retina where the edge of Snell's window falls (Munk 1970). The much greater brightness of the image falling on the ventral part of the retina compared to that on the dorsal one may also lead to problems from scattered light's degrading contrasts of the image of the scene below the fish focussed on the dorsal retina. Some species such as *Pantodon* and the Osteoglossidae have a septum derived from the falciform process that runs horizontally across the retina and may be useful in optically isolating the upper and lower hemispheres of the eye (Saidel and Braford in press).

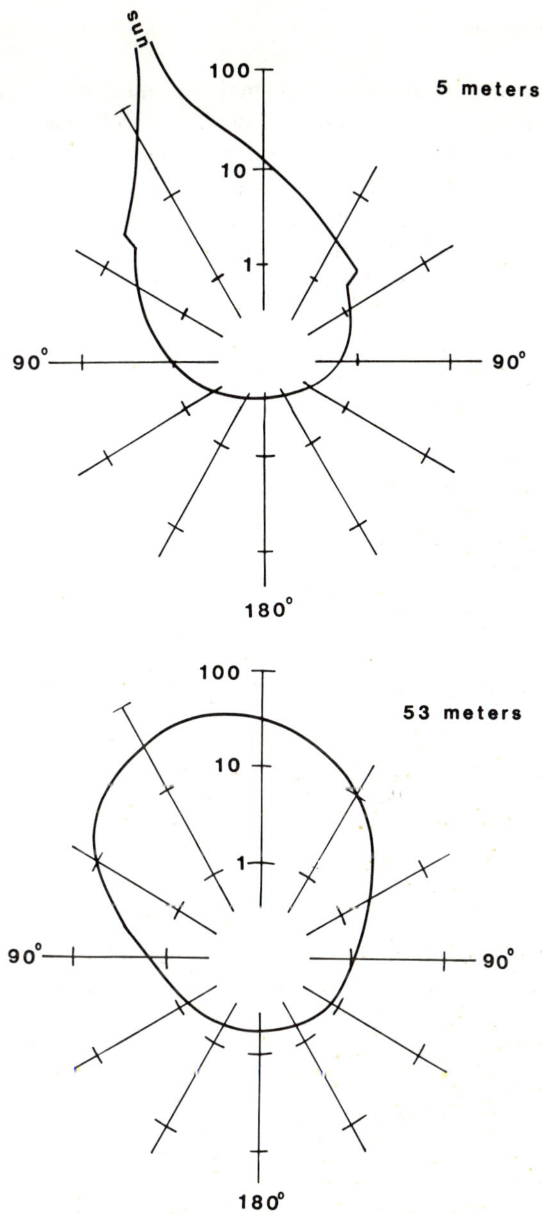


Figure 3.2. Angular distribution of light in open ocean water at depths of 5 m and 53 m. Scales on radii indicate relative radiance. At 5 m the direction of the sun is clearly visible, and the edge of Snell's window is marked by shoulders on radiance distribution curve at 48.5 degrees. At 53 m, influences of Snell's window and the sun are scarcely apparent, and there will be little difference in radiance distribution at greater depths. Measurements were made on a clear day. The wind was gusting to 25 knots, and waves were 1–1.5 m high. Spectral sensitivity of the instrument centered about 497 nm was approximately the wavelength of greatest water transparency (data from Smith 1974).

## Radiance Distribution

As the depth increases, the complex distribution of underwater light settles down into that shown in Fig. 3.2. Eventually a depth is reached at which the angular distribution of light does not change significantly with depth. In the clearest water this asymptotic state is reached around 100 m, whereas in the Baltic Sea it is reached at less than half that depth (Jerlov 1976). It would be difficult to exaggerate the importance of the angular distribution of light in the sea, since it influences both the techniques for camouflage used by marine animals and the visual adaptations employed by their predators to break it.

On the seabed, the principles of camouflage are those which have been so ably described by Cott (1940) for terrestrial animals. In mid-water, however, two further techniques of radiance matching are used. The first is mirror camouflage; the second, downward-directed bioluminescence. A flat vertical mirror suspended in mid-water is almost invisible, because the spacelight reflected in it is almost the same as the background spacelight. Pelagic fishes often have mirrorlike silvery flanks, but these are not flat, having instead the rounded contours of the fish. Each of the tiny reflecting guanine crystals in the scales is orientated vertically, rather than parallel to the surface of the flank; in this way, the whole fish behaves like a vertical mirror (Denton and Nicol 1965, 1966). There are also many silvery-sided fishes that live close to the bottom, and Ward (1919) has suggested that the purpose of the silveriness is to reflect features of the bottom and thus to achieve camouflage.

Mirror camouflage is made more effective by some countershading produced by the disposition of darker chromatophores on the dorsal surfaces, but no amount of mirror camouflage or countershading can be effective when the fish is viewed in silhouette directly against the brightest part of the spacelight field. Many mesopelagic fishes have upwardly directed eyes specifically designed to detect their prey when it is silhouetted against the bright downwelling light (for a review, see Herring 1977; Marshall 1979). The solution adopted by many mesopelagic animals, including fishes, crustacea, and cephalopods, is to arrange of downwardly directed photophores along the lower surfaces of the animal in such a way that the radiance of the photophores matches that of the downwelling light in intensity, color, and angular distribution (Denton and Nicol 1966). The fine tuning of this camouflaging device and the techniques used by predators to see through it are remarkable. For example, Warner, Latz, and Case (1979) have shown that the mesopelagic shrimp *Sergestes similis* changes the intensity of the downwardly directed light from the photophores to match the radiance of the downwelling daylight, thus adjusting for the natural changes in the brightness of the downwelling light that follow from the time of day or from clouds passing across the sun. In fact, the spectral distribution of the bioluminescence is only an approximate match to the downwelling daylight, especially at longer wavelengths. Muntz (1976) has suggested that the yellow lenses in the eyes of some mesopelagic fishes, such as *Argyropelecus affinis* and *Scopelarchus analis,* may be designed to pass most light in the wave band where the camouflaging bioluminescence of their prey is brighter than the downwelling daylight. This technique will make the bioluminescence of the prey species brighter than the background, and the camouflage will be broken.

## Image Transmission

The scatter and the absorption of light are the two main factors that limit the ability of aquatic animals to see through the water. Scattered light reduces the contrast of the image on the retina, and absorption reduces its brightness. With a purely physical approach it is possible to calculate the apparent contrasts of different elements of the image on the retina and, since we have some idea of the contrast perception of the eye, come to some realistic estimate of the distance a particular object might be seen in a particular underwater situation. This approach was pioneered by Duntley (1951, 1963), who used Middleton's (1952) equations for vision through the atmosphere. However, the biologist has to take into account the mechanisms of vision, and some of the ways that the physiological mechanisms of the eye are adapted to interpreting the retinal image are discussed in later sections. In particular, it is impossible to consider in isolation from each other such questions as contrast perception, acuity, the visibility of moving objects, and color vision.

The daylight that penetrates down through the water becomes progressively less bright and more restricted in spectral bandwidth with increasing depth. The rate at which this occurs and the spectral distribution of the light depend on the water itself (Fig. 3.3). Equation (3.1) describes the exponential reduction in the spectral intensity of daylight with depth:

$$I_d = Ie^{-Kd} \tag{3.1}$$

where $I$ = irradiance, $d$ = depth, and $K$ = diffuse attenuation coefficient. Irradiance is measured as the light falling on a flat surface and includes both light travelling a direct path from the surface and light that has been scattered from its original path by particles in the water. Where light is scattered from particles that are similar in size to the wavelength of light, scatter increases with the fourth power of the wavelength (Rayleigh scatter), whereas scatter from larger particles is about the same at all wavelengths (Mie scatter). The effect of this is that the upwelling light in very clear ocean water contains more deep-blue shortwave light than does the downwelling light (Tyler and Smith 1970).

In the situation envisioned in Figure 3.4, the light that travels the direct path from an object to the observer contains little scattered light because the light that is scattered is diverted from the direct path and does not enter the eye. The equation that describes this situation is

$$R_z = Re^{-\alpha z} \tag{3.2}$$

where $R$ = radiance, $z$ = visual range and $\alpha$ = narrow beam attenuation coefficient. The forms of equations (3.1) and (3.2) are similar, but the attenuation of light with path length is greater for radiance than irradiance. The inherent optical properties of pure water and the dissolved and suspended matter that are always

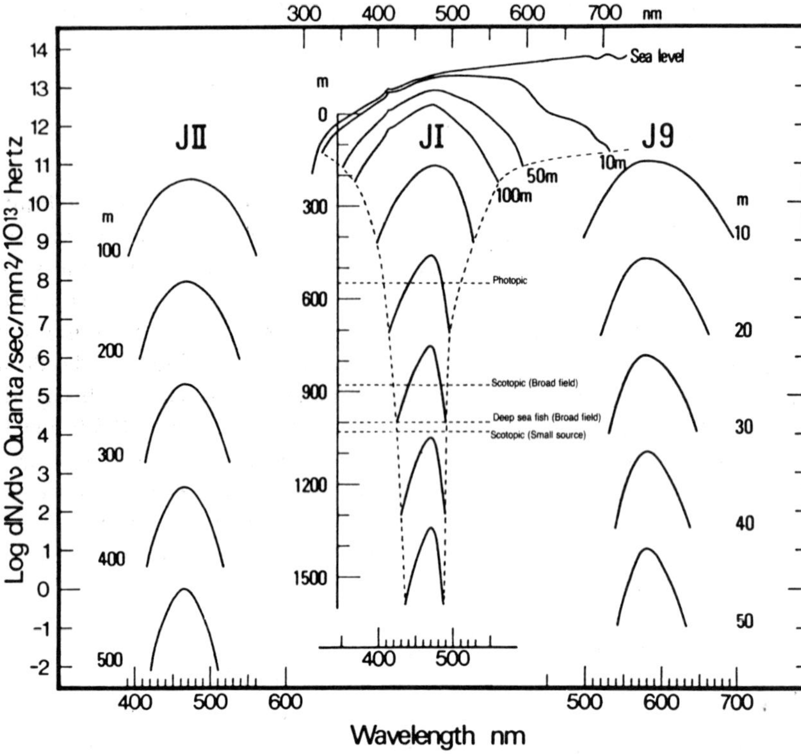


Figure 3.3. Spectral irradiance at various depths in two types of ocean water (Jerlov types II and I) and coastal water (Jerlov type 9). The irradiance is calculated in quanta (photons). These units are chosen to be directly relevant to estimates of visual sensitivity, and all curves are drawn in absolute relation to each other. Consequently, limits of vision indicated by interrupted lines drawn for Jerlov I water relate directly to other two water types by horizontal extension. Note how the waveband available for vision narrows rapidly in more shallow water and then settles down to steady asymptotic state as the depth increases. Radiance measurements on which these calculations were based (Jerlov 1968) were made for downwelling light and relate to fishes looking upwards (Dartnall 1975). Fishes looking downwards will have insufficient light, seeing approximately 200 m less deep (in clearest water) than when looking upwards.

present in natural water can be measured separately in the laboratory, and some values have been tabulated by Jerlov (1976).

If attenuation of image-forming light in its passage from the object to the observer were the only factor, the object would appear to get darker as the visual range increases, but the image contrast would not change. In fact, the apparent radiance of the object tends to equal that of the background space light as the visual range increases (Duntley 1963) because the background space light is derived from the (chiefly downwelling) daylight that is scattered by the water in the direction of the eye. In effect, this scattered "veiling" light is a source

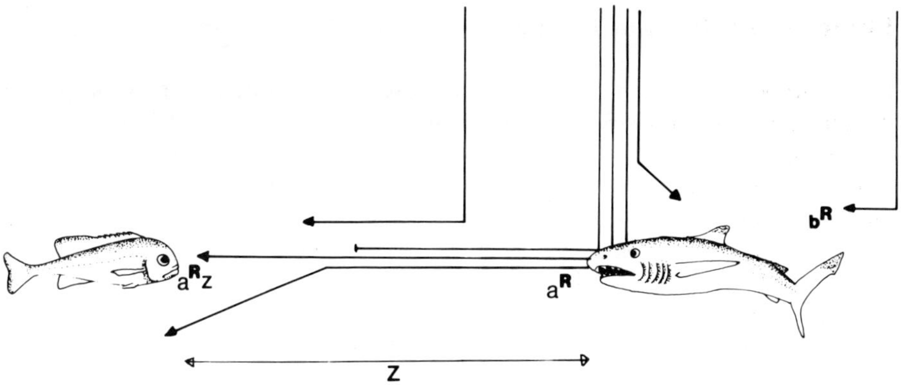


Figure 3.4. Three main factors that limit visibility in shallow water when light intensity is not limiting. Daylight is scattered by particles in the ocean in the direction of the eye, creating underwater space light and veiling light between the object and the eye. Light from the object is attenuated by combined effects of scatter and absorption in its passage between the object and the eye. Z is visual range, $_bR$ is background spacelight, $_aR$ is inherent radiance of the object (in this case the shark), and $_aR_Z$ is apparent radiance of the object at distance Z (see equations 2–5).

of environmental visual noise and probably the most important factor limiting underwater visibility in daylight.

The equation that describes the gain of veiling light at any given wavelength as path length increases is

$$_bR_z = {}_bR(l - e^{-\alpha R}) \tag{3.3}$$

where $_bR$ = background spacelight radiance.

The apparent radiance ($_aR$) of an object to an animal when both the loss of image-forming light (eq. [3.2]) and the gain in veiling light (eq. [3.3]) is taken into account is given in equation (3.4):

$$_aR_z = {}_aRe^{-\alpha z} + {}_bR(l - e^{-\alpha z}) \tag{3.4}$$

For the animal, it is the contrast between the object and its background that is important. Contrast ($C$) is conventionally described as the proportional difference between the target radiance and the background radiance or, in the case of an object seen against the water space light,

$$C_z = \frac{_bR_z - {}_aR_z}{_bR_z}. \tag{3.5}$$

By substituting the value for $_aR_z$ in equation (3.4) into equation (3.5), it is evident that the apparent contrast decreases with visual range as follows:

$$C_z = Ce^{-\alpha z} \tag{3.6}$$

where $C_z$ = apparent contrast at visual range $z$, and $C$ = inherent contrast presented by the object close to the eye.

## Photons and Retinal Image

Light can be looked on as a waveform whose energy is transferred to matter at discrete points in time and space, and these points of energy transfer are recorded as photons. Their number is proportional to the brightness of the light, but the exact time and place that individual photons appear is uncertain and can only be predicted statistically. Visual photoreceptors are photon counters, and it requires that fewer than 10 photons be absorbed by an integrated group of about 300 retinal rods within a period of 0.1 sec. for the (human) eye to register the sensation of light (Ripps and Weale 1976).

Because photoreceptors are photon counters and the exact arrival of photons is uncertain, the problems of vision are, at root, statistical. In Figure 3.5 is a representation of an image of a fish in dim light when it is close and when it is farther away through the water. Each white spot represents the absorption of an individual photon: the brighter the light, the more photons and the more information is contained in the image. The visibility of the more distant fish is much reduced, because photons that have been scattered into the eye by particles suspended in the water add veiling light, which is not image forming, to all parts of the retinal image and thus reduce all visual contrasts.

### The Probability of Photon Capture

A fundamental limit to the amount of information that an image contains is set by the number of photons that form it. The number of photons falling on the human retina limits spatial and temporal resolution well into ''daylight'' levels of illumination. In the failing light of late afternoon it becomes difficult to play a game such as tennis in which the ball is small and fast moving, but there is no such difficulty in playing football. In a very useful discussion of the statistical basis of the problem, Land (1980) remarks that at the lowest limits of vision, each rod receives one photon every 40 min. on average and, even in the light of the full moon, when it is bright enough to see some color and read large print, each rod receives a photon only once every second. The visual problem that is fundamental to all animals is whether a group of photoreceptors has captured a significantly different number of photons than have neighboring groups. It is, therefore, very relevant to know the brightness of the retinal image (how many photons per unit area and time), how large an area is covered by each grouping of photoreceptors, and how much time is taken for the collection of each sample.

### The Brightness of Retinal Image

The relationship between the anatomy of the eyes and the brightness of the retinal image has been discussed by Land (1980); and, as most aspects apply equally to terrestrial and aquatic animals, only a brief summary is given here. For point sources of light, such as bright distant specks of bioluminescence, the brightness of the image (the number of photons per unit time per unit area) is proportional to the pupil area; but for the more usual extended sources, the brightness of the

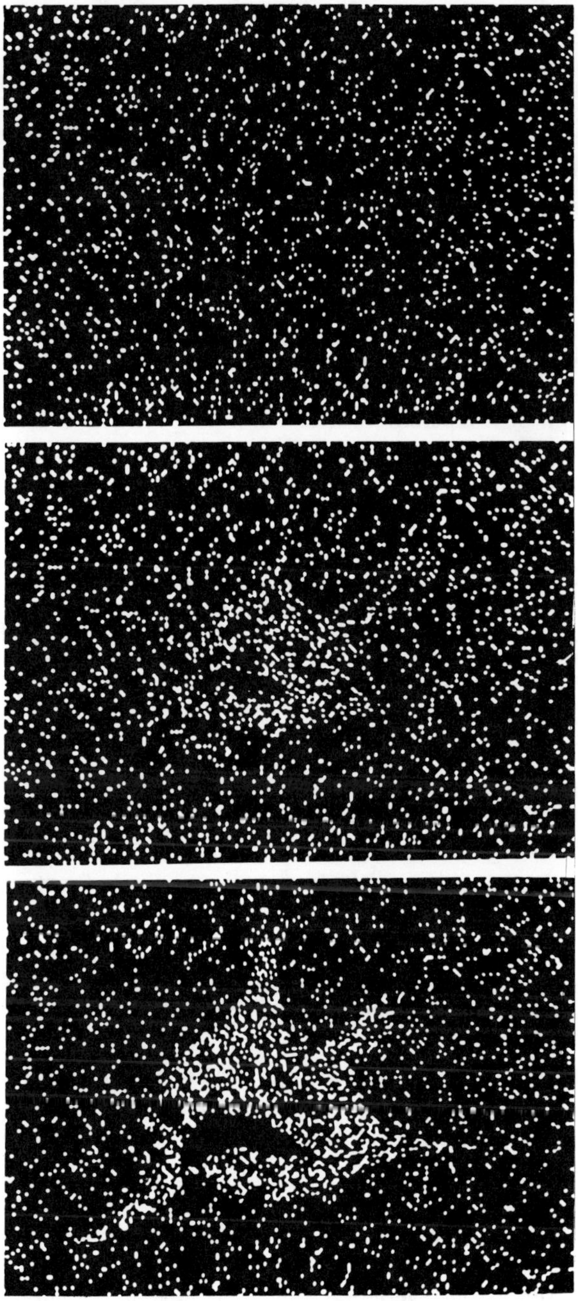

A

B

C

Figure 3.5. The number of photons here represented as white dots set fundamental limit to the information contained in an image. This figure illustrates the situation shown in Figure 3.4, in which a shark is swimming towards the observer. In (A) the shark cannot yet be seen, because image-forming light coming from it has been mostly absorbed by the water, and all that is to be seen is background spacelight between the shark and the observer. In (B) there is reduced veiling light because the thickness of water between the shark and observer is reduced. In (C) the shark is so close that little image-forming light has been lost, and there is negligible amount of veiling light. The brightness of the light is proportional to the number of photons. Note that the brightness of the background space light remains constant and it is the brightness of the image that changes.

image depends on the *f-number* of the eye (focal length/pupil aperture) and is the measure of image brightness equivalent to that used in photography. The single-lensed eye of aquatic animals is particularly interesting because the similarity in refractive index between the water and the fluids behind the cornea means that all the focussing power of the eye resides in the lens, which must be spherical to be able to focus sufficiently. The effective index of refraction of those fish lenses which have been measured is about 1.65 (Protasov 1970), and they have focal lengths about 2.55 times their radius (Matthiessen's ratio). Thus fishes in general may have very low f-numbers of about 0.78 (Munk and Frederiksen 1974). It is likely that cephalopod molluscs that also have round lenses may also have similarly low f-numbers (Land 1980).

The provision of a tapetum, which is a highly reflective layer at the back of the eye, redirecting light that has not been absorbed by the photoreceptors back through the photoreceptive layer, is another technique for increasing the effective brightness of the retinal image. Tapeta are a feature of many twilight-active animals, both terrestrial and aquatic. The gain in sensitivity may be achieved at the expense of image sharpness, because the tapetum is not a perfect mirror, but the many morphological types of tapeta in animals that need to see in dim light (Nicol, Arnott, and Best 1974) is a good example of convergent evolution and indicates that the gain in sensitivity, which cannot exceed one photographic stop, may be worthwhile.

## Integration Area

The area of the image that is sampled as a unit can be increased by harnessing the photoreceptors together so that they work as a single functional group. However, the improved perception of contrast that this brings is at the expense of the ability to see fine detail. There is some evidence that, in invertebrates, the cross-sectional area of individual rhabdomes is greater in animals living in dim light (Kirschfeld 1974; Blest and Land 1977), including the deep sea crustacean *Oplophorus* (Land 1980). In vertebrates, there is also an indication that the cones of crepuscular predators are unusually large (Munz and MacFarland 1977; Lythgoe 1979). In some mesopelagic fishes, such as *Scopelarchus guentheri, Scopelosaurus lepidus, Evermanella indica,* and *Ahliesaurus berryi* (Locker 1977; Munk 1977), groups of rods are contained within cups of reflecting tapetum, and it is difficult to see how these can act otherwise than as a single unit. Neural summation is at least as important as optical summation. In the human eye, some 300 rods may synapse together onto one ganglion cell (Pirenne 1967), and there is no doubt that nocturnal (terrestrial) vertebrates tend to have a greater degree of neural summation than diurnal ones (Walls 1942; Tansley 1965; Rodieck 1973).

## Integration Time

The photon count can be increased by prolonging the time over which each sample is collected, but this carries the penalty that moving parts of the image may be lost. At very low light intensity, a decrease in the intensity of a stimulus can be compensated for by a corresponding increase in the time of presentation up to a limit of 0.1 sec. For humans in very dim light the presentation time is about 100 msec and corresponds to the integration time (or retinal memory time), over which there is a reciprocal relationship between the time of exposure and the number of photons arriving in each unit area of the image (see Ripps and Weale 1976). The retinal integration time is related to the flicker fusion frequency, which is the highest frequency at which a flashing light can just be distinguished from a continuous one. In the human eye, this varies from about 55 Hz in bright light to as little as 10 Hz in very dim light, and similar values have been found in fishes (Blaxter 1970).

## Visual Noise

The term *noise* is used here to describe visual stimuli that do not form an image and that mask the information about the visual image that is useful to the animal. This noise can come either from physical sources in the environment or from the imperfect performance of the animal's own sensory system. Physiologists sometimes use the term *shot noise* to describe the irregularities in a sensory response due to the random fluctuations in the number of photons that arrive from moment to moment. This term is different from the *noise* described here, because it refers to real events that originate in the outside world.

### Environmental Noise

The most important source of environmental noise underwater is the scattering of light by suspended particles (Fig. 3.4). Reflections of the sky from the water surface must also be a troublesome source of noise for birds that need to see their prey in the water below. It is possible that the large complement of red oil droplets in the cones of some gulls represents a technique for increasing visual contrasts of objects floating at, or just below, the surface (Lythgoe 1979). Most of the light such objects reflect comes from the sun and sky and is relatively richer in long wave light than the upwelling light from the body of the ocean. A red-sensitive receptor will increase the contrast of the floating object by making it appear relatively brighter than the darker background of the ocean.

### Physiological Noise

It requires that only a single photon should be absorbed by one rhodopsin molecule to set off the cascade of events that results in a signal from a photoreceptor. Set against the fact that at low light levels a single rod receives only one photon every 40 min. and each rod contains $10^9$–$10^{10}$ rhodopsin molecules, it is apparent that rhodopsin molecules must be very stable. The absorption of a photon triggers an isomerization of the retinal group in the rhodopsin molecule from the 11-*cis* to the all-*trans* form. Similar isomerizations that are triggered by some other cause would substantially reduce the perception of small contrast differences.

Perhaps rhodopsins having a wavelength of maximum absorption near 500 nm are inherently more stable than those whose maximum is displaced to either longer or shorter wavelengths (see Lythgoe 1984 for a review). Stiles (1948) and Barlow (1957) have suggested that thermal energy from molecular movements might contribute some energy to the visual pigment molecule such that it combines with energy from an infrared quantum and together they supply enough energy for the isomerization of the chromophore. Both ideas are highly speculative, but both point to the extreme importance of reducing physiological noise at very low light intensities. Even so, the number of dark isomerizations will be proportional to the number of rhodopsin molecules present and may set a limit to the thickness of the functional rhodopsin layer in animals active in twilight.

## Visual Pigments

The spectral absorption curves of the visual pigments of the roach *Rutillus rutillus* are shown in Figure 3.6. The curves of all visual pigments have a similar shape and are chromoproteins consisting of a protein molecule (opsin) in which is embedded a chromophoric group that may be either retinal (vitamin $A_1$ aldehyde) or dehydroretinal (vitamin $A_2$ aldehyde). The retinal pigments are rhodopsins; the dehydroretinal pigments, porphyropsins. In the active state the chromophore

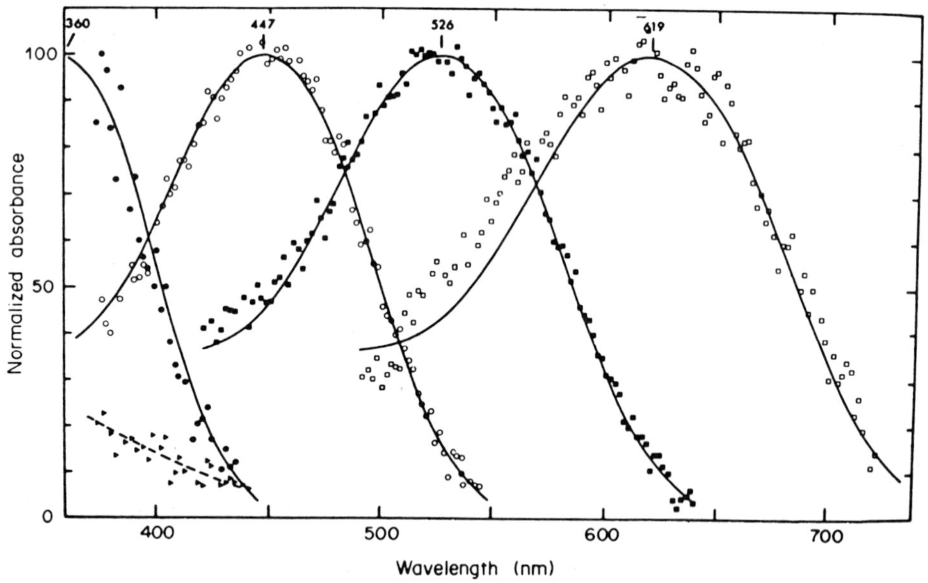


Figure 3.6. Spectral absorption curves of four cone visual pigments in roach *Rutillus rutillus* obtained by microspectrophotometry. Rods have mean wavelength of maximum absorption (λ max.) of 538 nm. It is possible that these cones contain some rhodopsin mixed with the porphyropsin. Maximum absorption of ultraviolet-sensitive pigment can be reliably determined from the position of long-waved limb of absorption curve (Bowmaker 1984).

is in the 11-*cis* isomeric configuration, but when a photon is absorbed it straightens to the all-*trans* form. Although the mechanism is not yet fully understood, it is likely that cyclic nucleotides and $Ca^{++}$ are mobilized, thus changing the permeability of the plasma membrane to the passage of ions, especially $Na^+$, and hence the polarity of the photoreceptor and the generation of a nerve impulse.

## Paired Pigment System

The wavelength of maximum absorption of porphyropsins is longer than that of rhodopsins, except possibly at wavelengths below about 450 nm, and the relationship is described by equation (3.7) (Dartnall and Lythgoe 1965):

$$\lambda_1 = .60\,\lambda_2 + 186 \tag{3.7}$$

where $\lambda_1$ = wavelength of maximum absorption ($\lambda_1$ max.) of the rhodopsin and $\lambda_2 = \lambda$ max. of its porphyropsin analogue. A shift from rhodopsin to porphyropsin results in a much greater shift for red-sensitive pigments than for green- or blue-sensitive ones, and presumably this has implications for color vision in animals that change the relative proportions of rhodopsins and porphyropsins in their retinal rods and cones.

Porphyropsins are frequently found in freshwater animals and only rarely in marine and terrestrial species. Most freshwater teleost fishes and larval amphibia have a substantial or overwhelming proportion of porphyropsin compared to rhodopsin. At least one freshwater turtle, *Pseudemys scripta elegans* (Liebman and Granda 1971), and the freshwater crayfish *Procambarus clarkii* (Suzuki and Makino-Tasak 1984) have retinal porphyropsin, but as yet, none has been recognized in any mammal or bird.

A significant feature of the rhodopsin-porphyropsin "paired pigment" system is that the proportions of the two pigments can vary with the season (Dartnall, Lander, and Munz 1961), with the developmental stage of the animal, and whether it is living in fresh water or the sea. Fishes that migrate between freshwater and the sea in their breeding migrations change their rhodopsin-porphyropsin ratio in the appropriate direction before actually entering their new environment (Table 3.1) (Knowles and Dartnall 1977; Beatty 1984). This suggests that the change is visually significant but is not triggered directly either by the water chemistry or by the spectral quality of the environment.

In the paired pigment species studied so far, including the freshwater crayfish studied by Suzuki, Arikawa, and Eguchi (1985), the proportion of porphyropsin increases in winter. The change appears to be triggered by changes in the light-dark cycle and is assisted by low temperatures. The actual relationship between day length and the rhodopsin-porphyropsin ratio varies from species to species (see Beatty 1984 for a review). Perhaps the variation in response to day length is not surprising, since most seasonal changes such as molting and gonadal development in birds are triggered weeks or months before they are complete (Menaker 1977). In all the fishes so far investigated, injections of thyroxin increase the rhodopsin-porphyropsin ratio; but pinealectomy of goldfish caused no change in it, and thus there is little evidence to suggest that the pineal-hypothalamic-pituitary

Table 3.1. Migratory Fishes Showing a Change in Rhodopsin/Porphyropsin Ratio

| Anadromous Species | Catadromous Species |
|---|---|
| *Oncorhynchus kisuch* | *Anguilla anguilla* |
| *O. tshawytscha* | *A. rostrata* |
| *O. keta* | |
| *O. gobuscha* | |
| *Salmo salar* | |
| *S. gairdneri* | |
| *Osmerus eperlanus mordax* | |
| *Petromyzon marinus* | |

*Source:* Beatty (1984).

pathway plays a part (Beatty, 1984). Indeed, Bridges and Yoshikami (1970) fitted an opaque contact lens to one eye of the rudd, *Scardinius erythrophthalmus,* and found that exposure to different light regimens induced the expected shift in the rhodopsin-porphyropsin ratio in the uncovered eye, but not in the one that no light could enter.

It is likely that an increase in the proportion of porphyropsin in the cones brings with it an increased sensitivity to longer wavelengths, and the effect is likely to be particularly marked for the long wave–sensitive cones, in which the λ max. shifts from about 565 nm to about 630 nm. It is also true that the ambient light in freshwater is relatively (though not absolutely) richer in long wavelengths than that in the sea. However, Muntz and Mouat (1984) measured both the seasonal change in rhodopsin-porphyropsin and the spectral irradiance underwater in natural populations of brown trout, *Salmo trutta,* in several Scottish lochs and found that the fish must have become less red sensitive as the underwater light became relatively richer in red light. There is a similar puzzle in fish such as the rudd that intensify the red of their fins in the breeding season. This occurs just when the sensitivity of their cones has shifted to shorter wavelengths (Loew and Dartnall 1976). Although it is difficult to know what emphasis to place on the observation, there does seem to be some correlation between shortwave sensitivity and life in shallow water (Muntz and Mouat 1984). Both rudd and trout appear to be more active in shallow water in the summer. Also, shallow-living fishes do seem to have more blue-sensitive cone classes and perhaps fewer red-sensitive cones (Fig. 3.8).

The paired pigment system may give considerable scope for variability in color processing. Seasonal changes have already been mentioned, and it is probably also true that the spectral quality and intensity of the light climate in fresh water varies more than in the ocean. There can also be great variability among fishes of the same species caught at the same time and place. For example, Bridges and Delisle (1974) found that, in a population of smelt, *Osmerus eperlanus mordax,* from eastern Canada, porphyropsin ratios at spawning time could vary from 10% to over 90% porphyropsin.

## Visual Pigments and Sensitivity

As long ago as 1936, it was realised that the color-filtering action of natural water might make it a convenient natural laboratory for the study of ways that visual systems might adapt to the color of the environmental light (Clarke 1936; Bayliss, Lythgoe, and Tansley 1936). The early predictions were that the homochromatic blue light that penetrates deep into the clear oceans would favor the use of shortwave-absorbing visual pigments. At that time, it was not yet known that pH had to be controlled in measuring visual pigment solutions, and no convincing correlations between water color and visual pigments were discerned. Later, with better extraction techniques, it became clear that the visual pigments of deep-sea fishes and invertebrates are indeed unusually blue sensitive and are precisely those which would be expected if the spectral absorption of the visual pigments were adjusted to maximize sensitivity (Fig. 3.7). In other environments, a trawl of animals all caught at the same depth and at the same time will also have visual pigments in their rods that all absorb in the same region of the spectrum. But in non-deep-sea environments the visual pigments do not absorb at long enough wavelengths to explain them on grounds of sensitivity alone. These calculations have to be made with some care, and a technique for doing this is given by Dartnall (1975).

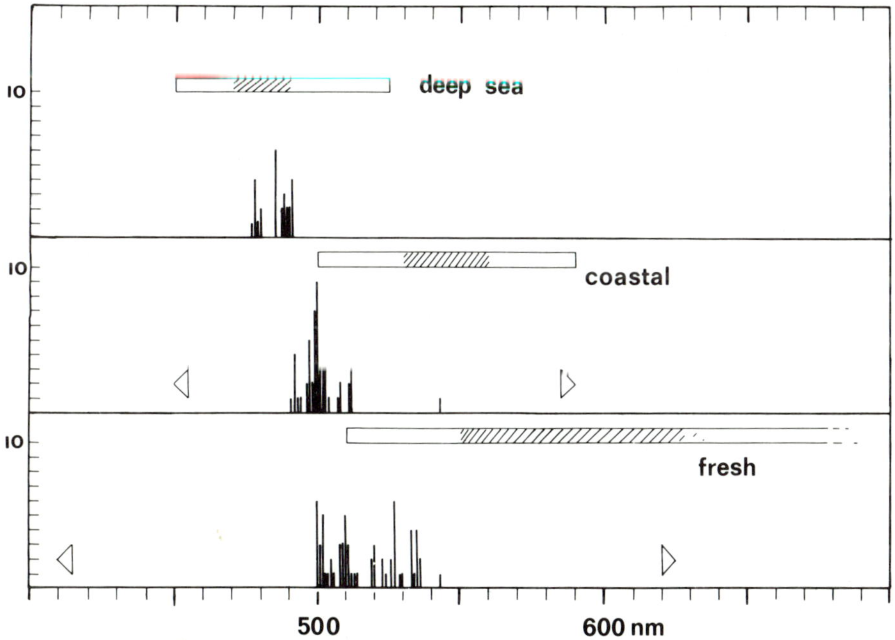


Figure 3.7. Histograms of numbers of fish species containing single rod pigments of known λ max. from different environments. Horizontal bars show possible range of visual pigment λ max. that would give maximum sensitivity in each water type; shaded areas show most likely range. (From Lythgoe 1980. Further data is contained in Crescitelli, McFall-Ngai, and Horwitz 1985.)

Sensitivity is important to a visual animal, because it needs a photon catch big enough to allow it to detect small contrasts. However, effective contrasts are often greater at wavelengths offset from those of greatest image radiance, and there may also be differences in the physiological noise introduced by different visual pigments (Lythgoe 1979, 1984). Factors such as these may help to explain why fishes that live in waters other than the deep oceans have visual pigments that are not sufficiently sensitive at the longer wavelengths to confer maximum sensitivity (Fig. 3.7).

## Visual Pigments and Visual Performance

Vision through water is often limited by the rapid reduction in contrasts brought about by the veil of scattered light between the eye and the object. It is evident from equation (3.6) that the maximum visible range of a large object depends on the inherent contrast of the object ($C$) at zero distance from the eye and the beam attenuation coefficient ($\alpha$) of the water. The inherent contrast of an object against any background is greatest if the object is either very dark or very bright, and all else being equal, either a dark or a light object will be seen at the greatest distance through the water. Contrasts reduce least rapidly when $\alpha$ is smallest (the water is most transparent), and it follows that an animal will experience the least reduction in apparent contrast if its photoreceptors are most sensitive to the wavelengths where the water is most transparent. An object becomes lost to sight when its apparent contrast falls below the contrast-perceiving threshold of the eye. This threshold depends in part, however, on the size and spatial frequency of the image on the retina, and therefore small objects will be lost to view sooner than larger ones. In very clear ocean water, a large dark object is unlikely to be visible farther than about 30 m, whereas in turbid water the visibility may be a few centimeters or less.

There is a special situation for white or gray objects that are brighter than their water background, because the inherent contrast depends on the wavelength. Such objects are chiefly illuminated by daylight that has travelled the direct path from the surface to the object and is then reflected directly into the eye. However, the object is seen against the water space light background, which is composed of scattered light that has travelled a longer path through the water than the light from above, which contributes most of the light that illuminates the object. The result is that spectral radiance curve of the object is broader than the spectral radiance of the background, and visual contrasts will be greater at wavelengths offset from the wavelength of maximum water transparency. This has been suggested as part of the reason that the visual pigments in the rods of fishes living in moderately shallow water do not confer the maximum sensitivity (Lythgoe 1968, 1979). An explanation of this type might also explain why species living much deeper where the downwelling light is homochromatic and there is little difference in the spectral radiance of light coming from different directions have visual pigments that are exactly those predicted on sensitivity grounds alone. The photoreceptor best suited to detecting a light gray object is thus most sensitive to wavelengths that are offset from the wavelength of maximum water transparency. The result is that the visual pigment required to detect a light gray object may be different from the one needed to detect a dark gray object. If an animal has two types of photoreceptor of different spectral sensitivity and the nervous system can compare the output from the two, the animal has the *potential* for color vision but does not necessarily have it.

Two radiances will be most easily distinguished if the photoreceptor is most sensitive to the wavelength at which they differ most. It is, however, a disadvantage to be most sensitive to the wavelengths at which the radiances are most similar. All else being equal, a photoreceptor with a narrow action spectrum centered in that part of the spectrum where the radiances differ most will enhance the signal-to-noise ratio. Put another way, if the non-image-forming "noisy" light is concentrated in a particular spectral region, it may be helpful to reduce sensitivity there.

The narrowing of the action spectrum of a photoreceptor can be achieved optically by means of cutoff filters in individual cones or neurally by lateral inhibition between cones containing visual pigments of different spectral sensitivity. The optical solution is exemplified by many reptiles and birds that have colored oil globules in the cone ellipsoid placed in such a way that the incoming light must traverse them before reaching the visual pigment in the outer segment. There is strong evidence that, in birds, the combination of at least four types of colored oil globules, with at least three types of rhodopsin, leads to at least seven types of cones when classified according to their action spectra (Bowmaker and Knowles 1977; Bowmaker 1984). The clear water marine green turtle, *Chelonia mydas,* has orange, two types of yellow, and colorless oil globules (Liebman and Granda, 1971); the freshwater *Pseudemys scripta* has red, orange, yellow, pale green, and colorless oil globules. In *Pseudemys* the four types of oil droplet are teamed with outer segments, each containing one of three visual pigments, to give six types of cone when classified by their spectral sensitivity curves (Ohtsuka and Teruya, 1985). It is also likely that color filters resembling cytochrome C are present in some cone classes in some fishes. These filters absorb blue light of 400–430 nm and would tend to increase contrasts in the blue-violet region of the spectrum (MacNichol et al. 1978).

## Color Vision

It is difficult to know which is more complex; the visual world that surrounds an animal or the brain that it uses to interpret it. Yet all the information available to the brain comes from a limited number of photoreceptor types containing between one and five visual pigments. A single photoreceptor, be it rod, cone, or retinular cell, can by itself signal no information about color, only about brightness. Thus, a photoreceptor that is most sensitive to green light might deliver an identical neural signal if stimulated by a dim green light or a strong orange one. The recognition of color requires that the retina and brain compare the outputs of photoreceptors of different spectral sensitivity, but serving the same element of the image.

Color vision gives an animal the ability to recognize the form of the spectral radiance curve of an object, with differently shaped curves giving the sensation

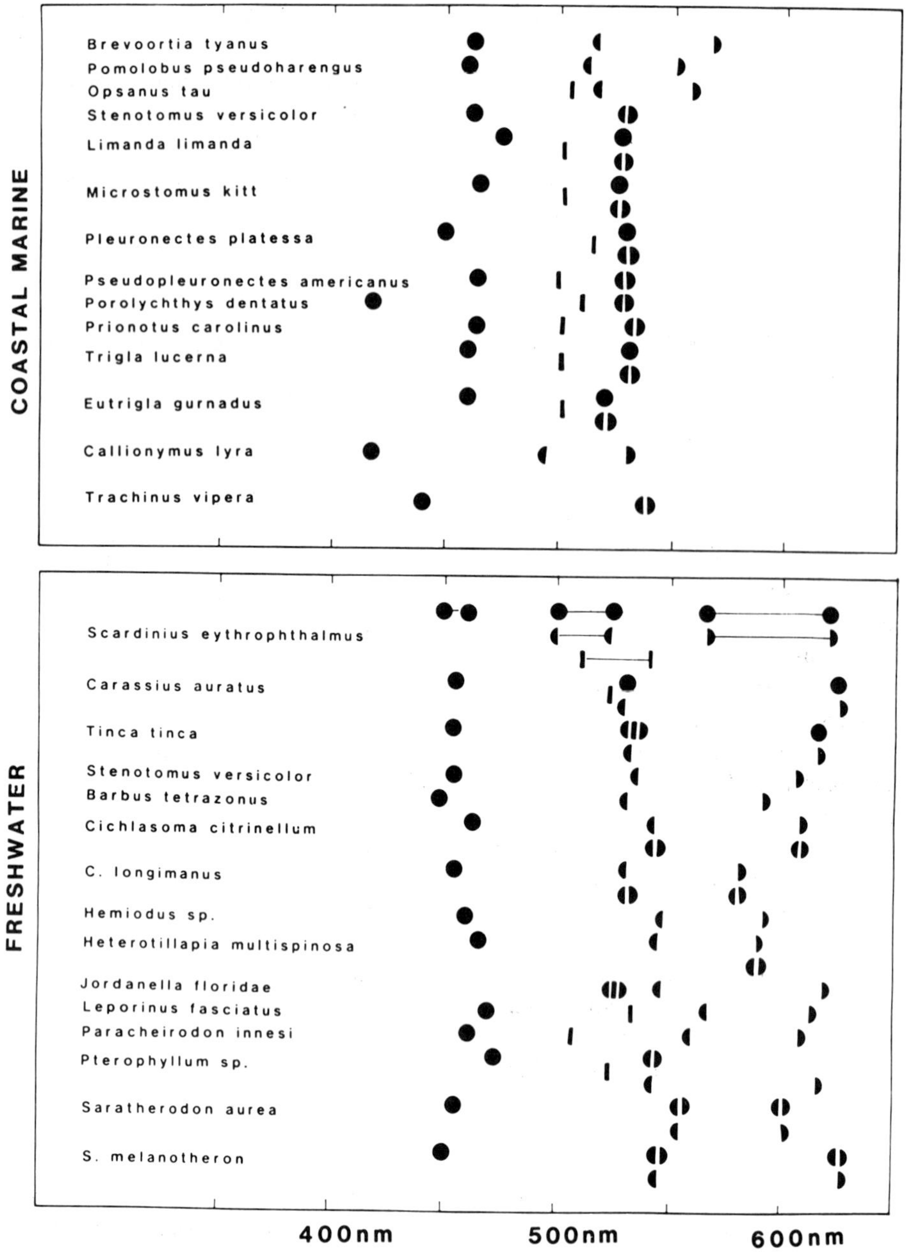


Figure 3.8. Absorption maxima of visual pigments in cones of fishes that live in coastal water, which is usually green, and those which live in freshwater, which is usually yellow green to brown. Greatest difference appears to be in paired cones. Absorption maxima of freshwater species are at longer wavelengths than those of marine species. All the fish present live at intermediate depths. Marine littoral and surface-living freshwater species are not included. Single circles are single cones. Half circles represent one cone of a pair. Where two half circles are adjacent, visual pigment in both cones of the pair are identical. Vertical bars represent rods. Horizontal lines indicate the extent of rhodopsin-porphyropsin shift in cones of *Scardinius* (data from Loew and Lythgoe 1978; Levine and MacNicholl 1979; Avery et al. 1983; Bowmaker 1984).

of particular colors. To do this, at least two types of receptor, each having a different spectral sensitivity curve, are needed. Because environmental light is stronger at some wavelengths than at others, it follows that not all the cones can be positioned in the spectrum to give maximum sensitivity; but it also follows that they are unlikely to be sensitive to wavelengths where there is little environmental light. In the human retina, there are three types of visual pigment contained separately in three types of cones (Bowmaker 1984). In some fishes, only two types of cones can be differentiated, and in some, perhaps only a single type is present (Loew and Lythgoe 1978; Levine and MacNichol 1979). However, there are known to be four types of cone in some fishes, such as the 14-spined stickleback, *Spinachia spinachia* (Loew and Lythgoe 1978); the roach *Rutillus rutillus* (Avery et al. 1983); and the Japanese dace, *Tribolodon hakonensis* (Harosi and Hashimoto 1983); and no doubt there are many others.

It is becoming clear that the visual pigments in the cones of marine fishes are broadly influenced by their habitat, but there are at present only rather vague indications that the behavior of the animal may also be important. In Figure 3.8, the visual pigments in the cones of fishes are arranged according to the color of the water in which they live. None of the species represented live exclusively in very deep or very shallow water and in that respect are broadly comparable. It is evident that freshwater species tend to possess a class of red-sensitive cones that are often paired and are not present in most marine and terrestrial animals. It is reasonable to suppose that this is a response to the long wave bias in the environmental light. There are not yet enough data to tell if the cones of fishes living in the green coastal waters of northern Europe broadly differ in λ max. from the cones of fishes living in the generally bluer water that surrounds coral reefs.

There are some interesting relationships between the depth at which freshwater species live and the visual pigments in their cones (Levine and MacNicholl 1979). In Figure 3.9 the fishes have been grouped according to life-style and depth. In freshwater it is likely to be the cones that absorb most strongly between about 550 nm and 620 nm or longer that are the most sensitive (Fig. 3.7), and it is evident that fishes that live in relatively deep water or are active at twilight possess cones matched to these wavelengths. These fishes also seem to lack blue-sensitive cones, meaning that they are sensitive to a rather narrow spectral bandwidth. This narrowing of spectral sensitivity must also be due to the narrowing of the wave band of the environmental light as the depth increases. Shallower-living fishes are sensitive to a much broader spectral wave band, and species such as the roach or dace can almost certainly see both shorter and longer wavelengths of light than we can. It is interesting that diurnal species living close to the surface appear to lack the long wave sensitivity of deeper-living species, and this may be due to the increased proportion of rhodopsin compared to porphyropsin in the retina. The occurrence of shortwave blue- and ultraviolet-sensitive cones in very shallow-living species has not been explained. The transparency of water to the near ultraviolet depends very much on the concentration of dissolved yellow substances it contains (Fig. 3.1B). At a depth of 1 or 2 m, there may be too little ultraviolet light to be useful, but in the top 100 mm of freshwater and

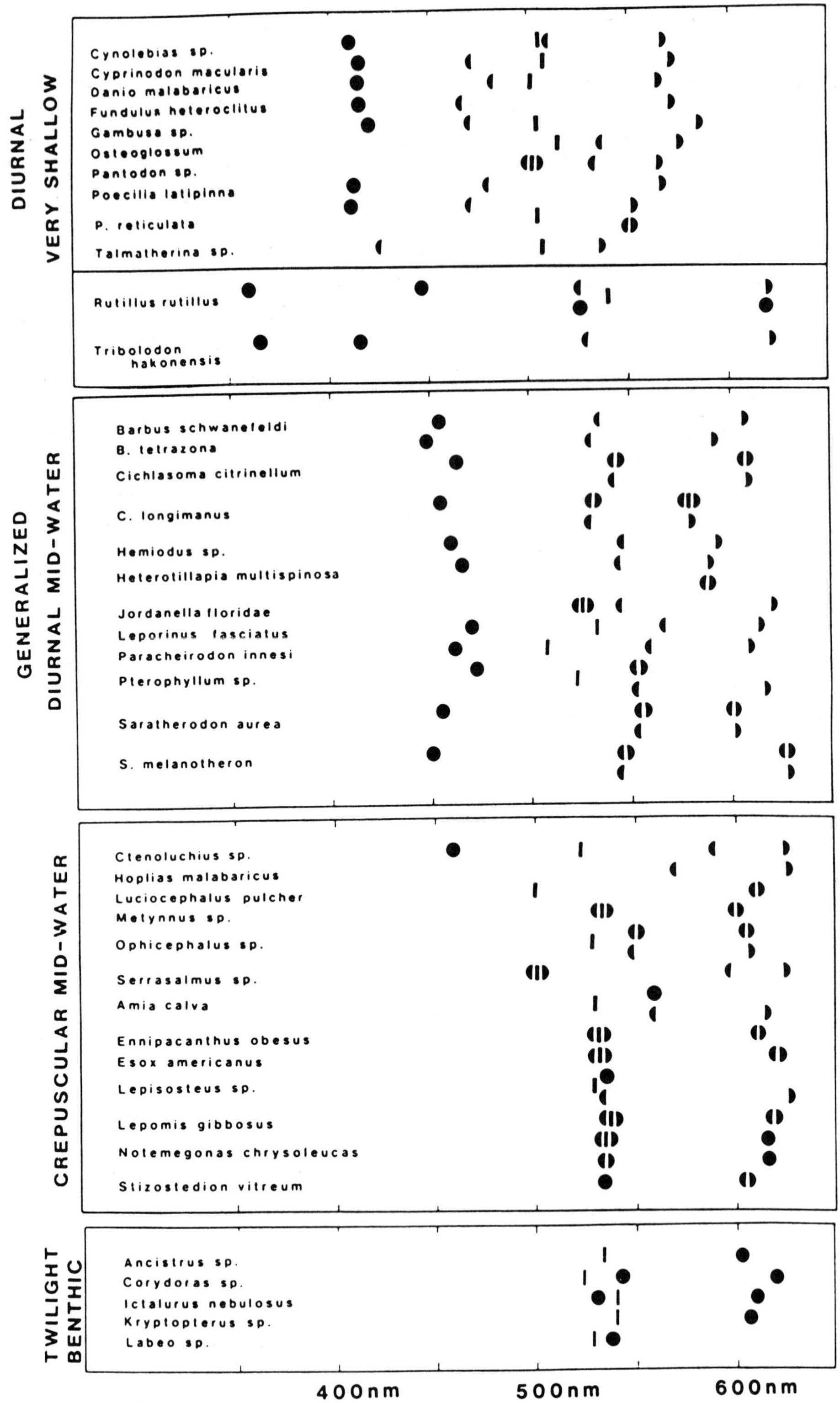
DIURNAL
VERY SHALLOW
Cynolebias sp.
Cyprinodon macularis
Danio malabaricus
Fundulus heteroclitus
Gambusa sp.
Osteoglossum
Pantodon sp.
Poecilia latipinna
P. reticulata
Talmatherina sp.
Rutillus rutillus
Tribolodon hakonensis
GENERALIZED
DIURNAL MID-WATER
Barbus schwanefeldi
B. tetrazona
Cichlasoma citrinellum
C. longimanus
Hemiodus sp.
Heterotillapia multispinosa
Jordanella floridae
Leporinus fasciatus
Paracheirodon innesi
Pterophyllum sp.
Saratherodon aurea
S. melanotheron
CREPUSCULAR MID-WATER
Ctenoluchius sp.
Hoplias malabaricus
Luciocephalus pulcher
Metynnus sp.
Ophicephalus sp.
Serrasalmus sp.
Amia calva
Ennipacanthus obesus
Esox americanus
Lepisosteus sp.
Lepomis gibbosus
Notemegonas chrysoleucas
Stizostedion vitreum
TWILIGHT
BENTHIC
Ancistrus sp.
Corydoras sp.
Ictalurus nebulosus
Kryptopterus sp.
Labeo sp.
400nm
500nm
600nm

Figure 3.9.

the top 10 m of very clear ocean water, there is likely to be enough ultraviolet light to be useful (Lythgoe 1984). This will be especially true for a fish looking upwards at angles between the vertical and about 48 degrees, where it will see floating objects silhouetted against the bright ultraviolet-rich sky.

Although we can probably predict the spectral sensitivity of individual photoreceptors, the complexity of neural processing in the retina and brain make it impossible in our present state of knowledge to infer much about the color vision of the animal. It is more feasible to understand what raw information is gathered by the animal at the photoreceptive level, and no amount of neural processing at higher levels can add information that has not been collected by the photoreceptors.

Neural interactions in the retina cause many horizontal and ganglion cells to have action spectra that differ from those of the individual cone types. These spectra may be narrower and displaced to longer wavelengths than either parent cone type (Sirovich and Abramov 1977). However, the exact shape and spectral location of the action spectra are variable and depend, in part, on the area of the receptive field and its position in the retina. One specific mechanism is for the tonic output from a ganglionic cell to be excited by cones from the center of its field and inhibited by cones, containing a visual pigment of different spectral sensitivity, from the peripheral field. For example, the red-green color opponent cells in the primate retina have an excitation maximum of about 620 nm and deepest inhibition about 500 nm, yet the visual pigments in the two types of cone have maximum sensitivities at 535 and 567 nm (Zrenner 1983). Similar examples of teleosts are known (Spekreijse, Mooij, and van de Berg 1981). For example, the second-order neurones in the dace *Tribolodon hakonensis*'s retina fed by the red- and blue-sensitive cones have a maximum stimulatory response at 700 nm and a deepest inhibitory response at 460 nm (Hashimoto and Inokuchi 1981). There are now known to be both a blue and an ultraviolet type of cone in dace retina (Harosi and Hashimoto 1983), but it is unlikely that the retina contains a visual pigment with a maximum absorption that exceeds 630 nm.

## Conclusions

In the underwater environment the combined effects of refraction, wavelength-selective absorption, and scatter cause the distance that objects can be seen through the water to be much reduced; the light is strongly directional and often restricted to a narrow spectral bandwidth. Below about 1,000 m, no daylight remains and

Figure 3.9. Absorption maxima of visual pigments in cones of freshwater fishes that live at various depths and are active at different times of day. Spectral band where the light is bright enough for vision gets narrower as the depth increases, and this is reflected in reduced spread of visual pigments in the spectrum in deeper-living species. Very shallow-living species appear to possess fewer long wave sensitive pigments. *Rutillus* and *Tribolodon,* which possess ultraviolet pigments, should probably be considered "generalized diurnal mid-water" species. Symbols as in Figure 3.8. (From Levine and MacNicholl 1979. Additional data for *Tribolodon:* Harosi and Hashimoto 1983; *Rutillus:* Avery et al. 1983; Bowmaker 1984; and *Scardinius:* Loew and Dartnall 1976.)

all light comes from bioluminescence. The biological consequences of these exacting conditions can be understood only by considering both the wave and the quantum nature of light.

The structure of fish eyes shows adaptations to the directional nature of the underwater light, and aquatic animals have unique techniques of camouflage. The spectral sensitivity of the retinal photoreceptors appears to depend on the color of the underwater light in ways that may give us an understanding of some of the primary mechanisms of color vision.

## References

Avery, J.A., J.K. Bowmaker, M.B.A Djamgos, and J.E.G. Downing. 1983. Ultraviolet sensitive receptors in a freshwater fish. *J. Physiol.* 334: pp 23.

Baker, K.S., and R.C. Smith. 1982. Bio-optical classification and model of natural waters. II. *Limnol. Oceanogr.* 27:500–509.

Barlow, H.B. 1957. Purkinje shift and retinal noise. *Nature* (*Lond.*) 179:255–256.

Bayliss, L.E., R.J. Lythgoe, and K. Tansley. 1936. Some new forms of visual purple found in sea fishes, with a note on the visual cells of origin. *Proc. R. Soc. Lond. B Biol. Sci.* 816:95–113.

Beatty, D.D. 1984. Visual pigments and the labile scotopic visual system of fish. *Vision Res.* 24:1563–1573.

Blaxter, J.H.S. 1974. "Light: Fishes." In *Marine Ecology,* vol. 1, by O. Kinney. London: Wiley Interscience.

Blest, A.D., and M.F. Land. 1977. The physiological optics of *Dinopus subrufus L.* Koch, a fish-eye lens in a spider. *Proc. R. Soc. Lond. B Biol. Sci.* 196:197–222.

Bowmaker, J.K. 1984. Microspectrophotometry of vertebrate photoreceptors. *Vision Res.* 24:1641–1650.

Bowmaker, J.K., and A. Knowles. 1977. The visual pigments and oil droplets of the chicken retina. *Vision Res.* 17:755–764.

Bridges, C.D.B., and C.E. Delisle. 1974. Post-glacial evolution of the visual pigments of the Smelt, *Osmerus eperlanus mordax. Vision Res.* 14:345–356.

Bridges, C.D.B., and S. Yoshikama. 1970. The rhodopsin-porphyropsin system in freshwater fishes. 2. Turnover and interconversion of visual pigment prosthetic groups in light and darkness: role of the pigment epithelium. *Vision Res.* 10:1333–1345.

Clarke, G.L. 1936. On the depth at which fishes can see. *Ecology* 17:452–456.

Cott, H.B. 1940. *Adaptive Coloration in Animals.* London: Methuen.

Crescitelli, F., M. McFall-Ngai, and J. Horwitz. 1985. The visual pigment sensitivity hypothesis: further evidence for fish of varying habitats. *J. Comp. Physiol. A Sens. Neural Behav. Physiol.* 157:323–333.

Dartnall, H.J.A. 1975. "Assessing the Fitness of Visual Pigments for Their Photic Environment." In *Vision in Fishes,* by M.A. Ali. NY: Plenum Press.

Dartnall, H.J.A., M.R. Lander, and F.W. Munz. 1961. "Periodic Changes in the Visual Pigment of a Fish." In *Progress in Photobiology,* by B.C. Christensen and B. Buckman. Amsterdam: Elsevier.

Dartnall, H.J.A., and J.N. Lythgoe. 1965. The spectral clustering of visual pigments. *Vision Res.* 5:81–100.

Denton, E.J., and J.A.C. Nicol. 1965. Studies on reflexions of light from silvery surfaces of fishes, with special reference to the bleak, *Alburnus alburnus. J. Mar. Biol. Assoc. U.K.* 45:683–703.

Denton, E.J., and J.A.C. Nicol. 1966. A survey of reflectivity in silvery teleosts. *J. Mar. Biol. Assoc. U.K.* 46:685–722.

Duntley, S.Q. 1951. The visibility of submerged objects. II. *Proc. Armed Forces–Nat. Res. Council Vision Comm.* 28:60.

Duntley, S.Q. 1963. Light in the sea. *J. Opt. Soc. Am.* 53:214–233.

Fernald, R.D. 1981. Chromatic organization of a cichlid fish retina. *Vision Res.* 21:1748–1753.

Harosi, F.I., and Y. Hashimoto. 1983. Ultraviolet visual pigment in a vertebrate: a tetrachromatic cone system in the dace. *Science* 222:1021–1023.

Hashimoto, Y., and M. Inokuchi. 1981. Characteristics of second order neurons in the dace retina: physiological and morphological studies. *Vision Res.* 21:1541–1550.

Herring, P.J. 1977. Bioluminescence of marine organisms. *Nature* 267:788–793.

Jerlov, N.G. 1968. *Optical Oceanography.* N.Y.: Elsevier.

Jerlov, N.G. 1951. Optical studies of ocean waters. *Rep. Swed. Deep-Sea Exped.* 3: 1–59.

Jerlov, N.G. 1976. *Marine Optics.* Amsterdam; Elsevier.

Kirschfeld, D. 1974. The absolute sensitivity of lens and compound eyes. *Z. Naturforsch.* 29c:592–596.

Knowles, A., and H.J.A. Dartnall. 1977. "The Photobiology of Vision." In *The Eye,* by H. Davson. N.Y.: Academic Press.

Land, M.F. 1980. "Optics and Vision in Invertebrates." In *Handbook of Sensory Physiology* VII/6B, by H. Autrum. Berlin: Springer-Verlag.

Levine, J.S., and E.F. MacNicholl, Jr. 1979. Visual pigments in teleost fishes: effects of habitat, microhabitat and behavior on visual system evolution. *Sensory Processes* 3:95–130.

Liebman, P.A., and A.M. Granda. 1971. Microspectrophotometric measurements of visual pigments in two species of turtle, *Pseudemys scripta* and *Chelonia midas. Vision Res.* 11:105–114.

Locket, N.A. 1977. "Adaptations to the Deep-sea Environment. In *Handbook of Sensory Physiology* VII/5, by F. Crescitelli. Berlin: Springer-Verlag.

Loew, E.R., and H.J.A. Dartnall. Vitamin $A_1/A_2$-based visual pigment mixtures in cones of the rudd. *Vision Res.* 16:891–896.

Loew, E.R., and J.N. Lythgoe. 1978. The ecology of cone pigments in teleost fish. *Vision Res.* 18:715–722.

Lythgoe, J.N. 1968. Visual pigments and visual range underwater. *Vision Res.* 8:997–1012.

Lythgoe, J.N. 1979. *The Ecology of Vision.* Oxford: Clarendon.

Lythgoe, J.N. 1980. "Vision in Fish: Ecological Adaptations." In *Environmental Physiology of Fishes,* by M.A. Ali. New York: Plenum.

Lythgoe, J.N. 1984. Visual pigments and environmental light. *Vision Res.* 24:1539–1550.

MacFarland, W.N., and E.R. Loew. 1983. "Wave Produced Changes in Underwater Light and Their Relation to Vision. In *Predators and Prey in Fishes,* by D.I.G. Noakes, D.G. Lindquist, G.S. Helfman, and T. Ward. The Hague: J.A. Junk.

MacNichol, Jr., E.F., Y.W. Kunz, J.S. Levine, F.I. Harosi, and B.A. Collins. 1978. Ellipsosomes: organelles containing a cytochrome-like pigment in the retinal cones of certain fishes. *Science* 200:549–552.

Marshall, N.B. 1979. *Explorations in Deep-sea Biology.* Poole, U.K.: Blandford.

Menaker, M. 1977. "Extra retinal photoreception." In *The Science of Photobiology,* by K.C. Smith. London: Plenum.

Middleton, W.E.K. 1952. *Vision through the Atmosphere.* University of Toronto Press.

Munk, O. 1970. On the occurrence and significance of horizontal band-shaped retinal areas in teleosts. *Vidensk. Medd. Dan. Naturhist. Foren.* 133:85–120.

Munk, O. and R.D. Frederiksen. 1974. On the function of aphakic apertures in teleosts. *Vidensk. Medd. Dan. Naturhist. Foren.* 137:65–94.

Munk, O. 1977. The visual cells and retinal tapetum of the foveate deep-sea fish *Scopelosaurus lepidus* (Teleostei). *Zoomorphologie* 87:21–49.

Muntz, W.R.A. 1976. On yellow lenses in mesopelagic animals. *J. Mar. Biol. Assoc. U.K.* 56:963–976.

Muntz, W.R.A., and G.S.V. Mouat. 1984. Annual variation in the visual pigments of brown trout inhabiting lochs providing different light environments. *Vision Res.* 24:1575–1580.

Munz, F.W., and W.N. MacFarland. 1977. "Evolutionary Adaptations of Fishes to the Photic Environment. In *Handbook of Sensory Physiology* by F. Crescitelli. Berlin: Springer-Verlag.

Nicol, J.A.C., H.J. Arnott, and C.G. Best. 1974. Tapeta lucida in bony fishes (Actinopterygii): a survey. *Can. J. Zool.* 59:61–81.

Ohtsuka, T. 1985. Relation of spectral types to oil droplets in cones of turtle retina. *Science* 229:874–877.

Pirenne, M.H. 1967. *Vision and the Eye.* London: Chapman & Hall.

Protasov, U.R. 1970. *Vision and near orientation of fish.* Translated by the Israel Program for Scientific Translation. Jerusalem.

Ripps, H., and R.A. Weale. 1976. "The Visual Stimulus." In *The Eye,* by H. Davson. New York: Academic Press.

Rodieck, R.A. 1973. *The Vertebrate Retina.* San Francisco: W.H. Freeman.

Saidel, W.M. and M.R. Braford, Jr. (In press.) Fundal variations in the eyes of osteoglossomorph fishes. Brain Behav. Evol.

Sirovich, L., and I. Abramov. 1977. Photopigments and pseudopigments. *Vision Res.* 17:5–16.

Smith, R.C. 1974. "Structure of Solar Radiation in the Upper Layers of the Sea. In *Optical Oceanography,* by N.G. Jerlov and E. Steeman Nielsen. London: Academic Press.

Spekreijse, H., T.E.M. Mooij, and T.J.T.P. van de Berg. 1981. Photopigments and carp ganglion cell action spectra. *Vision Res.* 21:1601–1604.

Stiles, W.S. 1948. "The Physical Interpretation of the Spectral Sensitivity Curve of the Eye. In *Transaction of the Optical Convention of the Worshipful Company of Spectacle Makers.* London: Spectacle Makers Company.

Suzuki, T., and M. Makino-Tasak. 1984. Rhodopsin-porphyropsin system in a crayfish. *Vision Res.* 24:1699.

Suzuki, T., K. Arikawa, and E. Eguchi. 1985. The effects of light and temperature on the rhodopsin-porphyropsin visual system of the crayfish, *Procambarus clarkii. Zool. Inst. Fac. Sci. Univ. Tokyo Annu. Rep.* 2:455–461.

Tansley, K. 1965. *Vision in Vertebrates.* London: Chapman & Hall.

Tyler, J.E., and R.C. Smith. 1970. *Measurements of Spectral Irradiance Underwater.* New York: Gordon and Breach.

Walls, G.L. 1942. *The Vertebrate Eye and Its Adaptive Radiation.* N.Y.: Hafner.

Ward, F. 1919. *Animal Life under Water.* London: Cassell.

Warner, J.A., M.I. Latz, and J.F. Case. 1979. Cryptic bioluminescence in a midwater shrimp. *Science* 203:1109–1110.

Zrenner, E. 1983. "Neurophysiological Aspects of Colour Vision Mechanisms in the Primate Retina." In *Colour Vision,* by J.D. Mollon and L.T. Sharp. New York: Academic Press.

# Chapter 4

# Hydrodynamic and Acoustic Field Detection

AD. J. KALMIJN *

## Introduction

Fishes have an impressive complement of hydrodynamic and acoustic sensors, commonly referred to as the lateral-line and inner-ear sense organs. The basic receptor elements are the *hair cells,* which detect the minute displacements imparted to their apical *ciliary bundles* (Fig. 4.1a). The directional sensitivity of the individual receptor cells is indicated by the asymmetric position of the single *kinocilium* relative to the several rows of *stereocilia.* Morphologically, the hair cells of the various sensory clusters are strikingly uniform. Their diversity in function is determined mainly by the peripheral structures coupling the ciliary bundles to the physical world that the animals inhabit.

The lateral-line organs, or *neuromasts,* are arranged in linear arrays on the head and along the trunk of the fishes. They connect to the water by gelatinous *cupulae* projecting from the surface of the animals (Fig. 4.1b). Many of the lateral-line organs are sunken into subcutaneous canals, which usually open to the outside by pores in the skin (Fig. 4.1c). The neuromasts respond to the motions of the water relative to the receptor epithelia. With regard to flows of external origin, the *lateral-line system* detects the *spatial derivative* of the flow field. Uniform flows are largely ineffective, since the nearly neutrally buoyant fish tend to move with the same velocity as the water, unless they rest on the substrate. (Cf. Coombs, Janssen, and Webb, Chapter 22; Denton and Gray, Chapter 23.)

The inner-ear *utricle, saccule,* and *lagena* feature sensory *maculae* loaded with solid calcareous *otoliths* or aggregates of *otoconia* (Fig. 4.1d,e). When the speed of a fish increases, the otoliths tend to lag behind by virtue of their *inertia,* thus deflecting the sensory hairs and stimulating the receptor cells. Naturally, the otolith organs also respond to gravitational forces. The inner-ear *semicircular canals* have large *cupulae,* resembling those of the lateral line (Fig. 4.1d,f).

* Scripps Institution of Oceanography, University of California, San Diego, La Jolla, CA 92093, USA

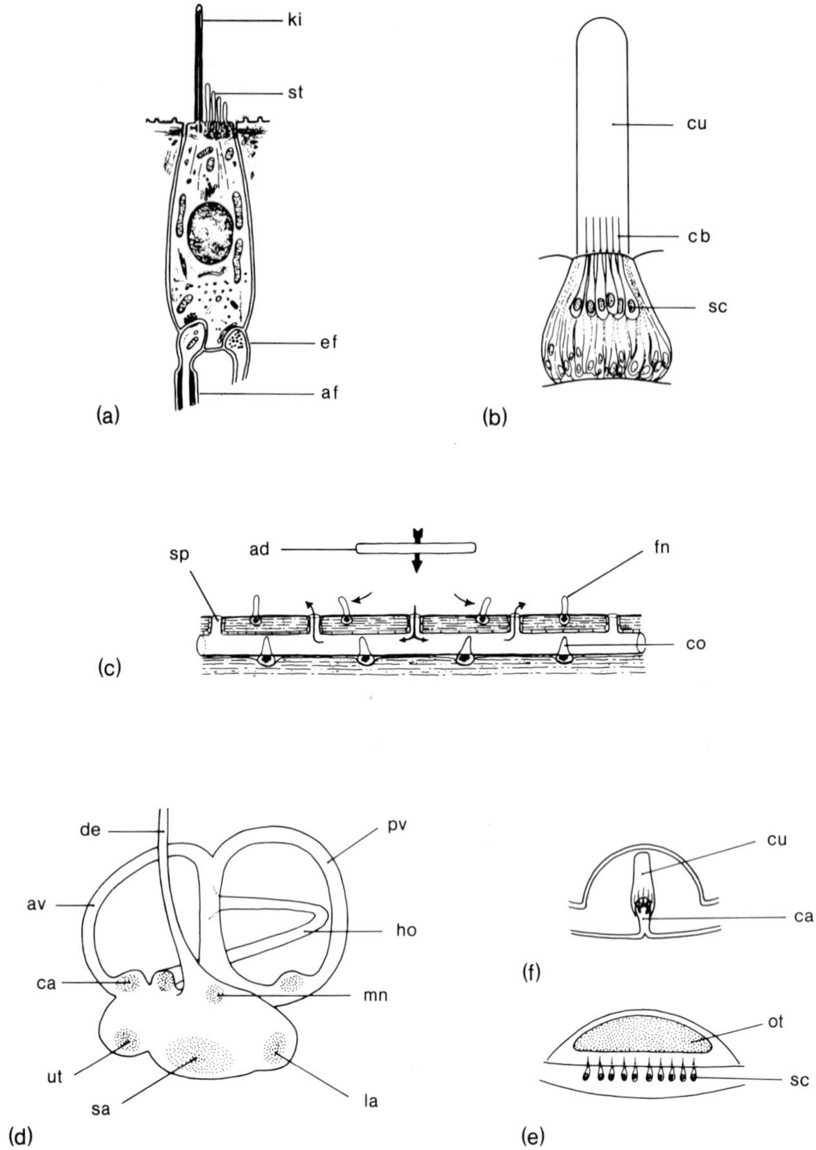


Figure 4.1. Lateral-line and inner-ear sense organs of fish. (*a*) Hair cell, *ki:* kinocilium, *st:* stereocilia, *af:* afferent nerve fiber, *ef:* efferent nerve fiber. (*b*) Lateral-line neuromast, *sc:* sensory cells, *cb:* ciliary bundles, *cu:* cupula. (*c*) Free neuromasts (*fn*) and lateral-line canal organs (*co*), *sp:* skin pores, *ad:* approaching disk. (*d*) Elasmobranch labyrinth, *ut:* utriculus, *sa:* sacculus, *la:* lagena, mn: macula neglecta, *ca:* crista ampullaris, *ho:* horizontal canal, *av:* anterior vertical canal, *pv:* posterior vertical canal, *de:* ductus endolymphaticus. (*e*) Utriculus, *ot:* otolithic mass, *sc:* sensory cells. (*f*) Ampulla of semicircular canal, *ca:* crista ampullaris, *cu:* cupula. (Figure 4.1a after Flock 1965, 4.1d after Roberts 1978, others after Dijkgraaf 1963a.)

When the head is rotated, the sensory hairs are deflected due to the *inertial moment* of the cupula and the canal fluid. Thus, the otolith organs and semicircular canals are driven by the linear and angular accelerations of the animal. (Cf. Platt, Chapter 31; Popper, Rogers, Saidel, and Cox, Chapter 27.)

Despite the lack of a cochlea, which is the organ of hearing in higher vertebrates, fishes commonly detect, and may even orient to, the acoustic fields of underwater objects. Might the utricle, saccule, lagena, or the macula neglecta, where present, support the auditory function? The *macula neglecta* (Fig. 4.1d) is an unloaded sensory patch that has been implicated in directional hearing, but without tenable physical basis (cf. p. 124). So far, the precise functions of the inner-ear organs have escaped identification, except perhaps for the semicircular canals. The lateral-line sense, in particular, has long been a source of controversy, mainly for a persistent lack of data on the natural stimulus, which some believe to be well above, and others well below, 20 Hz (cf. p. 109).

In the vicinity of underwater sound sources, acoustic fields have the nature of incompressible flows. The pressure–velocity wave that we usually think of as the sound we hear becomes apparent only at larger distances. Nevertheless, it would be inaccurate and misleading to suppose that the lateral line detects the *local hydrodynamic flow,* thus attending to the acoustic *near field,* and the inner ear detects the pressure variations or the fluid motions of the *propagating sound wave,* thus operating in the acoustic *far field.* Moreover, even though in many modern fishes the inner ear supports the detection of considerably higher frequencies than does the lateral line, a similarly clear distinction by frequency may not hold true for the basic inner-ear and lateral-line functions.

In fact, *with regard to the prevailing low-frequency sound sources, the lateral line covers only a small fraction of the near field, whereas the inner ear operates predominantly in the near, rather than the far, field.* To justify this assertion, I will discuss the role of low-frequency acoustic fields in the detection of natural sound sources, such as predator and prey. I will emphasize the relevance of multipoles over monopoles, of near fields over far fields, of fluid motion over acoustic pressure, of source velocity and acceleration over source displacement, and of low over high frequencies (cf. Rogers and Cox, Chapter 5). I thereby hope to reveal the original lateral-line and inner-ear functions and to trace the origin of vertebrate hearing, however futile the attempt may be.

The topics of acoustic communication, the detection of surface waves, the responses to stationary objects, and the control of posture and motion will not be discussed here (cf. Bleckmann, Chapter 24; Wilcox, Chapter 25). The issues whether the lateral line antedated the inner ear or whether the detection of gravity preceded the inertial function of the otoliths will also be left aside. I will *assume,* however, that aquatic vertebrates evolved their mechanical senses from direct contact, through the near field, into the far field, and started at low frequencies to extend their range to higher frequencies at a later stage. In short, I wish to elucidate the physical forces that have governed the functional evolution of the lateral-line and inner-ear senses.

The necessary physics has previously been introduced into the field of bioacoustics by Harris (1964). Admittedly, this chapter would not have been written without his and van Bergeijk's (1964) influential teaching. My presentation differs, however, in two respects. First, I have attempted to develop a conceptual understanding leading up to the acoustic equations, not vice versa. Secondly, on the basis of the same physics, I have come to drastically different conclusions, especially concerning the acoustic function of the inner ear. The most important issues Harris and van Bergeijk left unresolved are the response of the inner ear to the local flow fields of moving objects (cf. pp. 105–107) and the sensitivity of the lateral line to low-frequency accelerations (cf. pp. 113–117).

Adhering to the principle of biological significance, Dijkgraaf (1963a) has remained remarkably unscathed by the controversies of the past decades. Actually, Dijkgraaf's behavioral work and the biophysical studies of de Vries (1956) and co-workers form the cornerstones of the present chapter. In the first part (pp. 86–109), I will discuss the spatial aspects of the acoustic fields and address the problem of source localization. In the second part (pp. 109–125), I will discuss the temporal aspects of the fields and reevaluate the sensory biophysics of the lateral line and inner ear. The two sections may be consulted independently. A summary of their contents is given at the end of the chapter.

## Spatial Analysis

### Monopole Sources

In nearly all field studies on underwater hearing, high-quality monopole sources have been used. Monopoles are not only technically most efficient, but also produce the least complicated sound fields. Calibrated versions are readily available in the US Navy J-series transducers. Sinusoidally driven monopoles offer a good starting point for a discussion of the important concepts of acoustic near and far field, even though, I believe, monopoles are the wrong sources to employ for those who wish to discover the origin of vertebrate hearing. The use of sinusoidal signals also presents a convenient means of introducing the necessary physics, whether or not the animals operate predominantly in the time, rather than the frequency, domain (cf. p. 92).

The prototype of an acoustic monopole is the *pulsating sphere*. During expansion of the sphere, the water not only escapes the advancing surface radially, but also spreads laterally, occupying progressively larger, though thinner shells (Fig. 4.2a). Thus, in the immediate vicinity of a monopole source, the displacement and its derivatives, the velocity and acceleration, fall off steeply, as one over the distance squared, which constitutes the *local flow* of the acoustic *near field* (Fig. 4.3a). The displacement, velocity, and acceleration of the local-flow field of monopole or multipole sound sources are independent of frequency when referred to the corresponding quantities at the source. Whereas, in acoustics, the fields are usually expressed in terms of the source velocity, the animals may detect either the fluid velocity or fluid acceleration (cf. pp. 117–125).

* Since, by Newton's law, the negative of the pressure gradient relates to

---

* Paragraphs and sections marked with an asterisk may be omitted without loss of continuity. Pierce (1981) reminds us, though, that ". . . a deep understanding of acoustical principles is not acquired by superficial efforts . . ." (p. XI).

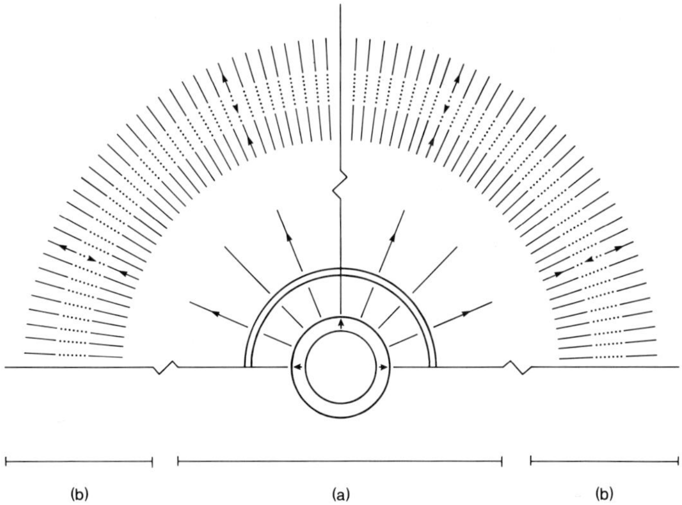


Figure 4.2. Monopole field of a pulsating sphere. (*a*) The local-flow field is centered on the source and radially symmetrical. When moving outward, the water occupies increasingly larger, but thinner shells. The particle velocities fall off as $1/r^2$. (*b*) The propagating sound wave is also centered on the source and radially symmetrical. The particle velocities fall off as $1/r$. *Arrows* indicate the sense of the particle velocities, *heavy shading* denotes compression, *light shading* denotes rarefaction. (Not to scale.)

the acceleration of the fluid particles, its spatial integral, the monopole local-flow pressure, falls off as one over the distance, rather than one over the distance squared (Fig. 4.3a). When referred to the velocity of the fluid particles, the local-flow acceleration, and thus the pressure and minus the pressure gradient, are directly proportional, whereas the local-flow displacement is inversely proportional to the frequency. Similarly, the acceleration, pressure, and minus the pressure gradient lead the velocity, while the displacement lags the velocity by 90 degrees. The ratio of the local-flow pressure and velocity amplitudes, the specific acoustic impedance, is proportional to both the frequency and the distance from the source, maintaining a phase angle of −90 degrees.

The fluid motions of the local-flow fields appear *delayed,* relative to the motions at the surface of the source, by the time required for the disturbance to travel the distance of separation. Except for this lag in time and the resulting phase shifts, the local flows about acoustic sources behave as ordinary incompressible flows and, because of their frequency independence, may, at any instant, be approximated by steady, DC *hydrodynamic flows* of corresponding source velocity. In the absence of vorticity, they furthermore qualify as frequency-independent *potential flows,* which render the local flows mathematically convenient to handle. Unfortunately, the effects of viscosity, predominantly arising from the sphere–water interface, often complicate the fields dramatically (cf. p. 125).

By advancing radially, the pulsating sphere also compresses the water to some extent. The locally increased pressure subsequently tends to relax, causing the water to expand against the adjacent parcel of water, which in turn is compressed, and so on (Fig. 4.2b). These variations in particle velocity and acoustic pressure,

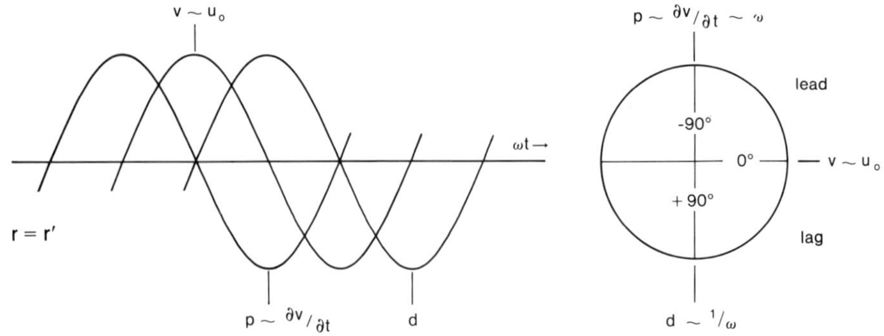


*Monopole Local Flow,* Adjusted for the Time of Travel, $ka \ll 1$, $kr \ll 1$

| | | |
|---|---|---|
| Velocity | $v$: | $+\frac{a^2}{r^2} U_0 \cos \omega t$ |
| Displacement | $\int v dt$: | $+\frac{a^2}{\omega r^2} U_0 \sin \omega t$ |
| Acceleration | $\frac{\partial v}{\partial t}$: | $-\frac{\omega a^2}{r^2} U_0 \sin \omega t$ |
| Pressure | $p$: | $-\frac{\rho \omega a^2}{r} U_0 \sin \omega t$ |
| Impedance | $\lvert p/v \rvert, p \rightarrow v$: | $\rho \omega r$, $-90$ degrees |

Note: $k = \omega / c = 2\pi / \lambda$, $u_0 = U_0 \cos \omega t$

$$-\frac{\partial p}{\partial r} = \rho \frac{\partial v}{\partial t}, \qquad p = -\int \rho \frac{\partial v}{\partial t} dr$$

Figure 4.3 (*a*). Time diagram (*top left*), phase diagram (*top right*), and table (*below*) of characteristic quantities and relations for the local-flow field of a pulsating sphere. Notations: $\omega$: angular frequency, $t$: time, $v$: particle velocity, $d$: particle displacement, $u_0$: radial source velocity, $U_0$: amplitude of radial source velocity, $a$: radius of pulsating sphere, $r$: radial distance, $p$: acoustic pressure, $\rho$: ambient density, $\lambda$: wavelength, $k$: wave number, $c$: speed of sound, $\lvert p/v \rvert$: specific acoustic impedance, $p \rightarrow v$: phase lag of $p$ relative to $v$. Diagrams apply to a point of given $r = r'$.

traveling outward from the source, constitute the familiar propagating *sound wave*. Even though near the source it is greatly exceeded by the local flow, in the *far field* the propagating sound wave dominates. Contrary to the local flow, the sound wave detaches itself from the source and is further controlled only by the physical properties of the medium, their gradients, and discontinuities (boundaries!). In natural waters, the speed of sound is nearly 1500 m/s or about $4\frac{1}{2}$ times faster than in air.

* In the propagating sound wave, the motions of the fluid particles are not only governed by Newton's law, stating that the inertial forces are proportional

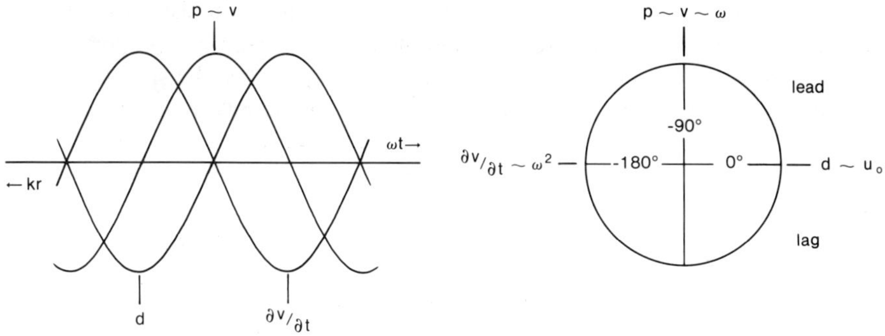


*Monopole Sound Wave,* Traveling, $ka \ll 1$, $kr \gg 1$

| | | |
|---|---|---|
| Velocity | $v$: | $-\dfrac{\omega a^2}{cr} U_0 \sin(\omega t - kr)$ |
| Displacement | $\int v\, dt$: | $+\dfrac{a^2}{cr} U_0 \cos(\omega t - kr)$ |
| Acceleration | $\dfrac{\partial v}{\partial t}$: | $-\dfrac{\omega^2 a^2}{cr} U_0 \cos(\omega t - kr)$ |
| Pressure | $p$: | $-\dfrac{\rho \omega a^2}{r} U_0 \sin(\omega t - kr)$ |
| Impedance | $\lvert p/v \rvert, p \rightarrow v$: | $\rho c$, 0 degrees |

Note: $k = \omega/c = 2\pi/\lambda$, $d = \int v\, dt$, $u_0 = U_0 \cos \omega t$

$$-\frac{\partial p}{\partial r} = \rho c^2 \frac{\partial^2 d}{\partial r^2} = \rho \frac{\partial^2 d}{\partial t^2}, \qquad p = \rho c^2 \frac{\partial d}{\partial r} = \int \rho \frac{\partial v}{\partial t} dr$$

Figure 4.3 (*b*). Time–space diagram (*top left*), phase diagram (*top right*), and table (*below*) of characteristic quantities and relations for the propagating sound wave emanating from a pulsating sphere. For notations, see Figure 4.3(a).

to the second derivative of the particle displacement with respect to time, but also by the volume–pressure relationships, stating that the pressure gradients due to the finite compressibility of the water are proportional to the second derivative of the particle displacement with respect to distance (Fig. 4.3b). It is this dynamic equilibrium between the inertial and compressional, rather than the inertial and source-imposed, forces that makes the propagating sound wave so different from the local flow. Nevertheless, the propagating sound wave and local flow remain physically inseparable components of one and the same acoustic field, ruled by one and the same wave equation (cf. p. 92).

* Since in the propagating sound wave of a monopole or multipole source the fluid particles behave as simple harmonic oscillators, their velocity is maximal when they rush through the equilibrium position in the direction of propagation (Fig. 4.3b). At that moment, the pressure gradient goes through zero to become

positive and the acoustic pressure is maximal. That is, in the undisturbed propagating sound wave, the particle velocity and acoustic pressure are proportional to, and in phase with, one another. The ratio of their amplitudes, the specific acoustic impedance, is independent of frequency and a function of only the physical properties of the medium. Hence, the constancy in specific acoustic impedance is often considered a criterion of true far-field conditions (cf. Rogers and Cox, Chapter 5).

While the acoustic intensity or energy flux of the propagating sound wave is proportional to the square of the particle-velocity amplitude or, equivalently, the square of the acoustic-pressure amplitude, by spreading three-dimensionally the monopole intensity diminishes as one over the distance squared. Consequently, both the particle velocity and acoustic pressure of the propagating sound wave fall off as one over the distance (Fig. 4.3b). Furthermore, when adjusted for the time of travel, the particle velocity and acoustic pressure of the sound wave are proportional to, and in phase with, the acceleration, rather than the velocity, at the source. Thus, for a monopole to produce a propagating sound wave, the advancing surface must have a nonzero acceleration.

In short, the monopole particle velocities of both the local flow and propagating sound wave are centered on the source. Near the source they fall off steeply, farther away more slowly. The local flow formally extends to infinity, but prevails only in the immediate vicinity of the source. The propagating sound wave begins right at the source, but is in full command only in the far field. Closer to the source, in the near field, the local flow is not negligible. The particle velocities of the local flow and the propagating sound wave are equal in strength at a distance of one wave length over $2\pi$. It is a dangerous practice, however, to regard all space within this distance as the near field, and all space beyond this distance as the far field, which has misled many a good scientist to infer far-field conditions barely beyond this magic boundary (cf. pp. 103–105).

If the fluid motions of monopole fields are all centered on the source, what might the near–far-field distinction mean to fishes insensitive to acoustic pressure? Pointing alternately towards and away from the source, the fluid velocities offer reliable directional cues. The remaining 180-degree ambiguity may, at short distance, be resolved by detecting the convergence of the fluid velocities or the gradient in their strengths. The issue is largely academic, however, because the majority of the velocity fields that fish encounter are not of a monopole nature. Monopole sources, like the pulsating sphere, are characterized by changes in volume, whereas most biologically relevant sound sources are of a nearly constant volume, as van Bergeijk noted already in 1964.

Swimbladders may be largely monopolar, but they are not ubiquitously present in fishes and need not be used to produce sound. Hence, I believe that the function of hearing evolved in response to acoustic fields of a more general nature and occurrence than the fields of monopole sources, and I consider the use of air cavities to produce and detect sound a later development. Sharks have a keen sense of hearing, but lack air cavities (cf. pp. 103–105). Swimbladders often play a vital role in acoustic communication (Hawkins and Myrberg 1983). When orienting to monopole sources, the "hearing specialists" among fishes may even compare the particle motion with the acoustic pressure (Buwalda 1981; Schuijf and Hawkins 1983; Popper et al., Chapter 27). Yet, in my opinion, we will not be able to ascertain the basic principles of acoustic source localization by applying solely monopole fields.

## * Monopole-Field Equations

Since, during the passage of the acoustic disturbance, (a) the mass of the medium remains conserved, (b) the fluid particles at all times obey Newton's law of motion, and (c) the pressure and density fluctuations are related to one another thermodynamically, it follows that the radially symmetric field of a simple or monopole source may be approximated, to first order, by the acoustic wave equation of the form (e.g., Lighthill 1980; Pierce 1981):

$$\frac{\partial^2(r\,\Phi_0)}{\partial t^2} = c^2\frac{\partial^2(r\,\Phi_0)}{\partial r^2} \tag{1}$$

where $\Phi_0$ is the monopole or zeroeth-moment velocity potential, expressed in spherical coordinates about the source, $r$ the radial distance to the fluid particle of interest, $t$ the time, and $c = \sqrt{\kappa/\rho}$ the speed of travel, with $\kappa$ the bulk modulus (one over the compressibility) and $\rho$ the density of the medium.

Although somewhat of a mathematical abstraction, the scalar velocity potential takes a central position in the description of the acoustic field, for its time derivative yields the negative of the excess or acoustic pressure, $p$, while its spatial gradient reproduces the fluid or particle velocity, $v$:

$$p = -\rho\frac{\partial\Phi_0}{\partial t} \qquad \text{and} \qquad v = \frac{\partial\Phi_0}{\partial r} \tag{2,3}$$

The solution of Eq. 1 for a small, sinusoidally *pulsating* sphere of a radius $a$ and a surface velocity $u_0 - U_0\,e^{-i\omega t}$ is:

$$\Phi_0 = -\frac{a^2}{r}U_0\,e^{-i(\omega t-kr)} \tag{4}$$

where $U_0$ is the amplitude of the radially oscillating source velocity, $(a^2/r)U_0$ the amplitude of the velocity potential, $\omega = 2\pi f$ the angular frequency, $k = 2\pi/\lambda = \omega/c$ the wave number, $\lambda$ the wavelength, and $i = \sqrt{-1}$. The radius of the source, $a$, is assumed to be negligibly small compared to the wavelength, $\lambda$, or $ka \ll 1$.

For the study of harmonic phenomena, the complex notation is the algebraic language of choice, as it keeps track of both the amplitude and the phase. The real part reproduces the actual physical expression, the square root of the summed real and imaginary parts squared yields the amplitude of the expression, and the imaginary over the real part yields the tangent of the phase angle.

The acoustic pressure, $p$, follows from Eqs. 2 and 4, giving in complex notation:

$$p = -\frac{i\rho c(ka)^2}{kr}U_0\,e^{-i(\omega t-kr)} \tag{5a}$$

or in real terms:

$$p = -\frac{\rho cka^2}{r} U_0 \sin(\omega t - kr) \tag{5b}$$

Thus, the acoustic pressure, $p$, is proportional to $1/r$ and to $k = \omega/c$, and leads the source velocity, $u_0 = U_0 \cos \omega t$, by 90 degrees, disregarding the time of travel. Moreover, there are no separate terms to be identified with the local flow and propagating sound wave.

Similarly, the particle velocity follows from Eqs. 3 and 4, giving in complex notation:

$$v = -i(ka)^2 \left(\frac{1}{kr} + \frac{i}{(kr)^2}\right) U_0 \, e^{-i(\omega t - kr)} \tag{6a}$$

or in real terms:

$$v = -\frac{ka^2}{r} U_0 \sin(\omega t - kr) + \frac{a^2}{r^2} U_0 \cos(\omega t - kr) \tag{6b}$$

The first terms of Eqs. 6a and 6b may be identified with the propagating *sound wave,* where $v$, like $p$, is proportional to $1/r$ and to $k = \omega/c$, and leads $u_0$ by 90 degrees. The second terms may be identified with the *local flow,* where $v$ is proportional to $1/r^2$, independent of $k = \omega/c$, and in phase with $u_0$. For $kr \gg 1$, the propagating sound wave dominates, which defines the monopole *far field.* Within this distance, the local flow is not negligible, which defines the monopole *near field.* For $kr = 1$ or $r = 1/k = \lambda/2\pi$, the local flow and propagating sound wave have equal velocity amplitudes, whereas for $kr \ll 1$ the local flow prevails. The velocity field is further proportional to the monopole moment, $4\pi a^2 U_0$, and hence to the *surface area* of the pulsating sphere.

However powerful, the monopole equations have their limitations and must be applied with the utmost care. Fortunately, most underwater sounds feature sufficiently small particle velocities and pressure variations that they are adequately described by the linear form of the acoustic wave equation and may be meaningfully expanded in terms of sinusoidal components. This justifies treating bioacoustics in the frequency domain, whether or not that is the method of preference or the method followed by the animals. Young herring, as a matter of fact, respond to a single cycle just as well as to a train of sinusoidal stimuli (Blaxter, Denton, and Gray 1981). At any rate, by thinking in terms of sinusoidal components and their sums, I have arrived at the conclusions of this chapter. Moreover, all frequency-domain phenomena have their time-domain counterparts, even though the transformations may not always be trivial.

The shape of the transducer is also crucial. In the derivation of the monopole formulas, the source was assumed to be radially symmetrical and small compared to the wavelength of the sound it produces. While the size of the source is usually no problem at the low frequencies of interest, the symmetry requirement tends to be more critical, certainly for the near field. None of the sources used for the study of underwater hearing, including those of the US Navy J-series, are radially symmetrical in design. Used at short range, they offer poor monopole sources, corrupted by higher-order moments and secondary flow, an often unrecognized form of vorticity (cf. p. 97).

If not stated otherwise, space usually is assumed to be unbounded. In experimental aquaria, however, reflections and interference have ruined many investigators' best intentions. Thus, in studies on underwater hearing, the effects of the walls, bottom, and the air–water interface must either be minimized satisfactorily or be fully understood, so that they can be taken into account. Actually, the presence of boundaries is quite natural (cf. Rogers and Cox, Chapter 5) and may even become the object of detection. The problem is least for local flows, which suffer mainly from distortion, rather than reflection, at rigid walls. Distortions are readily dealt with. Free surfaces, however, must either be kept at a sufficient distance or eliminated completely.

## Dipole Sources

If monopoles are not suited for the reproduction of the acoustic fields most relevant to the origin of underwater hearing, what other sources might be more adequate? Any source small compared to the wavelength of the sound it produces can be expanded mathematically into a multipole series, consisting of a monopole, a dipole, a quadrupole, and so forth. Such a representation has the illuminating feature that the local flows of subsequent terms fall off as one over increasing powers of the distance, so that sufficiently far from the source the lowest nonzero term of the series tends to override all higher terms. In the absence of a monopole moment, the dipole term tends to dominate, unless its moment also vanishes and the quadrupole or, perhaps, the octupole term takes over.

The prototype of a sinusoidally driven dipole source is a rigid, constant-volume *vibrating sphere*. The advancing sphere pushes the water ahead in front, while in the rear the water trails the sphere, which gives rise to the typical dipole configuration of the *local flow* (Fig. 4.4a), controlling the *inner part* of the *near field* (cf. p. 94). The axially symmetrical pattern reveals that the off-axis fluid motions of the local flow about a vibrating sphere are not centered on the source, as they are about the radially symmetrical monopole, but can have any direction with respect to the dipole axis. Moreover, the fluid displacement, velocity, and acceleration fall off as one over the third, rather than the second, power of the distance, and are, as in all local flows, independent of frequency when referred to the corresponding quantities at the source (Fig. 4.5a).

The dipole field of a vibrating sphere may be regarded as the equivalent of two equal-strength monopole fields of which the sources are separated only by a short distance and that pulsate 180 degrees out of phase. The required monopole moment would equal the strength of the dipole moment divided by the distance of separation. The steep cubic, rather than quadratic, decline of the dipole velocity field with distance is readily understood from the progressive cancellation of the two constituent monopole fields. Also, its frequency independence follows immediately from the frequency independence of the constituent monopole fields. The opposite-phase monopole pair furthermore offers a convenient practical implementation of the dipole source. Equivalently, a rigid, hollow sphere may be made to vibrate by a reciprocating mass inside its spherical shell.

* By Newton's law, the negative of the pressure gradient relates to the particle

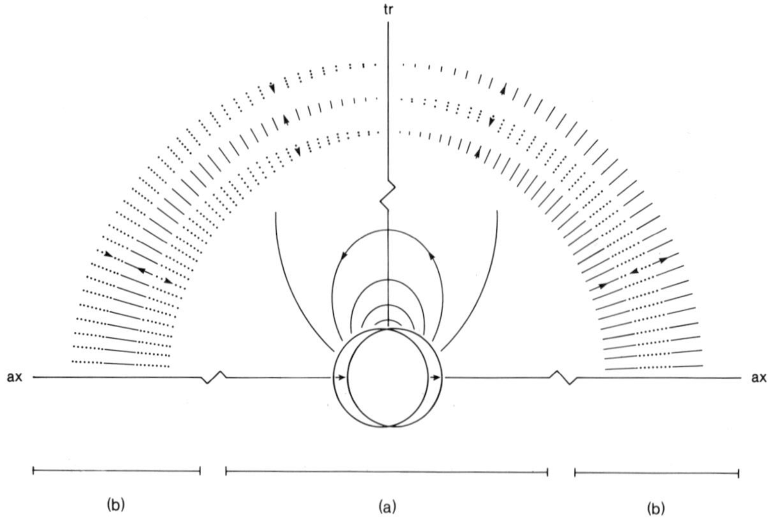


Figure 4.4. Dipole field of a vibrating sphere. (*a*) The local- and intermediate-flow fields are *not* centered on the source, except along the axis of vibration, and are *not* radially symmetrical. The field lines begin and end at the surface of the sphere. The particle velocities fall off as $1/r^3$ and $1/r^2$, respectively. (b) The propagating sound wave is centered on the source, but it is *not* radially symmetrical. The particle velocities fall off as $1/r$, being strongest along the axis of vibration and zero in the transverse plane. *Arrows* indicate the sense of the particle velocities, *heavy shading* denotes compression, *light shading* denotes rarefaction, *ax:* axis of vibration, *tr:* transverse plane. (Not to scale.)

acceleration, so that its spatial integral, the pressure itself, falls off as one over the distance squared, rather than one over the distance cubed (Fig. 4.5a). Furthermore, the local-flow acceleration, the pressure, and minus the pressure gradient are directly proportional to the frequency and lead the particle velocity by 90 degrees, whereas the displacement is inversely proportional to the frequency and lags the particle velocity by 90 degrees. By the same token, the specific acoustic impedance, the ratio between the pressure and velocity amplitudes, is proportional to both the frequency and the distance, while it maintains a phase angle of $-90$ degrees, as is characteristic for all local flows.

Before yielding to the pure far field, the fluid velocities about a dipole source tend to fall off less steeply, as one over the distance squared, than in the inner part of the near field. They also tend to become proportional to the frequency and, when adjusted for the time of travel, to lead the vibrating sphere by 90 degrees, so that the fluid velocity of this *intermediate flow* relates to the acceleration, rather than the velocity, at the source. Thus, between the regions dominated by the local flow and the propagating sound wave, the dipole field has an additional *near-field* region, that dominated by the intermediate flow. Interestingly enough, the fluid velocities of the intermediate flow *conform to the dipole configuration of the local flow,* even though they are frequency dependent and fall off less steeply with distance.

The particle velocities of the propagating *sound wave* emanating from a vibrat-

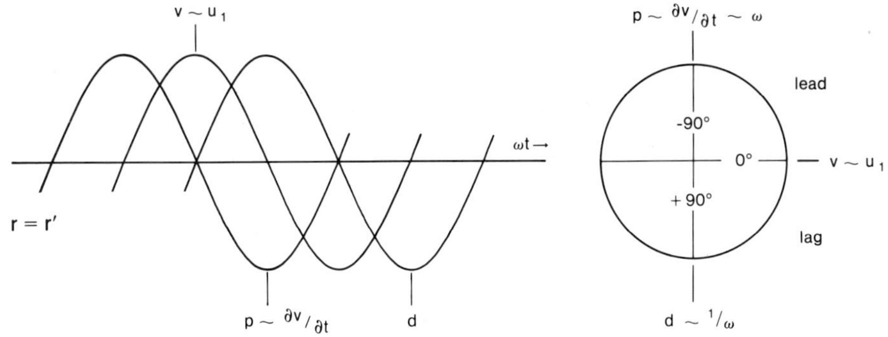


*Dipole Local Flow,* Adjusted for the Time of Travel, $ka \ll 1$, $kr \ll 1$

| | | | |
|---|---|---|---|
| Velocity | $v_{r,\theta}$: | $+\frac{a^3}{r^3} U_1 \cos\theta \cos\omega t,$ | $+\frac{a^3}{2r^3} U_1 \sin\theta \cos\omega t$ |
| Displacement | $\int v\,dt$: | $+\frac{a^3}{\omega r^3} U_1 \cos\theta \sin\omega t,$ | $+\frac{a^3}{2\omega r^3} U_1 \sin\theta \sin\omega t$ |
| Acceleration | $\frac{\partial v}{\partial t}$: | $-\frac{\omega a^3}{r^3} U_1 \cos\theta \sin\omega t,$ | $-\frac{\omega a^3}{2r^3} U_1 \sin\theta \sin\omega t$ |
| Pressure | $p$: | $-\frac{\rho\omega a^3}{2r^2} U_1 \cos\theta \sin\omega t$ | |
| Impedance | $\lvert p/v \rvert, p \rightarrow v$: | $\rho\omega r/2$, $-90$ degrees, along dipole axis | |

Note: $k = \omega/c = 2\pi/\lambda$, $\quad u_1 = U_1 \cos\omega t$

$$-\frac{\partial p}{\partial r} = \rho\frac{\partial v_r}{\partial t}, \quad -\frac{\partial p}{r\partial\theta} = \rho\frac{\partial v_\theta}{\partial t}, \quad p = -\int \rho\frac{\partial v_r}{\partial t}dr - \int \rho\frac{\partial v_\theta}{\partial t} r\, d\theta$$

Figure 4.5 (*a*). Time diagram (*top left*), phase diagram (*top right*), and table (*below*) of characteristic quantities and relations for the local-flow field of a vibrating sphere. Notations: $\omega$: angular frequency, $t$: time, $v$: particle velocity, $d$: particle displacement, $u_1$: axial source velocity, $U_1$: amplitude of axial source velocity, $a$: radius of vibrating sphere, $r$: radial distance, $\theta$: angle of radiation, $p$: acoustic pressure, $\rho$: ambient density, $\lambda$: wavelength, $k$: wave number, $c$: speed of sound, $\lvert p/v \rvert$: specific acoustic impedance, $p \rightarrow v$: phase lag of $p$ relative to $v$. Diagrams apply to a point of given $r = r'$.

ing sphere are directed radially and fall off as one over the distance, as they do in the instance of a monopole or any small multipole source (Fig. 4.4b). However, the *far field* about a dipole source is not radially symmetrical, for the amplitude of the particle velocity is a function of the angle of radiation. It is largest along the axis of vibration and tends to zero in the plane through the center of the sphere normal to the axis of vibration. Otherwise, at any point in the pure far field, the sound originating from a dipole source is indistinguishable from the sound originating from a monopole source. Ultimately, after freeing itself from the source, the spherically spreading wave depends only on the physical properties of the medium.

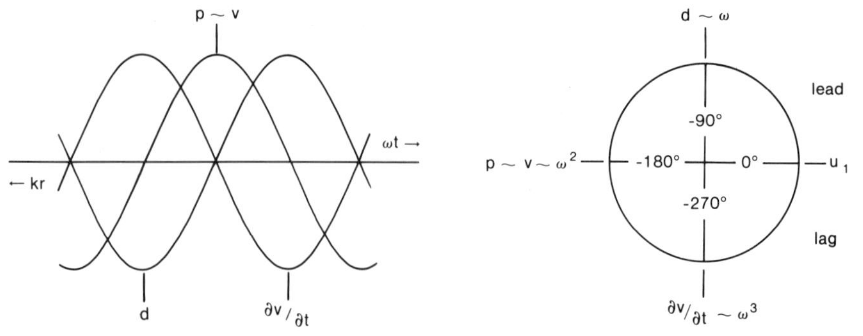


*Dipole Sound Wave,* Traveling, $ka \ll 1$, $kr \gg 1$

| | | |
|---|---|---|
| Velocity | $v$: | $-\dfrac{\omega^2 a^3}{2c^2 r} U_1 \cos\theta \cos(\omega t - kr)$ |
| Displacement | $\int v\,dt$: | $-\dfrac{\omega a^3}{2c^2 r} U_1 \cos\theta \sin(\omega t - kr)$ |
| Acceleration | $\dfrac{\partial v}{\partial t}$: | $+\dfrac{\omega^3 a^3}{2c^2 r} U_1 \cos\theta \sin(\omega t - kr)$ |
| Pressure | $p$: | $-\dfrac{\rho\omega^2 a^3}{2cr} U_1 \cos\theta \cos(\omega t - kr)$ |
| Impedance | $\lvert p/v \rvert, p \rightarrow v$: | $\rho c$, 0 degrees |

Note: $k = \omega/c = 2\pi/\lambda$, $\quad d = \int v\,dt$, $\quad u_1 = U_1 \cos \omega t$

$$-\frac{\partial p}{\partial r} = \rho c^2 \frac{\partial^2 d}{\partial r^2} = \rho \frac{\partial^2 d}{\partial t^2}, \qquad p = -\rho c^2 \frac{\partial d}{\partial r} = -\int \rho \frac{\partial v}{\partial t} dr$$

Figure 4.5 (*b*). Time–space diagram (*top left*), phase diagram (*top right*), and table (*below*) of characteristic quantities and relations for the propagating sound wave emanating from a vibrating sphere. For notations, see Figure 4.5(a).

* The particle velocity in the propagating sound wave of a dipole source is proportional to the frequency squared, when referred to the source velocity, and leads by 180 degrees after correction for the time of travel (Fig. 4.5b). Consequently, the far-field particle velocity relates not to the acceleration of the source, as in the monopole field, but to the derivative of the acceleration. Moreover, the far-field particle velocity is not proportional to the radius of the sphere squared over the speed of sound, as in the monopole field, but to the radius of the sphere cubed over the speed of sound squared. Thus, the lower the frequency and the smaller the source, the less favorably the vibrating sphere compares to the pulsating sphere in producing far-field sound. Near a dipole source, the local flow offers such a good escape for the water that the fluid compressions giving rise to the propagating sound wave remain only weak. This explains why dipole sources are such poor far-field sound projectors.

The distance at which the particle velocities of the local flow and the propagating sound wave are equal in magnitude depends on the angle of radiation. Measured along the axis of vibration, the equal-amplitude distance is approximately 1.4λ/

$2\pi$. At larger angles of radiation, it increases progressively, tending to infinity at 90 degrees, where the propagating sound wave vanishes. Although the concepts of local flow and propagating sound wave are crucial to the understanding of the lateral-line and inner-ear functions, the use of the equal-amplitude surface is of little help, since it indicates where the local flow and the propagating sound wave *do not* prevail, rather than where they do. Actually, at this magic boundary, the intermediate flow prevails (cf. pp. 94 and 98). For the problem of directional hearing, we need a more informative, biologically realistic approach (cf. pp. 99–102).

The most common biological dipole fields derive from the *whole-body* motions of swimming or coasting animals. Van Bergeijk (1964, 1967a) at first contrasted the monopole swimbladder pulsations with the undulations of the body and the oscillations of fins and tail, but later also called attention to the analogy between a gliding fish and the motion of a rigid sphere. It may seem strange to regard a fish moving at constant velocity as a source of "sound." Indeed, when the source frequency tends to zero, the propagating sound wave and, subsequently, the intermediate flow degenerate, and only a strong local flow is left. Yet, this usually is what the lateral line and inner ear detect, I believe (cf. pp. 105–107). Naturally, at the level of the sense organs, the stimulus varies in strength and direction with the change of relative position between the source and the recipient animal (cf. Kalmijn, Fig. 3 in Chapter 6).

Fluctuations in the forward speed of fishes due to their mode of propulsion result in source accelerations and their derivatives, thus producing true far-field sounds. For the predominantly low frequencies and low source strengths, however, the particle velocities of the propagating sound wave remain usually negligibly small compared to the local- and intermediate-flow velocities up to distances far beyond the range of detection in the often noisy, natural environment. Only when animals are startled and suddenly dart away to take flight might the far field become more relevant due to the presence of higher-frequency components and an elevated source strength. The relative motions of body parts cause higher-order moments as well (cf. pp. 102–103).

So far the effects of viscosity have been neglected. However, I wish to mention briefly one of its curious consequences, that of *secondary flow*. Without viscosity, the flow in front and in the rear of a moving sphere would be completely symmetrical. However, the slightest viscosity causes the trailing water to develop vortices, which are not nullified when the sphere reverses and moves in the opposite direction. Hence, the vortices grow with time, drawing the water in within the equatorial plane and expelling it in either direction along the axis of vibration. The presence of gradually increasing, steady flows may have plagued many, if not all, behavioral and neurophysiological experiments in which vibrating spheres were used. A discussion of other viscous effects arising from the boundary layer at the source will be deferred to a later section (pp. 124–125).

## * Dipole-Field Equations

The first-moment velocity potential of a doublet or dipole field may be obtained by summing the zeroeth-moment velocity potentials of two simple or monopole fields, the sources of which are only a short distance apart and pulsate in opposite phases (cf. p. 93), yielding for the velocity potential of a small, *vibrating* sphere of radius $a$ (e.g., Kinsler, Frey, Coppens, and Sanders 1982; Lamb 1931):

$$\Phi_1 = \frac{a^3}{2r}\left(ik - \frac{1}{r}\right) U_1 \cos\theta\, e^{-i(\omega t - kr)} \tag{7}$$

where $u_1 = U_1\, e^{-i\omega t}$ is the source velocity, $\theta$ the angle of radiation relative to the axis of vibration, and $ka \ll 1$ (cf. p. 91).

Thus, the acoustic pressure, $p$, is given by:

$$p = -\rho \frac{\partial \Phi_1}{\partial t}$$

$$= -\frac{\rho c (ka)^3}{2}\left(\frac{1}{kr} + \frac{i}{(kr)^2}\right) U_1 \cos\theta\, e^{-i(\omega t - kr)} \tag{8a}$$

$$= -\frac{\rho c\, k^2 a^3}{2r} U_1 \cos\theta \cos(\omega t - kr) \tag{8b}$$

$$- \frac{\rho c\, k a^3}{2r^2} U_1 \cos\theta \sin(\omega t - kr)$$

For $kr \gg 1$, the first terms of Eqs. 8a and 8b dominate and may be identified with the propagating sound wave, where $p$ is proportional to $1/r$ and to $k^2$, and leads the source velocity $u_1 = U_1 \cos \omega t$ by 180 degrees, disregarding the time of travel (cf. p. 87). For $kr \ll 1$, the second terms dominate, $p$ is proportional to $1/r^2$ and to $k$, and leads $u_1$ by 90 degrees. The acoustic pressure is further proportional to the cosine of the angle of radiation, $\theta$.

The radial particle velocity, $v_r$, and tangential particle velocity, $v_\theta$, are:

$$v_r = \frac{\partial \Phi_1}{\partial r} \quad \text{and} \quad v_\theta = \frac{1}{r}\frac{\partial \Phi_1}{\partial \theta}$$

$$v_r = -\frac{(ka)^3}{2}\left(\frac{1}{kr} + \frac{2i}{(kr)^2} - \frac{2}{(kr)^3}\right) U_1 \cos\theta\, e^{-i(\omega t - kr)} \tag{9a}$$

$$= -\frac{k^2 a^3}{2r} U_1 \cos\theta \cos(\omega t - kr) \tag{9b}$$

$$- \frac{k a^3}{r^2} U_1 \cos\theta \sin(\omega t - kr)$$

$$+ \frac{a^3}{r^3} U_1 \cos\theta \cos(\omega t - kr)$$

$$v_\theta = -\frac{(ka)^3}{2}\left(\frac{i}{(kr)^2} - \frac{1}{(kr)^3}\right) U_1 \sin\theta\, e^{-i(\omega t - kr)} \tag{10a}$$

$$= -\frac{k a^3}{2r^2} U_1 \sin\theta \sin(\omega t - kr) \tag{10b}$$

$$+ \frac{a^3}{2r^3} U_1 \sin\theta \cos(\omega t - kr)$$

The first terms of Eqs. 9a and 9b describe the propagating *sound wave,* where $v$, like $p$, is proportional to $1/r$ and to $k^2$, and leads $u_1$ by 180 degrees. The second terms represent the *intermediate flow,* where $v$ is proportional to $1/r^2$ and to $k$, and leads $u_1$ by 90 degrees. The third terms are those of the *local flow,* where $v$ is proportional to $1/r^3$, independent of $k$, and in phase with $u_1$. For $kr \gg 1$, the propagated sound wave dominates, which defines the dipole *far field.* Closer to the source, the local or intermediate flow is not negligible, which defines the dipole *near field.* The intermediate flow prevails when $kr \sim 1$, the local flow when $kr \ll 1$. The radial particle velocities of the dipole field are proportional to the cosine of the angle of radiation, $\theta$.

Equations 10a and 10b give the tangential components of the velocity field, with the first terms corresponding to the intermediate flow, and the second terms to the local flow. The tangential velocities of the local and intermediate flow are proportional to the sine of the angle of radiation. The propagating sound wave is strictly radial and lacks a tangential component. Both the local and intermediate flows feature the typical dipole configuration. (Quadrupole and higher-moment fields have increasing numbers of intermediate flows, all but the first of which differ in configuration from their respective local flows.) The particle velocities of the dipole local flow and propagating sound wave are equal in magnitude at about $1.4\lambda/2\pi$ along the axis of vibration and at infinity in the transverse plane through the sphere. The velocity field is further proportional to the dipole moment, $2\pi a^3 U_1$, and thus to the *volume* of the vibrating sphere.

The acoustic equations of this chapter are essentially the same as those given by Harris (1964). I have only written the complex exponent as $-i(\omega t - kr)$, rather than $ik\ (r - a - ct) = -i(\omega t - kr - ka)$, a simplification consistent with the approximations made in the derivation of the formulas (Pierce 1981). However, I have chosen not to adhere to the definitions of the near and far field given by Harris (1964) and van Bergeijk (1964), and later simplified by Siler (1969), since their use of the terms is not in keeping with the standard acoustic texts (Morse and Ingard 1968; Pierce 1981; Kinsler et al. 1982) and has led to unnecessary confusion in the field of bioacoustics (cf. Rogers and Cox, Chapter 5). Where Harris and van Bergeijk speak of the "near field," or Siler speaks of the "very-near field," they mean the region in which the local flow dominates, rather than the region in which either the local or intermediate flow is not negligible.

## Directional Analysis

In the propagating sound wave of a monopole or multipole source, the particle velocities are radially directed, as they are in the local flow of a monopole source. However, in the local *and* intermediate flow of a dipole source, the particle velocities feature the typical axial-symmetrical dipole configuration. Hence, for a dipole source, we may calculate where in space the particle-velocity field is approximately *dipolar,* and where in space it is more nearly *radial,* say, within plus or minus 10 degrees (Fig. 4.6). The region in which the velocity field is neither dipolar nor radial, within plus or minus 10 degrees, we will call the *transition* or *gray zone.* All three, the dipolar, transition, and radial zone, are of great biological interest, however different they are. The present chapter deals primarily with the dipolar zone.

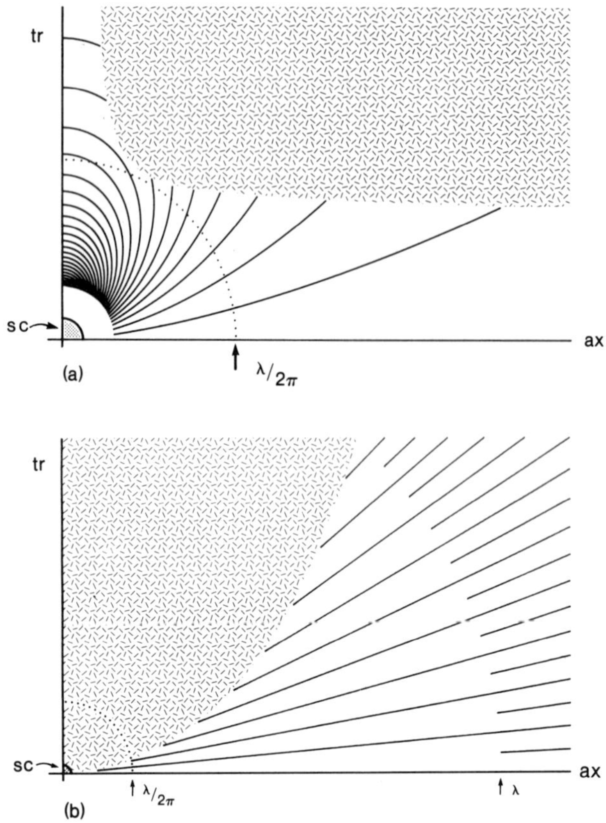


Figure 4.6. Particle-velocity field about a vibrating sphere. (*a*) Dipolar and nondipolar zones: the field is dipolar within 10 degrees up to a distance of at least λ/2π. (*b*) Radial and nonradial zones: the field is radial within 10 degrees in 30% of the solid angle only beyond a distance λ. The transition or gray zone is the region in which the field is neither dipolar, nor radial, i.e., the region in which the shaded areas of (*a*) and (*b*) overlap. Legends: *sc:* source, ax: axis of vibration, *tr:* transverse plane. Note the 1:3 difference in scale between (*a*) and (*b*).

The region in which the field is dipolar, within 10 degrees, extends to distances of at least λ/2π, or 24 m, for a frequency of 10 Hz. It stretches to infinity, not only along the axis of vibration, where the particle velocities of the local and intermediate flow are radial as they are in the propagating sound wave, but also in the transverse plane, where the particle velocities of the local and intermediate flow are tangential and the sound wave vanishes. The region in which the field is radial, within 10 degrees, increases with distance, but never covers the full 360 degrees. At a distance of one wavelength, or 150 m at 10 Hz, the field is radial only within about 45 degrees from the axis of vibration, or 30% of the solid angle. At shorter distances, this coverage diminishes rapidly and is not more than 6% at 1/π, and a mere 2% at 1/2π, of the wavelength. The boundaries between the three zones encroach upon the source as frequency increases.

The transition zone is readily recognized in the instance of a sinusoidally driven dipole source. Whereas the individual fluid particles generally trace imperceptibly curved trajectories in the local and intermediate flow, and perfectly straight trajectories in the propagating sound wave, in the transition zone they describe ellipses of various aspect ratios. For fields of arbitrary waveform, the boundaries of the three zones depend on the frequency components of interest. To be entirely in the dipolar zone, the recipient animal must be in the dipolar zone of the highest frequencies; to be entirely in the radial zone, it must be in the radial zone of the lowest frequencies. Receiving components in any two or in all three zones might, conceivably, allow the animals to interpret the particle velocities properly and to estimate the distance to the source, whether they do so in the time or in the frequency domain (cf. p. 92).

* In the local and intermediate flow of a dipole source, the imperceptibly curved trajectories make an angle $\beta$ with the radius vector, given by

$$\tan \beta = 1/2 \tan \theta \tag{11}$$

where $\theta$ is the angle between the radius vector of the point under consideration and the dipole axis. Equation 11 follows from Eqs. 9 and 10 for the radial and tangential components of the local- and intermediate-flow particle velocities.

* The elliptical shape of the particle trajectories in the transition zone results from the phase differences between the velocity contributions of the local flow, intermediate flow, and propagating sound wave. As the direction of the velocity field, I have taken the major axis of the ellipse, although only if it was, arbitrarily, at least five times longer than the minor axis. At each point of interest, the radial and tangential components of the velocity field were calculated numerically on the basis of Eqs. 9 and 10, and the ratio of their amplitudes was determined. Subsequently, the reference axes were rotated relative to the original coordinates until, at the angle $\beta$, the ratio of the velocity amplitudes along the new axes was maximal and, thus, the reference axes coincided with the major and minor axes of the ellipse.

* As may be seen by close inspection of Eqs. 9 and 10, the major axis of the velocity ellipse first turns to somewhat larger $\beta$ before swinging back and smoothly approaching the radial direction with increasing distance from the source. The total swing ranges from zero degrees along the axis of vibration to a limiting value of 90 degrees on nearing the transverse plane through the center of the sphere. Note that, during the gradual change in direction, the phase of the particle velocity advances by 180 degrees, whereas the phase of the acoustic pressure advances by only 90 degrees. However, fishes devoid of air cavities are not known to be sensitive to the acoustic pressure, and thus would lack a reference for judging the phase of the particle velocity.

In summary, the acoustic dipole field may be regarded to consist of the local flow, intermediate flow, and propagating sound wave (pp. 93–97). The far field is the region in which the propagating sound wave dominates, whereas the near field is the region in which the local and intermediate flow are not negligible (p. 98). The use of the near-field term for the region in which the local flow dominates should be abandoned in favor of the conventional engineering definitions (p. 99). I have furthermore introduced the terms dipolar, intermediate, and radial zones to highlight the directional information available in the particle-velocity field (this section). Their definitions are purely descriptive and have been chosen specifically for the discussion of directional hearing.

## Quadrupole and Higher-Order Moments

Mathematically, acoustic sources small compared to the wavelength of the sound they produce have unique monopole moments, characterized by a single, zero or nonzero, scalar quantity, independent of the choice of the reference point. Differences between sources of equal monopole moment are attributed to differences in their dipole and higher-order moments. Physically, monopole moments result from changes in the volume of the source, as in the instance of pulsating air bubbles, swimbladders, and similar air cavities.

In the absence of a monopole moment, acoustic sources have unique dipole moments, characterized by a single, zero or nonzero, vector quantity or, equivalently, by three scalar quantities, independent of the choice of the reference point. Differences between sources having equal dipole moments, but no monopole moment, are attributed to their quadrupole and higher-order moments. The dipole moment relates to the motion of the object's center of mass, which is particularly obvious during the whole-body movements of animals.

In contrast to the monopole moments, the dipole moments of moving animals seldom vanish, so that the quadrupole moments, as defined by two vector or six scalar quantities, are not unique, but depend on the choice of the reference point. More importantly, the local-flow quadrupole velocities fall off as one over the fourth power of the distance so that they are readily overridden by the dipole velocities, which fall off only as one over the third power of the distance. The higher-moment velocities fall off even more steeply.

A detailed discussion of the higher-order moments may not be particularly enlightening, but we should have some notion of their origin. Briefly, when the moving body is rigid, but not spherical, the higher-order moments are needed to complement the dipole moment so as to attain a complete field description. Furthermore, when the moving body is not rigid, its deformations generally lead to quadrupole and higher-order moments, whether or not it has a dipole moment. For a conceptual evaluation, it may be helpful to regard a quadrupole as two noncancelling dipoles, an octupole as two noncancelling quadrupoles, and so on.

Driven, reverberating gas bladders present biological monopole sources that often play an important role in animal communication and, at times, may even be used by predators to locate monopole prey. The motions of animals as a whole, regarded as rigid bodies, present the most common biological dipole sources. As proposed in this chapter, the basic acoustic functions of the lateral line and inner ear may have originated in the detection of the low-frequency, local and intermediate dipole flows. Most biological sources produce, in addition, quadrupole and higher-moment fields, which usually perturb the dipole configuration only in the immediate vicinity of the source.

A well-streamlined, coasting fish closely approaches the ideal of a dipole source, with only weak higher-order moments, certainly when observed from some distance in front or to the side. When the animal propels itself to overcome the viscous and inertial drag forces and, even more so, when it suddenly accelerates, as in escape, the velocity of the trailing water is reduced or may even reverse in direction. The resulting wake behind the fish clearly is not of a potential nature, as it shows abundant vorticity. Nevertheless, the fish emerges from the commotion with a rather clean velocity dipole field.

The tail beat of a swimming fish and the recoil motions of the head, moreover, form a connected system of transversely oscillating dipole sources, nearly a body length apart. In the vicinity of either the head or the tail, the two dipoles may be detected individually, but at larger distances they combine to form a quadrupole field. Fish that rhythmically move the pectoral fins to maintain a dynamic equilibrium, and struggling fish, making erratic translational and bending movements, may also create sizeable quadrupole and higher-moment fields, in addition to a great deal of vorticity.

The ventilatory flow of water, taken in through the mouth and expelled through the gills or the opercular slits, would also be largely dipolar, except that expelled water jets show ample vorticity and in that respect differ from the local-flow potential fields. It must be acknowledged that vorticity and turbulence are not considered in this chapter. This should not be construed, however, as a lack of appreciation for their biological importance.

Other biological sounds, caused by the grinding of teeth, spines, or skeleton parts, the snapping of the joints or the claws as in shrimp, the flow of water past the scales of fishes, and the breaking of the water surface, have been recorded and are usually given in the form of sonograms (Moulton 1960; 1963; Fish 1964; Tavolga 1964, 1971; Banner 1972). The heartbeat and the contractions of individual muscle fibers have also been suggested as possible sources of sound. However, neither the physical nature of the sources nor the configurations of the fields were identified in these studies. Moreover, all recordings were made by the use of pressure-sensitive hydrophones only.

## Acoustic-Field Detection

After these preliminaries, we will apply our directional analysis of the particle-velocity fields about monopole, dipole, and higher-moment sound sources to the problem of directional hearing. Extensive field data reveal that free-ranging *sharks are most readily attracted by sounds of very low frequency,* down to 10 Hz if not less (Nelson and Gruber 1963; Wisby, Richard, Nelson, and Gruber 1964; Myrberg, Gordon, and Klimley 1976). Sharks and the kindred skates and rays are devoid of air cavities and are, *I believe,* in their evolutionary development closer to the original function of underwater hearing than are the teleost fishes with swimbladders. Nevertheless, even though the latter may respond to sounds of several hundred, or even a few thousand, hertz, they also show clear reactions to the lower frequencies, down to a few hertz (Olson 1976).

Nelson and Gruber (1963), Nelson and Johnson (1972, 1976), and Banner (1972) found that sharks are particularly responsive to low-frequency sounds in spectral content and temporal pattern similar to the acoustic disturbances produced by struggling fish. Myrberg, Ha, Walewski, and Banbury (1972) furthermore observed the animals repeatedly approach a sound source from distances up to 400 m, that is, from well into the far field, even for the lowest frequencies applied. Despite the difficulties of working at sea, I have every reason to believe in the experimental validity of the data, highly respecting the original researchers and, equally so, the animals they worked with. Nevertheless, I have serious doubts about the biological validity of the results regarding the distance of response and the configuration of the fields that the animals oriented to.

In the shark experiments, *stronger-than-natural* fields were applied so as to attract the animals from sizeable distances and to ascertain whether they are truly capable of initiating their target-oriented movements in the acoustic far field (Myrberg, Banner, and Richard 1969; Myrberg 1978). Yet, struggling fish are not *expected* to produce low-frequency sounds detectable against the ambient noise at distances anywhere near 100 m, as pointed out by Richard (1968), and acknowledged by Myrberg et al. (1976) and Myrberg (1978). Moreover, only *monopole sources* were used, promoting the far field over the near field, even though in struggling fish the jerky, whole-animal motions and frantic flexing of the body are typically of a dipole and quadrupole nature. To generate monopole fields, the fish would have to show appreciable changes in volume as well.

When orienting to the low-frequency dipole fields of authentic prey, sharks operate, I believe, mostly in the dipolar, rather than the radial zone of the acoustic field. At 20 to 10 Hz, the dipolar zone extends to distances of at least 12 to 24 m, whereas the radial zone does not cover more than 30% of the solid angle up to distances of 75 to 150 m. Certainly, sharks may also detect the radial particle velocities of the monopolar swimbladder sounds of teleost fishes. Yet, swimbladder sounds, although reasonably common, remain occasional events as compared to the acoustic disturbances ubiquitously produced by all moving animals. Thus, I consider it unlikely that the swimbladder sounds provided the main driving forces for the development of the hearing function in sharks.

At not too large distances from a low-frequency dipole source, however, the animals are faced with the problem that the velocity vector generally is not radially directed. Nevertheless, all theories on directional hearing that I am aware of (cf. p. 105) tacitly assume the particle velocities to be centered on the source and, consequently, leave a triple-infinite ambiguity when applied to dipole or higher-moment fields. This leads to the unsettling conclusion that current theories on directional hearing, however ingenious, have not attended to the real challenge of acoustic source localization: the near-field detection of dipole and higher-moment sound sources. By the same token, the use of monopole sources for the study of directional hearing is bound to yield illusory results, unless applied only to the far field or the fields of monopole sources.

Note carefully that I have not argued against the orientation of fishes in the far field. At higher frequencies even sharks may be able to. Neither have I raised doubt about the orientation of fishes in monopole fields. I wish, however, to call attention to the question of directional hearing in the near fields of the common, low-frequency dipole and higher-moment sources. Actually, any source localization theory developed should be tested also on its validity for the detection of monopole sources (cf. pp. 107–109). So far, I have paid little regard to the detection of acoustic pressure, but instead have tried to discover what information fishes may derive from the particle motions of the acoustic near and far field. A few comments on earlier studies might be appropriate here.

An acute sense of hearing was found first in fishes that detect the pressure fluctuations in the sound wave, using the swimbladder as a pressure-to-motion converter (von Frisch 1938). Because of the scalar nature of the acoustic pressure, the animals were expected to lack the faculty of directional hearing, but would, from a short distance, be able to locate the sound source by the lateral line (von Frisch and Dijkgraaf 1935; van Bergeijk 1964). Later, however, also a direct sensitivity to the particle motions of the sound wave was inferred (Pumphrey 1950; de Vries 1956) and soon applied to the unresolved question of directional hearing (Dijkgraaf 1960; Sand and Enger 1974; Schuijf 1975).

Behavioral and physiological evidence has since confirmed the coexistence of acoustic-pressure and particle-motion channels in teleost fishes with swimbladders (Enger and Andersen 1967; Enger, Hawkins, Sand, and Chapman 1973; Sand and Enger 1973; Chapman and Hawkins 1973; Sand 1974; Fay and Popper 1974, 1975; Myrberg and Spires 1980; Buwalda 1981). In hindsight, the dual hearing thresholds observed in the earlier behavioral studies may be interpreted similarly, assuming that the particle motions were detected by the inner ear, instead of the lateral line (Tavolga and Wodinsky 1963; Cahn, Siler, and Wodinsky 1969). In sharks and teleost fishes without swimbladders, however, only a sensitivity to particle motion has been found (Banner 1967; Enger 1967; Chapman and Sand 1974).

Recently, these results have led to the hypothesis that fishes with a swimbladder relate the acoustic pressure to the particle motions, and fishes without a swimbladder relate the vertical to the horizontal particle motions, to remove the 180-degree ambiguity in the particle-velocity field (Schuijf 1975; Schuijf and Buwalda 1975, 1980; Buwalda, Schuijf, and Hawkins 1983; Popper et al., Chapter 27). Unfortunately, these authors did not state explicitly that their theories applied only to the far field or to the fields of monopole sources. In the instance of a dipole field, the mechanisms would suffer from an unsurmountable three-dimensional ambiguity at only slightly lower frequencies or slightly shorter distances from the source than those chosen for most tests. How might the animals have solved the problem, assuming they did?

## Hydrodynamic-Field Detection

Thus far, the discussion has been inspired by the increasing complexity of the monopole, dipole, and higher-order representations of the acoustic field. Next, I will endeavor to reconstruct the theory of underwater hearing, guided by the earlier inferred functional evolution, during which the detection range would have become progressively larger (cf. p. 85). I will begin with the most peripheral part of the system, the lateral line, and subsequently proceed to the organs of the inner ear. Similarly, I will first consider the detection and significance of the near field before advancing to the far field, in the hope of gaining a clear insight into the question of source localization and of revealing what might be the true origin of vertebrate hearing.

At the low frequencies of interest, the lateral line is stimulated only in the innermost part of the near field. At source distances large compared to the size of the recipient animal, the whole body, being nearly neutrally buoyant (Alexander 1966), tends to move with the water, thus leaving little differential motion for fish to detect (Sand 1981, 1984), except for the motions initiated by the fish itself. Formally, the lateral line detects the spatial derivative of the imposed local-flow field integrated along the lengths of the sensory arrays. Since in the monopole field the spatial derivative falls off as one over the third, and in the dipole field as one over the fourth, power of the distance, the lateral-line stimulus will be negligibly weak merely a few body lengths from the source. At larger distances, the whole-body motions deprive the lateral line of its stimulus.

But if the motion of the fish as a whole limits the efficiency of the lateral line, the animal could extend its detection range by monitoring just its motion as a whole. To accomplish this, fish have at their disposal three paired linear, and three paired angular, accelerometers in the form of the inner-ear otolith organs and the semicircular canals (Fig. 4.1). The inner-ear organs would be stimulated most effectively at the distance at which the lateral line loses its efficiency, usually well within the bounds of the acoustic near field. Thus, the difference between the inner-ear and lateral-line functions would not reside in the stimulus field, but in the particular field features that the sense organs attend to—the particle motions per se or their spatial derivatives—by virtue of their different physical coupling to the outside world.

This reasoning, however conjectural, has led me to the thesis that the otolith organs of fishes originally were, and in many cases still are, more concerned with the detection of the low-frequency local and intermediate flow of the acoustic near field than with the detection of the propagating sound wave of the acoustic far field. Naturally, the otolith organs would also detect the motions initiated by the animals themselves. At the low frequencies of interest, the near field constitutes, in my opinion, the most crucial part of a fish's world, the detection of which is a matter of life or death. The acoustic far fields of local sound sources would have been discovered only secondarily with increased sensitivity to the higher frequencies and the use of the swimbladder as a pressure-to-motion converter (van Bergeijk 1967a).

Certainly, we may speak of "hydrodynamic detection" in connection with the lateral line, since the flow fields that the lateral line detects are indistinguishable from incompressible hydrodynamic flows if we may ignore the minute time and phase differences (cf. p. 87). We also may speak of "hearing" when referring not only to the far-field, but also to the near-field function of the inner ear, since the local and intermediate flows about sound sources are legitimate parts of the acoustic field. Moreover, I see no reason to exclude the near field from the definition of "sound," neither do I feel inclined to apply the term "hearing" to the function of the lateral line. By these conventions, the inner ear of sharks would function as an organ of hearing, even when operating in the near field. Whether or not the animals detect acoustic pressure is a different question.

Pumphrey (1950) in his article on hearing came closest to the ideas presented here, although he did not mention the role of the inner ear in the near field. Yet, we all know from experience that the inner-ear organs are equally sensitive to the accelerations we initiate ourselves as to those we are subjected to, including the acceleration of gravity. Thus, it is generally understood that the inner ear of fishes, if sensitive enough, would also inertially respond to the particle motions of the propagating sound wave. However, few have alluded to the role of the inner ear in detecting the near field, perhaps because the lateral line was supposed to serve that function. Nevertheless, since the lateral line does so only for the innermost part, the remainder of the near field, I believe, is the original acoustic domain of the inner ear.

Curiously enough, van Bergeijk (1967a) categorically denounced the role of the inner ear in the detection of the near-field fluid motions, concluding that the "labyrinth appears to evolve purely as an inertial navigation system," but "is not very sensitive to near-field displacements, since it has been removed from the surface of the animal, and thus from contact with the environment," (p. 19). It almost seems that van Bergeijk speaks only about fishes tied to the substrate as in his and Harris' neurophysiological experiments. His lack of appreciation of the stimuli arising from the imposed whole-body motions becomes the more amazing when he resorts to the lateral line to explain the attraction of sharks from distances up to 600 yards from a 20-Hz monopole source.

Neurophysiological evidence for a sensitivity of the inner ear to near-field motions imposed upon free-moving fishes is overwhelming (Lowenstein and Roberts 1951; cf. p. 105). However, few attempts have been made to establish the function of the inner ear in the near field behaviorally under exclusion of both the lateral-line sense and the detection of acoustic pressure (see however Dijkgraaf 1963b). Moreover, the possible part played by the inner ear in the animals' responses commonly accredited to the lateral line also deserves further attention. Nevertheless, the function and biological significance of the lateral-line sense of fishes has been convincingly demonstrated by Dijkgraaf in his behavioral studies, leading from his early paper of 1934 till and beyond his splendid review article of 1963.

Related topics not discussed here are the role of the lateral line in schooling (Partridge and Pitcher 1980), in the detection of stationary objects (von Campenhausen, Riess, and Weissert 1981), and in the detection of surface waves (Bleckmann, Chapter 24). However, many of the basic ideas brought forward in the present chapter apply to those studies as well.

## Source-Localization Algorithms

The lateral-line arrays have the great advantage of providing instantaneous information about the spatial configuration of the fluid motions past the skin. Within a couple of body lengths from the source, the distribution of cupular deflections along the length of the arrays may suffice to locate the source of the field (Pumphrey 1950; van Bergeijk 1964). Consequently, the lateral-line response can be fast and precise, but takes place only within a short distance. This assumes that the fluid-velocity field is predominantly of a potential nature (cf. p. 87), like any well-behaved acoustic field, that is, relatively free of viscous effects, vortices, and turbulence. Indeed, the low-frequency near fields of moving fish are, to a first approximation, dipolar, except for the wake occupying a relatively narrow cone behind the animal (Kalmijn and Enger, p. 110; cf. p. 103).

The otolith organs, on the other hand, detect the fluid motions essentially at one point at a time only, since the distance between the right and left labyrinths is considered too short and the coupling at low frequencies too tight to expect appreciable differences in stimulation. That raises the question of how, by the use of the inner ear, fishes might orient to the source, not only in the radial, but also in the dipolar part of the field. As mentioned earlier (p. 104), all modern theories on directional hearing in fish tacitly imply that the fluid motions are centered on the source, but, in the near field, they generally are not, except in the near field of a monopole source (see however Rogers and Cox, Chapter 5; Popper et al., Chapter 27). For a discussion on the possible role of acoustic pressure in the dipole local-flow field, see Gray (1984).

One solution would be to assume that the animal probes the spatial configuration of the field while moving through it. By sampling the field over time, the inner ear would not respond as fast as the lateral line, but would enable the animal to detect the source from larger distances. For a radial field, the task of finding the source would be straighforward (Popper et al., Chapter 27), but for any other field, the problem of orienting to the source seems unsurmountable. The animal may sense the shape of the particle trajectories (cf. p. 101), or even the phase between the particle velocity and acoustic pressure (cf. p. 105), to find out where in the acoustic field it is, in the dipole, intermediate, or radial zone. Nevertheless, even if the field is known to be dipolar, it would still seem quite a demand on the part of the animal to calculate the position of the source from some distance.

How might directional hearing have developed with such a difficult, if not impossible start? The clue may be that, to attack its prey, a predator does not need to know where the target is in order to strike at it accurately and fast. From moment to moment, the animal has only to know in which direction to proceed so as to reach the source, until it is close enough to locate the prey, for instance, by the use of the lateral line. Thus, the animals would be guided to the source, rather than aim at it from a distance. The same problem arises in connection with the electric sense. The bioelectric fields of aquatic animals lack monopole moments and, from a distance, are also predominantly dipolar (Kalmijn, Chapter 6). By analogy, we may briefly consider two possible approach strategies, without strong claims that either one holds true for the inner ear.

Conceivably, the recipient animal follows the field lines of the particle motions in a direction in which the strength of the stimulus increases. This certainly would lead a predator to its prey, given that the prey produces a steady dipole field over a sufficient length of time. However, the attack would generally not be along the shortest path. Moreover, it would offer the prey a sure way of escaping only in front of the approaching predator. Whether the animals follow this strategy is readily verified, or rejected, experimentally. Actually, when locating their prey electrically, elasmobranch fishes tend to strike at their target by a more direct route, whereas the weakly electric fishes may indeed approach each other along the electric-organ field lines (Hopkins, Chapter 9).

Alternatively, the recipient fish could maintain a constant angle with the direction of the field by turning from the first moment of detection in such a way as to compensate for any apparent field rotation (Kalmijn, Chapter 6). When applied to fishes living in a horizontal plane—near the bottom for instance—this strategy would be surprisingly effective. Cueing in on the dipole fields of quietly swimming prey, the predator would be able to attack along realistically smooth paths without directional ambiguity. It would also allow the prey, if fast enough, to escape by turning in a direction opposite to the apparent field rotation effected by the predator. The electrically evoked feeding responses are consistent with this strategy, but experimental verification of the algorithm is still pending.

Whether fortuitous or real, the best evidence in favor of the conjecture that the animals approach the source area by keeping a constant angle with the particle-velocity field lines comes from the work of Myrberg et al. (1972). In the monopole fields that they applied, the proposed orientation mechanism would have led to an approach path in the form of a spiral. True enough, Myrberg and co-workers observed (pp. 939–941) that, in most instances, the sharks "did not veer suddenly and swim rapidly" in the direction of the source, but "moved unhurriedly," while approach paths in the form of "large arcs were noted in those sharks that had been moving in directions other than the correct one." Some day, I hope, we will see those tests repeated with dipole fields.

## Frequency Analysis

### A Frequency-Response Dilemma

For the lateral line to serve in the detection of prey, we might expect it to be particularly sensitive to the low-frequency fields that *quietly* moving animals produce. To the contrary, the neurophysiological response curves of the lateral-line canal organs feature "best frequencies" as high as 180 Hz in teleost fishes (Harris and van Bergeijk 1962; Muenz 1985), and around 64 Hz in the ray *Platyrhinoidis triseriata* (Goerner and Kalmijn, unpublished). The free neuromasts show slightly lower "best frequencies," up to 70 Hz, in both aquatic amphibia and teleost fishes (Kroese, van der Zalm, and van den Bercken 1978; Sand 1981; Muenz 1985). Under 10 Hz, only weak physiological responses are observed, although behavioral experiments indicate that the lateral-line sense plays a major role in the detection of just those very low frequencies (Dijkgraaf 1963a).

If the otolith organs of the inner ear begin their inertial task where the lateral-line system loses its grip on the physical stimulus, at a distance of a few body lengths or less (cf. p. 106), we would expect them to respond to rather low frequencies as well. The displacements of the local hydrodynamic flow field are frequency independent and, at low frequencies, a few body lengths means well within the near field. Even so, the displacement audiograms of teleost fishes without a swimbladder show "best frequencies" from a low of 100 to 200 Hz (cardiac response, Chapman and Sand 1974) up to as high as 1,500 Hz (saccular microphonics, Fay and Popper 1974), and a remarkably low sensitivity from 10 to 40 Hz, the frequencies most effective in attracting free-ranging sharks (cf. p. 103).

Nevertheless, I contend that these puzzling observations make perfect sense, as I will explain, first on the basis of the response curves as they were presented in the literature, and subsequently after replotting the curves in a more sensible way. Historically, the neurophysiological response curves and their so-called "best frequencies'' were referenced to stimuli of equal displacement amplitude, independent of frequency. The sense organs, however, evolved to detect the biologically relevant stimuli that the animals encounter in nature. Consequently, the frequency-response curves should, ultimately, be interpreted, not in absolute terms, but relative to the amplitude spectra of the natural fields known to elicit biologically meaningful behavioral responses.

Hence, Dr. Enger and I measured the near-field water motions about moving objects with a recently developed linear, DC-to-300-Hz underwater displacement transducer. Recordings were made of the local flow fields produced by fishes quietly approaching, and within a few centimeters passing, the transducer (Fig. 4.7a), and of the periodic water motions deriving from the fin and ventilatory movements of hovering fishes (Fig. 4.7b). Even though preliminary, the results show that the *displacement* spectra peak at frequencies well below 10 Hz and taper off steeply as frequency increases, thereby revealing a striking predominance of the lower frequencies. Frequency maxima in excess of 10 Hz were observed only in frantically struggling and shivering fish. For a comprehensive picture to emerge, however, substantially more research will be needed.

This leads us to our first interpretation. The frequency-response curves of the lateral line and inner ear *increase* just about as steeply as the amplitude spectra of the fluid displacements in the local-flow fields of quietly moving objects *decrease* over the same frequency range. Consequently, the detectability of the object, a dimensionless number defined by the stimulus amplitude times the sensitivity of the sense organ, remains approximately constant over the frequency range of interest (Fig. 4.8a). The sensory system would, in engineering terms, have a built-in prewhitening filter bringing all frequency components of the stimulus within a comfortable dynamic range, despite their large differences in amplitude. As the graphs show, the neurophysiological ''best frequency'' would signify merely the upper frequency limit of the biological range.

A more fundamental explanation, clearly revealing the man-wrought nature of the decades-old dilemma, presents itself when the sensitivities of the lateral line and inner ear are expressed in terms of the relevant physical quantities of the fields that the recipient animals detect. First of all, the physiologically adequate

Figure 4.7. Hydrodynamic flow field of a small goldfish, (*a*) approaching and passing a displacement transducer, and (*b*) hovering at a short distance from the transducer. 1. A single approach and pass, 0.65 s in duration, and one period of the regular, ventilatory movements, 1.14 s in duration, displacement versus time. 2. Continuous and discrete amplitude spectra of recorded fields, displacement versus frequency, linear–linear plots. 3. The same spectra, plotted double-logarithmically. 4. The same spectra expressed in terms of acceleration, rather than displacement, plotted double-logarithmically. The small wriggles in the trace after the fish passed the transducer, and, perhaps, the second, higher-frequency peaks in the amplitude spectra, represent the tail beat. The first, lower-frequency peaks are mainly due to the transient of the whole-body motion. The flow field of the hovering fish reflects the ventilatory movements and, to a lesser degree, the movements of the pectoral fins. The recordings were made with a Linear Electrolytic Displacement Transducer (LEDT) under development in the author's laboratory (Kalmijn and Enger, unpublished).

stimuli for the lateral-line and inner-ear hair cells are the displacements of the sensory hair bundles, parallel to the receptor epithelia, as has been convincingly demonstrated neurophysiologically for frequencies from virtually zero to 150 Hz by Hudspeth and Corey (1977). Hence, at the receptor level, the *hair cells* are essentially *displacement-sensitive* devices, the responses of which, at the lower frequencies of interest, are proportional to the shearing forces exerted on their apical hair bundles (cf. pp. 117–124).

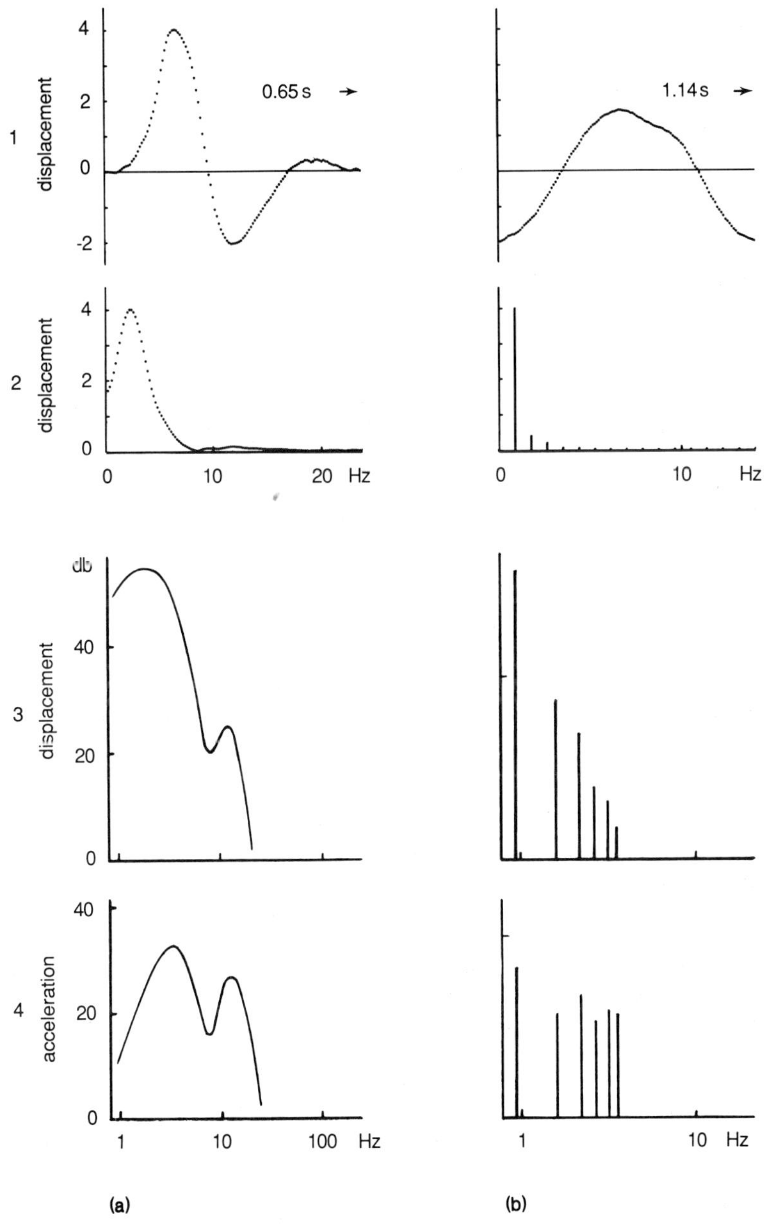

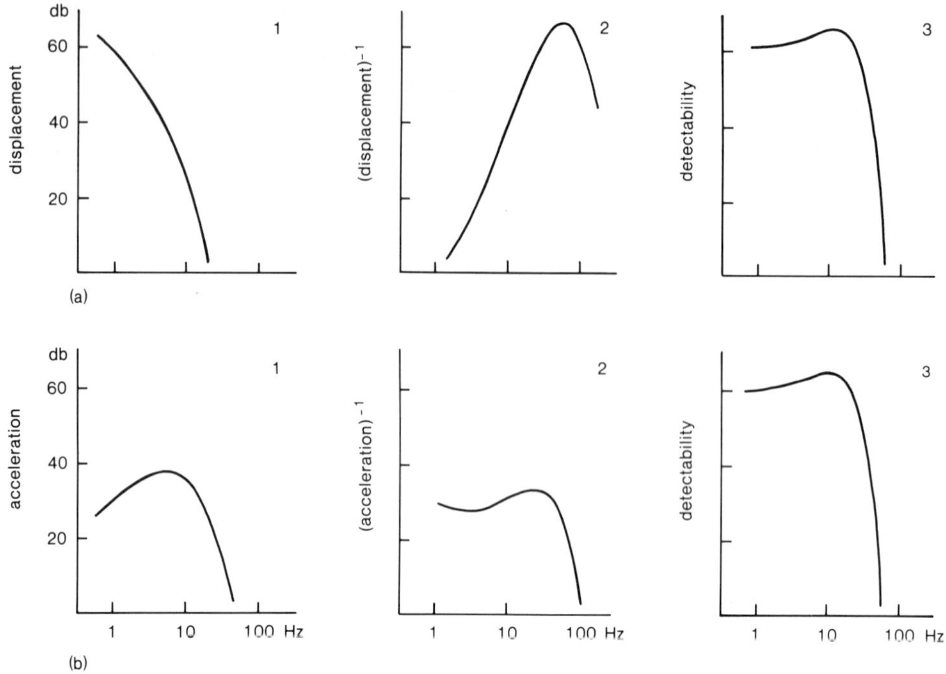


Figure 4.8. Hydrodynamic detectability of underwater objects. (*a*) Interpretation of literature data, presented as such: 1. displacement amplitude spectrum of the source field, 2. displacement frequency-response curve of the sensory system, 3. detectability of the object as a function of frequency. (*b*) Interpretation of literature data, replotted: 1. acceleration amplitude spectrum of the source field, 2. acceleration frequency-response curve of the sensory system, 3. detectability of the object as a function of frequency. The replotted curves express the data in terms of acceleration, which is the field quantity that the lateral-line canal and inner-ear otolith organs detect. Graphs were conceived for illustrative purposes only, uncalibrated.

The forces driving the lateral-line hair bundles usually derive from the viscous drag incurred by the cupula, which is about proportional to the velocity of the flow of water over the receptor epithelium (Jielof, Spoor, and de Vries 1952). Thus, even though the receptor cells are displacement detectors, for the lower frequencies, the *lateral-line neuromasts* are, at the organ level, *velocity-sensitive* devices. However, the velocity driving the cupula is not the same as the velocity of the hydrodynamic or acoustic flow field that the animal detects. In fact, for the canal organs, the velocity of the water driving the cupula is nearly proportional to the acceleration of the flow field only a short distance from the fish (Denton and Gray 1983; cf. pp. 120–124). Hence, for the lower frequencies, the *lateral-line canal organs* are, at the system level, *acceleration-sensitive* devices. For the free neuromasts, however, the velocity of the water driving the cupula is a function of both the velocity and the acceleration of the flow field only a short distance from the animal (Jielof et al. 1952, and pp. 120–124). Hence, for the lower frequencies, the *free neuromasts* are, at the system level, *partially velocity- and partially acceleration-sensitive* devices.

The mass-loaded sense organs of the inner ear, on the other hand, are driven by the inertial forces on the otoliths, or on the cupula and endolymph in the semicircular canals, forces that ensue from the linear or angular accelerations of the fish's head. Although displacement detectors at the receptor level, the *otolith organs* are, for the lower frequencies, *acceleration-sensitive* devices at the organ and at the system level. (For the semicircular canals, see pp. 117–120.)

In the light of these considerations, I have replotted the existing frequency-response curves as functions of the field quantities that the animals detect at the lower frequencies. The displacement curves for the lateral line are replotted, first, as a function of velocity to reveal the effect of the driving force and, subsequently, as a function of acceleration for the canal organs (Fig. 4.9a,b,c and Eq. 25, p. 122) and as a function of the geometric mean of the velocity and acceleration amplitudes for the free neuromasts (Fig. 4.10a,b,c and Eq. 24) to relate the receptor stimuli to the appropriate field quantities. Similarly, the displacement curves for the inner-ear otolith organs are replotted as a function of acceleration, which, in this instance, not only gives rise to the driving force, but also represents the relevant field quantity (Fig. 4.11a,b,c).

Which curve is most appropriate depends on the objective of the investigation. To study the hair-cell transduction process, the constant-displacement curves should be used. To study the physics of the detection mechanism, the velocity curves would be the proper choice for the friction-driven lateral-line neuromasts, whereas the acceleration curves would be more pertinent to the otolith organs of the inner ear. To study the biological significance of low-frequency natural stimuli, the acceleration curves are most relevant to the otolith organs of the inner ear and to the lateral-line canal organs, whereas the acceleration–velocity curves are most relevant to the free neuromasts. The usual choice of displacement, because velocity and acceleration would be conceptually more difficult to visualize, has led to more difficulties than it was meant to avoid.

## Old Mistakes and New Solutions

Inspection of the new curves presents a few intriguing surprises. When plotted against the velocity of the water driving the cupula, the lateral-line response reaches its maximum at about half the "best frequencies" of the original curves. Furthermore, when plotted against the acceleration, for the canal organs, or the geometric mean of the acceleration and velocity amplitudes, for the free neuromasts, the lateral-line response remains more or less constant as frequency increases, until it drops off steeply with or without overshoot. The once "best frequency" is now close to the actual cutoff point of the operating range or falls even beyond. Consequently, when plotted sensibly, the same data reveal that the lateral line indeed is exquisitely sensitive to the lower frequencies, down to 0.1 Hz or less, that elicited the observed behavioral responses (Dijkgraaf 1963a).

Historically, the justification for expressing the sensitivities in terms of displacement was, however, more than a matter of convenience. It all started with the lateral-line canal data on *Fundulus heteroclitus,* the killifish (Fig. 4.9a), which

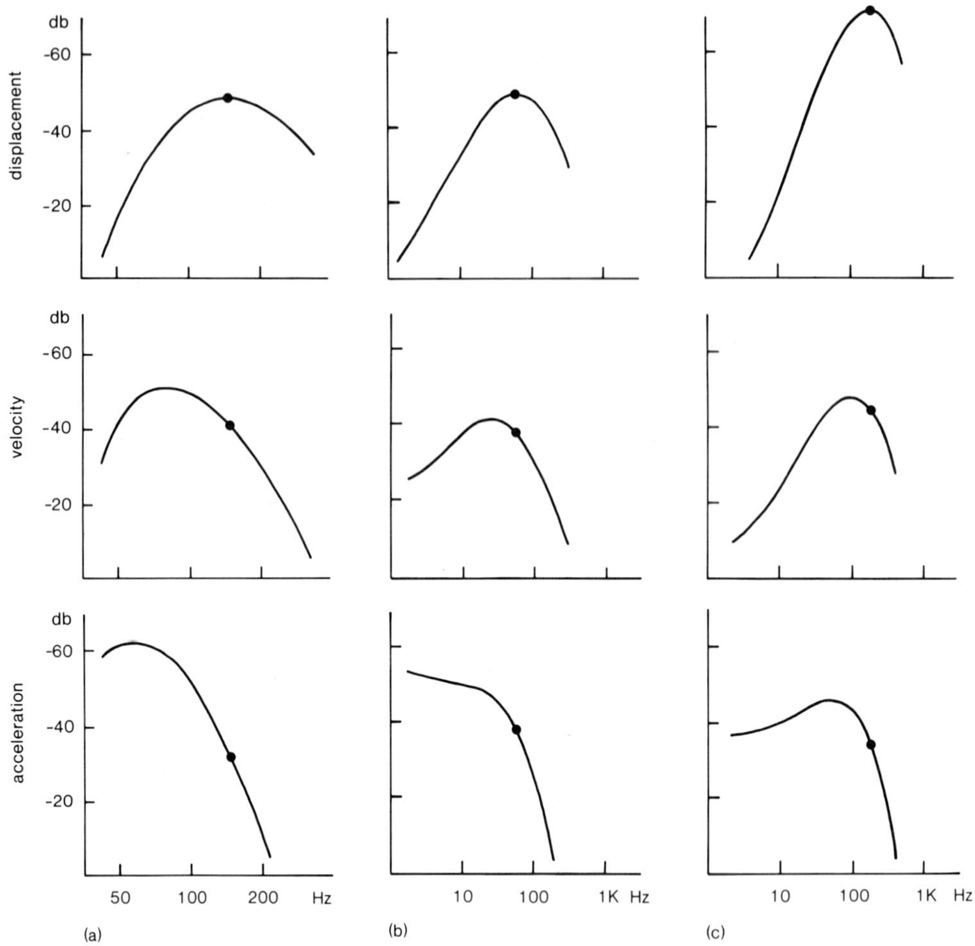


Figure 4.9. Frequency-response curves of lateral-line canal organs. (*a*) *Fundulus heteroclitus,* Harris and van Bergeijk (1962): microphonic potentials in response to a distant vibrating sphere, offering well-defined stimuli except for possible surface effects. (*b*) *Sarotherodon niloticus,* Muenz (1985): nerve action potentials in response to a tiny, nearby vibrating sphere, offering poor low-frequency stimulus definition due to unchecked boundary layer. (*c*) *Sarotherodon niloticus,* Muenz (1985): same as above, but for a different type of neuromast. Dots in curves indicate neurophysiological best frequencies. For the amplitude units, see the original literature.

are those of Harris and van Bergeijk (1962). The authors correctly introduced the vibrating sphere as the ideal local-flow source, but erroneously inferred that the lateral line would detect the displacement, not the velocity or acceleration, of the water near the fish (cf. van Bergeijk 1967b). Unfortunately, by disregarding all phase shifts that seemed to work in the wrong direction, they misinterpreted the microphonic potentials by 180 degrees (cf. pp. 120–124). Plotted properly, though, Harris and van Bergeijk's curves ''straighten out'' below the ''best frequencies'' and fully support Dijkgraaf's behavioral conclusions, leaving little basis for the heated arguments left on his doorstep.

Actually, the higher frequencies, not the lower, are attenuated by the physiologi-

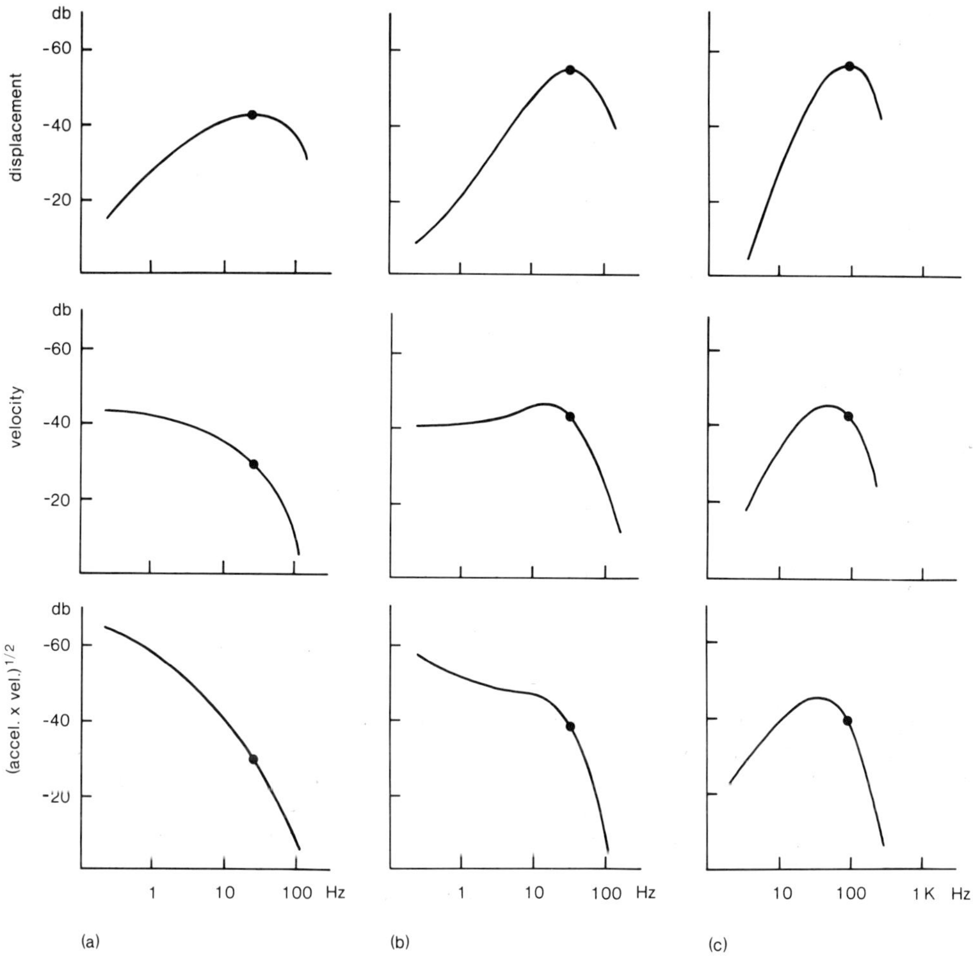


Figure 4.10. Frequency-response curves of free neuromasts. (*a*) *Xenopus laevis,* Kroese et al. (1980): microphonic potentials in response to a tiny, nearby vibrating sphere, offering poor stimulus definition. (*b*) *Xenopus laevis,* Kroese et al. (1978): nerve action potentials in response to a larger sphere, offering better stimulus definition. (*c*) *Sarotherodon niloticus,* Muenz (1985): nerve action potentials in response to a tiny, nearby sphere, low frequencies unduly suppressed for lack of boundary-layer control. Dots in curves indicate neurophysiological best frequencies. For the amplitude units, see the original literature.

cal and physical properties of the detection system. The "sharp tuning" suggested by the literature curves is purely artificial. What does it mean to be tuned to displacement, if it is the acceleration that the animal detects? The newly plotted curves, however, show low-pass characteristics instead, beginning their steep decline at 40 Hz or less for the free neuromasts and at 80 Hz or less for the canal organs. By comparison, the fast, Mauthner-initiated escape reflexes of fishes show single, biphasic tail flips, about 40 ms in duration, corresponding roughly to a fundamental frequency of 25 Hz (Eaton, Bombardieri, and Meyer 1977). (For a discussion of the phase relations and their sensitivity to experimental error, see pp. 120–125.)

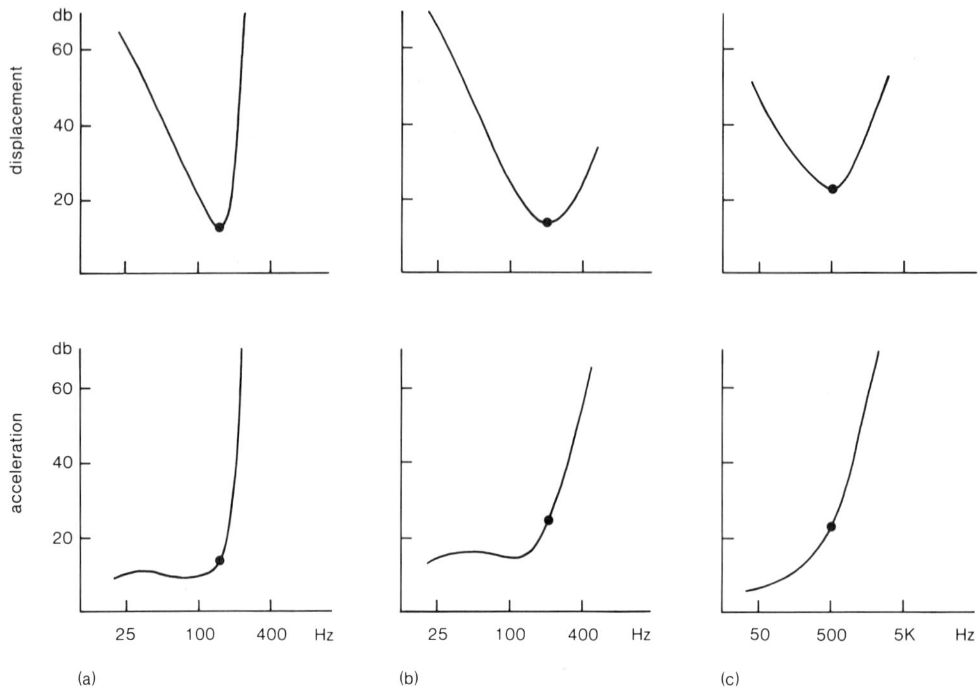


Figure 4.11. Audiograms of inner-ear particle motion detectors. (*a*) *Limanda limanda*, Chapman and Sand (1974): cardiac response in the near field of a monopole sound source. (*b*) *Salmo salar*, Hawkins and Johnstone (1978): cardiac response in the near field of a monopole sound source. (*c*) *Tilapia macrocephala*, Fay and Popper (1975): microphonic potentials in response to direct vibration of the fish's head. Note that in the detection of gravity the curves for the otolith organs extend to zero hertz. Dots indicate neurophysiological best frequencies. For the sensitivity units, see the original literature.

With regard to the detection of surface waves, the lateral-line-mediated responses of surface-feeding fishes are usually also referenced to the displacement of the stimulus (Bleckmann 1980), even though displacement is unlikely to be the proper field quantity. Specific data on the coupling between the free neuromasts and the natural stimulus are not available. Nevertheless, if the sense organs remain completely submerged (see Fig. 4 in Schwartz 1965), we may expect the considerations of previous paragraphs here to apply as well. Bleckmann (1980, p. 171) remarks that "the strong rise in threshold amplitudes from 30 Hz to 10 Hz suggests . . . a filter mechanism presumably located in receptive structures of the lateral line organ." After replotting the curves against the geometric mean of the velocity and acceleration amplitudes, however, the "strong rise" vanishes and with it the need of invoking a special filter mechanism.

Concerning the inner-ear curves, Enger (1966, p. 866) remarked that, ". . . if particle acceleration is the actual physical stimulus, the threshold would be, within generous limits, constant and independent of frequency . . .", up to about 100 Hz. Above that frequency, the inertial organs "degrade," first, to excellent velocity detectors, because of friction, and, at even higher frequencies, to superb displacement detectors, because of their inertia (cf. pp. 117–120).

Actually, the semicircular canals operate, by design, in the velocity mode, whereas the otolith organs, if involved in the detection of higher frequencies, would operate in the velocity, or even the displacement, mode. It is the astounding sensitivity of the hair-cell transduction mechanism that renders these sense organs so adaptive. Nevertheless, modern fishes have opted to recruit the swimbladder as a hearing aid to extend their sensitivity and frequency range even farther.

If the otolith and lateral-line organs respond to the acceleration or the geometric acceleration–velocity mean of the natural field, the stimulus spectra should also be expressed in those field quantities. Notably, after replotting the stimulus spectra, the higher frequencies stand out more clearly (Fig. 4.8b), and the sensitivities of the lateral line and inner ear appear once more well-matched to the frequency spectra of the natural stimuli. The prewhitening filter, originally invoked to explain the published displacement data as such, actually appears to be of a purely physical, not a physiological, nature. Studying the displacement curves, scientists may have been puzzled about an apparent lack of sensitivity to the lower frequencies or an apparent lack of the higher frequencies in the natural stimulus, but the animals did not suffer from such shortcomings.

But why should modern fishes be sensitive to the higher frequencies as well? Conceivably, the need arose when attempting to communicate with one another in a world already full of low frequencies. Note that the ambient pressure noise falls off steeply with frequency until it starts leveling off to reach a plateau or flat maximum between 200 and 500 Hz (Wenz 1964; Urick 1983). Moreover, when with increasing frequency the far field came within reach, the particle motions also began to fall off more slowly with distance. The real breakthrough, however, came with the advent of the swimbladder, which appeared to be highly useful not only to produce, but also to receive sounds of higher frequency. Let the "hearing specialists" among the fish "that come when you whistle" (von Frisch 1923) no longer mislead us in our attempts to trace the true origin of vertebrate hearing!

## Physics of Lateral-Line and Inner-Ear Sense Organs

The reasoning in the preceding discussion assumed a familiarity with the physical relations between the hair-bundle displacements and the natural lateral-line and inner-ear stimuli. Here, I will detail those relations and show that, whether caused by inertial or frictional forces, the displacements of the hair bundles may reflect any gradation of the acceleration, the velocity, or the displacement of the stimulus field, depending on the frequency of excitation and the physical properties of the detection system.

### * Inertial Detectors

The inertial detectors of the inner ear may, to a first approximation, be treated as simple, harmonic oscillators (Steinhausen 1931). Thus, in response to sinusoidal acceleration, the displacements of the mass-loaded sensory hair bundles are governed by the differential equation (de Vries 1956; Lewis 1984):

$$m_e \frac{d^2x}{dt^2} + b\frac{dx}{dt} + kx = -m_d\, a \cos \omega t \qquad (12)$$

where $x$ is the displacement of the hair bundle from its equilibrium position, measured relative to the sensory epithelium, $m_e$ the effective mass of the otolith and the entrained endolymph, $b$ the viscous drag force per unit velocity incurred by the moving mass, $k$ the restoring force per unit displacement due to the stiffness of the hair bundle, $m_d$ the driven mass of the otolith less the endolymph it displaces, $a \cos \omega t$ the acceleration *imparted to the fish,* and $-m_d a \cos \omega t$ the inertial force driving the otolith *in the coordinate system of the animal.*

For the semicircular canals, $x$ denotes the angular displacement (in radians) of the hair bundle and $a \cos \omega t$ the angular acceleration imparted to the fish, while $m_e = m_d$, $b$, and $k$ denote the moments of inertia, of the drag force per unit angular velocity, and of the restoring force per unit angular displacement, revealing the similarity between the linear and angular motions. The flow of endolymph is constrained by the viscous drag along the canal wall and by the stiffness of the cupula and the sensory hairs blocking the canal lumen (Fig. 4.1d,f).

Equation 12 is usually written in the form:

$$\frac{d^2x}{dt^2} + \gamma\frac{dx}{dt} + \omega_0^2 x = -\, na \cos \omega t \qquad (13)$$

where $\gamma = b/m_e$, $\omega_0^2 = k/m_e$, and $n = m_d/m_e$, the ratio of the driven over the effective mass (~1/2 for the otolith organs and unity for the semicircular canals).

Equation 13 may be solved in terms of the hair-bundle displacement $x(\omega)$ and the phase angle $\delta(\omega)$, defined as the angle by which the displacement lags the fictitious *inertial force* resulting from the oppositely directed acceleration imparted to the animal (e.g., Kinsler et al. 1982):

$$x(\omega) = -\frac{n}{\omega_0^2}\frac{\omega_0/\omega}{[(\omega_0/\omega - \omega/\omega_0)^2 + (\gamma/\omega_0)^2]^{1/2}}\, a \cos(\omega t - \delta) \qquad (14)$$

$$\text{where} \qquad \tan \delta(\omega) = \frac{\gamma/\omega_0}{\omega_0/\omega - \omega/\omega_0} \qquad (15)$$

The *natural* frequency of oscillation, in the absence of damping ($\gamma = 0$), is $\omega_0$ (in radians). For $\gamma < 2\,\omega_0$ the system still tends to oscillate, for $\gamma = 2\omega_0$ it is critically damped, for $\gamma > 2\omega_0$ it is overdamped (Fig. 4.12a,b,c,d). For $\omega_{1,2} = [(\gamma/2)^2 + \omega_0^2]^{1/2} \pm \gamma/2$ the two quadratic terms in the denominator of Eq. 14 are numerically equal and the phase angles $\delta_{1,2}$ given by Eq. 15 are $\pi/4$ and $3\pi/4$ radians or 45 and 135 degrees. The $\omega_{1,2}$ are the crossover frequencies delimiting the *modes of operation* for the various inertial detectors. Each mode is characterized by an approximate independence of frequency for either the acceleration, $a$, velocity, $v$, or displacement, $d$, of the natural stimulus.

For the lower excitation frequencies $\omega \ll \omega_1$:

$$x \cong -\, n/\omega_0^2\ a \cos(\omega t - \delta) \cong -\, m_d/k\ \, a \cos \omega t \qquad (16)$$

i.e., the hair-bundle displacement, $x$, is proportional, but opposite (−) to, the *acceleration, a* $\cos \omega t$, imparted to the animal and is in phase with the driving force ($\delta = 0$), as in the detection of gravity and low-frequency vibrations (Fig.

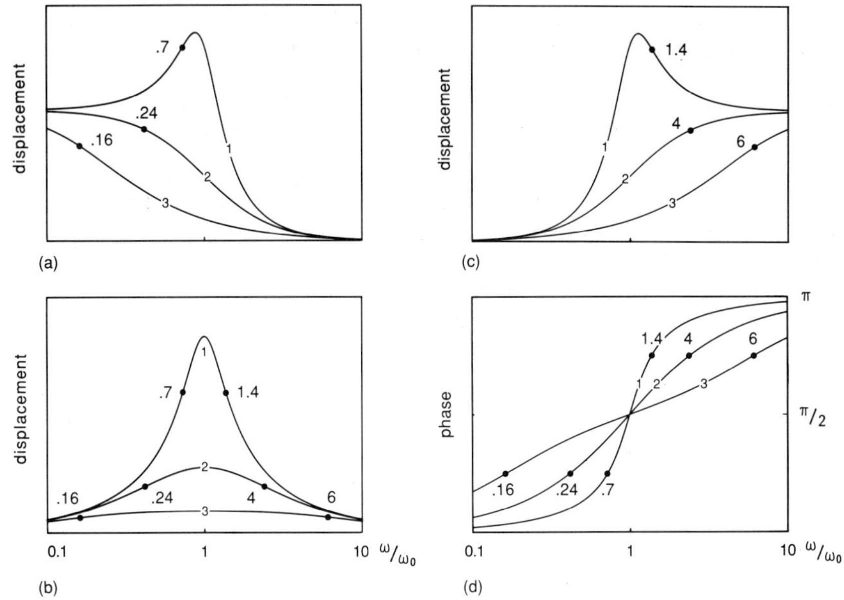


Figure 4.12. Amplitude and phase characteristics of lateral-line and inner-ear sense organs. (*a*), (*b*), and (*c*): Amplitude of the hair-bundle displacement, on a linear scale. (*d*): Phase lag of the hair-bundle displacement with respect to the driving force, in radians. Horizontal axis: frequency of the driving force, $\omega$, divided by the natural frequency, $\omega_0$, on a logarithmic scale. (1), (2), and (3): Curves for underdamped, $\gamma_1 = 2\omega_0/3$, critically damped, $\gamma_2 = 2\omega_0$, and overdamped, $\gamma_3 = 6\omega_0$, systems. The labeled dots denote the lower, $\omega_1$, and upper, $\omega_2$, crossover frequencies divided by the natural frequency, $\omega_0$. For the inertia-driven sense organs, governed by Eqs. 14 and 15: (*a*) At a constant acceleration amplitude of the driven mass, the curves are approximately flat below $\omega_1$: *acceleration mode* (direct reading accelerometer), e.g., the otolith organs at low frequency (curve 2). (*b*) At a constant velocity amplitude of the driven mass, the curves are approximately flat between $\omega_1$ and $\omega_2$: *velocity mode* (integrating accelerometer), e.g., the semicircular canal organs (curve 3) and otolith organs (curve 2) at intermediate frequencies. (*c*) At a constant displacement amplitude of the driven mass, the curves are approximately flat above $\omega_2$: *displacement mode* (double integrating accelerometer), e.g., the otolith organs at high frequencies (curve 2). For friction-driven sense organs, governed by Eqs. 21 and 22: (*a*) At a constant velocity amplitude of the water at the hair cells,* the curves are approximately flat below $\omega_1$: *velocity mode* (direct reading velocity meter), e.g., the lateral-line neuromasts (curve 2) and probably the macula neglecta at low frequency. (*b*) At a constant displacement amplitude of the water at the hair cells,* the curves are approximately flat between $\omega_1$ and $\omega_2$: *displacement mode* (integrating velocity meter), e.g., the lateral-line neuromasts (curve 2) and perhaps the macula neglecta at intermediate frequency. *Note that at low frequency, the fluid velocity driving the lateral-line canal organs is nearly proportional to the acceleration, whereas the fluid velocity driving the free neuromasts is approximately proportional to the geometric mean of the velocity and acceleration amplitudes of the water at only a short distance from the animal. Hence, curve (2) of (*a*) also gives the hair-bundle displacements of the canal organs in response to an external field of a constant acceleration amplitude, and of the free neuromasts in response to an external field of a constant geometric mean of the velocity and acceleration amplitudes.

4.12a, critically damped curve). In the acceleration mode, the system is predominantly controlled by the *stiffness, k,* of the restoring elements.

For the intermediate excitation frequencies $\omega_1 \ll \omega \ll \omega_2$:

$$x \cong -n/\gamma \ a/\omega \cos(\omega t - \delta) \cong -m_d/b \ v \cos(\omega t - \pi/2) \tag{17}$$

i.e., the hair-bundle displacement, $x$, is proportional, but opposite to, the angular or linear *velocity*, $v \cos(\omega t - \pi/2)$, of the animal and lags the driving force by 90 degrees, as in the detection of rotation by the semicircular canals during normal head movements (Fig. 4.12b, overdamped curve) and of intermediate-frequency vibrations (Fig. 4.12b, critically damped curve). In the velocity mode, the system is predominantly controlled by the *viscous drag, b,* incurred by the moving mass.

For the higher excitation frequencies $\omega \gg \omega_2$:

$$x \cong -na/\omega^2 \cos(\omega t - \delta) \cong -m_d/m_e \, d \cos(\omega t - \pi) \tag{18}$$

i.e., the hair-bundle displacement, $x$, is proportional, but opposite to, the *displacement*, $d \cos(\omega t - \pi)$, of the animal and lags the driving force by 180 degrees, as in the detection of high-frequency vibrations (Fig. 4.12c, critically damped curve). In the displacement mode, the system is predominantly controlled by the *inertia*, $m_e$, of the moving mass.

Before reaching the steady-state solution, the system usually shows a transient response, depending on the displacement and velocity of the sensing mass at the moment of stimulation. The transient is shortest and least disturbing when the system is critically damped, in which case $1/f_0 = 2\pi/\omega_0$ is the "time of indication" (de Vries 1950). The harmonic analysis applies to any arbitrary, physically realizable waveform, given a linear system response.

For the saccular otolith of the ruff, *Acerina* sp., de Vries (1950) inferred a $\gamma$ slightly less than $2\omega_0$ (nearly critical damping) and a natural frequency $f_0 = \omega_0/2\pi$ of 50 Hz, yielding a "time of indication," $1/f_0$, of ~ 20 ms and crossover frequencies $f_{1,2} = \omega_{1,2}/2\pi$ of ~ 20 Hz and ~ 120 Hz. For the smaller utricular otolith, assuming the same degree of damping, the values are: $f_0$ ~ 200 Hz, $1/f_0$ ~ 5 ms, $f_1$ ~ 80 Hz, and $f_2$ ~ 480 Hz. For the semicircular canals of the skate, *Raja* sp., Groen, Lowenstein, and Vendrik (1952) estimated a $\gamma$ of $35\omega_0$ (heavy overdamping), an $f_0$ of 0.16 Hz, an $f_1$ of 0.0045 Hz, and an $f_2$ of 5.5 Hz; covering the range of normal head movements. Although the presented physical models are overly simplistic and in dire need of refinement (Lowenstein 1972; Mayne 1974), they offer a realistic starting point for the interpretation of neurophysiological and behavioral data, if used with caution.

In summary, when detecting gravity, the otolith organs respond to acceleration, when detecting normal head rotations, the semicircular canals respond to angular velocity, and when detecting vibrations the otolith organs respond to acceleration at low frequencies, to velocity at intermediate frequencies, and to displacement at high frequencies.

## * Lateral-Line Detectors

The lateral-line neuromasts are coupled to the aquatic medium by the viscous drag forces exerted on the cupulae, as well as by the pressure differences developing across them. However, for cupulae that are elongated in the direction of sensitivity, the velocity-controlled drag forces dominate at the low frequencies of interest, and the differential equation governing cupular motion is:

$$m\frac{d^2x}{dt^2} + b\frac{dx}{dt} + kx = bv' \cos \omega t \tag{19}$$

where $x$ is the displacement of the hair bundle, the tip of which is embedded in the moving cupula, $m$ the effective mass of the cupula and the entrained fluid, $b$ the viscous force per unit velocity, tacitly assumed to have the same value for the velocity of the cupula and for the velocity of the water, $k$ the restoring force of the hair bundles per unit cupular displacement, and $v' \cos \omega t$ the velocity of the water driving the cupula, relative to the receptor epithelium.

Equation 19 may be written in the form:

$$\frac{d^2x}{dt^2} + \gamma\frac{dx}{dt} + \omega_0^2 x = \gamma v' \cos \omega t \tag{20}$$

with $\gamma = b/m$ and $\omega_0^2 = k/m,$ yielding a hair-bundle displacement $x(\omega)$ and a phase angle $\delta(\omega)$, the angle by which the displacement lags the driving *viscous force:*

$$x(\omega) = \frac{\gamma}{\omega_0^2}\,\frac{\omega_0/\omega}{[(\omega_0/\omega - \omega/\omega_0)^2 + (\gamma/\omega_0)^2]^{1/2}}\,v' \cos(\omega t - \delta) \tag{21}$$

$$\text{with} \qquad \tan \delta(\omega) = \frac{\gamma/\omega_0}{\omega_0/\omega - \omega/\omega_0} \tag{22}$$

Thus, the lateral-line *organs* respond to the velocity of the water *driving the cupula* (Eqs. 21 and 22) in a similar fashion as the otolith organs respond to the acceleration of the whole animal (Eqs. 14 and 15).

The analysis is incomplete, however, because the velocity $v'$ of the water driving the cupula differs from the velocity $v$ of the water only a short distance from the fish. In the water volume, the inertial forces dominate the viscous forces, and the *acceleration* of the fluid particles is proportional to, and in phase with, minus the pressure gradient. On the other hand, deep in the boundary layer on the skin and, even more so, in the lateral-line canals, the viscosity successfully competes with the inertia in controlling the fluid motions. Consequently, at the surface of the skin, it is a combination of the fluid *velocity and acceleration,* whereas, in the lateral-line canals, it is purely the fluid *velocity* that are proportional to, and in phase with, minus the pressure gradient at the frequencies of interest.

Since the inertial, acceleration-related forces increase more rapidly with frequency than do the viscous, velocity-related forces, the thickness of the boundary layer, defined by a one-percent-or-more influence of viscosity, is inversely proportional to the square root of the frequency (Schlichting 1979, p. 94). While the cupula of the free neuromasts has a half-length on the order of 50 μm, the boundary layer reaches a height of ~ 560 μm at 1 Hz, and of ~ 56 μm at 100 Hz, not counting the slime cover of the skin. Thus, at the lower frequencies, the cupula is driven by the velocity deep into the boundary layer, which is proportional to a combination of the velocity and the acceleration of the water volume. As frequency increases, the boundary layer becomes progressively thinner and the velocity driving the cupula approaches the velocity in the water volume. In most lateral-line canals, on the other hand, the viscosity dominates nearly completely, and the velocity driving the cupula is approximately proportional to, and in phase with, the acceleration of the water volume (Denton and Gray 1983).

For the lower excitation frequencies $\omega \ll \omega_1$:

$$x \cong \gamma/\omega_0^2 \; v' \cos(\omega t - \delta) \cong b/k \; v \; (\omega/\omega')^s \cos \omega t \tag{23}$$

where the displacement, $x$, of the hair bundle is proportional to, and in phase with, the velocity, $v' \cos \omega t$, of the water at, say, half the height of the cupula (cf. Eqs. 21 and 22). The amplitude of the velocity $v'$ in the boundary layer approximates $v(\omega/\omega')^s$, where $v$ is the amplitude of the velocity in the water volume, $\omega'$ a constant having the dimension of frequency, and $s$ a dimensionless number (after Schlichting 1979, p. 437). The values of $\omega'$ and $s$ reflect the inherent properties of the system under study.

For the free neuromasts, $f' = \omega'/2\pi$ is about 200 Hz, $s$ is slightly less than 0.5, and the phase lag, $\eta$, of the boundary layer velocity $v'$, relative to the volume velocity $v$, varies from nearly $-45$ degrees at 1 Hz to a lower value at 80 Hz, depending on the length of the cupula (see Fig. 4.13). Hence, Eq. 23 reduces to:

$$x \cong b/k \; v \; (\omega/\omega')^{1/2} \cos \omega t \cong b/k \; 1/(\omega')^{1/2} \, (va)^{1/2} \cos \omega t \tag{24}$$

i.e., the hair-bundle displacement, $x$, is approximately proportional to $v(\omega)^{1/2} = (v^2\omega)^{1/2} = (va)^{1/2}$, the geometric mean of the velocity and acceleration amplitudes outside the boundary layer, and leads the velocity or, equivalently, lags the acceleration, by a little less than 45 degrees.

For the canal organs, $f' = \omega'/2\pi$ is about 100 Hz, $s$ is nearly unity, and $\eta$ is close to $-90$ degrees for frequencies up to 80 Hz (Denton and Gray 1983), so that Eq. 23 reduces to:

$$x \cong b/k \; v \; \omega/\omega' \cos \omega t \cong b/k \; a/\omega' \cos \omega t \tag{25}$$

i.e., the hair-bundle displacement, $x$, is proportional to, and in phase with, the velocity $v'$ in the boundary layer, which in turn is proportional to, and in phase with, the *acceleration, a,* of the water volume. Thus, the lateral-line canal system responds to the acceleration of the water volume *relative to* the animal (Eq. 25) in a similar fashion as the otolith organs respond to the acceleration *of* the animal (Eq. 16).

The biophysics of lateral-line detection was first studied by Jielof et al. (1952). They described the sense organs as slightly underdamped with a natural frequency $f_0$ around 80 Hz (cf. Kuiper 1967), yielding crossover frequencies of 30 and 200 Hz (cf. p. 118). Jielof and co-workers already noticed the frequency dependence of the boundary layer for the free neuromasts. Recently, Denton and Gray (1983, Chapter 23) studied the hydrodynamics of the canal organs by direct observation of the fluid motions in the canals and established that the organs respond to the acceleration, rather than the velocity, of the outside medium.

In summary, at the low frequencies of interest, the free neuromasts operate in a mixed velocity–acceleration mode, whereas the canal organs operate in a nearly pure acceleration mode in detecting the natural fields. No doubt, the animals adhere to the rigor of physics, although they may not adhere to my understanding

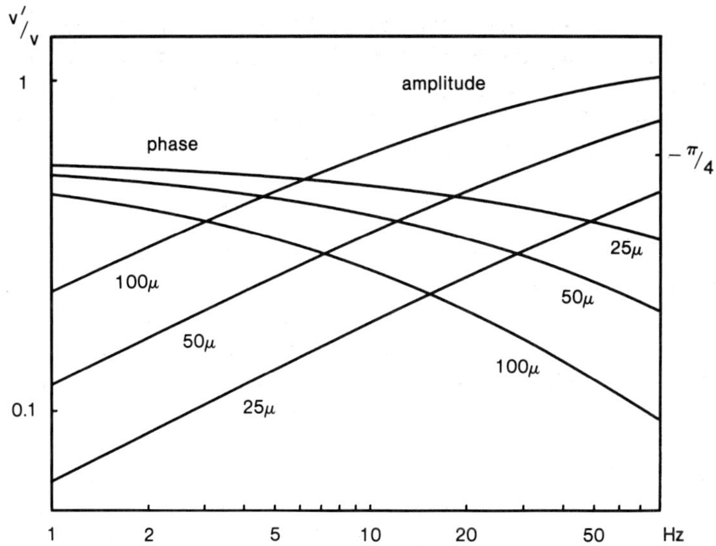


Figure 4.13. Amplitude and phase relations between the water velocity $v'$, driving the cupulae of the free neuromasts, and the velocity $v$ of the water outside the boundary layer, as a function of frequency. Left vertical axis: the ratio of $v'$ *over* $v$, on a logarithmic scale. Right vertical axis: the phase lag of $v'$ with respect to $v$, in radians. Horizontal axis: frequency in hertz on a logarithmic scale. The amplitude and phase curves were calculated for cupulae having half-lengths of 25, 50, and 100 μm, respectively, At low frequency, the slope of the amplitude curves is nearly 0.5 and the phase lag tends to −45 degrees. Both the amplitude and the phase angle decline with frequency, and do so more steeply the longer the cupula. Calculated from Schlichting (1979, p. 437).

of it. The detection of intermediate and higher frequencies will not be detailed here because of their lesser biological significance.

Thus far, the cupulae were assumed to be driven by viscous forces. In some fish, however, the cupulae seem to block the lateral-line canals more or less completely (Kroese, personal communication; Coombs et al., Chapter 22), and, driven by the pressure differential, would faithfully follow the motions of the canal fluid. In case the cupula is sufficiently stiff, the pressure difference between adjacent canal pores would control the displacement, rather than the velocity, of the canal fluid and the cupula, so that the end result would be the same. The velocity-driven cupulae may, however, be less liable to damage due to overstimulation.

For the sake of simplicity, I have, in the above, tacitly assumed that the flow in the boundary layer was laminar and perfectly parallel to the skin, and the pressure gradient along the skin was directly related to the acceleration of the water volume. Under those circumstances, a DC flow would develop a maximal boundary-layer thickness, and thus stimulate the free neuromasts minimally, and would develop a near-zero pressure gradient along the skin, and thus stimulate the canal organs even less. If, however, the flow impinges on the skin under a nonzero angle, the pressure distribution in the boundary layer is no longer determined only by the acceleration of the water volume, but also by the configuration of the flow field over the skin. Where the water is deflected by the presence of the animal, the pressure is elevated over that in the immediate surroundings, thus stimulating both the free neuromasts and the canal organs, also in the case of a steady DC flow. Both laminar and impinging flows arise from moving objects in the environment of the fish and from the animal's own motion through the water (Kuiper 1967). Further research in this direction is urgently needed.

## The Macula Neglecta

The macula neglecta of elasmobranch and a variety of bony fishes is an unloaded sensory patch, probably stimulated by the viscous force exerted by the endolymph on the gelatinous cupula. How this force comes about has been the subject of unrestrained speculation. The hypothesized sound funneling by the membranous duct in the wall of which the macula neglecta is located (Corwin 1981) defies the laws of physics, because the acoustic impedance of the structures involved does not significantly differ from that of seawater and their dimensions are negligibly small compared to the wave length of the sound detected. Also, the suggested Helmholtz resonance of the entire elasmobranch labyrinth (van den Berg and Schuijf 1983) remains not more than a technical term in the absence of any data on the rigidity of the cranium and the differential compressibility of cartilage and the inner-ear fluids. Inevitably, these fluids show complicated flow patterns, and the thin epithelia enclosing them undergo complex movements, not only because of the near field affecting the labyrinth through the roof of the cranium, but also because of the linear and angular inertial motions of the otoliths and the endolymph (Pumphrey 1950). Hence, the macula neglecta is bound to be affected by the stimuli acting on the other vestibular organs too. Nevertheless, even if the function is that suggested by either Corwin or van den Berg and Schuijf, it is not for the reasons given, I believe.

## The Boundary Layer Revisited

Although Hudspeth and Corey (1977) convincingly demonstrated that the hair cells of the saccule respond purely to the displacements of the sensory hair bundles up to frequencies of 150 Hz, other neurophysiologists, studying the free neuromasts and canal organs of the lateral line, felt compelled to invoke a sensitivity to the velocity of the hair bundles as well. Nevertheless, to explain the existing lateral-line data, the inferred velocity sensitivity is not needed if the frequency dependence of the boundary layer is taken into account. To substantiate this assertion, I will, in addition to my earlier comments on Harris and van Bergeijk's (1962) phase glitch (cf. p. 114), discuss some older and more recent neurophysiological studies, in which the attenuation and phase shifts effected by the boundary layer have also repeatedly escaped due attention.

Goerner (1963) applied uniform flow fields of constant velocity to the free lateral-line neuromasts of the clawed toad, *Xenopus* sp., so as to produce well-defined cupular displacements. He observed an unexpected transient overshoot in the nerve response at the onset of the velocity stimulus, which he attributed to a partially phasic nature of the transduction mechanism. Strelioff and Honrubia

(1978) reproduced Goerner's data and abided by his conclusions. Hydrodynamically, however, the initial acceleration of the water must have caused a similar overshoot of the fluid velocity monitored by the cupula during the time needed for the boundary layer to attain its full thickness. This would qualitatively explain Goerner's data and hence invalidate the evidence for the inferred partial velocity, rather than pure displacement, sensitivity of the hair-cell mechanism.

Kroese et al. (1978) determined the frequency response of the free neuromasts in *Xenopus* with a small vibrating sphere. Their data were consistent with the hypotheses of a velocity-driven cupula and a displacement-sensitive transduction mechanism except for a $0.2\pi$ phase lead, which, in hindsight, was to be expected from the boundary layer on the skin. For frequencies less than 5 Hz, they furthermore noticed a drastic advance in phase, relative to the stimulus, when the vibrating sphere was brought closer than 2 mm to the skin. Evidently, the boundary layer of the vibrating sphere had begun to exert control over the motion of the cupula. The effect was reported, but the suspicious data were not used until a later paper (Kroese et al. 1980), in which the authors ascribed the observed phase advances to a velocity sensitivity of the transduction mechanism, rather than to the coupling of the cupula to the sphere.

Muenz (1985) studied both the canal organs and the free neuromasts in the ciclid fish *Sarotherodon* sp., but used tiny vibrating spheres, after Kroese et al. (1980), at such short distances that the displacements applied were no longer those of the assumed inertia-controlled potential fields. The viscous boundary-layer forces inevitably rendered the coupling between the sphere and the cupulae of the free neuromasts highly frequency dependent. The 180-degree spread in the low-frequency phase data bears witness to this conjecture: directly under a horizontally vibrating sphere the water moves along with the source, whereas farther away it moves in the opposite direction. Consideration of the boundary layer phenomena, direct measurement of the fluid and cupular motion, or the use of a larger sphere at a greater distance could have solved the problem.

Ironically, nearly every article on the lateral line refers to Dijkgraaf's work, but few show an attempt to provide the sensory basis needed to explain the behavioral responses he observed. The lower frequencies often suffered from a neglect of the boundary layer or were shunned altogether. Moreover, as explained above, by plotting the data in terms of displacement, rather than the field quantities that the fish detect, the low-frequency response of the lateral line has been underestimated for years. Even the neurophysiologists' "best frequency" turned out to be merely the upper "cutoff frequency." Thus, instead of searching for the high-frequency signals to explain the *alleged* tuning properties of the sensory system, I have endeavored to reveal the low-frequency sensitivity *known* to mediate the animals' behavioral responses.

## Summary

A meaningful systems analysis requires a knowledge of the input and output signals of the process under study or, in biological terms, of the natural stimulus and the behavioral response. The peripheral sensory apparatus conditions the physical stimulus in preparation of its detection by the receptor cells, whereas the central nervous system processes the incoming data to provide the sensory basis of the animal's behavior. This chapter addresses the issues that have most seriously limited our understanding of the lateral-line and inner-ear senses of fishes: the lack of familiarity with (1) the biologically relevant hydrodynamic and acoustic fields and (2) the biophysics of the detection mechanisms.

In the first part, it is shown that, except for the monopole swimbladder fields of the more vocal fishes, most biological objects produce predominantly dipole fields, like all animals during swimming. The monopole moment is often negligibly small, while the dipole moment tends to override all higher-order moments. The lateral line responds to the local-flow field only within a few body lengths from the source, whereas the inner ear detects the remainder of the near field, as well as the particle motions or, via the swimbladder, the acoustic pressure of the propagating sound wave. The issues of source localization and the origin of vertebrate hearing are discussed.

The second part deals with the question of why behaviorally the lateral line and inner ear would be used to detect mostly low-frequency fields, whereas physiological data suggest a "best" response at considerably higher frequencies. To resolve this dilemma, the adequate physical stimuli for the receptor cells, for the sense organs, and for the animals are identified. It is shown that, although the data are routinely plotted in terms of fluid displacement, the animals respond to the velocity or acceleration. By plotting the data properly, the problem appears to vanish. In conclusion, the need of considering the boundary layer more seriously in lateral-line studies is emphasized.

*Acknowledgments.* The author has benefitted greatly from the intensive interactions following his conference presentation. The manuscript has seen many versions in response to constructive and destructive criticism, both of which have been instrumental in bringing the chapter to its present state. Dr. Jelle Atema has offered me the moral support to complete my task. Dr. D. Vance Holliday has helped me clarify the near- and far-field concepts. Drs. Charles S. Cox, Russ E. Davis, Walter H. Munk, and Peter H. Rogers have commented on the physics. Drs. James R. Enright, Arthur N. Popper, and William M. Saidel have suggested many improvements in the text. Drs. John Gray, Loren H. Haury, and Alfons B. A. Kroese have further scrutinized the manuscript. The discussions with Drs. Sven Dijkgraaf, Per S. Enger, Arthur A. Myrberg, and Olav Sand have been enjoyable. Ms. Georgia H. Greaves and my wife Vera have patiently assisted me in preparing the figures. I furthermore thank Drs. Eric O. Hartwig and Bernard J. Zahuranec of the Office of Naval Research for their scientific and financial support. The hydrodynamic measurements were conducted in collaboration with Dr. Per S. Enger, under the auspices of the National Science Foundation and the Norwegian Council for Science and the Humanities.

## References

Alexander, R.McN. (1966). Physical aspects of swimbladder function. Biol. Rev., 41:141–176.

Banner, A. (1967). Evidence of sensitivity to acoustic displacements in the lemon shark, *Negaprion brevirostris* (Poey). In: Lateral Line Detectors. Cahn, P.H. (ed.), Indiana University Press, Bloomington, pp. 265–273.

Banner, A. (1972). Use of sound in predation by young lemon sharks, *Negaprion brevirostris* (Poey). Bull. Mar. Sci., 22:251–283.

van Bergeijk, W.A. (1964). Directional and nondirectional hearing in fish. In: Marine Bio-Acoustics. Tavolga, W.N. (ed.), Pergamon Press, Oxford, pp. 281–299.

van Bergeijk, W.A. (1967a). The evolution of vertebrate hearing. In: Contributions to Sensory Physiology. Neff, W.D. (ed.), Academic Press, New York, pp. 1–49.

van Bergeijk, W.A. (1967b). Introductory comments on lateral line function. In: Lateral Line Detectors. Cahn, P.H. (ed.), Indiana University Press, Bloomington, pp. 73–81.

van den Berg, A.V. and Schuijf, A. (1983). Discrimination of sounds based on the phase difference between particle motion and acoustic pressure in the shark *Chiloscyllium griseum*. Proc. R. Soc. Lond., 218:127–134.

Blaxter, J.H.S., Denton, E.J., and Gray, J.A.B. (1981). Acousticolateralis system in clupeid fishes. In: Hearing and Sound Communication in Fishes. Tavolga, W.N., Popper, A.N., Fay, R.R. (eds.), Springer Verlag, New York, pp. 39–59.

Bleckmann, H. (1980). Reaction time and stimulus frequency in prey localization in the surface-feeding fish *Aplocheilus lineatus*. J. Comp. Physiol., 140:163–172.

Buwalda, R.J.A. (1981). Segregation of directional and nondirectional acoustic information in the cod. In: Hearing and Sound Communication in Fishes. Tavolga, W.N., Popper, A.N., Fay, R.R. (eds.), Springer-Verlag, New York, pp. 139–171.

Buwalda, R.J.A., Schuijf, A., and Hawkins, A.D. (1983). Discrimination by the cod of sounds from opposing directions. J. Comp. Physiol., 150:175–184.

Cahn, P.H., Siler, W., and Wodinsky, J. (1969). Acoustico-lateralis system of fishes: tests of pressure and particle-velocity sensitivity in grunts, *Haemulon sciurus* and *Haemulon parrai*. J. Acoust. Soc. Am., 46:1572–1578.

von Campenhausen, C., Riess, I., and Weissert, R. (1981). Detection of stationary objects by the blind cave fish *Anoptichthys jordani* (Characidae). J. Comp. Physiol., 143:369–374.

Chapman, C.J. and Hawkins, A.D. (1973). A field study of hearing in the cod, *Gadus morhua* L. J. Comp. Physiol., 85:147–167.

Chapman, C.J. and Sand, O. (1974). Field studies of hearing in two species of flatfish *Pleuronectes platessa* (L.) and *Limanda limanda* (L.) (Family Pleuronectidae). Comp. Biochem. Physiol., 47:371–385.

Corwin, J.T. (1981). Audition in elasmobranchs. In: Hearing and Sound Communication in Fishes. Tavolga, W.N., Popper, A.N., Fay, R.R. (eds.), Springer-Verlag, New York, pp. 81–105.

Denton, E.J. and Gray, J. (1983). Mechanical factors in the excitation of clupeid lateral lines. Proc. R. Soc. Lond., 218:1–26.

Dijkgraaf, S. (1934). Untersuchungen ueber die Funktion der Seitenorgane an Fischen. Z. Vergl. Physiol., 20:162–214.

Dijkgraaf, S. (1960). Hearing in bony fishes. Proc. R. Soc. Lond., 152:51–54.

Dijkgraaf, S. (1963a). The functioning and the significance of the lateral-line organs. Biol. Rev., 38:51–105.

Dijkgraaf, S. (1963b). Sound reception in the dogfish. Nature, 197:93–94.

Eaton, R.C., Bombardieri, R.A., and Meyer, D.L. (1977). The Mauthner-initiated startle response in teleost fish. J. Exp. Biol., 66:65–81.

Enger, P.S. (1966). Acoustic threshold in goldfish and its relation to the sound source distance. Comp. Biochem. Physiol., 18:859–868.

Enger, P.S. (1967). Effect of the acoustic near field on the sound threshold in fishes. In:

Lateral Line Detectors. Cahn, P.H. (ed.), Indiana University Press, Bloomington, pp. 239–248.

Enger, P.S. and Andersen, R. (1967). An electrophysiological field study of hearing in fish. Comp. Biochem. Physiol., 22:517–525.

Enger, P.S., Hawkins, A.D., Sand, O., and Chapman, C.J. (1973). Directional sensitivity of saccular microphonic potentials in the haddock. J. Exp. Biol., 59:425–433.

Fay, R.R. and Popper, A.N. (1974). Acoustic stimulation of the ear of the goldfish (*Carassius auratus*). J. Exp. Biol., 61:243–260.

Fay, R.R. and Popper, A.N. (1975). Modes of stimulation of the teleost ear. J. Exp. Biol., 62:379–387.

Fish, M.P. (1964). Biological sources of sustained ambient sea noise. In: Marine Bio-Acoustics. Tavolga, W.N. (ed.), Pergamon Press, Oxford, pp. 175–194.

Flock, A. (1965). The ultrastructure of the lateral line canal organ. Acta Oto-Laryngol., Suppl. 199:7–90.

von Frisch, K. (1923). Ein Zwergwels, der kommt, wenn man ihm pfeift. Biol. Zbl, 43:439–446.

von Frisch, K. (1938). The sense of hearing in fish. Nature, 141:8–11.

von Frisch, K. and Dijkgraaf, S. (1935). Koennen Fische die Schallrichtung wahrnehmen? Z. Vergl. Physiol., 22:641–655.

Gray, J. (1984). Interaction of sound pressure and particle acceleration in the excitation of the lateral-line neuromasts of sprats. Proc. R. Soc. Lond., 220:299–325.

Goerner, P. (1963). Untersuchungen zur Morphologie und Elektrophysiologie des Seitenlinienorgans vom Krallenfrosch (*Xenopus laevis* Daudin). Z. Vergl. Physiol., 47:316–338.

Groen, J.J., Lowenstein, O., and Vendrik, A.J.H. (1952). The mechanical analysis of the responses from the end-organs of the horizontal semicircular canal in the isolated elasmobranch labyrinth. J. Physiol., 117:329–346.

Harris, G.G. (1964). Considerations on the physics of sound production by fishes. In: Marine Bio-Acoustics. Tavolga, W.N. (ed.), Pergamon Press, Oxford, pp. 233–247.

Harris, G.G. and van Bergeijk, W.A. (1962). Evidence that the lateral-line organ responds to near-field displacements of sound sources in water. J. Acoust. Soc. Am., 34:1831–1841.

Hawkins, A.D. and Johnstone, A.D.F. (1978). The hearing of the Atlantic salmon, *Salmo salar*. J. Fish Biol., 13:655–673.

Hawkins, A.D. and Myrberg, A.A. (1983). Hearing and sound communication under water. In: Bioacoustics, a Comparative Approach. Lewis, B. (ed.), Academic Press, New York, pp. 347–405.

Hudspeth A.J. and Corey, D.P. (1977). Sensitivity, polarity, and conductance change in the response of the vertebrate hair cells to controlled mechanical stimuli. Proc. Natl. Acad. Sci., 74:2407–2411.

Jielof, R., Spoor, A., and de Vries, H1. (1952). The microphonic activity of the lateral line. J. Physiol., 116:137–157.

Kinsler, L.E., Frey, A.R., Coppens, A.B., and Sanders, J.V. (1982). Fundamentals of Acoustics, John Wiley & Sons, New York.

Kroese, A.B.A., van der Zalm, J.M., and van den Bercken, J. (1978). Frequency response of the lateral-line organ of *Xenopus laevis*. Pfluegers Arch., 375:167–175.

Kroese, A.B.A., van der Zalm, J.M., and van den Bercken, J. (1980). Extracellular receptor potentials from the lateral-line organ of *Xenopus laevis*. J. Exp. Biol., 86:63–77.

Kuiper, J.W. (1967). Frequency characteristics and functional significance of the lateral line organ. In: Lateral Line Detectors. Cahn, P.H. (ed.), Indiana University Press, Bloomington, pp. 105–121.

Lamb, H. (1931). The Dynamical Theory of Sound, Edward Arnold and Co, London.

Lewis, E.R. (1984). Inertial motion sensors. In: Comparative Physiology of Sensory Systems. Bolis, L., Keynes, R.D., Maddrell, S.H.P. (eds.), Cambridge University Press, Cambridge, pp. 587–610.

Lighthill, J. (1980). Waves in Fluids. Cambridge University Press, Cambridge.

Lowenstein, O. (1972). Physiology of the vestibular receptors. In: Progress in Brain Research, Vol. 37, Basic Aspects of Central Vestibular Mechanisms. Brodal, A., Pompeiano, O. (eds.), Elsevier, Amsterdam, pp. 19–30.

Lowenstein, O. and Roberts, T.D.M. (1951). The localization and analysis of the responses to vibration from the isolated elasmobranch labyrinth: a contribution to the problem of the evolution of hearing in vertebrates. J. Physiol., 114:471–489.

Mayne, R. (1974). A systems concept of the vestibular organs. In: Handbook of Sensory Physiology, Vol. VI/2. Kornhuber, H.H. (ed.), Springer-Verlag, New York, pp. 493–580.

Morse, P.M. and Ingard, K.U. (1968). Theoretical Acoustics: McGraw-Hill, New York.

Moulton, J.M. (1960). Swimming sounds and the schooling of fishes. Biol Bull., 119:210–223.

Moulton, J.M. (1963). Acoustic behaviour of fishes. In: Acoustic Behaviour of Animals. Busnel, R.G. (ed.), Elsevier, Amsterdam, pp. 655–693.

Muenz, H. (1985). Single unit activity in the peripheral lateral line system of the cichlid fish *Sarotherodon niloticus* L. J. Comp. Physiol., 157:555–568.

Myrberg, A.A. (1978). Underwater sound—its effect on the behavior of sharks. In: Sensory Biology of Sharks, Skates, and Rays. Hodgson, E.S., Mathewson, R.F. (eds.), U.S. Government Printing Office, Washington D.C., pp. 391–417.

Myrberg, A.A., Banner, A., and Richard, J.D. (1969). Shark attraction using a video-acoustic system. Mar. Biol., 2:264–276.

Myrberg, A..A., Gordon, C.R., and Klimley, A.P. (1976). Attraction of free ranging sharks by low frequency sound, with comments on its biological significance. In: Sound Reception in Fish. Schuijf, A., Hawkins, A.D. (eds.), Elsevier, Amsterdam, pp. 205–228.

Myrberg, A.A., Ha, S.J., Walewski, S., and Banbury, J.C. (1972). Effectiveness of acoustic signals in attracting epipelagic sharks to an underwater sound source. Bull. Mar. Sci., 22:926–949.

Myrberg, A.A. and Spires, J.Y. (1980). Hearing in damselfishes: an analysis of signal detection among closely related species. J. Comp. Physiol., 140:135–144.

Nelson, D.R. and Gruber, S.H. (1963). Sharks: attraction by low-frequency sound. Science, 142:975–977.

Nelson, D.R. and Johnson, R.H. (1972). Acoustic attraction of Pacific Reef sharks: effect of pulse intermittency and variability. Comp. Biochem. Physiol., 42:85–95.

Nelson, D.R. and Johnson, R.H. (1976). Some recent observations on acoustic attraction of Pacific Reef sharks. In: Sound Reception in Fish. Schuijf, A., Hawkins, A.D. (eds.), Elsevier, Amsterdam, pp. 229–239.

Olson, K. (1976). Evidence for localization of sound by fish in schools. In: Sound Reception in Fish. Schuijf, A., Hawkins, A.D. (eds.), Elsevier, Amsterdam, pp. 257–270.

Partridge, B.L. and Pitcher, T.J. (1980). The sensory basis of fish schools: relative roles of lateral line and vision. J. Comp. Physiol., 135:315–325.

Pierce, A.D. (1981). Acoustics, an Introduction to its Physical Principles and Applications. McGraw-Hill, New York.

Pumphrey, R.J. (1950). Hearing. Symp. Soc. Exp. Biol., 4:3–18.

Richard, J.D. (1968). Fish attraction with pulsed low-frequency sound. J. Fish. Res. Bd. Canada, 25:1441–1452.

Roberts, B.L. (1978). Mechanoreception and the behaviour of elasmobranch fishes with special reference to the acoustico-lateralis system. In: Sensory Biology of Sharks, Skates, and Rays. Hodgson, E.S., Mathewson, R.F. (eds.), U.S. Government Printing Office, Washington D.C., pp. 331–390.

Sand, O. (1974). Directional sensitivity of microphonic potentials from the perch ear. J. Exp. Biol., 60:881–899.

Sand, O. (1981). The lateral line and sound reception. In: Hearing and Sound Communication in Fishes. Tavolga, W.N., Popper, A.N., Fay, R.R. (eds.), Springer-Verlag, New York, pp. 459–480.

Sand, O. (1984). Lateral-line systems. In: Comparative Physiology of Sensory Systems. Bolis, L., Keynes, R.D., Maddrell, S.H.P. (eds.), Cambridge University Press, Cambridge, pp. 3–32.

Sand, O. and Enger, P.S. (1973). Evidence for an auditory function of the swimbladder in the cod. J. Exp. Biol., 59:405–414.

Sand, O. and Enger, P.S. (1974). Possible mechanisms for directional hearing and pitch discrimination in fish. Rheinisch-Westfaelische Akad. Wissensch., 53:223–242.

Schlichting, H. (1979). Boundary-Layer Theory. McGraw-Hill, New York.

Schuijf, A. (1975). Directional hearing of cod (*Gadus morhua*) under approximate free field conditions. J. Comp. Physiol., 98:307–332.

Schuijf, A. and Buwalda, R.J.A. (1975). On the mechanism of directional hearing in cod (*Gadus morhua* L.). J. Comp. Physiol., 98:333–343.

Schuijf, A. and Buwalda, R.J.A. (1980). Underwater localization—a major problem in fish acoustics. In: Comparative Studies of Hearing in Vertebrates. Popper, A.N., Fay, R.R. (eds.), Springer-Verlag, New York, pp. 43–77.

Schuijf, A. and Hawkins, A.D. (1983). Acoustic distance discrimination by the cod. Nature, 302:143–144.

Schwartz, E. (1965). Bau und Funktion der Seitenlinie des Streifenhechtlings (*Aplocheilus lineatus* Cuv. u. Val.). Z. Vergl. Physiol., 50:55–87.

Siler, W. (1969). Near- and farfields in a marine environment. J. Acoust. Soc. Am., 46:483–484.

Steinhausen, W. (1931). Ueber den Nachweis der Bewegung der Cupula in der intakten Bogengangsampulle des Labyrinthes bei der naturlichen rotatorischen und kalorischen Reizung. Pfluegers Arch., 228:322–328.

Strelioff, D. and Honrubia, V. (1978). Neural transduction in *Xenopus laevis* lateral line system. J. Neurophysiol., 41:432–444.

Tavolga, W.N. (1964). Sonic characteristics and mechanisms in marine fishes. In: Marine Bio-Acoustics. Tavolga. W.N. (ed.), Pergamon Press, Oxford, pp. 195–211.

Tavolga, W.N. (1971). Sound production and detection. In: Fish Physiology, Vol. 5. Hoar, W.S., Randall, D.J. (eds.), Academic Press, New York, pp. 135–205.

Tavolga, W.N. and Wodinsky, J. (1963). Auditory capacities in fishes: pure tone thresholds in nine species of marine teleosts. Bull. Am. Mus. Nat. Hist., 126:179–239.

Urick, R.J. (1983). Principles of Underwater Sound. McGraw-Hill, New York.

de Vries, Hl. (1950). The mechanics of the labyrinth otoliths. Acta Oto-Laryngol., 38:262–273.

de Vries, Hl. (1956). Physical aspects of the sense organs. In: Progress in Biophysics and Biophysical Chemistry. Butler, J.A.V. (ed.), Pergamon Press, Oxford, pp. 208–264.

Wenz, G.M. (1964). Curious noises and the sonic environment in the ocean. In: Marine Bio-Acoustics. Tavolga, W.N. (ed.), Pergamon Press, Oxford, pp. 101–119.

Wisby, W.J., Richard, J.D., Nelson, D.R., and Gruber, S.H. (1964). Sound perception in elasmobranchs. In: Marine Bio-Acoustics. Tavolga, W.N. (ed.), Pergamon Press, Oxford, pp. 255–268.

# Chapter 5

# Underwater Sound as a Biological Stimulus

PETER H. ROGERS* and MARDI COX*

Of all the sensory stimuli discussed in this volume, only sound allows long-range transmission of information underwater. This is a consequence of the extraordinarily low attenuation of sound in water and the ability of sound speed gradients in the ocean to channel sound so that it can propagate without interaction with the surface or bottom.

For example, as shown in Figure 5.1, at 500 Hz, which is in the middle of the hearing range for most fish, sound suffers only 1 dB of attenuation in 100 kilometers of propagation in seawater and 1 dB in 10,000 kilometers of propagation in freshwater. In comparison, at 500 Hz, electromagnetic radiation attenuates 1 dB in just 1 m, and blue-green light attenuates 1 dB in less than 3 m. The attenuation of sound in water is also several orders of magnitude lower than its attenuation in air, which is itself rather low.

The relationships presented in the following sections are fundamental to sound fields both near and far from the source. Although the fundamental theory presented in this chapter is applicable to all ranges and frequencies, the emphasis is on true "sound" (i.e., compressional waves), rather than "hydrodynamic" (i.e., essentially incompressible) flow, which may be the dominant stimulus at close range and low frequency (see Chap. 4 for a treatment emphasizing the latter stimulus). Because of the characteristics of underwater sound propagation, fish may simultaneously receive many signals from many distances. Understanding the acoustical stimulus to the fish requires the application of these fundamental relationships.

## The Nature of Sound

Sound is a longitudinal mechanical wave that propagates in a compressible medium. By *wave,* we simply mean a disturbance that propagates; by *longitudinal,*

* School of Mechanical Engineering, The Georgia Institute of Technology, Atlanta, GA 30332, USA

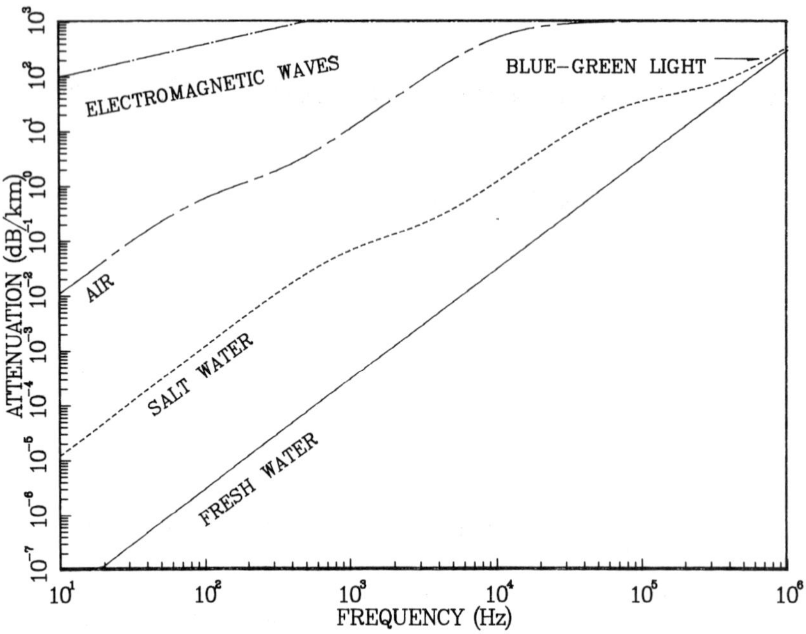


Figure 5.1. The attenuation of sound in freshwater, saltwater, and air compared with the attenuation of electromagnetic waves and blue-green light ($4.5 \times 10^{14}$ Hz) in water.

we mean that the displacement of the medium is predominantly parallel to the direction of propagation. An acoustical disturbance involves both motion of the medium and changes in the state variables. Thus, the acoustical field can be described in terms of the state variables—pressure, temperature, or density—or the motion variables—displacement, velocity, or acceleration. The state variables are scalars; the motion variables, vectors. The sound field can be completely characterized by any one of these quantities, either vector or scalar. Because a sound wave can be completely described by one of the scalar quantities, it is considered to be a *scalar* wave. It is incorrect to say that sound has scalar and vector "components," because any of the vectors associated with a sound wave can be expressed as the gradient of some scalar.

## Governing Equations

An acoustical disturbance propagates at the speed of sound ($c$), expressed in m/sec, which is related to the adiabatic compressibility of the medium:

$$c = \sqrt{\left(\frac{\partial p}{\partial \rho}\right)_S} \tag{5.1}$$

where $p$ = pressure, $\rho$ = density, $\partial$ = partial derivative, and the subscript (s) indicates that entropy remains constant. The "stiffer" the medium, the faster the speed of sound. The sound speed in water is approximately 1,500 m/sec. It is about 3% higher in seawater than in freshwater and increases almost linearly with temperature, salinity, and depth. The speed of sound in water is about four and a half times higher than the speed of sound in air and 150,000 times lower than the speed of light in water.

Two simple laws of physics govern the propagation of sound. The first is Newton's second law (force = mass × acceleration), which for an ideal fluid is expressed as Euler's equation:

$$\rho_o \frac{\partial \vec{v}}{\partial t} = -\vec{\nabla} p \tag{5.2}$$

where $\vec{v}$ = particle velocity, $t$ = time, and $\vec{\nabla}$ = gradient. This relationship shows that the gradient of the pressure exerts a force and the fluid responds by accelerating in the direction of the force. The second law is conservation of mass expressed by the continuity equation:

$$-\rho_o \vec{\nabla} \cdot \vec{v} = \frac{\partial \rho}{\partial t} = \frac{1}{c^2} \frac{\partial p}{\partial t} \tag{5.3}$$

This law states that any net flow of fluid into or out of a given volume causes the density of the fluid within it to change. That is, the divergence of the fluid velocity, which is a measure of mass flux at a point, is proportional to the time derivative of the density at that point. Since the fluid is compressible, any change in density is accompanied by a proportional change in pressure. From equation (5.1), we can see that the constant of proportionality is just $c^2$. Thus the divergence of the particle velocity is also proportional to the time derivative of the pressure. By adding the divergence of equation (5.2) to the time derivative of equation (5.3), we obtain the wave equation for sound:

$$\frac{1}{c^2} \frac{\partial^2 p}{\partial t^2} = \nabla^2 p \tag{5.4}$$

### Plane Waves

The wave equation is satisfied by any function of $(t - \vec{n} \cdot \vec{R}/c)$, where $\vec{n}$ = unit vector in the direction of propagation and $\vec{R}$ = propagation distance vector. This simple solution is the plane wave. Such waves are called "plane" because the surfaces of constant pressure are planes. The equations for a plane wave are

$$p(\vec{R}, t) = f\left(t - \frac{\vec{n} \cdot \vec{R}}{c}\right) \tag{5.5}$$

and

$$\vec{v}(\vec{R}, t) = \frac{\vec{n}}{\rho c} p(\vec{R}, t). \tag{5.6}$$

This mathematical form yields a pressure disturbance that propagates in the direction of $\vec{n}$ with speed $c$ and no change in shape. The particle velocity is in the direction of $\vec{n}$ and proportional to $p/\rho c$. Thus the wave is longitudinal, and the ratio of $p$ to $\vec{v}$ is positive for waves traveling in the positive $\vec{R}$ direction and negative for waves going in the opposite direction. Although no wave is ever truly ‘‘plane,’’ any wave will asymptotically behave like a plane wave at great distances. Thus, far away from the sound source, all waves approach plane waves.

## Spherical Waves

Another important solution to the wave equation is the ‘‘spherical wave,’’ for which the surfaces of constant pressure are spheres. Such waves are generated by any small pulsating source (i.e., monopole). This solution is given by

$$p(\vec{R}, t) = \frac{f(t - R/c)}{R} \tag{5.7}$$

and

$$\vec{v}(\vec{R}, t) = \left( \frac{p(\vec{R}, t)}{\rho c} + \frac{\int^{t} p(\vec{R}, \xi) d\xi}{\rho R} \right) \vec{e}_r \tag{5.8}$$

where $R$ = radial distance from the source, $\vec{e}_r$ is a unit vector in the radial direction, and $\xi$ is the dummy variable of integration. For a spherical wave, the pressure is $1/R$ times any function of $(t - R/c)$. The pressure waveform thus propagates radially with velocity, $c$, without any change in shape, but its amplitude decreases like $1/R$. This $1/R$ decay is necessary to conserve energy as the wave spreads.

The particle velocity is in the radial direction and consists of two terms. The first (like the plane wave solution) is just $p/\rho c$ and hence falls off like $1/R$. This part of the particle velocity is due to compression of the fluid by the source and hence is ‘‘true sound.’’ The second term is proportional to $1/R$ times the time integral of the pressure and falls off like $1/R^2$. This part of the velocity is essentially a ‘‘flow’’ term and would be present even if the fluid were incompressible.

### Nearfield vs. Farfield

The ‘‘flow’’ term will dominate the ‘‘sound’’ term close to the source. This region is called the *nearfield*. As $R$ increases, the pressure-to-velocity ratio approaches the plane wave value, and the spheres of constant pressure approach planes. This is the *farfield*.

Nearfields and farfields will exist for any kind of sound source. In general, we can define the farfield as the region where, to a good approximation, $p = \rho c v$. This requires $R$ to be much greater than the product of $c$ and some time that is characteristic of the waveform and also to be much greater than any source dimension (including images). The commonly used definition of the farfield as the region where $R$ is much greater than a wavelength (van Bergeijk 1964) is appropriate only for sinusoidal waves generated by point sources.

## Pressure–Velocity Ratios

Describing a sound field in terms of pressure–velocity ratios (p/v) is both difficult and dangerous. Obviously, since pressure is a scalar and velocity a vector, the expression has many interpretations. Does $v$ mean the magnitude of the velocity vector or one of its components? The ratio also depends on the nature of the source and the boundaries of its medium. Thus, formulae tabulated by Harris (1964) and Siler (1969) must be used with a thorough understanding of the assumptions made in their derivations.

In his consideration of sound production by fish, Harris (1964) tabulated component amplitudes of pressure and velocity, along with approximations for the farfield pressure and the nearfield radial velocity for monopole, dipole, and quadrupole radiation from a sphere with a radius much smaller than the wavelength of the emitted sound wave. Later, Siler (1969) simplified Harris's tabulation, this time with an interest in sound reception by fish.

Unfortunately, Harris and Siler's formulations have been misused in attempts to recast and interpret experimental data from settings in which the formulae have little or no meaning. Furthermore, in extending experimental results to enhance our understanding of the fish's true sound environment, we must realize that these formulations are for sinusoids of constant frequency generated by an idealized source. In general, the sound stimulating the fish ear will come from a variety of (nonideal) sources and will neither be sinusoidal nor have constant frequency.

The formulations by Harris and Siler are for an axisymmetrical spherical source in an unbounded, infinite medium. In contrast with the spherically symmetical case, the pressure and particle velocity now depend on the polar angle, as well as the radial distance from the source. Hence, except for the monopolar mode, the velocity has both radial and tangential components.

The $p/v$ ratios tabulated by Siler are for the magnitude of the radial component of the velocity vector. They show dependence neither on the size of the source nor on the polar angle, because it is the same for both the pressure and *radial* velocity. Moreover, Siler's ratios are not applicable to the usual laboratory experiments conducted in relatively small tanks and near an air–water interface. Even in the ocean, sound propagation is often channeled and these idealized relationships may not be applicable.

Finally, we note that the $p/v$ is the same for both plane and spherical waves in the farfield of the source and independent of frequency or source vibration mode.

## The Microvar

The *microvar,* a velocity unit invented specifically for marine biologists, is both misleading and confusing. Siler (1969) defines the microvar as the particle velocity associated with 1 microbar of sound pressure in the farfield. Since sound in the farfield of any source behaves like a plane wave with $p = \rho c v$, then 1 microvar

= 1 microbar/($\rho c$). Thus the microvar is really not a ''unit'' of velocity at all, since it depends on the properties of the fluid. Its only convenience is in making $p/v = 1$ in the far field.

This alleged unit presents a number of difficulties. First of all, $\rho c$ is not a constant. It can vary by as much as 1 dB with temperature and will change by 0.5 dB from freshwater to saltwater. Secondly, as we have already said, the far field itself is poorly defined. Thirdly, velocity is a vector and pressure a scalar. So, does the definition apply to the magnitude of the vector or to one of its components? Finally, as we shall see in the next section, the relevant biological stimulus is probably acceleration or displacement and not velocity anyway.

Using the microvar to report experimental results has no apparent advantage, because measurements usually are not made in the far field, but in whatever complex sound field is generated as a result of the boundaries of the experimental apparatus. To interpret these types of measurements, the fundamental relationships given by equations (5.2) and (5.3) must be applied. Furthermore, extrapolating experimental results obtained under less than fully understood conditions to far field conditions should not be attempted. Use of the microvar entails extra calculations and only adds a factor of confusion to analytical and experimental results.

Our recommendation is to banish the microvar and stick with the tried and true meter per second.

## Adequate Stimulus

A detailed description of the fish auditory system can be found in Chapter 27 and will not be repeated here. The receptors of sound in a fish's ear are the sensory hair cells, which respond positively when the ciliary bundles are bent toward the long kinocilium and negatively when bent away from the kinocilium (Flock 1971). No response occurs when the cilia are bent in the direction orthogonal to the kinocilium. Thus the hair cells respond to a single component of a displacement vector, and not directly to acoustic pressure. Sound can be detected only if it somehow induces the adequate stimulus: bending of the ciliary bundles.

This bending is believed to be caused by the relative motion between the otoliths and the rest of the fish's body, which has approximately the same density as water and, hence, in the presence of sound, moves with the acoustic particle velocity. The otoliths within the fish's ears have a density three times higher than water (i.e., more inertia) and thus move less. This results in relative motion between the otoliths and the rest of the ear.

The cilia are somehow attached to both the sensory epithelium and the otolith so that this relative motion induces bending, the adequate stimulus. The displacement of the otolith may be a direct response to the sound, with the measured quantity's being one component of the acoustical particle displacement; or, at higher frequencies, the displacement may be induced ''indirectly'' by motion of the swim bladder caused by the incident field.

Since the resonance frequency of the swim bladder is usually above the hearing range of the fish, the swim bladder acts like a spring at the frequencies of interest.

The displacement of the wall of the swim bladder is thus proportional to the incident pressure. Whether the motion of the swim bladder is coupled directly to the otolithic organs via the Weberian ossicles (as in the Otophysans) or transmitted acoustically through the intervening tissue, the resulting displacement of the otolith will be proportional to the displacement of the swim bladder and hence to the acoustic pressure. Thus the "indirect" path to the ear results in a signal proportional to the acoustic pressure, rather than to the acoustic displacement, as is the case for the "direct" path.

The size of the displacement stimulus at threshold is incredibly small. Many fish can detect an acoustic pressure of 0.01 Pa at 500 Hz. The acoustical particle displacement for such a wave is $2 \times 10^{-12}$ m. The displacement of the otolith relative to the fluid is approximately four-sevenths of this amount, and the induced displacement of the cilia must be smaller still. Therefore, we can conclude that the cilia are displaced by less than $10^{-12}$ m at threshold. The fish is detecting a displacement that is 1/100th of the diameter of a hydrogen atom!

## The Nature of the Coupling Between the Otolith and Cilia

In a direct coupling between the otolith and cilia, any displacement of the otolith will result in bending of the cilia. The displacement of the otolith is proportional to the acoustic displacement, which in turn is proportional to the quotient of the acoustic particle velocity and frequency. From equation (5.6), the displacement of the otolith and, hence, the response would be proportional to the *quotient of the acoustic pressure and frequency* (for farfield sound). Such a response is inconsistent with the observed decrease of 6 dB per octave in hearing threshold at low frequencies (Popper 1983). Instead, the hearing threshold data imply that the response should be proportional to the *product of the acoustic pressure and frequency*.

We suggest that the coupling between the otolith and cilia is not direct, but rather mediated by some elastic member, possibly the otolithic membrane. If the coupling is elastic, then at low frequency the cilia will not be bent, even if the amplitude of the displacement is large. As the frequency increases, bending begins and the amount increases quadratically with frequency. Thus, with elastic coupling, bending of the cilia is proportional not to the *displacement* of the otolith, but to its *acceleration,* which is proportional to the product of the square of the frequency and acoustical displacement (i.e., frequency times the acoustic velocity). Hence, from equation (5.6), the response will be proportional to the product of the acoustic pressure and frequency, which is in agreement with the low-frequency data.

## The Response of Hair Cells Without the Otolith

Even if there were no otolith, the hair cells would still respond to sound. The thresholds, however, would be higher and the directional response quite different.

Although it is not immediately obvious, a plane wave exhibits shear in spite of its longitudinal nature. As a result of this fluid shear, even without an overlying otolith the ciliary bundles would still be bent and hair cells respond. This effect is illustrated in Figure 5.2. The cilia are bent when the sound is incident at ±45 degrees, but not at 0 or 90 degrees. The bending switches direction between the fluid expansion and compression caused by the acoustic pressure.

The directional response of the hair cells without an overlying otolith is different than their response with an otolith. As shown in Figure 5.3, with the otolith the hair cell has dipole directivity; without the otolith, quadrupole directivity. Directional responses such as these are useful in determining the direction of a source and in increasing the detectability of a signal in the presence of high levels of the omnidirectional ambient noise commonly encountered in the ocean.

## Directional Cues

The ability of hair cells to detect a single component of the acoustic particle velocity and to indirectly detect pressure from the signal reradiated from the swim bladder provides fish with a means for discerning the direction of incidence of an acoustical signal. The mechanism by which the ear may sort out the individual components of the particle velocity and the pressure-related indirect signal is given in Chapter 27.

The process of localization seems simple enough. The fish shown in Figure 5.4(A) measures the $x$ and $y$ components of the acoustical particle velocity of a

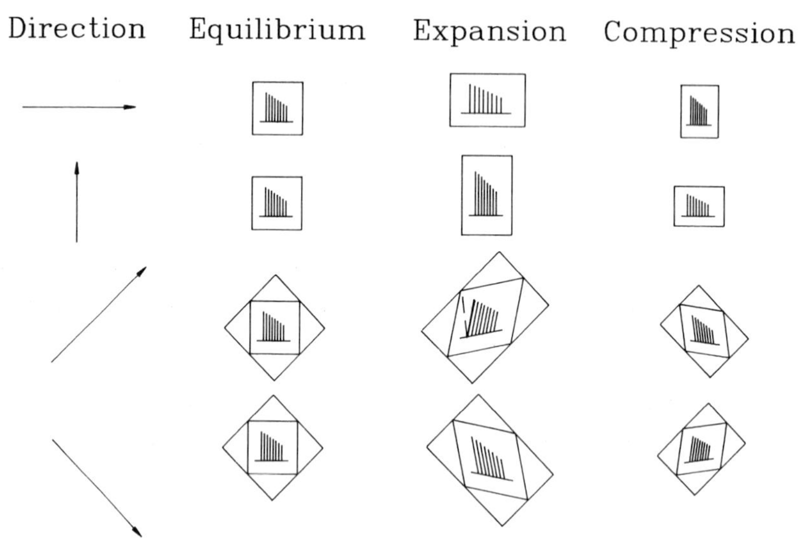


Figure 5.2. Uniaxial expansion and contraction of the fluid due to the incident plane wave transforming a fluid element with square cross section into one with rectangular cross section when the sound is incident at 0 or 90 degrees. At 45-degree incidence, this same expansion or contraction transforms the square into a rhombus, an obviously shearing motion. The effect on cilia is evident.

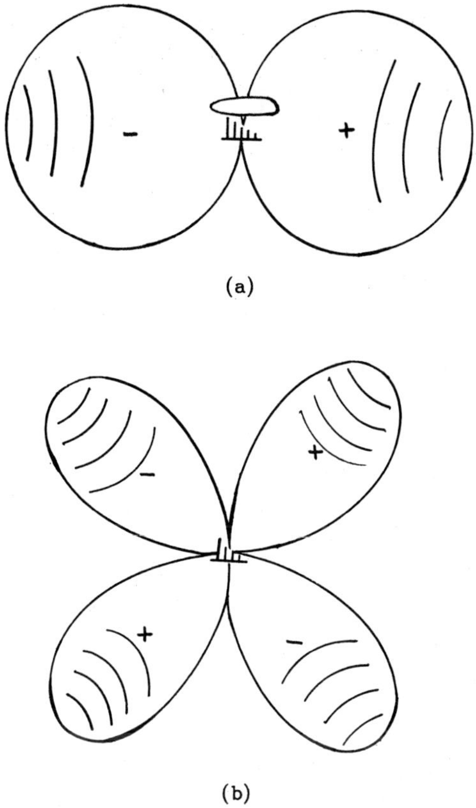


Figure 5.3. (a) Dipole directivity of hair cells with an overlying otolith. This response is just cos$\theta$ where $\theta$ is the polar angle in spherical coordinate system. (b) Quadrupole directivity of hair cells without overlying otolith. This response is cos$\theta$ sin$\theta$ cos$\phi$ where $\theta$ is the polar angle and $\phi$ is the azimuthal angle in spherical coordinates.

farfield source and determines the angle of incidence by the arc tangent of the ratio of the two components. The problem is that the tangent of the angle is equal to that plus 180 degrees. Thus, as shown in Figure 5.4(B), the fish cannot determine whether the signal is coming from 45 or 225 degrees—a potentially fatal ambiguity—which it resolves, as shown in Figure 5.4(C), with the pressure information obtained from the indirect signal. Knowing that the ratio of $p$ to $\rho cv$ is $-1$, rather than $+1$, eliminates the 45-degree possibility, and the fish can respond appropriately.

## Ambient Noise

*Noise* is unwanted sound. Signals of biological interest must be extracted from a background of ambient noise. High levels of more or less omnidirectional ambient noise exist underwater because of the excellent propagation of sound in

Figure 5.4(A) A fish may obtain sound source directional information from orthogonal components of the acoustic particle velocity. (B) A fish can tell the line of bearing of the source by using velocity components but cannot determine whether the sound is coming from its front or its rear. (C) "Indirect" signal from the stimulus is proportional to the acoustic pressure. If the pressure is out of phase with particle velocity ($p/\rho cv = -1$), then sound wave must be moving from the front to the rear of the fish. If they are in phase ($p/\rho cv = +1$), the wave is coming from opposite direction. Thus the fish has information that could enable it to decide how to respond.

the ocean and because it contains many sources of sound, including biological ones (e.g., marine mammals and snapping shrimp), ships, industrial work, rain, wind, surf, waves, current flow, earthquakes, volcanoes, and distant storms. In the very deep sea, surface waves at low frequency (<50 kHz) and thermal noise caused by the conversion of mechanical to thermal energy at the molecular level at high frequencies (>50 kHz) are the primary natural sources of ambient noise (Albers 1965).

Deepwater ambient noise was first studied during World War II. It is summarized by the Knudsen curves shown in Figure 5.5. These curves relate the spectrum of deepwater noise to sea state and shipping density. Sea states are characterized by the height of surface waves, ranging from sea state 0, with waves up to 1 ft high, to sea state 8, with waves more than 40 ft high.

Note that the ambient noise in the ocean over much of the fish's hearing range is dominated by man-made sources; in the absence of shipping, however, the sea state 0 noise level at frequencies around 500 Hz appears to agree well with the lowest hearing thresholds of several species of fish. This lends some evidence that the fish ear has evolved in accommodation to its environment; we find it somewhat puzzling, however, that fish are most sensitive to sound in the frequency range where the ambient noise is highest. Certainly, hearing in fish has been made difficult because of the recent advent of man-made underwater noise. Perhaps, in time, evolution will take care of this also.

Most fish live in the shallow coastal waters where the ambient noise level is

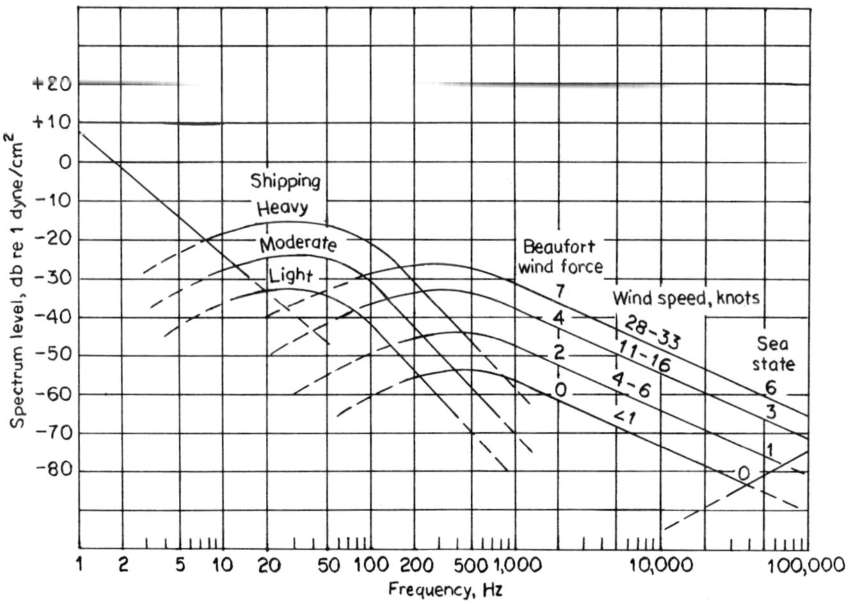


Figure 5.5. Average deepwater noise spectra characterized by sea state (Knudsen curves) and shipping density. (Reproduced with permission from Urick, R. J., *Principles of Underwater Sound for Engineers,* ©1967, McGraw-Hill Book Co.)

highest and has wide spatial and temporal variation. In this area, the primary noise sources are biological and man-made. In addition, intermittent sources, such as wind and rain, may increase the noise level by as much as 15 to 30 dB (depending on frequency). As we have previously stated, the directional response of the hair cells provides a 5-dB gain in the signal-to-noise ratio (see Chap. 27), which may help to explain the fish's ability to detect a sound source in the presence of such high levels of ambient noise.

## Underwater Sound Propagation

We now wish to review some aspects of the propagation of underwater sound. Underwater propagation is affected by absorption, reflection from the surface and seafloor, refraction, scattering, and water depth. As we have already mentioned, absorption is quite negligible in the hearing range of fish; it would be the limiting factor, however, in determining the propagation of the high-frequency signals used in the echolocation systems of many marine mammals.

### Reflection

Propagation of sound often involves reflection from the surface or bottom or both. Sound may also be reflected by any relatively large object that happens to be in the vicinity of a source. These reflections may give fish false information concerning the location of a source or destructively interfere with the direct signal.

The air–water interface at the surface is an excellent reflector at all frequencies; the reflection coefficient, however, is $-1$, which means that the reflected pressure signal has the same amplitude as the incident signal, but the opposite sign. Spatially then, the reflected signal is upside down and backward and tends to cancel the incident signal. This results in a weak signal with a rapid falloff with range when both the source and receiver are located near the surface.

The sea bottom is generally a poor reflector, with a reflection coefficient for normal incidence of less than 0.5; but for sound at the grazing incidence (angles of incidence less than the critical angle), it is a good reflector, with a reflection coefficient having a magnitude near unity. The critical angle, $\theta_c$ (measured with respect to the horizontal), depends on the sound speed of the sediment ($c_s$) and is given by

$$\theta_c = \arccos\left(\frac{c_w}{c_s}\right). \tag{5.9}$$

Since the speed of sound in ocean sediment is only slightly greater than the speed of sound in water ($c_w$), the critical angle tends to be rather small. For example, the critical angle is 9 degrees for a clayey silt bottom and 28 degrees for fine sand. Although the magnitude of the reflection coefficient is unity for critical reflection, it is a complex number. Hence, sound pulses that are critically reflected by the bottom undergo a drastic change in shape on reflection.

In addition to specular reflection (i.e., where the angle of reflection equals the angle of incidence), the surface and bottom may incoherently "scatter" sound in other directions owing to surface roughness.

## Scattering

*Scattering* is the alteration of an incident sound wave by some object. The scattered sound pressure is the pressure that would not be there if the scattering object were not present. Scattering at low frequencies will be significant for objects that are either large, like submarines, or highly compliant, like gas bubbles.

In the ocean, the main sources of scattering are fish, man-made objects, and local concentrations of gas bubbles, such as those that occur in the wakes of ships (Albers 1965). The gas-filled swim bladder is primarily responsible for the scattering from fish. Scattering is used commercially to get some measure of the size and number of fish in a particular area. Furthermore, as discussed previously, the scattered (i.e., reradiated) signal from a fish's own swim bladder may play some role in localization of sound sources.

## Refraction

*Refraction* is the bending of a propagating sound wave due to a change in sound speed. Refraction occurs when sound is transmitted through an interface between two mediums with differing sound speeds or is due to the presence of a sound speed gradient. A sound speed gradient bends the sound towards the direction of lower sound speed; consequently, the sound travels along a curved trajectory, instead of a straight line. Even though the total variation of sound speed throughout the height of the water column is very small, typically less than 6%, the effects of this variation are often profound. Refraction due to sound speed gradients in the ocean causes channeling of sound; regions of exceptionally good reception are called *caustics* or *convergence zones*, and regions where propagation is very poor are called *shadow zones*.

The long-range propagation of sound in deep water is controlled by the sound speed profile (Fig. 5.6), which depends on temperature, salinity, and pressure. At great depth the water is isothermal, so that the sound speed depends only on the hydrostatic pressure and hence increases linearly with depth. The temperature increases as the surface is approached, and eventually this effect overwhelms that of pressure so that the sound speed increases. At a depth of about 1,000 m the sound speed has a minimum, which is called the *SOFAR*, or channel, axis. The depth of the SOFAR axis depends on latitude. At northern latitudes, the axis is much closer to the surface and disappears entirely in arctic waters.

Near the surface, the sound speed profile varies seasonally. In the summer, the warm air temperature causes the sound speed to continue to increase right up to the surface. In the winter, convective and turbulent mixing of the water creates a surface layer of isothermal water up to 400 meters thick. Within this

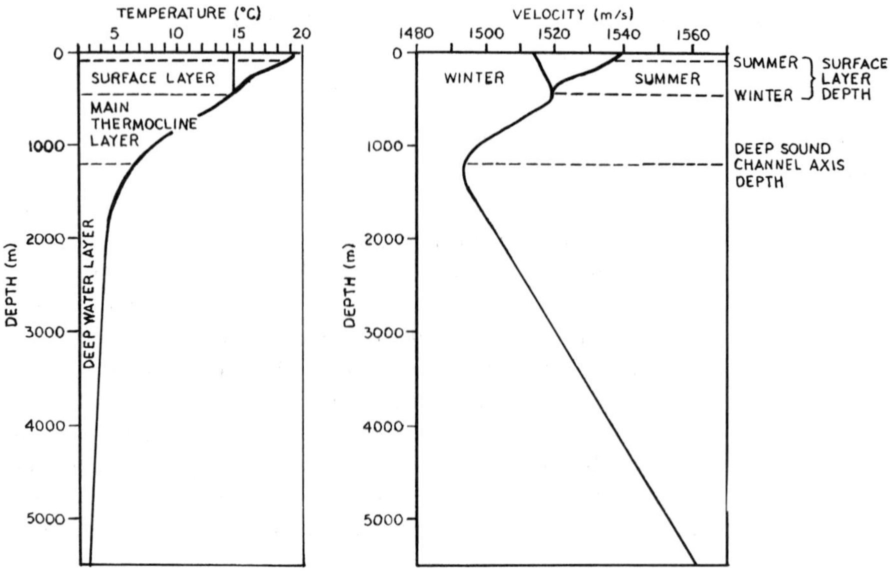


Figure 5.6. Typical summer and winter temperature and sound speed profiles of deep ocean at mid-latitudes.

layer, the sound speed decreases as the surface is approached because of the decrease in hydrostatic pressure.

In the summer, the propagational range for shallow sources and receivers may be limited by the downward refraction of sound. The fish in the upper right-hand corner of Figure 5.7(A) cannot hear the sound transmitted from the fish at the left because it is located in a shadow zone. Fish located somewhat closer to the source or somewhat deeper have no trouble hearing the sound. If the transmitting fish were located deeper, both listeners would be able to hear it.

In winter the isothermal water layer makes the situation quite different. As we see in Figure 5.7(B), if the transmitting fish is located within the surface layer, it will be heard by any listener located anywhere within the layer. The range of propagation to listeners below the layer, however, will be limited. In this case, the deeper fish is now in the shadow zone.

The propagation is excellent any time of the year for sources and receivers in the vicinity of the SOFAR axis (Fig. 5.7[C]). Within this axis, sound can propagate by refracted paths without ever contacting the surface or bottom. SOFAR propagation has been observed over distances of up to 10,000 miles.

## Shallow Water Propagation

Many aquatic species live in water that is very shallow, such as streams, ponds, tidal pools, and coral reefs. Sound propagation in shallow water is inherently

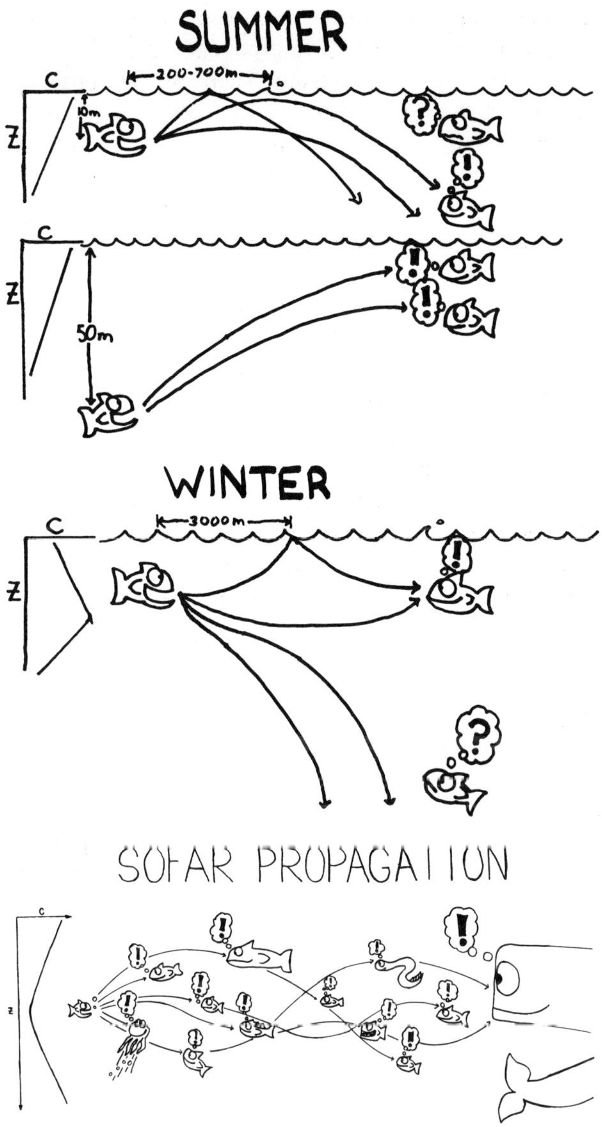


A

B

C

Figure 5.7(A) Effects of summer sound speed profile on propagational range and depth. (B) Effects of winter sound speed profile on propagational range and depth. (C) The propagation of sound in the neighborhood of SOFAR axis for any time of the year.

different than propagation in deep water, since sound can propagate only over distances greater than the water depth by repeatedly interacting with the surface and bottom.

Mathematically the problem must be treated as a boundary value problem. The solution is in terms of a finite sum of normal modes (Ferris 1972). In the vertical direction, each normal mode is a standing wave with a different number of nodes. Each mode propagates independently in the horizontal direction with its own frequency-dependent speed. The frequency-dependent wave speed (or dispersion) results in waveforms that change shape as they propagate.

Each mode has a certain characteristic cutoff frequency below which it cannot propagate. Most significantly, no sound can propagate at all at frequencies below the cutoff frequency for the lowest mode. This absolute cutoff frequency depends on the propagation speed in the sediment ($c_s$) and is inversely proportional to the depth of the water ($h$). The absolute cutoff frequency is given by

$$f_c = \frac{c_w/4h}{\sqrt{1 - c_w^2/c_s^2}} \tag{5.10}$$

where $c_w$ = speed of sound in water and $f_c$ is in Hz.

A plot of cutoff frequency versus depth for four different types of bottom is presented in Figure 5.8. Note that, in 10 cm of water, almost no sound that is within the hearing range of fish will propagate. In 1 or 2 m of water, the usable frequency range is still very limited, especially for slow bottoms. It would thus be quite advantageous for fish that live in shallow water to have an extended high-frequency hearing range. We find it significant to note that many Otophysans, which have such extended high-frequency capabilities, do live in shallow water.

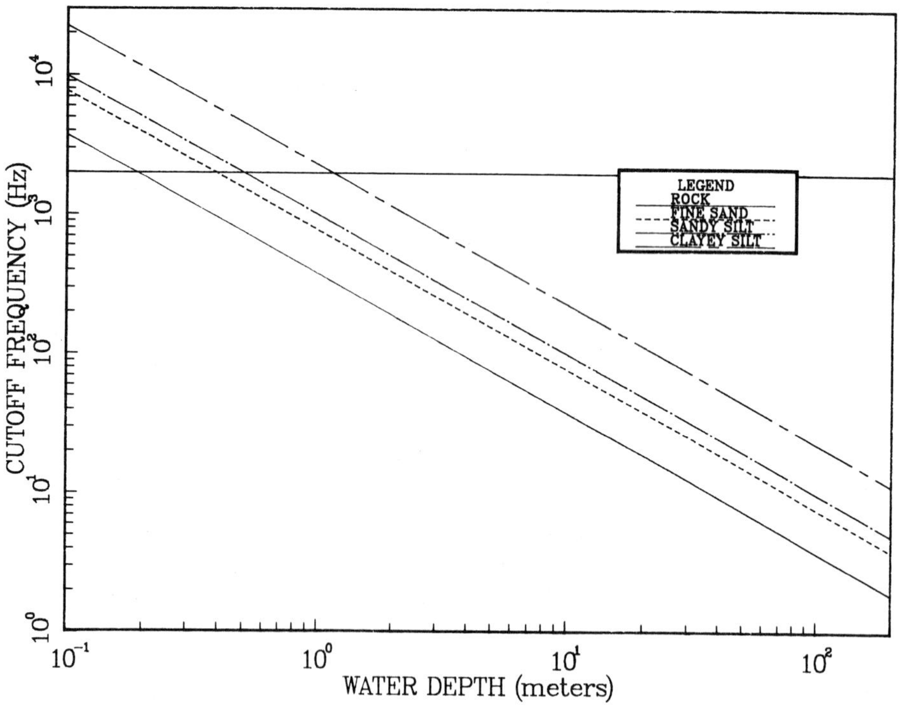


Figure 5.8. Cutoff frequency vs. water depth for different types of bottom sediment. Horizontal line passing through 2,000 Hz indicates typical upper limit of hearing frequency range for non-Otophysan species. Otophysans, however, usually hear up to approximately 6,000 Hz, indicating that they could detect sounds in water less than 0.5 m deep.

## Differences Between Underwater and Atmospheric Sound

We conclude by making some comparisons between atmospheric and underwater sound. These two acoustical environments are vastly different. It is important to be familiar with these differences so that we do not base our interpretations of underwater acoustical data on our own hearing environment.

These comparisons are summarized in Table 5.1. One of the major differences between air and water is the characteristic acoustical impedance ($\rho c$). The impedance for air is 400 kg/($m^2$-sec), while for water it is 1,500,000 kg/($m^2$-sec). This means that the acoustical particle velocity in air is about 4,000 times that in water for the same acoustical pressure. It is somewhat paradoxical then that terrestrial animals detect acoustical pressure while fish detect acoustical particle velocity.

In addition, the ambient noise level in the atmosphere is quite low compared to the underwater level. Any noise in the atmosphere is highly localized because propagation over any distance is rarely possible.

Finally, we note that sound is reciprocal in the ocean, but not in the atmosphere. This means that underwater, if a listener at some point *A* can hear a source located at point *B*, then a listener at *B* can hear a source at *A* equally well. This is not the case in the atmosphere, as experienced by our friends in Figure 5.9.

Table 5.1. Comparison of Underwater Acoustics with Atmospheric Acoustics

| | Air | Water |
|---|---|---|
| Sound speed | 340 m/sec<br>Less diffraction | 1,500 m/sec<br>More diffraction |
| Reflection | Ground helps ($R = +1$) | Surface hurts ($R = -1$) |
| Scattering | Air is compliant<br>No good scatterers | Water is stiff<br>Many good scatterers (e.g., swim bladders) |
| Sound speed profile | Large changes with altitude (up to 300%) due to winds and *large* temperature changes | Changes with depth are small (less than 6%) |
| Profile variability | Extremely variable due to winds and diurnal cycle | Variation tends to be seasonal and near the surface |
| Particle velocity | $\rho c = 400$<br>Particle velocity 4,000 times that of water for same pressure | $\rho c = 1{,}500{,}000$<br>Particle velocity is very small |
| Ambient noise | Low | High |
| Propagation | Long-range propagation is seldom possible | Long-range propagation is often possible |
| Reciprocity | No reciprocity, due to winds | Reciprocal (currents are always negligible) |

Figure 5.9. Much to the chagrin of the bull and his lady friend, sound is not reciprocal in the atmosphere, as it is underwater.

When the romantically inclined bull attempts to call his girl friend on the right, he is heard instead only by the cow on the left, because the propagation of sound is possible in just one direction. The wind shear causes sound propagating upwind to refract upwards over the head of the intended listener, while sound propagating downwind is refracted downwards. Note also that the cow on the left could not "reciprocate" by calling back to the bull. Reciprocity is destroyed in the atmosphere owing to the presence of winds. Winds in the upper atmosphere of over 100 m/sec are not uncommon, while ocean currents are generally only a few meters per second. Thus, the fluid velocity is comparable to the sound speed in air (340 m/sec), but very small compared to the sound speed in water (1500 m/sec). As a result, underwater sound propagation is always reciprocal—if you can hear, you can also be heard.

## Summary

We have briefly reviewed some aspects of the propagation and reception of sound underwater. Water provides an excellent medium for the propagation of sound, and fish have evolved a complex and competent system for detecting and processing it. The nature of underwater sound propagation, however, makes analysis of sound as a biological stimulus extremely difficult. We believe that it is best to apply the fundamental relationships given previously to the sound field being studied, rather than to generalize with nearfield or farfield approximations that may be totally inapplicable.

We are left, however, with the enigma of the biological relevance of hearing in fish. Although all species hear, only a relatively small number are known to vocalize. What are they listening to?

*Acknowledgments.* We would like to thank the Office of Naval Research for their continued support and David Rogers for the illustrations embellishing this chapter.

## References

Albers, V.M. (1965) Underwater Acoustics Handbook II, The Pennsylvania State University Press, University Park.

Clay, C.S., and Medwin, H. (1977) Acoustical oceanography: principles and applications, John Wiley and Sons, New York.

Ferris, R.H. (1972) Comparison of measured and calculated normal-mode amplitude functions for acoustic waves in shallow water, J. Acoust. Soc. Am., 52: 981–988.

Flock, A. (1971) Sensory transduction in hair cells, in Handbook of Sensory Physiology, Vol. 1, Lowenstein, W.R. (ed.), Springer-Verlag, Berlin, pp 396–411.

Harris, G.G. (1964) Considerations on the physics of sound production by fishes, in Marine Bio-acoustics I, Tavolga, W.N. (ed.), Pergamon Press, New York, pp. 233–247.

Pierce, A.D. (1981) Acoustics: An Introduction to Its Physical Principles and Applications, McGraw-Hill Book Co., New York.

Popper, A.N. (1983) Organization of the inner ear and auditory processing, in Fish Neurobiology, Vol. 1: Brain Stem and Sense Organs, Northcutt, R.G., and Davies, R.E. (eds.), University of Michigan Press, Ann Arbor, pp. 125–178.

Siler, W. (1969) Near- and farfields in a marine environment, J. Acoust. Soc. Am., 46: 483–484.

Urick, R.J. (1967) Principles of Underwater Sound for Engineers, McGraw-Hill Book Co., New York.

van Bergeijk, W.A. (1964) Directional and nondirectional hearing in fish, in Marine Bio-acoustics I, Tavolga, W.N. (ed.), Pergamon Press, New York, pp. 281–299.

# Chapter 6

# Detection of Weak Electric Fields

AD. J. KALMIJN*

## Introduction

Electric fields in natural waters present a wealth of sensory information. Bioelectric fields direct electrosensitive fishes to their prey, environmental fields provide important orientational cues, and the fields induced by the animals' motion through the earth's magnetic field offer oceanic species complete compass data. Particularly sensitive to electric fields are the marine sharks, skates, and rays, but the weakly electric fishes, the common catfishes, and several of the more primitive fishes are also known for their keen electric sense.

In detecting the fields of prey and orientational cues of external origin, the animals use the electric sense in the *passive* mode. Moreover, by deriving their magnetic compass heading from the electric fields they themselves induce, oceanic species may employ the electric sense in the *active* mode. Similarly, weakly electric fishes passively detect the electric-organ fields of each other and actively probe the environment with their own electrical discharges. The two modes of operation render the electric sense its great versatility.

The process of electroreception takes place in a series of steps leading from the occurrence of an electric field in nature to the behavioral response of the recipient animal. The peripheral sensory apparatus causes the physical stimulus to converge onto the receptor epithelia. Highly sensitive receptor cells encode the sensory message into physiological signals that convey the information to the central nervous system. Brain centers analyze the incoming signals and initiate the appropriate response. The animal's behavior may affect, or even produce, the physical stimulus, as in a feedback loop.

Successive events can be studied on their intrinsic merits or on their logical relations to one another. For example, the electroreceptive epithelia may be investigated to reveal the biophysical principles of electrical excitability or to determine the transfer functions connecting the physical stimuli to the afferent action-potential

* Scripps Institute of Oceanography, University of California, San Diego, La Jolla, CA 92093, USA

rates. For a full appreciation of the results, however, the different approaches must have as their objective an understanding of the sensory system as a whole. Only the entire sequence of events enables the animals to relate meaningfully to their physical environment.

This chapter discusses the electric fields occurring in nature, in particular the bioelectric fields, their discovery, and their physical features. The various functions of the electric sense are detailed and extended with the results of recent orientation studies at sea. Intermediate steps in the chain of events are evaluated for their parts in the process of sensory reception. The physics of the peripheral apparatus is treated in some depth, the topics of sensory transduction and nerve coding are touched upon briefly, and a new object detection algorithm is presented, suggesting central-processing requirements.

Sharks, skates, and rays have given us the first hints as to the function of an electric sense in fishes without electric organs. In deference to their astounding sensitivity and the great versatility of their electric sense, the elasmobranch fishes will receive considerable attention in the following pages. Other, mostly freshwater, electrosensitive species will be treated from a comparative point of view, especially where they have notably contributed to the fascinating tale of the electrical sensory modality. For further reading, see Bullock (1982), Hopkins (1983), and Bullock, Heiligenberg, Hopkins, Zakon, this volume.

## Electric Fields in Natural Waters

### The Discovery of Bioelectric Fields

The vital role of electricity in animal physiology, in particular in nerve conduction and muscle contraction, has been recognized since the contentious years of Galvani and Volta, during the late 1700s. Their pioneering work was antedated only by the painful, but supposedly invigorating, experiences with the strong electrical discharges of the marine electric ray and the freshwater electric eel and electric catfish (cf. Wu 1984).

In more recent years, Mathews (1902) observed steady electrical potential differences along cut, regenerating stems of the marine tubularian hydroid *Parypha*. He wondered, at the time, whether the recorded fields might take part in the functional polarization of individual cells and in the differentiation of the whole organism. Lund (1922, 1925, 1947) aptly suggested, on the basis of similar findings in *Obelia*, that the externally measured fields result from spatial variations in the potential difference across the living ectoendoderm (Fig. 6.1). He concurred with Mathews on the possible significance of the fields as directive forces during the process of growth.

Burr and Hovland (1937b) extended the investigations to salamander larvae and plotted the bioelectric fields as functions of the developmental stage and the distance from the animal. In young larvae, the potential difference between the head and tail had dropped from about 650 $\mu$V to less than 25 $\mu$V within 1 mm from the skin (cf. pp. 154–155). Burr (1941) also noted that the voltage pattern

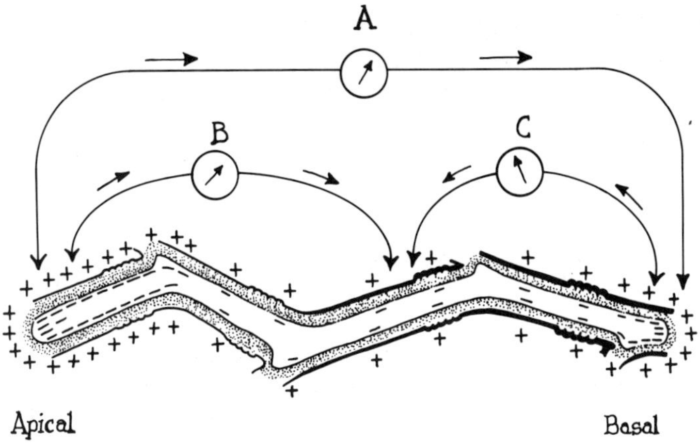


Figure 6.1. Bioelectric field of tubularian hydroid *Obelia*. The potential difference across the ectoendoderm is largest in the apical region, diminishes toward the middle region, and is higher again near the base of the stem. Steady electric currents flow within the stem in either the distal or proximal direction, closing their current paths through the seawater medium; *A*, *B*, and *C*, galvanometer readings. (Lund 1925.)

on the surface of frog eggs correlated with the future orientation of the developing embryo. Burr (1932, 1947) believed that the steady bioelectric fields of animals, though deriving from individual cells, functioned as important organizing and controlling agents and proposed that living organisms be regarded as morphological expressions of their electrodynamic fields [*sic*].

Searching for clues to explain the unusual electrical sensitivity of marine sharks and skates, Kalmijn (1966) found low-frequency bioelectric fields *at considerable distances* from the flounder *Pleuronectes platessa* (Fig. 6.2). The observed fields and the feeding responses they elicited from the sharks and skates would soon lead to a formal proof of the elasmobranchs' electric sense (pp. 160–162). Admittedly, when I first described the ventilatory potentials, I was unaware of their true origin. Actually, I was disturbed by the large "artifacts" appearing every time the salt-bridge electrodes were repositioned over the flounder, until it occurred to me that they were not artifacts, but manifestations of true *DC bioelectric fields* (Kalmijn 1969 [Dutch version], 1972 [English translation]).

In marine fishes, the electric fields appeared mainly to emanate from the mouth and the opercular slits, although other parts of the body, especially when lightly bruised, could be noticeably electrogenic as well. For instance, at a distance of 10 cm from the head of a small, undamaged flounder, I recorded DC fields of 0.2 μV/cm. Ventilatory and fin movements caused the fields to fluctuate rhythmically at rates from less than 1 to a few cycles per second. In the French Mediterranean, I measured the fields of an additional 60 vertebrate and invertebrate species, representing nine animal phyla (Kalmijn 1969, 1972, 1974). Among the invertebrates, the strongest fields were observed near wounded crabs, rendering them easy targets for the predatory sharks and skates. Octopuses, even when damaged, showed surprisingly weak electric fields.

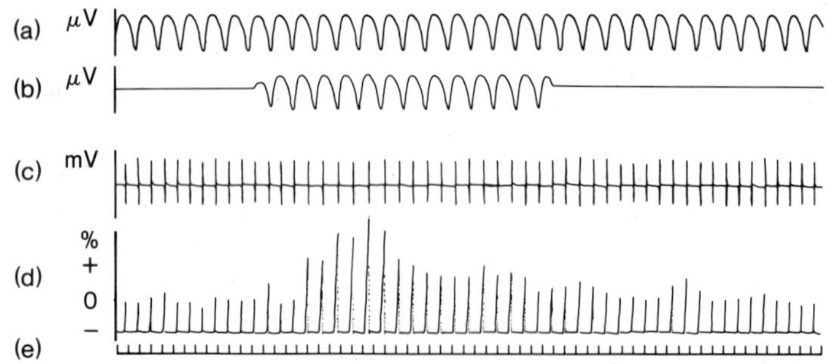


Figure 6.2. Cardiac reflex of skate *Raja clavata* in response to bioelectric field of flounder *Pleuronectes platessa*. (*a*) Ventilatory potentials measured in the vicinity of a flounder; (*b*) ventilatory potentials reproduced by a dipole source 10 cm in front of the skate; (*c*) heartbeat of the skate recorded with implanted electrodes; (*d*) instrumentally enhanced record of cardiac deceleration; (*e*) time marks, 2/s. (Kalmijn 1966.)

Besides the electric-organ discharges (Lissmann 1951), weak electric pulses, regarded as nerve or muscle action potentials, had been noticed in the vicinity of fishes before (Lissmann 1958, 1963; Kleerekoper and Sibakin 1956), but the often stronger DC and low-frequency electric fields had escaped attention. This should not come as a surprise, because DC fields were, and still are, difficult to measure in the aquatic medium. In those earlier years, the study of electrophysiology rested largely on the experimenter's proficiency in electronics, since suitable low-noise (vacuum-tube) amplifiers were not commonly available. The fabrication of stable DC electrodes has remained a matter of art.

## The Origin of Bioelectric Fields

In hindsight, the bioelectric fields could have been anticipated from the inevitable spatial variation in the potential difference across the boundary between the animal and the seawater medium, as Lund suggested for *Obelia*. Conspicuously diverse in structure and function are the protective integument, the respiratory and excretory epithelia of the gills, and the exposed mucous membranes of the oral cavity, the pharynx, and the digestive tract of fishes. Thus, in the mouth cavity and the pharynx, millivolt potentials of either polarity were measured relative to the outside medium by the use of cannula electrodes (Kalmijn 1969, 1972, 1974). The fields were often variable and, at times, even reversed polarity, depending on the physiological state of the animal.

At some distance from the animals, the bioelectric fields usually are of a purely electrical nature. However, when the measuring electrode approaches the skin to within about 0.5 mm, the potential gradient often changes more steeply with distance than one might naively expect. Evidently, the electrode then enters the *Nernst diffusion layer* over the surface of the animal, which is governed by ionic gradients and electrochemical potentials, rather than by purely electrical potentials. Thus, Burr and Hovland's (1937a, 1937b) measurements, made near or in contact with the skin, probably were largely electrochemical in nature, as judged from the steep, exponential drop in potential (cf. p. 152). The fields that I recorded at greater distance undoubtedly were strictly electrical.

Whereas the fields mentioned so far were measured with reversible electrodes, connected to the water by salt bridges, in pollution studies the ventilatory rhythm of fishes was routinely monitored by technically less refined means. The recorded potentials were commonly attributed to changes either in the polarization at the electrodes, caused by the ventilatory flow of water (Halsband and Halsband 1954) or, more vaguely, in the resistance of the water between the electrodes, resulting from the movements of the opercular covers (Roberts 1964). Thus, the recordings were interpreted as useful artifacts, which they might have been, rather than genuine biological effects. Pasztor and Kleerekoper (1962) also observed smoothly fluctuating fields "far greater in amplitude than any biopotentials," which they thought were generated by the respiratory currents.

Since the early measurements conducted expressly to reveal the electric sense of marine sharks and skates, DC and low-frequency bioelectric fields have been recorded also from various kinds of animals in the habitat of freshwater catfish (Peters and Bretschneider 1972; Roth 1972). Cardiac potentials were furthermore noticed in the vicinity of freshwater fish and salamanders (Goodman and Weinberger 1971; Lonsdale and Marshall 1973). Despite the many, often anecdotal, details, no fundamentally new ideas about the origin of the fields have been advanced since Lund's (1925) lucid interpretation, as is evident from the recent paper of Butsuk and Bessonov (1981), who mistakenly interpreted the electrochemical potentials near the skin in purely electrical terms (see above).

Any flow of water along an animal's skin creates streaming potentials, and any fin or whole body motions through the earth's magnetic field induce electric fields adding to the bioelectric fields proper (Kalmijn 1974). Although little or no data are available on the streaming potentials of animals, the induced electric fields are well understood and are believed to offer the physical basis of the elasmobranchs' magnetic compass sense (cf. pp. 159–160). However, neither the streaming, nor the induced, potentials contributed significantly to the reported bioelectric fields, since lightly anesthetized fishes had normal DC fields and showed typical ventilatory potentials when their opercular covers were moved experimentally, even though the flow of water was greatly reduced. The streaming potentials and induced electric fields of less bioelectrogenic species remain to be evaluated.

## The Representation of Bioelectric Fields

The configurations of bioelectric fields are conveniently expressed in multipole series, allowing a biologically meaningful interpretation of their spatial characteristics. Since aquatic animals lack monopole moments, the lowest nonzero term is that of the dipole moment, followed by a quadrupole, an octupole, and higher-order terms, all referred to a point within, and viewed from a point outside, the source region. The dipole contribution to the voltage gradient falls off as one over the third power of the distance, the quadrupole contribution as one over the fourth power, and so on. Consequently, at the distance at which sharks initiate their feeding responses, the dipole term usually suffices to simulate the field of arbitrary prey. Nearer the source, however, the higher-order moments must be included as well.

An estimate of the dipole strength may be obtained from the flounder data (Kalmijn 1974). A voltage gradient of 0.2 μV/cm at a distance of 10 cm in front of the fish yields, over a high-resistance bottom, a dipole moment of nearly 30 μA-cm at a seawater resistivity of 23 ohm-cm. Freshwater teleosts tend to have stronger voltage, but weaker current fields than marine teleosts of similar size and are electrically less loaded because of the high resistivity of the medium. It is not true, however, that in freshwater the bioelectric fields fall off less steeply with distance than in seawater, except perhaps in the thin Nernst diffusion layer (cf. p. 154).

Although they occur in the time domain, the bioelectric fields of prey are often more readily interpreted in the frequency domain. Periodic signals lead to discrete spectra, whereas transient signals lead to continuous spectra. A complete description in the frequency domain requires both amplitude and phase data. Energy spectra disregard the phase information and do not allow reconstruction of the original signal. Besides possible fluctuations in the source field, the relative motions of the predator and prey must be taken into account as well. Time series and amplitude spectra of the voltage gradients experienced by fishes passing through DC dipole fields are given in Fig. 6.3.

The analysis in terms of multipole series and frequency spectra provides scientists not only with concise descriptions of the fields and their sources, but also with an intuitive insight into the stimulus features that the receptor system can or cannot handle. The recipient fish, however, need not follow the engineering methods to acquire the desired sensory information. On the contrary, multipole expansion techniques require extensive spatial exploration, and transformation to the frequency domain takes at least one period of the lowest frequency of interest. In predation, the benefits of such methods would have to be substantial to outweigh the reduced distance, and increased time, of response. This trade-off between speed and accuracy may be decisive for the kind of data processing the recipient animal adopts.

## Environmental Electric Fields

In the oceans, electric fields induced by the flow of water through the earth's magnetic field prevail, whereas in freshwater the predominant fields are of electrochemical, rather than electromagnetic, origin. The oceanic fields were predicted by Faraday in 1832 and have since been studied extensively by physical oceanographers (Fig. 6.4). The freshwater fields are of diverse origin and usually have stronger voltage gradients, but are less well documented, than their oceanic counterparts. The electromagnetic and electrochemical fields, each in its own right,

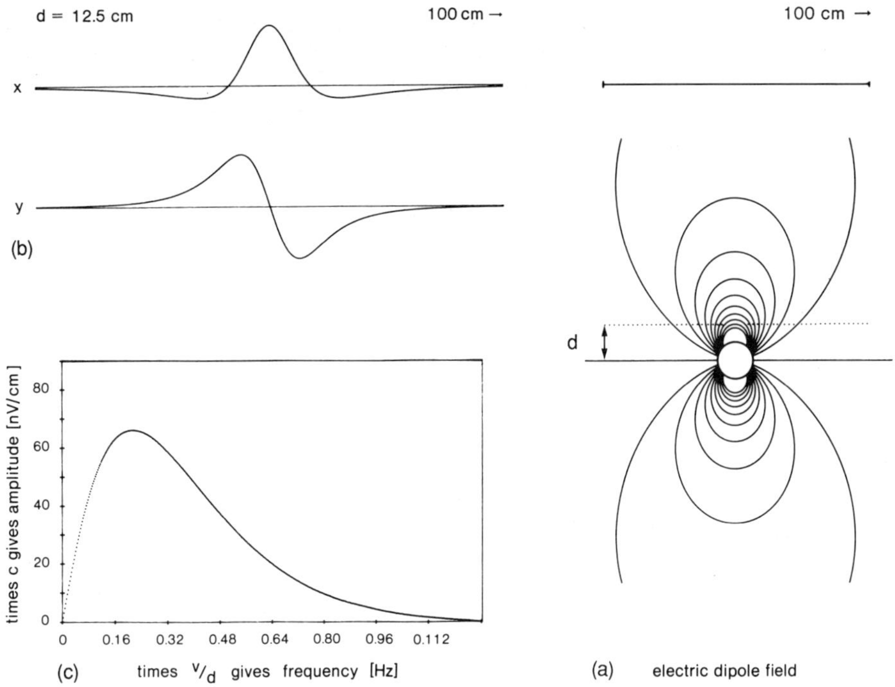


Figure 6.3. Field received by shark passing dipole source. (*a*) DC dipole field, simulating prey and trajectory of predator (*dotted line*); (*b*) longitudinal, *x*, and transverse, *y*, component of electric field along the path of travel; (*c*) normalized amplitude spectrum expressed in the speed of swimming, *v*, and the distance of passing, *d*. For a shark traveling at 50 cm/s and passing at 12.5 cm, $v/d$ = 4/s, or for a catfish traveling at 10 cm/s and passing at 0.5 cm, $v/d$ = 20/s, yielding amplitude maxima at about 0.8 and 4 Hz, respectively. The constant *c* is proportional to the dipole moment, the resistivity of the medium, and to $1/d^3$.

offer potentially important orientational cues, which are well appreciated by marine and freshwater electrosensitive fishes, as judged from the animals' trained behavioral responses (cf. pp. 165–167).

The global system of wind-driven océan currents and tidal flows induces a corresponding worldwide pattern of motional-electric fields ranging from less than 5 nV/cm to over 500 nV/cm (Longuet-Higgins, Stern, and Stommel 1954; von Arx 1962). Deelder (1952) speculated that migrating eels use the oceanic fields to detect their drift with the water. However, eels are not known to be particularly sensitive to electric fields (cf. Kalmijn 1984). The physical principles of the required orientation mechanism are, in the meantime, better understood. Moreover the demands on the detection system are adequately met by elasmobranch fishes (cf. pp. 168–174). Indeed, stingrays appreciate not only the direction, but also the polarity of uniform electric fields at voltage gradients as low as 5 nV/cm (cf. pp. 165–167).

The induced electric currents traverse the flow of water and complete their

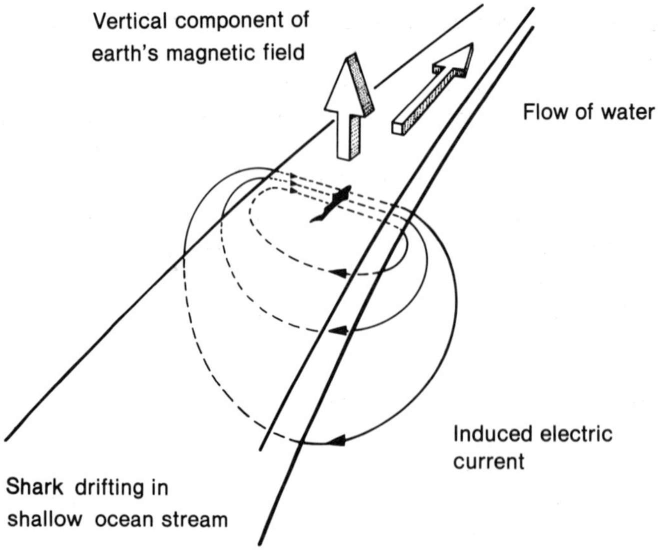


Figure 6.4. Motional-electric field of ocean current flowing through earth's magnetic field. The induced electric current gives rise to voltage gradients, which in the stream inform elasmobranch fishes about their drift with the water and outside the stream provide the animals with directional cues during their movements in familiar territory. (Kalmijn 1974.)

circuit through the ocean bottom and the water masses with respect to which the motion takes place, thereby invading the regions below and lateral to the stream (Fig. 6.4). Thus, in the Bahamas, the fringe fields of the nearby Gulf Stream penetrate the shallow Bimini waters and the funnel-shaped bay formed by the twin islands. At strengths of about 50 nV/cm and little affected by the tides, the fields furnish the local sharks, skates, and rays with directional cues during their daily movements (Kalmijn 1984, cf. pp. 165–167). The animals may distinguish the invading fields from the fields due to their drift by occasionally checking their velocity with respect to the bottom.

Fields of apparently electrochemical origin have been measured in freshwater bodies by Peters and Bretschneider (1972). Similar fields in the quiet glacial ponds of Cape Cod, Massachusetts, may provide the endemic North American catfish *Ictalurus nebulosus* with precious orientational cues on their nightly excursions (Kalmijn, unpublished). Captive catfish learned to home on the basis of uniform electric fields, as did weakly electric fishes, at a minimum voltage gradient of 1 μV/cm (cf. p. 166). The lesser sensitivity of the freshwater species, than of the marine elasmobranchs, agrees well with the greater strength of the electrochemical, than of the motional-electric, fields. However, the fields measured by Pals, Peters, and Schoenhage (1982) over clay bottoms in coastal waters were so strong and erratic that one wonders about their usefulness for marine elasmobranchs (Kalmijn 1984).

Regarding the fields in freshwater bodies and over coastal clay bottoms, a clear distinction should be made between *electrochemical potentials* due to differences in the ionic composition of the water at the electrodes and *purely electrical potentials from electrochemical sources*. The animals may well perceive the electrochemical potentials differently than we measure them with our electrodes (Kalmijn 1974). Electrochemical fields arising from spatial variations in the ionic content of the water at the surface of the recipient fish could conceivably be used by plankton-feeding elasmobranchs or by the freshwater paddlefish, *Polyodon spathula,* to detect food-rich patches. Detailed information on the fields in freshwater streams and lakes may be found in the earlier treatise by Krajew (1957) and the recent monograph by Alexandrov (1985, in Russian).

## Fields Induced by the Recipient Animal

Ships passing on the surface of the ocean (Faraday 1832), as well as sharks, skates, and rays traveling below (Kalmijn 1974), induce electromotive forces of which the gradients are perpendicular to both the magnetic field and the line of motion. For instance, when a shark swims in an easterly direction, the electric currents due to the animal's crossing of the horizontal magnetic component flow ventrodorsally across the body, closing their paths through the seawater medium (Fig. 6.5). The induced electric currents develop sizeable ohmic potential differences between the ventral and dorsal surfaces of the animal, thereby stimulating the electroreceptors (cf. pp. 168–171). At an ambient magnetic field of 0.5 gauss, a swimming speed of 1 cm/s suffices to produce a threshold stimulus of 5 nV/cm.

As the motionally induced fields depend on the animals' compass heading, they may serve as the physical basis for the elasmobranchs' magnetic sense (cf. pp. 167–168). If, for instance, the eastbound shark turns around to head west,

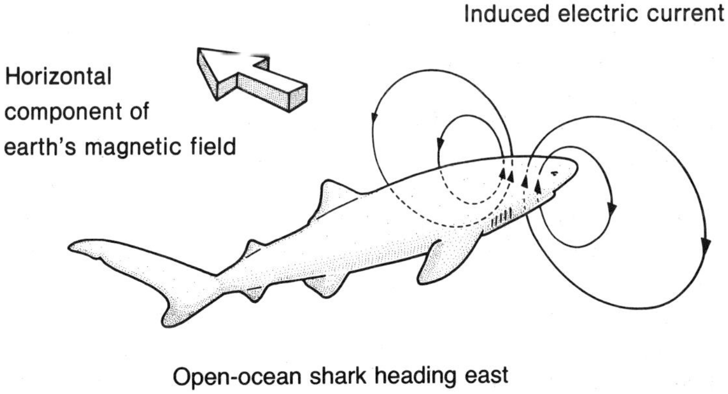


Figure 6.5. Motional-electric field of shark swimming through earth's magnetic field in easterly direction. The induced electric current leads to dorsoventral potential differences, which may inform the animal about its magnetic compass heading. (Kalmijn 1974.)

the ventrodorsal potential difference reverses in polarity to become dorsoventral instead. The strength of the field is furthermore a function of the angle of deviation from the east-west axis and is proportional to the fish's speed of travel. For an unambiguous interpretation of the induced electrical cues, the animal must probe the magnetic field in different directions, for which its swerving mode of swimming may suffice (Kalmijn 1981, 1984, and p. 175).

Although not apparent at first sight, the virtually uniform electric fields of ocean currents and the highly nonuniform electric fields resulting from the animals' own swimming motions produce similar potential distributions at the electroreceptive surfaces of the skin. This follows from the fact that a homogeneous ellipsoid body, placed in a uniform electric field, develops an internal field identical to the field it receives when moving rectilinearly through the earth's magnetic field (Kalmijn 1984, theorem of equivalence). That the tissues are not truly homogeneous does not greatly affect the outcome, as the electroreceptors detect the internal field *integrated* across the body (cf. pp. 168–170).

By crossing the vertical component of the earth's magnetic field, the horizontally swimming animal also induces transversely directed electromotive gradients, which in the northern hemisphere point to the left and in the southern hemisphere to the right of the animal. The ensuing ohmic electric fields differ from the predominantly horizontal fields of ocean currents, as they are independent of the direction of swimming. Whereas the horizontal component of the animal's field remains transverse at all times, the environmental field rotates with respect to the body axes when the animal turns left or right. Together, the transverse and dorsoventral electric fields may give the animal its magnetic latitude.

For a given speed of swimming and magnetic field strength, the animals' motional electric fields are at least 30 percent weaker in freshwater fishes than in the marine species, owing to the high resistivity of the freshwater medium (cf. pp. 168–170). More significantly, freshwater catfish and weakly electric fish are about two orders of magnitude less sensitive to low-frequency electric fields than are marine elasmobranchs (cf. pp. 165–167), while the prevailing freshwater electrochemical fields are just about that much stronger than the animals' motionally induced fields (cf. pp. 156–159). Consequently, I do not expect the electromagnetic compass mechanism to apply to freshwater electrosensitive species.

## Behavioral Responses to Weak Electric Fields

### A New Sensory Modality Revealed

Parker and van Heusen (1917) reported a remarkable sensitivity to galvanic fields in freshwater catfish, but did not recognize the biological significance of their findings, obviously unaware of the electric fields in the animals' natural habitat. Lissmann (1951, 1958), on the other hand, inferred the electric sense of weakly electric fishes purely on the basis of the animals' pulsed electric-organ discharges.

In behavioral experiments, he showed that the fishes not only respond to the electric-organ fields of each other, but also detect nearby objects as impedance inhomogeneities with their own electrical discharges.

In earlier studies on the sensory biology of small sharks in the Mediterranean, Dijkgraaf noticed that the animals were frightened by the approach of metallic rods (Dijkgraaf and Kalmijn 1962). Familiar with Parker and van Heusen's work on catfish, Dijkgraaf tentatively attributed the response to galvanic stimulation. In the years following Lissmann's discovery, Kalmijn offered direct evidence for the sharks' electrical sensitivity, applying purely electric fields of dipole and uniform configuration (Dijkgraaf and Kalmijn 1962). To evoke the startle response, voltage gradients of a mere 1–10 μV/cm sufficed. Nevertheless, the biological significance of the results remained unclear at the time.

After further diminishing the field strength, I observed distinct eye-blink reflexes in the shark *Scyliorhinus canicula* and hesitations in the rhythmic movements of the spiracle valves in the skate *Raja clavata* in response to uniform voltage gradients as low as 100 nV/cm (Dijkgraaf and Kalmijn 1962). The 5-Hz square-wave fields applied were symmetrical about zero, thus lacking a DC component. A frequency of 5 Hz was chosen, because 50-Hz fields were ineffective and 0.5-Hz fields resulted in individual eye blinks at each half cycle. In a later effort, I obtained even higher-sensitivity data by recording the heartbeat of skates, which, in a state of rest, exhibited transient cardiac decelerations at uniform voltage gradients *as low as 10 nV/cm* (Kalmijn 1966 and Fig. 6.2).

Besides the eye, spiracle, and cardiac reflexes, so far, only avoidance reactions had been observed. However, when I recorded the heartbeat in response to dipole fields, the skates occasionally approached and actually bit the hidden current electrodes, as in attack. Apparently, the applied dipole moments were now *sufficiently weak* to be biologically relevant. This observation and the measurements on the bioelectric fields of small flounder and other prey items (cf. pp. 152–154) suggested that elasmobranch fishes use their electrical sensitivity in predation (Kalmijn 1966). However, the idea met with skepticism and obviously required compelling, behavioral verification.

Hence, I presented small flounders to the bottom-feeding sharks and skates, first under a thin layer of coarse sand or seawater agar transparent to electric fields, which elicited strong feeding responses, and subsequently under agar covered by a thin plastic film screening the prey electrically, which left the flounders undetected. Apparent feeding responses were also observed when the prey fields were simulated by passing weak electric currents between two electrodes under agar or sand. In the final attack, the electric fields took precedence over any lingering odor cues. This concluded the *proof of the elasmobranchs' electric sense* and its role in predation (Kalmijn 1971 and Fig. 6.6).

Regarding the electroreceptive organs, Murray (1962) revealed that the ampullae of Lorenzini, in addition to their reputed mechanical and thermal sensitivity, are highly responsive to weak electric fields (p. 175). Dijkgraaf and Kalmijn (1963) demonstrated that selectively severing the ampullary nerves obliterated the eye-blink reflex only for stimuli applied to the affected skin areas. Kalmijn

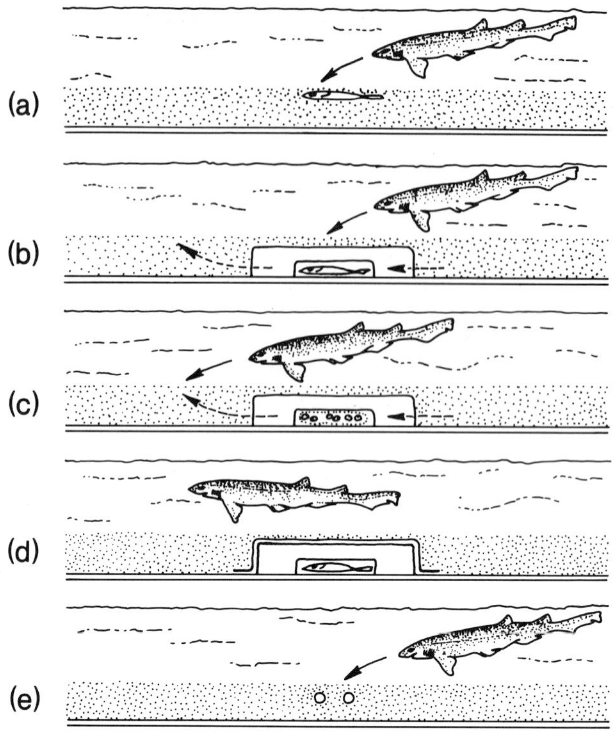


Figure 6.6. Responses of dogfish *Scyliorhinus canicula* to (*a*) a flounder buried under the sand, (*b*) a flounder in an electrically transparent agar chamber, (*c*) pieces of cut fish in an agar chamber, (*d*) a flounder in an agar chamber covered by an insulating plastic film, and (*e*) electrodes simulating the bioelectric field of a flounder; *solid arrows,* path of attack; *broken arrows,* flow of seawater. The sharks execute their well-aimed attacks in response to the electric fields of both real and simulated prey. (Kalmijn 1971.)

(1969, 1972) later recorded the reactions of the ampullary nerves, in resting sharks to the AC ventilatory potentials, and in swimming sharks to the DC bioelectric fields, of live prey. However, the *ampullae of Lorenzini* were not designated as *true electroreceptors* until the significance of the behavioral responses they mediate had been established (Kalmijn 1971).

The electrically evoked feeding behavior also offered a plausible explanation for the catfish's curious tendency to nibble at metallic rods (Parker and van Heusen 1917). The elasmobranch data moreover clarified the low-frequency sensitivity of weakly electric fish and catfish, which at first was thought to serve the detection of the electric-organ discharges (Machin and Lissmann 1960) or their possible precursors, the fields of muscle action potentials (Lissmann and Machin 1963). In weakly electric fish, many receptor units indeed respond to low-frequency fields, but remain unaffected by the electric-organ discharges (Bullock and Chichibu 1965). Moreover, denervation of the small pit organs, the electroreceptors of catfish, abolished the nibbling response (Dijkgraaf 1968).

## Prey Detection, Sensitivity, and Frequency Response

In the original studies on small *captive* sharks and skates, 30–60 cm in length, I observed apparent feeding responses to DC dipole fields of 20 μA-cm from distances up to 15 cm (Kalmijn 1971, cf. pp. 160–162). *At sea,* however, in the shallow waters off Cape Cod, the larger dogfish *Mustelus canis,* 90–120 cm in length, responded to dipole fields of 40 μA-cm from distances up to at least 40 cm (Kalmijn 1978b, 1982). The latter experiments revealed a *threshold sensitivity of 5 nV/cm,* rated conservatively. Although attracted by the scent of fish chum, the sharks bit solely at the electrodes, ignoring the nearby odor source. The swell shark, *Cephaloscyllium ventriosum,* waits in ambush and, when roused electrically, snaps at fishes coming close (Tricas 1982).

The 5-nV/cm threshold strength sufficed to elicit a complete feeding response. Because the sharks struck at the targets in fast, well-aimed moves without noticeable course corrections, the initial detection must have taken place from greater distance and at lower field strength than was estimated from the first visible manifestation of the attack. We shall refrain, however, from any unwarranted extrapolations. Actually, too large of a striking distance might strategically not be in the best interest of the predator, as it tends to alert the prey prematurely, facilitating escape (Kalmijn 1982).

Although the animals were clearly most responsive to DC and low-frequency fields (Kalmijn 1966, 1971), protracted series of experiments were conducted on the shark *Scyliorhinus canicula* and the skate *Raja clavata* to acquire more complete data (Kalmijn 1974). In summary, *DC dipole fields were most effective* in eliciting the feeding responses. The animals were almost equally sensitive to sine-wave fields of 1, 2, and 4 Hz as to DC fields. *At 8 Hz the threshold had increased by a factor of two.* At 16 Hz the sharks did not respond until the fields were so strong as to cause avoidance reactions. The skates, however, showed some responses at 16 and even 32 Hz, but only at increased field strengths. Within a factor of two, the threshold data were independent of the water temperature, which ranged from a low of 6°C to a high of 30°C.

The frequency-response curves derived from the shark's behavior with regard to local DC and sine-wave fields are not directly comparable with the amplitude spectra of the signals that the animals receive when swimming through DC dipole fields (cf. pp. 155–156 and Fig. 6.3). Nevertheless, neither the animals' sensitivity nor the prey fields lend strong support to the detection of frequencies much higher than 8 Hz, even when the motion of the predator and the fluctuations in the prey field are taken into account. Naturally, a lack of response need not mean a lack of detection. Paradoxically, stationary sharks are largely insensitive to DC fields, although in predation, DC fields elicit the most lively responses (cf. pp. 175–176). Whether the predators sense the condition or kind of prey from the temporal features of the fields remains to be determined. Given the short response times, an analysis in the frequency domain is unlikely (cf. pp. 155–156).

Peters and Evers (1985) reported a threshold curve for the ventilatory reflex of *Scyliorhinus* in response to uniform sine-wave fields, showing a minimum of

20 nV/cm from 0.1 to 1 Hz and a value of at least 200 nV/cm at 5 Hz. Compared with the local stimuli Dijkgraaf and Kalmijn (1963) applied in experiments designed to identify the shark's electroreceptors, the ventilatory reflex would be more sensitive than the eye-blink reflex. However, *local* dipole fields generally lead to weaker receptor stimuli than do *uniform* fields of the same maximal voltage gradient at the skin (cf. pp. 175–176). It therefore escapes me why Peters and Evers did not refer to my original uniform-field data revealing eye-blink, as well as ventilatory thresholds of 100 nV/cm at 5 Hz (Dijkgraaf and Kalmijn 1962). For the frequency dependence of the ventilatory reflex, see p. 173.

Because the fields of prey are predominantly of a dipole nature, they lack radial symmetry (cf. Fig. 6.3). Nevertheless, in bottom-feeding sharks, neither the accuracy nor the distance of response depends markedly on the angle of approach relative to the dipole axis. The animals are less skilled, however, in locating the prey higher in the water column. The oceanic blue shark, *Prionace glauca,* on the other hand, executed well-aimed attacks at electrically simulated prey, presented 5 m under the water surface, when tested at night in water 150 m deep off Nomansland, Cape Cod (Kalmijn 1982 and Fig. 6.7). In all experiments, the applied dipole moments were directed horizontally, as they probably are in most animals. A possible attack algorithm will be described below (pp. 177–179).

Electrically evoked feeding responses have also been observed in freshwater catfish (Peters and Meek 1973), weakly electric fish (Kalmijn 1974), and urodele amphibia (Himstedt, Kopp, and Schmidt 1982). The observed distances of response and the estimated sensitivities were, however, considerably less, about 5 μV/cm vs. 5 nV/cm and 5 cm vs. 40 cm, respectively, whereas the bandwidth extended to slightly higher frequencies (Machin 1962), than in the sharks. The higher-frequency sensitivity may be functionally significant, since smaller prey often produce faster field fluctuations, and near the source the fields vary more

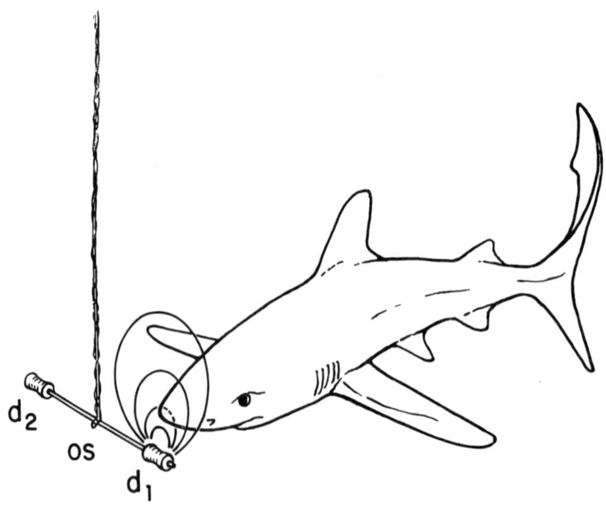


Figure 6.7. At night, the blue shark, *Prionace glauca,* relied on its keen electric sense in distinguishing between two visually identical targets, one passing a dipole current simulating prey, the other as a control. The observations were made at sea, under dim illumination, off Nomansland, Cape Cod. (Kalmijn 1982.)

steeply with distance (cf. pp. 175–176). In training experiments, Knudsen (1974) found a threshold of slightly less than 1 μV/cm in the weakly electric fish *Apteronotus albifrons* for frequencies of 5–15 Hz.

Although most renowned for sensing nearby objects with their electric-organ discharges, weakly electric fish have a low-frequency electric sense as well. To ascertain whether they can electrically locate their food without the use of the electric-organ discharges, I once more applied the method of selectively screening the prey (cf. pp. 160–162), this time with an additional silver-wire mesh between the predator and its target. By virtue of the frequency dependence of the silver–water interface, the wire mesh effectively shorted the predator's electric-organ discharges, while hardly affecting the prey's bioelectric fields. Nevertheless, persistent and well-aimed feeding attempts were observed both with respect to real and electrically simulated prey (Kalmijn 1974).

## Orientation to Environmental Fields

The electric fields of ocean currents not only inform elasmobranch fishes about their *drift with the water,* but also provide the animals with *local orientational cues* in protected waters (cf. pp. 156–159). Operating the electric sense in the *passive* mode, the fishes would sense both the direction and polarity of the virtually uniform fields. Thus, in a small, electromagnetically controlled, hut in the woods of Cape Cod, we trained the stingray *Urolophus halleri* to enter one, but not the other, of two enclosures on either side of the circular tank, solely on the basis of the imposed electric field. To prevent the use of alternative cues, the fields were applied in a random sequence of either normal or reversed polarity. The stingrays showed *definite orientation* even after the voltage gradient was lowered to *5 nV/cm,* confirming the threshold sensitivity derived from the feeding response (Kalmijn 1978b, 1982, cf. pp. 160–165).

The stingrays' orientational abilities were the more remarkable, since by moving through the earth's magnetic field the animals induced electric fields up to 20 times stronger than the imposed electric fields. Yet, cancelation of the magnetic field did not improve their orientational performances. The stingrays were obviously more comfortable with the intricate natural, than with the simpler artificial, field conditions. Pals, Valentijn, and Verwey (1982) failed to reproduce the food-reinforced, electric-field orientation tests in the shark *Scyliorhinus canicula,* perhaps because of the less relevant fields applied (cf. Kalmijn 1984). Interestingly, the ambient electric field appeared to enhance the feeding responses of the nurse shark, *Ginglymostoma cirratum* (Johnson, Scronce, and McManus 1984), in agreement with the low threshold values I obtained at sea (Kalmijn 1982).

Because captive stingrays learn so quickly to orient to electric fields similar to those in nature, it would be hard to believe that they forego such an amazing faculty in daily life. Nevertheless, whether or not they do remained a mere inference until verified behaviorally. Invited by Dr. Samuel H. Gruber, I therefore explored the possibility of conducting electric-field orientation tests in the Bahamas and, after recording the previously mentioned Gulf Stream fields, realized how excellent an opportunity this was (Kalmijn 1984, 1985, cf. pp. 156–159). Could it be that the local sharks and rays use their electric sense in navigating the otherwise featureless bay formed by the twin Bimini islands?

To answer this question, I decided to rotate the local environmental field slowly either to the right or left, by imposing the electric fields needed to yield the desired effect. Aided by Dr. Harry R. England, I tried the approach on the large stingray *Dasyatis americana* with astounding success. We made six near-ideal observations in which the rays came in on straight paths. When well within the test area, five of the animals *made their first turns in the proper directions within the 10 seconds that we rotated the field* by about 45 (Fig. 6.8). Although only a feasibility study, the results were unmistakable. In the nearby Gulf Stream, the same fields may inform elasmobranch fishes about their drift with the water (Kalmijn 1974). If so, the animals might be expected to correct their course of swimming in the open sea not only in response to field rotations, but also to changes in merely the strength of the field.

Freshwater catfish learned to discriminate between two hiding places on the basis of the polarity of nonuniform (Roth 1969), and uniform, electric fields (Kalmijn 1974; Peters and van Wijland 1974; Kalmijn, Kolba, and Kalmijn 1976a), and so did the weakly electric fish *Hypopomus* sp. (Kalmijn 1974 and unpublished). The lowest voltage gradient allowing orientation was 1 $\mu$V/cm. Catfish also learned to select one out of twelve hiding tubes evenly spaced along the perimeter of their circular tank, yielding a minimal angular discrimination

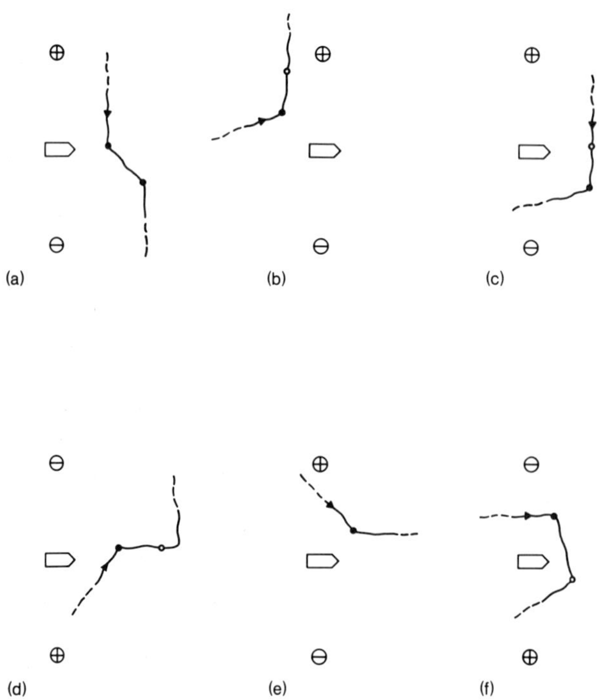


Figure 6.8. Responses of stingray *Dasyatis americana* to rotation of ambient electric field. The current passed between two remote electrodes, + and −, caused the ambient electric field to rotate within 10 s by about 45°, either to the right or left. *Solid dots* in the tracks indicate correct turns; *open circles* indicate wrong turns or a lack of response. In five of six trials, the animals turned in the expected directions during the initial field rotations.

of 30 degrees (Kalmijn, Kolba, and Kalmijn 1976b). The 1-μV/cm threshold was independent of the resistivity of the medium in the range of 2 to 40 kohm-cm, despite the 20-fold variation in current density. Thus, not only the receptors, but also the whole animals may be considered *voltage-sensitive* devices (Kalmijn 1984). The marine catfish *Plotosus anguillaris* has so far been trained to orient to fields as weak as 0.08 μV/cm, but we may not have reached the threshold yet.

## Detection of the Earth's Magnetic Field

In addition to the environmental fields, elasmobranch fishes may sense the electric fields they themselves induce when moving through the earth's magnetic field (cf. pp. 159–160). Operating the electric sense in the active mode, the animals would acquire *unambiguous compass information*. However, do elasmobranch fishes really orient to the earth's magnetic field and, if so, is their orientation mechanism of an electromagnetic nature? Convincing evidence for magnetic orientation is scarce. In no instance has the physical basis been determined, with the sole exception of the magnetic bacteria, which are simply constrained to swim along magnetic field lines (Blakemore 1975; Kalmijn and Blakemore 1978; Kalmijn 1981). Despite substantial data in support, the electromagnetic mechanism also remains a hypothesis, pending formal proof.

In behavioral experiments conducted at Woods Hole, the stingray *Urolophus halleri* was conditioned, this time, to enter an enclosure in the *magnetic east* of its circular tank and to avoid a similar enclosure in the *magnetic west* in the absence of any imposed electric fields. The magnetic fields were applied with the horizontal component randomly in either the normal or reversed direction by the aid of Helmholtz coils. Trials were carried out in northern-hemisphere fields at a vertical intensity of 0.42 gauss and a horizontal intensity of 0.26 gauss, in conformity with the magnetic conditions in the California region from where the animals were collected (Kalmijn 1978a, 1978b, 1981), *and* in strictly horizontal magnetic fields of 0.40 gauss simulating equatorial conditions (Kalmijn 1984). Magnetic orientation was established beyond doubt in both series.

Most significantly, the behavioral experiments, in particular the series conducted in *equatorial* fields, revealed that the stingrays not only detect the direction, but also the polarity, or the sense, of the magnetic field. Consequently, their magnetic sense is self-supporting, save for the detection of gravity. The proposed electromagnetic principle is inherently *vectorial* (cf. pp. 156–159), and the stingrays *are* capable of detecting the polarity of the fields (cf. pp. 165–167). In contrast, birds are believed to detect only the direction of the earth's magnetic field, not its polarity (Wiltschko 1972). It is unknown how birds resolve the inherent 180-degree ambiguity in equatorial regions, where elasmobranch fishes would be at their best. A polar sensitivity has recently also been found in salamanders (Phillips 1986).

Experiments designed to verify the electromagnetic principle of the elasmobranchs' magnetic compass sense are under way, following the approach outlined in Kalmijn (1984). Sharks, skates, and rays are expected to respond to vertically imposed *electric* fields as they would to horizontal *magnetic* fields if, and only if, the electromagnetic principle holds. The electric fields that sharks receive when crossing the horizontal component of the earth's magnetic field at cruising speeds of 50 cm/s range from 100 to 200 nV/cm, depending on the magnetic latitude, as compared with the 5–500-nV/cm environmental fields and the fishes' 5-nV/cm sensitivity. The validity of this comparison is based on the similar stimulus configurations resulting from the fields that the animals themselves induce and the imposed environmental fields (cf. p. 160).

Neurophysiological experiments by Andrianov, Brown, and Ilyinsky (1974), often referred to in connection with the electromagnetic principle of orientation, unfortunately were based on a misconception of the physics involved (Kalmijn 1978b, 1984). Brown and Ilyinsky (1978) later made more meaningful recordings, following the physics introduced by Kalmijn (1974), but left previous mistakes uncorrected (cf. p. 177). A magnetic sensitivity has also been claimed for eels (Tesch 1974), salmon (Quinn, Merrill, and Brannon 1981; Quinn and Brannon 1982), and tuna (Walker, Dizon, and Kirschvink 1982) without serious consideration of the electromagnetic mechanism. It is surprising, though, that a magnetic sense in teleost fishes seems so difficult to demonstrate. Concerning my doubts about the electromagnetic function of the inner-ear semicircular canals, entertained by Jungerman and Rosenblum (1980), see Kalmijn (1984).

The theory of electromagnetic orientation is most easily grasped when described by an observer at rest in the earth's frame of reference. When a shark *moves* on a straight course through the earth's magnetic field, it *induces* an electric field leading to appreciable potential differences across the receptor epithelia. However, in the reference frame of its own electroreceptors, the animal is *at rest* and, consequently, does *not* induce an electric field. Instead, it *receives* an electric field that appears to have its origin in the outside world. This dual description, phrased in technical, rather than biological, terms, led Einstein (1905) to the special theory of relativity, as recounted by Kalmijn (1985). The relativistic approach bears out the real challenge faced by the animals, for whom the issue is not merely academic, but a matter of survival.

## Physics of Peripheral Sensory Apparatus

### Resistivity of the External Medium

The low-frequency, ampullary electroreceptors are externally discernible by their skin pores, which connect to the deeper-situated ampullae proper by jelly- or fluid-filled canals. The walls of the ampullae proper consist mainly of the electroreceptive epithelia. The Lorenzinian ampullae of marine sharks, skates, and rays have canals from a few millimeters to at least several centimeters in length (Fig. 6.9). The ampullary receptors of freshwater fishes are microscopically small, having canals usually just long enough to traverse the skin.

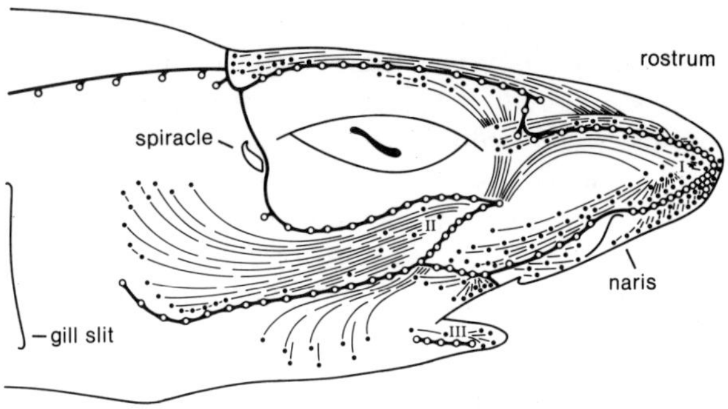


Figure 6.9. Electroreceptor and lateral-line system of the shark *Scyliorhinus canicula: open circles,* lateral-line pores; *heavy lines,* lateral-line canals; *solid dots,* electroreceptor pores; *broken lines,* ampullary canals. The ampullae proper are located in the rostral (*I*), infraorbital (*II*), and mandibular (*III*) capsules. (Kalmijn 1978b, after Dijkgraaf and Kalmijn 1963.)

The peripheral sensory apparatus comprises all tissues and structures contributing to the convergence of the physical stimulus onto the electroreceptive membranes. It passively conditions the natural field and enhances its biologically relevant features, dealing with such diverse stimuli as the dipole electric fields of prey and the uniform fields of ocean currents (cf. pp. 175–176). The electric fields due to the fishes' own motion through the earth's magnetic field are similar to the imposed environmental fields as far as the peripheral apparatus is concerned (theorem of equivalence, p. 160).

Averaged across the body, *marine elasmobranchs* are, with regard to *uniform electric fields,* nearly two orders of magnitude more resistive than is the seawater medium. Hence, the animals behave as relatively *good insulators,* contrary to common belief (cf. pp. 171–172). They distort the imposed fields favorably, strengthening the potential distribution at the skin as determined by the shape of the animal. Freshwater fishes are usually more resistive than marine elasmobranchs, but often less so than the freshwater medium, particularly in media of low ionic content. Consequently, *freshwater fishes* tend to load *uniform electric fields,* which reduces the electric potentials at the skin and, in terms of voltage sensitivity, renders the animals *inherently less efficient* (Kalmijn 1974, 1984).

With regard to *nearby dipole sources,* the animals appear more resistive than in uniform electric fields, because the stimulus current must, to enter and leave the body, pass through rather small skin areas of relatively high resistance. Thus, in the immediate vicinity of small dipole sources, both *marine and freshwater fishes* act as rather *good insulators,* which increases the effective stimulus at the receptor pores. These observations bear directly on the length of the ampullary canals, which, in turn, relates to the kinds of fields that the animals detect, the resistivity of the external medium, the noise of the gelatinous canal contents, and the useful sensitivity of the sensory system (cf. pp. 171–174).

The electric current induced by the fish's motion through the earth's magnetic field flows across the body and closes its path through the outside medium (cf. pp. 159–160 and Fig. 6.5). While the strength of this current depends on the total resistance of the circuit, the animal receives only the potential difference developed across the body (Kalmijn 1974, 1984). Thus, for the *electromagnetic compass sense* to be efficient, the resistance across the body must be high compared with the resistance of the return path through the outside medium. Ideally, the animal should be *electrically short-circuited* by the surrounding water, which to a high degree holds true for marine elasmobranchs.

## Potential Distribution Within the Animal

When marine elasmobranchs are subjected to uniform electric fields, about half the potential difference they receive appears across the skin on either side of the body, while the other half develops as voltage gradients in the body tissues (Kalmijn 1974). However, when they are exposed to fields of nearby dipole sources, almost the full potential difference appears across the skin in the areas through which the current enters and leaves the animal, while hardly any voltage gradients develop in the intervening body tissues (cf. pp. 168–170). Compared with the resistance of the body tissues, the *skin resistance* of marine elasmobranchs is only moderately high (Fig. 6.10). It nevertheless *plays an important role* in the functional organization of the sensory system.

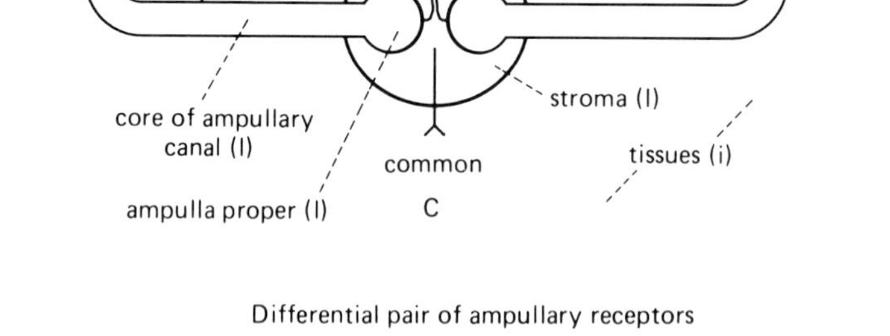


Figure 6.10. Differential design of elasmobranch ampullary system. The individual ampullae measure the pore potentials with respect to the common capsules. By subtracting the signals received from the various pore populations, the animal would sense the potential distribution over the skin, while suppressing the common-mode potentials of the capsules. The labels (*l*), (*i*), and (*h*) denote low, intermediate, and high resistance. (Kalmijn 1984.)

In freshwater electrosensitive fish, whether exposed to uniform electric fields or to the fields of nearby dipole sources, the stimulus appears largely as potential differences across the skin, while only weak voltage gradients develop in the body tissues. The skin resistance is evidently considerably higher than in marine elasmobranchs, presumably for osmotic reasons, and consequently plays a dominant role in the organization of the receptor system. In weakly electric fishes, the skin resistance is known to be particularly high (Bennett 1971a), whereas in freshwater catfishes and the freshwater ray *Potamotrygon* sp. it is not notably different from that in most freshwater teleosts (Kalmijn 1974).

Since the ampullary canals act as excellent low-frequency cables (cf. pp. 173–174), the sensory epithelia gather almost the full potential difference between the skin pores and the loose connective tissues in which the ampullae proper are embedded. This potential difference equals the voltage drop across the skin *plus* the voltage gradients in the body tissues integrated along the lengths of the canals (Kalmijn 1974). Thus, the ampullae with short canals, found in marine *and* freshwater species, rely almost entirely on the resistance of the skin, while those with long canals, found in marine species only, also collect the voltage gradients in the body tissues.

Owing to their high skin resistance, freshwater fishes remain internally nearly isopotential, whether subject to uniform or to dipole fields. Thus, the ampullary canals need only traverse the skin for the sense organs to receive almost the full potential difference available. Marine elasmobranchs remain internally nearly isopotential only in the proximity of dipole sources. They develop, however, appreciable voltage gradients in the body tissues when subjected to uniform electric fields. Instead, the elasmobranch ampullae proper are clustered in a few subcutaneous capsules, functioning as isopotential reference bodies, which connect to the widely scattered skin pores by canals of various lengths.

So far, only the resistive properties of the peripheral apparatus have been considered. Because of capacitive shunting, however, the skin loses its insulating properties at higher frequencies. Not only will the impedance of the whole animal decline, reducing the available signal, but the receptors also become less efficient, as they too depend on the impedance of the skin. Preliminary measurements indicate that the capacitive effects are insignificant at frequencies well below 10 Hz (Kalmijn 1974), although precise data are lacking. If not stated otherwise, our discussion will regard the sensory system to be purely resistive for the low frequencies of interest. The frequency response of the ampullary canals will be discussed below (pp. 173–174).

## Differential Operation and Common-Mode Rejection

The individual ampullary organs detect the electrical potentials at the skin pores relative to the common potentials of the loose subcutaneous or capsular tissues in which the ampullae proper are embedded and, thus, are single-ended input devices. The design of the peripheral apparatus suggests, however, that the electric sense *at a system's level* operates differentially, detecting the *potential distribution* over the electroreceptive skin areas, and disregards the ill-defined reference potentials of the subcutaneous or capsular tissues (Kalmijn 1974, 1984, and Fig. 6.9). Assuming that the pores provide sufficient skin coverage, the reference potentials would not yield any further information. On the contrary, they probably are the main cause of physiological self-noise.

I proposed the *differential mode of operation* and emphasized the importance of the pore distribution (Kalmijn 1974) to replace the widely accepted, but incorrect, notion that electric fields unimpededly invade marine elasmobranchs and that the direction and length of the individual ampullary canals would be the relevant parameters (Murray 1965; Bennett 1971b; cf. pp. 168–170). In fact, many of the canals are very short, and even the longer canals do not rely on their length for more than about half the signal strength (cf. pp. 170–171). In my opinion, the high sensitivity rests primarily on the *pore separation* and the intrinsic properties of the detection process. The length of the canals would make it possible to have the ampullae proper together in a few clusters so as to protect the system against interference from the animals' own, physiological fields.

An increase in skin resistance would not eliminate the need for the long canals and ampullary clusters of marine species. It might reduce the internal voltage gradients of the stimulus fields, but not the fishes' own physiological fields. The latter would render the subcutaneous tissues too noisy to serve as isopotential reference bodies. In freshwater species, including the stingray *Potamotrygon* sp., which lack long canals, the tissue noise is less of a problem, because the signals they receive are at least two orders of magnitude stronger than the oceanic fields. Conversely, in freshwater fishes, the resistance of long ampullary canals would be prohibitively high, unless the ionic strength of the canal cores were kept at a level well above that of the outside medium.

Although the ampullary capsules of marine elasmobranchs may present adequate isopotential reference bodies, the capsule stromata need not be electrically stable with respect to the outside medium. In particular, any changes in the capsular potentials resulting from the fishes' ventilatory movements or any physiologically induced variations in the potentials across the skin effectively stimulate the individual, single-ended ampullae (Dijkgraaf and Kalmijn 1966). The beauty of the inferred differential operation is that *common-mode signals,* by equally affecting all the ampullae of each capsule, *can be suppressed centrally,* given the required neural circuitry (Kalmijn 1974).

From this point of view, the capsular clusters with their receptor pores covering considerable skin areas, act as *paired subsystems, each with its own reference potential.* Although common-mode suppression must take place at the level of the subsystem, the various subsystems are readily integrated at higher nerve centers, since their pore distributions overlap generously or, along the midline of the animal, butt upon each other. These early inferences as to the differential mode of operation, made on the basis of the morphological and physical features of the peripheral apparatus, have recently received strong neurophysiological support (cf. pp. 179–181).

## Physical Properties of Ampullary Canals

In marine skates, the gelatinous content of the ampullary canals has nearly the same resistivity as the seawater medium (20–25 ohm-cm), while the specific resistance of the canal wall is the highest of any animal tissues known (6 Mohm-$cm^2$) (Waltman 1966). The specific capacity of the canal wall, however, is similar to that of other biological membranes (0.4 $\mu F/cm^2$). The resulting cable properties give rise to low-pass cut-off frequencies ranging from a few hundred hertz for the shorter canals to a mere three hertz for the longest canals. As these values were calculated for canals submerged in isopotential media, the in vivo cut-off frequencies might be slightly higher, because the internal stimulus fields would partially guard the canals.

According to Waltman (1966), the walls of the ampullae proper and thus also the sensory epithelia, would be so high-ohmic as to render the ampullary canals electrically open-ended. Bennett and Clusin (1978), however, came to the conclusion that the sensory epithelia were less insulating, so that the ampullary organs would suffer a noticeable, but still insignificant, DC loss, and have again somewhat higher cut-off frequencies than estimated before. Despite these uncertainties, the physical properties of the ampullary canals are impressive and fully in support of the electroreceptive function.

We recently measured the DC input resistance of the ampullary canals with sucrose-insulated, guarded pipet electrodes directly inserted into the external pores of 16-cm-long canals in lightly anesthetized specimens of the skate *Raja erinacea* (Fishman and Kalmijn 1985). The resistance values obtained (250 kohm) were consistent with the resistance of the canal wall (1 Mohm) calculated from Waltman's data and with the resistance of the blind ampulla proper (380 kohm) given by Bennett and Clusin. The technique will be further developed to acquire quantitative physical data under biologically valid conditions.

The animals' behavioral responses to uniform electric fields of threshold strength are expected to be mediated mainly by ampullae having their skin pores far apart. To reach the common capsules, these ampullae must have relatively long canals and, according to Waltman's data, filter out all but the lowest frequencies. In orienting to the weak, virtually uniform, environmental fields, the animals might benefit from such low-pass peripheral filters, as they help reduce the noise above the frequency range of interest (cf. pp. 175–176). Since during normal swimming the frequency of modulation is on the order of only 1 Hz, even the bandwidth of the ampullae with the longest canals would suffice.

Peters and Evers (1985) found a lowest threshold for the ventilatory reflex in response to uniform electric fields (cf. pp. 161) between 0.1 and 1 Hz, but did not think that the canals were responsible for the high-frequency cut-off. They argued, as did Montgomery (1984a), that in the small sharks and rays tested the canals were not as long as in Waltman's skates, but did not take the canal diameter into account. To remove any uncertainty, the physical and physiological filter properties of the ampullae with the longest canals shall be determined in the intact animal of the species under study.

Estimates of the thermal noise in the frequency range from DC to 8 Hz indicate that for threshold stimuli the ampullae with the longest canals operate at signal-to-noise ratios of about unity (Kalmijn 1984). The main sources of thermal noise are the gelatinous cores of the ampullary canals. Other major sources of predominantly low-frequency noise are the diffusion gradients at the fluid-membrane interfaces and the physiological processes in the membranes themselves. Whereas the canal noise acts coherently on the whole ampulla, the diffusion and physiological noise contributions average over a great number of individual receptor cells.

## Sensory Transduction and Nerve Coding

### Electroreceptor Mechanisms

Nonelectrical stimuli of extrinsic origin are commonly received at the apical structures of the receptor cells, where by the process of *sensory transduction* they give rise to graded *receptor potentials*. Spreading from the site of transduction to the basal, synaptic part of the cell, the receptor potentials control the release of neurotransmitter in the event of chemical mediation. The transmitter substance acts on the afferent nerve endings, causing them to produce the *generator potentials* that determine the rate of the afferent *action potentials*. The ampullary organs of marine and freshwater electrosensitive fishes present some interesting variations on this scheme, which I will discuss briefly (for further details, see Bennett and Clusin 1979; Obara and Sugawara 1984).

The elasmobranch ampullae of Lorenzini have pear-shaped receptor cells of which the small apical membranes have considerably higher resistance than do the large basal membranes (Waltman 1966; Obara and Bennett 1972). The apical membrane receives the physical stimulus, and the cell is excited by pore-negative fields. The receptor potentials spread to the basal, synaptic part of the cell, where they elicit the release of neurotransmitter. At a behavioral threshold of 5 nV/cm (cf. pp. 165–167), the potential differences received by the receptor epithelia of the small stingrays tested were at most 50 nV. The observed neurophysiological responses required at least 10 to 100 times stronger stimuli.

Conversely, in the ampullary receptors of teleost fishes, the apical membranes have larger surface areas and lower resistance than the basal membranes. Hence, the basal membrane receives the stimulus, and the cell is excited by pore-positive fields (Bennett 1971b). The stimulus itself, with or without the aid of a receptor potential (see below), causes the release of neurotransmitter. Here it is the stimulus that travels the height of the cell, whereas the receptor potential, where present, is the response of the same membrane that releases the neurotransmitter (Obara 1976). The lowest potential differences that the ampullary receptors of freshwater fishes detect are on the order of 10 μV. The receptors of the marine catfish *Plotosus* are 10 to 100 times more sensitive.

The astounding sensitivity of marine elasmobranchs rests on the process of *electrical-electrical transduction,* taking place at the small, 8-$\mu m^2$ apical membrane. However, the process is neither simple nor fully understood. Obara and Bennett (1972) and Bennett and Clusin (1978) suggested that the apical and basal membranes interact to form tiny voltage-dependent oscillators, giving the ampullae of Lorenzini their great sensitivity. However, Broun and Govardovskii (1984) offered an explanation without invoking oscillatory feedback. So far, the roles of the apical and basal membranes have been inferred only from whole-ampulla studies for want of intracellular recordings. Naturally, the useful sensitivity of the system depends also on the number of receptor cells per ampulla and the number of ampullae per animal (Kalmijn 1984).

In the freshwater catfish *Kryptopterus,* the physical stimulus does not give rise to detectable receptor potentials (Bennett 1971a), although in the marine catfish *Plotosus* it does (Obara 1976). In the freshwater case, the stimulus itself causes the release of neurotransmitter. The lesser sensitivity need not be a limitation of the simpler detection process, but may merely be a function of the high signal and noise levels in the natural habitat. The marine catfish *Plotosus* is definitely more sensitive to electric fields than are freshwater catfish (cf. p. 167), though it may not be as sensitive as the marine elasmobranchs. The freshwater stingray *Potamotrygon* is about as sensitive as are the freshwater teleosts and clearly less sensitive than the marine species (Kalmijn 1974).

## Detection of Temporal Field Characteristics

Murray (1962) observed clear changes in the action-potential rates of single ampullary fibers in response to pulsed electric fields of 1–2 μV/cm in the skate *Raja clavata,* from which he derived integration and differentiation time constants of 0.02 and 3 s, respectively (Murray 1962). The results were due to the physiological process of detection, as the ampullae studied had only short canals. Assuming linearity, the time constants translate to a bandwidth of 0.05–8 Hz, conforming to the results of behavioral tests (Kalmijn 1974, cf. pp. 163–165). The ampullary fibers of the thornback ray, *Platyrhinoidis triseriata,* (Montgomery 1984a) and the dogfish *Scyliorhinus canicula* (Peters and Evers 1985) showed a similar frequency dependence when tested with pure sine-wave fields.

When swimming through uniform electric fields or the fields of remote dipole sources, elasmobranch fishes receive approximately sinusoidal stimuli of about 1 Hz owing to the normal swaying of the head (cf. pp. 176–177). The high sensitivity required to detect the weaker fields may be attained by comparing the potential differences between far-separated pore populations, for even the longest canals are capable of handling the 1-Hz signal (cf. pp. 173–174). The stimuli that the animals receive when steadily cruising through the earth's magnetic field have a similar fundamental frequency (cf. pp. 159–160). The signals resulting from sudden turns and accelerations have broader amplitude spectra.

When the animal closes in on its prey, the field over the skin increases rapidly in strength, especially in the region of the head nearest the target. However, the stimulus received by the receptors does not increase as steeply as the field at the skin does, because, with the reduction of the affected skin area, the available stimulus diminishes owing to the smaller pore separation. For the shorter ampullae in the snout, detecting the fast-changing stimuli from nearby prey, the frequency response is unlikely to be limited by the cable properties of the canals (cf. pp. 173–174). For the amplitude spectrum of the signal received by a shark transecting a dipole field, see also pp. 155–156 and Fig. 6.3.

The function of a graded high-frequency attenuation, adjusted to the kind of signal to be detected, seems self-evident, but why the low-frequency cut-off? It is the DC-instability in the time domain or the low-frequency noise in the frequency domain that makes DC measurements unwise to attempt unless absolutely necessary. One solution is to shift the near-DC stimulus into a higher, less noisy, frequency band, for which the normal movements of the recipient animals may again suffice. Thus, the fishes' swaying motions could be considered either as the carrier for the stimulus or as the sampling function needed to realize the desired frequency shift. In predation, either the prey or the predator has to move. Without motion the prey can be neither detected nor caught.

In comparative studies, whole-nerve recordings from the catfish *Ictalurus nebulosus* revealed a minimum threshold of 1.5 μV/cm in the frequency range from 3 to 7 Hz (Peters and Buwalda 1972). Moreover, an electrical sensitivity to fields of 1–10 μV/cm has recently been demonstrated in the freshwater lamprey *Ichthyomyzon unicuspis,* featuring an afferent frequency response up to only 1 Hz (Bodznick and Preston 1983). The axolotl, *Ambystoma mexicanum,* another recent addition to the list of electrosensitive animals, showed its highest afferent sensitivity at 10 Hz in fields of 5 μV/cm (Muenz, Claas, and Fritszch 1984).

## Detection of Spatial Field Characteristics

The elasmobranch ampullary system, with its skin pores covering a closed surface area, seems well suited for the detection of the strength and direction of uniform electric fields (cf. pp. 171–172). However, because of the DC insensitivity of the receptors, the afferent nerve fibers do not signal the fields as such, but rather their spatial derivatives integrated over the arcs scanned by the animals' swaying or turning movements. Note that the integration is of the "leaky" kind, although it has a long time constant of decay (cf. pp. 175–176). In other words, the effective stimulus for the ampullary system is not the uniform field per se, but its *apparent rotation* relative to the animal's body axes. This, I believe, is the key to understanding the animals' electric-field orientation mechanisms (see below and pp. 177–179).

In horizontal fields, the signals ensuing from the fish's swerving mode of swimming are proportional to the strength of the field, the distance between the pores of the ampullae under consideration, the sine of the angle of swerving about the direction of swimming, and the sine of the angle between the line connecting the pores and the direction of the field. Consequently, the axis of the skin pores receiving the smallest signal gives the direction of the field, whereas the axes of the skin pores receiving the strongest signals define the isopotential plane, normal to the direction of the field. Even if the field is not purely horizontal, only the horizontal component is detected. The vertical component is filtered out by virtue of the system's DC insensitivity, assuming the fish to swim in a horizontal plane. (For the sake of clarity, the distortion of the field due to the animal's presence was not taken into account.)

The fields of nearby dipole or multipole sources, such as prey, vary significantly in strength and direction over distances comparable to the size of the recipient animal. Thus, near the source, the spatial configuration of the field becomes a viable cue as well, given that the animal appreciates the extent to which the field deviates from the purely uniform, clearly a second-order task. At the distance of response where the field is barely detectable, the nonuniformity is inconspicuous and would be difficult to detect (cf. pp. 177–179). When an animal approaches the source, however, the nonuniformity of the field becomes progressively more pronounced, until the stimulus distribution covers only a small skin area directly opposite the source.

The electric fields resulting from the animal's motion through the earth's magnetic field and their proposed role in magnetic orientation have been treated elsewhere (Kalmijn 1974, 1984). Here, I only wish to emphasize the importance of the relativity principle involved (Kalmijn 1985). To wit, a local change of magnetic flux imposed upon a stationary animal is definitely not equivalent to the change of magnetic flux that the animal receives when moving with respect to the water in the presence of the earth's magnetic field (see, however, Andrianov, Brown, and Ilyinsky 1974; Akoev, Ilyinsky, and Zadan 1976). In the first case, the induced currents form closed loops within the animal, whereas in the second case, they traverse the animal to loop around through the seawater medium (Kalmijn 1984).

Furthermore, a change of magnetic flux imposed upon just the recipient animal is not equivalent to a change in the ambient magnetic field. The latter produces regional, rather than local, electric fields, compared with the size of the animal (see, however, Brown and Ilyinsky 1978). Whether electrosensitive fishes might interpret the induced fields in terms of magnetic variations and why they would do so remain a matter of conjecture. For a functional interpretation of the neurophysiological data, only biologically meaningful stimuli should be applied. A good example is offered by the experiments in which Brown and Ilyinsky (1978) moved either the water or the fish, in conformity with the theory of electromagnetic orientation (Kalmijn 1974).

## Sensory Mechanisms and Central Processing

### A Prey-Detection Algorithm

At the distance of detection, the bioelectric fields of prey are similar in appearance to pure dipole fields (cf. pp. 155–156). Yet, bottom-feeding sharks aim their attacks directly at the source, whether they enter the field along the dipole axis, perpendicular to the dipole axis, or in an arbitrary direction (cf. p. 164). When on a tangential track, they veer left or right to approach the target along smoothly curved paths, without further course corrections. The need for such an impressive performance becomes obvious when one observes the dogfish *Mustelus canis* feeding at night on small fishes hovering over the ocean bottom in the shallow waters off Cape Cod (Kalmijn 1978b). Many prey, promptly alerted by the large predators, escape with great agility.

The most readily detected feature of the prey field, besides its mere presence, is the *direction* of the field, suitably averaged over the predator's electroreceptive skin area. Unfortunately, such a local average essentially pertains to only a single point in space and, in the case of a dipole field, does not carry any information about the position of the source. To locate the target, the animal could attempt to analyze the instantaneous potential distribution over the skin or the time series of the field directions detected while swimming. However, at the distance of response, the field deviates so little from uniform and the signal received is so near the noise of the receptor system that the precision required to infer the position of the source would be difficult to achieve.

The problem would be more tractable if it were possible for the animal to initiate its strikes without knowing the position of the target and to do so relying only on the local field direction. Being discouraged by other more sophisticated, but less conceivable, mechanisms, I recently discovered an amazingly simple *approach algorithm* satisfying the above requirements. The strategy would be as follows: when first sensing the presence of the prey field, the recipient fish fixes the direction of the field with respect to its body axes and, from then on, proceeds in such a fashion as to nullify, by feedback, any apparent rotation. This algorithm would guide the predators to their prey along approach paths *consistent with* two decades of *observation* (Fig. 6.11).

Applied to bottom-feeding predators and to prey having their dipole moments in a level plane, the proposed algorithm is surprisingly robust. It is insensitive to the angle of approach, to the polarity of the field, to temporal changes in the strength or direction of the field, and, consequently, to movements of the prey. In truly three-dimensional situations, the algorithm either needs further information or leaves some uncertainty in case the animal's path is not confined to a plane containing the dipole axis. Near the source, additional cues may be obtained from the nonuniformity of the field, for instance, from the curvature of the field lines or the gradient of the scalar field-line density (cf. pp. 175–176). A full analysis of the field configuration, however, is by no means so trivial a matter as merely detecting the direction of the field.

In physical dipole fields, the sharks were observed to strike at the *center* of the source, as expected from the proposed algorithm, even though the poles were separated by a distance of 5 cm. The fields of real prey, however, are no longer approximately dipolar when sensed from nearby. Hence, it would be more advantageous for the predator to follow the prescribed course only long enough to complete the remainder "ballistically." This would enable the animal to rush at the target, ignoring the more complicated inner part of the field. Actually, we have conducted experiments in which the source was electrically moved away

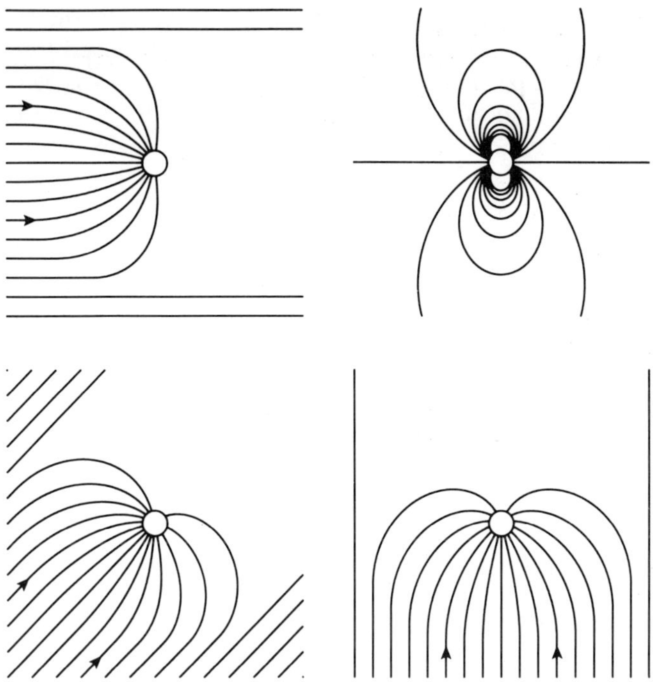

Figure 6.11. Approach paths calculated on the basis of newly proposed shark-attack algorithm. By tracing a course of constant field direction *with respect to its body axes,* a bottom-dwelling dogfish would invariably arrive at the source of a dipole field. *Top-right block,* dipole field simulating prey; *other three blocks,* approach paths of animals entering the field under an angle of respectively 0, 45, and 90° with respect to the dipole axis. The approach paths are all in the same plane as the dipole vector.

just after the animal had initiated the attack, with the result that it bit the sand at the original position of the source (Kalmijn and Weinger 1981).

The additional presence of environmental electric fields and of the fields that the fishes themselves induce by swimming through the earth's magnetic field complicates the situation to some extent. Yet, those fields help the animals to perform even better, not worse (cf. p. 165). By analogy with other sensory modalities, it is reasonable to assume that the anticipated signal is subtracted from the signal that the animals receive in order to identify the target field, a hypothesis calling for neurophysiological verification. The proposed detection algorithm offers a first attempt to explain existing behavioral observations, taking into account known physical and physiological data. Whether it will hold true or not, at least the simplicity of the new algorithm is appealing.

## Central Processing

From our knowledge of the physical stimulus and its biological relevance, some tentative inferences can be made about the central processing. Thus, the design of the peripheral sensory apparatus of marine elasmobranchs suggests that lower brain centers take the differences between the signals received from the various pore populations (cf. pp. 171–172). These expectations have recently been corroborated neurophysiologically by Montgomery (1984b) for the ray *Platyrhinoidis triseriata*. Furthermore, in order to reject the common-mode signals, the primary fibers must accommodate both the small differential *and* large common-mode signals, which explains their wide dynamic range. Only after rejection of the common-mode component would higher-order neurons be free to use their full dynamic range for the biologically relevant differential signals.

The DC insensitivity of the electroreceptors suggests that, in orienting to the DC uniform fields of ocean currents and to the DC dipole fields of remote prey, elasmobranch fishes rely in the first instance on the *apparent* changes in the direction of the field with respect to the body axes (cf. pp. 176–177). According to the prey-detection algorithm, they would actually null the *imposed* apparent field rotation by locomotory feedback (cf. pp. 177–179). In the environmental fields, the apparent field rotations would result, not from the nonuniformity of the field, but from the animals' own swerving movements (cf. pp. 175–178). These ideas have recently found their central correlate in the work of Schweitzer (1985), who studied the sensitivity of higher brain centers to uniform electric fields as a function of field direction in the thornback ray.

The medulla and higher brain centers of freshwater catfish also show an increase in sensitivity over the receptors (Roth 1975; McCreery 1977a; Tong 1982). Moreover, central units cover a wider frequency range, from less than 0.1 to 25 Hz as compared with 1 to 10 Hz for the receptors (Roth 1975; Knudsen 1976b). They may span the full frequency range or have narrower bandwidths with low, intermediate, or high center frequencies. In the medulla, mid- and low-frequency units prevail (Andrianov and Ilyinsky 1973, McCreery 1977a), whereas higher centers show a more even coverage (Knudsen 1976b). Receptor fields of medullar (McCreery 1977b) and higher-order units (Tong and Bullock 1982) have been reported, as has the representation of field direction (Knudsen 1976a, 1976b).

The increase in sensitivity of higher brain centers over the peripheral receptors suggests that also in freshwater species the effects of self-stimulation are eliminated at an early stage, whether by common-mode suppression, corollary discharge, or otherwise. Some authors have wondered about the response of central units to frequencies below 1 Hz and above 10 Hz (Roth 1975; Knudsen 1976b). A pure DC sensitivity may be impractical (cf. pp. 175–176), but a sensitivity to frequencies less than 1 Hz makes good sense in detecting the predominantly DC bioelectric and environmental fields. Moreover, the steep spatial gradients of prey fields would require a fast response in the time domain, corresponding to an extended range in the frequency domain (Knudsen 1978, cf. pp. 175–176).

However impressive, brain studies must be interpreted with the utmost caution. If meant to elucidate the normal operation of an animal, brain recordings should be made in response to biologically meaningful stimuli. What, for example, does it mean when the telencephalon of a skate shows neurophysiological responses to uniform electric fields 50 times stronger than needed to induce violent escape reactions and 100 to 10,000 times stronger than the oceanic fields (see, however,

Bodznick and Northcutt 1984). Similarly, if the evoked potentials recorded in the mesencephalon of a shark are negligibly weak at the low frequencies that the animals rely on for the detection of prey and to orient at sea, but are much stronger at higher frequencies, should we question the animals' behavior or, perhaps, the interpretation of the recorded potentials (see, however, Bullock 1979 and the explanation in Montgomery 1984a)?

The search for new categories of electric and magnetic signals in nature will continue, and each new one discovered will be evaluated for its biological significance. However, the responses to fields that the animals rarely, if at all, encounter in nature often are, in my opinion, merely *biological artifacts.* A meaningful sensory analysis seeks to explain the animals' behavioral responses to the physical stimuli they receive in the natural environment.

## Summary

The electric sense of lower aquatic vertebrates is approached from a systems point of view. After an introduction to the relevant electric fields in natural waters and the sensory information they offer, the animals' behavioral responses and their biological significance are discussed. On the basis of our knowledge of the physical input and the behavioral output, the intermediate physical and physiological steps are scrutinized as to their parts in the sequence of events that constitutes the animals' electric sense. Thus, the physics of the peripheral sensory apparatus, the process of sensory transduction, the coding of afferent nerve signals, and the central processing are analyzed to provide the groundwork for further research. In particular, new data on the orientation to weak electric fields at sea and a first prey-detection algorithm are presented.

### Acknowledgments

This chapter was prepared under contract with the Office of Naval Research, Oceanic Biology Program, Drs. Eric O. Hartwig, and Bernard J. Zahuranec, Program Managers. The studies in the Bahamas were conducted during three NSF-sponsored cruises on invitation of Dr. Samuel H. Gruber, Chief Scientist. I thank Drs. Charles S. Cox and James T. Enright for their helpful comments on the manuscript.

### References

Akoev, G.N., Ilyinsky, O.B., and Zadan, P.M. (1976). Responses of electroreceptors (Ampullae of Lorenzini) of skates to electric and magnetic fields. J. Comp. Physiol., 106:127–136.

Alexandrov, V.V. (1985). Fresh Water Electrophysics. Godrometeoizdat, Leningrad.

Andrianov, G.N., Brown, H.R., and Ilyinsky, O.B. (1974). Responses of central neurons to electrical and magnetic stimuli of the ampullae of Lorenzini in the Black Sea skate. J. Comp. Physiol., 93:287–299.

Andrianov, G.N. and Ilyinsky, O.B. (1973). Some functional properties of central neurons connected with the lateral-line organs of the catfish (*Ictalurus nebulosus*). J. Comp. Physiol., 86:365–376.

von Arx, W.S. (1962). An Introduction to Physical Oceanography. Addison-Wesley, Reading, MA.

Bennett, M.V.L. (1971a). Electrolocation in fish. Ann. N.Y. Acad. Sci., 188:242–269.

Bennett, M.V.L. (1971b). Electroreception. In: Fish Physiology. Hoar, W.S. and Randall, D.S. (eds.), Academic Press, New York, pp. 493–574.

Bennett, M.V.L. and Clusin, W.T. (1978). Physiology of the ampulla of Lorenzini, the electroreceptor of the elasmobranchs. In: Sensory Biology of Sharks, Skates, and Rays. Hodgson, E.S. and Mathewson, R.F. (eds.), U.S. Government Printing Office, Washington D.C., pp. 483–505.

Bennett, M.V.L. and Clusin, W.T. (1979). Transduction at electroreceptors: origins of sensitivity. In: Membrane Transduction Mechanisms. Cone, R.A. and Dowling, J.E. (eds.), Raven Press, New York, pp. 91–116.

Blakemore, R.P. (1975). Magnetotactic bacteria. Science, 190:377–379.

Bodznick, D. and Northcutt, R.G. (1984). An electrosensory area in the telencephalon of the little skate, *Raja erinacea*. Brain Res., 298:117–124.

Bodznick, D. and Preston, D.G. (1983). Physiological characterization of electroreceptors in the lampreys *Ichthyomyzon unicuspis* and *Petromyzon marinus*. J. Comp. Physiol., 152:209–217.

Broun, G.R. and Govardovskii, V.I. (1984). Electrical model of the electroreceptor of the ampulla of Lorenzini. Neurophysiology, 16:166–172.

Brown, H.R. and Ilyinsky, O.B. (1978). The ampullae of Lorenzini in the magnetic field. J. Comp. Physiol., 126:333–341.

Bullock, T.H. (1979). Processing of ampullary input in the brain: comparison of sensitivity and evoked responses among elasmobranch and siluriform fishes. J. Physiol. (Paris), 75:397–407.

Bullock, T.H. (1982). Electroreception. Ann. Rev. Neurosci., 5:121–170.

Bullock, T.H. and Chichibu, S. (1965). Further analysis of sensory coding in electroreceptors of electric fish. Proc. Nat. Acad. Sci., 54:422–429.

Burr, H. S. (1932). An electro-dynamic theory of development suggested by studies of proliferation rates in the brain of *Amblystoma*. J. Comp. Neur., 56:347–371.

Burr, H.S. (1941). Field properties of the developing frog's egg. Proc. Nat. Acad. Sci., 27:276–281.

Burr, H.S. (1947). Field theory in biology. Scientific Monthly, 64:217–225.

Burr, H.S. and Hovland, C.I. (1937a). Bio-electric potential gradients in the chick. Yale J. Biol. Med., 9:247–258.

Burr, H.S. and Hovland, C.I. (1937b). Bio-electric correlates of development in *Amblystoma*. Yale J. Biol. Med., 9:541–549.

Butsuk, S.V. and Bessonov, B.I. (1981). Direct current electric field in some teleost species: effect of medium salinity. J. Comp. Physiol., 141:277–282.

Deelder, C.L. (1952). On the migration of the elver (*Anquilla vulgaris* Turt.) at sea. J. Cons. Perm. Int. Explor. Mer, 18:187–218.

Dijkgraaf, S. (1968). Electroreception in the catfish, *Amiurus nebulosus*. Experientia, 24:187–188.

Dijkgraaf, S. and Kalmijn, A.J. (1962). Verhaltensversuche zur Funktion der Lorenzinischen Ampullen. Naturwissenschaften, 49:400.

Dijkgraaf, S. and Kalmijn, A.J. (1963). Untersuchungen ueber die Funktion der Lorenzinischen Ampullen an Haifischen. Z. Vergl. Physiol., 47:438–456.

Dijkgraaf, S. and Kalmijn, A.J. (1966). Versuche zur biologischen Bedeutung der Lorenzinischen Ampullen bei den Elasmobranchiern. Z. Vergl. Physiol., 53:187–194.

Einstein, A. (1905). On the electrodynamics of moving bodies. First published in Ann. Physik, 17:891–921, translated by Perrett, W. and Jeffery, G.B. (1952) in: The Principle of Relativity. Dover Publications, New York.

Faraday, M. (1832). Experimental researches in electricity. Phil. Trans. R. Soc. Lond., 122(1):125–194.

Fishman, H.M. and Kalmijn, A.J. (1985). Complex admittance of the ampullary canal of the skate *Raja erinacea*. Biol. Bull., 169:551.

Goodman, D.A. and Weinberger, N.M. (1971). Submerged electrodes in an aquarium: validation of a technique for remote sensing of behavior. Behav. Res. Meth. Instrum., 3:281–286.

Halsband, E. and Halsband, I. (1954). Untersuchungen ueber die Stoerungsschwellen im Stoffwechsel der Fische und Fischnaehrtiere nach Einwirkung verschiedener Abwassergifte. Arch. Fischerei, 5:119–132.

Himstedt, W., Kopp, J., and Schmidt, W. (1982). Electroreception guides feeding behaviour in amphibians. Naturwissenschaften, 69:552–553.

Hopkins, C.D. (1983). Functions and mechanisms in electroreception. In: Fish Neurobiology. Northcutt, R.G. and Davis, R.E. (eds.), Univ. of Michigan Press, Ann Arbor, pp. 215–259.

Johnson, C.S., Scronce, B.L., and McManus, M.W. (1984). Detection of DC electric dipoles in background fields by the nurse shark. J. Comp. Physiol., 155:681–687.

Jungerman, R.L. and Rosenblum, B. (1980). Magnetic induction for the sensing of magnetic fields by animals—an analysis. J. Theor. Biol., 87:25–32.

Kalmijn, A.J. (1966). Electro perception in sharks and rays. Nature, 212:1232–1233.

Kalmijn, A.J. (1969). In Dutch, for translation in English, see Kalmijn (1972).

Kalmijn, A.J. (1971). The electric sense of sharks and rays. J. Exp. Biol., 55:371–383.

Kalmijn, A.J. (1972). Bioelectric fields in sea water and the function of the ampullae of Lorenzini in elasmobranch fishes. Scripps Inst. Oceanogr. Ref. Ser., 72–83:1–21.

Kalmijn, A.J. (1974). The detection of electric fields from inanimate and animate sources other than electric organs. In: Handbook of Sensory Physiology, vol. III/3. Fessard, A. (ed.), Springer-Verlag, New York, pp. 147–200.

Kalmijn, A.J. (1978a). Experimental evidence of geomagnetic orientation in elasmobranch fishes. In: Animal Migration, Navigation, and Homing. Schmidt-Koenig, K. and Keeton, W.T. (eds.), Springer-Verlag, New York, pp. 347–353.

Kalmijn, A.J. (1978b). Electric and magnetic sensory world of sharks, skates, and rays. In: Sensory Biology of Sharks, Skates, and Rays. Hodgson, E.S. and Mathewson, R.F. (eds.), U.S. Government Printing Office, Washington D.C., pp. 507–528.

Kalmijn, A.J. (1981). Biophysics of geomagnetic field detection. IEEE Trans. Magnetics, 17:1113–1124.

Kalmijn, A.J. (1982). Electric and magnetic field detection in elasmobranch fishes. Science, 218:916–918.

Kalmijn, A.J. (1984): Theory of electromagnetic orientation: a further analysis. In: Comparative Physiology of Sensory Systems. Bolis, L., Keynes, R.D., and Maddrell, S.H.P. (eds.), Cambridge University Press, Cambridge, pp. 525–560.

Kalmijn, A.J. (1985). Electromagnetic orientation: a relativistic approach. In: Interactions between electromagnetic Fields and Cells. Chiabrera, A., Nicolini, C., and Schwan, H.P. (eds.), Plenum Press, New York, pp. 45–64.

Kalmijn, A.J. and Blakemore, R.P. (1978). The magnetic behavior of mud bacteria. In: Animal Migration, Navigation, and Homing. Schmidt-Koenig, K. and Keeton, W.T. (eds.), Springer-Verlag, New York, pp. 354–355.

Kalmijn, A.J., Kolba, C.A., and Kalmijn, V. (1976a). Orientation of catfish (*Ictalurus nebulosus*) in strictly uniform electric fields. I. Sensitivity of response. Biol. Bull., 151:415.

Kalmijn, A.J. and Weinger, M.B. (1981). An electrical simulator of moving prey for the study of feeding strategies in sharks, skates, and rays. Ann. Biomed. Eng., 9:363–367.

Kalmijn, V., Kolba, C.A., and Kalmijn, A.J. (1976b). Orientation of catfish (*Ictalurus nebulosus*) in strictly uniform electric fields. II. Spatial discrimination. Biol. Bull., 151:415–416.

Kleerekoper, H. and Sibakin, K. (1956). Spike potentials produced by the sea lamprey (*Petromyzon marinus*) in the water surrounding the head region. Nature, 178:490–491.

Knudsen, E.I. (1974). Behavioral thresholds to electric signals in high frequency electric fish. J. Comp. Physiol., 91:333–353.

Knudsen, E.I. (1976a). Midbrain responses to electroreceptive input in catfish: evidence of orientation preferences and somatotopic organization. J. Comp. Physiol., 106:51–67.

Knudsen, E.I. (1976b). Midbrain units in catfish: response properties to electroreceptive input. J. Comp. Physiol., 109:315–335.

Knudsen, E.I. (1978). Functional organization in electroreceptive midbrain of the catfish. J. Neurophysiol., 41:350–364.

Krajew, A.P. (1957). Grundlagen der Geoelektrik. VEB Verlag Technik, Berlin.

Lissmann, H.W. (1951). Continuous electrical signals from the tail of a fish, *Gymnarchus niloticus* Cuv. Nature, 167:201–202.

Lissmann, H.W. (1958). On the function and evolution of electric organs in fish. J. Exp. Biol., 35:156–191.

Lissmann, H.W. (1963). Electric location by fishes. Sci. Am. 208(3):50–59.

Lissmann, H.W. and Machin, K.E. (1963). Electric receptors in a non-electric fish (*Clarias*). Nature, 199:88–89.

Longuet-Higgins, M.S., Stern, M.E., and Stommel, H. (1954). The electrical field induced by ocean currents and waves, with applications to the method of towed electrodes. Pap. Phys. Oceanogr. Met., 13(1):1–37.

Lonsdale, E.M. and Marshall, W.C. (1973). Noncontact sensing of electric fields surrounding trout. ISA Trans., 12:82–87.

Lund, E.J. (1922). Experimental control of organic polarity by the electric current. II. The normal electrical polarity of *Obelia:* a proof of its existence. J. Exp. Zool., 36:477–494.

Lund, E.J. (1925). Experimental control of organic polarity by the electric current. V. The nature of the control of organic polarity by the electric current. J. Exp. Zool., 41:155–190.

Lund, E.J. (1947). Bioelectric Fields and Growth. Univ. of Texas Press, Austin.

Machin, K.E. (1962). Electric receptors. Symp. Soc. Exp. Biol., 16:227–244.

Machin, K.E. and Lissmann, H.W. (1960). The mode of operation of the electric receptors in *Gymnarchus niloticus*. J. Exp. Biol., 37:801–811.

Mathews, A.P. (1902). Electrical polarity in the hydroids. Am. J. Physiol., 8:294–299.

McCreery, D.B. (1977a). Two types of electroreceptive lateral lemniscal neurons of the lateral line lobe of the catfish *Ictalurus nebulosus:* connections from the lateral line nerve and steady-state frequency response characteristics. J. Comp. Physiol., 113:317–339.

McCreery, D.B. (1977b). Spatial organization of receptive fields of lateral lemniscus neurons of the lateral line lobe of the catfish *Ictalurus nebulosus*. J. Comp. Physiol., 113:341–353.

Montgomery, J.C. (1984a). Frequency response characteristics of primary and secondary neurons in the electrosensory system of the thornback ray. Comp. Biochem. Physiol., 79A:189–195.

Montgomery, J.C. (1984b). Noise cancellation in the electrosensory system of the thornback ray: common mode rejection of input produced by the animal's own ventilatory movement. J. Comp. Physiol., 155:103–111.

Muenz, H., Claas, B., and Fritzsch, B. (1984). Electroreceptive and mechanoreceptive units in the lateral line of the axolotl *Ambystoma mexicanum*. J. Comp. Physiol., 154:33–44.

Murray, R.W. (1962). The response of the ampullae of Lorenzini of elasmobranchs to electrical stimulation. J. Exp. Biol., 39:119–128.

Murray, R.W. (1965). Electroreceptor mechanisms: the relation of impulse frequency to stimulus strength and responses to pulsed stimuli in the ampullae of Lorenzini of elasmobranchs. J. Physiol. (Lond.), 180:592–606.

Obara, S. (1976). Mechanism of electroreception in ampullae of Lorenzini of the marine catfish *Plotosus*. In: Electrobiology of Nerve, Synaps, and Muscle. Reuben, J.P., Purpura, D.P., Bennett, M.V.L., and Kandel, E.R. (eds.), Raven Press, New York, pp. 129–147.

Obara, S. and Bennett, M.V.L. (1972). Mode of operation of ampullae of Lorenzini of the skate, *Raja*. J. Gen. Physiol., 60:534–557.

Obara, S. and Sugawara, Y. (1984). Electroreceptor mechanisms in teleost and non-teleost fishes. In: Comparative Physiology of Sensory Systems. Bolis, L., Keynes, R.D., and Maddrell, S.H.P. (eds.), Cambridge University Press, Cambridge, pp. 509–523.

Pals, N., Peters, R.C., and Schoenhage, A.A.C. (1982). Local geo-electric fields at the bottom of the sea and their relevance for electrosensitive fish. Neth. J. Zool., 32:479–494.

Pals, N., Valentijn, P., and Verwey, D. (1982). Orientation reactions of the dogfish, *Scyliorhinus canicula,* to local electric fields. Neth. J. Zool., 32:495–512.

Parker, G.H. and van Heusen, A.P. (1917). The responses of the catfish, *Amiurus nebulosus,* to metallic and non-metallic rods. Am. J. Physiol., 44:405–420.

Pasztor, V.M. and Kleerekoper, H. (1962). The role of the gill filament musculature in teleosts. Can. J. Zool., 40:785–802.

Peters, R.C. and Bretschneider, F. (1972). Electric phenomena in the habitat of the catfish *Ictalurus nebulosus* LeS. J. Comp. Physiol., 81:345–362.

Peters, R.C. and Buwalda, R.J.A. (1972). Frequency response of the electroreceptors ("small pit organs") of the catfish, *Ictalurus nebulosus* LeS. J. Comp. Physiol., 79:29–38.

Peters, R.C. and Evers, H.P. (1985). Frequency selectivity in the ampullary system of an elasmobranch fish (*Scyliorhinus canicula*). J. Exp. Biol., 118:99–109.

Peters, R.C. and Meek, J. (1973). Catfish and electric fields. Experientia, 29:299–300.

Peters, R.C. and van Wijland, F. (1974). Electro-orientation in the passive electric catfish, *Ictalurus nebulosus* LeS. J. Comp. Physiol., 92:273–280.

Phillips, J.B. (1986). Two magnetoreception pathways in a migratory salamander. Science, 233:765–767.

Quinn, T.P. and Brannon, E.L. (1982). The use of celestial and magnetic cues by orienting sockeye salmon smolts. J. Comp. Physiol., 147:547–552.

Quinn, T.P., Merrill, R.T., and Brannon, E.L. (1981). Magnetic field detection in sockeye salmon. J. Exp. Zool., 217:137–142.

Roberts, J.L. (1964). Metabolic responses of fresh-water sunfish to seasonal photoperiods and temperatures. Helgol. Wiss. Meeresunters., 9:459–473.

Roth, A. (1969). Elektrische Sinnesorgane beim Zwergwels *Ictalurus nebulosus* (*Amiurus nebulosus*). Z. Vergl. Physiol., 65:368–388.

Roth, A. (1972). Wozu dienen die Elektrorezeptoren der Welse? J. Comp. Physiol., 79:113–135.

Roth, A. (1975). Central neurons involved in the electroreception of the catfish *Kryptopterus*. J. Comp. Physiol., 100:135–146.

Schweitzer, J. (1985). Studies of the Elasmobranch Central Electrosensory System with Emphasis on Functional Organization of the Electroreceptive Midbrain in the Thornback Ray. Thesis, Univ. of California, San Diego.

Tesch, F.W. (1974). Influence of geomagnetism and salinity on the directional choice of eels. Helgol. Wiss. Meeresunters., 26:382–395.

Tong, S. (1982). The nucleus praeeminentialis: an electro- and mechanoreceptive center in the brainstem of the catfish. J. Comp. Physiol., 145:299–309.

Tong, S., and Bullock, T.H. (1982). Electroreceptive representation and its dynamics in the cerebellum of the catfish, *Ictalurus nebulosus* (Ictaluridae, Siluriformes). J. Comp. Physiol., 145:289–298.

Tricas, T.C. (1982). Bioelectric-mediated predation by swell sharks, *Cephaloscyllium ventriosum*. Copeia, 4:948–952.

Walker, M.M., Dizon, A.E., and Kirschvink, J.L. (1982). Geomagnetic field detection by yellowfin tuna. Oceans 82:755–758.

Waltman, B. (1966). Electrical properties and fine structure of the ampullary canals of Lorenzini. Acta Physiol. Scand., 66 (Suppl. 264):1–60.

Wiltschko, W. (1972). The influence of magnetic total intensity and inclination on directions preferred by migrating European robins (*Erithacus rubecula*). In: Animal Orientation and Navigation. Galler, S.R., Schmidt-Koenig, K., Jacobs, G.J., and Belleville, R.E., U.S. Government Printing Office, Washington DC, pp. 569–578.

Wu, C.H. (1984). Electric fish and the discovery of animal electricity. Am. Sci., 72:598–607.

Part II

# Behavior: Expectation, Communication, and Sensory Performance

The link between sensory performance and animal behavior is a strong double bond. Sensory information steers behavior generated by central neural and endocrine processes. Behavioral requirements demand specific sensory performance. To be competitive and to have offspring surviving in the process of natural selection, an organism not only needs immediate response capabilities, it also needs to know its physical and social environment; it needs capabilities to predict or anticipate short- and long-term future events; and it needs to learn. Therefore, sensory biology is driven by behavioral ecology and vice versa. The four chapters in this part deal with the second link of this double bond. The first link appears in the CNS chapters of Part 3.

Kamil (Chapter 7) uses the advances made recently in terrestrial behavioral ecology to point out the questions sensory biologists should consider whether they focus their research on terrestrial or aquatic systems. Some of the perceptual questions raised by Kamil continue to be explored by Saidel in Chapter 20.

Blaxter (Chapter 8) reviews the behavioral performance of fish and the widely varying characteristics of the sensory environment they occupy. Hopkins (Chapter 9) focuses on the specific demands of social communication where one often finds the most dramatic examples of sensory specialization. The principles of social communication in the context of electroreception and electric organ discharge are broadly applicable in other communication systems. Bullock (Chapter 10), again using electroreception as an example, discusses the ability of the central nervous system to acquire stimuli needed for behavioral decisions. Chapters 7 and 10 form an intellectual pair in which predictions of future events are explored from the vantage points of animal behavior and its neurobiological mechanisms. Many of the ideas and themes found in the chapters of this part are reflected in the physiological and morphological chapters in Part 3.

# Chapter 7

# Behavioral Ecology and Sensory Biology

ALAN C. KAMIL*

Zoologists use many different levels of analysis to study animals. Going from molecular to molar, these levels would include the neurochemical, neurophysiological, sensory biological, anatomical, behavioral, and ecological. Each of these represents part of the overall process of developing a full understanding of the structure, function, and evolution of animals. Therefore, no level of analysis exists in a vacuum; each is related to the others. In spite of this clear relationship, communication among scientists working at different levels of analysis is notoriously difficult (see Lehrman 1974, for an interesting and thoughtful discussion of this issue). In general, molar levels of analysis can provide direction to molecular levels, while molecular levels can provide information about mechanisms and constraints. In this chapter, I shall suggest some problems of sensory biology that are indicated by a more molar level of analysis, that of behavioral ecology.

Recent theory in behavioral ecology demonstrates that animals could increase their biological success by reacting to relatively subtle aspects of their environment. Research stimulated by this theory demonstrates that animals do, in fact, change their behavior in response to small environmental differences. This growing body of theory and research in behavioral ecology suggests that sensory biologists need to ask more subtle questions about how the senses of animals work. For example, much research in sensory biology concentrates on the specific cues and sensory apparatus that allow an animal to distinguish food from nonfood items (e.g., work on chemoreception that emphasizes the specific chemicals emanating from food items that attract foragers [see Chap. 14]). However, recent developments in the study of foraging behavior demonstrate that animals make many sophisticated decisions about foraging, showing sensitivity to subtle parameters (Kamil and Sargent 1981 and discussion below). Before discussing specifics, we need to discuss the relation between sensing and knowing.

In many cases, the more an animal knows about its environment, the greater

* Departments of Psychology and Zoology, University of Massachusetts, Amherst, MA 01003, USA

the biological success it can enjoy. What is meant by *knowing,* and what is the relationship between knowing and sensing? In general terms, an animal is said to know about some aspect of its environment when it changes its behavior adaptively to changes in that aspect. For example, an animal can be said to know about the density of its prey if it changes its behavior in response to changes in prey density. An animal cannot know anything, in this use of the term, unless it first senses something. Any consideration of what an animal knows inevitably has implications for sensory biology, but not all interesting cases of knowing pose interesting sensory problems. A behavioral response to environmental variation can be based on relatively trivial sensory biology. Nonetheless, many of the situations currently being studied by behavioral ecologists do pose interesting research questions for sensory biology, if only because the sensory bases for the information used by the behaving animal are unknown.

A functional approach to the study of organisms is implicit in this chapter. The functional approach is basic to behavioral ecology (see Maynard Smith 1978; Krebs and Davies 1978, 1981), but has also produced much controversy (Gould and Lewontin 1979). This is not the place to delve into this controversy in any detail. Suffice it to say that I believe the functional approach has contributed to a revolution in the study of animal behavior (Kamil and Yoerg 1982). Many fascinating phenomena have been discovered through functional reasoning, either formal or informal.

The functional approach is certainly not new to sensory biology. Von Frisch's (1954) discussion of how he discovered color vision in bees, for example, clearly reveals informal functional reasoning. He knew that the bright colors of flowers had to have some function. In many other cases, sensory biologists have worked on special systems of biological significance, such as echolocation in bats (Griffin 1958) or homing and migration in birds (Keeton 1974).

What does an animal need to know to succeed? At the simplest level, it needs to know three things: (1) how to find food, (2) how to avoid becoming food for another, and (3) with whom to mate. Each of these presents an animal with many more problems then might appear at first, however, and there are many related issues. I shall discuss some recent behavioral ecological results that bear on the problems of biological success. In each case, the potential implication for sensory biology is the same: What are the sensory bases for the knowledge demonstrated by the behavior?

## Discriminating Food from Nonfood Items

The first problem a foraging animal faces is detecting food (or, in other words, discriminating food from nonfood items). Most sensory biological research has concentrated on this aspect of the foraging problem—studying sensitivity to food-related stimuli. In the case of aquatic animals, much of this research has focussed on chemoreception (see Chap. 14).

Cryptic prey are, by definition, those which are hard to find, often because they resemble the backgrounds against which they must be detected. We have been studying cryptic prey detection in my laboratory for a number of years, using procedures in which blue jays (*Cyanocitta cristata*) are trained to search projected images for cryptic underwing (*Catocala*) moths (Pietrewicz and Kamil 1977). Birds can adjust several aspects of their behavior so that the efficiency of their prey detection is improved. Two of these behavioral adjustments pose particularly interesting questions for sensory biologists. When jays encounter the same kind of prey several times in a row, they become better at detecting it (Pietrewicz and Kamil 1979), an effect known as the *specific-searching image* (Tinbergen 1960; Croze 1970; Dawkins 1971). A similar finding has recently been reported for lobsters with chemical stimuli (Derby and Atema 1981), showing that the effect is not limited to vision.

Deborah Olson and I recently found another effect with sensory implications. *Catocala* moths achieve their crypticity by choosing resting substrates that they closely resemble (Sargent 1976). For example, *C. relicta* is white with black, disruptive markings. It actively selects trees with light-colored bark, such as white birch, on which to rest. *C. relicta* is much more cryptic on birch than on many other surfaces (Pietrewicz and Kamil 1977). The crypticity of this moth is dependent on both its own appearance and on that of the substrate on which it rests. We found that, when the substrate varied, the birds varied their speed of search. Surfaces on which the moths were most cryptic were searched most slowly. Since the surfaces were different birch trees that we had photographed, it is not clear how the birds discriminated between surfaces.

The variation in search speed did increase foraging efficiency. Several simulation analyses showed that the birds detected prey faster by varying their search speed than they could have with any fixed speed (Kamil, Lindstrom, and Peters 1985). It might initially appear that all a blue jay searching for cryptic moths needs to do is be able to discriminate the moth from its background. The bird knows more about its environment than this simple analysis suggests, however, and the jay uses this knowledge to improve its foraging success.

Similar phenomena occur in marine environments. Wicksten (1983) has shown that camouflage, including background matching, is found in many marine invertebrates and undoubtedly is a primary means of avoiding predation. Saidel's discussion of camouflage in flatfish (Chap. 20) provides an excellent example. Anderson (1984) has shown that largemouth bass (*Micropterus salmoides*) vary their search speed adaptively in response to changes in the degree of structure in their environment. The bass moved more slowly in the more structured, complex environments, a result similar to that obtained by Olson and me.

## Food: Which Food Types Should Be Eaten?

Recent developments in the study of foraging behavior show that detecting the presence of a prey is only the beginning of the problem for a forager. The foraging animal often needs to know a great deal more about the prey than whether it is present. Theoretical considerations show that foragers can dramatically increase their foraging efficiency if they know such parameters as prey value, density, and crypticity, among others. Before discussing these implications, however, a brief word about optimal-foraging theory is necessary.

First, there are some terminological problems. *Optimal-foraging theory* refers to a group of models of foraging behavior that takes a particular approach to model construction. These models develop a mathematical and/or graphical picture of a foraging problem that plot foraging efficiency as a function of some aspect(s) of behavior. If one assumes that selection has operated to maximize foraging efficiency, then the models can be used to make predictions about behavior. The behavior that maximizes foraging efficiency is usually a function of environmental parameters, such as prey density. Animals selected to maximize their foraging efficiency should vary their behavior when faced with such environmental variation. The major point is that optimal-foraging theory encompasses a variety of models. Different ones represent attempts to predict behavior in different foraging situations and often differ in the nature of their assumptions and in the "currency" presumed to be maximized. Therefore, it is incorrect to speak or write of the "predictions of optimal-foraging theory" in general terms. Rather, specific models must be tested and evaluated.

Second, optimal-foraging theory is a field undergoing rapid increases in complexity and breadth. There were two early models, the optimal diet model (MacArthur and Pianka 1966) and the marginal value theorem (Charnov 1976), that had an enormous influence and generated a large number of experiments (see Krebs, Stephens, and Sutherland 1983; Pyke 1984). Each model included many simplifying assumptions, and each was concentrated on a single decision that animals must make. These models have often been criticized for being too simple, omitting many important aspects of foraging. This criticism is somewhat misplaced, however, when generalized to optimal-foraging theory as a whole.

A new generation of models has begun to appear. These are often much more complex and incorporate many more factors of interest. This makes them more realistic, but also mathematically more difficult and perhaps more limited in application. Although it is too soon to judge the usefulness of these new models (see Gray 1987; Schoener 1987, for contrasting views), they may prove quite successful in predicting foraging behavior in complex situations.

In any event, optimal-foraging theory shows that foraging animals need to make many complex decisions to be efficient, and the research stimulated by optimal-foraging models shows that animals do make these decisions. In the course of their foraging, animals demonstrate that they know a great deal about the distribution and value of their prey (see Krebs, Stephens, and Sutherland 1983; Pyke 1984, for recent reviews). The sensory biology underlying these decisions is completely unknown.

Even the simplest optimal-foraging models strongly imply that animals make subtle discriminations among their prey. Take the first optimal-foraging model (MacArthur and Pianka 1966) as an example. The problem MacArthur and Pianka considered was that of a predator which encounters several different prey types sequentially and at random. Which of these should be included in the diet, i.e., which types should the predator always eat and which ignore? The answer turns out to be rather simple, given that the predator possesses considerable information about each prey type.

In the MacArthur and Pianka model, each prey type has a value (*E*/*H*) that is determined by its caloric value (*E*) divided by its handling time (*H*), where handling time includes the time required to pursue, capture, and consume the prey. If the forager can rank the types from best to worst in terms of *E*/*H*, the rule that maximizes rate of intake is to keep adding prey to the diet in ranked order until the *E*/*H* of the next prey is less than the average rate of intake being achieved without the addition of this type to the diet. This model makes several quantitative predictions: (1) under any given set of conditions, any prey type should either be taken every time it is encountered or never taken (no partial preferences should exist); (2) diets should become broader (include more prey types) as prey density decreases; and (3) whether a prey type is taken should depend not on its own density, but on that of higher-ranking prey types.

A number of studies have shown that foragers can rank their food types by *E*/*H*. For example, Elner and Hughes (1978) found that shore crabs (*Carcinus maenas*) prefer the size of mussel that maximizes E/H. Pulliam (1980) found a close relationship between *E*/*H* and preferences in chipping sparrows. How the animals achieve this ranking, which apparently requires integrating information about both food value and handling time, is not known.

There have been many tests of MacArthur and Pianka's model, in both the laboratory and the field. In general, the first prediction, that there are no partial preferences, has not been supported. The other predictions have been supported in many studies, at least qualitatively. For example, Krebs et al. (1977) presented large and small prey to great tits (*Parus major*). The prey were presented on a conveyor belt, to control the relative densities of the prey types. As the rate of encountering the large prey increased, the birds became more selective and rejected the smaller prey. This supports the prediction that the probability of taking the less valuable prey depends on the availability of more profitable types.

In an extensive series of laboratory and seminatural experiments, Werner and his associates (Werner and Hall 1974, 1976; Werner, Mittelbach, and Hall 1981; Werner et al. 1983) have investigated the foraging behavior of bluegill sunfish (*Lepomis macrochirus*) (reviewed by Werner and Mittelbach 1981). They have found that the theoretical framework provided by MacArthur and Pianka's (1966) model was useful in understanding prey and habitat selection, differences between age/size classes, and seasonal patterns of habitat use by bluegills. Thus, for example, bluegills of different sizes required different handling times for prey of different sizes. Knowledge of this relationship allowed the successful prediction of differences between age classes in both prey size selection and patterns of habitat use, since prey of different sizes tend to be found in different habitats.

## How to Avoid Being Eaten

One of the major criticisms of optimal-foraging theory has been that it concentrates exclusively on a single dimension—rate of intake. Although rate of intake is obviously important to biological success, it is only part of the story. Presumably, animals are selected on the basis of fitness, and not rate of intake alone: for example, if the foraging strategy that maximized rate of intake also maximized the risk of being eaten by another (predation), we should not expect the forager to maximize rate of intake. This criticism is literally true: in most optimal-foraging models, only the effects of different behaviors on foraging efficiency are considered; but the problem of trade-offs between foraging efficiency and other demands, such as avoiding predation risk, are not beyond the scope of theoretical or empirical investigation.

Milinski and Heller (1978) have investigated this trade-off in sticklebacks (*Gasterosteus aculeatus*). Normally, when faced with a swarm of daphnia, sticklebacks will attack the center of the swarm. When the presence of a predator was simulated by flying a model kingfisher over the experimental tank, the sticklebacks switched their attack to the edges of the swarm. Milinski and Heller interpreted this finding as showing that attacking the edge of a swarm made less demands on the sensory system of the sticklebacks, allowing the fish to attend more closely to the sky above the tank for the presence of a predatory bird. This interpretation may sound rather speculative, but it is supported by a variety of studies that have shown that animals will compromise between predation risk and foraging efficiency (Caraco 1979; Sih 1980; Cerri and Fraser 1983; Lima, Valone, and Caraco 1985).

Other experiments have found similar effects in other species of fish. For example, Dill and Fraser (1984) tested the foraging behavior of juvenile coho salmon (*Oncorhynchus kisutch*). They measured the distance from which the fish began to attack a fly in the absence of predators and after a model of a trout was presented. Following exposure to the predator model, reaction and attack distances were reduced. This supports the findings of Milinski and Heller (1978), since it also suggests that the presence of a predator causes foragers to give more attention to the environment, making them less sensitive to the presence of prey. Fraser and Emmons (1984) found that blacknose dace (*Rhinicthys atratulus*) react to the presence of predators by moving away from them, with a probable reduction in foraging efficiency, since the predators were in areas with high food availability.

These data clearly demonstrate that animals are very sensitive to the presence of potential predators. The sensory biology underlying this responsiveness is, in most cases, totally unknown. The study of sensitivity to predators has been productive for sensory biologists in the past. The most outstanding example is the study of the responses of noctuids to the hunting sounds of bats (Roeder and Treat 1961). I strongly suspect that the study of predator detection processes in other animals could yield rich rewards for the sensory biologist. The issue of trade-offs between attending to predators and attending to food distribution appears to be a particularly promising question.

## With Whom to Mate?

The first problem that an animal choosing a mate must face is to choose a member of the same species and of the opposite sex. In some instances, these decisions are known to be based on interesting and subtle (to human senses) cues, including visual and chemical stimuli. There are frequently more subtle decisions to be made, however. In particular, animals will have more successful offspring if they can choose a mate of maximum quality. A body of evidence is growing that such choices are often made, especially by females. But the sensory biology underlying such decisions is largely unknown.

For example, Andersson (1982) studied mate choice in widow birds (*Euplectes progne*). Male widow birds have long tails that are used in displays. Andersson experimentally altered the tail lengths of males in the field. He found that long-tailed males attracted more females than either short-tailed or control males. The cue on which mate choice was based is clear in this case. The example is a good one, because it clearly demonstrates active choice based on physical characteristics of the potential mate. This clarity is due to the use of good experimental design, including particularly the random assignment of birds to experimental conditions.

Another example of mate choice is seen in the breeding behavior of lemon tetras (*Hyphessobrycon pulchripinnis*) (Nakatsura and Kramer 1982). Male lemon tetras have limited daily fertility. After ten spawning acts in a single day, the fertilization rate for males showed significant decline, probably because of limited sperm supply. When female tetras were given a choice of males only one of which had spawned that day, the females showed a significant preference for the male that had not spawned and therefore had higher fertility. Although the mechanism is less clear in this case than in Andersson's (1982) widow bird research, the implications are interesting. What cues did the females use in choosing between males?

In many cases, the choice of a mate may not be based directly on physical quality, but on some other characteristic associated with the potential mate. In polygynous mating systems, some males have multiple mates while others have one or none. Studies of these mating systems indicate that the quality of the territory controlled by a male may influence female choice. The polygyny threshold model (Verner 1964; Verner and Willson 1966, Orians 1969) predicts that when there are large differences in quality between territories, a female may do as well to become a second female on the territory of an already mated male than to choose an unmated male whose territory is of lower quality. This requires females to discriminate territories of differential quality.

The model has received strong support from the work of Pleszczynska (1978), who studied lark buntings (*Calamospiza melanocorys*). Lark buntings are migratory. Males are the first to arrive on the breeding ground, where they establish territories. Females arrive later and over a period of time. The first females to arrive mate monogamously, whereas some later-arriving females mate bigamously. Pleszczynska found that a relatively simple measure of territorial quality, the amount of cover available there, successfully predicted which territories (with their associated males) would attract the most females. These data clearly suggest that female lark buntings discriminate among territories and choose their nesting territory quite systematically, but it is not at all clear what sensory evidence they employ to discriminate among territories.

## Social Relations

Although the choice of a mate is a crucial social decision, there are many other important social decisions animals must make. In particular, they must decide how to interact with a variety of conspecifics—whether to communicate, cooperate, or fight, for example. These issues have received considerable attention in behavioral ecology during the last 15 years. In each case, the evidence shows that animals know a great deal about the animals with which they interact.

The question of cooperation often includes questions of kinship. Kin selection theory (Hamilton 1964) holds that animals can significantly increase their inclusive fitness if they can discriminate kin from nonkin. Then they can help (behave altruistically towards) relatives. A wide variety of data shows that animals do treat kin differently from nonkin. For example, Sherman (1977) found that Belding's ground squirrels are much more likely to give alarm calls when kin are present than otherwise. In many birds, there are helpers at the nest that often feed young, guard against predators, or help maintain territorial boundaries. The helpers are usually previous offspring of one or both parents, so that the young they are helping to raise are siblings.

In many cases, kin recognition probably entails experience: for example, in most cases of avian helping at the nest, young from a previous nest help at the current one (Krebs and Davies 1981). Helping is not widely known among fishes, but Taborsky (1984) has recently studied the phenomenon in some detail in a cichlid, *Lamprologus brichardi.* He found that these fish often remain in their natal territory, assisting in brood care, territorial maintenance, and defense. Helpers grow more slowly than fish that leave the natal territory and join aggregations, but they also experience lower predation. This implies that the fish may be influenced in their decisions to stay or leave by food supply and predator pressure.

Bateson (1982) experimentally demonstrated the importance of early experience for kin selection. Japanese quail (*Coturnix coturnix japonica*) base their choice of mates on the appearance of the potential mate's plumage. Bateson found that quail which had been raised with siblings preferred to associate with cousins they had never seen than with either siblings or unrelated animals they had never seen. This preference for cousins, rather than siblings, may be related to the idea that it is best to breed with animals that are somewhat related, rather than with either close relatives or completely unrelated animals.

In other cases, early experience is not necessary for kin recognition. Wu et al. (1980) have shown that kin can be recognized even if they have never been seen before. They tested young pig-tailed macaques (*Macaca nemestrina*) for social preferences. They found that these monkeys preferred to spend time near a relative they had never seen, rather than an unrelated conspecific animal. Although visual cues appeared to be important in this study, the mechanism of kin recognition is unknown.

In other cases, individual recognition can be imporatnt. In a theoretical paper, Axelrod and Hamilton (1981) have shown that reciprocal altruism (the trading of altruistic behaviors by individuals at different times [Wilson 1975]) can be evolutionarily stable, provided animals can recognize others, so that they can remember their previous interactions and behave accordingly. Individual recognition is a well-known phenomenon, particularly among birds (Falls, 1982). Myrberg and Riggio (1985) recently demonstrated that bicolor damselfish (*Pomacentrus partitus*) can recognize their neighbors acoustically. They point out several important functions that this ability may serve, including territoriality and mate selection.

Animals also make many subtle decisions in determining when to contest resources. Recent theory in behavioral ecology suggests that many factors should affect the willingness of an animal to fight, including the relative strength and status of the contestants and the value of the resource being contested (Parker 1974, 1984).

A particularly nice example of research on these problems is provided by investigations of contests between hermit crabs (*Pagurus bernhardus*) for the shells in which they live (Dowds and Elwood 1983, 1985). These studies have shown that fighting behavior is affected by the relative size of the contestants, as well as the relative value of the shells. Perhaps the finding with the most fascinating sensory implications is that a crab's willingness to fight is based, in part, on the size of the shell of the other crab. The more closely the shell of the opponent matches body size, the more willing the crab is to display and fight. How this size, which must be judged relative to the body size of the crab, is assessed is not known.

The "selfish gene" approach to study of animal behavior (Dawkins 1976, 1982) emphasizes the concept that animals are machines whose purpose is to maximize the representation of their genes in future generations. This approach has led to a partial reevaluation of the concept of communication between animals (Krebs and Dawkins 1984). Whereas the classical ethological view emphasized the concept of the accurate transmission of information, the selfish gene view implies that animals often will avoid accurate communication.

Why should an animal avoid accurate communication? Consider the case of two animals contesting a resource. If one of them accurately transmits a limited willingness to fight, the other can use this information to win the contest. An animal should fight "all out" until it quits, because if it signals that it will soon quit, it will lose any chance of winning. Similar arguments can be offered for a male's trying to convince a female to mate with him. He should not communicate any weakness but attempt to be as impressive as possible to maximize his chances of successful courtship. This, in turn, suggests that animals should have evolved extremely accurate means of evaluating the signals and status of others.

Another idea with implications for sensory biology is what Krebs and Dawkins (1984) call the "conspiratorial whisper." When two communicating animals share a common goal for example, are closely related or are mates raising young— we might expect accurate communication; but communicating includes risks, such as attracting the attention of a predator. We might, therefore, expect that in such situations quiet communication would evolve. Although I know of no clear examples, loud, competitive signals would be more obvious to both animals and observers. There may be many social signals based on interesting sensory biology that are relatively quiet and difficult to detect and are yet to be discovered in the context of cooperating animals.

## Summary and Conclusions

My review of the literature in behavioral ecology has been selective and cursory (see Krebs and Davies 1978, 1981, 1984, for extensive reviews). Hopefully, it has demonstrated that animals make many decisions that must be based on relatively subtle environmental information. Thus they do not merely discriminate food from nonfood items, but decide which to consume among the available food types; do not merely select a member of the opposite sex of their species for mating, but choose the mate of highest quality or with the best territory; do not merely decide whether to fight for a resource, but base their decision on its value.

Since all these decisions must be based on information that comes through the senses, sensory biology plays a central role. It seems likely that, if sensory biologists would examine some of the implications of the sophisticated decisions animals make while foraging, avoiding predators, and engaging in social interactions, many new phenomena would be revealed.

In addition, increased information about the sensory biology of animal decisions will prove valuable to behavioral ecology. All animal behavior must be understood within the constraints placed on the animal by its structure and abilities. For example: if an animal cannot discriminate between two types of prey, it cannot be expected to eat one and ignore the other; if it cannot recognize its kin, kin selection cannot operate. Increased knowledge of the mechanisms underlying behavior, including the sensory ones, will be very important to developments in behavioral ecology.

*Acknowledgments.* The blue jay research reported in this chapter was supported by NSF grant BNS-8102335, and the preparation of the manuscript was supported by NSF grant BNS–8418721. I thank Arthur N. Popper, John Janssen, and Sonja Yoerg for their thoughtful comments on an earlier version of this chapter.

## References

Anderson, O. (1984) Optimal foraging by largemouth bass in structured environments, Ecology, 65:851–861.

Andersson, M. (1982) Female choice selects for extreme tail length in a widow bird, Nature, 299:818–820.

Axelrod, R. and Hamilton, W.D. (1981) The evolution of cooperation, Science, 211:1390–1396.

Bateson, P. (1982) Preferences for cousins in Japanese quail, Nature, 295:236–237.

Caraco, T. (1979) Time budgeting and group size: a theory, Ecology, 60:611–617.

Cerri, R.D. and Fraser, D.F. (1983) Predation and risk in foraging minnows: balancing conflicting demands, Am. Nat, 121:552–561.

Charnov, E.L. (1976) Optimal foraging: the marginal value theorem, Theor. Popul. Biol., 9:129–136.

Croze, H.J. (1970) Searching image in carrion crows, Z. Tierpsychol. (Supp.), 5:1–85.

Dawkins, M. (1971) Perceptual changes in chicks: another look at the "search image" concept, Anim. Behav., 19:566–574.

Dawkins, R. (1976) The Selfish Gene, Oxford University Press, Oxford.

Dawkins, R. (1982) The Extended Phenotype, W.H. Freeman, Oxford.

Derby, C.D. and Atema, J. (1981) Selective improvement in responses to prey odors by the lobster (*Homarus americanus*), J. Chem. Ecol. 6:1073–1080.

Derby, C.D. and Atema, J. (1982) The function of chemo- and mechanoreceptors in lobster (*Homarus americanus*) feeding behavior, J. Exp. Biol., 98:317–327.

Dill, L.M. and Fraser, A.H.G. (1984) Risk of predation and the feeding behavior of juvenile coho salmon (*Oncorhynchus kisutch*), Behav. Ecol. Sociobiol., 16:65–71.

Dowds, B.M. and Elwood, R.W. (1983) Shell wars: assessment strategies and the timing of decisions in hermit crab shell fights, Behav. 85:1–24.

Dowds, B.M. and Elwood, R.W. (1985) Shell wars. II. the influence on relative size on decisions made during hermit crab shell fights, Anim. Behav., 33:649–656.

Elner, R. and Hughes, R. (1978) Energy maximization in the diet of the shore crab *Carcinus maenas,* J. Anim. Ecol., 47:103–116.

Falls, J.B. (1982) Individual recognition by birds, in Acoustic Communication in Birds, vol. 1, Kroodsma, D. and Miller, E. (eds.), Academic Press, New York, pp. 237–278.

Fraser, D.F. and Emmons, E.E. (1984) Behavioral response of blacknose dace (*Rhinichthys atratulus*) to varying densities of predatory creek chub (*Semotilus atromaculatus*), Can. J. Fish. Aquat. Sci., 41:364–370.

Gould, S.J. and Lewontin, R.C. (1979) The spandrels of San Marco and the Panglossian paradigm: a critique of the adaptionist programme, Proc. R. Soc. Lond. B Biol. Sci., 205:581–598.

Gray, R. (1987) Faith and foraging: a critique of the "paradigm argument from design," in Foraging Behavior, Kamil, A.C., Pulliam, H.R., and Krebs, J.R. (eds.), Plenum, New York.

Griffin, D.R. (1958) Listening in the Dark, Yale University Press, New Haven.

Hamilton, W.D. (1964) The genetical evolution of social behavior, J. Theor. Biol. 7:1–52.

Kamil, A.C. and Sargent, T.D. (1981) Foraging Behavior: Ecological, Ethological and Psychological Approaches, Garland STPM Press, New York.

Kamil, A.C. and Yoerg, S. (1982) Learning and foraging behavior, in Perspectives in Ethology, vol. 5, Bateson, P.P.G. and Klopfer, P.H. (eds.), Plenum, New York, pp. 325–364.

Kamil, A.C., Lindstrom, F., and Peters, J. (1985) Foraging for cryptic prey by blue jays. I. The effects of travel time, Anim. Behav. 33:1068–1079.

Keeton, W.T. (1974) The orientational and navigational basis of homing in birds. Adv. Study Behav., 5:47–132.

Krebs, J.R. and Davies, N.B. (1978) Behavioural Ecology: An Evolutionary Approach, Sinauer Associates, Sunderland, Mass.

Krebs, J.R. and Davies, N.B. (1981) An Introduction to Behavioural Ecology, Blackwell Scientific Publications, Oxford.

Krebs, J.R. and Davies, N.B. (1984) Behavioural Ecology: An Evolutionary Approach, 2d ed., Sinauer Associates, Sunderland, Mass.

Krebs, J.R. and Dawkins, R. (1984) Animal Signals: Mind-reading and Manipulation, in Behavioural Ecology: An Evolutionary Approach, 2d ed., Krebs, J.R. and Davies, N.B. (eds.), Sinauer Associates, Sunderland, Mass., pp. 380–402.

Krebs, J.R., Erichsen, T.J., Webber, M.I., and Charnov, E.L. (1977) Optimal prey selection in the great tit (*Parus major*), Anim. Behav., 25:30–38.

Krebs J.R., Stephens, D., and Sutherland, W. (1983) Perspectives in optimal foraging, in Perspectives in Ornithology, Brush, A.H. and Clark, G.A., Jr. (eds.), Cambridge University Press, Cambridge, pp. 165–215.

Lehrman, D.S. (1974) Can psychiatrists use ethology? In Ethology and Psychiatry, White, N.F. (ed.), University of Toronto Press, Toronto, pp. 187–196.

Lima, S.L., Valone, T.J., and Caraco, T. (1985) Foraging efficiency–predation risk trade-off in the grey squirrel, Anim. Behav., 33:155–165.

MacArthur, R.H., and Pianka, E.R. (1966) An optimal use of a patchy environment, Am. Nat., 100:603–609.

Maynard Smith, J. (1978) Optimization theory in evolution, Annu. Rev. Ecol. Syst., 9:31–56.

Milinski, M. and Heller, R. (1978) Influence of a predator on the optimal foraging behavior of sticklebacks (*Gasterosteus aculeatus*), Nature, 275:642–644.

Myrberg, A.A., Jr. and Riggio, R.J. (1985) Acoustically mediated individual recognition by a coral reef fish (*Pomacentrus partitus*), Anim. Behav., 33:411–416.

Nakatsuru, K. and Kramer, D.L. (1982) Is sperm cheap? Limited male fertility and female choice in the lemon tetra (Pisces, Characidae), Science, 216:753–755.

Orians, G.H. (1969) On the evolution of mating systems in birds and mammals, Am. Nat., 103:589–603.

Parker, G.A. (1974) Assessment strategy and the evolution of fighting behaviour, J. Theor. Biol., 47:223–243.

Parker, G.A. (1984) Evolutionarily stable strategies, in Behavioural Ecology: An Evolutionary Approach, 2d ed., Krebs, J.R. and Davies, N.B. (eds.), Blackwell Scientific Publications, Oxford, pp. 30–61.

Pietrewicz, A.T. and Kamil, A.C. (1977) Visual detection of cryptic prey by blue jays (*Cyanocitta cristata*), Science, 195:580–582.

Pietrewicz, A.T. and Kamil, A.C. (1979) Search image formation in the blue jay (*Cyanocitta cristata*), Science, 204:1332–1333.

Pleszczynska, W.K. (1978) Microgeographic prediction of polygyny in the lark bunting, Science, 201:935–937.

Pulliam, H. (1980) Do chipping sparrows forage optimally? Ardea, 68:75–82.

Pyke, G.H. (1984) Optimal foraging theory: a critical review, Annu. Rev. Ecol. Syst., 15:523–575.

Roeder, K.D. and Treat, A.E. (1961) The detection and evasion of bats by moths, Am. Sci., 49:135–148.

Sargent, T.D. (1976) Legion of Night, University of Massachusetts Press, Amherst.

Schoener, T.W. (1987) A brief history of optimal foraging ecology, in Foraging Behavior, Kamil, A.C., Pulliam, H.R., and Krebs, J.R. (eds.), Plenum, New York.

Sherman, P.W. (1977) Nepotism and the evolution of alarm calls, Science, 197:1246–1253.

Sih, A. (1980) Optimal behavior: Can foragers balance two conflicting demands? Science, 210:1041–1043.

Taborsky, M. (1984) Broodcare helpers in the cichlid fish *Lamprologus brichardi:* their costs and benefits, Anim. Behav, 32:1236–1252.

Tinbergen, L. (1960) The natural control of insects in pinewoods. I. Factors influencing the intensity of predation by songbirds, Arch. Neerl. Zool., 13:265–343.

Verner, J. (1964) Evolution of polygamy in the long-billed marsh wren, Evolution, 18:252–261.

Verner, J. and Willson, M.F. (1966) The influence of habitats on mating systems of North American passerine birds, Ecology, 47:143–147.

von Frisch, K. (1954) Dancing Bees: An Account of the Life and Senses of the Honey Bee, translated by Dora Ilse, Methuen and Company, London.

Werner, E.E. and Hall, D.J. (1974) Optimal foraging and size selection of prey by the bluegill sunfish (*Lepomis macrochirus*), Ecology, 55:1042–1052.

Werner, E.E. and Hall, D.J. (1976) Niche shifts in sunfishes: experimental evidence and significance, Science, 191:404–406.

Werner, E.E. and Mittelbach, G. (1981) Optimal foraging: field tests of diet choice and habitat switching, Am. Zool. 21:813–829.

Werner, E.E., Mittelbach, G., and Hall, D.J. (1981) The role of foraging profitability and experience in habitat use by the bluegill sunfish, Ecology, 62:116–125.

Werner, E.E., Mittelbach, G., Hall, D.J., and Gilliam, J. (1983) Experimental tests of optimal habitat use in fish: the role of relative habitat profitability, Ecology, 64:1525–1539.

Wicksten, M.K. (1983) Camouflage in marine invertebrates, Mar. Biol. Annu. Rev. 21:177–193.

Wilson, E.O. (1975) Sociobiology, Harvard University Press, Cambridge.

Wu, H.M., Holmes, W.G., Medina, S.R., and Sackett, G.P. (1980) Kin preference in infant *Macaca nemestrina,* Nature, 285:225–227.

# Chapter 8

# Sensory Performance, Behavior, and Ecology of Fish

JOHN H.S. BLAXTER*

The interpretation of sensory performance requires some knowledge of the behavior of animals, the quality and amplitude of stimuli to which they respond under different conditions, the stimuli that they may ignore as having no "meaning," and the effect of motivation and habituation. Furthermore, the interface of sensory physiology with behavior, and so with ecology, is an essential area of study for explaining the mechanisms that regulate the responses of animals to their environment and their distribution in space and time.

This chapter deals with certain common themes that apply to vision, mechanoreception, and chemoreception of fish and their relation to behavior and ecology. These themes are (1) thresholds, (2) problems of background "noise", (3) directional perception, (4) enhancement of sensory awareness, and (5) behavior controlled by multiple stimuli. These themes are picked out of a much wider scenario; important aspects such as electroreception and polarized light reception are omitted. Some of these are dealt with in other chapters.

## Thresholds

The characteristics of stimuli and other factors that influence the sensitivity of animals are given in Table 8.1. They may operate at the receptor level and the level of the CNS, with some overlap. Because the receptors themselves have different sensitivities, as do individual animals, an intensity vs. response curve is usually sigmoid for a population of receptors (a sense organ) and a population of animals. At the lowest intensity the most sensitive receptors (or animals) respond, with a progressive "recruitment" of receptors or animals as the intensity increases. It may require an intensity increase of 2 log units (× 100) or more before the least sensitive receptors or animals respond. Thus the definition of the term *threshold* depends on several possible criteria, but in many cases it

* Dunstaffnage Marine Research Laboratory, Oban, Argyll PA34, 4AD, U.K.

Table 8.1. Stimulus Characteristics and Other Factors Influencing Thresholds

| | | |
|---|---|---|
| Wavelength or spectral composition | Receptor level | |
| Chemical structure | Receptor level | |
| Onset rise time | Receptor level | |
| Adaptation | Receptor level | CNS level |
| Duration | Receptor level | CNS level |
| Background noise | Receptor level | CNS level |
| Fatigue | | CNS level |
| Habituation | | CNS level |
| Motivational state (e.g., hunger or fear) | | CNS level |

refers to the amplitude of a stimulus at which 50% of the receptors (or animals) respond.

Thresholds also vary with wavelengths of light or sound and with chemical structure. Action spectra (threshold vs. stimulus quality relationships) show peaks or plateaux of maximum sensitivity to certain wavelengths or chemical compounds. The threshold is lower with sudden stimulus onset and stimuli of long duration and is higher during conditions of fatigue or high background noise (used in the sense of background stimuli that are often continuous and either contain no information or are of no value to the organism). Habituation (that is, the waning of a response caused by repeated stimulation in the absence of reinforcement) and low motivational state also increase the threshold. The direction of a signal is also important for the threshold of directional receptors. Few signals are pure, and so the bandwidth of a signal may influence the threshold, because receptors or the CNS may integrate intensity over a wide spectrum.

The human element also intrudes into the assessment of thresholds, which depend on the technique of measurement. Electrophysiological techniques are likely to yield the lowest thresholds, since they merely show whether information is passing between the receptor and the CNS. Such information may not be "meaningful" to the animal, and a higher intensity may be needed to elicit a behavioral response. Such behavior may be spontaneous or induced by training. Aversion conditioning may give a different threshold from behavior based on a reward, such as food, in response to which hunger and satiation play a role. Unconditioned behavior, such as a fright response, may give a different threshold than attraction to a food source.

Variations in action spectra as a result of different techniques are shown for vision in the herring, *Clupea harengus,* in Figure 8.1 and for hearing in the goldfish, *Carassius auratus,* in Figure 8.2. The herring shows differences in overall, as well as wavelength-dependent, sensitivity, depending on the criteria adopted. The audiograms of the goldfish have been grouped into three categories based on operant (usually reward) and classical (usually aversion) conditioning,

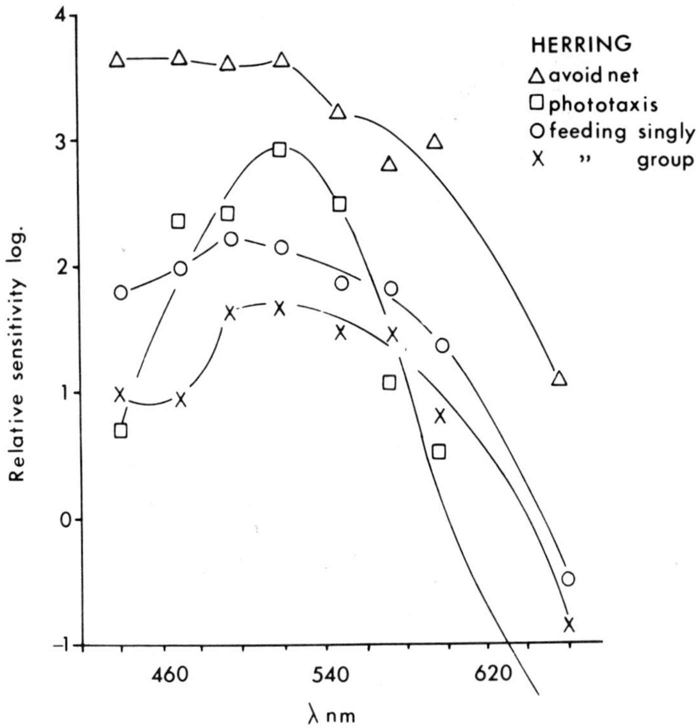


Figure 8.1. Spectral sensitivity curves of the herring, *Clupea harengus*, obtained by different behavioral techniques (from Blaxter 1964).

and on electrophysiological recordings from the medulla. The two figures indicate the influence that technique can have on the measurement of thresholds.

The ecological implications of thresholds are best followed up with vision and light-dependent behavior. In Figure 8.3 the range of light intensity over which feeding and schooling fall from a high level (near 100%) to a low one (near 0%) is shown for four groups of fish. Also shown are various other thresholds, especially for the process of dark adaptation obtained by histological examination of retinomotor responses of the rods, cones, and masking pigment in the retina. In general, the fall-off in behavioral performance matches the onset of dark adaptation. One may assume that light-adapted fish use cone vision to feed and school; at dusk the light intensity falls, the fish become dark adapted and cease feeding, and the schools disperse. At low light levels, rod-mediated vision may be involved in some phototactic responses, depth holding, and responses to movement, for example, of predators. At still lower levels, all vision ceases. During further discussion, it will be assumed that the change between light and dark adaptation and the cessation of light-dependent behavior occur at 0.05 $\mu W/cm^2$, as shown by the arrows in Figure 8.3. The lowest threshold for vision occurs between $10^{-5}$ and $10^{-8}$ $\mu W/cm^2$, and a theoretical maximum for deep-sea fish may be $10^{-9} - 10^{-10}$ $\mu W/cm^2$ (Clarke and Denton 1962). A summary

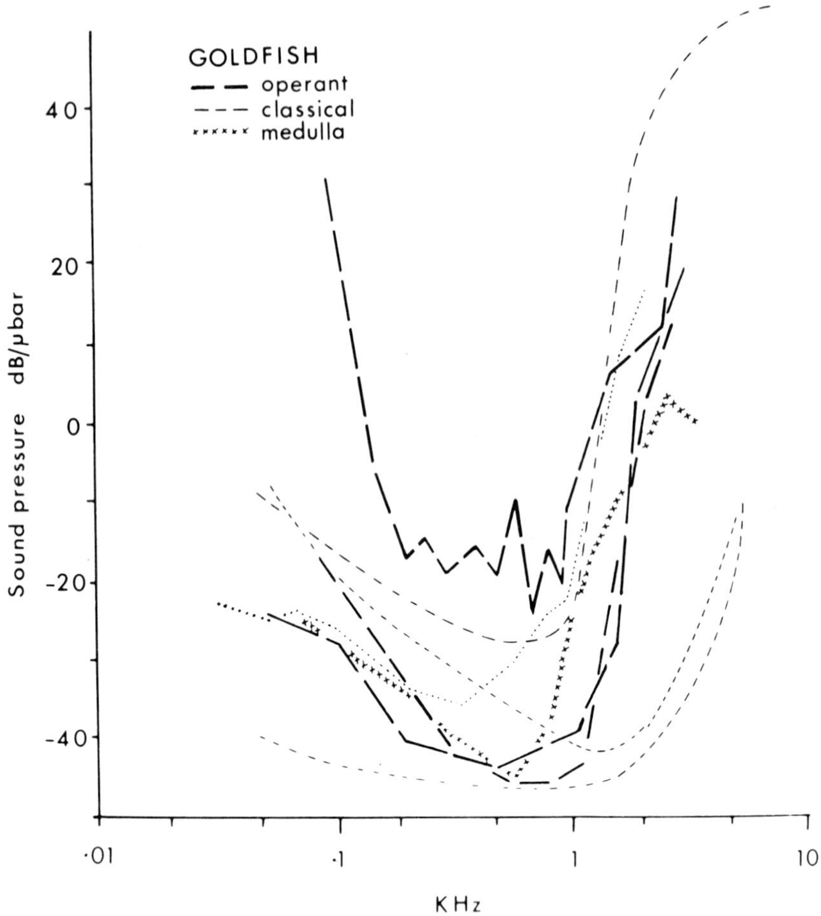


Figure 8.2. Audiograms of the goldfish, *Carassius auratus,* obtained by different techniques: *heavy lines,* operant conditioning; *thin lines,* classical conditioning; *medulla,* recording from brain stem (Redrawn from Hawkins 1981).

of the thresholds and their relation to astronomical conditions is given in Figure 8.4.

Clearly, many species of fish depend on sense organs other than the eye, but, where vision and light-dependent behavior predominate, the implications of these threshold values are far reaching. The time available for these behaviors will depend on day length and thus on season and latitude. In Figure 8.5 the number of hours per 24-hr day during which the irradiance remains above 0.05 $\mu W/cm^2$ at the water surface is plotted as isopleths against season and latitude. At high latitudes the seasonal effect is extreme, so that, in the polar summer, the fish have effectively a 24-hr day. In the polar winter, however, the light level is never high enough for light-dependent behavior to take place. In low latitudes the day length remains rather constant, nearly 13–14 hr, throughout

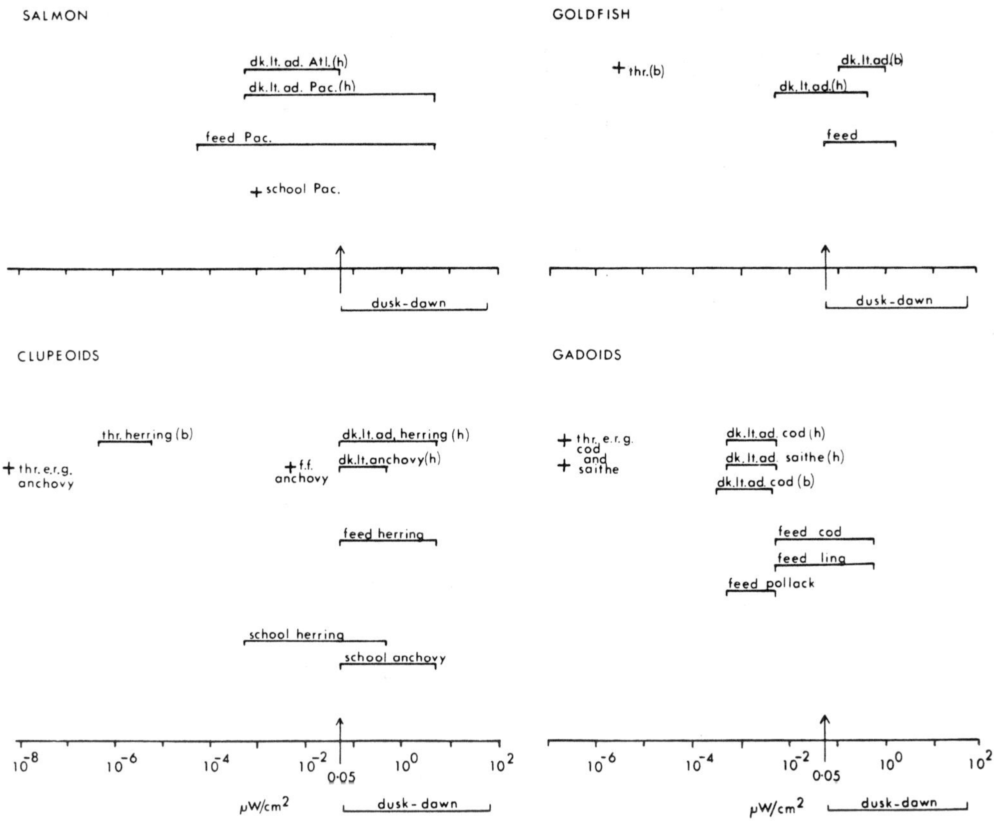


Figure 8.3. The range of light intensity (log. scale) over which various light-dependent behavioral and physiological functions cease. The figure is divided into four groups of fish: salmon, both Atlantic (*Atl.*), *Salmo salar,* and Pacific (*Pac.*), *Oncorhynchus* spp.; clupeoids, the herring, *Clupea harengus,* and the anchovy *Engraulis encrasicholus;* the goldfish, *Carassius auratus;* gadoids, the cod *Gadus morhua,* the coalfish, or pollack, *Pollachius virens,* and the ling *Molva* sp. The change of light intensity at dusk and dawn is given. Key: *dk.lt.ad.*, dark/light adaptation; (*h*), by histology of retina; (*b*), by behavioral techniques; *thr.*, threshold; *e.r.g.*, obtained by electroretinogram; *f.f.*, obtained by inflection in flicker fusion frequency relationship at different light intensities; *vertical arrow* at 0.05 $\mu W/cm^2$, average threshold used in further calculations (Data from Blaxter 1970; Tavolga 1977; and Anthony 1981).

the year. These differences in day length also influence the time and duration of spawning. There is a strong interaction with temperature, with direct effects on the fish through activity and energetics and with indirect effects through the timing of plankton blooms.

Below the water surface irradiance falls off exponentially as a result of the scattering and absorption of light. This further reduces the effective day length. Two extremes of light penetration (Sargasso and Baltic seas) are shown in Figure 8.6. Note the different scales of the ellipsoidal figures showing changes in the

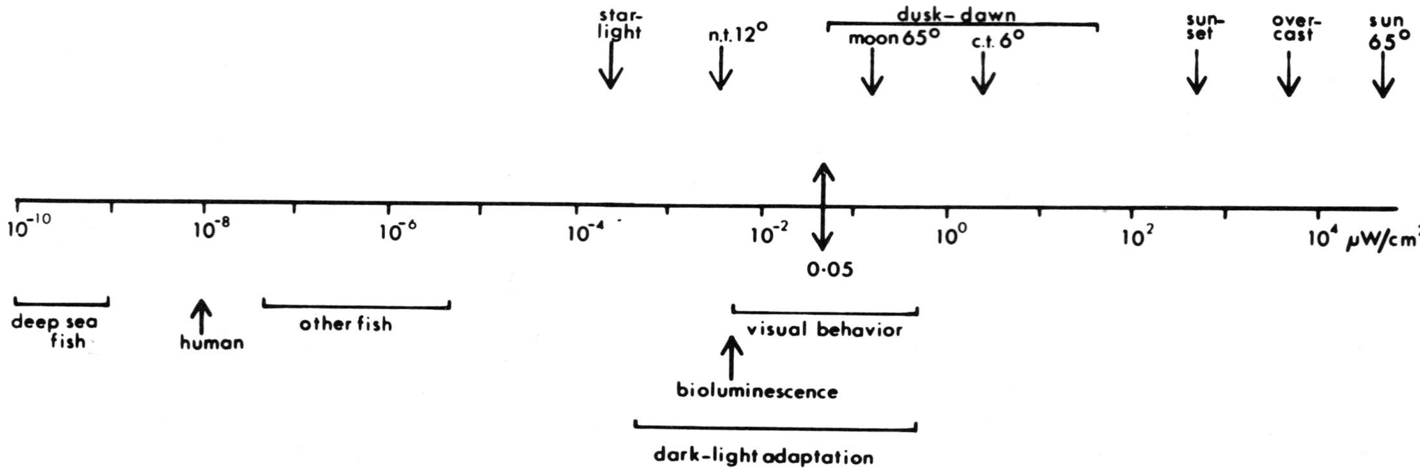


Figure 8.4. A summary of thresholds for various functions related to light intensity (log. scale) and astronomical conditions: *n.t.*, nautical twilight sun 12 degrees below horizon; *c.t.*, civil twilight sun 6 degrees below horizon, sun and moon altitude 65 degrees; 0.05 $\mu W/cm^2$, average threshold used in further calculations (see Fig. 8.5 and Table 8.2).

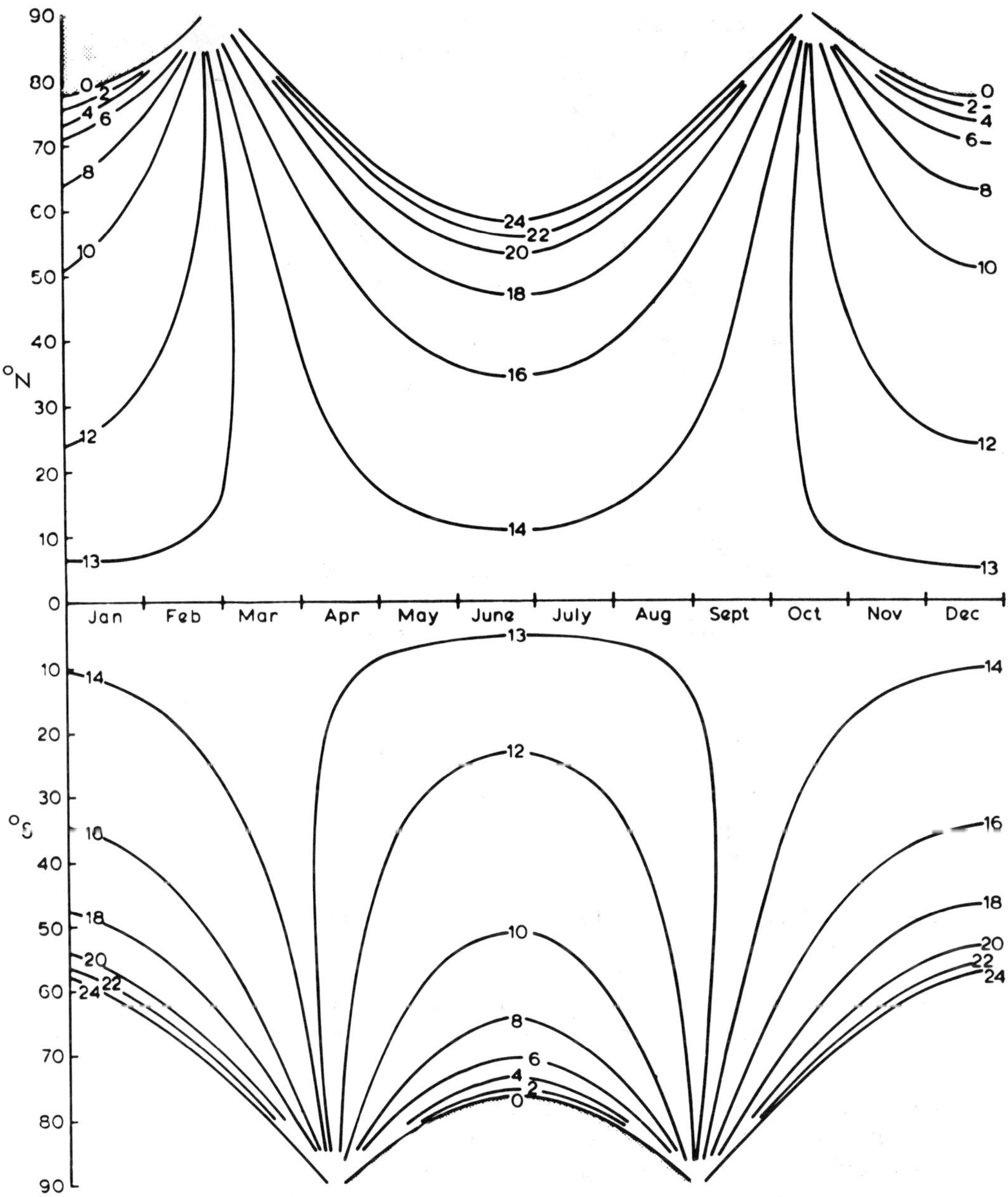


Figure 8.5. The number of hours per 24-hour day at different latitudes and seasons during which the irradiance at water surface is greater than 0.05 μW/cm$^2$ (From Blaxter 1970, by permission of Wiley Interscience.)

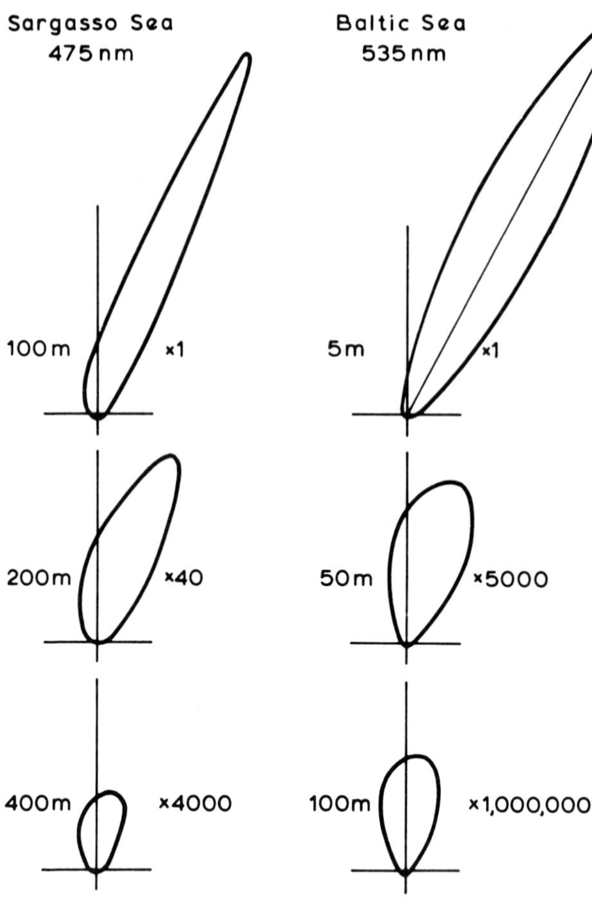


Figure 8.6. Polar diagrams of radiance distribution in arbitrary units for Sargasso Sea and northern Baltic Sea, showing the change with increase of depth to final state, where it is nearly symmetrical about the vertical. Ellipse axis on Baltic 5m diagram shows direction of maximum radiance. Note the difference in scales and wavelength of maximum penetration (From Jerlov 1970, by permission of Wiley Interscience).

polar distribution of radiance with depth. The wavelength of maximum penetration is shifted towards the red end of the spectrum in inshore waters, and light becomes more vertically distributed at greater depths. Using Jerlov's (1976) data on irradiance absorption in different types of seawater, it is possible to calculate the depth at which the irradiance, at noon, reaches the absolute threshold and the threshold for light-dependent behavior (Table 8.2). Even in the clearest oceanic (type I) water, light-dependent behavior cannot occur below 288 m unless mediated by bioluminescence. In coastal waters (200 m or less) the constraints can be severe, and only fish above 18–82 m depth will be at irradiance levels of more than 0.05 $\mu W/cm^2$. The absolute threshold for deep-sea fish (see Clarke and Denton 1962) is found at depths down to 700 m in oceanic waters, and the

Table 8.2. Maximum Depths in Meters for Vision (Solar Altitude 90 Degrees for Oceanic Water and 45 Degrees for Coastal Water)

| Water Type (Increasing Turbidity Moving Down Table) | Depth for Light-Dependent Behavior Applying a Threshold of 0.05 μW/cm² | Depth Applying Theoretical Deep-Sea Fish Threshold of $10^{-10}$ μW/cm² | Depth Applying Inshore Fish Threshold of $5 \times 10^{-7}$ μW/cm² |
|---|---|---|---|
| Oceanic | | | |
| I | 288 | 700 | |
| IA | 234 | 573 | |
| IB | 200 | 490 | |
| II | 164 | 402 | |
| III | 95 | 231 | |
| Coastal | | | |
| 1 | 82 | | 151 |
| 3 | 48 | | 87 |
| 5 | 28 | | 51 |
| 7 | 22 | | 40 |
| 9 | 18 | | 33 |

*Source:* Water types and irradiance attenuation data from Jerlov (1976).

absolute thresholds for other fish are found at depths down to 33–151 m inshore. A large proportion of the available oceanic space is thus effectively in darkness, even at noon. Fish have evolved in these dark habitats by developing bioluminescence or upgrading the relative roles of mechano- and chemoreception.

## "Noise"

Fish are subjected to "noise," or unwanted background stimulation, against which they have to discriminate stimuli that are important to them. Generally, noise causes increase of thresholds and reduces awareness of surroundings. Probably the most severe forms of noise are light-scattering and acoustical sea noise from wind-induced wave action. It is not clear whether noise of biological origin is ever completely superfluous. For example, in a school, information on the swimming of the nearest neighbors is obviously most important. Are stimuli from more distant fish merely a source of confusion, to be analyzed and discarded, or can they be used for early warning of large-scale changes in behavior, the presence of distant predators, or the approach to a source of food? We cannot answer such questions at this stage.

The general level of illumination as well as the size of an object and its color, pattern, and brightness in relation to the background determine the maximum sighting distance. The degree of turbidity also has a major influence, in that light scattering by particles in the water reduces the clarity of objects as viewed by the fish. Measured and calculated values rarely exceed 5–6 m (Protasov 1968;

Hester 1968) though Anthony (1981) calculates a maximum of 20 m for the cod *Gadus morhua* in conditions of high contrast and low turbidity. In the deep sea, where turbidity is especially low, bioluminescence can penetrate for relatively great distances. Nicol (1969) suggests a distance of 10–16 m for bathypelagic fish.

The influence of sea noise has been assessed in a number of fish species by measuring the threshold-to-noise (or signal-to-noise) ratios. This is the extent to which the amplitude of sound has to be raised above the background level (at a particular frequency) to be heard. In general, it is of the order of 20 dB, a factor of 10 (Buerkle 1968, 1969; Chapman 1973; Chapman and Johnstone 1974; Tavolga 1974, 1982). It is less if the source of the noise and the test signal have an angular separation of 10 degrees or more in azimuth or elevation. If noise and signal are separated in frequency the masking effect of the noise also decreases. (The *critical masking bandwidth* is defined as the frequency band of noise having a power equal to the signal at threshold.) Critical masking bandwidths increase with the frequency of the signal and vary with species; narrow critical bands are usually associated with fish having high sensitivity and a wide frequency range (see Chapter 5, 27, and 28).

One of the problems of determining acoustical thresholds in captivity is that background noise is always present, either in the laboratory or in sea cages. This may, in part, explain the great difference in audiograms reported for the goldfish (Fig. 8.2). In Figure 8.7, sound levels at various sea states are shown in relation to the audiogram for the cod obtained under quiet conditions. If the signal-to-noise (S/N) ratio obtained by experiment is added to sea state 3, it will be seen that the sound levels obtained exceed the audiogram thresholds. Thus, even moderate sea conditions can impair the sensitivity of cod and mask their awareness of their acoustical surroundings.

The chemoreceptors of fish are known to be affected by effluents such as heavy metals, especially copper, mercury, cadmium, and zinc, and by detergents, insecticides, and pulp mill effluents (Brown et al. 1982). At high concentrations these substances may be lethal; at lower concentrations the fish may make behavioral avoidance responses. The olfactory mucosa may also be damaged. It is not clear whether these effluents can mask the reception of chemicals of biological origin present in the normal life of fish. Although they may impair feeding or the action of pheromones, this may be caused by their destructive effect on the olfactory mucosa, rather than by a true (reversible) masking of one odor by another in which the receptor threshold is raised.

## Directional Perception

### Light

The perception of the source of stimulus is essential for all types of intra- and interspecific behavior. Within the sighting distance, fish can move directly towards

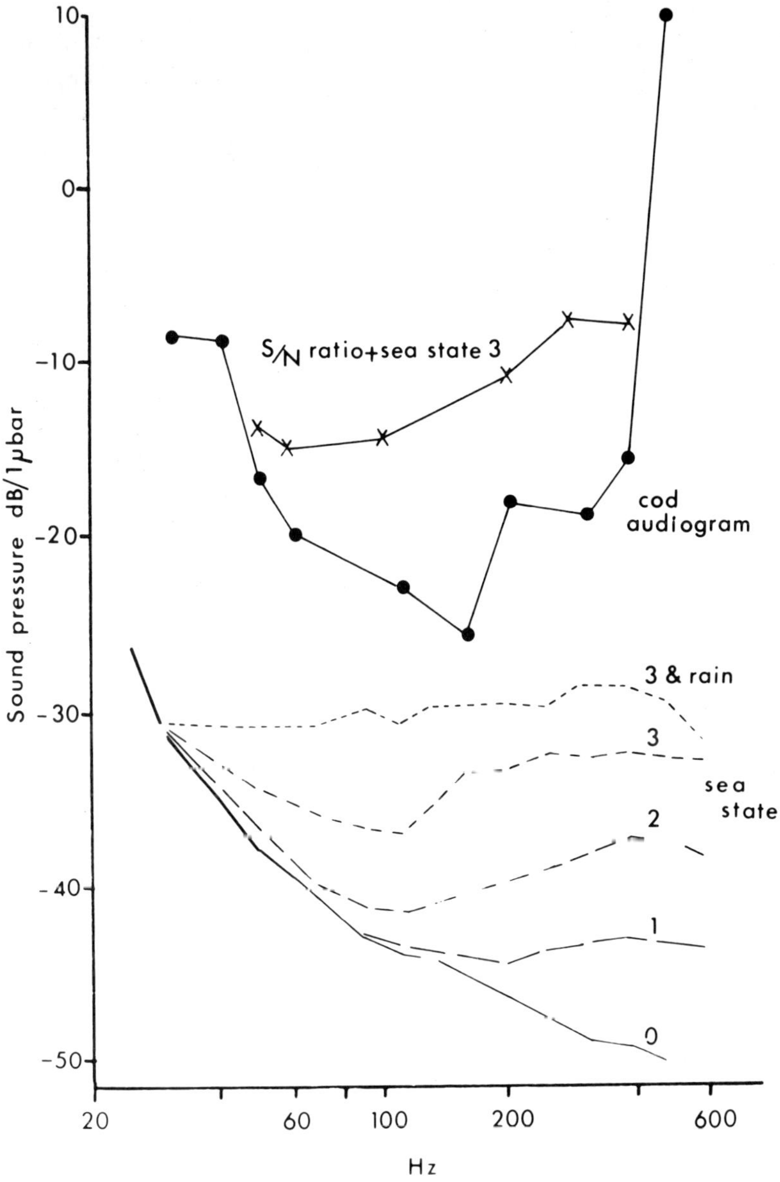


Figure 8.7. Ambient noise spectra for shallow water site at a depth of 15 m in Loch Torridon, Scotland, showing the effect of wind and weather (from Hawkins and Myrberg 1983) as related to the audiogram of cod (from Chapman and Hawkins 1973; Myrberg 1980b) with an additional relationship where the signal-to-noise (S/N) ratio is added to the spectral sound levels for sea state 3. Sea state defined on Beaufort scale in any nautical almanac.

or away from an object and, in many cases, probably make an estimate of its range, using stereoscopic vision or information on its size and clarity. The use of background light from the sky and sun as a reference presents different problems. It is well established that some species can use the sun as a reference point to make "sun compass" responses (see Quinn 1984; Stabell 1984, for recent reviews). The fish can not only swim at an angle to the sun, they can also allow for its movement through the day to maintain a constant compass heading.

Under water, azimuthal information from the sun is reduced by refraction at the surface, so that, even if the sun is at the horizon, the refracted angle is about 49 degrees to the vertical. Furthermore, especially in turbid water, the distribution of light becomes increasingly vertical with depth (Fig. 8.6). What are the problems for fish at different depths in using the sun as a reference point for migratory behavior? How does the azimuthal distribution of light change with depth, and what ability do fish have to discriminate brightness, and so how accurately could they locate the direction of the sun?

Some of the best data available are those of Tyler (1957–1960) from the clear water of Lake Pend Oreille. The angular distribution of light at three depths with the sun at an altitude of 57 degrees is shown in Figure 8.8. The upper figure shows the horizontal distribution and the lower figure the distribution at an angle of 30 degrees to the vertical (20 degrees at the greater depths) where the intensity is maximal in the azimuth of the sun. The ability of a fish to tell

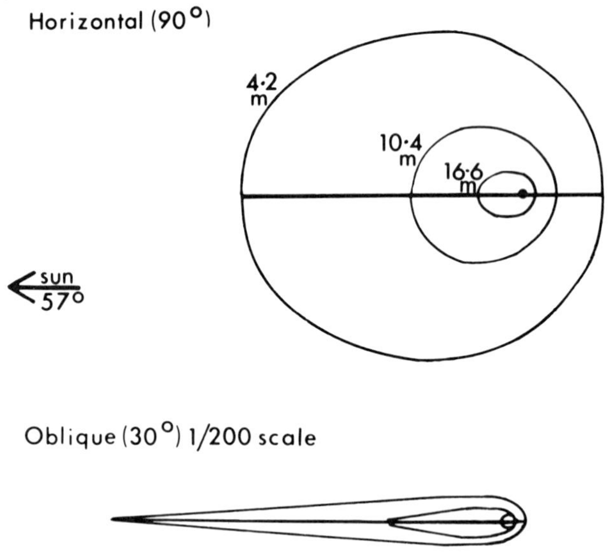


Figure 8.8. Polar diagrams of the radiance distribution in arbitrary units in the horizontal plane (*upper figure*) and at 30° to the vertical (at which the radiance in the direction of the sun was maximal, *lower figure*) at 4.2 m, 10.4 m, and 16.6 m below the surface of Lake Pend Oreille, Idaho (from Tyler 1957–60). Axes show direction of maximum and minimum radiance. Note the difference in scale of the *lower figure*. Sun altitude 57°.

the direction of the sun will depend on its visual axis (i.e., whether the fish looks horizontally or somewhat upwards) and its ability to discriminate the direction of maximum brightness. Clearly a fish looking upwards at 30 degrees to the vertical will be able to detect the sun's direction more accurately than one looking horizontally. With increasing depth, angular distribution becomes more symmetrical about the vertical, the sun appears more overhead, and its azimuth becomes more obscure.

The brightness discrimination of fish as determined by experiment is shown in Figure 8.9. A variety of techniques have been used, both behavioral and electrophysiological. In general, brightness discrimination improves at higher light reference levels. The Weber fraction, or the change in brightness discriminable as a percentage of the reference level (the so-called brightness discrimination limen) lies between 10% and 100% for several species and conditions, and these values will be applied to Tyler's data in Figure 8.8.

The angles discriminable from the sun's azimuth by fish with brightness limens of 10% and 100% looking horizontally (90 degrees) and in the direction of maximum brightness (20–30 degrees) are shown in Figure 8.10. Only near the surface and looking at the sun could fish detect its direction within 5–20 degrees. Bearing in mind that Lake Pend Oreille has unusually clear water, it seems unlikely that fish can use the sun as a reference point with much accuracy in more turbid inshore waters unless they are close to the sea surface.

It is not necessary to invoke an ability to swim at an angle to the sun to make good a particular compass course. A fish swimming towards the sun only in the morning will make a net southeasterly movement (in the northern hemisphere). Similarly, a fish swimming through the hours of daylight or only in the afternoon will make net southerly or southwesterly displacements, respectively (Fig. 8.11).

Evidence for fish making sun compass reactions under natural conditions is somewhat tenuous. For example, Winn et al. (quoted by Tesch 1975) displaced parrot fish *Scarus* spp. from their home range and marked them with small floats. They were found to return home only if the sun was visible. Groot (1965) studied the migrations of sockeye salmon, *Oncorhynchus nerka,* experimentally and during their migrations from lakes. He concluded that celestial navigation was being employed, probably from the sun and from polarized light, but that this species could switch to other clues in the absence of celestial information. Both Harden Jones (1968) and Quinn (1984) concluded that the return migration of Pacific salmon to the coast includes some degree of sun orientation, though other clues are probably available (see p. 223).

Brightness discrimination of a different sort must be associated with vertical migration, the regular migration of fish towards the sea surface at dusk and the seabed at dawn. The evidence strongly suggests that changes in light intensity not only trigger the movement, but control the rate of ascent and descent, in that the fish stay within a brightness "preferendum" (Blaxter 1975). Both the initial movement and the preferendum require that the fish remember a reference level of light intensity and move when the intensity changes from this level, either during the progression of dawn or dusk or by the fish moving up or down

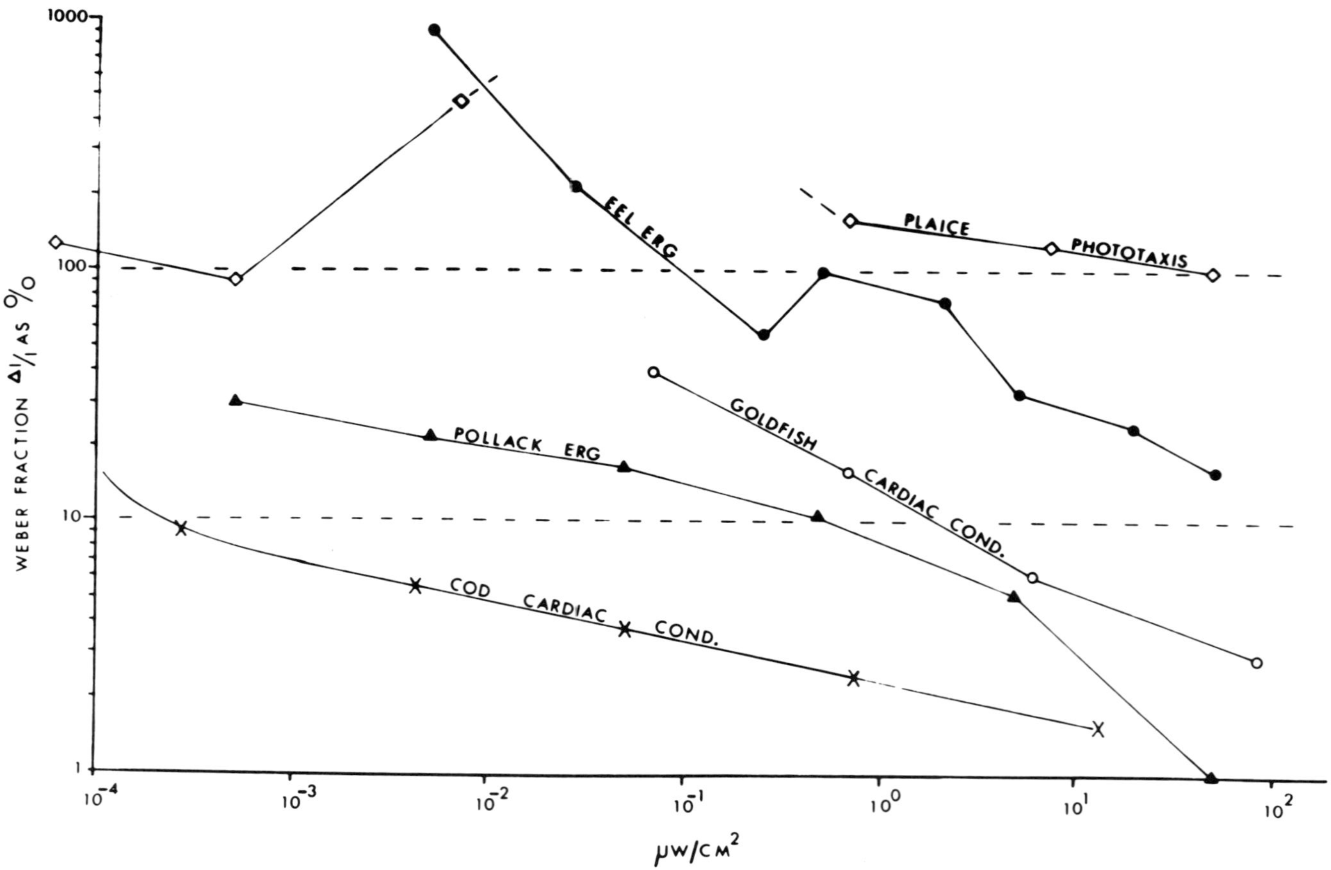


Figure 8.9. Brightness discrimination of white light as the Weber Fraction $\Delta I/I$ as percent (where $\Delta I$ is the increment of reference light intensity, $I$, discriminated) plotted against the reference intensity on a log. scale. Plaice, *Pleuronectes platessa* from Blaxter (1972), eel *Anguilla japonica* from Kobayashi (personal communication), goldfish *Carassius auratus* from Hester (1968), ''pollack'' (*Pollachius virens?*) from Protasov (1968), cod, *Gadus morhua* from Anthony (1981). Technique used is also given. ERG, electroretinogram; cardiac cond., cardiac conditioning. Note line for plaice is incomplete as no phototaxis occurs at intermediate intensities.

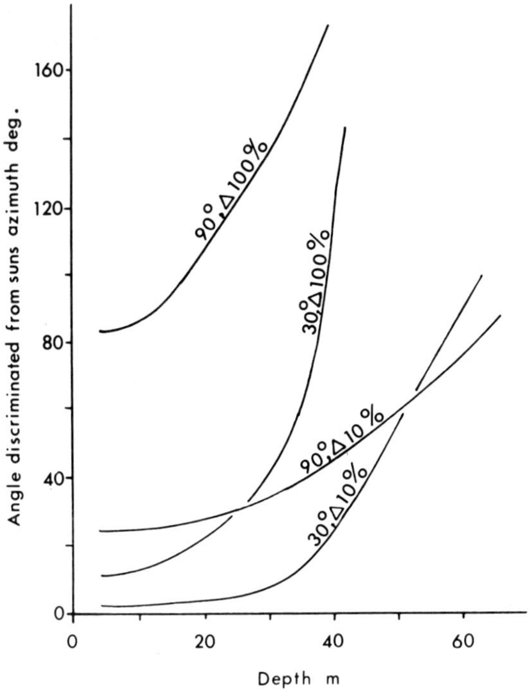


Figure 8.10. The accuracy with which fish with brightness discrimination limens (Δ) of 10% and 100% could locate the sun's azimuth expressed as the minimum angle discriminated from the true azimuth of the sun when looking in the direction of maximum radiance (30° to the vertical near the surface and 20° below about 40 m depth). Sun at 57° altitude in Lake Pend Oreille.

into brighter or dimmer light. Thus, in vertical migration the fish must discriminate a change in brightness over time, rather than make an instantaneous spatial discrimination to obtain the direction of the sun, as required in horizontal migration.

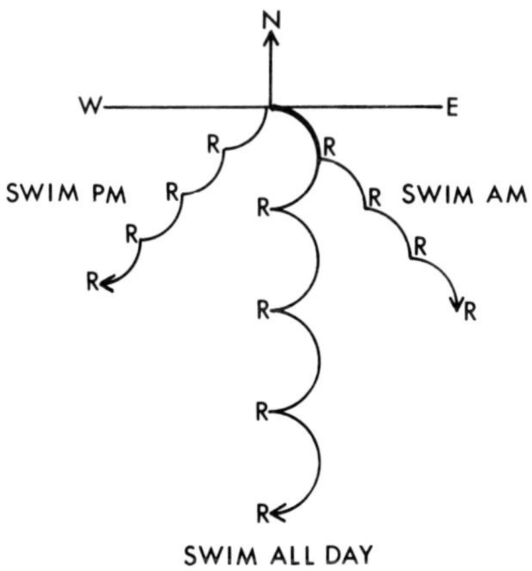


Figure 8.11. Fish swimming towards the sun it the morning (*AM*), during the daylight hours (all day), and in the afternoon (*PM*) and resting (*R*) at other times would make good net south-easterly, southerly, and southwesterly course (Adapted from Balchen 1977).

There are no experimental data available for temporal brightness discrimination. Using the instantaneous brightness limens of 10%–100% (Fig. 8.9), it is possible to calculate the distance that fish would have to move vertically to perceive a change in brightness in different types of seawater (Fig. 8.12). On the assumption that these limens are valid for temporal brightness discrimination, it is clear that fish can maintain a preferendum much more easily in turbid coastal waters, where changes in intensity of the order of 10%–100% occur over < ½–5 m.

## Sound

How fish scan a sound field to detect the direction of maximum amplitude (within the limits of their discrimination) is not known. Amplitude discrimination data are available for the goldfish (Fig. 8.13), and the discrimination limens are of the order of 10%–100%, as they are for brightness discrimination. Conditioning experiments in the sea show that cod can discriminate sound sources about 20 degrees apart in the horizontal plane and 16 degrees apart in the vertical plane and can make the "difficult" 180 degree discrimination between sounds coming from directly in front and directly behind them (Chapman 1973; Chapman and Johnstone 1974; Schuijf 1975; Schuijf and Buwalda 1975; Hawkins and Sand 1977).

The attracting and repelling effects of sound sources in the sea have been investigated by keeping fish in cages or by divers' observations of free-living fish, especially sharks (Table 8.3). The general conclusions are that sharks are attracted by low-frequency, broadband, irregular pulses. Most species are attracted from distances of only a few meters, and usually habituation is quite rapid if the stimuli are maintained. Fish may also become habituated to short-lived stimuli. Herring, for example, habituate to transient sounds that cause "fast-startle" (fright) responses. These are emergency responses to stimuli such as tail flips from predators or other herring that are too close. Continuous sound stimuli elicit nonhabituating, slower turns in herring. These sounds may be similar to tail beats of conspecifics, which are probably used to maintain interfish distances in the school, a situation in which habituation would be unacceptable.

## Chemicals

Various aspects of chemoreception are dealt with in the first two chapters. The directional perception of chemical sources by fish has been discussed by Kleerekoper (1969), Bardach and Villars (1974), Hara (1975), Atema (1980), and Jonsson (1980). Chemical stimulation often causes the general arousal of fish. In some species, such as elasmobranchs, anguillids, ictalurids, and salmonids, chemoreception plays a major role in orientation. The direction of the source may be obtained from comparison of bilateral intensity where the olfactory organs are well spaced, as in the hammerhead sharks (Sphyrnidae), where taste bands are dispersed over the body surface, or in sensory barbels, as in a catfish such as *Ictalurus* spp. More commonly, fish zigzag or adopt a more generalized search pattern to compare the intensity of an attractive chemical in space and time and

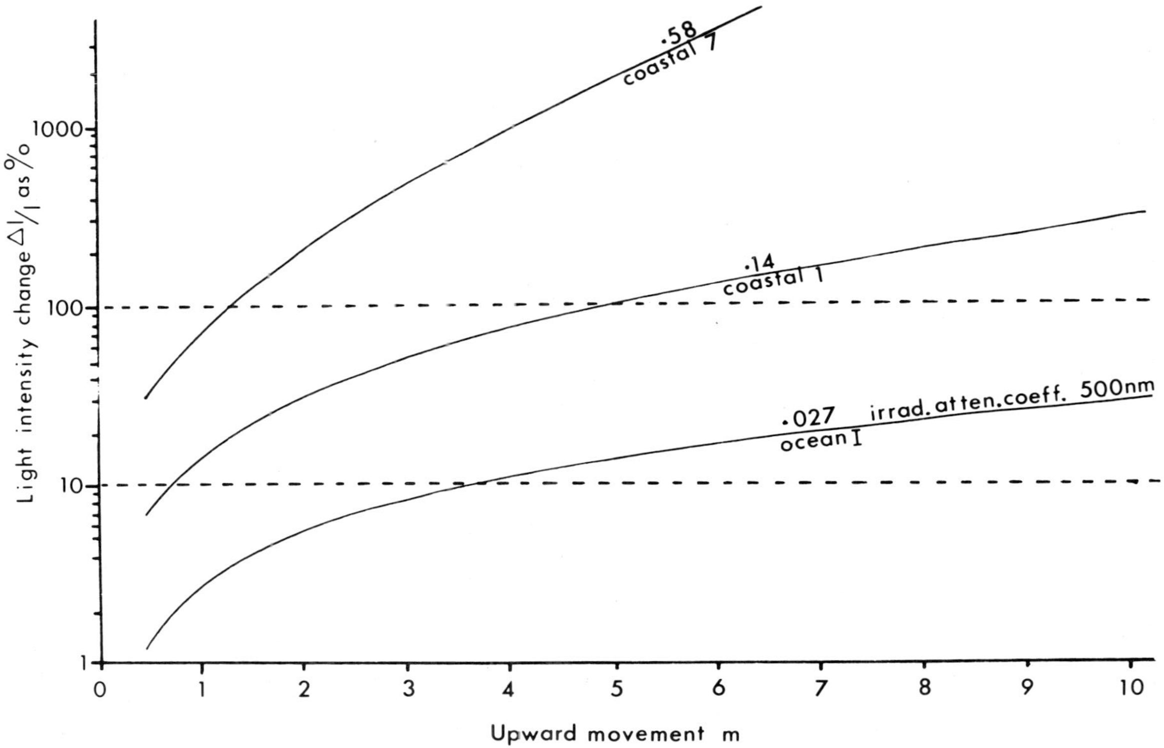


Figure 8.12. Light intensity (irradiance) increments as $\Delta I/I$, as percentage where $\Delta I$ = the change in irradiance during upward movements of different amounts related to reference irradiance ($I$) at the start of upward movement. Data are given for three types of seawater (Jerlov 1976; see also Table 8.2) with irradiance attenuation coefficients at a wavelength of 500 nm. *Horizontal dashed lines* show the range of brightness discrimination limens for white light from Figure 8.9. On the assumption that white light limens can be equated with irradiance change at 500 nm (near $\lambda_{max}$ of the rhodopsin for coastal fish) and on the assumption that the fish can memorize reference level $I$, it is possible to assess the extent of upward movement required before fish appreciate the change in irradiance.

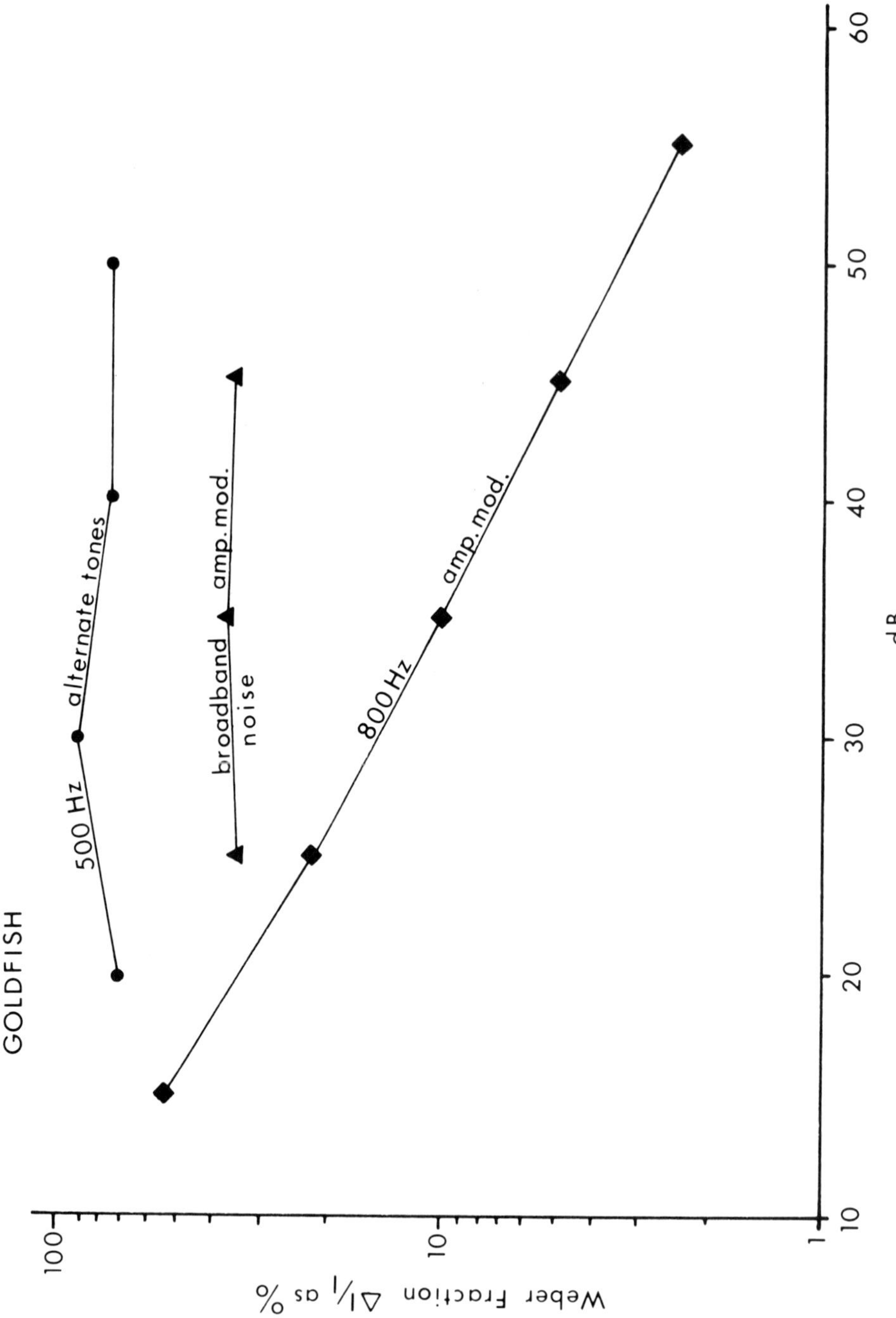


Figure 8.13. Amplitude discrimination limens for sound in goldfish *Carassius auratus*. Limens are given as Weber fraction $\Delta I/I$, as percentage of relative sound pressure level in dB. *Upper line* from Jacobs and Tavolga (1967); *lower lines* from Hall, Patricoski, and Fay (1981).

Table 8.3. Attraction or Repulsion by Sounds (Examples)

| Species | Conditions | Stimuli | Results | References |
|---|---|---|---|---|
| Menhaden | Cage in sea | 1 kHz–20 kHz (Continuous or amplitude modulated) | Move to area of higher intensity, but habituate | Moulton (1963) |
| Butterfish | Cage in sea | 2 kHz (Continuous or amplitude modulated) | Change distribution | |
| Various predators | Sea observations by TV | Pulses of 25–50 Hz | Attract some species | Richard (1968) |
| Lemon shark | Sea observations | Artificial and recorded biological sounds | Pulsed broadband sound attracts from 5-m range | Banner (1972) |
| Squirrelfish | Cage in sea, TV | Squirrelfish alarm calls | Attract from 2-m range | Popper, Salmon, and Parvulescu (1973) |
| Yellowtail, mackerel, jack | Sea observations by echo sounder | Playback of feeding sounds | Attracts | Maniwa (1976) |
| Silky shark, whitetip shark | Sea observations by diver | Pulses of 10–80 Hz | Attract by < 40-Hz broadband pulses; no response to pure tones | Myrberg, Gordon, and Klimley (1976) |
| Five shark species (four *Carcharhinus*) | Sea observations by diver | Broadband pulses of 25–100 Hz | Attract | Nelson and Johnson (1976) |
| Cod, saithe, lythe | Sea observations by diver and TV | 20–380 Hz, continuous or pulsed, broadband or pure tones | Attract by pure tones; pulsing probably better | Chapman (1976) |
| Saithe | Sea observations by TV | 150-Hz pure tone pulses | Attract from 80-m range | Olsen (1976) |
| Herring | Cage in sea, TV | 30 Hz–10 kHz from loudspeaker, sparker, and air gun | Repelled by low frequencies, but habituate | |
| Herring | Tanks | Transients 25–40 Hz | Repelled, but habituate | Blaxter, Gray, and Denton (1981) |
| Herring | Tanks | Continuously vibrating source of 20 Hz | Repelled; no habituation | Blaxter and Batty (1985) |

so move up the intensity gradient to the source. In a current, fish show rheotactic behavior and move upstream when they detect an attractive chemical.

Discrimination of intensity differences is required in all these conditions except rheotaxis. Only Johnsen and Teeter (1980) have measured discrimination limens. They trained the brown bullhead catfish *Ictalurus nebulosus* to discriminate different concentrations of an amino acid mixture passed across the taste receptors on the right and left barbels. The discrimination limens were similar to those for vision and hearing, 10%–30%, being lower at the low reference concentrations. Similarly Stewart, Bryant, and Atema (1979) found that a stimulus must be 5%–25% above an artificially raised background level to be detectable.

## The Enhancement of Sensory Awareness

Fish show a great range of specializations of their sensory systems to enhance their functions (Table 8.4). Perhaps this can largely be explained by the peculiar conditions of life in water. In particular, the effect of light scattering and absorption requires special adaptations of the eye to allow vision, especially in turbid waters and in the deep sea. The evolution of specialized retinae, and visual pigments matched to the wavelengths predominating underwater, improve sensitivity, whereas the light path to the retina may be adapted to improve contrast at the expense of sensitivity.

Because high-quality vision is impossible in so much of the deep ocean and in turbid inshore waters, the mechanoreceptors and chemoreceptors may assume great importance. The lateral line is peculiar to aquatic vertebrates (see Chaps. 22 and 23) and becomes particularly well developed in deep-sea fish. The Ostariophysi and clupeoids have evolved specialized sound pressure receptors to improve their hearing sensitivity and range. Chemoreception is enhanced by the increase in area of olfactory mucosa and proliferation of taste buds onto exploratory barbels or over the surface of the body.

Sensory awareness is also enhanced by the production of stimuli (Table 8.5). Stimuli such as visual images, swimming sounds, and excretory products, which fish produce incidentally, may be harmful by attracting predators. Such stimuli may also have evolved to perform an advantageous function, e.g., visual images in intraspecific signalling, such as courtship, hydrodynamic noise in maintaining interfish distances in schools, and odor for home stream or group recognition.

Fish have also evolved mechanisms for the "intentional" production of stimuli such as bioluminescence, sounds, and pheromones, which are employed in a wide variety of intra- and interspecific functions. These range from the recognition of others of the same species to the attraction of mates and the enhancement of predatory mechanisms (Table 8.5).

## Multiple Stimulation

It is simplistic to assume that behavioral patterns are determined by a single stimulus. Although many fish have evolved a particular sensory system to assume major importance for their life-styles, few species have allowed more than one sensory system to degenerate. The most common vestigial sense organ is the eye, which is of little importance in many deep-sea and cave species.

Table 8.6 sets out, in the most general terms, the wide range of stimuli that may be connected with particular behaviors. It is not being suggested that all the stimuli are linked to the particular behavior of any one species; but, in most species, behavior is likely to depend on more than one source of stimulation. Often the stimuli operate sequentially. For example, during the migration of adult salmonids from offshore feeding areas towards the coast, responses to the sun and to the earth's magnetic field may be used as rough orientating mechanisms (Tesch 1980; Quinn 1984). At the coast a memory of the topography of the seabed or chemoreception may become involved, and the fish may make rheotactic responses when they identify water of low salinity or the odor of the home stream (Johnsen 1984; Northcote 1984). Recently the role of imprinting of salmonid smolts to the odor of their home stream has been questioned and the importance of the genotype emphasized (Stabell 1984). When sockeye salmon smolts migrate from lakes to the sea, celestial clues from the sun and polarized light, as well as topographical features, may be involved (Groot 1965). Other mechanisms of migration have been established, especially tidal transport in plaice, *Pleuronectes platessa,* cod, *Gadus morhua,* and the eels *Anguilla vulgaris* (Creutzberg 1961; Harden Jones 1980; Arnold and Cook 1984). This allows the fish to make net movements in relation to the tide with minimum expenditure of energy. It is also possible that fish can respond to the direction of waves driven by the prevailing winds (Cook 1984) or keep track of compass heading by inertial clues (Harden Jones 1984).

Vertical migration is, however, dominated by light stimuli (Blaxter 1975) that trigger the movement and control its rates (see page 215). Hydrostatic pressure is important, especially in physoclistous (with closed swim bladder) fish, in that unrestricted upward movement could cause the swim bladder to burst. Pressure sensitivity, if it exists, probably depends on the activation of stretch receptors in the swim bladder wall as the swim bladder changes volume during ascent or descent. Since the swim bladder in physoclists is continually adapting by gas secretion or absorption, the existence of absolute pressure sensitivity is unlikely.

Other behavior, such as feeding, schooling, and reproduction, is usually mediated via a complex of stimuli. In feeding, arousal is often caused by chemical stimulation from the food or sounds of other fish feeding. The food may then be located by sight or by mechano- or chemoreceptors. After ingestion the palatability of the food is assessed by texture and taste within the mouth (Atema 1980). While aggregations of fish as unpolarized groups may be maintained by chemical contact, polarized school formation is maintained by a balance of visual-attracting and lateral line–repelling stimuli achieving a dynamic balance of nearest neighbor distance (Pitcher 1979; Partridge and Pitcher 1980). Contact aggregation, or pods, of fish are presumably held together by tactile stimulation. Where sensory processes in reproduction have been investigated, quite elaborate stimulus situations have been revealed, with initial attraction, courtship, mating, and parental care being mediated via all the senses (Myrberg 1980a), with the production of visual

Table 8.4. Enhancement of Sensory Awareness

| Level | Photoreception | Mechanoreception | Chemoreception |
|---|---|---|---|
| CNS | Increase brain size and so area for projection | | |
| Receptor | Summation increases sensitivity, but reduces acuity | | |
| | Duplex retina widens range of sensitivity and gives color vision<br>Pure cone *area* (clupeoids) improves acuity<br>Pure rod *fovea* improves binocular fixation (*Platytroctes*)<br>Tiered rod layers improve sensitivity<br>Visual pigment λ max. improves sensitivity when matched to ambient light; improves contrast if offset<br>Ultraviolet sensitivity (goldfish) | Hair cell orientation and number affect directional perception<br>Variation in number of cilia ? | Increased number of lamellae and sensory cells in macrosmatic fish (e.g., eel) |
| Intrasense organ | Retinomotor movements enhance rod-cone function<br>Tapetum (e.g., sharks) improves sensitivity<br>Tubular eyes (e.g., hatchetfish) improve acuity and binocularity<br>Yellow lenses enhance contrast, reduce chromatic aberration, but decrease sensitivity<br>Yellow lenses enhance bioluminescence (*Scopelarchus*)<br>Iridescent corneas enhance contrast, reduce intraocular flare | Specialization of utriculus (clupeoids)<br>Otolith size and shape may affect sensitivity and frequency discrimination | Enlargement of olfactory organ (eels, deep-sea fish) improve sensitivity and range of chemoreceptivity<br>Hydraulic devices to improve flow through nares (lamprey)<br>Intake tubes at nares (eels)<br>Grooves for entraining water into nares (skipjack) |

| | | | |
|---|---|---|---|
| Intersense organ | Binocularity enhances stereoscopy and range perception<br>(Separation of eyes in hammerhead shark and *Idiacanthus* larvae) | Binaurality (? value)<br>Weberian ossicles (Ostariophysi), otic bullae (clupeoids), and swim bladder diverticula enhance sensitivity and frequency range<br>Wide lateral lines and large neuromasts (macrourids) enhance sensitivity<br>Neuromasts on papillae or stalks (e.g., deep-sea snipe eels) | Paired chemoreceptors (? value)<br>Nares separated (hammerhead)<br>Chemoreceptors on barbels (catfish) or fin rays (gouramis) improve exploratory ability<br>Extension of taste buds over body (help to detect gradients ?) |

*Sources:* Data from Ali (1959); Atema (1980); Bardach and Villars (1974); Blaxter and Jones (1967); Hawryshyn and Beauchamp (1985); Jonsson (1980); Locket (1970, 1971, 1975); Lythgoe (1975, 1978, 1980); Marshall (1967, 1979); Munk (1966); Muntz (1975); Myrberg (1980a); Nicol (1975); Popper (1981); Popper and Coombs (1980); Yamamoto (1982).

Table 8.5. Enhancement of Sensory Awareness by Stimulus Production

| Level | Light | Sound/Mechanical | Chemicals |
|---|---|---|---|
| "Intentional" stimulus production<br>Intraspecific effects | Bioluminescence attracts mate, courtship displays, defines territory, agonistic | Sound production attracts mate, courtship (cichlids), defines territory (squirrelfish), induces maturity (toadfish)<br>Alarm calls<br>Individual recognition in coral fish (*Pomacentrus*)<br>Echolocation in sea catfish (*Arius*) | Chemicals (pheromones) elicit courtship (gobies), attract mates<br>Maintain aggregations (rudd)<br>Maintain parent-young contact (jewel fish)<br>Individual recognition in bullhead (*Ictalurus*)<br>Enhancement of home stream odor by conspecifics (salmonids)<br>Schreckstoff (Ostariophysi) prevents cannibalism, warns against predators |
| "Intentional" stimulus production<br>Interspecific effects | Bioluminescence attracts prey (anglerfish), illuminates food, camouflages against background light (hatchetfish), and warns or scares predators | Warning or scaring predators | Poisons reduce predation |
| Incidental stimulus production | Visual images attract conspecifics in schools, are involved in many visual behavior patterns (e.g., feeding, courtship).<br>Individual recognition in damselfish. Social facilitation in feeding, etc. | Hydrodynamic noise repels conspecifics in schools.<br>Feeding noises attract conspecifics and predators | Excretory products and damaged tissue may attract predators |

*Sources:* Data from Anctil (1975); Atema (1980); Bardach and Todd (1970); Fine, Winn, and Olla (1977); Hawkins and Myrberg (1983); Maniwa (1976); Marshall (1979); Myrberg (1980a, 1981); Myrberg and Riggio (1985); Nicol (1969); Pfeiffer (1962); Stabell (1984); and Tavolga (1976).

Table 8.6. Behavior Controlled by Multiple Stimulation

| Stimulus | Horizontal Migration | Vertical Migration, Depth Regulation | Feeding | Schooling | Reproduction |
|---|---|---|---|---|---|
| Light | Sun compass reaction<br>Polarized light<br>Rheotaxis requires visual clues<br>Tidal transport may require visual clues | Trigger response at dusk and dawn<br>Follow light intensity preferenda | Food location in daylight and clear water<br>Feeding act often visual | Vision attracts conspecifics | Courtship displays<br>Locating mates and mating |
| Mechanical | Inertial clues<br>Response to wind-driven waves | | Long-range location accurate to 20° | Conspecific repulsion | Courtship |
| Chemical | Home stream recognition<br>Salinity perception | | Arousal<br>Zigzagging and other searching<br>Taste during ingestion | Aggregations maintained in darkness | Females attract males<br>Pheromones |
| Tactile | Rheotaxis requires tactile clues | | Textural perception during ingestion | Pods (contact schools) | Courtship, mating |
| Hydrostatic pressure | Seabed topography related to depth | Pressure receptor—swim bladder system adapting in physoclists | | | |
| Electromagnetic | Earth's magnetic field | | Food location in electric fish | Aggregation in electric fish | Courtship in electric fish |

displays, sounds, and pheromones. The extent to which these complex situations occur in some of the marine species that assemble in large spawning aggregations is not clear. Where they have been investigated, e.g., in the haddock, *Melanogrammus aeglefinus*, pairing, sound production, and a visual display have been revealed (Hawkins, Chapman, and Symonds 1967).

## Summary and Conclusions

Fish are certainly a very versatile group, even colonizing, for brief periods, land and air. Within the aquatic habitat, they are found from relatively safe conditions in great depths, where high pressure, low temperature, and darkness prevail, to the stressful and dangerous conditions of the surf zone; from polar to tropical temperatures, and from freshwater to marine salinities. They have exploited to great effect the third (vertical) dimension, so readily available to them, by vertical migration and evolving adaptations to cope with the problems of changing light and pressure.

There is little in their structural, behavioral, and sensory adaptations that is unique. The swim bladder is a teleost characteristic, but gas-filled flotation mechanisms are found in many invertebrate groups; lateral lines are found in amphibians; the fins and myotomes of fish have close locomotory parallels in other vertebrates; bioluminescence is found in a wide range of invertebrates; the equivalent of schooling is extensively practiced on land. Perhaps the versatility of fish is best exemplified by bathypelagic species, in which the modification of sense organs, bioluminescence, loss of skeleton and increase of water content, reproductive adaptations, such as parasitical males, and bizarre modifications connected with feeding provide a wide-open field of research for the biologist.

*Acknowledgments.* I am grateful to Dr. J. Atema and Dr. R.R. Fay, who commented on this chapter in manuscript and made many helpful suggestions to improve it.

### References

Ali, M.A. (1959) The ocular structure, retinomotor and photobehavioral responses of juvenile Pacific salmon, Can. J. Zool., 37:965–996.

Anctil, M. (1975) Prospects in the study of interrelationships between vision and bioluminescence, in Vision in Fishes, Ali, M.A. (ed.), Plenum, New York, pp. 657–671.

Anthony, P.D. (1981) Visual contrast thresholds in the cod *Gadus morhua* L., J. Fish Biol., 19:87–103.

Arnold, G.P. and Cook, P.H. (1984) Fish migration by selective tidal stream transport: first results with a computer simulation model for the European continental shelf, in Mechanisms of Migration in Fishes, McCleave, J.D., Arnold, G.P., Dodson, J.J., and Neill, W.H. (eds.), Plenum, New York, pp. 227–261.

Atema, J. (1980) Chemical senses, chemical signals and feeding behavior in fishes, in Fish Behavior and Its Use in the Capture and Culture of Fishes, Bardach, J.E., Magnuson, J.J., May, R.C., and Reinhart, J.M. ICLARM, Manila, pp. 57–101.

Balchen, J.G. (1977) Modeling, prediction and control of fish behavior, technical note no. 93 from the University of Trondheim Institute of Technology.

Banner, A. (1972) Use of sound in predation by young lemon sharks *Negaprion brevirostris* (Poey), Bull. Mar. Sci. 22:251–283.

Bardach, J.E. and Todd, J.H. (1970) Chemical communication in fish, in Advances in Chemoreception, Johnston, J.W., Moulton, D.G., and Turk, A. (eds.), Appleton-Century-Crofts, New York, pp. 205–240.

Bardach, J.E. and Villars, T. (1974) The chemical senses of fishes, in Chemoreception in Marine Organisms, Grant, P.T., and Mackie, A.M. (eds.), Academic Press, London, pp. 49–104.

Blaxter, J.H.S. (1964) Spectral sensitivity of the herring *Clupea harengus* L., J. Exp. Biol., 41:155–162.

Blaxter, J.H.S. (1970) 2 Light 2.3 Animals 2.32 Fishes, in Marine Ecology, vol. 1, Part I, Kinne, O. (ed.), Wiley Interscience, New York, pp. 213–285.

Blaxter, J.H.S. (1972) Brightness discrimination in larvae of plaice and sole, J. Exp. Biol., 57:693–700.

Blaxter, J.H.S. (1975) The role of light in the vertical migration of fish—a review, in Light as an Ecological Factor II, 16th Symposium of The British Ecological Society, Evans, G.C., Bainbridge, R., Rackham, O. (eds.), Blackwell Scientific Publications, Oxford, pp. 189–210.

Blaxter, J.H.S. and Batty, R.S. (1985) Herring behaviour in the dark: responses to stationary and continously vibrating obstacles, J. Mar. Biol. Assoc. U.K., 65:1031–1049.

Blaxter, J.H.S. and Jones, M.P. (1967) The development of the retina and retinomotor responses in the herring, J. Mar. Biol. Assoc. U.K., 47:677–697.

Blaxter, J.H.S., Gray, J.A.B., and Denton, E.J. (1981) Sound and startle response in herring shoals, J. Mar. Biol. Assoc. U.K., 61:851–869.

Brown, S.B., Evans, R.E., Thompson, B.E., and Hara, T.J. (1982) Chemoreception and aquatic pollutants, in Chemoreception in Fishes, Hara, T.J. (ed.), Elsevier, Amsterdam, pp. 363–393.

Buerkle, U. (1968) Relation of pure tone thresholds to background noise levels in the Atlantic cod (*Gadus morhua*), J. Fish Res. Board Can., 25:1155–1160.

Buerkle, U. (1969) Auditory masking and critical band in Atlantic cod (*Gadus morhua*), J. Fish Res. Board Can., 26:1113–1119.

Chapman, C. J. (1973) Field studies of hearing in teleost fish, Helgol. Wiss. Meeresunters, 24:371–390.

Chapman, C.J. (1976) Some observations on the reactions of fish to sound, in Sound Reception in Fish, Schuijf, A. and Hawkins, A.D. (eds.), Elsevier, Amsterdam, pp. 241–255.

Chapman, C.J. and Hawkins, A.D. (1973) A field study of hearing in the cod *Gadus morhua* L., J. Comp. Physiol., 85:147–167.

Chapman, C. J. and Johnstone, A.D.F. (1974) Some auditory discrimination experiments on marine fish, J. Exp. Biol., 61:521–528.

Clarke, G.L. and Denton, E.J. (1962) Light and animal life, in The Sea, vol. 1, Hill, M.N. (ed.), John Wiley and Sons, New York, pp. 456–468.

Cook, P.H. (1984) Directional information from surface swell: some possibilities, in Mechanisms of Migration in Fishes, McCleave, J.D., Arnold, G.P., Dodson, J.J., and Neill, W.H. (eds.), Plenum, New York, pp. 79–101.

Creutzberg, F. (1961) On the orientation of migrating elvers (*Anguilla vulgaris* Turt.) in a tidal area, Neth. J. Sea Res., 1:257–338.

Enger, P.S. (1966) Acoustic threshold in goldfish and its relation to the sound source distance, Comp. Biochem. Physiol., 18:859–868.

Fay, R.R. (1969) Behavioral audiogram for the goldfish, J. Aud. Res., 9:112–121.

Fine, M.L., Winn, H.E., and Olla, B.L. (1977) Communciation in fishes, in How Animals Communicate, Sebeok, T.A. (ed.), Indiana University Press, Bloomington, pp. 472–518.

Groot, C. (1965) On the orientation of young sockeye salmon (*Oncorhynchus nerka*) during their seaward migration out of lakes, Behaviour, 14 (Suppl): 1–198.

Hall, L., Patricoski, M., and Fay, R.R. (1981) Neurophysiological mechanisms of intensity discrimination in the goldfish, in Hearing and Sound Communication in Fishes, Tavolga, W.N., Popper, A.N., and Fay R.R. (eds.), Springer-Verlag, New York, pp. 179–186.

Hara, T.J. (1975) Olfaction in fish, Prog. Neurobiol., 5:271–335.

Harden Jones, F.R. (1968) Fish migration, Arnold, London.

Harden Jones, F.R. (1980) The migration of plaice (*Pleuronectes platessa*) in relation to the environment, in Fish Behavior and Its Use in the Capture and Culture of Fishes, Bardach, J.E., Magnuson, J.J., May, R.C., Reinhart, J.M. (eds.), ICLARM, Manila, pp. 383–399.

Harden Jones, F.R. (1984) Could fish use inertial clues when on migration? In Mechanisms of Migration in Fishes, McCleave, J.D., Arnold, G.P., Dodson, J.J., and Neill, W.H. (eds.), Plenum, New York, pp. 67–78.

Hawkins, A.D. (1981) The hearing abilities of fish, in Hearing and Sound Communications in Fishes, Tavolga, W.N., Popper, A.N., and Fay, R.R. (eds.), Springer-Verlag, New York, pp. 109–137.

Hawkins, A.D. and Myrberg, A.A., Jr. (1983) Hearing and sound communcation under water, in Bioacoustics, a Comparative Approach, Lewis, B. (ed.), Academic Press, London, pp. 347–405.

Hawkins, A.D. and Sand, O. (1977) Directional hearing in the median vertical plane by the cod, J. Comp. Physiol. 122:1–8.

Hawkins, A.D., Chapman, C.J., and Symonds, D.J. (1967) Spawning of haddock in capitivity, Nature, 215:923–925.

Hawryshyn, C.W. and Beauchamp, R. (1985) Ultraviolet photosensitivity in goldfish: an independent U.V. retinal mechanism, Vision Res., 25:11–20.

Hester, F.J. (1968) Visual contrast thresholds of the goldfish (*Carassius auratus*), Vision Res., 8:1315–1335.

Jacobs, D.W. and Tavolga, W.N. (1967) Acoustic intensity limens in the goldfish, Anim. Behav., 15:324–335.

Jerlov, N.G. (1970) 2 Light 2.0 General Introduction, in Marine Ecology, vol. 1, Part I, Kinne, O. (ed.), Wiley Interscience, London, pp. 95–102.

Jerlov, N.G. (1976) Marine Optics, Elsevier, Amsterdam.

Johnsen, P.B. (1984) Establishing the physiological and behavioral determinants of chemosensory orientation, in Mechanisms of Migration in Fishes, McCleave, J.D., Arnold, G.P., Dodson, J.J., and Neill, W.H. (eds.), Plenum, New York, pp. 379–385.

Johnsen, P.B. and Teeter, J.H. (1980) Spatial gradient detection of chemical cues by catfish, J. Comp. Physiol., 140:95–99.

Jonsson, L. (1980) Chemical stimuli: role in the behavior of fishes, in Environmental Physiology of Fishes, Ali, M.A. (ed.), Plenum, New York, pp. 353–367.

Kleerekoper, H. (1969) Olfaction in Fishes, Indiana University Press, Bloomington.

Locket, N.A. (1970) Deep sea fish retinas, Br. Med. Bull., 26:107–111.

Locket, N.A. (1971) Retinal structure in *Platytroctes apus,* a deep sea fish with a pure rod fovea, J. Mar. Biol. Assoc. U.K., 51:79–91.

Locket, N.A. (1975) Some problems of deep-sea fish eyes, in Vision in Fishes, Ali, M.A. (ed.), Plenum, New York, pp. 645–655.

Lythgoe, J.N. (1975) The structure and phylogeny of iridescent corneas in fishes, in Vision in Fishes, Ali, M.A. (ed.), Plenum, New York, pp. 253–262.

Lythgoe, J.N. (1978) Fishes: vision in dim light and surrogate senses, in Sensory Ecology, Ali, M.A. (ed.), Plenum, New York, pp. 155–168.

Lythgoe, J.N. (1980) Vision in fishes: ecological adaptations, in Environmental Physiology of Fishes, Ali, M.A. (ed.), Plenum, New York, pp. 431–445.

Maniwa, Y. (1976) Attraction of bony fish, squid and crab by sound, in Sound Reception in Fish, Schuijf, A. and Hawkins, A.D. (eds.), Elsevier, Amsterdam, pp. 271–283.

Marshall, N.B. (1967) The olfactory organs of bathypelagic fishes, Symp. Zool. Soc. Lond., 19:57–70.

Marshall, N.B. (1979) Developments in deep-sea biology, Blandford Press, Poole, U.K.

Moulton, J.M. (1963) Acoustic orientation of marine fishes and invertebrates, Ergeb. Biol. 26:27–39.

Munk, O. (1966) Ocular anatomy of some deep-sea teleosts, Dana-Rep. Carlsberg Found., 70:1–62.

Muntz, W.R.A. (1975) The visual consequences of yellow filtering pigments in the eyes of fishes occupying different habitats, in Light as an Ecological Factor II, 16th Symposium of The British Ecological Society, Evans, G.C., Bainbridge, R., and Rackham, O. (eds.), Blackwell Scientific Publications, Oxford, pp. 271–287.

Myrberg, A.A., Jr. (1980a) Sensory mediation of social recognition processes in fishes, in Fish Behavior and Its Use in the Capture and Culture of Fishes, Bardach, J.E., Magnuson, J.J., May, R.C., and Reinhart, J.M. (eds.), ICLARM, Manila, pp. 146–178.

Myrberg, A.A., Jr. (1980b) Ocean noise and the behavior of marine animals: relationships and implications, in Advanced Concepts in Ocean Measurements for Marine Biology, Diemer, F.P., Vernberg, F.J., and Mirker, D.Z. (eds.), University of South Carolina Press, Columbia, pp. 461–491.

Myrberg, A.A., Jr. (1981) Sound communication and interception in fishes, in Hearing and Sound Communication in Fishes, Tavolga, W.N., Popper, A.N., and Fay, R.R. (eds.), Springer-Verlag, New York, pp. 395–425.

Myrberg, A.A., Jr. and Riggio, R.J. (1985) Acoustically mediated individual recognition by a coral reef fish *Pomatocentrus partitus,* Anim. Behav. 33:411–416.

Myrberg, A.A., Jr., Gordon, C.R., and Klimley, A.P. (1976) Attraction of free ranging sharks by low frequency sound, with comments on its biological significance, in Sound Reception in Fish, Schuijf, A. and Hawkins, A.D. (eds.), Elsevier, Amsterdam, pp. 205–228.

Nelson, D.R. and Johnson, R.H. (1976) Some recent observations on acoustic attraction of Pacific reef sharks, in Sound Reception in Fish, Schuijf, A. and Hawkins, A.D. (eds.), Elsevier, Amsterdam, pp. 229–239.

Nicol, J.A.C. (1969) Bioluminescence, in Fish Physiology vol. 3, Hoar, W.S. and Randall, D.J. (eds.), Academic Press, New York, pp. 355–400.

Nicol, J.A.C. (1975) Studies on the eyes of fishes: structure and ultrastructure, in Vision in Fishes, Ali, M.A. (ed.), Plenum, New York, 579–607.

Northcote, T.G. (1984) Mechanisms of fish migration in rivers, in Mechanisms of Migration in Fishes, McCleave, J.D., Arnold, G.P., Dodson, J.J., and Neill, W.H. (eds.), Plenum, New York, pp. 317–355.

Offutt, G.C. (1968) Auditory response in the goldfish, J. Aud. Res., 8:391–400.

Olsen, K. (1976) Evidence for localization of sound by fish in schools, in Sound Reception in Fish, Schuijf, A. and Hawkins, A.D. (eds.), Elsevier, Amsterdam, pp. 257–270.

Partridge, B.L. and Pitcher, T.J. (1980) The sensory basis of fish schools: relative roles of lateral line and vision, J. Comp. Physiol., 135:315–325.

Pfeiffer, W. (1962) The fright reaction of fish, Biol. Rev., 37:495–511.

Pitcher, T. (1979) Sensory information and the organization of behaviour in a shoaling cyprinid fish, Anim. Behav., 27:126–149.

Popper, A.N. (1981) Organization of the inner ear and auditory processing, in Fish Neurobiology and Behavior, Northcutt, R.G. and Davis, R.E. (eds.), University of Michigan Press, Ann Arbor, pp. 126–178.

Popper, A.N. and Coombs, S. (1980) Acoustic detection by fishes, in Environmental Physiology of Fishes, Ali, M.A. (ed.), Plenum, New York, pp. 403–430.

Popper, A.N., Salmon, M., and Parvulescu, A. (1973) Sound localization by the Hawaiian squirrel fishes *Myripristes berndti* and *M.argyromus,* Anim. Behav., 21:86–97.

Protasov, V.R. (1968) Vision and the near orientation of fish, Israel Program for Scientific Translations, Jerusalem.

Quinn, T.P. (1984) An experimental approach to fish compass and map orientation, in Mechanisms of Migration in Fishes, McCleave, J.D., Arnold, G.P., Dodson, J.J., and Neill, W.H., Plenum, New York, pp. 113–123.

Richard, J.D. (1968) Fish attraction with pulsed low-frequency sound, J. Fish Res. Board Can., 25:1441–1452.

Schuijf, A. (1975) Directional hearing of cod (*Gadus morhua*) under approximate free field conditions, J. Comp. Physiol., 98:307–332.

Schuijf, A. and Buwalda, R.J.A. (1975) On the mechanism of directional hearing in the cod (*Gadus morhua*), J. Comp. Physiol., 98:333–343.

Stabell, O.B. (1984) Homing and olfaction in salmonids: a critical review with special reference to the Atlantic salmon, Biol. Rev., 59:333–388.

Stewart, Ann, Bryant, B., and Atema, J. (1979) Behavioral evidence for two populations of amino acid receptors in catfish taste, Biol. Bull., 157:396.

Tavolga, W.N. (1974) Signal/noise ratio and the critical band in fishes, J. Acoust. Soc. Am., 55:1323–1333.

Tavolga, W.N. (1976) Acoustic obstacle detection in the sea catfish (*Arius felis*), in Sound Reception of Fish, Schuijf, A. and Hawkins, A.D. (eds.), Elsevier, Amsterdam, pp. 185–204.

Tavolga, W.N. (1977) Behavioural thresholds for diffuse illumination in the goldfish, J. Exp. Biol. 67:89–96.

Tavolga, W.N. (1982) Auditory acuity in the sea catfish (*Arius felis*), J. Exp. Biol., 96:367–376.

Tesch, F.W. (1975) 8. Orientation in space: animals, 8.2 Fishes, in Marine Ecology, vol. 2, Part II, Kinne, O. (ed.), John Wiley and Sons, London, pp. 657–707.

Tesch, F.W. (1980) Migratory performance and environmental evidence of orientation, in Environmental Physiology of Fishes, Ali, M.A. (ed.), Plenum, New York, pp. 589–612.

Tyler, J.E. (1957–1960) Radiance distribution as a function of depth in an underwater environment, Bull. Scripps Inst. Oceanogr. Univ. Calif., 7:363–407.

Yamamoto, M. (1982) Comparative morphology of the peripheral olfactory organ in teleosts, in Chemoreception in Fishes, Hara, T.J. (ed.), Elsevier, Amsterdam, pp. 39–59.

# Chapter 9

# Social Communication in the Aquatic Environment

CARL D. HOPKINS*

Many sense organs exhibit physiological adaptations designed to cope with a particular physical environment or to perform well for a specific behavioral function. Most of the chapters in this volume are concerned with the "sensory ecology" of the aquatic environment in one way or another. The present chapter will discuss adaptations for social communication within that environment.

Some sense organs behave essentially as "communication sensors" in that their sole function is the detection of communication signals. Other sense organs have more general functions and are used in sensory tasks not necessarily related to communication. It is useful to study the characteristics of some of these specialist communication sensors. They often exhibit conspicuous adaptations or exaggerated design characteristics that aid in our understanding of how the sense organ processes complex stimuli. This chapter will discuss a number of communication sensors.

The process of communication essentially involves three distinct processes: first, the generation of a signal by the sending organism; second, the transmission of that signal through the environment; and finally, its reception by the receiving organism. This chapter will discuss a number of physical properties of signal transmission in the aquatic world and how these factors might affect the choice of a modality, the design of the signals, and the nature of the sense organs. We need to know about background noise in the environment, about the speed of signal transmission, and about the laws of geometrical spreading. Sense organs evolve to cope with problems of detection and recognition in the real world, which is filled with other organisms that are communicating with, preying on, and escaping from other organisms. Rather than attempt an exhaustive review of animal communication in aquatic environments, this chapter will highlight examples in which we see important or unusual differences between aquatic and terrestrial environments.

*Section of Neurobiology and Behavior, Cornell University, Ithaca, N.Y. 14853, USA

## What Is Communication?

As a point of departure, we need to develop a working definition of *communication* that can be used to guide our thinking. Communication is a process that includes two participants: a sender and a receiver. Central to this process is the reception by one of these two of information conveyed in a signal produced by the other. Most ethologists include cases of signal exchanges between members of the same species, as well as those between members of different species; but they exclude exchanges in which the organism producing the stimulus suffers reduced fitness as a consequence of the informational exchange. Such exchanges (as in the case of the predator that attempts to conceal its presence, but nevertheless triggers an alarm response in its prey) have been called "interception" by Myrberg (1981). Communication may not always be of benefit to the receiver, as Dawkins and Krebs (1978) have pointed out. Indeed, in cases of mimicry, the signalling organism appears to benefit from producing the signal, while the recipient appears to suffer. The well-known anglerfish is a good example: common sense tells us that the anglerfish is communicating with its prey as it dangles an attractive lure in front of it, but clearly the approach response of the prey increases the fitness of the signaller, but not the receiver. Many common examples of communication result in a mutual benefit to both participants (see Wiley 1983).

Ethologists generally use an operational definition of communication: i.e., the detection of an observable response in one animal contingent on reception of a signal produced by a second. Such a response may be immediate or delayed. It may occur 100% of the time or less predictably. Curiously, the ethological literature is filled with studies describing signals produced by animals, but few studies are made of the responses these signals evoke. The chain of events leading up to a response includes many stages of sensory processing: peripheral stimulus filtering, central stimulus filtering, and learned recognition of complex stimuli. We know little about the complex central processing of communication signals that ultimately leads to the behavioral response, but we know of some intermediate stages.

There will always be some examples of behavior that fit an operational definition of communication but not the common sense one. The Jamming Avoidance Response (JAR) of New World electric fish is a case in point (Watanabe and Takeda 1963; Bullock, Hamstra, and Scheich 1972; see also Chap. 33). The JAR occurs when one electric fish with a wavelike electric discharge, like the gymnotiform *Eigenmannia,* approaches another of the same species, and the two have discharge frequencies within 1–5 Hz of one another. In a reflex-like response, one (or both) fish shifts its discharge frequency up or down away from that of the stimulus fish. Since one electric fish's signal evokes a response from a second one, this fits the operational definition of communication. Yet a number of careful behavioral studies (Heiligenberg 1977, for review) suggest that the principal function of the JAR is to prevent jamming of the electrolocation system, not communication.

In spite of the semantic difficulties here, what can we learn from the JAR? Sensory physiologists regard it as a unique opportunity to see how the nervous system processes a complex, biologically relevant stimulus from a conspecific animal. The sensory hardware used in this reflex is analogous to the hardware we expect to find for use in the analysis of communication displays by sensory systems (see Chapter 33 for further details on the JAR). In passing, we should note that the gymnotiform *Eigenmannia* has a rich variety of electrical signals that are used in agonistic, appeasement, and courtship behavior. Some of these electrical displays resemble the JAR itself (Hopkins 1974; Hagedorn and Heiligenberg 1985). The sensory/motor pathways mediating the JAR may well be similar, if not identical to those mediating the responses to these other, more traditional examples of social communication in agonistic and sexual behavior. Hopkins (1974) has speculated that the JAR may have given rise to the ritualized electrical "displays" used in social communication in much the same way that intention movements may give rise to communication displays in the visual modality. A sensory system may be adapted in only a general way for the detection of noncommunicative behaviors, but in highly specific ways for the detection of ritualized display behavior.

## Specialized Communication Sensors

We now turn to some examples of "communication sensors"—sense organs for which the sole function appears to be the reception of social communication signals.

The most clear-cut case concerns the submodalities of electroreceptors known for the electric fish of Africa, the Mormyridae. Three anatomically distinct classes of peripheral electroreceptors are found on the skin of all known mormyrids: ampullary receptors, mormyromasts, and Knollenorgans (anatomy and physiology described in Chap. 32; Szabo 1965, 1974; Bennett 1971b). Each of these three types of receptors project to different areas in the brainstem and remain relatively distinct as far as their pathways have been traced in the CNS (review in Bell 1979, in press). Three functions are ascribed to electroreception in mormyrids: passive electrolocation of DC or low-frequency electrical signals from prey or predators; active electrolocation of objects sensed as distortion in the electric field that is produced by the animal's own electric organ discharge; and electric communication. The Knollenorgan receptors and their central pathways appear to function solely in communication, and in this role they appear to function alone.

A number of adaptations exhibited by Knollenorgans point to their exclusive communication function. Firstly, Knollenorgans are quite sensitive to underwater AC electric fields, on the order of 0.1 mV/cm (Bennett 1971b; Hopkins 1981), so that, of all the tuberous electroreceptors, they appear to be best suited for detecting the distant and weak discharges of conspecific animals or other species (see Hopkins 1983b). Mormyromasts are of 1–2 orders of magnitude less sensitive and probably detect only an individual fish's own electrical discharge. Secondly, Knollenorgans behave as band-pass filters that are tuned to match the power spectrum of the species-specific electric organ discharges (EODs) of their own species. This is true in every species that has been examined (Hopkins 1981; Bass and Hopkins 1984). Ampullary receptors are tuned to a narrow range of frequencies below 30 Hz where the EODs have little energy; mormyromasts are tuned quite broadly over the frequency range where the EODs have energy but the match to the power spectrum of the EOD is poor (Hopkins 1981). Finally and most importantly, mormyrids appear unable to "hear" their own electrical discharge through the central pathways that are connected to Knollenorgan receptors even though Knollenorgans fire in the periphery in response to electrolocating signals from the fish's own electric organ. This is because the relay of sensory information to the midbrain is blanked just at the moment when a fish expects to receive its own EOD via the Knollenorgan pathway. Blanking is accomplished in the nucleus of the posterior lateral line lobe by inhibitory synapses that cover the relay cell's surfaces (Fig. 9.1). These synapses receive a "corollary discharge" input from the fish's EOD command circuitry. This discharge inhibits the transmission of all sensory input from Knollenorgans for a fraction of a millisecond around the time when the fish expects to receive input from its own discharge. Thus, mormyrids are essentially electrically "deaf" to themselves on this pathway. They can hear only the electrical signals of others on this sensory channel (Zipser and Bennett 1976a and b; Russell and Bell 1978).

---

Figure 9.1. The Knollenorgan electroreceptor of mormyrids is an example of a *communication sensor:* its sole function seems to be in the reception of electrical signals from others. (A) Typical mormyrid, *Gnathonemus petersii,* having Knollenorgans and two other types of electroreceptors distributed over the back, the belly, and the head (*lined area*); electric organ located in narrow caudal peduncle. (B) Physiological evidence that mormyrids are "deaf" to their own discharge on Knollenorgan pathway. Traces show extracellular spike potentials recorded from *g. mes. exterolateralis* in response to brief electrical pulse stimuli delivered across the skin at an intensity just sufficient to evoke Knollenorgan, but not mormyromast, responses. The beginning of each trace starts with fish's own EOD command signal recorded in the spinal cord (*bottom trace*) after the fish has been silenced electrically by curare. Stimulus latencies vary from 2.67 to 6 ms going from top to eighth trace after the beginning of the command. Upper traces show that the relay of sensory information to the mesencephalon is blanked at precise moment when the fish would normally have expected to receive input from its own EOD (i.e., *in starred traces 4 and 5*). Blanking is controlled by corollary discharge in the nucleus of lateral line lobe (from Russell and Bell 1978). (C) Circuit diagram of Knollenorgan pathway showing primary afferent neurons, running in posterior lateral line nerve and terminating in the nucleus of posterior lateral line lobe of brainstem on large spherical cells. These spherical cells send fibers bilaterally to *ganglion mesencephali exterolateralis pars anterior.* Two known synapses in this pathway, at the periphery and in the nucleus of posterior lateral line lobe, are electrotonic; hence the pathway is "fast conducting" (adapted with modifications from Enger, Libouban, and Szabo 1976). (D) Schematic drawing of large spherical cells in the nucleus of lateral line lobe, showing probable site of command-related inhibition that causes Knollenorgan system to blank its own EODs. Large cells receive large afferent terminals (club endings, *ce*) from primary afferents and small boutons that form inhibitory chemical (*ch*) or electrical synapses (*el*) on cell body and axon hillock. Small cells arise from an area near lateral lemniscus and probably carry corollary discharge to these cells to inhibit them. (From Bell and Szabo, in press.)

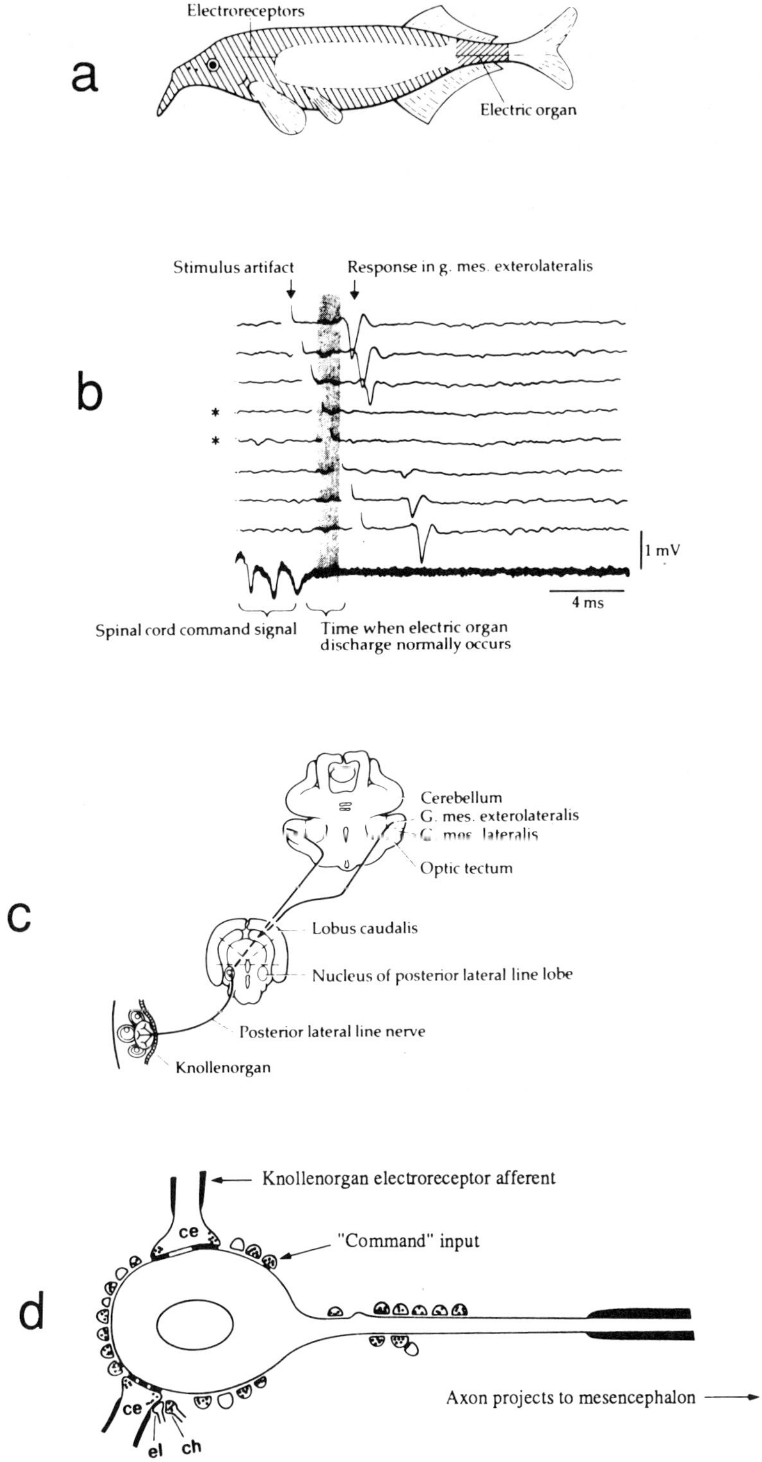
a
Electroreceptors
Electric organ
b
Stimulus artifact
Response in g. mes. exterolateralis
1 mV
4 ms
Spinal cord command signal
Time when electric organ discharge normally occurs
c
Cerebellum
G. mes. exterolateralis
Optic tectum
Lobus caudalis
Nucleus of posterior lateral line lobe
Posterior lateral line nerve
Knollenorgan
d
Knollenorgan electroreceptor afferent
ce
"Command" input
ce
el
ch
Axon projects to mesencephalon

Knollenorgan afferents project bilaterally to a midbrain nucleus peculiar to mormyrids (*ganglion mesencephali exterolateralis*). Bilateral lesions to this structure abolish communication responses of captive mormyrids (Moller and Szabo 1981), but the lesions have no effect on other behaviors requiring electroreceptors, such as active electrolocation.

Examples of communication sensors from other sensory modalities come mainly from terrestrial, rather than aquatic, animals. The antennae of many nocturnal insects carry olfactory receptors that are "tuned" in the periphery to sense-specific attractants (pheromones) used in communication. Schneider (1965, 1969) and Kaissling (1971) review the evidence for the specificity of olfactory receptors in the famed silkworm's pheromone detector cells. The auditory systems of many frogs may be considered as communication sensors much like the Knollenorgan system of mormyrids, though it is likely that functions besides communication can be ascribed to this structure. In one species after another, we see an extraordinary match between the frequency tuning of primary auditory afferents and the spectral properties of the male's mating call—a correspondence that is matched even in the tiny cricket frogs (*Acris crepitans*) of North America and that shows geographical variation in call spectrum with a corresponding match to auditory sensitivity (Capranica, Frishkopf, and Nevo 1973).

A most remarkable case of peripheral filter matching is seen in the "co-qui" frog from Puerto Rico (*Eleutherodactylus coqui*) (Narins and Capranica 1976, 1978). Male frogs make a co-qui call from their territories during the mating season but this two-part call appears destined for two classes of recipients. Males appear to direct the *co* part of the call (a 1-kHz tone) toward other males and the *qui* part (an upward FM sweep of about 2 kHz) toward females. Interestingly, males and females differ in the distribution of frequency sensitivities of auditory nerve fibers, with females being quite sensitive in the 2,000-Hz range typical of the *qui* note and males being sensitive to the 1,000 Hz *co* notes. Since it appears unlikely that differences in the acoustical ecology of the males and females can account for the sex-specific specializations in the auditory system of this species, it is likely that the differences have evolved to meet communication needs.

We recognize that a sense organ is specialized for communication when we see special purpose physiological effects, like the blanking seen in mormyrids or filtering designed to enhance certain well-known signals—or simply by the process of elimination: there may not be any other known function for a particular sense modality or sense organ.

We know much less about the sensory specializations for communication function in aquatic, as compared to terrestrial, organisms. Our understanding of the anatomy and physiology of the auditory systems of fish has improved dramatically in recent years (see Tavolga, Popper, and Fay 1981); specializations for a communication function in fish hearing have been difficult to establish, however. The goldfish *Carassius auratus,* an ostariophysan with its well-known Weberian ossicles, apparently does not use its auditory system for vocal communication. Clupeid fish have an inner ear air bladder extension of the swim bladder that enhances acoustical sensitivity and localization capabilities (Blaxter, Denton, and Gray 1981); yet clupeids also appear to be silent except for certain hydrodynamical noises they make while swimming in schools. The mormyrid fish are another group of auditory specialists (von Frisch 1938; Stipetić 1939). For a long time, they have been considered to be without a specific communicative function; but this auditory system must now be reexamined in the light of a recent discovery of sound production during breeding in *Pollimyrus isidori* (see p. 246). Damselfish (Pomacentridae), although not hearing specialists like the Ostariophysi, mormyrids, and clupeids, are active sound producers known to use "chirplike" sounds in courtship (Myrberg, Spanier, and Ha 1978). These sounds have very broad bandwidths with maximum energy at 200 to 1,100 Hz. Behavioral audiograms show a matching of auditory sensitivity to the frequency range of the sounds produced (Myrberg and Spires 1980). Such matching may be an adaptation for improving the signal-to-noise ratio of communication with sound at a distance (Myrberg 1980a). Fine (1981) has obtained preliminary evidence for a mismatch between the loud male calling signal of the toadfish *Opsanus tau* and the tuning curves of single nerve fibers in the sacculus; the behavioral audiogram for *Opsanus* obtained by Fish and Offutt (1972), however, indicates that there is a good behavioral sensitivity to the range of 200–400 Hz of the boat whistle call. Further study of the toadfish's auditory system is needed before we make a final conclusion about its specialization for communication.

The degree to which a given sensory system is specialized for receiving communication signals varies from modality to modality and from animal to animal. We are gaining increasing awareness of the types of social signals that an animal processes and of the environmental conditions under which it does so.

## Special Problems of Communication in Aquatic Environments

Animals living in aquatic environments communicate with each other by means of signals received by a variety of sensory modalities, including vision, audition, tactile and chemical senses, and electroreception. These modalities are already familiar to us from terrestrial organisms, except for electroreception, which occurs only in aquatic organisms. What special problems do aquatic organisms face when attempting to communicate that differ from those of their terrestrial counterparts?

### Speed of Signal Propagation

As a signal is transmitted from sender to receiver, it propagates at a velocity characterized by the physical properties of the medium. For some modalities of communication, the speed of signal propagation in water may be different from that in air, and these differences in velocity may alter the uses to which the modality is put for communication. Consideration of signal velocity is especially important for sound and chemical communication.

Sound travels about 5 times faster in water than in air: i.e., at $1.48–1.5 \times 10^3$ $m \cdot s^{-1}$ in water compared to 343 $m \cdot s^{-1}$ in air. Wavelengths of sound in water are 5 times that in air for a given frequency, and any wavelength-dependent effects, such as diffraction and sound shadowing, will be affected by this difference. The higher velocity of sound in water precludes the use of time-of-arrival cues for sound localization—cues that appear to be extremely important for sound localization in terrestrial organisms (Knudsen 1983).

Sounds are rarely propagated at a uniform velocity in most real environments. Temperature, salinity, and density gradients or inhomogeneities all affect the speed of sound propagation in water, and this can account for some unusual propagative phenomena such as "shadow zones," where sound cannot propagate at all, and SOFAR channel, where geometrical spreading is reduced and signals travel enormous distances. These effects, produced by temperature or salinity gradients or both of them, have been discussed in reviews by Hawkins and Myrberg (1983) and Payne and Webb (1971).

Chemical signals transmitted by diffusion will travel far more slowly in water than in air because the diffusion coefficients of water may be $10^4$ higher (see Chap. 2). It will take far longer in water for a chemical compound to diffuse out to the limit of its range, where the concentration has fallen to the threshold of a typical sensory receptor than it will in air. It will also take far longer for the signal to fade away once the source is discontinued. Bossert and Wilson's (1963) classical analysis of the transmission properties of chemical signals should be consulted for further details. One consequence of the long diffusion time is that the signalling animals must use water currents—either self-generated or natural—for the dispersion of their chemical messages (see Chap. 2).

Electrical signals are nonpropagating electrostatic fields which remove any time-of-arrival information that might have been used by the electrosensory system for localization or other purposes. The same consideration applies to optical signals, except that they often are transmitted across boundaries between different mediums and are thus refracted in proportion to the speed of transmission of light in the medium. We also see the dramatic effect of signal velocity as constraining the use of surface waves in communication (see Chap. 25). The highly frequency-dependent velocity leads to dispersion of surface wave signals as they travel over a distance (Markl 1983). Some surface-feeding fish can use information derived from a surface wave after dispersion to estimate the distance of a surface-floating target in prey capture (Hoin-Radkovsky, Bleckmann, and Schwartz 1984).

## Active Space

Of all the physical characteristics that can be measured for an animal communication system, the most important may well be the *range* of communication. The space, or volume, within which responses can be evoked by a signal has been called its *active space* by Bossert and Wilson (1963). It will vary from one modality to another and be different for different types of signals within the same modality. It is now recognized that active space is influenced by four different factors: (1) the rate of signal attenuation in the environment due to geometrical spreading, attenuation, and scattering; (2) the characteristics and magnitude of background noise in the environment; (3) the sensitivity of the animal receiver; and (4) the amplitude of the signal at the source (Brenowitz 1982).

## Signal Attenuation Due to Geometrical Spreading, Absorption, and Scattering

Acoustical signal range is influenced by geometrical spreading, absorption, and scattering in a complex way, which has recently been receiving increasing attention in the biological literature (see Michelsen 1978; Wiley and Richards 1978; Hawkins and Myrberg 1983; Payne and Webb 1971). The well-known phenomenon of geometrical spreading accounts for the most significant signal attenuation of a sound signal at close range. For monopolar sound sources in an infinite homogeneous medium (air or water) the sound intensity will decrease according to the inverse-square law. Since sound intensity levels are proportional to the square of pressure, sound pressure decreases according to the inverse first power of distance. Usually expressed in decibels, the inverse-square rate of geometrical spreading implies a 6-dB drop in sound pressure for each distance doubling. On log-log coordinates the inverse-square law for sound pressure defines a line with slope of $-1$, the equivalent of 6 dB per distance doubling (see Fig. 9.2). Sound pressure may fall off according to the inverse third power of distance when the source of the sound is dipole-like, rather than monopole-like, and the sound may become highly directional as well. This type of signal attenuation is most important in the near field of a dipole sound source. In the far field, sound pressure again follows the 6 dB per distance doubling. Where sound is confined by reflecting or refracting surfaces, such as in the thermocline/salinity–induced SOFAR channel in the ocean, sound may spread cylindrically, rather than spherically, and may attenuate at only 3 dB per distance doubling. Chapter 4 reviews some of the complexities of sound field geometry and geometrical spreading in signal attenuation.

In contrast to sound signals, electric currents generated by fish in water, which are well approximated by a dipolar field (Knudsen 1975; Heiligenberg 1977), decay according to the inverse cube of distance. For another comparison, Bossert and Wilson (1963) theorize that a chemical signal released steadily into still water or air will experience a decrease in concentration according to the inverse first power of distance (slope $= -1$ on Fig. 9.2).

If a signaller wishes to increase the active space of a given signal, it must increase the amplitude (or concentration) of the signal at the source. So, therefore, for chemical communication in still water, the law of geometrical spreading predicts that the signaller need only double the concentration at the source to double the radius of the active space. For sound communication, it must increase the sound pressure twice to double the radius. An electric fish must increase the voltage it produces eightfold. In the absence of other attenuating factors, the law of geometrical spreading will ultimately limit the active space of a communication modality. This may explain why electrical signalling is limited to a range of only a few meters, while acoustical signalling may operate over tens or hundreds of meters for the same group of organisms.

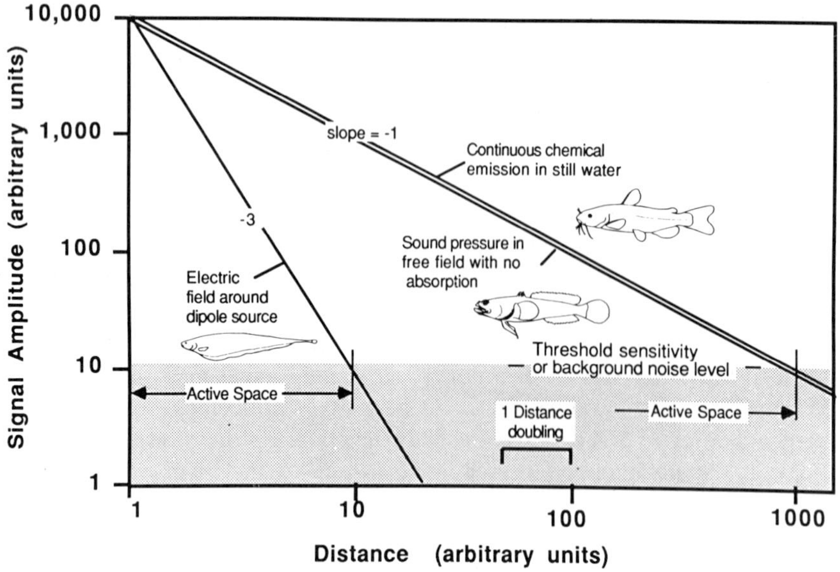


Figure 9.2. Effects of different geometrical-spreading laws on active space of communication. The abscissa shows the distance away from signal source measured in arbitrary units on log scale; the ordinate shows the amplitude of the signal, also on a log scale. Sounds propagate in either air or water according to the inverse square law, which means that the intensity varies in proportion to the inverse square of distance. Sound pressure, proportional to the square root of intensity, varies according to the inverse first power of distance (slope $= -1$). Chemical concentration decreases with the inverse first power of distance (slope $= -1$) around a chemical source diffusing into a still medium continuously. Electric fields generated by a dipole source decrease according to the inverse cube of distance (slope $= -3$). Active space is determined by the distance at which the signal decreases to the threshold of the receiver's sensory receptor, or the amplitude of the background noise of the communication channel. Whether limited by noise or by sensory threshold, the effect is the same. Active space may be increased by increasing the amplitude of the signal at the source, by decreasing the threshold of the receiver, or both. Another strategy would be to communicate under certain ecological conditions where the slope of the line is less than the ones shown here (see text).

Losses due to the absorption of signal energy during transmission differ from geometrical spreading, because the losses are directly proportional to distance, rather than distance doubling. For sound in air, for example, there is about a 1 dB loss of sound pressure per meter of transmission (at 1 kHz, 1 atm pressure, and 70% relative humidity) while in seawater, the loss due to absorption is only $5 \times 10^{-2}$ dB/m (see Griffin 1971; Payne and Webb 1971; Fisher and Simmons 1977; Piercy, Embleton, and Sutherland 1977). The effect of absorption is particularly important at long communication distances. For example, if the absorption of sound is 1 dB/m, then to double the distance of communication from 100 to 200 m would require not a 6-dB increase in sound intensity, as we would expect if geometrical spreading were the only factor, but a 106-dB increase to compensate for the absorption losses! Clearly the effect of absorption outweighs the importance of geometrical spreading at these distances. Therefore, it is especially relevant that the absorption of sound is of 2–3 orders of magnitude lower in water than in air (Rogers and Cox Chapter 5), a factor which contributes to the enormous importance of sound to underwater communication.

In visual communication, signal range is neither determined simply by geometrical spreading nor by the absorption characteristics of the aquatic environment except, perhaps, when a bioluminescent signaller is the sole source of light in an otherwise dark environment. Typically the light from a signalling organism originates with reflected ambient light from above, which may have already been filtered by the water column above the signalling organism. As such, the signal then competes with the background space light scattered from the water column between signaller and receiver (see Chap. 3).

## Background Noise and Receptor Sensitivity

As signal intensity decreases with distance from a source, it eventually becomes masked by signals and energy from other sources, both biological and nonbiological, which constitute the background noise for the communication channel. As with other physical properties of such a channel, it is important to understand the nature of this noise: its intensity, its source, how it changes throughout the day or the seasons, and its spectral characteristics.

Acoustic noise in marine environments has received the greatest attention, though most studies have been conducted in very deep water. In shallow marine environments the principal source of ambient acoustic noise comes from human marine traffic, and after that from wind and rain (review in Wenz 1962, 1964, 1972; Myrberg 1980a, 1980b). These acoustic noise sources tend to be of broad bandwidth, with a peak energy in the range of 100–1000 Hz. Ambient noise is highly predictable given local weather and sea state conditions. Added to this is noise from biological sources, which is highly variable in space and time, but can be a predominant source of noise in some environments—such as near populations of snapping shrimp, sea urchins, croakers, and other organisms. The level of acoustic noise in a particular environment may ultimately affect the absolute acoustic sensitivity of the organisms living there. To be more sensitive than the ongoing noise would mean constant sensory overstimulation. Myrberg (1980a), using acoustic sensitivity measurements conducted under low ambient noise, shows that behavioral audiograms of the shallow water damselfishes, *Eupomacentrus spp.* are pitched 20–26 dB above the ambient noise of their environment at sea state 1 (calm sea).

An interesting parallel can be noted for communication in the electrical modality, in which Hopkins (1973) has pointed out that the predominant source of nonbiological background noise in the bandwidth of 50–50,000 Hz comes from near and distant lightning storms. Although these storms may be infrequent and widely spaced, the noise from lightning may be nearly continuous, because the signals are so strong at the source and because they propagate so well in the atmosphere. Lightning flashes create wide-bandwidth electromagnetic waves that propagate long distances. At night, these signals reverberate between the earth and ionosphere to create "tweeks," "whistlers," and other characteristic signals (Hopkins 1980). In freshwater the amplitude of the lightning-generated noise tends to be 2 orders of magnitude greater than it is in seawater. Interestingly, there is a parallel increase in electroreceptor sensitivity in seawater, as can be seen by comparing behavioral thresholds of freshwater fish with other species living in seawater (see Hopkins 1973; Kalmijn 1974).

### Signal Source Localization

Communication receivers must not only detect signals at a distance, but must also be able to locate them in space. The question of "where" a signal is coming from may be crucial to its identification, and surely localization will be essential if a directed response such as approach or withdrawal is to occur. Localization may be accomplished by using a directional receiver. Visual signals in water are the most easily located, owing to the straight-line signal transmission. Olfactory signals are probably the least easily located and it may be accomplished only after searching for a chemical gradient.

Because sound travels so rapidly in water, travel times are too small to be detected by a bilaterally symmetrical nervous system, thereby eliminating the time-of-arrival cue considered essential for acoustical localization by most terrestrial animals. In addition, because the wavelengths of low-frequency sounds are so great and the density of an organism's body so close to that of water, an organism's body mass will have little shadowing effect on sound transmission that might impart a directionally dependent amplitude cue. While sound pressure, a scalar, may be impossible for fish to localize (van Bergeijk 1967), sound propagation sets particles of the medium into motion, and the direction of the motion, a vector, is determined by the direction toward the source, in the far field (see Chap. 4). It is now well established that fish can localize sound sources in all three planes, even in the far field (see Schuijf and Buwalda 1975); such capabilities may be mediated by the otolithic organs in the ear, which can be shown to respond directionally to the vector properties of particle displacement and are determined by the direction of the sound source (Hawkins and Horner 1981; see also Chap. 28).

New results on electrical signal localization in electric fish are discussed on page 253.

## Electrical and Acoustic Communication in Mormyrid Fish

We now turn to some specific illustrations of communication in aquatic organisms by reference to work in my own laboratory on the communication behavior of the mormyrid fish from Africa. The mormyrids are well known for the development of the electrosensory system and the production of electrical discharges, both of them used in electrical communication. Because the electrical sense is unfamiliar to us, as terrestrial vertebrates, I shall first provide a brief introduction to the physical properties of electrical signals in water.

## Signal Range

One of the major restrictions on electrical communication in water is that the active space for electrical signalling is apt to be rather small, because of the rapid rate of signal attenuation by geometrical spreading (see Fig. 9.2). Empirical measures of active space of mormyrid electrical signals were made in large aquaria by Squire and Moller (1982) and Moller and Bauer (1973) by slowly moving one tethered fish toward a second until it went electrically silent—a behavioral indication that it has detected the electrical signals from the first. One mormyrid, *Brienomyrus niger,* can detect another at a range of 135 cm as long as the signaller and recipient are favorably positioned with respect to one another, end to end and facing in the same direction. For other orientations, such as side to side antiparallel or perpendicular, the active space radius drops to about 50% of the maximum distance. When the water conductivity is raised, the active space shrinks dramatically, though the shape of the active space is relatively constant as an ellipsoid. This is because the electric organ behaves like an old battery with high internal resistance and the observed external peak voltage of the EOD is lowered when the external current path is shunted by high conductivity of the water (see Bell, Bradbury, and Russell 1976). There has never been a systematic field study in which active space has been measured under varying conditions of water depth, substrate composition, vegetation, signaller's size, receiver's size, or orientation of signaller with respect to receiver. Some electric fish produce signals that can be detected electronically in the field over a 5-m range, but we do not know at what range another mormyrid can detect the same signal.

Knudsen (1975) has measured the electric field around various gymnotid fish by holding them tethered at the surface of a pool 1 m deep and large enough to prevent boundary interference. As expected for a dipolar source, the electric field magnitude is proportional to both the inverse cube of the distance from the fish and the cosine of the angle of deviation from the longitudinal axis. By comparing electric field strength with behavioral sensitivity to electric fields, Knudsen concluded that the active space ranges between 50 cm and 1 m for gymnotids.

Active space for electric communication is short when compared with acoustic communication in terrestrial environments, where tens of meters is not uncommon. A range of 1 m may be comparatively good in water when one considers the alternatives of vision, which may be nil in some murky African streams and rivers, or sound, which may not propagate in shallow water. To increase the range of electrical signalling by increasing the amplitude of the signal at the electric organ may not be cost effective, since it will require an eightfold increase in amplitude to double the active space range. Compared to the size of most African river and lake environments, a 1–m range severely limits the electrical modality, but within the confines of a small stream, 1 m may be good enough to detect and locate other fish and superior to the range of visual, sonic, or chemical communication.

## Acoustic Signalling in *Pollimyrus isidori*

Because electrical signalling appears to be so range limited, it was of considerable interest to discover that one species of mormyrid, *Pollimyrus isidori,* also produces a rich repertoire of *acoustic* signals during its courtship (Crawford, Hagedorn, and Hopkins 1986). We now turn to a brief description of acoustical communication in *Pollimyrus,* a species of which we already know a good deal about electric-signalling behavior (Kirschbaum 1975; Kirschbaum and Westby 1975; Westby and Kirschbaum 1982).

*Pollimyrus isidori* can be induced to spawn in the laboratory by imitating conditions of the tropical rainy season in large community aquaria (see Kirschbaum 1975). The process begins as the males build a woven nest within a territory defended in a 50-cm diameter. Females do not establish territories, but rather remain relatively motionless in small areas between the territorial boundaries of adjacent males. Females set up a dominance hierarchy among themselves and jockey for the best positions. The dominant female's eggs ripen over a period of days or weeks as she becomes ready for spawning. Females are not tolerated in the male's territory except when she is ready for spawning, and then only after a period of courtship.

The male, being tied to his territory and nest, needs a way to attract an appropriate female from a distance. *Pollimyrus* does this with acoustic, not electric, signals. Males and females both discharge electrically during all stages of breeding, but we never detected patterned electrical discharges differing from electrolocation signals. Acoustically, however, we recorded a total of five different sounds. "Grunts," "moans," and "growls," (see Fig. 9.3) were given by adult males with a nest, but not yet mated. In addition to attracting females and perhaps to induce them to become ripe for spawning, these calls may repel other males. "Hoots" and "pops" were also produced by both males and females of this species during agonistical interactions (Fig. 9.4).

Male calling was stimulated by the nearby presence of a gravid female. We often observed a gravid female approaching the male's territory and noted that he would begin the grunt, moan, growl sequence immediately. Vigorous calling usually continued until the female left the territory, but males then continued to produce growls at a low rate.

On nights when a pair spawned, we observed intense calling by the territorial male, but, for the 2 to 3 hours when spawning actually took place, the male was acoustically silent. The motor acts during courtship are described in Crawford, Hagedorn, and Hopkins (1985, 1986). After spawning, the male guards the eggs and young for up to 2 weeks without producing these acoustical signals.

The "popping" sound is similar to the sharp impulsive "clicking" sound previously noted for *Gnathonemus petersii* by Rigley and Marshall (1973). It is

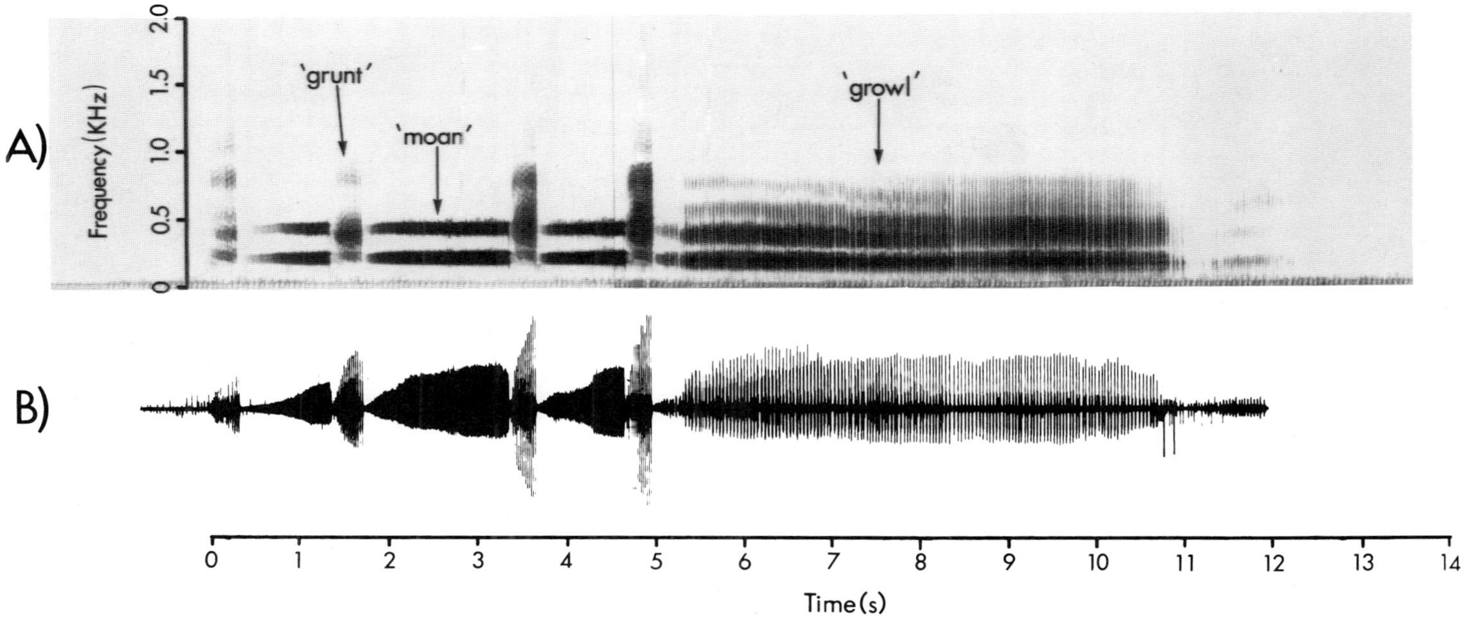


Figure 9.3. Acoustic courtship song in a mormyrid electric fish, *Pollimyrus isidori*, is composed of a sequence of grunts, moans, and growls produced by a male on its territory. Sonogram analysis in (A) has a bandwidth of 20 to 2000 Hz, and an analysis bandwidth of 75 Hz. The digitized waveform of the same signal (B) shows the signal waveform amplitude as a function of time. (From Crawford, Hagedorn, and Hopkins 1986.)

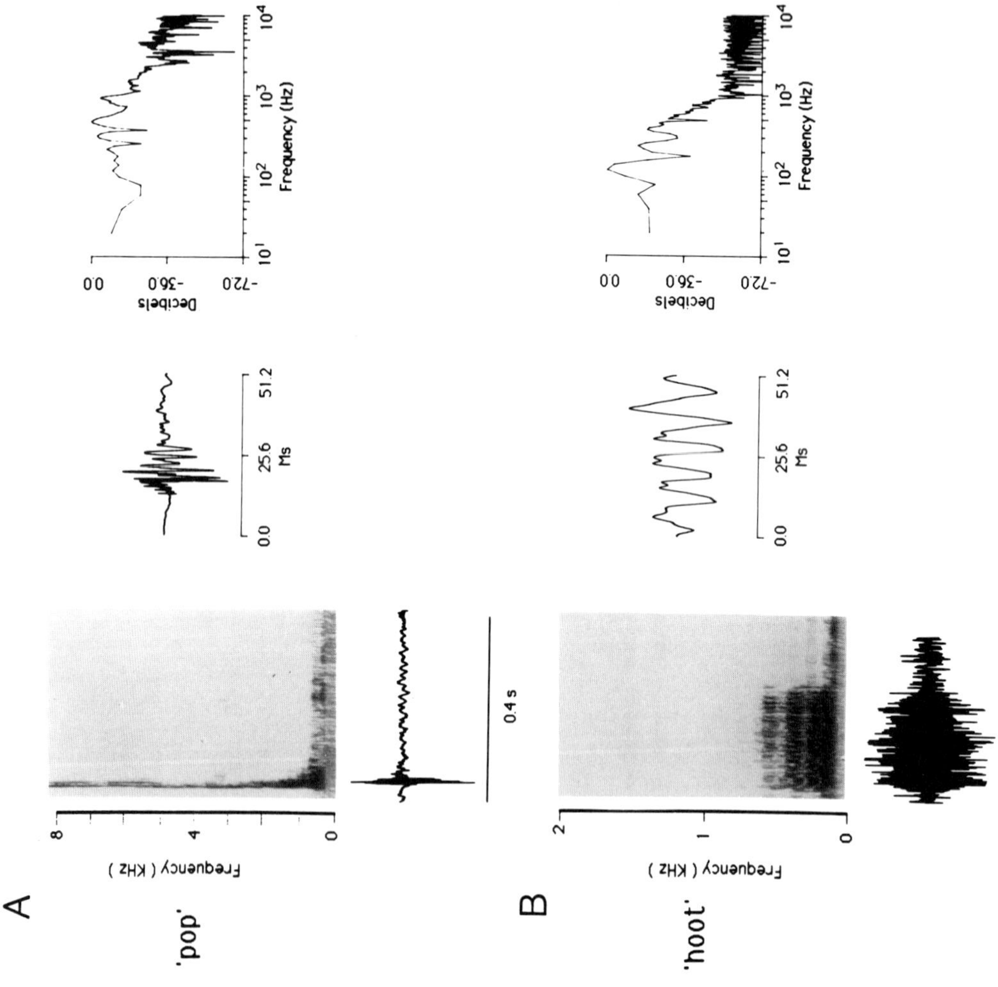
A
'pop'
Frequency ( KHz )
8
4
0
0.4 s
0.0
25.6
51.2
Ms
Decibels
0.0
-36.0
-72.0
$10^1$
$10^2$
$10^3$
$10^4$
Frequency (Hz)
B
'hoot'
Frequency ( KHz )
2
1
0
0.0
25.6
51.2
Ms
Decibels
0.0
-36.0
-72.0
$10^1$
$10^2$
$10^3$
$10^4$
Frequency (Hz)

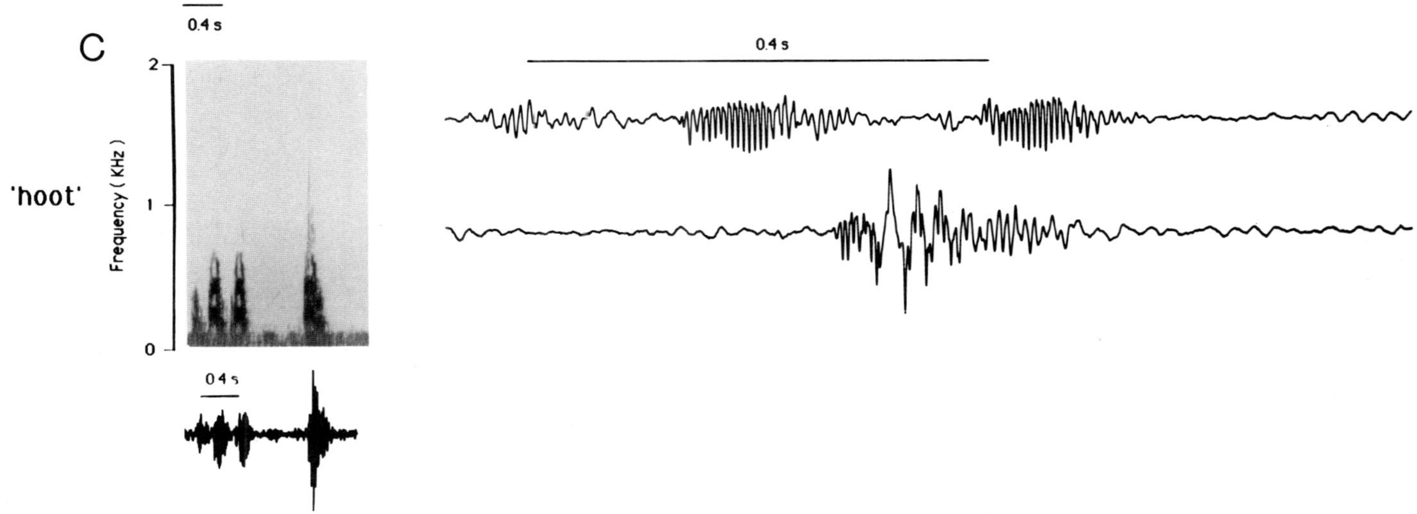


Figure 9.4. Sonograms, digitized waveforms, and pcwer spectra of pops (A) and hoots (B and C) from *Pollimyrus isidori*. These calls are given during agonistical interactions of either males or females.

produced during darting attack movements by dominant animals and territory owners and may be a mechanical sound produced by the sudden movement. Finally, the "hooting" call, which is extremely variable, is also associated with agonistical behavior.

Electrical discharge activity is reduced to a low rate during production of grunts, moans, and growls. Simultaneous electrical and acoustical recordings from the same individual fish show complete electrical silence during grunts and moans and only very slow pulsing during growls (Fig. 9.5).

## Audition in Mormyrids

Mormyrids have a highly specialized auditory system, with a tiny air bladder derived embryonically from swim bladder, located within the inner ear and attached to the saccular otolith (von Heusinger 1826; von Frisch 1938; Stipetić 1939). The air bladder is probably a transducer of sound pressure to displacement that gives these fish a sensitivity to sound pressure. The small size of the air bubble by comparison with the swim bladder may tune the sensitivity of the ear to higher frequencies (Werns and Howland 1976). McCormick and Popper (1984) have recently shown that the mormyrid, *Gnathonemus petersii,* has good sensitivity to sounds. The audiogram shows a maximum sensitivity at 400–600 Hz. Sensitivity drops by about 40 dB by 2,500 Hz on the high-frequency side and by 20 dB by 100 Hz on the low-frequency side. The overall behavioral sensitivity of this mormyrid appears comparable, on the high-frequency end, to acoustical specialists like the goldfish, though it is about 10 dB more sensitive than *Gnathonemus.* The air bubble in *Pollimyrus* resembles that of *Gnathonemus,* and we have no reason to believe that the auditory system is much different. There is a good match between the range of the behavioral audiogram of *Gnathonemus* and the acoustical call frequencies for *Pollimyrus,* as shown in Figures 9.3–9.5.

We are intrigued by the finding of sound production in these fishes and expect that sound communication may be important to this and other species of mormyrids. This example of acoustical signalling in fish is but one of many examples that can be cited from the literature. Fine, Winn, and Olla (1977), Myrberg (1981), and Hawkins and Myrberg (1983) present excellent reviews of acoustical signalling in a variety of groups and discuss the functions of acoustical communication.

Not all mormyrids have an acoustical courtship. Male *Brienomyrus brachyistius* (triphasic) produce an electrical call from their territory with no acoustical component. This species is typically found in streams 5–10 cm in depth, however, and under these conditions, sound propagation is poor (Chap. 5) and may offer no advantage for increasing active space compared with electric signalling. One is reminded of the classic study of Tavolga (1956, 1958) on the marine goby *Bathygobius soporator,* which compares different sensory modalities in communication. Visual, chemical, and acoustical signals were presented to territorial males, and courtship and combat were evoked in response. Different modalities of signals obviously served different functions, as they apparently do for electrical and acoustical signalling in mormyrids.

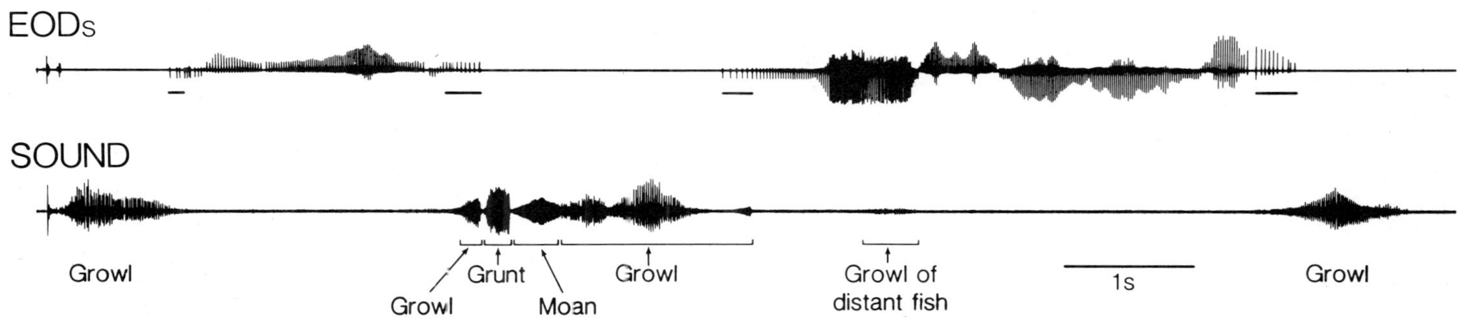


Figure 9.5. Electric organ discharges (EODs) appear to be suppressed during acoustic courtship calling of *Pollimyrus*. Oscillograph records show simultaneous recordings of EODs (*above*) and sounds (*below*) from one male. The EOD typically goes silent for the grunt and moan part of the call and then starts up again at reduced frequency during the growl.

## Nature of Electrical Signalling in Mormyrids

Mormyrid fish are best known not for their acoustic behavior, but for their abilities to produce and receive electric signals, which are produced in the electric organ in the caudal peduncle, which is composed of four columns of electrocytes, about 100 in each column. It is innervated by spinal electromotor neurons that receive input from the pacemaker in the medulla. The electric organ discharge (EOD) waveform is determined by the physiology and anatomy of the electric organ, which is relatively invariant for a given individual and species. The rhythm of discharges, encoded as a sequence of pulse intervals (SPI), is controlled from the pacemaker and highly variable from moment to moment.

One community of mormyrid fishes, particularly rich in species, is shown in Figure 9.6. All these species occur in the Ivindo River drainage system in Gabon, West Africa. The waveforms of the EODs are also shown on the same time base. We see that EODs are extremely diverse in terms of overall duration, number of peaks to the waveform, presence of inflection points within a peak, and the relative amplitudes of the peaks. We even see EODs with a reversed polarity compared to most species and several cases in which the two sexes differ in EODs. Bass (in press) has studied the anatomy of the electric organs of many of these species and discovered anatomical differences between them that can account for the differences in polarity between EODs and some of the complexities, such as inflection points and numbers of peaks. Bennett (1971a) also studied the electric organ of several species of mormyrids and found that the discharge of each individual electrocyte was similar to the waveform of the compound discharge of the whole electric organ. Bass and Hopkins (1983) demonstrated that several of the mormyrid species that have a natural sex difference in EODs can be induced to show it by androgen treatment. The species-specific and sex-specific EOD signatures potentially could provide recognition of species and sex. We examine this possibility on page 257.

In addition to EOD diversity, mormyrids produce SPIs in electrical communication. SPIs are difficult to study, because it is hard to determine which of a number of fish is producing a given electric signal. Bell, Myers, and Russell (1974) separated out EODs from two interacting individuals by attaching a fine electrical wire to one during interactions. Kramer (1974) observed interactions between fish of different species, where he could identify SPIs by following the species-specific EOD signatures. It is also possible to record SPIs from individuals that are interacting with a silenced partner where there is no confusion.

SPIs may be classified into four major categories based on signal form: single individuals can produce "discharge cessations," "frequency modulations," and "tonic discharge patterns." Pairs can interact with each other to produce "echoes" or "preferred avoidance intervals." All these patterns, along with their behavioral significance, are reviewed in Hopkins (in press).

One mormyrid, *Brienomyrus brachyistius* (triphasic) from Gabon, has EODs and SPIs worth mentioning here. This species was the first known to have a sex difference in the waveform and duration of the EOD (Figs. 9.6 and 9.7) (see Hopkins 1980; Hopkins and Bass 1981). Solitary (presumably unmated) males, recorded in the field, produce high-frequency bursts of EODs (sounding like "rasps") when ripe females pass their small stream territories. The same males discharge normally or go electrically silent in response to passing males or fish of other species. Rasps are detected only during the breeding (rainy) season. We sometimes detect a continuous series of rasps from a male that appears to be paired with a female. Females give occasional rasps (Fig. 9.8A and B).

*Rasps* are stereotyped bursts of EODs that start out at 150 pulses/sec and descend to 100–120 pulses/sec after a duration of about 200 to 400 msec. Rasps are repeated up to 10 times per minute. Females also produce rasps, usually briefer, and often in duet with male rasps, but this is observed only from females that have been paired with a male for several days running. Functionally the rasp display appears to serve much the same purpose as the acoustic calling in *Pollimyrus:* attraction of a female to the male's territory and possibly courtship. We have used the rasp response in playback experiments (see p. 257).

## Localization of Electric Signals by Mormyrids

The mechanism of localizing electric signals in water has received little attention to date except for excellent work on localization of objects by active electrolocation (review in Heiligenberg 1977). We expect that electric fish must be able to locate the source of a communication signal in order for the modality to be useful, and we know that fish do approach one another during social interactions and will approach electrical dipoles, but the mechanism of the localization is unknown and has not been studied. Furthermore, localization may require complex arrays of electroreceptors and complex computation by the CNS, and it is unclear whether the fish can make the necessary measurements and computations. The problem of localization of a signaller is simply the problem of determining distance and direction of a dipole source given measurements of the electric field at a distance. Electroreceptors in the receiver's skin sense the perpendicular component of the current and are therefore directional receivers. But though current is directional, a single point determination of that direction tells the receiver nothing about the distance and direction of the signal source, because the current lines are curved, not straight, as are sound waves or light signals. The source could be anywhere on the isocurrent circle. To reduce the ambiguity of the source position, the receiver would need an array of receivers designed to measure the divergence of the current line, if in fact it does diverge. (A similar problem is encountered in attempting to localize sound in the near field [see Chap. 4.]).

Recently, studies by Schluger and Hopkins (in preparation) suggest that mormyrids cannot locate the source of an electric signal in water, but they do have a simple, efficient strategy to enable them to approach such sources. Schluger filmed paths taken by individuals of *Brienomyrus* held in a large circular tank as they approached, in semidarkness, a dipole electrode playing synthetic electric pulses imitating a conspecific. The test fish started from a hiding place in the center of the arena and approached the electrodes on the edge. The electrical signal mimicked a conspecific, and it triggered an immediate approach and attack. Pathways

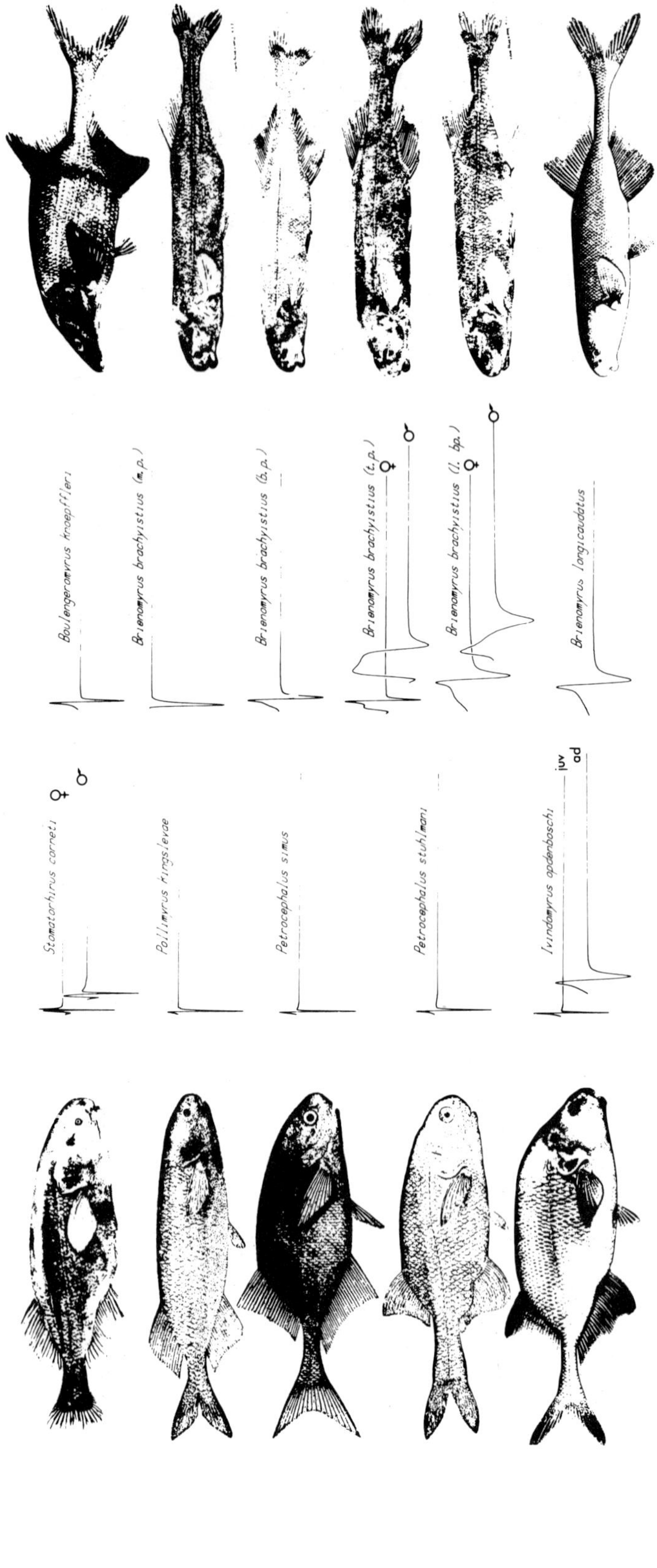
Boulengeromyrus knoepffleri
Brienomyrus brachyistius (m.p.)
Brienomyrus brachyistius (b.p.)
Brienomyrus brachyistius (t.p.)
♀
♂
Brienomyrus brachyistius (l. bp.)
♀
♂
Brienomyrus longicaudatus
Stomatorhinus corneti
♀
♂
Pollimyrus kingsleyae
Petrocephalus simus
Petrocephalus stuhlmanni
Ivindomyrus opdenboschi
juv
ad

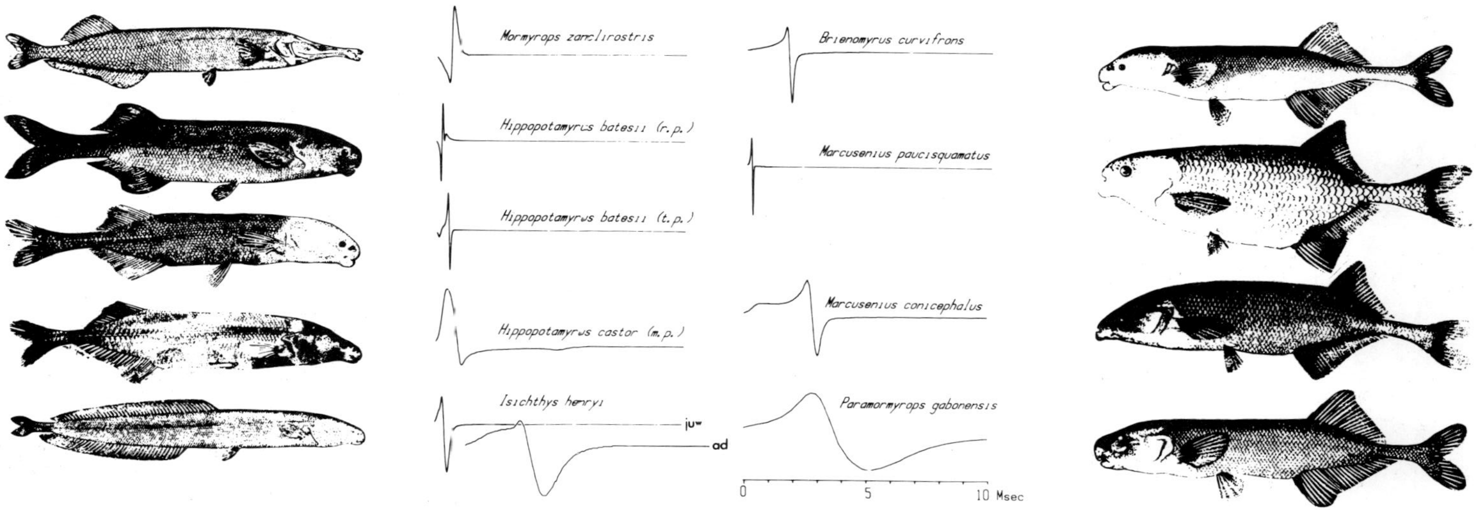


Figure 9.6. Twenty species of mormyrid electric fish from the community in Ivindo River of Gabon, along with their associated EODs, all of which are plotted on same time scale, with head positivity upward. Species identifications are tentative pending description of new species. The notation of *m.p.*, *t.p.*, *b.p.*, *l. bp.*, and *r.p.* refers to monophasic, triphasic, biphasic, long biphasic, and reverse polarity, which is characteristic of the EOD of the species.

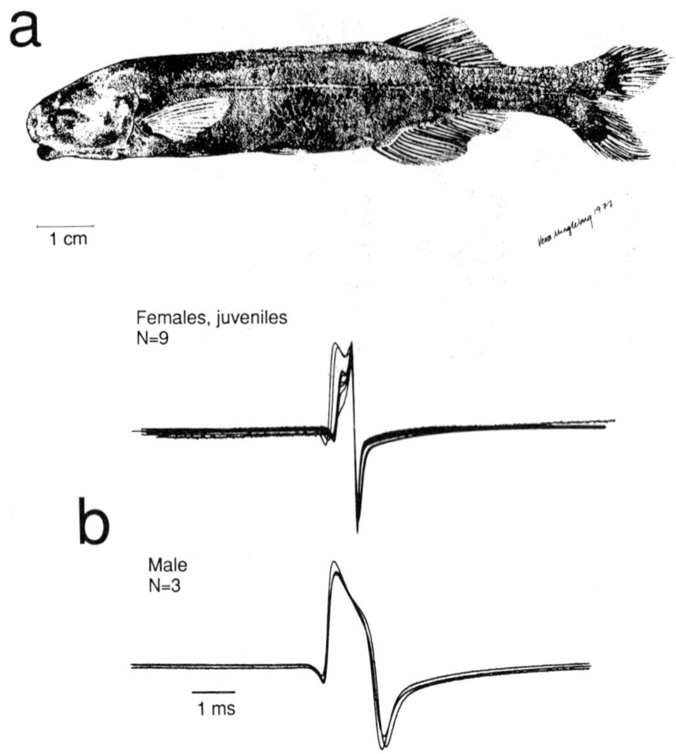


Figure 9.7. *Brienomyrus brachyistius* (triphasic) (A) is a mormyrid found in shallow streams in Gabon, West Africa. (B) EODs of adult males and females differ in duration and waveform. Superimposed EODs of 9 females and juveniles (*above*) are shorter in duration than those from 3 males (*below*). All EODs are scaled to the same peak-to-peak amplitude and centered around the point where the absolute value of the derivative of the EOD is maximum. Bass and Hopkins (1983b) were able to produce EODs of long duration in females and juveniles by administration of androgen. (Adapted from Hopkins 1983.)

of several such approaches are shown in Figure 9.9, along with vector diagrams of the electric fields around the stimulus electrodes. We oriented the electrodes in different ways to observe changes in the approach. In the left column in Figure 9.9, we see the approach to a dipolar field that is tangential to the edge of the tank (perpendicular to the radius). The field lines run through the shelter in a direction perpendicular to the angle between them. The paths of approach shown here arch outward away from the radius and then inwards to converge on the signal. In the center column of 9.9 we see the response to the same stimulus, now aligned along the radius of the circle. The fish swims towards the stimulus sometimes, but swims away just as often, in a seemingly random pattern. In the right column of Fig. 9.9, the stimulating electrodes are widely separated into a bipolar stimulus that spans the diameter of the tank to create a hemicircular bipole. The fish makes a straight approach to either of the two poles at random. Note that, in all three cases, the fish has its longitudinal axis parallel to the lines of current throughout the entire approach path. This sometimes takes it off in a long, arching approach or may even take it away, but all tracks eventually lead to the source. These simple observations suggest an orientation strategy for electric fish. First, the fish must turn and orient its body axis so that it is parallel to the direction of the local current flow. Second, it must swim forward while keeping the body parallel to the current lines. By following the current lines, the fish will be led to the source itself, though the pathway it takes will often be a curved one, not straight. Our tentative conclusion from these observations is that a fish cannot position an electrical signal in space unless it can move and make repeated electric field measurements. Since all field lines must converge on the source, this mechanism is adequate for approaching other electric fish.

## Signaller Identification and Signal Recognition

Receivers need to know "who" is signalling and "what" the signal means; hence they must be able to discriminate important signals from irrelevant noise—some of which may well be a "signal" for another species or individual or at another time. To make the necessary discriminations, receivers must have some prior knowledge about the nature of important communication signals, either innately or through learning. Hopkins and Bass (1981) and Hopkins (1983b) have studied sensory mechanisms involved in species and sex recognition in electric communication in the mormyrids.

We used the *Brienomyrus brachyistius* (triphasic) male's rasp response as an indicator of willingness to court passing females, and studied responses to recorded and computer-synthetic electric stimuli played back to the male on his territory at night (for description of methods, see Hopkins and Bass 1981; Hopkins 1983a). Signals were played back to males directly from a tape recorder or signal generator through electrodes placed near their territories.

Hopkins and Bass (1981) demonstrated that *Brienomyrus* could discriminate between EODs with identical power, but different phase, spectra. The natural female EOD was a preferred stimulus, and phase-shifted or time-reversed versions of this EOD were much less preferred (see Fig. 9.10). Hopkins and Bass (1981) concluded that EOD recognition is based on temporal cues in the waveform, not spectral cues.

The neural mechanism of this recognition is surprisingly simple. Knollenorgan receptors fire spikes in response to outside positive-going edges, but do not respond to negative-going edges. Like the excitable membrane itself, the Knollenorgan receptor behaves like a rectifier. When stimulated with the female EOD in the positive polarity, spikes cluster in at the transition between the first and second peaks to the waveform where the voltage transition is negative to positive. When stimulated with the same EOD with polarity inverted, the spikes shift to the transition between peak 2 and peak 3, because this is now the principal outside positive-going transition. For this EOD, though not for others that have been

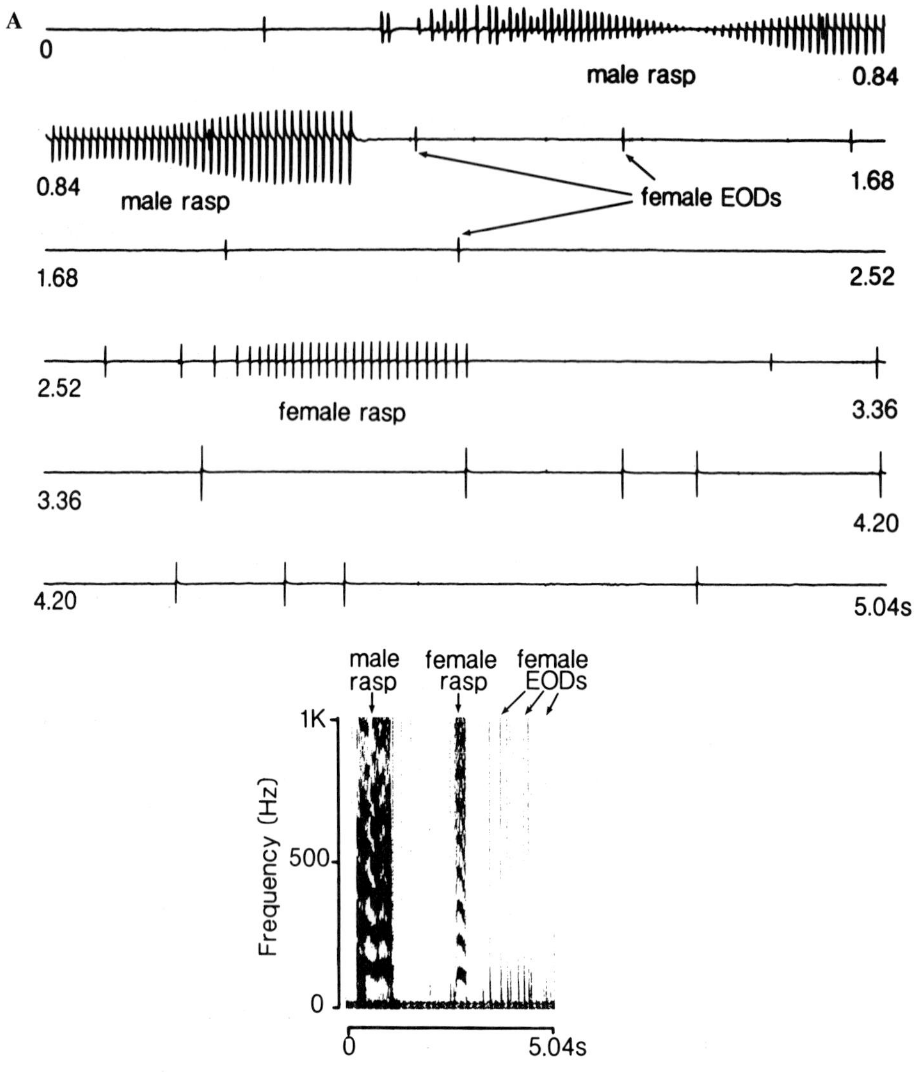


Figure 9.8(A) Oscillogram (*above*) and spectrogram (*below*) of the same rasps produced by *Brienomyrus brachyistius* (triphasic). Males have longer EODs than do females and can be distinguished from them by wider pulses in oscillogram plots from recordings made in the field. The male produces first rasp, a series of EOD pulses strung together at rates of more than 100/sec. As he turns with respect to electrodes, his EODs decrease in amplitude and then become inverted (*first line of oscillogram*). The female also produces a rasp, though shorter and at lower frequency. Numbers refer to time axis in seconds. Single female EODs at the end of the record show up on the spectrogram as broad bandwidth clicks.

tested, the transition back to the baseline from the third peak never evokes a spike, because the receptor cell is still refractory.

We examined the responses of Knollenorgans to both normal and inverted polarities of each EOD because, in nature, a fish experiences current flowing in

B

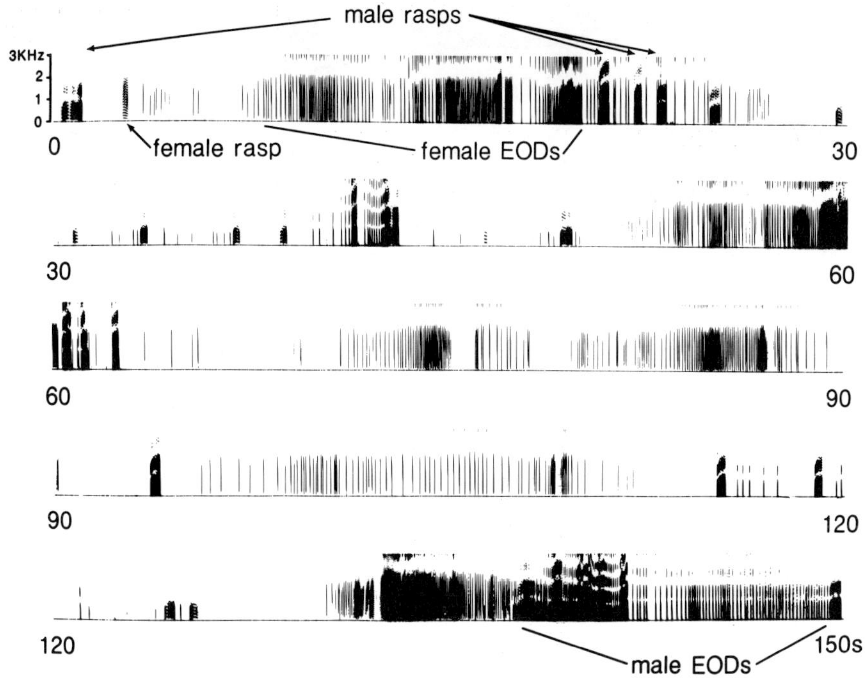


Figure 9.8(B) Male rasps are repeated at frequent intervals, as shown in this continuous sound spectrogram for 150 sec of field recording of one male and one female *Brienomyrus brachyistius* (triphasic). The female produces shorter, less frequent rasps, sometimes (not shown here) simultaneously with the male. Female's EODs are shorter than the male's and, when given singly, show up on the spectrogram with a peak about 1 kHz. The power spectrum of male's EOD has four distinct peaks in the range of 0–3 kHz shown. They can be distinguished easily from those of the female.

its body, in through the skin on one side and out the opposite, and Knollenorgans on opposite halves of the body "see" opposing polarities of the same stimulus in phase with each other. The brain presumably integrates the information from the two halves of the body to produce a composite temporal code from them. Figure 9.11. shows poststimulus time histograms in response to the female EOD and to the EODs of other species. The response to the positive waveform is plotted upwards and the response to the inverse polarity, downwards, to make a compound poststimulus time histogram (PSTH). The histogram of responses to the female *B. brachyistius* (triphasic) female EOD produces a unique PSTH when compared to all other species of mormyrids from the same habitat.

Hopkins and Bass (1981) successfully imitated the female EOD by presenting a rectangle wave stimulus to male fish in their territories at night. They reported the strongest responses to rectangles of 0.4-ms duration—an interval that matches the one between the peaks on the compound PSTH in Figure 9.11. Note that the onset of the rectangle evokes spikes from Knollenorgans on half the body, whereas the falling edge evokes spikes on the opposite side. By adjusting the

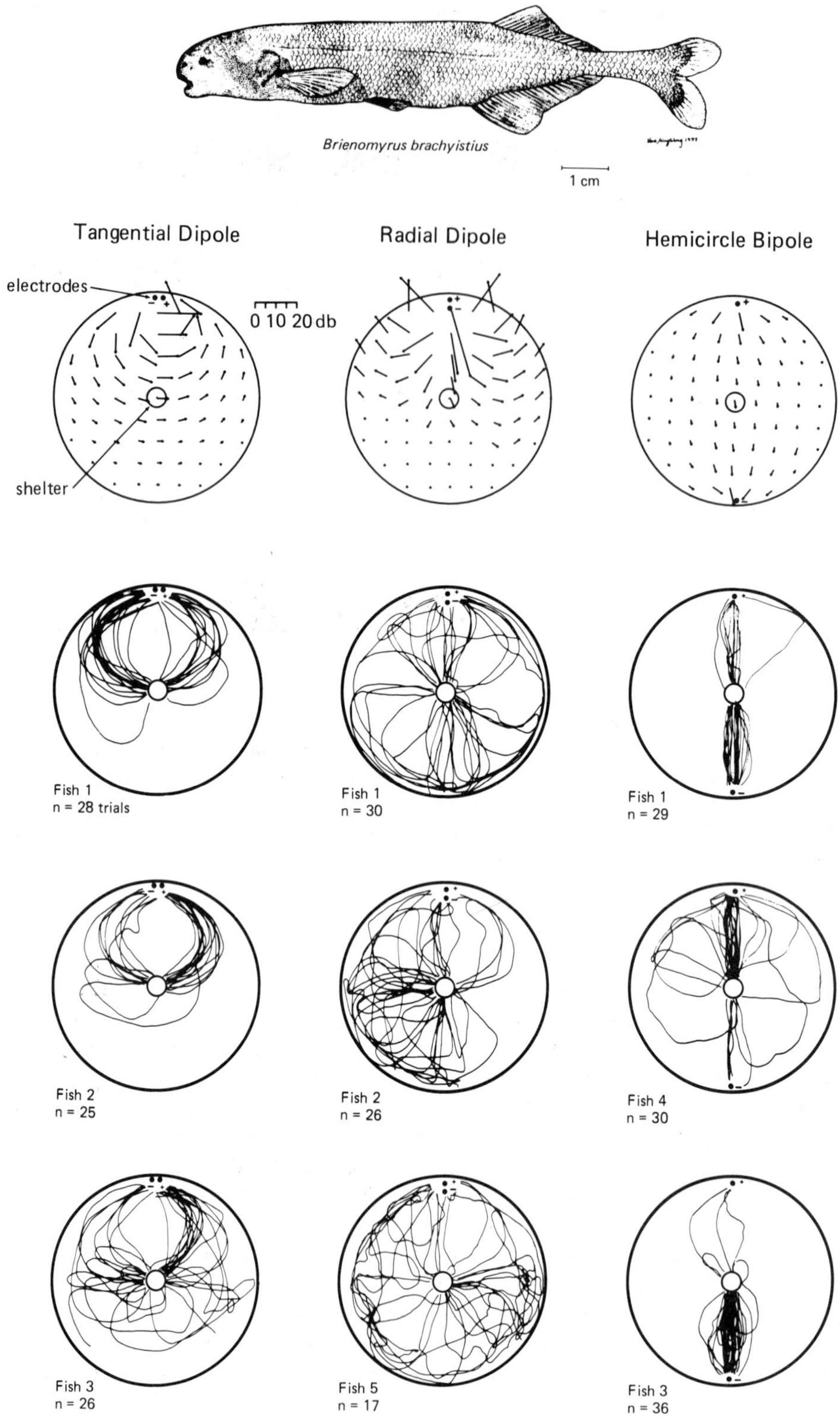
Brienomyrus brachyistius
1 cm
Tangential Dipole
Radial Dipole
Hemicircle Bipole
electrodes
0 10 20 db
shelter
Fish 1
n = 28 trials
Fish 1
n = 30
Fish 1
n = 29
Fish 2
n = 25
Fish 2
n = 26
Fish 4
n = 30
Fish 3
n = 26
Fish 5
n = 17
Fish 3
n = 36

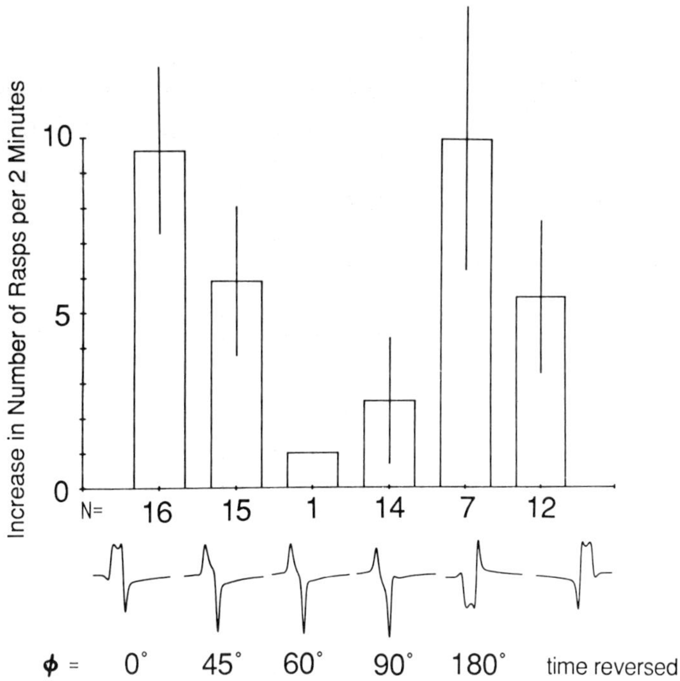


Figure 9.10. Unmated males of *Brienomyrus brachyistius* (triphasic) in their territories at night produce rasps in response to experimental playback of female EODs driven by scrambled SPIs. The ordinate in these histograms shows mean number of rasps produced during two minutes of EOD stimulation; the abscissa shows six different synthetic EODs derived from natural female EODs of the species. The first (0 degree) is an unmodified female EOD that elicits vigorous response from the male. Second to fifth histograms show responses to female EOD that has been filtered by unity gain, constant phase filter of 45, 60, 90, or 180 degrees, respectively. Final histogram shows responses to time-reversed female EOD. Males do not discriminate between 0- and 180-degree phase shifts, as might be expected. These shifts occur naturally as the female turns around. They do respond less strongly to phase-shifted and time-reversed EODs, however. Since all the stimuli have same power spectrum, this shows that the fish is "listening" for temporal, not spectral, cues in EOD signature when it makes EOD discrimination. (Adapted from Hopkins and Bass 1981.)

Figure 9.9. Electric fish do not appear able to locate another electric fish; instead, they follow local current lines until they reach the source (see text). In this experiment, a single mormyrid, *Brienomyrus brachyistius* (*top*), is held in a circular arena 1.1 m in diameter and provided with shelter in the center. In semidarkness, one pair of bipolar electrodes out of eight positioned every 45 degrees around the circumference of the tank is used to produce an electric signal mimicking a conspecific intruder. Active electrode is replotted here at the top of the circle (or along vertical axis in the case of hemicircle-bipolar electrode configuration). As resident fish approaches the active electrode, its path is traced on a video screen. Measurements of electric field amplitude and direction are shown as a vector diagram in the top row. Three electrode orientations are compared: tangential dipole (two electrodes spaced 4.5 cm apart tangential to the edge of the circle); radial dipole (electrodes aligned to the radius of the circle); and hemicircle bipole (electrodes widely separated along a diameter of the circle).

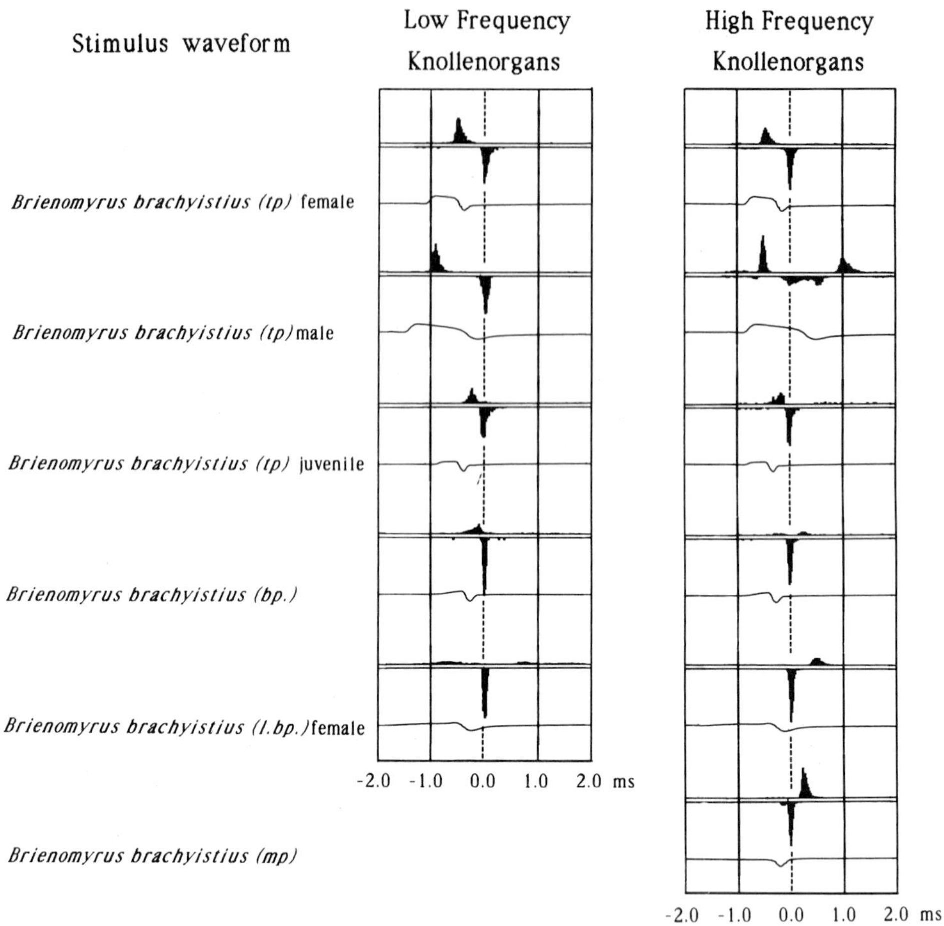


Figure 9.11. Knollenorgan electroreceptors translate EOD stimuli into a temporal pattern of spikes shown here as a series of compound PST histograms. Test species, *Brienomyrus brachyistius* (triphasic), has two "classes" of Knollenorgans, one tuned to 500–600 Hz (low frequency; responses shown in left column), others to 1,000 to 3,000 Hz (high frequency; responses shown in right column). Compound PST histograms are composed of two PST histograms: the response to normal pulse polarity is plotted upwards; the response to inverted polarity, downward. Stimulus EOD waveform is shown below each histogram and identified to the species at the left. Spikes tend to cluster into distinct peaks, and the interval between is characteristic for the stimulus waveform used.

duration of the rectangle, the interval between spikes from opposite sides of the body can be changed. Further testing with an array of rectangle stimulus durations (Fig. 9.12) demonstrates that males are tuned to respond behaviorally to 0.6-ms rectangles, which is slightly longer than reported originally. We conclude from these experiments that *B. brachyistius* (tp) recognizes the species-specific signature of the EOD by recognition of the temporal pattern of spikes evoked in the Knolle-

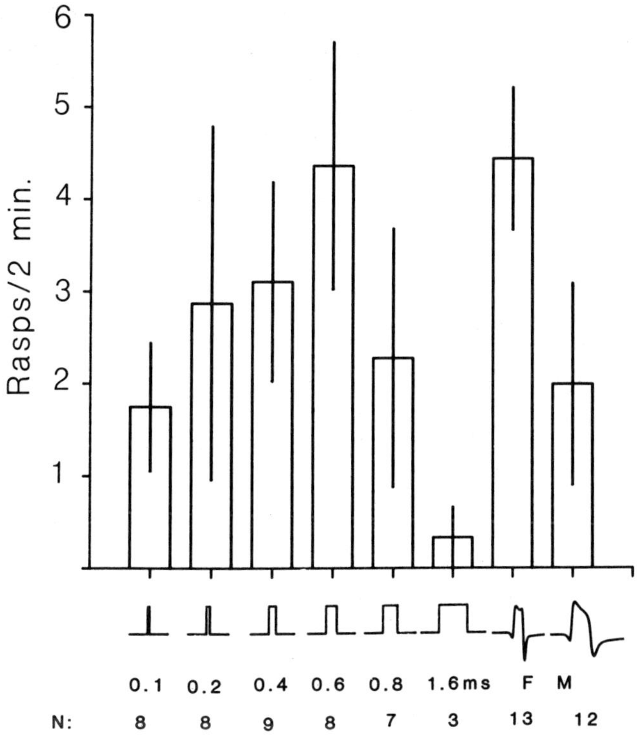


Figure 9.12. Rectangles are good "sign stimuli" for eliciting rasp response from male *B. brachyistius* (triphasic) in the field, and rectangles 0.6 ms in duration are an optimal sign stimulus. The ordinate shows average number of rasps elicited during 2-minute playback experiment; the abscissa shows stimulus waveform. The onset of the rectangle triggers spike response from Knollenorgans on one side of the body; the offset triggers spikes in Knollenorgans on opposite side. By adjusting the duration of the rectangle, one changes the interval between spikes elicited in two Knollenorgan populations. Males respond equally to their own female EODs and to rectangles 0.6 ms in duration and respond less to male EODs and to rectangles of longer or shorter duration. The 0.6-ms interval is slightly longer than the interval between peaks in PST histogram in response to the EOD shown in Figure 9.11.

norgan receptors over the body surface. This temporal pattern may require integration of inputs from different parts of the body that see opposite polarities of the same stimulus that are in phase with each other. Other species of mormyrids appear to have more complex temporal codes for their EOD signatures and may have more complex temporal recognition mechanisms as well.

To have species recognition rely on a simple neural code composed of only two spikes separated by an appropriate interval may be unique to the electrical communication modality. In this mode, signal waveforms are invariant, because the conduction of the signal is instantaneous, and there are no echoes and reverberations. Temporal cues, though important in the recognition of acoustic cues in species recognition, operate on a longer time scale and are probably far more complex. Another reason for the simplicity of the neural code for species recognition undoubtedly stems from the simplicity of the communication channel, which is used by only a few species of electric fish. By comparison, acoustic and chemical signalling is made complex by the sheer number of individuals and species using the same channel of communication.

## Summary

Social behavior between animals is mediated by communication through a variety of sensory modalities. Chemical, acoustic, tactile, visual, and electrical signals are used in aquatic environments. Some animals have specialized communication sensors for the detection of signals and displays of others. Signalling behavior and the receptors designed to receive them are under selection and adapted to the function and environment. Some signals are designed to carry great distances; others, for use at close range. Some signals are easily located, others less so. Some are species specific and unique in a given environment, other signals are less specific. Communication sensors are adapted to a variety of functions in communication: the initial detection of the signal in the presence of noise, the localization of the signal in space, and the recognition of the signal as one of a number of possible alternatives. This chapter attempts to illustrate a number of these adaptations with examples from work on electroreception in electrical communication.

### References

Bass, A.H. (1986) Electric organs revisited: evolution of a vertebrate communication and orientation organ, in Electroreception, Bullock, T.H. and Heiligenberg, W. (eds.), John Wiley and Sons, New York, pp. 13–70.

Bass, A.H. and Hopkins, C.D. (1983) Hormonal control of sexual differentiation: changes in electric organ discharge waveform, Science, 220:971–974.

Bass, A.H. and Hopkins, C.D. (1984) Shifts in frequency tuning of electroreceptors in androgen-treated mormyrid fish, J. Comp. Physiol., 155:713–724.

Bass, A.H. and Hopkins, C.D. (1985) Hormonal control of sex differences in the electric organ discharge (EOD) of mormyrid fishes, J. Comp. Physiol., 156:587–604.

Bell, C.C. (1979) Central nervous system physiology of electroreception, a review, J. Physiol. (Paris), 75:361–379.

Bell, C.C. (1986) Central physiology of the mormyrid electrosensory system, in Electroreception, Bullock, T.H. and Heiligenberg, W. (eds.), John Wiley and Sons, New York, pp. 423–452.

Bell, C.C., Myers, J.P., and Russell, C.J. (1974) Electric organ discharge patterns during dominance related behavioral displays in *Gnathonemus petersii*, J. Comp. Physiol., 92:210–228.

Bell, C.C., Bradbury, J., and Russell, C.J. (1976) The electric organ of a mormyrid as a current and voltage source, J. Comp. Physiol., 110:65–88.

Bell, C.C. and Szabo, T. (1986) Central structures and pathways of the mormyrid electrosensory system, in Electroreception, Bullock, T.H. and Heiligenberg, W. (eds.), John Wiley and Sons, New York, pp. 375–421.

Bennett, M.V.L. (1971a) Electric organs, in Fish Physiology, vol. 5, Hoar, W.S. and Randall, D.J. (eds.), Academic Press, New York, pp. 347–491.

Bennett, M.V.L. (1971b) Electroreception, in Fish Physiology, vol. 5, Hoar, W.S. and Randall, D.J. (eds.), Academic Press, New York, pp. 493–574.

Blaxter, J.H.S., Denton, E.J., and Gray, J.A.B. (1981) Acousticolateralis system in clupeid fishes, in Hearing and Sound Communication in Fishes, Tavolga, W.N., Popper, A.N., and Fay, R.R. (eds.), Springer-Verlag, New York, pp. 39–56.

Bossert, W.H. and Wilson, E.O. (1963) The analysis of olfactory communication among animals, J. Theor. Biol., 5:443–469.

Brenowitz, E. (1982) The active space of red-winged blackbird song, J. Comp. Physiol., 147:511–522.

Bullock, T.H., Hamstra, R.H., and Scheich, H. (1972) The jamming avoidance response of high-frequency electric fish. I. General features, J. Comp. Physiol., 77:1–22.

Capranica, R.R., Frishkopf, L.S., and Nevo, E. (1973) Encoding of geographic dialects in the auditory system of the cricket frog, Science, 182:1272–1275.

Crawford, J.D., Hagedorn, M.M., and Hopkins, C.D. (1985) Acoustic song in an electric fish, Abst. Soc. Neurosci., 11:270.

Crawford, J.D., Hagedorn, M.M., and Hopkins, C.D. (1986) Acoustic communication in an electric fish, *Pollimyrus isidori* (Mormyridae). J. Comp. Physiol. 159:297–310.

Dawkins, R. and Krebs, J.R. (1978) Animal signals: information or manipulation? in Behavioral Ecology, Krebs, J.R. and Davies, N.B. (eds.), Sinauer Associates, Sunderland, Mass., pp. 282–309.

Enger, P.S., Libouban, S., and Szabo, T. (1976) Fast conducting electrosensory pathway in the mormyrid fish, *Gnathonemus petersii,* Neurosci. Lett., 2:133–136.

Fine, M.L. (1981) Mismatch between sound production and hearing in the Oyster Toadfish, in Hearing and Sound Communication in Fishes, Tavolga, W.N., Popper, A.N., and Fay, R.R. (eds.), Springer-Verlag, New York, pp. 257–263.

Fine, M.L., Winn, H.E., and Olla, B.L. (1977) Communication in fishes, in How Animals Communicate, Sebeok, T. (ed.), Indiana University Press, Bloomington, pp. 472–518.

Fish, J.F. and Offutt, G.C. (1972) Hearing thresholds from toadfish, *Opsanus tau,* measured in the laboratory and field, J. Acoust. Soc. Am., 51:1318–1321.

Fisher, F.H. and Simmons, V.P. (1977) Sound in sea water, J. Acoust Soc. Am., 62:558–564.

Griffin, D.R. (1971) The importance of atmospheric attenuation for the echolocation of bats (Chiroptera), Anim. Behav., 19:55–61.

Hagedorn, M. and Heiligenberg, W. (1985) Court and spark: electric signals used during the courtship and mating of gymnotid electric fish, Anim. Behav., 33:254–265.

Hawkins, A.D. and Horner, K. (1981) Directional characteristics of primary auditory neurons from the cod ear, in Hearing and Sound Communication in Fishes, Tavolga, W.N., Popper, A.N., and Fay, R.R. (eds.), Springer-Verlag, New York, pp. 311–329.

Hawkins, A.D. and Myrberg, A.A., Jr. (1983) Hearing and sound communication under water, in Bioacoustics: A Comparative Approach, Lewis, B. (ed.), Academic Press New York, pp. 347–405.

Heiligenberg, W. (1977) Principles of Electrolocation and Jamming Avoidance in Electric

Fish. A Neuroethological Approach. Studies in Brain Function I, Springer-Verlag, New York.

Hoin-Radkovsky, I., Bleckmann, H., and Schwartz, E. (1984) Determination of source distance in the surface-feeding fish *Pantodon buchholzi* Pantodontidae, Anim. Behav., 32:840–851.

Hopkins, C.D. (1973) Lightning as background noise for communication among electric fish, Nature, 242:268–270.

Hopkins, C.D. (1974) Electric communication: functions in the social behavior of *Eigenmannia virescens,* Behaviour, 50:270–305.

Hopkins, C.D. (1980) Evolution of electric communication channels of mormyrids, Behav. Ecol. Sociobiol., 7:1–13.

Hopkins, C.D. (1981) On the diversity of electric signals in a community of mormyrid electric fish in West Africa, Am. Zool., 21:211–222.

Hopkins, C.D. (1983a) Neuroethology of species recognition in electroreception, in Advances in Vertebrate Neuroethology, NATO ASI Series A, Ewert, J.P., Capranica, R.R., and Ingle, D.J. (eds.), Plenum, New York, pp. 871–881.

Hopkins, C.D. (1983b) Functions and mechanisms in electroreception, in Fish Neurobiology, vol. 1, Northcutt, R.G. and Davis, R.E. (eds.), University of Michigan Press, Ann Arbor, pp. 216–259.

Hopkins, C.D. (1986) Behavior of Mormyridae, in Electroreception, Bullock, T.H. and Heiligenberg, W. (eds.), John Wiley and Sons, New York, pp. 527–576.

Hopkins, C.D. and Bass, A.H. (1981) Temporal coding of species recognition signals in electric fish, Science, 212:85–87.

Huber, F. and Markl, H. (eds.) (1983) Neuroethology and Behavioral Physiology: Roots and Growing Points, Springer-Verlag, New York.

Kaissling, K.E. (1971) Insect olfaction, in Handbook of Sensory Physiology, Beidler, L.M. (ed.), Springer-Verlag, New York, pp. 351–431.

Kalmijn, A.J. (1974) The detection of electric fields from inanimate and animate sources other than electric organs, in Handbook of Sensory Physiology, vol. 3, Fessard, A. (ed.), Springer-Verlag, New York, pp. 147–200.

Kirschbaum, F. (1975) Environmental factors control the periodical reproduction of tropical electric fish, Experientia, 31:1159–1160.

Kirschbaum, F. and Westby, G.W.M. (1975) Development of the electric discharge in mormyrid and gymnotid fish (*Marcusenius* sp. and *Eigenmannia virescens*), Experientia, 15:1290–1293.

Knudsen, E.I. (1975) Spatial aspects of the electric fields generated by weakly electric fish, J. Comp. Physiol., 99:103–118.

Knudsen, E.I. (1983) Space coding in the vertebrate auditory system, in Bioacoustics: A Comparative Approach, Lewis, B. (ed.), Academic Press, New York, pp. 311–344.

Kramer, B. (1974) Electric organ discharge interactions during interspecific agonistic behaviour in freely swimming mormyrid fish, J. Comp. Physiol., 93:203–235.

Markl, H. (1983) Vibrational communication, in Neuroethology and Behavioral Physiology, Huber, F. and Markl, H. (eds.), Springer-Verlag, New York, pp. 332–353.

McCormick, C.A. and Popper, A.N. (1984) Auditory sensitivity and psychophysical tuning curves in the elephant nose fish, *Gnathonemus petersii,* J. Comp. Physiol., 155:753–761.

Michelsen, A. (1978) Sound reception in different environments, in Sensory Ecology, Ali, M.A. (ed.), Plenum, New York, pp. 345–373.

Moiseff, A. and Konishi, M. (1981) Neuronal and behavioral sensitivity to binaural time differences in the owl, J. Neurosci., 1:40–48.

Moller, P. and Bauer, R. (1973) Communication in weakly electric fish, *Gnathonemus petersii* (Mormyridae). II. Interaction of electric organ discharge activities of two fish. Anim. Behav., 21:501–512.

Moller, P. and Szabo, T. (1981) Lesions in the nucleus mesencephali exterolateralis: effects on electrocommunication in the mormyrid fish, *Gnathonemus petersii* (Mormyriformes), J. Comp. Physiol., 144:327–333.

Myrberg, A.A., Jr. (1980a) Ocean noise and the behavior of marine animals: relationships and implications, in Advanced Concepts in Ocean Measurements for Marine Biology, Diemer, F.P., Vernberg, F.J., and Mirkes, D.Z. (eds.), University of South Carolina Press, Columbia, pp. 461–491.

Myrberg, A.A., Jr. (1980b) Fish bio-acoustics: its relevance to the "not so silent world," Environ. Biol. Fishes, 5:297–304.

Myrberg, A.A., Jr. (1981) Sound communication and interception in fishes, in Hearing and Sound Communication in Fishes, Tavolga, W.N., Popper, A.N., and Fay, R.R. (eds.), Springer-Verlag, New York, pp. 395–425.

Myrberg, A.A., Jr., Spanier, E., and Ha, S.J. (1978) Temporal patterning in acoustical communication, in Contrasts in Behavior, Reese, E.S. and Lighter, F.J. (eds.), John Wiley and Sons, New York, pp. 137–179.

Myrberg, A.A., Jr. and Spires, J. (1980) Hearing in damsel fishes: an analysis of signal detection among closely related species, J. Comp. Physiol., 140:135–144.

Narins, P.M. and Capranica, R.R. (1976) Sexual differences in the auditory system of the tree frog, *Eleutherodactylus coqui,* Science, 192:378–380.

Narins, P. and Capranica, R.R. (1978) Communicative significance of the two-note call of the treefrog, *Eleutherodactylus coqui,* J. Comp. Physiol., 127:1–9.

Payne, R. and Webb, D. (1971) Orientation by means of long range acoustic signalling in baleen whales, Ann. N.Y. Acad Sci, 188:110–141.

Piercy, J.E., Embleton, T.F.W., and Sutherland, L.C. (1977) Review of noise propagation in the atmosphere, J. Acoust. Soc. Am., 61:1403–1418.

Rigley, L. and Marshall, J.A. (1973) Sound production by the elephant-nose fish, *Gnathonemus petersii* (Pisces, Mormyridae), Copeia, 1973:134–135

Russell, C.J. and Bell, C.C. (1978) Neuronal responses to electrosensory input in mormyrid *valvula cerebelli,* J. Comp. Physiol. 41:1495–1510.

Schneider, D. (1965) Chemical sense communication in insects, Symp. Soc. Exp. Biol., 20:273–297.

Schneider, D. (1969) Insect olfaction: deciphering system for chemical messages, Science, 163:1031–1037.

Schuijf, A. and Buwalda, R.J.A. (1975) On the mechanism of directional hearing in cod (*Gadus morrhua* L.), J. Comp. Physiol., 98:333–343.

Squire, A. and Moller, P. (1982) Effects of water conductivity on electrocommunication in the weak-electric fish, *Brienomyrus niger* (Mormyriformes), Anim. Behav., 30:375–382.

Stipetić, E. (1939) Über das gehörorgan der Mormyriden, Z. Vgl. Physiol., 26:740–752.

Szabo, T. (1965) Sense organs of the lateral line system in some electric fish of the Gymnotoidae, Mormyridae, and Gymnarchidae, J. Morphol., 117:229–250.

Szabo, T. (1974) Anatomy of the specialized lateral line organs of electroreception, in Handbook of Sensory Physiology, vol. 3, Fessard, A. (ed.), Springer-Verlag, New York, pp. 13–56.

Tavolga, W.N. (1956) Visual, chemical, and sound stimuli as cues in the sex discriminatory behavior of the gobiid fish, *Bathygobius soporator,* Zoologica, 41:49–64.

Tavolga, W.N. (1958) The significance of underwater sounds produced by males of the gobiid fish, *Bathygobius soporator,* Physiol. Zool. 31:259–271.

Tavolga, W., Popper, A.N., and Fay, R.R. (1981) Hearing and Communication in Fishes, Springer-Verlag, New York.

van Bergeijk, W.A. (1967) The evolution of vertebrate hearing, Contrib. Sens. Physiol., 2:1–49.

von Frisch, K. (1938) The sense of hearing in fish, Nature, 141:8–11.

von Heusinger, C.F. (1826) Bemerkungen über das Gehörwerkzeug des *Mormyrus Cyprinoides, Gastroblecus Compressus* und *Pimelodus Synodontis,* Arch. Anat. Physiol. Meckel, 1:324–327.

Watanabe, A. and Takeda, K. (1963) The change of discharge frequency by A.C. stimulation in a weak electric fish, J. Exp. Biol., 40:57–66.

Wenz, G.M. (1962) Acoustic ambient noise in the ocean: spectra and sources, J. Acoust. Soc. Am., 34:1936–1956.

Wenz, G.M. (1964) Curious noises and the sonic environment in the ocean, in Marine Bio-Acoustics, Tavolga, W.N. (ed.), Pergamon Press, New York, pp. 101–119.

Wenz, G.M. (1972) Review of underwater acoustic research: noise, J. Acoust. Soc. Am., 51:1010–1024.

Werns, S. and Howland, H. (1976) Size and allometry of the saccular air bladder of *Gnathonemus petersii* (Pisces: Mormyridae): implications for hearing, Copeia, 1976:1200–1202.

Westby, G.W.M. and Kirschbaum, F. (1982) Sex differences in the waveform of the pulse-type electric fish, *Pollimyrus isidori* (Mormyridae), J. Comp. Physiol., 15:399–403.

Wiley, R.H. (1983) The evolution of communication: information and manipulation, in Animal Behaviour, vol. 2, Halliday, T.R. and Slater, P.J.B. (eds.), Blackwell Scientific Publications, pp. 156–189.

Wiley, R.H. and Richards, D.G. (1978) Physical constraints on acoustic communication in the atmosphere: implications for the evolution of animal vocalizations, Behav. Ecol. Sociobiol., 3:69–94.

Zipser, B. and Bennett, M.V.L. (1976a) Responses of cells of posterior lateral line lobe to activation of electroreceptors in a mormyrid fish, J. Neurophysiol., 39:693–712.

Zipser, B. and Bennett, M.V.L. (1976b) Interaction of electrosensory and electromotor signals in lateral line lobe of a mormyrid fish, J. Neurophysiol., 39:713–721.

# Chapter 10

# The Comparative Neurology of Expectation: Stimulus Acquisition and Neurobiology of Anticipated and Unanticipated Input

Theodore H. Bullock *

According to the plan of this book, this chapter is to ask what fish do in the process of stimulus acquisition for the particular modality of electroreception. In this chapter, I propose to point out some behaviors in fish associated with acquiring electrosensory stimuli and to reexamine the general neurobiological problem of expectation, especially the physiology of anticipated and unanticipated sensory input and some of its comparative neurological aspects. Besides its own interest, this will relate the purely aquatic to general and human biology.

I intend the term *expectation* here to mean an inferred state of the nervous system accounting for any of a wide variety of behavioral or physiological signs that a more or less specific sensory input is anticipated, rather than being a surprise or simply novel information. This is not a rigorous definition, but it will serve if we do not invoke the concept or infer an expectation unless the signs are rather clearly suggestive of such a state. I do not imply a conscious awareness of anticipation. Expectation includes stored information and, hence, the broad questions of how the CNS stores information, both innately and by learning.

A great body of important expectations is acquired genetically in most animals, representing the partly, or largely, innate reactions to one's own movements, food, partners, enemies, and environmental features. More or less innate expectations of specific input must follow many rhythmical and voluntary actions and sign and aversive stimuli.

Expectation is often a form of learning. Not all learning, however, entails expectation. Sensitization and nonspecific depressions need not include any specific expectation. Habituation used in the carefully defined sense that distinguishes it from other forms of response depression always does. Indeed, if the habituating stimuli were quite regular, there is likely to be a rather precisely timed anticipation so that stimuli on schedule are followed by habituated (depressed) responses;

* Neurobiology Unit, Scripps Institution of Oceanography and Department of Neurosciences, University of California, San Diego, La Jolla, CA 92093, USA

but either premature or delayed stimuli elicit larger responses, technically called *dishabituated.* Imprinting might be said to include a learned expectation of the stimulus set to which the animal has been imprinted and a phylogenetic expectation of the range and preference for stimuli to which the species can be imprinted; the same is true for birdsong in those species that depend on imitating a model. Associative learning always requires an expectation.

Using this definition, expectations are at least as widespread in the animal kingdom as habituation and associative learning. I do not know how widely distributed are those expectations that follow voluntary self-commands; presumably they are nearly universal, depending on the use of "command." One can properly call the vestibulo-ocular and crossed extension reflexes the results of expectations that certain well-defined inputs are due, namely, the retinal image shift normally associated with turning the head or the muscle loading normally caused on one side by lifting the other leg off the ground. The same is true of so-called space-constant receptive fields, such as those of higher-order visual units in crustaceans that compensate for body tilt (Wiersma and Yamaguchi 1967) and similar units in mammals (Horn and Hill 1969; Schlag et al. 1980). One more large category of expectations is that form of social feedback that is the normal consequence of emitting species-characteristic sign stimuli.

In light of this broad and important category of functions of central nervous systems, it is clear that at least a great deal of stimulus acquisition behavior must involve some degree of specific expectation. I will consider first some examples of stimulus acquisition behavior and central preparedness in electroreception and then turn to some comments on current concepts related to central expectations in general.

## Stimulus Acquisition in Electroreception

Among the wide variety of vertebrates with specialized peripheral and central mechanisms for electroreception, many examples of behavior associated with acquisition of stimuli can be discerned. Some, such as exploratory swimming, belong under appetitive behavior and are not obviously specific to this sensory modality. Particular body *postures,* however, can be connected to electroreception quite directly. *Eigenmannia* has been shown to govern its normal upright posture not only under the influence of vestibular input (gravity), but simultaneously under the influence of tonic input from the tuberous electroreceptor system, which both detects the substratum and computes its plane (Meyer, Heiligenberg, and Bullock 1976). By tilting an electrically opaque floor, one observes the hovering fish tilt to an angle between that of the floor and that normal to gravity if the fish is within about a centimeter of the floor. A visually opaque, but electrically transparent floor does not have this effect. The electroreceptive ventral substrate response (VSR), as this tendency to hover at right angles to the plane of the substratum was called, far outweighs any dorsal light response in these electric fish.

Electric fish of many species appear to *probe* novel objects with their tails. It is a striking sight to watch *Eigenmannia* investigating a novel object. The narrow, knifelike, pointed tail actively moves around the object while the fish backs into it. Knife fish in general tend to keep their bodies stiff and move forwards or backwards by adjusting the rapid undulations of their long ventral fins, perhaps to reduce relative motion between the electric field set up by the electric organ discharge and objects nearby. They can and do bend freely, as when probing, though this is mainly with the narrower part of the tail. Mormyrids, on the other hand, are in a different family and bend their wide fish tails more freely; they have a rich innervation in the narrow neck of the tail, presumably for accurate proprioceptive monitoring of the position of the tail.

Electric fish of many pulse species (as distinct from the less changeable wave species) increase the *rate of electric organ discharge* (EOD) under conditions suggestive of increased interest, as when food juice or mechanical, electrical, or social novelty is introduced; the increased rate may serve to increase input. Some mormyrids switch on or off episodically, a special form of EOD control called *echoing,* in which they discharge their brief pulse at a fixed, short interval after the discharge of a neighbor. Although we cannot assert the primary significance of echoing or even whether it is more important as a form of social signal or avoidance of jamming (interference with stimulus acquisition by the near coincidence of EOD from other fish), it is certainly a specialization in the realm of stimulus acquisition and one differentially developed in certain species.

Short-term control of *amplitude and form* of the EOD is essentially not used. A curious and apparently general feature of electroreceptive systems is that no *efferent control* of the receptors is present in any of the independently evolved electrical sense systems.

A proper knowledge of stimulus acquisition requires knowing the relevant stimuli and their biologically significant parameters, e.g., object characterization as to shape, size, conductance, inhomogeneity, localization, and motion. Although our knowledge is still inadequate, it seems possible that electroreception uses fewer features and parameters than do some other modalities, such as audition (at least in its more advanced forms), and therefore makes fewer kinds of discriminations. Electroreception and audition have been compared in other reviews (Bullock 1981, 1986b).

## Plastic Corollary Discharge

Some signs by which we can infer a central expectation (one that can be attributed to the CNS) are physiological, rather than behavioral. The African freshwater mormyriforms have a corollary discharge of the EOD command with a widespread central distribution. One of its properties is a plasticity that makes it perhaps the most interesting case of corollary discharge so far analyzed in any animal (see below). Assumed to be an accompaniment of all voluntary actions among vertebrates (H.-L. Teuber personal communication), the corollary discharge is believed to be a very common mechanism. Electrophysiological signs of neuronal activity believed to represent a corollary discharge have been reported in a number of cases (Richmond and Wurtz 1980; Evarts and Fromm 1981; Seal, Gross, and Bioulac 1982; Zaretsky 1982; Bell, Libouban, and Szabo 1983; Guthrie, Porter, and Sparks 1983; Stark and Bridgeman 1983; Bon, Corazza, and Inchingolo 1984; Hoffmann et al. 1984; Peck 1984a, 1984b, Toyama, Komatsu, and Shibuki 1984; Seal and Commenges 1985). In the African mormyrids, but not in the South American gymnotiform electric fish, a corollary discharge is found for each EOD and thought to be responsible for briefly suppressing the central effectiveness of input from a subset of receptors sensitive to other fish (electrocommunication) while relatively enhancing for a few milliseconds the input from electrolocation receptors (Bell, Libouban, and Szabo 1983).

The plastic consequence of the mormyrid corollary discharge, referred to above, was discovered by Bell's (1986) analyzing second-order afferent neurons in the medullary relay nucleus of the mormyrid *Gnathonemus*. He recorded from cells in the ampullary electroreceptive subsystem (equivalent to vestibular or cochlear nuclei) while monitoring the command to the motor neurons from the brainstem EOD command nucleus. With a curariform drug, he prevented the electric organ from firing. Now, triggering the oscilloscope sweep from the brainstem command signal, the activity in the medullary, second-order ampullary system neurons shows low-level spontaneous firing in the absence of any correlated input. If an artificial electric field is introduced into the water, triggered by the command signal, the recorded unit immediately shows a response, with peaks and troughs of characteristic latencies in the spike histogram. Their exact timing and sequence depend on the polarity, strength, duration, and delay of the stimulus. Maintaining the same stimulus for a few minutes, always triggered by the fish's EOD command signal, the response evoked in the second-order ampullary unit declines, and the histogram more or less flattens out. One may imagine some simple habituation mechanism but, on omitting the stimulus, one sees a clear mirror-image pattern of the initial response to the stimulus in the second-order neurons. This in turn disappears in a few minutes; reinstating the stimulus restores the initial pattern—for a minute or two. Reversing the polarity or changing the intensity, duration, or delay of the stimulus, within limits, alters the form of the initial response and therefore of the mirror-image response seen after a few minutes of stimulation followed by its omission. Evidently an expectation of a certain input is built up over about 3–5 min of maintained input, normally involving a few hundred command cycles. The neuronal expectation signal takes the form of the response to the input, but with the opposite sign and somewhat less amplitude, hence not quite cancelling the response to the expected input. Here are neurons with a stimulus-specific habituation, i.e., a response decrement specific to the polarity, intensity, and timing. In each new environment where the fish hovers, by this mechanism it will soon habituate to the ampullary stimulation. Only unexpected input gets through these cells.

The other principal group of electric teleosts, the gymnotiforms from South America, seem to do as well without a corollary discharge, using only input from an array of sense organs. Clearly, corollary discharge is not so universal as has been thought, but is an option in evolution.

Elasmobranchs are believed also to lack a corollary discharge, though to make a claim of absence convincing, substantially more effort will be necessary. Their highly sensitive ampullary electroreceptors are heavily influenced by their own movements, including those of inspiration and expiration, which swing all the receptors in the same direction. Inspiration in the thornback ray *Platyrhinoidis triseriata* strongly excites, and expiration nearly silences the first-order ampullary afferent fibers, therefore superimposing a large, cyclic interference on the signals of interest from novel objects. Montgomery (1984) found that these swings with ventilatory movements are eliminated in the activity of the second-order neurons of the medullary (dorsal octavolateral) nucleus, which are thus able to signal novelty without that interference. The evidence suggests that the mechanism amounts to *in-phase rejection* in the second-order cells. Presumably, convergence of two effectively equal input moieties occurs, one of which is excitatory and one inhibitory, from ampullae of opposite best orientations. If this is indeed the story in elasmobranchs, it is an elegantly simple solution to a common noise problem.

## Efference Copy vs. Corollary Discharge

I turn now to a brief comment on two long-standing concepts that concern anticipated input. Coming from the systems-analytic approach, which infers processes in the brain from behavioral measurements, the notions of efference copy and corollary discharge have had central positions for many years in discussions of central expectation. This is not the place for a full review, but it seems timely to point out that the field has moved, that a distinction that was originally real has been largely lost in usage, and that the term *efference copy* is inappropriate for studies on the physiology of the brain.

*Efferenzkopie* (von Holst and Mittelstaedt 1950; Mittelstaedt 1971) was a useful concept heuristically and became a favorite term in titles of articles, but it is rarely used in accordance with the original authors' specifications or in a sense distinct from the more permissive and physiological term *corollary discharge*. Efference copy referred to a highly specified expectation of sensory input caused by the execution of a command, precise in amplitude, direction, and latency. Fruitful in behavioral systems analysis, the term suffers in the physiological context from the lack of neuronally applicable criteria for recognizing the presumed message. The message left behind by the command until the reafference (afferent input due to one's own movements) arrives could hardly be a copy of the command, but was assumed to subtract from the expected reafference, so that any discrepancy signifies novelty. Only one or two cases were considered by Mittelstaedt (personal communication) to satisfy the criteria, that is, the analysis of the control circuit required the assumption of such a message; principally these were the voluntary lateral deviation of the eyes and possibly the visual accommodation movements in humans. The evidence, however, on which the analysis of space constancy in lateral deviation depends has been reinterpreted (Bridgeman 1979, 1981; Bridgeman et al. 1979; Nagle, Bridgeman, and Stark 1980; Stark and Bridgeman 1983; Bridgeman and Delgado 1984).

In contrast, *corollary discharge* (Sperry 1950; MacKay 1956, 1973; Teuber 1960; MacKay and Mittelstaedt 1974), which also refers to a message created by a voluntary command and left behind in the brain to influence the evaluation of consequent input, is a broad term. Teuber (personal communication) declared it must be a universal mechanism in voluntary actions, but we have just seen that it is not that general. The concept is physiological but at best imprecise; it does not make clear exclusions of nonmembers of the class or provide criteria for distinguishing among discharges, either in collaterals of the main motor pathway or following the first sign of a command.

These concepts have a historic and limited value currently. Physiologically, the need today is for detailed descriptions of the activity of relevant neurons or populations of them, following the initiation of effector actions that can alter sensory input. In the meantime, continued psychophysical and experimental behavioral studies can select among models and discriminate contexts, as MacKay (1973) and Bridgeman et al. (1979) did between sensory-motor coordination of eye movement and the perceptual stability of the world during eye movement.

## Spike and Wave Signs of Central Expectation and Fluctuations of Readiness

Progress in the neurobiology of central expectation has been spectacular, both in respect to findings on neuronal spikes and on slow wave field potentials. A large share of the responses of neurons in the brain can best be understood as representing expectations of what the body and the environment will do. In addition, it is relevant to our theme that temporal fluctuations in readiness to respond keep turning up, including unfamiliar short- and very short-term fluctuations. In the space available, only a selection of discoveries in each of these areas will be mentioned.

*Neurons and small clusters of them* have been found to develop an expectation with experience, that is, to become specially responsive to particular input or its absence. Bell's (1986) case in the mormyrid medullary nucleus, reviewed above, is the most elegant in respect to its being constantly readjusted with experience every few minutes. Computed maps, such as those for acoustical space in the owl's midbrain (Knudsen 1985; Konishi 1985) and the motor map for eye movement in the deep tectum of monkeys (Wurtz and Goldberg 1972; Sparks and Mays 1980), represent learned expectations of precisely where a stimulus will be found.

Ranck (1973, 1976) found cells in the rat hippocampus that fire when the rat goes to the place where the water bottle belongs and finds it not there, as well as other cells that fire when the water bottle is there. Chelidze (1975) found what were called "extrapolating cells" in the visual cortex of rabbits. After a number of light flashes at fixed intervals, in the range of 1–10 sec, the cell was said to begin to fire just before the next flash. When the light stimulus was discontinued, a spike discharge appeared in the background from one to three times with the same interval at which the flashes previously followed one another.

From the meager data presented, one must question the statistical significance of the claim, beyond a gradual recovery from light flash, which could take at least 10 sec and include facilitation of excitation, as well as inhibition, as shown in the same preparation by Skrebitsky (1975). Sokolov (1960, 1975), Vinogradova (1966, 1975), John (1967), and Vinogradova and Dudaeva (1972) point out that cells in the hippocampus act like comparator cells, firing only to novelty and soon habituating, but coming to life again if the expected stimulus is altered. In the model of Sokolov (1960, 1975) for the orienting response, there are expectation cells, which fire according to the expected input; sensory-reporting cells, which fire according to the actual stimulus; and comparator cells, which fire whenever there is a discrepancy between the other two. Cells of each kind have been found.

*Preparatory set* has been recorded in units of many parts of the brain (Evarts, Shinoda, and Wise 1984). Other neuronal data related to states of expectation following some kind of learning, as well as surprise (so-called orienting response), have been reported by many authors (Carew, Castellucci, and Kandel 1971; Buchwald and Humphrey 1972; Schaefer 1972; Barnebey and Carterette 1973; Wood 1973; Danilova 1975; Jacklet and Lukowiak 1975; Koopowitz 1975; Kupfermann 1975; Schaefer, Schott, and Meyer 1975; Farel and Thompson 1976; Murphey and Matsumoto 1976; Bernardo, Stone, and Koopowitz 1977; Bryan and Krasne 1977a, 1977b; Strumwasser 1976; Egger 1978; Sudakov and Zhuravlev 1981; Hawkins et al. 1983; Hawkins 1984; Maximova and Balaban 1984; Reinoso-Suarez and Ajmone-Marsan 1984; Voronin 1984; Goh, Lederhendler, and Alkon 1985; Hocherman and Yirmiya 1985).

*Acoustical reafference* due to one's own vocalizations is well known to cause middle ear muscle reflexes—as though the system expects a loud sound. Additional central processes may enhance or reduce sensitivity, according to which of these is adaptive ethologically, as noted above for mormyrid electric fish. Bullock and Ridgway (1972) found that the dolphin's auditory midbrain is better prepared for an echo to the animal's own voluntarily emitted click than to one imposed at an arbitrary time; the recovery time is less than half of that to the arbitrary click and already appreciable at 0.5 ms. It is not clear how similar is the preparedness for echoes in the midbrain of the bat. Margoliash (1983) found cells of the songbird's hyperstriatum ventrale pars caudale (HVC) that respond best to the bird's own song dialect and are suppressed during its own singing.

Waves representing field potentials of assemblies of cells are now well known to signal either expectation or its opposite, surprise. The extensive literature on (1) contingent negative variation (CNV), sometimes called an expectancy wave, (2) endogenous event-related potentials (ERP), (3) motor readiness potentials (Bereitsschaftpotentialen), (4) habituation (which is a form of learning that a certain stimulus is to be expected and need not be strongly responded to), and (5) arousal and attention (Picton 1973; Squires KC, Squires NK, and Hillyard 1975; Valle and Levine 1975; Kerkhof 1978; Begleiter 1979; Hari, Sams, and Jarvilehto 1979; Ruchkin, Sutton, and Stega 1980; Freeman 1983; Gevins et al. 1983; Hillyard and Kutas 1983; McCallum, Zappoli, and Denoth 1984; Simson, Ritter, and Vaughan 1985; Timsit-Berthier 1984; Tsuda 1984) all clearly show that humans, monkeys, cats, and other species have waves that can be averaged out of the ongoing background EEG, time-locked to such cognitive events. All these involve some form of expectation, and most studies concern the increment in activity with surprise or, if the stimulus is expected, it is not on an easily learned schedule ("There's one!"); clear responses, however, are reported in the absence of any stimulation if one was expected on a learned schedule (e.g., the omitted click in a series; in human subjects, "correct rejections," as well as false alarms).

## Fluctuations in the Expectation of Input

Fluctuations in anticipated input and stimulus acquisition behavior are of course large in annual cycles, migrations, sleep-wakefulness cycles, and other circadian rhythms (Laming 1983; Laming and Ewert 1984; Treisman 1984). Less familiar are other roughly periodical ups and downs. The so-called 90-minute cycle in humans and rhesus monkeys during daytime active states is well established (Bowden, Kripke, and Wyborney 1978; Kripke 1982) and manifested by a spontaneous change in the domain of interest or preoccupation of the subject. Still less known are minute rhythms—on the order of 1–3 min. Stebel and Sinz (1971) gave evidence of 2-min fluctuations in memory performance (errors in remembering strings of numbers) and in reaction times (to regular or stochastically presented mixed light and sound stimuli). Makeig, Galambos, and Stapells (1986) have new evidence from evoked potentials of ca. 1-min rhythms in human response to standard, monotonous, meaningless buzzes. It may not be too surprising that such fluctuations in brain state occur in view of findings such as those reviewed by Golenhofen (1970) that common to the entire smooth muscle system in mammals are rhythms of 0.5–2 min, even in isolated blood vessels; he considers it automatic myogenic activity; under normal conditions it is synchronized over some area, as well as modified by neural and hormonal influences.

*Still shorter-term changes in central expectation* or readiness for stimulus acquisition are manifested by behavioral and physiological fluctuations. Quantitative measurements have been made of the performances of skilled musicians in estimating time intervals, for example, in saying "That intercalated click in the 1 per sec. series was between a quarter and three-eighths notes," or in tapping out beats and fractions of beats. Without going into the numbers, one sample result is that, for small fractions, the perceptual judgment of the violinist and conductor Paul Zukofsky overestimates significantly but consistently (Sternberg and Knoll 1984).

Short-term changes in the responsivity of the brain to standard stimuli are known in evoked potential recording when large fluctuations in amplitude of evoked potentials occur to successive stimuli, generally on a pseudorandom basis. At least some of this variance is said to be systematic and correlated with the amplitude of the ongoing EEG just before the stimulus (Basar 1980). The EEG amplitude, evidently influential on responsivity, is most obviously manifested in the waxing and waning of each frequency band of the EEG and can be displayed as the envelope of the output of band-pass–filtered EEG. I have been impressed with the large component of this waxing and waning in the range of 1 Hz and lower frequencies (Bullock 1986a). In other words the given part of the brain is changing in its response to clicks or flashes every second and even more often, irregularly, but in part predictably.

## Strategies for New Research

The diversity of sources of information reviewed above underlines the importance of using all the windows and techniques we have—single unit recording, multichannel electrical and optical recording, compound potential recording, magneto-electroencephalography, 2-deoxyglucose, local blood flow, and any others that can reveal relevant activity.

Intriguing possibilities come to mind if we combine this knowledge with one or both of the following: First, there is the growing appreciation and, I predict, rapidly changing view of *built-in central mechanisms of recognition* of complex, naturally important stimuli. One thinks of face units in the temporal cortex of monkeys (Perrett and Rolls 1983), of pretectal worm and "antiworm" units of toads (Ingle 1983; Ewert 1984, 1985), and of units in the HVC of songbirds that discriminate the dialect or even the individual bird's own song (Margoliash 1983). It does not matter whether one wishes to emphasize the presumed requirement—at least for some recognitions—for a large, defined spatiotemporal pattern among a set of neurons activated by such stimuli or to emphasize the known, remarkable complexity of the stimulus requirements for firing in a small set of approximately equivalent neurons. These are not mutually exclusive ideas, as is often implied in remarks about alleged "grandmother cells" (Bullock 1986c). In either case, we are dealing with built-in mechanisms, either innate or learned or both, that express expectation of certain stimuli and can be more or less readied or suppressed, for example owing to the organism's own commands.

Second, there is much less appreciation of the potentialities for insight from the *comparative dimension*. It will make a great difference to our understanding if we learn which aspects of the variety of central physiological mechanisms I have alluded to—those related to anticipated and unanticipated sensory input—are general and equivalent in invertebrates or in lower and higher vertebrates and which are only developed in the more advanced taxa. Separate from that, one wonders whether some differences are characteristic for certain species or families in contrast to other species or families of the same class, as in the two orders of electric fish, and in such cases whether a relevant ethological correlate can be discerned. These two domains of comparative neurology—the differences between higher and lower taxonomic categories, orders or classes versus genera or species, are not always distinguished. They may have quite different evolutionary, ecological, and behavioral significance.

## Summary

The nervous system shows preparedness for expected input in a variety of ways. The modalities and submodalities of sense organs, their receptive fields, filter properties, distributions, and sensitivities, constitute a first-order set of adaptations for the broad range of biologically important stimuli. An array of behaviors in each species aids in stimulus acquisition, including control of posture, probing, and sampling rate. Examples are cited from electroreception in fish. The wide range of types of expectation includes short-term and long-term ones, those built in phylogenetically, and those due to learning. Not all expectations are learned, and some results of experience do not include expectation (sensitization and non-specific depression).

Physiological specializations at intermediate levels are differentially developed among taxa, as well as modalities. Efferent control of sense organs is absent in some, and well developed in other modalities or submodalities. Compensatory reflexes, like the vestibulo-ocular and crossed extension reflexes, are numerous and widespread, but probably none is universal, and their distribution and correlation with habit of life remain important comparative problems. The same is true of central compensations, such as space-constant, higher-order visual units that shift receptive fields with head tilt.

A wide variety of central neurons fire in consequence of commands to effectors (''corollary discharge'') and exert various effects on the evaluation of input, especially reafference. In one case, in electroreception in mormyrids, the effect is plastic; a mirror image of the actual input is built up in a few minutes and largely cancels the response at the first relay, compensating for the intensity, latency, and polarity of input. Corollary discharge is apparently absent in gymnotiform electric fish. It is, therefore, an evolutionary option, not a general property of equivalent commands.

Common mode or in-phase rejection is apparently used in the first electrosensory relay nucleus by elasmobranchs to improve the detection of signals in the face of certain kinds of noise.

Spike and wave signs of central expectation include extrapolating cells; comparator cells; cells responsive only to complex stimuli simulating natural, important events; habituated and conditioned cells; and waves dependent on stimulus contingencies or cognitive events. Fluctuations in responsiveness extend from circadian and slower rhythms to minute (ca. 60 sec) rhythms and even shorter-term changes, down to the range of seconds.

Opportunities for new research emphasize the comparative dimension and relative neglect of the search for complex recognition neurons, known in a few cases to be matched to important, expected input.

### References

Barnebey, A. and Carterette, E.C. (1973) On habituation in the cochlear nucleus, Science, 179:1250–1252.

Basar, E. (1980) EEG-Brain Dynamics, Elsevier, Amsterdam.

Begleiter, H. (1979) Evoked Brain Potentials and Behavior, Plenum, New York.

Bell, C.C. (1986) Central physiology of the mormyrid electrosensory system, in Electroreception, Bullock, T.H., and Heiligenberg, W.F. (eds.), John Wiley and Sons, New York.

Bell, C., Libouban, S., and Szabo, T. (1983) Pathways of the electric organ discharge command and its corollary discharges in mormyrid fish. J. Comp. Physiol., 216:327–338.

Bernardo, K., Stone, G.C., and Koopowitz, H. (1977) Primitive nervous systems: peripheral habituation in decerebrate polyclad flatworms, J. Neurobiol., 8:141–150.

Bon, L., Corazza, R., and Inchingolo, P. (1984) Oculomotor corollary discharge in the cat pretectum, Arch. Int. .Physiol. Biochem., 92:153–157.

Bowden, D.M., Kripke, D.F., and Wyborney, V.G. (1978) Ultradian rhythms in waking behavior of rhesus monkeys, Physiol. Behav., 21:929–933.

Bridgeman, B. (1979) Adaptation and the two-visual-systems hypothesis, Behav. Brain Sci., 2:64–65.

Bridgeman, B. (1981) Cognitive factors in subjective stabilization of the visual world, Acta Psychol., 48:111–1121.

Bridgeman, B. and Delgado, D. (1984) Sensory effects of eyepress are due to efference, Percept. Psychophys., 36:482–484.

Bridgeman, B., Lewis, S., Heit, G., and Nagle, M. (1979) Relation between cognitive and motor-oriented systems of visual position perception, J. Exp. Psychol. Hum. Percept. Perform., 5:692–700.

Bryan, J.S. and Krasne, F.B. (1977a) Protection from habituation of the crayfish lateral giant fibre escape response, J. Physiol. (Lond.), 271:351–368.

Bryan, J.S. and Krasne, F.B. (1977b) Presynaptic inhibition: the mechanism of protection from habituation of the crayfish lateral giant fibre escape response, J. Physiol. (Lond.), 271:369–390.

Buchwald, J.S. and Humphrey, G.L. (1972) Response plasticity in cochlear nucleus of decerebrate cats during acoustic habituation procedures, J. Neurophysiol. 35:864–878.

Bullock, T.H. (1981) Comparisons of the electric and acoustic senses and their central processing, in Hearing and Sound Communication in Fishes, Tavolga, W.N., Popper, A.N., and Fay, R.R. (eds.), Springer-Verlag, New York, pp. 525–571.

Bullock, T.H. (1986a) Compound potentials of the brain, ongoing and evoked: perspectives from comparative neurology, in Dynamics of Sensory and Cognitive Processing of the Brain, Basar, E. (ed.), Springer-Verlag, Berlin.

Bullock, T.H. (1986b) Significance of findings on electroreception for general neurobiology, in Electroreception, Bullock, T.H. and Heiligenberg, W. (eds.), John Wiley and Sons, New York, pp. 651–674.

Bullock, T.H. (1986c) Some principles in the brain analysis of signals: stimulus recognition and direction of evolution, Brain Behav. Evol., 28:145–156.

Bullock. T.H. and Ridgway, S.R. (1972) Evoked potentials in the central auditory system of alert porpoises to their own and artificial sounds, J. Neurobiol., 3:79–99.

Carew, T.J., Castellucci, V.F., and Kandel, E.R. (1971) An analysis of dishabituation and sensitization of the gill-withdrawal reflex in *Aplysia,* Int. J. Neurosci., 2:79–98.

Chelidze, L.R. (1975) Neuronal reactions of the rabbit's visual cortex as a function of the interval between flashes of light, in Neuronal Mechanisms of the Orienting Reflex, Sokolov, E.N. and Vinogradova, O.S. (eds.), English translation edited by Weinberger, N.E., Lawrence Erlbaum Associates, Hillsdale, N.J., and John Wiley and Sons, New York, pp. 52–62.

Danilova, N.N. (1975) The desynchronization reaction and its extinction in thalamic neuronal spike activity, in Neuronal Mechanisms of the Orienting Reflex, Sokolov, E.N. and Vinogradova, O.S. (ed.), Lawrence Erlbaum Associates, Hillsdale, N.J., pp. 200–213.

Egger, M.D. (1978) Sensitization and habituation of dorsal horn cells in cats, J. Physiol. (Lond.), 279:153–166.

Evarts, E.V. and Fromm, C. (1981) Transcortical reflexes and servo control of movement, Can. J. Physiol. Pharmacol., 59:757–775.

Evarts, E.V., Shinoda, Y., and Wise, S.P. (1984) Neurophysiological approaches to higher brain functions, John Wiley and Sons, New York.

Ewert, J.P. (1984) Tectal mechanisms that underlie prey-catching and avoidance behaviors in toads, in Comparative Neurology of the Optic Tectum, Vanegas, H. (ed.), Plenum, New York, pp. 247–416.

Ewert, J. P. (1985) Concepts in vertebrate neuroethology, Anim. Behav. 33:1–29.

Farel, P.B. and Thompson, R.F. (1976) Habituation of a monosynaptic response in frog spinal cord: evidence for a presynaptic mechanism, J. Neurophysiol., 39:661–666.

Freeman, W.J. (1983) The physiological basis of mental images, Biol. Psychiatry, 18:1107–1125.

Gevins, A.S., Schaffer, R.D., Doyle, J.C., Cutillo, B.A., Tannehill, R.S., and Bressler, S.L. (1983) Shadows of thought: shifting lateralization of human brain electrical patterns during brief visuomotor task, Science, 220:97–99.

Goh, Y., Lederhendler, I., and Alkon, D.L. (1985) Input and output changes of an identified neural pathway are correlated with associative learning in *Hermissenda,* J. Neurosci., 5:536–543.

Golenhofen, K. (1970) Slow rhythms in smooth muscle (minute rhythm), in Smooth Muscle, Bulbring, E., Brading, A.S., Jones, A.W. and Tomita, T. (eds.), Edward Arnold, London, pp. 316–342.

Guthrie, B.L., Porter, J.D., and Sparks, D.L. (1983) Corollary discharge provides accurate eye position information to the oculomotor system, Science, 221:1193–1195.

Hari, R., Sams, M., and Järvilehto, T. (1979) Auditory evoked transient and sustained potentials in the human EEG. I. Effects of expectation of stimuli, Psychiatry Res., 1:297–306.

Hawkins, R.D. (1984) A cellular mechanism of classical conditioning in *Aplysia,* J. Exp. Biol. 112:113–128.

Hawkins, R.D., Abrams, T.W., Carew, T.J., and Kandel, E.R. (1983) A cellular mechanism of classical conditioning in *Aplysia:* activity-dependent amplification of presynaptic facilitation, Science, 219:400–405.

Hillyard, S.A. and Kutas, M. (1983) Electrophysiology of cognitive processing, Annu. Rev. Psychol., 34:33–61.

Hocherman, S. and Yirmiya, R. (1985) Unit activity in the medial geniculate nucleus of a rhesus monkey reflects anticipation of acoustic signals, Neurosci. Lett. (Supp., 22):169.

Hoffmann, K.P., Bauer, R., Huber, H.P., and Mayr, M. (1984) Single cell activity in area 18 of the cat's visual cortex during optokinetic nystagmus, Exp. Brain Res., 57:118–127.

Horn, G. and Hill, R.M. (1969) Modifications of receptive fields of cells in the visual cortex occurring spontaneously and associated with bodily tilt, Nature (Lond.), 221:186–188.

Ingle, D.J. (1983) Prey selection in frogs and toads: a neuro-ethological model, in Handbook of Behavioral Neurobiology, Satinoff, E. and Teitelbaum, P. (eds.), Plenum, New York, pp. 235–261.

Jacklet, J.W. and Lukowiak, K. (1975) Neural processes in habituation and sensitization in model systems, in Progress in Neurobiology, vol. 4, Part I, Kerkut, G.A. and Phillis, J.W. (eds.), Pergamon Press, Oxford, pp. 1–56.

John, E.R. (1967) Mechanisms of Memory, Academic Press, New York.

Kerkhof, G.A. (1978) Decision latency: the P3 component in auditory signal detection, Neurosci. Lett. 8:289–294.

Knudsen, E.I. (1985) Auditory experience influences the development of sound localization and space coding in the auditory system, in Comparative Neurobiology: Modes of Communication in the Nervous System, Cohen, M.J. and Strumwasser, F. (eds.), John Wiley and Sons, New York, pp. 93–104.

Konishi, M. (1985) How auditory space is encoded in the owl's brain, in Comparative Neurobiology: Modes of Communication in the Nervous System, Cohen, M.J. and Strumwasser, F. (eds.), John Wiley and Sons, New York, pp. 335–350.

Koopowitz, H. (1975) Electrophysiology of the peripheral nerve net in the polyclad flatworm *Freemania litoricola,* J. Exp. Biol., 62:469–479.

Kripke, D.F. (1982) Ultradian rhythms in behavior and physiology, in Rhythmic Aspects of Behavior, Brown, F.M. and Graeber, R.C. (eds.), Lawrence Erlbaum Associates, Hillsdale, N.J., pp. 313–343.

Kupfermann, I. (1975) Neurophysiology of learning, Annu. Rev. Psychol., 26:367–391.

Laming, P.R. (1983) Relationships between the responses of visual units, EEGs and slow potential shifts in the optic tectum of the toad, in Advances in Vertebrate Neuroethology, NATO ASI Series A: Life Sciences, vol. 56, Ewert, J.P., Capranica, R.R and Ingle, D.J. (eds.), Plenum, New York, pp. 595–601.

Laming, P.R. and Ewert, J.P. (1984) Visual unit, EEG and sustained potential shift responses to biologically significant stimuli in the brain of toads *(Bufo bufo),* J. Comp. Physiol., 154:89–101.

MacKay, D.M. (1956) Towards an information-flow model of human behaviour, Br. J. Psychol., 47:30–43.

MacKay, D.M. (1973) Visual stability and voluntary eye movements, in Handbook of Sensory Physiology, Jung, R. (ed.), Springer-Verlag, Berlin, pp. 307–331.

MacKay, D.M. and Mittelstaedt, H. (1974) Visual stability and motor control (reafference revisited), in Cybernetics and Bionics, Proc. 5th Congr. Deutsche Gesellsch. f. Kybernetik, Nuremberg, 1973, R. Oldenbourg, Munich, pp. 71–80.

Makeig, S., Galambos, R., and Stapells, D. (1986) Studies of the auditory high-rates response; amplitude variations and their correlations with behavior, in Dynamics of Sensory and Cognitive Processing of the Brain, Basar, E. (ed.), Springer-Verlag, Berlin.

Margoliash, D. (1983) Acoustic parameters underlying the responses of song-specific neurons in the white-crowned sparrow, J. Neurosci., 3:1039–1057.

Maximova, O.A. and Balaban, P.M. (1984) Neuronal correlates of aversive learning in command neurons for avoidance behavior of *Helix lucorum* L., Brain Res., 292:139–149.

McCallum, W.C., Zappoli, R., and Denoth, F. (eds.) (1984) Proceedings 7th Congress on Event-Related Potentials, Elsevier, Amsterdam.

Meyer, D.L., Heiligenberg, W., and Bullock, T.H. (1976) The ventral substrate response. A new postural control mechanism in fishes, J. Comp. Physiol., 109:59–68.

Mittelstaedt, H. (1971) Reafferenzprinzip—Apologie und Kritik, in Vorträge der Erlanger

Physiologentagung 1970, Keidel, W.D. and Plattig, K.-H. (eds.), Springer-Verlag, Berlin, pp. 161–171.

Montgomery, J.C. (1984) Noise cancellation in the electrosensory system of the thornback ray: common mode rejection of input produced by the animal's own ventilatory movements, J. Comp. Physiol. 155:103–111.

Murphey, R.K. and Matsumoto, S.G. (1976) Experience modifies the plastic properties of identified neurons, Science, 191:564–566.

Nagle, M., Bridgeman., B., and Stark, L. (1980) Voluntary nystagmus, saccadic suppression, and stabilization of the visual world, Vision Res., 20:717–721.

Peck, C.K. (1984a) Saccade-related neurons in cat superior colliculus: pandirectional movement cells with postsaccadic responses, J. Neurophysiol., 52:1154–1168.

Peck, C.K. (1984b) Corollary discharge neurons in cat superior colliculus, Brain Res., 293:381–385.

Perrett, D.I. and Rolls, E.T. (1983) Neural mechanisms underlying the visual analysis of faces, in Advances in Vertebrate Neuroethology, NATO ASI Series A: Life Sciences, vol. 56, Ewert J.-P., Capranica, R.R. and Ingle, D.J. (eds.), Plenum, New York, pp. 543–568.

Picton, T.W. (1973) Evoked potential correlates of human auditory attention. Ph.D. thesis, University of California, San Diego.

Ranck, J.B., Jr. (1973) Studies on single neurons in dorsal hippocampal formation and septum in unrestrained rats, Exp. Neurol., 41:461–555.

Ranck, J.B., Jr. (1976) Behavioural correlates and firing repertoires of neurons in septal nuclei in unrestrained rats, in The Septal Nuclei, De France, J. (ed.), Plenum, New York, pp. 423–462.

Reinoso-Suarez, F. and Ajmone-Marsan, C. (eds.) (1984) Cortical Integration: Basic, Archicortical, and Cortical Association Levels of Neural Integration, Raven, New York.

Richmond, B.J. and Wurtz, R.H. (1980) Vision during saccadic eye movements. II. A corollary discharge to monkey superior cilliculus, J. Neurophysiol., 43:1156–1167.

Ruchkin, D.S., Sutton, S., and Stega, M. (1980) Emitted P300 and slow wave event-related potentials in guessing and detection tasks, Electroencephalogr. Clin. Neurophysiol., 49:1–14.

Schaefer, K.P. (1972) Neuronal elements of the orienting response. Microrecordings and stimulation experiments in rabbits, Bibl. Ophthalmol., 82:139–148.

Schaefer, K.P., Schott, D., and Meyer, D.L. (1975) On the organization of neuronal circuits involved in the generation of the orientation response (visual grasp reflex), Fortschr. Zool., 23:199–212.

Schlag, J., Schlag-Rey, M., Peck, C.K., and Joseph, J.P. (1980) Visual responses of thalamic neurons depending on the direction of gaze and the position of targets in space, Exp. Brain Res., 40:170–184.

Seal, J. and Commenges, D. (1985) A quantitative analysis of stimulus- and movement-related responses in the posterior parietal cortex of the monkey, Exp. Brain Res., 58:144–153.

Seal, J. Gross, C., and Bioulac, B. (1982) Activity of neurons in area 5 during a simple arm movement in monkeys before and after deafferentation of the trained limb, Brain Res., 250:229–243.

Simson, R., Ritter, W., and Vaughan, H.G., Jr. (1985) Effects of expectation on negative potentials during visual processing, Electroencephalogr. Clin. Neurophysiol., 62:25–31.

Skrebitsky, V.G. (1975) Plastic properties of the visual neurons in the alert rabbit, in

Neuronal Mechanisms of the Orienting Reflex, Sokolov, E.N. and Vinogradova, O.S. (eds.), English translation edited by Weinberger, N.E., Lawrence Erlbaum Associates, Hillsdale, N.J., and John Wiley and Sons, New York, pp. 94–100.

Sokolov, E.N. (1960) Neuronal models of the orienting reflex, in The Central Nervous System and Behavior, Brazier, M.A.B. (ed.), Josiah Macy, Jr., Foundation, New York, pp. 187–276.

Sokolov, E.N. (1975) Neuronal mechanisms of the orienting reflex, in Neuronal Mechanisms of the Orienting Reflex, Sokolov, E.N. and Vinogradova, O.S. (eds.), English translation edited by Weinberger, N.E., Lawrence Erlbaum Associates, Hillsdale, N.J., and John Wiley and Sons, New York, pp. 217–238.

Sparks, D.L. and Mays, L.E. (1980) Movement fields of saccade related burst neurons in monkey superior colliculus, Brain Res., 190:39–50.

Sperry, R.W. (1950) Neural basis of the spontaneous optokinetic response produced by visual inversion, J. Comp. Physiol. Psychol., 43:482–489.

Squires, K.C., Squires, N.K., and Hillyard, S.A. (1975) Decision-related cortical potentials during an auditory signal detection task with cued observation intervals, J. Exp. Psychol. Hum. Percept. Perform., 104:268–279.

Stark, L. and Bridgeman, B. (1983) Role of corollary discharge in space constancy, Percept. Psychophys., 34:371–380.

Stebel, J. and Sinz, R. (1971) On central nervous minute-periodicity and its coordination, J. Interdiscip. Cycle Res., 2:63–72.

Sternberg, S. and Knoll, R.L. (1984) Perception, production, and imitation of time ratios by skilled musicians, Ann. N.Y. Acad. Sci., 423:429–441.

Strumwasser, F. (1976) Properties of neurons and their relationship to concepts of plasticity, in Neural Mechanisms of Learning and Memory, Rosenzweig, M.R. and Bennett, E.L. (eds.), MIT Press, Cambridge, Mass., pp. 468–474.

Sudakov, K.V. and Zhuravlev, B.V. (1981) Burst pattern of unit discharges as a reflection of expectation of food reinforcement by hungry animals, Neurosci. Behav. Physiol., 11:155–158.

Teuber, H.-L. (1960) Perception, in Handbook of Physiology, Neurophysiology III, Field J. (ed.), Williams and Wilkins, Baltimore, pp. 1595–1668.

Timsit-Berthier, M. (1984) Variation contingente négative et composants endogènes du potentiel évoqué, Rev. Electroencephalogr. Neurophysiol. Clin. 14:77–96.

Toyama, K., Komatsu, Y., and Shibuki, K. (1984) Integration of retinal and motor signals of eye movements in striate cortex cells of alert cat, J. Neurophysiol., 51:649–665.

Treisman, M. (1984) Temporal rhythms and cerebral rhythms, Ann. N.Y. Acad. Sci., 423:542–565.

Tsuda, T. (1984) Contingent negative variation and the related thalamic activities. Observation on the responses evoked by ipsilateral and contralateral hand movements, Tokushima J. Exp. Med. (Japan), 31:29–32.

Valle, R.S. and Levine, J.M. (1975) Expectation effects in alpha wave control, Psychophysiology, 12:306–309.

Vinogradova, O.S. (1966) Dynamic classification of the reactions of hippocampal neurons to sensory stimuli, Fed. Proc. [Trans. Supp.] (U.S.), 25:397–403.

Vinogradova, O.S. (1975) The hippocampus and the orienting reflex, in Neuronal Mechanisms of the Orienting Reflex, Sokolov, E.N. and Vinogradova, O.S. (eds.), English translation edited by Weinberger, N.E., Lawrence Erlbaum Associates, Hillsdale, N.J., and John Wiley and Sons, New York, pp. 128–154.

Vinogradova, O.S. and Dudaeva, K.I. (1972) Comparator function of the hippocampus, (in Russian) Dokl. Akad. Nauk. S.S.S.R., 202:486–489.

von Holst, E. and Mittelstaedt, H. (1950) Das Reafferenzprinzip (Wechselwirkungen zwischen Zentralnervensystem und Peripherie), Naturwissenschaften, 37:464–476. English translation in Dodwell, P.C. (ed.), Perceptual Processing, Appleton-Century-Crofts, New York, pp. 41–71.

Voronin, L.L. (1984) Synaptic plasticity at the levels of the archicortex and neocortex, (in Russian) Neirofiziologiya, 16:651–665.

Wiersma, C.A.G. and Yamaguchi, T. (1967) Integration of visual stimuli by crayfish central nervous system, J. Exp. Biol., 47:409–431.

Wood, D.C. (1973) Stimulus specific habituation in a protozoan, Physiol. Behav., 11:349–354.

Wurtz, R.H. and Goldberg, M.E. (1972) Activity of superior collicular neurons in behaving monkey. III. Cells discharging before eye movements, J. Neurophysiol., 35:575–586.

Zaretsky, M. (1982) Quantitative measurements of centrally and retinally generated saccadic suppression in a locust movement detector neurons, J. Physiol. (Lond.), 328:521–533.

Part III

# Signal Extraction: Receptor Cells and Sensory Systems

In the context of physical constraints (Part I) and behavioral constraints (Part II), Part III treats the specific signal extraction characteristics of each different sensory system in its own subsection. Where appropriate, each subsection contains invertebrate and vertebrate examples indicating the importance of phylogenetic constraints. Phylogenetic diversity and constraint are treated specifically by Laverack (Chapter 11), Coombs, Janssen, and Webb (Chapter 22), Budelmann (Chapter 30) and Zakon (Chapter 32). In addition, phylogenetic diversity and constraints are treated in a global fashion in Part IV by Northcutt (Chapter 34).

## Part IIIA

# Chemoreception

Chemoreception is clearly the least understood of the major senses. Yet, in the aquatic environment chemoreception often plays a dominant role in animal behavior, particularly for many invertebrates. In harmony with the purpose of this book, there is no attempt to be complete in the coverage of chemoreceptive information. Instead, the chapters are chosen in their ability to generate much-needed conceptual advances. By necessity, this limits discussion to a few well-studied species. Examples of such species are catfish and lobsters. Other species, such as salmon, sharks, and mud snails exemplify similar and other principles in aquatic chemoreception. Some of the material learned from other species is incorporated in Chapters 1 and 2 by Carr and Atema and in Chapter 11 of this section.

In Chapter 11, Laverack illustrates the great diversity of known chemoreceptor systems, both at the level of receptor cells and receptor organs. Smell and taste, as known in terms referenced by humans, hardly suffice to describe the variety of chemoreceptor organs found even in one invertebrate species. Caprio (Chapter 12) describes the morphology and the physiological properties of smell and taste receptors in fish. The central pathways of smell and taste in fish are reviewed by Finger (Chapter 13), who includes a separate treatment of two other organ systems: the common chemical sense and the physiologically still quite unknown nervus terminalis system.

Derby and Atema (Chapter 14) discuss the recent developments in chemical signal extraction by the receptor cells of lobsters and crayfish, including the discovery of mixture suppression in receptor cells. Ache (Chapter 15) continues the chemosensory pathway of information processing in the lobster CNS where other forms of mixture suppression take place. Together, Chapters 12, 14, and 15 indicate that signal processing characteristics of fish and crustaceans have several common themes. This may well be an example of convergent evolution, presumably in response to yet unknown properties of the chemical stimulus environment.

# Chapter 11

# The Diversity of Chemoreceptors

MICHAEL S. LAVERACK*

Diversity is defined in the *Oxford English Dictionary* as "The condition or quality of being diverse, different or varied; difference, unlikeness."

In the present context, diversity can be qualified in a variety of ways.

The types of chemoreceptors in a range of animals may be many and varied in structure, and there may be concentration or dispersal of the organs across the body surface or both. Functionally there may be differences in sensitivity and in the nature of the stimulants. The resultant reflex responses or higher behavior may differ (e.g., acceptance or rejection; feeding or breeding). Total numbers may be few or many, and even the position of sensory cell bodies may be peripheral or central. The fact that there can be no specific rule at the organismic level may be reflected at the cellular level. We might ask if there is a general chemoreceptor cell type? Are all such cells the same throughout the animal kingdom? If not, what differences exist? Are chemoreceptors as conservative in structure as photoreceptors?

Chemoreceptors are composed of a number of elements, but the elements may be small or large in number. An organ is *usually* construed as being a collection of items forming a whole, rather than just individual cells, though in the protists, of course, organelles as part of a cell may have the function of several cells in a metazoan. It may help, therefore, to think about the fundamental requirements of animals in order to begin to organize thoughts about chemoreceptors.

See Chapters 1, 2, 12, 14, and 15 for information on specific attitudes towards the nature of chemoreception.

## Behavioral Diversity

Under this heading falls what most people immediately think of with regard to chemoreception, i.e., the impact on the animal of important and obvious chemical

* Gatty Marine Laboratory, University of St. Andrews, St. Andrews, Fife KY16 8LB, U.K.

cues in such matters as food detection, mate finding, and site selection (See Chap. 1).

## Functional Diversity

Functional diversity covers the physiological and biochemical responses of receptors to chemicals. The membrane of the receptor cell no doubt includes in its structure proteins or enzymes that alter in shape or charge under the influence of the chemical impact from outside and thus influence conductance across the membrane to establish the characteristic generator potential of the cell. There are real similarities between synaptic transmission and chemoreception, but though in the former the representative active chemicals seem at present to be relatively few in number, in the latter they are legion. There are likely to be several chemicals acting in common on both types of organ (e.g., glutamate or some short chain peptides), but the chemoreceptor has a far greater flexibility than the synapse.

### Generalists and Specialists

The phenomenon of special or general sensitivity has been discussed elsewhere (see Chap. 14). It covers the range from pheromonal detection (of one precise chemical) to the blunderbuss cell that responds to a number of compounds, possibly in the same chemical family.

### Dynamic or Static Endings

The effects of age (even short-lived animals age) are not always apparent but certainly exist. Turnover of receptor endings undoubtedly takes place, but how rapidly and for what reason is not clear. Do the endings become worn away or clogged up with chemicals that they cannot remove or degrade? In Crustacea, some of these events are obvious. The organs of the exoskeleton in the intermoult become abraded, broken, and worn down or covered with epiphytes and epizoites. In any holes and apertures, bacteria may gain access or algae settle, and thus impede the true function of the receptor. Blockage of any apertures may be significant, but, in a worn-down end organ, is it only the accessory structures that are affected, or are the internal dendrites also worn away? If hard structures are broken, what then of soft tissues? Do superficially placed ciliary structures become worn? Surely they must. Is the system dynamic, and are the cilia replaced? Surely they are.

### Internal and External Detection

The maintenance of life is not just a function of events external to the animal: it also demands some attention to internal conditions. Homeostasis requires that the variable modality be sensed and responded to; in the case of chemoreception, this may include blood chemistry or gaseous concentrations. In the blood, the significant feature may be the overall osmotic pressure or ion content, and it depends on the degree of hydration or dehydration, while gases ($O_2$, $N_2$, $CO_2$) may be influential in the reactions of respiratory centers and bring about fluctuations in breathing and ventilatory rates. So homeostatic detectors of body fluid and gases are vitally important and are chemoreceptors.

## Structural Diversity

Structural diversity covers a number of approaches, ranging from the gross morphological level to the finest cellular structure.

### Organ Systems

Human anthropocentric attitudes are misleading when one considers aquatic animals and their receptors. Chemoreceptors may be located anywhere on the animal body, and in some cases that means literally. While aggregations of chemoreceptors may form organs (rhinophores, osphradium, antennae), in many cases chemoreceptors seem scattered liberally all over the body surface. In Crustacea and in fish, it is possible that nearly every region of the body has chemosensors on it, including the tail (telson).

### Exposed or Protected Detectors

There are major differences in the arrangement of receptors and their exposure to the environment. Many are simply placed at the periphery, directly exposed to the bathing medium. They have endings that are open to the water, being, in the main, cilia, and without protection. Others, especially in arthropods, are heavily protected by accessory structures. There may be setae, canals, pouches, or depressions that hide the sensory organs.

### Few or Many?

It is not always clear why there is such a variation between two different organs in the numbers of neurons represented. There are two aspects of this. The organ (e.g., the antenna) may be composed of many individual branches and subdivisions, but, within each receptor, there may be just a few neurons, though the whole limb carries hundreds of thousands of receptors. In other cases, each seta contains numerous nerve cells (e.g., the lobster's aesthetasc hairs [Spencer 1986]). Elsewhere, chemoreceptive organs may possess either small numbers or intermediate clusters.

### Cilia

More and more descriptions show that receptor cells end in cilia. Many of these examples are known in chemoreceptors. One might reasonably ask whether there are any features that specify or distinguish a chemoreceptive cilium from other kinds. Twelve years ago Laverack (1974) tried to review this possibility, without reaching a conclusion. The variability possible in ciliary structure associated with sensors seems endless. The variations in chemoreceptors seem more extreme and numerous than in, say, mechanoreceptors, but there appears to be little that can be taken as a certain indicator of function: neither the number of total filaments nor the presence, absence, or multiplication of central filaments. This is disappointing, but makes the task of unravelling the puzzle even more challenging and interesting.

### Biochemical Structure

Information about the biochemical structure of individual chemoreceptor organs is now becoming available. Changes in genetic programming can no doubt alter the proteins of the receptor membranes, and minor alterations can occur. Human genetic variations, such as tasters and nontasters and people with a poor sense of smell, are known. Perhaps the same occurs in aquatic populations.

## Olfaction and Taste: Is There a Distinction?

Terrestrial and aerial animals have chemoreceptor populations that can conveniently be classified into those that respond to airborne substances, often at low concentrations and over long distances, as in olfaction, and those associated with food detection, which are usually more concentrated and presented directly by contact in a watery medium, as in the sensation of taste.

Molecules delivered in a watery medium probably surround the animal constantly, though more concentrated streams may originate from prey, mates, and other sources from time to time. These may lead to forms of behavior that culminate in locating the sources and then to further behaviors, such as mating and feeding.

In aquatic animals, it has long been a matter of debate as to what division, if any, can be made to discriminate these aspects. Are olfaction and taste separate modalities in water-living animals? Is the distinction worth making?

Laverack (1968) suggested that receptor organs responding to a high concentration of substances are often associated with tactile sensitivity and needing contact with material corresponding to taste, while receptors needing few molecules for stimulation and that may receive them in solution were effectively olfactory receptors. Atema (1977) proposed a functional division, with one class being concerned with food localization and needing local reflex activity, while a second class is concerned with the more complicated behavior observed in mate and prey detection. Altner and Prillinger (1980) stated that a "clear distinction between olfaction. . .and taste. . .is not convincing," and Ache (1982) believes that the separation is more than a simple physiological feature and quotes the complexity of known chemoreceptive pathways in the CNS as supporting evidence (see also Chaps. 2 and 15).

It is interesting, however, to note that Seelinger (1983) has shown clearly that in an aerial crustacean (the isopod *Hemilepistes reaumuri*) volatile substances are detected, a phenomenon analagous to olfaction, whereas waterborne substances are also detected, in a manner that may be equated with taste. This discrimination occurs at the sense cell level. Seelinger (1983) uses the terms *taste stimuli* for water-soluble compounds that require solution and contact with the organ and *odor stimuli* for those of higher vapor pressure that evaporate from the solvent and are transmitted by air. He suggests that the organs of *Hemilepistus* are often mixed in function but, nonetheless, identifies the division as a receptor function, rather than a central one.

The *Oxford English Dictionary* interestingly defines taste as the sense of "feeling or touch" and only subsequently as "the quality of a body brought *into contact* with certain organs of the mouth."

Acting on such clues, one might therefore suggest that, while aquatically borne substances surround an animal, there is still the prospect of a difference between olfaction and taste, based on what we now know of the structures involved (see p. 299 for further details). I would argue that, in both cases, input is likely to have far wider-ranging influence than we realize at present in terms of the effects on higher-level interneurons and behavioral events, and so the likelihood of a CNS neuronal network distinction being required is not really supported. What may now be evident on the basis of crustacean, insect, and molluscan anatomy is that there are some receptor organs having both mechano- and chemoreceptors together (e.g., molluscan rhinophores and crustacean funnel canals), whereas there are others in which only chemoreceptors are found (e.g., crustacean aesthetascs), even though neighboring sensilla may subserve the mechanical function (see also Keil and Steinbrecht 1984 for insects). Nerves that run to what is classified as a chemosensory neuropil in the CNS may well carry mechanoreceptor information as well, which may be integrated at the same or a different region.

I propose, therefore, that we consider a dual mechanical and chemical function as being the equivalent of taste and a more discrete chemoreceptor population as being concerned with olfaction. These attributes may have some correlation with threshold levels as well (Laverack 1968). In other words, the "contact" function is important as a component of taste. Investigation of other examples may confirm or deny such a proposition. A different view is held by other contributors to this volume (see Chaps. 1, 2, and 15).

**Proposed Definitions**

Taste: A combined sense, with both chemical and mechanical sensors active in the same end organ (e.g., funnel canals and "hedgehog hairs" in Crustacea). Contact is essential.

Olfaction: A sense in which chemical signals alone are received by a population of receptors (e.g., aesthetascs). Simultaneous contact is not essential.

## Diversity of Stimulatory Substances

The old classification of taste indicates a kind of diversity: salt, acid, bitter, and sweet. Now, and with particular respect to invertebrates, such a set of headings makes little sense. Even so, there are some distinct divisions of chemoreceptor populations, though few cases have been explored in sufficient detail to be definitive.

In most investigations, especially on new preparations, the general approach is to take some piece of possible prey or food (e.g., mussel or fish) and make an extract, pour this on the suspected appendage, and say whether the substance affects the preparation. The important step, however, is the next one, to stimulate the organ with a *known* substance, and then perhaps to combine several such known substances to determine what the rather complex stimulus of a natural prey might be.

The general thrust of these experiments has been to push into the background the old-fashioned anthropocentric view of the taste sense for a more modern approach based on the natural history of the animal at large. A lack of sensitivity toward sugars (e.g., sucrose) and bitters (e.g., quinine) in aquatic scavengers and predators might be expected (Mackie and Turner 1970), while a response to amines (found in many sources of food) and to water-soluble substances, such as saponins (Mackie, Lasker, and Grant 1968) should not surprise anyone. Of course, there are bound to be exceptions to such rules, but much work now points in the direction of using naturally occurring materials as sources of stimulants (see Chap. 1).

Table 11.1 is by no means exhaustive, but collects some of the information that has accrued in the last twenty years or so regarding stimulatory substances that affect various animals. This ignores materials such as pheromones that are very specific and probably exist in nearly all groups somehow, whether as specific lures for a mate or as general attractants for aggregation, or even as releasers that generate broadcast spawning in a population. Such materials are often highly specific, since any generally available chemical would tend to muddle any response; a sexually interesting and valuable substance (or mixture) ought to be limited to one species. Table 11.1 also tends to overlook the fact that many substances have not been assayed on particular preparations. Some authors use long lists of materials, but others very few, so that an omission from the list need not necessarily be exclusive of a possible effect.

Nonetheless, it does seem likely that such a table can give an indication of the types of material to which aquatic animals respond. It has become commonplace that amines and amino acids are important, yet this was not at all obvious twenty-five years ago. Similarly, attempts to stimulate with sugars tend to be without success, but of course natural foods and their extracts do tend to be successful.

Recourse to Table 11.1 suggests a number of things. Firstly, though it is an outmoded way of proceeding, there is no substitute for a knowledge of natural history: in particular, an examination of the natural food of the subject species. Trying to establish chemosensitivity of a carnivore when using plant extracts would not be very productive and vice versa. Even in highly visual animals, such as the cephalopods, long thought to be without much chemical sense at all, the presentation of certain dissolved substances leads to changes in respiratory behavior. These chemicals are not those associated with metabolism either, but are amines, such as glycine and proline, that might well be associated with natural prey such as bivalves (Boyle 1983, personal communication).

Frings and Frings (1965) showed that *Aplysia* is attracted by extracts of its natural food, the seaweed *Ulva*, and a concerted effort on chemosensitivity in this species might reveal the basic requirements of a herbivore. So far, it looks as though the osphradium of this animal is an osmo- cum mechanoreceptor (Stinnakre and Tauc 1969; Jahan-Parwar 1972), in contrast to that of a carnivorous prosobranch such as *Buccinum* (Bailey and Laverack 1966), which responds to a variety of amines.

Secondly, it appears that, in general, carnivores and scavengers respond to commonly occurring amines and amino acids arriving either in fluids originating close to the prey or from the prey tissues themselves. Little can be said about the proportionality of the components, and the results are often presented in such different ways that a comparison is difficult. Nonetheless, amino acids are important, and recently it seems that some less obvious cues might be equally significant: e.g., the noncyclic phosphatide AMP (Carr and Thompson 1983) and pyridine and its analogues (Hatt and Schmiedel-Jakob 1984). Imagination and good biochemical analyses of prey are invaluable tools of research into chemosensitivity.

Much of the work that has concentrated on electrophysiological analysis of chemoreceptors has simply indicated whether the units respond, but there is no way such an approach gives information on whether substances are behaviorally stimulatory or inhibitory. Even positive or negative receptor responses do not illustrate what the behavioral effect will be. This can only be seen by behavioral observations that show acceptance or rejection.

Some stimulatory chemicals, such as betaine and trimethylamine oxide, first tested by Laverack (1963), occur as natural products in dead tissues or normal components of tissue chemistry. Positive stimulation of a chemoreceptor population as determined by electrophysiological means is often taken as probably initiating acceptance by the animal, but, as Mackie and Mitchell (1982) have shown, trimethylamine depresses feeding in the sole *Solea solea*, while betaine stimulates feeding.

Substances that taste bitter to humans depress feeding in some fish (Hidaka, Ohsugi, and Kubomatsu 1978; Mackie 1981), and some naturally occurring compounds, such as anthroquinone and naphthypyrone sulfates, also deter fish feeding. A variety of unpalatable chemicals is released by marine invertebrates, and the mucous secretions from such organisms may all be loaded with such substances. Saponins, haloethers, and others may all be antagonistic toward potential predators.

It is not possible to determine, on electrophysiological or structural grounds,

Table 11.1. Correlation of Some Reports of the Sensitivity of Invertebrates to Chemical Substances with Selected Phyla

| Preparation | Alanine | Arginine | Aspartic acid | Aspartate | Asparagine | Citrulline | Cysteine | Glutamic acid | Glutamate | Glycine | Histidine | Hydroxyproline | Leucine | Lysine | Methionine | Ornithine | Proline | Phenylalanine | Serine | Taurine | Tyrosine | Tryptophan | Threonine | Valine | Betaine | Trimethylamine | Pyridine/Purine | Sucrose | Glucose | Trehalose | Lactic acid | GABA | Osmotic/salts | Natural products | Fish, mussels, etc. | Conspecifics | Authors |
|---|---|---|---|---|---|---|---|---|---|---|---|---|---|---|---|---|---|---|---|---|---|---|---|---|---|---|---|---|---|---|---|---|---|---|---|---|---|
| Behavior | | | | | | | | | | | | | | | | | | | | | | | | | | | | | | | | | | | | | |
| *Feeding* | | | | | | | | | | | | | | | | | | | | | | | | | | | | | | | | | | | | | |
| Polychaeta | | | | | | | | | | | | | | | | | | | | | | | | | | | | | | | | | | | | | |
| *Chaetopterus* | + | | + | | | | I | + | | + | | | | | I | | | | | + | | | | | | | | | | | | | | | | | Sumida and Case 1983 |
| Fanning | | | | | | | | | | | | | | | | | | | | | | | | | | | | | | | | | | | | | |
| *Aphrodite* | + | | | | | | | + | | ? | | | | | + | | ? | | | + | | | | | + | − | | | | | | | | | | | Lynes and Laverack unpublished |
| Pharynx | | | | | | | | | | | | | | | | | | | | | | | | | | | | | | | | | | | | | |
| Gastropoda | | | | | | | | | | | | | | | | | | | | | | | | | | | | | | | | | | | | | |
| *Bullia* | + | | | | | | + | + | | + | + | | − | | | | + | | + | | − | | | | | − | | − | − | | | | | | + | | Hodgson personal communication |
| Proboscis | | | | | | | | | | | | | | | | | | | | | | | | | | | | | | | | | | | | | |
| *Pleurobranchaea* | | | | | | | | | | | | | | | | | | | | | | | | | | | | | | | | | | | | | |
| Feeding | + | | + | | | | | − | + | + | | + | + | | + | | + | | | | | | | | | | | | | | | | | | | | |
| | | | | | | | | | | | | | | | | | | | + | | | | | | | | | | | | | | | | | | Bicker, Davis, and Matera 1982 |
| *Aplysia* | − | − | + | | + | | − | + | | − | | | − | − | + | | − | + | | | | | | | | | | | | | | | | | | | |
| | | | | | | | | | | | | | | | | | | | + | | + | − | − | − | | | | | | | | | | | + | | Frings and Frings 1965, |
| *Elysia* | | | | | | | + | + | + | | | | | + | + | | + | | | | | | | | | | | | | | | | | | | | Jahan-Parwar 1972 |
| Coelenterata | | | | | | | | | | | | | | | | | | | | + | | | | | + | | | | | | | | | | + | | Jensen 1982 |
| *Palythoa* | + | | | | | | | | | − | | + | | | + | | + | | | | | | | | | | | | | | | | | | | | |
| *Calliactis* | + | − | − | + | + | | − | + | | + | − | + | + | | + | | + | − | + | | | | | | | | | | | | | | | | | | Reimer 1971 |
| Crustacea | | | | | | | | | | | | | | | | | | | − | | − | + | − | − | | | | | | | | | | | | | Reimer 1973 |
| *Palaemonetes* | | | | | | | | | | | | | | | | | | | | | | | | | | | | | | | | | | | | | |
| *Homarus* | | | | | | | | | | | | | | | | | | | | | | | | | | | + | | | | | | | | | | Carr and Thompson 1983 |
| *Homarus* | + | + | − | | | + | | + | + | + | − | − | + | + | + | | + | + | | | | | | | | | | | | | | | | | + | + | McLeese 1973 |
| *Petrolisthes* | | | | | | | | + | | + | | | | | | | + | | − | | | + | − | − | + | | | | | | | | | | + | | McLeese 1970 |
| | | | | | | | | | | | | | | | | | | | | + | | | | | | | | + | + | | | | | | | | Hartmann & Hartmann 1977 |
| Echinodermata | | | | | | | | | | | | | | | | | | | | | | | | | | | | | | | | | | | | | |
| *Dendraster* | | | | | | | | | | | | | | | | | | | | | | | | | | | | | | | | | | | | + | Burke 1984 |
| *Marthasterias* | + | | | | | | | + | | + | | | + | + | + | | + | + | | | | | | | + | | | | | | + | | | | | | Valentincic 1979 |

| Preparation | Alanine | Arginine | Aspartic acid | Aspartate | Asparagine | Citruline | Cysteine | Glutamic acid | Glutamate | Glycine | Histidine | Hydroxyproline | Leucine | Lysine | Methionine | Ornithine | Proline | Phenylalanine | Serine | Taurine | Tyrosine | Tryptophan | Threonine | Valine | Betaine | Trimethylamine | Pyridine/Purine | Sucrose | Glucose | Trehalose | Lactic acid | GABA | Osmotic/salts | Natural products | Fish, mussels, etc | Conspecifics | Authors |
|---|---|---|---|---|---|---|---|---|---|---|---|---|---|---|---|---|---|---|---|---|---|---|---|---|---|---|---|---|---|---|---|---|---|---|---|---|---|
| *Ophiura brevicaudatum* |  |  | + |  | + | + | + |  | + |  |  |  | + | + | + | − | + | + | + | + | + | + | + | + | + |  |  |  |  |  |  |  |  |  |  |  | Reimer and Reimer 1975 |
| Respiration |  |  |  |  |  |  |  |  |  |  |  |  |  |  |  |  |  |  |  |  |  |  |  |  |  |  |  |  |  |  |  |  |  |  |  |  |  |
| Cephalopoda |  |  |  |  |  |  |  |  |  |  |  |  |  |  |  |  |  |  |  | + | − |  |  |  |  | + |  |  |  |  |  | + |  |  | + |  | Boyle 1983, pers. com. 1985 |
| *Eledone* | + | − |  |  |  |  |  |  |  | + |  |  |  |  |  |  |  | + |  |  |  |  |  |  |  |  |  |  |  |  |  |  |  |  |  |  |  |
| Electrophysiology |  |  |  |  |  |  |  |  |  |  |  |  |  |  |  |  |  |  |  |  |  |  |  |  |  |  |  |  |  |  |  |  |  |  |  |  |  |
| Gastropoda |  |  |  |  |  |  |  |  |  |  |  |  |  |  |  |  |  |  |  |  |  |  |  |  |  |  |  |  |  |  |  |  |  |  | + |  | Alkon, Akaike, and Harrigan |
| *Hermissenda* |  |  |  |  |  |  |  |  |  | + |  |  |  |  | + |  |  |  |  |  |  |  |  |  |  |  |  |  |  |  |  |  |  |  |  |  | 1978 |
|  |  |  |  |  |  |  |  |  |  |  |  |  |  |  |  |  |  |  |  |  |  |  |  |  |  | + | + |  |  |  |  |  |  |  | + |  | Bailey and Laverack 1966 |
| *Buccinum* |  |  | + |  |  |  |  | + |  |  |  |  |  |  |  |  |  |  |  |  |  |  |  |  |  |  |  |  | + |  |  |  |  | + | + |  | Bicker, Davis, and Matera |
| *Pleurobranchaea* | + |  |  |  |  |  |  |  |  | + |  |  |  |  |  |  |  |  |  |  |  |  |  |  |  |  |  |  |  |  |  |  |  |  |  |  | 1982 |
|  |  |  |  |  |  |  |  |  |  |  |  |  |  |  |  |  |  |  |  |  |  |  |  |  |  |  |  |  |  |  |  |  |  |  |  |  | Stinnakre and Tauc 1969 |
| *Aplysia* |  |  |  |  |  |  |  |  |  |  |  |  |  |  |  |  |  |  |  |  |  |  |  |  |  |  |  |  |  |  |  |  |  | + |  |  | Jahan-Parwar 1972 |
| Osphradium |  |  |  |  |  |  |  |  |  |  |  |  |  |  |  |  |  |  |  |  |  |  |  |  |  |  |  |  |  |  |  |  |  |  | + |  | Fredman and Jahan-Parwar |
| *Aplysia* |  |  |  |  |  |  |  |  |  |  |  |  |  |  |  |  |  |  |  |  |  |  |  |  |  |  |  |  |  |  |  |  |  |  |  |  | 1980 |
|  |  |  |  |  |  |  |  |  |  |  |  |  |  |  |  |  |  |  |  |  |  |  |  |  |  |  |  |  |  |  |  |  |  |  |  |  | Jahan-Parwar 1972 |
| Tentacle |  |  | + |  |  |  | + |  |  |  |  |  |  |  |  |  |  | + |  |  |  |  |  |  |  |  |  |  |  |  |  |  |  |  |  |  |  |
| Crustacea |  |  |  |  |  |  |  |  |  |  |  |  |  |  |  |  |  |  |  |  |  |  |  |  |  |  |  |  |  |  |  |  |  | + |  |  | Laverack 1975 |
| *Galathea* |  |  |  |  |  |  |  |  |  |  |  |  |  |  |  |  |  |  |  | + | + |  |  |  |  |  |  |  |  |  |  |  | − |  |  |  | Derby and Atema 1982 |
| *Homarus* |  | + |  | − |  | + |  |  | + | − |  |  | − |  |  |  | + |  |  |  | + |  | + |  |  |  |  |  |  |  |  |  |  |  |  |  | Ache 1972 |
| *Homarus* | + | + |  |  |  |  |  | + | − | + |  |  |  | + | − | + |  | + |  |  |  |  |  |  |  |  |  | + |  |  |  |  |  |  |  |  | Derby, Carr, and Ache 1984 |
| *Panulirus* |  |  |  |  |  |  |  |  |  |  |  |  |  |  |  |  |  |  |  | + |  |  | + |  | + |  | + |  |  |  |  |  |  |  |  |  | Shepheard 1974 |
| *Homarus* | + | + | + |  | + |  | + | + | + | + | − |  | + | + | + | + | + | + | + | + | + |  | − | + | + |  |  | + |  |  | − | + |  |  |  |  | Bauer, Dudel, and Hatt 1981 |
| *Austropotamobius* | + | + |  |  | + | − | + | + |  |  | + |  |  | + | + | + |  |  |  |  |  |  |  |  |  |  |  |  |  |  |  |  |  |  |  |  | Hatt and Bauer 1982 |
|  |  |  |  |  |  |  |  |  |  |  |  |  |  |  |  |  |  |  |  |  |  |  |  |  |  |  |  |  |  |  |  |  |  |  |  |  | Barber 1961 |
| Arachnida |  |  |  |  |  |  |  |  |  |  |  |  |  |  |  |  |  |  |  |  | + |  |  |  |  |  |  |  |  |  |  |  |  |  | + |  | Hayes and Barber 1982 |
| *Limulus* | + |  |  |  |  |  |  |  | + | + |  |  |  |  |  |  |  | + |  |  |  |  |  |  |  |  |  |  |  |  |  |  |  |  |  |  |  |
| Echinodermata |  |  |  |  |  |  |  |  |  |  |  |  |  |  |  |  |  |  |  | + |  |  |  |  | + | + |  |  |  |  | + |  |  |  |  |  | Moore and Cobb 1985 |
| *Ophiura ophiura* |  |  |  |  |  |  |  | + |  |  |  |  |  | + | + | − |  | − | + |  |  |  |  |  |  |  |  |  |  |  |  |  |  |  |  |  |  |

Notes: The list of chemicals is not exclusive of others that may have been used in a few circumstances, but it does include the commonest stimulatory materials assayed on such preparations. Absence of any sign (+ or −) only indicates that a compound has not been used, rather than that it is without effect. I shows inhibition; *?* indicates an ambiguous response.

the nature of acceptance or rejection of materials. This is a central phenomenon, and it is difficult to clarify what pathways lead to one or another response. It may be that some activities reported thus far are not associated with feeding or attraction, but rather with rejection. After all, both use the same population of receptors, similar pathways, and the same muscles, but the end result is very different.

Thirdly comes the question of trace materials. Fraenkel (1959) discussed the question of the evolution of secondary plant substances and their value in the lives of numerous insects. Dethier (1980), reflecting on the behavioral and electrophysiological responses of insects to the wide variety of secondary plant substances available, suggests that there are far more receptor types than the few typified in earlier work. He proposed that herbivorous insects evolved, from a common chemical sense, receptors sensitive to a wide variety of compounds. This I believe to be the essential truth for most animals, from the lowliest to the highest among aquatic organisms as well as terrestrial ones.

There can be little doubt that similar arguments can be made about many aquatic plants and their grazers and probably also about specific interrelations between prey and predator. It is highly likely that foraging rates, swimming speeds, homing movements, and so on can all be affected by very small amounts of selected chemicals.

The question of minor components being of importance as chemical cues is raised by biochemists every time they describe a new compound. For example, Rossi et al. (1985) have recently isolated 5-thiolhistidine from the eggs of *Octopus vulgaris*. The eggs of *Loligo vulgaris* contain a related substance, 1-methyl-5-thiolhistidine, which is also found in the urine of *Octopus* and *Loligo*. It is materials like these that could form important lures to others, but what about the various small organisms, such as protists, that wander about on *Octopus* eggs? There are various copepods that crawl around the surface of cephalopod eggs. Perhaps 5-thiolhistidine is a necessary amino acid for them, and they have receptors to detect it.

One might speculate on the sources of stimulatory materials and suggest that the variation of cues is at least as varied as the examined species. Could it be that sensitivity to purine derivatives (Derby, Carr, and Ache 1984) from nuclei and nucleic acids shows that other metabolic intermediates might be significant? Major biochemical pathways such as the Krebs cycle or hexose degradation give rise to a number of metabolic intermediates, such as succinate or malate or fumarate. These substances are commonplace in living tissues. The various components of urine, mucus, and other specific secretions from prey might be significant, as trace substances (secondary plant substances) are to herbivores (Fraenkel 1959). Chemically induced escape responses may thus be a function of specially adapted receptors that detect trace substances e.g., *Chlamys* escaping from *Asterias* (Mackie, Lasker, and Grant 1968). A 980 Dalton peptide from the sea urchin *Dendraster excentricus* promotes metamorphosis in larvae (Burke 1984).

The likelihood of demonstrating the reception of such chemicals depends on being able to locate, isolate, and identify them and to stimulate the sensors using them, but such a task often requires huge amounts of raw material (Jones 1979).

Alanine, one of the common amino acids, stimulates virtually every preparation tested from Cnidaria to Echinodermata. Alanine seems positively stimulatory everywhere except in the herbivore *Aplysia*. Other general stimulants with several positive results spread across several phyla are glutamic acid, glutamate, glycine, hydroxyproline, leucine, lysine, methionine, proline, serine, taurine, and betaine. Sugars, so important to many terrestrial animals, seem less so to aquatic representatives. Recently recognized sensitivity towards pyridines and purines (Hatt and Bauer 1982; Derby, Carr, and Ache 1984; Hatt 1984) may prove to be of more general relevance than is appreciated at present.

Peptides, proteins, and other such chemicals may be equally important, but the precise constitution of these longer-chain compounds can be highly varied and, consequently, specific. It may be that specific feeders (e.g., *Tritonia diomedea* on *Virgularia* sea whips and *Tritonia hombergi* on *Alcyonium digitatum*) may reflect not the common occurrence of amines and amino acids, but the more limited chemicals found in the prey only. Many such specific patterns of feeding exist and may be as limited as the possession of one or two proteins make them. It may also be true that it is the larva that is important in determining settlement and, hence, the adult food through choice of particular hosts.

Larval sensitivity towards chemical stimulants is not well known. Some work has been done on larval sensory structures (Annelida: Holborow and Laverack 1972; Urochordata: Cloney 1977; Mollusca: Bonar 1978; Chia 1982; Chia and Koss 1984). Lacalli (1984) describes several different types of sensory cells in the trochophore of the polychaete *Spirobranchus* and comments on the organization of the nervous system in this larva, indicating that, at an early stage in the life of the animal, the apparatus exists for a number of reflex activities, which might well include chemoreception. Even at 48 hr of age there are 36 cells in the larval nervous system and 16–18 recognizably different types of nerve cells. The apical tuft, an obvious candidate for sensory information, may well be a chemoreceptor (perhaps among other modalities) in many planktonic larval types (see Cantell, Franzen, and Sensenbaugh 1982).

What particular larvae respond to in the natural environment is not well known, but that the chemical nature of the substrate is crucial is unequivocal (see Crisp 1974; Burke 1983).

## Numbers and Location of Receptors

It is fairly clear in many groups of aquatic animals that the total number of chemoreceptive cells is large, possibly larger than all other sensory cells put together.

One might speculate on why this is so. Is it that the opportunities for capturing certain molecules are not statistically so great as that of capturing photons? Is it that the surrounding water comes from all directions and hence brings chemicals to all parts of the body, and that this is more likely than in aerial animals? Cephalization normally brings about the location of receptors at the head, yet the bodies of aquatic animals frequently have numerous chemoreceptors on other parts of the body (for fish: Bardach and Villars 1974; Lane and Whitear 1982; for Crustacea: Ache 1982).

Why are chemoreceptive axons more closely packed in nerves, and why are they often so much smaller in diameter?

It is not easy to answer these questions. General sensitivity is a function of all animal bodies, while localization comes with certain kinds of specialization. Areas around the mouth are likely to be well endowed with chemoreceptors; buccal cavities no doubt also have chemoreceptors, as do gill chambers and areas where water flow is chaneled for some particular reason; if there are feeding appendages, chemoreceptors are obviously appropriate in such places. It is less obvious what requirements lead to placement of receptors along the sides of the body.

The number of receptors also varies considerably within organs and for no very obvious reason. In the decapod Crustacea, the number of neurons on the prime chemoreceptor organs, the antennules, are extremely high. Each seta, or aesthetasc, may contain up to 300 or more dendrites, and each may be subdivided (Laverack and Ardill 1965; Ghiradella, Case, and Cronshaw 1968; Ache 1982; Spencer 1986). There are many aesthetascs, and the total population may be hundreds of thousands, representing a most important olfactory source (Ache 1982). Large numbers are also reported in the aesthetascs of isopods (Mead, Gabouriant, and Corbière-Tichanée 1976; Heimann 1984). By contrast, the chelate pereiopod "hedgehog hairs" (Shelton and Laverack 1970; Derby 1982) contain 8 chemosensory neurons and 2 mechanoreceptors (for the crayfish *Austropotamobius torrentium,* Altner, Hatt, and Altner 1983; for lobster *Homarus gammarus,* Laverack unpublished), as do similar setae of the freshwater shrimp *Atya innocous* (Felgenhauer and Abele 1983). There are several thousand possible receptors on the chelate legs, including funnel canals (Luther 1930; Shelton and Laverack 1968; Gnatzy, Schmidt, and Rombke 1984), which are located at the tips of the legs, and these have been thought of as mechanosensors or chemoreceptors or both. It is now known that they subserve both modalities (for mechanoreception: Shelton and Laverack 1968; Barth 1980; for chemoreception; Gnatzy, Schmidt, and Rombke 1984). There are 2 mechanically sensitive units and up to 20 or so chemically responsive units (see p. 302).

In the esophagus, there are perhaps 500–600 chemoreceptor organs, each with 2 or 3 (rarely 4) neurons, giving a few more thousand sensors in total (Allen 1894; Robertson and Laverack 1979; Altner pers. comm.; Marshall and Laverack unpublished).

There are almost certainly other types of setae carrying chemoreceptor dendrites (e.g., plaques on the chelae and cones on the gills), but internal details are not known for these so far. In a medium-sized lobster, I would estimate that the total chemoreceptor population is approximately 1,000,000. Small wonder that the diameters of the sensory axons that run to the brain or to the ventral chain ganglia are very small (ca. 0.1 μm) and numerous. Limbs such as the maxillipeds, mandibles, and other mouthparts are all heavily supplied with neurons of small diameter. Each may be capable of individual sensitivity, providing many more inputs than is possible with a few large units.

## Sensory Cilium

The cilium is known to be an important organelle in sensors, as well as in locomotion, but if one observes the surface of animals with the SEM, there are no clues to which cilia subserve which modality. Some may well be stiff and mechanically sensitive, and some may be discocilia (paddle cilia), but they are chemoreceptors; and some just look like cilia and could be anything. One looks at annelids and sees patches of cilia, and frequently they are said to be in hillocks that are sensors; but they look like motile cilia and have little specialization (Fig. 11.1[A]).

A dual function of the cilia has been suspected for some time, and if sensors

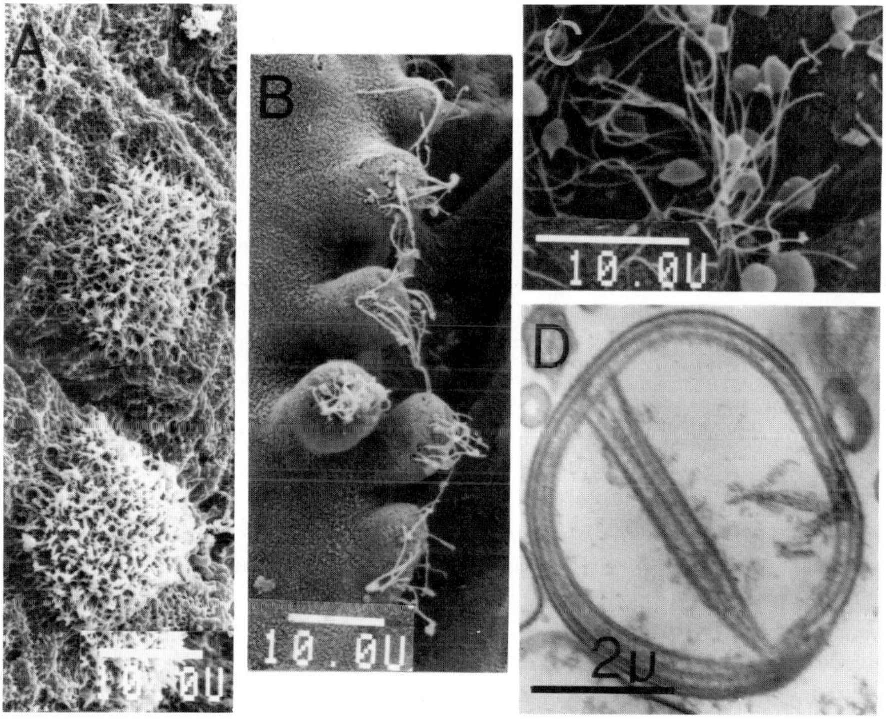


Figure 11.1 Ciliated structures from various sites. (A) From the prostomium of oligochaete *Allolobophora longa.* Cilia are clustered in clumps, but are externally indistinguishable from other cilia. This is the seat of considerable chemosensitivity in earthworms, but organelles are similar to motile cilia in external appearance. (B) From ventral cirrus of polychaete *Spinther miniaceus.* Cilia are mounted on short papillae. Some appear straight and filamentous, while others have rounded distensions (Miller unpublished). (C) From another site on the body of *Spinther miniaceus.* These are typical discocilia (paddle cilia) comparable to those reported for molluscs and suggested to be chemoreceptors (Davis and Matera 1982; Haszprunar 1985; Miller unpublished). (D) Internal structure of a discocilium, but this example is taken from a statocyst of *Arenicola marina* (Polychaeta) (Laverack unpublished). These are probably mechanoreceptors.

evolved from such organelles, it is perhaps not surprising if examples show both movement and sensitivity. Stiff cilia are certainly known to beat from time to time, but more significant is the finding that the cilia may beat in reverse when mechanically agitated (this has been shown by Thurm 1968 in mussels; Strathmann 1971 in echinoderm larvae; and Strathmann 1982 in Bryozoa). The implication of chemical sensitivity in ciliary reversal has been made also by Strathmann (1984) in plutei in which inert plastic spheres coated with algal extracts brought about a change in ciliary beat. It is clear that the sensitivity of the cilia is a very basic characteristic and that it is involved in all forms of stimulus perception.

The participation of cilia in sense organs is now well established. They are the only locations for cilia in groups not renowned for possessing such organelles, such as the arthropods and nematodes. Their ubiquity in sensors is apparent, yet the various modifications of basic ciliary structure gives us few clues as to why they are so important. The alterations in the standard 9 + 2 configuration of locomotory cilia found in sensors has received much attention, but without any rules being defined as to the significance of the changes.

One problem about characterizing chemosensors lies in the often multiple functions of the organs in which they are found; distinguishing between a chemosensitive cilium and a neighboring mechanosensor may not be simple on structural grounds, though accessory structures may assist in the identification (Fig. 1[B], [C], and [D]). It seems, therefore, that there is not a single feature one can point to that permits certain identification of a chemoreceptor on the basis of its ultrastructure, though various authors have attempted this. It also appears that the internal structures are not critical to its function, but perhaps the extended membrane of the projection is important.

Several authors (Davis and Matera 1982; Haszprunar 1985) have recently suggested that, at least in molluscs, discocilia, which are flared and expanded at the tip, may be typical chemoreceptive organelles. This is based on findings from the osphradium of prosobranchs and the rhinophores of opisthobranchs, but such cilia may not be specifically located on receptors and may also be mechanoreceptive (Laverack unpublished, on the statocysts of *Arenicola* Polychaeta).

Similarly, in molluscs, peripheral ciliary bunches exist on tentacles and feeding structures, as well as the osphradium. The osphradium has been shown physiologically to be chemically sensitive (Bailey and Laverack 1966; Jahan-Parwar, Smith, and van Baumgarten 1969; Stinnakre and Tauc 1969; Jahan-Parwar 1972). It is covered in ciliary endings (Welsch and Storch 1969; Crisp 1981). Tentacles have been shown behaviorally and electrophysiologically to be chemosensitive (Bicker, Davis, and Matera 1982; Chase 1982) and ciliated (Phillips 1975, 1976, 1979; Davis and Matera 1982; Hackney, McCrohan, and Hawkins 1983). Chia and Koss (1984), working on the cephalic sense organs of the nudibranch *Rostanga* have three types of cells labelled as receptors and suggest that one or more of them may be chemically sensitive, but also recognize that there are no shortcuts to certainty based on ultrastructure alone.

We might conclude, therefore, that there is a great diversity of form, that many sensors are ciliated, and that ciliary structure may be varied; but there are few, if any, definite clues as to what modality is served by any individual example without supporting physiological evidence, which is often difficult to acquire.

The difficulties of identifying chemoreceptors from among a population of receptors is nowhere better exemplified than in echinoderms. The sensory hillock of pedicellariae seems to possess one type of ciliated ending, yet both mechanical and chemical responses are shown (Campbell and Laverack 1969). The tube feet suckers carry numerous populations of small sensilla. Some are short, just protruding above the surface, while others are rather long, stout, and surrounded by a palisade of microvilli (Cobb, personal communication, 1986). Both are ciliary in nature, but the internal structure is highly varied, and many or no filaments may be seen. Which modality is served by which type of sensor is unknown and, given the difficulties of working with echinoderms, is likely to be for some time. Electrophysiology of central neurons is possible, but sensory information *per se* is not yet available (see Moore and Cobb 1985). Sloan and Campbell (1982) give an overall review of chemoreception in echinoderms.

## Crustacean Chemoreceptors

To complete this survey of diversity in chemoreceptors, I would like to discuss specific examples from Crustacea. This is intended to show that (1) the numbers of neurons differ markedly from organ to organ, (2) all have ciliated endings, (3) distinction may be made between those with an integral mechanoreceptor and those without, and (4) these distinctions reflect what are believed to be olfactory and taste receptors. On Crustacea other organs exist that are probably chemoreceptive, but for which evidence is incomplete.

### Aesthetasc Hairs

The antennule of Crustacea carries a large battery of thin-walled, untanned, cuticular setae, which have been described by numerous authors (Laverack and Ardill 1965; Ghiradella, Cronshaw, and Case 1968; Snow 1973; Mead, Gabouriant, and Corbière-Tichanée 1976; Ache 1982; Heimann 1984; Spencer 1986) (Fig. 11.2). These aesthetascs represent the largest aggregation of chemoreceptors in these animals. Each hair usually contains a large number of dendrites, up to 300, and each ends in a cilium, which may be divided (Ghiradella, Case and Cronshaw et al 1968; Heimann 1984). No mechanoreceptor is associated with the seta.

Much physiological work has been carried out on these organs, and Chaps. 2 and 15 show how important these structures are as sources of information on chemoreception. They undoubtedly form an organ of olfaction.

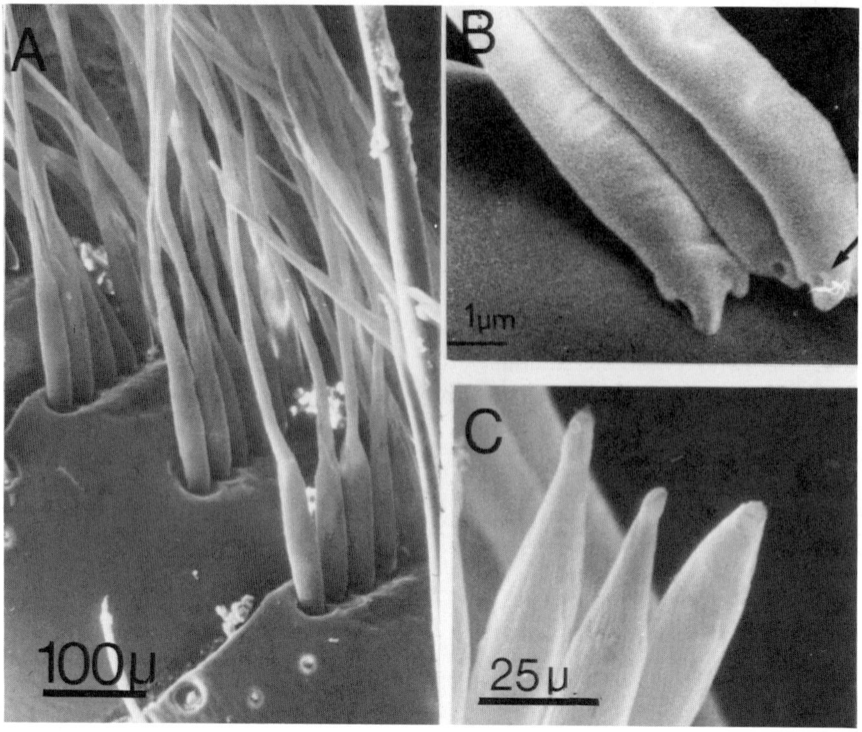


Figure 11.2. Aesthetasc setae. (A) Bases and shafts of aesthetascs of lobster *Homarus gammarus*, showing swollen basal region, which contains dendrites. (B) The tip of aesthetascs of cumacean *Diastylis*, which are perforated by a subterminal pore and should be compared with (C). (C) Tips of equivalent hairs of *Homarus* (newly molted), which do not carry a pore.

## Esophageal Chemoreceptor in *Homarus*

The mouthparts, labrum, and paragnath of *Homarus* all appear to be provided with chemoreceptors, and any rejection or acceptance of food objects must be a function of their activity. When food is presented to the mouth, however, and enters the alimentary tract, there is still a set of receptors to be negotiated, and these lie in two separate groups, anterior and posterior, on the wall of the esophagus.

These patches of receptors have been known for many years (Allen 1894). They appear at the surface as a series of small pores, in some places solitary and elsewhere in clusters (Fig. 11.3[A]). The anterior and posterior groups are slightly different. Perhaps the important thing to note about them is that they are mounted more or less flush with the cuticular surface and are usually in an area without setae, yet may be surrounded by flimsy thin-walled setae and are open to the lumen at the ends (Fig 11.3[A]). Their ultrastructure is not completely worked out, though the dendrites are certainly modified cilia that run up a cuticular

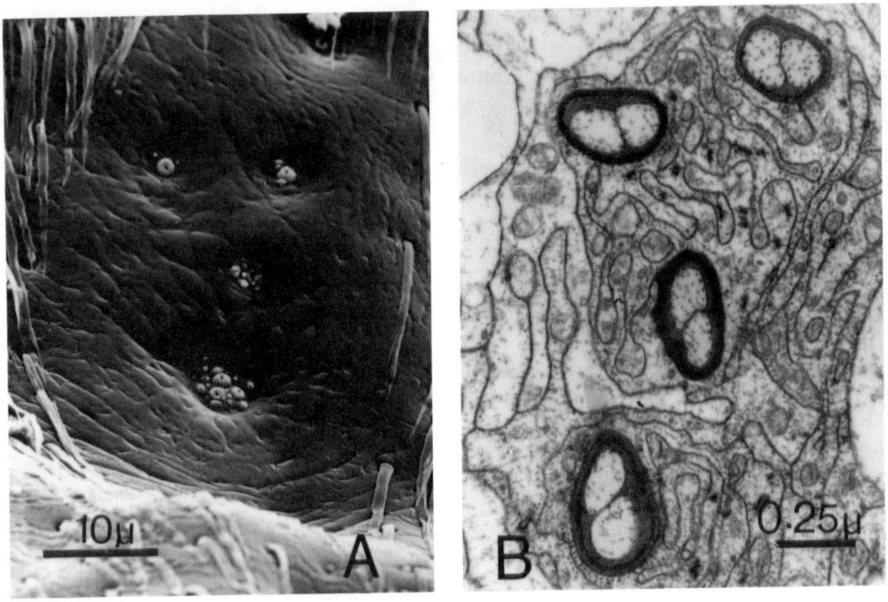


Figure 11.3. Esophageal chemoreceptors of *Homarus gammarus*. (A) The appearance of groups of chemoreceptor endings on esophageal lumen. They are in areas bare of setae and may occur in small aggregations or singly. (B) Each receptor is innervated by two dendrites enveloped in cuticular tube extending inward from the surface. One of these ends at lower level than the other. They are derived from nerve cells that probably carry ciliary structures at lower level. They are typically double, but three or four may occur in some examples.

tube (Fig. 11.3[B]). The number is not absolute, since 2, 3, or 4 dendrites have been observed in one capsule. Of these dendrites, one is probably shorter than the other(s), since one disappears before the surface is reached.

Homologous receptors exist in the crayfish *Austropotamobius torrentium* (Altner, Hatt, and Altner personal communication), in which three dendrites are the norm with one of these believed to be a mechanoreceptor. This would be in accord with the notion that this organ within the alimentary canal is effectively an organ of taste. It is known to be chemically sensitive.

## Funnel Canals in Dactyls of Decapod Crustacea

Luther (1930) described small canals that penetrate the dactyl tips of the crab *Carcinus maenas* and proposed that they were chemoreceptors. Shelton and Laverack (1968) reinvestigated them and concluded that they were mechanoreceptors, a finding that was confirmed by Barth (1980). Recently, however, an ultrastructural study by Schmidt and Gnatzy (1984) and Gnatzy, Schmidt, and Rombke (1984) has shown that the organ is really a mixed modality sensory. There are a number of neurons serving each canal, and of these, two have a different structure than the remainder. These are interpreted as the mechanoreceptors recorded by the earlier workers, and the remainder (between 3 and 23) are believed to be chemoreceptors (Fig. 11.4). Both types of receptors are modified ciliary endings. The surface is covered with a thin membrane that protects the dendritic tips.

The chemical sensitivity of these cells is difficult to quantify. It is possible that these cells are involved in the responses recorded by Laverack (1963) and Case and Gwilliam (1961) from the tips of decapod legs, but since other organs were also included in these preparations, this is not absolutely sure. The inclusion of mechanoreceptors in the same organ indicates that these are receptors of taste.

## Hedgehog Hairs

These setae were first described by Shelton and Laverack (1970) as short, stout projections from the interior faces of the chelae of the pereiopods on the second and third walking legs of *Homarus gammarus*. Single examples of these setae may be observed on other walking legs of lobsters close to the tip of the dactyl. They earned the soubriquet "hedgehog hairs" from Derby (1982) as an indication of their appearance. They are stiff, robust setae with a small pore at the tip (Fig. 11.5[A] and [B]). The fringing setules are entire and represent the fragmented ends of the cuticular laminae that form the surface. The terminal pore is linked to the interior of the main shaft by a canal, which extends from the apex to the

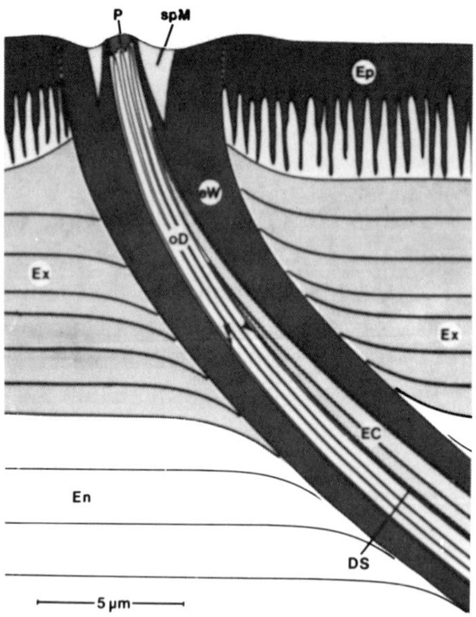


Figure 11.4. Endings of funnel canals of pereiopods of *Carcinus maenas*, containing both mechanoreceptors (2) and chemoreceptors (up to 23 or so) and ending in small membrane-covered depression at the surface: *DS*, dendritical sheath; *EC*, enveloping cells; *En*, endocuticle; *Ep*, epicuticle; *Ex*, exocuticle; *eW*, epicuticular wall; *oD*, outer dendritical segments; *P*, plug; *spM*, spongy material; *arrowheads* show the end of dendritical sheath. (From Schmidt and Gnatzy 1984, with permission of Springer-Verlag.)

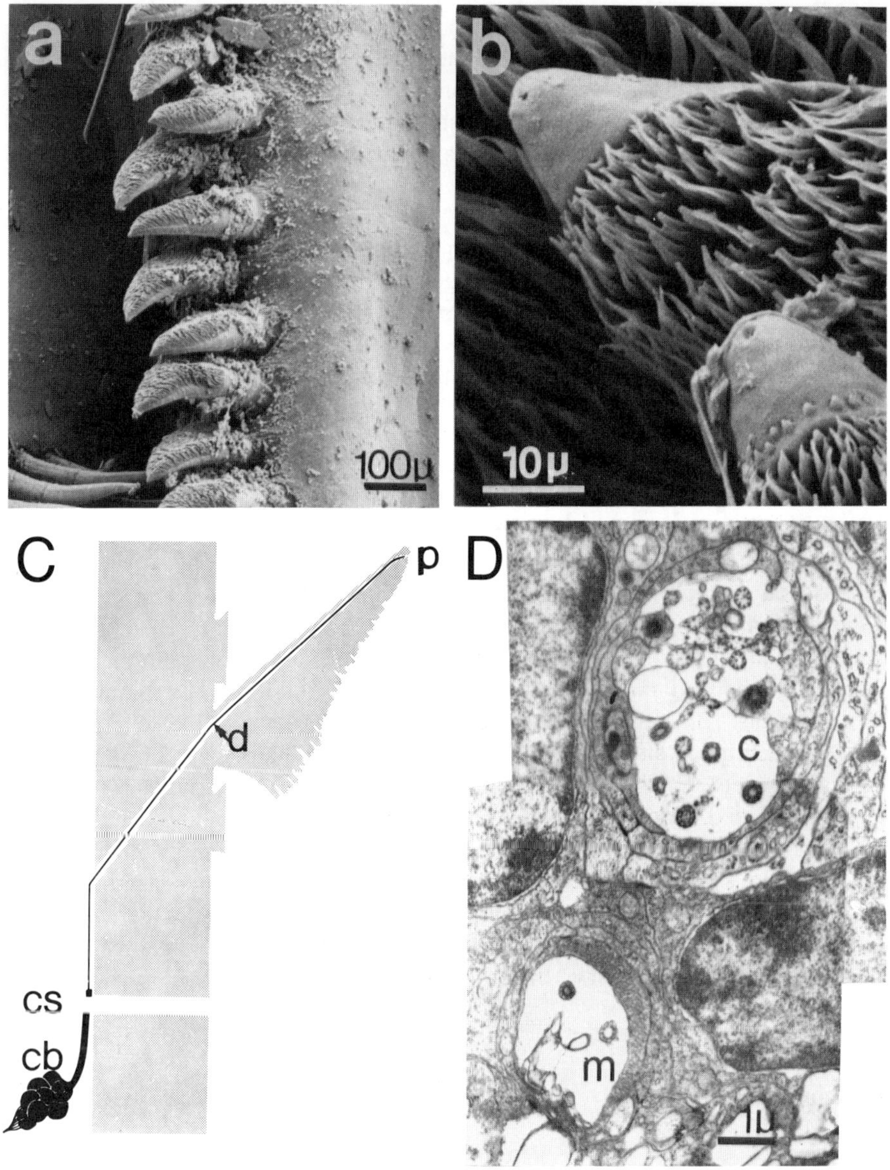


Figure 11.5. "Hedgehog hairs" (Derby 1982), or fringed setae (Altner, Hatt, and Altner 1983) from the chelate pereiopods of Macrura. (A) A line of setae from second pereiopod of *Homarus gammarus* dactyl. (B) Tips of such setae, showing narrow aperture close to the tip (Laverack unpublished). (C) The reconstruction of internal innervation for crayfish *Austropotamobius*. Dendritical strand consisting of 8 dendrites of chemoreceptors runs to the tip of the hair, but another pair of neurons terminates in scolopale sheath close to the base. Both types are ciliated: *cb,* cell bodies; *cs,* ciliary; *d,* dendrites; *p,* pore. (Altner, Hatt, and Altner 1983, with permission of Springer-Verlag.) (D) Dendritical cluster of single seta close to the origin at cell body. Mechanoreceptors (*m*) are surrounded by the scolopale, and chemoreceptors (*c*) lie in open (fluid-filled) space (Laverack unpublished).

base (Fig. 11.5[C]). Within the lumen lies a small collection of dendrites that originate from cell bodies located in the main portion of the limb (either dactyl or propus). Eight dendrites are all that are found within each massive seta (Altner, Hatt, and Altner 1983; Laverack unpublished). At the base, however, are two further dendrites, surrounded by scolopale material, that represent mechanoreceptor units (Fig. 11.5[C]).

These setae are placed where they are involved in feeding, with the food being picked up by the two sets of pereiopods and presented to the mouth. Therefore a tactile sense built into the organ is significant. Their range of sensitivity has been tested by Hatt in a series of papers (Hatt and Bauer 1982; Hatt 1984; Hatt and Schmiedel-Jakob 1984).

These setae fall into the category of taste receptors according to the definition on page 289.

## Dorsal Organ of Decapod Larvae

The dorsal surface of the late larva (megalopa) of decapod Crustacea bears a unique organ, the "dorsal organ" (Laverack and Barrientos 1985; Barrientos unpublished). This organ is arranged in the form of a central aperture, with four bordering pits and nipples (Fig. 11.6[A]). Transmission electron microscopic investigations show the central pore to be glandular, while the four peripheral pits are each innervated by four cilia arranged in two pairs (Fig. 11.6[B]). These appear to be the endings of two sensory cells, and the dendrites terminate in a highly modified cuticle that is quite distinct from the surrounding one.

The function of this structure is unknown, and the adult animal does not bear it. It would seem a reasonable proposal that the sensors are chemoreceptors, bearing a branched or double cilium, as do aesthetasc dendrites (Crouau 1978; Heimann 1984). No sensor that could be interpreted as a mechanoreceptor has been seen. This organ may, therefore, represent a larval olfactory organ of a special type.

## Conclusion

I have used the Crustacea as an example to indicate that (1) the number of neurons may vary from 2 or 3, as in the esophageal receptors, through 8 in the hedgehog hairs, up to 22–23 in the funnel canals, and to hundreds in the aesthetasc hairs; (2) organs coming into contact with food, such as those on leg ends, in the gut, and on food-manipulating limbs, carry a dual innervation of chemoreceptors and mechanoreceptors and are in situations in which taste is the appropriate sense, while antennules (aesthetascs) and dorsal organs have only chemoreceptors and are therefore olfactory organs; (3) all sensory cells are ciliated, some being divided or multiplied (olfactory: aesthetascs and dorsal organ) while others are not (taste).

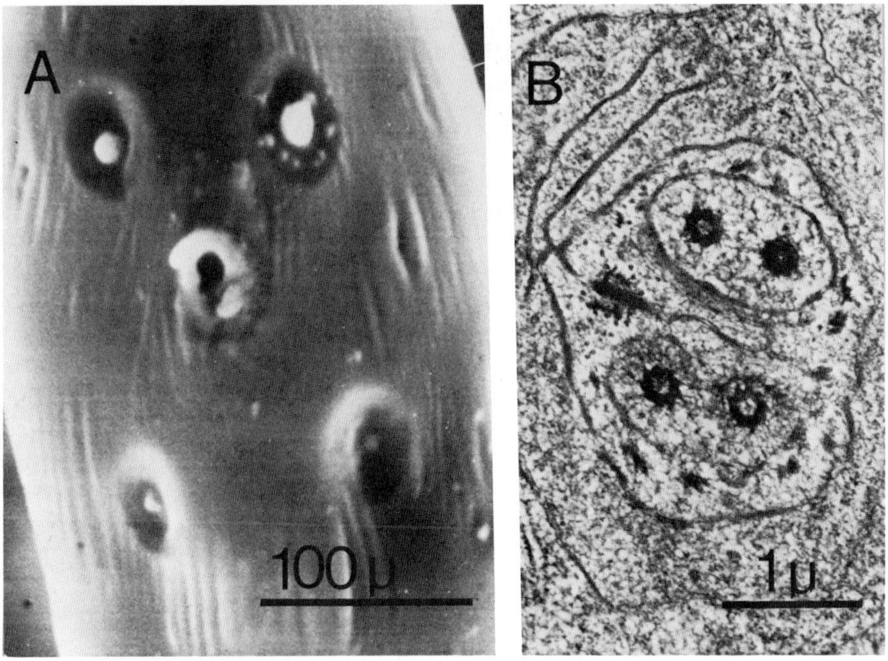


Figure 11.6. Dorsal organ complex of the larva of *Hyas cornutus*. (A) External appearance of the organ with central pore (glandular opening) and 4 ''nipples'' located in depressions at corners of a rectangle around the gland. (B) Each ''nipple'' is composed of unusual cuticle and is the ending of a pair of dendrites of ciliary type. Each pair comes from single cell and indicates double-ciliated ending, as in common in other chemoreceptors (Laverack and Barrientos 1985, with permission of the Royal Society of Edinburgh).

Whether such conclusions can be applied to other phyla, especially the vertebrates, is a matter for argument.

## Summary

This has been an attempt to demonstrate the diversity in chemoreceptors, and it appears that there is no easy way to determine and identify receptors without several modes of evidence being available. The organs that have evolved to receive chemical information from a watery environment seem as numerous as the species that bear them.

The diversity of organs falls into several categories, and each may be permutated with others. The differences between them may be structural; ultrastructural; enzymatic; open or closed to the environment; generalist or specialist in response; or concerned with feeding, breeding, homing, and settlement. They may have few or many neurons and be localized or widespread. Many, if not all, are based on ciliary principles, yet there is no standard structure.

The problems of establishing the chemosensory nature of a cell are legion and so far has been definitively achieved for very few examples. We have a long way to go before being able to design a chemoreceptor that takes into account all the possible combinations.

*Acknowledgments.* I am extremely grateful to Yolande Barrientos Chacon, Catherine Gill, Jim Cobb, Alan Hodgson, and Peter Boyle for permission to use as yet unpublished information and to Sandy Edwards and Charles Roemmélé for unstinting technical advice and help.

## References

Ache, B.W. (1972) Amino acid receptors in the antennules of *Homarus americanus*, Comp. Biochem. Physiol., 42A:807–811.

Ache, B.W. (1982) Chemoreception and thermoreception, in Biology of Crustacea, vol. 3, Bliss, D.H., Atwood, H.L. and Sandeman, D.C., (eds.), Academic Press, New York, pp. 369–398.

Alkon, D.L., Akaike, T., and Harrigan, J. (1978) Interaction of chemosensory, visual and statocyst pathways in *Hermissenda crassicornis*, J. Gen. Physiol., 71:177–194.

Allen, E.J. (1894) Studies on the nervous system of Crustacea. II. The stomatogastric system of *Astacus* and *Homarus*, Q.J. Microsc. Sci., 36:483–496.

Altner, H. and Prillinger, L. (1980) Ultrastructure of invertebrate chemo- thermo- and hygroreceptors and its functional significance, Int. Rev. Cytol., 67:69–139.

Altner, I., Hatt, H., and Altner, H. (1983) Structural properties of bimodal chemo- and mechano-sensitive setae on the pereiopod chelae of the crayfish *Austropotamobius torrentium*, Cell Tissue Res., 228:357–374.

Atema, J. (1977) Functional separation of smell and taste in fish and Crustacea, Olfaction Taste Proc. Int. Symp. 6:165–174.

Bailey, D.F. and Laverack, M.S. (1966) Aspects of the neurophysiology of *Buccinum undatum* L. (Gastropoda). I. Central responses to stimulation of the osphradium, J. Exp. Biol., 144:131–148.

Barber, S.B. (1961) Responses of *Limulus* chemoreceptors to amino acid stimulation, Am. Zool., 1:435.

Bardach, J. and Villars, T. (1974) The chemical senses of fishes, in Chemoreception by Marine Organisms, Grant, P.T. and Mackie, A.M. (eds.), Academic Press, London, pp. 49–104.

Barth, F.G. (1980) Campaniform sensilla; another vibration receptor in the crab leg, Naturwiss enschaften 67:201.

Bauer, U., Dudel, J., and Hatt, H. (1981) Characteristics of single chemoreceptive units sensitive to amino acids and related substances in the crayfish leg, J. Comp. Physiol., 144:67–74.

Bicker, G., Davis, W.J., and Matera, M. (1982) Chemoreception and mechanoreception in the gastropod mollusc *Pleurobranchaea californica*, J. Comp. Physiol., 149:235–250.

Bonar, D.B. (1978) Ultrastructure of a cephalic sensory organ in larvae of the gastropod *Phestilla sibogae* (Aeolidacea, Nudibranchia), Tissue Cell, 10:153–165.

Boyle, P.R. (1983) Ventilation rate and arousal in the octopus, *Octopus vulgaris*, J. Exp. Mar. Biol. Ecol., 69:129–136.

Burke, R.D. (1983) The induction of marine invertebrate larvae; stimulus and response, Can. J. Zool., 61:1701–1719.

Burke, R.D. (1984) Pheromonal control of metamorphosis in the Pacific sand dollar, *Dendraster excentricus,* Science, 225:442–443.

Campbell, A.C. and Laverack, M.S. (1969) The responses of pedicellariae from *Echinus esculentus,* J. Exp. Mar. Biol. Ecol., 2:191–214.

Cantell, C.E., Franzen, A., and Sensenbaugh, T. (1982) Ultrastructure of multiciliated collar cells in the pilidium larva of *Lineus bilineatus,* Zoomorphologie, 101:1–15.

Carr, W.E.S. and Thompson, H. W. (1983) Adenosine 5′-monophosphate, an internal regulatory agent, is a potent chemoattractant for a marine shrimp, J. Comp. Physiol. A, 153:47–53.

Case, J.F., and Gwilliam, G.F. (1961) Amino-acid sensitivity of the dactyl chemoreceptors of *Carcinides maenas,* Biol. Bull., 127:428–446.

Chase, R. (1982) The olfactory sensitivity of snails, *Achatina fulica,* J. Comp. Physiol., 148:225–235.

Chia, F.S. (1982) Fine structure of larval rhinophores of the nudibranch *Rostanga pulchra* with emphasis on sensory receptor cells, Cell Tissue Res. 225:235–248.

Chia, F.S. and Koss, R.F. (1984) Fine structure of the cephalic sensory organ in the larva of the nudibranch *Rostanga pulchra* (Mollusca, Opisthobranchia, Nudibranchia), Zoomorphologie, 104:131–139.

Cloney, R.A. (1977) Larval adhesive organs and metamorphosis in Ascidians. 1. Fine structure of the everting papillae of *Distaplia occidentalis,* Cell Tissue Res., 183:423–444.

Crisp, D.J. (1974) Factors influencing the settlement of marine invertebrate larvae, in Chemoreception in Marine Animals, Grant, P.T. and Mackie, A.M. (eds.) Academic Press, New York, pp. 177–205.

Crisp, M. (1981) Epithelial sensory structures of trochids, J. Mar. Biol. Assoc. U.K., 61:95–106.

Crouau, Y. (1978) Organes sensoriels d'un Mysidace sous terrain anophthalme, *Antromysis juberthiei:* étude ultra-structurale des aesthetascs, Bull. Mus. Nat. Hist. Nat. Paris, 3; Zoologie, 352:165–175.

Davis, W.J. and Matera E.M. (1982) Chemoreception in gastropod molluscs; electron microscopy of putative receptor cells, J. Neurobiol., 13:79–84.

Derby, C.D. (1982) Structure and function of cuticular sensilla of the lobster, *Homarus americanus,* J. Crustacean Biol., 2:1–21.

Derby, C.D. and Atema, J. (1982) Narrow-spectrum chemoreceptor cells in the walking legs of the lobster *Homarus americanus;* taste specialists, J. Comp. Physiol., 146:181–189.

Derby, C.D., Carr, W.E.S. and Ache, B.W. (1984) Purinergic olfactory cells of crustaceans: response characteristics and similarities to internal purinergic cells of vertebrates, J. Comp. Physiol. A, 155:341–349.

Dethier, V.G. (1980) Evolution of receptor sensitivity to secondary plant substances with special reference to deterrents, Am. Nat., 115:45–66.

Felgenhauer, B.E. and Abele, L.G. (1983) Ultrastructure and functional morphology of feeding and associated appendages in the tropical freshwater shrimp *Atya innocous* (Herbst) with notes on its ecology, J. Crustacean Biol., 3:336–363.

Fraenkel, G.S. (1959) The raison d'être of secondary plant substances, Science, 129:1466–1470.

Fredman, S.M. and Jahan-Parwar, B. (1980) Processing of chemosensory and mechanosensory information in identifiable *Aplysia* neurons, Comp. Biochem. Physiol., 66A:25–34.

Frings, H. and Frings, G. (1965) Chemosensory bases of food-finding and feeding in *Aplysia juliana,* Biol. Bull., 128:211–217.

Ghiradella, H., Case, J.F., and Cronshaw, J. (1968) Structure of aesthetascs in selected marine and terrestrial decapods: chemoreceptor morphology and environment, Am. Zool., 8:603–621.

Ghiradella, H., Cronshaw, J., and Case, J.F. (1968) Fine structure of the aesthetasc hairs of *Pagurus hirsutiusculus* Dana, Protoplasma, 66:1–20.

Gnatzy, W., Schmidt, M., and Rombke, J. (1984) Are the funnel-canal organs the "campaniform sensilla" of the shore crab *Carcinus maenas* (Crustacea: Decapoda)? I. Topography, external structure and basic organization, Zoomorphologie, 104:11–20.

Hackney, C.M., McCrohan, C.R., and Hawkins, S.J. (1983) Putative sense organs on the pallial tentacles of the limpet *Patella vulgata* (L), Cell Tissue Res. 231:663–674.

Hartmann, H.B. and Hartmann, M.S. (1977) The stimulation of filter feeding in the porcellain crab *Petrolisthes cinctipes* Randall by amino acids and sugars, Comp. Biochem. Physiol. 56A:19–22.

Haszprunar, G. (1985) The fine morphology of the osphradial sense organs of the Mollusca. I. Gastropoda, Prosobranchia, Philos. Trans. R. Soc. Lond. B Biol. Sci., 307:457–496.

Hatt, H. (1984) Structural requirements of amino acids and related compounds for stimulation of receptors in crayfish walking leg, J. Comp. Physiol., 155:219–231.

Hatt, H. and Bauer, U. (1982) Electrophysiological properties of pyridine receptors in the crayfish walking leg, J. Comp. Physiol., 148:221–224.

Hatt, H. and Schmiedel-Jakob, I. (1984) Electrophysiological studies of pyridine-sensitive units on the crayfish walking leg. I. Characteristics of stimulatory molecules, J. Comp. Physiol., 154:855–863.

Hayes, W.F. and Barber, S.B. (1982) Peripheral synapses in *Limulus* chemoreceptors, Comp. Biochem. Physiol., 72A:287–293.

Heimann, P. (1984) Fine structure and moulting of aesthetasc sense organs on the antennules of the isopod, *Asellus aquaticus* (Crustacea), Cell Tissue Res., 235:117–128.

Hidaka, I., Ohsugi, T., and Kubomatsu, T. (1978) Taste receptor stimulation and feeding behaviour in the puffer, Chem Sens. Flavour, 3:341–354.

Holborow, P.L. and Laverack, M.S. (1972) Presumptive photoreceptor structures of the trochophore of *Harmothoe imbricata,* Mar. Behav. Physiol., 1:139–156.

Jahan-Parwar, B. (1972) Behavioural and electrophysiological studies on chemoreception in *Aplysia,* Am. Zool., 12:525–537.

Jahan-Parwar, B., Smith, M., and van Baumgarten, R. (1969) Activation of neurosecretory cells in *Aplysia* by osphradial stimulation, Am. J. Physiol., 216:1246–1257.

Jensen, K.R. (1982) Chemoreception as a factor in food location of *Elysia cauze* Marcus (Opisthobranchia, Ascoglossa), Mar. Behav. Physiol., 8:205–218.

Jones, D.A. (1979) Chemical defense; primary or secondary function? Am. Nat. 113:445–451.

Keil, T.A. and Steinbrecht, R.A. (1984) Mechanosensitive and olfactory sensilla of insects, in Insect Ultrastructure vol. 2 King, R.C. and Akai, H. (eds.), Plenum, pp. 477–516.

Lacalli, T.C. (1984) Structure and organization of the nervous system, in the trochophore larva of *Spirobranchus,* Philos. Trans. R. Soc. Lond. B Biol. Sci., 306:79–135.

Lane, E.B. and Whitear, M. (1982) Sensory structures at the surface of fish skin. I. Putative chemoreceptors, Zool. J. Linn. Soc., 75:141–152.

Laverack, M.S. (1963) Aspects of chemoreception in Crustacea, Comp. Biochem. Physiol., 8:141–151.

Laverack, M.S. (1968) On the receptors of marine invertebrates, Oceanogr. Mar. Biol. Annu. Rev., 6:249–324.

Laverack, M.S. (1974) Structure of chemoreceptor cells, in Chemoreception in Marine Organisms, Grant, P.T., and Mackie, A.M. (eds.), Academic Press, New York, pp. 1–49.

Laverack, M.S. (1975) Properties of chemoreceptors in marine Crustacea, Olfaction Taste, Proc. Int. Symp. 5:141–146.

Laverack, M.S. and Ardill, D.J. (1965) The innervation of the aesthetasc hairs of *Panulirus argus,* Q. J. Microsc. Sci., 106:45–60.

Laverack, M.S. and Barrientos, Y. (1985) Sensory and other superficial structures in living marine Crustacea, Trans. R. Soc. Edinb. Earth Sci., 76:123–136.

Luther, W. (1930) Versuche über die Chemorezeption der Brachyuren, Z. Vgl. Physiol., 12:177–205.

Mackie, A.M. (1981) Identification of the gustatory feeding stimulants, in Chemoreception in fishes, Hara, T.J., (ed.), Elsevier, North Holland, pp. 275–291.

Mackie, A.M., Lasker, R., and Grant, P.T. (1968) Avoidance reactions of a mollusc *Buccinum undatum* to saponin-like surface-active substances in extracts of the starfish *Asterias rubens* and *Marthasterias glacialis,* Comp. Biochem. Physiol., 26:415–428.

Mackie, A.M. and Mitchell, A.I. (1982) Further studies on the chemical control of feeding behaviour in the dover sole, *Solea solea,* Comp. Biochem. Physiol., 73A:89–93.

Mackie, A.M. and Turner, A.B. (1970) Partial characterization of a biologically active steroid glycoside isolated from the starfish *Marthasterias glacialis,* Biochem. J., 117:543–550.

Matera, E.M. and Davis, W.J. (1982) Paddle cilia (discocilia) in chemosensitive structures of the gastropod mollusc *Pleurobranchaea californica,* Cell Tissue Res. 222:25–40.

McLeese, D.W. (1970) Detection of dissolved substances by the American lobster (*Homarus americanus*) and olfactory attraction between lobsters, J. Fish Res. Board Can., 27:1371–1378.

McLeese, D.W. (1973) Olfactory responses of lobsters (*Homarus americanus*) to solutions from prey species and to seawater extracts and chemical fractions of fish muscle and effects of antennule ablation, Mar. Behav. Physiol., 2:237–249

Mead, F., Gabouriant, D.T., and Corbière-Tichanée, G., (1976) Structure de l'organe sensoriel apical de l'antenne chez l'isopode terrestre *Metaponorthus sexfasciatus* Budde-Lund (Crustacea: Isopoda), Zoomorphologie, 53:253–269.

Moore, A. and Cobb, J.L.S. (1985) Neurophysiological studies on the detection of amino acids in *Ophiura ophiura,* Comp. Biochem. Physiol., 82A:395–399.

Phillips, D.W. (1975) Localisation and electrical activity of the distance chemoreceptors that mediate predator avoidance behaviour in *Acmaea limatula* and *Acmaea scutum* (Gastropoda, Prosobranchia), J. Exp. Biol., 63:403–412.

Phillips, D.W. (1976) A scanning electron microscope study of sensory tentacles on the mantle margin of the gastropod *Acmaea (Notacmaea) scutum,* Veliger 19:266–271.

Phillips, D.W. (1979) Ultrastructure of sensory cells on the mantle tentacles of the gastropod *Notacmaea scutum,* Tissue & Cell, 11:623–632.

Reimer, A.A. (1971) Chemical control of feeding behaviour in *Palythoa* (Zoanthidae, Coelenterata), Comp. Biochem. Physiol. 40A:19–38.

Reimer, A.A. (1973) Feeding behaviour in the sea anemone *Calliactis polypus* (Forskal 1975), Comp. Biochem. Physiol., 44A:1289–1301.

Reimer, A.A. and Reimer, A. (1975) Chemical control of feeding in four species of tropical ophiuroids of the genus *Ophioderma,* Comp. Biochem. Physiol., 51A:917–927.

Robertson, R.M. and Laverack, M.S. (1979) Oesophageal sensors and their modulatory influence on oesophageal peristalsis in the lobster, Proc. R. Soc. Lond. B Biol. Sci., 206:235–263.

Rossi, F., Nardi, G., Palumbo, A., and Prota, G. (1985) 5-thiolhistidine, a new amino acid from eggs of *Octopus vulgaris,* Comp. Biochem. Physiol., 80B:843–845.

Schmidt, M. and Gnatzy, W., (1984) Are the funnel canal organs the ''campaniform sensilla'' of the shore crab, *Carcinus maenas* (Decapoda: Crustacea)? II. Ultrastructure, Cell Tissue Res., 237:87–93.

Seelinger, G. (1983) Response characteristics and specificity of chemoreceptors in *Hemilepistus reaumuri* (Crustacea: Isopoda), J. Comp. Physiol., 152:219–229.

Shelton, R.G.J. and Laverack, M.S. (1968) Observations on a redescribed crustacean cuticular sense organ, Comp. Biochem. Physiol., 25:1049–1059.

Shelton, R.G.J. and Laverack, M.S. (1970) Receptor hair structure and function in the lobster *Homarus gammarus,* J. Exp. Mar. Biol. Ecol., 4:201–210.

Shepheard, P. (1974) Chemoreception in the antennule of the lobster, *Homarus americanus,* Mar. Behav. Physiol., 2:261–273.

Sloan, N.A. and Campbell, A.C. (1982) Perception of food, in Echinoderm Nutrition, Jangoux, M. and Lawrence, J.M. (eds.), Balkema, Rotterdam, pp. 3–23.

Snow, P.J. (1973) Ultrastructure of the aesthetasc hairs of the littoral decapod, *Paragrapsus gaimardii,* Z. Zellforsch. Mikrosk. Anat., 138:489–502.

Spencer, M. (1986) The innervation and chemical sensitivity of single aesthetasc hairs, J. Comp. Physiol., 158:59–68.

Stinnakre, J. and Tauc, L. (1969) Central neuronal response to the activation of osmoreceptors in the osphradium of *Aplysia,* J. Exp. Biol., 51:347–361.

Strathmann, R.R. (1971) The feeding behaviour of planktotrophic echinoderm larvae: mechanisms, regulation and rates of suspension feeding, J. Exp. Mar. Biol. Ecol., 6:109–160.

Strathmann, R.R. (1982) Cine films of particle capture by an induced local change of beat of lateral cilia of a bryozoan, J. Exp. Mar. Biol. Ecol., 62:225–236.

Strathmann, R.R. (1984) Effect of flavour and size on selection of food by suspension-feeding plutei, Limnol. Oceanogr., 29:357–361.

Sumida, B.H. and Case, J.F. (1983) Food recognition by *Chaetopterus variopedatus* (Renier): synergy of mechanical and chemical stimulation, Mar. Behav. Physiol., 9:249–274.

Thurm, U. (1968) Steps in the transducer process of mechanoreceptors, Symp. Zool. Soc. Lond., 23:199–216.

Valentincic, T. (1979) Structure-activity relationships of amino acids in *Marthasterias glacialis* chemoreception, Proc. Europ. Colloq. Echinoderms, Balkema, Rotterdam, pp. 337–341.

Welsch, U. and Storch, V. (1969) Über das Osphradium der prosobranchen Schnecken *Buccinum undatum* L und *Neptunea antiqua* (L), Z. Zellforsch. Mikrosk. Anat., 95:317–330.

# Chapter 12

# Peripheral Filters and Chemoreceptor Cells in Fishes

JOHN CAPRIO *

Olfaction in vertebrates is the sense used in the processing of chemical information that is detected and transmitted to the central nervous system by bipolar neurons that compose the olfactory nerve (cranial nerve I). Gustation is the sense used in the processing of chemical information that is detected by taste cells and transmitted centrally by facial (cranial nerve VII), glossopharyngeal (IX), or vagal (X) nerves. Receptor cells of both systems must discriminate relevant chemical stimuli from background chemical "noise" that exists in the immediate environment of all vertebrates. This ability of chemoreceptors to act as "peripheral filters" by detecting and preferentially passing information concerning biologically important chemical stimuli occurs through the evolutionary development of different receptor molecules with different chemospecificities built into the membranes of the receptor cells. This chapter will review the general organization of the olfactory epithelium and taste bud structure in fishes and summarize recent physiological findings as they relate to the sensitivities and specificities of the "filtering elements" (i.e., transduction processes and types of receptor sites) of the receptive olfactory and gustatory receptor cells. For books specific to chemoreception in aquatic organisms, see Kleerekoper (1969), Mackie and Grant (1974), Hara (1982), and Hasler and Scholz (1983).

Although this chapter is concerned specifically with the olfactory and gustatory systems, it is important to note here that branches of the trigeminal nerve (V) terminate as free nerve endings in both the nasal and gustatory regions of vertebrates and are also capable of detecting chemicals (Silver and Maruniak 1981). Furthermore, spinal chemosensitivity mediated by free nerve endings may be rather well developed in certain fishes. For information on these nonolfactory-nongustatory chemosenses (sometimes termed "the common chemical sense") in fish, see Sheldon (1909), Finger (1982), Belousova, Devitsina, and Malyukina (1983), and Silver and Finger (1984).

---

* Department of Zoology and Physiology, Louisiana State University, Baton Rouge, LA 70803, USA

## Olfactory Organ and Receptor Cell Distribution

Although considerable variation occurs among bony fishes, generally each of the paired olfactory organs is located within a blind pit located on the dorsal side of the head (Burne 1909). In elasmobranchs, however, the olfactory organs are situated on the ventral side of the snout (Sheldon 1909). An incurrent naris (located more rostrally) and an excurrent naris provide pathways of inlet and exit, respectively, for the water bathing the olfactory organ. In bony fishes termed "isosmates," a continuous water current through the olfactory capsule is generated by ciliary action of apical cells in the nonsensory epithelium (Døving et al. 1977). In the "cyclosmates," the compression and extension of accessory olfactory sacs due to respiratory movements of the fish create a pumping action that provides a pulsatile flow of water through the olfactory capsule, somewhat analogous to the "sniffing" action of terrestrial vertebrates (Døving et al. 1977). The olfactory organ of the few sharks studied appears to be more complex than in bony fishes owing to an overlying nasal flap with an attached septum covering apposing "valvelike" structures that descend into the olfactory capsule proper (Tester 1963; Zeiske, Caprio, and Gruber 1986). Ventilation of the organ is hypothesized to result from a pressure differential between inlet and outlet chambers caused by the forward motion of the animal (Zeiske, Caprio, and Gruber 1986).

In the majority of fish studied, the floor of the nasal cavity forms a series of folds, the olfactory lamellae (rosette), in which the sensory epithelium resides [Fig. 12.1(A)]. Each lamella is composed of two layers of epithelium enclosing a thin stromal layer, within which the olfactory receptor axons course towards the olfactory bulb (Caprio and Raderman-Little 1978; Yamamoto 1982). In some species, such as the topminnow (*Aplocheilus lineatus*), lamellae do not occur, and the sensory and nonsensory epithelia are distributed on the relatively flat floor of the nasal chamber (Zeiske et al. 1976). The geometry and number of olfactory lamellae within each capsule and the organization of the sensory and nonsensory epithelia across each lamella vary widely among the different species (Burne 1909; Yamamoto 1982). The sensory epithelium is columnar and pseudostratified, containing receptor, supporting, and basal cells. In addition, a few goblet cells, which, along with the supporting cells, produce mucus that covers the olfactory surface, may be found scattered in the apical region. The mucous sheet, however, is thinner than that produced in tetrapods, primarily by the subepithelial Bowman's glands, which are lacking in fish. Olfactory receptor cell density is quite variable and has, for different species, been reported to range between 46,000 and 500,000 $mm^{-2}$ (Teichman 1954; Zeiske et al. 1976; Yamamoto 1982). The total number of unmyelinated axons, whose individual diameters average 0.14–0.2 $\mu m$, in a single olfactory nerve of a fish is between $4.2 \times 10^4$ and $1.3 \times 10^7$ (Gemne and Døving 1969; Easton 1971; Kreutzberg and Gross 1977; Yamamoto 1982). Basal cells within the sensory epithelium of vertebrates differentiate into olfactory neurons during normal receptor turnover and in response to olfactory receptor injury (Graziadei and Metcalf 1971; Moulton 1975; Cancalon 1982). The nonsensory epithelium is stratified and composed of a surface cell

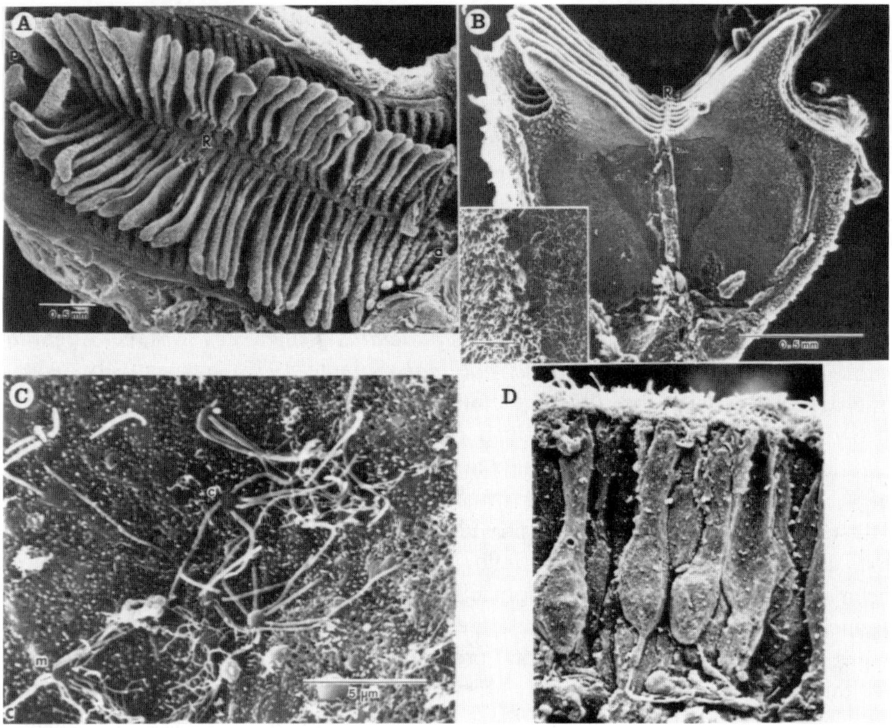


Figure 12.1. Olfactory organ and receptor cells of channel catfish *Ictalurus punctatus*. (A) Scanning electron micrograph (SEM) of dorsal surface of olfactory rosette. (B) Transection of olfactory organ between adjacent lamellae. Central (darker) sensory epithelium is easily distinguished from surrounding indifferent epithelium. *Inset:* Distinct interface between sensory and indifferent epithelia is seen at higher magnification. (From Erickson and Caprio 1984.) (C) SEM of intermediate region of sensory epithelium showing dendritical endings of both ciliated (*c*) and stellate microvillous (*m*) olfactory receptor neurons. (D) SEM of transverse fracture face within sensory region of a lamella showing three prominent olfactory receptor neurons. One axon is shown leaving receptor cell proper. (From Caprio and Raderman-Little 1978.)

layer, several intermediate layers, and a basal layer; goblet cells are also located in this region. Both the sensory and nonsensory epithelia are separated from the underlying loose connective tissue by a basal lamina.

The sensory regions of the olfactory epithelium of the majority of teleosts studied contain at least two types of bipolar olfactory neurons, ciliated and microvillous, which are classified on the basis of their apical dendritical processes (Fig. 12.1C) (Ichikawa and Ueda 1977; Yamamoto 1982; Cancalon 1983; Muller and Marc 1984). In elasmobranchs, however, only microvillous olfactory receptors are thought to occur (Reese and Brightman 1970; Bronshtein 1976; Zeiske, Caprio, and Gruber 1986). Generally, the knoblike dendritical ending of the ciliated olfactory neuron contains 4–12 cilia that are 3–8 μm long and 0.2–0.3 μm in diameter, showing a (9 + 2) or (9 + 0) microtubular pattern in cross section (Theisen et al. 1980; Yamamoto 1982). Microvillous cells found in the sensory region of the olfactory mucosa of teleosts give rise to either 30–80, 2–5-μm-long microvilli originating from a protruding apical ending (i.e., "tufted" cells) (Thommesen 1982; Yamamoto 1982) or 15–25, 3–6-μm-long microvilli radiating laterally from a flat, disclike apical surface (i.e., "stellate" cell) (Cancalon 1982; Yamamoto 1982; Erickson and Caprio 1984). The microvilli of both tufted and stellate cells are approximately 0.1 μm in diameter. Both types of microvillous cells have been identified in differing proportions in the same species (Schulte 1972; Cancalon 1982). Direct evidence that these types of microvillous cells are bipolar primary neurons comes from observations of stellate and tufted cells possessing axonlike processes in the channel catfish *Ictalurus punctatus* (Cancalon 1983), eel *Anguilla anguilla*, and piranha *Serrasalmus hollandi* (Schulte 1972), respectively. Also, microvillous cells in the olfactory epithelium of the brown trout (*Salmo trutta*) (Rowley and Moran 1985), goldfish (*Carassius auratus*), and channel catfish (Muller and Marc 1984) accumulated horseradish peroxidase (HRP) after labelling of the peripheral cut end of the olfactory nerve and HRP injections into the olfactory bulb, respectively. This provided the evidence that axons of the microvillous cells project to the olfactory bulb. Although microvillous neurons may have been derived phylogenetically from ciliated receptors (Pyatkina 1976; Delfino, Bianchi, and Ercolini 1981), these neurons are presently independent cellular types and not merely developmental precursors of the ciliated receptor. Evidence for this comes primarily from the observations that both types renew concurrently following zinc sulfate–induced degeneration of olfactory receptors (Cancalon 1982) and that ciliated olfactory neurons are most numerous during larval development (Pyatkina 1976; Breucker, Zeiske, and Melinkat 1979).

An incomplete segregation of the ciliated and microvillous olfactory receptors occurs across the surface of the sensory region of the lamellae in four species of teleosts studied quantitatively by scanning electron microscopy. The dorsomedial portion of the sensory region of the lamellae located in the central one-third of the rosette in the channel catfish *Ictalurus punctatus* contains significantly more stellate microvillous (20/600 $\mu m^2$) than ciliated (10/600 $\mu m^2$) olfactory neurons, whereas the ventrolateral sensory epithelium is composed almost exclusively of ciliated receptors (28 ciliated: 1 microvillous cell/600 $\mu m^2$) (Erickson and Caprio 1984). In three species of salmonids (*Salmo alpinus, S. fontinalis,* and *S. gairdneri*), the tufted microvillous cells are more uniformly distributed than the stellate cells of the catfish; but, like the catfish, the density of microvillous cells increases towards the medial portions of the olfactory lamellae. In both the channel catfish and the three species of salmonids, the frequency of ciliated cells is greatest in the lateral lamellar corridors (Thommesen 1983).

Although it is now established that olfactory receptor neurons exist in at least two anatomically distinct forms, ciliated and microvillous, and that they may be differentially distributed across the sensory regions of the olfactory mucosa, the question of their physiological differences remains. Only two reports currently exist for which electrophysiological recordings were obtained from regions of the sensory epithelium having different relative proportions of ciliated and microvillous neurons. The tufted microvillous olfactory neurons of three species of salmonids were reported to be more sensitive to amino acids as compared to bile salts, whereas the reverse was indicated for the ciliated neurons (Thommesen 1982). It is hypothesized that bile salts may be important pheromonal cues during the spawning migration of salmonids (Døving, Selset, and Thommesen 1980; Selset and Døving 1980). In the nonmigratory channel catfish, however, a differential specificity to amino acids and bile salts in olfactory regions rich in stellate microvillous and ciliated receptors, respectively, was not observed (Erickson and Caprio 1984).

An additional type of cell, the "rod cell," was proposed to be an olfactory receptor neuron (Bannister 1965; Schulte 1972). Recent evidence suggests, however, that they may be degenerating ciliated receptors (Theisen et al. 1980; Muller and Marc 1984) or artifacts of the histological fixation process (Eller, Rowley, and Moran 1985).

## Taste Bud Anatomy and Distribution

Taste cells in fishes, as in other vertebrates, are primarily contained within pear-shaped structures termed *taste buds,* though cells purported to be isolated taste cells have been suggested (Reutter 1974; Lane and Whitear 1982). Taste buds are located in a stratified squamous epithelium (epidermis) and are separated from the underlying corium by a basal membrane (Reutter 1982). Taste buds of fishes vary between 30 and 80 μm in length and 20 and 50 μm in width (Hirata 1966; Crisp, Lowe, and Laverack 1975; Grover-Johnson and Farbman 1976; Kiyohara, Yamashita, and Kitoh 1980) and are generally distributed within the oral cavity and pharynx. In some fishes, however, taste buds may extend to the external body surface. Those taste buds on the surface, lips, and rostral palate are innervated by the facial nerve (VII), whereas the glossopharyngeal (IX), and vagus (X) nerves innervate only oropharyngeal taste buds (Herrick 1901). The wide variation in distribution of the oral and extraoral taste buds is seen in a comparison of the ictalurid catfish and Japanese minnow. In the yellow bullhead catfish, *Ictalurus natalis,* the extraoral taste buds predominate (90%) (Fig. 12.2A and B) (Atema 1971), whereas in the Japanese minnow, *Pseudorasbora parva,* the majority (86%) of taste buds are located within the oral cavity (Fig. 12.2[C] and [D]) (Kiyohara, Yamashita, and Kitoh 1980). The Japanese minnow (6 cm in length) has fewer total taste buds (8,000) than the catfish (>175,000; body length 25 cm), but the Japanese minnow has a higher taste bud density (approx. 140 $mm^{-2}$ on lips, palatal organ, and gill rakers) than the catfish (>50 $mm^{-2}$ on the gill arches and >25 $mm^{-2}$ on maxillary barbels). Presently, there is no evidence to suggest that differences in total numbers or in the distribution of taste buds between species of fishes relate to gustatory sensitivity. For example, the Japanese minnow (Kiyohara, Yamashita, and Harada 1981), ictalurid catfish (Caprio 1975, 1982, 1984), and the Japanese eel (*Anguilla japonica*) (Yoshii et al. 1979) have similar gustatory sensitivities to amino acids measured electrophysiologically, though the eel lacks taste buds on its external body surface. Numbers and distribution patterns of taste buds, however, are implicated in the ability of

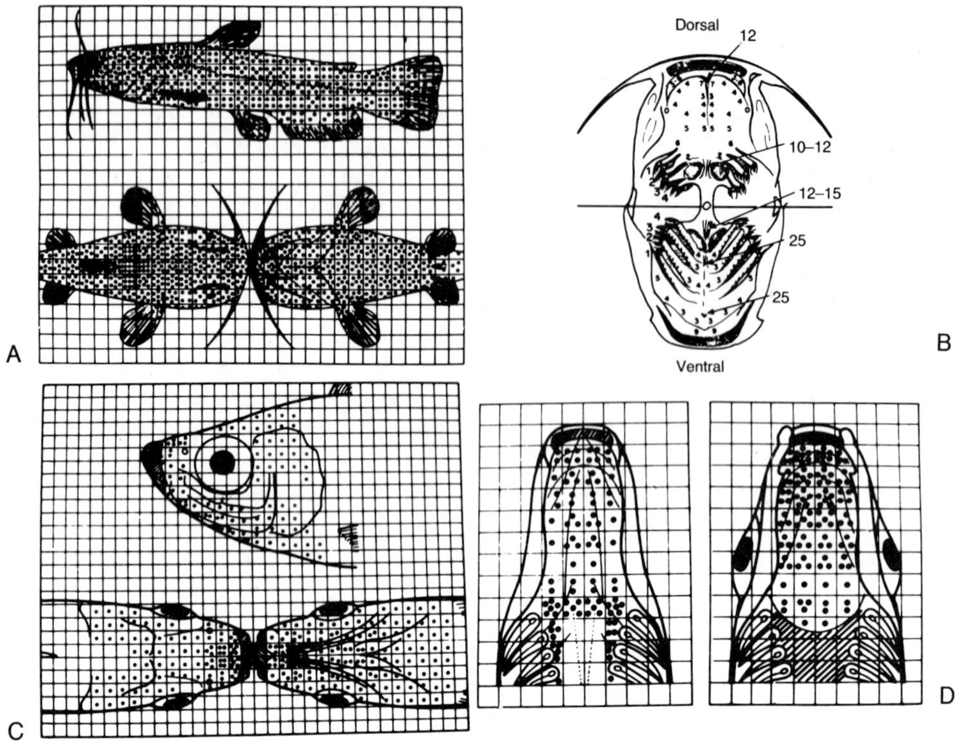


Figure 12.2. Distribution of taste buds in yellow bullhead catfish, *Ictalurus natalis,* and Japanese minnow, *Pseudorasbora parva: each square* = 1 mm$^2$. (A) The distribution of extraoral taste buds in adult catfish: *each dot* = 1 taste bud; *solid black* = more than 15 taste buds; *top,* lateral view; *bottom left and right,* dorsal and ventral views. (B) The distribution of oral taste buds in the catfish. High concentration on ridges is shown by *arrows and attached numbers*. Gill arches are numbered 1–4. Upper and lower portions of the figure are dorsal and ventral views inside oropharyngeal cavity. (From Atema 1971.) (C) The distribution of extraoral taste buds in adult minnow 6 cm in length. Each *solid dot* and *open dot* represents 5 and 2 taste buds, respectively. *Oblique lines* = more than 60 taste buds. *Top:* lateral view; *bottom left and right:* dorsal and ventral views. (D) The distribution of oral taste buds in the minnow: *each dot* = 15 taste buds: *oblique lines* on lips and palatal organ = more than 140 taste buds. Large parts of branchial apparatus are removed. *Left and right:* ventral and dorsal view inside the mouth. (From Kiyohara, Yamashita, and Kitoh 1980, with permission from Pergamon Press, Ltd.)

a fish to locate a food source effectively, (Bardach, Todd, and Crickmer 1967). The difference in distribution and density of taste buds between the Japanese minnow and ictalurid catfish may reflect differences in their respective feeding behaviors. The minnow possibly uses vision more than the catfish does in locating a food source and is therefore not dependent on a profuse extraoral supply of taste buds (Kiyohara, Yamashita, and Kitoh 1980). However, this difference may also reflect, in part, the overall size differential of the two species. An extensive external body distribution of taste buds may not be so critical to a fish only 6 cm long as it is to a fish that may reach a body length in excess of 1.5 m. The reduced length of the minnow may limit the effectiveness of a gustatory tropotactic (spatial discrimination of chemical cues) analysis of concentration gradients, which was indicated for the larger catfish (Johnsen and Teeter 1980).

Two types of microvillous cells of ectodermal origin and a type of neuroectodermal basal cell are routinely described within the taste buds of teleosts (Fig. 12.3) (Hirata 1966; Grover-Johnson and Farbman 1976; Reutter 1978). Terminology for the two types of microvillous cell varies. They are termed *light and dark cells,* based on light microscopic descriptions after formaldehyde fixation, or *t-cells* (i.e., tubule-containing cells, analogous to the light taste cell) and *f-cells* (i.e., fibril-containing cells, analogous to the dark taste cell) (Crisp, Lowe, and Laverack 1975). The light cell, which terminates at its apical (proximal) surface in a large (1.5–3 μm) club-shaped microvillus, is generally accepted as a gustatory receptor cell because of identified synaptic contacts with nerve processes within the taste bud (Hirata 1966; Crisp, Lowe, and Laverack 1975; Grover-Johnson and Farbman 1976; Reutter 1978). The dark cell, which contains numerous small (0.5–1.0 μm) microvilli, encloses the light cell with lobed processes in the distal regions of the taste bud and is possibly a supporting cell (Hirata 1966; Crisp, Lowe, and Laverack 1975; Grover-Johnson and Farbman 1976).

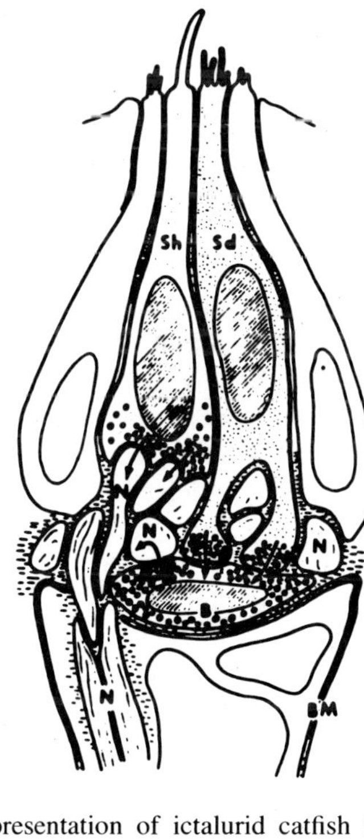


Figure 12.3 Schematic representation of ictalurid catfish taste bud: *Sh,* light cell; *Sd,* dark cell; *B,* basal cell; *N,* nerve fibers of cranial nerves VII, IX, or X; *BM,* basement membrane. (From Reutter 1971.)

However, Reutter (1978) indicated that dark cells in the ictalurid catfish also form synapses with nerve terminals within taste buds (Fig. 12.3).

Gustatory stimuli may have a more direct access to the receptor membranes of fish than to those of mammals, which are submerged in an apical pore filled with a mucous barrier (Murray 1973). Only a thin mucous layer overlays the generally poreless taste buds of fish, possibly allowing the large microvillous receptor process of the light cells to penetrate the mucus, as scanning electron micrographs of freeze-dried specimens suggest (Reutter 1980). In both fish (Raderman-Little 1979) and mammals (Beidler and Smallman 1965; Farbman 1980), light and dark cells undergo renewal from postmitotic daughter cells of epithelial cells surrounding the taste bud. The time required for taste cell renewal (time of degeneration of 50% of labelled cells) in the channel catfish at 30°C and rats and mice at 37°C is remarkably similar, taking 10–12 days.

The monoamine-containing basal cells, which generally number 2–5/taste bud, are located at the base of the bud, resting against the basal lamina (Grover-Johnson and Farbman 1976; Reutter 1978; Toyoshima, Nada, and Shimamura 1984). These authors suggest that the basal cells in teleosts are derived from Schwann cells, form afferent synapses with neurons penetrating into the taste bud, and may function as mechanoreceptor cells. Reutter (1982) also makes the proposal, based on transmission electron micrographs of catfish taste buds, that besides forming synapses with nerve terminals, both light and dark taste bud cells also synapse with basal cells, thus providing the currently controversial possibility that basal cells may function as modifers of gustatory receptor activity.

## Receptor Mechanisms

Our knowledge of specific transduction events in vertebrate olfactory and gustatory receptor cells is extremely limited. The binding of chemical stimuli to the apical membrane of an olfactory receptor cell leads to the production of a generator potential that depolarizes the cell, resulting in impulse initiation in the axon. The molecular interaction of chemical stimuli on a gustatory receptor cell initiates a receptor potential that triggers the release of a yet unknown neurotransmitter, leading to the production of action potentials in the first-order neuron. Voltage-dependent $Na^+$ and $Ca^{++}$ channels, indicated in the bullfrog *Rana catesbeiana* taste cells, may have a possible role in the electrical events linking receptor membrane depolarization and neurotransmission (Kashiwayanagi, Miyake, and Kurihara 1983). Further, taste bud cells in isolated lingual mucosa of the mud puppy *Necturus maculosus* were shown to be capable of evoking regenerative impulses in response to either electrical (Roper 1983) or chemical (0.1 M–1.0 M KCl) (Roper and McPheeters 1984) stimulation, whereas epithelial cells lying outside the taste buds were not electrically excitable. Whether this phenomenon occurs in taste buds in vivo and is a general phenomenon of taste cell transduction that was previously overlooked remains to be verified. Presently, the regenerative electrical activity of taste bud cells in the mud puppy is stimulating a renewed interest in taste cell physiology.

Successful intracellular recordings of olfactory and presumed gustatory receptor cells in fish were reported only in the lamprey *Entosphenus japonicus* (Suzuki 1977, 1982) and brown bullhead catfish, *Ictalurus nebulosus* (Teeter 1974), respectively. Resting membrane potentials averaged −52 mV in the lamprey olfactory receptor cell and −26 mV in the taste bud cells of the catfish. In contrast to taste bud cells of the mud puppy, the electrical properties of epithelial cells of the catfish adjacent to, but lying outside, the taste bud were not significantly different from those located within the bud, suggesting that both taste and nontaste epithelial cells possess similar ionic mechanisms for responding to certain chemical stimuli (salt, acid, and amino acid). The response of these cells to a chemical stimulus is a graded depolarization and decreased membrane resistance (generally) or hyperpolarization and increased membrane resistance. Reversal potentials for the depolarizing response were near 0 for the olfactory receptor and ranged from +4 to +53 mV for the taste bud cell, indicating ionic current carriers.

Experimental alterations of the ionic environment over the olfactory and gustatory receptors should influence the electrical activity of the receptor cells and define their ionic requirements. Integrated electroencephalographic recordings (EEG) from the olfactory bulb of the carp (*Cyprinus carpio*), rainbow trout (*Salmo gairdnerii*), and bullfrog showed that olfactory responses to amino acids were reduced or eliminated with the assumed experimentally induced removal of ions from the surface of the olfactory epithelium and responses recovered with their replacement (Yoshii and Kurihara 1983a). Replacement with any one of 11 different inorganic cations or particular, individually presented organic cations supported the olfactory responses to amino acids of normal magnitude. Further, divalent inorganic cations were effective at a lower concentration than the monovalent ions. A study of the ionic requirements of gustatory receptors in the eel *Anguilla japonica* to amino acids resulted in similar findings (Yoshii and Kurihara 1983b). Some discrepancy, however, exists between the previous findings and those reported by Suzuki (1978, 1982) for single olfactory receptor responses in the lamprey to altered ionic environments. Suzuki indicated that responses to L-arginine were blocked when a small amount of calcium-chelating agent (EGTA) was added to the calcium-free solution bathing the receptors. Because of an earlier finding (Suzuki 1980) that the binding activity of arginine increased with increasing calcium ion concentrations in the range between $10^{-7}$M and $10^{-4}$M, but decreased gradually at higher concentrations, it was suggested that calcium ions act only as modifiers of receptor cell activity. These studies collectively indicate that cations do not act at the apical membrane surface of chemoreceptor cells as current-carrying ions, but that the binding to the chemoreceptive membranes played an essential role in the sensory process. Possible results of calcium binding are the maintenance of the stimulus-receptor complex in the active state and an increase in affinity of the receptor site for the stimulus. Thus, separate sites may exist in the receptor cell for binding chemical stimuli and initiating permeability changes. How these different membrane areas are coupled by additional receptor events, be they electrical, chemical, or physical, is unknown. Future voltage-clamp and patch-clamp studies of chemoreceptor cells should add to our limited understanding of the transduction process.

## Sensitivity and Specificity to Amino Acids

The emphasis placed in this chapter on amino acid stimuli is indicative of the number of recent studies of aquatic organisms that have tested these compounds. Other chemicals implicated as possibly important stimuli in fish include bile salts (Døving, Selset, and Thommesen 1980; Hara et al. 1984), nucleotides (Hidaka, Kiyohara, and Oda 1977), and fatty acids (Sutterlin and Sutterlin 1970; Yoshii et al. 1979). However, to date, no other single class of compounds that stimulate the olfactory and gustatory systems of fish has yielded a body of consistent electrophysiological, biochemical, and behavioral data, like that obtained with amino acids.

Possible reasons for amino acids' being potent chemosensory stimuli include their water solubility and presence in the mucus that covers the external body of fish (Uskova and Chaykovskaya 1971; Stabell and Selset 1980), which may serve as a source for conspecific recognition. Amino acids are released from cellular stores to the environment from both living aquatic invertebrates (Johannes and Webb 1970) and decaying organic matter, thus providing a chemical trail for orientation to potential food sources. Importantly, background levels of dissolved free amino acids in the water column are relatively low ($< 10^{-7}$ M) (Pocklington 1972; Gardner and Lee 1975; North 1975), which is probably due to the highly efficient amino acid uptake mechanisms of bacteria and algae (Gardner and Lee 1975). This diminishes the possibility of sensory adaptation to these compounds and supports the hypothesis that amino acids may serve as excellent signal molecules in fish.

Unsubstituted L-alpha amino acids are highly stimulatory to both the olfactory and gustatory systems of fishes (Tucker 1983; Caprio 1984). The primary exceptions are that both alpha- and beta-alanine are equally stimulatory to the palatal gustatory system of the Japanese eel, *Anguilla japonica* (Yoshii et al. 1979), and that primary carboxyl esters of some amino acids are either equally (Caprio 1978; Marui, Harada, and Kasahara 1983) or more (Kiyohara, Yamashita, and Harada 1981) stimulatory than the parent free amino acid. Peptides are always less effective chemosensory stimuli to fish tested electrophysiologically than the most stimulatory amino acid residue in the molecule. Further, the overall sensitivity to amino acids of both systems is also highly similar. Estimated electrophysiological thresholds for the more effective stimuli generally range between $10^{-7}$ and $10^{-9}$ M, though for the gustatory systems of a few species, amino acid thresholds extend below nanomolar concentrations (see Caprio 1984 for a review).

The above studies established conclusively that olfactory and gustatory receptors of fish are highly sensitive to amino acids. The two chemosensory systems are not redundant, however, as the major physiological distinction between olfaction and gustation in the same species is the differential specificity for the amino acid side chains (Table 12.1). On electrophysiological comparisons of the amino acid response spectrum in different species, it is evident that the population of olfactory receptors respond best to amino acids consisting of linear, unbranched, and uncharged side chains (e.g., methionine, ethionine, glutamine, norvaline, norleucine, alanine, serine, and cysteine). For the most part, neutral amino acids

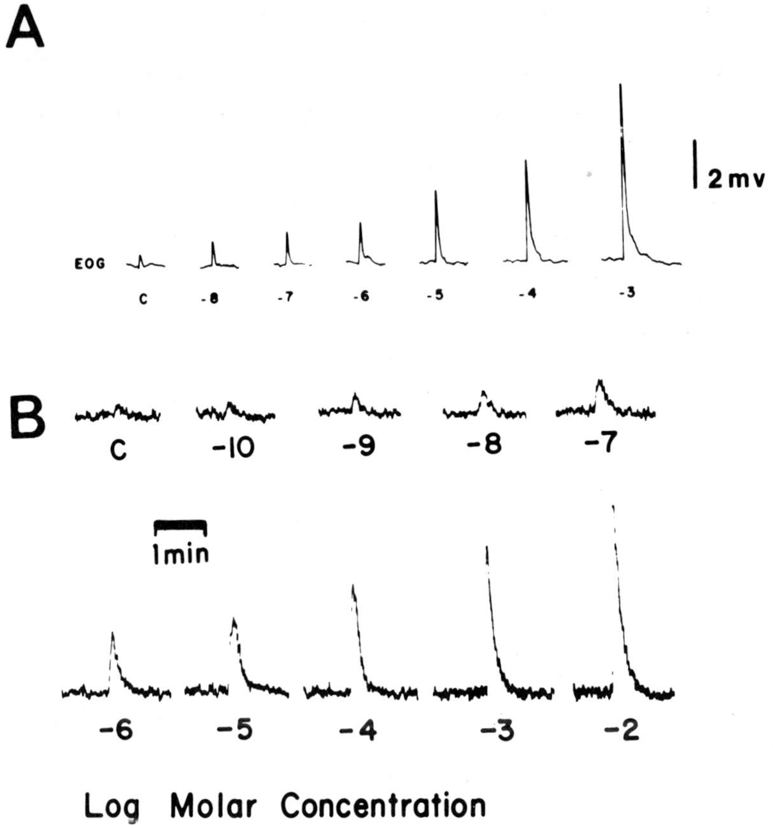


Figure 12.4. Olfactory and gustatory responses of ictalurid catfish to different concentrations of amino acids. (A) Olfactory (*EOG*) responses of channel catfish *Ictalurus punctatus* to increasing concentration of L-arginine. (From Byrd and Caprio 1982.) (B) Integrated (rise time 0.5 sec.) taste responses from facial (VII) nerve innervating maxillary barbel of spotted bullhead catfish, *Ictalurus serracanthus,* to increasing concentrations of L-alanine: *C,* response to control well water.

with the longer side chains are more stimulatory (Table 12.1). The amino acid specificities of the gustatory systems of fishes, however, show more variability than the olfactory system. Electrophysiological recordings from the facial taste systems in nine species of fish indicate that neutral amino acids with short (containing 0 or 1 carbon atom) side chains (e.g., glycine, alanine, or serine) are more potent stimuli than those with longer side chains, (Table 12.1); other examples are the Japanese minnow, *Pseudorasbora parva* (Kiyohara, Yamashita, and Harada 1981) and the puffer *Fugu pardalis* (Hidaka 1982). This finding is in direct contrast to that for olfaction. Further, basic amino acids (L-arginine and L-lysine, pH adjusted to nearly neutral), an imino acid (L-proline), and an amino acid derivative (betaine) show considerable variability in their ability to stimulate gustatory receptors in different species of fish effectively. Each of these three types

Table 12.1. Electrophysiologically Potent Amino Acid Stimuli

| Animal | Olfaction | Gustation |
|---|---|---|
| Channel catfish, *Ictalurus punctatus* | CysH, Met, Gln, nVal, nLeu[a] | Ala, Arg[a], Pro[b] |
| Sea catfish, *Arius felis* | CysH, Met, Gln[c] | Ala, Arg[d] |
| Japanese sea catfish, *Plotosus anguillaris* | nLeu, Met, Gln, CysH[e] | Pro, Betaine, Arg, Gly[d] |
| Mullet, *Mugil cephalus* | Gln, Met, Ala[f] | Arg, Lys, Ala, Ser, Betaine[f] |
| Sea bream, *Chrysophrys major* | Gln, Met, Ala[f] | Ala, Gly, Arg, Ser, Lys[f] |
| Rainbow trout, *Salmo gairdneri* | Gln, Met, Leu, Asn, Ala[g] | Pro, OHPro, Betaine, Leu, Ala[h] |
| Carp, *Cyprinus carpio* | (Lys, Arg, Glu), Gln, Met, Ala, Leu[i] | Pro, Ala, CysH, Glu;Na, Betaine[j] |
| Japanese eel, *Anguilla japonica* | —[k] | Arg, Gly, Ala, β-Ala, Lys[l] |
| American eel, *Anguilla rostrata* | CysH, Gln, Met, Ala[m] | —[k] |

[a] Caprio 1978; Byrd and Caprio 1982
[b] Kanwal and Caprio 1983
[c] Caprio 1980
[d] Caprio unpublished
[e] Silver unpublished
[f] Goh and Tamura 1980
[g] Hara 1975
[h] Marui, Harada, and Kasahara 1983
[i] Goh and Tamura 1978; ( ) possible pH effect
[j] Marui, Evans, and Hara 1983
[k] Not tested
[l] Yoshii et al. 1979
[m] Silver 1982

of stimuli are highly effective for some fishes and rather nonstimulatory to others. Acidic amino acids (pH adjusted to nearly neutral) and their sodium salts are relatively poor olfactory and gustatory stimuli. The single exception is monosodium glutamate (MSG), which is a potent stimulus to the facial taste system of the carp *Cyprinus carpio* (Marui, Harada, and Kasahara 1983). It is also important to recognize that the present statements on the specificity of olfactory and gustatory receptor cells of fish must be tempered with the fact that only an extremely limited number of species have been tested in comparison with the number that inhabit both the marine and freshwater habitats of the world.

Since taste buds in various bodily locations may be innervated by different cranial nerves (VII, IX, or X), it is necessary to determine, in the same species of fish, if amino acid specificities vary across these nerves. Presently, in only a single species, the channel catfish (*Ictalurus punctatus*), is that information available. In the channel catfish, the facial (Caprio 1975, 1978; Davenport and Caprio 1982), glossopharyngeal, and vagal (Kanwal and Caprio 1983) gustatory systems show a similar amino acid response spectrum, though thresholds were generally lower for the facial system. This is consistent with the finding that the facial gustatory system is used by the catfish to orient to a distant food source, whereas the glossopharyngeal and vagal gustatory systems are in contact with higher concentrations of materials within the oropharyngeal cavity and serve as the afferent limb of the swallowing and rejection reflexes (Atema 1971; Morita and Finger 1985a, 1985b).

## "Peripheral Filters": Transduction Processes and Receptor Site Types

Rank orders of effective amino acids obtained from electrophysiological recordings from populations of receptor cells do not directly indicate the number of different types of binding sites that are present in the membranes of these cells, but they do provide important clues. Different patterns of chemospecificity, as shown for olfaction and gustation, along with distinct species variations in the gustatory spectrum of sensitivity to specific amino acids, provide an initial insight into the possible types of amino acid–binding sites that may exist. Analysis of the chemospecificity patterns of single olfactory receptor cells and single gustatory units (a single gustatory fiber and the taste cells it innervates) would provide additional information on receptor site types; at present, however, there are few such published reports.

Theoretically, the number of different receptor "site types" could be more than the number of different amino acids that are detected by the particular chemoreceptive system, or it could be as few as a single site type having broad specificity that interacts with all the commonly occurring amino acids. Major problems with the latter include (1) a background of any amino acid in the water column that could effectively reduce or eliminate the ability of the chemosensory system to respond to other amino acids owing to sensory adaptation; (2) the variation in the side-chain structure of the different amino acids that may preclude the development of a broadly tuned, highly sensitive binding site; and (3) a single site type that may prevent the organism from distinguishing different amino acids (quality coding) from differences in concentration of a single amino acid (quantity coding). On the other hand, the necessity to build multiples of 20 or more different types of amino acid–binding sites per receptor cell may not be economical in terms of either the cellular energy or genetic requirements involved for the benefit gained (i.e., high specificity and little problem with adaptation).

Most of the electrophysiological information concerning amino acid receptor site types in fish is derived primarily from only a few published cross-adaptation experiments in olfaction (Caprio and Byrd 1984; Ohno, Yoshii, and Kurihara 1984) and gustation (Yoshii et al. 1979; Caprio 1982; Davenport and Caprio 1982; Kanwal and Caprio 1983). The experimental protocol of these studies entails the continuous presentation of an "adapting" stimulus to the receptor surface, followed by the individual application of "test" stimuli, once the phasic response to the adapting stimulus declines to baseline or to an ongoing tonic level of activity. The depression in the phasic response of the system, compared with its unadapted value, is used as a measure of the cross-reactivity of the compounds. A test stimulus that is normally stimulatory in the unadapted state, but fails to produce a phasic response or whose response is greatly attenuated in the presence of the adapting stimulus, is considered to be cross-adapted and to interact with the same "transduction process" (and possibly the same binding site) as the adapting stimulus. Since electrophysiological experiments cannot distinguish between two compounds interacting at the same site or at independent sites that are intricately linked to membrane events leading to the production of generator (olfactory) and receptor (gustatory) potentials, the term *transduction process* is used to include either case. For a test stimulus that shows relatively little reduction in response magnitude (and possibly even an enhancement of the response) (see Yoshii et al. 1979) in the presence of a background stimulus, these compounds are assumed to interact with different transduction processes and thus with different binding sites.

Results of cross-adaptation studies in the channel catfish *Ictalurus punctatus* indicate that the olfactory receptors possess different transduction processes for the acidic (A), basic (B), and neutral amino acids (Fig. 12.5) (Caprio and Byrd 1984). In addition, at least two partially interacting neutral transduction processes exist, one for the hydrophilic, neutral amino acids containing short side chains (SCN) and the second for the more hydrophobic, neutral amino acids containing long side chains (LCN) (Fig. 12.5[B]). Thus, at least in this species, as few as

A

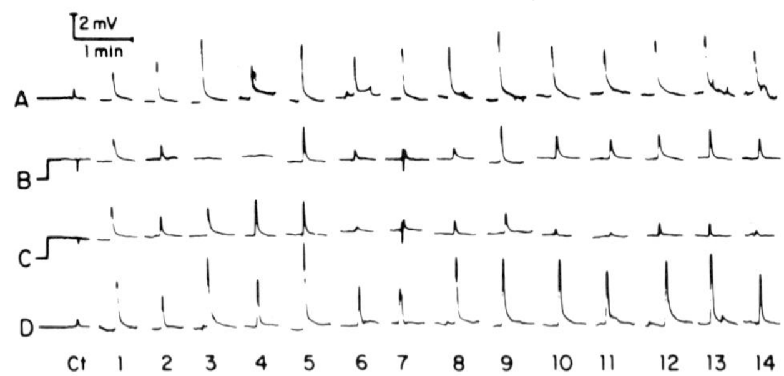


B

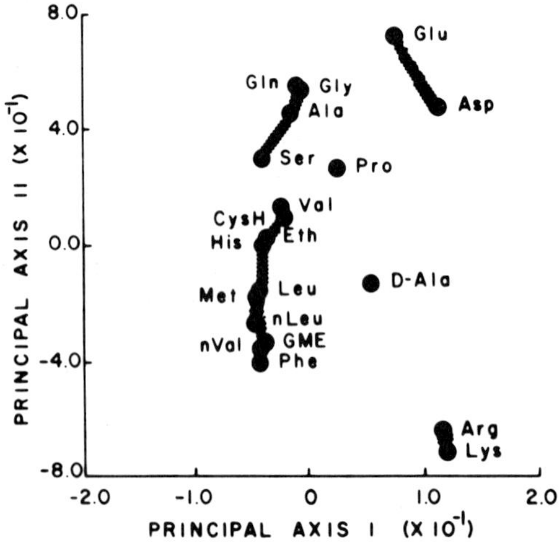

4 different amino acid receptor site types may be all that is necessary for the efficient detection of the common L-amino acids in the water column. In the carp, *Cyprinus carpio,* the results of amino acid cross-adaptation studies (Ohno, Yoshii, and Kurihara 1984) are somewhat similar to that indicated for the catfish. In the carp, as in the catfish, acidic, basic, and neutral amino acids interact with different transduction processes; in the carp, however, the evidence further suggests that the basic amino acids, arginine and lysine, interact with their own respective binding sites. Although binding sites for the neutral amino acids are different from those for the acidic and basic amino acids, at present there are insufficient data to determine, in the carp, the probable number of different transduction processes that exist for this class of amino acids.

Although Ohno, Yoshii, and Kurihara (1984) conclude that there are many different olfactory receptor site types for amino acids in the carp, this statement is based on the authors' interpretation of three types of results enumerated by Caprio and Byrd (1984) in the catfish and common to cross-adaptation experiments in both species: (1) all the adapting compounds affect, to varying degrees, the olfactory responses to the test stimuli; (2) full reciprocal cross-adaptation does not always occur for particular pairs of amino acids thought to share the same transduction process; and (3) specific adapting regimens affect differently the responses to amino acids assumed to interact with the same transduction process. Interpretation of these results may be based on a simple model of olfactory stimulus-receptor interaction, as being consistent with the existence of numerous and

---

Figure 12.5. (A) Representative EOG records obtained from single channel catfish to 14 of the 20 "test" amino acids presented in a background of well water (*A* and *D*), $10^{-4}$M L-arginine (*B*), or $10^{-5}$M L methionine (*C*). Phasic EOG responses in *A* and *D* rise from baseline activity, whereas in *B* and *C* the EOG records rise from ongoing tonic level of activity to respective adapting stimulus. Amino acids are arranged in the order of acidic (*1,* L-aspartic acid; *2,* L-glutamic acid); basic (*3,* L-arginine; *4,* L-lysine); and neutral compounds with short side chains (SCN) (*5,* glycine; *6,* L-alanine; *7,* D-alanine; *8,* L-serine; *9,* L-cysteine) and neutral compounds with long side chains (LCN) (*10,* L-methionine; *11,* L methionine; *12,* L-leucine; *13,* L-norleucine; *14,* L-glutamic acid -γ- methyl ester). *Ct* indicates control response to well water injected into either continuous flow of well water (*A* and *D*) or respective adapting solutions (*B* and *C*) bathing olfactory mucosa. The depression of *Ct* seen in *B* and *C* indicates a decline in ongoing tonic response levels caused by water blanks. Responses *3B* and *10C* served as controls during respective adapting regimens. Note in *B* that arginine-adapting solution selectively eliminated phasic EOG response to itself (*3B*, adaptation) and L-lysine (*4A,* cross adaptation). Note in *C* that methionine-adapting solution reduced responses to LCN test amino acids (*10C–14C*) to greater degree than those to other amino acid groups, leaving responses to acidic (*1,2*) and basic (*3,4*) amino acids. Breaks in records indicate 4-min. interstimulus intervals. (B) Locations of each of 20 test amino acids in two-dimensional space obtained from principal components ordination. This multivariate statistical technique relates each test amino acid to 19 others based on similarities of EOG responses to test stimuli across 8 different adapting regimens (adapting simuli included amino acids 2, 3, 4, 6, 9, 10, 13, 14 listed in A). (Reproduced from the Journal of General Physiology, 1984, 84:403–422, by copyright permission of the Rockefeller University Press.)

somewhat overlapping types of amino acid–binding sites. The first observation, however, may only be evidence of the presence of multiple binding sites for different amino acids on the same olfactory receptor cell (i.e., different binding sites sharing the same "neural channel"). Olfactory receptor neurons would, therefore, be partially depolarized by an adapting stimulus and unable to respond fully to other stimuli. Further, the second and third observations are compatible with an allosteric model of odorant-receptor interaction, in that the binding of one amino acid to a site may allosterically affect the binding properties of other sites on the receptor molecule, even though the actual binding sites are relatively independent. This would result in significantly fewer different receptor site types required for interpretation of the data. Still unknown for any species of fish and an important topic for future research, is the distribution of these transduction processes per individual olfactory receptor cell, which defines the specificity of an individual neural channel.

For the gustatory system, current electrophysiological evidence indicates the presence of three different amino acid transduction processes (and thus at least three different binding sites) in the channel catfish, one each for L-alanine (including other neutral amino acids), for L-arginine, and for the imino acid, L-proline (Fig. 12.6) (Kanwal and Caprio 1983). Further, the alanine- and arginine-binding sites are somewhat segregated on different taste cells, as multi-unit and single unit recordings indicate the existence of "alanine-best" and "arginine-best" gustatory fiber types in the facial (Davenport and Caprio 1982), glossopharyngeal, and vagal nerves (Kanwal and Caprio 1983). In the Japanese eel, both SCN and B amino acids appear to interact with the same transduction process, but this is distinct from the relatively independent binding sites for L-proline, betaine, and L-histidine, respectively (Yoshii et al. 1979). The independence of the proline and histidine sites were further indicated by protease (papain and pronase E) treatment of the palatal taste buds. Thus the number of different gustatory transduction processes that exist for amino acids in fish may also be relatatively few.

Specific information concerning possible binding site types for amino acid stimuli in fish is derived from biochemical binding experiments with presumed olfactory (Cancalon 1978; Novoselov, Krapivinskaya, and Fesenko 1980; Brown and Hara 1981; Rhein and Cagan 1983) and gustatory (Cagan 1981; Cagan and Boyle 1984) receptor cell membranes. These studies have been subject to some criticism because of the observed slow kinetics of stimulus-receptor association, which are not in accord with physiological response times, and because of possible involvement with tissue amino acid transport processes independent of olfactory and gustatory mechanisms (Brown and Hara 1981). Although some discrepancies exist between the biochemical and electrophysiological data, the similarities are striking. For example, the SCN, LCN, B, and A olfactory transduction processes identified electrophysiologically in the channel catfish have support from biochemical studies performed on other species (see Caprio and Byrd 1984 for further discussion). Also, different gustatory binding sites for alanine (and other neutral amino acids) and arginine in the channel catfish have also been identified biochemically in the same species (see Cagan 1981).

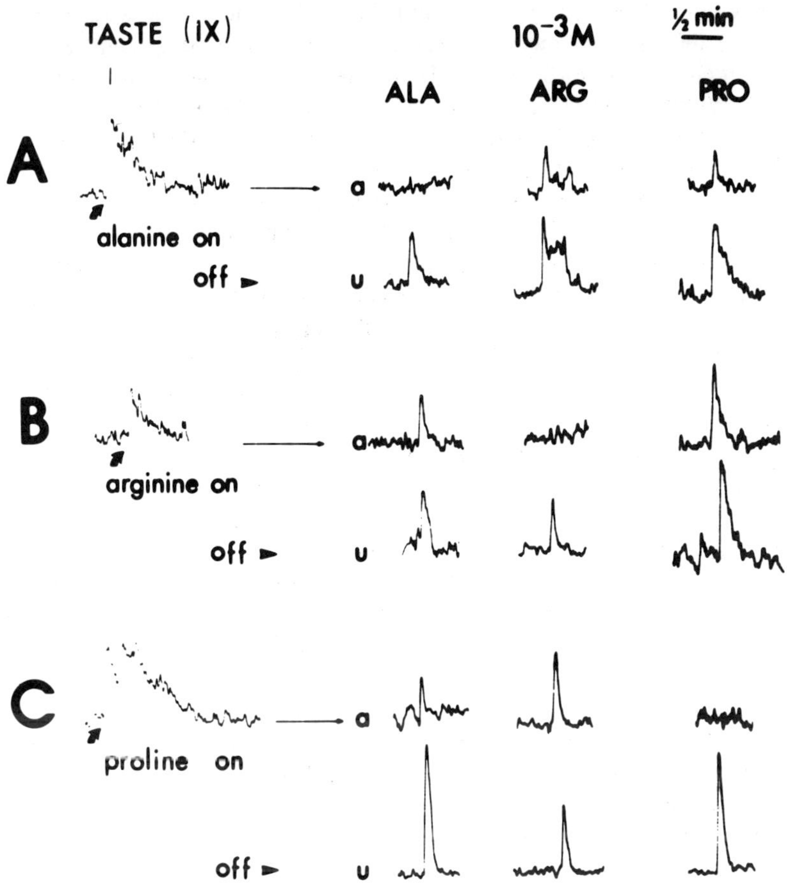


Figure 12.6. Reciprocal gustatory cross-adaptation from glossopharyngeal (IX) nerve of channel catfish, indicating relatively independent transduction processes for L-arginine, L-alanine, and L-proline. Similar results were obtained from vagus (X) nerve. (A) Continuous presentation of alanine to oropharyngeal taste buds eliminates subsequent responses to alanine, whereas responses to arginine and proline are maintained: *a*, adapted state; *u*, unadapted state following cessation of adapting stimulus. (B) Adaptation with arginine eliminates taste responses to itself, but not to alanine or proline. (C) Adaptation with proline eliminates taste response to itself, but not to alanine and arginine. *Off* refers to removal of adapting stimulus and resumption of well water flow. (Reprinted from Kanwal and Caprio 1983, with permission from Springer-Verlag.)

## Effects of Chemical Mixtures

In contrast to the conditions in most experimental studies, olfactory and gustatory receptor cells rarely encounter sequentially spaced single chemicals under natural circumstances. The normal chemical world is a mixture of substances varying in concentration and potency. Thus, it is critical to the basic understanding of the processes involved in olfaction and gustation to determine how these senses respond to the simultaneous presentation of compounds comprising, at least initially, simple mixtures. Studies indicate that the response to a mixture cannot generally be predicted from knowledge of the responses to individual components (Carr, Netherton, and Milstead 1984; Derby and Ache 1984; Zimmer-Faust et al. 1984). This is attributed, in many cases, to the occurrence of both mixture suppression and synergism among different components in the mixture. Recent evidence from olfactory (EOG) recordings in the channel catfish to binary (Dudek and Caprio 1985) and trinary (Dudek and Caprio unpublished) mixtures of amino acids, however, clearly shows that responses to these mixtures are predictable (at least in the catfish) with knowledge of the respective transduction processes involved (Caprio and Byrd 1984). In these studies, all stimuli were adjusted in concentration to provide for approximately equal response magnitude, which is an effective control for inherent differences in stimulus potency. The results show that the responses to amino acids that interact with different (i.e., ''across group'') transduction processes are significantly larger than those produced by a mixture of amino acids constituting a single transduction process (i.e., ''within group''). For the latter the magnitude of response to the mixture is similar to that produced by a higher concentration of any one of the components (i.e., ''stimulus substitution'' model based on the dose-response functions of the stimuli) (Hyman and Frank 1980; Rifkin and Bartoshuk 1980) (Fig. 12.7[A]), whereas the response to the former is significantly larger and may approach additivity (Fig. 12.7[B]).

Dose-response functions of chemoreceptive systems to amino acids in aquatic organisms are much less than unity on log–log coordinates (Caprio 1984; Derby and Ache 1984). Thus, because of (1) the ability of the population of receptor cells to respond to such a tremendously wide range (>100,000-fold change) of stimulus strengths and (2) the physiological limits on the magnitude of the maximum response attainable by a single receptor cell, logarithmic increments in stimulus concentration produce progressively smaller relative response increments (i.e., sensory compression) (Bartoshuk 1975). The previously described studies (Dudek and Caprio 1985) on the effects of amino acid mixtures on the olfactory receptor responses of the catfish indicate that sensory stimulation of chemoreceptor cells by a finite number of stimulus molecules is enhanced more by the activation of different types of receptor sites than by stimulus interaction at a single site. This finding is critically important in accurately predicting the response to mixtures and has not been taken into account in previous studies. Furthermore, responses to mixtures that are significantly greater than those predicted on a ''stimulus substitution'' model may not be examples of synergism, but only of the simultaneous activation of different receptor site types. For example, in the olfactory mixture experiments in the catfish (Dudek and Caprio 1985), the term *synergism* should be limited to responses to (1) within group mixtures that result in significantly greater responses than those predictions based on the stimulus substitution model and (2) across group mixtures that result in significantly greater responses than those predictions based on an additive model. Neither occurred nor was there evidence of mixture suppression. This suggests the possibility that some of the previous examples of this suppression may be explained by simple competitive

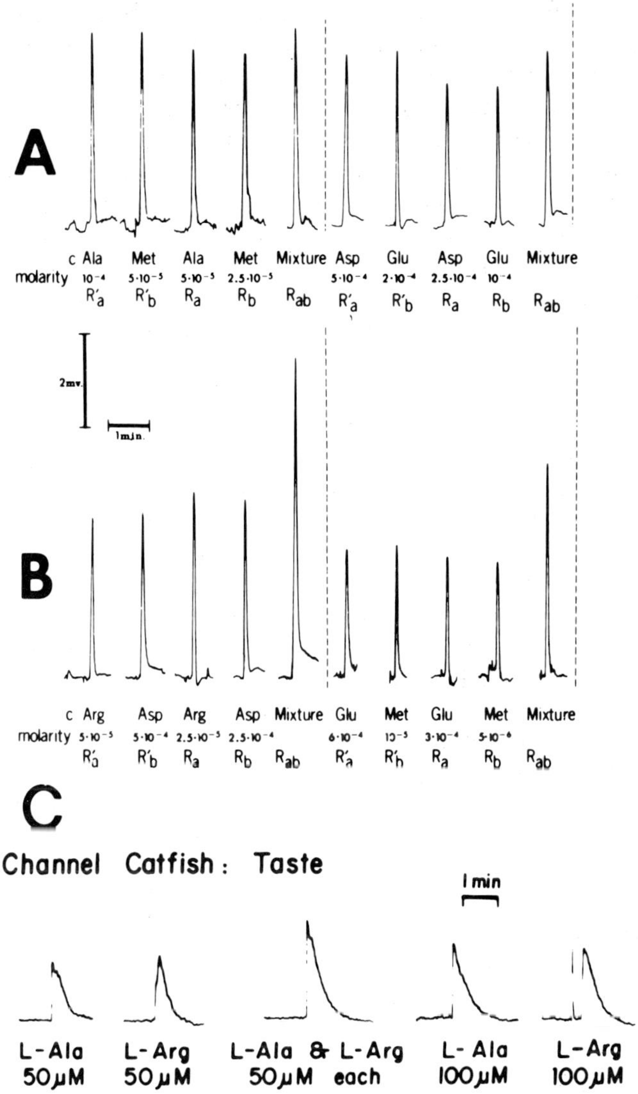


Figure 12.7. Olfactory (A and B) and gustatory (C) responses to binary mixtures of amino acids and to their individual components. Concentrations of component compounds are adjusted to provide approximately equal response magnitude. (A) Olfactory (EOG) responses to binary mixtures composed of amino acids that interact with same transduction process (i.e., neutral + neutral, acidic + acidic). (B) Olfactory (EOG) responses to binary mixtures composed of amino acids that interact with different transduction processes (i.e., acidic + basic, acidic + neutral): $R'_a$, $R'_b$, the response to each amino acid presented alone at the concentration used to create binary mixture (mixture formed from equal volumes of *a* and *b*); $R_a$, $R_b$, the response to each amino acid presented alone at its resulting concentration in the mixture; $R_{ab}$, the response to binary mixture. (C) Integrated (rise time 0.5 sec.) gustatory responses to binary mixture of alanine and arginine and to individual components of the mixture.

binding among stimuli with differing potencies that share the same transduction process.

Only a single quantitative report exists on the electrophysiological responses of the gustatory system of a fish to mixtures of amino acids (Hidaka, Nyu, and Kiyohara 1976). In that report, both integrated and single unit gustatory responses of the puffer *Fugu pardalis* were enhanced by particular binary mixtures. Betaine mixed with glycine, alanine, or serine resulted in enhanced responses, whereas binary mixtures of glycine with alanine or serine resulted in a reduction in response from that obtained to glycine alone at the same total amino acid concentration. Although cross-adaptation studies to indicate how these compounds are sorted into particular transduction processes were not performed, these results indicated that the increased responsiveness of the taste system was possibly due to the simultaneous activation of more than one receptor site type (i.e., a site for neutral amino acids with short side chains and one for betaine). From unpublished experiments (Caprio) on the channel catfish, amino acids that interact with different gustatory transduction processes (e.g., alanine and arginine) (Fig. 12.7[C]) result in greater neural activity than that produced by an equal number of amino acid molecules of either one of the two compounds (i.e., greater than that predicted by the stimulus substitution model). Future studies will attempt to determine whether chemoresponses of both the olfactory and gustatory systems to complex mixtures are similarly predictable with knowledge of both the relative potencies of the component stimuli and their respective organization into transduction processes.

## Summary

Olfactory receptor anatomy and taste bud structure in fish are generally similar to those found in other vertebrates. For aquatic organisms, however, both olfactory and gustatory receptors are stimulated by water-soluble compounds. Volatility, an essential characteristic for olfactory stimulants of air breathers, is unimportant to aquatic animals. Further, the sensitivity of olfactory and gustatory receptors of certain species of fish to stimuli such as amino acids is similarly high, thus resulting in both chemosensory systems' serving as "distant" sensors, a characteristic of olfaction alone in air breathers. L-α-Amino acids are potent olfactory and gustatory stimulus compounds to almost all fish tested, but the relative responses of the olfactory and gustatory systems to specific amino acids differ. Whereas the olfactory system across species of fish is most responsive (at equal stimulus strength) to the neutral amino acids, the gustatory system is more variable. Knowledge of the number and specificities of the different transduction processes (i.e., receptor site types) that exist and their distribution in receptor cells will provide better understanding of the responses of the two chemosensory systems to more "natural" stimuli, such as complex mixtures.

*Acknowledgments.* Supported by grants NIH NS14819 and NSF BNS8418299. I thank Mr. J. Erickson for Figures 12.1(A) and (C) and Mr. John Dudek for Figures 12.7(A) and (B). I also thank Ms. Lynnda Halbrook for secretarial and Mr. Ron Bouchard for photographic assistance.

## References

Atema, J. (1971) Structures and functions of the sense of taste in catfish (*Ictaluris natalis*), Brain Behav. Evol., 4:273–294.

Bannister, L.H. (1965) The fine structure of the olfactory surface of teleostean fishes, Q. J. Microsc. Sci., 106:333–342.

Bardach, J.E., Todd, J.H., and Crickmer, R. (1967) Orientation by taste in fish of the genus *Ictalurus*, Science, 155:1276–1278.

Bartoshuk, L.M. (1975) Taste mixtures: Is mixture suppression related to compression? Physiol. Behav., 14:634–649.

Beidler, L.M. and Smallman, R.L. (1965) Renewal of cells within taste buds, J. Cell Biol., 27:263–272.

Belousova, T.A., Devitsina, G.V., and Malyukina, G.A. (1983) Functional peculiarities of fish trigeminal system, Chem. Sens., 8:121–130.

Breucker, H., Zeiske, E., and Melinkat, R. (1979) Development of the olfactory organ in the rainbow fish *Nematocentris maccullochi* (Atheriniformes, Melanotaeniidae), Cell Tissue Res., 200:53–68.

Bronshtein, A.A. (1976) Some peculiarities of fine structure of the olfactory organ in elasmobranchs, Z. Evol. Biokhim. Fiziol., 12:63–67.

Brown, S.B. and Hara, T.J. (1981) Accumulation of chemostimulatory amino acids by a sedimentable fraction isolated from olfactory rosettes of rainbow trout (*Salmo gairdnerii*), Biochem. Biophys. Acta, 675:149–162.

Burne, R.H. (1909) The anatomy of the olfactory organ of teleostean fishes, Proc. Zool. Soc. London, 2:610–637.

Byrd, R.P., Jr. and Caprio, J. (1982) Comparison of olfactory receptor (EOG) and bulbar (EEG) responses to amino acids in the catfish, *Ictalurus punctatus*, Brain Res., 249:73–80.

Cagan, R.H. (1981) Recognition of taste stimuli at the initial binding interaction, in Biochemistry of Taste and Olfaction, Cagan, R.H. (ed.), Academic Press, New York, pp. 175–203.

Cagan, R.H. and Boyle, A.G. (1984) Biochemical studies of taste sensation. XI. Isolation, characterization and taste ligand binding activity of plasma membranes from catfish taste tissue, Biochem. Biophys. Acta 799:230–237.

Cancalon, P. (1978) Isolation and characterization of the olfactory epithelial cells of the catfish, Chem. Sens. Flavour, 3:381–396.

Cancalon, P. (1982) Degeneration and regeneration of olfactory cells induced by $ZnSO_4$ and other chemicals, Tissue Cell, 14:717–733.

Cancalon, P. (1983) Receptor cells of the catfish olfactory mucosa, Chem. Sens., 81:203–209.

Caprio, J. (1975) High sensitivity of catfish taste receptors to amino acids, Comp. Biochem. Physiol., 52A:247–251.

Caprio, J. (1978) Olfaction and taste in the channel catfish: an electrophysiological study of the responses to amino acids and derivatives, J. Comp. Physiol., 123:357–371.

Caprio, J. (1980) Similarity of olfactory receptor responses (EOG) of freshwater and marine catfish to amino acids, Can. J. Zool., 58:1778–1784.

Caprio, J. (1982) High sensitivity and specificity of olfactory and gustatory receptors of catfish to amino acids, in Chemoreception in Fishes, Hara, T.J. (ed.), Elsevier, New York, pp. 109–134.

Caprio, J. (1984) Olfaction and taste in fish, in The Comparative Physiology of Sensory Systems, Bolis, L., Keynes, R., and Maddrell, S.H.P. (eds.), Cambridge University Press, Cambridge, pp. 257–283.

Caprio, J. and Byrd, R.P., Jr. (1984) Electrophysiological evidence for acidic, basic, and neutral amino acid olfactory receptor sites in the catfish, J. Gen. Physiol., 84:403–422.

Caprio, J. and Raderman-Little, R. (1978) Scanning electron microscopy of the channel catfish olfactory lamellae, Tissue Cell, 10:1–9.

Carr, W.E.S., Netherton, J.C., III, and Milstead, M.L. (1984) "Chemoattractants of the shrimp, *Palaemonetes pugio:* variability in responsiveness and the stimulatory capacity of mixtures containing amino acids quaternary ammonium compounds, purines and other substances," Comp. Biochem. Physiol., 77A:469–474.

Crisp, M., Lowe, G.A., and Laverack, M.S. (1975) On the ultrastructure and permeability of taste buds of the marine teleost *Ciliata mustela,* Tissue Cell, 7:191–202.

Davenport, C.J. and Caprio, J. (1982) Taste and tactile recordings from the ramus recurrens facialis innervating flank taste buds in the catfish, J. Comp. Physiol., 147:217–229.

Delfino, G., Bianchi, S., and Ercolini, A. (1981) On the olfactory epithelium in cyprinids: a comparison between hypogean and epigean species, Monitore Zool. Ital. II (N.S.) Supp., 14:153–180.

Derby, C.D. and Ache, B.W. (1984) Electrophysiological identification of the stimulatory and interactive components of a complex odorant, Chem. Sens., 9:201–218.

Døving, K.B., Dubois-Dauphin, M., Holley, A., and Jourdan, F. (1977) Functional anatomy of the olfactory organ of fish and the ciliary mechanisms of water transport, Acta Zool. (Stockh.), 58:245–255.

Døving, K.B., Selset, R., and Thommesen, G. (1980) Olfactory sensitivity to bile acids in salmonid fishes, Acta Physiol. Scand., 108:123–131.

Dudek, J. and Caprio, J. (1985) "Additional" evidence for acidic, basic and neutral amino acid olfactory receptor sites in the catfish, Assoc. Chemorecept. Sci. Abst., 7:49.

Easton, D.M. (1971) Garfish olfactory nerve: easily accessible source of numerous long, homogeneous nonmelinated axons, Science, 72:952–955.

Eller, P.A., Rowley, J.C., III, and Moran, D.T. (1985) The "rod cell" in trout olfactory epithelium: fact or artifact, Assoc. Chemorecept. Sci. Abst., 7:451.

Erickson, J.R. and Caprio, J. (1984) The spatial distribution of ciliated and microvillous olfactory receptor neurons in the channel catfish is not matched by a differential specificity to amino acid and bile salt stimuli, Chem. Sens., 9:127–141.

Farbman, A.I. (1980) Renewal of taste bud cells in rat circumvallate papillae, Cell Tissue Res., 13:349–357.

Finger, T.E. (1982) Somatotopy in the representation of the pectoral fin and free fin rays in the spinal cord of the sea robin, *Prionotus carolinsus,* Biol. Bull., 163:154–161.

Gardner, W.S. and Lee, G.F. (1975) The role of amino acids in the nitrogen cycle of Lake Mendota, Limnol. Oceanogr., 20:379–388.

Gemne, G. and Døving, K.B. (1969) Ultrastructural properties of primary olfactory neurons in fish (*Lota lota* L), Am. J. Anat., 126:452–476.

Goh, Y. and Tamura, T. (1978) The electrical responses of the olfactory tract to amino acids in carp, Bull. Jpn. Soc. Sci. Fish, 44:341–344.

Goh, Y. and Tamura, T. (1980) Olfactory and gustatory responses to amino acids in two marine teleosts—red sea bream and mullet, Comp. Biochem. Physiol., 66C:217–224.

Graziadei, P.P.C. and Metcalf, J.F. (1971) Autoradiographic and ultrastructural observations in the frog's olfactory mucosa, Z. Zellforsch. Mikrosk. Anat., 116:305–318.

Grover-Johnson, N. and Farbman, A.I. (1976) Fine structure of taste buds in the barbel of the catfish, *Ictalurus punctatus,* Cell Tissue Res., 169:395–403.

Hara, T.J. (1975) Olfaction in fish, in Progress in Neurobiology, vol. 5, Kerkut, G.A. and Phillips, J.W. (eds.), Pergamon Press, Oxford, pp. 271–335.

Hara, T.J. (1982) Chemoreception in Fishes, Elsevier Press, Amsterdam.

Hara, T.J., MacDonald, S., Evans, R.E., Marui, T., and Arai, S. (1984) Morpholine bile acids and skin mucus as possible chemical cues in salmonid homing: electrophysiological re-evaluation, in Mechanisms of Migration in Fishes, McCleave, J.D., Arnold, G.P., Dodson, J.D., and Neill, W.H. (eds.), Plenum, New York, pp. 363–378.

Hasler, A.D. and Scholz, A.T. (1983) Olfactory imprinting and homing in salmon, Springer-Verlag, New York.

Herrick, C.J. (1901) The cranial nerves and cutaneous sense organs of the North American siluroid fishes, J. Comp. Neurol., 11:177–249.

Hidaka, I. (1982) Taste receptor stimulation and feeding behavior in the puffer, In Chemoreception in Fishes, Hara, T.J. (ed.), Elsevier, New York, pp. 243–257.

Hidaka, I., Nyu, N., and Kiyohara, S. (1976) Gustatory response in the puffer. IV. Effects of mixtures of amino acids and betaine, Bull. Fac. Fish Mie Univ., 3:17–28.

Hidaka, I., Kiyohara, S., and Oda, S. (1977) Gustatory response in the puffer. III. Stimulatory effectiveness of nucleotides and their derivatives, Bull. Jpn. Soc. Sci. Fish., 43:423–428.

Hirata, Y. (1966) Fine structure of the terminal buds on the barbels of some fishes, Arch. Histol. Jpn., 26:507–523.

Hyman, A.M. and Frank, M.E. (1980) Effects of binary taste stimuli on the neural activity of the hamster chorda tympani, J. Gen. Physiol., 76:125–142.

Ichikawa, M. and Ueda, K. (1977) Fine structure of the olfactory epithelium in the goldfish, *Carassius auratus.* A study of retrograde degeneration, Cell Tissue Res., 183:445–455.

Johannes, R.E. and Webb, K.L. (1970) Release of dissolved organic compounds by marine and fresh water invertebrates, in Symposium on Organic Matter in Natural Waters, Hood, D.W. (ed.), Alaska Institute of Marine Science, University of Alaska, pp. 257–274.

Johnsen, P.B. and Teeter, J. (1980) Spatial gradient detection of chemical cues by catfish, J. Comp. Physiol., 140:95–99.

Kanwal, J.S. and Caprio, J. (1983) An electrophysiological investigation of the oropharyngeal (IX–X) taste system in the channel catfish, *Ictalurus punctatus,* J. Comp. Physiol., 150:345–357.

Kashiwayanagi, M., Miyake, M., and Kurihara, K. (1983) Voltage-dependent $Ca^{+2}$ channel and $Na^{+}$ channel in frog taste cells, Am. J. Physiol., 244:C82–C88.

Kiyohara, S., Yamashita, S., and Kitoh, J. (1980) Distribution of taste buds on the lips and inside the mouth in the minnow, *Pseudorasbora parva,* Physiol. Behav., 24:1143–1147.

Kiyohara, S., Yamashita, S., and Harada, S. (1981) High sensitivity of minnow gustatory receptors to amino acids, Physiol. Behav., 26:1103–1108.

Kleerekoper, H. (1969) Olfaction in Fishes, Indiana University Press, Bloomington.

Kreutzberg, G.W. and Gross, G.W. (1977) General morphology and axonal ultrastructure of the olfactory nerve of the pike, *Exos lucius*. Cell Tissue Res., 181:443–457.

Lane, E.B. and Whitear, M. (1982) Sensory structures at the surface of fish skin. I. Putative chemoreceptors, Zool. J. Linn. Soc., 75:141–151.

Mackie, A.M. and Grant, P.T. (1974) Chemoreception in Marine Organisms, Academic Press, New York.

Marui, T., Evans, R.E., and Hara, T.J. (1983) Gustatory responses of the rainbow trout (*Salmo gairdnerii*) palate to amino acids and derivatives, J. Comp. Physiol., 153:423–433.

Marui, T., Harada, S., and Kasahara, Y. (1983) Gustatory specificity for amino acids in the facial taste system of the carp, *Cyprinus carpio* L., J. Comp. Physiol., 153:299–308.

Morita, Y. and Finger, T. (1985a) Reflex connections of the facial and vagal gustatory systems in the brainstem of the bullhead catfish, *Ictalurus nebulosus*, J. Comp. Neurol., 231:547–558.

Morita, Y. and Finger, T. (1985b) Topographic and laminar organization of the vagal gustatory system in the goldfish, *Carassius auratus*, J. Comp. Neurol., 238:187–201.

Moulton, D. (1975) Cell renewal in the olfactory epithelium of the mouse, in Olfaction and Taste V, Denton, D.A. and Coghlan, J.P. (eds.), Academic Press, New York, pp. 111–114.

Muller, J.F. and Marc, R.E. (1984) Three distinct morphological classes of receptors in fish olfactory organs, J. Comp. Neurol., 222:482–495.

Murray, R.G. (1973) The ultrastructure of taste buds, in The Ultrastructure of Sensory Organs I, Friedman, I. (ed.), North Holland Publishing Co., Amsterdam, pp. 1–81.

North, B.B. (1975) Primary amines in California coastal waters: utilization by phytoplankton, Limnol. Oceanogr., 20:20–27.

Novoselov, V.I., Krapivinskaya, L.D., and Fesenko, E.E. (1980) Molecular mechanisms of odor sensing. V. Some biochemical characteristics of the alanineous receptor from the olfactory epithelium of the skate *Dasyatis pastinaca*, Chem. Sens., 5:195–203.

Ohno, T., Yoshii, K., and Kurihara, K. (1984) Multiple receptor types for amino acids in the carp olfactory cells revealed by quantitative cross-adaptation model, Brain Res., 310:13–21.

Pocklington, R. (1972) Determination of nanomolar quantities of free amino acids dissolved in North Atlantic ocean waters, Analyt. Biochem., 45:403–421.

Pyatkina, G.A. (1976) Receptor cells of various types and their proportional interrelation in the olfactory organ of larvae and adults of acipenserid fishes (in Russian with English summary), Tsitol., 18:1444–1449.

Raderman-Little, R (1979) The effect of temperature on the turnover of taste bud cells in catfish, Cell Tissue Kinet., 12:269–280.

Reese, T.S. and Brightman, M.W. (1970) Olfactory surface and central olfactory connections in some vertebrates, in Taste and Smell in Vertebrates, Ciba Foundation Symposium, Wolstenholme, G.E.W. and Knight, J. (eds.), Churchill, London, pp. 115–143.

Reutter, K. (1971) Die Geschmacksknospen des Zwergwelses *Amiurus nebulosus* (Lesueur). Morphologische und histochemische Untersuchungen, Z. Zellforsch, Mikrosk. Anat., 120:280–308.

Reutter, K. (1974) Cholinergic innervation of scattered sensory cells in fish epidermis, Cell Tissue Res., 149:143–146.

Reutter, K. (1978) Taste organ in the bullhead (Teleostei), Adv. Anat. Embryol. Cell Biol., 55:1–98.

Reutter, K. (1980) SEM-study of the mucus layer on the receptor field of fish taste buds, in Olfaction and Taste VII, van der Starre, H. (ed.), IRL Press, Washington, D.C., p. 107.

Reutter, K. (1982) Taste organ in the barbel of the bullhead, in Chemoreception in Fishes, Hara, T.J. (ed.), Elsevier, New York, pp. 77–91.

Rhein, L.D. and Cagan, R.H. (1983) Biochemical studies of olfaction: binding specificity of odorants to a cilia preparation from rainbow trout olfactory rosettes, J. Neurochem., 41:569–577.

Rifkin, B. and Bartoshuk, L.M. (1980) Taste synergism between monosodium glutamate and disodium 5′-guanylate, Physiol. Behav., 24:1169–1172.

Roper, S. (1983) Regenerative impulses in taste cells, Science, 220:1311–1312.

Roper, S., and McPheeters, S.M. (1984) Chemosensory stimulation in mudpuppy: KCl evokes impulses in taste cells, Assoc. Chemorecept. Sci. Abst., 6:10.

Rowley, J.C. and Moran, D.T. (1985) HRP applied to transected trout olfactory nerve fills ciliated receptors, microvillar receptors and some basal cells, Assoc. Chemorecept. Sci. Abst., 7:157.

Schulte, E. (1972) Studies of the regio olfactoria in the eel, *Anguilla anguilla*, Z. Zellforsch, Mikrosk. Anat., 125:210–228.

Selset, R. and Døving, K.B. (1980) Behavior of mature anadromous char (*Salmo alpinus* L) towards odorants produced by smolts of their own population, Acta Physiol. Scand., 108:113–122.

Sheldon, R.E. (1909) The reactions of the dogfish to chemical stimuli, J. Comp. Neurol. Psychol., 19:273–311.

Silver, W.L. (1982) Electrophysiological responses from peripheral olfactory system of the American eel, *Anguilla rostrata*, J. Comp. Physiol., 148:379–388.

Silver, W.L. and Finger, T.E. (1984) Electrophysiological examination of a non-olfactory, non-gustatory chemosense in the sea robin, *Prionotus carolinsus*, J. Comp. Physiol., 154:167–174.

Silver, W.L. and Maruniak, J.A. (1981) Trigeminal chemoreception in the nasal and oral cavities, Chem. Sens., 6:295–305.

Stabell, O.B. and Selset, R. (1980) Comparison of mucus collecting methods in fish olfaction, Acta Physiol. Scand., 108:91–96.

Sutterlin, A.M. and Sutterlin, N. (1970) Taste responses in Atlantic salmon (*Salmo salar*) parr, J. Fish Res. Board Can., 27:1927–1942.

Suzuki, N. (1977) Intracellular responses of lamprey olfactory receptors to current and chemical stimulation, in Food Intake and Chemical Senses, Katjuki, Y., Sato, M., Takagi, S.F., and Oomura, Y. (eds.), University of Tokyo Press, Tokyo, pp. 13–22.

Suzuki, N. (1978) Effects of different ionic environments on the responses of single olfactory receptors in the lamprey, Comp. Biochem. Physiol., 61A:461–467.

Suzuki, N. (1980) Binding activities of radioactively labeled amino acids to the lamprey olfactory tissue and its fractions, in Proceedings of the 14th Japanese Symposium on Taste and Smell, Tomota, H. (ed.), pp. 33–36.

Suzuki, N. (1982) Responses of olfactory receptor cells to electrical and chemical stimulation, in Chemoreception in Fishes, Hara, T.J. (ed.), Elsevier, New York, pp. 93–100.

Teeter, J. (1974) Electrical properties of taste bud cells and surrounding epithelial cells in catfish and mudpuppies, Ph.D. Thesis, University of Pennsylvania, Philadelphia.

Teichman, H. (1954) Vergleichende Untersuchungen an der Nase der Fische 2, Morphol. Okol. Tiere., 43:171–212.

Tester, A.L. (1963) Olfaction, gustation and the common chemical sense in sharks, in Sharks and Survival, Gilbert, P.W. (ed.), Heath, Boston, pp. 255–282.

Theisen, B., Breuker, H., Zeiske, E., and Melinkat, R. (1980) Structure and development of the olfactory organ in the garfish *Belone belone* (L.) (Teleostei, Atheriniformes), Acta Zool., 61:161–170.

Thommesen, G. (1982) Specificity and distribution of receptor cells in the olfactory mucosa of the char (*Salmo alpinus*), Acta Physiol. Scand., 115:47–56.

Thommesen, G. (1983) Morphology, distribution, and specificity of olfactory receptor cells in salmonid fishes, Acta Physiol. Scand., 117:241–250.

Toyoshima, K., Nada, O., and Shimamura, A. (1984) Fine structure of monoamine containing basal cells in the taste buds on the barbels of three species of teleosts, Cell Tissue Res., 235:479–484.

Tucker, D. (1983) Fish chemoreception: peripheral anatomy and physiology, in Fish Neurobiology: Brainstem and Sense Organs, Vol. I, Northcutt, R.G., and Davis, R.E. (eds.), University of Michigan Press, Ann Arbor, pp. 311–349.

Uskova, Y.T. and Chaykovskaya, A.V. (1971) The amino acid content of mucus from the skin of various marine fishes, Hydrobiologia, J7:79–80.

Yamamoto, M. (1982) Comparative morphology of the peripheral olfactory organ in teleosts, in Chemoreception in Fishes, Hara, T.J. (ed.), Elsevier, New York, pp. 39–59.

Yoshii, K. and Kurihara, K. (1983a) Role of cations in olfactory reception, Brain Res., 274:239–248.

Yoshii, K. and Kurihara, K. (1983b) Ion dependence of the eel taste response to amino acids, Brain Res., 280:63–67.

Yoshii, K., Kamo, N., Kurihara, K., and Kobatake, Y. (1979) Gustatory responses of eel palatine receptors to amino acids and carboxylic acids, J. Gen. Physiol., 74:301–317.

Zeiske, E., Caprio, J., and Gruber, S.H. (1986) Morphological and electrophysiological studies on the olfactory organ of the lemon shark, *Negaprion brevirostris*. (Poey) (Elasmobranchii: Carcharhinidae), in Second International Conference on Indo-Pacific Fishes, Uyeno, T., Arai, R., Taniuchi, T., and Matsuura, K. (eds.), The Ichthyological Society of Japan, Tokyo, pp. 381–391.

Zeiske, E., Melinkat, R., Breucker, H., and Kux, J. (1976) Ultrastructural studies on the epithelia of the olfactory organ of cyprinodonts (Teleostei, Cyprinodontoidea), Cell Tissue Res., 172:245–267.

Zimmer-Faust, R.K., Tyre, J.E., Michel, W.C., and Case, J.F. (1984) Chemical mediation of appetitive feeding in a marine decapod crustacean: the importance of suppression and synergism, Biol. Bull., 167:339–353.

## Chapter 13

# Organization of Chemosensory Systems Within the Brains of Bony Fishes

Thomas E. Finger *

The chemical senses in all vertebrates can be separated according to end organ, mode of innervation, and primary terminus within the CNS. According to these criteria, three exteroceptive chemical senses can be distinguished: olfaction, gustation, and the common chemical sense (Parker 1922; Atema 1980; Finger 1983b). The olfactory sense is mediated via olfactory receptors, themselves neurons with axons extending into the CNS and terminating in the olfactory bulb, the rostralmost extension of the forebrain (Fig. 13.1). In contrast, gustation is mediated through specialized axonless neuroepithelial receptors called taste buds. The chemosensory elements of the taste buds synapse on the distal processes of ganglionic cells of either the facial, glossopharyngeal, or vagal nerves, which terminate centrally in the special visceral sensory column of the medulla (Fig. 13.1). The common chemical sense is the least characterized, but appears to be mediated by either isolated chemosensory cells (Lane and Whitear 1982) innervated by the somatic sensory (trigeminal and spinal) nerves or by free nerve endings of these same nerves. In most, and especially aquatic, vertebrates, these chemical senses mediate different sorts of behavior (Atema 1980). The olfactory sense is associated with functions such as intraspecific social interactions or being alerted to the presence of food, whereas gustation is used in food localization and determination of palatability. In general, stimulation of the common chemical sense results in avoidance, but in at least some specialized species, such as sea robins, this sense is used for food localization (Parker 1922; Silver and Finger 1984).

Despite the clear-cut separability according to end organ and pattern of innervation, smell and taste are intermingled sensations for air-breathing vertebrates in which the nasal cavity communicates with the oral cavity. Being air-breathing vertebrates, we often confuse the separability of the various chemical senses. When we place food in our mouths, we sense the chemical composition not only through our taste buds, but also through our noses. In fact, most of the sensation we term flavor is mediated through olfactory receptors. Thus, linguisti-

* Department of Cellular and Structural Biology, B-111, Denver, CO 80262, USA

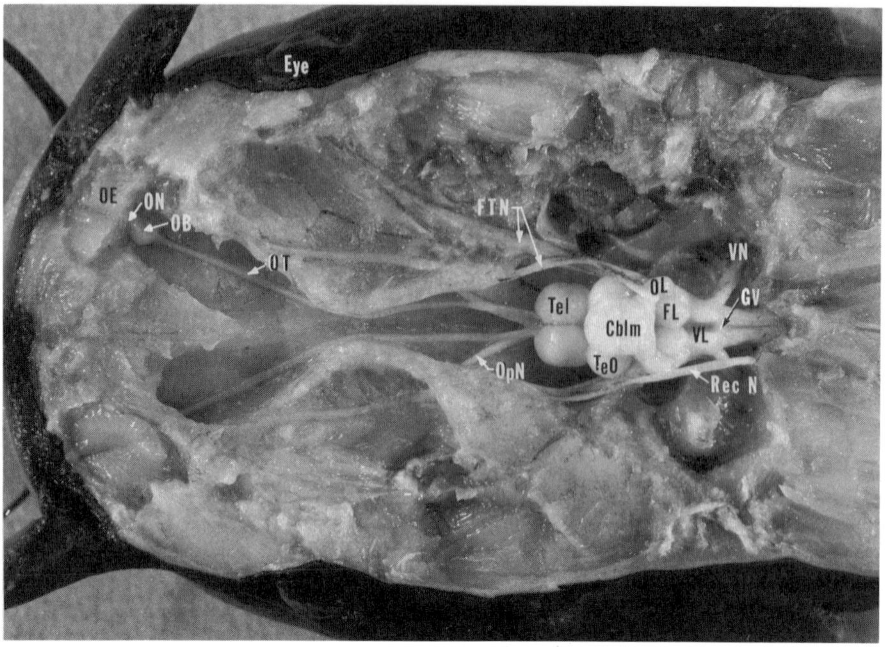


Figure 13.1. Dorsal view of dissected head of *Ictalurus,* showing the brain and many of the cranial nerves involved in chemoreception. Olfactory epithelium and surrounding bone on right side has been cut away to show olfactory nerve and bulb. Anterior to the left: *Cblm,* cerebellum; *FL,* facial lobe; *FTN,* facial-trigeminal nerve (mixed nerve to barbels); *GV,* general visceral nucleus; *OB,* olfactory bulb; *OE,* olfactory epithelium; *OL,* octavolateral area; *ON,* olfactory nerve; *OpN,* optic nerve; *OT,* olfactory tract; *Rec N,* recurrent branch of facial nerve (which innervates taste buds on fins, flank, and tail); *Tel,* telencephalon; *TeO,* optic tectum; *VL,* vagal lobe; *VN,* vagus nerve. (Dissection courtesy of Y. Morita.)

cally, there is an intermingling of meaning for words associated with cranial chemical senses. In this chapter, the term *gustation* denotes that chemosensory system mediated by taste buds, while *olfaction* denotes the system mediated by the bipolar receptors of the olfactory epithelium (see Chap. 12).

## Olfactory System

As in other vertebrates, the olfactory receptors in teleosts are bipolar neurons. At least two types of receptors, ciliated and microvillous, both with axons reaching into the brain, are evident in most teleosts (Muller and Marc 1984; Eller, Rowley, and Moran 1985). Whether these two receptor types are responsive to different kinds of chemical stimuli is still a matter of debate (see Thommesen 1983; Erickson and Caprio 1983; see also Chap. 12). In many mammals, microvillous receptors are segregated into a separate receptor organ called the *vomeronasal organ,* while the olfactory organ comprises mostly ciliated receptors. Bertmar (1981) suggested that the vomeronasal organ developed from the microvillous receptors of fish, whereas the main olfactory organ evolved from the ciliated receptors. This neat hypothesis is somewhat less attractive now that microvillous receptors have been found in the main olfactory organ of mammals with vomeronasal organs. To date, no clear cases have been presented of fish with vomeronasal organs.

The olfactory organs of all vertebrates, including fish, are associated with the terminal nerve—a nonolfactory nerve that directly connects the periphery with the brain. This nerve is not connected to olfactory receptors and may, in fact, not even be a sensory nerve at all (see below).

## Olfactory Nerve

The axons of the olfactory nerve emerge from the base of the receptor cells to gather into fascicles that penetrate the bone underlying the olfactory mucosa. The nerve runs caudally to terminate in the superficial layers of the olfactory bulb (Fig. 13.2). In some fish, such as *Carassius* and *Ictalurus,* the olfactory nerve is quite short, the olfactory bulb being close to the epithelium, but relatively remote from the brain. In other fish, such as *Salmo,* the nerve is comparatively long with the olfactory bulb directly opposed to the rostral end of the telencephalon. In other species, the bulb lies in an intermediate position.

## Olfactory Bulb

The olfactory nerve fibers contact the olfactory bulb at its rostral end and spread across the surface of the bulb to terminate in the superficial glomerular layer. Various types of small juxtaglomerular cells are located in the vicinity of the glomeruli, while most of the larger efferent bulbar cells lie deep to the glomerular layer. Deeper still is the granule cell core of the olfactory bulb.

The olfactory bulb in teleosts appears to have essentially the same organization as that in other vertebrates (Oka, Ichikawa, and Ueda 1982). This basic organization is illustrated semischematically in Figure 13.3[A]. The olfactory nerve fibers terminate on the dendrites of the juxtaglomerular cells and efferent bulbar cell types. The deep granule cells are predominantly axonless, with apical dendrites extending outwards to form dendrodendritic synapses with some of the efferent cell populations. In *Carassius auratus* (goldfish), several types of efferent cells are recognizable. One class of such cells includes typical mitral/tufted types (Fig. 13.3[B]); another class is the ruffed cell (Fig. 13.3[C]) (Kosaka and Hama 1979), which may be a specialized form of the mitral/tufted cell population.

In experimental material prepared by labelling the olfactory mucosa of goldfish with horseradish peroxidase, two sorts of olfactory nerve terminals have been noted (Finger and Fernald unpublished observation). One type of terminal comprises fine processes with small varicosities (Fig. 13.4[B]); the second type of terminal arborization comprises thicker nerve processes with coarser varicosities

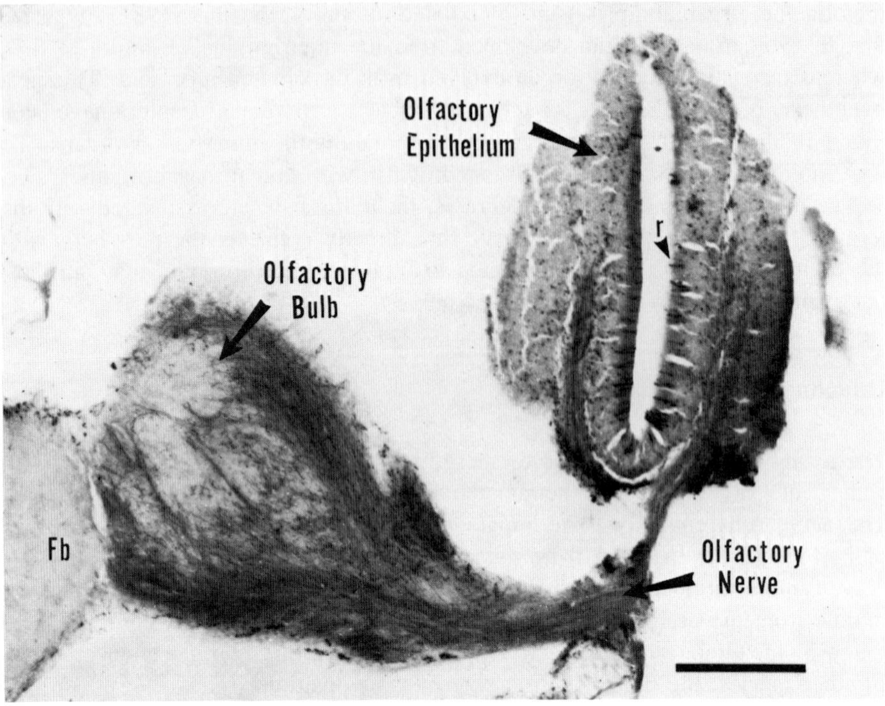


Figure 13.2. Low-power photomicrograph of longitudinal section through olfactory epithelium, nerve, and bulb of small specimen of *Carassius auratus*. Because of small size of the animal, olfactory bulb lies close to the telencephalon. In larger specimens, the olfactory bulb lies several millimeters from the telencephalon, as shown in Figure 13.1 for *Ictalurus*. In this specimen, olfactory receptor cells, and hence their axons in the nerve and bulb, have been labelled with marker horseradish peroxidase (HRP), which shows up darkly in this photomicrograph. Olfactory nerve can be seen to spread out across the surface of the bulb to terminate in glomerular layer: *Fb*, Forebrain; *r*, labelled receptor cell in the epithelium. Scale bar = 250 μm.

Figure 13.3. (A) Semischematic diagram illustrating the organization of olfactory bulb. Primary olfactory input enters via olfactory nerve fibers situated in olfactory nerve layer (*Olf. Nerve L.*) at the surface of the bulb. Olfactory axons turn inwards to synapse in glomeruli, which are complexes of neuropil comprising dendritic processes of two output cell classes, mitral cells and ruffed cells. Not shown are interneurons that probably participate in glomeruli. Dendrites of mitral cells make reciprocal dendrodendritic synapses (*dd*) with apical dendrites of granule cells. Apical dendrites of granule cells also form reciprocal and unidirectional synapses with initial segment of ruffed cell axon. Granule cells also have short, deep dendrites that receive synaptic input from mitral cell collaterals and centrifugal fibers. *Small arrows* indicate the direction of transmission at different synapses within olfactory bulb: *Ext. Plex. L*, external plexiform layer; *L*, layer; *Olf. Nerve L*, olfactory nerve layer. (B) Photomicrograph of retrogradely labelled mitral cell following an injection of HRP into olfactory terminal areas of the telencephalon of a goldfish (*Carassius auratus*). Scale bar = 50 μm. (C) Photomicrograph of retrogradely labelled ruffed cell from same preparation and magnification as in (B).

(Fig. 13.4[A]). These two types of terminal arborization occur in different glomerulii in different parts of the bulb. Whether these two terminal types correspond to bulbar functional domains (Satou et al. 1984) remains to be determined.

## Central Connections of Olfactory Bulb

In all species of fish examined to date, the olfactory bulb projects to a limited portion of the telencephalon (Scalia and Ebbesson 1971; Finger 1975; Murakami, Morita, and Ito 1983; Northcutt and Davis 1983; von Bartheld et al. 1984; Prasada Rao and Finger 1984). Even higher order olfactory connections appear to involve a limited area of the telencephalon (Murakami, Morita, and Ito 1983). Thus the

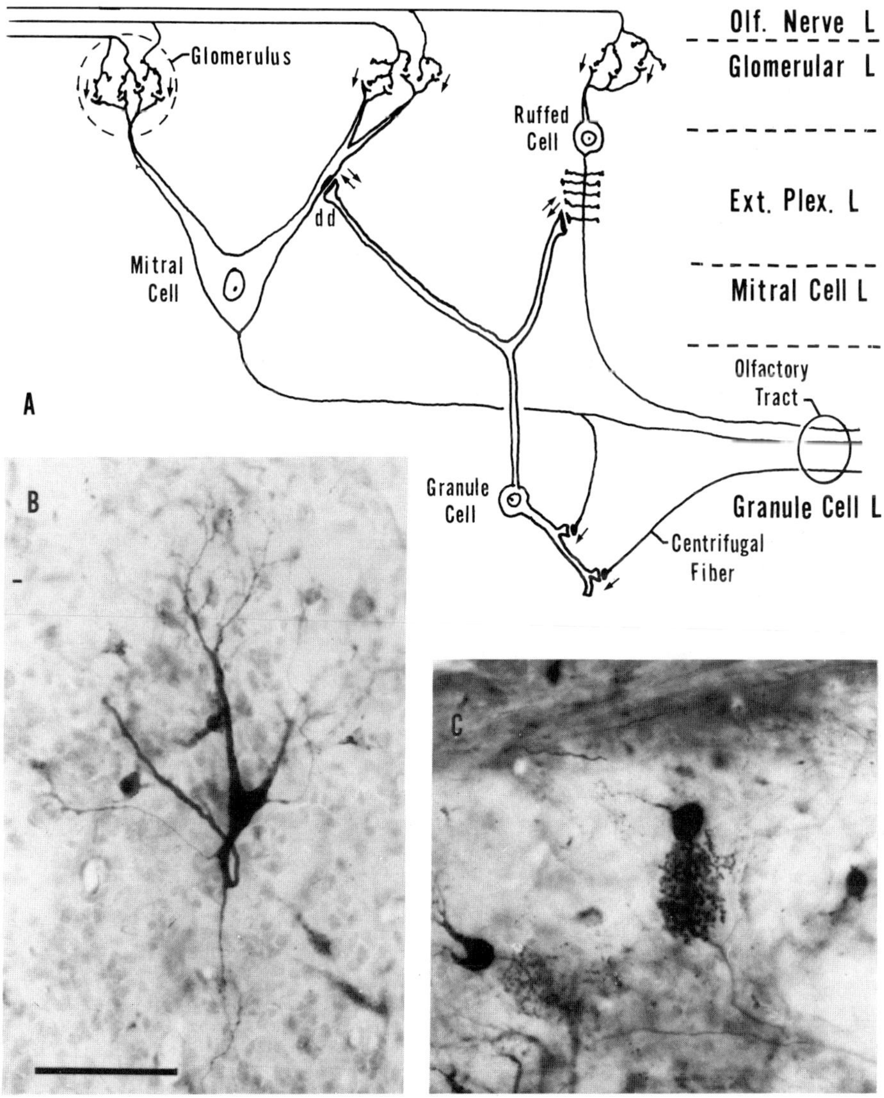

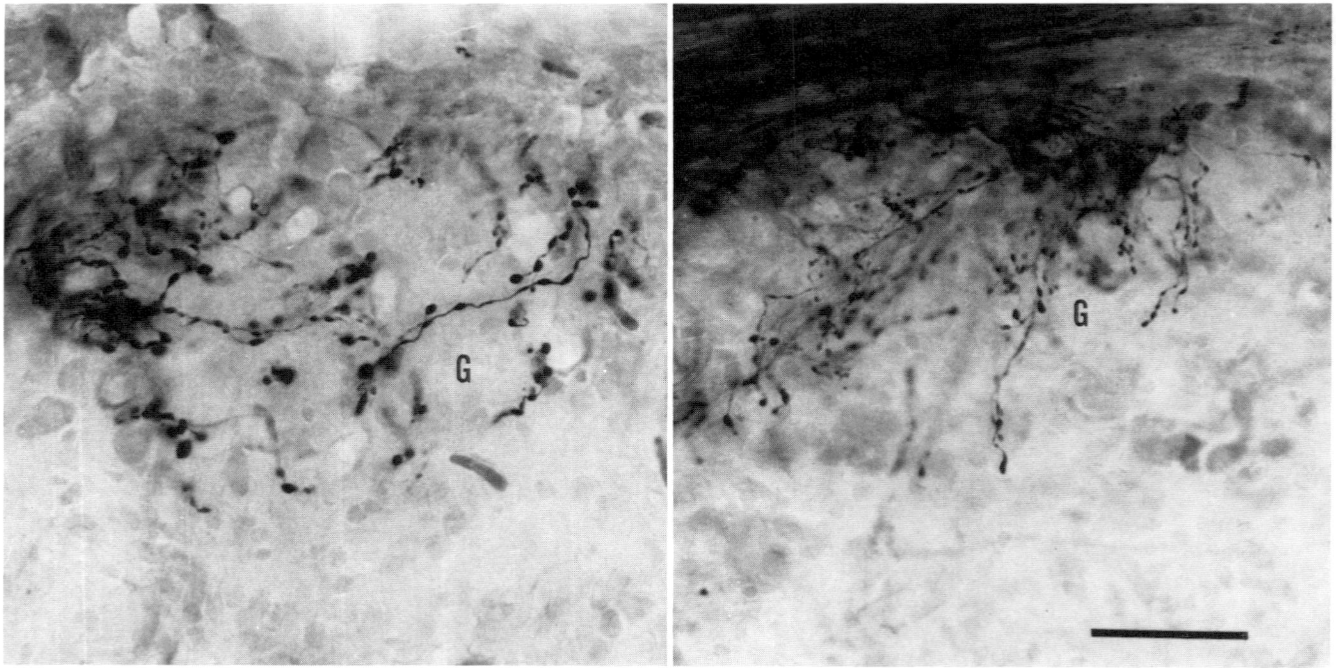


Figure 13.4. A comparison of olfactory nerve terminals in olfactory bulb of *Carrassius auratus*. Primary olfactory nerve fibers have been filled with HRP, as in Figure 13.2. (A) Coarse terminal arbor. (B) Fine terminal arbor: *G*, glomerulus. Scale bar = 50 μm.

concept of the telencephalon in fishes as being entirely devoted to the olfactory sense (e.g., Ariens Kappers, Huber, and Crosby 1936; Romer 1970) is outdated and should be abandoned.

The centripetal fibers leave the olfactory bulb in two main bundles, the medial and lateral olfactory tracts. The two tracts differ slightly in their central distributions, but substantial overlap exists between the two systems. The olfactory tract fibers from the olfactory bulb on each side terminate bilaterally, crossing the midline in the anterior commissure and, in some species, the habenular commissure as well (see von Bartheld et al. 1984 for more detailed interspecific comparisons). The ipsilateral terminal fields are always larger than their contralateral counterparts. The olfactory tracts terminate mainly in two broad areas of the telencephalon (Fig. 13.5). These areas are largely confluent in *Lepisosteus* (Northcutt and Davis 1983), but are quite distinct in fishes such as *Salmo* (Northcutt and Davis 1983), *Ictalurus* (Finger 1975) and *Polypterus* (von Bartheld and Meyer 1986a).

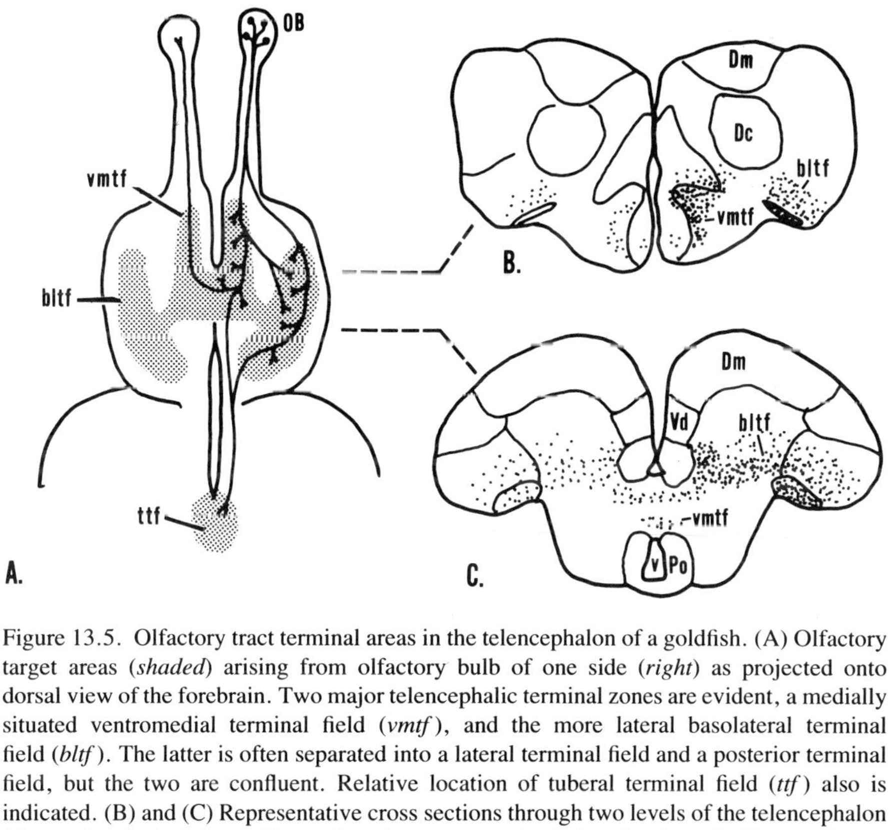


Figure 13.5. Olfactory tract terminal areas in the telencephalon of a goldfish. (A) Olfactory target areas (*shaded*) arising from olfactory bulb of one side (*right*) as projected onto dorsal view of the forebrain. Two major telencephalic terminal zones are evident, a medially situated ventromedial terminal field (*vmtf*), and the more lateral basolateral terminal field (*bltf*). The latter is often separated into a lateral terminal field and a posterior terminal field, but the two are confluent. Relative location of tuberal terminal field (*ttf*) also is indicated. (B) and (C) Representative cross sections through two levels of the telencephalon (shown by *dashed lines*) illustrating the extent and relative density of olfactory inputs arising from olfactory bulb of right side: *bltf,* basolateral terminal field; *Dc,* telencephalic area dorsalis centralis; *Dm,* telencephalic area dorsalis medialis; *OB,* olfactory bulb; *Po,* preoptic area; *ttf,* tuberal terminal field; *v,* ventricle; *Vd,* telencephalic area ventralis dorsalis; *vmtf,* ventromedial terminal field.

The larger telencephalic terminal zone comprises the basolateral portions of the dorsal telencephalon (*bltf,* Fig. 13.5), including portions of pallial areas (dorsalis centralis (DC) and dorsalis posterior (DP) in teleosts). This terminal zone includes the posterior and lateral terminal fields described by various investigators (Scalia and Ebbesson 1971; Finger 1975; Ebbesson, Meyer, and Scheich 1981; von Bartheld et al. 1984; Levine and Dethier 1985).

The second major telencephalic olfactory target zone involves a ventromedial column of nuclei (*vmtf,* Fig. 13.5) extending rostrally from the region just caudal to the anterior commissure. Most of the ventral tier of telencephalic nuclei, including, in teleosts, the intermediate (Vi), dorsal (Vd), supracommissural (Vs), and, perhaps, ventral (Vv) nuclei, receive input from the olfactory bulb. In addition, the olfactory tracts provide a major contribution to the fibers crossing in the anterior commissure; these decussating fibers give rise to small terminal fields within the zone around the anterior commissure.

Relatively minor components of the olfactory tract terminate in the posterior tuberal region of the ventral diencephalon. This diencephalic terminal field has been referred to as the hypothalamic terminal field (Finger 1975), though it is not clear that the posterior tuber region is in any way equivalent to the hypothalamus (Braford and Northcutt 1983). The olfactory bulb of one side of the brain also has a minor connection to the olfactory bulb on the contralateral side (Finger 1975; Bass 1981).

Of the various olfactory target areas, only the connections of the dorsalis posterior (DP) region of the basolateral terminal zone have been studied by experimental methods (Murakami, Morita, and Ito 1983). This study shows that the DP region has limited intratelencephalic connections, projecting only to a small region on the medial wall of the dorsal telencephalon (ventral region of the dorsalis medialis [vDM]). The vDM itself has limited intratelencephalic connections mostly involving the contralateral vDM and the region of the commissural ridge. These findings of limited olfactory connectivity reinforce the concept that the telencephalon of teleosts is *not* dedicated mainly to olfactory processing.

## Olfaction in Flatfish: Lopsided Smelling

The size of the olfactory system in different fish varies a great deal according to habitat and lineage (see Yamamoto 1982 for a recent review on the diversity of the olfactory organ in teleosts). In all cases except the flatfish species, the olfactory organs of the two sides of the animal are equal in size, and the central connections of the olfactory bulbs are symmetrical. In the flatfish, the olfactory organ on the upward-facing side is considerably larger than the one on the downward-facing side. The olfactory centers in the CNS are similarly asymmetrical with the olfactory bulbs, and even the telencephalon of the upward side being larger than the contralateral counterparts (Fig. 13.6) (Prasada Rao and Finger 1984). Despite the marked asymmetry in the size of the olfactory bulbs, the bulb on each side maintains a projection to the contralateral telencephalon.

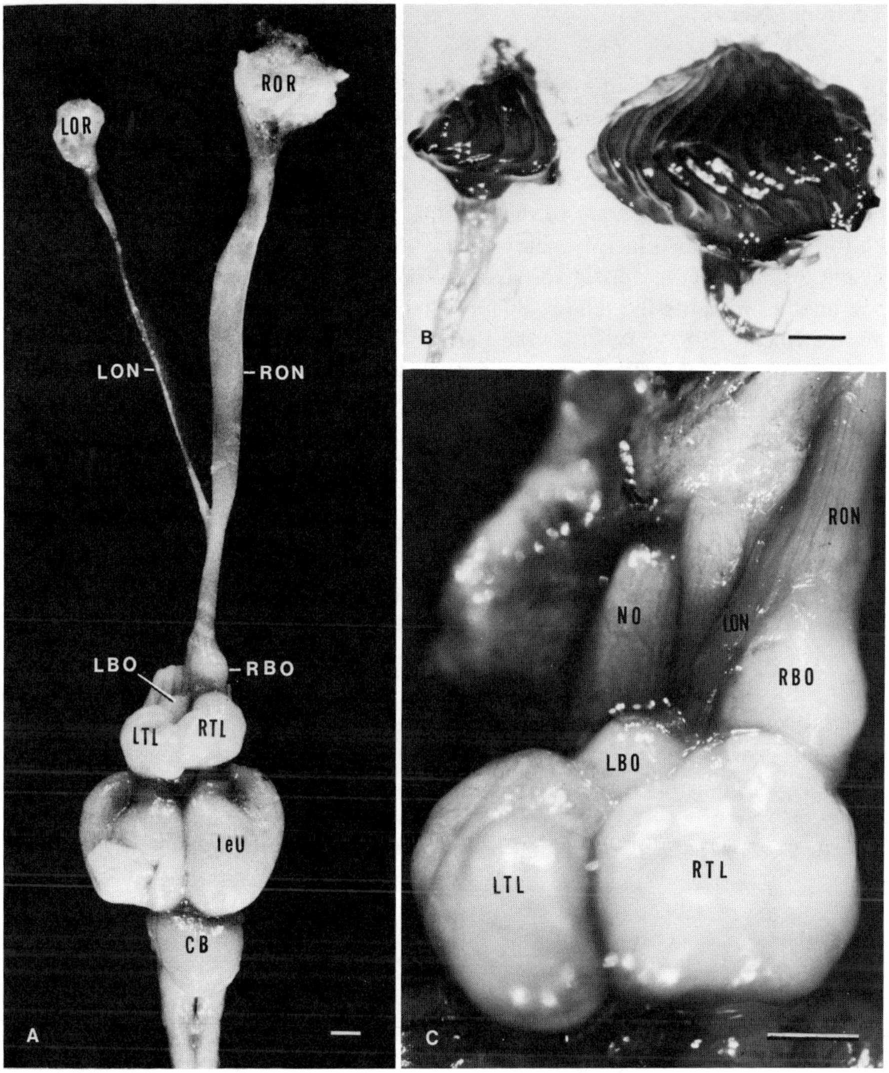


Figure 13.6. Olfactory system of flounder *Pseudopleuronectes americanus*. (A) Dorsal view of the brain, olfactory bulbs (*RBO, LBO*), olfactory nerves (*RON, LON*), and olfactory organs (*ROR, LOR*). Note much larger development of olfactory structures on the right (*RBO, RON, ROR*) than on the left (*LBO, LON, LOR*). (B) Comparison of olfactory organs. (C) Higher-power view of olfactory bulbs and telencephalic hemispheres (*LTL, RTL*). Both olfactory bulb and telencephalon on right side (*RBO, RTL*) are larger than on the left: *CB*, cerebellum; *LBO*, left olfactory bulb; *LON*, left olfactory nerve; *LOR*, left olfactory organ; *LTL*, left telencephalon; *NO*, optic nerve; *RBO*, right olfactory bulb; *RON*, right olfactory nerve; *ROR*, right olfactory organ; *RTL*, right telencephalon; *TeO*, optic tectum.

## Terminal Nerve

In all vertebrates, including bony fishes, a separate group of fibers emerges from the telencephalon or olfactory bulb to join the olfactory nerve in entering the olfactory capsule. This distinct nerve has been called the *terminal nerve* (Pinkus 1984) and is usually considered a separate cranial nerve, distinct from the olfactory nerve. The neuronal somata of the terminal nerve are situated near the olfactory bulb or among the olfactory nerve fibers. At least some of these ganglion cells of the terminal nerve appear to have one process directed towards the periphery and another towards the telencephalon (Stell et al. 1982; Springer 1983; Demski and Northcutt 1983). Whether the peripherally directed process is sensory and whether the peripheral process reaches the surface of the olfactory epithelium are matters open to dispute. In any event, the central processes of the terminal nerve ganglion cells are distributed primarily to the preoptic region and the commissural ridge (Bass 1981; Munz et al. 1982; Demski and Northcutt 1983; Levine and Dethier 1985; von Bartheld and Meyer 1986b). In addition, the central processes of some of the ganglion cells of the terminal nerve extend into the telencephalon, turn anteriorly into the optic tracts, and terminate within the synaptic layers of the retina (Fig. 13.7). Those terminal nerve ganglion cells which project to the retina contain LHRH (luteinizing hormone–releasing hormone) and have been named the *nucleus olfactoretinalis* (Munz et al. 1982).

In goldfish, the large ganglion cells of the terminal nerve receive direct somatic input from a small number of catecholaminergic axons arising from the locus coeruleus (Fernald and Finger 1984). This connection may indicate that this population of terminal nerve cells is involved in a behavioral or attentional, rather than sensory, process. One hypothesized role for the terminal nerve is in the control of some aspects of mating behavior (Demski and Northcutt 1983).

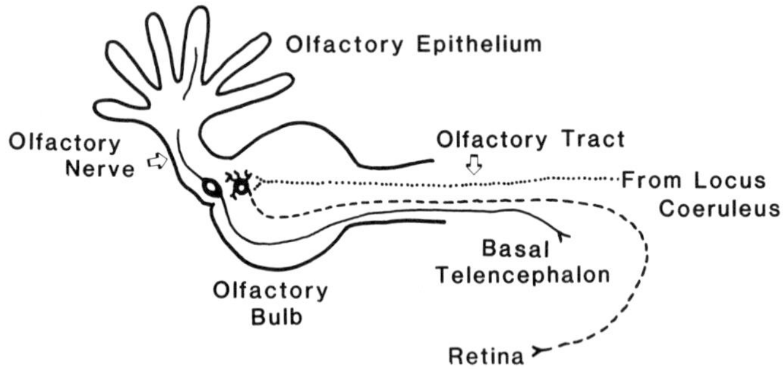


Figure 13.7. Semischematic diagram of the organization and position of terminal nerve ganglion in *Carassius*. Two types of cells are present (1) bipolar type, which has peripherally directed process and centrally directed process that reaches only as far as the basal telencephalon, and (2) multipolar type, with short dendrites and which projects to the retina, as well as to basal telencephalon. Multipolar cell is immunoreactive for LHRH. At least some multipolar terminal nerve ganglion cells receive centrifugal input from the vicinity of locus coeruleus (von Bartheld, Rickmann and Meyer 1986).

## Gustatory System

Taste buds always are innervated by fibers derived from either the facial (VII), glossopharyngeal (IX), or vagal cranial nerves (X). This is true whether the taste buds are within the orobranchial cavity or spread across the animal's body surface, as in Ictaluridae and many Cyprinidae (see Chap. 12). The external taste buds are innervated by branches of the facial nerve; in many species (e.g., *Ictalurus*), a recurrent branch of the facial nerve exits from the skull caudally to innervate the taste buds on the flank, fins, or tail. This recurrent branch is different from, and does not anastomose with, the lateral line nerves, though the two nerves may be in proximity. The recurrent facial nerve does, however, distribute peripherally with fibers from the dorsal root ganglia of the spinal cord, which provide the general cutaneous innervation to the flank, fins, and tail (Herrick 1901, 1905).

A small intraoral field of taste buds at the anterior end of the mouth also is innervated by branches of the facial nerve. The glossopharyngeal nerve (cranial nerve IX) in most fish innervates only a relatively minor intraoral gustatory field, while the vagus nerve (X) provides the bulk of the oropharyngeal innervation. Large numbers of vagally innervated taste buds are located on the gill arches and palate, which have evolved into elaborate feeding organs in some cyprinid species (Sibbing 1984), some clupeids, and an African fish, *Heterotis* (Braford 1986).

### Gustatory Systems in Central Nervous System

All the gustatory nerves terminate centrally in the visceral sensory column of the rostral medulla. This column of nuclei is situated at the ventricular surface of the medulla, laterally adjacent to the sulcus limitans (Fig. 13.8). Thus the primary gustatory nuclei lie medial to the octavolateral nuclei and lateral to the visceral motor nuclei of the rostral medulla.

#### Primary Gustatory Nuclei

In species like *Gnathonemus* and *Prionotus,* which lack peripheral gustatory specializations, such as a palatal organ or barbels, the gustatory nuclei form a rather undifferentiated column. The facial nerve gustatory inputs terminate in the anterior end of this column, the vagal gustatory inputs in the posterior end, and the glossopharyngeal gustatory inputs in an intermediate position (see Fig. 13.8) (Maler, Karten, and Bennett 1973a, 1973b; Finger unpublished observations). In species with peripheral gustatory specializations, portions of the primary gustatory nuclear column can be greatly elaborated to become lobular or laminated, as shown in Fig. 13.9 (Finger 1976; Marui and Caprio 1982; Braford 1986).

At least in those species with an elaborate gustatory apparatus, the gustatory lobes may exhibit a relatively fine-grain somatotopical representation of the end organs. For example, each barbel is represented in a unique lobule of the facial lobe (Finger

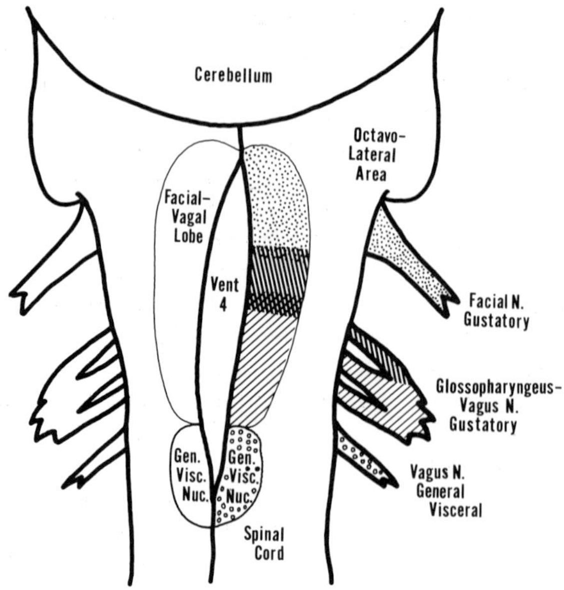


Figure 13.8. Semischematic drawing of dorsal view of gustatory and general visceral nuclei in the medulla of unspecialized teleost. Gustatory roots of facial, glossopharyngeal, and vagal nerves terminate in rostrocaudal sequence in gustatory lobes. General visceral inputs terminate in separate, more caudal general visceral nucleus, including commissural nucleus of Cajal (not shown separately): *Gen. Visc. Nuc.*, general visceral nucleus and commissural nucleus of Cajal; *N.*, nerve; *Vent 4*, fourth ventricle.

1976; Marui and Caprio 1982). Electrophysiological studies demonstrate further that the neurons within a given lobule have receptive fields restricted to a small portion of the given barbel (Marui and Caprio 1982). Similarly, anatomical (Morita and Finger 1985a) and electrophysiological (McGlone 1977; Kanwal 1986) studies demonstrate that the orobranchial cavity is represented topographically (orotopically) within the vagal lobe.

Since most of the work on higher order connections of the gustatory systems in fishes has been carried out on ictalurid catfish (Finger 1978a, 1983b; Morita and Finger 1985b) and cyprinids (Finger 1983b; Morita, Murakami, and Ito 1983), the remainder of this section will describe the findings from studies on these groups. My own unpublished results with *Gadus* and *Prionotus*, however, suggest that the previously published studies are generally applicable to most teleost orders. Thus, what is described below about the term *facial lobe* is probably applicable to the rostral end of the gustatory column in fish lacking gustatory specializations, and what is described about the vagal lobe probably applies to the caudal end of the gustatory column.

## Second-Order Projections

The output from the primary gustatory nuclei forms three major output systems: reflex-reticular, descending, and ascending (Fig. 13.10). In addition, the primary gustatory nuclei themselves comprise a large number of local interneurons. Thus the output of the primary gustatory nuclei, though sometimes referred to as second order, may not actually represent the output from neurons that receive synapses from the primary sensory nerve fibers.

Each of the gustatory lobes projects to the subjacent reticular formation and, in the case of the vagal lobe, to the subjacent vagal motor nuclei (Morita and Finger 1985b). Thus the vagal lobe is intimately associated with reflex swallowing as mediated by the vagal motor nuclei (Atema 1971).

Each of the gustatory lobes, but especially the facial lobe, maintains a distinct descending projection system terminating in a spinotrigeminal funicular nucleus, as well as in the dorsal horn of the spinal cord (Finger 1978a; Morita and Finger 1985b). For those fish with extensive external taste systems, this pathway may provide an avenue for the correlation of somatosensory and gustatory inputs from similar areas of the body (Herrick 1905, 1906).

The major efferent fiber system of the gustatory lobes assumes a rostral trajectory to reach the superior secondary gustatory nucleus in the pons (metencephalon, i.e., that portion of the brain lying between the medulla and midbrain). The superior secondary gustatory nucleus is relatively large in most teleost species. Originally this nucleus was mistaken for a deep cerebellar nucleus (Ariens Kappers, Huber, and Crosby 1936), but experimental methods demonstrate that this large pontine nucleus is associated with the gustatory system and not the cerebellum (Herrick 1905; Finger 1978a, 1978b). In fish with large gustatory systems, such as ictalurids, many cyprinids, and gadids, the superior secondary gustatory nucleus represents a significant proportion of the cross-sectional area of the brain stem at pontine levels.

The superior secondary gustatory nucleus contains a population of large cells whose dendrites extend throughout large areas of the nucleus (Finger 1978a) These cells would be expected to exhibit larger receptive fields than those of the primary nucleus and, indeed, this appears to be the case (Marui and Caprio 1982).

Some of the ascending second-order gustatory fibers, and especially those originating in the facial lobe, bypass the superior secondary gustatory nucleus to terminate directly in the ventral diencephalon. These direct medullodiencephalic fibers terminate mostly in a nuclear complex situated in the ventrolateral thalamus. Some of these fibers also terminate in the vicinity of large neurons that constitute the nucleus lobobulbaris, a nucleus that projects directly back to the facial and vagal lobes (Finger 1978a).

## Higher Order Ascending Connections

The major ascending central gustatory pathway emerges from the superior secondary gustatory nucleus to terminate throughout much of the ventral diencephalon, including the ventral thalamus and lateral part of the inferior lobe (Fig. 13.11). Portions of the inferior lobe in teleosts appear to be similar to the lateral hypothalamus in amniote vertebrates. These connections have been studied carefully only in the catfish *Ictalurus* (Finger 1976; Kanwal 1986) and *Carassius* (Morita, Ito, and Masai 1980); and the connections in these related otophysan species are

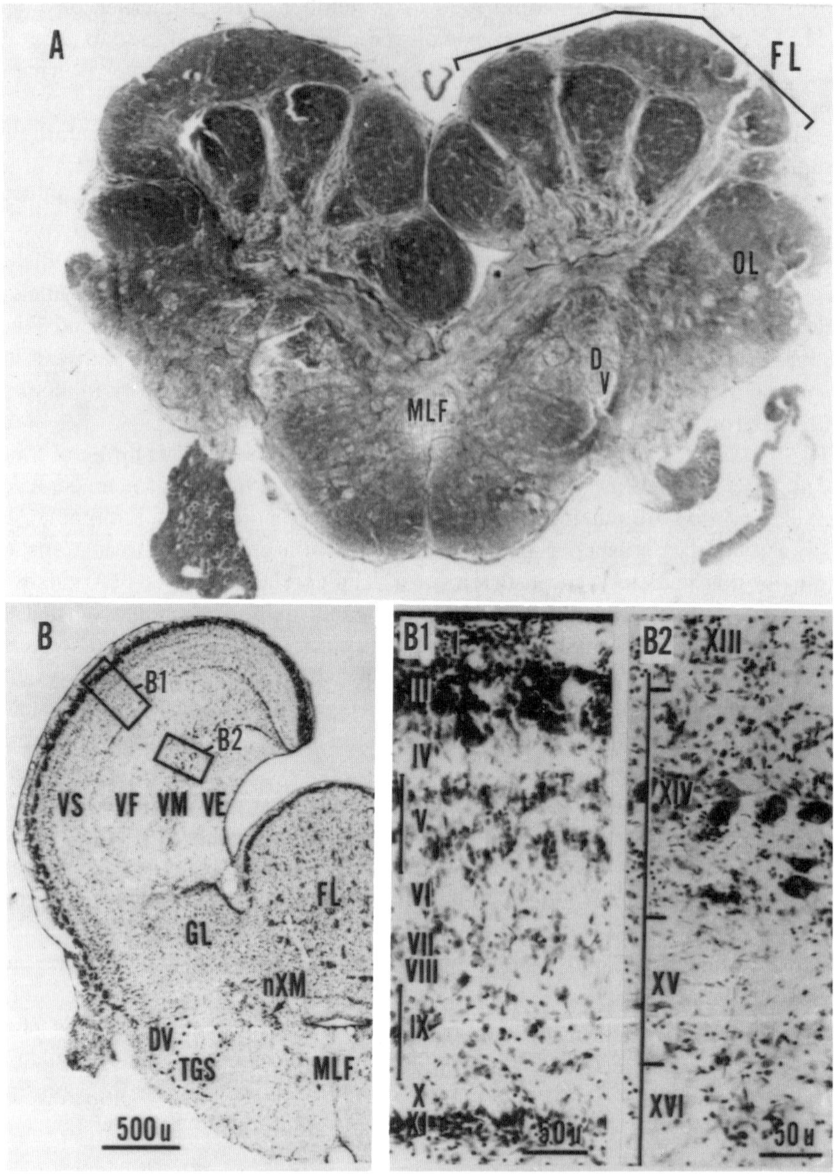


Figure 13.9. Fish having specializations of central gustatory nuclei corresponding to specializations of peripheral gustatory apparatus. (A) Facial lobes of catfish (in this case, Japanese sea catfish, *Plotosus*) divided into lobules by fascicles of fibers. Each lobule contains neural representation of single barbel. (Photograph courtesy of T. Marui.) (B) Vagal lobe of *Carassius* is rigidly laminated system in which the oral cavity is mapped onto the lobe. Other gustatory lobes in these species (*GL* and *FL*) are not highly laminated. Enlargements B1 and B2 illustrate the pattern of lamination, respectively, in sensory and motor layers.

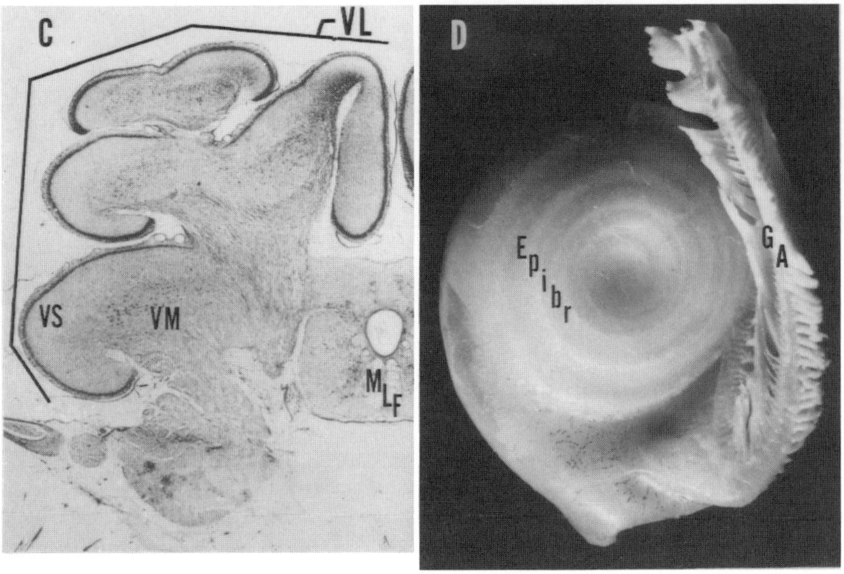


Figure 13.9. (C) and (D) Extreme example of the correspondence between CNS structures and peripheral gustatory elaboration in *Heterotis*, in which spiraled epibranchial organ is matched by spiraled portion of vagal lobe. (Photographs courtesy of M. Braford.) (C) Cross section through spiraled vagal lobe. Note that overall organization of vagal lobe of *Heterotis* is similar to that of *Carassius*. In both species, outer sensory layer contains superficial layer of small, densely packed neurons, and motor layer lies deep to sensory layer. (D) Photomacrograph of epibranchial organ dissected free of the animal. One gill arch (*GA*) remains attached: *DV*, descending trigeminal nerve root; *Epibr*, epibranchial organ; *FL*, facial lobe; *GA*, gill arch; *GL*, glossopharyngeal lobe; *MLF*, medial longitudinal fasciculus; *nXM*, somatic motor nucleus of vagus nerve; *OL*, octavolateral area; *TGS*, sensory nucleus of descending trigeminal nerve root; *VE*, ependymal layer of vagal lobe; *VF*, central fiber layer of vagal lobe; *VL*, vagal lobe; *VM*, motor layers of vagal lobe; *VS*, sensory layers of vagal lobe.

similar. Whether similar patterns of connectivity exist for other teleostean species is unknown.

In *Ictalurus*, a recent study shows that the output of the superior secondary gustatory nucleus reaches widespread areas of the lateral part of the inferior lobe and the ventral thalamus (Kanwal 1986). The thalamic gustatory nuclei in these fish have been named the "posterior thalamic nucleus" (Finger 1978a), though this nucleus does not appear to be equivalent to what has been termed the *posterior thalamic nucleus* in other species of teleosts (e.g., Morita, Ito, and Masai 1980). The posterior thalamic nucleus described for *Ictalurus* may be equivalent to a portion of the preglomerular complex, but further study of this is required before this equivalence can be drawn with any certainty.

Morita and co-workers (Morita, Ito, and Masai 1980; Morito, Murakami, and Ito 1983), following the terminology of Peter and Gill (1975), describe the ascending gustatory inputs as terminating in the nucleus glomerulosus, as well as in

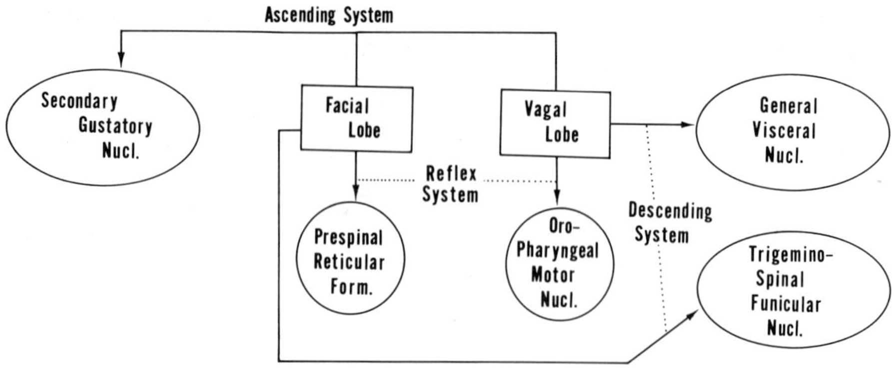


Figure 13.10. Schematic diagram indicating connections of gustatory lobes in teleosts: *Form.*, formation; *Nucl.*, nucleus.

the inferior lobe. The nucleus glomerulosus of cyprinid species as described by these workers does not, however, appear equivalent to the more obvious nucleus glomerulosus evident in perciform species (Sakamato and Ito 1982), which is involved in visual processes and has quite different connections (see also Braford and Northcutt 1983). Since the two nuclei are not homologous, the term *nucleus glomerulosus* as applied to the gustatory nucleus in cyprinids should be abandoned. Positionally and hodologically, the tertiary gustatory nucleus in cyprinids appears equivalent to a component of the nucleus preglomerulosus. Evidence for this contention comes from studies on the sea robin *Prionotus carolinus,* in which I have found a second-order gustatory projection reaching a portion of nucleus preglomerulosus (Finger 1983a). This nucleus appears to be a multimodal complex, rather than a single nucleus, and receives inputs from ascending spinal, as well as gustatory systems. The multimodal input to the nucleus preglomerulosus is not overlapping; i.e., one portion of the nucleus receives the spinal, while

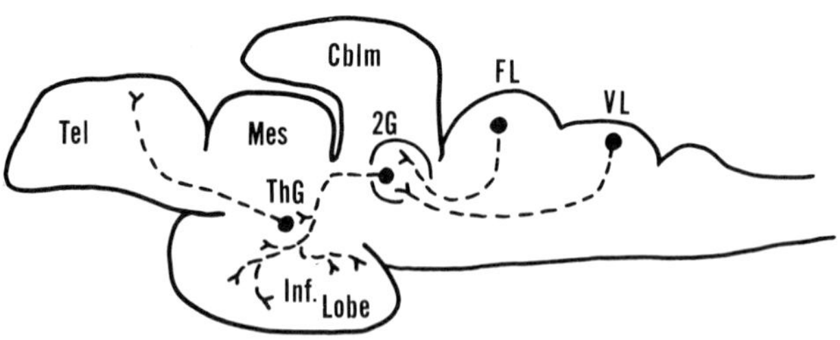


Figure 13.11. Gustatory input to the forebrain as indicated on lateral view of the brain of a catfish. Major gustatory target areas lie in ventral diencephalon, including lateral and dorsal areas of inferior lobe. Small telencephalic gustatory target zone also is present: *Cblm,* cerebellum; *FL,* facial lobe; *Inf. Lobe,* inferior lobe; *Mes,* mesencephalon; *Tel,* telencephalon; *ThG,* thalamic gustatory complex; *VL,* vagal lobe; *2G,* superior (pontine) secondary gustatory nucleus.

another part receives the gustatory, input. In this respect, the nucleus preglomerulosus is similar to the ventrobasal complex of the thalamus in mammals.

Another major recipient group for ascending gustatory information is the lobobulbar nucleus. The large cells of this nucleus project back onto the primary gustatory lobes (Finger 1978a). In contrast, smaller neurons in this same region relay gustatory information to the telencephalon (Kanwal 1986). The telencephalic gustatory recipient zone lies along the ventral portion of telencephalic zone dorsalis medialis (Dm). Single unit recording from this area demonstrates the gustatory nature of the telencephalic units (Kanwal 1986).

## Common Chemical Sense

The common chemical sense is best developed in the genus *Prionotus* (sea robins and gurnards), though all fish may exhibit this sense to some degree (Parker 1922). In sea robins, the three most anterior pairs of fin rays of the pectoral fins are specialized exploratory sensory organs that are chemosensory (Silver and Finger 1984). These free fin rays lack taste buds and olfactory receptors and so, by definition, are elements of the common chemical sense as defined by Parker (1922).

The transducing elements of the common chemical sense as occurring on the fin ray are not definitively established, though Lane and Whitear (1982) give ultrastructural evidence for their being isolated chemosensory cells in the epithelium. Other workers suggest that free nerve endings within the epithelium may directly serve the function of chemosensory transduction. Similar isolated, presumed chemosensory cells have been described in the epithelium of the dorsal fin of the rockling *Gaidropsarus* (Kotrschal, Whitear, and Adam 1984). In the case of this species, however, the isolated chemosensory cells may be innervated by the facial nerve—the nerve that in other species, innervates external taste buds. This situation may be indicative of the manner in which taste buds evolved from a more generally chemosensitive epithelium.

### Primary Sensory Nerves and Their Site of Central Termination

The fin ray chemoreceptors in *Prionotus* are innervated solely by spinal nerves; no cranial nerve components are involved. This contrasts sharply with the situation of cods and gouramis (*Trichogaster*) in which the specialized fin appendages contain taste buds and are innervated by recurrent branches of the facial nerve.

The spinal nerves that innervate the fin rays in sea robins arise from a large dorsal root ganglion (Fig. 13.12) that appears to be a fused ganglionic mass containing elements derived from two or three spinal roots (Morrill 1895). The neurons of this fused ganglion are bipolar (Finger unpublished observations), as are the ganglionic neurons of most teleosts.

The central processes of these ganglion cells terminate within the enlarged dorsal horn of the first few spinal segments (Herrick 1907; Finger 1982). Virtually no primary sensory processes ascend to the spinomedullary level. Therefore the dorsal root fibers of this system do not form a primary dorsal column system as found in the spinal cords of most amniotes.

The rostral spinal cord of sea robins is modified from the typical teleostean pattern by the presence of four paired enlargements, termed *accessory spinal lobes,* which extend dorsally from the surface of the cord (Fig. 13.12) (Morrill 1895; Finger 1982). The rostralmost accessory lobe on each side is smaller than the caudal three, and so the former is called the *minor accessory lobe,* in contradistinc-

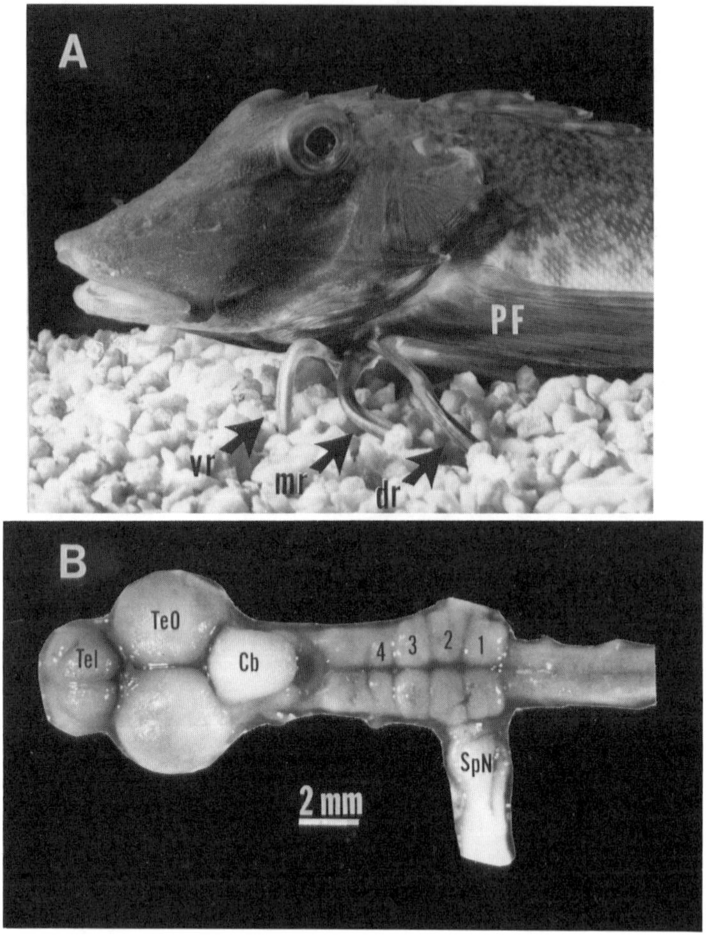


Figure 13.12. (A) Sea robin *Prionotus carolinus.* Note pectoral appendages (*vr, mr, dr*), which, though modified fin rays, are freely mobile and unattached to webbed pectoral fin (*PF*). (B) Dorsal view of the brain and rostral spinal cord of this species. Fin rays are innervated by spinal nerves (*SpN*), which enter rostral spinal cord to terminate in accessory lobes (*1–4*): *Cb,* cerebellum; *dr,* dorsal fin ray; *mr,* middle fin ray; *PF,* pectoral fin; *SpN,* spinal ganglion and nerve roots that innervate fin rays; *Tel,* telencephalon; *TeO,* optic tectum; *vr,* ventral fin ray; *1–4,* spinal accessory lobes.

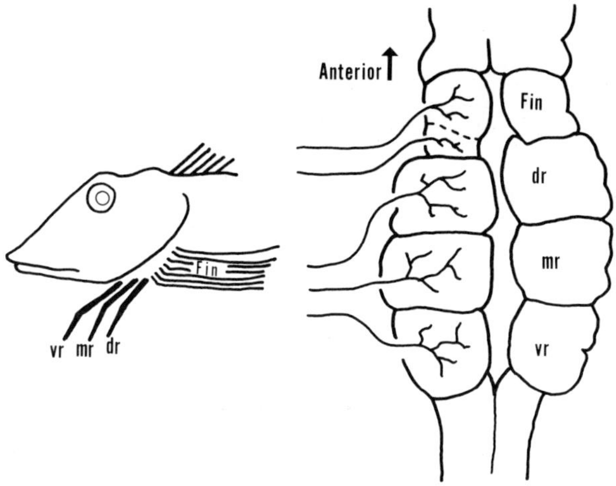


Figure 13.13. Semischematic drawing of somatotopic representation of pectoral fin and fin rays in rostral spinal cord. *Left,* Lateral view of the fish to show the position of fin rays and pectoral fin. *Right,* Dorsal view of rostral part of spinal cord, indicating which appendage is represented in which accessory spinal lobe. Note counterintuitive arrangement; i.e., ventral fin ray (*vr*), which is anatomically most anterior, is represented in most posterior accessory lobe: *dr,* dorsal fin ray or its spinal representation; *mr,* middle fin ray or its spinal representation; *vr,* ventral fin ray or its spinal representation.

tion to the latter, called *major accessory lobes* (Finger 1982). Anatomical (Finger 1982) and electrophysiological (Finger and Silver unpublished observations) experiments reveal that the three free pectoral fin rays are each represented in a major accessory lobe, while the webbed pectoral fin is represented in the minor accessory lobe (Fig. 13.13). Thus a somatotopical arrangement is evident, but would appear counterintuitive, since the most rostral fin ray is represented in the most caudal lobe. This arrangement, however counterintuitive, appears related to the rotation of the fin during embryogenesis (Kuntz and Radcliffe 1918; Finger 1982; Finger and Kalil 1985).

## Central Pathways

The output of the spinal accessory lobes reaches two principal destinations, the cerebellum and a funicular nucleus situated at the spinomedullary junction (Finger 1981). The spinocerebellar system terminates within a restricted portion of the corpus cerebelli and arises from large neurons situated deep in the accessory lobes. The spinofunicular system terminates in the lateral funicular nucleus and arises from smaller neurons scattered throughout the accessory lobes. The lateral funicular nucleus itself comprises two divisions, a lateral and a medial portion. The medial portion is smaller and projects to the spinal area of the cerebellum, which thus receives both a direct and an indirect spinocerebellar input. The major lemniscal type of ascending spinal sensory system arises from the lateral portion of the lateral funicular nucleus.

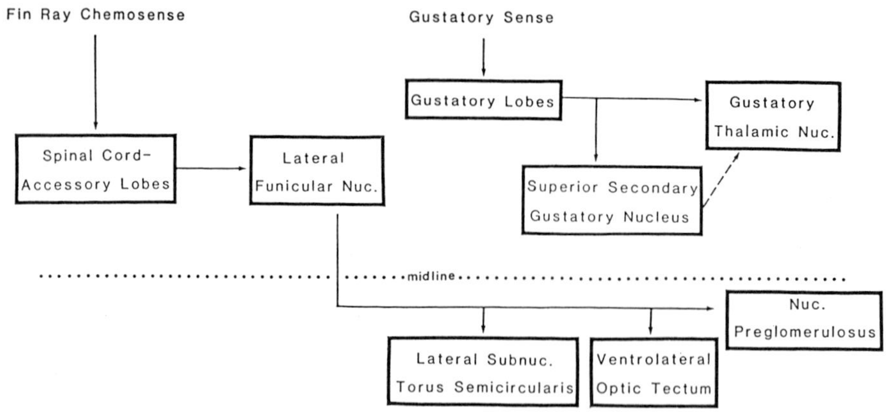


Figure 13.14. Schematic diagram of brain of *Prionotus* comparing connections of fin ray chemosense and gustatory sense in this sea robin. Anterior is to the right. Note that ascending gustatory pathway remains ipsilateral, while ascending spinal pathway crosses the midline in caudal medulla. Thus chemosensory inputs on one side of the animal that stimulate both chemosensory systems (such as amino acids would be expected to do) end up on opposite sides of the brain: *Nuc.*, nucleus.

Efferent fibers leave the lateral funicular nucleus in a ventromedial direction to cross the midline in midmedulla. The ascending fibers assume a superficial position along the ventral margin of the rostral hindbrain. The fibers gradually move dorsally as the tract ascends through the mesencephalon. At the level of the torus semicircularis, some of the fibers turn laterally from the main bundle to terminate in a specialized lateral subnucleus of the torus semicircularis, as well as in the optic tectum. The bulk of the ascending spinal lemniscal tract continues rostrally past the midbrain to terminate in a part of the nucleus preglomerulosus of the diencephalon (Fig. 13.14). The ascending spinal (probably including chemosensory) system has virtually no overlap with the ascending gustatory system in *Prionotus* (Finger 1983a), though both sensory modalities project to adjacent portions of the preglomerular complex.

## Summary

Three chemosensory systems can be recognized in teleost fish, as in other vertebrates: olfaction, gustation, and the common chemical sense. These modalities are distinct on the bases of receptor morphology and innervation, behaviors controlled by each system, and organization of each system within the CNS (Table 13.1).

The olfactory receptor neurons have axons that terminate within the olfactory bulb. The output cells of the bulb project, in turn, to restricted areas of the basolateral telencephalon. The gustatory receptors are axonless neuroepithelial cells innervated by either the facial, glossopharyngeal, or vagal nerves. These

Table 13.1. Comparison of Three Major Chemoreceptive Systems in Fish

| Olfactory Sense | Gustatory Sense | Common Chemical Sense |
|---|---|---|
| Olfactory epithelium | Taste buds | Receptors |
| ↓ | ↓ | ↙ CN V ↘ Spinal nerves |
| Olfactory bulb | CN VII, IX, X | ? Dorsal horn |
| ↙ ↘ | ↓ | ↓ |
| Ventrobasal and ventromedial telencephalon; Tuberal zone (hypothalamus?) | Medullary gustatory lobes | Funicular relay nucleus |
| ↓ (from ventrobasal and ventromedial telencephalon) | ↓ | ↓ |
| Medial wall dorsal telencephalon | Superior secondary gustatory nucleus | Diencephalic somatosensory nucleus (including nucleus preglomerulosus) |
| | ↓ | |
| | Diencephalic gustatory nucleus | |
| | ↓ | |
| | Dorsomedial central telencephalon | |

Note: Little overlap exists for the three systems at any level of the neuraxis studied to date. One might expect, however, on the basis of behavior, that the chemoreceptor systems ultimately converge on some central structure. *CN V*, cranial nerve 5 (trigeminal nerve); *CN VII, IX, X*, cranial nerves 7 (facial), 9 (glossopharyngeal), and 10 (vagal).

nerves terminate centrally within the gustatory sensory column situated along the dorsomedial surface of the medulla. The primary gustatory nuclei project rostrally to a pontine second-order gustatory nucleus and to a group of nuclei situated in the ventral diencephalon. One or more of these diencephalic nuclei project to a telencephalic gustatory area in the dorsomedial telencephalon. The common chemical sense is mediated via general cutaneous nerves that terminate centrally in the spinal cord or trigeminal nuclei. In sea robins, which possess a highly developed fin ray chemosense, the rostral portion of the spinal cord includes four dorsally situated enlargements, called accessory spinal lobes. These lobes project rostrally to the lateral funicular nucleus situated at the spinomedullary junction. This funicular nucleus relays the ascending spinal input to the contralateral midbrain and ventral diencephalon. The ventral diencephalic area devoted to the common chemical sense in this species is, however, separate from the gustatory nuclei lying nearby. Thus the olfactory, gustatory, and common chemical senses maintain distinct channels within the CNS.

## References


Ariens Kappers, C.U. 1906 The structure of the teleostean and selachian brain. J. Comp. Neurol. 16:1–109.

Ariens Kappers, C.U., Huber, G.C., and Crosby, E. (1936) The comparative anatomy of the nervous system of vertebrates, including man, Hafner Publishing Co., New York.

Atema, J. (1971) Structures and functions of the sense of taste in the catfish, *Ictalurus natalis,* Brain Behav. Evol. 4:273–294.

Atema, J. (1980) Smelling and tasting underwater, Oceanus, 23:4–18.

Bass, A.H. (1981) Olfactory bulb efferents in the channel catfish, *Ictalurus punctatus,* J. Morphol., 169:91–111.

Bertmar, G. (1981) Evolution of vomeronasal organs in vertebrates, Evolution, 35:359–366.

Braford, M.A., Jr. (1986) De Gustibus non disputandum: a spiral center for taste in the brain of the teleost fish, *Heterotis niloticus,* Science, 232:489–491.

Braford, M.A., and Northcutt, R.G. (1983) Organization of the diencephalon and pretectum of the ray-finned fishes, in Fish Neurobiology, vol. 2, Higher Brain Areas and Functions, Davis, R.E. and Northcutt, R.G. (eds.) University of Michigan Press, Ann Arbor, pp. 117–164.

Demski, L.S. and Northcutt, R.G. (1983) The terminal nerve: a new chemosensory system in vertebrates? Science, 220:435–437.

Ebbesson, S.O.E., Meyer, D., and Scheich, H. (1981) The olfactory system of the black piranha, *Serrasalamus natterei,* Cell Tissue Res., 216:167–180.

Eller, P., Rowley, J.C. III, and Moran, D.T. (1985) The "rod cell" in trout olfactory epithelium: fact or artefact? Chem. Sens., 10:392.

Erickson, J., and Caprio, J. (1983) Similar neural responses from regions rich in ciliated and microvillus cells, respectively in the catfish olfactory epithelium, Abst. Assoc. Chem. Sci., 5:34.

Fernald, R.D. and Finger, T.E. (1984) Catecholaminergic neurons of locus coeruleus project to the ganglion cells of the nervus terminalis (NT) in goldfish, Abst. Soc. Neurosci. 10:50.

Finger, T.E. (1975) Distribution of the olfactory tracts in the bullhead catfish, *Ictalurus nebulosus*, J. Comp. Neurol., 161:125–142.

Finger, T.E. (1976) Gustatory pathways in the bullhead catfish. I. Connections of the anterior ganglion, J. Comp. Neurol. 165:513–526.

Finger, T.E. (1978a) Gustatory pathways in the bullhead catfish. II. Facial lobe connections, J. Comp. Neurol., 180:691–706.

Finger, T.E. (1978b) Efferent neurons of the teleost cerebellum, Brain Res., 153:608–614.

Finger, T.E. (1981) Fish that taste with their feet: spinal sensory pathways in the searobin, *Prionotus carolinus*, Biol. Bull., 161:343.

Finger, T.E. (1982) Somatotopy in the representation of the pectoral fin and free fin rays in the spinal cord of the sea robin, *Prionotus carolinus*, Biol. Bull., 163:154–161.

Finger, T.E. (1983a) Comparison of gustatory and finray chemosensory pathways in the CNS of the sea robin, *Prionotus carolinus*, Abst. Assoc. Chemorecept. Sci., 5:38.

Finger, T.E. (1983b) The gustatory system in teleost fish, in Fish Neurobiology, vol. 1, Northcutt, R.G. and Davis, R.E. (eds), University of Michigan Press, Ann Arbor, pp. 285 310.

Finger, T.E. and Kalil, K. (1985) Organization of motor neuronal pools in the rostral spinal cord of the sea robin, *Prionotus carolinus*, J. Comp. Neurol., 239:384–390.

Herrick, C.J. (1901) The cranial nerves and cutaneous sense organs of the North American siluroid fishes, J. Comp. Neurol., 11:177–247.

Herrick, C.J. (1905) Central gustatory paths in brains of bony fishes, J. Comp. Neurol., 15:375–456.

Herrick, C.J. (1906) On the centers for taste and touch in the medulla oblongata of fishes, J. Comp. Neurol., 16:403–439.

Herrick, C.J. (1907) The tactile centers in the spinal cord and brain of the sea robin, *Prionotus carolinus L.*, J. Comp. Neurol. Psychol., 17:307–327.

Kanwal, J. (1986) The functional organization of the gustatory system in the brains of Ictalurid catfish: a combined electrophysiological and neuroanatomical study, Ph.D. Thesis, Louisiana State University.

Kosaka, T. and Hama, K. (1979) Ruffed cell: a new type of neuron with a distinctive initial unmyelinated portion of the axon in the olfactory bulb of the goldfish (*Carassius auratus*). I. Golgi impregnation and serial thin sectioning studies, J. Comp. Neurol., 186:301–320.

Kotrschal, K., Whitear, M., and Adam, H. (1984) Morphology and histology of the anterior dorsal fin of *Gaidropsarus mediterraneus* (Pisces Teleostei), a specialized sensory organ, Zoomorphologie, 104:365–372.

Kuntz, A. and Radcliffe, L. (1918) Notes on the embryology and larval development of twelve teleostean fishes, Bull. U.S. Bur. Fish., 35 (1915–1916):87–134.

Lane, E.B. and Whitear, M. (1982) Sensory structures at the surface of fish skin. I. Putative chemoreceptors, Zool. J. Linn. Soc., 75:141–151.

Levine, R.L. and Dethier, S. (1985) The connections between the olfactory bulb and the brain in the goldfish, J. Comp. Neurol., 237:427–444.

Maler, L., Karten, H.J. and Bennett, M.V.L. (1973a) The central connections of the posterior lateral line nerve of *Gnathonemus petersii*, J. Comp. Neurol., 151:57–66.

Maler, L., Karten, H.J., and Bennett, M.V.L. (1973b) The central connections of the anterior lateral line nerve of *Gnathonemus petersii*, J. Comp. Neurol., 151:67–84.

Marui, T. and Caprio, J. (1982) Electrophysiological evidence for the topographical arrangement of taste and tactile neurons in the facial lobe of the channel catfish, Brain Res., 231:185–190.

McGlone, F. (1977) The gustatory system of goldfish *Carassius auratus L.*, Ph.D. Thesis, University of Sussex, U.K.

Morita, Y. and Finger, T.E. (1985a) Reflex connections of the facial and vagal gustatory systems in the brainstem of the bullhead catfish, *Ictalurus nebulosus*, J. Comp. Neurol, 231:547–558.

Morita, Y. and Finger, T.E. (1985b) Topographic and laminar organization of the vagal gustatory system in the goldfish, *Carassius auratus*, J. Comp. Neurol., 238:187–201.

Morita, Y., Ito, H., and Masai, H. (1980) Central gustatory paths in the crucian carp, *Carassius carassius*, J. Comp. Neurol., 191:119–132.

Morita, Y., Murakami, T., and Ito, H. (1983) Cytoarchitecture and topographic projections of the gustatory centers in a teleost, *Carassius carassius*, J. Comp. Neurol., 218:378–394.

Morrill, A.D. (1895) The pectoral appendages of *Prionotus* and their innervation, J. Morphol., 11:177–192.

Muller, J.F. and Marc, R.E. (1984) Three distinct morphological classes of receptors in fish olfactory organs, J. Comp. Neurol., 222:482–495.

Münz, H., Claas, B., Stumpf, W.E., and Jennes, L. (1982) Centrifugal innervation of the retina by luteinizing hormone releasing hormone (LHRH)—immunoreactive telencephalic neurons in teleostean fishes, Cell Tissue Res., 222:313–332.

Murakami, T., Morita, Y., and Ito, H. (1983) Extrinsic and intrinsic fiber connections of the telencephalon in a teleost, *Sebasticus marmorato*, J. Comp. Neurol., 216:115–131.

Northcutt, R.G. and Davis, R.E. (1983) Telencephalic organization in ray-finned fishes, in Fish Neurobiology, vol. 2, Davis, R.E. and Northcutt, R.G. (eds.), University of Michigan Press, Ann Arbor, pp. 203–236.

Oka, Y., Ichikawa, M., and Ueda, K. (1982) Synaptic organization of the olfactory bulb and central projection of the olfactory tract, in Chemoreception in Fishes, Hara, T.J. (ed.), Elsevier, New York, pp. 61–76.

Parker, G.H. (1922) Smell, taste and allied senses in the vertebrates, Lippincott, Philadelphia.

Peter, R.E., and Gill, V.E. (1975) A stereotaxic atlas and technique for forebrain nuclei of the goldfish, *Carassius auratus*, J. Comp. Neurol., 159:69–102.

Pinkus, F. (1894) Über einen noch nicht beschriebenen Hirnnerv des *Protopterus annectens*, Anat. Anz., 9:562–566.

Prasada Rao, P.D. and Finger, T.E. (1984) Asymmetry of the olfactory system in the brain of the winter flounder, *Pseudopleuronectes americanus*, J. Comp. Neurol., 225:492–510.

Romer, A.S. (1970) The vertebrate body, 4th ed. W.B. Saunders Co., Philadelphia.

Sakamato, N., and Ito, H. (1982) Fiber connections of the corpus glomerulosum in a teleost, *Navodon modestus*, J. Comp. Neurol., 205:291–298.

Satou, M., Fujita, I., Ichikawa, M., Yamaguchi, K., and Ueda, K. (1983) Field potential and intracellular potential studies of the olfactory bulb in the carp: evidence for a functional separation of the olfactory bulb into lateral and medial subdivisions, J. Comp. Physiol., 152:319–333.

Scalia, F. and Ebbesson, S.O.E. (1971) The central projections of the olfactory bulb in a teleost, Brain Behav. Evol., 4:376–399.

Sibbing, F.A. (1984) Food handling and mastication in the carp, *Cyprinus carpio L.*, PhD Thesis, Agricultural University of Wageningen, The Netherlands.

Silver, W.L. and Finger, T.E. (1984) Electrophysiological examination of the chemoreceptors on the free fin rays of the sea robin, *Prionotus carolinus*, J. Comp. Physiol., 154:167–174.

Springer, A.D. (1983) Centrifugal innervation of goldfish retina from ganglion cells of the nervus terminalis, J. Comp. Neurol., 214:404–416.

Stell, W.K., Chohan, K.S., Lam, D.M.K., and Kozlowski, G.P. (1982) Luteinizing hormone releasing hormone (LHRH) immunoreactive fibers in goldfish retina, Invest. Opthalmol. & Visual Sci., 22:278.

Thommesen, G. (1983) Morphology, distribution and specificity of olfactory receptor cells in salmonid fishes, Acta Physiol. Scand., 117:241–249.

von Bartheld, C.S. and Meyer, D.L. (1986a) Central connections of the olfactory bulb in the bichir, *Polypterus palmas*, reexamined, Cell Tissue Res., 244:527–535.

von Bartheld, C.S. and Meyer, D.L. (1986b) Tracing of single fibers of the nervus terminalis in the goldfish brain, Cell Tissue Res., 245:143–158.

von Bartheld, C.S., Meyer, D.L., Fiebig, E., and Ebbesson, S.O.E. (1984) Central connections of the olfactory bulb in the goldfish, *Carassius auratus*, Cell Tissue Res., 238:475–487.

von Bartheld, C.S., Rickmann, M.J., and Meyer, D.L. (1986) A light- and electron-microscopic study of mesencephalic neurons projecting to the ganglion of the nervus terminalis in the goldfish, Cell Tissue Res., 246:63–70.

Yamamoto, M. (1982) Comparative morphology of the peripheral olfactory organ in teleosts, in Chemoreception in Fishes, Hara, T.J. (ed.), Elsevier, New York, pp. 39–59.

# Chapter 14

# Chemoreceptor Cells in Aquatic Invertebrates: Peripheral Mechanisms of Chemical Signal Processing in Decapod Crustaceans

CHARLES D. DERBY * and JELLE ATEMA **

## Introduction

### Stimulus Filtering

The sensory capabilities of any animal are determined by a sequential set of physical and biological filters that regulate which environmental disturbances will stimulate the receptive surface of the animal. The environment itself is the first filter as it transmits physical or chemical disturbances from one or another source. The animal's sense organs contain additional physical filters that select—each with its own degree of specificity—which part of the disturbances will have best access to the receptor membrane. Subsequent filtering occurs at the level of receptor cells and at each next level of the CNS. Environmental disturbances that alter the activity of receptor cells are called stimuli; they include disturbances that trigger molecular mechanisms, such as adaptation in the cell without necessarily causing it to fire nerve impulses or other means of information coding. Subsequent interneuronal steps of information processing are all based on neural signals that code certain aspects of the original stimuli.

With respect to the chemical senses of aquatic animals, the nature of the chemicals themselves, such as the degree of solubility, determines which chemical disturbances are potential chemical stimuli (Chap. 1). Most biologically significant disturbances will be substances released from plants and animals, such as tissue fluids, metabolites, and pheromones, in addition to environmental "signposts" such as home streams, territories, and trails. Actions can selectively bring stimuli to the receptor cells. These actions include (1) ventilating the gills and olfactory organs by fish; and (2) exopoditic fanning, gill bailing, and pleopod fanning by

* Department of Biology, Georgia State University, Atlanta, GA 30303, USA
** Boston University Marine Program, Marine Biological Laboratory, Woods Hole, MA 02543, USA

crustaceans (Atema 1985 and Chap. 2). Additionally, the method of packaging receptor cells into sensory organs can affect the nature of the message encoded by these cells. An example of this is the packaging of antennular chemoreceptor cells of some crustaceans into dense tufts of sensilla and the subsequent flicking of the antennules. The result is an improvement in the spatial- and temporal-resolving powers of the system, both physiologically (Schmitt and Ache 1979) and behaviorally (Reeder and Ache 1980; Devine and Atema 1982).

This chapter will review the recent advances in our knowledge of chemoreceptor cells of aquatic invertebrates. Knowing the physiological properties of chemoreceptor cells will allow us to determine which environmental disturbances are chemical stimuli for the species in question. This information is also required for studies of signal processing in the CNS and studies of the molecular mechanisms that operate in chemoreceptor cells.

## Crustacean Chemoreceptor Cells

Our aquatic invertebrate focus will be narrowed to crustaceans and, more specifically, to the Florida spiny lobster (*Panulirus argus*), the American lobster (*Homarus americanus*), and the European crayfish (*Austropotamobius torrentium*), since most of our knowledge of the filtering properties of receptor cells of aquatic invertebrates comes from these three species of decapod crustaceans. In large part, this is the result of our ability to record the activity of single chemoreceptor cells in these species. These cells, be they in the antennules, antennae, legs, or mouthparts, have common anatomical features (Ache 1982; Ache and Derby 1985), for example, (1) they are modified ciliary cells; (2) they are bipolar, with somata in the periphery; and (3) they have axons projecting into the CNS, apparently without forming synapses in the periphery (Fig. 14.1).

## General Constraints on Receptors

In this chapter, we shall examine how chemoreceptor cells of aquatic crustaceans handle two general problems that are faced by virtually all sensory systems: resolving type (quality) and intensity (quantity) of the stimulus.

The first task that sensory systems must accomplish is to determine the stimulus quality, enabling the animal to identify shapes, sounds, or smells emanating from potential competitors, mates, predators, or prey. In receptor organs containing cells with narrow and nonoverlapping response spectra, the identity of a specific stimulus can be made on the basis of the activation of those cells specifically sensitive to that stimulus: this is referred to as a *labelled-line code*. Pheromone-detecting systems of some insects (Kaissling 1979; Boeckh 1980) and auditory systems with critical bands (Ehret and Capranica 1980) are two possible examples. However, most sensory systems are thought to code stimulus quality by an *across-neuron pattern code* (Erickson 1978). In these systems which are often based on cells having broad and overlapping response profiles, the identity

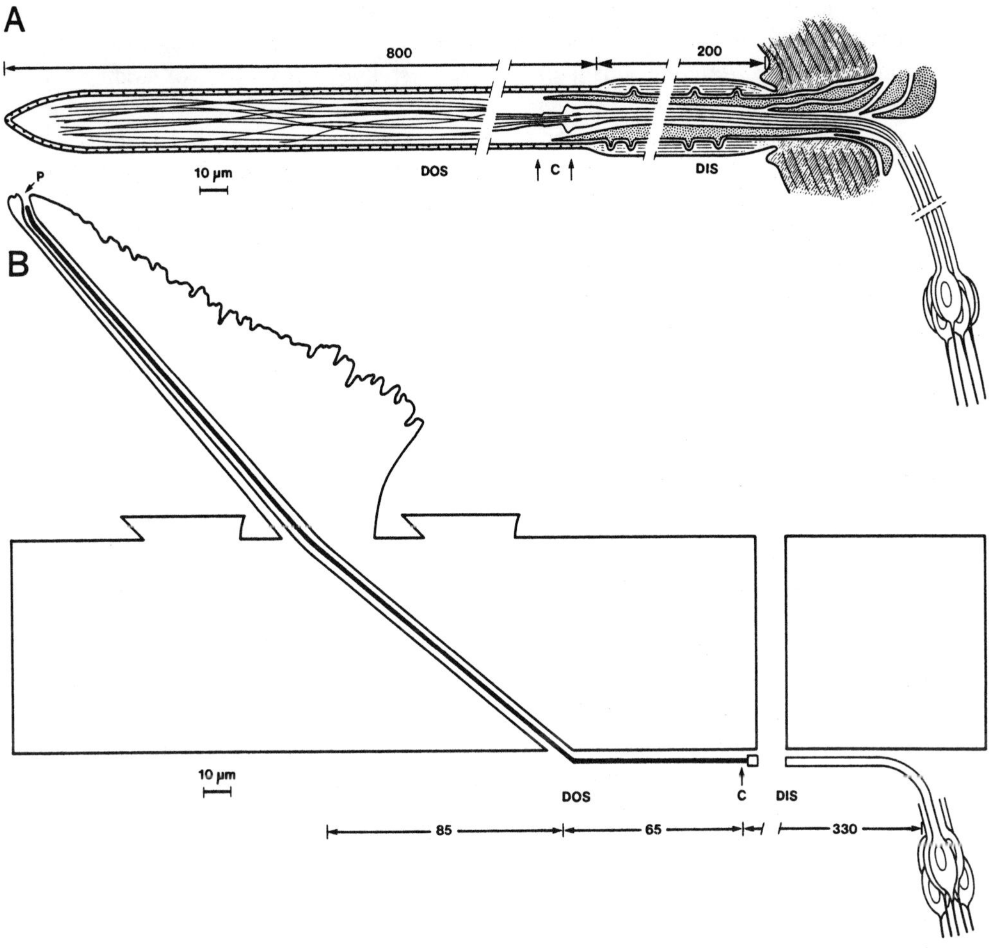


Figure 14.1. Schematic representation of two major chemoreceptive crustacean sensilla and their receptor cells. (A) Aesthetasc sensillum on the antennule of the lobster *Homarus americanus*. Entire sensillum is about 1 mm long and 30 μm in diameter. Somata of some of approximately 400 receptor cells are shown. Their long dendrites innervate the sensillum, coursing through a sheath of supporting cells that extend into proximal thick-walled part of the sensillum. Dendrites (one shown greatly enlarged) sprout cilia, which soon branch and extend the length of the thin-walled distal part of the sensillum. Chemical stimuli are believed to reach dendrites by passing through spongy, thin-walled cuticle. (After Atema 1980.) (B) Fringed seta on the leg of crayfish *Austropotamobius torrentium*. Note again peripheral location of somata and bipolar nature of these modified ciliary cells, with their dendrites extending in one direction into the lumen of the sensillum and their axons beginning their course toward CNS. Chemical stimuli are believed to enter through subapical pore. (After Altner, Hatt, and Altner 1983.) *DOS*, outer (ciliary) segment of sensory dendrite; *DIS*, inner segment of sensory dendrite; *c*, ciliary junction; *p*, pore near the tip of the seta. Distances and scale in μm.

of a particular stimulus is thought to be encoded by a distinct pattern of activity across a population of cells, and not by the activity in any one class of cells. Color vision in vertebrates is perhaps the best known example of coding by across-neuron patterns.

The second problem a receptor organ faces is to determine the intensity of the stimulus. Although not all animals may always need it, an ideal sensory system should maintain high resolving powers over a large range of stimulus intensities. There is a limit to the resolving powers of single cells, however. The critical parameters are the slope of the stimulus-response curve and the response variance (Maes 1984). Some investigators have estimated that single sugar receptor cells of insects may be able to discriminate no more than 3–4 discrete intensity levels (Smith, Bowdan, and Dethier 1983). A whole sensory organ, however, could improve on this situation by employing "range fractionation," in which individual receptor cells restrict the linear portion of their stimulus-response curve to a fraction of the entire dynamic range of natural stimuli. By employing cells with different thresholds, the receptor organ as a whole could cover a wide working range.

A sensitive detecting system also should be able to maintain its resolving power in various stimulus backgrounds (i.e., environmental noise). The sensitivity of some receptor cells can be adjusted through adaptation, either by efferent feedback onto receptor cells or by internal changes in the receptor cells themselves. Examples include the hair cells in auditory and lateral line systems (Flock and Russell 1976; Art 1984), muscle spindles (Hunt and Kuffler 1951), crustacean muscle receptor organs (Kuffler and Eyzaguirre 1955), tactile receptors (Kennedy, Calabrese, and Wine 1974), cone cells (Normann and Werblin 1974; Dowling 1977), and invertebrate photoreceptor cells (Chap. 17). No efferent control of invertebrate or vertebrate chemoreceptor cells is known.

Although clarity of presentation and initial experimental approaches demand a separation of quality and quantity coding, real sensory systems, even at the level of receptor cells, probably code these parameters simultaneously and by using the same neural elements. Ultimately, the demands on the sensory system will be to extract variable and complex stimuli from a variable and complex background and to determine which stimuli must be treated as important signals and which rejected as noise.

## The Determination of Stimulus Quality

Because of the wave properties of visual and auditory stimuli, the stimuli within each of these modalities form a continuous spectrum along a frequency scale. Chemical stimuli, on the other hand, have different and discrete molecular shapes and properties and thus form a discontinuous spectrum. This makes it impossible to ever truly know the tuning characteristics of a chemoreceptor cell, since one cannot test a cell with all the molecules of its natural environment, let alone with all known chemical substances. In addition, molecular mixtures presented to a single receptor cell often cause responses that are unpredictable from the cell's responses to single compounds, almost as if the mixtures represent stimuli as unique as single compounds (see pp 373–378). Nevertheless, each receptor cell does not respond to all chemical substances. On the contrary, when presented with a small series of different molecules, most crustacean chemoreceptor cells will respond to only one or a few (see next section). Therefore, in practice, one can coarsely determine the "tuning" characteristics of a cell by choosing a handful of chemical substances of particular interest. This interest can be that of the animal, by considering what environmental chemicals it responds to behaviorally, or that of the investigator, by probing the cell with structural analogues of known stimuli or compounds stimulatory to other animals.

## Narrow Tuning

Applying such methods of reasoning, chemoreceptor cells of most animals tend to be "broadly tuned," responding to several different compounds with different, but comparable, intensity. "Narrowly tuned" cells have been found in specialized cases, such as insect sex pheromone receptors, in which—one might argue—the demands of species-specific communication have shaped both signal production and reception to extremes of specificity. Pheromone receptor cells of many insects respond best to one of the pheromone components, considerably less to a few chemical analogues, and not at all to other compounds tested (Boeckh 1980).

It was a great surprise when many crustacean chemoreceptor cells were found to be narrowly tuned to one best compound from a test series of substances comprising not a pheromone mixture, but a mixture based on their food. When presented with single compounds, both smell and taste receptor cells showed strong responses to one or another compound, but rarely to more than one. This holds for chemoreceptor cells in the antennules of *Panulirus argus* (Fig. 14.2) (Fuzessery, Carr, and Ache 1978; Derby and Ache 1984a; Derby, Carr, and Ache 1984), the legs of *Homarus americanus* (Fig. 14.3) (Derby and Atema 1982b; Johnson et al. 1984), and the antennules of *H. americanus* (Johnson and Atema 1983). When cells are grouped by their response to a series of compounds, populations of receptors emerge, with each population tuned primarily to one compound (Fig. 14.2). Tuning breadth varies within a population, usually with a predominance of very narrowly tuned cells (Fig. 14.3). Broadly tuned cells exist but are less common.

The crayfish *Austropotamobius torrentium* represents a somewhat different picture. It has three known classes of cells in its walking legs, each class responsive to amino acids, pyridines, or amines (Bauer and Hatt 1980; Bauer, Dudel, and Hatt 1981; Hatt and Bauer 1982). The response profile of the class of amino acid–sensitive cells is shown in Figure 14.4. In each class, the excitatory response profiles are much broader than those of most cells of *P. argus* and *H. americanus*. They are, nonetheless, restricted in their sensitivity, with little overlap between classes. Molecular models, binding affinity, and efficacy have been studied in *A. torrentium* chemoreceptor cells for amino acids and pyridines (Hatt 1984; Hatt and Schmiedel-Jakob 1984, 1985). Hatt and Schmiedel-Jakob propose that

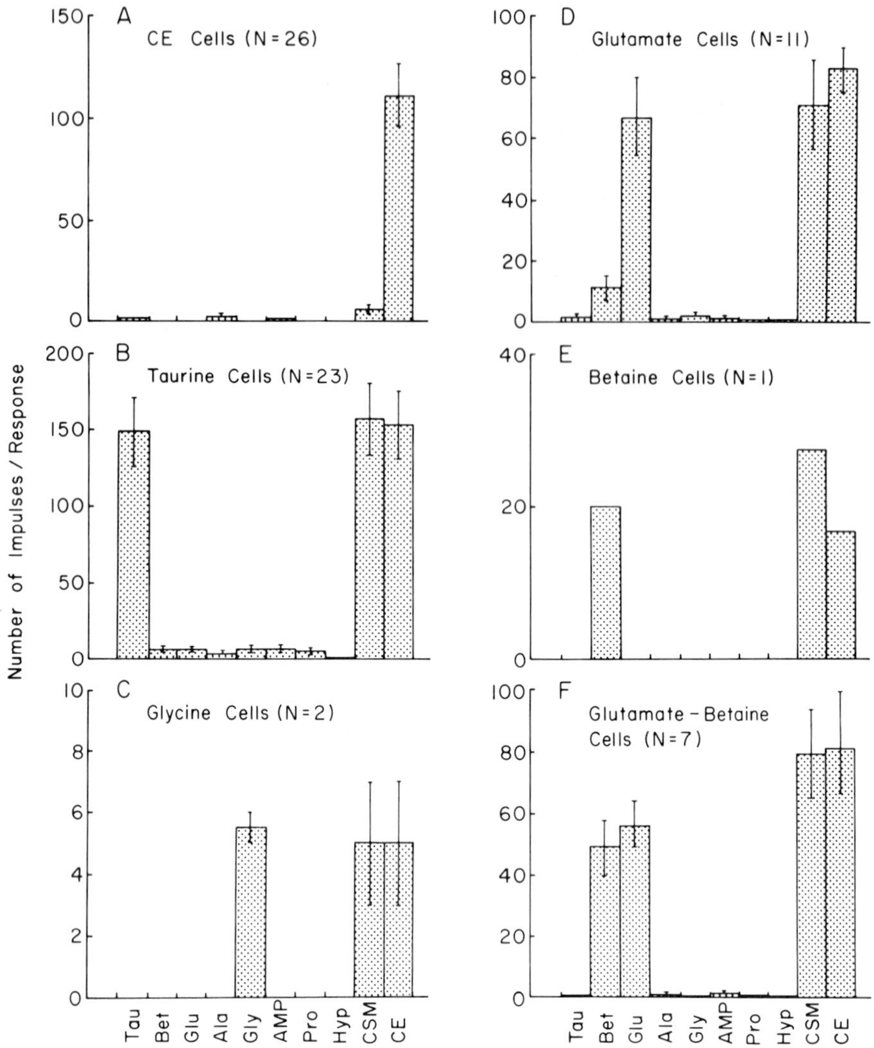


Figure 14.2. Narrow excitatory response profiles for six types of chemoreceptor cells in antennules of spiny lobster, *Panulirus argus*. Responses are to taurine, betaine, L-glutamate, L-alanine, glycine, adenosine-5′-monophosphate, L-proline, hydroxy-L-proline, 23-component artificial crab mixture (CSM), and crab muscle extract (CE). Response values are means ± SEM, with *N* representing sample size for each cell type. Each individual compound was tested at the concentration in which it occurs in the mixture and extract. Cell types were identified by cluster analysis. (From Derby and Ache 1984a.)

competition for one major type of receptor site is typical for these cells. This conclusion fits the result that each cell in *A. torrentium* legs responds predictably to a series of amino acids. Such a predictable series of excitatory chemicals for all cells of a given class does not appear to be the case for chemoreceptor cells in the legs of *H. americanus* (Johnson et al. 1984) and the antennules of *P.*

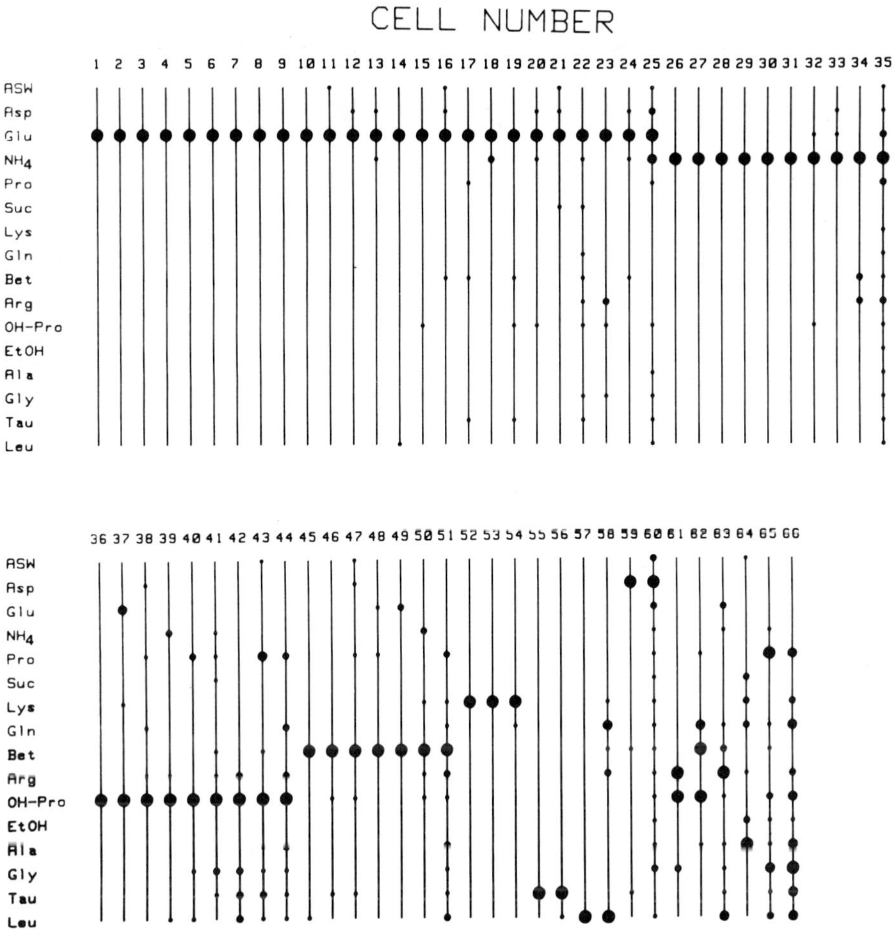


Figure 14.3. Response spectra of 66 chemoreceptor cells from legs of *Homarus americanus*. Responses in each cell are normalized to its response to best stimulus. Continuous line indicates no response; smallest dots indicate less than 30% of maximum response; next larger dots, greater than 30%, but less than 70%; third larger dots, greater than 70%, but less than 100%; and largest dots, 100%. Responses are to artificial seawater, L-aspartate, L-glutamate, ammonium chloride, L-proline, sucrose, L-lysine, L-glutamine, betaine, L-arginine, hydroxy-L-proline, ethanol, L-alanine, glycine, taurine, and L-leucine, each at $3 \times 10^{-4}$ M. Response values are means ± SEM. (From Johnson et al. 1984).

*argus* (Derby and Ache 1984a, unpublished data), in which the second-best compound, third-best compound, etc., vary from cell to cell within a class. This difference in the chemoreceptor cells of *A. torrentium* and the two species of lobsters may reflect different molecular receptor mechanisms, but it is not yet known if these differences result in different mechanisms of resolving chemical mixtures.

The excitatory response spectra of these crustacean receptor cells, especially

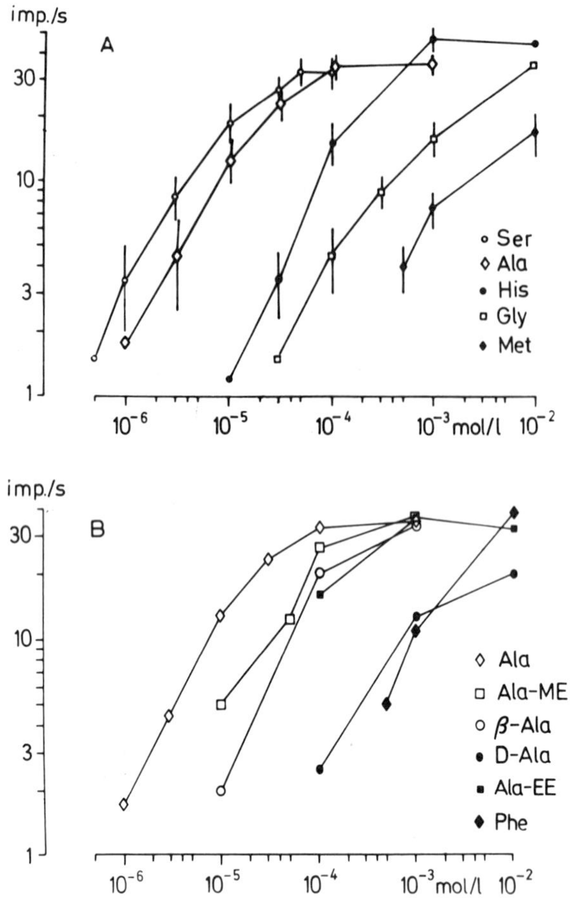


Figure 14.4. Excitatory response profiles of 29 amino acid–sensitive receptor cells in legs of crayfish *Austropotamobius torrentium*. (A) Responses to different amino acids: L-serine, L-alanine, L-histidine, glycine, and L-methionine. Response values are means ± SEM for 29 cells. (B) Responses to L-alanine and its structural analogues: L-alanine methyl ester, β-alanine, D-alanine, L-alanine ethyl ester, L-phenylalanine. Response values are means for same cells as in (A). (From Bauer, Dudel, and Hatt 1981.)

of the lobster species, are much narrower than those of food-sensitive chemoreceptor cells of most other animals. This forces us to consider the function of these narrowly tuned cells. If only the excitatory response spectra of these receptor cells are considered, one might infer that certain individual chemicals or groups of chemicals (those for which narrowly tuned cells with nonoverlapping response spectra exist) could be discriminated by the next neuronal level on the basis of a labelled-line code. But narrow tuning of receptor cells does not necessarily dictate that the receptor cells are treated as labelled lines by subsequent neuronal levels. The information carried by the narrowly tuned cells might be integrated in parallel with other receptor cells, possibly even those more broadly tuned cells, such that the substance is identified by an across-neuron pattern code.

Indeed, Derby, Hamilton, and Ache (1984) and Derby and Ache (1984a) found that higher order neurons of *P. argus* were generally more broadly tuned than the receptor cells (see Chap. 15). Theoretically, the receptor cell populations of both *H. americanus* (Johnson et al. 1984) and *P. argus* (Derby and Ache 1984a; Girardot, Fine, and Derby 1986) could distinguish between some chemicals based on an across-neuron pattern code. Such a code would obviously be most important for differentiating among chemicals for which narrowly tuned cells do not exist.

If either coding scheme seems possible at the receptor level, at least for certain stimuli, then we might ask the question, why do the narrowly tuned cells exist? One likely role of these cells is to pick out key compounds from complex chemical mixtures. A second possible role is to make a special contribution to the across-neuron patterns for that chemical or structurally similar chemicals. In support of this idea is the observation that in the antennular system of *P. argus,* no one class of narrowly tuned cells is sufficient by itself to establish different across-neuron patterns for all compounds tested. However, a single class of narrowly tuned cells can make a significant contribution toward establishing distinct across-neuron patterns (Girardot, Fine, and Derby 1986). As an example, the ability of this antennular chemosensory system to differentiate glutamate from other chemicals is best if receptor cells of all classes are included in the analysis. Discrimination based on information from glutamate-sensitive cells alone is better than that based on information from all other cell groups combined, however.

## Mixture Effects

Animals in their natural habitat are usually barraged simultaneously with many chemicals. To understand how chemoreceptor systems function in natural conditions, we must consider the effects of chemical mixtures on receptor cells. In addition, the effects of chemical noise background must be considered (see p 381).

The responses of many crustacean chemoreceptor cells to a mixture cannot be predicted from responses to the individual components of the mixture. These events are called *mixture interactions*—either suppression or synergism. Mixture suppression has been observed in representatives of most of the populations of receptor cells in antennules of *H. americanus* (Johnson and Atema 1984, 1985; Johnson, Borroni, and Atema 1985) and in some of, but not all, the cell types in the antennules of *P. argus* (Derby and Ache 1984a; Derby, Ache, and Kennel 1985; Gleeson and Ache 1985; Ache, Gleeson, and Thompson 1986). The mixture suppression is not restricted to the periphery, since it is common in central neurons as well (Derby and Ache 1984a, 1984b; Derby, Ache, and Kennel 1985; see also Chap. 15). In some cells, the suppression can be caused by any of several single compounds, as well as by mixtures of compounds (Johnson and Atema 1984, 1985; Gleeson and Ache 1985; Ache, Gleeson, and Thompson 1986). In other cells that are suppressed by a mixture of compounds, any one of the compounds by itself is insufficient to elicit suppression (Johnson and Atema 1984, 1985). Mixture synergism has also been observed in some receptor cells of *H. americanus;* this effect is of less magnitude than suppression and tends to occur only at low stimulus concentrations (Johnson and Atema 1984, 1985; Johnson,

Borroni, and Atema 1985). If mixture synergism occurs in cells of the olfactory pathway of *P. argus,* its effect is much smaller than is that of mixture suppression (Derby and Ache 1984b; Derby, Ache, and Kennel 1985; Gleeson and Ache 1985; Ache, Gleeson, and Thompson 1986). These mixture effects are always fully reversible.

The mixture interactions in lobster chemoreceptor cells are detailed in Figures 14.5 and 14.6. The parallel shifts to the right seen in the stimulus-response functions of Figures 14.5 and 14.6(A) seem to suggest competitive inhibition as hypothesized for the *A. torrentium* leg receptors (Hatt and Schmiedel-Jakob 1985). Many other molecular models could explain the shifts and changes in slope obtained in lobsters, however. Since the mixture interactions recorded thus far are varied and complex, it is not productive to speculate beyond the hypothesis that most of these receptor cells of lobsters probably contain sites for several different compounds, with a predominance of one or another site, to give the cell its observed narrow tuning to a best compound. This does not exclude the possibility of competitive inhibition at any of the sites, making the response of any one cell difficult to predict accurately.

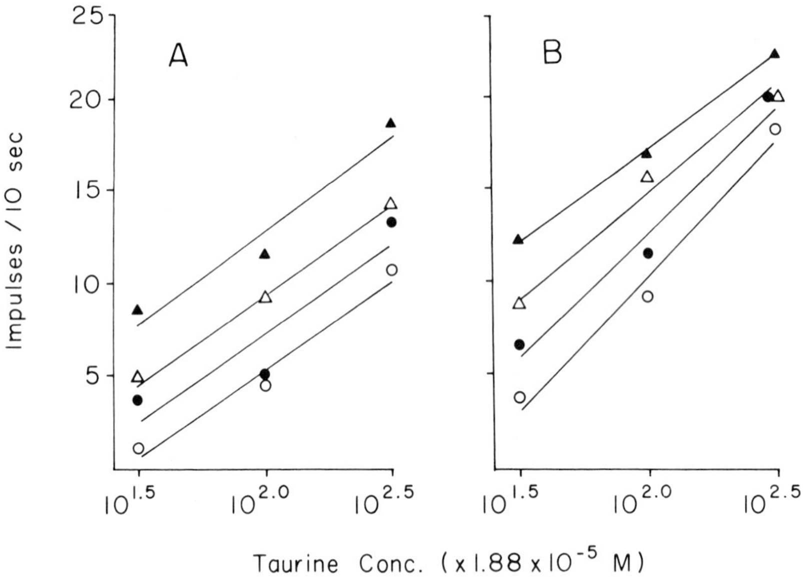


Figure 14.5. Mixture suppression in taurine-sensitive cells in antennules of spiny lobster, *Panulirus argus.* (A) Mean stimulus-response functions for 8 cells stimulated with taurine alone (*closed triangles*) and binary mixtures of taurine plus glycine. Glycine concentrations in binary mixtures were $1.13 \times 10^{-4}$ M (*open triangles*), $3.59 \times 10^{-4}$ M (*closed circles*) and $1.13 \times 10^{-3}$ M (*open circles*). (B) Mean stimulus-response functions for 6 cells stimulated with taurine alone (*closed triangles*) and binary mixtures of taurine plus L-alanine. Alanine concentrations in binary mixtures were $1.13 \times 10^{-4}$ M (*open triangles*), $3.59 \times 10^{-4}$ M (*closed circles*), and $1.13 \times 10^{-3}$ M (*open circles*). (Reprinted with permission from Gleeson RA and Ache BW. *Brain Res.* 335:99–107, 1985.)

It has been suggested that mixture suppression is a form of adaptation, more or less instantaneously setting the sensitivity of the cell to the background concentration of those chemicals which can interact with the cell (Bartoshuk 1975). This seems to be a reasonable function of mixture suppression for more broadly tuned cells. As a hypothetical example, assume there is a broadly tuned cell that is substantially excited by taurine, betaine, glutamate, or AMP, and consider this cell's ability to detect the presence and determine the amount of taurine in a mixture of these four compounds. At certain high concentrations, a mixture of the three components other than taurine will lead to response saturation, even when the response to any one component is not saturated. When this occurs, the addition of taurine to the mixture will elicit no increase in response; but through mixture suppression, in which the compounds other than taurine effectively decrease the sensitivity of the cell to taurine by shifting the working range of the stimulus-response curve for taurine to the right, the cell is more likely to detect the additional effect of taurine when it is added to the mixture.

For narrowly tuned cells, whose function we have suggested as being to detect and identify a single compound in a noisy chemical background, it is hard to imagine such a need for, or function of, mixture suppression. As an example, consider a narrowly tuned cell, excited only or principally by taurine, and consider its ability to detect, as in the previous example, taurine in a mixture of taurine and several other compounds. Since the chemicals other than taurine do not excite this cell, at no concentration would they saturate this cell's response. Thus, suppression of the response to the best chemical by nonexcitatory chemicals would be unnecessary in these narrowly tuned cells.

For receptor organs based on cells with narrow excitatory response spectra, we suggest two functions of mixture suppression. Both are related to improved coding of chemical quality by feature or contrast enhancement. The first is that mixture suppression could make the information about stimulus quality carried by single cells even more distinct. As mentioned previously, some of the narrowly tuned cells of lobsters, while having responses dominated by a single compound, still respond to a limited degree to other compounds (see Fig. 14.3). If mixture suppression decreases the response not only to the best stimulus, but also to these other stimuli, then mixture suppression could actually result in narrower excitatory response spectra. Such an effect is, in many ways, similar to two-tone suppression in audition (Rhode, Geisler, and Kennedy 1978; Sellick and Russell 1979; Ehret, Moffat, and Capranica 1983). Auditory hair cells and fibers in frogs and mammals respond to some range of frequencies, with a maximal response at a specific frequency. When a tone of best frequency is presented simultaneously with a second tone (often at a frequency just above or below the best) or with noise, there is a reduction in the response to the best frequency and, in some cases, in the response to nearby frequencies as well. The behavioral function of two-tone suppression is still uncertain, but in some cases, it has the effect, as does mixture suppression, of narrowing the tuning of the cells, thus improving the contrast between stimulus and background.

A second and related function of mixture suppression in narrowly tuned cells may be to create greater contrast between across-neuron patterns generated by

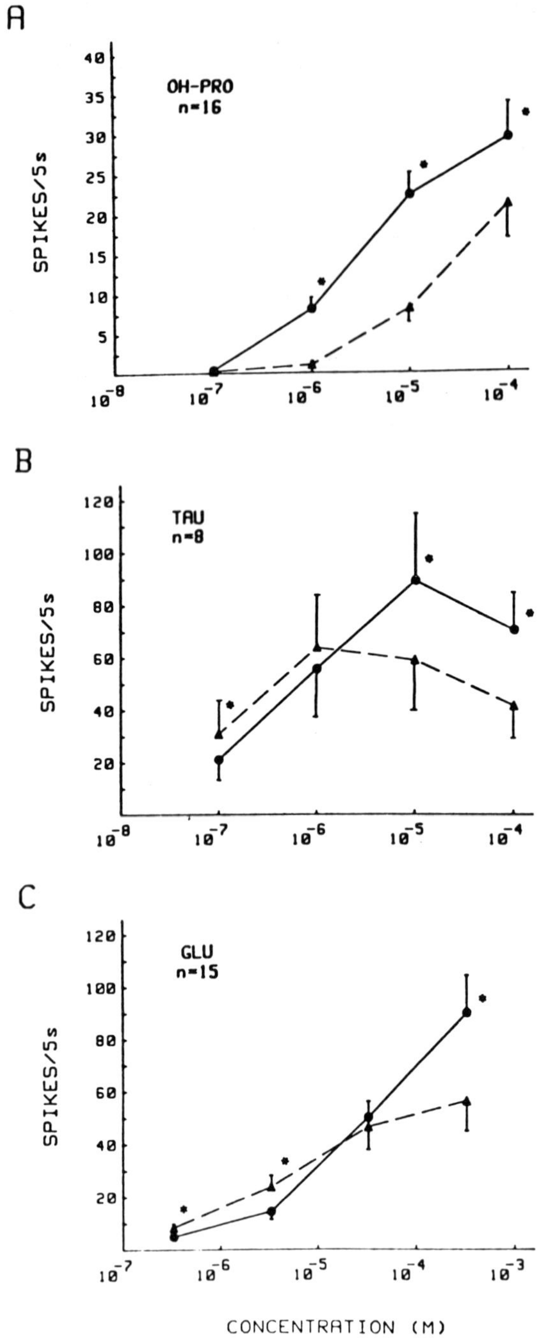
A
OH-PRO
n=16
SPIKES/5s
B
TAU
n=8
SPIKES/5s
C
GLU
n=15
SPIKES/5s
CONCENTRATION (M)

different mixture or single compound stimuli. As an example, consider the system presented in Figure 14.7, which is a greatly simplified representation of the antennular system of *P. argus*. Shown are the responses of six cell types: (A) represents betaine-sensitive cells, (B) a class of taurine-sensitive cells that give high-intensity responses to taurine and do not show mixture suppression, (C) a class of taurine-sensitive cells that give low-intensity responses to taurine and do show mixture suppression, (D) glutamate-sensitive cells, (E) AMP-sensitive cells, and (F) glycine-sensitive cells. This figure shows the across-neuron patterns for four stimulus conditions: taurine, glycine, taurine plus glycine without mixture suppression, and taurine plus glycine with mixture suppression. Consider the ability of these populations of cells to distinguish the mixture of taurine plus glycine from a component of the mixture, taurine. For this model we assume that, with regard to any two across-neuron patterns, the more elements (cells) whose responses differ, the more distinct will be the patterns. This figure shows that the addition of glycine to taurine without suppression causes a change in the response of only one cell (*cell F*); but the addition of glycine with suppression causes a change in the response of two cells (*cells C and F*). Therefore the pattern for the mixture of taurine plus glycine becomes more distinct from the pattern for taurine alone as a result of suppression. The implication of this analysis is that mixture suppression results in mixtures taking on qualities of their own, distinct from those of the components. This suggests that the quality of chemical mixtures might be coded by a "synthetic," rather than an "analytic," method (Erickson 1978). In *synthetic processing*, when stimuli of different qualities are mixed, a new quality, distinct from that of the component stimuli, is formed. An example of synthetic processing is color vision: when two wavelengths representing two different colors are mixed, the colors fuse, resulting in a single color from which the component colors cannot be detected. *Analytical processing* occurs if stimuli of different qualities when mixed do not fuse, but rather remain separate. An example of analytical processing is audition, in which the two tones in a chord can be distinguished as distinct.

In behavioral studies, complex chemical mixtures are usually found to be more stimulatory than are their individual components. Recent examples of this come from the grass shrimp *Palaemonetes pugio* (Carr and Derby 1986) and *Homarus americanus* (Borroni, Handrich, and Atema 1986). In fact, mixture synergism has been demonstrated behaviorally in the case of *P. pugio*. Exceptions to this generalization do exist. For the common gulf-weed shrimp *Leander tenui-*

Figure 14.6. Mixture suppression in chemoreceptor cells in American lobster, *Homarus americanus*. Stimulus-response functions for populations of single cells to single compounds (*circles*) and 15-component mixture (*triangles*) that contains these three plus other compounds. (A) Responses of 16 antennular cells to hydroxy-L-proline and to the mixture. (B) Responses of eight antennular cells to taurine and to the mixture. (C) Responses of 15 walking leg cells to L-glutamate and to the mixture. Response values are means ± SEM of spikes per 5 sec. Asterisk indicates that responses to the single compound and the mixture are significantly different. (From Johnson, Borroni, and Atema 1985.)

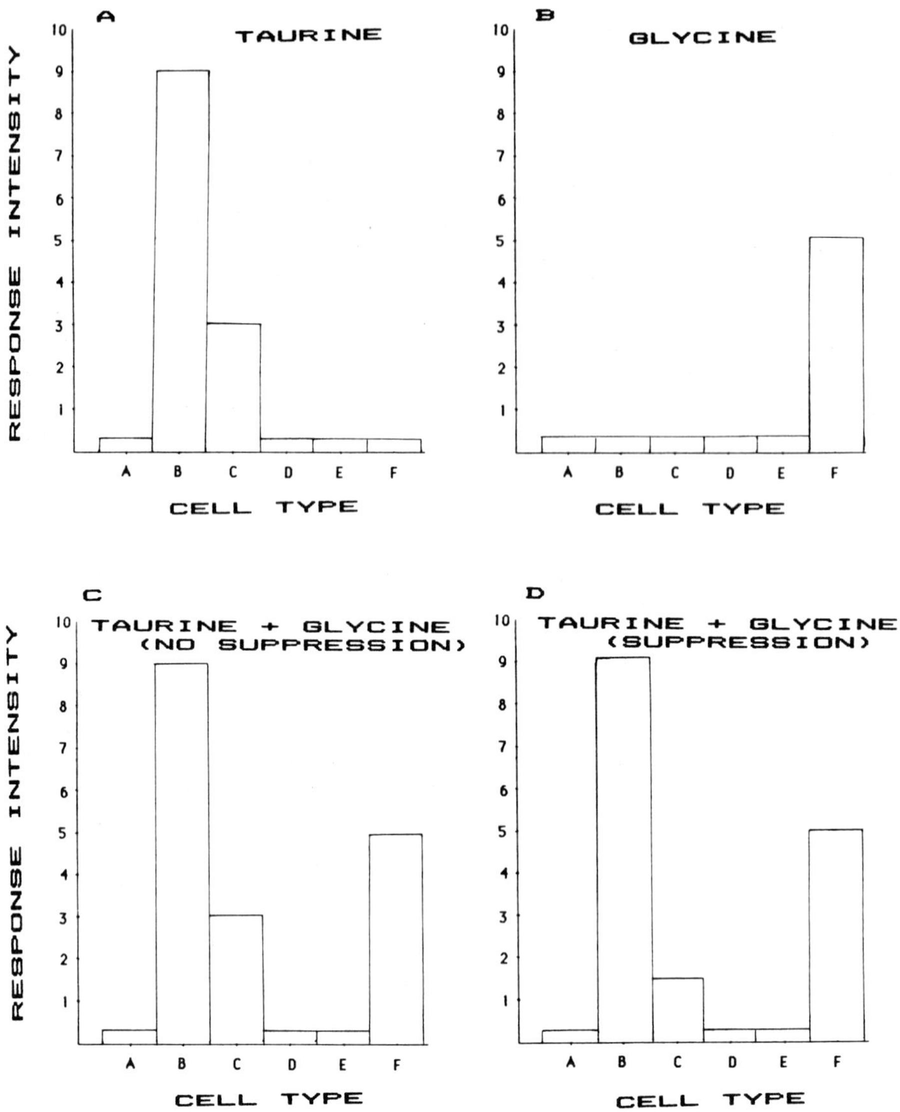


Figure 14.7. Effect of mixture suppression on quality coding. Represented is response profile of simplified version of antennular chemosensory system of spiny lobster, *Panulirus argus*. Shown are responses of 6 types of cells (*A* through *F*) to 4 stimulus conditions. This figure shows that suppression results in across-neuron pattern for the mixture becoming more distinct from patterns for any of mixture's components. See the text for explanation.

*cornis,* a single component (taurine) of a mixture of amino acids is as stimulatory as the mixture itself (Johnson and Atema 1986), and for the spiny lobster, *Panulirus interruptus,* a single component (glycine) can be more effective than a mixture containing it (Zimmer-Faust et al. 1984).

## The Determination of Stimulus Quantity

### Behavioral Thresholds

Chemosensory systems of aquatic crustaceans can operate over an extremely wide range of chemical concentrations, as has been shown in behavioral studies. One example of acute sensitivity is the ability of blue crabs *Callinectes sapidus* to detect food-related chemicals. By measuring the flicking of antennules, Pearson and Olla (1977) found that these crabs, as a population, can detect concentrations of freeze-dried clam extract as low as $10^{-15}$ g/L (containing an estimated concentration of individual amino acids such as taurine on the order of $10^{-13}$ M) and have a dynamic working range of at least 10 orders of magnitude. Similar examples of high sensitivity in different aquatic species suggest the importance of chemoreception to aquatic animals in general.

### Physiological Thresholds

To understand how these animals maintain such impressive behavioral thresholds and working ranges, one must investigate first the peripheral chemosensory system. One of the best examples is found in the responses of a specific population of taurine-sensitive receptor cells in the antennules of *P. argus*. As a population, these cells showed statistically significant responses to taurine at concentrations as low as $10^{-13}$ M, with thresholds probably well below that level (Thompson and Ache 1980). Under these conditions, the working range of this population of cells was at least 6 orders of magnitude. Other types of chemoreceptor cells of lobsters have population thresholds in the micromolar to nanomolar concentration range and working ranges of 3 to 6 or more orders of magnitude (Derby and Atema 1982a; Derby, Carr, and Ache 1984; Johnson, Borroni, and Atema 1985). Pyridine receptor cells of *A. torrentium* seem to fall into two categories, one with a limited working range (2 orders of magnitude) and one with an extensive working range (7 orders of magnitude) (Hatt and Bauer 1982).

### Range Fractionation

Range fractionation (see p 368) has been demonstrated for taurine-sensitive cells (Thompson and Ache 1980) and AMP-sensitive cells (Derby unpublished data) in the antennules of *P. argus* and glutamate-sensitive cells in the legs of *H. americanus* (Fig. 14.8) (Derby, unpublished data). The result of range fractionation is that the slopes of the stimulus-response curves for many of the individual cells are steeper than the average slope for the population (Fig. 14.8[A-1] and [B-1]). This arrangement has the advantage over one in which all the cells of a population have broad working ranges by allowing better dynamic signal resolution, since a steeper slope results in a greater change in response for any given

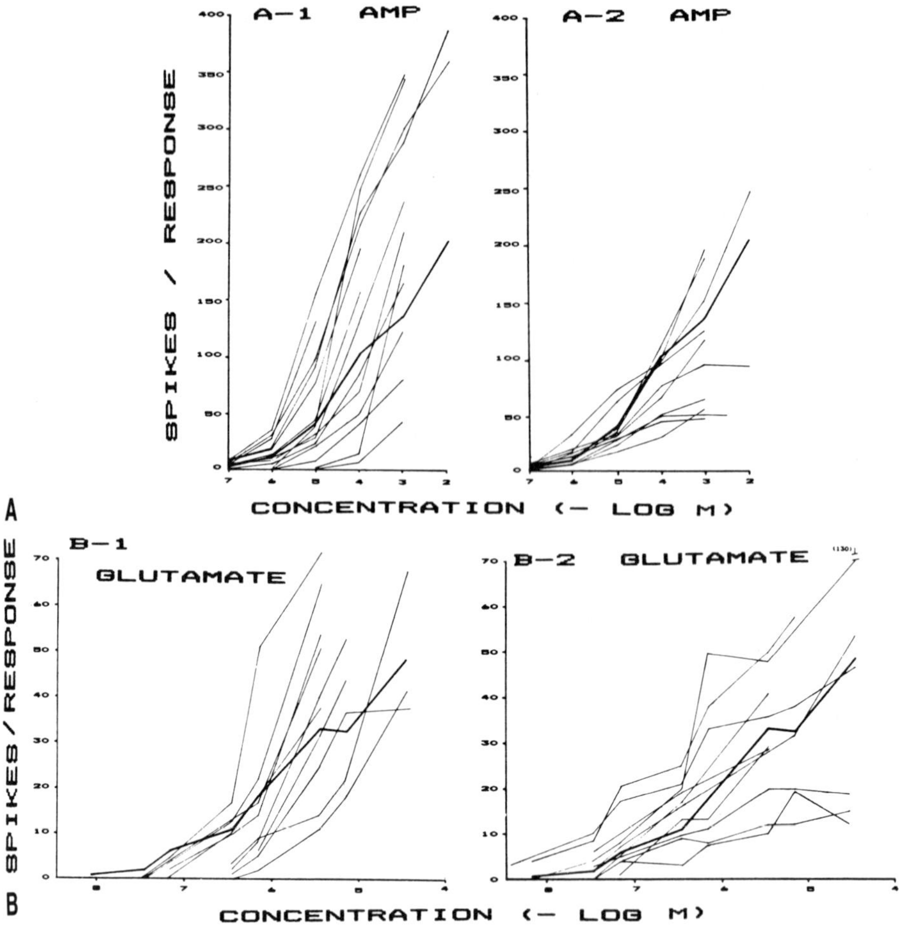


Figure 14.8. Range fractionation in populations of crustacean chemoreceptor cells. (A) Responses of AMP-sensitive cells in antennules of *Panulirus argus*. (B) Responses of glutamate-sensitive cells in walking legs of *Homarus americanus*. *Light lines* represent responses of individual cells: (A) $n = 24$; (B) $n = 20$. *Dark lines* represent average responses of all 24 AMP cells (A) and all 20 glutamate cells (B). In both (A) and (B), cells exist (*A-1* and *B-1*) that divide the range covered by the population into smaller ranges represented by steeper slopes. Cells whose stimulus-response functions more closely follow that of the average of the population also exist for both populations (*A-2* and *B-2*). Concentrations are expressed in molarity. (Derby, unpublished data.)

change in stimulus intensity. The disadvantage of range fractionation is that there is a potential for poorer sensitivity, since only a fraction of the neurons of a population are operating in the dynamic portion of their working range at any one time. This may be less of a problem, however, in the lobster's chemosensory systems, in which single sensory appendages, such as antennules and legs, can have tens to hundreds of thousands of chemoreceptor cells (see Ache 1982).

Mixture effects that result in a flattening of the stimulus-response function (e.g., taurine cells of Fig. 14.6[B]) could greatly decrease the ability of a cell to code for differences in stimulus intensity.

### Detection in Chemical Noise

In addition to being able to resolve differences in stimulus intensity over a large range of intensities, an effective chemosensory system should have the ability to operate effectively when the quality or quantity of stimuli in the background fluctuates widely—that is, in the presence of chemical noise (Atema 1985). Chemoreceptor cells of lobsters have working ranges that are not fixed but, rather, vary depending on the chemical background. For example, in ammonium-sensitive cells in the legs of *H. americanus,* self-adaptation with ammonium causes a parallel shift to the right in the stimulus-response curve for ammonium (Borroni and Atema 1985). The dynamic working range of the stimulus-response function does not change in extent, only in location. The advantage of adjusting the working range of these receptor cells to match the intensity of the background chemical noise is that more receptor cells are optimally responsive at any background level and thus should maintain a high resolving power under different environmental conditions. Further studies on the dynamic nature of such adaptation (Voigt and Atema 1985) lead to two additional conclusions: (1) adaptation occurs fairly quickly, suggesting that in the animal's natural environment, where chemical noise abounds, most of these chemoreceptor cells are operating under some degree of adaptation; and (2) the time and magnitude of self-adaptation depend more on the signal-to-noise ratio (i.e., the ratio of concentrations of stimulus and background) than on the absolute concentrations of the signal or the noise. This is in addition to the generally suppressing effects of mixtures described earlier (see pp 373–378).

## Interspecific Differences in Sensitivity

One would expect that environmental constraints and behavioral requirements would have led to substantial differences in spectral sensitivity between species. Behavioral and physiological studies of marine crustaceans, especially lobsters and shrimp, indicate that amino acids, nucleotides, and quaternary ammonium compounds are usually the most stimulatory components of food extracts (Derby and Ache 1984b; Carr and Derby 1986; see also Chap. 1). In contrast, behavioral data on freshwater crayfish (Ashby and Larimer 1965) and semiterrestrial crabs (Robertson, Fudge, and Vermeer 1981; Trott and Robertson 1984) show that sugars are among the most stimulatory chemicals and are usually more effective than amino acids. Some interspecific differences in sensitivity may be related to feeding habits and types of food in their diets. For example, crayfish and semiterrestrial crabs are more herbivorous than are lobsters and shrimp, which might account for the differences in chemosensitivities. Supporting this idea is the observation that a marine herbivorous crab, *Pugettia producta,* shows a greater response to some sugars than to many (but not all) amino acids (Zimmer-Faust, Cook, and Case 1979).

Nonetheless, some of the presumed interspecific differences in chemosensitivity may simply be due to lack of data. For example, the sensitivity of crayfish to pyridines is supposed to aid in detecting plant materials (Hatt and Bauer 1982), which crayfish readily eat. Lobsters, which are more carnivorous, might not be expected to have pyridine receptors; but when searched for, cells sensitive to pyridines were found in *P. argus* (Hatt and Ache unpublished data). Similarly, both the smell and taste receptor cells of *H. americanus* responded to secondary compounds of terrestrial plants (a class of compounds often used as feeding deterrents), though no significant behavioral reactions to these compounds could be observed and lobsters are not known to use plants as an important food source (Derby, Reilly, and Atema 1984).

Despite the lack of data, we conclude that there are some differences in chemical sensitivity among crustaceans, and that many of these animals may have physiological sensitivity that is not expressed behaviorally. The latter would represent "potential" for behavioral sensitivity.

## Summary

The chemoreceptor cells of aquatic crustaceans clearly are not simply passive filters, always allowing some information through their sensory window, while always excluding others. The ability of these cells to sample their environment spatially and temporally is improved by movements of their sensory appendages. The response properties of these cells range from very narrowly to more broadly tuned and from being unaffected by the presence of other compounds to being suppressed by other compounds. Nonetheless, a major focus on one best compound is the rule for each cell. Thresholds vary from cell to cell within a tuned population, as well as between populations. Population differences may be related to the natural background levels of the compounds in question. Range fractionation may improve the ability of crustacean chemosensory systems to resolve changes in stimulus concentration. This would be particularly useful in these systems, wherein mixture effects can limit or distort quantity-resolving capabilities of single cells by changing their stimulus-response functions to slopes near 0 or to complex, nonlinear shapes. Natural stimulus mixtures would be represented in the receptor cells by patterns of firing across "mixture-suppressed" or "enhanced" cells, with each cell contributing information primarily or exclusively about its best stimulus compound in the mixture. At higher stimulus concentrations, more cells will be recruited. Both at higher stimulus levels and at higher background levels, cells will self-adapt and cross-adapt, allowing room for detection and resolution of yet stronger stimuli and of different stimulus qualities. Our current state of knowledge indicates that at the level of receptor cells, more effort is invested in quality discrimination than in resolution of quantity differences. Various mechanisms discussed in this chapter point to quality (spectral) contrast enhancement at the cost of quantity resolution. This may well reflect the realities of the distribution of chemical signals in natural environments, where information of subtle intensity may be of limited use.

## References


Ache, B.W. (1982) Chemoreception and thermoreception, in The Biology of Crustacea, vol. 3, Neurobiology: Structure and Function, Atwood, H.L. and Sandeman, D.C. (eds.), Academic Press, New York, pp. 369–398.

Ache, B.W. and Derby, C.D. (1985) Functional organization of olfaction in crustaceans, Trends Neurosci., 8:356–360.

Ache, B.W., Gleeson, R.A., and Thompson, H.D. (1986) Mechanism of interaction between odorants at olfactory receptor cells, Chem. Sens. 11:575.

Altner, I., Hatt, H., and Altner, H. (1983) Structural properties of bimodal chemo- and mechanosensitive setae on the pereiopod chelae of the crayfish, *Austropotamobius torrentium,* Cell Tissue Res., 228:357–374.

Art, J.J. (1984) Auditory function at the receptor level in reptiles, in Comparative Physiology of Sensory Systems, Bolis, L., Keynes, R.D., and Maddrell, S.H.P. (eds.), Cambridge University Press, Cambridge, U.K., pp. 47–66.

Ashby, E.A. and Larimer, J.L. (1965) Modification of cardiac and respiratory rhythms in crayfish following carbohydrate chemoreception, J. Cell. Comp. Physiol., 65:373–380.

Atema, J. (1980) Smelling and tasting underwater, Oceanus, 23:4–18.

Atema, J. (1985) Chemoreception in the sea: adaptations of chemoreceptors and behaviour to aquatic stimulus conditions, Symp. Soc. Exp. Biol., 39:387–423.

Bartoshuk, L.M. (1975) Taste mixtures: is mixture suppression related to compression? Physiol. & Behav., 14:643–649.

Bauer, U., Dudel, J., and Hatt, H. (1981) Characteristics of single chemoreceptive units sensitive to amino acids and related substances in the crayfish leg, J. Comp. Physiol., 144:67–74.

Bauer, U. and Hatt, H. (1980) Demonstration of three different types of chemosensitive units in the crayfish claw using computerized evaluation, Neurosci. Lett., 17:209–214.

Boeckh, J. (1980) Neural basis of coding of chemosensory quality at the receptor cell level, in Olfaction and Taste, vol. 7, Van der Starre, H. (ed.), IRL Press, London, pp. 113–122.

Borroni, P.F. and Atema, J. (1985) Adaptation of primary chemoreceptor cells: parallel shift of dose-response functions and reduced variance, Soc. Neurosci. Abst., 11:1219.

Borroni, P.F., Handrich, L.S., and Atema, J. (1986) The role of narrowly tuned taste cell populations in lobster (*Homarus americanus*) feeding behavior, Behav. Neurosci., 100:206–212.

Carr, W.E.S. and Derby, C.D. (1986) Chemically stimulated feeding behavior in marine animals: the importance of chemical mixtures and the involvement of mixture interactions, J. Chem. Ecol., 12:989–1011.

Derby, C.D. and Ache, B.W. (1984a) Quality coding of a complex odorant in an invertebrate, J. Neurophysiol., 51:906–924.

Derby, C.D. and Ache, B.W. (1984b) Electrophysiological identification of the stimulatory and interactive components of a complex odorant, Chem. Sens., 9:201–218.

Derby, C.D., Ache, B.W., and Kennel, E.W. (1985) Mixture suppression in olfaction: electrophysiological evaluation of the contribution of peripheral and central neural components, Chem. Sens. 10:301–316.

Derby, C.D. and Atema, J. (1982a) Chemosensitivity of walking legs of the lobster *Homarus americanus:* neurophysiological response spectrum and thresholds, J. Exp. Biol., 98:303–315.

Derby, C.D. and Atema, J. (1982b) Narrow-spectrum chemoreceptor cells in the walking

legs of the lobster *Homarus americanus:* taste specialists, J. Comp. Physiol., 146:181–189.
Derby, C.D., Carr, W.E.S., and Ache, B.W. (1984) Purinergic olfactory cells of crustaceans: response characteristics and similarities to internal purinergic cells of vertebrates, J. Comp. Physiol., 155:341–349.
Derby, C.D., Hamilton, K.A., and Ache, B.W. (1984) Processing of olfactory information at three neuronal levels in the spiny lobster, Brain Res., 300:311–319.
Derby, C.D., Reilly, P.M., and Atema, J. (1984) Chemosensitivity of the lobster *Homarus americanus* to secondary plant compounds: unused receptor capabilities, J. Chem. Ecol., 10:879–892.
Devine, D.V. and Atema, J. (1982) Function of chemoreceptor organs in spatial orientation of the lobster, *Homarus americanus:* differences and overlap, Biol. Bull., 163:144–153.
Dowling, J.E. (1977) Receptoral and network mechanisms of visual adaptation, Neurosci. Res. Program Bull., 15:1–12.
Ehret, G. and Capranica, R.R. (1980) Masking patterns and filter characteristics of auditory nerve fibers in the green treefrog (*Hyla cinerea*), J. Comp. Physiol., 141:1–12.
Ehret, G., Moffat, A.J.M., and Capranica, R.R. (1983) Two-tone suppression in auditory nerve fibers of the green treefrog (*Hyla cinerea*), J. Acoust. Soc. Am., 73:2093–2095.
Erickson, R.P. (1978) Common properties of sensory systems, in Handbook of Behavioral Neurobiology, vol. 1, Sensory Integration, Masterton, R.B. (ed.), Plenum, New York, pp. 73–90.
Flock, A. and Russell, I.J. (1976) Inhibition by efferent nerve fibres: action on hair cell and afferent synaptic transmission in the lateral line canal organ of the burbot *Lota lota,* J. Physiol. (Lond.), 257:45–62.
Fuzessery, Z.M., Carr, W.E.S., and Ache, B.W. (1978) Antennular chemosensitivity in the spiny lobster *Panulirus argus:* studies of taurine sensitive receptors, Biol. Bull., 154:226–240.
Girardot, M.-N., Fine, J.B., and Derby, C.D. (1986) Coding of odorant quality by the olfactory system of the lobster: behavioral and neural analysis of discrimination of quality of single chemicals and chemical mixtures, Soc. Neurosci. Abst., 12:1352.
Gleeson, R.A. and Ache, B.W. (1985) Amino-acid suppression of taurine-sensitive chemosensory neurons, Brain Res., 335:99–107.
Hatt, H. (1984) Structural requirements of amino acids and related compounds for stimulation of receptors in crayfish walking legs, J. Comp. Physiol., 155:219–231.
Hatt, H. and Bauer, U. (1982) Electrophysiological properties of pyridine receptors in the crayfish walking leg, J. Comp. Physiol., 148:221–224.
Hatt, H. and Schmiedel-Jakob, I. (1984) Electrophysiological studies of pyridine-sensitive units on the crayfish walking leg. I. Characteristics of stimulatory molecules, J. Comp. Physiol., 154:855–863.
Hatt, H. and Schmiedel-Jakob, I. (1985) Specific antagonists at the pyridine receptor: evidence from electrophysiological studies with acetylpyridines, Chem. Sens., 10:325–332.
Hunt, C.C. and Kuffler, S.W. (1951) Further study of efferent small-nerve fibers to mammalian muscle spindles. Multiple spindle innervation and activity contraction, J. Physiol., 108:315–339.
Johnson, B.R. and Atema, J. (1983) Narrow-spectrum chemoreceptor cells in the antennules of the American lobster, Neurosci. Lett., 41:145–150.
Johnson, B.R. and Atema, J. (1984) Mixture suppression in primary olfactory receptor cells in the lobster, Soc. Neurosci. Abst., 10:861.

Johnson, B.R. and Atema, J. (1985) Mixture effects in primary olfactory cells in the lobster, Chem. Sens., 10:395.

Johnson, B.R. and Atema, J. (1986) Chemical stimulants for a component of feeding behavior in the common gulf-weed shrimp, *Leander tenuicornis,* Biol. Bull., 170:1–10.

Johnson, B.R., Borroni, P.F., and Atema, J. (1985) Mixture effects in primary olfactory and gustatory receptor cells from the lobster, Chem. Sens., 10:367–373.

Johnson, B.R., Voigt, R., Borroni, P.F., and Atema, J. (1984) Response properties of lobster chemoreceptors: tuning of primary taste neurons in walking legs, J. Comp. Physiol., 155:593–604.

Kaissling, K.-E. (1979) Recognition of pheromones by moths, especially in saturnids and *Bombyx mori,* in Chemical Ecology: Odour Communication in Animals, Ritter, F.J. (ed.), Elsevier, Amsterdam, pp. 43–56.

Kennedy, D., Calabrese, R.L., and Wine, J.J. (1974) Presynaptic inhibition: primary afferent depolarization in crayfish neurons, Science, 186:451–454.

Kuffler, S.W. and Eyzaguirre, C. (1955) Synaptic inhibition in an isolated nerve cell, J. Gen. Physiol., 39:155–189.

Maes, F.W. (1984) A neural coding model for sensory intensity discrimination, to be applied to gustation, J. Comp. Physiol., 155:263–270.

Normann, R.A. and Werblin, F.S. (1974) Control of retinal sensitivity. I. Light and dark adaptation of vertebrate rods and cones, J. Gen. Physiol., 63:37–61.

Pearson, W.H. and Olla, B.L. (1977) Chemoreception in the blue crab, *Callinectes sapidus,* Biol. Bull., 153:346–354.

Reeder, P.B. and Ache, B.W. (1980) Chemotaxis in the Florida spiny lobster, *Panulirus argus,* Anim. Behav., 28:831–839.

Rhode, W.S., Geisler, C.D., and Kennedy, D.T. (1978) Auditory nerve fiber responses to wide-band noise and tone combinations, J. Neurophysiol., 41:692–704.

Robertson, J.R., Fudge, J.A., and Vermeer, G.K. (1981) Chemical and live feeding stimulants of the sand fiddler crab, *Uca pugilator* (Bosc), J. Exp. Mar. Biol. Ecol., 53:47–64.

Schmitt, B.C. and Ache, B.W. (1979) Olfaction: responses of a decapod crustacean are enhanced by flicking, Science, 205:204–206.

Sellick, P.M. and Russell, I.J. (1979) Two-tone suppression in cochlear hair cells, Hear. Res., 1:227–236.

Smith, D.V., Bowdan, E., and Dethier, V.G. (1983) Information transmission in tarsal sugar receptors of the blowfly, Chem. Sens., 9:81–101.

Thompson, H. and Ache, B.W. (1980) Threshold determination for olfactory receptors of the spiny lobster, Mar. Behav. Physiol., 7:249–260.

Trott, T.J. and Robertson, J.R. (1984) Chemical stimulants of cheliped flexion behavior by the Western Atlantic ghost crab *Ocypode quadrata* (Fabricius), J. Exp. Mar. Biol. Ecol., 78:237–252.

Voigt, R. and Atema, J. (1985) Self-adaptation of receptor cells in the lobster: effects of elevated backgrounds on responses to repeated stimulation, Chem. Sens., 10:395–396.

Zimmer-Faust, R.K., Cook, D.P., and Case, J.F. (1979) Chemosensory induced bradycardia in the kelp crab, *Pugettia producta* (Randall), J. Exp. Mar. Biol. Ecol., 38:135–150.

Zimmer-Faust, R.K., Tyre, J.E., Michel, W.C., and Case, J.F. (1984) Chemical mediation of appetitive feeding in a marine decapod crustacean: the importance of suppression and synergism, Biol. Bull., 167:339–353.

# Chapter 15

# Integration of Chemosensory Information in Aquatic Invertebrates

BARRY W. ACHE *

Previous chapters in this volume have considered the molecular nature and distribution of chemical signals in aquatic environments (Chap. 1), the behavior of aquatic invertebrates associated with stimulus acquisition (Chap. 2), and the structure (Chap. 11) and functional organization of chemoreceptors in aquatic invertebrates (Chap. 14). The present chapter continues this logical progression from stimulus to response by considering the fate of the information the chemoreceptor cells of aquatic invertebrates send to the CNS. This is a bittersweet topic, in that our understanding of chemosensory processing in invertebrates is just beginning, even though numerous species of invertebrates have served as important animal models in neurobiology and their motor organization is well understood. In fact, the best-characterized motor networks typically underlie behaviors such as feeding that are naturally triggered by chemical stimuli. Understanding chemosensory processing in these organisms would be a major step towards deciphering the "black box" in which sensory input is coupled to motor output, and ultimately to behavior.

The chemical senses of the leech, *Aplysia,* and other aquatic invertebrates of neurobiological interest are beginning to be studied in more detail (Croll 1983; Elliott 1984), but much of our present knowledge of chemosensory processing in invertebrates comes from studies of the olfactory pathway of arthropods. The arthropods are well represented in aquatic environments by the crustaceans, which therefore will be the focus of this chapter. Much of what we know about how arthropods process chemosensory information, however, comes from studies with insects. There is no a priori reason to expect sensory processing to be markedly different in these two groups of arthropods; so some reference to insects is made where it enhances our understanding of the general arthropod situation. Any differences that ultimately do emerge between chemosensory processing in crustaceans and insects are of interest, however, since they may reflect specific adaptations to chemical signal processing in aquatic environments.

---

* C. V. Whitney Laboratory, University of Florida, St. Augustine, FL 32086, USA

Both aquatic and terrestrial arthropods are faced with the problem of deciphering the four basic parameters of their odor environment—quality, quantity, space, and time—if they are going to exploit the full informational capacity of the modality. After briefly describing the neuroanatomy of the arthropod olfactory pathway, this chapter will consider what is known about how arthropods process chemosensory information relative to each of these four basic stimulus parameters.

## Structural Organization

A striking feature of the olfactory brain of arthropods is the anatomical similarity of its gross organizational plan to that of the vertebrate olfactory pathway (Fig. 15.1). This feature was appreciated in the emerging years of "simple systems" neurobiology by Maynard (1967), who pioneered using the arthropod brain, in particular that of the spiny lobster, *Panulirus argus,* as a model to study neural processing in complex neuropil.

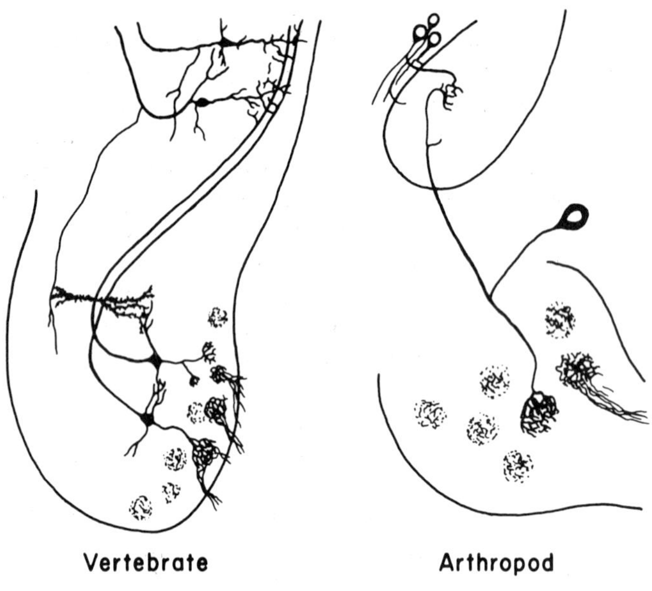


Figure 15.1. Simplified, schematic representation of central olfactory connections in a vertebrate and an arthropod, used by Maynard (1967) to illustrate their similar organizational plans. Axons of many thousands of primary olfactory receptor cells enter from the right in each case and converge upon and ramify in glomerularly organized neuropil. In the vertebrate, mitral cells, first-order interneurons, receive this input and project to pyriform cortex and other regions of the brain; in the arthropod, similar interneurons project to corpora pedunculata or analogous structures. Anatomical similarities in terms of specific dendritical distributions become more obscure. (Reprinted with permission from Maynard [1967], organization of central ganglia, in *Invertebrate Nervous Systems,* Wiersma (ed.), University of Chicago Press.)

Axons from several hundred thousand primary bipolar receptor cells, an estimated 400,000 in the spiny lobster, *Panulirus argus* (Laverack and Ardill 1965), project from the olfactory organ, a tuft of aesthetasc sensilla borne on the lateral filament of the antennule, to the brain via the antennular nerve. No morphological or physiological evidence to date refutes the idea that these receptors project as independent elements. A similar situation characterizes the olfactory periphery in insects, where primary bipolar receptors innervate several types of sensilla borne on the antenna—the homologue of the crustacean antennule—and project to the brain via the antennal nerve (Altner and Prillinger 1980).

In crustaceans and most insects, axons from the peripheral receptors project to the ipsilateral deutocerebrum, specifically to the paired antennal lobes in insects and the paired olfactory lobes in crustaceans. In some dipterans, they also cross over to the contralateral side via the supraesophageal commissure (Boeckh et al. 1984), but this appears to be atypical of the general arthropod situation. In the deutocerebrum, the primary afferents converge onto many fewer first-order interneurons. One class of deutocerebral interneurons, the extrinsic, or projection, neurons, sends a major process external to the lobe (Fig. 15.2[A]) (Matsumoto and Hildebrand 1981; Ernst and Boeckh 1983). From the number of these projection fibers relative to the number of afferent fibers, convergence ratios in the cockroach have been estimated to be about 760:1 for "food" receptors and as high as

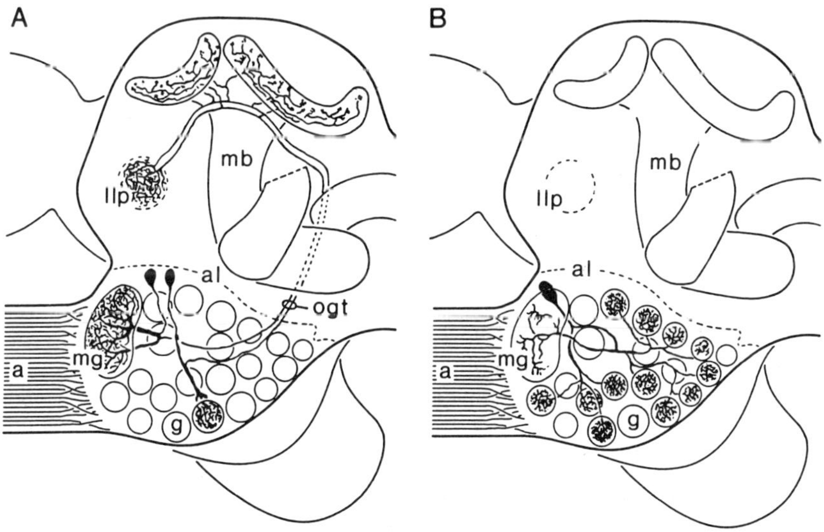


Figure 15.2. Diagrammatical representation of horizontal sections of the brain of the cockroach showing camera lucida drawings of two types of interneurons in antennal lobes of the deutocerebrum. (A) Projection interneurons whose axons project from single glomerulus (*g*, *mg*) in antennal lobe (*al*) via olfactory-globular tract (*ogt*) to higher levels of olfactory pathway, including calyces of mushroom bodies (*mb*) and lateral lobe of the protocerebrum (*llp*). (B) Local interneurons that lack an axon and ramify in many glomeruli. Primary afferents enter antennal lobe via antennal nerve (*a*). (From Ernst and Boeckh 1983.)

4,000-1 for pheromone-specific receptors (Boeckh et al. 1984). Comparable ratios are not available for crustaceans but presumably are equally high.

Characteristically the neuropil at the first synaptical level of the olfactory pathway is organized into discrete clusters known as *glomeruli,* where all synaptical interactions of the primary afferents are thought to occur. This type of organization also characterizes the first synaptical level of the vertebrate olfactory pathway (Shepherd 1977), suggesting that there must be some fundamental advantage of this particular type of structured neuropil for processing olfactory information. The number of glomeruli can vary widely among species. Boeckh et al. (1984), for example, cites several thousand glomeruli per antennal lobe in the locust vs. only 120 in the cockroach, insects with not totally dissimilar habits. There is no evidence to date that any of the glomeruli in crustaceans are morphologically unique, but in insects, a large "macroglomerulus" (*mg* in Fig. 15.2[A]) appears to be uniquely associated with pheromone sensitivity (Boeckh and Boeckh 1979; Matsumoto and Hildebrand 1981). The development of these structures, interestingly, is in part influenced by contact with the ingrowing axons of pheromone-specific receptors (Schneiderman, Matsumoto, and Hildebrand 1982; Tolbert, Matsumoto, and Hildebrand 1983). What is beginning to emerge is the idea that individual glomeruli are identifiable and functionally specific (Rospars and Chambille 1981; Stocker et al. 1983), and that they serve as functional units for processing olfactory information.

The glomeruli of the olfactory lobes of the crayfish are arranged radially in tapered columns. Here, many small, presumably sensory, axons synapse onto interneurons, some of which loop down from the periphery, branch, and then return to the periphery and others of which enter from the center of the lobe and branch into the columns (Sandeman and Luff 1973). The glomeruli of insects, more spherical than those of crustaceans, show evidence of complex synaptic interaction, including several types of neuritic profiles, synapses with different ratios of pre- to postsynaptic elements and serial synapses (Tolbert and Hildebrand 1981). Numerous candidate neurotransmitters have been localized to this region of the brain (Barker et al. 1972; Maxwell, Tait, and Hildebrand 1978), but the actual transmitters and their mode of action in the lobe are just beginning to be resolved (see Christensen, Waldrop, and Hildebrand 1985).

A second class of deutocerebral interneurons, *local neurons,* lack an axon and have all their processes confined to the antennal lobe, where they ramify among many glomeruli and are thought to mediate interactions among the glomeruli (Fig. 15.2[B]) (Matsumoto and Hildebrand 1981; Ernst and Boeckh 1983). Local interneurons tend to be numerous compared to projection neurons. Although local interneurons have yet to be described in the olfactory neuropil of crustaceans, they are well known from other crustacean ganglia and include both spiking and nonspiking neurons (Reichert et al. 1982; Reichert, Plummer, and Wine 1983).

In crustaceans, the olfactory lobe is intimately associated with another glomerularly organized neuropil, the paired *accessory lobes*. The accessory lobe glomeruli, which may number many thousands, in contrast to one hundred to several hundreds in the olfactory lobes, consist of spherical clusters of small processes surrounding one larger element that is presynaptic to the smaller processes (Sandeman and

Luff 1973). Because the accessory lobes are not thought to receive primary afference and are connected to the olfactory lobes via interneuronal tracts, they have been considered to be secondary processing centers for olfactory information (Sandeman and Luff 1973; Sandeman 1982).

Axons of the projection interneurons (including, in crustaceans, those from the accessory lobes) ascend the olfactory-globular tract to the protocerebrum. In stalk-eyed crustaceans, such as lobsters, the target of these interneurons is thought to be the two major neuropils of the most proximal of the eyestalk ganglia, the medulla terminalis. More is known of the neuroanatomy of this level of the olfactory pathway in insects, in which the structural equivalent to the medulla is thought to be the corpora pedunculata, or mushroom bodies. The paired mushroom bodies (see Fig. 15.2) consist of a parallel array of intrinsic fibers that arborize in the calyx, or "cup," of the structure and project through the pedunculus, or "stalk," of the structure to the α and β lobes, terminal bifurcations of the pedunculus. The projection neurons from the deutocerebrum arborize abundantly within the calyx and represent a major, though not the only, input to this neuropil. In contrast to the convergent organization seen in the antennal lobes, this synaptical level is thought to show divergence. Boeckh et al. (1984) calculated divergence ratios of 1-700–1-1,400 in the calyces of the cockroach. Extrinsic neurons that project into the α and β lobes connect the mushroom bodies with other parts of the protocerebrum and are thought to represent the output fibers of the mushroom bodies (Homberg 1984; Schildberger 1984). This neuropil is not thought to receive primary afference, though Strausfeld (1976) reports fibers from long olfactory sensilla project directly to the mushroom bodies of the fly.

Little can yet be said of the neuroanatomy of subsequent levels of the olfactory pathway. It is known, however, that interneurons carry olfactory information from the brain to the ventral nerve cord by way of the circumesophageal connectives. Staining these neurons in the spiny lobster indicates that they are a mixed population of fibers, but that none synapse in the olfactory/accessory lobe complex or in the medulla terminalis, suggesting that they are, in fact, higher-order elements in the olfactory pathway (Hamilton and Ache 1983). These neural elements provide a useful place to monitor the "output" of the olfactory system for experimental analyses of integrative mechanisms in olfaction.

## Functional Organization

A real advantage of most invertebrate models for neurobiological research is that particular neurons can be characterized (morphologically, physiologically, biochemically, or all three) and subsequently relocated in each individual organism, the "identified neuron" concept. The small size of most neurons in the olfactory pathway of arthropods precludes working with identified cells, but it is often possible to assign physiologically characterized neurons to particular morphological classes. Rigorous correlation of structure and function is just beginning to emerge, however, and much of our present understanding of the functional organization of the arthropod olfactory pathway has come from cells whose detailed morphology is as yet unknown.

## The Discrimination of Odor Quality

Determining exactly what constitutes a qualitatively different odor to a lower organism is not a simple matter. This can be done by conditioning experiments on the assumption that compounds so discriminated represent functionally different qualities. This tack has been used to some extent with insects (Borst 1983), but not crustaceans. Another approach has been to consider that the tissue constituents or exudates of naturally occurring stimulus sources, such as in prey organisms, define sets of compounds that organisms must decipher as an entity at any given instant. The latter approach has been used to investigate the question of quality discrimination in crustaceans. In fact, one advantage of invertebrate models, especially aquatic invertebrates, in chemoreception is that many and diverse behavioral contexts in which chemical signals act have been described, and the behaviors typically have one or more stereotyped components that can be used as a bioassay to dissect the naturally occurring signal into its active molecular constituents. As detailed in Chapter 1, feeding cues in aquatic invertebrates tend to be mixtures of common organic substances such as amino acids, amines, organic acids, and nucleotides. These mixtures are not only easily identified, reconstructed, and delivered to the preparation as solutions but, because of the broad biological activity of many of the component chemicals, much is already known about how the components act at the cellular level.

Natural odors consist of complex mixtures of these chemicals that collectively constitute the ''gestalt'' of the stimulus. Because of this, they coactivate many parallel inputs, allowing for the possibility that interactions occur among the inputs. Quality coding must be studied with this in mind. In the spiny lobster, such interactions among coactivated olfactory neurons can be observed physiologically, primarily as mixture suppression, in which one or more of the components of a complex odor suppress the stimulatory capacity of another component (Derby and Ache 1984a). Mixture suppression in the spiny lobster is the result of at least two phenomena that occur at different levels of the olfactory pathway (Derby, Ache, and Kennel 1985) (Fig. 15.3). First, mixture suppression occurs in the periphery, possibly the result of competitive inhibition between or among receptor sites on individual receptor cells (Gleeson and Ache 1985). The odor-evoked responses in some interneurons, however, are suppressed even when the stimulatory and suppressive components of a mixture are applied to different parts of the receptor population (antennular filaments), thereby eliminating the possibility of interaction at the receptor level. The mechanism(s) underlying this central suppression are unknown.

Behaviorally, there is evidence for mixture synergism, as well as mixture suppression in crustaceans. Mixture synergism occurs in shrimp, in which synthetic mixtures based on extracts of natural tissues tended to elicit much greater responses than would be predicted by summing the responses to the individual com-

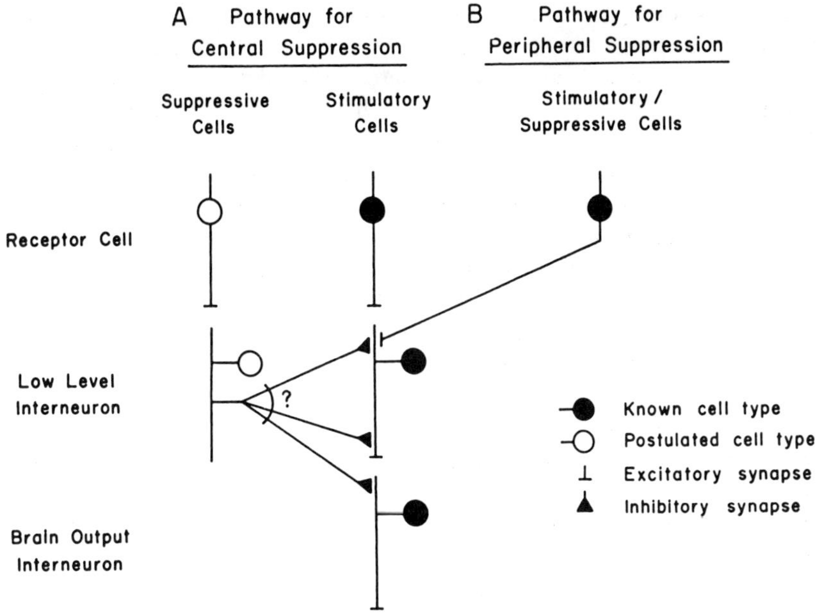


Figure 15.3. Possible model to explain neural mechanisms underlying mixture suppression observed in CNS nervous of spiny lobster. (A) Centrally mediated suppression occurs owing to inhibitory interaction of inputs from two types of receptor cells, those excited by suppressive compounds in complex mixture and those excited by stimulatory compounds. Suppressive receptor cells make excitatory connections on low order interneuron(s) otherwise excited by input from stimulatory receptor cells. (B) Peripherally mediated suppression also occurs owing to yet another type of receptor cell. These are also excited by stimulatory compounds, but their level of excitation is reduced when they are coactivated by the suppressive compounds. This type of receptor cell makes excitatory connections within the CNS. (From Derby, Ache, and Kennel 1985).

ponent of the mixtures (Carr and Derby 1986). If the olfactory system of shrimp is similar to that of the spiny lobster, then mixture synergism must be weakly expressed at the level of single neurons and therefore not readily detected, or it is an emergent property across populations of neurons that drive behavior, or it is the result of input from receptor populations other than the antennules. In support of the first alternative is the observation that mixture synergism accompanies suppression in olfactory receptor cells in the clawed lobster *Homarus americanus* (Johnson, Borroni, and Atema 1985). Mixture suppression has been demonstrated behaviorally in the California spiny lobster, in which feeding responses to single, low-molecular-weight components of a potential food organism are eliminated when the same components are presented to the animal as part of multicomponent mixtures of these compounds (Zimmer-Faust et al. 1984).

How might information about odor quality be encoded? Although this question is the focus of much research in the chemical senses, it remains unanswered. While the receptor cells of crustaceans exhibit various degrees of tuning, they do not appear to be tuned to particular arrays of compounds that are likely to constitute naturally occurring "qualities." Thus, quality coding must be largely the result of the central processing of information derived from the population of receptor cells. Studies with crustaceans have perhaps contributed some novel information to our understanding of how this happens. The action spectrum or bandwidth of some single crustacean olfactory receptor cells can be extremely narrow, such that even minor alterations in the native molecule's structure markedly reduce or eliminate its effectiveness as a stimulant. Input from these cells could serve as an unambiguous labelled line for the compound to which the cell is tuned. This type of cell occurs along with others that are more broadly tuned to a variety of different compounds. Although the input from any one of the more broadly tuned cells would be ambiguous, stimulus quality could be represented over the entire population of receptor cells as particular across-neuron patterns of activity. Since crustaceans appear to have these two different classes of olfactory receptor cells, each class with its characteristically different response profile, the mechanism for quality coding in olfaction may not be unlike that proposed for coding different tastes in vertebrates by Smith et al. (1983). They proposed that the activity in more than one group of cells is necessary for the unambiguous discrimination of different taste qualities, while particular types of cells are critical for generating similarities in the neural codes for tastants of similar quality.

No evidence reported to date suggests that a complex, feature-detecting type of organization exists in the crustacean olfactory pathway. As a population, olfactory interneurons are more broadly tuned than are the receptor cells (Fig. 15.4). They cannot be grouped into discrete, highly correlated cell types, as can the receptors, though some interneurons have specificities as narrow as receptor cells (Derby and Ache 1984b). Across-neuron pattern codes are therefore probably used at the interneuronal levels to characterize stimulus quality of single chemicals. Additional evidence against interneurons' being feature detectors is that they are not uniquely triggered by complex stimuli (Hamilton and Ache 1983). In fact, a single chemical can be as or more stimulatory for interneurons than mixtures containing that chemical. Since not all complex stimuli activate these interneurons equally, it is likely that, as for single components, features of the mixtures are extracted across populations of these fibers. This proposition, of course, requires experimental verification.

If different odorants are to be reliably discriminated under a variety of conditions, quality coding must be independent of stimulus quantity, at least over some range of concentrations. Experimental results with the crustacean olfactory system (Derby and Ache 1984b) indicate that this is true for several compounds, at least at relatively low concentrations. Mixture interactions potentially could also have a confounding effect on quality coding, especially since not all classes of receptors responsive to a particular chemical show equal degrees of interaction. As a result, the across-neuron pattern of activity generated by that chemical may, in part, be a function of which other chemicals are simultaneously presented and in which concentration and, therefore, depend on the precise composition of the mixture.

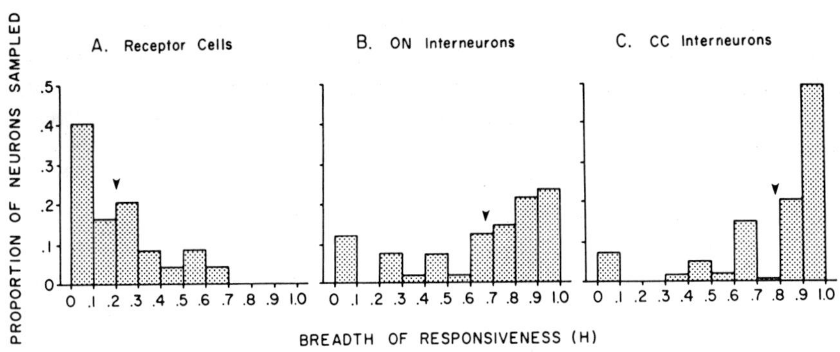


Figure 15.4. Response spectra of neurons at three different levels of olfactory pathway of spiny lobster. The breadth of tuning is quantified here with H metric, diversity measure based on the equation for entropy from information theory. *H* values can vary from 0–1, with H–0 indicating neurons that respond to only one of stimulants in test group and H–1 indicating neurons that respond equally to all. Plotted are the distribution of *H* values for (A) 26 *receptor cells*, (B) 43 low-order *interneurons* that project from the olfactory–accessory lobe complex via olfactory-globular tract in optic nerve (*ON*) to medulla terminalis (see p 391 for anatomical description), and (C) 41 higher-order *interneurons* that project from the brain via circumesophageal connectives (*CC*) to ventral nerve cord. H values were determined at all levels for a group of 3 stimuli: taurine, glutamic acid, and betaine. Mean H value (*arrows*) increases at higher levels, indicating an increase in the breadth of tuning at higher levels of olfactory pathway. (From Derby, Hamilton, and Ache 1984.)

## The Discrimination of Odor Quantity

The olfactory system is probably more concerned with detecting changes in stimulus concentration than in detecting the absolute concentration of the stimulus. A key element in discriminating changes in odor concentration is at the periphery and results from the intermittent exposure most olfactory organs have to the odor environment. In vertebrates, typically, this comes from tidal flushing of the olfactory chamber, but intermittent exposure also occurs in crustacean olfaction as well (Schmitt and Ache 1979). This phenomenon, called *flicking*, is considered elsewhere (Chap. 2), but of interest here is that the receptors discharge as a series of phasic bursts, each reflecting the net change in concentration that occurred during the interflick interval. The intensity of each burst is maximized, because the receptors are protected from adaptation to any stimulus changes occurring during the interval and because the onset of the stimulus is rapid (rising in about 100 ms in the spiny lobster).

The basis of coding of odor quantity at the receptor level appears to be a simple function of the frequency of discharge of the receptor cells (see Chap. 14). There is no evidence that any new variable is used in the CNS to code concentration besides that used in the periphery. For example, interneurons in the circumesophageal connectives of crustaceans, shown to be higher-order neurons projecting from the brain to the ventral nerve cord (Hamilton and Ache

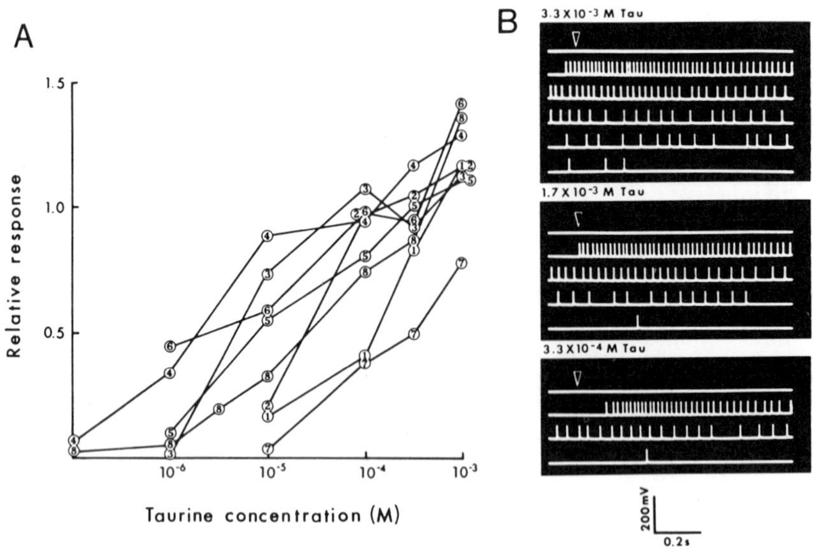


Figure 15.5. Concentration dependency of the response of interneurons that descend circumesophageal connectives in spiny lobster. (A) Dose-response relationships of 8 interneurons to ascending series of concentrations of taurine (3.3 times the molarity indicated) applied to olfactory receptors. (B) Intracellular traces of actual response to one such interneuron. *Arrows* indicate stimulus introduction (approx. 1 sec. is required for the stimulus to reach the preparation postintroduction). (From Hamilton and Ache 1983.)

1983), exhibit power functions and patterns of discharge similar to those of the receptor cells (Fig. 15.5).

On conceptual grounds, one would assume that the massive convergence at the first synaptic level of the olfactory pathway would enhance quantity discrimination. Only limited data address this possibility in arthropods, but comparison of the sensitivity of antennal receptor cells to deutocerebral interneurons responsive to sex pheromones in insects suggests that as much as orders of magnitude increase in absolute sensitivity is gained as a result of convergence in the antennal lobes (Boeckh and Selsam 1984).

## The Discrimination of Spatial and Temporal Aspects of Odors

Although chemical stimuli are inherently weak in their spatial and temporal properties, both aquatic and terrestrial arthropods rely on such information to get about in their odor environment (see review by Bell and Tobin 1982). While oriented movements are frequently based on multimodal sensory information, behavioral studies indicate that crustaceans orient to distant odor sources, at least in part, by antennular chemoreception (Reeder and Ache 1980; Devine and Atema 1982; Tierney, Thompson, and Dunham 1984) and do so by making both instantaneous (simultaneous sampling of two spaced points, as is possible with paired, widely spaced receptor organs) and temporal (successive sampling) comparisons of their odor environment (Fig. 15.6).

An important component of spatial-temporal integration is at the periphery, where as mentioned in the previous section, the intermittent exposure to odorants produced by flicking markedly improves quantitative resolution of the stimulus. The two olfactory organs are not necessarily flicked together, suggesting that successive averaging or some other time-dependent process in the CNS is also involved in the integration process.

Central components of spatial-temporal integration have yet to be described, however. In crustaceans, olfactory interneurons with bilateral receptor fields coexist with those having unilateral fields, even at higher levels of the olfactory pathway (Ache and Sandeman 1980; Derby and Ache 1984b. Differential responses of these two classes of neurons would provide the organism with spatial information about its odor environment. Whether bilateral contrast is enhanced

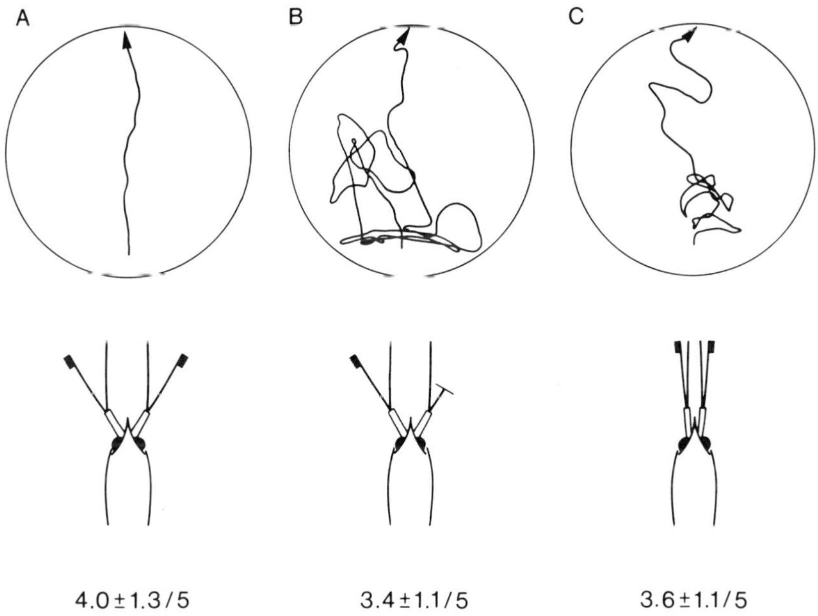


Figure 15.6. Tracings of typical videotaped paths of spiny lobsters walking towards point source of odor (shrimp extract). The odor enters test arena at the top (12 o'clock) of each figure; lobsters start walking from the bottom (6 o'clock). (A) Path of intact lobster. (B) Path of a lobster with right lateral filament of the antennule (right olfactory organ) ablated to interfere with instantaneous (simultaneous bilateral) comparisons of odor field. (C) Path of a lobster with both antennules immobilized in forward-pointing position to interfere with temporal (successive) comparisons of odor field. Ratios below each figure indicate average number of times (X ± SD) out of five trials each of 5 lobsters reached criterion (odor inlet) within test interval (4 min.). Note that both procedures (B) and (C) reduce, though do not eliminate, the ability of subjects to reach criterion. (From Reeder PB and Ache BW. *Anim. Behav.* 28:831–839, 1980. Reproduced with permission of Baillière Tindall.)

by inhibition from the side that is stimulated more strongly or earlier or both, as is known to occur in the mammalian olfactory system (Leveteau and Macleod 1969), is as yet unknown. One class of output interneurons in the brain of insects "flip-flop" their rates of discharge in response to changes in the strength of pheromonal stimulation of the antenna (Olberg 1983). Such patterns of discharge are not known at the level of receptors, suggesting that this patterned output, of potential relevance to the orientation of odor, is generated by specific connections in the CNS.

## Summary

Our understanding of how aquatic invertebrates integrate chemosensory information is, at best, limited. This is not surprising, since neural integration is certainly the most complex aspect of any stimulus-response chain and since we still have much more to learn about the nature of the chemosensory input to the CNS from the periphery. Some progress has been made in understanding integration in the olfactory system of arthropods, but it represents only a beginning. The past performance of arthropods as neurobiological models and the importance of chemosensory stimuli to arthropod behavior, however, suggest a promising future for this line of investigation and, for similar reasons, for research on the other species of aquatic invertebrates whose chemical senses are beginning to be studied in more detail.

Both morphological and physiological data, though limited, suggest that the basic structural and functional organization of the olfactory system of crustaceans does not differ significantly from that of insects, allowing that there may be few unique adaptations for processing chemosensory information in aquatic environments. Any differences that ultimately do emerge from a comparison of insects and crustaceans, however, may be phylogenetic, rather than habitat dependent. It should be possible to resolve this problem by determining if integrative mechanisms vary among those species of brachyuran and anomuran crustaceans that show graded degrees of independence from the water, including species, such as *Coenobita* and *Gecarcinus,* that except for their reproduction, are essentially terrestrial.

Because the basic organizational plan of the olfactory system of insects and crustaceans is not totally dissimilar to that of vertebrates, it is reasonable to expect that principles of chemosensory integration described in arthropods will have more general applicability. There are hints that this may be true both at the periphery (e.g., discontinuous sampling) and in the CNS (e.g., a code for quality that uses both narrowly and broadly tuned neurons). Certainly much more work needs to be done before any similarities in how vertebrates and invertebrates process chemosensory information can be rigorously evaluated. If so, however, understanding how aquatic invertebrates such as crustaceans process chemosensory information will extend beyond its immediate implications to chemical sensing in the aquatic environment and contribute, symbiotically, to an overall understanding of the chemical senses. The numerous technical advantages of working with aquatic invertebrates suggest they may play an important role in this symbiosis.

*Acknowledgments.* I should like to thank my colleagues Drs. Kathryn Hamilton and Charles Derby for their contributions to much of the work described here, which was supported by NSF awards BNS-8308120 and BNS-8511256.

## References

Ache, B.W. and Sandeman, D.C. (1980) Olfactory-induced central neural activity in the Murray crayfish, *Euastacus armatus,* J. Comp. Physiol., 140:295–301.

Altner, H. and Prillinger, L. (1980) Ultrastructure of invertebrate chemo-, thermo-, and hygroreceptors and its functional significance, Int. Rev. Cytol., 67:69–139.

Barker, D.L., Herbert, E., Hildebrand, J.G., Kravitz, E.A. (1972) Acetylcholine and lobster sensory neurones, J. Physiol., 226:205–229.

Bell, W.J. and Tobin, T.R. (1982) Chemo-orientation, Biol. Rev., 47:209–260.

Boeckh, J. and Boeckh, V. (1979) Threshold and odor specificity of pheromone-sensitive neurons in the deutocerebrum of *Antheraea Pernyi* and *A. polyphemus* (Saturnidae), J. Comp. Physiol., 132:235–242.

Boeckh, J., Ernst, K.D., Sass, H., and Waldow, U. (1984) Anatomical and physiological characteristics of individual neurones in the central antennal pathway of insects, J. Insect Physiol., 30:15–26.

Boeckh, J. and Selsam, P. (1984) Quantitative investigation of the odour specificity of central olfactory neurones in the American cockroach, Chem. Sens., 9:369–380.
Boeckh, J., Ernst, K.D., Sass, H., and Waldow, U. (1984) Anatomical and physiological characteristics of individual neurones in the central antennal pathway of insects, J. Insect Physiol., 30:15–26.
Borst, A. (1983) Computation of olfactory signals in *Drosophila melanogaster,* J. Comp. Physiol., 152:373–383.

Carr, W.E.S. and Derby, C.D. (1986) Behavioral chemoattractants for the shrimp, *Palaemonetes pugeo:* identification of active components in food extracts and evidence of synergistic mixture interactions, Chem. Sens., 11:49–64.

Christensen, T.A., Waldrop, B.R., and Hildebrand, J.G. (1985) GABA mediated inhibition in the antennal lobes of the moth *Manduca sexta,* Soc. Neurosci. Abst., 1:163.

Croll, R.P. (1983) Gastropod chemoreception, Biol. Rev., 58:293–319.

Derby, C.D. and Ache, B.W. (1984a) Electrophysiological identification of the stimulatory and interactive components of a complex odorant, Chem. Sens., 9:201–218.

Derby, C.D. and Ache, B.W. (1984b) Quality coding of a complex odorant in an invertebrate, J. Neurophysiol., 51:906–924.

Derby, C.D., Ache, B.W., and Kennel, E.W. (1985) Mixture suppression in olfaction: electrophysiological evaluation of the contribution of peripheral and central neural components, Chem. Sens., 10:301–316.

Derby, C.D., Hamilton, K.A., and Ache, B.W. (1984) Processing of olfactory information at three neuronal levels in the spiny lobster, Brain Res., 300:311–319.

Devine, D.V. and Atema, J. (1982) Function of chemoreception organs in spatial orientation of the lobster, *Homarus americanus:* differences and overlap, Biol. Bull., 163:144–153.

Elliott, E.J. (1984) Chemoreception in the leech, Neurosci. Soc. Abst., 10:656.

Elofsson, R. (1983) 5-HT-like immunoreactivity in the central nervous system of the crayfish, *Pacifastacus leniusculus,* Cell Tissue Res., 232:221–236.

Ernst, K.D. and Boeckh, J. (1983) A neuroanatomical study on the organization of the central antennal pathways in insects, Cell Tissue Res., 229:1–22.

Gleeson, R.A. and Ache, B.W. (1985) Amino acid suppression of taurine-sensitive chemosensory neurons, Brain Res., 335:99–107.

Hamilton, K.A. and Ache, B.W. (1983) Olfactory excitation of interneurons in the brain of the spiny lobster, J. Comp. Physiol., 150:129–140.

Homberg, U. (1984) Processing of antennal information in extrinsic mushroom body neurons of the bee brain, J. Comp. Physiol., 154:825–836.

Johnson, B.R., Borroni, P.F., and Atema, J. (1985) Mixture effects in primary olfactory and gustatory receptor cells from the lobster, Chem. Sens., 10:367–373.

Laverack, M.S. and Ardill, D.J. (1965) The innervation of the aesthetasc hairs of *Panulirus argus,* Q. J. Microsc. Sci., 106:45–60.

Leveteau, J. and Macleod, P. (1969) Reciprocal inhibition at glomerular level during bilateral olfactory stimulation, Proc. 3rd Int. Symp. Olfaction Taste, 3:212–215.

Matsumoto, S.G. and Hildebrand, J.G. (1981) Olfactory mechanisms in the moth *Manduca sexta:* response characteristics and morphology of central neurons in the antennal lobes, Proc. R. Soc. Lond., B213:249–277.

Maxwell, G.D., Tait, J.F., and Hildebrand, J.G. (1978) Regional synthesis of neurotransmitter candidates in the CNS of the moth *Manduca sexta,* Comp. Biochem. Physiol., 61C:109–119.

Maynard, D.M. (1967) Organization of central ganglia, in Invertebrate Nervous Systems, Wiersma, C.A.G. (ed.), University of Chicago Press, Chicago, pp. 231–255.

Olberg, R.M. (1983) Pheromone-triggered flip-flopping interneurons in the ventral nerve cord of the silkworm moth, *Bombyx mori,* J. Comp. Physiol., 152:297–307.

Reeder, P.B. and Ache, B.W. (1980) Chemotaxis in the Florida spiny lobster, *Panulirus argus,* Anim. Behav., 28:831–839.

Reichert, H., Plummer, M.R., Hagiwara, G., Roth, R.L., and Wine, J.L. (1982) Local interneurons in the terminal abdominal ganglion of the crayfish, J. Comp. Physiol., 149:145–162.

Reichert, H., Plummer, M.R., and Wine, J.J. (1983) Identified nonspiking local interneurons mediate nonrecurrent, lateral inhibition of crayfish mechanosensory interneurons, J. Comp. Physiol., 151:261–276.

Rospars, J.P., Chambille, I. (1981) Deutocerebrum of the cockroach *Blaberus craniifer* Burm. Quantitative study and automated identification of the glomeruli, J. Neurobiol., 12:221–247.

Sandeman, D.C. (1982) Organization of the central nervous system, in The Biology of Crustacea, Atwood, H.L. and Sandeman, D.C. (eds.), Academic Press, New York, pp. 1–61.

Sandeman, D.C. and Luff, S.E. (1973) The structural organization of glomerular neuropile in the olfactory and accessory lobes of an Australian freshwater crayfish, *Cherax destructor,* Z. Zellforsch. Mikrosk. Anat., 142:37–61.

Schildberger, K. (1984) Multimodal interneurons in the cricket brain: properties of identified extrinsic mushroom body cells, J. Comp. Physiol., 154:71–79.

Schmitt, B.C. and Ache, B.W. (1979) Olfaction: responses of a decapod crustacean are enhanced by flicking, Science, 205:204–206.

Schneiderman, A.M., Matsumoto, S.G., and Hildebrand, J.G. (1982) Trans-sexually grafted antennae influence development of sexually dimorphic neurones in moth brain, Nature, 298:844–846.

Shepherd, G.M. (1977) The olfactory bulb: a simple system in the mammalian brain, in *Handbook of physiology*, Brookhard, J.M., and Montcastle, V.B. (eds.), American Physiological Society, Maryland, pp. 945–968.

Smith, D.V., Van Buskirk, R.L., Travers, J.B., and Bieber, S.L. (1983) Coding of taste stimuli by hamster brain stem neurons, J. Neurophysiol., 50:541–558.

Stocker, R.F., Singh, R.N., Schorderet, M., and Siddiqi, O. (1983) Projection patterns of different types of antennal sensilla in the antennal glomeruli of *Drosophila melanogaster,* Cell Tissue Res., 232:237–248.

Strausfeld, N.J. (1976) Atlas of an Insect Brain, Springer-Verlag, New York.

Tierney, A.J., Thompson, C.S., and Dunham, D.W. (1984) Site of pheromone reception in the crayfish *Orconectes propinquus* (Decapoda, Cambaridae), J. Crustacean Biol., 4:554–559.

Tolbert, L.P. and Hildebrand, J.G. (1981) Organization and synaptic ultrastructure of glomeruli in the antennal lobes of the moth *Manduca sexta:* a study using thin sections and freeze-fracture, Proc. R. Soc. Lond. B. Biol. Sci., 213:279–301.

Tolbert, L.P., Matsumoto, S.G., and Hildebrand, J.G. (1983) Development of synapses in the antennal lobes of the moth *Manduca sexta* during metamorphosis, J. Neurosci., 3:1158–1175.

Zimmer-Faust, R.K., Tyre, J.E., Michel, W.C., and Case, J.F. (1984) Chemical mediation of appetitive feeding in a marine decapod crustacean: the importance of suppression and synergism, Biol. Bull., 167:339–353.

# Part IIIB

# Vision

Our human bias favors vision and it is not surprising to see that both the stimulus and its receptors have been rather well described. Underwater, a great variety of photic environments are rather well known, as described by Lythgoe in Chapter 3, thus allowing us to see matching of receptor and environment in some detail. These adaptations involve optics, retinal morphology, receptor physiology, and circadian rhythms, among others. In terms of general principles of signal extraction and environmental adaptation, our knowledge of vision offers many models for other sensory systems.

The six chapters in this part include two on invertebrates and two on vertebrates; in each case, a general chapter is followed by a more specialized topic. The last two chapters deal with principles of signal extraction and central processing.

Cronin (Chapter 16) reduces the staggering variety of marine invertebrate photoreceptor organs to a few basic designs. The variety seems to have phylogenetic as well as environmental origins. The subject is treated from the level of chemical relationships among rhodopsins to structural features of whole eyes. Barlow, Powers, and Kass (Chapter 17) then put one of the invertebrate eyes to work in the mating behavior of *Limulus*. This is a specific example of how circadian rhythms in structure and function of retina culminate in peak visual sensitivity at night when behavioral demands are great.

Fernald (Chapter 18) discusses the critical morphological and physiological features of fish eyes in light of the physics of their particular photic environment. Special attention is given to the visual consequences of eye growth during development. Adaptation to the aquatic environment is seen most dramatically in the optics of amphibious animals, a topic treated by Sivak in Chapter 19. Amphibians make the air–water transition slowly compared to diving birds, where we can observe optic adaptations to the aquatic environment taking place in a fraction of a second.

In a play on figure and ground, Saidel (Chapter 20) continues Kamil's ecological discussion of crypticity (Chapter 7) but Saidel uses the phenomenon of camouflage to deduce certain fundamental signal extraction features of visual systems. Northcutt and Wullimann (Chapter 21) give a phyletic analysis of the visually related centers and pathways in fishes with particular emphasis on areas for future research.

# Chapter 16

# Vision in Marine Invertebrates

THOMAS W. CRONIN *

The marine metazoan invertebrates exhibit a variety in eye design as great as their own phyletic diversity. Although this diversity can be conceptually reduced to a large number of variations on relatively few basic themes, as I shall demonstrate in this chapter, the variations, nonetheless, present an initially bewildering array. This can be traced, in part, to the multiplicity of photic environments inhabited by the huge assemblage of organisms conveniently lumped by one mammalian species as ''invertebrates,'' but it also reflects the large number of roughly parallel evolutionary lines, each based on a different ancestral eye type. Contributing to the complexity has been the independent evolution of radically different ocular organs in single lineages, such as the bivalve molluscs. The situation stands in sharp contrast to that of vertebrates, in which all extant eyes represent radiation from a single ancestral design.

Another feature of invertebrate vision that presents problems for the reviewer is that the photoreceptor organs do not always turn up where they might be expected. Although eyes frequently are located in the cephalic or anterior regions, particularly in the larger and more active marine invertebrates, noncephalic eyes are also common. Some of these are not simple eyespotlike organs, but are of optically elegant design. Furthermore, extraocular photoreceptors occur widely in invertebrate phyla, and these may assume some of the functions normally assigned to eyes (see reviews of Bennett 1979 and Yoshida 1979). The types of information concerning the ambient light that may be gathered by eyes (and extraocular photoreceptors) are these: (1) intensity; (2) temporal changes, such as motion, flicker, or photoperiod; (3) spatial distribution; (4) spectral distribution; and (5) polarization pattern. Even the simplest invertebrate photoreceptor is suitable for discriminating light intensity and duration, while some marine invertebrate (e.g., crustacean compound) eyes are able to measure all the given features of a light stimulus.

---

* Department of Biological Sciences, University of Maryland, Baltimore County, Catonsville, MD 21228, USA

As the foregoing discussion implies, presenting a proper, comprehensive review of marine invertebrate vision in the space available is impossible. Fortunately, such a review is not necessary. Three recent volumes of the *Handbook of Sensory Physiology* have been devoted to invertebrate vision, including extensive and generally excellent coverage of marine species (Autrum 1979,1981a,1981b), and Ali (1984) has edited a fine collection of papers on the same topic. Moreover, a number of shorter papers dealing with various aspects of invertebrate vision, but limited by taxonomical area (Shaw and Stowe 1982; Cronin 1986a: crustacea) or visual function (Cronin 1986b: photoperiodism) are available. These works largely describe the current status of our understanding of invertebrate visual design, function, and evolution, so I will treat these topics with relative brevity here. When possible, however, I will point out how marine invertebrates actually use their eyes and how eye design in various groups is adaptive for the lives these animals lead.

## Invertebrate Visual Pigments

It appears that all animals pack their light-sensitive receptor molecules, or visual pigments, into the plasma membranes of specialized sensory cells. To enhance sensitivity, these cells typically have a great proliferation of their plasma membranes, permitting an increased content of visual pigment and thus a greater light absorption. The membrane is most frequently arranged perpendicular to the direction that light normally enters the receptor. Photoreceptor cells may or may not derive their receptor membrane from ciliary membranes; those which do are known as *ciliary photoreceptors,* and all other types are called *rhabdomeric photoreceptors* (Eakin 1972). The term *microvillous* is sometimes used synonymously with *rhabdomeric,* but some ciliary types also have microvilli. The classification may have some functional correlates: rhabdomeric photoreceptor cells depolarize on illumination, while ciliary types hyperpolarize (see Land 1984a). Nevertheless, there is such taxonomical and functional variation among animals that generalizations about either function or evolution of photoreceptor cells are difficult to conceive (see Westfall 1982).

In contrast to the confusion about the origins of photoreceptor cells, there is a clear homology among the visual pigments, and all go under the generic name of "rhodopsin." The absorption spectra of invertebrate rhodopsins can be fit by the Dartnall nomogram curve (Ebrey and Honig 1977) or similar empirical functions that describe the shapes of vertebrate absorption spectra (Fig. 16.1). All known rhodopsins consist of a chromophore, derived from a molecule of the vitamin A family, linked to a single polypeptide chain or opsin. Recent sequence analysis of opsin from *Drosophila melanogaster* reveals considerable similarity to mammalian opsins (O'Tousa et al. 1985; Zuker, Cowman, and Rubin 1985). Sequence homology between such disparate groups as insects and mammals implies that most or all opsins may derive from a single ancestral molecule. Ironically, the chromophore of *D. melanogaster* rhodopsin, as well as that of several other dipteran and lepidopteran species, turns out not to be the expected retinal$_1$,

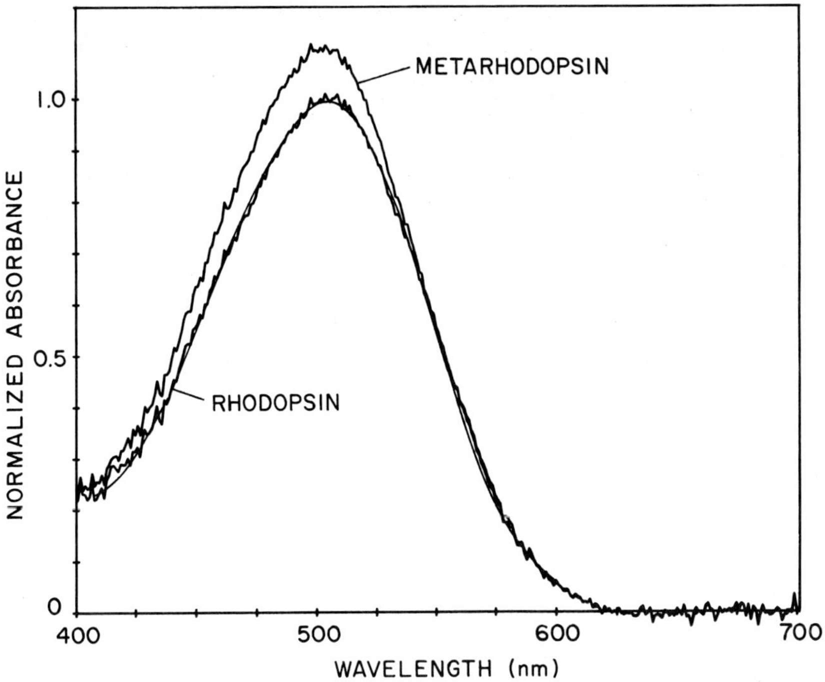


Figure 16.1. Absorption spectra for stable states of visual pigment of stomatopod crustacean *Squilla empusa*, obtained by microspectrophotometry of isolated rhabdoms (microvillar photoreceptors). The spectrum for the rhodopsin was obtained by averaging difference spectra for photobleaching of initially dark-adapted rhabdoms ($n = 20$), whereas photoproduct's absorption spectrum was computed with data from same rhabdoms, as discussed in Cronin (1985). Rhodopsin spectrum is fit well by Dartnall nomogram curve computed for a 507-nm vertebrate visual pigment (smooth curve) (see Ebrey and Honig 1977). This figure is modified from Cronin (1985).

derived from vitamin $A_1$, but instead the novel 3-hydroxyretinal (Vogt 1983, 1984). Although the significance of this finding is unclear, other unsuspected carotenoids may function as chromophores for various invertebrate rhodopsins.

The invertebrate visual pigments that have been studied are unlike those of vertebrates in that each forms a thermally stable photoproduct following photon capture and photoisomerization. While vertebrate rhodopsins release their chromophores and literally "bleach" after photoisomerization, the invertebrate photoproducts have distinctive, stable absorption spectra (Fig. 16.1) and may be reconverted to the original rhodopsin state on subsequent irradiation. This recycling of visual pigment is biochemically parsimonious when compared to the vertebrate situation, but it produces special problems for sensitivity control and the regeneration of rhodopsin in the dark. Many marine invertebrates apparently solve these problems by altering the amount of photoreceptor membrane, with its visual pigment content (Brandenberger and Eakin 1980; Kataoka and Yamamoto 1981, 1983; Chamberlain and Barlow 1984; Cronin and Goldsmith 1984).

The only marine invertebrates proved to possess more than a single spectral type of rhodopsin are the crustaceans (Cummins and Goldsmith 1981). They have a rhodopsin absorbing principally in the near-ultraviolet or violet in a minority population of photoreceptor cells and a blue-green or green-absorbing rhodopsin in the rest. Also, only in crustaceans has color vision (i.e., the ability to discriminate hues independently of their intensities) unequivocally been demonstrated by behavioral methods (Hyatt 1975). This probably reflects the relatively limited experimentation in this area more than the capabilities of marine invertebrates in general.

## Eye Design in Marine Invertebrates

Photoreceptors may be specialized for any of a number of functions. For the purposes of this chapter, I will consider an eye to be a photoreceptor organ with at least a minimal provision for restricting the directions from which it may be excited by incoming light. By this definition, eyes are photoreceptor organs having some type of optical arrangement (Land 1981a); they may be cephalic or located elsewhere on (or in) the body. The function of the optical system is to enable at least a crude determination of the spatial distribution of light; to make use of the optics, receptors must be placed in an organized array—the retina. This is usually spread out over two dimensions, forming a somewhat spherical surface, but rarely it may be one-dimensional (a line of receptors) or even zero-dimensional (a small receptor group), with a rasterlike sweep through the other dimension(s). Optical systems known to exist in marine invertebrates are diagrammatically illustrated in Figures 16.2 and 16.3.

### Invertebrate Optical Systems

It is useful to group eyes into two well-defined classes. Either the entire retina receives light through a single optical system (simple eye) (Fig. 16.2[A] to [C]), or it is illuminated by light's passing through a number of optically independent units (compound eye) (Figs. 16.2[D], and 16.3). In marine invertebrates, simple eyes image light onto the retina in one of three ways: (1) by simple aperture restriction (pigment-cup eye) (Fig. 16.2[A]); (2) by refracting light with a spherical lens or, less frequently, multiple aspherical lenses (lens eye or camera eye) (Fig. 16.2[B]); or (3) by reflecting off a concave mirror, possibly adding a correcting lens if the mirror is spherical (mirror eye) (Fig. 16.2[C]). Pigment-cup eyes are the most widely occurring type of eye in spite of their transparent functional limitations, while lens eyes achieve the greatest light-gathering and resolution capacity of any. These eye designs will be discussed in some detail later.

Compound eyes function in one of two fundamentally different ways. The most common type operates by illuminating each retinal unit through a separate aperture or lens (apposition eye) (Figs. 16.2[D] and 16.3[A]). The two alternative designs combine light passing through separate optical elements to form an image

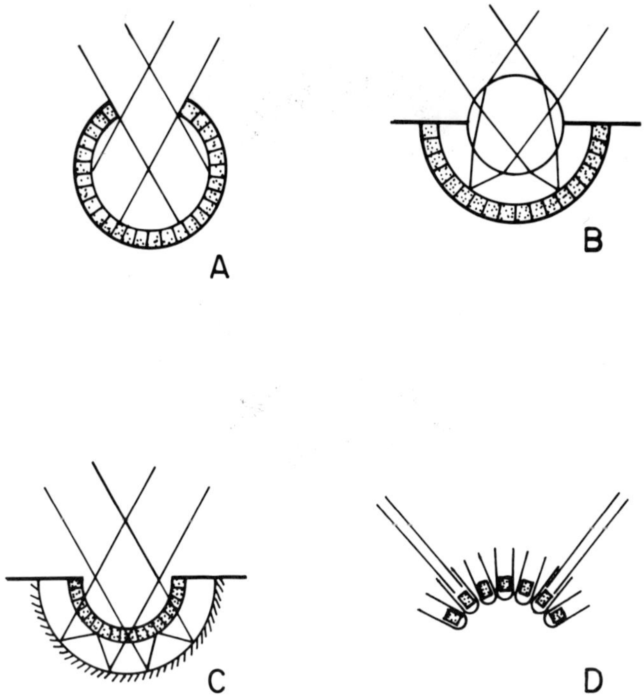


Figure 16.2. Major mechanisms of image formation in marine invertebrate eyes other than crustacean compound eyes. In each case, light enters the eye from above. Retinal region of the eye is indicated by *stippling* and divided to show individual receptor units diagrammatically. (A) Pigment-cup eye. (B) Lens eye. (C) Mirror eye. (D) Compound eye, consisting of multiple pigment tubes. (From Land 1981a.)

on the retinal surface, using either refraction (refracting superposition eye) (Fig. 16.3[B]) or reflection (reflecting superposition eye) (Fig. 16.3[C]). Among marine invertebrates, compound eyes only attain prominence in crustaceans.

## Pigment-Cup Eyes

Pigment-cup eyes occur in virtually every invertebrate phylum; a hydromedusan example (*Spirocodon saltatrix*) is illustrated in Figure 16.4[A]. They are often called *ocelli,* but since this term also refers to some lens eyes or corneal eyes, it is better avoided. Pigment-cup eyes have independently appeared numerous times (Burr 1984a) and may have ciliary, rhabdomeric, or both types of photoreceptor cells. When present, they are generally the only eye type in the animal. They vary in number from two in chaetognaths and some platyhelminths, such as *Dugesia* and other polyclads, as well as rhabdocoels (Fournier 1984), to many thousand in *Onithochiton* and other chitons. (I know of no metazoan with a single eye, though on occasion several pigments cups may fuse to form one cluster.) They may occupy a cephalic (or at least anterior) location, as in many

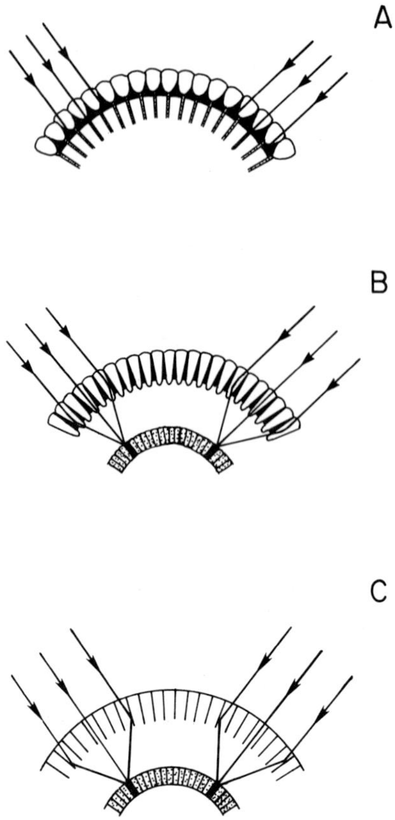


Figure 16.3. Mechanisms of image formation in crustacean compound eyes. (A) Apposition design, in which ommatidia are optically isolated and the aperture for each receptor is generally small. (B) Refracting superposition design, whereby each receptor receives refracted light through a number of corneal units. (C) Reflecting superposition design, operating analogously to refracting type but using mirrors instead of refractile units. (From Cronin 1986a, after Land 1981a.)

platyhelminths, rotifers, some annelids, crustaceans, many gastropods, and chaetognaths. Alternatively, they may occur in fixed, but noncephalic, locations, particularly in radially symmetrical animals like cnidarians or echinoderms; or they can be spread out over the body surface, as in many annelids and the polyplacophorans. In some groups, they are present only in certain life stages; the pigment-cup eyes of some bryozoan larvae are lost on metamorphosis.

Because pigment-cup eyes have a partial pigment screen for the retina, they provide some directional sensitivity to light. Animals possessing them can orient in a light field and exhibit phototaxis (Burr 1984b). They also are adequate for initiating shadow responses, whereby an animal performs rapid withdrawal or closure reflexes when the light intensity decreases rapidly. In some animals, their directional sensitivity, coupled with their multiplicity, permit a crude spatial perception of the object producing the shadow and an ability to detect movement, so that the most appropriate evasive response can be initiated.

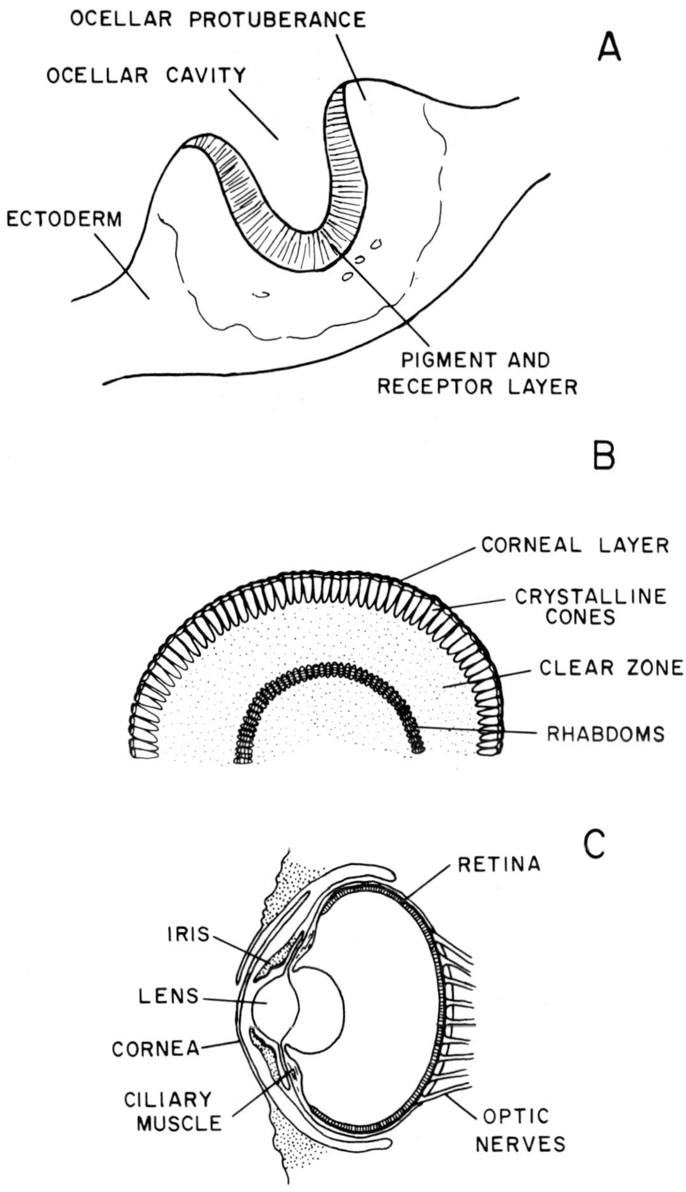


Figure 16.4. Examples of most common marine invertebrate eye designs. (A) Pigment-cup eye from hydromedusan *Spirocodon saltatrix,* consisting of a layer of photoreceptor cells lining pigment-enclosed pit. The opening of the cup is about 50 μm across. (From a photograph in Toh, Yoshida, and Tateda 1979.) (B) Refracting superposition eye of euphausiid crustacean *Nyctiphanes simplex*. Eye diameter is 0.9 mm, and principal optical components are indicated. This eye operates exactly like one diagrammed in Figure 16.3(B). (From a photograph in Kampa 1965.) (C) Cephalopod lens eye, from *Octopus*. High-quality lens has radial gradient in refractive index and forms sharp and corrected image on the retina. Retinal illumination is controlled by changes in pupillary area. The eye is about 1 cm in diameter. (From Wells 1962.)

Optics of pigment-cup eyes are easy to understand. In the simplest case of all, the photoreceptor(s) overlies a pigment backing in a flat sheet, and nothing like an image can form because there is no aperture. As the receptors become confined in a literal "cup," where light must enter through some opening, vision becomes increasingly directional, however, and a blurred image can form (Fig. 16.2[A]). The hydromedusan eye of Figure 16.4[A] could potentially resolve two points of light having a large angular separation. Further restriction of the aperture forms a pinhole eye, as in the molluscs *Haliotis* and *Nautilus* (Land 1981a), having reasonably good resolution, but poor light-gathering ability.

In many pigment-cup eyes, a lenslike body overlies the retina. Its location is such that it cannot conceivably form a focussed image, especially since its outer surface contacts water (Land 1981a), but it could enhance both directional and absolute sensitivity even without a sharp image. Since their proximal borders generally contact the retina, these lenses could function in water only if they possessed a radial gradient of refractive index, but this has not been studied.

It is easy to see that, with the movement of the "lens" away from the retina, the enlargement of the retina, and the addition of accessory structures for control of the aperture, focus, and eye orientation, an optically excellent lens eye could evolve directly from a pigment-cup eye—indeed, this is a fair description of the ontogeny of a vertebrate eye. Other ways to enhance the effectiveness of pigment-cup eyes are to limit their aperture and group them into assemblages with each member sampling a different region of the visual field or to segregate groups of receptors on the retina and provide each group with its own aperture. Either of these pathways produces a compound eye. The evolution of high-quality eyes from pigment cups may thus be straightforward, and this partially explains the often-noted independent emergence of convergent eye designs among animals.

Pigment-cup eyes are optically simple, but their receptor physiology could be as advanced as in the photoreceptor cells of more complex eyes. For instance, in the pigment cups of asteroid echinoderms, there is a daily cycle of photoreceptor microvillar membrane reminiscent of the membrane cycles of arthropod compound eyes (compare the review of Yoshida, Takasu, and Tamotsu 1984 with Shaw and Stowe 1982). Physiological properties of pigment-cup photoreceptor cells are poorly known but of special interest for comparative studies and deserve attention. In summary, these eyes may be of basic design, but they perfectly meet the needs of sessile or slow-moving animals with simple nervous systems. They provide a design from which the complicated eyes of more active animals can readily arise.

## Compound Eyes

One normally associates compound eyes with arthropods, but this eye type has arisen independently in one annelid family (Sabellidae) and a molluscan family (Arcidae). In these cases, eye anatomy is much simpler than in arthropods; in sabellid polychaetes (e.g., *Protula*) the simplest design is an arrangement of pigmented tubes with receptors at the bottom of each (see Fig. 16.2[D]). The design is elaborated in other sabellids (e.g., *Branchiomma*) and in ark clams (e.g., *Barbatia*) by the addition of a lens at the distal tip of the tube. Such eyes may be regarded as a collection of pigment cups, organized only in that their optical axes are radially oriented, and they are of the apposition type. Their function is certainly to act as broadfield shadow or motion detectors; in both sabellids and ark clams, high-speed withdrawal reflexes are well known.

In contrast to the eyes just described, modern crustacean compound eyes have departed so far from their ancestral type that their early evolution is largely concealed from us. Their ommatidia, the individual elements of compound eyes, consist of a number of cells, including photoreceptors, supporting cells, pigment-containing cells, and cells forming optical components. Light is brought to a focus in crustacean compound eyes by the dioptric apparatus of individual ommatidia. Each of the three mechanisms used requires a particular arrangement of ommatidial elements (Fig. 16.3). In the apposition eye, each ommatidium independently views a region of visual space, the light being focussed on its receptor unit by optics of the same ommatidium. This eye type occurs in some anomurans, all true crabs, and most nondecapod crustaceans having compound eyes, excepting the mysids and euphausiids. It offers two design advantages: ease of obtaining a relatively high resolution (up to the diffraction limits of the optics), and a plastic arrangement of optical units that permits specialization of areas of the eye looking into particular regions of visual space. Regional specialization reaches its extreme in the Stomatopoda (mantis shrimp), in which various eye regions have striking anatomical differences and up to three different parts of each eye regard a single point in space (Horridge 1978).

The apposition eye suffers a serious functional limitation: each receptor unit can only be illuminated by light entering a necessarily small collecting aperture; therefore retinal luminance and absolute sensitivity are low. A solution to this problem has appeared in many nocturnal or deep-sea crustaceans; it is to form a retinal image by superimposing the separate images focussed by each ommatidium's optical apparatus. This greatly expands the aperture collecting light for each retinal unit. Recent work with some diurnal insects (Land 1984b) has shown, surprisingly, that the resolution of such an eye can approach the diffraction-limited case, which is the best any eye can do. To achieve this, the ommatidia must be aligned with great precision. Since crustaceans, unlike insects, continue eye growth for many molts after metamorphosis, it is not clear whether they could maintain this order of structural perfection. Another reason that crustacean compound eyes might not achieve their theoretical limits is that they usually operate at larger apertures than the ones Land studied, and superposition eyes suffer from a unique spherical aberration at larger apertures (Kunze 1979; Land, Burton, and Meyer-Rochow 1979).

There are two ways to form a superposition image. These are discussed by Kunze (1979), Land (1980a, 1981a), and Cronin (1986a) and these publications may be referred to for a more thorough treatment than here presented. The first way is to refract light as it passes through each optical unit, so that it leaves at the same angle to the unit's optical axis as it had on entering. Refracting superposition eyes (Fig. 16.3[B]) are found in euphausiids (an example is illustrated in Fig. 16.4[B]), and they probably occur in mysids. Alternatively, the light can be reflected in each unit, again causing it to leave at the same angle at which it entered (this is how all plane mirrors behave), thus forming the reflecting superposition eye (Fig. 16.3[C]). These eyes are characteristic of shrimp, lobsters, and some anomurans, and are easily recognized because they are the only crustacean compound eyes with square facets.

While compound eyes are capable of gathering much more information about the visual world than are pigment cups, they continue to be limited in resolution, and the apposition type has poor light-gathering ability. Why then have they persisted? Several replies to the question exist.

In the first place, among marine invertebrates, compound eyes are found in a group of animals that generally have slow-moving, scavenging life-styles, for which the demands on an eye are not severe, or that live in the deep sea, where the light-gathering qualities of superposition eyes can be exploited and low ambient light levels preclude outstanding resolution in any type of eye. This answer is not wholly satisfying, since fast-moving or predatory crustaceans, like crabs or mantis shrimp, do well with apposition eyes. Furthermore, flying insects (including the predatory dragonflies) all have compound eyes.

As noted before, the apposition eye readily lends itself to regional specializations. There are the additional benefits of packing huge visual fields (nearly 360°) and great depth of field into a very small package. For animals having nervous systems that are small and mainly hard wired, these advantages, coupled with the geometrical regularity and simplicity of the compound eye, may outweigh its low resolution.

A final, and more tentative, answer to the question "why compound eyes" is that they arose early in crustacean evolution and it may have been more convenient to stay with them than to start over. Although this is perhaps true for some crustacean taxa, it is untenable as a complete explanation, because compound eyes arose independently in another marine arthropod taxon (xiphosurans) and again twice in terrestrial arthropods (myriapods and insects). Compound eyes might be naturally suited to animals with exoskeletons, though it is not yet possible to be certain why.

Crustaceans use these eyes for many distinctive visual tasks (see Cronin 1986a). Planktonic forms, including cladocerans and decapod larvae, visually orient to the underwater light field, and the larvae have well-developed predator-evasive photic responses (Forward 1977). Many deep-sea crustaceans also orient and direct vertical migrations visually. In addition, hyperiid amphipods and euphausiids apparently locate visual targets (e.g., prey) against downwelling light, and euphausiids orient their photophores under visual control (Land 1980b, 1981b). The largest crustaceans, stomatopods and decapods, have multifunctional eyes used for orientation and navigation, communication, predator detection, and tracking and locating prey for capture. Crabs are known to discriminate forms (Langdon and Herrnkind 1985). The compound eyes of these animals reach the greatest size and optical specialization of any in existence.

### Lens Eyes

Eyes of marine animals that form extended images with a single optical path must use at least one refractive surface behind the cornea because of the weak refractive power at the cornea-water interface. This differs from the case of most terrestrial animals, for which the corneal surface is the principal refractive interface.

Small lens eyes are widespread among marine invertebrates. The best examples are in the gastropod molluscs and copepod crustaceans. Periwinkles of the genus *Littorina* have well-corrected eyes that function only in air (Hamilton, Ardizzoni, and Penn 1983), but other gastropods have eyes that work under water (see reviews by Messenger 1981 and Land 1984a). Copepods have evolved several exotic eye designs employing multiple elements, scanning retinas, and severe sexual dimorphism (Land 1984c); behavioral observations to explain the ocular specializations are lacking.

One group of gastropods, the heteropods, have smallish, highly modified lens eyes. Their unusual feature is that the retina is not a spherical surface, but instead comprises a curved strip many receptors long, but only a few wide (Land 1982). The strip of retina scans through the image produced by the lens, and the eye thus functions like a linear-diode-array digitizing TV camera. The advantages are the same for both the natural and artificial design: fewer high-quality receptors are required, and the associated neuronal/electronic circuitry is much simplified. To stretch the analogy further, both need ''software'' so that a central processor, knowing the current position of the retina, can locate the visual target.

Large and highly competent eyes of the camera type (Fig. 16.4[C]) have independently arisen at least three times among animals; today they are found in alciopid polychaetes, cephalopod molluscs, and vertebrates. The two invertebrate groups having them are fast-moving, raptorial predators. Certainly the need to detect, track, and locate prey precisely contributed to the perfection of these animals' eyes. Alciopid eyes (see Wald and Rayport 1977) are no doubt used to find the coelenterates on which the polychaetes prey, though no observational data exist to confirm this. Coleoid cephalopods (octopuses, squids, and cuttlefish) are of course well known for their rapid pounces or strikes upon prey.

Cephalopods have invested impressively in their eyes. Messenger (1981) notes that cuttlefish eyes can constitute 50% of the body mass, while eyes of giant squids may reach the size of a beach ball ($\approx$40 cm)! Besides the general optical plan, other parallels with vertebrate eye evolution are noteworthy. The improvement of the eye is accompanied by expansion of the optic lobes of the brain; and, in both groups, the means of moving the eyes and specializing them for vision in the deep sea are quite similar. Land (1984a) discusses these last points nicely.

Nevertheless, fundamental differences can be found between cephalopod and vertebrate eyes. In cephalopods, the photoreceptor cells are rhabdomeric, as opposed to vertebrate ciliary cells. The photoreceptor membranes are not stacks of plates or discs, but well-oriented microvilli, which in a given cell run either horizontally or vertically with regard to gravity (Fig. 16.5). This retinal design permits analysis of the polarization pattern of light (Moody and Parriss 1960; Saidel, Lettvin, and MacNichol 1983), a capability that could be useful for distinguishing counterilluminated or silvered prey against background light. In this aspect, cephalopod photoreceptors function analogously to the better-known ones of the decapod crustaceans (Waterman and Horch 1966), though their polarized

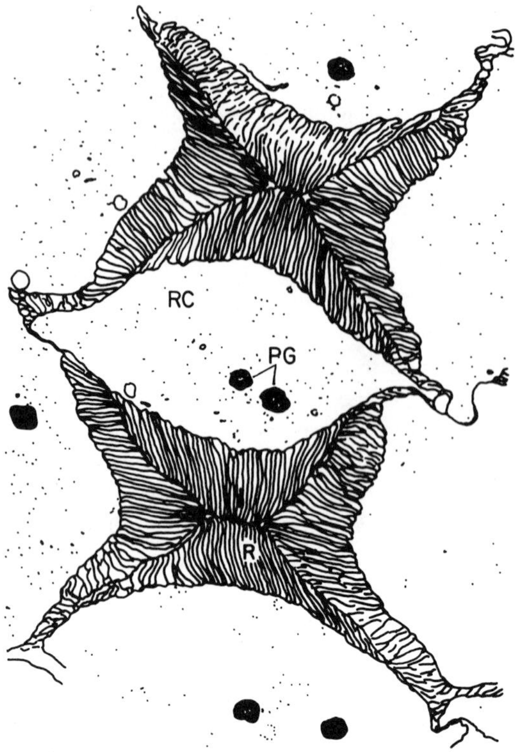


Figure 16.5. Cross section of photoreceptor outer segments in light-adapted retina of cephalopod *Octopus ocellatus*. Each receptor cell (*RC*) gives rise to microvilli along two opposed borders (like a 2-sided toothbrush), thus contributing to 2 rhabdoms (*R*). Although rhabdoms consist of 4 sets of orthogonally oriented microvilli, single receptor cells have microvilli that are all parallel and oriented either horizontally or vertically with respect to visual world. Since absorbing dipoles of rhodopsin molecules tend to be parallel to microvillar axis (Goldsmith and Wehner 1977), this retinal anatomy provides structural basis for polarized light vision. Pigment granules (*PG*) are involved in dark and light adaptation. (From a photograph in Kataoka and Yamamoto (1983).

light vision is perhaps offset by their inability to discriminate hues (Messenger 1981).

Another difference between the cephalopod and vertebrate retinal design is that, in cephalopods, the outer segments of the receptors form the innermost layer of the retina and are thus the first absorbing elements to receive incoming light. Unlike the vertebrate case, their vision is not degraded by overlying blood vessels and layers of higher-order neurons, and there is no blind spot. Cephalopods need offer no apologies for their eyes—they are among the best ever made. Diamond (1985) observes that if the Creator bestowed his own ocular design on an earthly species, He must have been a squid.

## Eye Designs Using Mirrors

Besides the eyes already discussed, a few invertebrate eyes form images or at least concentrate light onto the retina, using mirror optics. The "mirrors" are multilayer interferential reflectors typically optimized for intermediate (blue to green) wavelengths. Mirrors are commonly used as reflecting tapeta in many eyes and, as previously mentioned, form superposition images in some crustacean compound eyes. The eyes discussed next, however, are something different. In describing them, I shall draw on the work of Land (1981a, 1984a, 1984c), who has also written a very readable account of reflecting eyes for the nonspecialist (Land 1978).

As illustrated in Figure 16.2[C], mirrors can be used to concentrate light and focus it onto a retina. It is hardly surprising that some animals use this principle; after all, it is the one used by every major optical astronomical telescope, partly because of its optical simplicity. The only animal definitely known to use pure mirror optics is the large deep-sea ostracod *Gigantocypris*. It has 3-mm wide, paired eyes looking out through transparent patches of carapace and resembling nothing more than a pair of flashlight reflectors (but operating in reverse!). The eyes have the optically correct parabolic profile in the horizontal plane but are spherical in the vertical plane, and they thus form poor images on the centrally located retina. Vision is further degraded because light rays from a single source pass through many receptors. Resolution must be poor, but the eyes have the greatest light-concentrating power known. Their use is undescribed, but they might be used for orientation or for locating luminescent sources in the deep sea.

A few species of small invertebrates have eyecuplike photoreceptors that appear to operate somewhat like the *Gigantocypris* type. These include the cerebral photoreceptors of some rotifers (Clément and Wurdak 1984), and also the pigment-cup eyes of some archiannelids (Verger-Bocquet 1984) and some flatworms (Fournier 1984). In all these, the rhabdomeric microvillar mass is placed in front of a small spherical cup of layered platelets of appropriate dimensions to form a multilayer interferential reflector. If the layered structure were in reality a reflector, these eyespots would have an increased sensitivity due to a double light path and possibly an increased collecting aperture; the receptors, however, appear to lie nearer the center of the sphere than to its focus. These eyes would be somewhat directionally sensitive, because the mirror works only for light entering the open part of the cup.

The eyes of the bivalve molluscs known as scallops also have spherical reflectors, which produce well-focussed images on the distal (ciliary) retina (there is a rhabdomeric proximal retina receiving an unfocussed image). The focussing is possible because light first enters the eye through a lens that corrects for the spherical aberration generated by the mirror. The eye can thus operate effectively with a wide aperture, though the path of the light, traversing the centrally located retina while both entering and leaving, means that contrast is degraded. Sensitivity must also be fairly low, for, if the retina were an excellent absorber, little light would reach the mirror in the first place. Scallop eyes are movement detectors and stimulate a rapid shell-closing response. Several dozen eyes exist in a single animal; their input would seem to offer the scallop much to ponder.

## Summary

In this chapter, I have attempted at least to touch on much of the diversity that characterizes marine invertebrate vision, while concentrating on the most prevalent and apparently successful designs. Although the scale of ocular organization in these animals ranges from the simplest imaginable to highly advanced types, it is always useful to remember that the animals possessing any type of eye are successfully competing within their own niches. Obviously the type arrived at by each species meets its needs.

A second point to be made in summary is that, in spite of the huge number of marine invertebrate species, occupying the greatest diversity of photic habitats in any biogeographical realm and with their unimaginably complex evolutionary convergences and divergences, only three types of eye are common: eyecups, compound eyes, and lens eyes. There appear to be only a few basic designs that work, without being evolutionary dead ends, and it may well be that we see all of them in modern marine invertebrates.

*Acknowledgments.* I thank T. Ford and F. Baldwin for assistance in preparing the figures used in this paper. Some of this material is based on research supported by Grant BNS-8311567, National Science Foundation.

### References

Ali, M.A. (1984) Photoreception and vision in invertebrates, Plenum, New York.

Autrum, H. (ed.) (1979) Invertebrate Photoreceptors: Handbook of Sensory Physiology, vol. VII/6A, Springer-Verlag, Berlin.

Autrum, H. (ed.) (1981a) Invertebrate Visual Centers and Behavior: Handbook of Sensory Physiology, vol. VII/6B, Springer-Verlag, Berlin.

Autrum, H. (ed.) (1981b) Invertebrate Visual Centers and Behavior: Handbook of Sensory Physiology, vol. VII/6C, Springer-Verlag, Berlin.

Bennett, M.F. (1979) Extraocular light receptors and circadian rhythms, in Invertebrate Photoreceptors: Handbook of Sensory Physiology, vol. VII/6A), Autrum, H. (ed.), Springer-Verlag, Berlin, pp. 641–663.

Brandenberger, J.L. and Eakin, R.M. (1980) Cytochemical localization of acid phosphatase in ocelli of the seawater *Patiria miniata* during recycling of photoreceptoral membranes. J. Exp. Zool., 214:127–140.

Burr, A.H. (1984a) Evolution of eyes and photoreceptor organelles in the lower phyla, in Photoreception and Vision in Invertebrates, Ali, M.A. (ed.), Plenum, New York, pp. 131–178.

Burr, A.H. (1984b) Photomovement behavior in simple invertebrates, in Photoreception and Vision in Invertebrates, Ali, M.A. (ed.), Plenum, New York, pp. 179–215.

Chamberlain, S.C. and Barlow, R.B. (1984) Transient membrane shedding in *Limulus* photoreceptors: control mechanisms under natural lighting, J. Neurosci., 4:2792–2810.

Clément, P. and Wurdak, E. (1984) Photoreceptors and photoreceptions in rotifers, in Photoreception and Vision in Invertebrates, Ali, M.A. (ed.), Plenum, New York, pp. 241–288.

Cronin, T.W. (1985) The visual pigment of a stomatopod crustacean, *Squilla empusa,* J. Comp. Physiol., 156:679–687.

Cronin, T.W. (1986a) Optical design and evolutionary adaptation in crustacean compound eyes. J. Crustacean Biol., 6:1–23.

Cronin, T.W. (1986b) Photoreception in marine invertebrates, Am. Zool., 26:403–415.

Cronin, T.W. and Goldsmith, T.H. (1984) Dark regeneration of rhodopsin in crayfish photoreceptors, J. Gen. Physiol., 84:63–81.

Cummins, D.R. and Goldsmith, T.H. (1981) Cellular identification of the violet receptor in the crayfish eye, J. Comp. Physiol., 142:199–202.

Diamond, J. (1985) Voyage of the overloaded ark, Discover, 6 (6):82–92.

Eakin, R.M. (1972) Structure of invertebrate photoreceptors, in Photochemistry of Vision: Handbook of sensory physiology, vol. VII/1, Autrum, H. (ed.), Springer-Verlag, Berlin, pp. 625–684.

Ebrey, T.G. and Honig, B. (1977) New wavelength dependent visual pigment nomograms, Vision Res., 17:147–151.

Forward, R.B., Jr (1977) Occurrence of a shadow response among brachyuran larvae, Mar. Biol., 39:331–342.

Fournier, A. (1984) Photoreceptors and photosensitivity in platyhelminthes, in Photoreception and Vision in Invertebrates, Ali, M.A. (ed.), Plenum, New York, pp. 217–239.

Goldsmith, T.H. and Wehner, R. (1977) Restrictions on rotational and translational diffusion of pigment in the membranes of a rhabdomeric photoreceptor, J. Gen. Physiol., 70:453–490.

Hamilton, P.V., Ardizzoni, S.C., and Penn, J.S. (1983) Eye structure and optics in the intertidal snail, *Littorina irrorata,* J. Comp. Physiol., 152:435–445.

Horridge, G.A. (1978) The separation of visual axes in apposition compound eyes, Philos. Trans. R. Soc. Lond., B285:1–59.

Hyatt, G.W. (1975) Physiological and behavioral evidence for color discrimination by fiddler crabs (Brachyura, Ocypodidae, genus *Uca*), in Physiological ecology of estuarine organisms, Vernberg, F.J. (ed.), University of South Carolina Press, Columbia, pp. 333–365.

Kampa, E.M. (1965) The euphausiid eye—a re-evaluation, Vision Res., 5:475–488.

Kataoka, S. and Yamamoto, T.Y. (1981) Diurnal changes in the fine structure of photoreceptors in an abalone, *Nordotis discus,* Cell Tissue Res., 218:181–189.

Kataoka, S. and Yamamoto, T.Y. (1983) Fine structure and formation of the photoreceptor in *Octopus ocellatus,* Biol. Cell, 49:45–54.

Kunze, P. (1979) Apposition and superposition eyes, in Invertebrate Photoreceptors: Handbook of Sensory Physiology, vol. VII/6A, Autrum, H. (ed.), Springer-Verlag, Berlin, pp. 441–502.

Land, M.F. (1978) Animal eyes with mirror optics, Sci. Am., 239(12):126–134.

Land, M.F. (1980a) Compound eyes: old and new optical mechanisms, Nature, 287:681–686.

Land, M.F. (1980b) Eye movements and the mechanism of vertical steering in euphausiid crustacea, J. Comp. Physiol., 137:255–265.

Land, M.F. (1981a) Optics and vision in invertebrates, in Invertebrate Visual Centers and Behavior: Handbook of Sensory Physiology, vol. VII/6B, Autrum, H. (ed.), Springer-Verlag, Berlin, pp. 471–592.

Land, M.F. (1981b) Optics of the eyes of *Phronima* and other deep-sea amphipods, J. Comp. Physiol., 145:209–226.

Land, M.F. (1982) Scanning eye movements in a heteropod mollusc, J. Exp. Biol., 96:427–430.

Land, M.F. (1984a) Molluscs, in Photoreception and Vision in Invertebrates, Ali, M.A. (ed), Plenum, New York, pp. 699–726.

Land, M.F. (1984b) The resolving power of diurnal superposition eyes measured with an ophthalmoscope, J. Comp. Physiol., 154:515–534.

Land, M.F. (1984c) Crustacea, in Photoreception and Vision in Invertebrates, Ali, M.A. (ed.), Plenum, New York, pp. 401–438.

Land, M.F. Burton, F.A., and Meyer-Rochow, V.B. (1979) The optical geometry of euphausiid eyes, J. Comp. Physiol., 130:49–62.

Langdon, J.W. and Herrnkind, W.F. (1985) Visual shape discrimination in the fiddler crab, *Uca pugilator,* Mar. Behav. Physiol., 11:315–326.

Messenger, J.B. (1981) Comparative physiology of vision in molluscs, in Invertebrate Visual Centers and Behavior: Handbook of Sensory Physiology, vol. VII/6C, Autrum, H. (ed.), Springer-Verlag, Berlin, pp. 93–200.

Moody, M.F. and Parriss, J.R. (1960) The discrimination of polarized light by *Octopus,* Nature, 186:839–840.

O'Tousa, J.E., Baehr, W., Martin, R.L., Hirsh, J. Pak, W.L., and Applebury, M.L. (1985) The *Drosophila* ninaE gene encodes an opsin, Cell, 40:839–850.

Saidel, W.M., Lettvin, J.Y., and MacNichol, E.F., Jr. (1983) Processing of polarized light by squid photoreceptors, Nature, 304:534–536.

Shaw, S.R., and Stowe, S. (1982) Photoreception, in Neurobiology: Structure and Function: The Biology of Crustacea, vol. 3, Sandeman, D.C. and Atwood, H.L. (eds.), Academic Press, New York, pp. 291–367.

Toh, Y., Yoshida, M., and Tateda, H. (1979) Fine structure of the ocellus of the hydromedusan, *Spirocodon saltatrix.* I. Receptor cells, J. Ultrastruct. Res., 68:341–352.

Verger-Bocquet, M. (1984) Photoréception et vision chez les Annélides, in Photoreception and Vision in Invertebrates, Ali, M.A. (ed.), Plenum, New York, pp. 289–334.

Vogt, K. (1983) Is the fly visual pigment a rhodopsin? Z. Naturforsch., 38c:329–333.

Vogt, K. (1984) The chromophore of the visual pigment in some insect orders, Z. Naturforsch., 39c:196–197.

Wald, G. and Rayport, S. (1977) Vision in annelid worms, Science, 196:1434–1439.

Waterman, T.H. and Horch, K.W. (1966) Mechanism of polarized light perception, Science, 154:467–475.

Wells, M.J. (1962) Brain and behavior in cephalopods, Stanford University Press, Stanford, Calif.

Westfall, J.A. (1982) Visual cells in evolution, Raven, New York.

Yoshida, M. (1979) Extraocular photoreception, in Invertebrate Photoreceptors: Handbook of Sensory Physiology, vol. VII/6A, Autrum, H. (ed.), Springer-Verlag, Berlin, pp. 581–640.

Yoshida, M., Takasu, N., and Tamotsu, S. (1984) Photoreception in echinoderms, in Photoreception and Vision in Invertebrates, Ali, M.A. (ed.), Plenum, New York, pp. 743–771.

Zuker, C.S., Cowman, A.F., and Rubin, G.M. (1985) Isolation and structure of a rhodopsin gene from *D. melanogaster,* Cell, 40:839–850.

# Chapter 17

# Vision and Mating Behavior in *Limulus*

Robert B. Barlow, Jr.,* Maureen K. Powers,** and Leonard Kass ***

Modulation of sensory function can provide a key for understanding the organization of sensory systems. This is indeed the case for the visual system of the horseshoe crab *Limulus polyphemus*. Large fluctuations in sensitivity occur every day in the lateral eyes of this animal. They are generated by nerve signals transmitted from a circadian clock located in the animal's brain (Barlow, Bolanowski, and Brachman 1977; Kaplan and Barlow 1980; Barlow 1983). The clock's signals can increase lateral eye sensitivity by as much as 100,000 times at night (Barlow, Bolanowski, and Brachman 1977; Barlow, Chamberlain, and Kass 1984).

Modulation of visual function via efferent input is not unique to *Limulus*. The scorpion *Androctonus australis* (Fleissner and Fleissner 1978), orb-spider, *Argiope bruennichii* (Yamashita and Tateda 1981), and rabbit (type American-Dutch 4) (Brandenburg, Bobbert, and Eggelmeyer 1981) all share this property. In each case, retinal sensitivity is modulated by efferent signals transmitted from a circadian clock located more centrally in the nervous system. The retinal sensitivity of the green sunfish, *Lepomis cyanellus,* also exhibits a circadian rhythm, but, in this case, efferent signals from the brain modulate rather than generate the retinal rhythms (Dearry and Barlow 1987). Efferent signals appear to modulate the firing patterns of retinal ganglion cells in the triggerfish *Hemibalistes chrysopterus* (Sandeman and Rosenthal 1976), as well as the excitability of semicircular canal afferents of the toadfish *Opsanus tau* (Highstein and Baker 1985); but, in these cases, the efferent signals are generated by somatosensory and vestibular stimulation and not by a circadian clock. Modulation of sensory function by efferent nerve signals has also been observed in the cat (Fex 1962), chick (Miles 1970), pigeon (Pearlman and Hughes 1976), and turtle *Pseudemys scripta elegans* (Art et al. 1982), though the mechanism for generating the efferent signals is not fully understood in these and other cases (Klinke and Galley 1974). Efferent

---

Marine Biological Laboratory, Woods Hole, MA 02543, USA Permanent addresses:
* Institute for Sensory Research, Syracuse University, Syracuse, NY 13244–5290, USA
** Department of Psychology, Vanderbilt University, Nashville, TN 37240, USA
*** Department of Biology, University of Maine, Orono, ME 04473, USA

pathways are known to exist in a number of other organisms (see Chap. 30), but their function is not known. Finally, in at least one organism, the electric fish, sensory function can be modulated in the apparent absence of an efferent pathway (Chap. 10).

Modulation of sensory function, regardless of its origin or specific effector mechanism, may serve to "tune" a particular sensory modality for optimal operation under specific environmental conditions. Aquatic animals are especially interesting in this respect because many of them live in environments that change periodically with the time of day, month, and year. Such conditions place serious constraints on sensory function (Chaps. 3, and 8).

The intertidal zone is unique among aquatic environments because it is periodically transformed from terrestrial to aquatic by movements of the earth, sun, and moon. Animals inhabiting this zone must cope all or part of their lives with its dramatic fluctuations. *Limulus* has successfully exploited the intertidal zone for the important biological function of mating. In the spring, animals move in from deep water along the northern Atlantic coast to build nests at the water's edge (Lockwood 1870; Barlow et al. 1986). During the mating season, males use visual cues to help them locate and mate with females (Barlow, Ireland, and Kass 1982; Barlow et al. 1983; Barlow et al. 1984).

Is the endogenous modulation of *Limulus* visual sensitivity related to the physical constraints imposed by the dynamic environment of the intertidal zone? In this chapter we will examine *Limulus* mating in the intertidal zone in terms of its relationship to circadian changes in visual sensitivity as measured physiologically (Barlow, Bolanowski, and Brachman 1977; Kaplan and Barlow 1980; Barlow, 1983) and in visual performance as measured behaviorally (Powers and Barlow 1985). Sensory and behavioral thresholds need not be the same, and we may learn something about the organization of a sensory system by analyzing it at different levels. *Limulus* provides an excellent opportunity for such an analysis. We will describe here the close correlations between the physiological and behavioral functions in *Limulus* and how they relate to the environmental conditions.

## Circadian Rhythms in Retinal Structure and Function

When *Limulus* is held in constant darkness, the lateral and median eyes produce larger electroretinographic responses (ERGs) at night than during the day. Figure 17.1 (*top*) shows the ERGs recorded over a 3-day period from the lateral eye in situ. The animal was maintained in a natural light-dark cycle before being placed in darkness. On day 1, in darkness, the amplitude of the ERG began to increase at 3:30 P.M., reached a maximum at 6 P.M., remained high until 2:30 A.M., and then decreased gradually to a low daytime level. This cycle continued on days 2 and 3 with a period of 24.8 hr. After 2 weeks in darkness (data not shown), the rhythm had drifted completely out of phase with the solar day.

Figure 17.1 (*bottom*) shows that the endogenous rhythm persists without attenuation for periods of up to a year in darkness. The recording procedure and test flash intensity was the same as in Figure 17.1 (*top*) but the animal was not. The circadian rhythm of this animal had a period of 23.4 hr. Experiments with other

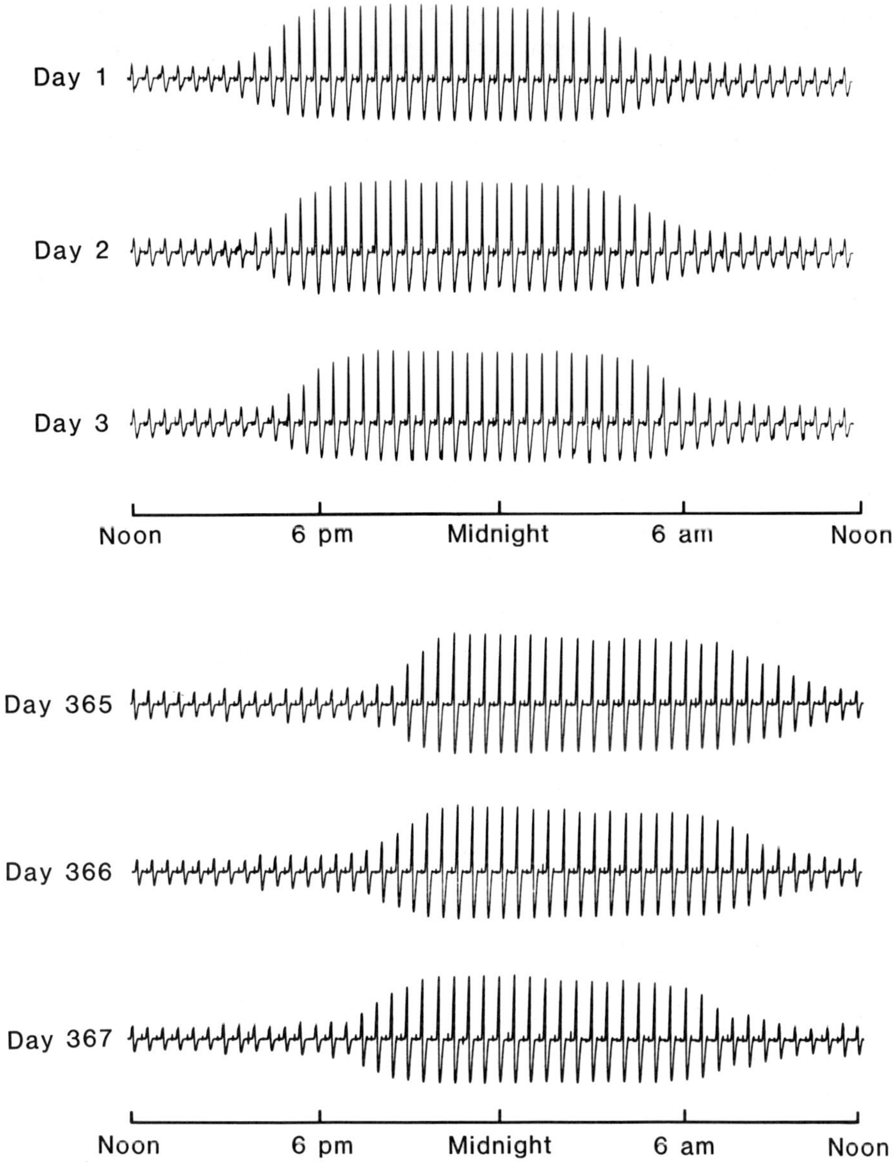


Figure 17.1. Circadian rhythms in the amplitude of the ERG of *Limulus* lateral eye. (*Top*) Continuous 3-day chart recording started at noon of day 1 when the recorder was gated on for 700 ms, during which time a 100-ms-flash was presented to the eye. The sequence was repeated every 30 min. for 3 days with a constant flash intensity of $10^6$ photons/ommatidium. Peak-to-peak amplitude was 210 μV during the night of day 1. (*Bottom*) Continuous chart recording of ERG responses from lateral eye of an animal maintained in darkness for 1 year. Recording procedure and test flash intensity were the same as above. Response amplitude was 130 μV during the night of day 365. (From Barlow 1983.)

animals yielded periods ranging from 22.2 to 25.5 hr, with a mean value of 23.9 ± 0.7 hr. Circadian changes in visual sensitivity corresponding to the changes in ERG amplitude in Figure 17.1 can be determined from the intensity-response function (Barlow 1983). The changes in ERG amplitude for both experiments in Figure 17.1 correspond to approximately a 25-fold increase in sensitivity at night. Other experiments yielded 20- to 100-fold increases in nighttime sensitivity (Barlow 1983).

Efferent optic nerve activity mediates the circadian rhythms shown in Figure 17.1. If the optic nerve is cut at midnight, the ERG amplitude will decline to the daytime level, following an exponential time course, with a time constant of 0.78 hr. If the optic nerve is cut during the day, the ERG amplitude will remain at the daytime level. Recordings from the proximal stump of the cut optic nerve reveal efferent activity having a circadian rhythm in synchrony with that of the ERG amplitude. Delivering current pulses to the distal stump of the cut optic nerve mimics the effects of endogenous efferent activity on the ERG amplitude. Thus it can be concluded that efferent fibers in the optic nerve trunk transmit the signals that mediate the circadian rhythms in visual sensitivity (Barlow 1983).

The nighttime increase in ERG amplitude in Figure 17.1 results primarily from changes in the retinal structure. Figure 17.2 gives anatomical reconstructions of dark-adapted ommatidia from lateral eyes fixed during the day and night. The longitudinal reconstructions show that an ommatidium is composed of several photoreceptors (retinular cells) tightly clustered around the dendrite of an eccentric cell. Light collected by the corneal lens passes through an aperture formed by surrounding pigment cells and enters the photosensitive rhabdom of the retinular cells (Chamberlain and Barlow 1977; Barlow and Chamberlain 1980). During the day, the narrow aperture (17 μm in diameter) of a dark-adapted ommatidium restricts the light entering the retinular cells. In addition, the elongated, light-sensitive rhabdom and associated pigment granules reduce the effectiveness of oblique rays of light. Both structural features contribute to the relatively narrow daytime field of view plotted at the top of Figure 17.2.

At night, structural changes increase the field of view from 6 to 13 degrees (angular width at half maximum sensitivity) (Barlow, Chamberlain, and Levinson 1980). The wider acceptance angle allows photoreceptors to absorb more photons at night, which in turn increases the amplitude of the ERG. Over 95% of the nighttime increase in ERG amplitude can be attributed to an increase in photon catch (Barlow 1983).

The wider field of view at night also reduces spatial resolution. The resolution of compound apposition eyes, such as the *Limulus* lateral eye, is determined by the degree of overlap of the fields of view of adjacent ommatidia. Since the nighttime acceptance angle of 13 degrees is about twice the angle separating the axes of adjacent ommatidia, the fields of view of adjacent ommatidia overlap considerably (Barlow, Chamberlain, and Kass 1984). High daytime acuity is, therefore, exchanged for high nighttime photon catch under the control of a circadian clock.

Table 17.1 summarizes the circadian rhythms we have detected in the retinal properties of the *Limulus* eye. All are mediated by the *efferent input* from a circadian clock located in the brain, first entry in the table. The next three—

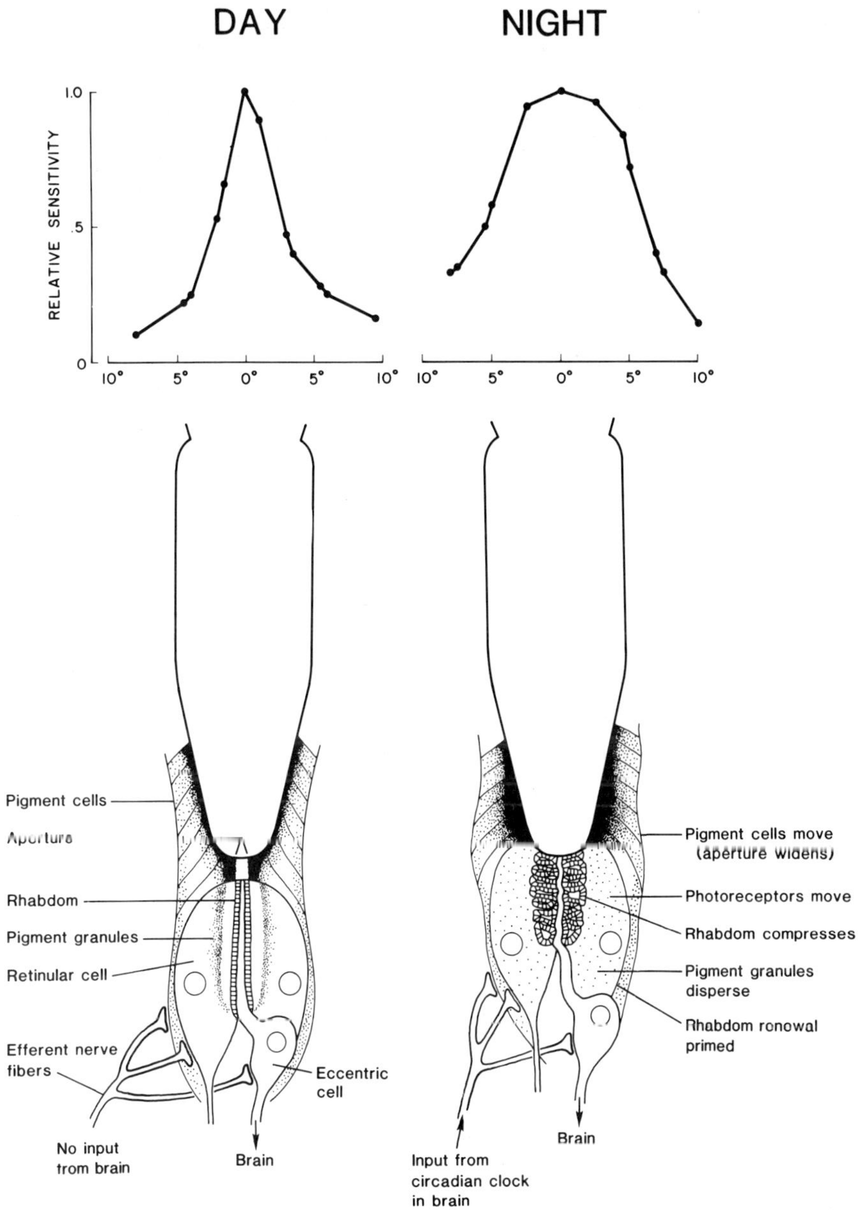


Figure 17.2. Circadian changes in the structure and field of view of a *Limulus* ommatidium. Longitudinal reconstructions are based on micrographs of serial cross sections of dark-adapted ommatidia fixed in situ during the day and night. Schematic efferent terminals are included to show that each type of retinal cell receives synaptic input from circadian clock in the brain. The diameter of corneal lens is 200 μm. Curves at the *top* give relative sensitivity of the discharge of a single optic nerve fiber as a function of the angle of incidence of light flashes delivered from a point source. The width of an acceptance angle at 0.5 sensitivity was 6 degrees during the day and 13 degrees at night. (From Barlow, Chamberlain, and Levinson 1980.)

Table 17.1. Circadian Rhythms in the *Limulus* Retina

| Retinal property | Day | Night |
|---|---|---|
| Efferent input | Absent | Present |
| Gain | Low | High |
| Noise | High | Low |
| Frequency response | Fast | Slow |
| Photon catch | Low | High |
| Aperture | Constricted | Dilated |
| Acceptance angle | 6° | 13° |
| Cell position | Proximal | Distal |
| Pigment granules | Clustered | Dispersed |
| Membrane turnover | Trigger | Prime |
| Lateral inhibition | Strong | Weak |

*gain, noise,* and *frequency response*—are physiological properties that originate in photoreceptor cells (Kaplan and Barlow 1980; Batra 1983; Renninger, Kaplan, and Barlow 1984; Kaplan, Renninger, and Barlow 1986; Barlow et al. 1987). Circadian changes in structure (*aperture* and *cell position*) contribute to rhythms in *photon catch* and *acceptance angle*. We do not know what effect changes in *pigment granule distribution* and *membrane turnover* may have on retinal sensitivity (Chamberlain and Barlow 1979, 1984). *Lateral inhibition* is an integrative mechanism of the retina (Hartline 1972). Reducing its strength at night increases retinal responses (Batra and Barlow 1982). Circadian rhythms in the properties in Table 17.1 can increase retinal sensitivity as much as 100,000 times at night (Barlow, Chamberlain, and Kass 1984).

## Laboratory Studies of Circadian Rhythms in Visual Sensitivity

Do circadian changes in the sensitivity of the lateral eyes influence the animal's behavior? To answer this question, we measured reflexive movements of the telson (tail) in response to lateral eye illumination. Tail movements were detected by attaching a thermistor to the tail and recording changes in the voltage drop across the thermistor when the tail moved in response to a brief series of flashes delivered with a light pipe to one lateral eye. We recorded the ERG from the same eye for a convenient measure of retinal sensitivity during the reflexive response (Powers and Barlow 1985).

Figure 17.3 plots the amplitude of the ERG and the probability of tail movements as a function of the time of day for one animal while it remained in darkness. Tail movements occurred more frequently during the animal's subjective night when the ERG was elevated. Both responses changed systematically with the time of day and were highly correlated with one another ($r = 0.84$; $p < 0.001$). In addition to these reflexive measures, we developed a classical conditioning paradigm for assessing the magnitude of circadian changes in visual sensitivity (Powers and Barlow 1985). We found that the animals maintained in

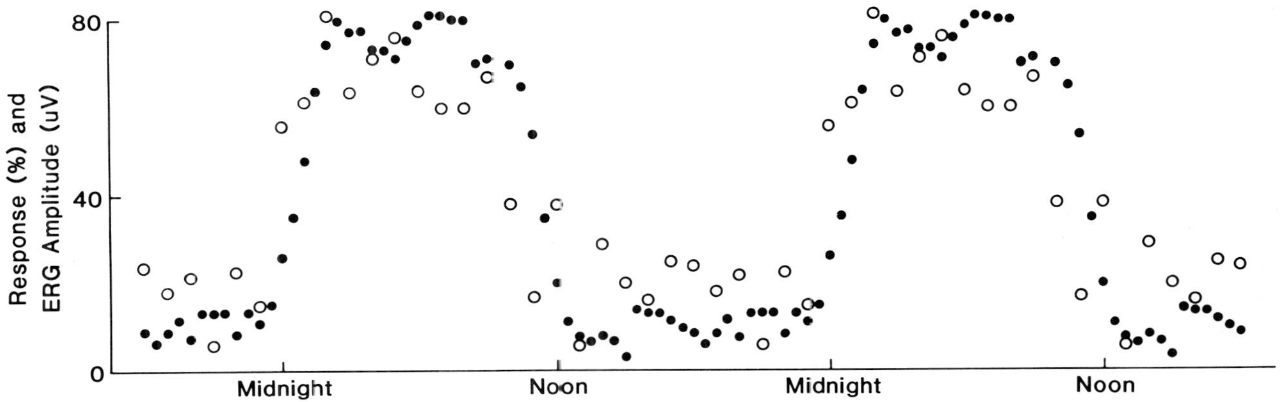


Figure 17.3. Circadian rhythm in reflexive movements of the tail elicited by visual stimuli (*open circles*). The probability of tail movement correlated with the circadian rhythm in ERG recorded simultaneously from the lateral eye (*filled circles*). Light stimulus was a constant intensity flash delivered every 15 min. to the lateral eye. (From Powers and Barlow 1985.)

constant darkness were about 10 times more sensitive during the subjective night than during the subjective day. This day-night change in visual sensitivity is consistent with that measured physiologically (Barlow, Chamberlain, and Levinson 1980). Thus, under controlled laboratory conditions, changes in specific behavioral responses appear to reflect circadian changes in retinal sensitivity.

## The Role of Vision in *Limulus* Mating Behavior

In 1928 Hartline noted that the lateral eye of *Limulus* was an admirable preparation for research in visual physiology (Hartline 1928, 1972). Extensive studies of the lateral eye since that time have provided a wealth of information on the basic mechanisms in vision. Until recently, however, nothing was known about the usefulness of vision to the animal in its natural habitat. Several years ago, we began investigating mating behavior to determine whether this behavior involves vision. In this section, we will briefly describe the behavior itself—when it occurs and other characteristics of interest—and then we will describe our experiments on vision.

*Limulus* on the northeast coast of North America mate in the spring (Lockwood 1870; Barlow et al. 1986). Male-female pairs and single males move in periodically from deeper water to lay eggs and fertilize them near the water's edge (see Fig. 17.4, *top*). During mating activity, the single males swim parallel to the beach within about 10 m of the water's edge. Pairs also swim, but the female eventually digs into the sand to deposit her eggs. The single males gather at these nest sites and release sperm along with the paired male, presumably to increase their individual chances of fertilizing the eggs. The ratio of males to females during mating activity is high. In the 1984 season, we counted a total of 14,995 animals at our research site on Mashnee Dike, Cape Cod, Massachusetts. Only 4,278 of them were females, and all but six were in amplexus. This surplus of males, apparently in search of mating partners, was the subject pool for the visual experiments described below.

The migration of the animals into the intertidal zone for mating is correlated with the height of the tide and the phase of the moon (Barlow et al. 1986). At our research site, *Limulus* (males and females) migrated only during high tides; the greatest numbers appearing the spring high tides of the new and full moons. Also two tide cycles occur each day at this site, and the heights of consecutive high tides are not equal. Our field studies show that the animals prefer to mate during the higher of the two high tides and that they use the first light of dawn as a cue to shift from one high tide to the other at the fortnightly transition of the tidal inequality (Barlow et al. 1986). In sum, the mating behavior of this species at Mashnee Dike is regulated by sunlight, tide height, and lunar phase.

To test whether vision has a role in mating behavior, we observed the behavior of the single males in the vicinity of cement castings placed in the nesting area during high tide. The castings were made from an adult female carapace, a hemisphere, and a cube, all having equally exposed surface areas. Each casting was painted black, gray, or white, giving relative contrasts underwater against

Figure 17.4. (*Top*) A pair of horseshoe crabs nesting near the water's edge at night on Mashnee Dike, Cape Cod, Massachusetts. (*Bottom*) A male *Limulus* mounted on a submerged black cement casting of a female carapace in the normal position for egg fertilization.

sand of −0.70, +0.1, and +0.75, respectively. The *Limulus* carapace has an average contrast against sand of $-0.33 \pm 0.03$. Relative contrast is defined as the logarithm of the ratio of reflectances. Figure 17.4 (*bottom*) shows a male *Limulus* mounted on the dorsal part of a black cement casting of the female carapace. This is the usual position for egg fertilization. The male's behavior is indistinguishable from that of normal mating, as shown in Figure 17.4 (*top*). Both photographs were taken on the same night near the water's edge at high tide on Mashnee Dike, Cape Cod.

The attraction of males to a specific casting depended on its form and contrast. We counted the number of males near each submerged casting during eight high tides in May and June 1984, four at night and four during the day (Barlow, Ireland, and Kass 1982). Figure 17.5 (*top*) shows that the number of males was greatest around the female models, moderate around hemispheres, and lowest around cubes. Relative numbers of animals did not change from day to night. Male *Limulus* can thus discriminate form. Such discrimination may result from tactile or visual inputs or both. Figure 17.5 (*bottom*) shows that the greatest number of animals congregated around the black castings (40%), regardless of the time of day. Males were attracted in almost equal numbers to the gray castings during the day, but not at night. The contrasts of black and gray castings bracket that of the female carapace. Both were visible during the day, but at night the low relative contrast of the gray castings against the sand (+0.1) may have made them less visible, since fewer animals were observed there. Indeed, we could not see the gray castings at night but had no difficulty in detecting the white and black castings by illumination from either a new or full moon. When visibility is not a limiting factor, males are apparently attracted to objects similar in shape to the female carapace. These results, together with those showing that blinded animals were not attracted to the castings (Barlow, Ireland, and Kass 1982), indicate that vision has a role in mating behavior.

To test the relative attraction of specific forms, we placed black silhouettes of an adult female carapace, a hemisphere, and a cube 1.1 m in front of the opening of a submerged holding pen. The three silhouettes were located at 45, 90, and 135 degrees with respect to the pen's opening. Male *Limulus* were placed in the pen and allowed to exit via the opening of their own accord. Most of the animals leaving the pen (255 of 349) approached and contacted one of the three silhouettes. The orientation behavior did not appear to be influenced by the specific shape of the silhouettes (Barlow et al. 1983). We therefore concluded that these objects were equally attractive.

Our field observations are consistent with the following scheme: Males (1) approach all objects they can see, (2) explore the objects by tactile contact, and (3) attempt to mate with those having the appropriate shape. Thus, both vision and taction appear to be involved. We should note that the animal has a well-developed chemosensory system (Wyse 1971); but, as yet, no evidence shows that this system is involved in mating.

What limits *Limulus* vision underwater? The data in Figure 17.5 show that contrast is a significant factor. To investigate other possible factors, we analyzed the approach behavior of males in the vicinity of dark, submerged objects. Figure

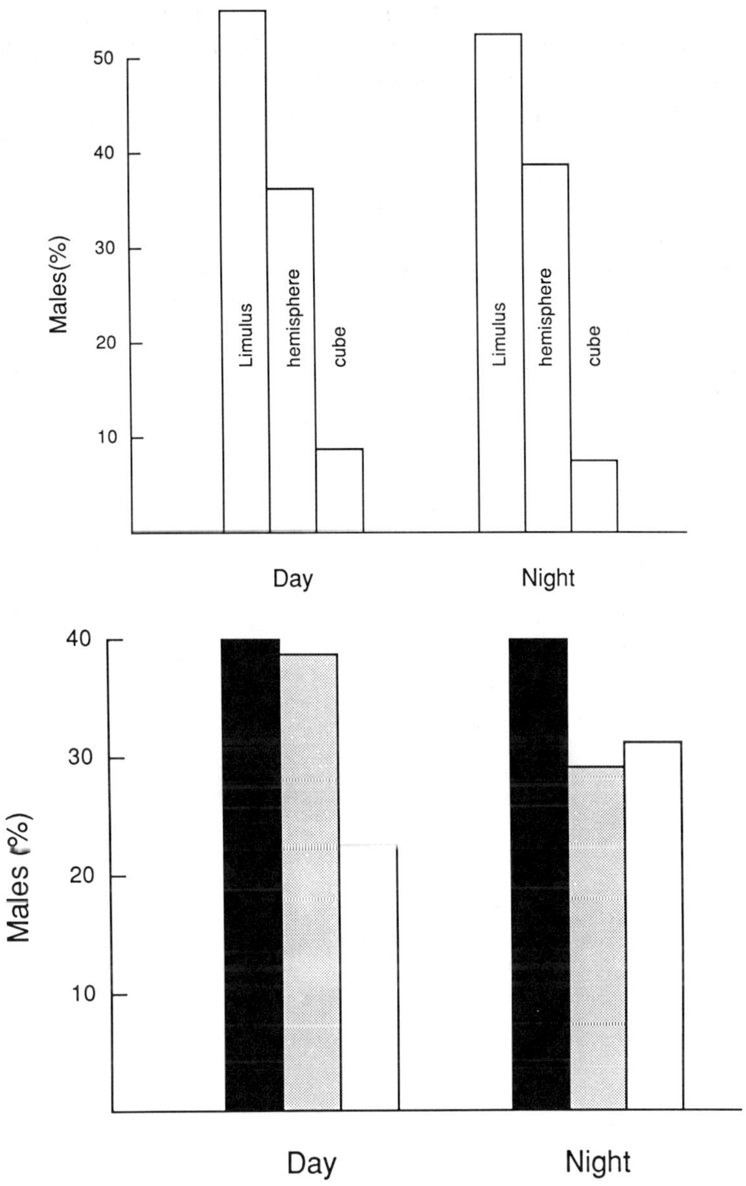


Figure 17.5. The attraction of male *Limulus* to cement castings during the day and at night. (*Top*) Average percentage of males observed at each cement form during four daytime and four nighttime observation periods. Castings of the female carapace attracted the greatest percentage of males both day and night. (*Bottom*) The percentage of males in contact with black, gray, and white castings during the same eight observation periods. Black castings attracted the greatest number of males regardless of time of day. The preference for gray castings was almost equal to that for black during the day, but not at night. Results are based on a total of 6,988 males in contact with eighteen castings. (From Barlow, Ireland, and Kass 1982.)

17.6 shows the experimental setup at the water's edge on the south side of Mashnee Dike, Cape Cod. A black cement hemisphere (29-cm diameter) was placed 5 m from the water's edge at high tide. The hemisphere, which was one of the objects used in the study described above, is about the size of an adult female *Limulus*. We recorded the behavior of males near the hemisphere on videotape, using a battery-operated infrared illumination and TV system.

Figure 17.7 shows how we analyzed the videotapes. Male behavior was divided into two categories. We scored males that made contact with the black hemisphere as ''hits'' and measured the distance from the hemisphere at which the animals oriented toward it. We scored males that did not make contact as ''misses'' and measured the distance between the hemisphere and the animal's point of closest approach. In a preliminary study with this technique, we observed the behavior of 523 males during seven daytime and six nighttime high tides of the mating season in 1984 (Barlow et al. 1984).

Figure 17.8 gives the number of hits and misses during the day (*solid line*) and at night (*broken line*). All hits are grouped together at 0 on the abscissa. All misses are divided into bins indicating their distance of closest approach. Note that during the day many animals passed by the hemisphere at distances >1.2 m and did not orient toward it. Relatively few animals passed within 1.0 m without hitting it. In fact, all animals passing within 0.1 to 0.3 m of the

Figure 17.6. Field setup for observing *Limulus* behavior in the vicinity of a submerged black cement casting. Animals are detected with a battery-powered infrared TV system that includes lamps attached to an overhead boom for nighttime work and a video cassette recorder for data collection. Field station was on the south side of Mashnee Dike, Cape Cod, Massachusetts.

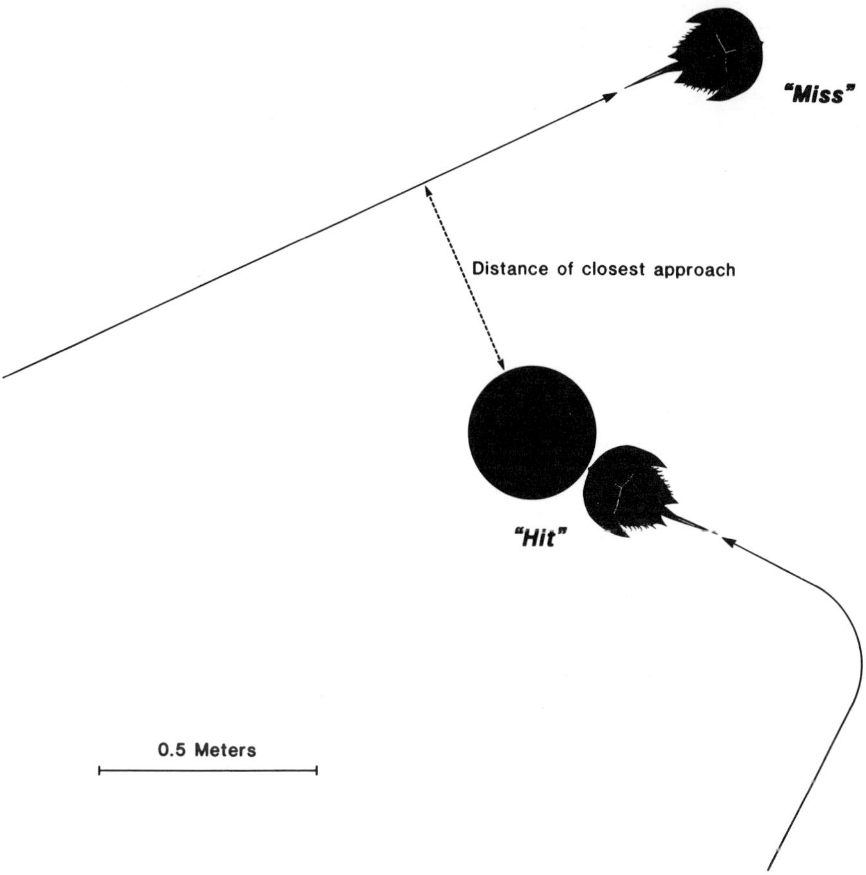


Figure 17.7. A diagram of camera view from Figure 17.6. Animals passing by the black hemisphere were scored as "misses," and their distance of closest approach was measured by playing back videotape recordings. Animals making contact with the cement hemisphere were scored as "hits."

hemisphere changed direction and hit the casting. Control experiments carried out with the hemisphere removed show that the density of animals is uniform throughout the test area; i.e., all distances of animal movement with respect to an arbitrary reference point in the test area were equally probable.

The distance at which the number of animals missing the hemisphere declines can be used as a measure of the "sighting distance," because it is the greatest distance at which animals begin to orient towards the hemisphere. Figure 17.8 shows that the sighting distance was 1.2 m during the day and 1.0 m at night, suggesting that the animals must be about 0.2 m closer to the hemisphere to see it at night. The base of the hemisphere subtended visual angles of 14 and 17 degrees, respectively, for the day- and night-sighting distances (Barlow et al. 1984). For all males that hit the hemisphere, the average distances at which orientation occurred were $0.97 \pm 0.04$ m and $0.53 \pm 0.04$ m, respectively, for day and night.

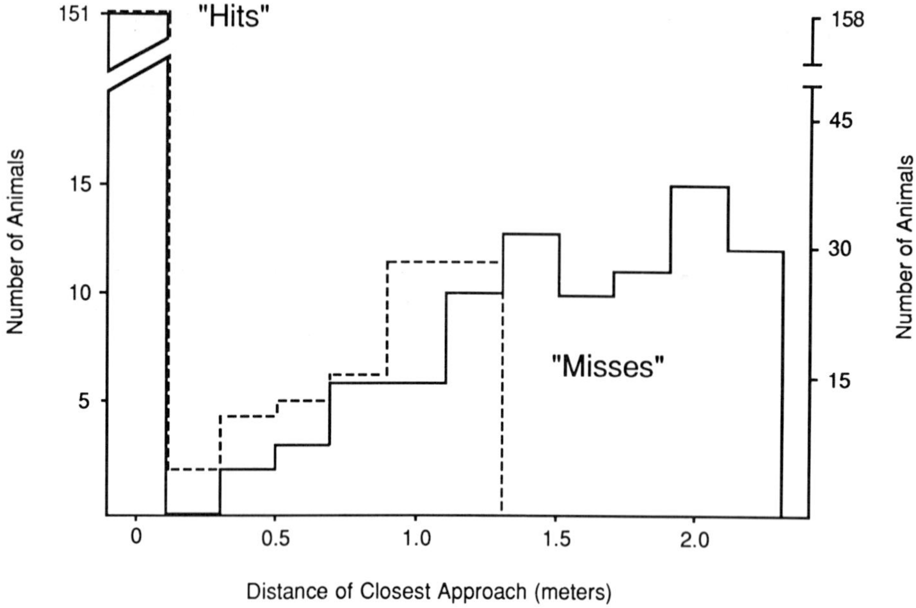


Figure 17.8. The distance of closest approach of males to the black hemisphere during the day (*solid line, left ordinate*) and at night (*broken line, right ordinate*). Hits are scored at zero on the abscissa. Results are based on a total of 523 males observed in seven daytime and six nighttime sessions at the field station on Mashnee Dike. (From Barlow et al. 1984.)

Sighting distances yield information on factors that limit *Limulus* vision underwater. For example, the visual angles subtended by the hemisphere for day and night orientation may be related to the optical properties of the lateral eye. In the central portion of the eye, the optic axes of adjacent ommatidia diverge by about 6 degrees (Barlow, Chamberlain, and Kass 1984), which is about equal to the width of the field of view of an ommatidium during the day and half that at night (Barlow, Chamberlain, and Levinson 1980). Calculations show that objects subtending visual angles of 14–17 degrees can be seen by no more than four ommatidia both day or night. Detection by so few ommatidia suggests that the animal may be operating near the limit of the optical properties of its compound lateral eyes. Although acuity is sacrificed somewhat at night to increase photon catch, the loss does not appear significant.

## Summary

*Limulus* vision at night in its natural habitat appears to be limited by at least three factors: (1) retinal sensitivity, (2) acuity, and (3) target contrast. The animals rely on vision to find mates, and much of this behavior occurs at night. The animals have adapted to the low levels of illumination from the nighttime sky by evolving elegant cellular mechanisms that increase retinal sensitivity and thus enable them to detect targets of sufficiently high contrast. The endogenous increase in retinal sensitivity at night occurs without a significant reduction in acuity. Further underwater studies of *Limulus* vision with targets of different size and contrast may well allow us to analyze the animal's behavior in terms of specific physiological and anatomical characteristics of its visual system.

*Acknowledgments.* Research was supported by NIH Grants EY-00667, EY-03352, EY-00246, and EY-05443, by NSF Grant BNS-8320315, and by a grant from the University Research Council of Vanderbilt University. For their assistance and contributions we thank Mark D. Camara, Robert W. Fiordalice, Heidi A. Howard, Vivian Mancini, and Janice L. Pelletier.

## References

Art, J.J. Crawford, A.C. Fettiplace, R., and Fuchs, P.A. (1982) Efferent regulation of hair cells in the turtle cochlea, Proc. R. Soc. Lond. Biol. Sci. 216:377–384.

Barlow, R.B., Jr. (1983) Circadian rhythms in the *Limulus* visual system, J. Neurosci., 3:856–870.

Barlow, R.B., Jr., Chamberlain, S.C., and Kass, L. (1984) Circadian rhythms in retinal function, in Molecular and Cellular Basis of Visual Acuity, Hilfer, S.R. and Sheffield, J.B. (eds.), Springer-Verlag, New York, pp. 31–53.

Barlow, R.B., Jr. and Chamberlain, S.C. (1980) Light and a circadian clock modulate structure and function in *Limulus* photoreceptors, in The Effects of Constant Light on Visual Processes, Williams, T.P. and Baker, B.N. (eds), Plenum, New York, pp. 247–269.

Barlow, R.B., Jr., Bolanowski, S.J., Jr., and Brachman, M.L. (1977) Efferent optic nerve fibers mediate circadian rhythms in the *Limulus* eye, Science, 197:86–89.

Barlow, R.B., Jr., Chamberlain, S.C., and Levinson, J.Z. (1980) *Limulus* brain modulates the structure and function of the lateral eyes, Science, 210:1037–1039.

Barlow, R.B., Jr., Ireland, L.C., and Kass, L. (1982) Vision has a role in *Limulus* mating behavior, Nature, 296:65–66.

Barlow, R.B. Jr., Kass, L., Mancini, V., and Pelletier, J.L. (1983) Vision in *Limulus* mating behavior: tests for detection and form discrimination, Biol. Bull., 165:539.

Barlow, R.B., Jr., Powers, M.K., Kass, L., Fiordalice, R.W., Camara, M.D., and Howard, H.A. (1984) Vision in *Limulus* mating behavior during the day and at night, Biol. Bull., 167:522.

Barlow, R.B., Jr., Powers, M.K., Howard, H., and Kass, L. (1986) Migratory behavior of *Limulus* for mating: Relation to lunar phase, tide height, and sunlight. Biol. Bull., 171:310–329.

Barlow, R.B., Jr., Kaplan, E., Renninger, G.H., and Saito, T. (1987) Circadian rhythms in *Limulus* photoreceptors. I. Intracellular recordings. J. Gen. Physiol., 89:353–370.

Batra, R. (1983) Efferent control of visual processing in the lateral eye of the horseshoe crab, Ph.D. Thesis, Syracuse University, Syracuse, New York.

Batra, R. and Barlow, R.B. Jr. (1982) Efferent control of pattern vision in *Limulus*, Soc. Neurosci. Abst., 9:849.

Brandenburg, J., Bobbert, A.C., and Eggelmeyer, F. (1983) Circadian changes in the response of the rabbit's retina to flashes, Behav. Brain Res., 7:113–123.

Chamberlain, S.C. and Barlow, R.B. Jr. (1977) Morphological correlates of efferent circadian activity and light adaptation in the *Limulus* lateral eye, Biol. Bull., 153:418–419.

Chamberlain, S.C. and Barlow, R.B., Jr. (1979) Light and efferent activity control rhabdom turnover in *Limulus* photoreceptors, Science, 206:361–363.

Chamberlain, S.C. and Barlow, R.B. Jr. (1984) Transient membrane shedding in *Limulus* photoreceptors: control mechanisms under natural lighting, J. Neurosci., 4:2792–2810.

Dearry, A., and Barlow, R.B., Jr. (1987) Circadian rhythms in the green sunfish retina. J. Gen. Physiol., 89 (in press).

Fex, J. (1962) Auditory activity in centrifugal and centripedal cochlear fibers in cat, Acta Physiol. Scand. 55 (supp.), 189:2–68.

Fleissner, G. and Fleissner, G. (1978) The optic nerve mediates the circadian pigment migration in the median eyes of the scorpion, Comp. Biochem. Physiol. (A), 61:69–71.

Hartline, H.K. (1928) A quantitative and descriptive study of the electrical response to illumination of the arthropod eye, Am. J. Physiol., 83:466–483.

Hartline, H.K. (1972) Visual receptors and retinal interaction, in Nobel Lectures: Physiology or Medicine 1963–1970, Elsevier, New York, pp. 269–288.

Highstein, S.M. and Baker, R. (1985) Action of the efferent vestibular system on primary afferents of toadfish, *Opsanus tau,* J. Neurophysiol., 54:370–384.

Kaplan, E. and Barlow, R.B., Jr. (1980) Circadian clock in *Limulus* brain increases response and decreases noise of retinal photoreceptors, Nature, 286:393–395.

Kaplan, E., Barlow, R.B., Jr., and Renninger, G.H. (1986) The circadian clock in the *Limulus* brain modifies the electrical properties of the photoreceptor membrane. Biol. Bull., 171:495.

Klinke, R. and Galley, N. (1974) Efferent innervation of vestibular and auditory receptors, Physiol. Rev., 54:316–357.

Lockwood, S. (1870) The horse foot crab, Am. Nat., 4:257–274.

Miles, F.A. (1970) Centrifugal effects in the avian retina, Science, 170:992–995.

Pearlman, A.L. and Hughes, C.P. (1976) Functional role of efferents to the avian retina. II. Effects of reversible cooling of the isothmo-optic nucleus, J. Comp. Physiol., 166:123–132.

Powers, M.K. and Barlow, R.B., Jr. (1985) Behavioral correlates of circadian rhythms in *Limulus* visual system, Biol. Bull., 169:578–591.

Renninger, G.H. Kaplan, E., and Barlow, R.B., Jr. (1984) A circadian clock increases the gain of photoreceptor cells of the *Limulus* lateral eye, Biol. Bull., 167:501–541.

Sandeman, D.C. and Rosenthal, N.P. (1976) Efferent axons in the fish optic nerve and their effect on the retinal ganglion cell, Brain Res., 68:41–54.

Wyse, G.A. (1971) Receptor organization and function in *Limulus* chelae, Z. Vgl. Physiol., 73:249–273.

Yamashita, S. and Tateda, H. (1981) Efferent neural control in the eyes of orb weaving spiders, J. Comp. Physiol. (A), 143:477–483.

# Chapter 18

# Aquatic Adaptations in Fish Eyes

RUSSELL D. FERNALD *

> To suppose that the eye with all its inimitable contrivances for adjusting the focus to different distances, for admitting different amounts of light, and for the correction of spherical and chromatic aberration, could have been formed by natural selection, seems, I freely confess, absurd in the highest degree.
>
> (Charles Darwin, *The Origin of Species,* 1859, p. 168)

Of all the sense organs, eyes have probably attracted the most attention because of both their central importance and intricate construction. Darwin knew that such "organs of extreme perfection and complication" posed a crucial test of his theory because they seemed *too* good to have been shaped by natural selection (Darwin [1859] 1958). Since eyes must obey the optical laws of physics, fundamental physical constraints on their structure provide an important analytical basis for understanding adaptive ocular specializations. In light of these physical constraints, inferences about the selective forces that have shaped eye design can be made with some confidence, particularly in the study of aquatic eyes.

Fish eyes offer an unusually interesting and accessible opportunity to ask general questions about visual evolution, development, and function (Fernald 1984). Although a great deal of work has been done on the domestic goldfish, this narrow experimental focus obscures the fact that bony fish form the largest class of vertebrates, comprising over 25,000 species. Evidence of their success can be found in the adaptive radiation of forms occupying an extraordinary range of habitats. Fish eyes and brains variously reflect success in mastering different life history strategies with the unsurprising general rule that the favored sensory modality occupies disproportionately more brain volume (see Geiger 1956).

There are several summaries of information about the fish eye available, so here I will present primarily recent findings. Since fish eyes have a relatively standard construction, with some deviations resulting from adaptations to specialized situations, I will first discuss underlying physical constraints that apply generally to the evolution of eyes. Then I will analyze various structures that have evolved to meet these constraints in teleost fish and, finally, present some new evidence about novel solutions to problems posed by regulation of fish eye growth.

---

* Institute of Neuroscience, University of Oregon, Eugene, OR 97403, USA

## Physical Constraints on Eye Design

In striking contrast to the enormous variety of eye types that have evolved among invertebrate species (Land 1981; see also Chap. 16), there exists essentially a single type of image-forming vertebrate eye. Because the vertebrate eye first evolved in water it is sensitive to only a narrow band of wavelengths of electromagnetic radiation, which are transmitted through water without significant attenuation (Fig. 18.1). The only other band with comparably low transmitted attenuation ($< 10^3$ Hz) has been exploited by weakly electric fish (see Chaps. 9 and 33). At low frequencies, however, the very long wavelengths make image formation impossible, since any detector must be large relative to the wavelengths detected. As a consequence, electric fish have detectors distributed all along their bodies from whose response they ''compute'' an image.

The exploitation of light for vision has proven a powerful selective force in all organisms (see Chap. 3). Light travels in straight lines, can be reflected, and varies in both wavelength (color) and intensity. Many of the design principles and even apparent errors we recognize in eyes reflect constraints arising directly from the physical properties of light. Natural selection cannot be expected to produce an error-free eye, but rather one just good enough for the animal's particular circumstances (Darwin 1859, 187). To inform their owners about the visual world, image-forming eyes must both *capture light* and *resolve it into images*. These two requirements for vision have shaped the evolution of eyes throughout the animal kingdom.

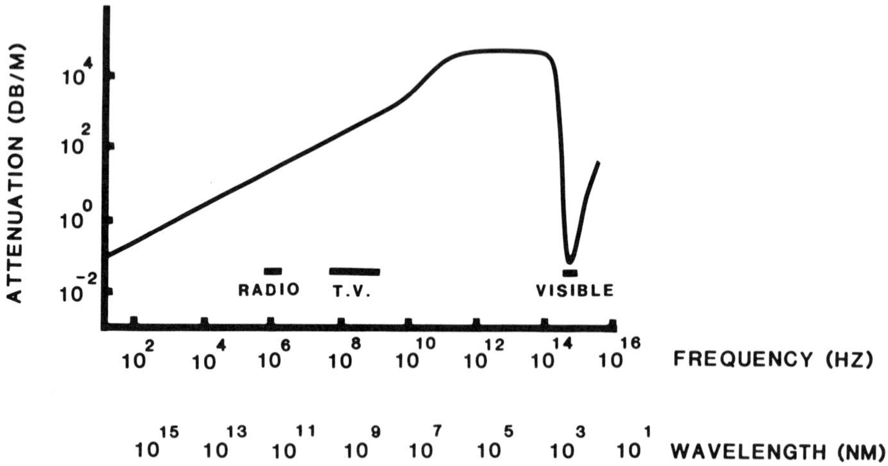


Figure 18.1. The attenuation (dB/m) of electromagnetic radiation in seawater plotted as a function of frequency (Hz) and wavelength (nm) of that radiation. Bands of electromagnetic energy used for radio and television transmission are shown, as is the narrow band corresponding to visible light.

## Light Collection

The sensitivity of an eye depends on the brightness of the image reaching the retina. In viewing a point source of intensity, $I$ (in candelas), in which all the light entering the eye comes from a single point, the flux $F$ (in lumens) in the eye is given by

$$F = (I \times P_a)/D^2 \quad (1)$$

where $P_a$ = pupillary area and $D$ = distance from the source to the plane of the pupil (Land 1981). Thus the brightness of the image increases in proportion to both the intensity of the source and the area of the pupil and decreases in proportion to the square of the distance from the source to the plane of the pupil. For deep-sea fish, there may well be point sources of light, such as bioluminescent organisms or organs of other fish; so light-gathering ability was likely a selective force leading to the large eyes in these creatures (see below).

Most animals living in higher levels of ambient daytime light near the surface of the water, however, usually view scenes illuminated by extended sources for which the retinal illuminance $E$ (in lumen/m$^2$) is

$$E = (L \times P_a)/f^2 \quad (2)$$

where $L$ = luminance of the extended surface source, $P_a$ = pupillary area, and $f$ = focal length (Land 1981).

The retinal illuminance can be related to the well-known f-number of a lens

$$\text{f-number} = \text{focal length/aperture diameter} \quad (3)$$

because, in most fishes, *both* the focal length and aperture diameter are related to the lens diameter. The focal length divided by the lens radius equals a constant, *m*, called *Matthiessen's ratio* (Matthiessen 1882; see also Table 18.1):

$$f/r = m. \quad (4)$$

The effective aperture diameter is the lens diameter (Fernald, Wright and Shelton 1985 in prep.), since the iris does not cover any part of the lens. So, for fish, where $m$ is a constant equalling Matthiessen's ratio,

$$\text{f-number} = (m \times r)/(2 \times r) = m/2. \quad (5)$$

Since the pupillary area equals $\pi r^2$ where $r$ = lens diameter, the retinal illuminance can be expressed in terms of f-number by combining equations (2) and (3):

$$E = (\pi/4) \times L \times (1/\text{f-number})^2. \quad (6)$$

Substituting equation (5) into (6), the retinal illuminance $E$ of an extended image can be expressed as

Table 18.1. Matthiessen's Ratio

| Species | Matthiessen's Ratio | λ (nm) | Reference |
|---|---|---|---|
| *Haplochromis burtoni* (African cichlid) | 2.25 | 494 | Fernald and Wright 1983 |
| *Salmo gairdnerii* | 2.26 | 435.8 | Scroczyński 1975a |
| (trout) | 2.32 | 546 | |
| | 2.35 | 640.2 | |
| *Esox lucius* | 2.45 | | Matthiessen 1882 |
| (pike) | 2.40 | 546 | Scroczyński 1976 |
| *Rutilus rutilus* (roach) | 2.23 | 546 | Scroczyński 1977 |
| *Carassius auratus* (goldfish) | 2.35 | Incandescent | Charman and Tucker 1973 |
| *Cyprinus carpio* (carp) | 2.4 | | Matthiessen 1882 |
| Marine | | | |
| *Gadus callarias* | 2.8 | | Matthiessen 1882 |
| | 2.4 | | Matthiessen 1882 |
| *Gadus mochua* (cod) | 2.56 | Tungsten | Sadler 1973 |
| *Gadus virens* (saithe) | 2.43 | Tungsten | Sadler 1973 |
| *Pseudopleuronectes* | | | Sivak 1982 |
| *americanus* | 2.17 (center) | Helium- | |
| (flounder) | 2.47 (edge) | neon laser | |

Note: Mathiessen (1882) first noted that the focal length of the fish lens divided by its radius was a constant value. He reported this constant as 2.55, though his 10 measurements ranged from 2.4 to 2.8 and varied with species. He cited Soemmering (1818) as measuring five eyes with a range of values from 2.1 to 2.5. As can be seen here, the value varies by about 30% among species. The f-number of fish thus lies in the range of 1.1 to 1.3 (Eq. 5), which is greater than that of the cat (0.89) (Vakkur and Bishop 1963) and less than that of the human (2.1) (Martin 1977).

$$E = \frac{\pi}{4} \times L \times (2/m)^2 = (\pi/m^2) \times L, \tag{7}$$

which is *independent* of eye size. So the retinal illuminance is directly porportional to the intensity of the source and inversely proportional to the square of Matthiessen's ratio for the species in question.

Thus, when viewing an extended source, larger fish eyes do *not* have a brighter retinal image resulting from a proportionally larger pupillary aperture. This result is particularly interesting because most fish eyes do grow larger as the fish grows, suggesting that, for fish living in well-illuminated habitats, increasing the image brightness was not an important selective force for growing a larger eye.

**Resolution**

A second important constraint on the evolution of eyes has been the necessity to ensure that enough resolving power is present to deliver a focused image onto the retina. The resolution depends on lens quality, receptor size, and distance between receptors, while the focusing depends on accommodative mechanisms available to move the lens.

Assuming a perfect image is delivered to the retina, the resolving power of an eye is

$$\text{Resolving power} = f/(2 \times p) \tag{8}$$

where $f$ = *focal length and* $p$ = receptor separation (Land 1981). A larger eye can therefore produce improved resolving power by virtue of increased focal length as long as the receptor spacing and image quality remain constant. If, however, receptor spacing, as well as focal length, increase in direct proportion to the increase in eye size, resolving power remains approximately constant.

A second possible limit on the resolving power is given by the diffraction limitation of small apertures. This limit, derived from equation (8) is

$$\text{f} - \text{number} = (2 \times p)/\lambda. \tag{9}$$

So, since the minimum useful receptor diameter is approximately 1μm, owing to the wavelength of light (Land 1981), at 500 nm the f-number should be less than 4 to prevent diffraction from being limiting. From equation (5) and Table 18.1 it is clear that teleost eyes have evolved so that diffraction is not a limiting factor.

Since resolution generally does increase with eye size, it is tempting to assert that larger eyes will always be selected in the course of evolution. Kirschfeld (1976) has suggested instead that it may be important to achieve a certain "subjective resolution" that is matched to the animals' height. The factor he proposed, the product of interreceptor angle and body height, is surprisingly constant across a wide range of terrestrial animal species (see Kirschfeld 1976, Fig. 1). For aquatic animals, however, additional factors must be taken into account to understand selective forces at work on eye size.

## The Structure of Fish Eyes

The structure of teleost fish eyes (Fig. 18.2) is superficially quite similar to that of terrestrial vertebrates (Walls 1942). As seen in Figure 18.2, the axial length of the fish eye is short and the apparent hemispherical shape of the eye obvious. The actual eye shape depends on the amount of accommodation possible in the species, since the lens moves within the globe to achieve different states of focus. In fish species without significant powers of accommodation, the retina is approximately hemispherical. In fish that do accommodate, the axis along

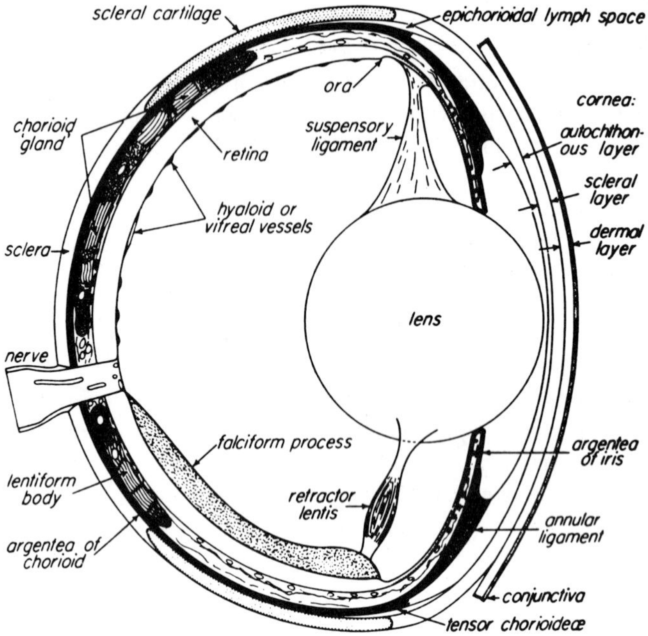


Figure 18.2. Diagrammatical vertical section of generalized teleost eye from Walls (1942). Some structures shown may not be present in some eyes. For example, either lentiform body and falciform process or hyaloid vessels are present. (Reprinted by permission from Walls, 1942.)

which accommodation occurs is normally considerably longer to allow lens movement. Typically, this is the nasal-temporal axis (see below).

The optics of animals living underwater are significantly different from those of animals living in air in one major respect: underwater there is no air-cornea interface (see Chap. 19 for solutions by amphibians, which inhabit both worlds). This interface is an important element in the optics of land-living vertebrates. In humans, for example, the air-cornea interface serves as an important refractory element of the eye, providing about 43 diopters of refractive power, with the lens responsible for an additional 13 diopters (Westheimer 1968). Underwater eyes have no optical benefit from the cornea because its refractive index is nearly identical to that of water. Thus, in all underwater eyes, the refractive power of the cornea is neutralized and the lens, which provides the only dioptric power for image formation, needs to be stronger. In addition, lenses of fish need a short focal length to minimize eye size. Together these constraints have been resolved through the evolution of a spherical lens with very high refractive power in most fish. The exceptions to this rule include skates and other rays, which have proportionally larger eyes and presumably, therefore, aspherical lenses, as well as a few species of fish that live at the air-water interface, including the "four-eyed" minnow, *Anableps* sp. and the amphibious fish *Dialommus fuscus* (see below).

Although the cornea does not contribute to the optical power of the aquatic eye, shallow-living diurnal fish often have yellow corneas and occasionally yellow lenses. These yellow optical filters are thought to serve at least two purposes (Walls 1942; Tansley 1965; Lythgoe 1979 and Chap. 3). First, in deeper-living animals, the yellow filter serves to counter camouflage coloration of animals swimming above. Second, the yellow filter reduces the amount of shortwave light, which is most scattered within the eye. Degradation of the image due to this scattering of shorter wavelengths is thus reduced. Since any filter necessarily reduces the sensitivity of the eye, some fish species have a differential distribution of corneal pigment (Moreland and Lythgoe 1968), and others have corneal chromatophores, providing a filter in the day, but becoming virtually transparent at night (Orlov and Gamburtzeva 1975).

Exceptions to the general description given here are the eyes of deep-sea fishes. In these animals, several adaptations can be found, including some that allow greater photon capture and others that increase visual field size (see Locket 1977). Some deep-sea fish have evolved retinae with multiple banks of receptors, including a disproportionate number of rods (see Vilter 1953). In at least one species, these numerous rods are located in a fovea-like pit at the temporal pole of the retina (Locket 1985). This suggests that these multiple layers of rods act like a photon detector that can be pointed at sources of illumination. In addition, numerous deep-sea fish have a highly reflective tapetum that scatters light among the photoreceptive outer segments, presumably to increase sensitivity, though at the cost of acuity. Many mesopelagic fish species have evolved assymetrical tubular eyes, because photon capture is extremely important and space considerations exclude the possibility of enlarging a spherical eye. This type of eye has been found in 30 of the 750 mesopelagic species, distributed among 11 families (Marshall 1971), suggesting that it may have evolved several times. In these eyes, which appear like sections of a larger eye, the lens is spherical and often fills the dorsal part of the eye. The main retina lies at the end of the ''tube'' and satsifies Matthiessen's ratio, meaning that it is at a distance appropriate for a typical fish eye lens. There is an accessory retina, which extends up the medial wall of the eyecup, from the choroid fissure to the dorsal region of the eye, often until it nearly touches the lens. In some fish, this accessory retina has light directed onto it by a bundle of transparent fibers (Locket 1977). These tubular eyes, similar to those which have evolved in some species of owls, are mostly immobile. The second adaptation to the deep-sea environment is the appearance of retinal diverticula presumed to increase the visual field (Munk 1966; Frederiksen 1973). These diverticula project from the eyecup laterally through a slitlike opening. They presumably receive light via reflecting crystals located in a well-developed argenteum located laterally near the primary lens. Retinae in these fish contain rodlike photoreceptors that are significantly longer than are normal rods.

The eyes of deep-sea fish clearly have evolved to optimize photon capture (see Eq. [1]). As described above, the tubular eyes have the advantages of a larger eye without the space requirements. In these eyes, the cost of having a reduced visual field is at least partially compensated for by the evolution of accessory retinae and diverticula. Surface-dwelling fish, however, have evolved eyes that clearly weigh the advantages of photon capture with other constraints, which will be discussed below.

## Focus and Resolution

### Lens Quality

Spherical lenses have fascinated scientists since early attempts to understand how light is refracted by raindrops to produce rainbows. In the 1300s, Persian scientists followed light rays through a water-filled sphere (Fig. 18.3) to understand how colors arose from white light, thereby discovering the problem of spherical aberration (Devons 1985). Synthetic spherical lenses could be subject to spherical or any of five other primary aberrations characteristic of optical systems (Fincham 1959). Of these, only chromatic aberration is a function of the material of the lens itself, while the rest depend on the structure of the lens and the shape of its focusing surfaces. Most of these aberrations represent failures of the lens to focus an image onto a flat plane. Since retinae are curved, this is not a meaningful constraint in physiological optics. In spherical fish lenses, however, spherical or chromatic aberration or both might adversely affect image quality.

Spherical aberration occurs when light rays entering at different distances from

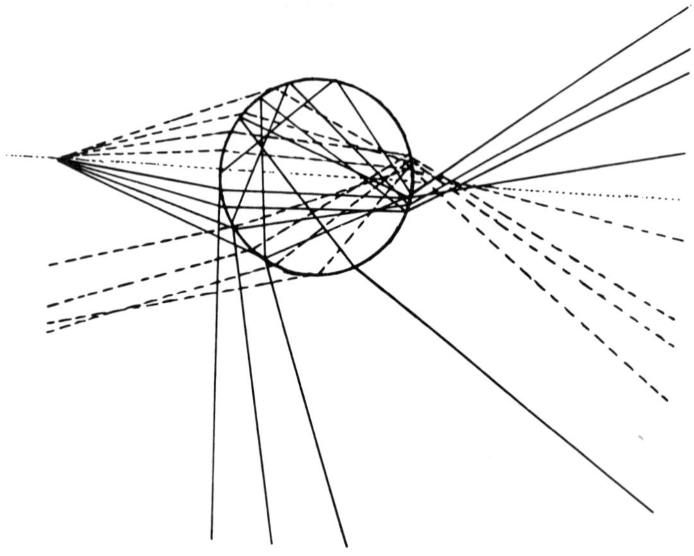

Figure 18.3. Ray-tracing in a transparent sphere by Kamal al-din al-Farisi (ca. 1304–1310). Rays arising from a point source at the left are either refracted or reflected by the sphere. The extreme spherical aberration is evident in the failure of the incident beams to converge on a single point. (Reprinted by permission from Devons, 1985.)

the optic axis are focused at different distances from the lens. Chromatic aberration occurs when light of different wavelengths focuses at different distances from a lens along its main axis (longitudinal chromatic aberration) or different distances off the main axis of the lens (lateral chromatic aberration).

How good is the quality of an image viewed through a spherical fish lens? Teleost lenses deliver surprisingly good images right up to their edges (Fig. 18.4). In most fish, the lens protrudes through the pupil, so that the iris cannot limit images to the central area of the lens. Thus, to be useful, the image *must* be of high quality all across the lens. This is in contrast to the lenses found in most vertebrate eyes, which evolved for aerial vision, in which a pupil restricts vision to the central portion of the lens. It has long been presumed that the spherical fish lens is of high quality because there exists a refractive index gradient within it. For this gradient to correct spherical aberration, it would require a high value in the center of the lens, decreasing continuously and symmetrically with the radius in all directions. Maxwell (1854) allegedly first postulated this gradient while contemplating his breakfast herring (cited in Pumphrey 1961), though the idea arose independently in the writings of Young (1801), Maxwell (1854), and Matthiessen (1886). Brewster (1816) describes his attempts at producing a spherical lens with a refractive index gradient to test this hypothesis.

Luneberg (1944) computed the general function describing an ideal refractive index gradient that results in a spherical lens without aberration. Such ideal spherical lenses are now often referred to as "Luneberg lenses." Later, Fletcher, Mur-

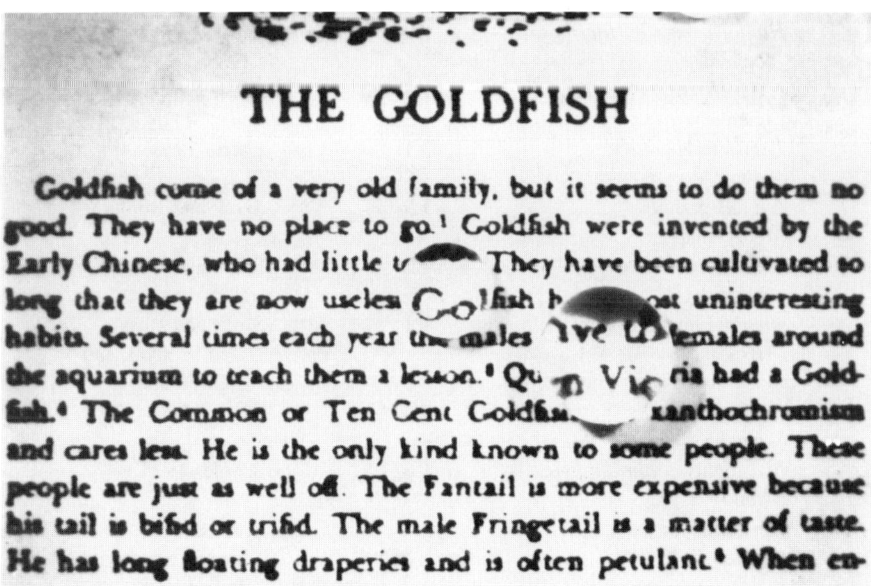


Figure 18.4. The text from Cuppy (1941, 8) imaged by fish lens (*left*) and glass bead (*right*). Note that the image through fish lens is of high quality up to the edge, in contrast to the image through glass bead.

phy, and Young (1954) again found this function, using a slightly different method explicitly to examine spherical biological lenses. It is important to note that the problem of producing good images with a spherical fish lens is made much more complicated by the fact that fishes grow throughout life requiring that some mechanism must exist to modify any refractive index gradient as the eye enlarges.

Pumphrey (1961) claimed that the fish lens was aberration-free and hence must have the design first formally proposed by Maxwell (1854). Since then, investigators have shown that the fish lens is not perfect, suffering from both chromatic and spherical aberration.

The chromatic aberration of teleost lenses has been measured by numerous investigators and ranges from 2% to 5% of the focal length (see Table 18.2). In some cases, these are overestimates, because the methodology introduced additional chromatic dispersion (Sivak and Bobier 1978). Eberle (1968) proposed the appealing idea that, if cones sensitive to longer wavelengths were physically longer, this could compensate for chromatic error of the lens, since light of longer wavelength is focussed farther from the lens. This hypothesis depends on two critical factors; (1) the magnitude of the difference in distance from the lens to the focal plane where light of different wavelengths is focussed, and (2) the location of the region of photon capture within the cone photoreceptors and their

Table 18.2. Chromatic Aberration in Teleosts

| Species | Percent of Focal Length | References |
|---|---|---|
| *Haplochromis burtoni* (cichlid) | 1.9 | Fernald and Wright 1985a |
| *Haplochromis elegans* (cichlid) | 1.8–2.2 | Otten 1981 |
| *Lebistes reticulatus* (guppy) | 2.1 | Eberle 1968 |
| *Lutjaus griseus* (gray snapper) | 3.9 | Sivak and Bobier 1978 |
| *Chilomycterus schoepfi* (burrfish) | 4.0 | Sivak and Bobier 1978 |
| *Ambloplites rupestris* (rock bass) | 4.1 | Sivak and Ross, cited in Sivak and Bobier 1978 |
| *Perca fluvialtis* (perch) | 4.5 | Scroczyński 1979 |
| *Scardinius erythrophalmus* (rudd) | 4.6 | Scholes 1975 |
| *Carassius auratus* (goldfish) | 5.3 | Sivak and Ross, cited in Sivak and Bobier 1978 |

Note: Details of the measurement technique vary and are described in each individual case. The difference in focal length at the C and F Fraunhofer lines (485 and 656 nm, respectively) is expressed as a percentage of the focal length of the lens.

chromatic sensitivity. Figure 18.5 summarizes these data for the African cichlid fish *Haplochromis burtoni* (see Fernald and Wright 1985a for more details). In short, since ''blue'' cones are shorter and blue light focusses closer to the lens, the difference in cone sizes is in the right direction; the differences are not large

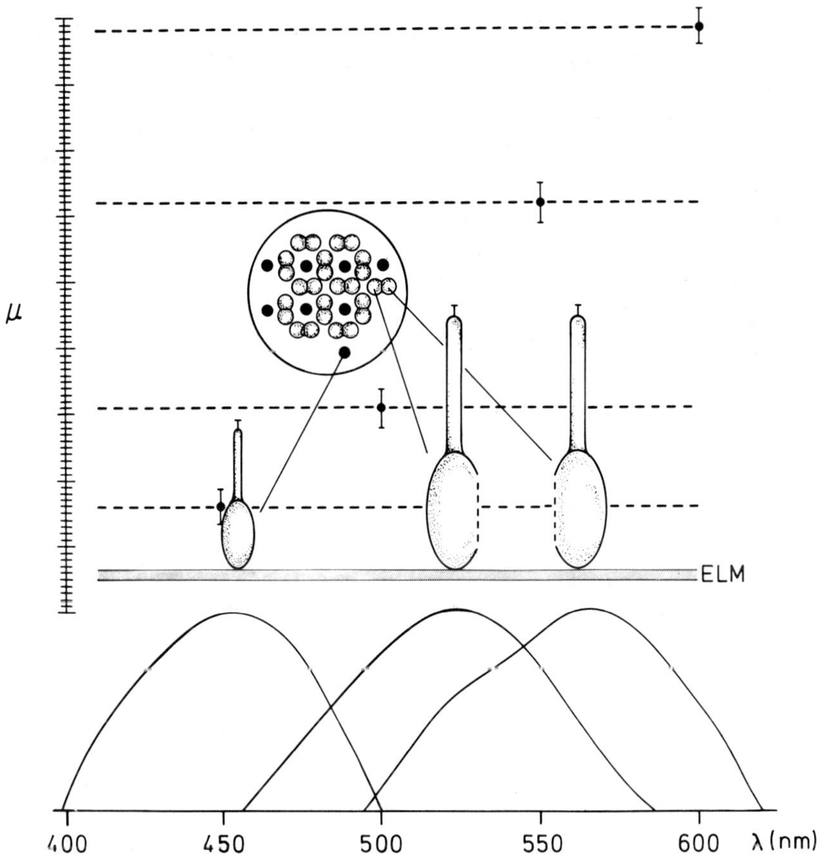


Figure 18.5. Schematic representation of several important optical features of the eye of *H. burtoni*. Smaller single cone and two members of the pair of twin cones are located at horizontal positions corresponding to their peak optical absorbance as measured microspectrophotometrically (Fernald and Liebman 1980). Optical densities (normalized) are plotted below each cone type along the abscissa. Actual spatial relationship among cone types (*in transverse section*) is shown in the *insert*. The retina is schematized so that the outside of the eye is upward from external limiting membrane (*ELM*) in the figure. Cones are drawn to scale with innermost parts of their inner segments aligned on the ELM, as they are in light-adapted retina. Planes of focus measured for four wavelengths in this representative lens are shown by dotted horizontal lines and standard deviations by bars. These focal planes, though fixed in relationship to each other, will move relative to the retina as the lens moves. They have been arbitrarily positioned in the figure so that light at optimal wavelength (454 nm) for small blue cone would be focussed on its inner segment boundary (see text).

enough to compensate for measured chromatic error, however. Scholes (1975) computed similar results for *Scardinius erythrophthalmus*.

What then are the visual consequences of chromatic error in the lens? If we assume that the plane of focus of a single wavelength is in the receptor layer of the retina, then the effect of any other wavelength would be to make a colored fringe around the images projected onto the retina. For these fringes to be visible, the width of the fringe would have to extend over the separation of at least two cones. For *H. burtoni,* the computed blur circle produced by 650-nm light produced when a 2-mm lens is focussed for 500-nm light would be too small to see (see Fernald and Wright 1985a for details). Clearly, however, the potential problem is proportionately worse for fish with smaller lenses and may have been one of several selective forces for eye growth in the course of evolution.

The optical resolution of the fish lens has been measured less often than chromatic aberration, possibly because the lens appears so good when compared with a glass bead of similar size (Fig. 18.6; see also Fig. 18.4). We have measured the optical resolution of the lens as a function of lens size and compared it with the possible anatomical resolution based on the measurement of cone photoreceptor spacing (Fig. 18.7) (Fernald and Wright 1985a). In other species, such as humans, the optical resolution of the lens approximately matches that of the retina. The fish lens, in contrast, has an angular resolution about 10 times better than that of the retina. Only at very small lens sizes (i.e., the size of the lens when the animal first becomes free swimming) does the theoretical limit of the lens resolution match retinal resolution. This suggests that, for fish that do not live in the deep sea, growth of the eyes serves primarily to increase the optical resolution and perhaps to improve contrast detection.

How does the fish preserve the optical quality of its spherical lens despite continual growth? As discussed above, if a spherical lens has no spherical aberration at all and the refractive index a continuous value from the center to the edge, there is a single unique refractive index gradient that can satisfy this requirement. We have recently estimated the refractive index gradient by the direct measurement of the optical properties of a successively surgically reduced lens and have proposed that there is a central hard core surrounded by a gelatinous outer cortex. In all sizes of lenses, the radius of the core is 0.67% of the radius of the whole lens (Fernald and Wright 1983). This core appears to have different optical properties than does the entire lens, suggesting that there might be a discontinuous distribution of refractive index across the lens, rather than a continuous one. More recent theoretical calculations, however, have suggested that such a discontinuous gradient might not produce the very small spherical aberration that has been measured (Campbell and Sands 1984). To resolve this question, we need more accurate direct measurement of the refractive index gradient within individual lenses.

It is worth noting that some fish species do not have spherical lenses. In the elasmobranchs, for example, the lens is subspherical (Walls 1942). In two other genera of fish, specialized optical systems reflect selection for good vision at the interface of air and water. The amphibious fish *Dialommus fuscus* is probably unique in that its eyes have flat corneal windows that form a prism enabling it to see sharply in both air and water (Walls 1942; Stevens and Parsons 1980).

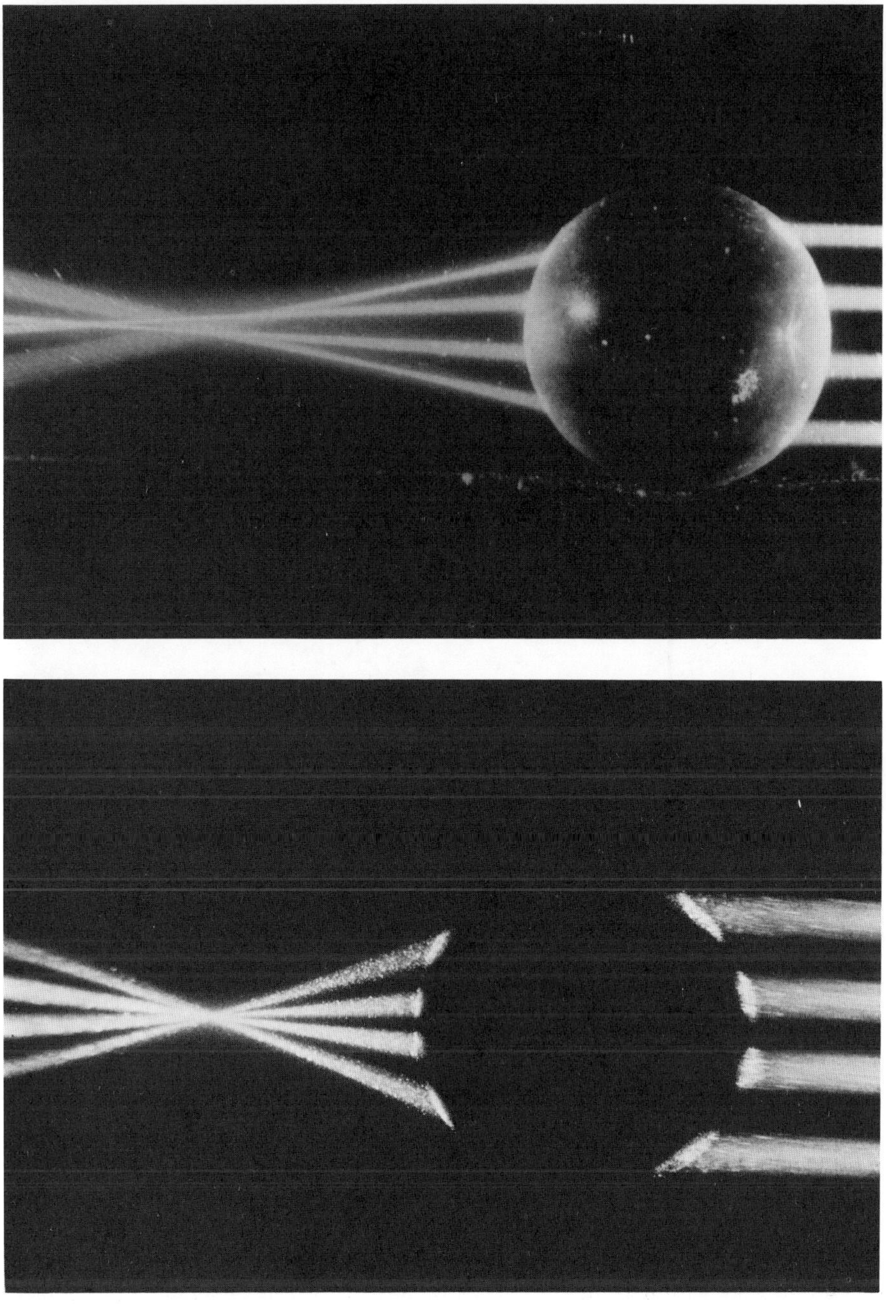

Figure 18.6. (*Top*) Glass sphere of uniform refractive index (n $\simeq$ 1.53) focussing four laser beams, illustrating longitudinal spherical aberration. Argon laser beam is $\lambda$ = 494 nm (Coherent Radiation). (*Bottom*) Freshly excised *H. burtoni* lens suspended by its ligament in oxygenated fish Ringer's solution and illuminated as *above*. Finely focused cone of light is typical of fresh fish lenses. (Reprinted by permission from Fernald, 1985a.)

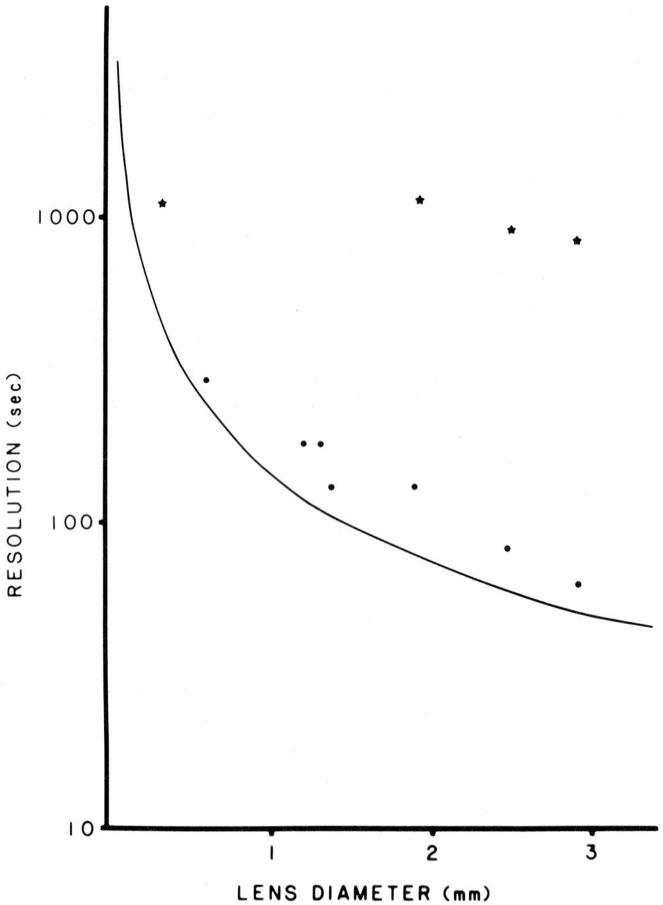


Figure 18.7. Measured resolving power of seven lenses (*solid circles*) in 604-nm light and computed retinal acuity (*stars*) as a function of lens diameter. Solid line is the graph of the function that describes diffraction-limited resolution of round aperatures (Airy's disc). Note that the resolution of fish lens is nearly diffraction limited.

This small animal, native to the Galápagos islands, lives primarily in the tide pools and spends about 50% of its time out of the water, sitting on lava rocks waiting to jump back into tide pools, either to escape predators or to feed. According to an analysis of the optics of the *D. fuscus* eye (Stevens and Parsons 1980), while the animal is out of the water the prismatic cornea probably produces a double image of the world in a single eye. The speculation that binocular information might exist within a single eye is extremely interesting and should be pursued.

*Anableps* sp. called the "four-eyed minnows," swim at the surface of the water in such a way that half the eye looks down into the water and half looks up into the air. Their egg-shaped lens (Walls 1942) appears to have evolved to serve each of its two visual worlds appropriately. Images arriving through the air are refracted by both the cornea and lens and pass through the latter along its flatter side. In contrast, rays arriving through the water are imaged only by the lens, and these pass through the point of the lens, where the radius of curvature is very small and the lens essentially spherical. *Anableps* appears to have evolved its unique optical system in response to an increasingly murky habitat.

## The State of Focus

The visual image available to an animal depends not only on the quality of the lens, but also on the state of the focus produced by the accommodative apparatus. In contrast to land vertebrates, which focus their eyes by adjusting the refractive strength of the lens by modifying its shape, accommodation in fish, when it occurs, results from the movement of the spherically shaped lens within the globe. In most fish examined, the lens moves approximately parallel to the plane of the pupil along an axis of pupillary eccentricity (see below). This means that different parts of the eye are focussed at different distances from the eye on different parts of the visual world simultaneously. For example, in the relaxed accommodative state in fish, the lens lies nearer to the nasal pole of the eye. Thus the temporal part of the retina (nasal visual field) is focussed for near vision and the nasal pole retina (temporal visual field) for far vision (Fernald and Wright 1985b).

Accommodative state is described by where incident parallel rays are focussed. For seeing images clearly, both accommodative state and location of the object of interest are important. Emmetropia is the focus of parallel rays in the plane of the photoreceptive elements and results in sharp focus of objects lying from an intermediate distance to infinity. Myopia is focus of the parallel rays between the lens and retina and results in sharp focus of objects lying from an intermediate distance to close up. Hyperopia is the focus of parallel rays behind the retina and results in no focussed images.

Investigators have used electrical stimulation, as well as drugs, to induce accommodative change in fish to allow measurement of the location of image focus. Beer (1894) first measured the state of refraction of fish eyes, concluding that the fish eye is slightly myopic. It is safe to say, however, that a discussion of accommodation in fish based on experience with the human visual system may not be appropriate, for at least two reasons. The first is that, from an anthropocentric perspective, investigators have proposed emmetropia as the desired state in teleost visual systems. Since underwater visibility is limited by backscatter and turbidity, however, myopic vision may be quite appropriate. The second difficulty has been that investigators often have not taken into account the fact that the state of accommodation is a function of the angle of view, because the lens moves within the globe. Thus, measures of accommodation may not be comparable among species and occasionally are not comparable within a single study. In addition, since lens movement varies with fish size (Fernald and Wright 1985b), the size of the specimen used will affect the detectability and measurement of accommodative state.

The retinoscopic measurement of refractive state is difficult in fish eyes because they are small. Moreover, because the measurements are made underwater, the vergence of the retinoscopic reflection must be corrected for in the air-water interface, or the degree of ametropia in either myopic or hyperopic eyes will be overestimated (Hueter and Gruber 1980).

Perhaps the most difficult issue is in regard to the source of the reflection in the eye. In most fish studied, there are no retinal blood vessels to use as landmarks, since the choroid body provides oxygen to the eye. Estimating or assuming a particular reflecting surface can lead to sizeable errors, particularly in small eyes, because the retinal thickness is nearly constant over a wide range of eye sizes. A greater error can arise, however, because of the chromatic aberration of the optic system (Nuboer and Genderen-Takken 1978).

We have recently measured the lens movement during accommodation and the corresponding refractive state of a variety of sizes of eyes of *H. burtoni*, using two different methods (Fernald and Wright 1985b). The magnitude of the lens movement responsible for accommodation increases with eye size, but with increasing lens diameter, the power of the lens decreases so that the net effect is nearly neutralized (Fig. 18.8). That is, as the animal gets larger, its accommodative mechanism is appropriately scaled to body size. The consequence of the slight loss of accommodative amplitude with growth means that the distance to the nearest focal point increases from about 2 cm to 4.5 cm as the animal grows. Since the fish's length is increasing as well, the near point maintains its same position relative to the body.

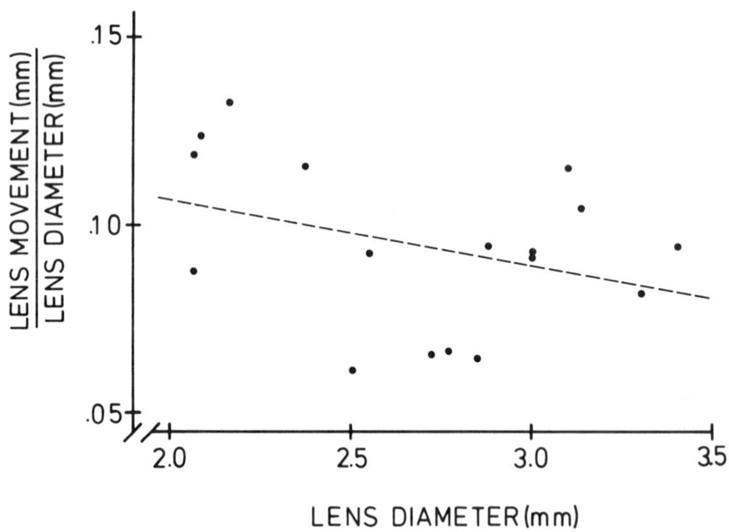


Figure 18.8. The relationship between total magnitude of electrically induced lens movement (normalized with respect to lens diameter, *ordinate*) as a function of lens diameter (*abscissa*). The magnitude of lens movement increases as lens size increases. (Reprinted by permission from Fernald RD and Wright SE. *Vis. Res.* 25:158–167, 1985. Copyright © 1985 by Pergamon Press.)

We have also computed the accommodative amplitude as a function of retinal location for two different sizes of fish. This comparison illustrates two points: (1) larger animals have a smaller dynamic range of accommodative power, and (2) active accommodation can only be effective for images falling onto either the nasal or temporal poles of the retina. Our two methods of estimating refractive state gave somewhat different results, but we concluded that fish are probably emmetropic to slightly myopic. This agrees with the results of recordings from the optic tectum of marine bass, which also suggested an emmetropic state (Meyer and Schwassmann 1970).

## Optic Field and Binocular Field

Since the teleost lens is a nearly incompressible sphere that protrudes through the pupil, it allows the fish periscopic vision, so that each eye has a field of view larger than 180 degrees. In many fish species, the pupillary opening is slightly eccentric, with a small aphakic space nasal to the spherical lens. The major axis of this eccentricity is colinear with the axis of lens movement and has been hypothesized to enlarge the field of binocular vision (Kahmann 1936). We have measured the optic field size in the African cichlid fish *Haplochromis burtoni* and found it to be radially symmetrical and 181° ± 1.1°, independent of both fish size and state of accommodation (Fernald, Wright, and Shelton 1986). From this evidence, it appears that the pupillary eccentricity probably exists to allow lens movement and not to increase optic field size. This hypothesis is made more plausible, since the aphakic space is minimized during lens movements by iris movements (Fernald, Wright, and Shelton 1986). In the goldfish, which has a round pupil and apparently little or no accommodative power, the optic field size is approximately the same as that measured here and is also independent of fish size (Easter, Johns, and Baumann 1977). Binocular field size is about 28 degrees, independent of fish size (Fernald, Wright, and Shelton 1986). It is interesting that a binocular field exists, even though there is no known brain site that receives binocular retinal projection (Fernald 1982a).

The constancy of optic field size, despite a significant increase in retinal size, means that individual retinal cells view different portions of the world as the animal grows. This has two important consequences: First, it means that the photopic (daytime or bright light) visual resolution should increase with eye size, since a greater number of cones view the same solid angle. This has been found to be true (see below). Second, since a larger fish eye viewing an extended source will have the same retinal illuminance as a small eye (see Eq. 7 above), scotopic (nightime or low-light) visual threshold will remain constant only if the number of rods per unit area remains constant. Indeed, the number of rods per unit area does remain constant (see Müller 1952; Johns and Easter 1977; Fernald 1983) and the threshold remains constant as well (see Powers and Bassi 1981; Allen and Fernald 1985). The novel regulatory processes that enable this are described below.

## Retinal Function During Growth

### Retinal Growth

Fish have a gelatinous, thin, transparent retinal structure characteristic of all vertebrates. Fish retinae differ fundamentally from those of other vertebrates, however, because they continue to grow throughout the life of the animal, both by adding new neurons and by stretching existing retinal tissue. In most vertebrates, postnatal eye growth is due solely to the stretching of existing neural tissue, though in some species of birds and mammals, there is also some retinal cell addition in a period immediately after birth. Amphibians add retinal neurons past metamorphosis, but it is unknown exactly how long this continues. In fish, the rate of eye growth relative to body growth depends on the importance of vision to the animal, and in some species, the individual growth rate may be under social control (see Fraley and Fernald 1982). The proportion of retinal area that increases owing to cell addition, as compared with that due to retinal stretching, varies greatly among species. In all species examined, cells are added from a germinal zone located all around the retinal margin as originally described (Müller 1952).

In fish, the absolute amount of growth can be quite large. For example, the eye of the highly visual African cichlid *Haplochromis burtoni* may triple in radius during the first two months of life (Fernald 1985a). In considering the consequences of retinal growth, three determinants of visual function must be considered: (1) the image made available by the optics, (2) the photopic visual acuity, and (3) the scotopic sensitivity. Regarding the image presented to the retina, as shown above, the lens improves its resolution significantly with size, and chromatic error appears to be undetectable. So both theoretically (see Eq. 8) and actually, the optics improve with lens (eye) size, and in addition, a larger eye will have a proportionally larger image to analyze.

### Visual Acuity and Retinal Growth

Photopic visual acuity depends on the solid angle viewed by the eye, on the cone photoreceptor spacing across the retina, and on the convergence of the cones onto higher-order processing cells (ganglion), since in bright light only the cones participate in phototransduction (see above). Since the solid angle viewed by the eye is independent of eye size, the same volume of space projects onto a larger retinal surface as the eye grows. So, to maintain the visual acuity, the density of cone photoreceptors per degree of visual angle must remain the same, as must the ratio of cones to higher-order cells. In fact, because new cones are added throughout life, the cone density per visual angle increases slightly with eye size, and with that increase the best theoretically achievable acuity increases as well. For example, as *Haplochromis burtoni* grows from eight weeks to four months and its eye increases in size from a radius of 0.85 mm to 2.4 mm, the cone density increases from 6.9 cones/degree to 14.8 cones/degree (Fernald 1985a). Because the cell addition accounts for only 40% of the increase in retinal area (Fernald 1983), the cone density *per unit area* decreases, while the cone density per visual angle increases (see Fernald 1985a for details). Behavioral studies of the sunfish have confirmed that visual acuity improves with fish size (Hairston, Li, and Easter 1982).

## Visual Sensitivity and Retinal Growth

Scotopic visual sensitivity depends on capture of photons by rod photoreceptors, since cones are moved out of the plane of focus in low light (see below). The behaviorally measured scotopic visual threshold is constant over a wide range of eye sizes (Allen and Fernald 1981; Powers and Bassi 1981), meaning that the fish can detect very low levels of light energy despite its enlarging eye. As Müller (1952) first showed, the number of rods per unit area remains constant as the fish grows. Since the number of ganglion cells per visual angle, like the cones, increases slightly, the solid angle of visual space per ganglion cell is slightly smaller in larger animals. Clearly these ganglion cells receive their input from an increasingly large number of rod photoreceptors, and how they preserve their response properties during growth is unknown.

When first describing the preservation of rod density, Müller (1952) also noted that mitotic figures associated with cell division only at the retinal margin. He suggested, therefore, that rod photoreceptor density was maintained by migration of rods from the marginal zone into the retina. Since the rod photoreceptor has a long outer segment ($\simeq$ 50 $\mu$) connected by an ellipsoid to a cell body, lateral migration would appear to cause disruption, particularly since the number of rods and the distances involved are large (Fernald 1984). Nonetheless, Johns (1977) presented evidence from $^3$H-thymidine–labelled goldfish in support of Müller's original hypothesis. We have replicated these experiments using the African cichlid fish *Haplochromis burtoni* with quite different results.

As expected, following treatment with $^3$H-thymidine, we found heavily labelled cells in all retinal layers around the margin of the retina. In addition, we found significant numbers of labelled rod nuclei in the outer nuclear layer all across the retina. These appear to be newly generated rods produced by cell division within the retina (Fernald and Johns 1980). Repeating the experiments with the goldfish has confirmed that new retinal rods are produced similarly in this teleost (Johns and Fernald 1981). The failure to find this result in earlier experiments may have been due to the significantly slower growth rate of the goldfish, though labelled rods are visible in published autoradiographs from fish retinae (Scholes 1976), and several workers had previously described mitotically active cells in the outer nuclear layer (e.g., Baburina 1955; Lyall 1957a).

## The Regulation of Retinal Growth

Thus, in the teleost retina, new retinal tissue is being added all around the iris margin at the ora serrata. In addition, the existing retina is being stretched, accommodating the intercalation of newly generated rods exclusively. These two processes preserve visual sensitivity by maintaining constant rod density while adjusting the angular spacing of cones and hence visual acuity to increasing eye size. As a consequence of these essentially competing processes, there is a critical regulatory problem at the margin of the eye (Fernald and Scholes 1985a, 1985b). Since the density of rods is constant at all sizes of fish, while the density of all other cell types decreases, the ratio of the number of rods to the number of other cells is increasing. For example, at hatching, *H. burtoni* has ca. 1 rod/cone, whereas at six months the ratio may be 4–6 rods/cone. In central retina, rod addition is distributed and is responsible for incrementing the rod-cone ratio throughout life. At the margin of the eye, however, the germinal zone is obliged to produce an increasing ratio of rods to other neurons, depending on the size of the fish. How does the germinal zone "know" the current ratio of rods to cones, particularly since the individual growth rates can vary enormously, depending on rearing conditions (Fernald 1983)? We have recently discovered a second exclusive rod germinal zone near the margin of the eye that achieves this regulatory feat (Fernald and Scholes 1985a, 1985b).

Examining the outer nuclear layer (ONL) near the retinal margin in fish of different species and widely different sizes, we found that there is a single sheet of cells with the rods arranged in a sparse square lattice among the cones, which are arranged in their typical array. Nearer the center of the retina, however, rod nuclei become much more numerous, packed confluently in a separate layer vitread to the cone nuclei, more than one deep. The density of rods, shown as a function of distance from the retinal margin in Fig. 18.9, increases abruptly from ca. $2.5 \times 10^4/\text{mm}^2$ to ca. $9 \times 10^4/\text{mm}^2$, a value typical in *Haplochromis burtoni* of all sizes. The increase begins about 100 μm from the margin. The density towards the center of other retinal neurons declines somewhat in this region of the retina but is uniform elsewhere in the retina at any given age.

Corresponding spatially with the sharp increase in rod density, we found numerous extra cell profiles concentrated in a narrow annulus in the ONL, supernumerary to the rods and cones. We have called these *pleomorphic cells* (Fernald and Scholes 1985a) because of their variable shapes, which include occasional mitotic figures (Fig. 18.10). After injecting $^3$H-thymidine and examining the tissue after short survival times (24 hr.), we found labelled clustered over pleomorphic cells. After intermediate survival times (4 d.), pleomorphic cells were comparatively weakly labelled, whereas heavy clusters of grains were seen over differentiated rod nuclei. After long a survival time (10 d.), only rod nuclei were labelled.

These findings suggest a novel solution to the problem of generating a variable ratio of rods to cones at the margin of the eye during the life of the animal. The pleomorphs operate independently of the primary retinal germinal zone, continually moving outwards through the ONL to colonize new tissue made at the

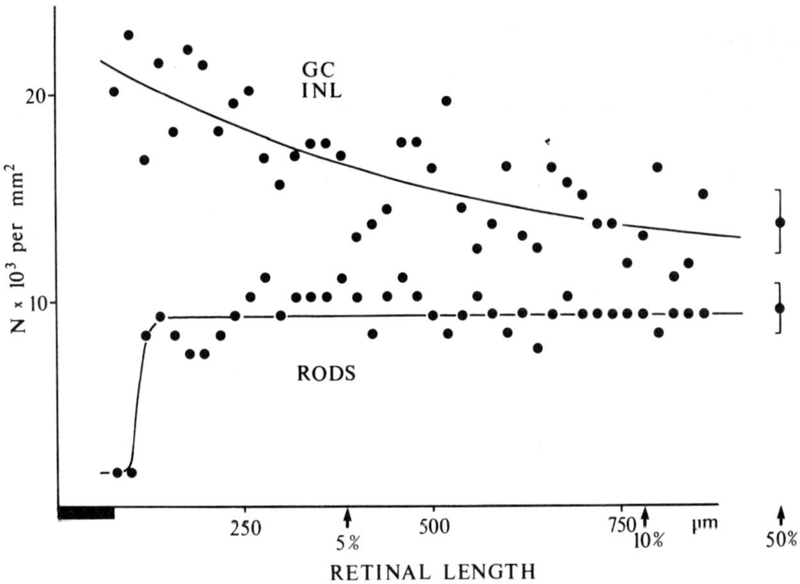


Figure 18.9. Retinal cell density plotted against retinal length (expressed in µm and as %), with *black bar* signifying marginal retinal blastema. The density of rod nuclei (*RODS)* increases 80–100 µm inside the margin, contrasting with inner nuclear layer (*INL*) and ganglion cell density (*GC*), which falls in peripheral 10% of retinal length but therafter remains constant throughout, like rod density. Values are corrected cell counts from photomicrographs using 20-µm sampling bins over first ca. 12% of retinal length and then 100-µm bin at the center of the retina (retinal length = 50%), where standard deviations are also shown.

margin during growth in the adult to produce rods. Moreover, this same zone appears to populate the newly generated retina with the precursors that are responsible for inserting rods later as the tissue stretches (Fernald and Scholes 1985a, 1985b).

An important feature of this second, exclusive rod germinal zone is that the stem cells (pleomorphs) themselves appear to increase in number, maintaining a constant density as the eye grows, whereas the primary germinal zone appears to decrease in density with growth. This crucial difference allows the pleomorphs to accumulate a numerical advantage over the marginal stem cells so that the secondary rod production in new retinal tissue can produce enough rods to match the existing retina.

## Teleost Photoreceptors

For animals dependent on vision, the evolution of color vision was extremely important, since it allowed features of the environment to be distinguished on the basis of spectral, as well as intensity, differences. On both environmental

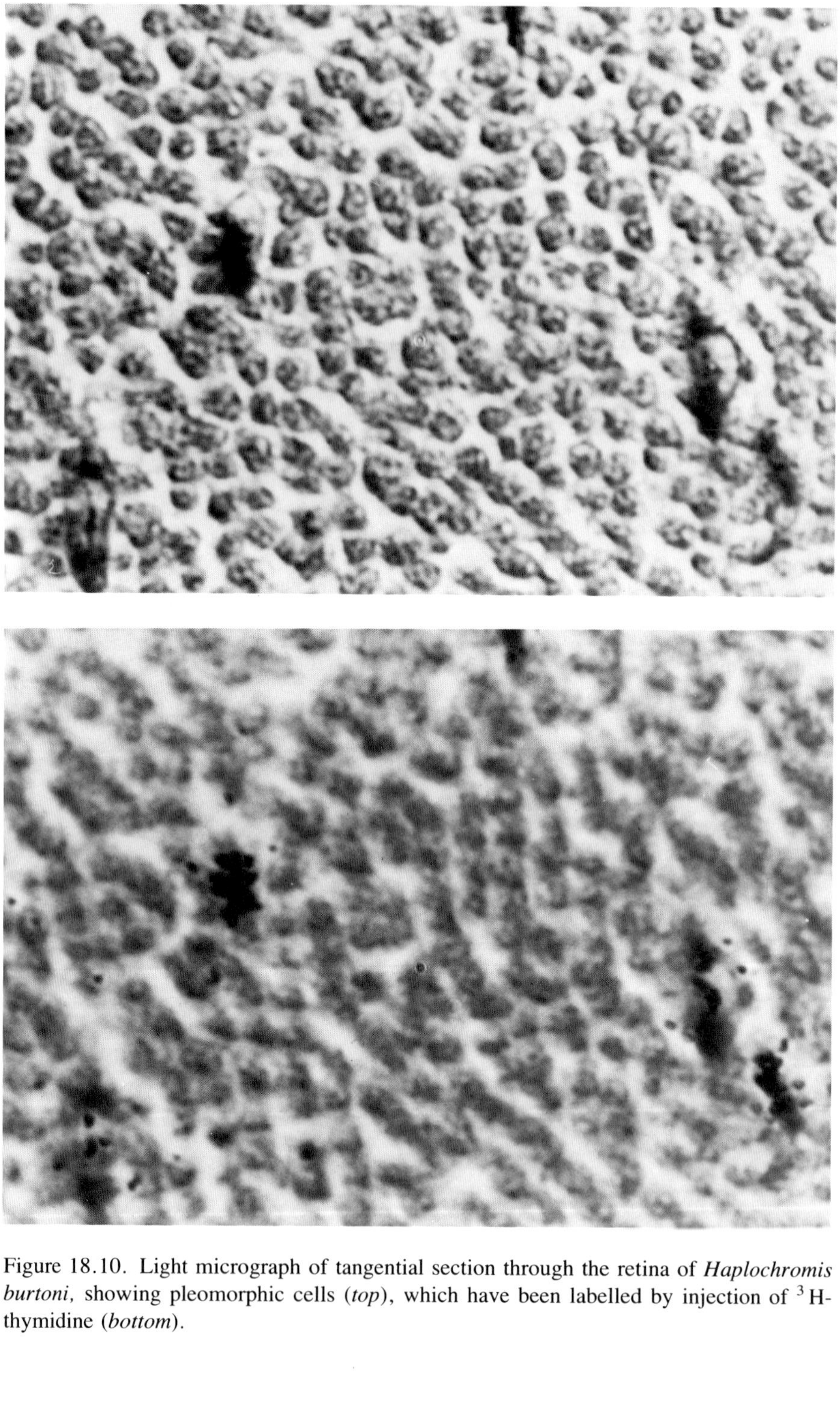

Figure 18.10. Light micrograph of tangential section through the retina of *Haplochromis burtoni*, showing pleomorphic cells (*top*), which have been labelled by injection of $^{3}$H-thymidine (*bottom*).

and behavioral grounds, Munz and McFarland (1977) postulated that color vision may have evolved in fish over 400 million years ago. Color vision requires that chromatic differences in the light reflected from objects be detected and information about the differences be transmitted to the brain. Physiological substrates for both these processes have been found in a number of other species. Detecting different colors has been achieved in most species through the evolution of retinal photoreceptors that are maximally sensitive to different wavelengths of light. In a few cases, however, filters positioned in front of photopigments offer another solution (see Kong, Fung, and Wasserman 1980; Ohtsuka 1985). The brain is informed about the detected chromatic differences via color-opponent neurons (see Daw 1967). Teleost fish species have been used as experimental subjects in many of the important experiments and observations related to color vision.

## Structure

Photoreceptors in fish, as in other vertebrates (see Borwein 1981), are elongated structures lying along the axis parallel to that of incident light and subdivisible into obvious regions. Most sclerad is the outer segment, where the light-sensitive photopigments are located in discs lying perpendicular to the axis of incident light. Cones have outer segments shaped like truncated cones, with a slight taper, while rods are long, slender cylinders. Vitread lies the barrel-shaped inner segment connected to the outer segment by a thin cilium. The scleral end of the inner segment, called the *ellipsoid,* contains a collection of mitochondria, while the vitread end, called the *myoid,* contains the machinery responsible for moving the photoreceptors during retinomotor movements (see below). The nuclei lie vitread, with the sparser, lighter-staining cone nuclei in a row along the outer nuclear layer and the more numerous, darkly staining, smaller rod nuclei in several rows more vitread. These nuclei connect via thin fibers to their synaptical terminals, cone pedicles, and rod spherules. The terminals contain synaptical vesicles and synaptical ribbon(s) and are responsible for connecting with bipolar and horizontal cells and other photoreceptors. The discs containing photopigment are constantly shed and renewed (Young 1967). Visual pigment proteins are synthesized in the myoid, travel through the cilium, and are incorporated into the outer segments. In rods, the incorporation proceeds in an orderly fashion from the base of the outer segment, whereas, in cones, new material is incorporated in a diffuse pattern (see Borwein 1981 for a review).

Most fish species have double or twin cones, which may or may not both have the same photopigment (Lowe and Lythgoe 1978). Such directly apposed photoreceptors are widespread in the animal kingdom (Walls 1942), but there is little agreement about their adaptive value. In addition, as noted above, some benthic fish have multiple tiered banks of rod outer segments most likely to optimize photon capture.

## Function

The photosensitive chemicals that absorb light to initiate a neural response are chemically related, as seen in their nearly identically shaped absorbance spectrum. They consist basically of a chromophoric group based on vitamin $A_1$(retinol) or vitamin $A_2$(dehyroretinol) joined to a protein (opsin) molecule. Early workers measured wavelength sensitivity of retinal pigments extracted from the retina chemically. They could ascertain the absorbance peaks of the numerically dominant photoreceptor species, but only under special circumstances could they distinguish among different less numerous ones. With the development of the microspectrophotometer (Liebman and Entine 1964), absorbance could be measured as a function of wavelength in individual photoreceptors, allowing hypotheses to be formulated about the relationship between peak absorbance of the photopigment and the habitat of the animal.

It was originally proposed that the visual sensitivity of fish living in deeper water would be shifted to shorter wavelengths when compared to that of fish inhabiting brighter, spectrally broader surface waters, since at greater depths the predominate wavelength shifts as well to shorter wavelengths (Bayliss, Lythgoe, and Tansley 1936; Clarke 1936). For deep-sea fish, absorption maxima of the visual pigments matched almost exactly the wavelength of maximum water transparency (Denton and Warren 1957; Munz 1958). Because extraction methods were used, the photopigments measured in these experiments were presumably primarily rod photopigments. In contrast to these results from deep-sea fish, fish living in green coastal water and in freshwater are not matched to the background but shifted away from the peak transmissivity (Lythgoe 1979). Lythgoe has argued that this slight spectral shift will be the most efficient contrast detector and hence has been selected during evolution.

Munz and McFarland (1977) recognized that, because of the scattering of light in water, fish usually detect objects against a background of space light that has a relatively constant color along any particular line of sight. The color will be different depending on the direction of gaze: looking up it will appear green, straight ahead yellow, and looking down red. Thus, for various microhabitats of water, different absorbance spectra can be predicted. The hypotheses of Munz and McFarland have broadly been confirmed (e.g., Lowe and Lythgoe 1978), suggesting that, since multiple pigment systems allow both dark and bright objects to be discriminated against background space light, contrast sensitivity may have been a major force in the evolution of color vision.

## Spatial Organization

Cone photoreceptors in most teleost retinae are arranged in a highly regular mosaic pattern (e.g., Eigenmann and Shafer 1900; Lyall 1957b). Two basic types of patterns have been described: a row pattern and a square pattern. In both, a row of double or twin cones alternates with a row of single cones. The patterns differ in that in one case the twin or double cones are arranged so that lines through the centers of each pair are at right angles with respect to one another (the square pattern), whereas in the other case the axes of the double cones are parallel to one another (the row pattern) (see Fernald 1981). The square mosaic is generally found in fish species that are highly dependent upon vision. In some species, both patterns exist in different regions of the adult retina or occur at different times during development. In every case, rods when present are interspersed among the cone array throughout the extent of the retina, as if perfectly ordered bowling pins (cones) sat in a field of alfalfa (rods).

Cones with different photopigments must be distributed systematically across the retina to be effective as contrast or color detectors. To discover the pattern of pigment distribution in the cone mosaic of teleosts, Marc and Sperling (1976) used a vital staining technique with nitroblue tetrazolium di-formazin (NBT-DF), which is deposited in the mitochondria-filled ellipsoids of photoreceptors whose outer segments have been exposed to light. If the absorbance spectrum of visual pigments from microspectrophotometric measurements are known, cones can be selectively stimulated and thereby stained (Fig. 18.11).

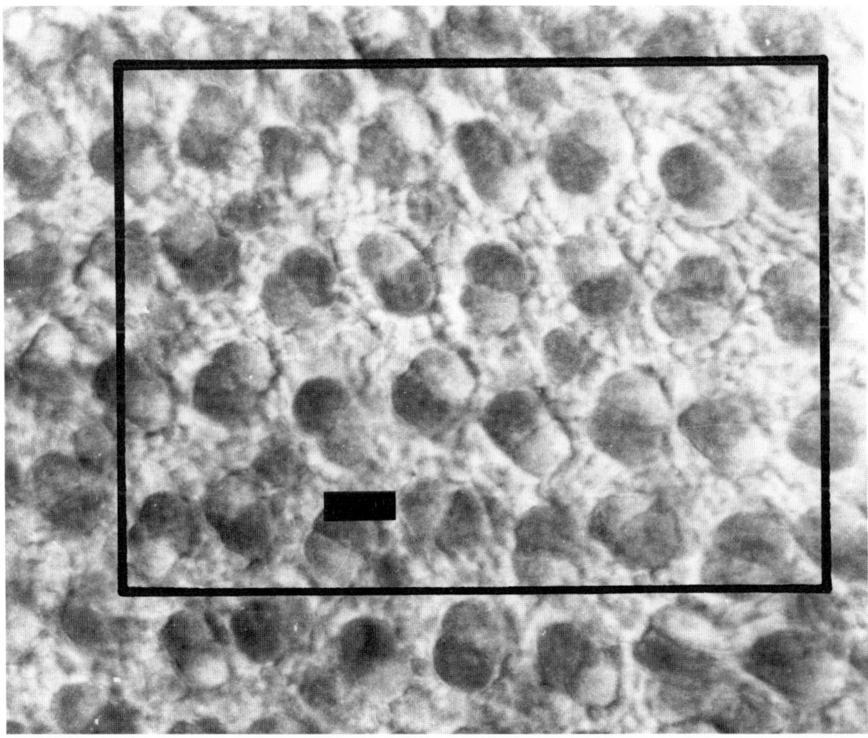

Figure 18.11. Light micrograph of whole mounted retina, showing cones that were stimulated by 487-nm light ($9 \times 10^6$ photons/$\mu m^2 -$ s) and then incubated in NBT (see Fernald 1981 for details). Alternating symmetry is evident in labelling pattern of twin cones. *Bar* is 10 μm. (Reprinted with permission from Fernald RD. *Vis. Res.* 21:1751, 1981. Copyright © 1981, Pergamon Press.)

In the cichlid *H. burtoni* photopigments with twin cones have an alternating symmetry about each single cone: that is, if the single cone is taken as the center, the twin cones are arranged like square dance partners, in this case with alternating 523- and 562-nm absorbance peak pigments around the central cone. Since twin cones are shared by neighboring single-centered cones, the sense of alternating symmetry is reversed in neighboring cone arrays. Formally, this alternating symmetry is a fourfold rotational symmetry about the single cone, so that rotating the entire pattern 90 degrees about any blue cone results in the same pattern. This distribution pattern places the cones with different photopigment types uniformly across the retina, being the optimal distribution for the ratio of photoreceptors available. The symmetrical pattern of retinal photoreceptors described here suggests that patterns of conductivity among neurons in other retinal layers may also reflect the alternating symmetry.

Recent analysis of the retinae of representatives of two genera of holosteans has revealed surprisingly similar cellular mechanisms for color vision to those described here for teleosts (Burkhardt et al. 1983). Trichromatic photoreceptor systems and opponent color-coding mechanisms appear to have evolved in ancient fish before teleosts arose, supporting the view that vertebrate color vision originated in an ancient fish (Munz and McFarland 1977).

## Retinomotor Movements

### Description

In most lower vertebrates, the retinal photoreceptors and retinal pigmented epithelium move radially in response to changes in ambient light (Boll 1877; Kühne 1887). In some animals, these changes, known as *retinomotor* or *photomechanical movements,* exhibit an endogenous circadian rhythm. In the light, myoids on cone photoreceptors contract, moving the cones so that the inner (vitreous) end of the inner segments lie along the external limiting membrane, while myoids of the rods elongate, moving the rods away from the outer limiting membrane. Pigment granules of the retinal pigmented epithelium disperse so that rod outer segments are completely immersed in the dark black pigmented epithelium. In the dark, these movements reverse. Pigment aggregates in the pigmented epithelium to expose rod outer segments, and rods move into position adjacent to the external limiting membrane. Cone myoids lengthen, and the cones move out away from the external limiting membrane.

Retinomotor movements are maximal in teleost fish with duplex retinae (see review by Ali and Wagner 1975). In these species, photoreceptor movement can be as much as 40–80 μm, and the pigmented epithelium may move as much as 100 μm into the tips of the apical projections near the outer limiting membrane. The retinomotor movements are quite slow, lasting from twenty to seventy minutes. In the laboratory, movements resulting from the light-to-dark transition are slower than those from the reverse, though similar differences have not been found under natural circumstances (Douglas 1982). Coincident with this retinomotor activity at dawn and dusk is a quiescent period in the behavior of the animal, suggesting that the photoreceptors may not be useful during their movement (Hobson 1972; Munz and McFarland 1977).

In fish with duplex retinae, retinomotor movements first occur when both types of photoreceptors have developed. Cones generally develop first in young fish, followed by rods. In embryonic animals, retinomotor movements begin as soon as both types of photoreceptors are present and their outer segments have become distinct (Ali and Wagner 1975; Kunz and Ennis 1983). It has been suggested that during retinomotor movements the cone matrix changes regularly though this has not been confirmed in cichlids (Fernald 1982b).

### Function

Generally, proposals about the adaptive advantages of retinomotor movements fall into three main classes. (1) Photoreceptor position is arranged so that the outer segments occupy the full capture area for appropriate lighting conditions: cones for photopic light and rods for scotopic light. This results essentially in two separate functional retinae (Herzog 1905). (2) In the light, dispersed pigmented epithelium may contribute to photopic acuity by absorbing scattered rays refracted out of the cone outer segments (Garten 1907). (3) Migration of pigmented epithelium shields rod outer segments from bleaching during bright light (Herzog 1905).

A fourth possible reason for the evolution of retinomotor movements in the fish might be to maximize retinal acuity, particularly in small animals. As illustrated above (Fig. 18.7), at this early age, the lens is probably quite adequate for resolution of most important images. When *H. burtoni* becomes free-swimming at fourteen days of age, its lens diameter is about 0.3 mm, with a focal length of 0.34 mm. The diffraction-limited acuity for a lens of this size would be about 600″ of arc, and in fact the lens must be somewhat worse, say 850″ or 14.2′ of arc. The retinal acuity is considerably less, however, based on measurement of the distance between cones in the retinal cone mosaic (see Fernald and Wright 1985a for details). We estimate the acuity to be between 1.5 and 2 degrees of arc. Without retinomotor movements, the rods would be interspersed among the cones, increasing cone spacing still more. This would cause a decrease in acuity by a factor of about 2, simply because interdigitated rods oblige the cone separation to be increased. Retinomotor movements offer a neat solution to the problem of packing cone photoreceptors to minimize the distance between them during their time of optimal use. Similarly, by having a uniform field of rod photoreceptors during periods of low light intensity, threshold detection will be enhanced, since fewer photons will fall onto low-sensitivity cones.

### Control

All experiments with teleost retinomotor movements are consistent with the idea that the mechanisms of force production reside with each cell type, though the regulation of the movement may depend on interaction among the cells (see Burnside and Nagle 1983 for a review). In rods, cones, and pigmented epithelium, the force responsible for movement following the dark-to-light transition is actin dependent (Burnside and Nagle 1983). Movement following the light-to-dark transition depends on actin in the rods and microtubules in the cones and is unknown in the pigmented epithelium. Retinomotor movement is initiated locally, and in at least one case, the action spectrum of cone contraction matches the absorbance spectrum of cones measured microspectrophotometrically (Douglas and Wagner 1984). In contrast, movement of the pigmented epithelium appears to be mediated by rod photon absorption (Liebman, Carroll, and Laties 1969). The magnitude of endogenous rhythms exhibited by retinomotor movements varies greatly among the species studied and depends on both the activity patterns in the natural life of the animal and the immediately prior maintainence conditions (Douglas and Wagner 1984).

## Summary

Fundamentally limited by the physics of light, fish have eyes exquisitely adapted to a variety of visual environments, as can be seen in the structure of their eyes. For example, deep-sea fish have evolved large, specialized eyes largely in response to photon collection as the primary selective force. Surface-dwelling fish, on the other hand, have large eyes to achieve greater acuity and contrast sensitivity. Numerous developmental and functional processes serve the ultimate purpose of providing a good image in sharp focus on receptors appropriate for the habitat. Even as we learn more about the "inimitable contrivances" that Darwin puzzled over, these insights reveal only another layer of complexity to be understood.

*Acknowledgments.* This work was supported by NIH Grants EY-05051, F06-TW-00832, EY-05693, the Whitehall Foundation, and the Medical Research Foundation of Oregon. I should like to thank Drs. J. Scholes and A. Fernald for helpful discussion and Nancy Hirata and Linda Shelton for important technical help.

### References

Ali, M.A. and Wagner, H.H. (1975) Distribution and development of retinomotor responses, in Vision in Fishes, Ali, M.A. (ed.), Plenum, New York, pp. 369–396.

Allen, E.E. and Fernald, R.D. (1981) Scotopic visual threshold in the African cichlid fish, *Haplochromis burtoni*, Soc. Neuro. 7:270.

Allen, E.A. and Fernald, R.D. (1985) Spectral sensitivity of the African cichlid fish, *Haplochromis burtoni*, J. Comp. Physiol. 157:247–253.

Baburina, E.A. (1955) The eye of the retina in the caspian shad, Dokl. Akad. Nauk. S.S.S.R., 100(6):1167–1170.

Bayliss, L.E., Lythgoe, R.J., and Tansley, K. (1936) Some new forms of visual purple found in sea fishes, with a note on the visual cells of origin, Proc. R. Soc. B, 816:95–113.

Beer, T. (1894) Die Accommodation des Fischauges, Pfluegers Archiv. Gesamte Physiol. Menschen Tiere, 58:523–650.

Boll, F. (1877) Zur Anatomie und Physiologie der Retina, Arch. Anat. Physiol., 4:783–787.

Borwein, B. (1981) The retinal receptor: a description, in Vertebrate Photoreceptor Optics, Enoch, J.M. and Tobey, F.L.,Jr (eds.), Springer-Verlag, New York, pp. 11–81.

Brewster, D. (1816) On the structure of the crystalline lens in fishes and quadrupeds, as ascertained by its action on polarised light, Philos. Trans. R. Soc. Lond., 311–317.

Burkhardt, D.A., Gottesman, J., Levine, J.S., and MacNichol, E.F.,Jr. (1983) Cellular mechanisms for color-coding in holostean retinas and the evolution of color vision. Vision Res., 23:1031–1041.

Burnside, B. and Nagle, B. (1983) Retinomotor movements of photoreceptors and retinal pigment epithelium: mechanisms and regulation, in Progress in Retinal Research, vol. 2, Osborne, N. and Chader, G. (eds.), Pergamon Press, New York, p. 67–109.

Campbell, M. and Sands, P.J. (1984) Optical quality during crystalline lens growth, Nature, 312:291–292.

Charman, W.N. and Tucker, J. (1973) The optical system of the goldfish eye, Vision Res., 13:1–8.

Clarke, G.L. (1936) On the depth at which fishes can see, Ecology, 17:452–456.

Cuppy, W. (1941) How to Become Extinct, University of Chicago Press.

Darwin, C. (1859) The Origin of Species, New American Library Edition (1958), p. 187.

Daw, N.W. (1967) Goldfish retina: organization for simultaneous color contrast, Science, 158:942–944.

Denton, E.J. and Warren, F.J. (1957) The photosensitive pigments in the retinae of deep-sea fish, J. Mar. Biol. Assoc. U.K., 36:651–652.

Devons, S. (1985) Optics through the eyes of the medieval churchmen, in Science and Technology in Medieval Society, Long, P.O. (ed.), Ann. N.Y. Acad. Sci., pp. 205–224.

Douglas, R.H. (1982) The function of the photomechanical movements in the retina of rainbow trout (*Salmo gairdnerii*), J. Exp. Biol., 96:389–403.

Douglas, R.H. and Wagner, H-J. (1984) Action specturm of photomechanical cone contraction in the catfish retina, Invest. Ophthalmol. Visual Sci., 25:534–538.

Easter, S.S., Johns, P.R., and Baumann, L.R. (1977) Growth of the adult goldfish eye. I Optics, Vision Res., 16:469–476.

Eberle, H. (1968) Zapfenbau, Zapfenlänge und Chromatische Aberration im Auge von *Lebistes reticulatus* (Peters Guppy), Zool. Jb. Physiol., 74:121–154.

Eigenmann, C.H. and Shafer, G.E. (1900) The mosaic of single and twin cones in the retinas of fishes, Am. Nat., 34:109–118.

Fernald, R.D. (1980) Optic nerve distention in a cichlid fish, Vision Res., 20:1015–1019.

Fernald, R.D. (1981) Chromatic organization of a cichlid fish retina, Vision Res., 21:1749–1753.

Fernald, R.D. (1982a) Retinal projections in the African cichlid fish, *Haplochromis burtoni,* J. Comp. Neurol., 206:379–389.

Fernald, R.D. (1982b) Cone mosaic in a teleost retina: no difference between light and dark adapted states, Experentia, 38:1337–1338.

Fernald, R.D. (1983) Neural basis of visual pattern recognition, in Advances in Vertebrate Neuroethology, Ewert, J-P., Capranica, R.R., and Ingle, D.J. (eds.), Plenum, New York, pp. 569–580.

Fernald, R.D. (1984) Vision and behavior in an African cichlid fish, Am. Sci., 72(1):58–65.

Fernald, R.D. (1985a) Growth of the teleost eye: novel solutions to complex constraints, Environ. Biol. Fishes, 13:113–123.

Fernald, R.D. (1985b) Eye movements in the African cichlid fish, *Haplochromis burtoni,* J. Comp. Physiol., 156:199–208.

Fernald, R.D. and Johns, P.R. (1980) Retinal structure and growth in the cichlid fish, *Haplochromis burtoni,* Invest. Ophthalmol. Visual Sci. (supp.) 69.

Fernald, R.D. and Liebman, P. (1980) Visual receptor pigments in the African cichlid fish, *Haplochromis burtoni,* Vision Res., 20:857–864.

Fernald, R.D. and Scholes, J. (1985a) A zone of exclusive rod neurogenesis in the teleost retina, Soc. Neuro. Abst., 11:810.

Fernald, R.D. and Scholes, J. (1985b) Retinal neurogenesis in teleosts: a second germinal zone, Submitted.

Fernald, R.D. and Wright, S. (1983) Maintenance of optical quality during crystalline lens growth, Nature, 301:618–620.

Fernald, R.D. and Wright, S. (1985a) Growth of the visual system of the African cichlid fish, *H. burtoni:* optics, Vision Res., 25(2):155–161.

Fernald, R.D. and Wright S. (1985b) Growth of the visual system of the African cichlid fish, *H. burtoni:* accommodation, Vision Res., 25(2):163–170.

Fernald, R.D., Wright, S., and Shelton, L.C. (1986) Growth of the visual system of the African cichlid fish, *H. burtoni:* optic field and retinal field, (in preparation).

Fincham, W.H.A. (1959) Optics, Hatton Press, London.

Fletcher, A., Murphy T., and Young, A. (1954) Solutions of two optical problems, Proc. R. Soc. Lond. A., 223:216–225.

Fraley, N.B. and Fernald, R.D. (1982) Social control of developmental rate in the African cichlid fish, *Haplochromis burtoni,* Z. Tierpsychol. 60:66–82.

Frederikson, R.D. (1973) On the retinal diverticula in the tubular-eyed opisthoproctid deep-sea fishes *Macropinna microstoma* and *Dolichopteryx longipes.* Vidensk, Medd. Dan. Naturhist. Foren., 136:233–244.

Garten, S. (1907) Die Veränderungen der Netzhaut durch Licht, Graefe-Saemisch Handbuch der gesamten Augenheilkunde, Leipzig, pp. 250–280.

Geiger, W. (1956) Quantitative Untersuchungen über das Gehirn der Knochenfische, mit besonderer Berücksichtigung seines relativen Wachstums, Acta Anat. 26:121–163; 27:324–350.

Hairston, N.G., Jr. Li, K.T., and Easter, S.S., Jr. (1982) Fish vision and the detection of planktonic prey, Science, 218:1240–1242.

Herzog, H. (1905) Experimentelle Untersuchungen zur Physiologie der Bewegungsorgange in der Netzhaut, Arch. Anat. Physiol. (Physiol. Abst.), 516:413–464.

Hobson, E.S. (1972) Activity of Hawaiian reef fishes during evening and morning transitions between daylight and darkness, U.S. Fish. Bull. 70:715–740.

Hueter, R.E. and Gruber, S.H. (1980) Retinoscopy of aquatic eyes, Vision Res., 20:197–200.

Johns, P.R. (1977) Growth of the adult goldfish eye. III. Source of the new retinal retinal cell number, J. Comp. Neurol., 176:331–342.

Johns, P.R. and Fernald, R.D. (1981) Genesis of rods in teleost fish retina, Nature, 293:141–142.

Kahmenn, H. (1936) Über das foveale sehen der Wirbeltiere. I. Über die Fovea centralis und die Fovea lateralis bei einigen Wirbeltieren. Albrecht von Graefe's Arch. Ophthalmol., 135:265–276.

Kirschfeld, K. (1976) The resolution of lens and compound eyes, in Neural Principles of Vision, Zettler, F. and Weiler, R. (eds.), Springer-Verlag, Berlin, pp. 354–369.

Kong, K.L., Fung, Y.M., and Wasserman, G.S. (1980) Filter mediated color vision with one visual pigment, Science, 207:783–786.

Kühne, W. (1887) Fortgesetzte Untersuchungen über die Retina und die Pigmente des Auges, Untersuch. Physiol. Inst. Univ. Heildelberg, 2:89–109.

Kunz, Y. and Ennis, S. (1983) Ultrastructural diurnal changes of the retinal photoreceptors in the embryo of a viviparous teleost (*Poecilia reticulata P.*), Cell. Differ. 13:115–123.

Land, M.C. (1981) Optics and vision in invertebrates, in Handbook of Sensory Physiology, vol. VII 6B, Autrum, H.J. (ed.), Springer-Verlag, Berlin, pp. 472–592.

Liebman, P.A. and Entine, G. (1964) Sensitive low-light level microspectrophotometer: detection of photo-sensitive pigments of retinal cones, J. Opt. Soc. Am., 54:1451–1459.

Liebman, P.A., Carroll, S., and Laties, A. (1969) Spectral sensitivity of retinal screening pigment migration in the frog, Vision Res. 9:377–384.

Locket, N.A. (1977) Adaptations to the deep-sea environment, in Handbook of Sensory Physiology, vol. VII/5, Crescitelli, F. (ed.), Springer-Verlag, Berlin, pp. 67–192.

Lowe, E.R. and Lythgoe, J.N. (1978) The ecology of cone pigments in teleost fishes, Vision Res., 18:715–722.

Luneberg, R.K. (1944) Mathematical Theory of Optics, Brown University Press, Providence, R.I., pp. 208–213.

Lyall, A.H., (1957a) The growth of the trout retina, Q. J. Microsc. Sci., 98:101–110

Lyall A.H. (1957b) Cone arrangements in teleost retinae. Q. J. Microsc. Sci., 98:189–209.

Lythgoe, J.N. (1979) The Ecology of Vision, Clarendon Press, Oxford.

Marc, R.E. and Sperling, H.G. (1976) Color receptor identities of goldfish cones, Science, 191:487–489.

Marshall, N.B. (1971) Explorations in the life of fishes, Harvard University Press, Cambridge, Mass.

Matthiessen, L. (1882) Über die Beziehungen, welche zwischen dem Brechungsindex des Kernzentrums der Krystalllinse und den Dimensionen des Auges bestehen, Pflügers Arch ges Physiol, 27:510–523.

Matthiessen, L. (1886)Über den physikalisch-optischen Bau des Auges der Cetacean und der Fische, Pflügers Archiv. Gesamte Physiol., Menschen Tierre, 38:521–528.

Maxwell, J.C. (1854) Some solutions of problems, Cambridge & Dublin Math. J., 1:76–78.

Meyer, D.L. and Schwassmann, H.O. (1970) Electrophysiological method for determination of refractive state in fish eyes, Vision Res., 10:1301–1303.

Moreland, J.D. and Lythgoe, J.N. (1968) Yellow corneas in fishes, Vision Res., 8:1377–1380.

Müller, H. (1952) Bau und Wachstum der Netzhaut des Guppy (*Lebistes reticulatus*), Zool. Jb. Allgemein Zool. Physiol. Tier, 63:275–324.

Munk, O. (1966) Ocular anatomy of some deep-sea teleosts, Dana-Rep Carlsberg Found., 70:1–62.

Munz, F.W. (1958) Photosensitive pigments from the retinae of certain deep sea fishes, J. Physiol., 140:220–225.

Munz, F.W. and McFarland, W.N. (1977) Evolutionary adaptations of fishes to the photic environment, in Handbook of Sensory Physiology, vol. VII/5, Crescitelli, F. (ed.), Springer-Verlag, Berlin, pp. 193–274.

Nuboer, J.F.W. and van Genderen-Takken, H. (1978) The artifact of retinoscopy, Vision Res., 18:1091–1096.

Ohtsuka, T. (1985) Relation of spectral types to oil droplets in cones of turtle retina, Science, 229:874–976.

Orlov, O.Y. and Gamburtzeva, A.G. (1975) Dynamics of corneal colorations in fish, *Hexagrammos octagrammus,* Biofizika, 21:362–365.

Otten, E. (1981) Vision during growth of a generalized *Haplochromis* species: *H. Elegans* Trewavas 1933 (Pisces, Cichlidae), Neth. J. Zool., 31:650–700.

Powers, M.K. and Bassi, C.J. (1981) Absolute visual threshold is determined by the proportion of stimulated rods in the growing goldfish retina, Neurosci. Abst., 7:541.

Powers, M.K. and Easter, S.S., Jr. (1983) Behavioral significance of retinal structure and function in fishes, in Fish Neurobiology, Northcutt, R.G., and Davis, R.E. (eds.), University of Michigan Press, Ann Arbor, pp. 377–404.

Pumphrey, R.J. (1961) Concerning vision, in The Cell and the Organism, Ramsay, J.A. (ed.), Cambridge University Press, pp. 193–208.

Sadler, J.D. (1973) The focal length of the fish eye lens and visual acuity, Vision Res., 13:417–423.

Scholes, J.H. (1975) Colour receptors and the synaptic connexions in the retina of a cyprinid fish, Philos. Trans. R. Soc. B Lond., 270:61–118.

Scholes, J.H. (1976) Neuronal connections and cellular arrangement in the fish retina, in Neural Principles of Vision, Zettler, F., and Weiler, R. (eds.), Springer-Verlag, Berlin, pp. 354–369.

Scroczyński, S. (1975a) Die sphärische Aberration der Augenlinse der Regenbogenforelle (*Salmo gairdnerii Rich*), Zool. Jb. Physiol., 79:204–212.

Scroczyński, S. (1975b) Die sphärische Aberration der Augenlinse des Hechts (*Esox lucius L.*), Zool. Jb. Physiol., 79:547–558.

Scroczyński, S. (1977) Spherical aberration of crystalline lens in the roach *Rutilus rutilus L.*, J. Comp. Physiol., 121:135–144.

Sivak, J.G (1982) Optical characteristics of the eye of the flounder, J. Comp. Physiol., 146:345–349.

Sivak, J.G., and Bobier, W.R. (1978) Chromatic aberration of the fish eye and its effect on refractive state, Vision Res., 18:453–455.

Soemmering, D.W. (1818) De Oculorum Hominis Animaliumque Secone Horizontali Commentatio, Vandenhoeck and Ruprecht, Göttingen.

Stevens, J.K. and Parsons, K.E. (1980) A fish with double vision, Nat. Hist., 89:62–67.

Tansley, K. (1965) Vision in Vertebrates, Chapman and Hall, London.

Vilter, V. (1953) Existence d'une rétine à plusieur mosaîques photoréceptrice chez un poisson abyssal bathypelagique, *Bathylagus benedicei,* C. R. Soc. Biol. (Paris), 147:1937–1939.

Walls, G.L. (1942) The vertebrate eye and its adaptive radiation, Hafner, New York [1962].

Westheimer, G. (1968) "The eye," in Mountcastle, V.B. (ed.), Medical Physiology, 12th ed., Mosby, St. Louis, pp. 1532–1553.

Young, R.W. (1967) The renewal of photoreceptor outer segments, J. Cell Biol., 33:61–72.

Young, T. (1801) On the mechanism of the eye, Philos. Trans., 92:23–88.

# Chapter 19

# Optics of Amphibious Eyes in Vertebrates

JACOB G. SIVAK*

## Vision in Air and Water

The change from an aquatic to a terrestrial life-style is the factor of paramount importance in the phylogenetic development of the optics of the vertebrate eye (Sivak 1980). The fundamental point concerns the realization that the cornea of the eye is not a significant refractive structure when the eye is underwater.

The cornea is a transparent structure consisting of two curved and almost parallel surfaces, so that its refractive function is dependent on the existence of media of unequal refractive index in front and in back. The refractive index of the cornea itself is almost irrelevant, since the refractive effect of the front surface is neutralized by the negative effect of the posterior surface. Clearly the aquatic habitat of ancestral vertebrates limited the original function of the cornea to that of a transparent window. The typical aquatic cornea is relatively flat in shape, presumably an adaptation related to the need for streamlining in a dense medium (Jamieson 1971).

Although the earliest vertebrate eye is no longer available for study, existing aquatic species, be they fishes (Agnatha, cartilaginous or bony) or mammals (cetaceans or sirenians), possess an eye characterized by the presence of a spherical or nearly spherical crystalline lens (Fig. 19.1) of rigid consistency and elevated refractive index (Franz 1934; Walls 1942; Duke-Elder 1958). The lens, as the only refractive structure of these eyes, is responsible for accommodative changes in refractive power. The various accommodative mechanisms proposed for a variety of aquatic species always involve lens movement (Beer 1894; Franz 1934; Walls 1942; Sivak 1980).

Both the lens and the cornea are transparent, avascular structures that permit light to enter the eye and reach the retina. The cornea is mainly acellular (collagen), however, and its principal cellular component, the epithelium, forms the external

* School of Optometry and Department of Biology, University of Waterloo, Waterloo, Ontario N2L 3G1, Canada

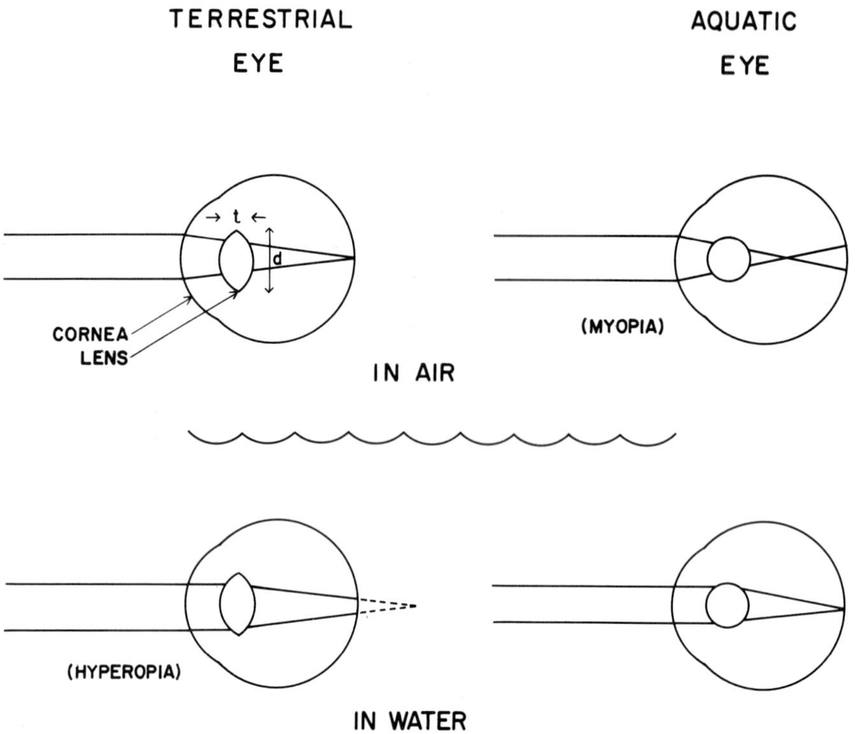


Figure 19.1. Schematic representation of optical effects of moving aquatic eye into air and terrestrial eye into water. Lens diameter (*d*) and thickness (*t*) are indicated.

surface of the eye (Hogan, Alvarado, and Weddell 1971). The process of cell multiplication therefore, does not affect corneal structure or optical function.

The lens, on the other hand, is a completely cellular structure located inside the eye and within an acellular capsule. Cellular multiplication is restricted to the peripheral equator, so that the older lens material is compressed towards the center as the number of lens cells increases throughout life (Duke-Elder 1958). In addition, the lens is surrounded by the humors of the eye, which have a refractive index ranging from 1.334 to 1.336 (Hughes 1977). The lens must have a refractive index that is substantially higher than the media surrounding it in order to focus light on the retina. Since protein content is directly related to refractive index, all lens tissue shows unusually high protein concentrations (Spector 1982). In the case of the typical fish lens, an equivalent index of 1.65 is usually quoted (Walls 1942; Sivak and Kreuzer 1983). The difficulty in maintaining a clear and functional lens due to location within the eye and continued growth is reflected in the prevalence and variety of lens opacities, with current evidence indicating that as many as 95% of human lenses in persons over 65 years of age show some cataractous change (Kinoshita 1974; Duncan 1981; Worgul 1982).

In air, the cornea assumes a dominant refractive role because of the substantial refractive index separation between air and aqueous humor (1.0–1.335). In addition, a flattened cornea is no longer a significant streamlining factor, and the terrestrial cornea usually has a radius of curvature that is smaller than that of the rest of the eye (Walls 1942). A more convex shape may be important architecturally in maintaining the smooth optical surface required of the cornea once it is exposed to air (Bito pers. com.). Whatever its ultimate shape, the terrestrial cornea is frequently the principal refractive structure of the eye. For example, its power in the case of the human eye is approximately double that of the unaccommodated lens (Bennett and Francis (1962).

The adult human lens is markedly elliptical in shape (diameter > thickness), relatively soft in consistency and relatively low in refractive index (equivalent refractive index = 1.41) (Bennett and Francis 1962). It is interesting to note that descriptions of human lens development typically state that the lens becomes more flattened during growth from infancy to childhood because of the equatorial origin of new lens cells (secondary lens fibers) and their tapered anterior—posterior extensions toward the lens poles (Fig. 19.1) (Tripathi and Tripathi 1983). A spherical shape may be difficult to maintain once the hollow embryological lens vesicle is filled by the primary lens fibers, but there is little information available concerning the cellular growth factors controlling lens shape, particularly where the spherical lens is concerned.

Accommodation remains a function of the lens in terrestrial eyes. When it exists, it usually involves changes in lens shape brought about by a variety of intraocular mechanisms (Sivak 1980). This difference seems to fit with the change toward a softer and flattened lens.

The foregoing is a generalized comparison that represents an effort to understand the effect of air and water on the overall refractive plan of the vertebrate eye. There are a number of notable exceptions. For example, among the cartilaginous fish, the lens is not always spherical, with a substantially flattened lens being apparent in certain stingrays (Sivak 1976). By comparison, the lens of nocturnal mammals (e.g., rats) is often nearly spherical in shape (Fig. 19.2), an adaptation designed to enhance the light-gathering ability of the eye (Walls 1942). Nevertheless, the lens of the eye of newborn rats is as elliptical as that of newborn humans (Sivak and Dovrat 1984). Snakes are believed to accommodate by moving the lens, as do fish (Michel 1932; Sivak 1977). The snake eye departs from the typical terrestrial plan in many respects, however, including the existence of a spherical lens, perhaps owing to a subterranean evolutionary history (Walls 1940, 1942). The accommodative response of the cat eye may include lens movement, though the literature is not clear on this point (Lovasik 1979). Despite exceptions already reported or still to be found, the air-water refractive dichotomy of the vertebrate (and simple invertebrate) eye will not be altered.

In illustrating the optical consequences of the air-water transition, two vertebrate groups stand out as being of particular interest, amphibians and diving birds. Both groups are amphibious: the former by metamorphic permanent changes from water to air and the latter by rapid and frequent transitions between the two media when they dive. Amphibians are living examples of the evolutionary step from water to air. Unless a species is neotenic and remains in the water,

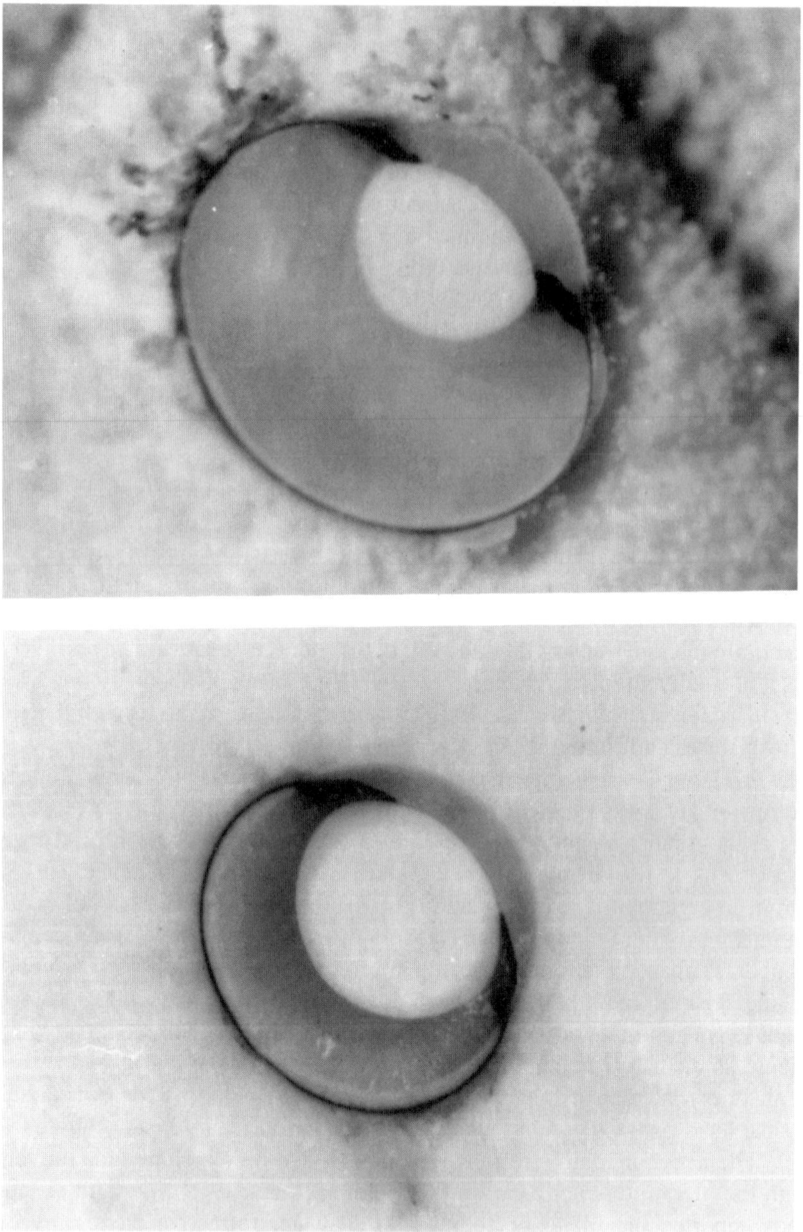

Figure 19.2. Internal appearance of eyes of diurnal ground squirrel (*top*) and nocturnal rat (*bottom*) eyes after freeze-sectioning to geometrical center of the lens. Note the difference in relative size and shape of diurnal and nocturnal lenses. (From Gur and Sivak 1979.)

the amphibian visual system must adapt to the change in the medium contacting the eye as the animals metamorphose. Birds are of interest because of the large number of species that depend on the marine or freshwater or both habitats for food. If the life history of a particular species indicates that good visual resolution is needed in air and in water, then it must cope with the loss of corneal refraction that takes place as soon as the eye enters water. Interesting solutions to these problems will be discussed in the following sections. Because less is known in other amphibious groups (i.e., aquatic reptiles and mammals), these will not be discussed here.

## The Metamorphosis of Amphibian Ocular Optics

A series of studies has examined the rate and extent of optical change undergone by the eyes of a number of amphibians during metamorphosis (Sivak and Warburg 1980, 1983; Sivak et al. 1985). The approach involves the use of a rapid freezing and sectioning technique to photograph and measure gross ocular morphology at various stages before, during, and after metamorphosis. Attention is paid to lens and corneal shape and to their positions relative to themselves and to the retina. In addition, refractive errors (myopia or hyperopia, expressed in diopters) were determined retinoscopically in vivo in the appropriate medium (air or water), and at various metamorphic stages. (The retinoscope is a clinical instrument that can be used to measure ocular refractive errors objectively, with the test eye in air or water, while one *diopter* is defined as the refractive power required to focus parallel light on a point located 1 m away.)

The results indicate that, while the eye of the larval amphibian is essentially identical in optical plan to the typical teleost eye (flattened cornea and spherical lens) the eye of the adult varies widely, both in terms of rate and extent of optical change (Table 19.1). Larval refractive states in water vary from emmetropia (zero refractive error) to mild hyperopia (5 diopters) in *Salamandra salamandra* and from 9 to 10 diopters of hyperopia in *Pelobates syriacus* and *Rana catesbeiana*. Despite the relative flatness of the cornea, corneal refractive power in air would be considerable. Values of 280, 180, and 78 diopters have been calculated from sample measurements of *Salamandra salamandra, Pelobates syriacus,* and *Rana catesbeiana,* respectively. Clearly, such eyes would be of little use as resolving instruments in air.

The precision of refractive state determinations is confounded by two factors: (1) uncertainty regarding the reflecting layer of the retina and (2) the effect of ocular aberrations, particularly longitudinal chromatic aberration. If the reflected light used with objective measures such as retinoscopy emanates from the vitreous surface, rather than the receptor layer of the retina, a spurious quantity of hyperopia will be measured. The amount will depend of the refractive equivalent of retinal thickness as a proportion of the overall refractive power of the eye. Since the larval amphibian eye is small and has large retinal cells (Brown, Gibbons, and Wald 1963; Grusser-Cornehls and Himstedt 1976), this factor may be important. On the other hand, the low amounts of hyperopia measured (0 to 10 diopters)

Table 19.1. Sample Intraocular Dimensions (to ± 0.05 mm or ± 0.01 mm) of Eyes of Amphibians before and after Metamorphosis

| Species and Stage | Eye Diameter[a] Anteroposteriorly | Cornea to Anterior Lens | Corneal Radius of Curvature | Lens Thickness | Lens Diameter | % Difference in Lens Dimensions[b] Diameter vs. Thickness (N and Range)[c] |
|---|---|---|---|---|---|---|
| *Pelobates syriacus* | | | | | | |
| Tadpole (early) | 2.80 | 0.35 | 1.80 | 1.15 | 1.15 | 0 (N = 9, range = 0%–3.5%) |
| Tadpole (late) | 3.50 | 0.60 | 2.00 | 1.30 | 1.50 | 13.3 (N = 7, range = 10.0%–19.2%) |
| Juvenile adult | 4.20 | 0.65 | 1.70 | 1.80 | 2.30 | 21.7 (N = 8, range = 24.3%–28.3%) |
| *Salamandra salamandra* | | | | | | |
| Larva (early) | 1.36 | 0.12 | 1.18 | 0.61 | 0.61 | 0 (N = 4, range = 0%–2%) |
| Larva (late) | 2.22 | 0.28 | 1.22 | 0.89 | 0.93 | 4.2 (N = 3, range = 4.3%–7.3%) |
| Juvenile adult | 2.22 | 0.40 | 1.21 | 1.08 | 1.25 | 13.6 (N = 4, range = 11.4%–15.7%) |
| *Rana catesbeiana* | | | | | | |
| Tadpole (early to late) | 5.76 | 0.73 | 3.74 | 2.57 | 2.57 | 0 (N = 3, 0 range) |
| Adult | 6.78 | 1.13 | 2.95 | 3.28 | 3.75 | 12.5 (N = 3, range = 11.8%–12.5%) |
| *Notophthalmus viridescens* | | | | | | |
| Larvae | 0.92 | 0.11 | 0.49 | 0.59 | 0.60 | 1.7 (N = 6, range = 0%–3.2%) |
| Eft (land) | 1.30 | 0.23 | 0.92 | 0.71 | 0.76 | 6.6 (N = 5, range = 4.8%–8.5%) |
| Adult | 1.97 | 0.26 | 1.27 | 0.95 | 0.93 | 2.2 (N = 5, range = 0%–2.5%) |

*Source:* Reproduced with permission from Sivak et al. 1985.
[a] Eye diameter from anterior corneal surface to retina-vitreous border.
[b] % difference in lens dimensions refers to the lenses of the individual examples.
[c] The number of individuals studied and the range values for % difference in lens diameter are included in parentheses.

relative to the short focal length of the eyes (1.5–3.0 mm) suggest otherwise. This point merits more intensive study in amphibians.

The refractive power of all vertebrate eyes varies somewhat with wavelength, with short wavelengths (blue) refracted more than long (red) ones (Sivak 1974). Marine and freshwater environments can be restricted to one region of the visible spectrum, depending on the presence or absence of a variety of impurities or the absorptive nature of water, which favors the transmittance of short (blue and blue-green) wavelengths, or both of them (Lythgoe 1972; Munz and MacFarland 1977). Thus, an eye in which focal length is too long (hyperopic), as measured with normal (tungsten or halogen) retinoscopic illumination, may be emmetropic in a naturally blue habitat.

The above considerations assume that refractive error and its effect on the resolving ability of the eye is the key visual concern. In fact, the relation between small eyes and hyperopia was pointed out by Walls (1942), who noted that it is often associated with nocturnal species, at least in mammals. In these cases, sensitivity to light, and not resolution ability, is the key visual concern, and findings of hyperopia may simply reflect indifference to optical quality.

As mentioned earlier, the uniform optical appearance of the larval amphibian eye can be contrasted with the widely differing metamorphic changes found in a variety of adults. For example, the postmetamorphic lens of *Pelobates syriacus,* an anuran, is considerably flattened (Fig. 19.3), with diameter exceeding thickness (Fig. 19.1) by 21.7%. The eye gets progressively more hyperopic over a long 20-week period, until the tail of the animal is completely resorbed, at which point it leaves the water. In air the metamorphosed eye is slightly (1–3 diopters) hyperopic (Sivak and Warburg 1983). The larval form of *Pelobates* is herbivorous, and vision may be less important before and during metamorphosis than in the terrestrial juvenile and adult carnivorous states (Degani 1982). A similar slow change was noted from sections of fixed eyes of *Rana temporaria* (de Jongh 1967).

In the bullfrog *Rana catesbeiana* the change in lens shape is less (diameter 12.5% greater than thickness), but the major part of the change takes place more rapidly, perhaps over a 24–48-hour period just before the animal moves onto land (Sivak et al. 1985). The difference between *Rana* and *Pelobates* may be related to *Rana*'s carnivorous feeding habits, both before and after metamorphosis (Barbour 1971).

Another anuran, *Xenopus laevis,* provides a third example (Fig. 19.4). Here the lens is virtually unchanged in shape (spherical or slightly flattened), persumably because water is the dominent habitat of the adult (Sivak et al. 1985).

Variations are found among the urodeles as well (Table 19.1). The lens dimensions of juvenile *Salamandra salamandra* differ by 13.6% (Sivak and Warburg 1980). Furthermore, most of this change takes place during a critical 24-hour period before movement to land. Before this time, the larvae are slightly hyperopic (5 diopters) in water. The juveniles are somewhat hyperopic (1–6 diopters) in air. *Notophthalmus viridescens* is an animal that spends two or three years on land as an adult before returning to the water. Here the change is slower and less dramatic (6.6% difference in lens dimensions). The lens returns to a spherical shape when the land (eft) stage is over (Sivak et al 1985).

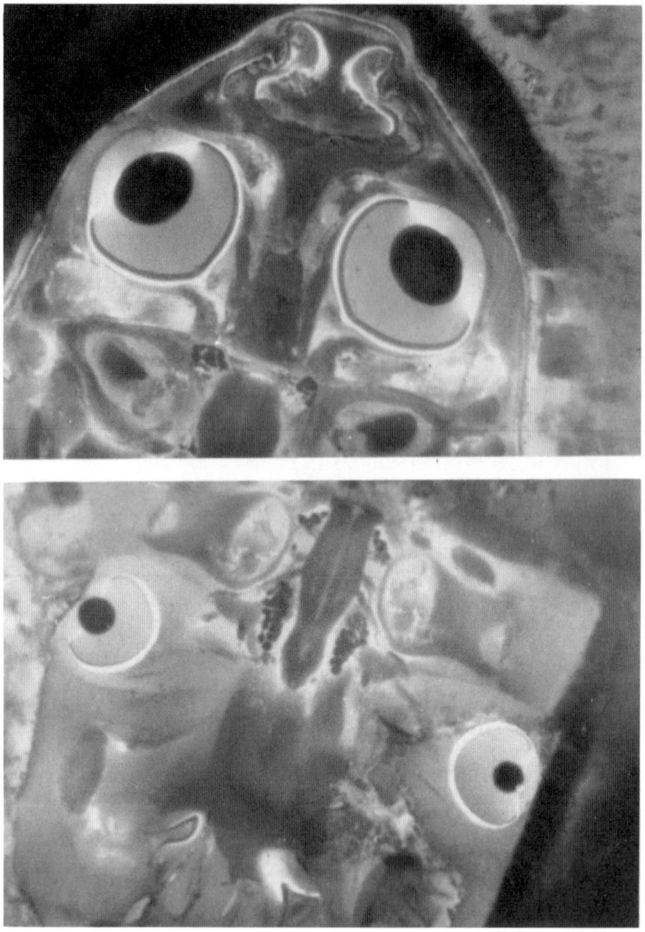

Figure 19.3. Hemisections of larval (*top*) and adult (*bottom*) heads and eyes of *Pelobates syriacus*. Note the metamorphosis of lens shape. (Approx. 2.75×.) (Modified from Sivak and Warburg 1983.)

Metamorphic alterations of lens shape are accompanied by changes in relative corneal curvature and laterality of eye position. In general, the adult cornea is more prominent than the larval one. In *Pelobates*, for example, the radius of curvature of the tadpole cornea (1.8 mm) is larger than the radius of the whole eye (1.4 mm). In the juvenile, corneal radius (1.7 mm) is smaller than eye radius (2.1 mm) (Sivak and Warburg 1983).

Metamorphosis for aerial vision is usually associated with increased binocularity. For example, the angle separating the ocular axes is reduced from 170° to 95° in *Pelobates* and from 120° to 103° in *Salamandra*. By contrast, the eyes of *Xenopus* maintain their lateral and dorsal position (ocular axes 120° apart) in the adult (Sivak et al. 1985). It appears that binocular vision is more important in air than in water, though the significance of this finding is not clear. In the

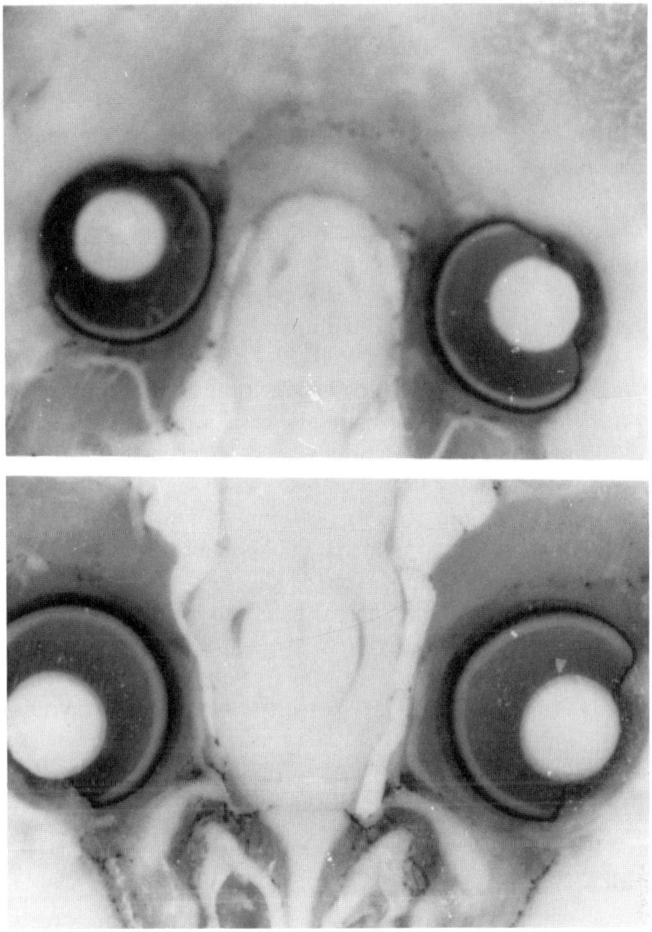

Figure 19.4. Frozen sections of the head and eyes of *Xenopus laevis* in larval (*top*) and adult (*bottom*) stages. Note virtual absence of change in lens shape. Eye diameters are approximately 2 mm. (From Sivak et al. 1985.)

case of *Pelobates,* this change may reflect the change from herbivorous to carnivorous feeding habits.

These descriptions emphasize the relationship between visual optics and environment. Obviously, the degree of adaptation to air or water is dependent on the extent to which one medium predominates in the behavioral repertoire of a particular species.

## Super Accommodation in Eyes of Diving Birds

A number of avian species appears to be able to cope with the loss of corneal refractive power when diving into water. In contrast to the slow, at least 24-

hour, metamorphic period of the amphibian eye, the bird eye must be able to adapt to a rapid (less than one-second) change in external media. As with the amphibians, however, behavioral indications of visual need are of prime importance in the choice of experimental species and in the interpretation of results. Already in 1910, Hess described an accommodative mechanism in cormorants capable of drastically increasing the refractive power of the lens when the eye is in water.

The avian accommodative mechanism and its reptilian precursor are the most effective means known of varying ocular focal power among terrestrial vertebrates. The accommodative apparatus includes a large and striated ciliary muscle, the contractile force of which is transmitted directly to the crystalline lens as a result of a series of structural adaptations (Walls 1942; Pumphrey 1948). These include (1) the existence of large ciliary folds between the lens and ciliary body, (2) an annular pad made up of collumnar epithelial cells around the lens equator, which increases its diameter, and (3) a pronounced corneoscleral sulcus due to the existence of a ring of ossicles in the limbal sclera (Fig. 19.5).

To add perspective to the role of accommodation in amphibious vision of some diving birds, it is useful to note that the human eye would have to accommodate by an amount roughly equivalent to the loss of corneal power in water (i.e., 40–45 diopters), an amount approximately four times the amount of accommodation available to the young eye (Grosvenor 1982).

In determining the accommodative range of the cormorant, Hess (1910) recorded a refractive state shift of 40–50 diopters (emmetropia to myopia) in response to the artificial stimulation of the eye. Hess assumed that (1) an accommodative change of this amount had to be related to the need for acute vision underwater

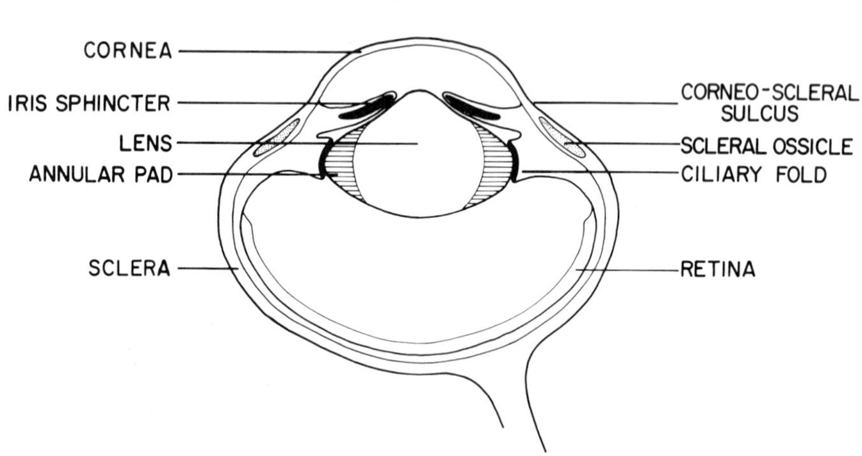


Figure 19.5. Schematic drawing of avian eye showing components of accommodative apparatus and accommodative lens changes suggested for the cormorant in the description by Hess (1910). The action of ciliary muscle is transmitted directly to the lens. In addition, massive sphincter muscle of the iris deforms anterior lens surface to produce a bulge of short radius of curvature. Combined effects markedly increase refractive power of the lens to compensate for refractive loss of the cornea in water.

and (2) that both the iris sphincter and ciliary body deform the lens to produce the change in curvature required. A mechanism of this type had been proposed earlier by Beer (1898) for certain aquatic snakes and turtles.

Subsequent workers have used the same approach to estimate whether the accommodative abilities of a variety of avian species are great enough to be considered a means of overcoming the optical barrier to high visual resolution underwater. Goodge (1960) has shown, for example, that the dipper *Cinclus mexicanus* can accommodate as much as 48 diopters when the eye is electrically stimulated. By retinoscopy, the eye is 40 diopters hyperopic underwater; so this range is sufficient to compensate for the optical loss of the cornea in water.

A series of more recent refractive studies, carried out on three pairs of aquatic species with varying underwater visual needs, emphasizes the relationship between feeding habits and accommodation (Sivak, Bobier, and Levy 1978). Underwater retinoscopic measurements carried out in vivo indicate that species such as the double-crested cormorant (*Phalacrocorax auritus*), the black guillemot (*Cepphus grylle*), and the hooded merganser (*Mergus cucullatus*), can accommodate to achieve a refractive condition of emmetropia or near emmetropia when in water. All are carnivorous species that actively pursue their prey underwater. The brown pelican (*Pelecanus occidentalis*) is a fish feeder, but it does not pursue its prey underwater. Rather, pelicans plunge almost perpendicularly to the water surface on a hit-or-miss basis. The herring gull (*Larus argentatus*) is a scavanger that feeds mainly on the surface, while the common mallard (*Anas platyrhynchos*) is herbivorous and merely ''tips up'' when feeding from the water. The last three species become very hyperopic (>45 diopters) when the eye is in water. Additional work, using the techniques of retinoscopy and photorefraction, has shown that penguins also have an exaggerated accommodative ability (Howland and Sivak 1984). The penguin cornea is relatively flat in comparison to that of other birds (Howland and Sivak 1984; Martin and Young 1984), however, so that the amount of accommodation needed for underwater vision is less.

Comparative anatomists have known for some time that the avian intraocular muscles (ciliary and iridic) are striated (Walls 1942; Duke-Elder 1958). In species such as the hooded merganser and black guillemot, the iris sphincter is very large, in both absolute and relative terms. In these birds, the sphincter occupies almost all the cross-sectional area of the iris (Fig. 19.5) with little nonmuscular tissue in evidence (Sivak and Vrablic 1982). An ultrastructural comparison of mallard and merganser muscles shows that the merganser sphincter includes a mitochondria-rich type of cell, perhaps indicating a concentration of myoglobin for aerobic metabolism and sustained activity (Sivak and Vrablic 1982).

That the iris is definitely used in merganser accommodation was shown by Levy and Sivak (1980) in a freeze-sectioning study of the eye with and without drug-induced accommodation (Fig. 19.6). When the accommodative mechanism is stimulated with nicotine sulphate, thickness of the lens increases, and its anterior portion bulges through the iris to form an anterior lenticonus (Fig. 19.6) of a short radius of curvature. Although Hess and subsequent authors, such as Walls, suggested that the iris actively squeezes the anterior surface of the lens, it is possible that, in fact, the iris forms a rigid annulus through which the relatively soft lens is pushed by the ciliary folds.

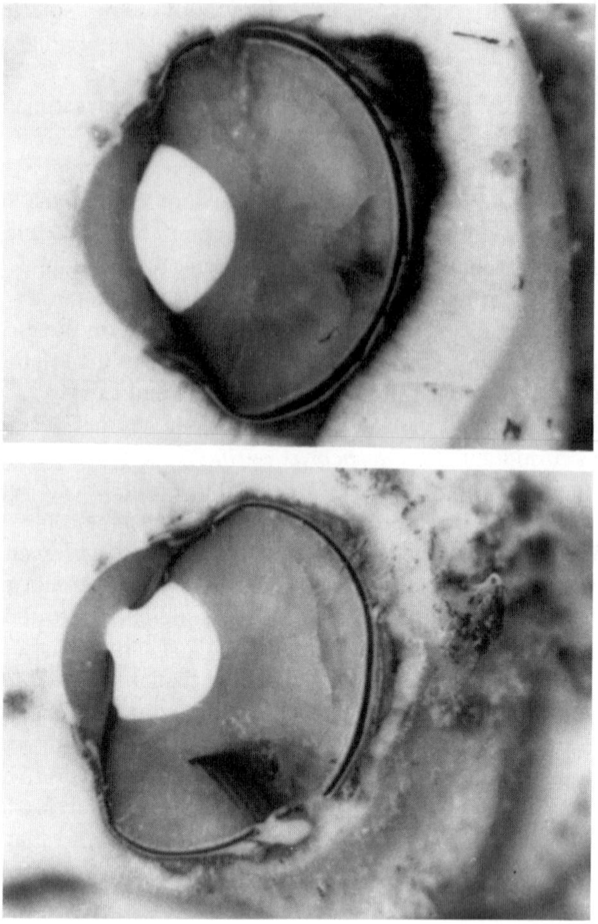

Figure 19.6. Frozen hemisections of the eye of hooded merganser (*Mergus cucullatus*) when accommodation is relaxed (*top*) and after stimulation with nicotine sulphate (*bottom*). (Axial eye diameter is approximately 20 mm.) (Reproduced with permission from Levy and Sivak 1980.)

The accommodative dynamics of a variety of diving and nondiving ducks was examined in vitro as follows: Excised eyes were placed in a saline bath, and a portion of the posterior sclera was removed. Fine, parallel, low-powered helium-neon laser beams (two or three) were projected through the pupil into the eye. Since the eye was in water, the cornea was not refractive, and the incident beams exited through the hole in the sclera to focus at a point well beyond the posterior pole of the eye (Fig. 19.7). The effect of accommodation on the lens was timed and measured from videotapes of the movement of the focal point toward the retina in response to electrical stimulation of the ciliary body (Sivak, Hildebrand, and Lebert 1985).

The results demonstrate that divers such as the merganser and the goldeneye

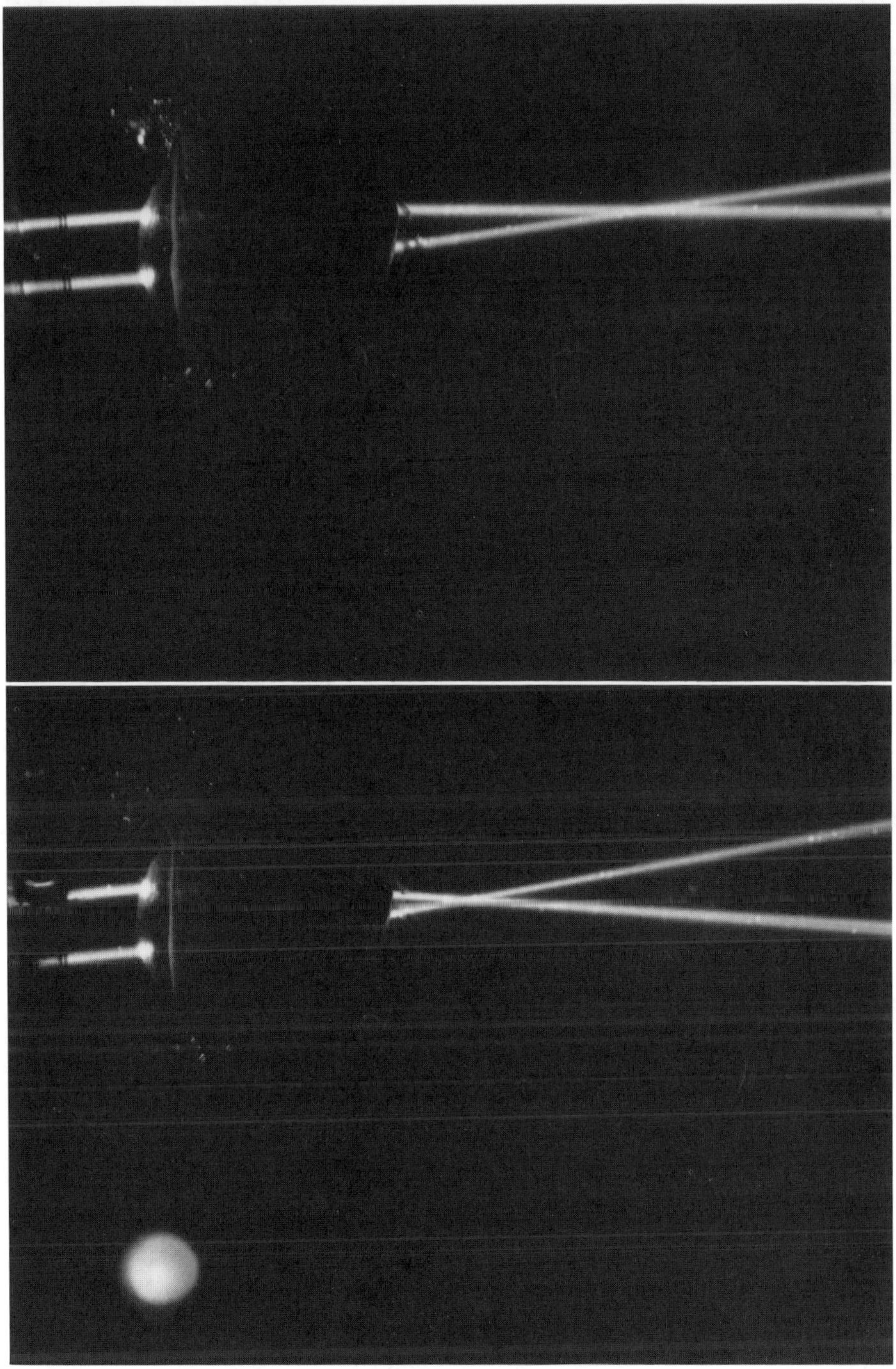

Figure 19.7. Parallel laser beams entering hooded merganser eye in Ringer's solution, before (*top*) and during (*bottom*) electrical stimulation of ciliary body. Marked change in focal point position toward posterior pole of the eye is brought about by accommodative lens change. (Approximately 4.8×.)

(*Bucephala clangala*) can bring the focal point to the vicinity of the retina (equivalent to the 70–80 diopter refractive loss of the cornea in water) in about one-third of a second (Fig. 19.8). A study of the effect of continuous stimulation indicates that this amount of accommodation can be sustained for about a minute, the duration of typical dives (Johnsgard 1975). Video records show the lens being pushed through the pupil in accommodation.

Herbivorous species that do not dive, such as the common mallard and the wood duck (*Aix sponsa*), demonstrate much lower magnitudes of accommodation (3 and 6 diopters, respectively). The redhead duck (*Aythya americana* ) is somewhat in between (16 diopters), a finding consistent with this species's omnivorous feeding habits.

It would be interesting to examine the dynamics of accommodation in a penguin eye by the same approach to see whether a sustained accommodative response

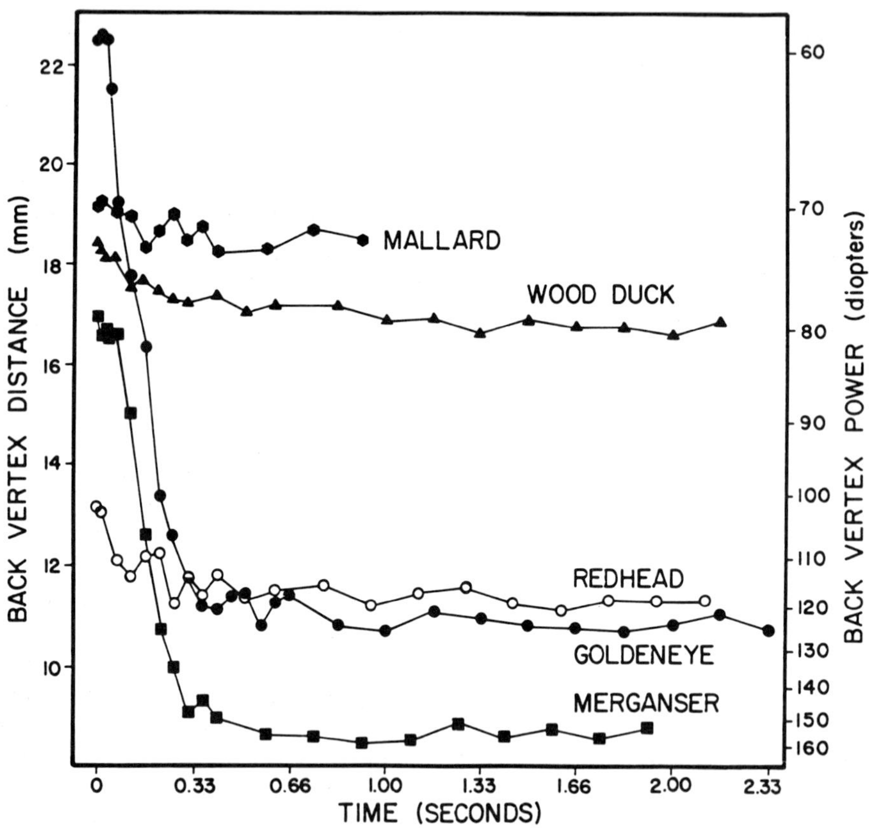


Figure 19.8. Change in back vertex (focal) distance and refractive power of eyes of a variety of waterfowl having varying need for acute underwater vision. Focal changes are measured from sample video recordings of focal change brought about by ciliary body stimulation. Time is determined from the number of video frames. (Modified from Sivak, Hildebrand, and Lebert 1985.)

can be maintained for the much longer dives common in this group. On a relative scale, the sphincter of the iris of the Adélie penguin is as massive as that of the hooded merganser (Sivak and Vrablic 1982). Perhaps the penguin's accommodative response can be maintained longer because the amount required is minimized by the flatter cornea.

## Conclusion—Questions Asked, Answers Needed

The research approaches described here, involving two vertebrate groups, leave many unanswered questions and pose many new ones. For example, the change in shape of the amphibian lens in association with metamorphosis emphasizes that lens shape is not simply a result of the packing arrangement of old and new cells that takes place as the lens grows and develops. Rather, lens shape is in some way regulated to fit the overall refractive plan of the eye. How is a change in lens shape brought about? Is the signal for refractive change initiated by a certain degree of retinal defocus, as seems to be the case in the development of myopia in chicks and other species when a transluscent goggle is worn during the developmental periods (Hodos and Kuenzel 1984)?

What happens at the cellular level? Since the metamorphosed amphibian eye is usually larger than the larval one, the change in lens shape may result from an acceleration of equatorial cell multiplication, provided the newer cells have a tapered shape appropriate for a nonspherical lens. The fact, however, that lens metamorphosis takes place rapidly (24–48 hrs) in some species and that the metamorphosed lens may not be larger (e.g., *Notophthalmus viridescens*) suggests that a change in cell shape or intercellular connections of existing cells or both are possible.

The amphibian eye has been used extensively for the study of various aspects of ocular anatomy, physiology, and embryology. Little attention, however, has been directed to metamorphosis in the context of optical change from an aquatic to a terrestrial eye. It would seem that the amphibian eye would be a useful model for the study of factors controlling ocular growth or lens morphology or both of them. From a behavioral point of view, the structure of the eye in terms of aquatic or terrestrial optics would be a simple method of determining whether air or water is the critical habitat for a given species.

From physiological studies in excised eyes, it appears that an exaggerated accommodative ability, capable of neutralizing the refractive loss of the cornea in water, exists in certain diving birds. Additional study to show that such responses take place in vivo, perhaps by means of slit lamp photography of lens shape or ultrasound measurements or both, is an obvious necessity. The need to establish the relative contributions of the iris and ciliary body to changes of lens shape is also required. Careful comparisons of the rates of pupil closure and focal changes may indicate whether the role of the iris in producing the accommodative lenticonus is passive or active.

A further point of interest concerning the iris is the question of pupil size at lower light levels. Even completely diurnal species may dive into water in which light is poorly transmitted. If contraction of the iris sphincter is required for adequate accommodation, then pupil dilation is not possible. Conversely, if dilation of the pupil takes place because of the decreased availability of light, then accommodation will be insufficient to neutralize the loss of corneal power in water.

A final note of interest concerns the neurophysiology of accommodation in birds. Classically (e.g., Duke-Elder 1958), accommodation in mammals is described as a parasympathetic–smooth muscle system. In fact, a number of studies have shown that the ciliary muscle of higher mammals has striate anatomical and physiological features (Ishikawa 1962; Ripps, Siegel, and Getz 1962), and the existence of an accommodation-related synapse in the ciliary ganglion has been questioned (Westheimer and Blair 1973). Although it has been known for a long time that avian intraocular muscles are striated, little attention has been directed to their neural control. Some evidence indicates that the avian ciliary ganglia contain synapses in which pre- and postsynaptic elements are electrically, rather than chemically, coupled (Martin and Pilar 1963a, 1963b).

Future workers should make an effort to tie together the different aspects of avian accommodation, both in diving and nondiving birds. The information required includes intraocular muscle anatomy, rate and magnitude of intraocular muscle response, pharmacological response of intraocular muscles, ciliary-ganglionic morphology, and CNS origins of accommodative neural control.

## Summary

An important environmental factor to consider in the adaptive radiation of the vertebrate eye is the effect of air and water on visual optics. Although the cornea is normally the main refractive structure of the terrestrial eye, it is of little or no refractive importance when the eye is in water. This is not a result of the close similarity between the refractive indices of water and the cornea, but rather of the fact that the cornea is bounded by parallel or nearly parallel surfaces exposed to air or water in front and an aqueous medium (aqueous humor) behind. As a result, the terrestrial eye differs morphologically, physiologically, and optically from it aquatic precursor.

Two vertebrate groups, the amphibians and birds, are of particular interest—the former because of environmental change associated with metamorphosis from larval to adult stages, the latter because of the dependence of many birds on aquatic environments for feeding. The changes in visual optics that accompany metamorphosis are described in detail for a variety of urodeles and anurans. Although the eyes of all larval amphibians are typically aquatic, the eyes of adults show varying degrees of terrestrial characteristics, depending on the extent to which the life history of a particular species frees it from dependence on water. For example, the lens of urodele *Salamandra salamandra* changes from a spherical shape to a markedly flattened one during a relatively brief critical period just before metamorphosis. On the other hand, the lens of *Xenopus laevis*, an anuran that remains aquatic as an adult, is more or less spherical at both stages.

Evidence showing an exaggerated accommodative ability in certain diving birds is reviewed. This response is restricted to carnivorous species that pursue their prey underwater. Recent measurements of its rate and magnitude indicates that it is a response designed to compensate for the refractive loss of the cornea when the eye moves underwater.

*Acknowledgments.* The author's research described here is supported by the Natural Sciences and Research Council of Canada. The author gratefully acknowledges the drafting and typing assistance of A. Weber, J. MacLean, and D. Clermont.

## References

Barbour, R.W. (1971) Amphibians and reptiles of Kentucky, University Press of Kentucky, Lexington.

Beer, T. (1894) Die Accommodation des Fishauges, Pfluegers Archiv. Gesanite. Physiol. Menschen Tiere, 58:523–650.

Beer, T. (1898) Die Accommodation des Auges bei den Reptiles, Pfluegers Archive. Gesanite. Physiol. Menschen Tiere, 69:507–568.

Bennett, A.G. and Francis, J.L. (1962) The eye as an optical system, in The Eye, vol. 4, Visual Optics and the Optical Space Sense, Davson, H. (ed.), Academic Press, New York, pp. 101–132.

Brown, P.K., Gibbons, I.R., and Wald, G. (1963) The visual cells and visual pigment of the mudpuppy, *Necturus*, J. Cell Biol., 19:79–106.

Degani, D. (1982) Amphibian tadpole interaction in a winter pond, Hydrobiologia, 96:3–7.

Duke-Elder, S. (1958) System of ophthalmology, vol. 1, The Eye in Evolution, Henry Kimpton, London.

Duncan, G. (1981) On classifying human cataractous lenses, in: Mechanisms of Cataract Formation in the Human Lens, Duncan, G. (ed), Academic Press, London, pp. 1–5.

Franz, V. (1934) Höhere Sinnesorgane, in Handbuch der vergleichenden Anatomie der Wirbeltiere, vol. 2, Part III, Bolk, L. Goppert, E., Kallius, E. and Lubosch, W. (eds.), Urban & Schwartzenberg, Berlin, pp. 989–1292.

Goodge, W.R. (1960) Adaptations for amphibious vision in the Dipper (*Cinclus mexicanus*), J. Morphol., 107:79–91.

Grosvenor, T.P. (1982) Primary care optometry, Porfessional Press, Chicago.

Grusser-Cornehls, U. and Himstedt, W. (1976) The urodele visual system, in The Amphibian Visual System, a Multidisciplinary Approach, Fite, K.V. (eds.), Academic Press, New York, pp. 203–266.

Gur, M. and Sivak, J.G. (1979) Refractive state of the eye of a small diurnal mammal: the ground squirrel, Am. J. Optom. Physiol. Op., 56:689–695.

Hess, C. (1910) Die Akkommodation bei Tauchervögeln, Archiv. Vl. Ophthalmol., I:163–164.

Hodos, W. and Kuenzel, W.J. (1984) Retinal image degradation produces ocular enlargement in chicks, Invest. Ophthalmol. & Visual Sci., 25:652–659.

Hogan, M.J., Alvarado, J.A., and Weddell, J.E. (1971). Histology of the human eye, Saunders, Philadelphia.

Howland, H.C. and Sivak, J.G. (1984) Penguin vision in air and water, Vision Res., 24:1905–1909.

Hughes, A. (1977) The topography of vision in mammals of contrasting life style: comparative optics and retinal organization, in Handbook of Sensory Physiology, vol. II/5, The Visual System in Vertebrates, Crescitelli, F. (ed.), Springer-Verlag, Berlin, pp. 613–756.

Ishikawa, T. (1962) Fine structure of the human ciliary muscle, Invest. Opthalmol. & Visual Sci., 1:587–608.

Jamieson, G.S. (1971) The functional significance of corneal distortion in marine mammals, Can. J. Zool., 49:421–423.

Johnsgard, P.A. (1975) Waterfowl of North America, Indiana University Press, Bloomington.

deJongh, H.J. (1967) Relative growth of the eye in larval and metamorphosing *Rana temporaria,* Growth, 31:93–103.

Kinoshita, J.H. (1974) Mechanisms initiating cataract formation, Invest. Ophthalmol. & Visual Sci., 13:713–724.

Levy, B. and Sivak, J.G. (1980) Mechanisms of accommodation in the bird eye, J. Comp. Physiol., 137:267–272.

Lovasik, J.V. (1979) Neural control mechanisms of ocular accommodation in cat, Ph.D. thesis, University of Waterloo, Waterloo, Ontario.

Lythgoe, J.N. (1972) The adaptation of visual pigments to the photic environment, in Handbook of Sensory Physiology, vol. VII/1, Photochemistry of Vision, Crescitelli, F. (ed.), Springer-Verlag, Berlin, pp. 566–603.

Martin, A. R. and Pilar, G. (1963a) Dual mode of synaptic transmission in the avian ciliary ganglion, J. Physiol. (Lond.), 168:443–463.

Martin, A.R. and Pilar G. (163b) Transmission through the ciliary ganglion of the chick, J. Physiol. (Lond.), 168:464–475.

Martin, G.R. and Young, S.R. (1984) The eye of the humboldt penguin, *Spheniscus humboldti:* visual fields and schematic optics, Proc. R. Soc. Lond. B Biol. Sci., 223:197–222.

Martof, B.S., Palmer, W.M., Bailey, J.R., and Harrison, J.R., III (1980) Amphibians and reptiles of the Carolinas and Virginia, University of North Carolina Press, Chapel Hill.

Michel, K. (1932) Die Akkommodation des Schlangenauges, Jena A. Naturwiss., 66:577–628.

Munz, F.A. and McFarland, W.N. (1977) Evolutionary adaptations of fishes to the photic environment, in Handbook of sensory physiology, vol. VII/5, The Visual System of Vertebrates, Crescitelli, F. (ed.), Springer-Verlag, Berlin, pp. 193–274.

Pumphrey, R.J. (1948) The sense organs of birds, Ibis, 90:171–199.

Ripps, H., Siegel, I.M., and Getz, W.B. (1962) Functional organization of ciliary muscle in the cat, Am. J. Physiol., 203:851–859.

Sivak, J.G. (1974) Refractive error of the fish eye, Vision Res., 14:209–213.

Sivak, J.G. (1976) The accommodative significance of the "ramp" retina of the stingray, Vision Res. 16:945–950.

Sivak, J.G. (1977) The role of the spectacle in the visual optics of the snake eye, Vision Res., 17:293–298.

Sivak, J.G. (1980) Accommodation in vertebrates: a contemporary survey, in Zadunaisky, J.A., and Davson, H. (eds.), Current Topics in Eye Research, vol. 3, Academic Press, New York, pp. 281–330.

Sivak, J.G., Bobier, W.R., and Levy, B. (1978) The refractive significance of the nictitating membrane of the bird eye, J. Comp. Physiol., 125:335–339.

Sivak, J.G. and Dovrat, A. (1984) Aging and the optical quality of the rat crystalline lens, Invest. Ophthalmol & Visual Sci., 24:1162–1166.

Sivak, J.G., Hildebrand, T., and Lebert, C. (1985). Magnitude and rate of accommodation in diving and nondiving birds, Vision Res., 25:925–933.

Sivak, J.G. and Kreuzer, R.O. (1983) Spherical aberration of the crystalline lens, Vision Res., 23:59–70.

Sivak, J.G., Levy, B., Weber, A.P., and Glover, R.F. (1985). Environmental influence on the shape of the crystalline lens: the amphibian example, Exp. Biol. Med., 44:29–40.

Sivak, J.G. and Vrablic, O.E. (1982) Ultrastructure of intraocular muscles of diving and nondiving ducks, Can. J. Zool., 60:1588–1606.

Sivak, J. and Warburg, J. (1980) Optical metamorphosis of the eye of *Salamandra salamandra,* Can. J. Zool., 58:2059–2064.

Sivak, J.G. and Warburg, M. (1983) Changes in optical properties of the eye during metamorphosis of an anuran, *Pelobates syriacus,* J. Comp. Physiol., 150:329–332.

Spector, A. (1982) Aging of the lens and cataract formation, in Aging and Human Visual Function, Sekuler, R., Kline, D., and Dismukes, K. (eds.), Liss, New York, pp. 27–43.

Tripathi, R.C. and Tripathi, B.J. (1983) Lens morphology, aging and cataract, Gerontology, 38:259–270.

Walls, G.L. (1940) Ophthalmological implications for the early history of snakes, Copeia, 1–8.

Walls, G.L. (1942) The vertebrate eye and its adaptive radiation, Cranbrook Institute of Science, Bloomfield Hills, Mich.

Westheimer, G. and Blair, S.M. (1973) The parasympathetic pathways to the internal eye muscles, Invest. Ophthalmol & Visual Sci. 12:193–197.

Worgul, B.V. (1982) Lens, in Ocular Anatomy, Embryology and Teratology, Jakobiec, F.A. (ed.), Harper and Row, Philadelphia, pp. 355–389.

# Chapter 20

# How to Be Unseen: An Essay in Obscurity

WILLIAM M. SAIDEL*

## The Problem Defined

Camouflage as a concept may be either an attribute (a disguise) or an act (to disguise oneself). Ethological texts tend to use the word *crypsis* to describe the condition of being "camouflaged to resemble part of the environment" (Edmunds 1974, 7), leaving unclear which (or both) meaning(s), is being considered. In the discussion of a particular instance of animal camouflage, which concept is being examined must be clearly stated, because different biological processes are involved with each definition. The designs of camouflage (patterns, a noun) act to camouflage (a verb) an animal only if a viewer can appreciate its optical effects. Understanding the processes of camouflage (verb!) depends on understanding how the designs of camouflage (noun!) hide the animal. This essay is an inquiry into some visual processes that may be implicated in the evolution and design of a particular camouflage (noun)—the camouflage of flatfish.

One fundamental fact will emerge. At no stage in the discussion of any of the different aspects of flatfish camouflage will the camouflaged animal use feedback of any sort to measure the appropriateness of its appearance. In fact, little of the body is even seen by their retina (Gulley, Cochran, and Ebbesson 1975; Luckenbill-Edds and Sharma 1977). Appropriateness is measured only by the predator, for ". . . predation on the least well concealed individuals has been the selective factor responsible for perfecting the crypsis" (Edmunds 1974, 16).

### Seeing Objects

Under normal circumstances, we humans visually attend to one object at a time. Our attention to any particular object may change with great rapidity, but at all times, our retina views far more of the world than one single attended object.

*Department of Zoology, The University of Maryland, College Park Campus, College Park, MD 20742, USA

Humans preferentially attend to objects seen in the fovea. What are all the other stimuli that are sensed by the retina, but not attended to? They are visual texture. Texture is always presented to an entire retina, but analysis of texture by a specialized retinal area (e.g., fovea, area centralis, area temporalis) or central brain region (the "Cyclopean" eye of Julesz 1971) is preferentially used for object definition. The finer grain of a specialized area may permit the acute measure of local differences in texture that is required for object identification.

The problem for a camouflaged animal is to avoid detection when surrounded by objects in the environment, that is, not to be seen as a distinct object. A general solution to this problem is a disguise that is seen as sufficiently similar to the background that a viewer cannot distinguish between the animal and its surround. The disguise then must act like a malleable mirror having a reflection that is interpreted only in relation to the immediate environment.

Thayer (1909) and Cott, in his monumental volume, *Adaptive Coloration in Animals* (1957), were the first to emphasize the theory of protective resemblance ". . . in terms of the vision of potential enemies and prospective prey" (Cott 1957, 173). As Cott posed the theory, camouflage evolved to conceal the effects of directional illumination on solid objects (such as animals) that are seen by an eye capable of recognizing them. He identified a limited set of designs used repeatedly throughout the animal kingdom that minimized the visibility of a directionally illuminated animal (Cott 1957, 1971). These designs are based on four principles: (1) color resemblance or agreement; (2) obliterative shading (a graded lightening and darkening on the surface whose effect diminishes the appearance of roundness, or relief); (3) disruptive coloration (a pattern of colors and tones that divides the continuous body outline and surface into sections); and (4) shadow elimination.

His analysis is limited to camouflage as a set of designs. These four principles describe the patterns of disguise most adequately, but leave unanswered the key question about their effects on the viewer's vision.

Directional illumination provides an observer with two distinctive cues: highlights and shadows (Cott's principles 2 and 4), as well as providing the medium for the chromatic appearance. Flatfish, except when swimming, lie directly upon, or partially buried within, the ground. Thus, the appearance of camouflage in these fish limits consideration of camouflage primarily to the problem of texture discrimination (Cott's principle 3) without the additional consequences of directional illumination.

## Abbreviated Natural History of Flatfish and Their Changes in Appearance

Flatfish undergo a unique metamorphosis. In the first two to three months of life, the larvae metamorphose with one eye rotating through or around a cartilaginous cranium to rest on the other side, leaving one side with a pair of eyes and the other with none (Agassiz 1879). This is accompanied by a rotation of the body axis of 90°. The postmetamorphic fish subsequently lie on a genetically determined side (Policansky 1982), either their embryonic left side (families Bothidae and Cynoglossidae), their right side (families Pleuronectidae and Soleidae), or either side (families Psettodidae and Citharidae). The skin directly against the bottom typically appears white in nearly all flatfish, whereas the chromatic and textural pattern of the premetamorphic larvae remains on the visible side.

The study of animal color changes began with the origin of recorded natural history (see Parker 1948, 8ff). When the change in an animal's appearance is in accord with the local environment, the change is termed *adaptive camouflage*. This ability is found in many fishes, but appears to reach a pinnacle with flatfish. This group of fish has evolved the ability to change not just their skin tone, but also their skin texture, and this ability is triggered by visual cues.

For the first twenty years of the 20th century, two views of this pattern-adaptative ability by flatfish were argued within the scientific community. Loeb, Pitkin, and others suggested that the skin of flatfish adapted to reproduce the local ocean bottom as seen by the fish's retina (Mast 1914, 179). Sumner (1911) and Mast (1914) identified the change in appearance as changes in tone and appearance of local skin areas that mimicked the local environmental patterns. Sumner and Mast commented that the skin patterns copied the surround only within the possibilities allowed by the morphological elements within the skin.

For both views, however, an accurate description of the nature of the adaptive changes in appearance proved difficult to enunciate. Various comments illustrated an analysis of matching by a visually dependent judgement and described its nature in comparative terms. For instance, "The way in which this [WMS, a change in the fish's appearance] was brought to pass is difficult to analyze in detail, but the results are indicated in the figures" (Sumner 1911, 422), or, "Our evidence shows conclusively that . . . there is a configuration produced in the skin which resembles that of the background. This is clearly shown in the photographs . . ." (Mast 1914, 188). The analysis was one of similarities and differences with the background, and the comparison was usually made in terms of environmental structures found around the fish (e.g., sand, rocks, pebbles, and algae).

Subsequently, the analysis of flatfish camouflage focused primarily on physiological mechanisms of the control of skin chromatophores (see reviews in Parker 1948; Fujii 1969; Bagnara and Hadley 1973) until Lanzing (1977) applied ethological notions to the various adaptations of two flatfish species, the turbot *Scopthalmus maximus* and the plaice *Pleuronectes platessa*. He focused on (1) the permanency of the chromatophore markers, (2) the symmetry in position of the markers, and (3) the limitations in appearance that a flatfish could assume on different backgrounds. Lanzing (1977, 221) concluded that "a limited range of preset chromatophore stimulation patterns is available from which the fish 'selects' the one closest to the current background patterning." Moreover, he inferred that the fish probably has separate patterned and unpatterned chromatophore control systems, reminiscent of those described by Baerends and Baerends-van Roon (1950). Thus, the analysis of flatfish camouflage in the twentieth century had moved from an exact reproduction of the fish's background to a restricted representation of the background to a set of modal or fixed action patterns that are elicited by seeing the background.

These analyses of flatfish camouflage asked the question, "How does a fish respond to defined substrates?" The results assumed that the observer passively viewed the fish in its surround. The studies avoided examining an explicit relationship between the camouflaged fish and the observer, even though the latter is integral to the phylogenetically successful designs.

With only a passive relationship between the fish and its observers, there should be no restrictions on the appearance of a flatfish, because the relationship between the designs of camouflage and the environment is open looped. As far as the camouflager is concerned, it responds to its environment without seeing its own skin. It is the viewer, e.g., a predator, that forms a link between the camouflage designs and the environment by extracting those appearances that remain visible. After all, the appearances of survivors are the viewer's criteria of successful camouflage (see Edmunds 1974). This closure resulting from a predator's visual perception probably accounts for the basic similarity of morphological markers for diverse pleuronectids (Hewer 1931). Thus, the perception of the pattern is most important in the evolution of flatfish camouflage patterns.

This chapter is an inquiry into the perception of flatfish patterns. It is structured to identify changes in a flatfish's appearance, using some material from original studies, and the physiological control of those changes. This information is then used to infer a step in the visual perception of the observer that closes the loop between the designs of camouflage and the environment, thereby establishing a cryptic state.

## Adaptive Behavior and Its Structural Basis

### Adaptive Parameters: A Note on Methodology

The traditional macroscopic analysis of flatfish camouflage has been to describe subjectively the appearances of a fish or photographs of it on a set of different backgrounds (Sumner 1911; Mast 1914; Osborn 1939; Chinarina 1960; Scott 1965; de Groot, Norde, and Verheijen 1969; Lanzing 1977). A descriptive analysis of changes in appearance that has evolved to confuse just such an analysis is tautological.

To avoid this problem, I resorted to photometric measurements of skin reflectance or reflectance from photographs to analyze the adaptive changes in the appearance of two flatfish, *Pseudopleuronectes americanus* (the winter flounder) and *Paralichthys lethostigma* (the southern flounder). A 4- or 8-level gray scale was also photographed on the same roll of film, either separately or simultaneously with the fish, to facilitate accurate tonal reproduction. Fish were photographed about thirty minutes after being exposed to a particular background.

The f-stop and exposure time necessary to reproduce the gray scale were used to print the image of the fish. No reference was made to the fish's appearance.

A print was glued to a glass plate and placed at the object plane of a compound microscope onto which a photomultiplier tube had been mounted at the image plane of one eyepiece. The other eyepiece was used to monitor the experiment visually. The amount of light reflected from a path that passed though the different skin markers on the fish image was measured. The measures of reflectance from the same path of a single fish adapted to different backgrounds were compared to obtain a measure of the variations in adaptation. The average reflectance of a particular adaptation or background was obtained by planimetric integration of the measured reflectance.

### Tonal Responses to the Background

In the first example, the adaptations to backgrounds of different uniform reflectances were measured (Fig. 20.1). Three points are immediately noticeable: (1) The average reflectance of the skin responded to that of the background without equalling it; that is, for a group of backgrounds whose reflectance varied, the darkest average reflectance was obtained on the darkest background; the lightest reflectance, on the lightest background, etc. (2) An intrinsic pattern corresponding to the morphological markers remained apparent under all adaptations. (3) The total range of reflectances available to the fish was less than the range measured from the various backgrounds. Figure 20.1 illustrates the responses by *Paralichthys. Pseudopleuronectes* was less vivacious in its adaptations, and the range in variation of reflectances to a variety of backgrounds was substantially less than the range demonstrated by *Paralichthys* (Saidel 1977a).

### Responses to Different Textures

In Figure 20.2, an individual fish was placed on two backgrounds that were chosen to illustrate uniform and contrasting backgrounds. The average reflectance of the fish on the different patterns shows that this fish is lighter when on the uniform background than on the contrasty one, even though the brightest areas on either background are equal in reflectance. The decrease is about 12%. Thus, the fish skin adapts to an average of the background reflectance from a relatively large area of its visual field (see Mast 1914).

However, the contrast (ratio of reflectance from adjacent skin areas) has increased. The maximal contrast has gone from 14% on the uniform background to over 70% on the contrasty background. The increase in contrast resulting from adaptation on the contrasty background occurred even though the dynamic range between the lightest and darkest areas of the skin had decreased. Importantly, every border between adjacent skin features in the photometric tracing of one adaptation is visible in the other adaptation.

In Figure 20.3, the contrasts across an edge of the backgrounds are identical, but the overall reflectances differ. (The ratios of dark to light are 3:1 (*left*) and 1:3 (*right*), respectively.) The minimum reflectances (from the darkest areas) are nearly equal in both adaptations, while the average reflectance on the lighter background is about 16% lighter elsewhere. It is as if a 0.2 neutral density filter

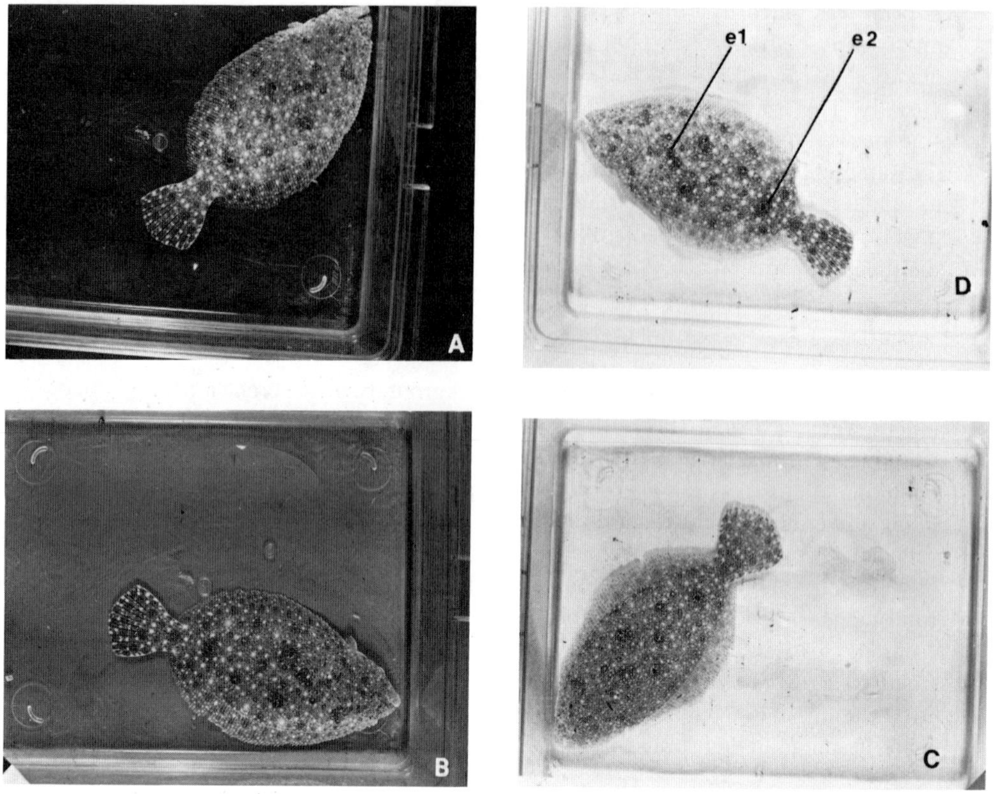
e1
e2
A
B
C
D

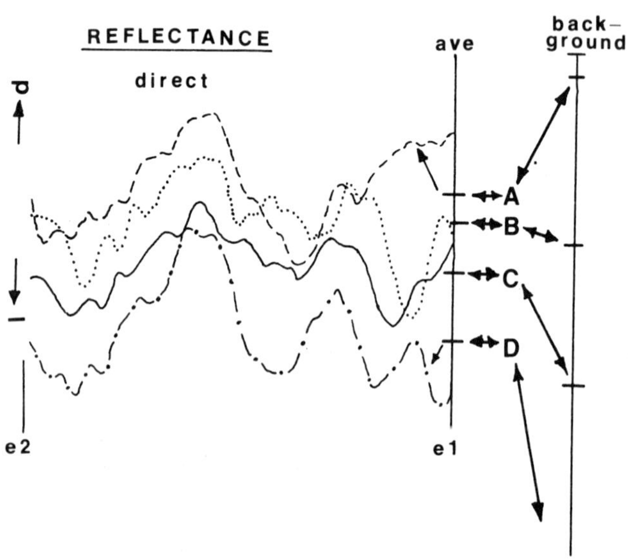
REFLECTANCE
direct
ave
back-
ground
d
l
e2
e1
A
B
C
D

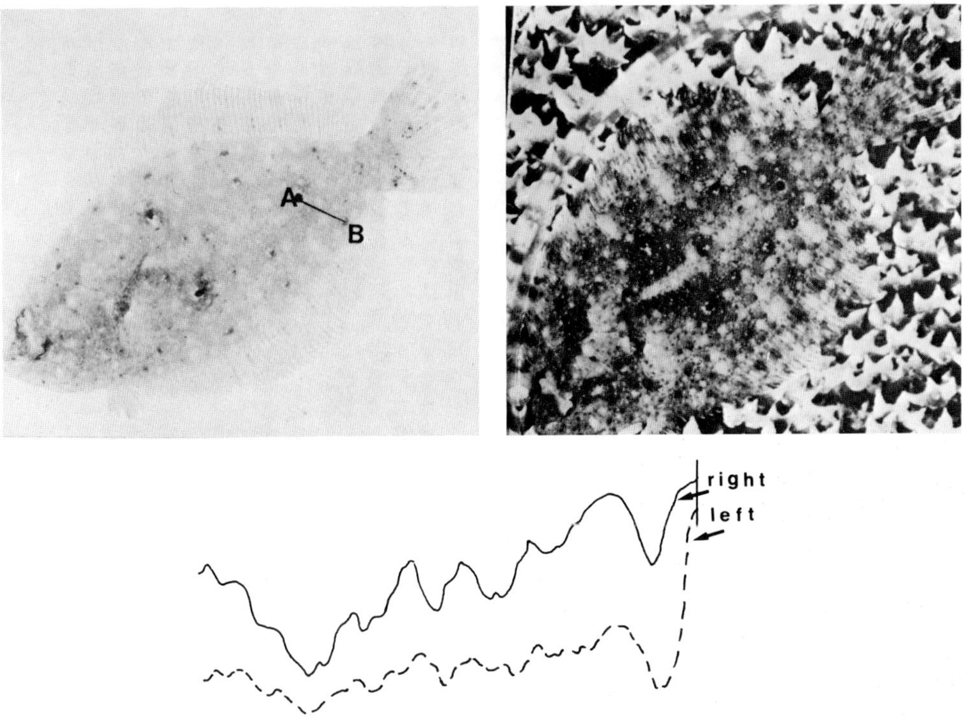


Figure 20.2. *Above*, illustrations of adapted changes to a white (*left*) and patterned (*right*) background. *Below*, graphs measuring the reflectance from the path indicated *above* in left figure as running from caudal central spot [*morphological marker 1* (*A*) to the edge of a fin through *morphological marker 4* (*B*)].

Note that the contrast is greater in patterned response, though average reflectance is obviously greater in white adaptation.

were placed over the skin adapted to the lighter background to construct the adaptation to the darker background.

## Checkerboard Story

An adapatation to a checkerboard-like background is a dramatic textbook demonstration of cryptic behavior (Fig. 20.4A and B). The same test was included in

Figure 20.1. *Above*, the adaptation of *Paralichthys lethostigma* to uniform backgrounds of different reflectances. *Below*, photometric scans of a path indicated in (D) by *e1* and *e2* from all four adaptations. The ordinate represents variations in the reflectance measured across the path. Increasing reflectance is indicated in downward direction.

The area under a curve was measured to determine its average reflectance. This average is indicated by black bar on right-hand side (*ave*), while background reflectance is indicated on far right (*background*). (Note that the reflectance measured from white background exceeded the scale.)

The convention of these scans is used for Figures 20.2 and 20.3.

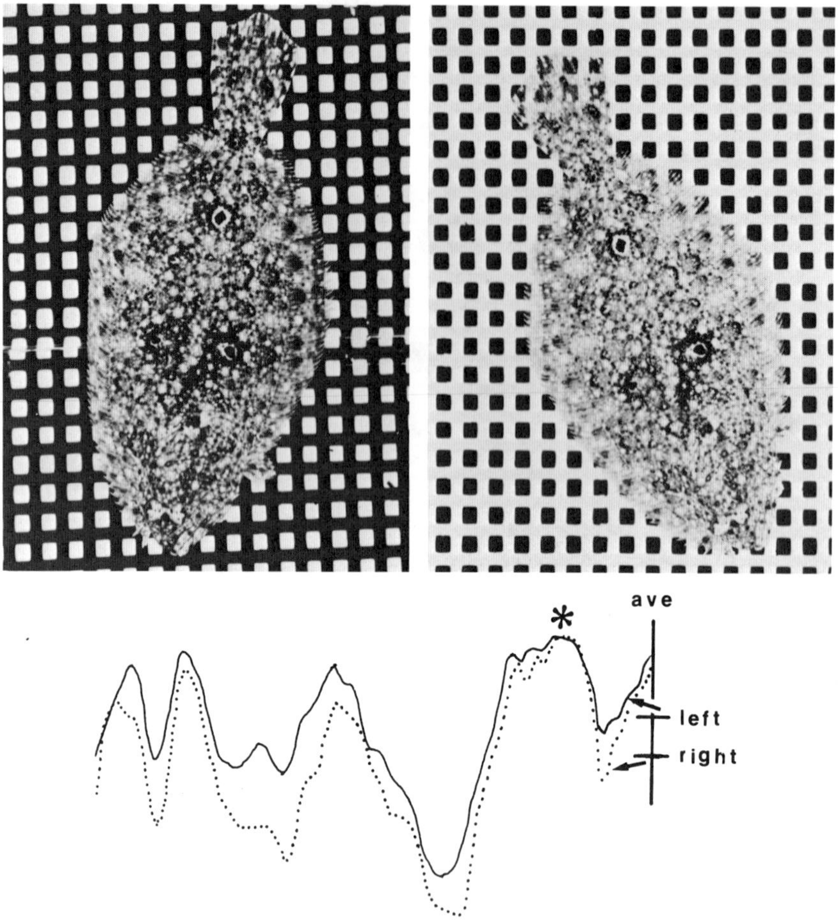


Figure 20.3. *Above,* adaptation *of P. lethostigma* to backgrounds whose average reflectances differ by 1:3 (*left*) and (*right*), while contrasts across any edge in either background are identical. *Below,* both average reflectance and the contrast across a border is reduced from left compared with right illustration, suggesting that overall decrease in skin reflectance is involved in the establishment of contrast. However, reduced contrast from areas containing high concentrations of epidermal melanophores are nearly identical (*asterisk*). This result is equivalent to placing a 0.2 neutral density figure over right adaptation to produce the adaptation on the left (see text).

this set of experiments with the controls as described in the section on methodology (Fig. 20.5A). One way to determine the adaptive change in appearance of a fish is to remove the background (Fig. 20.5B). The justification for this procedure is contained on pp. 504–506. In all tests of this sort, *Paralichthys* appeared far more similar to each other than one would expect from Figure 20.4A or B.

Similar background removals of Figures 20.4A and B are illustrated in Figures 20.4C and D. It is apparent that, whatever adaptations occurred, they are consistent with changes in the average contrast of some visual field seen by the fish. However, assessing the details of the changes in these illustrations is difficult. As shown in Figures 20.1–20.3, *all* features of skin texture are present for *all* adaptive changes. Thus, the lack of detail in the white areas of Figure 20.4A and B suggests that these photographs do not accurately represent the entire skin appearance. Secondly, small flatfish (<8 cm) are translucent (p.o.). Hence, one cannot assess what degree of the match in texture seen in Figures 20.4A and B results in part from seeing the background through the body (also see pp. 505–506 below regarding fin transparency).

## Design Materials

The designs in a fish skin are like a nineteenth century pointillist painting. Seurat, for example, painted the color of an area not with broad strokes that form a sheet of color, as did van Gogh, but by filling an area with many dots or flicks of paint. The viewer's eye integrates these dots into a perceived color, using the basic principles of color addition or subtraction. Just as the chromatic component of a pointillist painting is a consequence of integrating many dabs of color, the appearance of a fish depends on the visual integration of many microscopic skin chromatophores. That a flatfish looks like a set of patches suggests that chromatophores are nonuniformly distributed.

### Chromatophore Types

Chromatophores are embryologically related because they are derived from the same population of neural crest cells (Du Shane 1935; Bagnara 1983). Different types of these pigment-containing cells are embedded within the different layers of the skin (epidermis, suprascalar dermis, and subscalar dermal layers) (see Brown and Wellings 1970; Junqueira, Toledo, and Porter 1970; Roberts, Young, and Milne 1971; Burton 1978). These cells may be stellate or ovoid, and the types of chromatophores differ within the various skin layers. One type of stellate cell, a melanocyte, is found in the subdermis and epidermis of flatfish (and most fish) skin. Both stellate-shaped, pigment-containing cells and ovoid light-scattering cells are found in the suprascalar dermis.

Pigment is sequestered within cytoplasmic vesicles or granules. In the stellate chromatophores, these granules migrate during adaptations to different environments (Parker 1948; Green 1968). The pigment is said to be *dispersed* when the granules are located in the cell processes and *aggregated* when in the central spherical area surrounding the nucleus.

The color of a chromatophore depends on the chemical nature of the pigment: melanin-containing vesicles (melanosomes) are black or brown (dermal melanophores and epidermal and subdermal melanocytes), and pteridine- and carotenoid-containing vesicles (pterinosomes and carotenoid vesicles) appear red (erythrophores) or yellow (xanthophores) (Bagnara and Hadley 1973). Leucophores, or

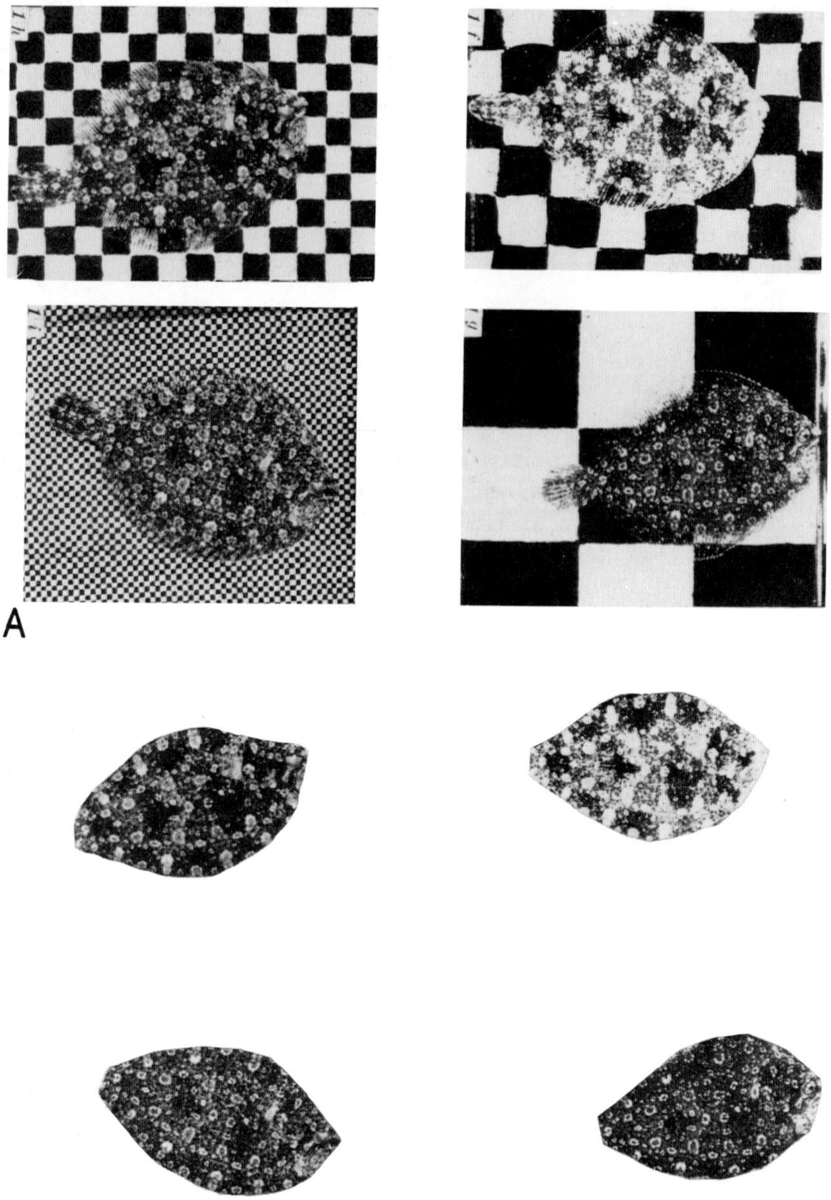

Figures 20.4 and 20.5. Adaptations to checkerboard backgrounds of high local contrast. Figure 20.5, but neither 20.4(A) nor (C) were performed under controlled conditions (see description in text). Figure 20.4(A) is reprinted from Sumner 1911 (with permission); Figure 20.4(B) is reprinted from Mast 1914. The background and fins of 20.4(A) and (B), and 20.5(A) were masked to make 20.4(C) and (D), and 20.5(B), respectively.

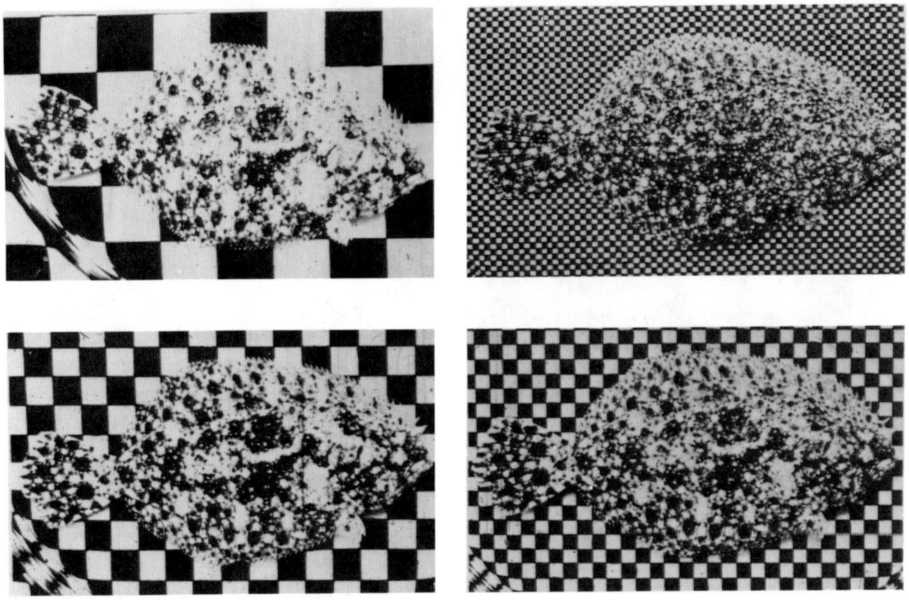

B

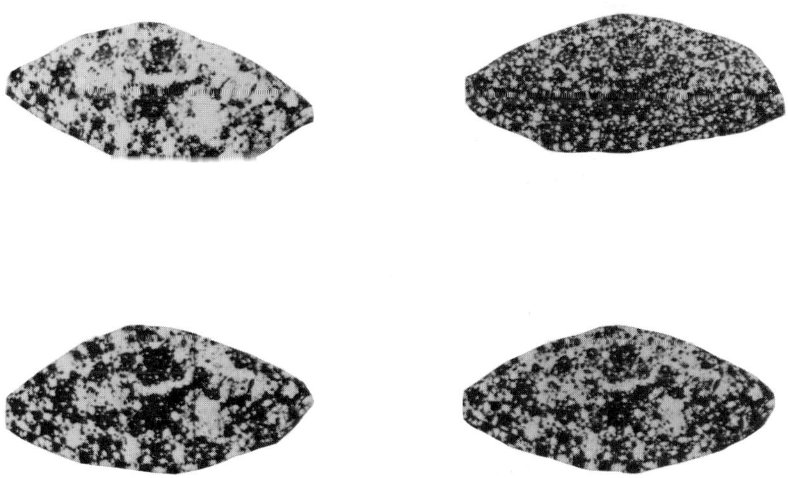

D

Under controlled conditions (Fig. 20.5[B]), adaptation to these backgrounds were minimal, presumably because the fish adapted to average reflectance of the background, and the relevant area (as seen by the fish) included nearly equal areas of white and black. (See pp. 493–495 for a discussion of ''apparent'' checkerboard adaptations of Figure 20.4.)

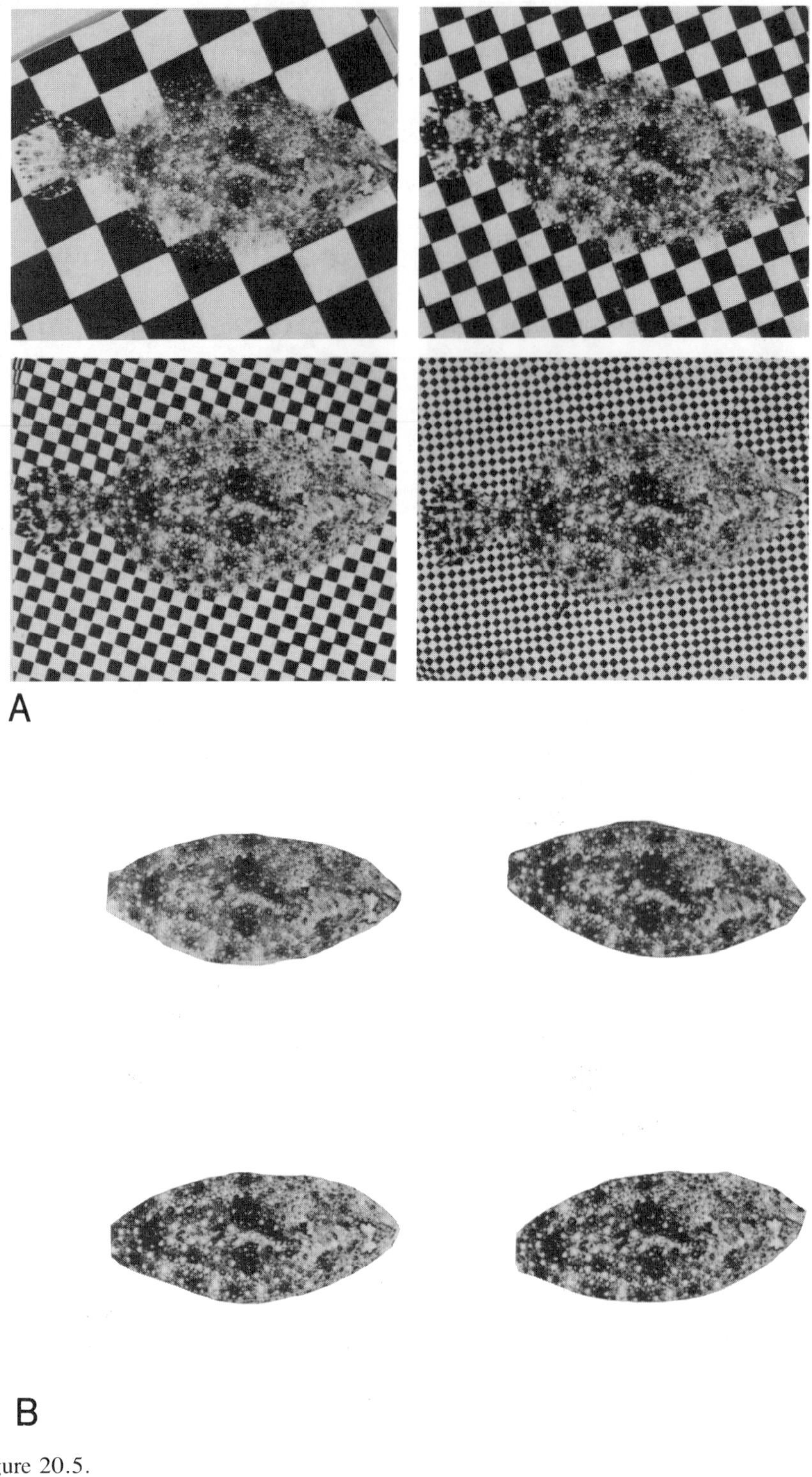

Figure 20.5.

white-reflecting cells, contain an electron-dense material within randomly oriented, reflecting vesicles (Menter et al. 1979).

Iridophores, the ovoidal cells, contain groups of oriented platelets (guanophores or refractosomes). Iridophores and leucophores reflect or scatter light by reflection or interference (Land 1972), whereas chromatophores with colored pigment produce colors by selective filtration of the incident and reflected illumination (Bagnara and Hadley 1973; Lythgoe 1979).

The color of an area of skin seen by a viewer depends on addition and subtraction of the colors of overlapping chromatophores (or dermal chromatophore units) (Bagnara, Taylor, and Hadley 1968; Hawkes 1974). For example, a cell body of a dermal melanophore may be located beneath iridophores, with its processes extending above and over them. When the melanophore pigment is aggregated, a melanophore provides a backing to the overlying iridophores and absorbs light not scattered from them. The bluish appearance of the iridophores will not be desaturated by the addition of white light reflected from the underlying tissues. When melanophore pigment is dispersed, melanosomes absorb the light reflected or scattered from iridophores, which leads to a darker appearance. Similar chromatophore associations may occur between leucophores and melanophores or xanthophores and iridophores.

Epidermal melanocytes play a different role in the formation of visible patterns. When their pigment is dispersed, they act like an opaque filter. Thus, skin areas containing relatively high concentrations of these cells are the darkest areas. In fact, the lowest reflectances measured in Figures 20.1B. 20.2B, and 20.3B are of areas with the highest concentrations of these cells.

## Morphological Markers

Differing concentrations of chromatophores define morphological markers (distinct patches of skin). In *Pseudopleuronectes,* there are five such markers: (1) a black central spot; (2) a surround of this spot; (3) a white marginal spot; (4) a black marginal spot; and (5) an interspot, or background area (Fig. 20.6A). This characterization of morphological markers is found in all pleuronectid flatfish species, with variations primarily in their spatial distribution (Hewer 1931).

Each of these markers has its own characteristic chromatophore densities (cell type/unit area) (Saidel 1977a; Burton 1980). Markers 1 and 4 contain the highest density of epidermal melanophores (Fig. 20.6B). Epidermal melanophores are absent from the white marginal areas (marker 3), whereas the reflecting cells (iridophores or leucophores or both) have the highest density in this area. The spot surround (marker 2) and background (marker 5) areas are intermediate in density for each of these cell types. The distribution of dermal reflecting cells (iridophores and leucophores) is inverse to that of the epidermal melanocytes (Fig. 20.6B), while xanthophore and epidermal melanocyte densities are similar (Burton 1980). In contrast, dermal melanophores are nearly uniformly distributed throughout the skin (Fig. 20.6B).

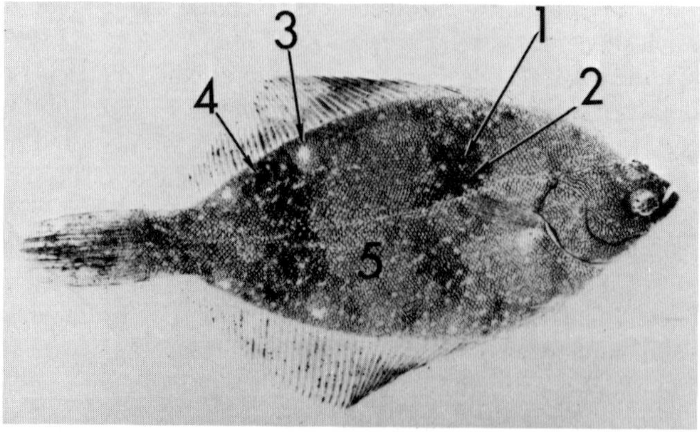


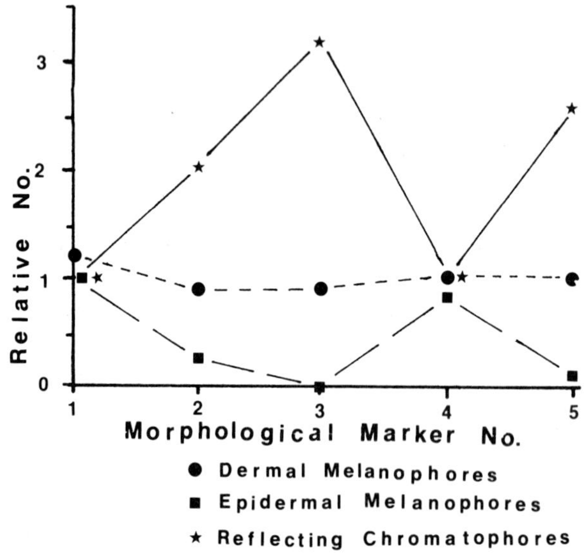


Figure 20.6. *Above,* morphological skin markers of *Pseudopleuronectes americanus.* The number of each area corresponds to morphological marker number in *lower graph. Below,* relative densities of three chromatophores plotted in this graph to illustrate the *pattern* of distribution. Note that (1) dermal melanophores are nearly uniformly distributed and (2) epidermal melanophores and dermal reflecting cells are inversely distributed.

Dermal melanophores were normalized to the density of those found in *marker 5;* and other two chromatophores were normalized to respective densities in *marker 1.*

## Control Processes

The tonal adaptations described on pp. 491–495 result primarily from pigment granule redistribution in melanophores. This section emphasizes the relationship between a fish's vision, its nervous system, and control of melanosome position within a melanophore. (See Byers and Porter 1977; Porter and McNiven 1982;

Schliwa and Euteneuer 1983; and Stearns 1984 for cell biological aspects of pigment granule migrations.)

Vision is a necessary stimulus for an adaptive texture change. Blinded fish darken and do not lighten when placed on light, dark, or contrasting backgrounds. Keeble and Gamble (1904) proposed that control of skin tone depended on the *albedo* (a measure of the ratio of the light reflected from the substrate compared to the incident, or overhead, illumination) (Sumner 1911; Sumner and Keys 1929; though see Mast 1914 for contrary results). Several others argued that stimuli from separate retinal areas controlled darkening or lightening of the skin (de Groot, Norde, and Verheijen 1969). In the sole *Solea solea,* de Groot, Norde, and Verheijen (1969) argued that local stimulation of the dorsal part of the retina (whose visual field included the ground) led to changes in skin patchiness, not just tone; but this result might also follow from a uniform change in the adapted state of the nonuniformly distributed chromatophores.

Over time, a blinded fish may slowly change its overall tone as visual functions recover during optic nerve regeneration, as nonretinal photoreception converges on the visual pathway (Ekstrom et al. 1983; Ekstrom 1984), or as incident light directly affects melanosome position. Melanophores of several fish and larval anurans respond to direct illumination (see Weber 1983). For example, melanophores of the adult cyprinid *Zacco* disperse in a graded manner to direct photostimulation and aggregate to a decrease in light flux, to nerve stimulation and to catecholamine stimulation (Iga and Takabatake 1983). The direct photoeffect appears to be independent of the nervous system, since the effect is not blocked by α-receptor antagonists, whereas such agents block nerve and catecholamine control of aggregation (see the next section).

Coloration may also be partly subject to direct control by incident illumination. The wavelength of light scattered from iridophores is directly modulated by the incident illumination, because photoreception by an opsin-based photopigment located within or adjacent to the iridocyte leads to a change in the spacing of adjacent "guanosomes" (Lythgoe, Shand, and Foster 1984). Given the common embryonic origin of chromatophores (Bagnara 1983), a direct response to illumination may play a more important role in setting the position of pigment within chromatophores in accordance with the ambient illumination than has been realized.

## Nervous Pathway for Melanophore Control: Peripheral Aspects

Melanophore aggregation is graded (Fujii and Novales 1969). A single peripheral nerve fiber innervates several melanophores, and each melanophore is innervated by several nerves (see Parker 1948, 133; Bagnara and Hadley 1973). The evidence is strong that the peripheral nerve fibers innervating individual melanophores and other chromatophores are part of the sympathetic nervous system (Pouchet 1876) and are only catecholaminergic (with only the silurid catfish as exceptions! [Fujii and Miyashita 1976; Fujii, Miyashita, and Fujii 1982]). The evidence is based on microscopic observations of catecholaminergic fluorescence in the nerves around melanophores (Jacobowitz and Laties 1968; Falk, Muntzing, and Rosengren 1969); the absence of visually induced skin adaptations following injections of 6-hydroxydopamine (6-OHDA) (Follenius 1971); the complete denervation of dermal melanophores by 6-OHDA (Saidel 1977a; Junqueira and Salles 1978); and many physiological and pharmacological experiments in which responses to catecholamine agonist and antagonist agents were assessed by changes in melanosome position (e.g., Fujii 1961, 1969; Scott 1965). In all physiological and pharmacological experiments (except with the silurids), catecholamine stimulation of α-adrenergic receptors resulted in skin lightening "in vivo" and melanophore aggregation "in vitro" (Ruffin, Reed, and Finnin 1969; Reed and Finnin 1972; Fernando and Grove 1974a, 1974b; Fujii and Miyashita 1975; Burton 1985).

Only two exceptions to the general rule of catecholaminergic innervation of dermal melanophores have been found, and both belong to the catfish family Siluridae, *Parasilurus* and *Kryptopterus*. In the silurid but not ictalurid catfish, the melanophore-innervating fibers appear to be cholinergic (Fujii and Miyashita 1976; Fujii, Miyashita, and Fujii 1982).

## Nervous Pathway for Melanophore Control: Central Aspects

A locus in the medulla contains cells whose descending axons exit the spinal cord (at a species-specific vertebra) (see Healey 1951, 1954; Scott 1965; Burton 1969; Wilhelm 1969; Iwata and Fukuda 1973), enter the autonomic chain ganglia, and exit via peripheral nerves to innervate melanophores (von Frisch 1911). (This entire pathway has not been confirmed. It is not known whether the axons of these medullary cells directly innervate melanophores, as von Frisch formulated it, which would be unique for a vertebrate sympathetic pathway, or if these cells synapse within the sympathetic chain ganglia.) Direct stimulation of these cells results in skin lightening (von Frisch 1911; Parker 1948; Scott 1965; Fujii 1969; Fujii and Novales 1969; Bauer and Demski 1980); the absence of stimulation to this medullary "nucleus" is followed by skin darkening. A second neuronal pathway for a melanophore-dispersing response was explored by von Gelei (1942), but many have since cast serious doubt upon its existence (see Gray 1956; Iwata and Fukuda 1973; and the next section).

For vision-mediated skin lightening, the optic tectum is necessary (Dijkgraaf 1949; Gentle 1971a, 1971b). Iwata and Fukuda (1973) took advantage of the topography between the tectum and the visual field to examine the albedo hypothesis. Electrical stimulation of the tectal area receiving a representation of the dorsal visual field led to skin darkening. Stimulation of tectal areas receiving the representation of the ventral part of the visual field elicited lightening of the body. These experiments assumed that the unstimulated tectal region still signalled an unchanged measure of "albedo." Lesions of either area led to darkening, suggesting that the medullary "lightening" cells required competitive inputs from the entire visual field that are mediated by competing excitation and inhibition from the optic tectum.

Most convincingly, Iwata and Fukuda (1973) demonstrated that the degree of lightening of a fish was directly correlated with the rate of action potentials in the sympathetic chain and that this rate could be modulated by stimulation of the optic tectum. The more action potentials recorded in the sympathetic chain, the lighter the fish; the fewer potentials, the darker. Iwata and Fukuda concluded (as others before) that a separate neural pathway for darkening was unlikely to exist.

## Melanophore Dispersion: How Does It Happen?

Since the spike rate in the sympathetic chain correlates positively with the degree of body lightening (Iwata and Fukuda 1973), activity in the melanophore-innervating fibers is consistent with an innervation that controls the degree to which the melanosomes aggregate, and it does so via α-adrenergic receptors. Melanosomes disperse primarily because they are not being aggregated (Gray 1956; Saidel 1977a, 1977b). Inhibition of ATP production also aggregates melanosomes (Junqueira, Raker, and Porter 1974; Saidel 1977a, 1977b), suggesting that nerve activity inhibits an intracellular process that continually promotes melanosome disperson. Thus, any experiment to which a direct exogenous stimulation of pigment dispersion is attributed must contend with the automatic nature of dispersion. On the other hand, β-adrenergic receptors have been suggested as cellular mediators for dispersion (Reed and Finnin 1972; Miyashita and Fujii 1975). Melanophores of *Pseudopleuronectes* do not appear to possess β-adrenergic receptors, however (Burton 1985).

Experiments with a variety of fish species have shown the influences of pituitary-derived hormones on the position of chromatophores (Parker 1948; Bagnara and Hadley 1973). A family of peptidergic melanophore-stimulating hormones (e.g. MSH, α-MSH) has been located in the intermediate lobe of the pituitary. One MSH-producing cell type in the pars intermedia exhibits changes in several morphological characters when a fish is adapted to a white or black background (van Eys 1980). Thus, ambient exogenous MSH may also be influenced by environmental factors.

This influence may be mediated by a fiber tract that connects the nucleus lateralis tuberis pars lateralis (NLT-pl) with the pars intermedia. Cells in the NLT-pl exhibited a high level of immunocytochemical staining to 27-S-methylglucagon in dark-adapted fish and negligible staining in light-adapted ones (Terlou, Gielen, and Notenboom 1985). Heavy staining indicated storage of a melanophore-concentrating hormone (MCH), whereas light staining suggested a release of MCH. MCH affects melanophores "in vitro" (Kawauchi et al. 1983). Thus, hormonal influences on chromatophores may depend on a competitive relationship between MCH and MSH.

### Peripheral Control of Other Chromatophores

Neural control of pigment granule position in subdermal and epidermal melanocytes is less well understood. These cells are probably not innervated (Brown and Wellings 1970; Roberts, Young, and Milne 1971; Imaki and Chavin 1975; Lamer and Chavin 1975; Miyashita and Fujii 1980). Like dermal melanophores, they aggregate to catecholamine stimulation, so pigment adaptation in these cells may be controlled only by nonneural processes.

Innervation of dermal erythrophores, leucophores (Fujii and Miyashita 1979), and xanthophores has been established, but has been poorly characterized compared to that of dermal melanophores. However, xanthophores and erythrophores aggregate their pigment granules to "in vitro" catecholamine stimulation, though, in *Pseudopleuronectes,* xanthophores are less sensitive than dermal melanophores (Burton 1985). Leucophores (alternately called iridocytes) respond to catecholamines (Iga, Yamada, and Iwakiri 1977), direct nerve stimulation (Fujii and Miyashita 1979; Oshima and Fujii 1984), and ionically manipulated salines (Miyoshi 1952) by dispersing their pigment granules, which is opposite to the response of melanophores.

## Camouflage as Visual Processing of Pattern and Texture

In analyzing the perceived image of a camouflaged animal, one is inexorably led to the general problem of distinguishing a figure from its background. At the heart of the problem are questions relating to visual identification: What are the steps in texture discrimination, and what relationships are involved in seeing one textured object embedded within a differently textured field? Let us ask, for the example of flatfish, how this problem is exacerbated by basically simple changes of a fixed skin pattern. The following ideas, from a variety of sources, present a glimmer of a solution.

Detection and identification of one textured field embedded within a differently textured one (analogous to detecting a camouflaged animal in its surround) is a two-step process (Bergen and Julesz 1983; Julesz 1984; Sagi and Julesz 1985). An observer requires up to 100 milliseconds to determine that something is spatially different in one part of the visual field. A few hundred more milliseconds are then required to identify the location of the difference (Bergen and Julesz 1983). These two stages of recognition are called *preattentive texture discrimination,* which occurs instantly across the entire visual field, and *attentive texture discrimination,* or *scrutiny,* which requires local comparisons of texture within a small area of the visual field. Identifying an object that statistically resembles its surround is more difficult and requires more time than identifying a dissimilar object. These observations reveal that limitations of visual processing can be discovered as a function of an image's structure.

Detection of an object may be compromised by the masking effects of the surrounding environment. A masking stimulus is a second stimulus that is given simultaneously or displaced in time at the same spatial location as a test object. The strength of the masking depends upon some relationship between the test texture and the masking texture. This relationship has been described in statistical terms as spatial frequency components or as geometrical elements (see Julesz 1971, 1975, 1984), and it may be possible to relate texture to the fractal dimensionality of the object and its surround (Mandelbrot 1983). For individual instances of animal camouflage, the most suitable analysis remains to be clarified.

The results described in the previous section on the structure of adaptation show that the flatfish skin markers change their reflectances to different backgrounds. These changes are seen as adaptations in the average reflectance of the skin as a whole and the contrast across borders of adjacent morphological markers. The invisibility of a flatfish seen monocularly or binocularly and the time required for preattentive texture discrimination suggest that the physiological process relating to this masking is retinal.

Thus, the phenomenology of flatfish camouflage suggests a dependancy on preadapting the viewer's visual system for certain textural characteristics, and the particular characteristics are defined by the surrounding local environment. The consequential adaptations in the viewer's visual system decreases sensitivity to the textural characteristics of flatfish skin, so the adaptive skin behavior of flatfish need not be designed to equal the background at all. The adaptations of the skin need only change the relevant characteristics of skin texture to a level below that necessary for detection by the processes of preattentive texture discrimination. Thus, the notion of "adaptive matching" of the environment to produce a cryptic state does not enter into the equation between the camouflager, the environment, and the viewer.

With scrutiny, an observer whose visual field includes a flatfish will eventually see the fish. Consciously or unconsciously, the observer behaves as if the predominant textural aspects of the surround are protracted onto the fish, so that the texture of a flatfish skin is interpreted as part of the surrounding environmental texture (Sumner 1911; Mast 1914). One might argue that the uniqueness of the flatfish's skin pattern is its ability to accept the imposition of the surrounding texture as the dominant aspect of its skin appearance.

As Cott and many others have pointed out (see Breder 1955), boundaries or edges are the features that establish a separation of one form from another or from a background. Boundaries are a textural discontinuity. They contain a disproportionate representation of high spatial frequencies compared with adjacent areas. The structural transparency of flatfish fin skin aids in the elimination of its boundaries. Fin skin is transparent between chromatophore markers. Consequently, one sees through them a combination of the ground and skin textures. This mixture forms a gradient that ameliorates the difference between skin and environment textures and results in the elimination of a distinctive edge to the fish. In the presence of a distinct boundary, the perceptive process that imposes the texture of the environment onto the fish skin could not operate. Behaviorally, flatfish further diminish textural differences by a bedding reaction in which sand and rocks are fluffed onto their bodies when they settle.

By eliminating discrete boundaries between the flatfish and the background, the textural similarities between the fish and its surround are emphasized and the differences diminished. Otherwise, if the object cannot be recognized as different from the background, then the object must be a part of the background and is indiscriminable. One might say that the upper surface of a flatfish looks like an ''exquisite imitation of the surface of the sand bed; or it is marked with variegated pebble-patterns, or with the two kinds in combination, or with seaweed colors and patterns, as the case may be'' (Thayer 1909, 165). Such a relational style of description is exactly that used by previous researchers to describe the adaptive appearance of a flatfish.

Note one futher point. This discussion has been phrased as if the location of a flatfish is known in advance. This is akin to viewing with the most sensitive part of a retina. If one forces oneself to gaze with peripheral vision, identifying the textural differences becomes much more difficult, possibly because the grain of the less sensitive area of the retina (or peripheral retina in mammals) is coarser than the grain necessary for preattentive texture perception. However, this situation is also the more natural one.

The realization that camouflage patterns of flatfish did not evolve from human predation places certain constraints on this hypothesis, but two arguments support its basic elements. Firstly, the scale of measurement inherent in human foveal vision exceeds that of most nonfoveal aquatic vertebrates. This basic relationship has evolved from the biophysical limits of acuity by photoreceptors. In air, diffraction is the limiting factor in acuity, while in water, light scatter degrades resolution (see Chap. 3). It follows that a camouflaged appearance that is difficult for human foveal vision to make out will likely be more difficult for nonfoveal or aquatic vision or both of them.

Secondly, the retinal circuitry of all vertebrates (with few exceptions) is comparable. (Differences may be found in certain aspects of retinal structure (photopigments, convergence ratios of photoreceptors to bipolar cells, local cell densities, to name a few), but such differences may also be found among species of the same order.) Thus, the rules for visual operations found in one species are not likely to be qualitatively different from those of any other. Seeing camouflaged animals is difficult with a human retina. Seeing camouflaged animals with any other retina is apt to be equally or more difficult.

## Afterword

Can behavioral complements of texture adaptation be identified in fish, and can a similar but independently evolved strategy for texture adaptation be found somewhere in the natural world? Both of these questions can be answered in the affirmative. Although anecdotal, the answer to the first question can be illustrated with two accounts of behavior in the wild; the second, by considering the camouflage patterns of cephalopods.

To appreciate the success of flatfish in camouflage, they must be observed in their own environment. (The following observations on *Pseudopleuronectes* were made while SCUBA diving near Cohasset Harbor, Cohasset, Massachusetts.) A flatfish swims in bursts just above the ocean bottom. When it ceases swimming after being chased, it darts to the bottom. Invariably, the dart is angled away from its path of swimming. This behavior strongly hinders tracking because, when settled to the bottom, the fish is no longer the subject of visual attention. All the factors of its cryptic appearance make relocating it difficult.

Secondly, the predatory behavior of the trumpet fish *Aulostomus maculatus* cannot be explained without resorting to arguments of the sort presented above. Its skin has a set of lines the length of its body and a variety of colorational patterns. The trumpet fish preys on smaller species such as *Chromis* from a head-down vertical position amidst staghorn corals while exposing its longitudinal stripes (Randall 1968; Kaufman 1976). Since staghorn corals rise from the ocean floor as a set of vertical lines, in this assumed posture a trumpet fish appears stationary, moving only with the local water flow like a branch of the surrounding coral head. For a trumpet fish to turn its body lengthwise in accordance with the direction of these lines suggests that *Chromis* do not see such a fish, camouflaged by its skin and behavior, as different from the coral. Rather, this behavior, in conjunction with the skin texture, appears to a prey with similar textural characteristics as the surround. (See Cott 1957, 131–139, for other examples of posture as one aspect of cryptic behavior.)

Cephalopods have independently evolved from vertebrates (Packard 1972). When one considers the camouflage of cephalopods, a sense of déjà vu rapidly develops, because the physiologically different camouflage system of cephalopods appears to use principles in the designs of its texture and chromatic adaptations similar to those of flatfish.

The appearance of a cephalopod's skin is due to two skin characteristics: coloration changes from chromatophore organs and active control of skin papillae. Chromatophore pigments may be red, yellow, black, or brown. Skin colors result from the expansion of sacs of pigment by radial muscle fibers. Skin papillae are raised by the activities of other skin muscles. Under nervous control, the radial muscle fibers contract against the elastic membrane of the sac that expands the size of the pigment spot. Individual muscle fibers are under independent nerve control, and each nerve may innervate several muscles (Florey 1969; Cloney and Brocco 1983; Dubas and Boyle 1985). Another cell type, the leucophore, is seen as white, while still another, the iridophore, produces structural coloration by reflection or scattering. The latter two types are not unlike their analogues in fish skin.

Chromatophore organs are spatially associated into morphological and physiological units (Maynard 1967; Packard and Hochberg 1977). When a motor neuron soma in the brain or its axon is stimulated, a collection (a "physiological" unit) of associated chromatophores exhibits change, and this collection is immutable (Maynard 1967; Packard and Hochberg 1977; Hanlon and Messenger in preparation). Nonuniform distributions of chromatophores allow for different-sized and different-appearing physiological units that are used in a remarkable display of behavioral and textural adaptations (octopods: Packard and Hochberg 1977; *Loligo:* Hanlon 1982; *Sepia:* Hanlon and Messenger in preparation). In situations requiring camouflage to different substrates, these animals respond by controlling for brightness, which is another way of describing tone, and by expressing disruptive patterns, which is another way of generating contrasting spatial relationships. The responses, however, are only within the limitations of the appearance of physiological units, which is not unlike the variations in appearance of the morphological markers of flatfish skin. Again similar to flatfish, the neural pathway involved in adaptive changes of a cephalopod's skin pattern involves no visual feedback to determine the appropriateness of the pattern (Boycott 1953, 1961; Young 1971).

When taken together, the camouflage patterning of flatfish and those aspects of the chromatophore system of cephalopods relating to camouflage suggest that these two groups exhibit a convergence of strategy for textural adaptation. This convergence is likely due to the similarity in visual texture discrimination capabilities of their respective vertebrate predators (see Packard 1972, 287–297, for a discussion on the coevolution of cephalopods and fish). The patterns we now see in individual flatfish and octopuses or squids are sufficient because predation would have eliminated the patterns more distinguishable to their predators. This story thus concludes by ultimately revealing in two distinct groups of animals a convergence in the strategies for active camouflage that have been shaped from a common source, the limitations of vertebrate texture discrimination.

## Summary

Macroscopically, the fundamental adaptive changes seen in flatfish texture adaptation can be explained by adaptations in (1) skin reflectance and (2) contrast between discrete skin areas. These changes occur on the mural provided by the distribution of different chromatophores—the uniformly distributed dermal melanophores and the spatially distinct distributions of melanocytes and dermal reflecting cells. Detailed modulation is not required for changes in appearance because skin structure restricts the available textural adaptations. Therefore, visual analysis of environmental texture by a fish may be surprisingly minimal. Psychophysically, the skin patterns on which texture adaptations occur conform to limitations in the perceptual processes used to discriminate between textures, and it is this relationship that has led to the extant patterns within the skin of flatfishes.

Microscopically, increased activity in the fibers innervating melanophores leads to a graded inhibition of the intracellular process promoting dispersion. Rapid changes of pigment position are controlled by the nervous system, but this neuronal control may be superimposed on a level of pigment dispersion relating to competing hormonal effects.

*Acknowledgments.* Some of the original work in this chapter were performed in fulfillment of Ph.D. requirements at the Massachusetts Institute of Technology and presented at the annual meeting of the American Society of Zoologists (1978; see Saidel, W. M., Am. Zool. 18: abst. 36). I wish to acknowledge an everlasting debt to Dr.

J. Y. Lettvin of M.I.T. and especially thank Joan C. Feldman-Saidel for her continuous and unfailing support and Drs. R. Fernald, E. Gruberg, and R. Hanlon for reviewing a previous version of this chapter.

## References

Agassiz, A. (1879) On the young stages of bony fishes, Proc. Am. Acad. Arts Sci., 14:1–25.

Baerends, G.P. and Baerends-van Roon, J.M. (1950) An introduction to the study of the ethology of cichlid fishes, Behavior (supp), 1:1–242.

Bagnara, J.T. (1983) Developmental aspects of vertebrate chromatophores, Am. Zool., 23:465–478.

Bagnara, J.T., and Hadley, M.E. (1973) Chromatophores and Color Change, Prentice-Hall, Englewood Cliffs, N.J.

Bagnara, J.T., Taylor, J.D., and Hadley, M.E. (1968) The dermal chromatophore unit, J. Cell Biol., 38:67–79.

Bauer, D.H. and Demski, L.S. (1980) Vertical banding evoked by electrical stimulation of the brain in anaesthetized green sunfish, *Lepomis cyanellus,* and bluegills, *Lepomis macrochirus,* J. Exp. Biol., 84:149–160.

Bergen, J.R. and Julesz, B. (1983) Parallel versus serial processing in rapid pattern discrimination, Nature, 303:696–698.

Boycott, B.B. (1953) The chromatophore system of cephalopods, Proc. Linn. Soc. Lond., 164:235–240.

Boycott, B.B. (1961) The functional organization of the brain of the cuttlefish *Sepia officinalis,* Proc. R. Soc. Lond. B Biol. Sci., 153:503–534.

Breder, C.M., Jr. (1955) Special features of visibility reduction in flatfishes, Zoologica, 40:91–98.

Brown, G.A. and Wellings, S.R. (1970) Electron microscopy of the skin of the teleost, *Hippoglossoides elassodon,* Z. Zellforsch. Mikrosk. Anat., 103:149–169.

Burton, D. (1969) The colour change of the minnow, *Phoxinus phoxinus,* with particular reference to the effect of spinal lesions, J. Zool. Lond., 157:169–185.

Burton, D. (1978) Melanophore distribution within the integumentary tissues of two teleost species, *Pseudopleuronectes americanus* and *Gasterosteus aculeatus* form *leiurus,* Can. J. Zool., 56:526–535.

Burton, D. (1980) A cellular analysis of chromatophore patterning in winter flounder (*Pseudopleuronectes americanus* Walbaum), Comp. Biochem. Physiol., 67A:453–457.

Burton, D. (1985) Differential *in vivo* sensitivity of melanophores and xanthophores to catecholamines in winter flounder (*Pseudopleuronectes americanus* Walbaum) integumentary patterns, J. Exp. Biol., 114:649–659.

Byers, H.R. and Porter, K.R. (1977) Transformations in the structure of the cytoplasmic ground substance in erythrophores during pigment aggregation and dispersion. I. A study using whole-cell preparations in stereo high voltage electron microscopy, J. Cell Biol., 75:541–558.

Chinarina, A.D. (1960) Changes in the coloration of marine flounders (*Platessa platessa* L.) and capelin (*Mallotus villosus villosus* Muller) under experimental conditions, Dokl. Akad. Nauk SSSR, 135:1543–1545.

Cloney, R.A. and Brocco, S.L. (1983) Chromatophore organs, reflector cells, iridocytes and leucophores in cephalopods, Am. Zool., 23:581–592.

Cott, H.B. (1957) Adaptive Coloration in Animals, 2d ed., Methuen, London.
Cott, H.B. (1971) Animal form in relation to appearance, in Aspects of Form, Whyte, L.L. (ed.), Indiana University Press, Bloomington, pp. 121–156.
de Groot, S.J., Norde, R., and Verheijen, F.J. (1969) Retinal stimulation and pattern formation in the common sole *Solea solea* (L.) (Pisces: Soleidae), Neth. J. Sea Res., 4:339–349.
Dijkgraaf, S. (1949) Lokalizationversuche am Fischgehirn, Experientia, 5:44–48.
Dubas, F. and Boyle, P.R. (1985) Chromatophore motor units in *Eledone cirrhosa* (Cephalopoda: Octopoda), J. Exp. Biol., 117:415–431.
Du Shane, G.P. (1935) An experimental study of the origin of pigment cells in amphibia, J. Exp. Zool., 72:1–31
Edmunds, M. (1974) Defense in animals: a survey of anti-predator defences, Longman Group, Essex, U.K.
Ekstrom. P. (1984) Central neural connections of the pineal organ and retina in the teleost *Gasterosteus aculeatus* L., J. Comp. Neurol., 226:321–335.
Ekstrom, P., Borg, B., Paulson, G., Reschke, M., and van Veen, Th. (1983) Central nervous connections of the pineal organ in the teleost, *Carassius carassius* L., Neurosci. Lett. (supp), 14:S104.
Falk, B., Muntzing, J., and Rosengren, A.M. (1969) Adrenergic nerves to the dermal melanophores of the rainbow trout, *Salmo gairdneri, Z. Zellforsch. Mikrosk. Anat.*, 99:430–434.
Fernando, M.M. and Grove, D.J. (1974a) Melanophore aggregation in the plaice (*Pleuronectes platessa* L.). I. Changes in *in vivo* sensitivity to sympathomimetic amines, Comp. Biochem. Physiol., 48A:711–721.
Fernando, M.M. and Grove, D.J. (1974b) Melanophore aggregation in the plaice (*Pleuronectes platessa* L.), II. *In vitro* effects of adrenergic drugs, Comp. Biochem. Physiol., 48A:723–732.
Florey, E. (1969) Ultrastructure and function of cephalopod chromatophores, Am. Zool., 9: 429–442.
Follenius, M.E. (1971) Action de la 6 OH dopamine sur les mélanophores et sur l'adaptation chromatique chez le poisson téléostéen *Gasterosteus aculeatus,* C.R. Acad. Sci., 272D:733–736.
Fujii, R. (1961) Demonstration of the adrenergic nature of transmission at the junction between melanophore-concentrating nerve and melanophore in bony fish, J. Fac. Sci. Univ. Tokyo, Section IV, 9:171–196.
Fujii, R. (1969) Chromatophores and pigments, in Fish Physiology, vol. 3, Hoar, W.S. and Randall, D.J. (eds.), Academic Press, New York, pp. 307–353.
Fujii, R. and Miyashita, Y. (1975) Receptor mechanisms in fish chromatophores. I. Alpha nature of adrenoceptors mediating melanosome aggregation in guppy melanophores, Comp. Biochem. Physiol., 51C:171–178.
Fujii, R. and Miyashita, Y. (1976) Receptor mechanisms in fish chromatophores. III. Neurally controlled melanosome aggregation in a siluroid (*Parasilurus asotus*) is strangely mediated by cholinoceptors, Comp. Biochem. Physiol., 55C:43–49.
Fujii, R. and Miyashita, Y. (1979) Photoelectric recording of motile responses of fish leucophores, Annot. Zool. Jpn., 52:87–94.
Fujii, R., Miyashita, Y., and Fujii, Y. (1982) Muscarinic cholinoceptors mediate neurally evoked pigment aggregation in glass catfish melanophores, J. Neurol. Trans., 54:29–39.
Fujii, R. and Novales, R.R. (1969) The nervous mechanism controlling pigment aggregation in *Fundulus* melanophores, Comp. Biochem. Physiol., 29:109–124.

Gentle, M.J. (1971a) The central nervous control of colour change in the minnow (*Phoxinus phoxinus* L.). I. Blinding and the effects of tectal removal on normal and blind fish, J. Exp. Biol., 54:83–91.

Gentle, M.J. (1971b) The central nervous control of colour change in the minnow (*Phoxinus phoxinus* L.). II. Tectal ablations in normal blind fish, J. Exp. Biol., 54:93–102.

Gray, E.G. (1956) Control of the melanophores of the minnow (*Phoxinus phoxinus* [L.]), J. Exp. Biol., 33:448–459.

Green, L. (1968) Mechanisms of movements of granules in melanocytes of *Fundulus heteroclitus,* Proc. Nat. Acad. Sci. USA, 59:1179–1186.

Gulley, R. L., Cochran, M., and Ebbesson, S.O.E. (1975) The visual connections of the adult flatfish, *Achirus lineatus,* J. Comp. Neurol., 162:309–320.

Hanlon, R.T. (1982) The functional organization of chromatophores and iridescent cells in the body patterning of *Loligo plei* (Cephalopoda: Myopsida), Malacologia, 23:89–119.

Hawkes, J.W. (1974) The structure of fish skin. II. The chromatophore unit, Cell Tissue Res., 149:159–172.

Healey, E.G. (1951) The colour change of the minnow (*Phoxinus laevis* AG.). I. Effects of spinal section between vertebrae 5 and 12 on the responses of the melanophores, J. Exp. Biol., 28:298–319.

Healey, E.G. (1954) The colour change of the minnow (*Phoxinus laevis* AG.). II. Effects of spinal section between vertebrae 1 and 15 and of anterior autonomic section on the responses of the melanophores, J. Exp. Biol., 31:473–490.

Hewer, H.R. (1931) Studies in colour-changes in fish. V. The colour-patterns in certain flat-fish and their relation to the environment, J. Linn. Soc. Lond. (Zool.), 37:493–513.

Iga, T. and Takabatake, I. (1983) Melanophores of *Zacco temmincki* (Teleostei) are light sensitive, J. Exp. Zool., 227:9–14.

Iga, T., Yamada, K., and Iwakiri, M. (1977) Adrenergic receptors mediating pigment dispersion in leucophores of a teleost, *Oryzias latipes,* Mem. Fac. Lit. and Sci. Shimane Univ. (Nat. Sci.), 11:63–72.

Imaki, H. and Chavin, W. (1975) Ultrastructure of the integumental melanophores of the South American lungfish (*Lepidosiren paradoxa*) and the African lungfish (*Protopterus* sp.), Cell Tissue Res., 158:375–389.

Iwata, K.S. and Fukuda, H. (1973) Central control of color changes in fish, in Responses of Fish to Environmental Changes, Chavin, W. (ed.), Charles C Thomas, Springfield, Illinois, pp. 316–341.

Jacobowitz, D.M. and Laties, A.M. (1968) Direct adrenergic innervation of a teleost melanophore, Anat. Rec. 162:501–504.

Julesz, B. (1971) Foundations of Cyclopean Perception, University of Chicago Press, Chicago.

Julesz, B. (1975) Experiments in the visual perception of texture, Sci. Am., 232:34–44.

Julesz, B. (1984) A brief outline of the texton theory of human vision, Trends Neurosci., 1984:41–45.

Junqueira, L.C.U., Raker, E., and Porter, K.R. (1974) Studies on pigment migration in the melanophores of the teleost *Fundulus heteroclitus* (L), Arch. Histol. Jpn., 36:339–366.

Junqueira, L.C.U. and Salles, L.M.M. (1978) Effects of 6-hydroxydopamine and 5-hydroxydopamine on the teleost melanophores' innervation, J. Fish Biol., 13:415–419.

Junqueira, L.C., Toledo, A.M.S., and Porter, K.R. (1970) Observations on the structure of the skin of the teleost *Fundulus heteroclitus* (L), Arch. Histol. Jpn., 32:1–15.

Kaufman, L. (1976) Feeding behavior and functional coloration of the Atlantic trumpetfish, *Aulostomus maculatus,* Copeia, 377–378.

Kawauchi, H., Kawazoe, I., Tsubokawa, M., Kishida, M., and Baker, B.I. (1983) Characterization of melanin-concentrating hormone in chum salmon pituitaries, Nature, 305:321–323.

Keeble, F. and Gamble, F.W. (1904) The colour-physiology of higher Crustacea, Philos. Trans. R. Soc. Lond. B Biol. Sci., 196:295–388.

Lamer, H.I. and Chavin, W. (1975) Ultrastructure of the integumental melanophores of the coelacanth, *Latimeria chalumnae,* Cell Tissue Res., 163:383–394.

Land, M.F. (1972) The physics and biology of animal reflectors, Prog. Biophys. Mol. Biol., 24:75–106.

Lanzing, W.J.R. (1977) Reassessment of chromatophore pattern regulation in two species of flatfish (*Scophthalmus maximus; Pleuronectus platessa*), Neth. J. Sea Res. 11:213–222.

Luckenbill-Edds, L. and Sharma, S.C. (1977) Retinotectal projection of the adult winter flounder (*Pseudopleuronectes americanus*), J. Comp. Neurol., 173:307–318.

Lythgoe, J.N. (1979) The Ecology of Vision, Clarendon Press, Oxford.

Lythgoe, J.N., Shand, J., and Foster, R.G. (1984) Visual pigment in fish iridocytes, Nature, 308:83–84.

Mandelbrot, B. (1983) The Fractal Geometry of Nature, 2d Ed., W.H. Freeman, San Francisco.

Mast, S.O. (1914) Change in shade, color, and pattern in fishes, and their bearing on the problems of adaptation and behavior, with especial reference to the flounders *Paralichthys* and *Ancylopsetta,* Bull. U.S. Bur. Fish., 34:173–232.

Maynard, D.M. (1967) Organization of central ganglia, In Invertebrate Nervous Systems, Wiersma, C.G.A. (ed.), University of Chicago Press, Chicago, pp. 231–255.

Menter, D.G., Obika, M., Tchen, T.T., and Taylor, J.D. (1979) Leucophores and iridophores of *Fundulus heteroclitus:* biophysical and ultrastructural properties, J. Morphol., 160:103–120.

Miyashita, Y. and Fujii, R. (1975) Receptor mechanisms in fish chromatophores. II. Evidence for beta adrenoceptors mediating melanosome dispersion in guppy melanophores, Comp. Biochem. Physiol., 51C:179–187.

Miyashita, Y. and Fujii, R. (1980) Observations of the fine structure of catfish epidermal melanophores with due consideration to their functions, Annot. Zool. Jpn., 53:174–181.

Miyoshi, S. (1952) Response of iridocytes in isolated scales of the medaka (*Oryzias latipes*) to chlorides, Annot. Zool. Jpn., 25:21–29.

Osborn, C.M. (1939) The physiology of color change in flatfishes, J. Exp. Zool., 81:479–514.

Oshima, N. and Fujii, R. (1984) A precision photoelectric method for recording chromatophore responses in vitro, Zool., Sci., 1:545–552.

Packard, A. (1972) Cephalopods and fish: the limits of convergence, Biol. Rev., 47:241–307.

Packard, A. and Hochberg, F.G. (1977) Skin patterning in *Octopus* and other genera, Symp. Zool. Soc. Lond., 38:191–231.

Parker, G.H. (1948) Animal Colour Changes and Their Neurohumours, University Press, Cambridge, U.K.

Policansky, D. (1982) The asymmetry of flounders, Sci. Am., 246:116–122.

Porter, K.R. and McNiven, M.A. (1982) The cytoplast: a unit structure in chromatophores, Cell, 29:23–32.

Pouchet, G. (1876) Des changements de coloration sous l'influence des nerfs, J. Anat. Physiol., 12:1–90, 113–165.

Randall, J.E. (1968) Caribbean Reef Fishes, TFH Publications, Jersey City, N.J.

Reed, B.L. and Finnin, B.C. (1972) Adrenergic innervation of melanophores in a teleost fish, in Pigmentation: its genesis and biologic control, Riley, V. (ed.), Appleton-Century-Crofts, New York, pp. 285–294.

Roberts, R.J., Young, H., and Milne, J.A. (1971) Studies on the skin of plaice (*Pleuronectes platessa* L.). I. The structure and ultrastructure of normal plaice skin, J. Fish Biol., 4:87–98.

Ruffin, N.E., Reed, B.L., and Finnin, B.C. (1969) Pharmacological studies on teleost melanophores, Eur. J. Pharmacol., 8:114–118.

Sagi, D. and Julesz, B. (1985) "Where" and "what" in vision, Science, 228:1217–1219.

Saidel, W.M. (1977a) Physiological and psychological aspects of flatfish camouflage, Ph.D. thesis, Massachusetts Institute of Technology, Cambridge, Mass.

Saidel, W.M. (1977b) Metabolic energy requirements during teleost melanophore adaptations, Experientia, 33:1573–1574.

Schliwa, M. and Euteneuer, U. (1983) Comparative ultrastructure and physiology of chromatophores, with emphasis on changes associated with intracellular transport, Am. Zool., 23:479–494.

Scott, G.T. (1965) Physiology and pharmacology of color change in the sand flounder *Scopthalamus aquosus,* Limnol. Oceanogr. 10:230–246.

Stearns, M.E. (1984) Cytomatrix in chromatophores, J. Cell Biol., 99:144s–151s.

Sumner, F.B. (1911) The adjustment of flatfishes to various backgrounds. A study of adaptive color change, J. Exp. Zool., 10:409–505.

Sumner, F.B. and Keys, A.B. (1929) The effects of differences in the apparent source of illumination upon the shade assumed by a flatfish on a given background, Physiol. Zool., 2:495–504.

Terlou, M., Gielen, J.T., and Notenboom, C.D. (1985) Nucleus lateralis tuberis-pars lateralis and background adaptation in the rainbow trout, *Salmo gairdneri,* Cell Tissue Res., 240:185–190.

Thayer, G.H. (1909) Concealing Coloration in the Animal Kingdom. An exposition of the laws of disguise through color and pattern, Macmillan Co., New York, pp. xix and 160.

van Eys, G.J.J.M. (1980) Structural changes in the pars intermedia of the cichlid teleost *Sarotherodon mossambicus* as a result of background adaptation and illumination. I. The MSH-producing cells, Cell Tissue Res., 208:99–110.

von Frisch, K. (1911) Beiträge zur Physiologie der Pigmentzellen in der Fischaut, Pflüegers, Archiv. Gesamte. Physiol. Menschen. Tiere., 138:319–387.

von Gelei, G. (1942) Zur Frage der Doppelinnervation der Chromatophoren, Z. Vgl. Physiol., 29:532–540.

Weber, W. (1983) Photosensitivity of chromatophores, Am. Zool., 23:495–506.

Wilhelm, R. (1969) Vergleichende Untersuchungen über den nervös gesteuerten Farbwechsel der Fische, Z. Vgl. Physiol., 65:153–190.

Young, J.Z. (1971) The Anatomy of the Nervous System of *Octopus vulgaris,* Clarendon Press, Oxford.

# Chapter 21

# The Visual System in Teleost Fishes: Morphological Patterns and Trends

R. GLENN NORTHCUTT* and MARIO F. WULLIMANN*

Information on the organization of the nervous system in vertebrates has increased tremendously during the past twenty years, primarily spurred by new experimental techniques for tracing pathways (Nauta and Ebbesson 1970; Cowan and Cuenod 1975; Heimer and RoBards 1981). One focus of this experimental flurry has been the visual system of teleost fishes (Table 21.1). However, most of these studies have focused on cypriniform teleosts, specifically the common goldfish, *Carassius auratus,* whose visual system has been adopted as a model in the analysis of general problems of development and plasticity. Alternatively, one or another species has been the focus of an opportunistic analysis, in which some particularly well-developed visual character facilitates recognition of its biological role. Phyletic analyses, representing a third approach, have been infrequent and usually lack rigor. Such analyses should include the following steps: (1) an examination of the characters in a number of taxa; (2) recognition of the patterns of character distribution; (3) determination of the polarity of character transformations (i.e., recognizing homologous characters and determining which are primitive—plesiomorphic—and which are derived—apomorphic); (4) formulation and testing of hypotheses regarding these transformations and their underlying mechanisms. The theory and practice of phylogenetic systematics (Hennig 1966; Eldredge and Cracraft 1980; Wiley 1981; Northcutt 1984a, 1985) appear to provide the most rigorous criteria for such analyses.

A phyletic analysis not only allows one to recognize character patterns and trends, but by posing hypotheses regarding their history, one can frequently reveal structural and functional relationships not otherwise apparent and reveal gaps in our knowledge of the characters being examined, as well as related characters, that alter the course of our research (Northcutt 1986). For these reasons, we attempt to review the extensive literature on the primary visual projections in teleost fishes within the framework of a phyletic analysis and emphasize what additional studies are needed. This review considers aspects of the optic nerve and retinofugal and retinopetal projections of the retina, as well as higher-order visually related centers and pathways, to the extent that such centers and pathways

*Neurobiology Unit, Scripps Institute of Oceanography, University of California, San Diego, La Jolla, CA 92093, USA

Table 21.1. Experimental Retinofugal Studies of Actinopterygian Fishes

| Reference | Method[a] | Species Described[b] |
|---|---|---|
| Cladistia | | |
| Repérant, Rio, and Amouzou 1979 | A | *Polypterus senegalus* |
| Repérant et al. 1981 | A,D | *Polypterus senegalus* |
| Braford and Northcutt 1983 | A | *Polypterus palmas* |
| Chondrostei | | |
| Repérant et al. 1982 | A,D,H | *Acipenser guldenstadti* |
| Ginglymodi | | |
| Northcutt and Butler 1976 | A,D | *Lepisosteus osseus* |
| Braford and Northcutt 1983 | A | *Lepisosteus osseus* |
| Teleostei | | |
| Osteoglossomorpha | | |
| Lázár, Libouban, and Szabo 1984 | A,D,Cl,H | *Gnathonemus petersii* |
| Elopomorpha | | |
| Ebbesson 1968 | D | *Gymnothorax funebris** |
| Ekström 1982 | C | *Anguilla anguilla* |
| Euteleostei | | |
| Esocae | | |
| Repérant et al. 1976 | A | *Esox lucius** |
| Ostariophysi | | |
| Characiformes | | |
| Voneida and Sligar 1976 | A,D | *Astyanax hubbsi* |
| | | *Astyanax mexicanus* |
| Ebbesson and Ito 1980 | A,D | *Serrasalmus niger* |
| Cypriniformes | | |
| Roth 1969 | D | *Carassius auratus** |
| | | *Idus melanotus** |
| Sharma 1972 | D | *Carassius auratus* |
| Repérant and Lemire 1976 | A,D, | *Rutilus rutilus* |
| | | *Cyprinus macrolepidotus** |
| | | *Cyprinus carpio** |
| Repérant et al. 1976 | A | *Scardinius erythrophthalmus* |
| | | *Alburnus alburnus** |
| | | *Cyprinus carpio** |
| | | *Phoxinus phoxinus** |
| | | *Rutilus rutilus** |
| | | *Tinca tinca** |
| Peyrichoux et al 1977 | H | *Scardinius erythrophthalmus* |
| | | *Alburnus alburnus** |
| | | *Leuciscus rutilus** |
| | | *Rutilus rutilus** |
| | | *Tinca tinca** |
| Springer and Gaffney 1981 | Cl | *Carassius auratus* |
| Braford and Northcutt 1983 | A | *Carassius auratus* |
| Fraley and Sharma 1984 | H | *Carassius auratus* |
| Siluriformes | | |
| Ebbesson and O'Donnel 1980 | D | *Malapterurus electricus* |
| Meyer and Ebbesson 1981 | A,D,H | *Synodontis nigriventris* |
| Prasada Rao and Sharma 1982 | A,H | *Ictalurus punctatus* |

Table 21.1. (*Continued*)

| Reference | Method[a] | Species Described[b] |
|---|---|---|
| Salmonidae | | |
| Pinganaud and Clairambault 1975 | D | *Salmo irideus* |
| Lemire and Repérant 1976 | A | *Salmo irideus* |
| Repérant et al. 1976 | A | *Salmo irideus** |
| Pinganaud and Clairambault 1979 | A,D | *Salmo irideus* |
| Paracanthopterygii | | |
| Ebbesson 1968 | D | *Opsanus tau* |
| Percomorpha | | |
| Gasterosteiformes | | |
| Ekström 1984 | H | *Gasterosteus aculeatus* |
| Scorpaeniformes | | |
| Ito et al. 1984 | H | *Sebastiscus marmoratus* |
| Perciformes | | |
| Roth 1969 | D | *Aequadens portalogensis**<br>*Astronotus ocellatus**<br>*Cichlasoma meeki** |
| Vanegas and Ebbesson 1973 | D | *Eugerres plumieri* |
| Anders and Hibbard 1974 | D | *Cichlasoma biocellatum* |
| Repérant et al. 1976 | A | *Perca fluviatilis** |
| Butler and Northcutt 1981 | A | *Lepomis cyanellus** |
| Fernald 1982 | D,H | *Haplochromis burtoni* |
| Finck 1984 | H | *Coris julis* |
| Presson, Fernald, and Max 1985 | Cl,H | *Haplochromis burtoni* |
| Springer and Mednick 1985d | A,Cl | *Astronotus ocellatus* |
| Tetraodontiformes | | |
| Ito et al. 1984 | H | *Navodon modestus* |
| Pleuronectiformes | | |
| Gulley, Cochran, and Ebbesson 1975 | D | *Achirus lineatus* |
| Beryciformes | | |
| Campbell and Ebbesson 1969 | D | *Holocentrus sp.* |

[a] A = autoradiography; D = degeneration; H = horseradish peroxidase; C = cobalt; Cl = cobaltous lysine.
[b] * = described but not illustrated.

resolve discrepancies, but not details of the retina nor the development and plasticity of these visual centers.

While an accepted, uniform nomenclature does not exist for the nervous system in ray-finned fishes, we shall use the nomenclature developed by Northcutt (1983) for the optic tectum; Braford and Northcutt (1983) for the diencephalon; and Northcutt and Davis (1983) (modified from Nieuwenhuys 1963) and Northcutt and Braford (1980) for the telencephalon, with modifications as necessitated by new observations.

A phyletic analysis is strongly affected by the genealogical relationships of the taxa involved and thus frequently entails distinguishing among hypotheses of such relationships. The present analysis assumes that the hypotheses regarding

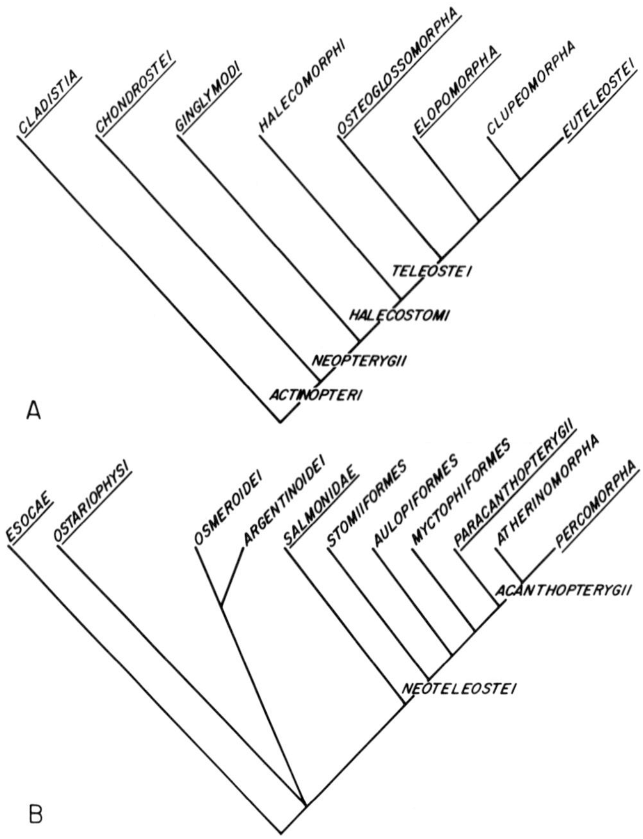


Figure 21.1. The actinopterygian (A) and euteleost (B) fishes. Experimental retinofugal studies exist only for members of underlined taxa.

the genealogy of ray-finned fishes proposed by Lauder and Liem (1983a, 1983b) and illustrated in Figure 21.1 are the most corroborated. Visually related characters that are possible synapomorphies (shared derived characters) that would support or invalidate these hypotheses are noted.

## Optic Nerve

The detailed structure of the optic nerve in teleosts has been examined in an ostariophysan, *Carassius* (Rusoff and Easter 1980; Bunt 1982; Easter, Rusoff, and Kish 1981) and the percomorphs *Aequidens, Astronotus,* and *Tilapia* (Scholes 1979), *Eugerres* (Tapp 1973, 1974), *Haplochromis* (Presson, Fernald, and Max 1985), *Helostoma,* and *Lepomis* (Rusoff 1984). In *Carassius* and *Eugerres* the optic nerve consists of approximately 200,000 axons, one-fourth of which are unmyelinated. The optic nerve in many teleosts appears to add axons throughout life by the addition of new ganglion cells in the retina, and the unmyelinated fibers appear to be the axons of the youngest retinal ganglion cells. In *Carassius* the youngest optic axons, arising from the periphery of the retina, occupy the center of the head of the optic nerve, whereas the oldest axons occupy the periphery. As the optic nerve is traced centrally, the fibers are rearranged so that the axons of the central and peripheral parts of the optic head form distinct ventral and dorsal groups, respectively, within the optic nerve. As the optic nerve is traced toward the optic chiasm, the compact optic nerve of *Carassius* divides into a series of loosely connected fascicules, forming an indistinct, transversely folded, band, with the younger fibers growing over the surface of the older so that the older fibers are ventromedial within each nerve. In many percomorph teleosts, ganglion cells are added to the peripheral retina, but—unlike in *Carassius*—the younger axons immediately occupy a ventral position in the optic head. Furthermore, the optic nerve in many percomorph teleosts is not subdivided into separate fascicules, as in *Carassius,* but forms a distinctly pleated ribbon beginning immediately beyond the head of the optic nerve.

Examination of the optic nerve in a number of ray finned fishes (Table 21.2) reveals that both fasciculated and complexly pleated optic nerves occur in all radiations except ginglymodes, halecomorphs, and clupeomorphs, where the optic nerve appears to be always pleated. A cladistic analysis of these characters suggests one of two scenarios: (1) if a fasciculated optic nerve was the plesiomorphic condition for ray-finned fishes, pleated nerves must have arisen at least twice, and reversals must have occurred four times independently; (2) if a pleated optic nerve was the plesiomorphic condition for ray-finned fishes, then a fasciculated nerve must have arisen five times independently. Assuming that gains and losses are equally probable and that the lowest number of character transformations is most parsimonious, a pleated nerve would be assumed to be the plesiomorphic condition for ray-finned fishes. An out-group analysis of other vertebrates, including lampreys and cartilaginous fishes, suggests that a fasciculated nerve is the plesiomorphic condition for all vertebrates and that a pleated optic nerve is a synapomorphy for ray-finned fishes. Given the large number of teleost groups whose optic nerves have not been examined and the presence of a pleated nerve in ginglymodes and halecomorphs, it is probable that a pleated nerve was present in ancestral teleosts and became fasciculated independently in each teleost radiation. It should be possible to determine the polarity of this transformation by a more detailed examination of the distribution of these characters in each teleost radiation and by comparing the development of the optic nerve in several teleost species.

It is claimed that a pleated optic nerve is an adaptation for maintaining constant electrical properties in the nerve in spite of extensive movements of the eye (Fernald 1980). Loose folds in the optic nerve would allow the nerve to lengthen during extensive rotation of the eye, without stretching individual optic axons, thus avoiding transient distortions in the electrical properties of the fibers. Although a pleated optic nerve may be so adapted, other possibilities should be considered. A pleated optic nerve may be exaptive (Gould and Vrba 1982; Northcutt, Chapter 34), and its effect may be due to other historical reasons. For example, fasciculated

Table 21.2. Variation in Visually Related Characters

| Species Examined | Fasciculated (f) vs. Pleated (p) Optic Nerve | PSp | PSm | PSi | PSp/p | Nucleus Corticalis | Nucleus Glomerulosus |
|---|---|---|---|---|---|---|---|
| Cladistia | | | | | | | |
| *Polypterus palmas* | p | + | – | – | – | – | – |
| *Erpetoichthys calabaricus* | f | + | – | – | – | – | – |
| Chondrostei | | | | | | | |
| *Polyodon spathula* | p | + | – | – | – | – | – |
| *Scaphirhynchus platyrynchus* | f | + | – | – | – | – | – |
| Ginglymodi | | | | | | | |
| *Lepisosteus osseus* | p | + | + | – | – | – | – |
| Halecomorpha | | | | | | | |
| *Amia calva* | p | + | + | – | – | – | – |
| Teleostei | | | | | | | |
| Osteoglossomorpha | | | | | | | |
| *Osteoglossum bicirrhosum* | p | + | + | +? | + | ? | – |
| *Pantodon buchholzi* | f | + | + | – | – | – | – |
| Elopomorpha | | | | | | | |
| *Elops saurus* | p | + | + | +? | + | – | – |
| *Gymnothorax moringa* | p | – | + | – | – | – | – |
| *Anguilla rostrata* | f | + | + | ? | – | – | – |
| Clupeomorpha | | | | | | | |
| *Clupea harengus* | p | + | + | + | + | – | – |
| *Brevoortia tyrranus* | p | + | + | + | – | – | – |
| *Engraulis mordax* | p | + | + | + | + | – | – |
| *Chirocentrus dorab* | p | + | + | + | + | – | – |
| Euteleostei | | | | | | | |
| Esocidae | | | | | | | |
| *Umbra limi* | f | + | + | – | + | – | – |
| *Esox lucius* | f | + | + | – | + | – | – |
| Ostariophysi | | | | | | | |
| Cypriniformes | | | | | | | |

| | | | | | | | |
|---|---|---|---|---|---|---|---|
| *Carassius auratus* | f | + | + | − | − | − | − |
| Characiformes | | | | | | | |
| *Crenuchus spilurus* | p | + | + | +? | + | +? | +? |
| *Serrasalmus nattereri* | p | + | + | +? | − | +? | +? |
| Gymnotoidei | | | | | | | |
| *Eigenmannia virescens* | f | −? | + | − | − | − | − |
| Siluriformes | | | | | | | |
| *Ictalurus melas* | f | − | + | − | − | +? | − |
| *Sorubim lima* | f | − | + | − | − | − | − |
| Salmonidae | | | | | | | |
| *Salmo gairdneri* | p | + | + | + | + | − | − |
| Paracanthopterygii | | | | | | | |
| *Opsanus tau* | f | + | + | ? | − | + | + |
| Atherinomorpha | | | | | | | |
| *Leuresthes tenuis* | p | + | + | + | + | + | + |
| *Strongylura notata* | p | + | + | + | + | + | + |
| Percomorpha | | | | | | | |
| *Lepomis cyanellus* | p | + | + | + | + | + | + |
| *Toxotes jaculatrix* | p | + | + | + | + | + | + |
| *Trachinotus falcatus* | p | + | + | + | + | + | + |
| *Astronotus ocellatus* | p | + | + | + | + | + | + |
| *Oreochromis mossambica* | ? | + | + | + | + | + | + |
| *Hemichromis bimaculatus* | p | + | + | + | + | + | + |
| *Cichlasoma erythraeum* | ? | + | + | + | + | + | + |
| *Coris julis* | p | + | + | + | + | + | + |
| *Embiotoca jacksoni* | p | + | + | + | + | + | + |
| *Cymatogaster aggregata* | p | + | + | + | + | + | + |
| *Astroscopus y-graeceum* | f | + | + | + | − | + | + |
| *Sphyraena baraccuda* | p | + | + | + | + | + | + |
| *Anabas testudineus* | p | + | + | + | + | + | + |
| *Ophicephalus africanus* | p | + | + | + | + | + | + |
| *Rhinecanthus aculeatus* | p | + | + | + | + | + | + |

NOTE: +, character present; −, character absent; ±?, probable condition; ?, condition unknown; PSp, PSm, PSi, pars parvocellularis, pars magnocellularis and pars intermedius, respectively, of the superficial pretectal nucleus; Psp/p, pleated pars parvocellularis.

lated and pleated nerves appear to be correlated, respectively, with punctate and elliptical optic nerve heads (S. S. Easter pers. com.). An elliptical optic head may be an efficient configuration for packing a large number of retinal axons, while minimizing reduction of the visual field. Thus, pleating of the optic nerve may be a mechanical consequence of an elliptical optic head, which, in turn, may be an adaptation to minimize reduction of the visual field in species under strong selection to increase the number of retinal receptors. In this case, a pleated nerve may effect electrical constancy in retinal axons, but could not be said to be adaptive. These hypotheses, as well as others, can be tested by examining the distribution and correlations of the relevant anatomical and functional characters among various fishes (see Northcutt, Chapter 34).

## Connections of the Retina

Anatomical (Ito and Murakami 1984; Ito et al. 1984) and physiological (Vanegas, Essayag-Millán, and Laufer 1971; Vanegas, Amat, and Essayag-Millán 1973; Sandemann and Rosenthal 1974; Schmidt 1979; Vanegas 1983) studies indicate that the optic nerve in teleosts comprises both retinofugal and retinopetal fibers. Six classes of different-sized retinal ganglion cells have been identified in the percomorphs *Navodon* and *Sebastiscus,* and the diameter of an optic nerve axon appears to be closely linked to its cell class: type I cells, with small round or oval somata and small dendritic fields in the inner plexiform layer, give off thin axons (mean diameter 0.8 μm), as do type II cells, with small somata and dendrites that arborize in the innermost region of the inner plexiform layer; type III cells, with medium-sized round somata and extensively branching dendrites, give off medium-sized axons (mean diameter 1.3 μm), as do type IV cells, with large somata and dendrites in the innermost part of the inner plexiform layer; type V cells, with large, variously shaped, somata, usually dislocated in the inner plexiform or granular layer, give off thick axons (mean diameter 2.5 μm), as do type VI cells, which have the largest somata, also dislocated.

Anatomical studies of *Navodon* and *Sebastiscus* (Ito et al. 1984) demonstrate that axons of all three size groups include retinofugal fibers that terminate in various diencephalic and mesencephalic visual centers. However, electrophysiological evidence suggests that the largest axons include retinopetal fibers that arise from the optic tectum and, possibly, other neural centers (Vanegas, Amat, and Essayag-Millán 1973; Schmidt 1979).

The occurrence of three classes of retinofugal fibers and at least one class of retinopetal fibers is probably widespread among euteleosts, given the occurrence of these classes in cyprinids and percomorphs. However, it is presently impossible to determine if this is a plesiomorphic or apomorphic condition for teleosts or ray-finned fishes, as other groups have not been examined. Given considerable variation in the retinorecipient laminae of the optic tectum (Northcutt 1983) and pretectum (Braford and Northcutt 1983) in ray-finned fishes, it is particularly important to determine the physiological properties of the different classes of optic nerve fibers and their phylogeny.

## Retinofugal Projections

The extensive literature covering experimental studies of retinofugal projections in ray-finned fishes (Table 21.1) and reviews of same (Braford and Northcutt 1983; Northcutt 1983; Vanegas 1983; Vanegas and Ito 1983) report from seven (Vanegas and Ito 1983) to fourteen (Springer and Gaffney 1981) distinct centers, including the optic tectum, that receive contralateral retinofugal projections. Reports vary because of several factors: different experimental methods, different interpretations of the number and boundaries of various cell groups, and differences between species. A consensus does exist, however, regarding retinofugal terminal fields, and most of the differences in interpretation appear to relate to the number of cell groups, particularly in the thalamic and pretectal regions, and their homologues among various species of teleosts. Figure 21.2 summarizes the primary targets of the contralateral retinofugal projections in a cyprinid, *Carassius* (Sharma 1972; Springer and Gaffney 1981; Braford and Northcutt 1983; Fraley and Sharma 1984) and a centrarchid, *Lepomis* (Butler and Northcutt 1981), whose retinofugal projections are comparable to other percomorphs such as *Astronotus* (Springer and Mednick 1985a), *Haplochromis* (Fernald 1982; Presson, Fernald, and Max 1985), *Gasterosteus* (Ekström 1984), *Navodon* (Ito et al. 1984) and *Sebastiscus* (Ito et al. 1984). The nomenclature used is modified from Braford and Northcutt (1983). Differing anatomical names used for several of these retinofugal targets in other studies are noted, and some higher-order visual pathways that bear on these nomenclatural differences and their phylogenetic implications are discussed.

### Preoptic Centers

The fibers of the optic nerve in teleosts, as in other anamniotes, form a lateral marginal and a medial axial optic tract caudal to the optic chiasm. As the optic fibers decussate in the chiasm, some of them turn dorsally to terminate in the suprachiasmatic nucleus (Fig. 21.2A, B, and D; *hypothalamic optic nucleus* of Meyer and Ebbesson 1981). Similarly, dorsomedially directed optic fibers also arise from the axial optic tract and innervate the caudal portion of the suprachiasmatic nucleus. Presson, Fernald, and Max (1985) claim that the suprachiasmatic nucleus of *Haplochromis* receives retinal projections only from the axial optic tract, which they believe consists solely of axons of ganglion cells located in the central portion of the retina, dorsal to the optic papilla. Fraley and Sharma (1984) similarly report that the axial optic tract, which they term the *medial fascicle of the optic tract,* innervates the suprachiasmatic nucleus in *Carassius,* but they claim that the axial tract consists of ganglion axons from all retinal quadrants that terminate in the suprachiasmatic nucleus, with the ventral retinal projections, not the central ones, as in *Haplochromis,* being the most extensive. Furthermore, Springer and Gaffney (1981) noted, in the preoptic region in *Carassius,* additional optic fascicles that innervate not only the suprachiasmatic nucleus, but also more dorsally located cellular fields that Braford and Northcutt (1983) termed the *nucleus preopticus parvocellularis posterioris* (Fig. 21.2A, B, D, and E) and a rostral tuberal region (Fig. 21.2E) of the hypothalamus. After contributing fibers to the suprachiasmatic nucleus, the axial optic tract continues dorsally to contribute fibers to the medial optic tract (Fig. 21.3B) and then joins the marginal optic tract.

## Thalamic and Periventricular Pretectal Centers

As fibers of the marginal optic tract are traced dorsally at a level comparable to that of Figure 21.2D, fibers turn medially to form a medial optic tract (*fasciculus dorsalis tractus opticus* of Springer and Gaffney 1981), which turns caudally to

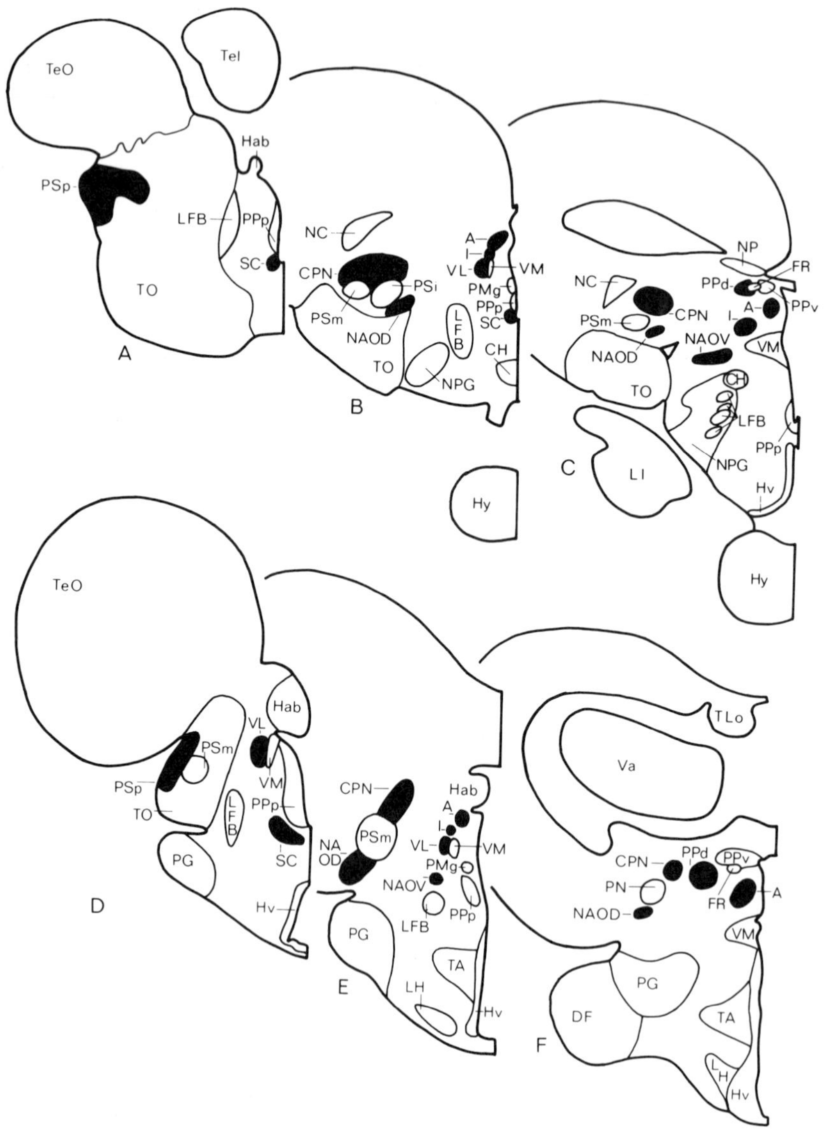

innervate the anterior, intermediate, and ventrolateral thalamic nuclei (Figs. 21.2B–F, 21.3A –C, 21.5, and 21.6; collectively termed the *dorsomedial optic nucleus* by Meyer and Ebbesson 1981, the *dorsolateral and ventrolateral optic nucleus* by Springer and Gaffney 1981, and the *dorsolateral thalamic nucleus* by Ito et al. 1984), the pars dorsalis of the periventricular pretectal nucleus (Fig. 21.2C and F; termed the *caudal part of the dorsomedial optic nucleus* by Meyer and Ebbesson 1981, the *caudal part of area pretectalis* Ekström 1984, and the *nucleus of the posterior commissure* by most other authors) before terminating in the optic tectum. In *Haplochromis* the medial optic tract is composed of axons arising from ganglion cells of all retinal quadrants, but no retinotopic termination is apparent in any of the terminal fields of this tract (Presson, Fernald, and Max 1985). However, Fraley and Sharma (1984) report that retinotopic terminations of axons of the medial optic tract, which they term the *dorsomedial optic tract,* exist in *Carassius.* Rostal (*nucleus opticus dorsolateralis*) and caudal (*nucleus of the posterior commissure*) terminal fields were recognized with axons of the dorsal and ventral retinal quadrants projecting onto the dorsorostral and ventrocaudal portions, respectively, and the axons of the nasal and temporal retinal quadrants projecting onto the dorsocaudal and ventrorostral portions, respectively, of the *dorsolateral optic nucleus.* Retinotopic projections to the *nucleus of the posterior commissure* were not so clear. Axons of the temporal and ventral retinal quadrants appear to innervate the caudal and rostral portions of the nucleus, respectively, but there is no clear topography for the termination of axons of the dorsal and nasal retinal quadrants.

Braford and Northcutt (1983) recognized the periventricular zone adjacent to the posterior commissure as a periventricular pretectal nucleus (*caudal dorsolateral optic nucleus* or *nucleus of the posterior commissure* of most studies) and subdivided this region into a pars dorsalis and a pars ventralis (Fig. 21.2C, and F). The pars dorsalis is the main site of the termination of the caudally coursing

Figure 21.2. Schematic drawings of transverse sections through one-half of the diencephalon and mesencephalon of a percomorph, *Lepomis* (A–C), and a cyprinid, *Carassius* (D–F), illustrating the position and relative development of primary retinofugal terminal fields (nuclei in black); *A,* nucleus anterioris thalami; *CH,* commissura horizontalis; *CPN,* nucleus pretectalis centralis; *DF,* nucleus diffusus lobi inferiores; *FR,* fasciculus retroflexus; *Hab,* habenula; *Hy,* hypophysis; *Hv,* ventral zone of periventricular hypothalamus; *I,* nucleus intermedius thalami; *LFB,* lateral forebrain bundle; *LH,* nucleus lateralis hypothalami; *LI,* lobus inferior; *NAOD,* nucleus accessorius opticus dorsalis; *NAOV,* nucleus accessorius opticus ventralis; *NC,* nucleus corticalis; *NPG,* nucleus preglomerulosus; *PG,* preglomerular nuclear complex; *PMg,* nucleus preopticus magnocellularis pars gigantocellularis; *PN,* nucleus pretectalis posterioris; *PPd,* nucleus pretectalis periventricularis pars dorsalis; *PPp,* nucleus preopticus parvocellularis posterioris; *PPv,* nucleus pretectalis periventricularis pars ventralis; *PSi,* nucleus pretectalis superficialis pars intermedius; *PSm,* nucleus pretectalis superficialis pars magnocellularis; *PSp,* nucleus pretectalis superficialis pars parvocellularis; *SC,* nucleus suprachiasmaticus; *TA,* nucleus tuberis anterior; *Tel,* telencephalon; *TLo,* torus longitudinalis; *TO,* tractus opticus; *TeO,* tectum opticum; *Va,* valvula cerebelli; *VL,* nucleus ventrolateralis thalami; *Vm,* nucleus medialis thalami.

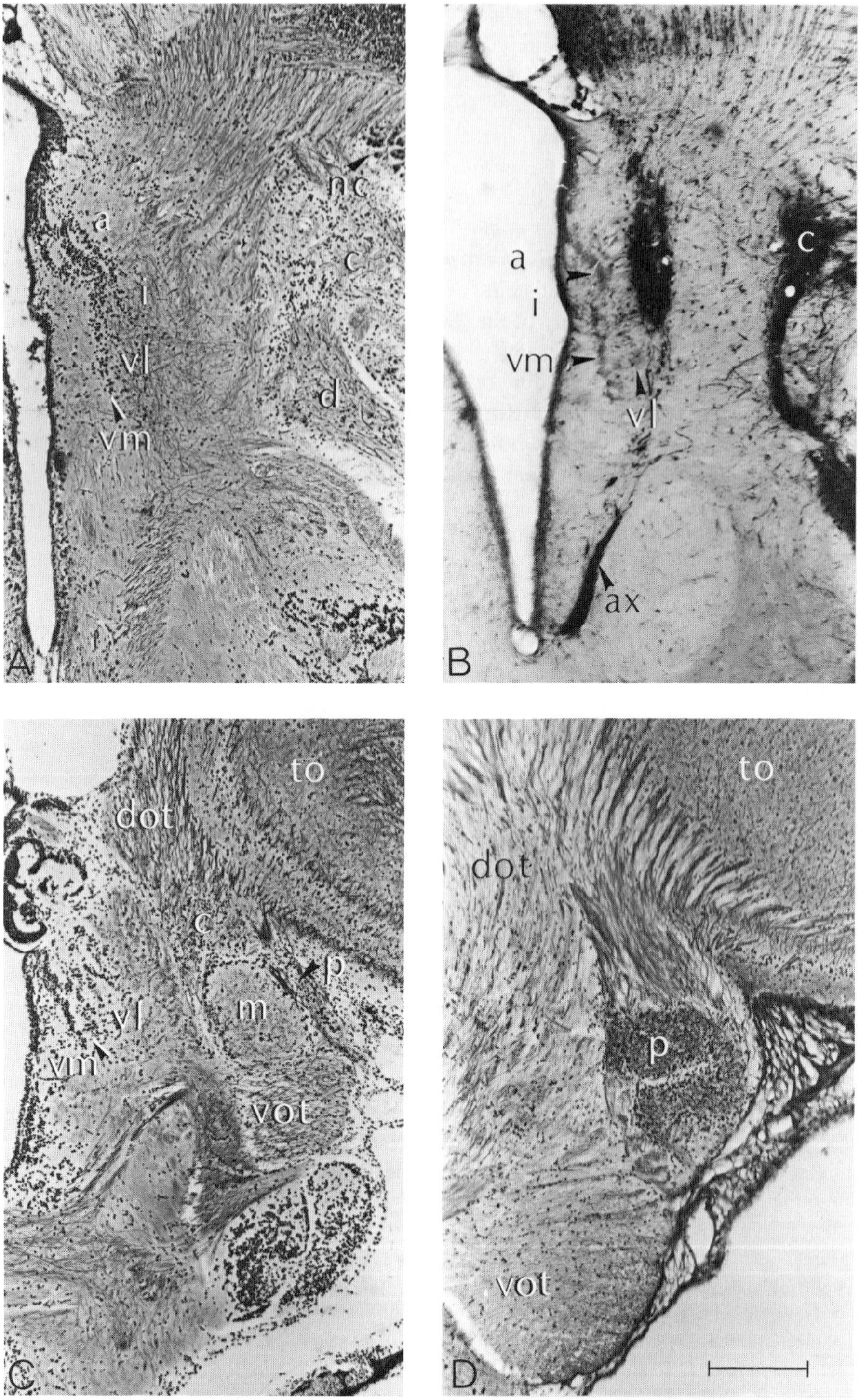
nc
a
c
i
vl
d
vm
A
c
a
i
vm
vl
ax
B
to
dot
c
p
m
vl
vm
vot
C
to
dot
p
vot
D

medial optic tract; however, cells of pars ventralis may also receive a retinal input if their dendrites extend laterally. Pars ventralis is a major target of ascending cerebellar efferents in *Carassius* (Braford 1971) and in *Lepisosteus* (Braford and Northcutt 1983), and the cells of the periventricular pretectal nucleus (Fig. 21.7B and 21.8B) appear to project to the optic tectum (Luiten 1981; Northcutt 1982a, 1982b; Fiebig, Ebbesson, and Meyer 1983) and cerebellum (Finger 1978a; Luiten 1981; Northcutt 1982c; Wullimann and Northcutt 1985).

There is substantial evidence that the thalamic region generally identified as the *dorsolateral optic* or *thalamic nucleus* consists of at least four nuclei rostrally (the anterior, intermediate, ventrolateral, and ventromedial thalamic nuclei of Braford and Northcutt 1983). The thalamic wall immediately beneath the rostral habenular nuclei consists of obliquely oriented cellular laminae that Braford and Northcutt (1983) termed the *ventrolateral and ventromedial nuclei of the ventral thalamus* (Fig. 21.3C). The ventrolateral nucleus is distinguished from the ventromedial nucleus by its more loosely organized cellular laminae and by a dense neuropil composed of retinofugal and tectal terminals. The ventromedial thalamic nucleus may also receive retinofugal terminals, as the dendrites of its cells extend laterally into the neuropil of the ventrolateral thalamic nucleus. It is these rostral portions of the ventrolateral and ventromedial thalamic nuclei that many studies describe as the rostral pole of the *dorsolateral thalamic nucleus*. However, the cells of this rostral thalamic region (Fig. 21.6), like the more caudally and ventrally located cells described as the *ventrolateral optic nucleus* by Springer and Gaffney (1981), the *nucleus preopticus* by Luiten (1981), and the *dorsomedial optic nucleus* by Fiebig, Ebbesson, and Meyer (1983), receive a dense cerebellar input throughout their rostrocaudal extent (Braford 1971; Braford and Northcutt 1983), whereas the anterior and intermediate thalamic nuclei do not. Although the ventrolateral and ventromedial thalamic nuclei project to the optic tectum (Grover and Sharma 1981; Luiten 1981; Fiebig, Ebbesson, and Meyer 1983), as does the anterior thalamic nucleus (Figs. 21.5 and 21.6), the last-named also projects to the telencephalon (Echteler and Saidel 1981; Northcutt 1981a). Projections to the telencephalon from a cell group identified as a *ventromedial thalamic nucleus* have been reported in *Sebastiscus* (Murakami, Fukuoka, and Ito 1986), but other studies (Braford and Northcutt 1983; Echteler 1984, 1985) have identified these cells as

Figure 21.3. Photomicrographs of transverse sections through the diencephalon and optic tectum of *Lepomis* (A, B, and D) and *Carassius* (C) illustrating relative development of retinofugal recipient nuclei of the rostral thalamus and pretectum. A, C, and D are Bodian-stained preparations; B illustrates the course and termination of retinofugal fibers labeled with horseradish peroxidase. Dorsal and lateral surfaces in A–D are to the *top* and *right*, respectively. Bar scale equals 200 um; *a*, nucleus anterioris thalami; *ax*, axial optic tract; *c*, nucleus pretectalis centralis; *d;* nucleus accessorius opticus dorsalis; *dot*, dorsal optic tract; *i*, nucleus intermedius thalami; *m*, nucleus pretectalis superficialis pars magnocellularis; *nc*, nucleus corticalis; *p*, nucleus pretectalis superficialis pars parvocellularis; *to*, tectum opticum; *vl*, nucleus ventrolateralis thalami; *vm*, nucleus ventromedialis thalami; *vot*, ventral optic tract.

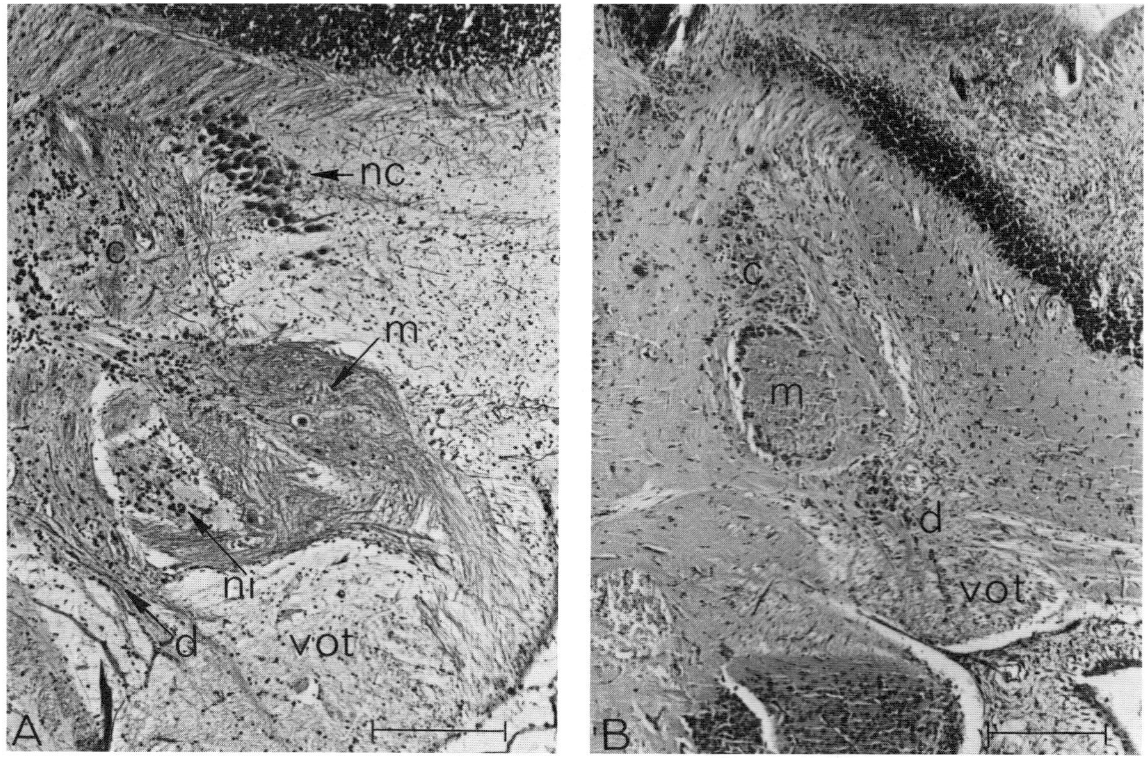


Figure 21.4. Photomicrographs of transverse Bodian-stained section through the pretectum of *Lepomis* (A) and a Nissl-stained section through the pretectum of *Carassius* (B). Dorsal and lateral brain surfaces are to the *top* and *right*, respectively. Bar scales equal 200 um; *c*, nucleus pretectalis centralis; *d*, nucleus accessorius opticus dorsalis; *m*, nucleus pretectalis superficialis pars magnocellularis; *ni*, nucleus pretectalis superficialis pars intermedius; *nc*, nucleus corticalis; *vot*, ventral optic tract.

part of the posterior central nucleus, a more caudal and dorsal division of the dorsal thalamus.

## Pretectal Centers

At approximately the same level that the medial optic tract arises from the medial edge of the marginal optic tract, the main portion of the marginal optic tract divides into dorsal and ventral optic tracts (Fig. 21.3C and D) that continue caudally to innervate the rostromedial and caudolateral portions of the optic tectum. Bifurcation of the marginal optic tract is approximately coincident with the rostral border of the pretectum, and a number of pretectal nuclei also receive retinofugal input. The most superficial of these pretectal centers is the pars parvocellularis (*nucleus geniculatus lateralis* of most studies) of the superficial pretectal nucleus, which, in *Carassius* (Figs. 21.2D and 21.3C) and many other cyprinids, consists of a sphere folded dorsoventrally with the lips of the folds directed rostrally. In cichlids and many other teleosts (Table 21.2), the pars parvocellularis (Figs. 21.2A and 21.3D) is located more rostrally in the marginal optic tract and consists of multiple folds similar to a pleated optic nerve. A cladistic analysis of the occurrence and degree of folding of the pars parvocellularis (Table 21.2) suggests that the superficial pretectal nucleus of cladistians and chondrosteans is homologous to the pars parvocellularis of neopterygians, and that an nonfolded or simply folded pars parvocellularis is the plesiomorphic condition in neopterygians. Furthermore, a complexly folded pars parvocellularis appears to be plesiomorphic for teleosts, and the reduction or loss of the nucleus in several groups of teleosts (Table 21.2) is an independently occurring apomorphy.

Retinofugal projections to the pars parvocellularis of the superficial pretectal nucleus are known to be retinotopically organized in *Carassius* (Fraley and Sharma 1984; Springer and Mednick 1985b) and *Haplochromis* (Presson, Fernald, and Max 1985), with the dorsal and ventral retinal quadrants terminating laterally and medially, respectively; the nasal and temporal quadrants, caudally and rostrally, respectively. There is a slight discrepancy in the sites of termination of the projections of the nasal and temporal quadrants in *Carassius* and *Haplochromis*, but these are likely due to differences in the degree of folding of this nucleus in these two genera. The pars parvocellularis also receives input from the optic tectum (Sligar and Voneida 1976; Grover and Sharma 1979a; Northcutt and Butler 1980; Luiten 1981; Striedter and Northcutt 1986) and from the ipsilateral nucleus isthmi (Striedter and Northcutt 1986). Although the pars parvocellularis has frequently been termed the *lateral geniculate nucleus,* it does not project to the telencephalon (Ito and Kishida 1978; Grover and Sharma 1979b; Northcutt and Braford 1984), as would be expected if it were homologous to the lateral geniculate nucleus in amniotes. Instead, pars parvocellularis in *Lepomis* (Fig. 21.7) projects bilaterally to another pretectal center, nucleus intermedius of Brickner (1929), and to the inferior raphe nucleus (Striedter and Northcutt 1986). A projection to the retina has been reported in *Tetraodon* and *Julidochromis* (Ebbesson and Meyer 1981), but has not been observed in other teleosts (Münz and

Claas 1981; Springer and Gaffney 1981; Uchiyama, Sakamoto, and Ito 1981; Ekström 1984; Springer and Mednick 1985a).

In cyprinids, such as *Carassius,* an additional large spherical nucleus, termed *nucleus rotundus* by Schnitzlein (1962), is closely associated with pars parvocellularis and is easily recognized by its extensive central neuropil ringed by medium-sized cells (Figs. 21.2D, and 21.3C). Schnitzlein (1962) assumed that this nucleus occupied a central position within the dorsal thalamus and homologized it to the same-named nucleus in birds and reptiles, in which it receives bilateral input from the optic tectum and projects to a subdivision (dorsal ventricular ridge) of the telencephalic pallium. Although Sharma (1972) reported that this cell group receives primary retinal input in *Carassius,* subsequent studies (Springer and Gaffney 1981; Braford and Northcutt 1983) have failed to confirm this observation. Braford and Northcutt (1983) argued that, although the *nucleus rotundus* of Schnitzlein does receive a bilateral tectal input (Grover and Sharma 1979a; Northcutt and Butler 1980; Luiten 1981), its topographical position is incompatible with that of the nucleus rotundus in amniotes and it should be considered a pretectal nucleus, which they termed *pars magnocellularis* of the superficial pretectal nucleus because of its close association with the smaller-celled pars parvocellularis in ginglymodes and halecomorphs, as well as in cyprinids. Subsequently, the pars magnocellularis in cyprinids has been demonstrated (Northcutt and Braford 1984) to project ipsilaterally to a caudal hypothalamic nucleus (the so-called mammillary body) and to the nucleus lateralis valvulae (Fig. 21.8B). Thus both its topography and its efferent projections are incompatible with its' being the homologue of nucleus rotundus in amniotes.

Although a *nucleus rotundus,* or pars magnocellularis, has been recognized in percomorphs (Fernald 1982; Springer and Mednick 1985a), the comparable pretectal region of these fishes is more complex (Fig. 21.4A). In many percomorphs, such as *Lepomis,* there are two distinct neuropils, one of which contains medium-sized cells and the other large cells, rather than a single neuropil ringed by cells, as in *Carassius.* The dorsolateral neuropil contains cells as large as those of nucleus corticalis (Fig. 21.4A) and is easily distinguished from those of the finer-fibered ventromedial neuropil, which contains smaller neurons. We term these two neuronal populations the pars magnocellularis and pars intermedius, respectively, of the superficial pretectal nucleus (Fig. 21.7). The pars magnocellularis (*nucleus geniculatus posterior pars ventralis* of Ebbesson and Vanegas 1976 and *nucleus pretectalis* of Ito et al. 1981; Ito, Sakamoto, and Takatsuji 1982) receives ipsilateral input from the optic tectum (Ebbesson and Vanegas 1976) and projects bilaterally to nucleus isthmi (Ito et al. 1981; Ito, Sakamoto, and Takatsuji 1982; Striedter and Northcutt 1986), whereas pars intermedius (*nucleus intermedius* of Brickner 1929 and Sakamoto and Ito 1982) receives a bilateral input from pars parvocellularis (Striedter and Northcutt 1986) and projects ipsilaterally to nucleus glomerulosus (Sakamoto and Ito 1982). We believe that the pars magnocellularis is homologous in cyprinids and percomorphs, because in each case the nucleus occupies a comparable topographical position, consists of similar neurons, and receives tectal input (Figs. 21.7 and 21.8). However, there is no evidence that the pars magnocellularis of cyprinids projects to nucleus isthmi (Northcutt and Braford 1984) nor that the pars magnocellularis of percomorphs projects to the mammillary body and nucleus lateralis valvulae (Ito, Sakamoto, and Takatsuji 1982; Sakamoto and Ito 1982). The known efferents of the pars magnocellularis in the percomorph *Navodon* were determined by injecting horseradish peroxidase (HRP) into nucleus isthmi and not into pars magnocellularis. It is thus possible that there are other efferents from pars magnocellularis in percomorphs and that projections exist to the mammillary body and nucleus lateralis valvulae. Nucleus intermedius may not exist in cyprinids because these teleosts do not possess a nucleus glomerulosus (Sakamoto and Ito 1982; Braford and Northcutt 1983) and yet they may still exhibit a pars parvocellularis owing to its projections to conserved targets. Similarly, a projection to nucleus isthmi from pars magnocellularis may not exist in *Carassius,* as nucleus isthmi in that genus is poorly differentiated relative to that in most percomorphs. Because of these differences in efferent projections of the proposed homologues, pars magnocellularis in cyprinids and in percomorphs, it is necessary to consider several alternative hypotheses: (1) pars magnocellularis in cyprinids is a field homologue of pars intermedius and pars magnocellularis of percomorphs; (2) pars intermedius or pars magnocellularis in percomorphs is a de novo nucleus; and (3) both nuclei in percomorphs possess separate homologues, not yet identified, in cyprinids. Although additional experimental studies are needed to evaluate these hypotheses, a cladistic analysis of the distribution of a recognizable pars intermedius and pars magnocellularis among ray-finned fishes (Table 21.2) suggests that a pars magnocellularis arose with the origin of neopterygians and that a pars intermedius is a plesiomorphic character of teleosts or possibly only of clupeomorphs and euteleosts. If this hypothesis is correct, the absence of a pars intermedius in cypriniform, gymnotoid, and siluriform teleosts is an apomorphic condition rather than a plesiomorphic one.

At least three additional nuclei (the central pretectal nucleus and dorsal and ventral accessory optic nuclei) occur within the pretectal region in teleosts (Fig. 21.2) but have been differently named in various studies. A distinct population of medium-sized neurons occurs within the marginal optic tract or begins caudally in the dorsal optic tract immediately after bifurcation of the marginal optic tract; this cell population has been termed the central pretectal nucleus (Northcutt and Butler 1976; Braford and Northcutt 1983; Presson, Fernald, and Max 1985). The central pretectal nucleus (Figs. 21.3C, 21.4, 21.7B, and 21.8B) receives bilateral tectal input (Ebbesson and Vanegas 1976; Grover and Sharma 1979a; Northcutt and Butler 1980) and projects to the optic tectum (Luiten 1981; Northcutt 1982b; Fiebig, Ebbesson, and Meyer 1983) and the corpus of the cerebellum (Finger 1978a; Finger and Karten 1978; Luiten 1981; Northcutt 1982c; Wullimann and Northcutt 1985). This nucleus has been identified as a pretectal (P1) nucleus in *Carassius* and *Ictalurus* (Finger and Karten 1978), as the *pretectal nucleus* in *Serrasalmus* (Fiebig, Ebbesson, and Meyer 1983), as the *pars dorsalis of nucleus* and *area opticus pretectalis dorsalis* in *Carassius* (Springer and Gaffney 1981), as the *dorsal pretectal area and nucleus* in *Carassius* (Fraley and Sharma 1984), as *nucleus pretectalis medialis* in *Astronotus* (Springer and Mednick 1985a), and as *pars dorsalis of area pretectalis* in various teleosts (Vanegas and Ito

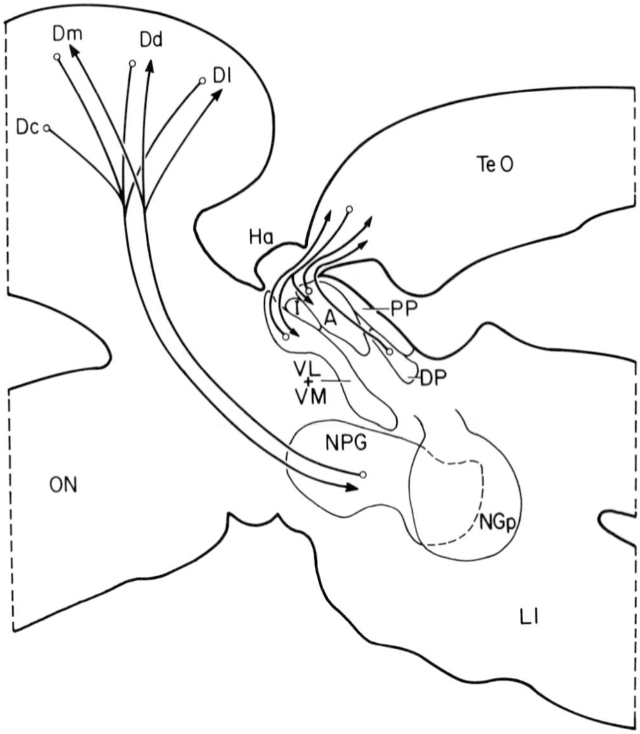


Figure 21.5. Diagram of some of the afferent and efferent projections of thalamic and preglomerular nuclei of *Lepomis* viewed in sagittal plane; *A*, nucleus anterioris thalami; *Dc*, central zone of area dorsalis telencephali; *Dd*, dorsal zone of area dorsalis telencephali; *Dl*, lateral zone of area dorsalis telencephali; *Dm*, medial zone of area dorsalis telencephali; *DP*, nucleus dorsalis posterioris thalami; *Ha*, habenula; *I*, nucleus intermedius thalami; *LI*, lobus inferior; *NGp*, nucleus glomerulosus pars posterior; *NPG*, nucleus preglomerulosus; *ON*, optic nerve; *PP*, nucleus pretectalis periventricularis; *Te O*, tectum opticum; *VL*, nucleus ventrolateralis thalami; *VM*, nucleus ventromedialis thalami.

1983). Presson, Fernald, and Max (1985) report that the central pretectal nucleus in *Haplochromis* receives retinofugal projections solely from the ventral hemiretina; while Fraley and Sharma (1984) report that the comparable retinofugal terminal field in *Carassius* is richly innervated by the ventral hemiretina, they also claim that this area receives projections from the dorsal hemiretina. A central pretectal nucleus apparently occurs in all radiations of ray-finned fishes, and no striking apomorphy has been noted among the species examined (Table 21.2).

A dorsal accessory optic nucleus occurs in *Carassius* (Figs. 21.2E and F and 21.4B) and *Lepomis* (Figs. 21.2B and C and 21.3A), adjacent to the dorsomedial border of the ventral optic tract, and a ventral accessory optic nucleus (Fig. 21.2C, E) occurs in both genera slightly more caudally and receives retinofugal fibers that diverge medially from the ventral optic tract as a distinct bundle termed the *accessory optic tract* by Springer and Gaffney (1981). The dorsal accessory

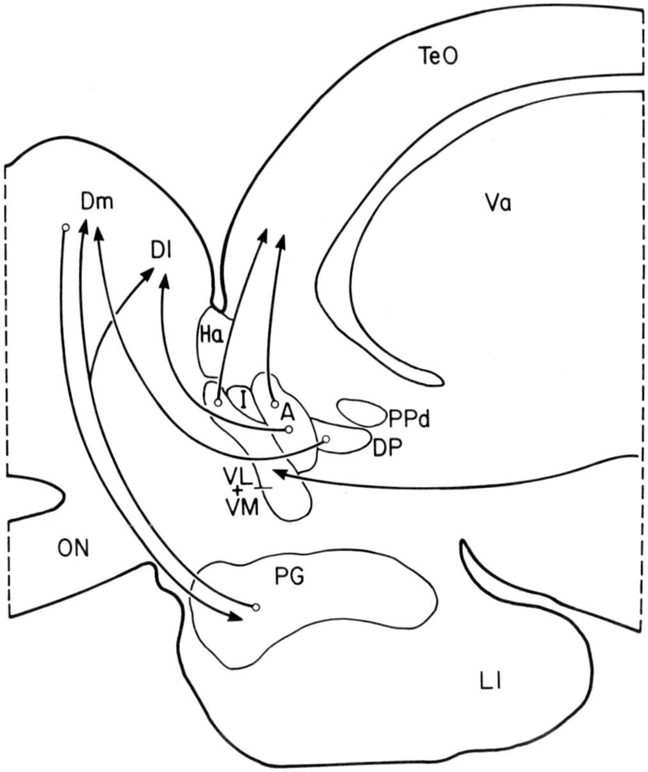


Figure 21.6. Diagram of some of the afferent and efferent connections of the thalamus and preglomerular complex of *Carassius* viewed in sagittal plane. Ascending brain stem tract to *VL* + *VM* is the so-called mesencephalocerebellar tract, which carries efferents to the thalamus; *A*, nucleus anterioris thalami; *Dl*, lateral zone of area dorsalis telencephali; *Dm*, medial zone of area dorsalis telencephali; *DP*, nucleus dorsalis posterioris thalami; *Ha*, habenula; *I*, nucleus intermedius thalami; *LI*, lobus inferior; *ON*, optic nerve; *PG*, preglomerular nuclear complex; *PPd*, nucleus pretectalis periventricularis pars dorsalis; *Te O*, tectum opticum; *Va*, valvula cerebelli; *VL*, nucleus ventrolateralis thalami; *VM*, nucleus ventromedialis thalami.

optic nucleus has been identified as the *nucleus of the ventral optic tract* in *Lepisosteus* (Northcutt and Butler 1976), the *area opticus pretectalis pars ventralis* and *nucleus corticalis* (Springer and Gaffney 1981), the *basal optic nucleus* (Braford and Northcutt 1983), a pretectal (P2) nucleus in *Carassius* (Finger and Karten 1978), the *anterior glomerular nucleus* in *Ictalurus* (Finger 1978a), the medial *ventral pretectal nuclei* in *Astronotus* (Springer and Mednick 1985a), and the *area pretectalis pars ventralis* in various other teleosts (Vanegas and Ito 1983). The ventral accessory optic nucleus has been identified as the *accessory optic nucleus* in *Acipenser* (Repérant et al. 1982), *Polypterus* (Repérant et al. 1981; Braford and Northcutt 1983), *Lepisosteus* (Northcutt and Butler 1976), and *Carassius* (Springer and Gaffney 1981; Braford and Northcutt 1983), and as the *area*

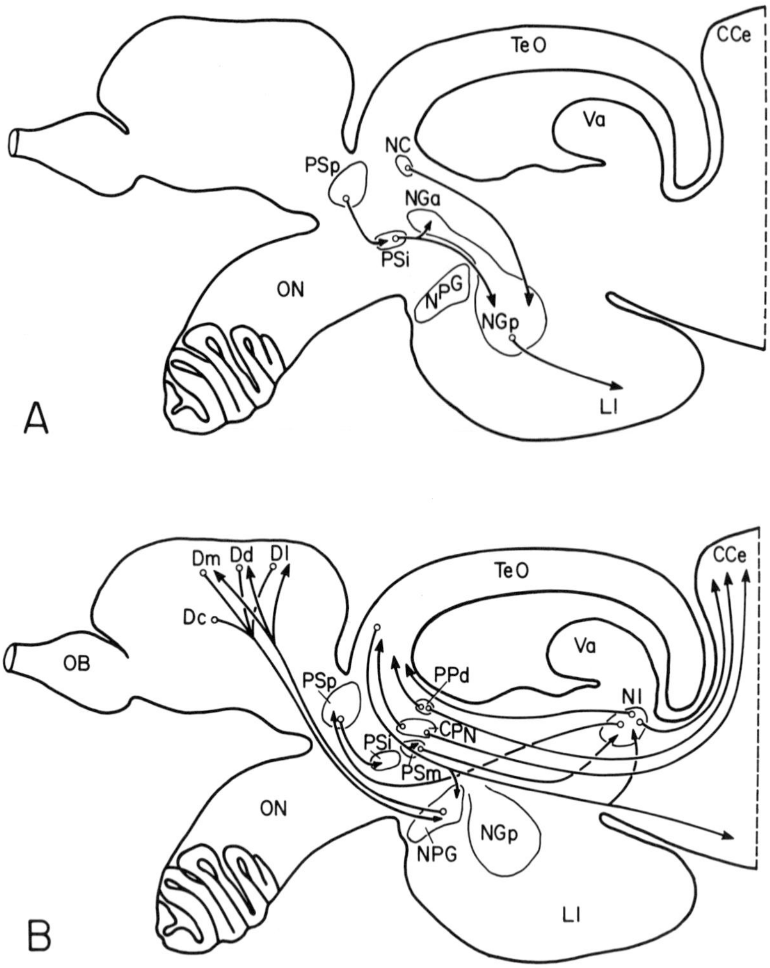


Figure 21.7. Diagram of some of the afferent and the efferent connections of nucleus glomerulosus (A) and pretectum (B) of *Lepomis* viewed in sagittal plane; *CCe,* corpus cerebelli; *CPN,* nucleus pretectalis centralis; *Dc,* central zone of area dorsalis telencephali; *Dd,* dorsal zone of area dorsalis telencephali; *Dl,* lateral zone of area dorsalis telencephali; *Dm,* medial zone of area dorsalis telencephali; *LI,* lobus inferior; *NC,* nucleus corticalis; *NGa,* nucleus glomerulosus pars anterior; *NGp,* nucleus glomerulosus pars posterior; *NI,* nucleus isthmi; *NPG,* nucleus preglomerulosus; *OB,* olfactory bulb; *ON,* optic nerve; *PPd,* nucleus pretectalis periventricularis pars dorsalis; *PSi,* nucleus pretectalis superficialis pars intermedius; *PSm,* nucleus pretectalis superficialis pars magnocellularis; *PSp,* nucleus pretectalis superficialis pars parvocellularis; *Te O,* tectum opticum; *Va,* valvula cerebelli.

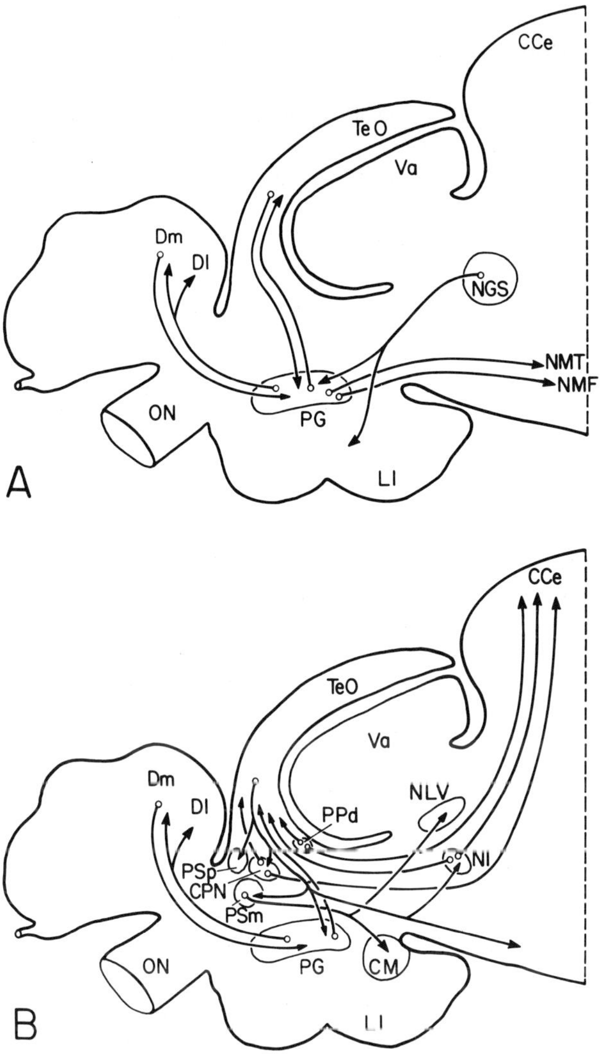


Figure 21.8. Diagram of some of the afferent and efferent connections of preglomerular complex (A) and pretectum (B) of *Carassius* viewed in sagittal plane; *CCe,* corpus cerebelli; *CM,* corpus mammillare; *CPN,* nucleus pretectalis centralis; *Dl,* lateral zone of area dorsalis telencephali; *Dm,* medial zone of area dorsalis telencephali; *LI,* lobus inferior; *NGS,* nucleus gustatorius secundarius; *NI,* nucleus isthmi; *NLV,* nucleus lateralis valvulae; *NMF,* nucleus motorious nervi facialis; *NMT,* nucleus motorius nervi trigemini; *ON,* optic nerve; *PG,* preglomerular nuclear complex; *PPd,* nucleus pretectalis periventricularis pars dorsalis; *PSm,* nucleus pretectalis superficialis pars magnocellularis; *PSp,* nucleus pretectalis superficialis pars parvocellularis; *Te O,* tectum opticum; *Va,* valvula cerebelli.

*ventrolateralis* in various other teleost species (Vanegas and Ito 1983). Only the dorsal accessory optic nucleus of *Ictalurus* has been shown to project to the corpus of the cerebellum and to the oculomotor nucleus (Finger and Karten 1978), but both dorsal and ventral accessory optic nuclei project to the cerebellar corpus in *Lepisosteus* (Northcutt 1982c) and *Lepomis* (Wullimann and Northcutt 1985), but apparently not in *Carassius* (Finger and Karten 1978; Grover and Sharma 1981) or *Cyprinus* (Luiten 1981).

Dorsal and ventral accessory optic nuclei appear to occur in all radiations of ray-finned fishes except, perhaps, for a dorsal accessory optic nucleus in cladistians. The ventral accessory optic nucleus is well developed in highly visual percomorphs, such as *Astronotus* (Springer and Mednick 1985a) and *Navodon* (Ito et al. 1984), but too few species have been examined to determine whether a well-developed ventral accessory optic nucleus is a percomorph or euteleost apomorphy. However, other highly visual teleosts (elopomorphs and clupeomorphs) do not exhibit a well-developed ventral accessory optic nucleus (Ekström 1982; Wullimann and Northcutt unpubl. obs.).

Acanthopterygian teleosts exhibit a distinct large-celled pretectal nucleus termed nucleus corticalis (Figs. 21.2B and C, 21.4A, and 21.7A) that probably receives retinal projections. The somata of nucleus corticalis are among the largest in the pretectum and can easily be confused with the somata of the mesencephalic trigeminal nucleus, particularly in osteoglossomorphs and elopomorphs. The cell bodies of nucleus corticalis form a mediolaterally elongated plate adjacent to the dorsolateral border of the central pretectal nucleus and pars magnocellularis of the superficial pretectal nucleus (Figs. 21.2C and 21.4A), and the dendrites of the corticalis cells extend into the superficial and central gray zone of the optic tectum, where they may form synapses with the entering retinotectal fibers (Sakamoto and Ito 1982). Retinofugal projections are claimed to terminate on the cell bodies of nucleus corticalis in *Carassius* (Sharma 1972; Springer and Gaffney 1981) and other cyprinids (Repérant and Lemire 1976; Repérant et al. 1976); but Springer and Mednick (1985a) and Presson, Fernald, and Max (1985) reported that retinofugal fibers pass through nucleus corticalis without forming synapses in *Astronotus* and *Haplochromis,* respectively. These results are rather surprising, as there is no question in the literature that percomorphs exhibit a distinct nucleus corticalis, whereas Sharma (1972) identified nucleus corticalis in *Carassius* as an indistinct population dorsal to the central pretectal nucleus, and Springer and Gaffney (1981) and, more recently, Fraley and Sharma (1984) identified the nucleus as caudal and ventral to pars magnocellularis of the superficial pretectal nucleus. Braford and Northcutt (1983) failed to identify a nucleus corticalis in *Carassius,* and the present survey (Table 21.2) suggests that a nucleus comparable to the nucleus corticalis of acanthopterygian teleosts does not exist in cypriniform and gymnotoid teleosts. Horseradish peroxidase injections into the optic tectum of the percomorphs *Navodon* (Sakamoto and Ito 1982) and *Lepomis* (Northcutt 1982b) label dendrites of nucleus corticalis, whose axons project ipsilaterally, in association with fibers of the horizontal commissure, before terminating in nucleus glomerulosus (Fig. 21.7A). Tectal injections in ostariophysans (Sligar and Voneida 1976; Grover and Sharma 1981; Luiten 1981), including

*Carassius,* fail to demonstrate a projection to the "glomerular nucleus," but tectal efferents do project via another tract (transverse commissure) to the preglomerular complex, as in many other species (Ebbesson and Vanegas 1976; Northcutt and Butler 1980; Luiten 1981). In fact, a distinct nucleus glomerulosus has not been identified in cyprinids (Braford and Northcutt 1983; Vanegas and Ito 1983). These observations and a cladistic analysis of the distribution of a recognizable nucleus corticalis and nucleus glomerulosus (Table 21.2) suggest that both nuclei are clearly synapomorphic for paracanthopterygians, atherinomorphs, and percomorphs. This conclusion is consistent with Mazzi's observation (1953) that other neoteleosts, in addition to those listed in Table 21.2, exhibit a well-developed nucleus glomerulosus. The possibility that these nuclei may also occur in characiforms suggests that two additional hypotheses must be considered: (1) nucleus corticalis and nucleus glomerulosus may have arisen with euteleosts and may subsequently have been lost in some ostariophysans; or (2) similar nuclei have evolved independently in some ostariophysans and neoteleosts. These hypotheses can be evaluated only after extensive descriptive and experimental examination of the various euteleost clades.

Nucleus glomerulosus (*nucleus rotundus* of Ito and Kishida 1975, 1977) of many percomorph teleosts, such as *Lepomis* and *Navodon,* is one of the most distinct diencephalic nuclei and can be recognized as a bilobed neural aggregate oriented obliquely in the caudal plane (Fig. 21.7). This nucleus exhibits rostral and caudal subdivisions (Brickner 1929; Sakamoto and Ito 1982; Wullimann 1985), which we term pars anterior and pars posterior. The phyletic distribution of a bilobed nucleus glomerulosus is unclear. It may characterize only acanthopterygians, or it may be a neoteleostean apomorphy. In *Navodon* the nucleus glomerulosus consists of four laminae arranged concentrically around a central core of fibers in the following centripetal order: peripheral capsule of fibers and large neurons, small-celled layer, fibrous layer, and glomerular layer (Sakamoto and Ito 1982). Nucleus glomerulosus receives a bilateral input from nucleus corticalis and an ipsilateral input from pars intermedius of the superfical pretectal nucleus; the axons of corticalis neurons terminate exclusively in glomeruli, whereas the axons of intermedius neurons terminate in the small-celled layer of pars anterior or the fibrous layer of pars posterior (Sakamoto and Ito 1982). Both the large and small cells of nucleus glomerulosus are efferent neurons and project to the ipsilateral nucleus diffusus of the inferior lobe (Fig. 21.7A) of the hypothalamus (Sakamoto and Ito 1982). Given these connections of nucleus glomerulosus, it is very likely that it is an important link in visually related circuits from the pretectum and optic tectum to the hypothalamus (Wullimann in press).

## Optic Tectum, Preglomerular Complex, and Nucleus Isthmi

The course of entering retinotectal fibers and their termination in the various tectal laminae have been reviewed in detail for teleosts (Meek 1983; Vanegas 1983; Vanegas and Ito 1983), as well as nonteleost ray-finned fishes (Northcutt 1983). The plesiomorphic pattern of lamination of retinotectal fibers in ray-finned fishes appears to consist of a dorsal lamina located deep in the marginal zone, a central lamina spanning the superficial white and gray zone and extending down into the central zone, and a deep lamina at or immediately below the dorsal border of the deep white zone (Northcutt 1983). Both cyprinids (Repérant and Lemire 1976; Schmidt 1979; Springer and Gaffney 1981) and percomorph teleosts (Vanegas and Ebbesson 1973; Anders and Hibbard 1974; Landreth et al. 1975; Fernald 1982; Springer and Mednick 1985a) exhibit an additional retinotectal terminal lamina located immediately above the periventricular gray zone. It is presently uncertain whether the deep terminal lamina of nonteleost ray-finned fishes is the homologue of the same-named lamina in teleosts or the deeper periventricular lamina or is a field homologue of both. Clearly, teleosts exhibit an apomorphic pattern of retinotectal termination in comparison to other ray-finned fishes, but additional studies are needed to explain its functional significance.

Efferent projections of the optic tectum in ray-finned fishes are known for *Lepisosteus* (Northcutt and Butler 1980), three genera of ostariophysan teleosts (*Astyanax,* Sligar and Voneida 1976; *Carassius,* Grover and Sharma 1979a; *Cyprinus,* Luiten 1981), and two genera of percomorph teleosts (*Eugerres* and *Holocentrus,* Ebbesson and Vanegas 1976). All these studies report ascending projections to the pretectum and diencephalon, projections from one tectal lobe to the other (intertectal), and descending projections to the brain stem. The details of these studies have been reviewed elsewhere (Vanegas and Ito 1983; Northcutt 1983) and will not be repeated here. Instead, we shall focus on the details of two visually related circuits that involve the optic tectum and which may be apomorphic for ray-finned fishes.

In some percomorph teleosts the optic tectum projects ipsilaterally to a large kidney-shaped nucleus located in the rostral diencephalon, sandwiched between the marginal optic tract and the lateral forebrain bundle (Ebbesson and Vanegas 1976; Ito and Vanegas 1983). This nucleus, identified as *nucleus prethalamicus* (Meader 1934), does not receive retinofugal projections in *Holocentrus* (Campbell and Ebbesson 1969) and has been shown to project ipsilaterally to the telencephalon in *Adioryx* (Ito and Vanegas 1984), *Holocentrus* (Ebbesson 1980a; Ito et al. 1980), and *Myripristis* (Ito et al. 1980) and specifically to the lateral zone of area dorsalis telencephali in *Holocentrus* (Ito and Vanegas 1983) and *Sebastiscus* (Murakami, Morita, and Ito 1983). All these studies have noted the similarity in the afferent and efferent connections of *nucleus prethalamicus* of percomorphs and nucleus rotundus of amniotes. *Nucleus prethalamicus,* like nucleus rotundus, does not receive retinofugal projections, but does receive tectal input and projects to the telencephalon. Basing their conclusion on these similarities in connections, but failing to note or comment on the differences in topographical position of *nucleus prethalamicus* and nucleus rotundus (nucleus rotundus is located in the dorsal thalamus of birds and reptiles, whereas *nucleus prethalamicus* is located in the preoptic area of the hypothalamus of percomorphs) Ebbesson (1980a), Ito et al. (1980), and Ito and Vanegas (1983) proposed that *nucleus prethalamicus* of percomorphs is homologous to nucleus rotundus of amniotes. Interpreting the homology of *nucleus prethalamicus,* however, appears to be a small part of a larger problem. The diencephalon of most neopterygians is characterized by a series of nuclei termed the *preglomerular complex* (Braford and Northcutt 1983) that extends through most of the length of the basal diencephalon (Fig. 21.7 and 21.8). Various subdivisions of the preglomerular complex receive higher-order gustatory (Finger 1978b; Morita, Ito, and Masai 1980), lateral line (Finger 1980; Echteler 1984; Murakami, Fukuoka, and Ito 1986), and tectal (Ebbesson and Vanegas 1976; Northcutt and Butler 1980) inputs (Figs. 21.7B and 21.8) and project to the telencephalon (Ito and Kishida 1978; Finger 1980; Echteler and Saidel 1981) and optic tectum (Luiten 1981). Given the close association of *nucleus prethalamicus* to the preglomerular nuclei and the histological properties of *nucleus prethalamicus* (Ito and Vanegas 1983), it may be an apomorphic character of the preglomerular complex of beryciform and scorpaeniform percomorphs. Clearly, additional species among these percomorph categories should be examined for the distribution of a distinct *nucleus prethalamicus,* as well as possible transitional stages.

There are obviously two schools of thought concerning *nucleus prethalamicus* and the preglomerular complex. Members of one school (Ebbesson 1980a; Murakami, Morita, and Ito 1983; Vanegas and Ito 1983) interpret these nuclei as dorsal thalamic, primarily or solely on the basis of connections, and they have not addressed a topography that is peculiar to dorsal thalamic nuclei nor how the dorsal thalamus can be recognized when other diencephalic nuclei receive ascending sensory inputs and project to the telencephalon (Riley and Moore 1981; Ricardo 1981; Köhler, Haglund, and Swanson 1984; Shammah-Lagnado, Negrao, and Ricardo 1985). For example, the zona incerta in mammals, a subthalamic nucleus, receives tectal input (Shammah-Lagnado, Negrao, and Ricardo 1985) and projects to the globus pallidus (Ricardo 1981). The argument would be equally reasonable, based solely on connections, that *nucleus prethalamicus* of teleosts or nucleus rotundus of birds is homologous to the zona incerta of mammals. Proponents of the second school (Northcutt 1981b; Echteler and Saidel 1981; Braford and Northcutt 1983) interpret the nuclei of the preglomerular complex as possible derivatives of the posterior tuberculum, a more ventrally located major division of the diencephalon, which also receives ascending sensory inputs and projects to the telencephalon in other vertebrates. This interpretation is based on (1) the topographical position of the preglomerular complex; (2) the occurrence of other nuclei in the topographical position usually interpreted to be dorsal thalamus in other vertebrates (Braford and Northcutt 1983); (3) the projection of these more dorsally located nuclei, like dorsal thalamic nuclei in other vertebrates, to the telencephalon (Echteler and Saidel 1981; Northcutt 1981a; Murakami, Fukuoka and Ito, 1986); and (4) revealed by examination of cladistians, chondrosteans, and neopterygians, a morphocline from an unmigrated periventricular posterior tuberculum to a posterior tuberculum with complexly migrated nuclei (Braford and Northcutt 1983). The differing views of these two schools can be tested by further hodological and immunohistochemical studies of the nuclei identified as anterior, posterior dorsal, and posterior central thalamic by Braford and Northcutt (1983), as well as hodological experiments on the nucleus of the torus lateralis of cladistians and chondrosteans. The nucleus of the torus lateralis may be a field homologue of both the preglomerular complex and torus lateralis of neopterygians, as cladistians and chondrosteans do not exhibit any migrated nuclei in the rostral basal diencephalon. If experiments on the connections of the nucleus of the torus lateralis do not reveal similarities to those of the preglomerular complex of neopterygians, the simplest interpretation is that the preglomerular complex of neopterygians is an autapomorphy.

There may be similar problems concerning nucleus isthmi, a visually related nucleus of the caudal tegmentum in many teleosts. Nucleus isthmi is a spherical structure consisting of an outer shell of medium-sized neurons surrounding an inner noncellular neuropil (Sakamoto, Ito, and Ueda 1981). In the cyprinids *Carassius* and *Cyprinus,* nucleus isthmi (Fig. 21.8B) receives input from the ipsilateral optic tectum (Grover and Sharma 1979a; Luiten 1981) and projects back onto the same tectal lobe (Grover and Sharma 1981; Luiten 1981), as well as ipsilaterally to the corpus of the cerebellum (Luiten 1981). Identical reciprocal connections between the optic tectum and nucleus isthmi (Fig. 21.7B) have been reported in the percomorph *Navodon* (Sakamoto, Ito and Ueda 1981), and an additional input from pars magnocellularis of the superficial pretectal nucleus is reported in *Navodon* and in another percomorph, *Lepomis* (Striedter and Northcutt 1986), but does not appear to exist in *Carassius* (Northcutt and Braford 1984). Efferents to the ipsilateral corpus of the cerebellum also exist in *Lepomis* (Wullimann and Northcutt 1985). Nucleus isthmi, at least in percomorphs, receives a direct tectal input and an indirect input via pars magnocellularis and projects to the cerebellum, as back onto the optic tectum. At present it is impossible to determine whether the indirect pathway in percomorphs is plesiomorphic or apomorphic.

A similarly named tegmental nucleus occurs in tetrapods, with reciprocal connections to the optic tectum, but additional pretectal or cerebellar connections have not been reported (Gruberg and Udin 1978; Northcutt 1984b). Nucleus isthmi does not appear to exist in cartilaginous fishes (Smeets, Nieuwenhuys, and Roberts 1983), and a distinct nucleus isthmi has not been described in cladistians (Nieuwenhuys 1983; Nieuwenhuys and Oey 1983); chondrosteans have not been examined. It is thus possible that nucleus isthmi arose with the origin of bony fishes and that the same-named nuclei in teleosts and tetrapods are homologous; it is also possible that these nuclei are homoplasous and have evolved independently in actinopterygians and sarcopterygians.

## Retinopetal Projections

Although numerous studies document retinopetal projecting nuclei in various piscine radiations, no clear picture of the plesiomorphic pattern of retinopetal cell groups or their phylogeny has emerged. In *Lampetra* bilateral retinopetal projections with a heavier contralateral component arise from the *medial M5 cell group of Schober,* and a more laterally located *mesencephalic reticular cell group* also projects to the contralateral retina (Vesselkin et al. 1980, 1984). Similar bilateral retinopetal projections are reported from medial and lateral tegmental cell groups termed the *pars superficialis* and *profunda,* respectively, of the isthmo-optic nucleus of the cladistian *Polypterus* (Meyer et al. 1983). Although an isthmo-optic nucleus has not been described in teleosts, it is possible that a retinopetally projecting tegmental nucleus is the plesiomorphic condition for anamniotes; but the phyletic distribution of this nucleus is unknown, as cartilaginous, chondrostean, ginglymodian, and halecomorph fishes have not been examined.

Retinopetal projections have been described for a number of teleosts (Ebbesson and Meyer 1981; Münz and Claas 1981; Münz, Stumpf, and Jennes 1981; Münz et al. 1982; Meyer and Ebbesson 1981; Meyer, Fiebig, and Ebbesson 1981; Uchiyama, Sakamoto, and Ito 1981; Gerwerzhagen et al. 1982; Stell et al. 1982, 1984; Crapon de Caprona and Fritzsch 1983; Demski and Northcutt 1983; Springer 1983; Finck 1984; Ekström 1984; Ito et al. 1984; Uchiyama and Ito 1984; Springer and Mednick 1985a), but the reported projections include a highly variable number of nuclei, ranging from cells in the olfactory bulb to cells in the optic tectum, and no obvious phyletic pattern has been discerned.

In cyprinids and silurids, fibers of the nervus terminalis whose cell bodies are located ventral to or among the axons of the olfactory nerve course through the ventral wall of the forebrain to enter the optic nerve and terminate within the retina (Bass 1981; Demski and Northcutt 1983; Springer 1983). In *Carassius* many of the ganglion cells and retinopetal fibers of the nervus terminalis are luteinizing hormone–releasing hormone (LHRH)–immunoreactive. Similar LHRH-positive cells, termed the *nucleus olfactoretinalis* (Münz, Stumpf, and Jennes 1981; Münz et al. 1982), occur in the olfactory bulb and area ventralis of the telencephalon of many poecilid, centrarchid, and cichlid species, and many of these cells possess axons that terminate in the retina (Ebbesson and Meyer 1981; Ekström 1984; Ito et al. 1984; Springer and Mednick 1985a). The neurons of the ganglion of the nervus terminalis are believed to arise ontogenetically from the olfactory placode (Locy 1899, 1903) and to migrate centrally along the course of the axons of the olfactory nerve. It is thus possible that the cells of *nucleus olfactoretinalis* and nervus terminalis are homologous and have been identified as different populations because of differences in the degree of their central migration.

Retinopetally projecting neurons have also been identified at the level of the optic chiasm as *nucleus intrachiasmaticus* in *Pantodon* (Gerwerzhagen et al. 1982) and the *preoptic retinopetal nucleus* in *Navodon* (Uchiyama, Sakamoto, and Ito 1981; Ito et al. 1984; Uchiyama and Ito 1984), as well as in thalamic and pretectal nuclei (Ebbesson and Meyer 1981; Meyer and Ebbesson 1981; Meyer, Fiebig, and Ebbesson 1981; Ekström 1984; Ito et al. 1984; Springer and Mednick 1985a) and in the optic tectum (Schmidt 1979; Ebbesson and Meyer 1981; Meyer and Ebbesson 1981; Meyer, Fiebig, and Ebbesson 1981). All retinopetal projections in teleosts reportedly terminate in the contralateral retina, and—as in nonteleost fishes—ipsilateral terminations have been reported also in *Astronotus* (Springer and Mednick 1985a) and in *Navodon* and *Sebastiscus* (Ito et al. 1984).

Most of the retinopetally projecting neurons of the thalamus and pretectum are located adjacent to various divisions of the optic tract and have been identified as components of an assortment of nuclei (Ebbesson and Meyer 1981; Meyer and Ebbesson 1981; Ekström 1984) or as a single *nucleus thalamoretinalis* (Springer and Mednick 1985a). Furthermore, there is considerable controversy regarding the sites of some retinopetally projecting neurons in the diencephalon and tectum, as most studies have utilized the retrograde transport of horseradish peroxidase (HRP) to visualize these neurons, and interpretations may be confounded by gliosis following optic nerve transection or transynaptic transport of HRP. The latter is particularly likely when lysolecithin, dimethyl sulfoxide, or both are used to facilitate the uptake of HRP (Ebbesson and Meyer 1981; Meyer and Ebbesson 1981). However, the existence of retinopetally projecting neurons located in the ganglion of nervus terminalis in *Astronotus* and *Carassius* and *nucleus thalamoretinalis* in *Astronotus* has been corroborated by use of the cobaltous lysine technique.

Although the data on the location of retinopetally projecting neurons are scanty, the broad phylogenetic distribution of a nervus terminalis (Demski and Northcutt 1983; Bullock and Northcutt 1984) allows us to pose several hypotheses regarding the phylogeny of these projections: (1) the plesiomorphic pattern for jawed fishes comprised only a retinopetally projecting nervus terminalis; (2) the plesiomorphic pattern comprised a retinopetally projecting nervus terminalis and a more caudally situated isthmo-optic nucleus, as in lampreys and *Polypterus;* or (3) the plesiomorphic pattern comprised a retinopetally projecting nervus terminalis, one or more diencephalic and pretectal nuclei, such as the so-called thalamoretinal nucleus, and an isthmo-optic nucleus. The first hypothesis assumes that the nervus terminalis projects to the retina in members of other gnathostome radiations, as it does in teleosts, that the retinopetally projecting cells of the diencephalon and pretectum are migrated ganglion cells of the nervus terminalis, and that an isthmo-optic nucleus is not homologous in lampreys, cladistians, and tetrapods. The second hypothesis depends on the same assumptions regarding the nervus terminalis and thalamoretinal nucleus and further assumes that an isthmo-optic nucleus is widespread phyletically and homologous. The third hypothesis again assumes that the nervus terminalis projects to the retina in many gnathostome radiations, but also assumes that the thalamoretinal and isthmo-optic nuclei are distinct and homologous phyletic characters across many gnathostome radiations. These hypotheses can be tested only when additional data exist on the embryology and immunohistochemistry of the nervus terminalis and thalamoretinal nucleus, as well as additional experimental studies of retinopetal projections in more phyletically diverse gnathostome species.

In this context, it is surprising that Ebbesson and collaborators cite the distribution of retinopetal projections as evidence supporting the phyletic hypothesis of parcellation (Ebbesson 1980b). Ebbesson and Meyer (1981) suggest that the presence of multiple retinopetal centers in teleosts is a plesiomorphic pattern for vertebrates and that reduction or loss characterizes the so-called advanced vertebrates. However, Meyer et al. (1983) note the occurrence of a single retinopetal center in the cladistian *Polypterus,* as in lampreys (Vesselkin et al. 1980). Unless Ebbesson and collaborators wish to argue that both lampreys and *Polypterus* exhibit degenerate retinopetal projections, they would logically conclude that a single isthmal retinopetal center is plesiomorphic for gnathostomes and that teleosts exhibit an "advanced" condition owing to parcellation. Even this conclusion requires the assumption that all retinopetal projections are homologous among different gnathostome radiations, an assumption clearly not substantiated at present. This example, however, demonstrates the futility of attempts to discern evolutionary processes when the homology and polarity of characters are unknown and when taxa, rather than characters, are assumed to be plesiomorphic or apomorphic.

## Conclusions

Although a considerable number of neurobiologists have focused on the visual system of ray-finned fishes during the past twenty years, it is clear that most of our attention has been directed towards a very few species. Most frequently, cyprinids or cichlids have been examined, and their visual circuitry has been used as a general model for the analysis of axon guidance and other developmental problems, or details of their visual pathways and centers have been compared to those for members of other vertebrate radiations to support contentions regarding the origin and subsequent evolution of these neural characters, without regard for evolutionary diversity and its basis within radiations. These trends are clearly reflected by the paucity of data on numerous groups of ray-finned fishes. There are no descriptive or experimental studies for one of the teleost clades (clupeomorphs) and only one and two experimental studies, respectively, of species in two of the other three clades (osteoglossomorphs and elopomorphs). In each case the species chosen exhibits large numbers of apomorphic characters, which increase the probability that many of the visual characters are also apomorphic. In the case of euteleosts, the largest single clade, less than half of the sister groups have been examined (Fig. 21.1B), and it is difficult or impossible to evaluate the polarity of the visual characters reported in the other euteleost groups.

The literature on visual centers and pathways would support the contention that there are two basic patterns of visual organization among teleosts: a percomorph pattern (Fig. 21.2A–C) and a cyprinid pattern (Fig. 21.2D–F). The percomorph pattern can be characterized by a complexly pleated optic nerve and pars parvocellularis of the superficial pretectal nucleus, an extensive and highly differentiated pars intermedius and magnocellularis of the superfical pretectal nucleus, a distinct nucleus corticalis and bilobed nucleus glomerulosus, and a highly differentiated ventral accessory optic nucleus and nucleus isthmi. In contrast, the cyprinid pattern can be characterized by a simple fasciculated optic nerve and an nonfolded or simply folded pars parvocellularis of the superficial pretectal nucleus, a small or nonexistent pars intermedius of the superficial pretectal nucleus, a small or absent nucleus corticalis and nucleus glomerulosus, and a poorly developed ventral accessory optic nucleus and nucleus isthmi. Given the usual propensity for believing that simple characters are plesiomorphic, one might conclude that the cyprinid pattern is plesiomorphic and the percomorph pattern apomorphic. In fact, neither pattern is plesiomorphic, and a cladistic analysis indicates that both patterns exhibit a mixture of plesiomorphic and apomorphic characters.

As noted earlier, a complexly pleated optic nerve and folded pars parvocellularis of the superficial pretectal nucleus are most parsimoniously interpreted as plesiomorphic characters for actinopterygians and teleosts, respectively, and a fasciculated optic nerve and nonfolded or simply folded pars parvocellularis would then be considered apomorphic characters for cypriniforms, if not ostariophysans. A conclusion regarding ostariophysans is presently impossible, as the visual characters of anotophysans, the sister group of all ostariophysans examined, have not been described. Similarly, the presence of a distinct pars intermedius of the superficial pretectal nucleus is most parsimoniously interpreted as a plesiomorphic character for clupeomorphs and euteleosts, if not teleosts, and its reduction or loss is apomorphic for cyprinids.

The absence or presence of only an indistinct nucleus corticalis and nucleus glomerulosus cannot be evaluated because of the difficulty of recognizing such poorly differentiated nuclei without experimental evidence. In this case, a phyletic analysis reveals gaps in our knowledge and the necessity for new studies of a number of neglected groups of ostariophysans, as well as other teleosts.

The presence of a small ventral accessory optic nucleus in cyprinids is most parsimoniously interpreted as the retention of an actinopterygian plesiomorphic character, whereas the presence of a small, poorly differentiated nucleus isthmi in cyprinids is most parsimoniously interpreted as an apomorphic character, and a well-developed nucleus isthmi appears to be a plesiomorphic character of teleosts. However, there is insufficient evidence to decide whether a nucleus isthmi occurs in other ray-finned fishes; thus it is presently impossible to determine whether the nucleus isthmi of teleosts is homologous to the same-named nucleus in tetrapods. Consideration of many of the apomorphic visual characters of the cyprinid pattern (fasciculated optic nerve, simply folded pars parvocellularis, indistinct or absent pars intermedius, and a poorly developed nucleus isthmi) suggests that cyprinids arose from a more visual ancestor. It is of interest whether reduction in vision occurred with the origin of ostariophysans, the origin of otophysans (was reduction in vision correlated with increase in acoustical abilities?) or with the origin of cypriniforms. Many gymnotoids and siluroids are characterized by relatively small eyes, although some siluroids possess relatively large eyes, so that reduction in vision may characterize the origin of otophysans or may have occurred a number of times independently. Thus the ostariophysan fishes may be an appropriate group in which to examine the effects of reduction and the constraints on redevelopment of this particular sensory system.

Although the present cladistic analysis is hampered by the lack of information on visual characters in a large number of groups of teleosts, it is clear that more than two patterns of visual organization occur (Table 21.2). A complexly folded pars parvocellularis of the superficial pretectal nucleus is not always correlated with a pleated optic nerve. The optic nerves of *Esox* (Schilling and Northcutt 1986) and *Umbra* are fasciculated, but each possesses a complexly folded pars parvocellularis. Conversely, the optic nerves of *Amia, Brevoortia,* and *Gymnothorax* are pleated but each possesses an nonfolded or simply folded pars parvocellularis. In percomorph teleosts, both nucleus intermedius and nucleus corticalis project to a large, well-differentiated nucleus glomerulosus (Fig. 21.7A); on the other hand, clupeomorph teleosts appear to possess a well-developed nucleus intermedius, but a nucleus corticalis and nucleus glomerulosus cannot be recognized (Table 21.2). The efferents of nucleus intermedius presently known are based on HRP injections into nucleus glomerulosus (Sakamoto and Ito 1982). Thus a phyletic analysis would suggest that nucleus intermedius may possess additional efferents that are presently unknown, but it is also possible that we have misidentified this superficial pretectal nucleus in clupeomorphs. Similarly, *Ictalurus* may possess a nucleus corticalis (Table 21.2) but there is no recognizable nucleus intermedius or nucleus glomerulosus. Again, it is possible that nucleus corticalis has additional efferents in many teleosts, that efferents have evolved *de novo* in *Ictalurus,* or that we have misidentified the nucleus. Another notable exception to the suspected distribution of nucleus corticalis occurs in *Osteoglossum.* This genus exhibits a strikingly extensive population of large neurons in the pretectum in a position comparable to that of nucleus corticalis. If this population does represent nucleus corticalis, it would clearly alter our understanding of the origin and evolution of nucleus corticalis. However, it is also possible that this population in *Osteoglossum* is an expanded mesencephalic trigeminal nucleus associated with the spectacular feeding behavior of these animals. It is equally interesting that our survey reveals large "pretectal" nuclei of uncertain identity in *Osteoglossum,* clupeomorphs, esocids, and salmonids. In each case, a well-differentiated nucleus occurs in a position comparable to a pars intermedius, a pars magnocellularis, or an anterolateral portion of the preglomerular complex. In each case, the histology of the mysterious nucleus is sufficiently different from that of the structures described in most euteleosts that it cannot be identified with any certainty. In fact, the histology varies sufficiently among the above taxa to make it unlikely that these nuclei are homologous. We therefore suspect that additional nuclei, of which we are unaware, may exist in this region or that other pretectal nuclei that do not receive retinofugal fibers in percomorphs, such as the posterior pretectal nucleus (Fig. 21.2F), may be more extensive in other teleost groups. In each of the above cases, the accuracy of our interpretations regarding connections and of our speculations regarding homology is really secondary to the realization that phyletic analysis reveals correlations among neural characters, as well as new characters not previously recognized. Regardless of the final details regarding these nuclei and their distribution, it is clear that the pretectum in ray-finned fishes results from, and likely continues to reflect, a series of unique natural experiments in the processing of visual information.

*Acknowledgments* Original research reported herein was supported in part by grants NS-11006 and EY-02485, National Institutes of Health, to R. Glenn Northcutt; Mario Wullimann has been supported by the Janggen-Pöhn-Stiftung and the Swiss National Science Foundation. Mary Sue Northcutt assisted in many phases of the research and in the preparation of the manuscript. We gratefully acknowledge our indebtedness to our colleagues Robert Bernhardt and Stephen Easter for discussing with us many of the topics covered herein. Finally, for providing ceaseless stimulation in the course of many related investigations of vision in fishes, we thank other laboratory members: Richard Puzdrowski, Thomas Schilling, Jiakun Song, and Georg Striedter.

## References

Anders, J.J., and Hibbard, E. (1974) The optic system of the teleost *Cichlasoma biocellatum,* J. Comp. Neurol., 158: 145–154.

Bass, A.H. (1981) Organization of the telencephalon in the channel catfish. *Ictalurus punctatus*. J. Morphol., 169: 71–90.

Braford, M.R., Jr. (1971) Efferent projections of the corpus cerebelli in the goldfish, thesis, Case Western Reserve University, Cleveland.

Braford, M.R., Jr., and Northcutt, R.G. (1983) Organization of the diencephalon and pretectum of the ray-finned fishes, in Fish Neurobiology, vol. 2, Davis, R.E., and Northcutt, R.G. (eds.), University of Michigan Press, Ann Arbor, pp. 117–163.

Brickner, R.M. (1929) A description and interpretation of certain parts of the teleostean midbrain and thalamus, J. Comp. Neurol., 47: 225–282.

Bullock, T.H., and Northcutt, R.G. (1984) Nervus terminalis in dogfish (*Squalus acanthias,* Elasmobranchii) carries tonic efferent impulses, Neurosci. Lett., 44: 155–160.

Bunt, S.M. (1982) Retinotopic and temporal organization of the optic nerve and tracts in the adult goldfish, J. Comp. Neurol., 206: 209–226.

Butler, A.B., and Northcutt, R.G. (1981) Retinal projections in the green sunfish: a case of homoplasy, Soc. Neurosci. Abstr., 7: 84.

Campbell, C.B.G., and Ebbesson, S.O.E. (1969) The optic system of a teleost: *Holocentrus* re-examined, Brain Behav. Evol., 2: 415–430.

Cowan, W.M., and Cuenod, M. (1975) The Use of Axonal Transport for Studies of Neuronal Connectivity, Elsevier, Amsterdam.

Crapon de Caprona, M.D., and Fritzsch, B. (1983) The development of the retinopetal nucleus olfacto-retinalis of two cichlid fish as revealed by horseradish peroxidase, Dev. Brain Res., 11: 281–301.

Demski, L.S., and Northcutt, R.G. (1983) The terminal nerve: a new chemosensory system in vertebrates? Science, 220: 435–437.

Easter, S.S., Rusoff, A.C., and Kish, P.E. (1981) The growth and organization of the optic nerve and tract in juvenile and adult goldfish, J. Neurosci., 1: 793–811.

Ebbesson, S.O.E. (1968) Retinal projections in two teleost fishes (*Opsanus tau* and *Gymnothorax funebris*): an experimental study with silver impregnation methods, Brain Behav. Evol., 1: 134–154.

Ebbesson, S.O.E. (1980b) The parcellation theory and its relation to interspecific variability in brain organization, evolutionary and ontogenetic development and neuronal plasticity, Cell Tissue Res., 213: 179–212.

Ebbesson, S.O.E. (1980a) A visual thalamo-telencephalic pathway in a teleost fish (*Holocentrus rufus*), Cell Tissue Res., 213: 505–508.

Ebbesson, S.O.E., and Ito, H. (1980) Bilateral retinal projections in the black piranha (*Serrasalmus niger*), Cell Tissue Res., 213: 483–495.

Ebbesson, S.O.E., and Meyer, D.L. (1981) Efferents to the retina have multiple sources in teleost fish, Science, 214: 924–926.

Ebbesson, S.O.E., and O'Donnel, D. (1980) Retinal projections in the electric catfish (*Malapterurus electricus*), Cell Tissue Res., 213: 497–503.

Ebbesson, S.O.E., and Vanegas, H. (1976) Projections of the optic tectum in two teleost species, J. Comp. Neurol., 165: 161–180.

Echteler, S.M. (1984) Connections of the auditory midbrain in a teleost fish, *Cyprinus carpio,* J. Comp. Neurol., 230: 536–551.

Echteler, S.M. (1985) Organization of central auditory pathways in a teleost fish, *Cyprinus carpio,* J. Comp. Physiol. A, 156: 267–280.

Echteler, S.M., and Saidel, W.M. (1981) Forebrain connections in the goldfish support telencephalic homologies with land vertebrates, Science, 212: 683–685.

Ekström, P. (1982) Retinofugal projections in the eel, *Anguilla anguilla* L. (Teleostei), visualized by the cobalt-filling technique, Cell Tissue Res., 225: 507–524.

Ekström, P. (1984) Central neural connections of the pineal organ and retina in the teleost *Gasterosteus aculeatus* L., J. Comp. Neurol., 226: 321–335.

Eldredge, N., and Cracraft, J. (1980) Phylogenetic Patterns and the Evolutionary Process, Columbia University Press, New York.

Fernald, R.D. (1980) Optic nerve distention in a cichlid fish, Vision Res., 20: 1015–1019.

Fernald, R.D. (1982) Retinal projections in the African cichlid fish, *Haplochromis burtoni,* J. Comp. Neurol., 206: 379–389.

Fiebig, E., Ebbesson, S.O.E., and Meyer, D.L. (1983) Afferent connections of the optic tectum in the piranha (*Serrasalmus nattereri*), Cell Tissue Res., 231: 55–72.

Finck, W. (1984) Zum optischen System von *Coris julis* (L.) (Labridae, Perciformes), thesis, University of Basel.

Finger, T.E. (1978a) Cerebellar afferents in teleost catfish (Ictaluridae), J. Comp. Neurol., 181: 173–182.

Finger, T.E. (1978b) Gustatory pathways in the bullhead catfish. II. Facial lobe connections, J. Comp. Neurol., 180: 691–706.

Finger, T.E. (1980) Nonolfactory sensory pathway to the telencephalon in a teleost fish, Science, 210: 671–673.

Finger, T.E., and Karten, H.J. (1978) The accessory optic system in teleosts, Brain Res., 153: 144–149.

Fraley, S.M., and Sharma, S.C. (1984) Topography of retinal axons in the diencephalon of goldfish, Cell Tissue Res., 238: 529–538.

Gerwerzhagen, K., Rickmann, M.J., Meyer, D.L., and Ebbesson, S.O.E. (1982) Optic tract cells projecting to the retina in the teleost, *Pantodon buchholzi,* Cell Tissue Res., 225: 23–28.

Gould, S.J., and Vrba, E.S. (1982) Exaptation—a missing term in the science of form, Paleobiology, 8: 4–15.

Grover, B.G., and Sharma, S.C. (1979a) Tectal projections in the goldfish (*Carassius auratus*): a degeneration study, J. Comp. Neurol., 184: 435–454.

Grover, B.G., and Sharma, S.C. (1979b) A retino-thalamo-telencephalic system in goldfish, Invest. Ophthalmol. Suppl., 18: 156.

Grover, B.G., and Sharma, S.C. (1981) Organization of extrinsic tectal connections in goldfish (*Carassius auratus*), J. Comp. Neurol., 196: 471–488.

Gruberg, E.R., and Udin, S.B. (1978) Topographic projections between the nucleus isthmi and the tectum of the frog *Rana pipiens,* J. Comp. Neurol., 179: 487–500.

Gulley, R.L., Cochran, M., and Ebbesson, S.O.E. (1975) The visual connections of the adult flatfish, *Achirus lineatus,* J. Comp. Neurol., 162: 309–320.

Heimer, L., and RoBards, M.J. (1981) Neuroanatomical Tract-Tracing Methods, Plenum, New York.

Hennig, W. (1966) Phylogenetic Systematics, translated by Davis, D.D., and Zangerl, R., University of Illinois Press, Urbana.

Ito, H., and Kishida, R. (1975) Organization of the teleostean nucleus rotundus, J. Morphol., 147: 89–108.

Ito, H., and Kishida, R. (1977) Synaptic organization of the nucleus rotundus in some teleosts, J. Morphol., 151: 397–418.

Ito, H., and Kishida, R. (1978) Telencephalic afferent neurons identified by the retrograde HRP method in the carp diencephalon, Brain Res., 149: 211–215.

Ito, H., Morita, Y., Sakamoto, N., and Ueda, S. (1980) Possibility of telencephalic visual projection in teleosts, Holocentridae, Brain Res., 197: 219–222.

Ito, H., and Murakami, T. (1984) Retinal ganglion cells in two teleost species, *Sebastiscus marmoratus* and *Navodon modestus*, J. Comp. Neurol., 229: 80–96.

Ito, H., Sakamoto, N., and Takatsuji, K. (1982) Cytoarchitecture, fiber connections and ultrastructure of nucleus isthmi in a teleost (*Navodon modestus*) with special reference to degenerating isthmic afferents from optic tectum and nucleus pretectalis, J. Comp. Neurol., 205: 299–311.

Ito, H., Tanaka, H., Sakamoto, N., and Morita, Y. (1981) Isthmic afferent neurons identified by the retrograde HRP method in a teleost, *Navodon modestus*, Brain Res., 207: 163–169.

Ito, H., and Vanegas, H. (1983) Cytoarchitecture and ultrastructure of nucleus prethalamicus, with special reference to degenerating afferents from optic tectum and telencephalon, in a teleost (*Holocentrus ascensionis*), J. Comp. Neurol., 221: 401–415.

Ito, H., and Vanegas, H. (1984) Visual receptive thalamopetal neurons in the optic tectum of teleosts (Holocentridae), Brain Res., 290: 201–210.

Ito, H., Vanegas, H., Murakami, T., and Morita, Y. (1984) Diameters and terminal patterns of retinofugal axons in their target areas: an HRP study in two teleosts (*Sebastiscus* and *Navodon*), J. Comp. Neurol., 230: 179–197.

Köhler, C., Haglund, L., and Swanson, S.W. (1984) A diffuse ÁMSH-immunoreactive projection to the hippocampus and spinal cord from individual neurons in the lateral hypothalamic area and zona incerta, J. Comp. Neurol., 223: 501–514.

Landreth, G.E., Neale, E.A., Neale, J.H., Duff, R.S., Braford, M.R., Jr., Northcutt, R.G., and Agranoff, B.W. (1975) Evaluation of [$^3$H]proline for radioautographic tracing of axonal projections in the teleost visual system, Brain Res., 91: 25–42.

Lauder, G.V., and Liem, K.F. (1983a) The evolution and interrelationships of the actinopterygian fishes, Bull. Mus. Comp. Zool., 150: 95–197.

Lauder, G.V., and Liem, K.F. (1983b) Patterns of diversity and evolution in ray-finned fishes, in Fish Neurobiology, vol. 1, Northcutt, R.G., and Davis R.E. (eds.), University of Michigan Press, Ann Arbor, pp. 1–24.

Lázár, G., Libouban, S., and Szabo, T. (1984) The mormyrid mesencephalon. III. Retinal projections in a weakly electric fish, *Gnathonemus petersii*, J. Comp. Neurol., 230: 1–12.

Lemire, M., and Repérant, J. (1976) Analyse radio-autographique des projections visuelles primaires chez la truite, *Salmo irideus* Gibb (comparaison avec quelques Téléostéens d'eau douce), C.R. Acad. Sci. Paris, 283 (Sér. D): 951–954.

Locy, W.A. (1899) New facts regarding the development of the olfactory nerve, Anat. Anz., 16: 273–290.

Locy, W.A. (1903) A new cranial nerve in selachians, in Mark Anniversary Volume, Parker, G.H. (ed), Henry Holt, New York, pp. 39–55.

Luiten, P.G.M. (1981) Afferent and efferent connections of the optic tectum in the carp (*Cyprinus carpio* L.), Brain Res., 220: 51–65.

Mazzi, V. (1953) Il corpo glomerulare del talamo degli attinotterigi, Publ. Stn. Zool. Napoli, 24: 373–433.

Meader, R.G. (1934) The optic system of the teleost, *Holocentrus*. I. The primary optic pathways and the corpus geniculatum complex, J. Comp. Neurol., 60: 361–407.

Meek, J. (1983) Functional anatomy of the tectum mesencephali of the goldfish, Brain Res. Rev., 6: 247–297.

Meyer, D.L., and Ebbesson, S.O.E. (1981) Retinofugal and retinopetal connections in the upside-down catfish (*Synodontis nigriventris*), Cell Tissue Res., 218: 389–401.

Meyer, D.L., Fiebig, E., and Ebbesson, S.O.E. (1981) A note on the reciprocal connections between the retina and the brain in the puffer fish *Tetraodon fluviatilis*, Neurosci. Lett., 23: 111–115.

Meyer, D.L., Gerwerzhagen, K., Fiebig, E., Ahlswede, F., and Ebbesson, S.O.E. (1983) An isthmo-optic system in a bony fish, Cell Tissue Res., 231: 129–133.

Morita, Y., Ito, H., and Masai, H. (1980) Central gustatory paths in the crucian carp *Carassius carassius*, J. Comp. Neurol., 191: 119–132.

Münz, H., and Claas, B. (1981) Centrifugal innervation of the retina in cichlid and poecilid fishes: a horseradish peroxidase study. Neurosci Lett 22: 223–226.

Münz H., Claas B., Stumpf W.E., Jennes L. (1982) Centrifugal innervation of the retina by luteinizing hormone releasing hormone (LHRH)–immunoreactive telencephalic neurons in teleostean fishes, Cell Tissue Res., 222: 313–323.

Münz H., Stumpf W.E., Jennes L. (1981) LHRH systems in the brain of platyfish. Brain Res 22: 1–13.

Murakami, T., Fukuoka, T., and Ito, H. (1986) Telencephalic ascending acousticolateral system in a teleost (*Sebastiscus marmoratus*), with special reference to the fiber connections of the nucleus preglomerulosus, J. Comp. Neurol., 247: 383–397.

Murakami, T., Morita, Y., and Ito, H. (1983) Extrinsic and intrinsic fiber connections of the telencephalon in a teleost, *Sebastiscus marmoratus*, J. Comp. Neurol., 216: 115–131.

Nauta, W.J.H., and Ebbesson, S.O.E. (1970) Contemporary Research Methods in Neuroanatomy, Springer-Verlag, New York.

Nieuwenhuys, R. (1963) The comparative anatomy of the actinopterygian forebrain, J. Hirnforsch., 6: 171–192.

Nieuwenhuys, R. (1983) The central nervous system of the brachiopterygian fish *Erpetoichthys calabaricus*, J. Hirnforsch., 24: 501–533.

Nieuwenhuys, R., and Oey, P.L. (1983) Topological analysis of the brain stem of the reedfish, *Erpetoichthys calabaricus*, J. Comp. Neurol., 213: 220–232.

Northcutt, R.G. (1981b) Evolution of the telencephalon in nonmammals, Annu. Rev. Neurosci., 4: 301–350.

Northcutt, R.G. (1981a) Localization of neurons afferent to the telencephalon in a primitive bony fish, *Polypterus palmas*, Neurosci. Lett., 22: 219–222.

Northcutt, R.G. (1982a) Localization of neurons afferent to the optic tectum in longnose gars, J. Comp. Neurol., 204: 325–335.

Northcutt, R.G. (1982b) Cells of origin of pathways afferent to the optic tectum in the green sunfish, *Lepomis cyanellus*, Invest. Ophthalmol. & Visual Sci. Suppl., 22: 245.

Northcutt, R.G. (1982c) Cerebellar afferents in the longnose gar (Holostei), Soc. Neurosci. Abstr., 8: 762.

Northcutt, R.G. (1983) Evolution of the optic tectum in ray-finned fishes, in: Fish Neurobiology, vol. 2, Davis, R.E., and Northcutt, R.G. (eds.), University of Michigan Press, Ann Arbor, pp. 1–42.

Northcutt, R.G. (1984a) Evolution of the vertebrate central nervous system: patterns and processes, Am. Zool., 24: 701–716.

Northcutt, R.G. (1984b) Anatomical organization of the optic tectum in reptiles, in Comparative Neurology of the Optic Tectum, Vanegas, H. (ed.), Plenum, New York, pp. 547–600.

Northcutt, R.G. (1985) Brain phylogeny: speculations on pattern and cause, in Comparative Neurobiology: Modes of Communication in the Nervous System, Cohen, M.J., and Strumwasser, F. (eds.), Wiley, New York, pp. 351–378.

Northcutt, R.G. (1986) Strategies of comparison: how do we study brain evolution? Verh. Dtsch. Zool. Ges., 79:91–104.

Northcutt, R.G., and Braford, M.R., Jr. (1980) New observations on the organization and evolution of the telencephalon of actinopterygian fishes, in Comparative Neurology of the Telencephalon, Ebbesson, S.O.E. (ed.), Plenum, New York, pp. 41–98.

Northcutt, R.G., and Braford, M.R., Jr. (1984) Some efferent connections of the superficial pretectum in the goldfish, Brain Res., 296: 181–184.

Northcutt, R.G., and Butler, A.B. (1976) Retinofugal pathways in the longnose gar, *Lepisosteus osseus* (Linnaeus), J. Comp. Neurol., 166: 1–16.

Northcutt, R.G., and Butler, A.B. (1980) Projections of the optic tectum in the longnose gar, *Lepisosteus osseus,* Brain Res., 190: 333–346.

Northcutt, R.G., and Davis, R.E. (1983) Telencephalic organization in ray-finned fishes, in Fish Neurobiology, vol. 2, Davis, R.E., and Northcutt, R.G. (eds.), University of Michigan Press, Ann Arbor, pp. 203–236.

Peyrichoux, J., Weidner, C., Repérant, J., and Miceli, D. (1977) An experimental study of the visual system of cyprinid fish using the HRP method, Brain Res., 130: 531–537.

Pinganaud, G., and Clairambault, P. (1975) Sur l'architecture des voies et centres optiques primaires de la truite (*Salmo irideus* Gibb), C.R. Acad. Sci. Paris, 281 Sér. D: 411–414.

Pinganaud, G., and Clairambault, P. (1979) The visual system of the trout *Salmo irideus* Gibb: a degeneration and radioautographic study, J. Hirnforsch., 20: 413–431.

Prasada Rao, P.D., and Sharma, S.C. (1982) Retinofugal pathways in juvenile and adult channel catfish, *Ictalurus* (*Ameiurus*) *punctatus*: an HRP and autoradiographic study, J. Comp. Neurol., 210: 37–48.

Presson, J., Fernald, R.D., and Max, M. (1985) The organization of retinal projections to the diencephalon and pretectum in the cichlid fish, *Haplochromis burtoni,* J. Comp. Neurol., 235: 360–374.

Repérant, J., and Lemire, M. (1976) Retinal projections in cyprinid fishes: a degeneration and radioautographic study, Brain Behav. Evol., 13: 34–57.

Repérant, J., Lemire, M., Miceli, D., and Peyrichoux, J. (1976) A radioautographic study of the visual system in fresh water teleosts following intraocular injection of tritiated fucose and proline, Brain Res., 118: 123–131.

Repérant, J., Rio, J.P., and Amouzou, M. (1979) Analyse radio-autographique des projections rétiniennes chez le poisson osseux/primitif *Polypterus senegalus* C, C.R. Acad. Sci. Paris, 289 Sér. D: 947–950.

Repérant, J., Rio, J.P., Miceli, D., Amouzou, M., and Peyrichoux, J. (1981) The retinofugal pathways in the primitive African bony fish *Polypterus senegalus* (Cuvier, 1829), Brain Res., 217: 225–243.

Repérant, J., Vesselkin, N.P., Ermakova, T.V., Rustamov, E.K., Rio, J.P., Palatnikov, G.K., Peyrichoux, J., and Kasimov, R.V. (1982) The retinofugal pathways in a primitive actinopterygian, the chondrostean *Acipenser guldenstadti:* an experimen-

tal study using degeneration, radioautographic and HRP methods, Brain Res., 251: 1–23.
Ricardo, J.A. (1981) Efferent connections of the subthalamic region in the rat. II. The zona incerta, Brain Res., 214: 43–60.
Riley, J.N., and Moore, R.Y. (1981) Diencephalic and brainstem afferents to the hippocampal formation of the rat, Brain Res. Bull., 6: 437–444.
Roth, R.L. (1969) Optic tract projections in representatives of two fresh-water teleost families, Anat. Rec., 163: 253–254.
Rusoff, A.C. (1984) Paths of axons in the visual system of perciform fish and implications of these paths for rules governing axonal growth, J. Neurosci., 4: 1414–1428.
Rusoff, A.C., and Easter, S.S. (1980) Order in the optic nerve of goldfish, Science, 208: 311–312.
Sakamoto, N., and Ito, H. (1982) Fiber connections of the corpus glomerulosum in a teleost, *Navodon modestus,* J. Comp. Neurol., 205: 291–298.
Sakamoto, N., Ito, H., and Ueda, S. (1981) Topographic projections between the nucleus isthmi and the optic tectum in a teleost, *Navodon modestus,* Brain Res., 224: 225–234.
Sandemann, D.C., and Rosenthal, N.P. (1974) Efferent axons in the fish optic nerve and their effect on the retinal ganglion cells, Brain Res., 68: 41–54.
Schilling, T.F., and Northcutt, R.G. (1986) A cladistic analysis of the euteleost pretectum, Soc. Neurosci. Abstr., 12:104.
Schmidt, J.T. (1979) The laminar organization of optic nerve fibres in the tectum of goldfish, Proc. R. Soc. Lond. Biol. Sci., 205: 287–306.
Schnitzlein, H.N. (1962) The habenula and the dorsal thalamus of some teleosts, J.Comp. Neurol., 118: 225–268.
Scholes, J.H. (1979) Nerve fibre topography in the retinal projection to the tectum, Nature, 278: 620–624.
Shammah-Lagnado, S.J., Negrao, N., and Ricardo, J.A. (1985) Afferent connections of the zona incerta: a horseradish peroxidase study in the rat, Neuroscience, 15: 109–134.
Sharma, S.C. (1972) The retinal projections in the goldfish: an experimental study, Brain Res., 39: 213–223.
Sligar, C.M., and Voneida, T.J. (1976) Tectal efferents in the blind cave fish *Astyanax hubbsi,* J. Comp. Neurol., 165: 107–124.
Smeets, W.J.A.J., Nieuwenhuys, R., and Roberts, B.L. (1983) The central nervous system of cartilaginous fishes, Springer-Verlag, Berlin.
Springer, A.D. (1983) Centrifugal innervation of goldfish retina from ganglion cells of the nervus terminalis, J. Comp. Neurol., 214: 404–415.
Springer, A.D., and Gaffney, J.S. (1981) Retinal projections in the goldfish: a study using cobaltous-lysine, J. Comp. Neurol., 203: 401–424.
Springer, A.D., and Mednick, A.S. (1985a) Retinofugal and retinopetal projections in the cichlid fish *Astronotus ocellatus,* J. Comp. Neurol., 236: 179–196.
Springer, A.D., and Mednick, A.S. (1985b) Topography of the retinal projections to the superficial pretectal parvicellular nucleus of goldfish: A cobaltous-lysine study, J. Comp. Neurol., 237:239-250.
Stell, W.K., Chohan, K.S., Lam, D.M.K., and Kozlowski, G.P. (1982) Luteinizing hormone–releasing hormone (LHRH)–immunoreactive fibers in goldfish retina, Invest. Ophthalmol. & Visual Sci., Suppl. 22: 278.
Stell, W.K., Walker, S.E., Kuldip, S., Chohan, K.S., and Ball, A.K. (1984) The goldfish nervus terminalis: a luteinizing hormone–releasing hormone and a molluscan cardioex-

citatory peptide immunoreactive olfactoretinal pathway, Proc. Nat. Acad. Sci. USA, 81: 940–944.

Striedter, G., and Northcutt, R.G. (1986) Connections of the superficial pretectum and nucleus isthmi in the sunfish *Lepomis cyanellus,* Soc. Neurosci. Abstr., 12:103.

Tapp, R.L. (1973) The structure of the optic nerve of the teleost: *Eugerres plumieri,* J. Comp. Neurol., 150: 239–252.

Tapp, R.L. (1974) Axon numbers and distribution, myelin thickness, and the reconstruction of the compound action potential in the optic nerve of the teleost: *Eugerres plumieri,* J. Comp. Neurol., 153: 267–274.

Uchiyama, H., and Ito, H. (1984) Fiber connections and synaptic organization of the preoptic retinopetal nucleus in the filefish (Balistidae, Teleostei), Brain Res., 298: 11–24.

Uchiyama, H., Sakamoto, N., and Ito, H. (1981) A retinopetal nucleus in the preoptic area in a teleost, *Navodon modestus,* Brain Res., 222: 119–124.

Vanegas, H. (1983) Organization and physiology of the teleostean optic tectum, in Fish Neurobiology, vol. 2, Davis, R.E., and Northcutt, R.G. (eds.), University of Michigan Press, Ann Arbor, pp. 43–90.

Vanegas, H., Amat, J., and Essayag-Millán, E. (1973) Electrophysiological evidence of tectal efferents to the fish eye, Brain Res., 54: 309–313.

Vanegas, H., and Ebbesson, S.O.E. (1973) Retinal projections in the perch-like teleost *Eugerres plumieri,* J. Comp. Neurol., 151: 331–358.

Vanegas, H., and Ebbesson, S.O.E. (1976) Telencephalic projections in two teleost species, J. Comp. Neurol., 165: 181–196.

Vanegas, H., Essayag-Millán, E., and Laufer, M. (1971) Response of the optic tectum to stimulation of the optic nerve in the teleost *Eugerres plumieri,* Brain Res., 31: 107–118.

Vanegas, H., and Ito, H. (1983) Morphological aspects of the teleostean visual system: a review, Brain Res. Rev., 6: 117–137.

Vesselkin, N.P., Ermakova, T.V., Repérant, J., Kosareva, A.A., and Kenigfest, N.B. (1980) The retinofugal and retinopetal systems in *Lampetra fluviatilis:* an experimental study using radioautographic and HRP methods, Brain Res., 195: 453–460.

Vesselkin, N.P., Repérant, J., Kenigfest, N.B., Miceli, D., Ermakova, T.V., and Rio, J.P. (1984) an anatomical and electrophysiological study of the centrifugal visual system in the lamprey (*Lampetra fluviatilis*), Brain Res., 292: 41–56.

Voneida, T.J., and Sligar, C.M. (1976) A comparative neuroanatomic study of retinal projections in two fishes: *Astyanax hubbsi* (the blind cave fish), and *Astyanax mexicanus,* J. Comp. Neurol., 165: 89–106.

Wiley, E.O. (1981) Phylogenetics, Wiley, New York.

Wullimann, M.F. (1985) Zum Gehirn von *Rhinecanthus aculeatus* (Linnaeus, 1758) (Balistidae, Tetraodontiformes, Teleostei), thesis, University of Basel.

Wullimann, M.F. (in press) The hypothalamic, ventricular channel-system and its phylogenetic distribution among fishes, Proc. V Congr. Europ. Ichthyol.

Wullimann, M.F., and Northcutt, R.G. (1985) Cerebellar afferents in the green sunfish, *Lepomis cyanellus,* Soc. Neurosci. Abstr., 11: 1310.

Part IIIC

# Mechanoreception

For most animals, the environment is in constant motion. Not only do almost all animals at one time or another move in a fluid medium, they also cause fluid motion. In addition, the medium itself moves at many size scales. Fluid motion is characterized by flow and compression. The relationship of flow and compression in space and time is described mathematically by the wave equation by Kalmijn (Chapter 3) and by Rogers and Cox (Chapter 4). What is commonly called ''sound'' is a special aspect of the propagation of mechanical energy in a medium; it depends on elastic properties of the medium. Hydrodynamic flow fields are described by ignoring these elastic properties. Hydrodynamics and acoustics can be considered separately under many conditions of biological importance. Different receptor organs exist in fish and aquatic invertebrates to respond preferentially to one or another aspect of the mechanical disturbance.

Three distinct physical conditions are (1) *local shearfields* that can be measured as differential flow by an array of mechanoreceptors across the body surface; (2) *local laminar flow* causing whole-animal movements that can be measured—depending on the detector's frequency response—as displacement, velocity, or acceleration by inertial detectors; and (3) *compressions* of the medium such as hydrostatic pressure and sound. For detection principles, far-field displacement components of acoustic compression waves can be considered as (oscillating) laminar flow. In fishes, the relevant receptor organs include the mechanoreceptive lateral line for shearfield detection, the inner ear for body movement detection, and the swimbladder (when present) for pressure detection. In all cases, elaborate specializations are known among different species. In addition, animals organize their movements and position in space with equilibrium organs. Such organs extract the direction of the gravity force and help the animal coordinate its body movements. In fishes, equilibrium functions are ascribed to the semicircular canals and the otolith organs of the inner ear, while the semicircular canals detect body movements with rotational components. Functionally and morphologically similar organs are found for each of these functions across the invertebrates.

For practical and historical reasons, Part IIIC is divided into three subsections dealing with hydrodynamic, acoustic, and equilibrium senses, respectively. Despite the physical and functional differences in these major mechanoreceptor systems, in vertebrates all three are based on a single morphological and physiological unit, the sensory hair cell. Invertebrates use hair cells as well, or they use functionally similiar mechanoreceptive cuticular hair sensilla.

Part IIIC-1

# Hydrodynamic Reception

In fishes, the lateral line consists of the free neuromast system and the canal system. In Chapter 22, Coombs, Janssen, and Webb organize the bewildering variety of canals and free neuromast distributions found in the diverse class of fishes into a small number of general patterns. They focus on phylogenetic relationships as a constraint on ecological adaptation.

While the physiology of hair cells is rather well studied in some sensory systems, the adequate stimulus for the lateral line organ is still poorly known. Denton and Gray (Chapter 23) use some examples of lateral line specialization to provide new insights into the mechanical coupling of hydrodynamic flow fields and the motion of the cupula and the hair cells. Their chapter complements the physical descriptions of Kalmijn (Chapter 3) and Rogers and Cox (Chapter 4). Another specialized function of the lateral line is the detection of surface waves. Bleckmann (Chapter 24) addresses this case and makes a comparison between fish and semiaquatic spiders. Wilcox (Chapter 25) analyzes surface wave detection and compares aquatic insects, fishes, and amphibians. Wiese (Chapter 26) reviews CNS processing of hydrodynamic information in crayfish, one of the best known aquatic models. Sensory principles of feature extraction described for crayfish may be useful in future studies of lateral line information processing. Lateral line projections in the CNS are described by McCormick and Braford (Chapter 29).

# Chapter 22

# Diversity of Lateral Line Systems: Evolutionary and Functional Considerations

SHERYL COOMBS*, JOHN JANSSEN,* and JACQUELINE F. WEBB**

The lateral line system is the only sensory system unique to aquatic vertebrates, being restricted to fish and some aquatic amphibians. It contains at least two, and probably three, different receptor classes: ciliated mechanoreceptors, electroreceptors, and a largely unknown group of receptors called the *vesicles of Savi,* found only in certain elasmobranch fishes (Garman 1888; Coggi 1891; Norris 1932). While all these organs are innervated by branches of the lateral line nerve and may have common embryological origins (Coggi 1891; Norris 1932; Northcutt 1985), it is quite clear that ciliated mechanoreceptors and electroreceptors respond to different stimuli and that their CNS projections are segregated at least at the level of the brain stem (McCormick 1983; Northcutt 1985). Thus, the concept of a single lateral line system has given way to one of at least two different sensory systems, with emerging evidence of anatomical functional variability within each of these.

Variation in the structure of lateral line mechanoreceptive organs has been documented in numerous fields of biology, ranging from systematics to embryology. Not since Dijkgraaf (1962), however, has there been a major synthesis of this variation, as there has for electroreceptive organs (Bullock 1982; Hopkins 1983). Consequently, little is known about the evolutionary and functional significance of variation in the mechanoreceptive lateral line system. The primary goals of this chapter are to (1) describe the basic anatomy and distribution pattern of mechanoreceptive lateral line organs in fish, (2) identify major dimensions of anatomical variability, and (3) place this variation in both an evolutionary and functional context.

---

* Parmly Hearing Institute and Department of Biology, Loyola University, Chicago, IL 60626, USA
** Department of Biology, Boston University, Boston, MA 02215, USA

## Basic Anatomy of Mechanoreceptive Lateral Line System

The neuromast is the most typical mechanoreceptive end organ found in the lateral line system of all fish, but a single spiracular organ restricted to the spiracular gill is also found in subteleostean fishes (Jarvik 1980). Although both end organs are characterized by a hair cell epithelium, the spiracular organ has an organization and function believed to be significantly different from that of the neuromast (Barry and Boord 1984) and will not be considered further in this chapter.

The hair cell epithelium of the neuromast is surrounded by two types of nonsensory cells: support cells dispersed between the hair cells and mantle cells at the periphery (Münz 1979) (Figs. 22.1 and 22.2). Overlying the neuromast is a bilayered gelatinous cupula (Fig. 22.2C), with an inner core presumably secreted by support cells and an outer core by mantle cells (Münz 1979). Projecting into the cupula are ciliary bundles from the apical ends of hair cells (Figs. 22.1 and 22.2D). Each ciliary bundle consists of several stereocilia of graded lengths, arranged along the cell from shortest to tallest, with the tallest being adjacent to a single kinocilium characterized by a 9 + 2 mictrotubule configuration (Flock

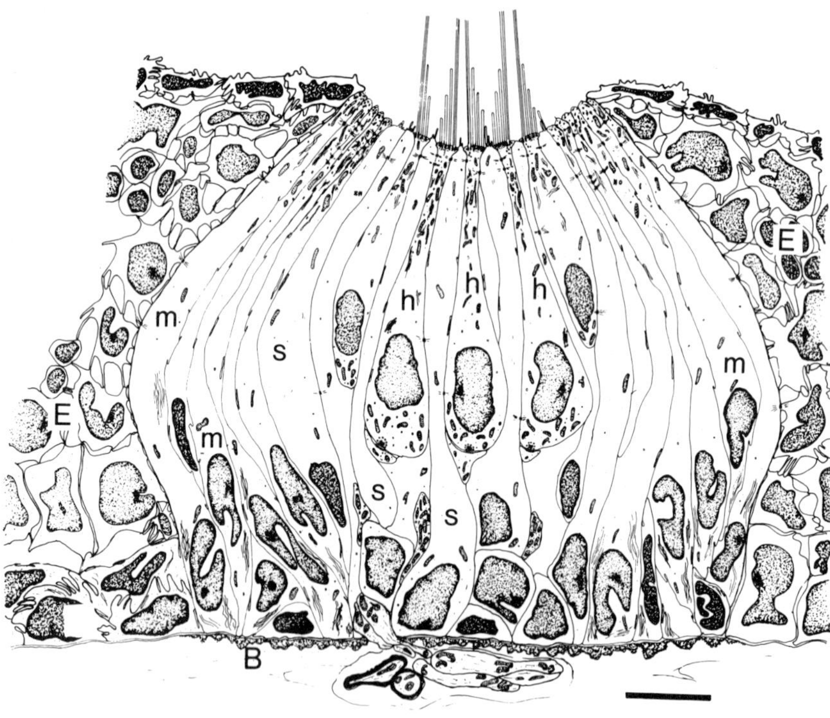


Figure 22.1. Schematic cross section through lateral line neuromast from cichlid *Sarotherodon niloticus* showing the epidermis (*E*); mantle cells (*m*); support cells (*s*); hair cells (*h*); and basement membrane (*B*). Scale bar: 5μm. (From Münz 1979.)

1965). The directional response property of the hair cell corresponds to this anatomical polarization such that bending of the stereocilia towards the kinocilium results in depolarization of the cell, whereas bending in the opposite direction results in hyperpolarization (Flock 1965). In all neuromasts that have been examined, the anatomical polarization of hair cells is in one of two opposing directions (Flock 1965; Best and Gray 1982; Janssen et al. in press), resulting in a single axis of best sensitivity for any given neuromast (Fig. 22.2D).

## Variations in Canal Structure

Neuromasts are distributed along the head and trunk of fish, either superficially on the skin, or in grooves and canals (Fig. 22.2D). The structure and size of head canals varies considerably among fishes. In most elasmobranch fishes, head canals generally lie in shallow troughs in the neurocranium and are embedded in the compact layer of the dermis (Tester and Kendall 1969). In the Holocephali, cartilaginous rings support open, groovelike canals (Fig. 22.3C), which are quite broad and deep relative to trunk canals (von Lubitz 1981). In the Osteichthyes, head canals are partially or completely enclosed in dermal bone.

Among the bony fishes, head canal diameters vary from approximately 100 μm to 7 mm (Chap. 23), and canal width is correlated with various structural specializations. In general, narrow canals on the head are completely enclosed by bone perforated periodically by pores (Jakubowski 1966b, 1967b). Wider canals, found in a wide variety of teleostean taxa (Fig. 22.4 and Table 22.1), are usually covered by an elastic connective tissue, reinforced by bone only in the area of each neuromast (Jakubowski 1965, 1967a). Compartmentalization of wide canals is frequently achieved via narrow bony constrictions or other, nonbony structures, such as membranous lamellae (Fig. 22.4) or ridges protruding from the roof of the canal in the area of the neuromast (Moore and Burris 1956; Hoyt 1972; Schemmel 1977).

In most groups of fish, trunk canals are contained within or between modified lateral line scales, and among teleosts, lateral line scales modified to contain trunk canals are of several types. Scales in the steelhead trout, *Salmo gairdneri* (Salmonidae) have tubes that enclose a portion of the canal. The tubes have two suprascalar openings, one of which overlaps an infrascalar opening on the adjacent scale, to form continuous support for the canal, and the other of which opens to the environment (Fig. 22.3B). Other lateral line scales may be reduced to simple tubes that abut one another (Fig. 22.3D). In small scaled forms, like some pricklebacks (Stichaeidae), the trunk canal is not contained within a single scale but is enveloped by modified scales on two sides (Makushok 1961). Other scale modifications on the trunk include heavy sculptured scales called *scutes,* which support the caudal portion of the canal in tuna and mackerel (Scombridae) and cardioid scales in pike (Esocidae), which form an open groove for trunk neuromasts (Fig. 22.3E). Although the absence of the trunk canal has been correlated with the absence of scales (Webb in preparation), a trunk canal and its development has

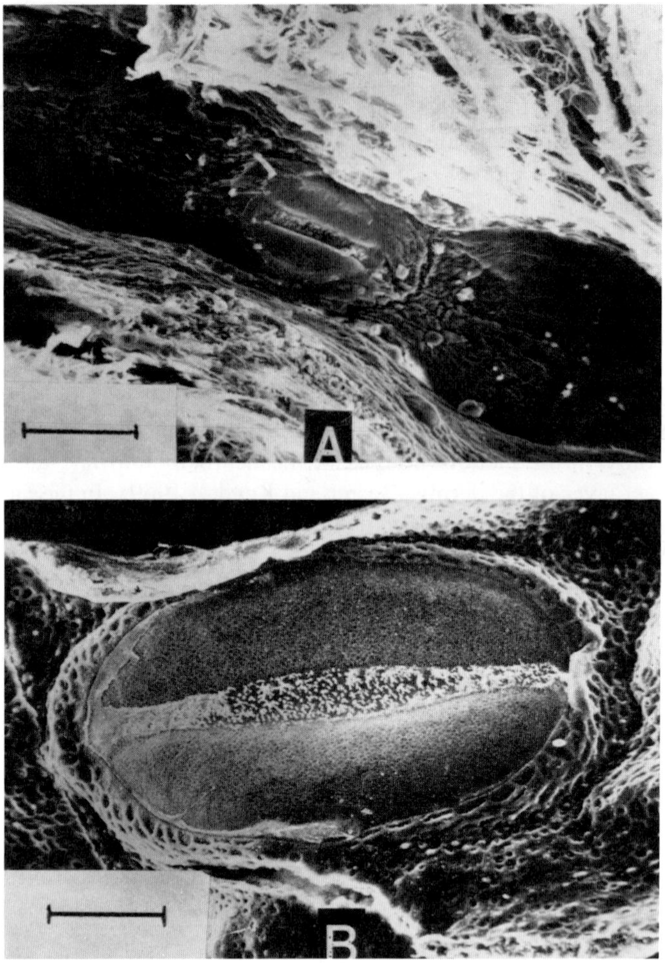


Figure 22.2 Scanning electron micrographs of neuromasts in *Cottus bairdi* (Cottidae) (from Janssen et al. in preparation) showing (A) the position of a neuromast within trunk canal; (B) the strip of hair cell epithelium flanked by mantle cells on both sides; (C) partial cupula; and (D) the orientation of ciliary bundles along canal axis. Scale bar: 100 μm (A–C), 0.1 μm (D).

recently been documented in the scaleless *Cottus bairdi* (Cottidae) (Fig. 22.5) (Janssen et al. in press).

In most fishes, lateral line canals open to the environment through a series of pores, which vary in shape and size and may be nearly flush with the surface of the canal or sit at the end of tubules or canaliculi (Jakubowski 1966c) extending from the canal. In carcharinid sharks, tubules are located beneath adjacent crowns of scales notched to frame openings for pores at the tips of tubules (Tester and

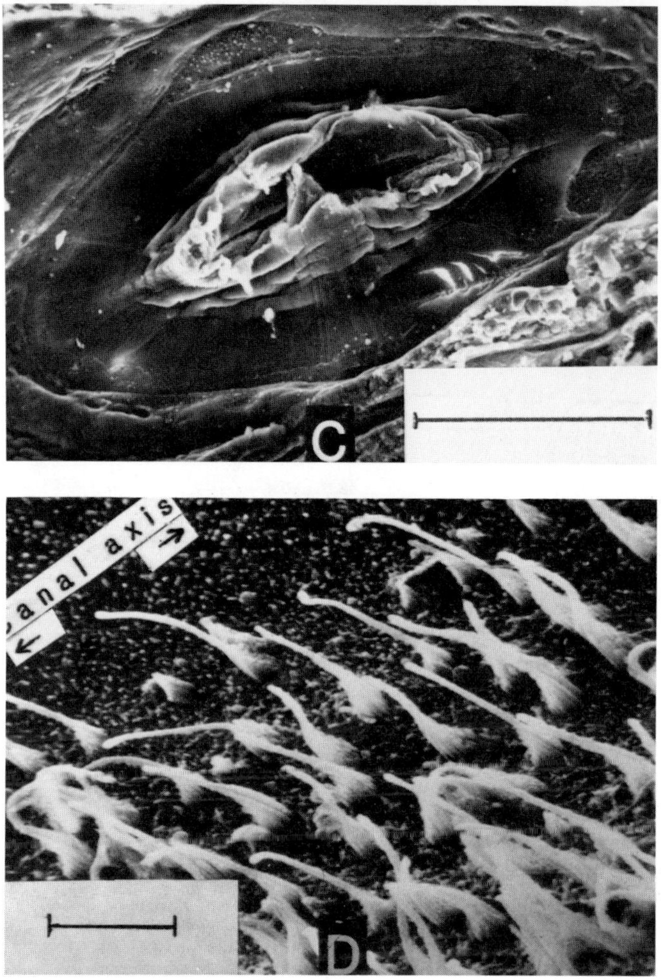


Kendall 1969). In teleosts such as *Salmo gairdneri,* having tube-containing scales, the pore is simply a hole on the suprascalar side of the tube (Fig. 22.3B).

Tubules also vary in size, shape, and number. Those associated with widened canals in teleostean fish tend to form baglike diverticula (Jakubowski 1966a, 1966b, 1966c) (Fig. 22.4). Secondary and tertiary branching of tubules occurs in nearly all elasmobranchs, most primitive actinopterygians, and many higher teleosts (Fig. 22.6 and Table 22.I). Pores are usually found on the tips of terminal branches; thus a primary tubule that branches into several secondary and tertiary ones may lead to several openings at the ends of tertiary tubules.

Neuromasts are always found in the canals and never in tubules. The typical arrangement in bony fishes is one neuromast between every two pores or primary tubules, but in some teleosts such as *Prionotus* (Triglidae) and *Anoplogaster* (Anoplogasteridae), in which tubules are absent and canals covered with membranes, one neuromast is located between what appear to be groups of pores in

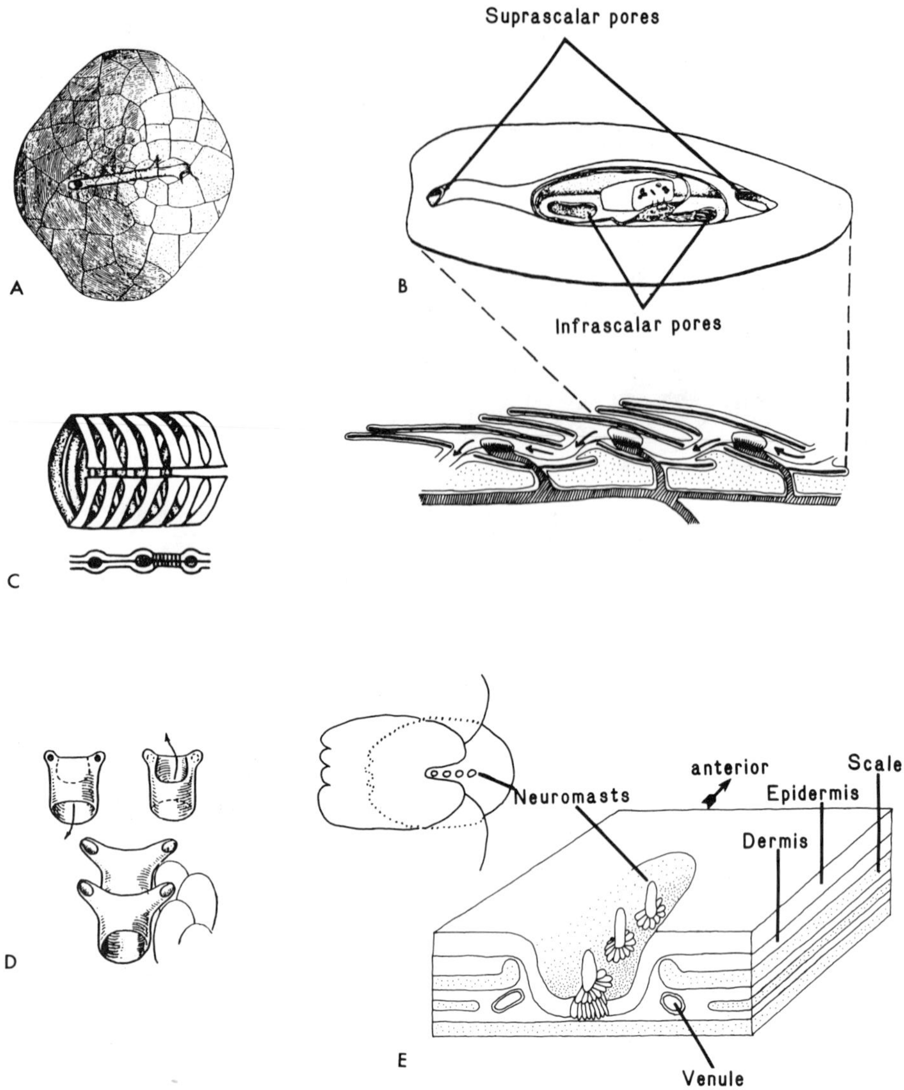


Figure 22.3. Canal support systems. (A) Shinglelike scales in *Osteoglossum bicirrhosum* (Osteoglossidae) (from Lauder and Liem 1983 after Taverne 1977); (B) shinglelike scales in cross section showing the interconnection of tubes via supra- and infrascalar pores in *Salmo gairdneri* (Salmonidae) (from Weber and Schiewe 1976); (C) cartilaginous rings that support canals between neuromast positions in *Chimaera* (from Chu and Wen 1979); (D) lateral line scales reduced to tubes in a stichaeid (from Makushok 1961); and (E) cardioid scales in *Esox americanus*. Neuromasts sit in grooves in posterior edge of each scale.(From Merrilees and Crossman 1973a.)

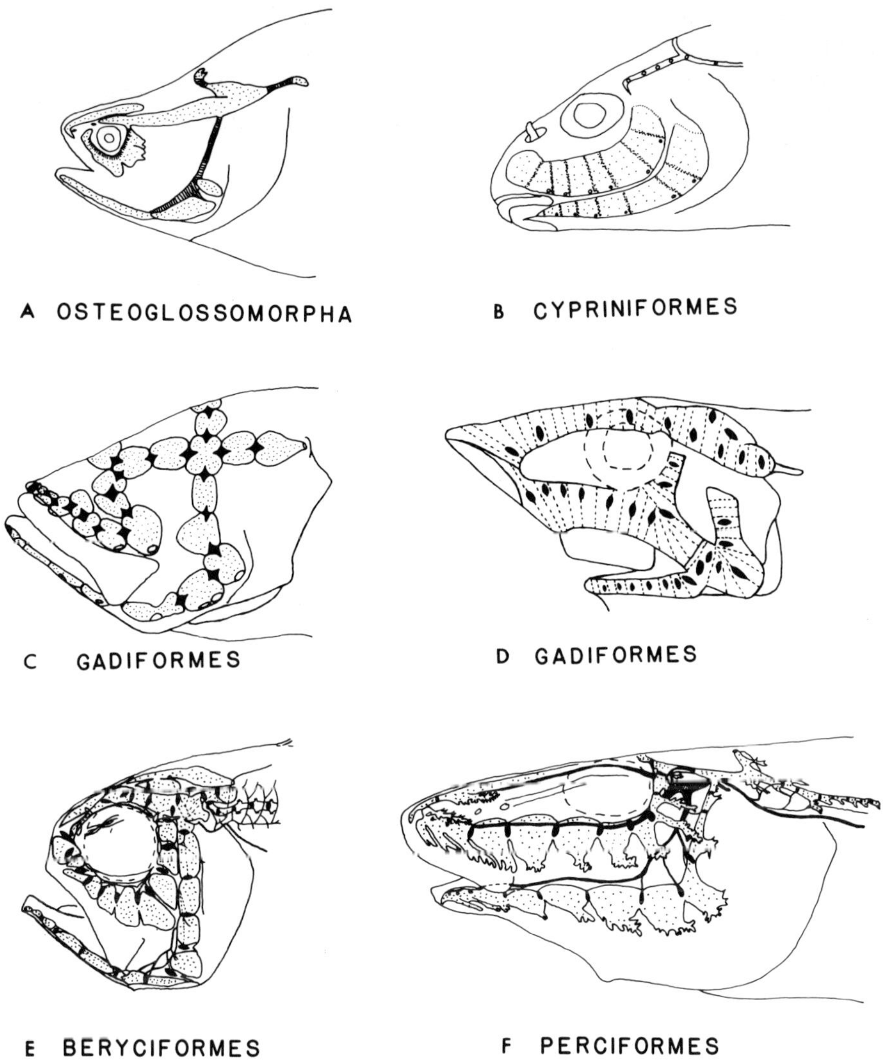


Figure 22.4. Examples of widened cephalic canals in various fish taxa: (A) *Notopterus chitala* (Notopteridae) (Reprinted with permission from *Acta Zool.* 65:77–91, 1974. Copyright © 1974, Pergamon Press.) (B) *Ericymba buccata* (Cyprinidae) (from Hoyt 1972); (C) *Typhliasina pearsei* (Brotulidae) (from Schemmel 1977); (D) *Coelorhynchus flabellispinis* (Macrouridae) (Reprinted with permission from *Deep Sea Res.* 12:299–322, 1965. Copyright © 1965, Pergamon Press.) (E) *Hoplostethus mediterraneus* (Trachichthyidae) (from Jakubowski 1974); and (F) *Aspro zingle* (Percidae) (from Jakubowski 1967b). Darkened triangular structures represent neuromast positions in C and D. Canal pores are absent in A, not depicted in E and F, and indicated by open circles in B and C.

the membrane (Janssen and Coombs pers. obs., Chap. 23). In other taxa having membranous canal coverings, like mormyrids and some percids, the number of pores have been reduced (Pehrson 1945; Jakubowski 1967a); and in the Notopteroidei, the absence of pores is characteristic of the group (Greenwood 1973). In

Table 22.1. Incidence of Widened Head Canals (*A*), Branched Tubules (*B*), and Head Canal Loss (*C*)

| Taxon | A | B | C | Taxon | A | B | C |
|---|---|---|---|---|---|---|---|
| Agnatha | | | | | | | |
| Petromyzontidae | — | — | 36 | Argentinoidei | | | |
| Myxinidae | — | — | 42 | Searsiidae | 41 | — | — |
| Elasmobranchiomorpha | | | | Aulopiformes | | | |
| Holocephali | 7, 18, 64 | — | — | Evermannellidae | 41 | — | — |
| | | | | Myctophiformes | | | |
| Selachii | — | 7, 18 | — | Myctophidae | 41 | — | — |
| Batoidei | — | 7, 18 | — | Paracanthopterygii | | | |
| Osteichthyes | | | | Aphredoderidae | 45 | — | — |
| Cladistia | — | — | — | Gadidae | 17, 24, 28 | — | — |
| Polypteridae | — | 10 | — | | | | |
| Chondrostei | | | | Macrouridae | 40 | — | — |
| Polyodontidae | — | 11 | — | Ophidiidae | 40, 58 | — | — |
| Ascipenseridae | — | 10 | — | Amblyopsidae | — | — | 12 |
| Ginglimodi | | | | Batrachoididae | — | — | 8 |
| Lepisosteidae | — | 9, 59 | — | Chaunacidae | 40 | — | — |
| Halecomorpha | | | | Ceratoididae | — | — | 42 |
| Amiidae | — | 1 | — | Ogocephalidae | 40 | — | — |
| Osteoglossomorpha | | | | Atherinomorpha | | | 52 |
| Hiodontidae | — | 47 | — | Cyprinodontidae | — | — | 14, 63 |
| Pantodontidae | 20 | — | — | Poeciliidae | — | — | 57 |
| Notopteridae | 32, 33, 50 | — | — | Scomberesocidae | — | 53 | — |
| | | | | Belonidae | — | 53 | — |
| Mormyridae | 54 | — | — | Hemirhamphidae | — | 53 | — |
| Elopomorpha | | | | Exocoetidae | — | 53 | — |
| Elopidae | — | 46 | — | Percomorpha | | | |
| Megalopidae | — | 46 | — | Centrarchidae | 5 | — | 5 |
| Halosauridae | 40 | — | — | Percidae | 29 | — | 15, 26, 35, 51 |
| Muraenidae | 2 | — | — | | | | |
| Nemichthyidae | — | — | 34 | Gobiidae | — | — | 43, 44, 55, 61 |
| Eurypharyngidae | — | — | 42 | | | | |
| Clupeomorpha | | | | Gunnellichthyidae | — | — | 43 |
| Denticepitidae | — | 60 | — | Stichaeidae | — | 39 | 39 |
| Chirocentridae | — | 4 | — | Soleidae | — | — | 3 |
| Clupeidae | — | 21, 60 | — | Holocentridae | 30 | — | — |
| Engraulidae | — | 48, 49 | — | Berycidae | 30 | — | — |
| Esocae | | | | Monocentridae | 30 | — | — |
| Umbridae | — | — | 47 | Trachichthyidae | 30 | — | — |
| Esocidae | — | — | 47 | Melamphaidae | 41 | — | — |
| Galaxiidae | — | — | 47 | Cottidae | 31 | — | — |

Table 22.1 *continued*

| Taxon | A | B | C | Taxon | A | B | C |
|---|---|---|---|---|---|---|---|
| Ostariophysi | | | | Cichlidae | 38 | — | — |
| Plotosidae | — | 37 | — | Sparidae | — | 27, 62 | — |
| Cyprinidae | 22 | — | 23, 19, | Mullidae | — | 27 | — |
| | | | 56 | Serranidae | — | 27, 62 | — |
| Cobitidae | — | — | 25, 37 | Stromateidae | — | 13 | — |
| Characidae | — | — | 65, 66 | Coryphaenidae | 16 | — | — |
| | | | | Sciaenidae | 6 | — | — |

*Sources:* 1, Allis (1889); 2, Allis (1903); 3, Applebaum and Schemmel (1983); 4, Bardack (1965); 5, Branson and Moore (1962); 6, Chao (1978); 7, Chu and Wen (1979); 8, Collette (1966); 9, Collinge (1893a); 10, Collinge (1893b); 11, Collinge (1895); 12, Cox (1905); 13, Deng, Xiong, and Zhan (1981); 14, Denny (1937); 15, Disler and Smirnov (1977); 16, Fange, Larsson, and Lidman (1972); 17, Flock (1965); 18, Garman (1888); 19, Gosline (1974); 20, Hoin-Radkovsky, Bleckmann, and Schwartz (1984); 21, Hoss and Blaxter (1982); 22, Hoyt (1972); 23, Illick (1956); 24, Jakubowski (1965); 25, Jakubowski (1966a); 26, Jakubowski (1966b); 27, Jakubowski (1966c); 28, Jakubowski (1967a); 29, Jakubowski (1967b); 30, Jakubowski (1974); 31, Janssen et al. (unpubl.); 32, Kapoor (1964); 33, Kapoor (1966); 34, Karmovskaya (1981); 35, Kuiper (1967); 36, Lane and Whitear (1982); 37, Lekander (1949); 38, Liem (1981); 39, Makushok (1961); 40, Marshall (1965); 41, Marshall (1971); 42, Marshall (1979); 43, Masuda et al. (1984); 44, Miller and Wongrat (1979); 45, Moore and Burris (1956); 46, Nelson (1969); 47, Nelson (1972); 48, Nelson (1983); 49, Nelson (1984); 50, Omarkan (1948); 51, Page (1977); 52, Parenti (1981); 53, Parin and Astakhov (1982); 54, Pehrson (1945); 55, Pinchuk (1981); 56, Reno (1971); 57, Rosen and Mendelson (1960); 58, Schemmel (1977); 59, Song (1984); 60, Stephens (1985); 61, Takagi (1957); 62, Tretyakov (1944); 63, van Bergeijk and Alexander (1962); 64, von Lubitz (1981); 65, Weitzman and Fink (1983); 66, Weitzman and Fink (1985).

elasmobranchs, several neuromasts appear to run together under a continuous cupula that may be associated with few or no pores (Johnson 1917; Tester and Kendall 1969; Chap. 23).

## Variations in Neuromast Morphology

In general, superficial neuromasts, on the skin, are remarkably different from those enclosed in canals. Superficial neuromasts are generally smaller ($< 100$ μm in diameter) than canal neuromasts, have fewer hair and support cells (Münz 1979; Janssen et al. in press), and may differ from canal neuromasts in their development (see p. 576) and orientation of hair cells (see p. 582). Moreover, superficial neuromasts can be distinguished from canal neuromasts by the extent of myelination on the afferent neurons (Sato 1956; Flock and Wersall 1962; Jakubowski 1965, 1966b; Münz 1979; Münz and Claas 1983) and possibly by the absence of efferent innervation (Münz 1979; Hama 1978).

### Canal Neuromasts

Canal neuromasts, including their cupulae (Chap. 23) vary in size, shape, topography, and orientation within the canal; and these factors are usually correlated

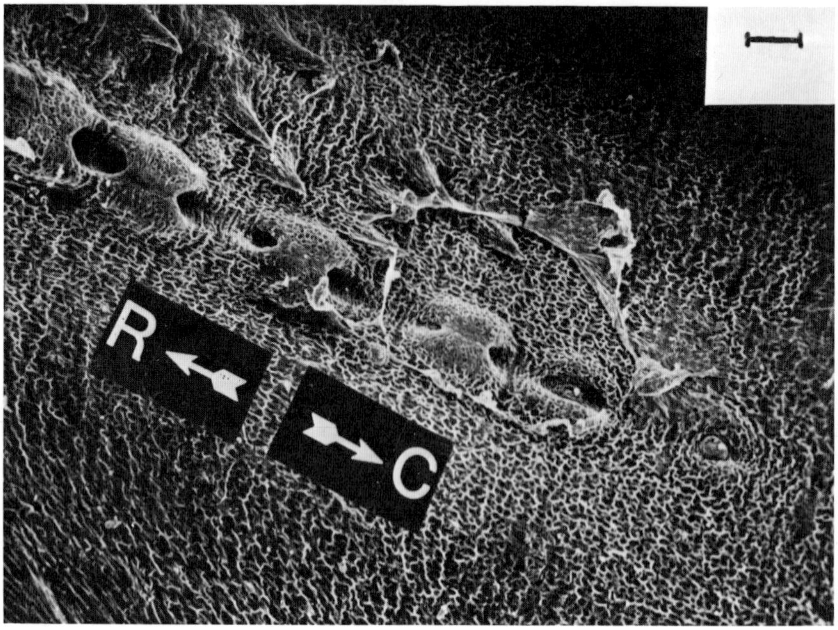


Figure 22.5. Developing trunk canal on 2-cm mottled sculpin, *Cottus bairdi* showing progressive invagination and canal and pore formation in caudal-to-rostral direction (from Janssen et al. in press). Scale bar: 100 μm.

with canal diameter. The canal neuromast shown for the mottled sculpin (Fig. 22.2A) is typical of those described for the sprat (Clupeidae) (Best and Gray 1982) and other teleosts with relatively narrow canals (e.g., various percids, Jakubowski 1967a). The neuromast is located in a local constriction of the canal with the long axis of the neuromast parallel to the canal axis. Mantle cells are arranged in moundlike protuberances on each side of the hair cell epithelium to form a narrow trough in which the sensory strip lies (fig. 22.2A and B). Kinocilia are aligned along the canal axis so that the sensory tissue is physiologically polarized to respond maximally to water movements along this axis (Fig. 22.2D). Furthermore, the major axis of the overlying cupula is also parallel to the canal axis (Fig. 22.2C).

In contrast, neuromasts found in widened head canals, like those of the pirate perch, *Aphredoderus sayanus* (Percidae). (Fig. 22.7A), and the clown knifefish, *Notopterus chitala* (Fig. 22.7D), have a shape different from those characteristic of narrow canals. The mantle cells do not form a trough, and the hair cells are more widely distributed on the neuromast. The major axis of hair cell orientation is still parallel to the axis of the canal, but the longest aspect of the neuromast and cupula appears to be transverse to the canal (Fig. 22.7B and C). In addition, specializations like the one shown in Figure 22.7C for *Aphredoderus* often help to compartmentalize wide canals in the vicinity of the neuromast (see also p. 555).

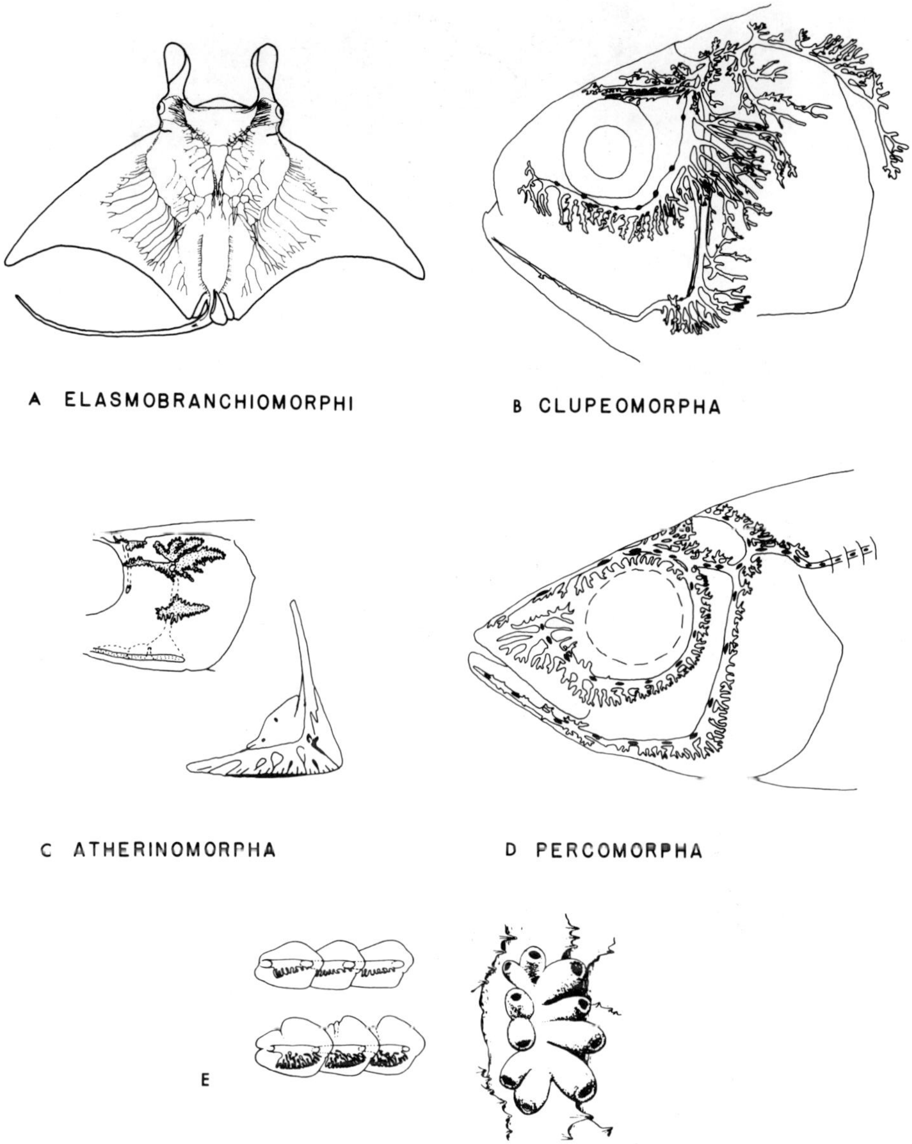


Figure 22.6. Examples of branched tubules found in various fish groups: (A) *Manta birostris* (Mobulidae) (from Chu and Wen 1979); (B) *Brevoortia tyrannus* (Clupeidae) (Hoss and Blaxter 1982, adapted from orignal drawing by C.M.R. Hoss); (C) branching tubules on preopercular canal in *Hyporhamphus neglectissimus* (*left*) and *Oxyporhamphus micropterus* (*right*) (Hemirhamphidae) (from Parin and Astakhov 1982); (D) *Spicara smaris* (Mullidae) (from Jakubowski 1966c); and (E) branched tubules on trunk scales of beloniform fish (*left*) (from Parin and Astakhov 1982) and of *Solea vulgaris* (Soleidae) (*right*). (From Applebaum and Schemmel 1983.)

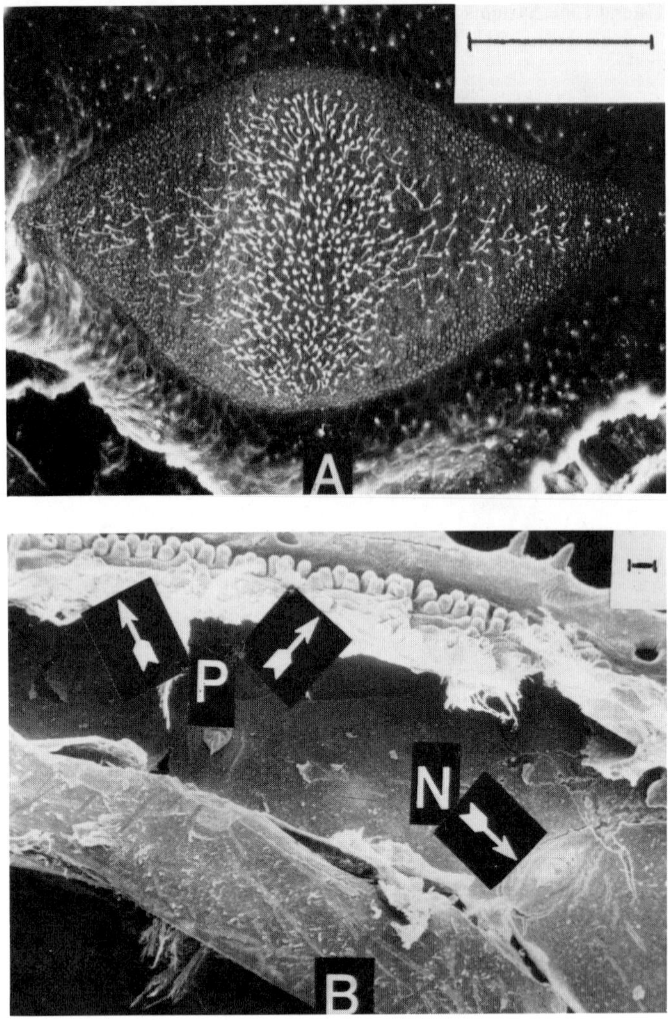


Figure 22.7. Scanning electron micrographs of neuromasts in *Aphredoderus sayanas* (Percidae) showing: (A) the distribution of hair cells on the neuromast and (B) the position of the neuromast (*N*) within the canal and accessory papillate neuromasts (*P*) accompanying the canal; in (C), the neuromast with cupula has been torn away from the canal, revealing liplike projection (*L*) that hangs over the neuromast; (D) a neuromast from *Notopterus chitala* (Notopteridae) shows different-shaped hair cell patch with mantle flanges that spread across the canal. *Scale bar:* 100 μm.

## Superficial Neuromasts

Like canal neuromasts, there is significant variation in the structure of superficial neuromasts and, unfortunately, in the terminology used to describe this variation. To standardize the terminology, we have devised a classification scheme, similar to that used by Srivastava and Srivastava (1968), that is based on the relationship between the neuromast, the dermal and epidermal layers, and the basement membrane between them. Several types of superficial neuromasts are recognized,

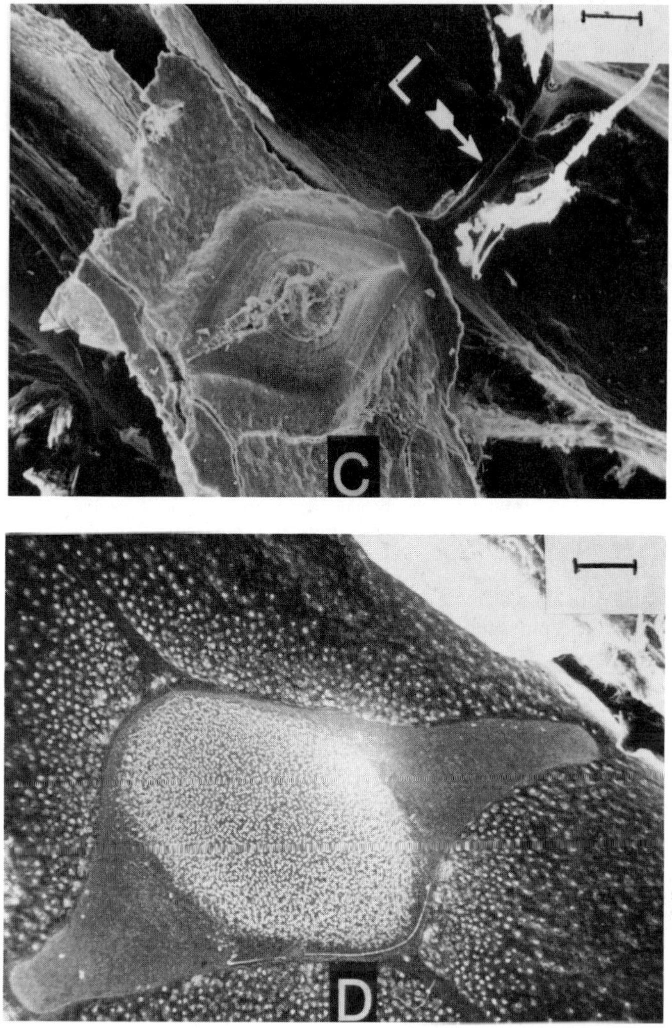


though some types may represent ontogenetic stages in the development of adult neuromast morphology, as well as an adult condition (Fig. 22.8).

*Flush neuromasts* sit above a relatively straight basement membrane, and the top of the neuromast is nearly flush with the top of the surrounding epidermal layer. Called type A by Srivastava and Srivastava (1968), this superficial neuromast is the general type found in larval fish (Pehrson 1945; Srivastava and Srivastava 1968). Neuromasts that appear flush in larval animals may later sink into the dermis to become *canal neuromasts* (Allis 1889; Johnson 1917; Janssen et al. in preparation) (Fig. 22.4). Alternatively, they may remain flush in adult fish (Münz 1979; Janssen et al. in press) or develop into one of the other types of superficial neuromasts (Fig. 22.8) (Allis 1889).

Several types of superficial neuromasts sit in depressions. *Dermal pit* organs, like the ones shown in Figure 22.3E for *Esox* (Esocidae), sit on the bottom of a

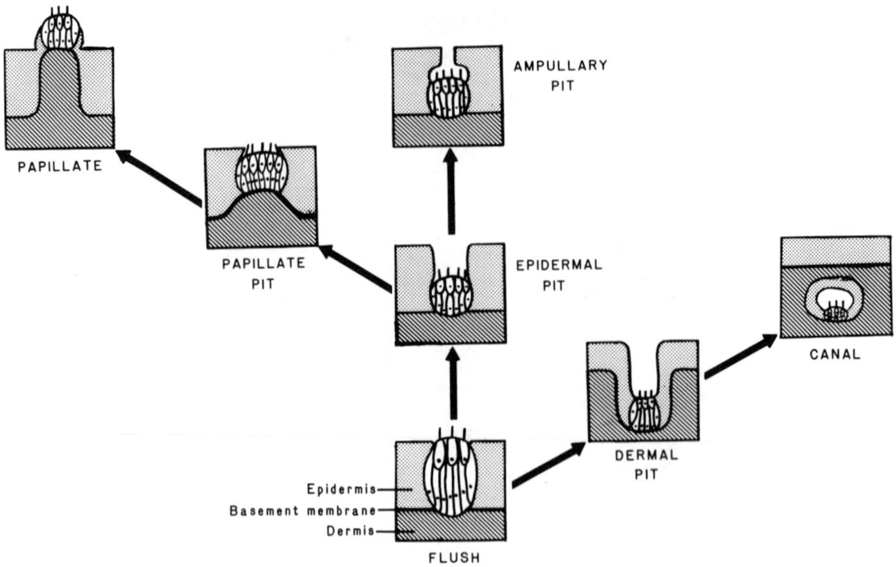


Figure 22.8. Classification scheme for superficial neuromast types based on the relationship of the neuromast to epidermal and dermal layers. Possible ontogenetic sequences are shown by *arrows*.

pit or groove formed by the invagination of the basement membrane into the dermis. These have been called type B (Srivastava and Srivastava 1968), *cryptes sensorielles* (Budker 1958), and *fossettes sensorielles* (Szabo 1963) and probably represent an ontogenetic stage in canal neuromast formation in larval fish (Allis 1889; Johnson 1917) (Fig. 22.5) or an arrested developmental (paedomorphic) state in adult fish (Nelson 1969). *Epidermal pit* organs sit in pits formed by a gap in the epidermis and are morphologically similar to large pit organs in *Ictalurus* (*Ameiurus*) (Herrick 1901; Sato 1955), conical hillocks (Merkel 1880), type D neuromasts in adult *Cirrhina* (Cyprinidae) (Srivastava and Srivastava 1968), and pit line organs in adult *Amia* (Allis 1889). *Ampullary pit* organs are superficially similar to ampullary electroreceptive organs found in silurids (Herrick 1901, 1903; Mullinger 1964; Bretschneider 1979) and in *Polypterus* (Jorgensen 1982). Like the electroreceptive ampullary organs, originally termed "small pit organs" (Herrick 1901, 1903), the mechanoreceptive counterpart sits on a relatively straight basement membrane at the bottom of an ampulla within the epidermis. This type of superficial neuromast has been described for cobitids (Jakubowski 1966a) and the flatfish *Peltorhamphus novaezeelandiae* (Roper 1981) (Fig. 22.9A).

In contrast to pit organs, *papillate neuromasts* sit on a dermal stalk formed by the evagination of the basement membrane. Figure 22.7B shows a series of papillate organs that follow the course of the preoperculomandibular canal in *Aphredoderus sayanas* (Aphredoderidae) (Moore and Burris 1956). Stalked neuromasts like these have also been described for gobies (Srivastava and Srivastava 1968), amblyopsid blind cavefish (Poulson 1963), ceratioid anglerfishes (Marshall

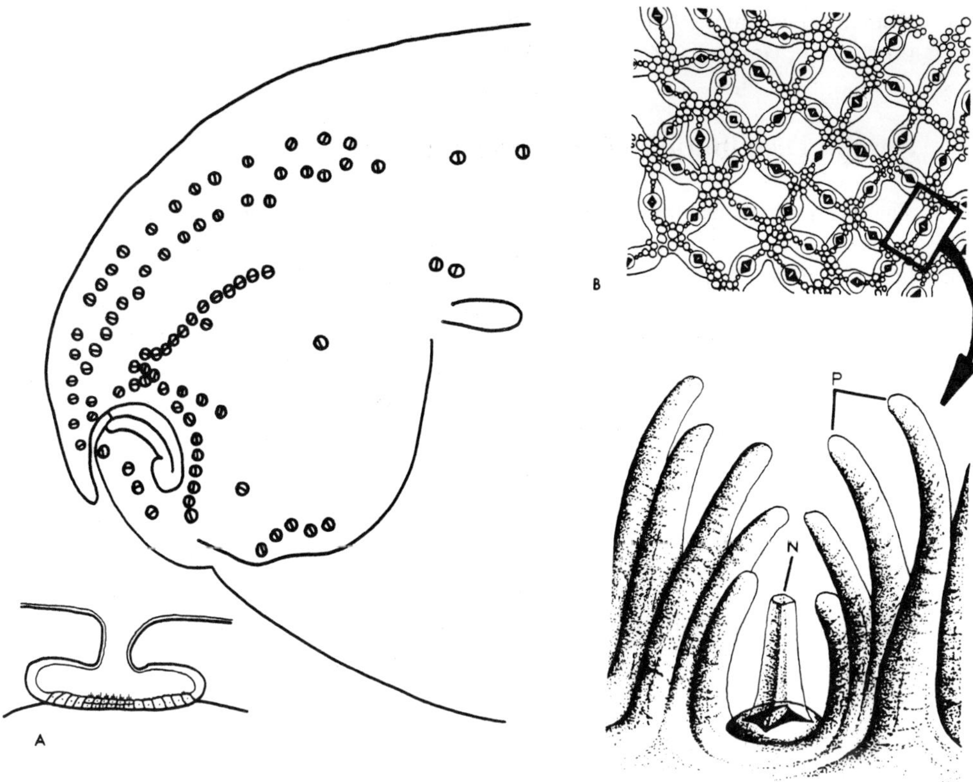


Figure 22.9. Two examples of adaptations for directing flow over superficial neuromasts. (A) Slitlike openings to mechanoreceptive ampullary pit organs distributed in various directions on abocular side of *Peltorhamphus novaezeelandiae* (Soleidae) (from Roper 1981). Bottom insert depicts schematic cross section of single ampullary pit organ. (B) *Top*, mosaic arrangement of superficial neuromasts and nonsensory papillae; *bottom*, close-up showing the arrangement of one neuromast (*N*) and papillae (*P*) that surround it in *Solea vulgaris* (Soleidae). (From Applebaum and Schemmel 1983.)

1971), and certain deep-sea gadoids (Schemmel 1977). The *papillate pit* organ is elevated by a slight evagination of the dermis, but also sits in an intraepidermal pit. This type has been described for several species of adult lamprey (Yamada 1973; Lane and Whitear 1982), the Japanese sea eel, *Conger myriaster* (Congridae) (Hama 1978) and the burbot, *Lota lota* (Gadidae) (Jakubowski 1965).

In some cases, specialized epidermal structures associated with the neuromast may assist in channeling water in particular directions past the neuromast. In *Peltorhamphus novaezeélandiae* (Soleidae), the hair cell sensory strip of ampullary pit organs is parallel to a narrow slitlike opening above the strip (Roper 1981) (Fig. 22.9A). Hair cells lying in the strip are presumed to be maximally sensitive to water flow parallel to the slit. Similarly, papillae surrounding neuromasts on the abocular side of *Solea vulgaris* (Soleidae) form canal-like structures (Applebaum & Schemmel 1983) (Fig. 22.9B) not unlike those formed by setae in sergestid shrimp (Marshall 1979; Chap. 23).

## Spatial Distribution of Receptor Organs

The terminology used to describe the distribution of canals and superficial neuromasts in different taxa is extremely varied. A standard terminology based on homologies has been difficult to establish (see Jarvik 1980), partly because various criteria have been used, including the bones with which neuromasts are associated, the neurogenic origin and rate of neuromast development (Allis 1889; Holmgren and Pehrson 1949; Lekander 1949; Pehrson 1949; Srivastava and Srivastava 1969), and the peripheral innervation pattern (Allis 1889; Holmgren & Pehrson 1949; Lekander 1949; Jakubowski 1966b, 1966c, 1967a; Northcutt 1985). Although the organization of projections from various regions of the lateral line system in the CNS would help in determining homologies, it is largely unknown. Peripheral innervation patterns presently constitute the most widely available source of information for determining homologies and have been used here to establish a standard terminology for describing the distribution of canal and superficial neuromasts across taxa. Although we have treated the distribution patterns for canal and superficial neuromasts separately, their distribution overlaps considerably.

### Basic Distribution Pattern of Canals

The distribution and innervation of canal neuromasts generally follow a basic plan in all fish (Fig. 22.10). The *supra- and infraorbital canals* course above and below the eye and join behind it to form a *main body canal,* which extends onto the trunk. Several canals may lie on the lower jaw and in the head region, the most typical being the *mandibular canal.* There are also several commissural lines, including the widely distributed *supratemporal commissure,* which runs dorsally to connect the two main body canals. The supraorbital, infraorbital, and mandibular canals are innervated by separate rami of the anterior lateral line nerve: the superficial opthalmic, buccal, and mandibular rami, respectively. The trunk portion of the main body canal is innervated by the main trunk ramus of the posterior lateral line nerve. In most actinopterygian fish, the anterior portion of the main body canal can be further divided into a series of sections caudal to

Figure 22.10. The distribution of canals in representatives of major fish taxa: Holocephali (*Chimaera cubana*) (after Bigelow and Schroeder 1953); Elasmobranchii (*Chlamydoselachus*) (after Holmgren and Pehrson 1949) and (*Scyliorhynchus*) (after Dikgraaf 1967 reprinted with permission from Indiana University Press); Halecomorphi (*Amia calva*) (after Jarvik 1980); Teleostei (*Hiodon alosoides*) (after Nelson 1972); Dipnoi (*Neoceratodus forsteri*) (Holmgren and Pherson 1949) and Actinistia (*Latimeria chalumnae*). (After Jarvik 1980.)

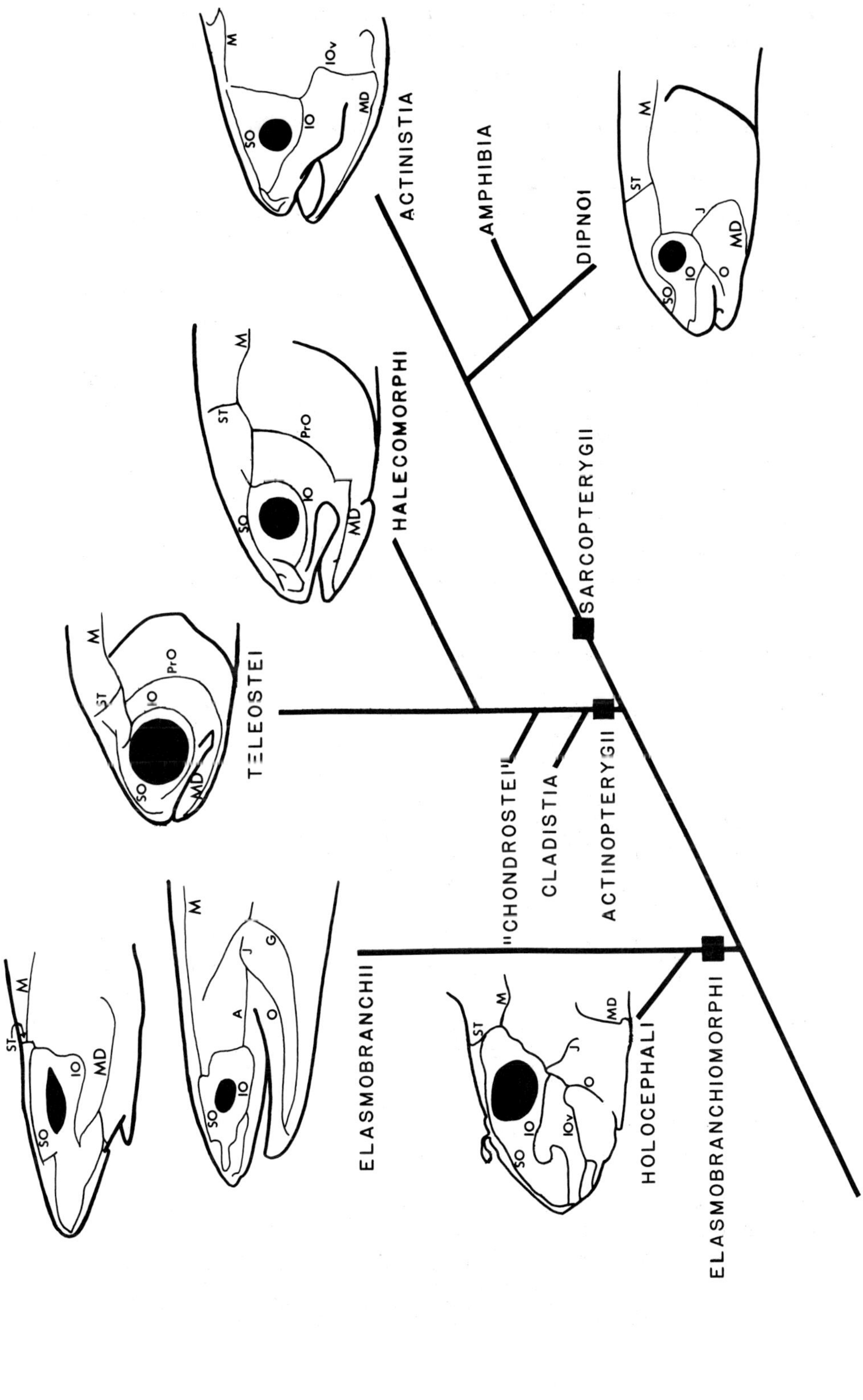
ACTINISTIA
AMPHIBIA
DIPNOI
HALECOMORPHI
SARCOPTERYGII
TELEOSTEI
"CHONDROSTEI"
CLADISTIA
ACTINOPTERYGII
ELASMOBRANCHII
HOLOCEPHALI
ELASMOBRANCHIOMORPHI

where the orbital canals join (Northcutt 1985) (Fig. 22.11). These include (1) the *otic canal,* which is innervated by the otic ramus of the anterior lateral line nerve; (2) the *post-otic canal,* innervated by the squamosal ramus (or lateralis component of the glossopharyngeal nerve); and (3) the *temporal canal* and the *supratemporal commissure,* innervated by the supratemporal-dorsal accessory ramus of the posterior lateral line nerve.

## Variation in the Distribution of Canals

In dipnoans, holocephalans, and primitive elasmobranchs, the mandibular ramus innervates a number of canals in addition to the mandibular canal, including the jugal, oral, gular, and angular canals (Fig. 22.10) (Cole 1896; Allis 1923; Jarvik 1980; Northcutt 1985). Fewer canals are innervated by the mandibular ramus in advanced elasmobranchs and actinopterygian fishes (see Fig. 22.11), but corresponding lines of superficial neuromasts may represent the evolutionary replacement of these canals (see p. 576). Similarly, the buccal ramus innervates both the infraorbital canal and a canal ventral to that in Holocephali and Actinistia (Fig. 22.10). The homologue of this ventral infraorbital canal is either absent in actinopterygians or has been replaced by a line of superficial neuromasts frequently found below the infraorbital canal (Fig. 22.11).

Among teleosts, trunk canal extent and placement can be clearly categorized

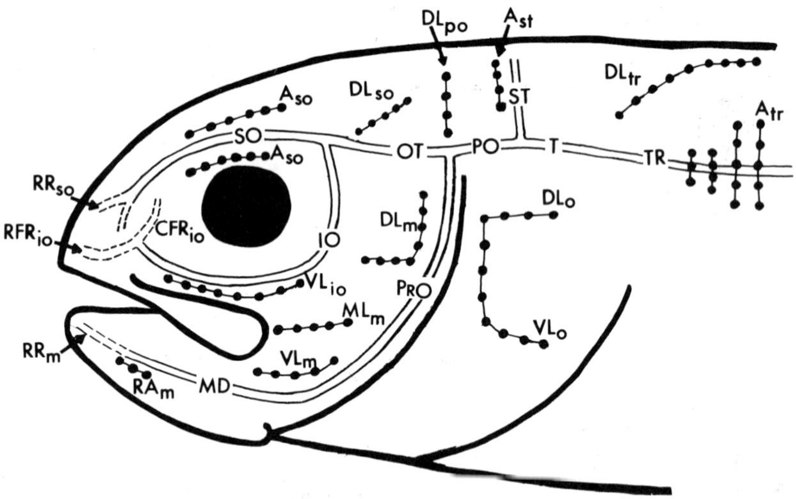


Figure 22.11. Common distribution patterns for lateral line canals and superficial neuromasts in actinopterygian fish. *Double solid lines* represent locations of canals, *dashed lines* the location of canals or superficial neuromasts thought to replace canals, and *dots connected by single lines* the location of other types of superficial neuromasts. Abbreviations for superficial neuromast lines are outlined in Tables 22.3 and 4. Canals: *SO,* supraorpital canal; *IO,* infraorbital canal; *OT,* otic canal; *PrO,* preopercular canal; *MD,* mandibular canal; *PO,* postotic canal; *ST,* supratemporal division of temporal canal (*T*); *TR,* trunk canal.

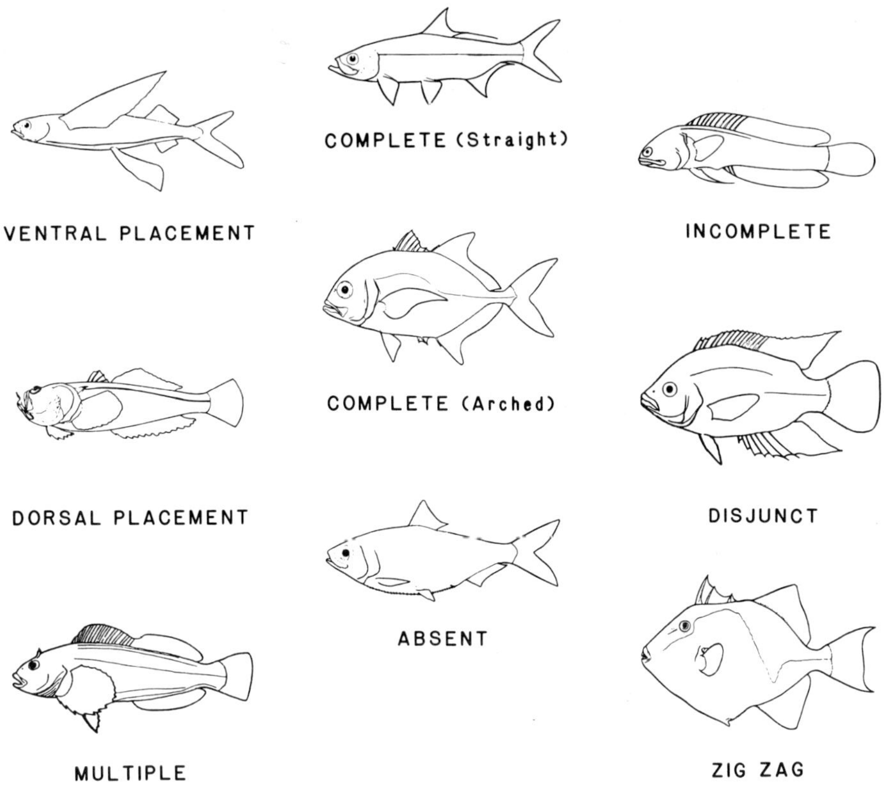


Figure 22.12. Trunk canal patterns in teleost fish (figures from Nelson 1984).

(Webb in preparation) (Fig. 22.12 and Table 22.2). It can be (1) *complete,* from the head to the end of the caudal peduncle; (2) *incomplete,* ending anterior to the caudal peduncle; (3) *absent;* (4) *caudally extended* onto the caudal fin rays; (5) *disjunct,* with an interruption about three-fourths down the flank; and (6) *multiple,* with up to five parallel trunk lines. In addition, the trunk canal may be either straight or arched and lie in a dorsal, midlateral, or ventral position.

Relatively little is known about the innervation of trunk neuromasts. In paracanthopterygian and atherinomorph fish, the main trunk nerve branches into several dorsal and ventral segmented rami (Freihofer 1972) (Fig. 22.13), which presumably innervate both superficial and canal neuromasts. The ventral rami apparently innervate ventrally placed canal organs where they exist in atherinomorphs, but both ventral and dorsal rami are reduced or lacking in higher teleosts (Freihofer 1972). Freihofer (1972) has suggested that ventral and dorsal trunk canals have arisen secondarily in teleosts and thus are not homologous to dorsal and ventral canals reported in primitive bony fish (Watson 1937; Stensio 1947). Until more is understood about the innervation and development of these canals, however, homologies are uncertain.

Just as there is reduction in the number of canals innervated by the buccal,

Table 22.2. Incidence of Trunk Canal Patterns in Teleosts

| Taxon | Complete | Absent | Incomplete | Disjunct | Multiple | Reference |
|---|---|---|---|---|---|---|
| Osteoglossomorpha | | | | | | |
| Osteoglossoidei | X | | | | | |
| Notopteroidei | X | | | | | |
| Elopomorpha | | | | | | |
| Elopidae | X | | | | | |
| Megalopidae | X | | | | | |
| Anguilliformes | X | | | | | |
| Clupeomorpha | | | | | | |
| Denticepitoidei | X | | | | | |
| Clupeoidei | | X | | | | |
| Esocae | X | | | | | |
| Ostariophysi | | | | | | |
| Gonorhynchiformes | X | | | | | |
| Cypriniformes | X | X | X | | | |
| Characiformes | X | | X | | | Roberts (1967) |
| Siluriformes | X | X | | | | Mees (1974) |
| Argentinoidei | X | | | | | |
| Osmeroidei | X | X | X | | | |
| Salmonidae | X | | | | | |
| Stomiiformes ?[a] | | | | | | |
| Aulopiformes ? | | | | | | |
| Myctophiformes | X | X | | | | |

| | | | | | | | |
|---|---|---|---|---|---|---|---|
| Paracanthopterygii | | | | | | | |
| Percopsiformes | | X | X | X | | | |
| Gadiformes | | X | | | | X | Nelson (1984) |
| Batrachoidiformes | | X | X | | | X | Collette and Russo (1981) |
| Lophiiformes | ? | | X | | | | |
| Gobiesosiformes | ? | | | | | | |
| Atherinomorpha | | | | | | | |
| Cyprinodontiformes | | X | X | | | | |
| "Atherinoids" | | X | X | | | | |
| Beloniformes | | X | X | | | | |
| Percomorpha | | | | | | | |
| Lampridiformes | | X | | | | | |
| Gasterosteiformes/ Pegasiformes | | X | X | | | | |
| Dactylopteriformes | ? | | | | | | |
| Beryciformes | | X | X | | | | |
| Zeiformes | | X | | | | X | |
| Synbranchiformes | ? | | | | | | |
| Channiformes | ? | | | | | | |
| Scorpaeniformes | | X | X | | | X | Rutenberg (1962) |
| Perciformes | | X | X | X | X | X | Makushok (1961); Jakubowski (1966b); Winterbottom (1982); Collette (1983) |
| Pleuronectiformes | | X | X | | | X | |
| Tetraodontiformes | | X | X | | | X | |

*Source:* All data from Nelson, J.S. (1984) unless otherwise indicated.
[a] ? no data available.

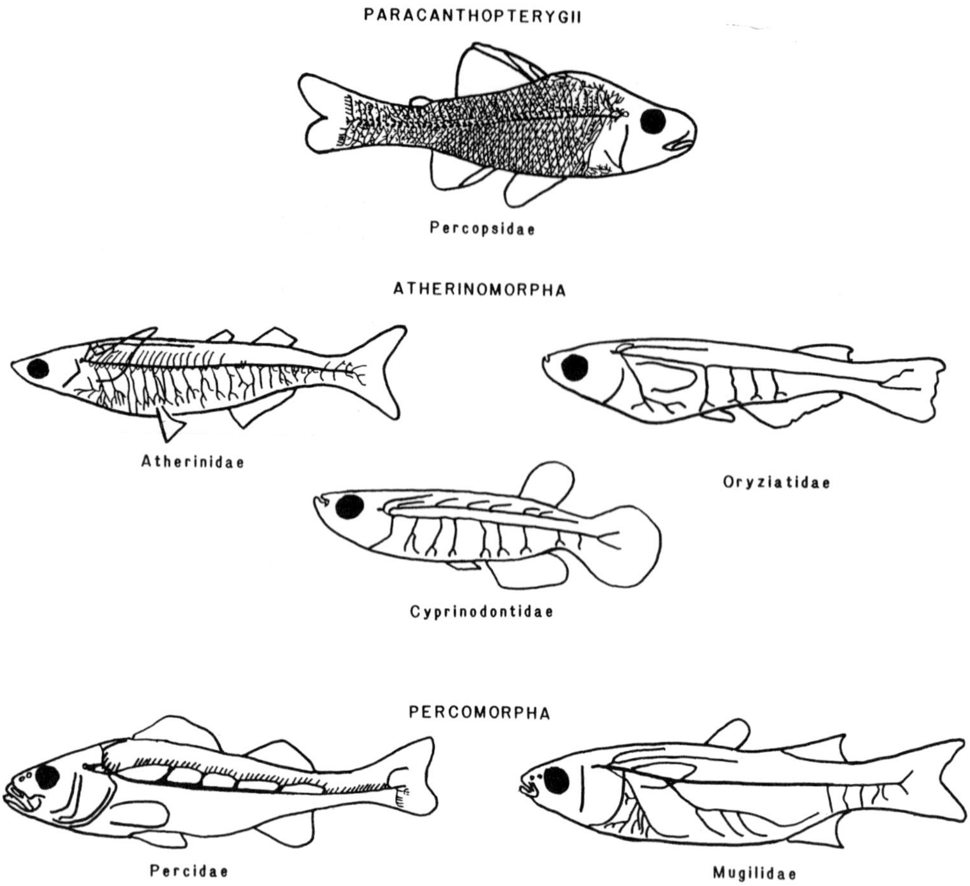


Figure 22.13. Examples of posterior lateral line nerve patterns in Paracanthopterygii, Atherinomorpha, and Percomorpha. (From Freihofer 1972.)

mandibular, and trunk rami in certain fish groups, there is also loss of portions of, and in some cases entire, canals (Fig. 22.14 and Table 22.1). Canals are entirely absent in extant Agnatha (Yamada 1973; Lane and Whitear 1982) and in many teleosts, such as certain gobies (Pinchuk 1980) and amblyopsids (Poulson 1963). In cyprinids (Lekander 1949; Illick 1956) and other taxa, the degree of cephalic lateral line canal loss varies.

In rarer instances, there may be an apparent extension of existing canals or an addition of a new one, such as the premaxillary canal found in some beloniform fishes (Parin & Astakhov 1982) or the dorsal portion of the preopercular canal hypothesized by Holmgren and Pehrson (1949) to be an addition found only in actinopterygian fishes.

Finally, there is also variability in the location of head canals relative to the inner ear of fish. In the Clupeomorpha, all head canals radiate from a central enlarged sinus called the *lateral recess* (Lauder and Liem 1983). Positioned in

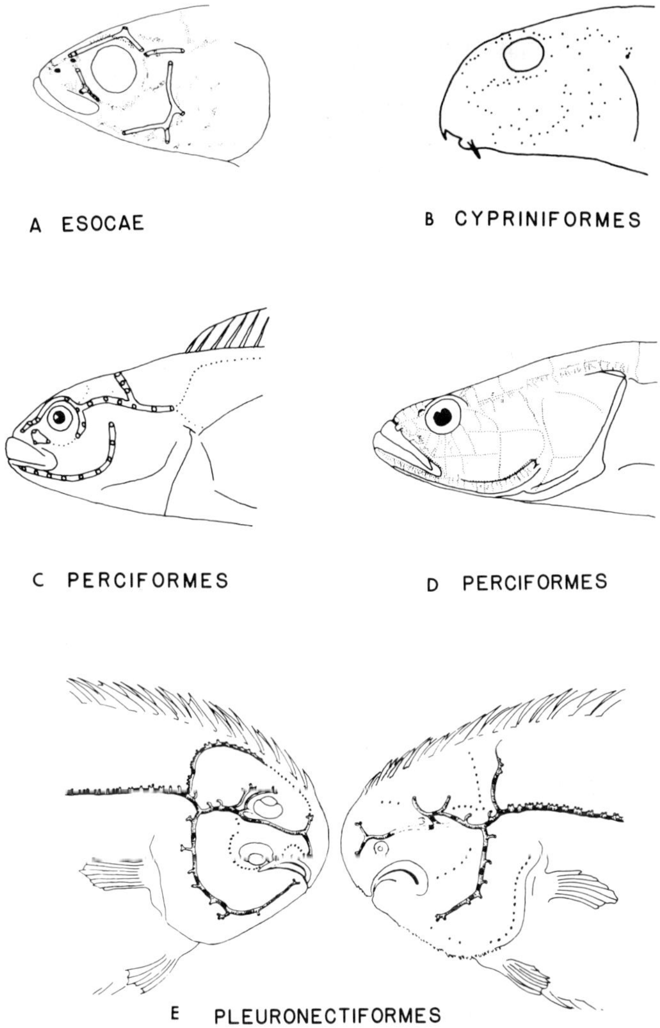


Figure 22.14. Examples of cephalic lateral line canal loss and replacement by superficial neuromasts in (A) *Umbra limi* (Umbridae) (from Nelson 1972); (B) *Cobitis taenia* (Cobitae) (from Lekander 1949); (C) *Opisthocentrus dybowskii* (Stichaeidae) (from Makushok 1961); and (D) *Eleotris pisonis* (Gobiidae) (from Miller and Wongrat 1979); and (E) *Solea vulgaris* (Soleidae). (From Applebaum and Schemmel 1983.)

the wall of the skull at the back of the lateral recess is a compliant membrane in contact with inner ear fluids. This mechanical linkage enables the lateral line system, usually thought to respond to relative movements between the fish and surrounding water (Chap. 4), to respond to the same propagated pressure changes as does the ear by virtue of a gas-filled cavity near the ear (Denton and Blaxter 1976; Denton, Gray, and Blaxter 1979). In other groups, such as the Notopteroidei, for which the morphology suggests a similar link between ear and lateral line, gas cavities are also present near the ear (Kapoor 1964; Greenwood 1973). Although the clupeomorphs represent the only fish group for which a mechanical relationship between ear and lateral line has been demonstrated, the possibility of similar relationships should be explored in other fish groups, especially those known to have air cavity–ear associations, including members of the Elopomorpha, Osteoglossomorpha, Ostariophysi, and various other teleosts (Popper and Coombs 1982).

## Basic Distribution of Superficial Neuromasts

The terminology for the distribution and homology of superficial neuromasts is even more variable than that for canals (Holmgren and Pehrson 1949; Lekander 1949; Jarvik 1980). Furthermore, the notion that superficial neuromasts in actinopterygian fish may arise from a number of different developmental processes complicates the task. In *Amia,* cyprinids, and certain other teleosts, primary superficial neuromasts develop in the embryo from epidermal thickenings called *placodes,* whereas other secondary superficial neuromasts appear later during larval and juvenile stages, either by budding off from primary neuromasts or by arising in situ from the basal layer of the epidermis, presumably via neural induction (Allis 1889; Lekander 1949; Disler 1960; O'Connell 1981). According to Lekander (1949), some primary neuromasts become enclosed in canals, and others remain free; but secondary neuromasts always remain superficial. This has led to the suggestion that primary neuromasts remaining superficial are homologous to canal neuromasts (Lekander 1949). However, it is not clear to what extent developmental patterns differ across major taxa (Pehrson 1949; Srivastava and Srivastava 1969; Jarvik 1980).

Although a classification of superficial neuromasts into primary and secondary types may not be useful for all fish groups, the distribution of superficial neuromasts among actinopterygian fishes is somewhat consistent with this dichotomy. Most superficial neuromasts tend to follow either the general course of existing canals as *accessory lines* (Fig. 22.7B) or that of canals in closely related or more primitive taxa, presumably as *replacement lines* for canals lost in evolution (Nelson 1972) (Fig. 22.14). Replacement lines may be represented by either a single row of neuromasts (e.g., infraorbital replacement on *Opistocentrus dybowskii* [Fig. 22.14C]) or multiple lines or groupings (e.g., infraorbital replacement neuromasts on *Umbra limi* [Fig. 22.14A]). Similarly, supernumerary papillate neuromasts are clearly accessory to the preoperculomandibular canal of *Aphredoderus sayanus* (Fig. 22.7B), and transverse rows and groups of neuromasts are frequently seen to accompany the trunk canal on many actinopterygian fishes (Fig. 22.11) (Allis 1889; Münz 1979). In some teleost fishes, superficial neuromasts are so numerous and widely distributed (Fig. 22.14D) (Schemmel 1977) that it is difficult to discern distinct patterns or the precise relationship between canals and superficial neuromasts.

Despite these difficulties of interpretation, several distinct lines or patterns of superficial neuromasts, defined by their topographical position and innervation, can be recognized across diverse actinopterygian taxa and are shown schematically in Figure 22.11. Because the names of these lines are so varied in the literature, we have devised a standard terminology based on the canal with which the superficial line is associated and the type of association (i.e., replacement vs. accessory). Presumed homologues in various fish groups are summarized in Tables 22.3 and 22.4 by the names originally used to describe them. In cases in which the distinction between accessory and replacement lines could not be made, the simple designation of ''line'' is used, and these cases are explained below.

The *dorsal supraorbital line,* or anterior pit line (Allis 1889), appears to form a posterior continuation of the supraorbital canal. In certain cyprinids, the homologue of this line is apparently a canal (Lekander 1949). Since this line in primitive actinopterygian fish and its putative homologue in dipnoans (Pehrson 1949) is superficial, it is likely that the superficial condition is primitive for bony fish.

Table 22.3. Lines of Superficial Neuromasts Innervated by the Superficial Opthalmic (*1*), Buccal (*2*), Mandibular (3), Opercular (*4*), and Undescribed (*U*) Rami of the Anterior Lateral Line Nerve

| Ramus | Superficial Line | Fish Taxon | References |
|---|---|---|---|
| | | Dorsal Supraorbital Line ($DL_{so}$) | |
| 1 | Anterior series | Lepidosteidae | Song (1984) |
| 1 | Anterior pit line | Amiidae | Allis (1889) |
| 1 | Anterior head line | Gymnarchidae | Pehrson (1945) |
| U | Anterior line | Elopidae | Nelson (1972) |
| U | Anterior line | Salmonidae | Nelson (1972) |
| U | Anterior line | Osmeridae | Nelson (1972) |
| U | Anterior line | Umbridae | Nelson (1972) |
| U | Anterior line | Esocidae | Nelson (1972) |
| 1 | Posterior portion of supraorbital canal[a] | Cyprinidae | Lekander (1949) |
| 1 | Parapostorbital line | Percidae | Jakubowski (1966b) |
| | | Supraorbital Rostral Replacement ($RR_{so}$) | |
| 1 | Tectal pit line | Polypteridae | Jarvik (1947) |
| U | Nasal pit line | Polypteridae | Pehrson (1947) |
| 1 | Vestigial anterior commissure | Amiidae | Allis (1889) |
| 1 | Nasal line | Cyprinidae | Lekander (1949) |
| 1 | Nasal line | Cobitidae | Lekander (1949) |
| 1 | Rostral line | Gadidae | Jakubowski (1967a) |
| 1 | Rostral line | Percidae | Jakubowski (1966b) |
| | | Supraorbital Accessory ($A_{so}$) | |
| 1 | Medial and lateral supraorbital pit lines | Cyprinidae<br>Cobitidae | Lekander (1949)<br>Lekander (1949) |
| 1 | Parasupraorbital | Gadidae | Jakubowski (1967a) |

Table 22.3 *continued*

| Ramus | Superficial Line | Fish Taxon | Reference |
|---|---|---|---|
| | Caudal Fork Replacement of the Infraorbital Canal ($CFR_{io}$) | | |
| 2 | Posterior commissure[a] | Amiidae | Allis (1889) |
| 2 | Antorbital canal[a] | Engraulidae | Nelson G.J. (1984) |
| 2 | Postnasal line | Esocidae | Pehrson (1944) |
| 2 | Antorbital line | Esocidae | Nelson (1972) |
| 2 | Antorbital line | Umbridae | Nelson (1972) |
| U | — | Gymnarchidae | Pehrson (1945) |
| 2 | Antorbital pit line | Cyprinidae | Lekander (1949) |
| 2 | Antorbital pit line | Cobitidae | Lekander (1949) |
| 2 | Paranasal line | Gadidae | Jakubowski (1967a) |
| 2 | Paranasal line | Percidae | Jakubowski (1966b) |
| | Rostral Fork Replacement of the Infraorbital Canal ($RFR_{io}$) | | |
| U | Ethmoidal commissure[a] | Polypteridae | Traquair (1871) |
| 2 | Ethmoidal commissure[a] | Amiidae | Allis (1889) |
| 2 | Rostral canal[a] | Clupeidae | Wolfhart (1937) |
| 2 | Rostral organ (in part)[a] | Engraulidae | Nelson G.J. (1984) |
| U | Supraethmoidal commissure | Esocidae | Pehrson (1944) |
| U | Ethmoidal line | Esocidae | Nelson (1972) |
| U | Ethmoidal line | Umbridae | Nelson (1972) |
| 2 | Rostral pit line | Gymnarchidae | Pehrson (1945) |
| 2 | Rostral pit commissure | Siluridae | Lekander (1949) |
| 2 | Rostral pit commissure | Cyprinidae | Lekander (1949) |
| 2 | Rostral pit commissure | Cobitidae | Lekander (1949) |
| U | Ethmoidal Commissure[a] | Salmonidae | Pehrson (1944) |
| 2 | Supralabial line | Gadidae | Jakubowski (1967a) |
| 2 | Rostral commissure | Percidae | Jakubowski (1966b) |
| 2 | Rostral canal[a] | Mullidae | Jakubowski (1966c) |
| | Ventral Infraorbital Line ($VL_{io}$) | | |
| 2 | Infraorbital pit line | Esocidae | Pehrson (1944) |
| 2 | Accessory infraorbital line | Cyprinidae | Lekander (1949) |
| 2 | Accessory infratorbital line | Siluridae | Lekander (1949) |
| 2 | — | Gadidae | Jakubowski (1967a) |
| 2 | — | Percidae | Jakubowski (1966b) |
| | Dorsal Mandibular Line ($DL_m$) | | |
| 3 | Horizontal line | Amiidae | Allis (1889) |
| 3 | Horizontal cheek series | Lepisosteidae | Song (1984) |
| 3 | Horizontal cheek line | Esocidae | Pehrson (1944) |
| 3 | Horizontal line | Cobitidae | Lekander (1949) |
| 3 | Horizontal line | Siluridae | Lekander (1949) |

Table 22.3 *continued*

| Ramus | Superficial Line | Fish Taxon | Reference |
|---|---|---|---|
| | | Median Mandibular Line ($ML_m$) | |
| 3 | Vertical line | Amiidae | Allis (1889) |
| 3 | Vertical cheek series | Lepisosteidae | Song (1984) |
| 3 | S-line | Esocidae | Pehrson (1944) |
| 3 | Mandibular line of pits | Gymnarchidae | Pehrson (1945) |
| 3 | Vertical line | Cyprinidae | Lekander (1949) |
| 3 | Oral lines | Gadidae | Jakubowski (1967a) |
| 3 | Buccal line | Percidae | Jakubowski (1966b) |
| | | Ventral Mandibular Line ($VL_m$) | |
| 3 | Mandibular line of pits | Amiidae | Allis (1889) |
| U | Mandibular line of pits | Esocidae | Pehrson (1944) |
| 3 | Accessory preopercular line | Siluridae | Lekander (1949) |
| 3 | Accessory preopercular line | Cyprinidae | Lekander (1949) |
| 3 | Accessory preopercular line | Cobitidae | Lekander (1949) |
| 3 | Medial line | Gadidae | Jakubowski (1967a) |
| | | Rostral Mandibular Accessory ($RA_m$) | |
| 3 | Anterior mandibular line | Gymnarchidae | Pehrson (1945) |
| 3 | Mandibular line | Esocidae | Pehrson (1944) |
| 3 | Symphysial line | Cyprinidae | Lekander (1949) |
| | | Rostral Mandibular Replacement ($RR_m$) | |
| U | Mentomandibular line | Esocidae | Nelson (1972) |
| U | Mentomandibular line | Umbridae | Nelson (1972) |
| 3 | Anterior mandibular line | Cyprinidae | Lekander (1949) |
| | | Ventral Opercular Line ($VL_o$) | |
| 4 | Opercular line of pits | Gymnarchidae | Pehrson (1945) |
| 4 | Vertical opercular line and lower horizontal line | Cyprinidae<br>Cobitidae | Lekander (1949) |
| 4 | Opercular line and lower opercular line | Gadidae | Jakubowski (1967a) |
| 4 | Opercular line | Percidae | Jakubowski (1966b) |
| U | Transverse row and inferior longitudinal row (in part) | Gobiidae | Pinchuk (1980) |

Table 22.3 *continued*

| Ramus | Superficial Line | Fish Taxon | Reference |
|---|---|---|---|
| | | Dorsal Opercular Line ($DL_o$) | |
| 4 | Upper horizontal line | Cobitidae<br>Cyprinidae | Lekander (1949) |
| 4 | Upper opercular line | Gadidae | Jakubowski (1967a) |
| 4 | Inferior longitudinal row (in part) | Gobiidae | Pinchuk (1980) |

[a] Line exists as a canal.

Similarly the *dorsal postotic line,* commonly called the middle pit line, does not appear as a canal in any extant actinopterygian fish, and its presumed homologue in dipnoans is also superficial (Pehrson 1949).

Several lines of superficial neuromasts, named here according to their relative positions in *Amia,* occur near the preoperculomandibular canal in actinopterygian fishes. The *median and ventral mandibular lines* are innervated by mandibular ramus twigs (presumably the cutaneous mandibular externus of Harrison 1981) that course rostrodorsally to the main ramus. One or both of these twigs may be homologous to the internal mandibular ramus that innervates the oral canal in dipnoans (Northcutt 1985). Holmgren and Pehrson (1949) claim that these lines in actinopterygian fish are homologous to the jugal and secondary oral canals of dipnoans. The *dorsal mandibular line,* also innervated by the mandibular ramus, runs horizontally from the preopercular canal to the infraorbital canal in *Amia* (Allis 1889), but in various teleosts runs more vertically along the edge of the preopercular canal (Lekander 1949). This line is thought to be homologous to the secondary jugal line in dipnoans (Holmgren and Pehrson 1949). There is also an opercular ramus, which appears to branch off from the mandibular ramus to innervate the *dorsal and ventral opercular lines*. Opercular lines have been reported in many extant fish groups (Holmgren and Pehrson 1949), but it is unclear whether these lines are superficial in all cases.

Finally, the innervation and distribution of superficial neuromasts on the trunk are less clear than those on the head. A *dorsal trunk line* extending from above the gills to the first dorsal fin is characteristic of many actinopterygian (Table 22.4 and Fig. 22.12) and elasmobranch fishes (Tester and Kendall 1967). In actinopterygian fishes, this line is innervated by a branch of the supratemporal dorsal accessory ramus (Northcutt 1985), which is called the dorsal longitudinal ramus by Freihofer (1972) and the dorsalis ramus by Jakubowski (1966b). Apparently this line extends along the entire body in the elasmobranch *Mustelus* (Johnson 1917; Tester and Kendall 1967), and a similar dorsal line is formed by a row of short stitches in dipnoan fishes (Northcutt 1985). The evolutionary relationship between dorsal trunk lines found in extant actinoptergian, elasmobranch, and dipnoan fishes is unclear, however, as is the relationship of these superficial

Table 22.4. Lines of Superficial Neuromasts Innervated by the Squamosal (*5*), Supratemporal Dorsal Accessory (*6*), Trunk (*7*), and Undescribed (*U*) Rami of the Posterior Lateral Line Nerve

| Ramus | Superficial Line | Fish Taxon | Reference |
|---|---|---|---|
| | | Dorsal Post-otic Line ($DL_{po}$) | |
| 5 | Middle pit line | Amiidae | Allis (1889) |
| 5 | Middle series | Lepisosteidae | Song (1984) |
| U | Middle line | Hiodontidae | Nelson (1972) |
| U | Middle line | Elopidae | Nelson (1972) |
| U | Middle line | Esocidae | Nelson (1972)<br>Pehrson (1944) |
| U | Middle line | Umbridae | Nelson (1972) |
| 5 | Medial extrascapular line | Cyprinidae | Lekander (1949) |
| 5 | Middle pit line | Ictaluridae | Herrick (1901) |
| 5 | Medial line | Gadidae | Jakubowski (1967a) |
| | | Supratemporal Accessory ($A_{st}$) | |
| 6 | Posterior pit line | Amiidae | Allis (1889) |
| 6 | Posterior series | Lepisosteidae | Song (1984) |
| | — | Cyprinidae | Lekander (1949) |
| 6 | Parasupratemporal line | Gadidae | Jakubowski (1967a) |
| 6 | Parasupratemporal line | Percidae | Jakubowski (1966b) |
| | | Dorsal Trunk Line ($DL_{tr}$) | |
| 6 | Dorsal line of pits | Amiidae | Allis (1889) |
| U | Dorsolateral series | Lepisosteidae | Song (1984) |
| U | Dorsolateral line | Cyprinidae | Strivastava and Strivastava (1969) |
| 6 | — | Ictaluridae | Herrick (1901) |
| 6 | — | Atherinidae | Freihofer (1972) |
| 6 | — | Cyprinodontidae | Freihofer (1972) |
| 6 | — | Oryziatidae | Freihofer (1972) |
| 6 | — | Percidae | Freihofer (1972) |
| 6 | Predorsal fin line | Percidae | Jakubowski (1966b) |
| | | Trunk Accessories ($A_{tr}$) | |
| 7 | — | Amiidae | Allis (1889) |
| 7 | Dorsal, adjacent series | Lepisosteidae | Song (1984) |
| U | Trunk accessories | Cyprinidae | Strivastava and Strivastava (1969) |
| U | — | Ictaluridae | Herrick (1901) |
| 7 | — | Percopsidae | Freihofer (1972) |
| 7 | — | Atherinidae | Freihofer (1972) |

lines to dorsal trunk canals reported for primitive bony fishes (Watson 1937; Stensio 1947).

Variations in Superficial Neuromast Distribution

In those elasmobranch fishes which have been examined, superficial neuromasts on the head appear to be restricted to a pair in front of the endolymphatic pore, a mandibular row, and in some cases an umbilical row (Boord and Campbell 1977; Tester and Kendall 1967). On the trunk, the pattern of superficial neuromasts is more similar to that found in actinopterygian fishes (Tester and Kendall 1967). It is clear from Table 22.3 and Figure 22.11 that superficial neuromasts on the heads of most primitive actinopterygian fishes and many teleosts are not randomly placed and tend to occur in association with canals. Proliferation of superficial neuromasts in many teleostean groups may lead to considerable variation in distribution patterns, however, especially on the trunk.

Unlike canal neuromasts, which have their hair cells oriented along the canal axis, superficial neuromasts may be oriented in various directions, but tend to form lines that run either parallel or perpendicular to nearby canal axes (Dijkgraaf 1962; Merrilees and Crossman 1973a; Hama 1978; Münz 1979). Preliminary work on *Cottus bairdi* (Cottidae) (Janssen et al. in press) suggests that accessory superficial neuromasts may be oriented either parallel or perpendicular to the axis of the canal they accompany, whereas superficial replacement neuromasts align themselves along the axis of the canal they have replaced.

## Morphological Constraints in the Evolution of Lateral Line Systems

In general, superficial neuromasts are more labile in their development, distribution, and hair cell orientation than are canal neuromasts. Neuromasts are known to have morphogenetic influences on bone and scale formation (Merrilees and Crossman 1973b; Graham-Smith 1978), and it is probable that bone and scale in turn place certain morphological constraints on canal neuromast growth and distribution. The shape of canal neuromasts is clearly correlated with canal diameter (see p. 562), suggesting that canals constrain neuromast shape. Moreover, developing canal neuromasts are clearly capable of budding before becoming enclosed in canals (Allis 1889; Lekander 1949; O'Connell 1981), whereas multiplication after canal formation has not been documented. Although superficial neuromasts are known to increase in number and change their pattern of organization in the growing anchovy *Engraulis mordax* (O'Connell 1981), it is not clear to what extent this occurs in adult fish.

The apparent ability of superficial neuromasts to multiply, though not carefully studied, is reminiscent of the budding process described for tuberous and ampullary electroreceptors in *Sternopygus* (Gymnotidae) (Zakon 1984). Other characteristics shared by superficial neuromasts and electroreceptors include similar innervation and distribution patterns in association with canals, ampullary morphology of pits (Fig. 22.8), and the possibility that superficial neuromasts, like electroreceptors, lack efferent innervation (Hama 1978; Münz 1979). In this regard, it is interesting to note that superficial neuromasts are few or nonexistent in some silurids (Lekander 1949) and mormyrids (Harder 1968), suggesting that superficial neuromasts in these freshwater electroreceptive fish have been replaced by or are homologous to electroreceptive ampullary organs (Lekander 1949; Northcutt 1985). Similarly the ampullae of Lorenzini and vesicles of Savi are restricted to the head of elasmobranch fishes (Boord and Campbell 1977), where superficial neuromasts are fewer in number relative to the trunk. Coggi (1891) and Garman (1888) have suggested that the vesicles of Savi evolved from lateral line canal organs, but the notion that they, along with the ampullae of Lorenzini, are more closely related to superficial neuromasts has received no attention.

## Convergent Trends

Although a few major taxa can be defined by characteristics of the lateral line system (e.g., Notopteroidei and Clupeomorpha), several major trends are convergent in many diverse taxa (Webb in preparation), including (1) widening of the cephalic lateral line canals, (2) branching of tubules, (3) loss of canals and their replacement by superficial neuromasts, and (4) placement and extent of canals.

Differences in the placement and extent of lateral line organ systems seem to be related to other morphological changes. Among gnathostomes, modifications in the distribution of lateral line organs have accompanied rostral elongation of the body, the development of gill slits and covers, and changes in the directions of the visceral arches (Jarvik 1980). Variations in the head and body shape of fish are also reflected in the course and extent of canals. In the elasmobranchs, the flattening of the body in the skates and rays is correlated with the extension of the hyomandibular canal onto the greatly enlarged pectoral fin, just as the elongation of the sawfish (*Pristis*) rostrum is correlated with an extension of the supraorbital canal (Chu and Wen 1979). The dramatic morphological changes during flatfish metamorphosis are also accompanied by modifications in the number, type, and distribution of lateral line organs on the ocular and abocular sides (Fig. 22.14E) (Roper 1981; Applebaum and Schemmel 1983).

Finally, the arching of the trunk line is correlated with the evolutionary migration of the pectoral fin onto the flank of the Paracanthopterygii and Acanthopterygii (Marshall 1971). Since swimming patterns and maneuverability are altered by the change in fin position (Webb 1984), arching may reduce the hydrodynamical interference of the pectoral fin (Marshall 1971). Similarly, for trichurids, which swim by undulating their dorsal fin, it has been hypothesized that the ventral location of the trunk canal is an adaptation for removing the hydrodynamical sensors from the source of self-generated noise (Bone 1971).

Habitat preference and feeding behaviors are also correlated with the position of the lateral line system. According to Marshall (1971), dorsally placed trunk canals are found in sixteen families of benthic-associated fishes, such as Uranoscopidae, whereas ventrally placed trunk canals are characteristic of surface-dwelling fish, such as members of the Cyprinodontiformes and Beloniformes. Similarly the lateral line system in the benthic skates and rays is much more elaborated on the dorsal than on the ventral surface when compared to the pelagic species (Chu and Wen 1979). Interestingly, many surface-feeding fish, like the butterflyfish *Pantodon* (Pantodontidae) and the topminnow *Aplocheilus* (Cyprinodontidae) have cephalic specializations of the lateral line system on the dorsal surface of the head (Schwartz 1965; Hoin-Radkovsky, Bleckmann and Schwartz 1984), but a ventrally placed trunk canal. Thus, though the cephalic system is apparently specialized for surface feeding (Bleckmann 1985), the trunk system may serve a different function, perhaps predator detection.

Canal loss and replacement with superficial neuromasts occurs repeatedly in the Euteleostei (Fig. 22.14 and Table 22.1) and seems to represent a derived paedomorphic state in which primary superficial neuromasts fail to develop into canal neuromasts in the adult. This tends to occur in small species like the neon tetras (Characidae, Weitzman and Fink 1983), suggesting that there are size constraints on canal function

Branched tubules are characteristic of most elasmobranchs and subeuteleostean fishes (Table 22.1 and Fig. 22.6), but are lacking in *Polypterus* (Collinge 1893b). This primitive condition appears to have been lost and reevolved in a few advanced teleosts (Webb in preparation). The number and complexity of secondary branches apparently increases with the size of the fish (Allis 1889; Stephens 1985), indicating that branching may be a mechanism for sampling signals over larger areas as the fish grows.

Finally, widened head canals and their associated specializations (see pp. 562) appear to be derived conditions that have arisen independently in a number of euteleostean groups (Table 22.1 and Fig. 22.4).

## Functional Implications of Morphological Variation

The many examples of convergent evolution outlined above suggest that various morphological patterns of the lateral line system represent functional specializations that have evolved many different times under similar selective pressures. Figure 22.15 illustrates three major morphological variations of the lateral line system that have emerged from our comparative survey. The biggest variation is between canal and superficial neuromasts. Although the latter may range from pit to papillate, they differ as a group from canal systems along many anatomical dimensions (see p. 561). Evidence for a functional distinction comes from studies on the African cichlid *Sarotherodon niloticus,* which show major differences in the frequency response and absolute sensitivity of canal and superficial neuromasts in the same species (Münz 1985).

On the basis of relative canal diameter and other anatomical specializations, canal systems can be further subdivided into narrow and wide subsystems. In narrow canals, the detecting element (neuromast + cupula) is located in a local canal constriction with its long axis parallel to the canal axis (Figs. 22.2A and 22.15). This maximizes water flow past the sensor, suggesting that neuromasts

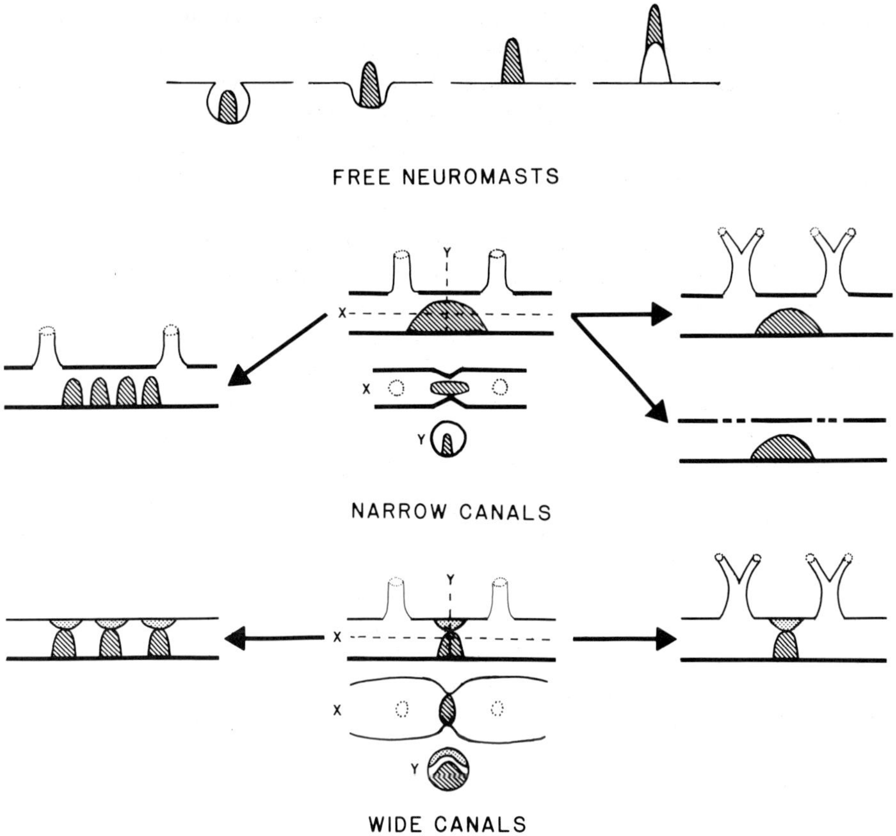


Figure 22.15. Diagrammatical representation of superficial (free) neuromast types (*top panel*) and two major types of lateral line canals (*middle and bottom panels*). All structures are shown in sagittal section with dorsal (*X*) and cross-sectional (*Y*) views of basic narrow and wide canal types. Neuromasts are represented as *hatch-marked structures,* whereas canal outlines are represented by heavy lines if bone and thin lines if membranous. *Stippled areas* represent structural specializations for canal compartmentalization. Variations in these three basic types are shown to the *left and right of center.*

in narrow canals are velocity detectors, as has been shown for clupeids by Gray (1983). The canal is enclosed in either bone or scale, so that its walls are relatively rigid and the only stimulation pathway is through relatively small pores. Several pores may have an effect on a single neuromast; conversely, several neuromasts may be affected by only a few pores.

In contrast, widened canals have greatly enlarged tubules and stretches of tissue that may function as flexible, tympanum-like interfaces with the environment. In some cases, there is a tendency toward the loss of pores, resulting in a single membrane stretched across several neuromasts (i.e., Notopteroidei) (Fig. 22.3). The long axis of the detecting element is perpendicular to the canal (Fig. 22.7). Partitions are located at each neuromast (Figs. 22.4 and 22.7C), but it is unclear to what extent the canal is further occluded by the cupula. Partial occlusion would maximize water velocity in the area of the neuromast, as local constrictions apparently do in narrower canals. Complete occlusion would mean that the detecting element is displacement-sensitive, with the cupula serving as the only release point for the incompressible canal fluids when the overlying membrane is depressed.

It is quite clear from work by Münz (1985) and Denton and Gray (Chap. 23) that the frequency response properties of some of these morphologically different types of sensors differ, but whether they have different biological functions remains to be seen. Correlations between sensor types and the habitat and life-style of fishes suggest that ambient noise levels provide a major selection pressure for the evolution of lateral line systems. Canal loss and replacement with superficial neuromasts are associated with fish living in relatively still waters or having sedentary behaviors (Dijkgraaf 1962; Marshall 1971; Merrilees and Crossman 1973a). Similarly, Jakubowski (1967b) has shown that percid species with narrow canals and parallel neuromasts live in strong currents, whereas those with widened canals and transverse neuromasts live in relatively stagnant environments. Furthermore, it is not unusual to find widened canals that are accompanied by a proliferation of papillate superficial neuromasts (Moore and Burris 1956; Schemmel 1977). It may be that low-noise enviroments favor the evolution of both widened canals and supernumerary superficial neuromasts. Obviously, information about biologically relevant signals in addition to background noises is needed before a complete understanding of the function and evolution of these systems can be reached.

## Summary and Conclusions

Several dimensions of variability have been identified for the lateral line systems of fish, including (1) the spatial distribution of canals and superficial neuromasts; (2) the structure of canals and their openings (pores and tubules); and (3) the size, shape, topography, and orientation of superficial and canal neuromasts. A few fish taxa can be defined by unique features of the lateral line system, but there are also major trends that are convergent across taxa. These include (1) secondary branching of tubules, (2) loss of canals and their replacement with superficial neuromasts, (3) widening of cephalic canals, and (4) placement and extent of canals on the bodies of fish. On the basis of convergent trends and shared morphological features, the lateral line system has been categorized into three major functional classes: superficial neuromasts and narrow and wide canal systems. In some instances, these broad functional classes can be correlated with specific life-styles or habitat, with ambient noise level being a common denominator. The distribution of canal organs is remarkably similar in all fish groups, but the distribution of superficial neuromasts, particularly on the trunk, is more variable. It is suggested that superficial neuromasts, being under fewer morphological constraints than canal neuromasts, are more labile, in both a developmental and evolutionary sense.

*Acknowledgments.* Original research for this chapter was supported by Grant 1-PO1-NS-2197, National Institues of Health. Many thanks to Denise Hoekstra for her technical assistance in many phases of the research and manuscript preparation and to Robert Zeller and Phillip Newmark for their help in preparing figures. We also want to thank Jiakun Song for fruitful discussions during the early stages of preparing the manuscript and Dr. Eric Denton, Sir John Gray, Mr. Rick Puzdrowski, Dr. R. Glen Northcutt, Dr. Catherine H. McCormick and Dr. Arthur N. Popper for many helpful suggestions on how to improve the chapter.

## References

Allis, E.P. (1889) The anatomy and development of the lateral line system in *Amia calva,* J. Morphol. 2 (3):463–542.

Allis, E.P. (1903) The lateral sensory system in the Muranidae, Int. Monatsschr. Anat. Physiol., 20 (4–6):1–48.

Allis, E.P. (1923) The cranial anatomy of *Chlamydoselachus anguineus,* Acta Zool. (Stockh.) 4:123–221.

Applebaum, S., and Schemmel, C. (1983) Dermal sense organs and their significance in the feeding behavior of the common sole *Solea vulgaris,* Mar. Ecol. Prog. Ser., 13:29–36.

Bardack, D. (1965) Anatomy and evolution of chirocentrid fishes, Univ. Kans. Paleontol. Contrib. 10:1–88.

Barry, M.A. and Boord, R.L. (1984) The spiracular organ of sharks and skates: anatomical evidence indicating a mechanoreceptive role, Science, 226:990–992.

Best, A.C. and Gray, J.A.B. (1982) Nerve fibre and receptor counts in the sprat utriculus and lateral line, J. Mar. Biol. Assoc. U.K. 62:201–213.

Bigelow, H.B., and Schroeder, W.C. (1953) Chimeroids, in Fishes of the Western North Atlantic, Tee-Van, J. (ed.), Sears Foundation for Marine Research, pp. 515–562.

Bleckmann, H. (1985) The water surface as signal transmission channel: How surface waves are used for prey identification, prey localization and intraspecific communication, in Ottoson, D. (eds.), Progress in Sensory Physiology, Springer-Verlag, New York.

Bone, Q. (1971) On the scabbard fish, *Aphanopus carbo,* J. Mar. Biol. Assoc. U.K., 51:219–225.

Boord, R.L. and Campbell, C.B.G. (1977) Structural and functional organization of the lateral line system of sharks, Am. Zool., 17:431–441.

Branson, B.A. and Moore, G.A. (1962) The lateralis components of the acoustico-lateralis system in the sunfish family Centrarchidae, Copeia, 1:1–108.

Bretschneider, F. (1979) Aspects of electroreceptor functioning in the catfish, *Ictalurus nebulosus,* Ph.D. thesis, University of Utrecht.

Budker, P. (1958) The cutaneous sensory organs of Selachians, Grasse's Traité de Zoologie, 13:1033–1062.

Bullock T.H. (1982) Electroreception, Annu. Rev. Neurosci., 5:121–170.

Chao, L.N. (1978) A basis for classifying Western Atlantic Sciaeidae, (Teleostei: Perciformes), NOAA Tech. Rep. NMFS Circ., pp. 415:64.

Chu, Y.T. and Wen, M.C. (1979) A study of the lateral line canals system and that of Lorenzini ampullae and tubules of elasmobranch fishes of China, Monog. Fish. China, 2:132.

Coggi, A. (1891) Le vesciole di Salvi e gli organi della linea laterale nella torpedini, Atti. Accad. Lincei. Roma, 4, ser. 7:2.

Cole, E.J. (1896) On the cranial nerves of *Chimaera monstrosa* with a discussion of the lateral line system and the morphology of the chorda tympani, Trans. R. Soc. Edinb., 38:631–680.

Collette, B.B. (1966) A review of the venomous toadfishes, subfamily Thalassophryninae, Copeia, 1966 (4):846–864.

Collette, B.B. (1983) Recognition of two species of double lined mackerels (Grammatorcynus: Scombridae), Proc. Biol. Soc. Wash., 96 (4):715–718.

Collette, B.B. and Russo, J.L. (1981) A revision of the scaly toadfishes, genus *Batrachoides,* with descriptions of the two new species from eastern Atlantic, Bull. Mar. Sci., 31 (2):197–233.

Collinge, W.E. (1893a) The lateral canal system of *Lepidosteus osseus*. Proc. Birmingham Philos. Soc., 8:263–272.

Collinge W.E. (1893b) Note on the lateral canal system of *Polypterus,* Proc. Birmingham Philos. Soc., 8:255–262.

Collinge, W.E. (1895) The sensory canal system of fishes. I. Ganoidei, Q. J. Microsc. Sci., 36:499–537.

Cox, V.O. (1905) A revision of the cave fishes of North America, Appendix to the Report of Commission of Fisheries to the Secretary of Commerce and Labor for the year ending June 30, 1904, pp. 377–393.

Deng, S., Xiong, G., and Zhan, H. (1981) A comparative study of the lateral line canal system of Chinese Stromateidae, Acta Zool. Sin., 27:232–239.

Denny, M. (1937) The lateral line system of the teleost *Fundulus heteroclitus,* J. Comp. Neurol., 68 (1):49–65.

Denton, E. and Blaxter, J. (1976) The mechanical relationships between the clupeid swimbladder, inner ear and the lateral line, J. Mar. Biol. Assoc. U.K., 56:787–807.

Denton, J., Gray, J.A.B., and Blaxter, J.A.S. (1979) The mechanics of the clupeid acoustico-lateralis system: frequency responses, J. Mar. Biol. Assoc. U.K. 59:27–47.

Dijkgraaf, S. (1962) The functioning and significance of the lateral line organs, Biol. Rev., 38:51–105.

Dijkgraaf, S. (1967) Biological significance of the lateral line organs, in Lateral Line Detectors, Cahn, P. (ed.), Indiana University Press, Bloomington, pp. 83–95.

Disler, N.N. (1960) Lateral line sense organs and their importance in fish behavior, Acad. Sci. USSR Severtsov Institute of Animal Morphology (translated from Russian: Israel Program for Scientific Translations, Jerusalem, 1971, U.S. Department of Commerce, National Technical Information Service, Springfield, Va. 22151).

Disler, N.N. and Smirnov, S.A. (1977) Sensory organs of the lateral line canal system in two percids and their importance in behavior, J. Fish Res. Board Can., 34:1492–1503.

Fange, R., Larsson, A., and Lidman, U. (1972) Fluids and jellies of the acousticolateralis system in relation to body fluids in *Coryphaenoides rupestris* and other fishes, Mar. Biol., 17:180–185.

Flock, A. (1965) Electron microscopic and electrophysiologial studies on the lateral line canal organ, Acta Oto-laryngol., 199:1–90.

Flock, A. and Wersall, J. (1962) A study of the sensory hairs of the receptor cells in the lateral line organ of fish with special reference to the function of the receptors, J. Cell Biol., 15:19–27.

Freihofer, W. (1972) Trunk lateral line nerves, hyoid arch, gill rakers, and olfactory bulb location in atheriniform, mugilid, and percoid fishes, Occas. Pap. Calif. Acad. Sci., 95:31.

Garman, S. (1888) On the lateral canal system of the Selachia and Holocephala, Bull. Mus. Comp. Zool., 17:57–120.

Gosline, W.A. (1974) Certain lateral-line canals of the head in Cyprinid fishes with particular references to the derivation of North American forms, Jpn. J. Ichthyol., 21 (1):9–15.

Graham-Smith, W. (1978) On the lateral lines and dermal bones in the parietal region of some crossopterygian and dipnoan fishes, Philos. Trans. R. Soc. Lond. Biol. Sci., 282:41–105.

Gray, J. A. B. (1983) Interaction of sound pressure and particle acceleration in the excitation of lateral line neuromasts of sprats, Proc. R. Soc. Lond. Biol. Sci., 220:299–325.

Greenwood, P.H. (1973) Interrelationships of osteoglossomorphs, in Interrelationships of Fishes, Greenwood, P.H., Miles, R.S., and Patterson, C. (eds.), Academic Press, London, pp. 307–332.

Hama, K. (1978) A study of the fine structure of the pit organ of the common Japanese sea eel *Conger myriaster,* Cell Tissue Res., 189:375–388.

Harder, W. (1968) Die Beziehungen zwischen Elektrorezeptoren, elektrischem Organ, Seitenlinienorganen und Nervensystem bei den Mormyridae (Teleostei, Pisces), Z. Vgl. Physiol., 59:272–318.

Harrison, G. (1981) The cranial nerves of the teleost *Trichiurus lepturus,* J. Morphol. 167:119–134.

Herrick, C.J. (1901) The cranial nerves and cutaneous sense organs of the North American siluroid fishes, J. Comp. Neurol., Vol. II 177–249.

Herrick, D.J. (1903) On the morphology and physiological classification of the cutaneous sense organs of fishes, Am. Nat., 37:313–318.

Hoin-Radkovsky, I., Bleckmann, H., and Schwartz, E. (1984) Determination of source distance in the surface-feeding fish *Pantodon buchholzi,* Anim. Behav., 32:840–851.

Holmgren, N. and Pehrson, T. (1949) Some remarks on the ontogenetical development of the sensory lines on the cheek in fishes and amphibians, Acta Zool. Stockh., 30:249–314.

Hopkins, C.D. (1983) Functions and mechanisms in electroreception, in Fish Neurobiology, vol. I, Brainstem and Sense Organs, Northcutt, R.G., and Davis R.E. (eds.), University of Mich Press, Ann Arbor.

Hoss, D.E. and Blaxter, J.H.S. (1982) Development and function of the swimbladder-inner ear–lateral line system in the Atlantic menhaden, *Brevoortia tyrannus* (Latrobe), J. Fish Biol., 20:131–142.

Hoyt, R.D. (1972) Anatomy and osteology of the cephalic lateral line system of the silverjaw minnow, *Ericymba buccata,* Copeia, 1972:812–816.

Illick, H.J. (1956) A comparative study of the cephalic lateral-line system of the North American Cyprinidae, Am. Midl. Nat., 56 (1):204–223.

Jakubowski, M. (1965) Cutaneous sense organs of fishes. II. The structure of lateral line organs in the burbot (*Lota lota*) and pike (*Esox lucius*), Acta Biol. Cracov., 8:87–99.

Jakubowski, M. (1966a) Cutaneous sense organs of fishes. III. The lateral line organs in some *Cobitidae,* Acta Biol. Cracov. Ser. Zool., 9:71–80.

Jakubowski, M. (1966b) Cutaneous sense organs of fishes. IV. The lateral-line organs in the perch-like *Lucioperca lucioperca* and perch *Perca fluviatilis,* their topography, innervation, vascularization, and structure, Acta Biol. Cracov. Ser. Zool., 9:138–149.

Jakubowski, M. (1966c) Cutaneous sense organs of fishes. Canal system of lateral-line organs in *Mullus barbatus ponticus* Essipov and *Spicara smaris* L. (topography, innervation, structure), Acta Biol. Cracov., 9:225–237.

Jakubowski, M. (1967a) Cutaneous sense organs of fishes. VI. The structure, topography, and innervation of lateral line organ in the burbot (*Lota lota*), Acta Biol. Cracov. Ser. Zool., 10:39–47.

Jakubowski, M. (1967b) Cutaneous sense organs of fishes. VIII. The structure of the system of lateral-line canal organs in the *Percidae,* Acta Biol. Cracov. Ser. Zool., 10:69–81.

Jakubowski, M. (1974) Structure of the lateral line canal system and related bones in the berycoid fish *Hoplostethus mediterraneus* Cuv. et Val. (Trachichthyidae, Pisces), Acta Anat., 87:261–274.

Janssen, J., S. Coombs, D. Hoekstra, C. Platt. In press. Anatomy and differential growth of the lateral line system of the mottled sculpin, *Cottus balvili* (Scorpaeniformes: Cottidae). Brain Behav. Evol.

Jarvik, E. (1947) Notes on the pit lines and dermal bones on the head in *Polypterus,* Zool. Bidr. Upps., 25:70–78.

Jarvik, E. (1980) Basic Structure and Evolution of Vertebrates, Vols. 1 and 2, Academic Press, New York, pp. 337 and 575.

Johnson, S.E. (1917) Structure and development of the sense organs of the lateral canal system of selachians (*Mustelus canis* and *Squalus acanthias*), J. Comp. Neurol., 28:1–74.

Jorgensen, J.M. (1982) Fine structure of the ampullary organs of bichir (*Polypterus senegalus)* with some notes on the phylogenetic development of electroreceptors, Acta Zool. Stockh., 63:211–217.

Kapoor, A.S. (1964) Functional morphology of latero-sensory canals in the Notopteridae (Pisces), Acta Zool. Stockh. 65:77–91.

Kapoor, A.S. (1966) Cephalic lateral line system in *Notopterus chitala,* Copeia, 1966 (2):352–353.

Karmovskaya, E.S. (1981) The systematics and distribution of eels of the genus *Labichthys.* (F. Nemichthyidae), J. Ichthyol., 19 (6):31–39.

Kuiper, J.W. (1967) Frequency characteristics and functional significance of the lateral line organ, in Lateral Line Detectors, Cahn, P.H. (ed.), Indiana University Press, Bloomington.

Lane, E.B. and Whitear, M. (1982) Sensory structures at the surface of fish skin. II. Lateralis system, Zool. J. Linn. Soc., 76:19–28.

Lauder, G.V. and Liem, K.F. (1983) The evolution and interrelationships of the Actinopterygian fishes, Bull. Mus. Comp. Zool., 150 (3):95–197.

Lekander, B. (1949) The sensory line system and the canal bones in the head of some Ostariophysi, Acta Zool. (Stockh.), 30:1–131.

Liem, K.F. (1981) A phyletic study of the Lake Tanganyika cichlid genera *Asprotilapia, Ecotodus, Lestradea, Cunningtonia, Opthalmochromis and Opthalmotilapia,* Bull. Mus. Comp. Zool. 149:191–214.

Makushok, V.M. (1961) Some peculiarities in the structure of the seismosensory system of the northern Blenniids (Stichaeoidae, Blennoideii, Pisces), Tr. Inst. Okeanol. Akad. Nauk USSR 43:225–269.

Marshall, N.B. (1965) Systematic and biological studies of the macrourid fishes (Anacanthini-teleostei), Deep-Sea Res., 12:299–322.
Marshall, N.B. (1971) Explorations in the Life of Fishes, Harvard University Press, Cambridge, pp. 204.
Marshall, N.B. (1979) Deep Sea Biology Developments and Perseptives, Garland PSTPM Press, New York, 556 pp.
Masuda, H., Amaoka, K., Araga, C., Uyeno, T., and Yoshino, T. (eds.) (1984) Fishes of the Japanese Archipelago, Tokai University Press, 437 pp. and plates.
McCormick, C.A. (1983) Organization and evolution of the octavolateralis areas of fishes, in Fish Neurobiology, vol. 1; Brain Stem and Sense Organs, Northcutt, R. Glen, and Davis, R.E. (eds.), University of Michagan Press, Ann Arbor.
Mees, G.E. (1974) The Auchenipteridae and Pimelodidae of the Suriname (Pisces: Nematognathi), Zool. Verh. (Leiden), 132:256 pp.
Merkel, F. (1880) Über die Endigungen der sensiblen Nerven in der Haut der Wirbelthiere, Rostock.
Merrilees, M.J. and Crossman, E.J. (1973a) Surface pits in the family Esocidae. I. Structure and types, J. Morphol., 141:307–320.
Merrilees, M.J. and Crossman, E.J. (1973b) Surface pits in the family Esocidae. II. Epidermal-dermal interaction and evidence for aplasia of the lateral line sensory system, J. Morphol. 141 (3):321–343.
Miller, P.J. and Wongrat, P. (1979) A new goby (Teleostei: Gobiidae) from the South China Sea and its significance for gobioid classification, Zool. J. Linn. Soc., 67:239–257.
Moore, G.A. and Burris, W.E. (1956) Description of the lateral line system of the pirate perch, *Aphredoderus sayanus,* Copeia, 1956:18–20.
Mullinger, A.M. (1964) A fine structure of ampullary electic receptors in *Ameirus,* Proc. R. Soc. Lond. B. Sci., 160:345–359.
Münz, H. (1979) Morphology and innervation of the lateral line system in *Sarotherodeon niloticus,* Zoomorphologie, 93:73–86.
Münz, H. (1985) Single unit activity in the peripheral lateral line system of the cichlid fish *Sarotherodon niloticus* L., J. Comp. Physiol. A, 157:555–568.
Münz, H. and Claas, B. (1983) The functional organization of neuromasts in the lateral line system of a cichlid fish, in Advances in Vertebrate Neuroethology, Ewert, J.P., Capranica, R.R., and Ingle, D.J. (eds.), Plenum, New York.
Nelson, G.J. (1969) Infraorbital bones and their bearing on the phylogeny and geography of Osteoglossomorph fishes, Am. Mus. Novit. 2394:1–37.
Nelson, G.J. (1972) Cephalic sensory canals, pitlines, and the classification of Esocoid fishes, with notes on Galaxiid and other teleosts, Am. Mus. Novit., 2492:1–49.
Nelson, G.J. (1983) *Anchoa argentivittata* with notes on other Eastern Pacific anchovies and the Indo-Pacific genus *Encrasicholina,* Copeia, 1983 (1):48–54.
Nelson, G.J. (1984) Notes on the rostral organ of anchovies, (Family Engraulidae), Jpn. J. Ichthyol., 31:86–87.
Nelson, J.S. (1984) Fishes of the World, John Wiley and Sons, New York.
Norris, H.W. (1932) The laterosensory system of *Torpedo marmorata,* innervation and morphology, J. Comp. Neurol., 56:169–178.
Northcutt, R.G. (1985) Electroreception in non-teleost bony fishes, in Electroreception, Bullock, T.H., and Heiligenbergieds, W. (eds.), John Wiley and Sons, New York.
O'Connell, C.P. (1981) Development of organ systems in the northern anchovy, *Engraulis mordax* and other teleosts, Am. Zool., 21:429–446.

Omarkhan, M. (1948) The lateral sensory canals of larval *Notopterus*, Proc. Zool. Soc. London, 118:938–972.

Page, L.M. (1977) The lateralis system of darters (Etheostomatini), Copeia, 1977 (3):472–475.

Parenti, L. (1981) A phylogenetic and biogeographic analysis of cyprinodontiform fishes (Teleostei, Atherinomorpha), Bull. Am. Mus. Nat. Hist., 168 (4):335–557.

Parin, N.V. and Astakhov, D.A. (1982) Studies on the acousticolateralis system of beloniform fishes in connection with their systematics, Copeia, 1982 (2):276–291.

Pehrson, T. (1944) The development of latero-sensory canal bones in the skull of *Esox lucius*, Acta Zool., 25:135–157.

Pehrson, T. (1945) The system of pit organ lines in *Gymnarchus niloticus*. Acta Zool., 26:1–8.

Pehrson, T. (1947) Some new interpretations of the skull in *Polypterus*, Acta Zool. (Stockh.) 28: 399–455.

Pehrson, T. (1949) The ontogeny of the lateral line system in the head of dipnoans, Acta Zool. (Stockh.), 30:153–182.

Pinchuk, V.I. (1980) The lateral line system of *Caspiosoma caspium* (Kessler 1877) and the systematic status of *Asra tyrcomanus*, Iljin 1941 J. Fish Biol., 17:231–233.

Popper, A.N., and Coombs, S. (1982) The morphology and evolution of the ear in Actinopterygian fishes, Am. Zool., 22:311–328.

Poulson, T. (1963) Cave adaptation in amblyopsid fishes, Am. Midl. Nat., 70:257–290.

Reno, H.W. (1971) The lateral-line system of the silverjaw minnow, *Ericymba baccata* Cope, Southwest. Nat., 15 (3):357–358.

Roberts, T.R. (1967) *Virilia*, a new genus of sexually dimorphic characid fishes from West Africa, with remarks on characoids having an incomplete lateral line, Stanford Ichthyol. Bull., 8(4):251–259.

Roper, D.S. (1981) Superficial neuromasts of the flatfish *Peltorhamphus novaezeelandiae*, J. Fish Biol., 18:753–758.

Rosen, D.E., and Mendelson, J.R. (1960) The sensory canals of the head in poeciliid fishes (Cyprinodontiformes) with reference to dentitional types, Copeia, 1960 (3):203–210.

Rutenberg, E.P. (1962) Survey of the fishes of family Hexagrammidae, in Greenlings, Rass, T.S. (ed.), Trans. Inst. Oceanol., 1–103, transposed from Russian, Israel Prog for Scientific Translation, 1970, Jerusalem.

Sato, M. (1955) Studies on the pit organs of fishes. I. Histological structure of the large pit organs, Jpn. J. Zool., 11:443–452.

Sato, M. (1956) Studies on the pit organs of fishes. IV. The distribution, histological structure, and development of the small pit organs, Anno. Zool. Jpn., 9:207–212.

Schemmel, C. (1967) Vergleichende Untersuchungen aus den Hautsinnesorganen ober- und unterirdisch lebender *Astyanax*-Foramen. Ein Beitrag zur Evolution der Cavernicolen, Z. Morphol. Tiere, 61:255–316.

Schemmel, C. (1977) Zur Morphologie und Funktion der Sinnerorgane von *Typhliasina pearsei* (Hubbs) (Ophidioidea, Teleostei), Zoomorphologie, 87:191–202.

Schwartz, E. (1965) Bau und Funktion der Seitenlinie des Streifenhechtlings (*Aplocheilus lineatus* Cuv. Val.), Z. Vgl. Physiol., 50:55–87.

Song, J. (1984) The distribution of lateral line receptors in garfish (Holostei), Am. Zool., 24:(3):134A.

Srivastava, C.B.L. and Srivastava, M.D.L. (1968) Lateral line organs in some teleosts: Cirrhina mrigala Ham. Buch., *Ophicephalus punctatus* Bloch and *Gobius striatus* Day, J. Comp. Neurol., 134:339–352.

Srivastava, M.D.L. and Srivastava, C.B.L. (1969) The development of neuromast in *Cirrhina mrigala*. Ham Buch. and *Ophicephalus punctatus* Bloch, J. Morphol. 122:321–344.

Stensio, E.A. (1947) The sensory lines and dermal bones of the cheek in fish and amphibians, Kungl. Sven. Vet Akad. Handl., 24:1–195.

Stephens, R.R. (1985) The lateral line system of the gizzard shad, *Dorosoma cepedianum* (Pisces: Clupeidae), Copeia, 1985 (3):540–555.

Szabo, T. (1963) Les fossettes sensorielles du *Gymnarchus niloticus*, Arch. Anat. Microsc. Morphol. Exp., 52:507–510.

Takagi, K. (1957) Description of some new gobioid fishes of Japan, with a proposition on the sensory line system as a taxonomic character, J. Tokyo Univ. Fish., 43 (1):98–126.

Taverne, L. (1977) Ostéologie, phylogenèse et systématique des Téléosteens fossiles et actuels du super-ordre des Osteoglossomorphes I, Acad. R. Belg. Mem. Cl. Sci. 42 (3):1–235.

Tester, A.L. and Kendall, J.I. (1967) Innervation of free and canal neuromasts in the sharks *Carcharhinus, Menisorrah* and *Sphyrna lewini,* in Lateral Line Detectors, Cahn, P.H. (ed.), Indiana University Press, Bloomington.

Tester, A.L. and Kendall, J.I. (1969) Morphology of the lateralis canal system in the shark genus, *Carcharhinus,* Pac. Sci., 23:1–16.

Traquair, R.H. (1871) On the cranial osteology of *Polypterus,* J. Anat. Physiol., 5:166–183.

Tretyakov, D.K. (1944) Outlines of the phylogeny of fishes, Acad. Sci. Ukranian S.S.R. Kiev.

van Bergeijk, W.A. and Alexander, S. (1962) Lateral line canal organs on the head of *Fundulus heteroclitus,* J. Morphol. 110:333–346.

von Lubitz, D.K.J. and Ekstrom (1981) Ultrastructure of the lateral line sense organ of the ratfish, Cell Tissue Res. 215:651–665.

Watson, D.M.S. (1937) Acanthodian fishes, Philos. Trans. R. Soc. Lond. B Biol. Sci., 228:49–146.

Webb, P.W. (1984) Body form, locomotion and foraging in aquatic vertebrates, Am. Zool., 24 (1):107–120.

Weber, D.D. and Schiewe, M.H. (1976) Morphology and function of the lateral line of juvenile steelhead trout in relation to gas-bubble disease, J. Fish Biol., 9:217–233.

Weitzman, S.H. and Fink, W.V. (1983) Relationships of the neon tetras as a group of South American freshwater fishes (Teleostei, Characidae) with comments on the phylogeny of New World Characiforms, Bull. Mus. Comp. Zool., 150 (6):339–395.

Weitzman, S.H. and Fink, S.V. (1985) Xenurobryconin phylogeny and putative pheromone pumps in glandulocaudine fishes (Teleostei: Characidae), Smithson. Contrib. and Zool., 421:1–121.

Winterbottom, R.R. (1982) A revision of the congrogadid fish genus *Halidesmus* (Pisces: Perciformes) with the description of a new species from Kenya and a list of the species included in the family, Can. J. Zool., 60:754–763.

Wolfhart, T.A. (1937) Anatomische untersuchungen über die Seitenkanäle der Sardine (*Clupea pilchardus* Walb), Z. Morphol. Oekol. Tiere, 33:381–411.

Yamada, Y. (1973) Fine structure of the ordinary lateral line organ. I. The neuromast of lamprey, J. Ultrastruct. Res., 43:1–17.

Zakon, H.H. (1984) Postembryonic changes in the peripheral electrosensory system of a weakly electric fish: addition of receptor organs with age, J. Comp. Neurol., 228:557–570.

# Chapter 23

# Mechanical Factors in the Excitation of the Lateral Lines of Fishes

ERIC J. DENTON* and JOHN A. B. GRAY*

The effective stimulus to a neuromast lying inside a typical lateral line system will be one that produces a movement of the cupula of the neuromast *relative* to the wall of the lateral line canal. The force moving the cupula must derive from liquid movement in the canal, and such motion arises from the *net* pressure gradient along the short stretch of canal carrying the neuromast. This gradient can be measured in terms of the local accelerations of the medium adjacent to the canal *relative* to the surface of the fish. If the medium adjacent to the lateral line moves in exactly the same way as the surface of the fish carrying the lateral line, the sense organs of the lateral line will not be stimulated (Denton and Gray 1983; see also Chap. 23.4).

It is widely recognized that the most likely natural stimuli to lateral line sense organs are the following:

(1) Local pressure gradients produced by a fish's own swimming movements. Thus, when a fish is gliding at constant velocity, the static pressure differences over its head and body can be very large (DuBois, Cavagna, and Fox 1974). There will also be large, varying pressure gradients over the head and body when it swims, especially when it changes direction quickly.

(2) Mechanical disturbance of the lateral line system caused by distortion in the fish's tissues arising from its movements.

(3) Local pressure gradients produced by external sources (e.g., by neighboring animals in mid-water, by disturbances at the surface, or by the flow of water around rocks in streams. The fields around vibrating bodies are discussed in Chapters 4 and 28.

We have only made observations with external vibrating sources in mid-water. However, the ability to detect and evaluate stimuli of all the types described above under (1) and (3) could be useful to fish. For all such stimuli the nature of the mechanical relations between local pressure gradients and cupular displace-

* Marine Biological Association of the United Kingdom, The Laboratory, Plymouth, PL1 2PB, UK

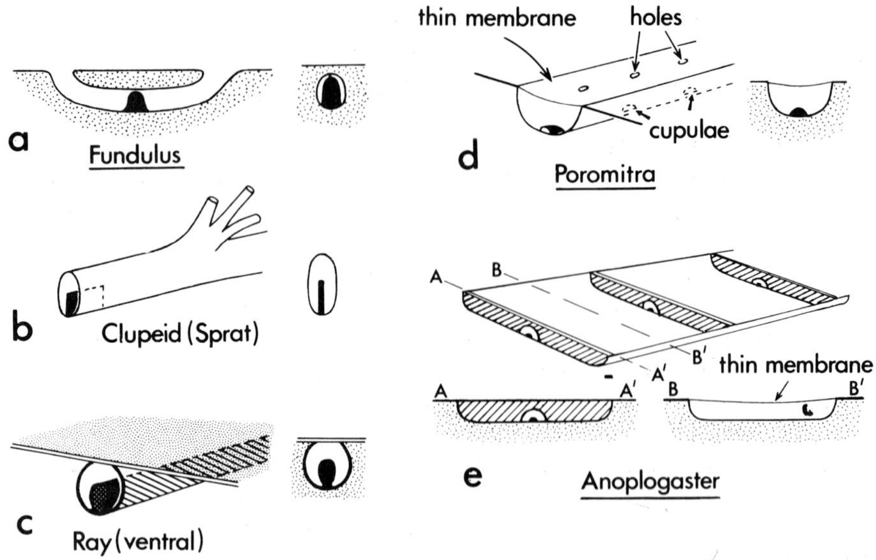


Figure 23.1. Diagrams illustrating some of the variety of lateral line systems that can be found among fish. (Cupulae of neuromasts are shown as black objects; in nature they are almost transparent). For each animal, there is a drawing of a cross section through a canal at a position where there is a neuromast. The figure shows parts of lateral lines from (a) *Fundulus*. In this part of lateral line system the cupula almost completely blocks lateral line canal. (After van Bergeijk and Alexander 1962 and Harris and van Bergeijk 1962.) (b) A clupeid. For fuller description see Denton and Gray (1983). (c) A ray. Here we show part of canal system from ventral surface of the animal. The cupula runs the length of the canal, which has rigid walls and compliant external surface. (d) *Poromitra*. Part of lateral line system of the head. The canal is covered by thin, flexible membrane, which has holes. Cupulae are low structures that obstruct only small part of the lumen of the canal. (e) *Anoplogaster*. Part of lateral line system of the head. The canal is covered by thin, continuous membrane. It is divided into compartments by partitions. Cupulae lie in openings in partitions. Lines AA′ and BB′ show the position of sections.

ments (these are discussed below) must be significant in determining such properties as sensitivity, directionality, and frequency response.

The structures of lateral line systems vary greatly between different species of fish, and we should expect the properties, particularly the mechanical properties, of the lateral line systems of different fish to differ greatly from each other. There are, indeed, sometimes enormous differences between different parts of the lateral line system in the same fish (e.g., between the head and body lateral lines of macrourid fish and between the dorsal and ventral lateral lines of rays. Such differences are sometimes so great that they suggest very different functions for the different parts of the lateral line system on the same animal.

The main objectives of the work described in the present chapter are (1) to find experimentally, for a few fish, the relations between particular structures in lateral lines and the mechanical performances of these lateral lines; (2) to develop, from an analysis of the results of (1) and by the study of model lateral lines, ways of predicting the likely performances of the lateral lines of fish in general.

Although we have examined the anatomy, including the fine anatomy, of the lateral lines of a number of fish, the types to which we shall give the most attention are the following:

(1) Fish with reduced bony head canals and cupulae almost completely filling their lumens, e.g., *Fundulus heteroclitus* (Atheriniformes; Cyprinodontidae) (Fig. 23.1a).

(2) Fish with canals confined to the head and freely open to the sea through pores, with the cupulae blocking only a small fraction of the lumens of the canals. Examples of such fish are *Sprattus sprattus* L. (Clupeiformes; Clupeidae) (sprat) (Fig. 23.1b) and *Stolophorus purpureus* (Clupeiformes; Engraulidae) (anchovy).

(3) Fish with long lateral lines along the side of the body, the lateral line canal opening periodically to the sea, with neuromasts lying in the canal between openings (pores). Canals of this type are often formed by overlapping scales. The cupulae in some fish almost fill the lumen of the canal, whereas in other fish the cupulae block only a small part of it. Examples of such fish are *Trachurus trachurus* (Perciformes; Carangidae) (horse mackerel) and *Apogon fragilis* (Perciformes; Apogonidae) (cardinal fish).

(4) Fish with large head canals that are covered with thin membranes, the neuromasts of which, though large, do not fill the lumens of the canals, e.g., *Poromitra* sp. (Beryciformes; Melamphaeidae) (Fig. 23.1d).

(5) Fish with large head canals that are covered with thin membranes and divided into compartments by bony partitions. The neuromasts lie in openings in these partitions. An example of such a fish is *Anoplogaster* (Beryciformes; Anoplogasteridae) (Fig. 23.1e).

Figure 23.1 shows some lateral lines diagrammatically. This gives only an indication of some of the structural variety that can be found. For an excellent comprehensive description of lateral line systems in fish, see Chapter 22).

In discussing sources of stimuli, we will confine our attention to external vibrating sources. For such sources we have analyzed the relations between the vibrations of the source and the pattern of neural activity in the whole population of neuromasts in a lateral line into five successive steps.

(1) The relation between the acceleration of the surface of a source of vibration and the net pressure gradients (measured as accelerations in the medium) adjacent to the lateral line system
(2) The relation of the net local acceleration of the medium over a short section of lateral line to the motion of liquid inside the lateral line canal
(3) The relation of the motion of the liquid in the canal to the forces acting on the cupulae
(4) The relation of the forces on a cupula to the activity of the nerve fiber populations from the neuromast
(5) The relation of the position of the source with respect to the fish, to the pattern of activity among the neuromasts of the whole lateral line system

## Methods

The general approach was to use sinusoidal mechanical stimuli and measure the displacements of the fish or of liquid inside lateral line canals by observing the displacement of light-scattering particles through a dissecting microscope. The apparatus used has been described by Denton and Gray (1983). Many direct observations of structures were made on freshly killed or anesthetized fish, in particular on the positions and sizes of the cupulae; these usually stain intensely with Janus green B. Nerve activity was recorded from whole neuromast nerves, and the stimuli needed for constant responses were found (Gray 1984).

Models of lateral lines, some examples of which are shown in Figure 23.2, were made from Plexiglas. Each model consisted of two parts: (1) A base in which a groove or grooves had been made. Sometimes obstacles (corresponding to partitions or cupulae in lateral line systems) were placed in these grooves. (2) A lid in which holes or slits had been made. The combined base and lid was filled with a liquid containing light-scattering particles. The model lateral line was fixed to a platform in a chamber filled with the same liquid as in the model lateral line. The displacements of particles inside the "lateral line" and in the medium around it were observed and measured when the medium around the model was made to vibrate in various ways. In some models a piece of thin

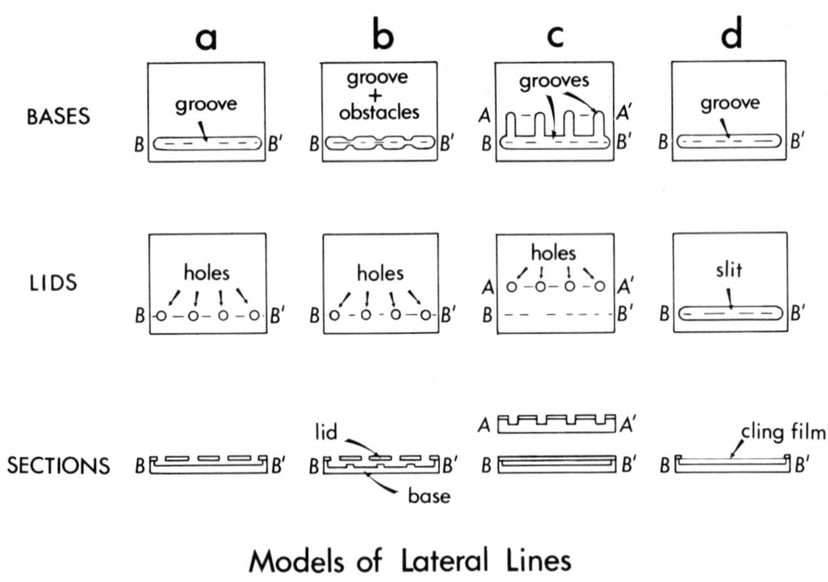


Figure 23.2. Some examples of model lateral lines. A particular "lateral line" was formed by combining a base within which grooves had been cut with a lid containing appropriate holes or slits. Bases were 3 mm thick and grooves mostly cut to a depth of 1 mm. Lids were 1 mm thick. To mimic lateral lines covered with a membrane, a piece of thin plastic, slightly stretched, could be placed between a lid and a base. An example of such a combination is shown in *Model d*. Lines AA′ and BB′ show positions of sections.

plastic film was clamped between the base and the lid. This corresponded to the membranes that cover some lateral lines.

## Steps in Excitation

The great bulk of our experimental biological work has been carried out on clupeids, especially the sprat, which is the only species on which we have experimental evidence connected with all the steps in the excitation of lateral line neuromasts listed on page 597. Most of this work, structural and experimental, has already been published (Best and Gray 1982; Denton and Gray 1982; Blaxter, Gray, and Best 1983; Denton and Gray 1983; Gray 1984). It forms the base on which we have made a comparative study of lateral lines by the use of physical and mathematical models, anatomy, and a limited amount of experimental work on other species. This work is referred to in each of the following sections as appropriate. Following our consideration of these steps, we include a short paragraph to summarize our ideas on the biological importance of the lateral line system to sprats.

### Step 1. The Source to Net Acceleration of the Medium Adjacent to the Lateral Line

We showed earlier in experiments with four species of fish (Denton and Gray 1983) that a fish suspended in the field of a vibrating source behaves as a rigid body along its length. Points along a fish always moved in an identical manner along the fish's long axis, and the amplitude of the motion could be predicted from the difference of pressure between snout and tail. The same principle held for other axes for rigid structures like heads. With a simple source whose properties are known, pressures and pressure gradients (these are proportional to accelerations) can be calculated at all relevant points in the field (see, for example, Harris 1964). The magnitude and direction of the acceleration of the fish can be calculated from the pressures at its snout and tail, modified when appropriate, by acceleration in other axes in which there is rigidity. The acceleration of the water outside the canal in the region of each neuromast can be calculated, and hence the magnitude and direction of the net acceleration between fish and water at any point can be found.

A simple situation is one in which a fish is pointing exactly towards a vibrating sphere (Fig. 23.3). The acceleration of the fish along its long axis will have some value between the relatively large acceleration of the water around the snout and the lower values found around the more distant tail. Acceleration in these conditions decays in the water inversely with distance cubed. Near the snout the accelerations of the water will be much greater than that of the fish; this gives large net accelerations. At some point in the center of the fish the two accelerations will be equal, giving a net acceleration of zero. Towards the tail the fish will move more than the water, producing at any given instant a net

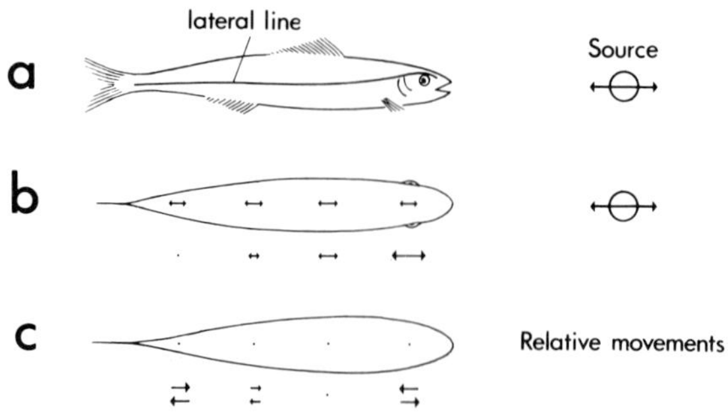


Figure 23.3. A diagram illustrating the way in which longitudinal rigidity of a fish will affect the pattern of stimulation of its lateral line system. (a) A fish facing a source of vibration—here vibrating sphere. (b) The fish seen from above. *Arrows* represent amplitudes of displacement. For the fish, these amplitudes are constant along its whole length. For the medium, displacements fall rapidly as the distance from the source increases. They are much greater near the head of the fish than near its tail. (c) Relative movement of the fish and the medium at various positions along the fish.

acceleration in the direction opposite to that at the snout. With a fish in this position the net accelerations between water and fish decline as distance to the fourth power. Patterns of net acceleration can be worked out for any fish in any calculable sound field if the axes of rigidity have been established; some such patterns are given by Denton and Gray (1983).

Important consequences arise from the motion of the fish in the field of a vibrating source. The first is that the pattern of net accelerations on the different neuromasts in the lateral line system changes dramatically with the position or angle of a fish relative to a source of sound. The second is that, when the distance of the source is great relative to the length of a fish (this can happen well within the near field), the stimuli given to the lateral line system become extremely weak. Long fish will, with respect to the lateral line, have an advantage over short fish in detecting vibrating sources.

### Step 2. From Net Motions in the Medium Close to Lateral Lines to Motions inside Lateral Line Canals

We shall now consider the relation between the motions of liquid inside canals (both model and real) to those in the external medium adjacent to these canals. We shall mostly confine our discussions to sinusoidal motions. The simplest model of a canal is a straight capillary tube of uniform cross section and open at its ends. Experiments show that the motion inside the capillary relative to that outside is proportional to the cosine of the angle ($\alpha$) between the direction of the motion in the medium and the long axis of the capillary. The frequency response measured as the ratio $x_{in}/x_{out}$ of the maximum "inside" displacements (near the center of the capillary) and the "outside" displacements (measured in

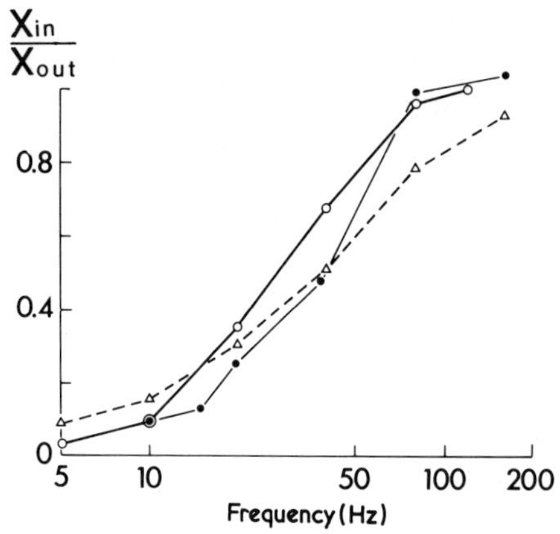


Figure 23.4. The ordinate giving the ratio ($x_{in}/x_{out}$) of amplitudes inside and outside canal system under study; the abscissa, the frequency (Hz) plotted on logarithmic scale. *Curves* are: ●–● for sprat lateral line canal; ○–○ for capillary tube of a radius of 140 μm; △ △ for theoretical relation, $j\omega I/(j\omega I + R)$, for capillary tube of a radius of 140 μm (see text for explanation). (From Denton and Gray 1983.)

the medium) are shown in Figure 23.4 for a capillary with an internal radius of 140 μm when the angle $\alpha = 0°$. At higher frequencies (> 80 Hz), $x_{in}/x_{out}$ becomes a little greater than 1. At frequencies below about 80 Hz, the ratio falls at approximately the inverse of the frequency. Denton and Gray (1983) give a simple theory that approximately explains the relationship. To produce sinusoidally changing displacements in the medium, there must be sinusoidally changing pressure gradients. The impedance relating the displacements to the pressure gradients in the medium will depend on only the frequency and the mass involved. Within the capillary the impedance will have a similar inertial, but also a resistive, term. The ratio $x_{in}/x_{out}$ of displacements inside and outside the capillary tube will be given approximately by the equation

$$\frac{x_{in}}{x_{out}} = \frac{j\omega I}{j\omega I + R} \tag{23.1}$$

where $I$ = inertance (mass divided by the square of the cross-sectional area of the tube), $R$ = frictional (viscous) resistance calculated for steady (Hagen-Poiseuille) flow conditions, $j$ = complex operator and $\omega = 2\pi \times$ frequency. ($I$ and $R$ are calculated for the same length of tube, and hence the ratio of equation (23.1) is independent of this length).

A curve calculated from this equation giving $x_{in}/x_{out}$ plotted against frequency for a capillary tube with a radius of 140 μm is given in Figure 23.4, where it is shown to agree reasonably well with experimental results found for such a capillary. An equally good agreement between theory and experiment was found for a range of capillaries of different radii. We may note that, for frequencies for which $R \gg \omega I$, the values of $R$ calculated for steady flow will be a good approximation to the truth, because the flow pattern across the tube will be parabolic and the oscillating flows in the tube will everywhere be in phase. At "high" frequencies (i.e., where $\omega l \gg R$), the flow pattern is more complicated. Velocities in the tube can exceed those in the medium, and there can be phase differences between different points across the tube. (Both of these effects were seen in our experiments on capillaries and lateral lines.) At such frequencies the flow pattern across the tube will no longer be parabolic, and the velocity gradients that are perpendicular to the wall of the tube will be greater than would be expected from a parabolic flow pattern. (A good account of the flow patterns described above is given by Schlichting 1968.)

We should note that, though the amplitudes of vibration used in our experiments were greater than those found in lateral lines in nature, the flow patterns were seen to be laminar, and the ratio $x_{in}/x_{out}$ was independent of the amplitude of $x_{out}$, Reynold's numbers calculated for the maximum velocities of flow were almost always less than one, indicating that we can safely apply the results we obtained in our experiments to the much lower amplitudes of stimulation found in nature. The phase relations between the flows inside capillary tubes and in the medium were close to those expected from the simple theory given above.

Further experiments were made with model lateral lines more closely approximating fish lateral lines. In some models in Figure 23.2c, the side grooves arising from the main groove (canal) were made of very different lengths, so that the line of holes *AA′* was no longer parallel to the line of the main "canal" *BB′*. Experiment showed that the greatest response in the main canal was given when the axis of vibration in the medium was parallel to the line of holes, not to the main canal itself. In some models the grooves were made in a V pattern. Observations again showed that the different sections of the Vs behaved independently of each other in their directionality. Quite shallow grooves (bases without lids) had marked directional properties. Further results and calculations on models are given on page 610.

We have made more observations on the sprat lateral line system than on the lateral lines of other fish. The anatomy of this system is described in some detail by Denton and Gray (1983), who give references to earlier studies. The behavior of liquid inside the main canal system is, in frequency response, much like that of a capillary tube of approximately the same cross-sectional area (see Figure 23.4). Again $x_{in}/x_{out}$ varies approximately as the cosine of the angle between the axis of the canal and the axis of vibration in the medium. A similar cosine law was found by Harris and van Berjeijk (1962) on *Fundulus heteroclitus* and by Sand (1981) on *Rutilus rutilus* (Cypriniformes; Cyprinidae). Displacements in the canal were advanced in phase relative to those in the medium, and these phase advances were a little larger than those found in capillaries. There was little mechanical coupling between neighboring parts of the main canals, confirming Sand's observations on *Rutilus*.

In Figure 23.5, we show results on frequency response for parts of the lateral line systems of two fish, both having lateral lines with pores spaced along the canal. One (*Harengula ovalis;* Clupeiformes; Clupeidae) is for a canal of greater cross-sectional area; the other (*Apogon fragilis*) is for a canal of smaller cross-sectional area than the sprat. The curves are similar to those of the capillary

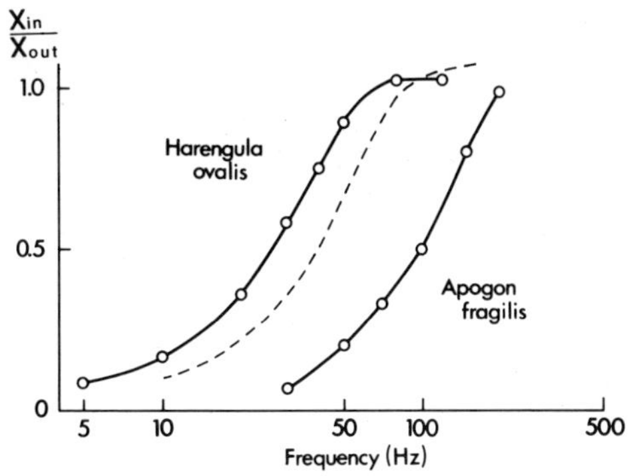


Figure 23.5. Similar to Figure 23.4 for the sprat, but giving experimental results for two other fish. For one, *Apogon fragilis,* the part of lateral line canal studied was smaller in cross-sectional area than the one studied in the sprat. For the other, *Harengula ovalis,* the part of lateral line canal studied was larger in cross-sectional area than the one studied in the sprat. The curve for the sprat is shown by *dashed line.*

tube and the sprat (Fig. 23.4), but are displaced in the sense expected from the simple theory, the curve being shifted towards higher frequencies for the narrower canal. Another pair of fish (*Aphenopus carbo* and *Lepidopus caudatus;* both Perciformes; Trichuridae) gave a similar result; their canals do not have branches, but slits between overlapping plates.

We have dealt above with models and fish lateral lines in which the "cupulae" obstructed the canals to only a limited extent and for which there were no partitions dividing the canal into compartments. We shall now consider what the effect of obstacles will be. Since the cupulae themselves are obstacles and cupulae are, in some fish, found in parts of the canal system where the canal is narrowed (e.g., as is shown for *Anoplogaster* in Fig. 23.1), the frictional forces bending the cupulae will generally depend on the flows in the narrower parts of the canal, rather than those in the broader parts.

Figure 23.6 shows a comparison of results obtained on a model of the *Type a* with those obtained for a model of *Type b*. In the figure, we again plot $x_{in}/x_{out}$ against frequency. It may be seen that placing obstacles in the canals changes the shape of the frequency response greatly. It diminishes the responses at "lower" frequencies and increases those at "higher" frequencies. There is a further important consequence of narrowing the canal, which we shall discuss in the next section.

So far we have discussed only systems in which compliant elements play a negligible role in determining the flows of liquid within the canals. There are at least two types of canals for which these flows will depend greatly on compliant structures. The first type is one in which the cupula almost fills the lateral line canal, as it does in the short sections of canal of *Fundulus* shown in Figure 23.1. Here the displacements of the liquid within the canal and of the cupula itself must be, especially at higher frequencies, approximately equal. The system can, therefore, be treated as one of a resistance, mass, and compliance in series.

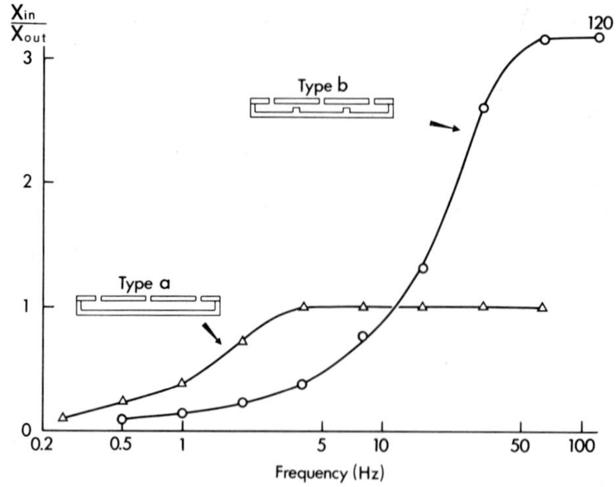


Figure 23.6. A comparison of results obtained on models of Types *a* and *b* of Figure 23.2. For the curve ○–○, which was obtained with model canal with obstructions (*Type b*), values of $x_{in}$ were measured above obstacles. Abscissa: frequency in Hz (logarithmic scale). Ordinate: ratio $x_{in}/x_{out}$.

Harris and van Bergeijk (1962) have analyzed their results on *Fundulus* and found their theoretical and experimental results in accord. They decided that the cupulae are nearly critically damped. Similar conclusions were reached by Kuiper (1967) on *Acerina,* i.e., *Gymnocephalus cernus* (Perciformes; Percidae) and by Flock (1965) on *Lota vulgaris* (Gadiformes; Gadidae).

A second type of lateral line for which compliances are important is one in which the lateral line canals are covered with a membrane: e.g., in the ventral lateral line canals of the rays (e.g., *Raja clavata;* Rajiformes; Rajidae) and in the head canals of *Poromitra, Anoplogaster,* and *Hoplostethus* sp. (Beryciformes; Trachichthyidae).

We have studied models of the kind shown in Figure 23.2d. Here a thin, slightly stretched membrane is placed over the canal. The effect can be (as might be expected) to make the system show resonant properties with greatly diminished responses (i.e., low ratios of $x_{in}/x_{out}$) below the resonant frequency and greatly enhanced "responses" around this frequency. We shall discuss the behavior of lateral line canals in *Poromitra* and *Anoplogaster* in more detail later.

## Step 3. From Motions of Liquid Inside Lateral Line Canals to Forces on Cupulae

When considering the forces acting on cupulae, there are two anatomical types of canal that are relatively easy to analyze. The first is one in which the cupulae almost completely fill the lumen of the canal. The second is one in which the cupulae take up only a small fraction of the cross-sectional areas of the canals;

examples are found (1) in the sprat, where the cupulae are thin flat structures with their broad surfaces parallel to the axes of the canals in which they lie and (2) in *Poromitra* and *Stolophorus,* where the cupulae rise from the internal surfaces of the canals, but project only a relatively short distance into them, by no means obscuring their lumens. Here we shall discuss mainly the lateral line systems for which we would expect the forces on the cupulae to be mainly frictional.

Some of the factors determining such frictional forces will be discussed by reference to Figure 23.7. This shows, diagrammatically, velocity profiles (1) for unobstructed parts of a canal; (2) alongside a *central plate* (corresponding to the cupulae of the sprat); and (3) alongside an *obstacle blocking half of the cross-sectional area of the canal* (we shall suppose that here a low cupula projects from the inner surface of the obstacle into the lumen of the canal). The velocity profiles shown are for low frequencies (in which the $R$ terms of Equation (23.1) are appreciably greater than the $\omega l$ terms), and these profiles are parabolic in shape. The frictional forces will depend on the velocity gradients perpendicular to the wall of the surface being considered. Since the instantaneous total flows must be constant along the canal, the maximum velocities will be slightly increased where the central plate blocks a small part of the lumen and doubled where the obstacle blocks half the lumen. The frictional forces on the central plate will be twice those found against the wall of the main canal, because the velocity gradients perpendicular to the plate will be doubled. Furthermore, the plate has two sides on which these forces can act. The frictional force on the surface of the obstacle facing the center of the canal will be increased fourfold by the narrowing of the canal (here half of the increase comes from a doubling of the maximum velocities

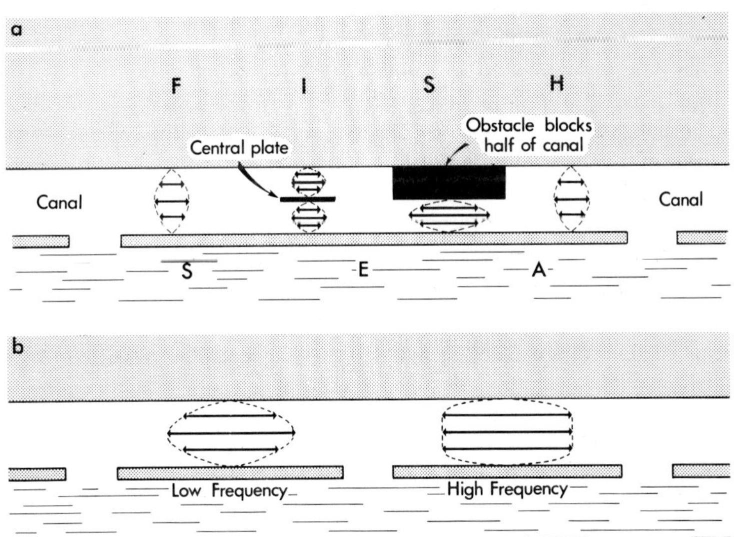


Figure 23.7. A diagram showing velocity profiles in "lateral line canals." (a) For low frequency, such a canal containing central plate and an obstacle blocking half of cross-sectional area. (b) A comparison of profiles for low and high frequencies.

of flow). By an extension of these arguments, it is easy to see that different dispositions of canals and cupulae will produce very different magnification factors of this kind.

The simple arguments used above have concerned only situations in which the profiles of flow across the tubes are parabolic. This is, as Schlichting (1968) shows, the case when the values for the dimensionless group $k = r(\omega/\upsilon)^{1/2}$ are low (i.e., $\ll 5$), where $r$ = radius of the tube, $\omega = 2\pi \times$ frequency, and $\nu$ = kinematical viscosity. For higher values of $k$ the flow patterns are more complicated. For our present purposes, however, it is enough to recognize (1) that velocities in the central parts of the tube are substantially constant across the tube and everywhere half a period out of phase with respect to the exciting force and (2) that, even if the maximum velocities in the central parts of the tube remain constant, the velocity gradients perpendicular to the wall of the tube will continue to increase to some extent with frequency as the boundary layer against the wall becomes progressively steeper (Fig. 23.7b).

On Figures 23.3, 23.5, and 23.6, we have plotted ratios of displacements inside the canal to those in the medium ($x_{in}/x_{out}$) against frequency. These show, as we noted for the sprat earlier, that for lower frequencies, the displacements of liquid inside the canal will be proportional to velocities in the medium adjacent to the canal. If the cupulae are driven mainly by frictional forces and the velocity profile across the canal does not change with frequency, then these forces will be proportional to the velocities in the canal and, hence, to net *accelerations* in the medium. This relationship will be to some extent maintained at higher frequencies, since as frequency rises, two opposing factors come into play. Firstly, $x_{in}/x_{out}$ becomes more independent of frequency, and secondly, the velocity profile across the canal changes in shape, becoming steeper against the wall of the canal and the surfaces of the cupulae (Figure 23.7).

In the above argument we have not taken into account the way in which the flow of liquid along a canal will divide itself between different compartments alongside and above the cupulae. In general, the effect of this will be to diminish the forces exerted on the cupulae at lower frequencies relative to those at higher frequencies.

We may note that, in addition to calculating forces acting on the cupulae, we have made some (unpublished) experiments in which the displacements of the cupulae in sprat lateral line canals were measured directly. These showed that these cupulae were relatively stiff structures that gave displacements approximately proportional to the velocities of seawater in the lateral line canals.

### Step 4. The Force on the Cupula to the Activity in the Nerve

There have been many reports of recordings of microphonics, or nerve impulses, from neuromasts and their nerves. These show that neuromasts contain two populations of hair cells (Jielof, Spoor, and de Vries 1952) and that each responds to the displacement of the hairs in one of two opposite directions (Flock 1971). For other references on microphonics, see Kuiper (1967), Flock (1971), and Sand (1981). In the context of this section, we are concerned with the relation between the force on the cupula and the pattern of activity in each population of nerve fibers from the neuromast. We only consider the sprat and, from the arguments given above, assume that the force is roughly proportional to the velocity of the water past the cupula. Records have been obtained from nerves of neuromasts in the sprat's infraorbital canal (Gray 1984). The nerve from each neuromast contains about 90 fibers (7–9 cm fish) and the receptor cell–fiber ratio in about 7 (Best and Gray 1982). The results show that, as in other fish, there are two populations of fibers. Single transients or individual peaks of a sine wave evoke volleys in these populations, and the number of fibers active in such a volley is related to the amplitude of the transient or peak.

In the sprat, the velocities of water in the lateral line canals depend not only on the differences of acceleration of the water and the fish, as discussed above (see Fig. 23.3), but also on water movements set up by the motion of the auditory bulla's membrane as the pressure changes (Blaxter, Denton, and Gray 1981). The displacements of water in the canals are proportional to the pressure on the bulla, and hence, the velocities in the canals are related to the rate of change of the pressure, which in turn is proportional to the rate of change of the acceleration of the source surface. Thus, responding as it does to water acceleration, i.e., acceleration of the source surface, in combination with a stimulus related to the rate of change of this acceleration, the system appears to be designed to detect the earliest signs of change (Gray 1984).

In the sprat, the mechanics are arranged to ensure instantaneous detection of the earliest changes. The neural response is arranged to signal the amplitude and polarity of each change. In other fish in which the canals are designed for higher sensitivity with a sacrifice in time resolution, the coding in the nerve may be different.

### Step 5. Patterns Between Neuromasts

A sprat could obtain information about the position, or at least change of position (including distance), of a source relative to itself if it were able to collate the information about amplitude and polarity signalled from each neuromast. In the case of the sprat, we have sufficient information to calculate the velocities in the canal at the position of each neuromast in the infraorbital canal with transient accelerations in either direction of a neighbor's tail, at most positions around the "receiving" fish. Such calculations show that the system is very sensitive to changes of position or angle. Figure 23.8 shows families of patterns, each of which includes a standard pattern (the same for all four sets) and one each for an increase and decrease of the value of one parameter. It can be seen that quite small changes in a parameter can produce major changes in pattern.

### The Importance of Lateral Line to the Sprat

If a function of the lateral line system is to provide information about external sources, since the effective stimulus falls rapidly with distance (see step 1), only

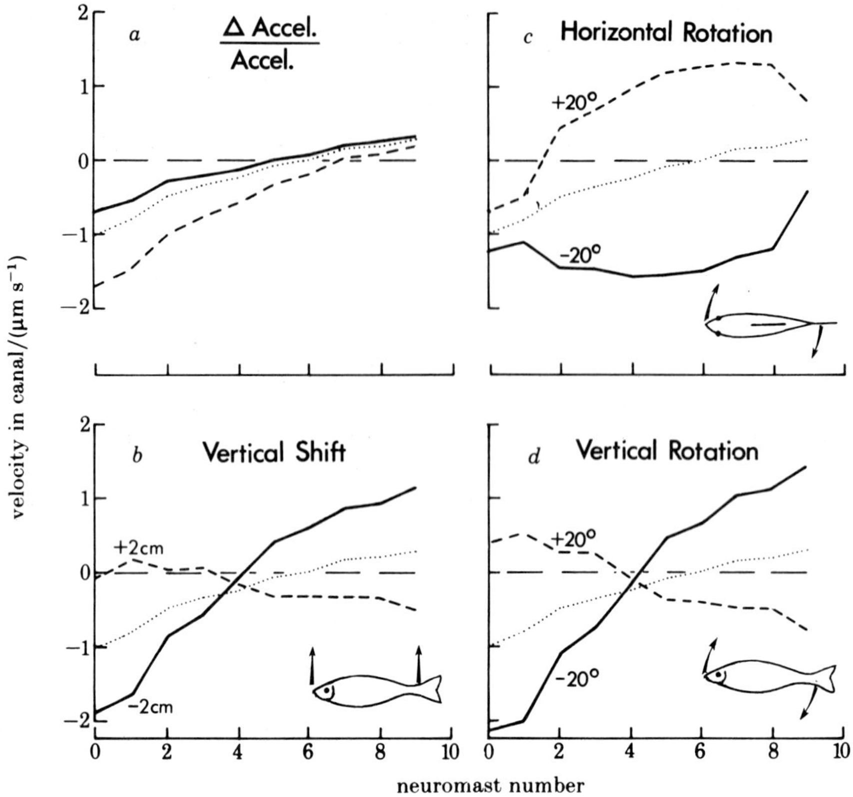


Figure 23.8. For the sprat, four sets of graphs showing patterns of the stimulation of lateral line neuromasts by vibrating source. Each set includes a curve for standard position (.....) (the same for all sets); a curve (-----) with the value of particular quantity increased; and another (——) with same quantity decreased. The ordinate indicates the velocity in the canal in $\mu m\ s^{-1}$. The abscissa indicates the neuromast in infraorbital canal numbered in sequence forward from lateral recess. In all cases the fish was 22 cm away from the source in horizontal plane and 8 cm ahead of the axis of vibration. (a) Altering the ratio of peak rate of the change in the acceleration of source surface to its peak acceleration: the standard is 100 $s^{-1}$; increased value, 250 $s^{-1}$; decreased value, 50 $s^{-1}$. (b) Altering the height above or below the axis of vibration: the standard is 0; others are + and − 2 cm. (c) Altering the value of fish heading in horizontal plane: the standard is at right angles to the axis of vibration; others are + and − 20°. (d) Altering the value of fish heading in vertical plane: the standard is 0; others are + and − 20°. (After Gray 1984.)

sources within a few fish lengths need be considered. The most sensitive range of frequency of the organs, in terms of velocity in the canal to give a constant response, lies between 20 and 80 Hz (Gray 1984). It must be remembered, however, that it is the acceleration of a transient peak that is signalled and accelerations can be high in movements that have low repeat frequencies or in slow movements of large amplitude. The most obvious external source fitting all these constraints is a neighbor's tail. A station-keeping device needs to detect the earliest changes in acceleration of neighbors and any small changes of position relative to a neighbor, including changes in distance. The sprat lateral line does transmit information about all these things. There is no direct proof that it uses the information from this receptor system, but sprats swim close together in schools, make rapid changes of direction, and hardly ever collide. Partridge and Pitcher (1980) have shown the importance of lateral lines to schooling in other fish.

## The Behavior of Other Types of Lateral Line

For this chapter, we have picked out, from the variety of animals on which we have some information, three examples that illustrate particular points. The simple mathematical model that we used to explain the results, such as those for the sprat's lateral line and straight capillary tubes, is clearly inadequate for more complicated lateral line systems. We have, therefore, made a model with sufficient components to allow it to simulate a number of the lateral line systems that we have studied. This "lateral line" consisted of a tube of cross-sectional area $a_1$, containing a narrower section of cross-sectional area $a_2$, with the lengths of the wider and narrower parts being $1_1$ and $1_2$, respectively. The ends of the tube are covered by membranes of acoustic compliance $C$, and each membrane had an opening, or pore, of area $a_3$ and length $1_3$.

If we designate by the ratio $x_{in}/x_{out}$ the relative amplitudes of displacements in the narrow part of the tube and those in the medium adjacent to the tube, we find that

$$\frac{x_{in}}{x_{out}} = \frac{\text{impedance outside the tube}}{(\text{impedance inside the tube}) + (\text{impedance of membrane})}$$

$$= \frac{j\omega I_{out}}{(j\omega I_{in} + R_{in}) + \left(\dfrac{2}{j\omega C + \dfrac{1}{R_p}}\right)} \tag{23.2}$$

where

$$I_{out} = \frac{\rho(l_1 + l_2)}{a_2}$$

$$I_{in} = \rho\left(\frac{l_1}{a_1} + \frac{l_2}{a_2}\right)$$

$$R_{in} = 8\pi\eta\left(\frac{l_1}{a_1^2} + \frac{l_2}{a_2^2}\right)$$

$$R_p = \frac{8\pi\eta l_3}{a_3^2}$$

where $\rho$ = density of the medium and $\eta$ its viscosity. Acoustical compliance $C$ is related to tension $T$ in a circular membrane by the equation $C = \frac{\pi r}{8T}$, where $r$ = radius of the membrane.

Equation (23.2) reduces to the earlier Equation (23.1) used on the sprat when $a_1 = a_2 = a_3$ and $\frac{1}{C} = 0$.

The resistances were calculated for steady flow. They did not reflect the changes in resistance that occur with the change of frequency on the spatial distribution of the flow (see Schlichting 1968), and the discrepancies seen in Figure 23.4 can, in large part, be attributed to this factor. Equation (23.2) fits the results on the model of the simple canal with pores (Fig. 23.2a) reasonably well. Figure 23.9 shows calculated and observed results for a model with small pores for media of two different viscosities. Increasing the size of the pores shifted the curves slightly to the left, and within experimental error, the experimental and calculated curves shifted to the same extent. There was not such good agreement when we tried to calculate the consequences of having a narrow section in the tube. The general pattern was correct, but the experimental curves were shifted farther to the right, were steeper, and the "amplification" at the high frequency was less than that predicted. We had no quantitative means of checking the predictions of the calculations with a compliant membrane, but there was qualitative agreement with experiments. These models were simplified not only by the use at all frequencies of resistance calculated for steady flow, but also because distributed impedance elements were lumped and end effects neglected. Although these models give only a first approximation to the truth we have found them useful in our attempts to understand complicated lateral lines.

## Poromitra

*Poromitra* sp. have large canals (7 mm wide in a 50-mm fish) on the head, and these canals occupy most of the surface of the head (Fig. 23.10). The cupulae are low and do not obstruct the lumens of the canals appreciably. Each canal is

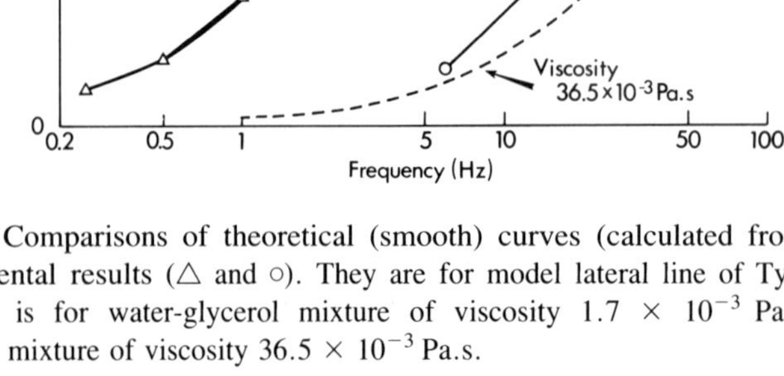


Figure 23.9. Comparisons of theoretical (smooth) curves (calculated from Eq. [23.2]) with experimental results (△ and ○). They are for model lateral line of Type *a* of Figure 23.2. (△–△) is for water-glycerol mixture of viscosity $1.7 \times 10^{-3}$ Pa.s. ○–○ is for water-glycerol mixture of viscosity $36.5 \times 10^{-3}$ Pa.s.

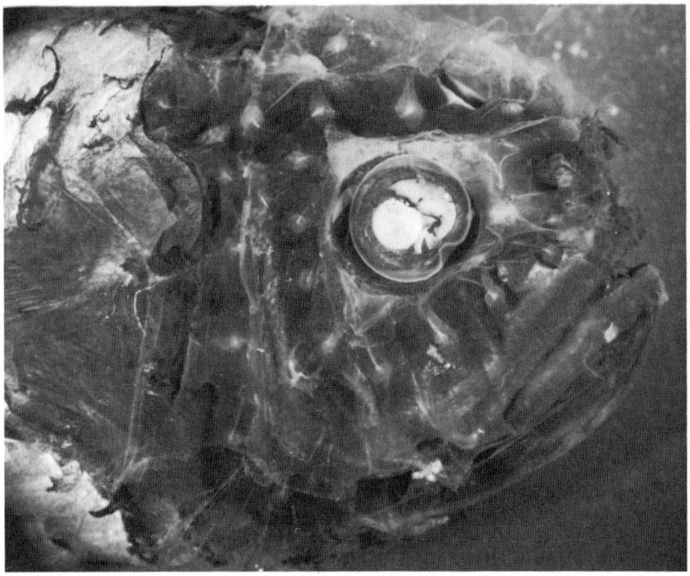

Figure 23.10. Right side of the head of *Poromitra.* Membranes covering lateral line canals have been removed to show neuromasts. Membranes had openings (pores) between neuromasts.

covered by a soft membrane, which sags when the fish is lifted out of water. The membrane has pores of about 0.4 mm in diameter, and the neuromasts lie on the floor of the canal between pores. All the measurements necessary to allow us to use Equation (23.2) except the value of the compliance were made on several fish. These were adjusted linearly to a standard fish of 50-mm length and means found. After such adjustments, the scatter in measurements was small. Figure 23.11 gives the results of calculations based on Equation (23.2). The two *dashed lines* represent the effects of assuming (*on the right*) a rigid membrane and (*on the left*) no membrane. The two *solid lines* give the results of calculations made with two values of compliance, one 2.5 times the other, chosen to give curves near the experimental points. There are uncertainties in converting these acoustic compliances to tensions in the membrane, but they are certainly minuscule, on the order of very few tens of mN/m, and so comparable to the surface tensions of liquids. Such low tensions are probably all that is required to allow large canals and chambers to behave as mechanical units, rather than as parts of the open sea. Without pores in the membranes the resonant peaks would be considerably higher. The *triangles* in Figure 23.11 are means of uncorrected results of three experiments (it was not possible to get readings at higher frequencies). The *circles* are for one of the experiments after correction for the angle between the canal and the outside motion. These results suggest that, below about 10 Hz, the ratio ($x_{in}/x_{out}$) increases more than proportionally with frequency. The velocity in the canal may, at these low frequencies, be a function of something approaching the rate of change of acceleration of the source and so designed to

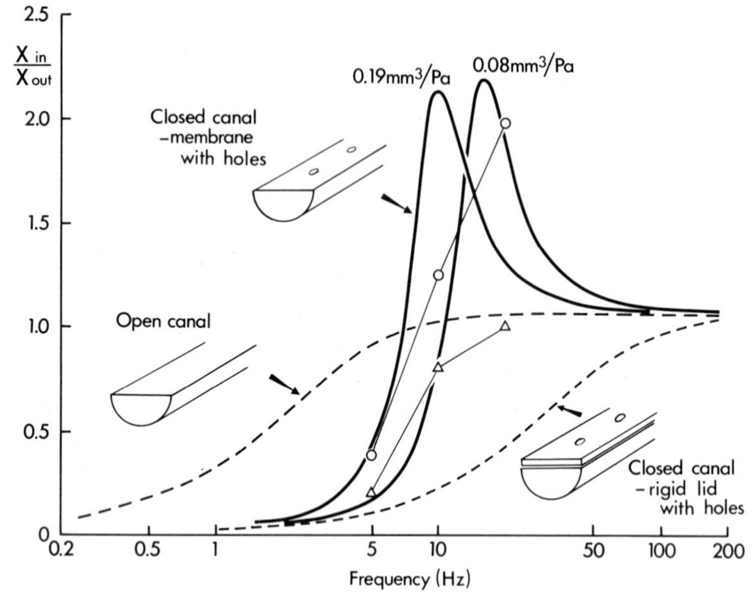


Figure 23.11. Curves shown by heavy *continuous and dashed lines* based on calculated ratios $x_{in}/x_{out}$ for a range of frequencies for a part of lateral line system of *Poromitra:* ---- left-hand curve, open canal; right-hand curve, the canal covered by rigid lid with holes; —— a canal covered by thin, flexible membrane pierced by pores. Two values for acoustic compliance were taken: these were 0.19 and 0.08 $mm^3$/Pa. Curves shown by thin lines give experimental results: (△–△) mean of three experiments (uncorrected, see text); (○–○) one experiment (corrected).

detect the earliest signs of a change in motion of a neighboring source. As the frequency rises above 10 Hz, the velocity in the canal more and more approximates the velocity outside.

The cupulae are shallow domes with an elliptical base with (in a 50-mm fish) a length of 0.5 mm along the canal and a width of 0.4 mm across it. The height of the cupula is 0.1 mm. Short of more direct evidence, this might suggest that the force on the cupula is proportional to the velocity in the canal. Calculations suggest that the velocity gradient perpendicular to the length of the canal will, in a canal of this size, be large close to the wall and small across the rest of it. The low profile of the cupulae may therefore be adapted to receive a maximum force for a given mean flow.

The number of nerve fibers in one fish 44 mm long was approximately 100 per neuromast and the ratio of hair cells to fibers approximately 20 (cf. the sprat above). The neuromast area is large, and by our counts (these will give minimum figures), the density of hair cells (about 1 for every 20 $\mu m^2$) is approximately the same as that found for an anchovy and about a fifth of that found for the sprat. The total number of hair cells per neuromast is, however, appreciably larger for *Poromitra*. It is about 15 times greater than for a sprat of equal length. Let us suppose that the stiffness of a cupula (for small displacements, such as are found in nature) increases with the number of hair cells; then, in the ''design'' of these systems, there must be a payoff between the advantages of having more hair cells and the advantage of having fewer cells, but with each one stimulated to a greater extent for a given force on the cupula. (On the simplest theory, this would be by a factor inversely proportional to the number of hair cells).

At 10 Hz, the velocity in a canal of *Poromitra* will, for a given acceleration in the external medium, be about eight times greater than the corresponding velocity in the sprat. If spontaneous activity is a limiting factor, then sensitivity could be increased by averaging from a large number of receptors, and the greater number of hair cells in *Poromitra* would, assuming a square root law, give it an advantage of 3–4 times. A further increase of sensitivity can be expected from the steepness of the velocity gradient near the wall of a large canal. It is possible that, overall, *Poromitra* might be as much as 100 times more sensitive in terms of pressure gradient than a sprat in the frequency range of 5–15 Hz; this would be $10^4$ in terms of power. These fish are meso- to bathypelagic and feed on small crustacea at depths that are lightless and silent; they move with their pectoral fins (Professor N.B. Marshall personal communication). These canals may indeed be well adapted for catching such prey.

## Anoplogaster

*Anoplogaster* sp. has head canals that occupy most of the surface area of the head. These, unlike those of the melamphids, are shallow chambers (about 1 mm deep) separated from each other by bony partitions; in one 100-mm fish, the surface areas of the chambers ranged from 4.5 to 105 $mm^2$. The chambers are covered with thin, soft membranes, and there does not seem to be any regular system of pores. Small holes of variable shape and position were found in the specimens that we studied, but these holes seem likely to have been artifacts. The partitions are pierced by openings about 2 mm wide and 0.6 mm deep, and in these openings are the neuromasts, surmounted by cupulae 1.2 mm along the axis of the opening, 0.4 mm across the width of the opening, and 0.4–0.5 mm high. Hair cell numbers are smaller than for *Poromitra* of the same body length. One neuromast had about 1,200 hair cells, and the nerve from the same neuromast had 65 fibers, giving a receptor cell-fiber ratio of 18.6, while another nerve had 120 fibers. The expected motion in the canal was calculated for the dimensions measured, assuming a tension in the very soft membrane of 10 mN/m. We have been unable to do any experiment to check this. Calculation suggests that the high-frequency asymptote might have an inside-outside ratio ($x_{in}/x_{out}$) of 4, and there might be a resonance at about 8 Hz. If these results give even a rough approximation to the truth, it seems that this lateral line will have a greatly increased sensitivity over that found for the sprat, particularly at low frequencies. The sensitivity will probably be further increased because of the high-velocity gradients perpendicular to the direction of flow that occur when a cupula of the size found here divides the flow through a small opening. Like the melamphids, these are meso- to bathypelagic fish living in dim light and relative silence.

They feed on small fish, squid, and decapod crustaceans, and these might well produce disturbances for which such a canal system might be appropriate.

## Penaeidae (Crustacea; Decapoda; Natantia)

The last example is drawn, not from a fish, but from certain pelagic shrimp of the families Sergestidae and Penaeidae (Denton and Gray 1985). These animals have long antennal flagella. Each flagellum has a proximal part held approximately at right angles to the body and a distal trailing part, which is held parallel to the body (Fig. 23.12a). The trailing parts of these flagella have curved setae that form a "canal" (Fig. 23.12b) and another type of setae spaced along the inside of this canal (Fig. 23.12c) that are parts of mechanoreceptors (Foxton 1969; Ball and Cowan 1977). We have shown that the proximal parts are laterally flexible and that the trailing distal parts follow the water movements at frequencies above about 1 Hz at least to 100 Hz. These distal parts of the flagella resemble two lateral line systems operating independently and held nearly two body lengths apart. We were able to observe the motions of the tips of the mechanoreceptors and to find, over a range of frequencies, their displacements as fractions of the

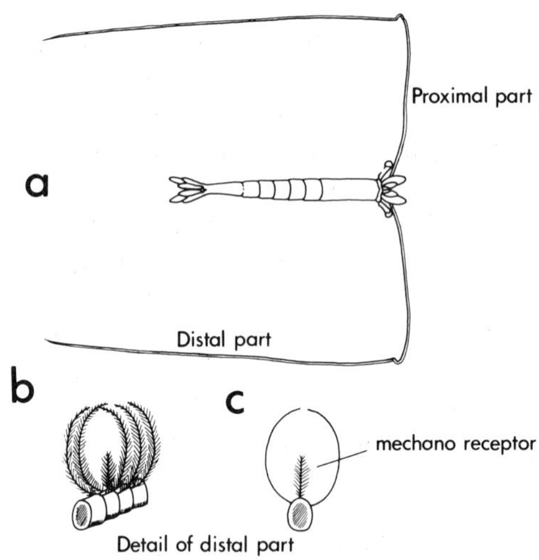


Figure 23.12. (a) Diagram to show the way in which second antennae (flagella) of crustaceans discussed in this chapter are held in nature. (b) Diagram of short length of distal part of a flagellum. This shows how curved setae with many setules, arising from the segments of the flagellum, form a canal enclosing other setae that are parts of mechanoreceptors. Such canals have radii up to about 0.3 mm. (c) Diagram perpendicular to the axis of distal part of a flagellum. It shows specialized seta that is part of a mechanoreceptor and projects (like the cupula of fish lateral line neuromast) into a "canal." Diagrams (from Denton and Gray 1985) are based on the work of Foxton (1969) and Ball and Cowan (1977) and our own observations.

outside displacements in the medium. The mechanoreceptor moved freely, approximately following the displacement of the seawater in the canal.

We have calculated patterns of stimulation for various relative positions of a shrimp and a source of vibrations. These calculations show that, as might be expected, such patterns change even more dramatically with position than they do in fish (Fig. 23.8).

## A Discussion

Advances in our understanding of the acousticolateral systems of fish are difficult because these systems are, in many ways, more sophisticated than our own sense of hearing. There is the additional complication that different parts of an acousticolateral system measure different aspects of the mechanical disturbance to which the fish is exposed. Thus, in the sprat, the free neuromasts might measure something approaching the *local net velocity* of the medium against the skin (see Liff and Shamres 1972 and Oman 1972); the neuromasts in the lateral line system, the *local net acceleration* of the medium against the skin; the otolith-carrying organs (including the utricle), the *absolute accelerations* of the head; the utricle, the pressure and also, by a special adaptation, changes in depth of the animal. In many fish, there is also a much greater emphasis on the analysis of transients and much less on the analysis of frequency than in human hearing. We know far too little about the natural stimuli to which fish are exposed, and the experiments described in Chapter 4 are particularly welcome. Even with good knowledge of the possible stimuli, we should still need to know which aspects of the stimuli are significant in the lives of fish. We would like to have, for other situations, the kind of knowledge that we have of the ways in which an animal like *Xenopus* can locate objects at the surface (see Chapters 24 and 25).

For most lateral lines, there is a consensus that lateral line systems are concerned with low frequencies (< 200 Hz) and with disturbances that originate either close to the fish or from its own movements. We believe that, when dealing with external sources of vibration, it is particularly important to study the interactions of the displacements of the fish in relation to the displacements of the medium around the fish. Because of such interactions, the patterns of stimulation of the lateral line system change much more dramatically with the position of the source than might be expected (see Fig. 23.8).

A common type of lateral line consists of a canal with regularly spaced pores and neuromasts in the canal between the pores. In different lateral lines, the cupulae block the canal to a greater or lesser extent. In the limiting cases of a cupula that blocks the canal and one in which the cupula is low and hardly obstructs the canal at all, the mechanical performances of these systems are much more alike than might be expected. Thus, at lower frequencies (*low* is defined on p. 601), the forces on the cupulae of both are approximately proportional to the local net accelerations of the external medium close to the lateral line (or to express this in another way, proportional to frequency squared). The "design" of a lateral line can allow it to select particular features of a transient stimulus or to act as an attenator for some frequencies and an amplifier for others. Thus, to achieve higher sensitivies for slow transients or low frequencies, large lateral lines are necessary and, with these, it is possible to increase the sensitivity for particular bands of frequencies by having very compliant membranes covering the canals. The great variety of structure in lateral lines makes us suppose that they are specialized for different purposes in different animals and sometimes even in different parts of the same animal. Thus, the sprat's lateral line, through its connection with the auditory bulla, seems to be specialized to deal with transients and to detect the earliest sign of change of behavior of a neighboring animal. In contrast, the systems found in such fish as *Anoplogaster* or *Gymnocephalus cernus* (the ruffe) probably have lateral lines with resonant properties allowing them to be extremely sensitive to particular frequency bands, but with poor resolution in time.

In this chapter we have considered a number of factors influencing the response of lateral line sense organs to stimulation by external sources. These have included the structure and dimensions of the canals, the properties of the cupula, and the number of receptors and nerve fibers. We have suggested how some of these may in some cases be of importance to the animal. It is important to remember that such interpretations require assumptions about the sources that matter and the information the fish may require. For example, a large canal could mean higher sensitivity at low frequency or a requirement to determine the velocity of the source independent of frequency over a wider range of frequency.

*Acknowledgments.* The Marine Biological Association of the U.K. is generously supported by a grant from the Natural Environment Research Council. J.A.B.G. has a grant from the Medical Research Council. Some of the work described was done by E.J.D. while visiting Australia as a Senior Queen's Fellow. For all this support we are very grateful.

## References

Ball, E.E. and Cowan, A.N. (1977) Ultrastructure of the antennal sensilla of *Acetes* (Crustacea, Decapoda, Natantia, Sergestidae), Philos. Trans. R. Soc. Lond. Biol. Sci., 277:429–457.

Best, A.C.G. and Gray, J.A.B. (1982) Nerve fibre and receptor counts in the sprat utriculus and lateral line, J. Mar. Biol. Assoc. U.K., 62:201–213.

Blaxter, J.H.S., Denton, E.J., and Gray, J.A.B. (1981) Acousticolateralis system in clupeid fishes, in Tavolga, W.N., Popper, A.N., and Fay, R.R. (eds.), Hearing and sound communication in fishes, Springer-Verlag, New York, pp. 39–59.

Blaxter, J.H.S., Gray, J.A.B., and Best, A.C.G. (1983) Structure and development of the free neuromasts and lateral line system of the herring, J. Mar. Biol. Assoc. U.K., 63:247–260.

Denton, E.J. and Gray, J.A.B. (1982) The rigidity of fish and patterns of lateral line stimulation, Nature, 297:679–681.

Denton, E.J. and Gray, J.A.B. (1983) Mechanical factors in the excitation of clupeid lateral lines, Proc. R. Soc. Lond. Biol. Sci., 218:1–26.

Denton E.J. and Gray, J.A.B. (1985) Lateral-line-like antennae of certain of the Penaeidae (Crustacea, Decapoda, Natantia), Proc. R. Soc. Lond. Biol. Sci., 226:249–261

DuBois, A.B., Cavagna, G.A., and Fox, R.S. (1974) Pressure distribution on the body surface of swimming fish, J. Exp. Biol. 60:581–591.

Flock, Å. (1965) Electron microscopic and electrophysiological studies on the lateral line organ, thesis, University of Stockholm, cited by Kuiper, A. (1967).

Flock, Å. (1971) Sensory transduction in hair cells, in Handbook of Sensory Physiology, vol. 1, Principles of Receptor Physiology, Loewenstein, W.R. (ed.), Springer-Verlag, New York, pp. 396–441.

Foxton, P. (1969) The morphology of the antennal flagellum of certain of the Penaeidae (Decapoda, Natantia), Crustaceana, 16:33–42.

Gray, J.A.B. (1984) Interaction of sound pressure and particle acceleration in the excitation of the lateral-line neuromasts of sprats, Proc. R. Soc. Lond. Biol. Sci., 220:299–325.

Harris, G.G. (1964) Considerations on the physics of sound production by fishes, in Marine Bioacoustics, Tavolga W.N., (ed.), Oxford, Pergamon Press, pp. 233–247.

Harris G.G. and van Bergeijk, W.A. (1962) Evidence that the lateral-line organ responds to near-field displacements of sound sources in water, J. Acoust. Soc. Am., 34:1831–1841.

Jielof, R., Spoor, A., and de Vries, H.L. (1952) The microphonic activity of the lateral line, J. Physiol., 116:137–157.

Kuiper, J.W. (1967) Frequency characteristics and functional significance of the lateral line organ, in Lateral Line Detectors, Cahn, P.H. (ed.), Indiana University Press, Bloomington, pp. 105–121.

Liff, H.J. and Shamres, S. (1972) Structure and motion of cupulae of lateral line organs in *Necturus maculosus*. III. A technique for measuring the motion of free-standing lateral line cupulae, Quarterly Progress Report, Research Laboratory of Electronics, M.I.T., 104:331–336.

Oman, C.M. (1972) Structure and motion of cupulae of lateral line organs in *Necturus maculosus*. IV. Preliminary model for the dynamic response of the free-standing lateral line cupula, based on measurements of cupula stiffness, Quarterly Progress Report, Research Laboratory of Electronics, M.I.T., 104:336–342.

Partridge, B.L. and Pitcher, T.J. (1980) The sensory basis of fish schools: relative roles of lateral line and vision, J. Comp. Physiol. 135:315–325.

Sand, O. (1981) The lateral line and sound reception, in Hearing and Sound Communication in Fishes, Tavolga, W.N., Popper, A.N., and Fay, R.R. (eds.), Springer-Verlag, New York, pp. 459–480.

Schlichting, H. (1968) Boundary-layer theory, McGraw-Hill, New York, pp. 419–421.

van Bergeijk, W.A. and Alexander, S. (1962) Lateral line canal organs on the head of *Fundulus heteroclitus*, J. Morphol., 110:333–346.

# Chapter 24

# Prey Identification and Prey Localization in Surface-feeding Fish and Fishing Spiders

Horst Bleckmann*

Surface-feeding fish and fishing spiders prey on terrestrial insects trapped at the water surface. In addition, semiaquatic spiders feed on small fish, frogs, and tadpoles (Gudger 1922). The stimuli involved are surface waves such as those generated by insects trapped at the water–air boundary or aquatic vertebrates contacting the water surface to feed or breath (Schwartz 1965, 1971; Bleckmann and Barth 1984; Bleckmann and Lotz 1987). Both surface-feeding fish and fishing spiders determine the direction and distance to a concentric wave source. They can do this even under open loop conditions (i.e., even if the wave generator is turned off before or at the moment the fish or spider starts to turn to the wave source) (Bleckmann and Schwartz 1982; Bleckmann and Barth 1984). In addition, these animals are faced with the problem of distinguishing between prey and nonprey stimuli. The present chapter does not attempt to review the pertinent literature on surface-dwelling animals (see Bleckmann 1985a and Chap. 25). Instead, it focuses on two questions; namely, how can surface-feeding fish and fishing spiders determine the source distance, and how can they distinguish between prey and nonprey wave stimuli?

## Physical Principles

The study of sensory biology requires a thorough knowledge of not only the physics of the biologically relevant stimuli, but also that of the transmission channel through which these stimuli are propagated. Water surface waves are boundary waves caused by moving a solid object through the water, as well as by shock, wind, and all similar forces that generate and maintain waves. Like all physical waves, water surface waves can be treated as temporal and spatial

* Department of Neurosciences, School of Medicine and Neurobiology Unit, Scripps Institution of Oceanography, University of California at San Diego, La Jolla, California 92093, USA

oscillations, unequivocally characterized in either the time domain by their amplitude function or the frequency domain by their amplitude and phase spectrum.

## Characteristics of Prey and Nonprey Wave Stimuli

Only wave stimuli generated by the following sources are of importance for this chapter: (1) biotic stimuli caused by aquatic, semiaquatic, and terrestrial insects that have fallen into the water; (2) biotic stimuli caused by frogs, tadpoles, and fish contacting the water-air interface (e.g., to feed or breathe); (3) abiotic wave stimuli caused, for instance, by a falling leaf, seed, or twig; and (4) permanent wind-generated waves that occur as additive background noise in all larger bodies of water (Fig. 24.1) (see also Chap. 25).

### Time Domain

With respect to the stimulus's time structure, three different types of waves can be found. These are (1) concentrical waves that are more regular in the time course; (2) concentrical waves that are more irregular in the time course, and (3) permanent wind-generated waves that have no center of origin. More regular concentrical waves are often evoked by semiaquatic and aquatic insects and vertebrates. These waves also include ''noise'' waves generated by falling leaves, twigs, and seeds. Waves of this type consist of a single wave train whose duration rarely exceeds 1.5 sec. If more wave stimuli are set free, they usually arise from different loci with time intervals that are longer than the average train duration (Lang 1980; Bleckmann 1985b).

More irregular concentrical waves are caused by kicking leg movements of terrestrial insects struggling at the water surface. These waves often have a duration that exceeds 10 sec., with values of 60 or more sec. common. If more wave stimuli are set free, they arise from the same locus, with time intervals often shorter than that of the stimulus duration (Lang 1980; Bleckmann 1985b).

The amplitude of wave stimuli caused by these sources is quite variable. It depends not only on the size of the wave emitter, but also on the impact velocity and source distance. Some typical values for the mean peak-to-peak displacement amplitude of wave stimuli are 2–80 μm (terrestrial insects struggling at the water surface [Lang 1980]), 106 μm ∓ 34 (falling leaf of *Salix alba*), 1,239 μm ∓ 57 (falling piece of wood), and 22–1,320 μm (aquatic and semiaquatic vertebrates contacting the water surface) (Bleckmann 1985b). Even if the wind speed is only 6–12 km/hr, wind-generated waves measured in the natural biotope of the fishing spider *Dolomedes triton* can have a height of at least 40 mm (Bleckmann and Rovner 1984). Thus we cannot distinguish between the different wave types by amplitude alone (Lang 1980; Bleckmann 1985b).

### Frequency Domain

In the frequency domain, the relevant wave stimuli can be divided into roughly three categories (for a more refined distinction, see Lang 1980): (1) wave stimuli that contain frequencies (f) above about 50 Hz (i.e., stimuli caused by terrestrial

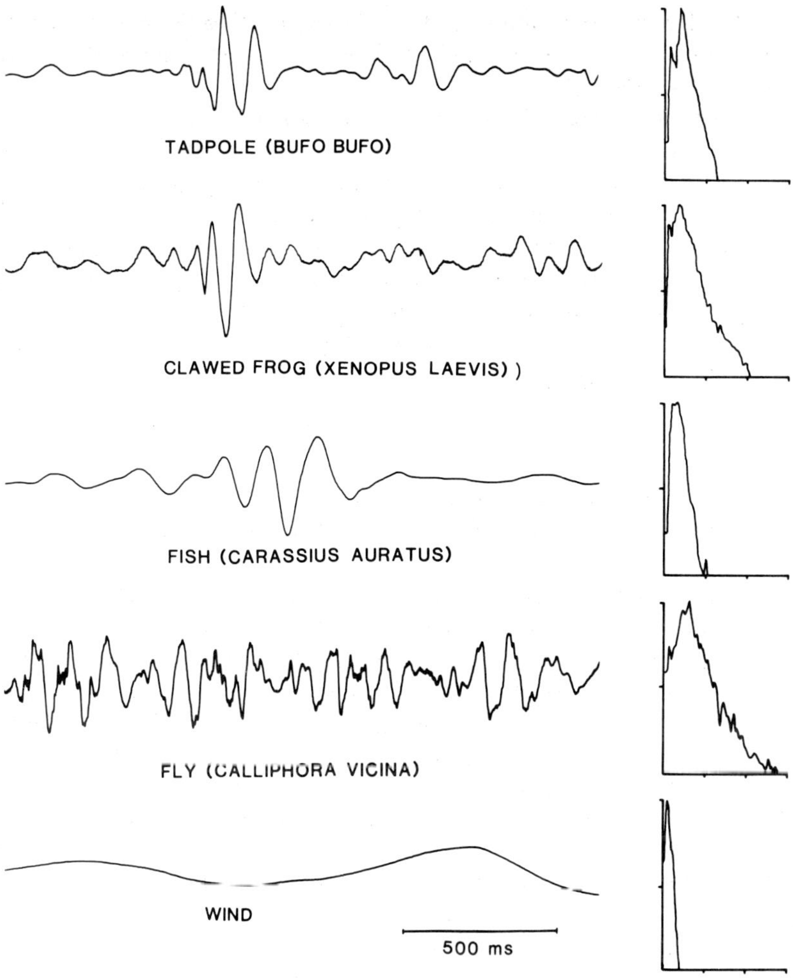


Figures 24.1. Examples of water surface waves caused by different biotic sources and by wind. Amplitudes are not to scale. *Right:* relative spectra of corresponding wave stimuli. In all cases, at least 10 wave spectra were averaged. Highest power value was set equal to 0 dB. *x−axis,* 0–60 Hz; *y−axis,* 0 to −60 dB.

insects' struggling at the water surface); (2) wave stimuli with an upper frequency limit of about 20–50 Hz (i.e., waves caused by fish, frogs, and tadpoles, as well as abiotic waves, caused by falling leaves, seeds, or twigs; and (3) wind-generated wave stimuli that do not contain frequencies above about 10 Hz (Fig. 24.1) (Lang 1980; Bleckmann and Rovner 1984; see also Chap. 25).

## Water Surface as Stimulus Transmission Channel

To find out which stimulus parameters might be useful for determining the wave source distance, some knowledge about the physical properties of the water surface is essential. During propagation, surface waves are strongly modified. They not only decrease in amplitude, but also change in frequency composition, duration, and time course (Fig. 24.2).

The propagation (phase) velocity ($C_{ph}$) of water surface waves depends on gravitational acceleration (gravity-induced waves: $\leq$ 4 Hz), the surface tension (capillary-induced waves: $\geq$ 46 Hz), or both (capillary gravity-induced waves: 4 Hz $<$ f $<$ 46 Hz) (Sommerfeld 1970). As a result, surface waves with amplitudes much smaller than the water depth (h) are spread by dispersion; i.e., propagation velocity decreases with decreasing wave length ($\lambda$) until it reaches a minimum (23.1 cm/sec. at 13 Hz, $\lambda$ = 1.7 cm) and then increases continuously with decreasing $\lambda$ (e.g., 40.37 cm/sec. at 140 Hz, $\lambda$ = 0.2888 cm) (Fig. 24.2A). The exact relation between $\lambda$ (and hence f) and $C_{ph}$ is

$$C_{ph} = \sqrt{\frac{g \cdot \lambda}{2\pi} + \frac{2\pi\eta}{\xi \cdot \lambda}} \qquad (24.1)$$

where $g$ = gravitational acceleration (981 cm/s$^2$), $\eta$ = surface tension (72 dyn/cm), and $\xi$ = density of water (0.998 g/cm$^3$) (all values refer to 20°C) (Sommerfeld 1970).

Most biologically relevant surface waves consist of a mixture of contiguous frequency components. Owing to their frequency-dependent propagation velocity, the high- and low-frequency components of a single short wave train are separated more and more during stimulus propagation. This can easily be demonstrated by dipping a rod (glued, for instance, to the coil of a modified loudspeaker) once into the water. The resulting stimulus, called a *click,* is always frequency-modulated downward to a degree that depends on the source distance, but is nearly independent of the kind of wave production or initial stimulus intensity (Fig. 24.2C and D). If both the mean frequency of the initial part of a wave stimulus and the relative frequency change around this local frequency are known,

Figure 24.2. (A) Propagation (phase) velocity of water surface waves with different frequencies and wave lengths. (B) The attenuation of water surface waves associated with stimulus frequency and the distance from vibration source. *Dotted lines,* measured damping curves for frequencies 20 and 140 Hz. *Inset,* the change of spectral composition of water surface waves during stimulus propagation. Source distances at which wave stimuli were measured are 3, 8, 13, 18, and 23 cm: *x-axis,* 0–160 Hz; *y-axis,* 0 to −60 dB. Curves in A and B were calculated according to Eqs. (24.1.) and (24.2). (C) The nature of click stimulus as it passes points 5, 10, and 15 cm from the source: *time bar,* 100 ms; *Inset,* clicks of various amplitudes (0.06–2.5 μm, amplitudes not to scale) measured at a distance of 10 cm. Note that independent of stimulus amplitude, constant frequency modulation appears at certain source distance. (D) Time intervals $t_1$—$t_8$ (*y-axis*) of first 8 *x-intercepts* (i.e., points where the stimulus crosses the *x-axis*) of wave stimuli produced either by air blast or by rods of different diameters (2.5, 5, and 8 mm) in relation to the kind of stimulus production and source distance. All observed differences of frequency modulation occurred within *shaded area.* Between *dashed lines,* 5 cm; *solid lines,* 10 cm; and *dotted lines,* 15 cm. (From Bleckmann and Schwartz 1982; Bleckmann 1985a.)

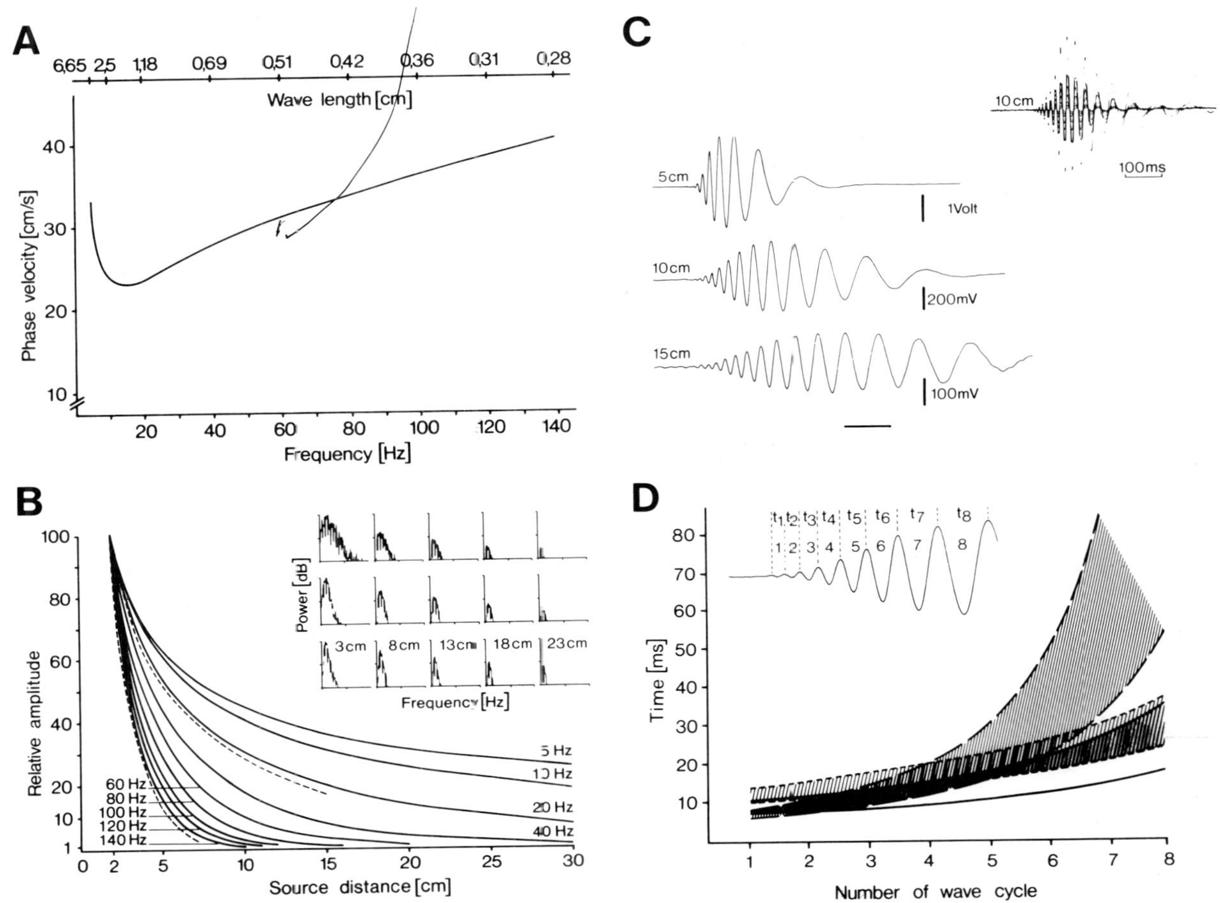
A
6,65 2,5 1,18 0,69 0,51 0,42 0,36 0,31 0,28
Wave length [cm]
Phase velocity [cm/s]
Frequency [Hz]
B
Relative amplitude
Source distance [cm]
Power [dB]
3cm 8cm 13cm 18cm 23cm
Frequency [Hz]
5 Hz
10 Hz
20 Hz
40 Hz
60 Hz
80 Hz
100 Hz
120 Hz
140 Hz
C
5cm
1Volt
10cm
200mV
15cm
100mV
10 cm
100ms
D
Time [ms]
Number of wave cycle

the wave source distance (D) unequivocal can be calculated (Käse and Bleckmann in press).

Owing to geometrical spreading and the physical properties (viscosity) of the water, concentrical surface waves are strongly attenuated during propagation. Close to the wave source, the slope of the attenuation depends on both f and the distance (D) the stimulus has already traveled (Fig. 24.2B). For D $\geq$ 2 cm (for technical reasons, the wave stimuli could not be measured closer to their source), the damping of water waves can be approximately described by the formula

$$A = A_i \sqrt{\frac{R_i}{R}} \cdot e^{\left(\frac{16\, æ\, \pi^2}{\lambda^2 \cdot C_{ph}}\right)(R_i - R)} \qquad (24.2)$$

where $A_i$ = amplitude of the wave stimulus at distance $R_i$, $A$ = actual wave amplitude at distance $R$, and æ = kinematical viscosity of water (=0.01 g/cm·sec·ξ at 20°C) (Bleckmann 1985a). According to geometrical spreading (eq. 24.2), the attenuation of water waves occurs mostly in the near vicinity of a wave source (Fig. 24.2B). In addition, attenuation increases strongly with f (decreasing λ), but is independent of the stimulus amplitude. Thus, owing to the complex filter characteristics of the water surface, a polyfrequency wave stimulus loses mainly its high-frequency components during propagation (Fig. 24.2B, *inset*).

### Sound Waves

A splash made by the entry of any object into water radiates not only surface, but also sound waves in both air and water. At present, there is no available information about the frequency and amplitude content of underwater sound generated by insects struggling at the water surface. For a first approximation, however, the pressure waves resulting from a falling droplet might be useful. For example, an 11-mg waterdrop hitting the water surface (impact velocity = 2 m/sec.) produces sound frequencies in the range of 100 Hz–10 kHz. For the frequencies that fall into the hearing range of many fish (about 50 Hz–1 kHz), the average sound energy reaches up to $10^{-6}$ ergs (Franz 1959). At a source distance of 10 cm, this is approximately −19 dB re 1 dyne/cm $^2$. Even if we take into account that the pressure waves correlated with the initial impact of the droplet are very short (about 10 ms), this may be above the hearing threshold of many teleosts (Fay and Popper 1980). Compared to surface waves, the propagation velocity of pressure (sound) waves in water is very high (1,450 m/sec. at 20°C). Therefore, in theory at least, the difference in arrival time between pressure waves and surface waves can be used to calculate D (Hoin-Radkovski, Bleckmann, and Schwartz 1984).

## Useful Stimulus Parameters for Distance Determination

From the information given in the preceding two sections, it is obvious that the following five stimulus parameters or mechanisms, are useful in determining the wave source distance under open loop conditions:

1. The curvature of the wave front
2. The relative amplitude decrease per unit of distance
3. The frequency modulation of the initial part of the wave stimulus
4. The amplitude spectrum
5 The difference in arrival time between surface and pressure waves

Note that only mechanism 1 works independently of both stimulus amplitude and frequency. Owing to the frequency-dependent damping and phase velocity of water surface waves, all other mechanisms require the additional knowledge of the frequency content of the wave stimulus. It should be stressed that, in waves lacking higher frequencies, the amplitude spectrum (mechanism 4) can give some information about D only if compared with a commonly experienced standard. A stimulus that contains high-frequency components, however, in general can be expected to have traveled only a short distance. Electrophysiological and behavioral experiments indicate that surface-feeding fish have the sensory and neural capabilities to use at least three of the five parameters proposed above.

## Distance Determination in Surface-feeding Fish

### Electrophysiological Results

In surface-feeding fish, the sense organs used in water surface wave perception are the lateral line organs located on the fish's head and back (Schwartz 1970). To date, few electrophysiological experiments (microphonic potentials and recordings from primary lateral line afferents) have been done on surface-feeding fish. Recordings from higher levels of the neuraxis are completely lacking. The electrophysiological results, however, show that the neural responses of single head neuromasts of the Indian freshwater fish *Aplocheilus lineatus* (Cyprinodontidae) and the African butterfly fish *Pantodon buchholzi* (Pantodontidae) are phase-coupled to the stimulus (Unbehauen 1980; Topp 1983; Müller 1984). Although the variability of these responses has not yet been determined, the calculation of stimulus frequency on the basis of interspike time intervals may provide a suitable mechanism for frequency and phase discrimination.

Electrophysiologically obtained threshold values (for examples, see open squares in Fig. 24.3) demonstrate that both *A. lineatus* and *P. buchholzi* are especially sensitive to high- and single-frequency (sf) wave stimuli (Bleckmann and Topp 1981; Topp 1983; Müller 1984). In connection with the phasic-tonic response characteristics of the head neuromasts (Topp 1983; Müller 1984), this selective sensitivity guarantees that the initial low-amplitude, high-frequency part of a wave stimulus is overrepresented in the neural response. In *Aplocheilus*, the increase of neural response with increasing amplitude and *slew rate* (defined as steepness of amplitude modulation) is more pronounced if low-amplitude, high-frequency sf-stimuli are presented (Fig. 24.4).

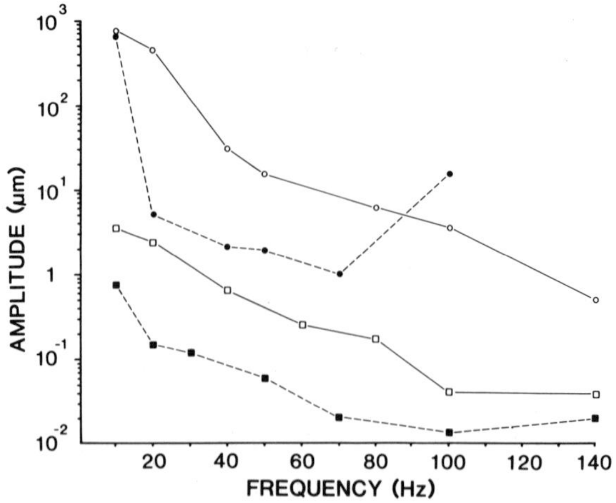


Figurer 24.3. Vibration sensitivity of surface-feeding fish *A. lineatus* and fishing spiders *Dolomedes triton,* determined in behavioral (*filled symbols*) and electrophysiological (*open symbols*) experiments: *squares, A. lineatus; circles, D. triton. Aplocheilus* recordings were made from a branch of truncus supraorbitalis innervating head neuromast II.2. In *Dolomedes,* responses of single slits of metatarsal organ of first pair of legs were recorded. For stimulation, electrodynamic vibrator was brought into contact with the tarsus. (From Bleckmann and Topp 1981; Bleckmann and Barth 1984.)

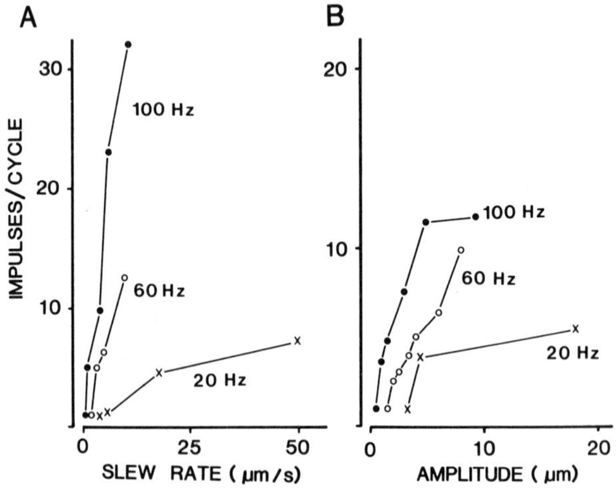


Figure 24.4. Phasic response of single head neuromast of *A. lineatus* to the onset (A) and tonic response to the plateau (B) of surface wave stimulus. In all cases, the number of spikes per stimulus cycle were taken as a measure. (From Topp 1983.)

## Behavioral Results

### Vibration Threshold

The striking sensitivity of the head lateral line system of surface-feeding fish to high-frequency sf-stimuli is confirmed by behavioral experiments. In terms of displacement, sensitivity rises from 10–100 Hz (surface-feeding fish rarely respond to frequencies below 10 Hz) by about 12 dB/octave (Fig. 24.3). This indicates that these fish respond to the acceleration component of a wave stimulus (if the displacement amplitude of an sf-stimulus is held constant, the acceleration amplitude of the stimulus increases at 12 dB/octave). Thus, with respect to the filter properties of the water surface (Eq. 24.2), surface-feeding fish are well adapted to perceiving the weak-amplitude, high-frequency components within a wave train.

### Integration Time

The basic frequency of a wave stimulus and its frequency modulation and amplitude spectrum can be obtained only if the stimulus is integrated over at least a few wave cycles. If a click is presented, the first stimulus-evoked action potentials in the red swimming muscles of *A.lineatus* can be recorded only after about 8.5 wave cycles have passed the head of the fish (Bleckmann and Schwartz 1981). This holds true even if the clicks presented vary in displacement amplitude and slew rate up to a factor of 148 and 240, respectively (Bleckmann 1980). Thus, the reaction time of *A. lineatus* to water surface waves is mainly a function of f, but not of either stimulus amplitude or slew rate. Neither the threshold values of sf-stimuli nor the reaction time to clicks increases as long as the swimming speed of *A. lineatus* does not exceed 10 cm/sec. (Bleckmann 1982).

### Stimulus Discrimination

Tests to demonstrate that surface-feeding fish can distinguish complex wave stimuli have not been done. Behavioral studies show, however, that *A. lineatus* can discriminate between stimuli differing in only one parameter—frequency, amplitude, or slew rate. For instance, in the range of 10–150 Hz, *A. lineatus* can discriminate water wave frequencies to a resolution of 10% (Bleckmann, Waldner, and Schwartz 1981). As expected from the electrophysiological results (Fig. 24.4), slew rate and amplitude discrimination are best when high-frequency, low-amplitude wave stimuli are presented. To give a numerical example: *A. lineatus* can discriminate between wave stimuli (peak-to-peak amplitude 1.7 μm, duration 900 ms) whose slew rate differs by 2.5 μm/sec. (carrier frequency 70 Hz), 10 μm/sec. (40 Hz), and 12 μm/sec. (20 Hz). The corresponding values for amplitude discrimination (carrier frequency 40 Hz, signal duration again 900 ms) are 4 μm (amplitude of reference stimulus 21 μm) and 1 μm (0.15 μm) (Waldner 1981). Thus, *A. lineatus* can discriminate wave stimuli using all three types of cues.

## Distance Determination

As we have seen, important sensory prerequisites for determining D by some or all the five mechanisms proposed above are fulfilled by surface-feeding fish. By testing their orientation behavior (i.e., by measuring the swimming distance as a function of D) to artificially manipulated wave stimuli, we learn which cues are used for distance determination. In order to avoid visual cues, all tests were performed with blinded animals. For localization tests, artificial wave stimuli can be generated most elegantly by blowing a defined airstream on the water surface. In this way, unwanted cues (e.g., the meniscus around a rod contacting the water surface) that give information about the position of the wave generator are excluded (Bleckmann 1980).

*Responses to Clicks.* One might think that the ability of surface-feeding fish to determine D is based on the simple assumption (due to Eq. [24.2]) that a high-amplitude wave stimulus with a high slew rate and upper frequency limit has traveled, on an average, a shorter D than a low-amplitude stimulus with a low slew rate and upper frequency limit. That is, however, not the case. If we generate click stimuli that differ at a certain source distance in amplitude, slew rate, and upper frequency limit, the relative localization errors of *A. lineatus* and *P. buch-*

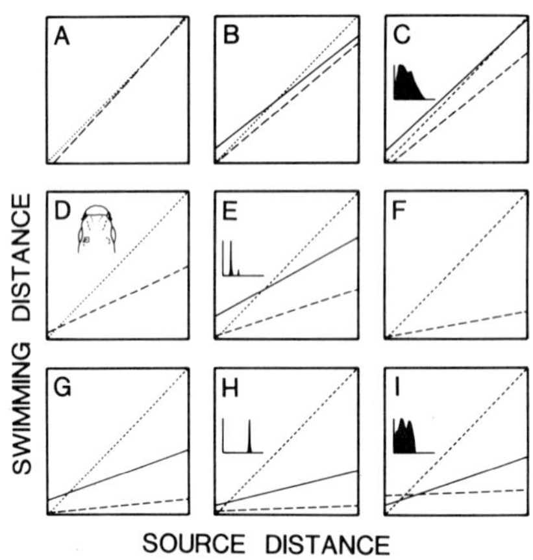


Figure 24.5. Mean swimming distance of *A. lineatus* (----) and *P. buchholzi* (———) to click stimuli of 0.5-, 1.5-, and 4-V intensity (A–C), to sf-stimuli of 30 Hz (*Aplocheilus*) and 35 Hz (*Pantodon*) (E), 50, 70, and 100 Hz (F–H), and to upward frequency-modulated wave stimulus (I). In D, responses of *A. lineatus,* in which all, except one head lateral line organ, were destroyed (see *inset*), are shown: *dotted line,* line of perfect localization (slope 1): the *x-axis* (source distance) and *y-axis* (swimming distance) are, in all cases, 3–18 cm. The number of fish for both species and all experiments is $\geq 3$. *Insets:* Relative power spectra obtained at source distance of 5 cm (C, E, and H) and 7 cm (I). In all cases, highest power value was set equal to 0 dB. *x-axis,* 0–150 Hz; *Y-axis,* 0 to −60 dB. (From Bleckmann 1980; Bleckmann and Schwartz 1982; Müller and Schwartz 1982; and Hoin-Radkovski, Bleckmann, and Schwartz 1984.)

*holzi* are independent of both the type of click stimulus presented and the source distance (Fig. 24.5A–C) (Bleckmann and Schwartz 1982; Hoin-Radkovski, Bleckmann, and Schwartz 1984). In *A.lineatus* the accuracy of distance determination is independent of swimming speed (tested up to 14 cm/sec.) (Bleckmann 1982); thus, even a fast-swimming fish is able to use lateral line information for the detection and localization of prey.

*Responses to Simulated Click.* Experiments with a simulated click give clear evidence that distance estimation is based first of all on the evaluation of the frequency modulation of a wave stimulus. If, for example, *A. lineatus* and *P. buchholzi* are confronted with a stimulus at a D of 7 cm whose frequency modulation has been artificially modified so that it closely resembles that of a click at 15 cm (Fig. 24.6A), the fish swim on an average of 3–5 cm beyond the wave source (Fig. 24.6B), even though all other distance-dependent wave parameters (see above) are still available. It is clear that the fish were not completely deceived, since the standard deviation of the responses is significantly greater for the artificially altered click as compared to a normal click from 15 cm. The use of frequency modulation for distance determination is further supported in experiments with *A. lineatus* in which all but one head lateral line organ have been destroyed. Although these animals have no way to determine the curvature of the wave stimulus or the amplitude decrease per unit of distance, they still show an increase of swimming distance with source distance if clicks are presented (Müller and Schwartz 1982). Distance determination in ablated fish, however, is somewhat impaired (Fig. 24.5D) (Müller and Schwartz 1982), thus indicating again that, besides frequency modulation, these fish use other parameters for distance determination.

*Responses to Single-Frequency Stimuli.* That the evaluation of both the curvature of the wave front and the amplitude spectrum is used in distance determination can be seen in experiments in which sinusoidal sf-stimuli, which do not contain any frequency modulation at all, are presented. Although *A. lineatus* and *P. buchholzi* now tend to underestimate D if it exceeds 6–8 cm (Fig. 24.5E–H), there is still a positive correlation between D and swimming distance (for *A. lineatus* this holds true only up to frequencies of 50 Hz [Bleckmann 1980]). Summarizing all the results obtained with sf-stimuli, we learn that the relative localization error at a given D rises with f, while at a given f it rises with D (Bleckmann 1980; Hoin-Radkovski, Bleckmann, and Schwartz 1984). Assuming that the curvature of the wave front is determined through arrival time differences of the wave front at different neuromasts, this is exactly what is to be expected. Theoretical calculations (Hoin-Radkovski, Bleckmann, and Schwartz 1984) suggest that the accuracy of distance determination from sf-stimuli should decrease with both increasing source distance (because the curvature becomes more and more flattened) and increasing frequency of the stimulus (because an increase in frequency results in an increase of phase velocity and thus, again, in a decrease of arrival time differences). At a given D, surface-feeding fish swim progressively shorter distances in response to sf-stimuli at higher frequencies. This may indicate

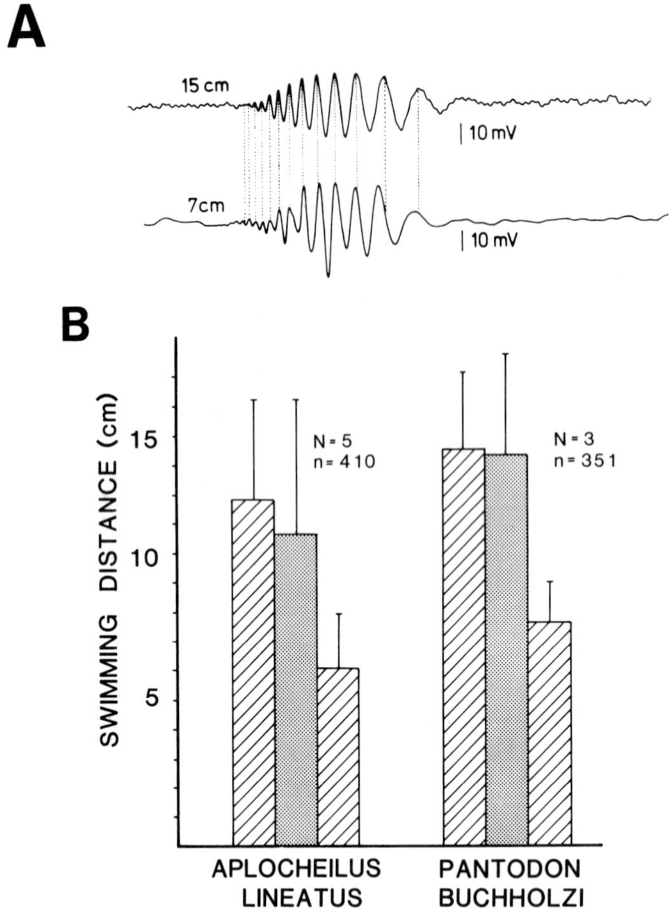


Figure 24.6. (A) Wave stimulus of 1.5-V click at 15 cm and simulated click at 7 cm. (From Bleckmann and Schwartz 1982; Hoin-Radkovski, Bleckmann, and Schwartz 1984.) (B) Mean swimming distance towards wave source for *A. lineatus* and *P. buchholzi*. For each species from left to right: response to a click at 15 ± 1.5 cm source distance, to simulated wave stimulus at 7 ± 0.5 cm source distance, and to a click at 7 ± 0.5 cm source distance. Here and in Figures 7 and 8, N = number of animals and *n* = number of trials.

that the amplitude spectrum is also evaluated for distance determination (i.e., if no other cues are available, a high-frequency stimulus is ''expected'' to have traveled a shorter D than a low-frequency one). This point of view is further supported in experiments with artificial frequency-upward modulated wave stimuli containing high-amplitude, high-frequency components (Fig. 24.5I, *inset*) (Bleckmann and Schwartz 1982). These stimuli in both species tested also cause an underestimation of D (Fig. 24.5I). Considering the filter properties of the water surface, this probably reflects a good localization strategy.

From the results given above, it is obvious that the stimulus parameters of frequency and frequency modulation are the most important ones for distance determination. The importance of the frequency parameter in contrast to amplitude is also stressed by the following observation: The mean swimming distance of *A. lineatus* to the source of a 50-Hz wave stimulus decreases by only about 20% if the displacement amplitude of the stimulus or its slew rate vary by a factor of 25 (Bleckmann 1980). Thus the strong decrease in swimming distance observed with increasing frequency of sf-stimuli is probably due to a frequency discrimination, not to a discrimination of these stimuli on the basis of subjective intensity differences (as has been speculated by Elepfandt, Seiler, and Aicher (1985). The assumption that surface-feeding fish indeed discriminate frequencies is also supported by the observation that prey capture reactions ( a sudden vigorous jump in the direction of the wave source) to clicks, independent of amplitude and slew rate, occur only close to the wave source ($\leq$ 10 cm) (Bleckmann 1980). In contrast, in high-frequency ($\geq$ 70 Hz) sf-stimuli, prey capture responses in *A. lineatus* can be released at least up to a distance of 14 cm (Bleckmann 1980). This demonstrates again that high-frequency sf-stimuli lead to an underestimation of the wave source distance.

Lateral line organs are directionally sensitive. The magnitude of the neural response approximates a cosine function with respect to the organs' long axis (Flock 1965; Unbehauen 1980; Bleckmann 1986; see also Chap. 22). It can be shown that there exists an unequivocal relationship between the output differences of different neuromasts and the curvature of the wave front (Hoin-Radkovski, Bleckmann, and Schwartz 1984). If surface-feeding fish determined D with the aid of organ output differences, a better performance for high-frequency sf-stimuli would be expected because the differences in the neural output of single lateral line organs are more pronounced if high-frequency stimuli are presented (Fig. 24.4). Because the orientation behavior of surface-feeding fish contradicts this assumption (Fig. 24.5E–H), we can assume that a localizing mechanism based on the organ's directional sensitivity is, at most, of minor importance.

## Sound

A carefully conditioned *A. lineatus* responds slightly to water surface waves even when all dorsally located lateral line organs are eliminated (Müller and Schwartz 1982). This may be due to more ventrally located lateral line organs. Another possibility is that surface-feeding fish may be able to hear sound waves correlated with the air-to-water entry of an object (Franz 1959) or with the water surface wave itself (Görner 1976). However, the remarkable change of swimming distance depending on the kind of wave stimulus presented and the reduction in the localization accuracy in intact fish at all frequencies tested make it unlikely that the arrival time difference between sound and surface waves is used for a calculation of D.

## Some Speculative Considerations

The morphology of the mechanosensitive lateral line of fish shows a marked, and as yet not well understood, diversity that may be due to phylogeny or different ecological and behavioral tasks the fish must face or both (see Chaps. 22 and 23). Like many other fish, surface feeders vary in their degree of cephalic lateral line canal reduction (Schwartz 1970). *A. lineatus* has eighteen free-standing neuromasts on its head, which differ in size (about 0.2–0.5 mm) and orientation (Schwartz 1965). In the head region of *P. buchholzi,* there are numerous small (about 0.05 × 0.03 mm) free neuromasts, but its head lateral line also contains large sacs of connective tissue located between the epidermis and the bones of the skull (Schwartz 1970; Müller 1984). These sacs contain ten large (up to 2.15 mm in length) curved neuromasts that have up to 3,000 hair cells (Müller 1984). Irrespective of these differences, however, their ability to orient to water surface waves and their underlying mechanisms are quite similar in both fish species. This may support the hypothesis that superficial neuromasts and widened canals are alternative adaptations to a low-noise environment, like the surface of ponds and pools (see Chap. 22).

Some speculations regarding the innervation and function of single neuromasts in surface-feeding fish might be useful. As in *Aplocheilus* (Münz pers. com.), the ratio between the number of hair cells in a neuromast and the number of afferent nerve fibers in *Pantodon* is about 3:1 (Müller 1984). We can therefore assume that the main function of this large number of hair cells is not to improve the signal-noise ratio. From non-surface-feeding fish we know that different neuromasts and afferent nerve fibers innervating a single neuromast can be tuned to different frequencies (Münz 1986). There are some indications that the same holds true for the head lateral line system of *A. lineatus* (Sann 1984, quoted by Müller 1984). Intrinsic mechanisms of hair cell frequency tuning have been proposed for electroreceptive (Meyer and Zakon 1982) and auditory hair cells (Crawford and Fettiplace 1981). It is conceivable that the intrinsic mechanism of hair cell tuning may also be operating within the mechanosensitive lateral line system and that the ability to analyze the frequency amplitude content of a complex wave stimulus on the receptor level is as sophisticated in the mechanosensitive lateral line as in the acoustical and electroreceptive parts of the octavolateralis system.

## Stimulus Discrimination in Fishing Spiders

All organisms are faced with the problem of differentiating between important and unimportant stimuli. In predators, especially, distinction at a distance between prey, nonprey, and conspecific animals is of importance. It may help to increase prey capture success, conserve motor energy, and minimize the risk of the predator's becoming prey itself. The ability to discriminate between complex wave stimuli has been studied in the semiaquatic spider *Dolomedes triton* (Pisauridae).

Before we look at this spider's behavior, let us again view the relevant wave stimuli with respect to possible and relevant differences.

## Useful Stimulus Parameters for Prey and Nonprey Wave Discrimination

From the information given in the section on physical principles, it is obvious that only wave stimuli generated by the kicking of terrestrial insects are easy to recognize as prey waves. These waves combine several prey-indicating characteristics because they are irregular in time course and long lasting and have frequency components above 50 Hz. In addition, in most cases, more than one wave train is set free from the same locus. In contrast, an unequivocal distinction between wave stimuli caused by aquatic and semiaquatic insects and vertebrates and those caused by leaf, seed, and twig fall is not possible. These stimuli are similar with respect to their short duration and their low upper frequency limit. Wind waves, on the other hand, are clearly distinguishable from all other types of waves because their upper frequency limit is restricted to 10 Hz (Bleckmann and Rovner 1984). Whether and how fishing spiders distinguish between different wave types can only be tested in behavioral experiments with defined wave stimuli. These experiments, which were performed on blinded animals to eliminate visual cues, show that the semiaquatic spider *D. triton* does discriminate between wave stimuli and that this discrimination is based on the evaluation of several wave parameters.

## Behavioral Results

Spiders have very sensitive vibration receptors (slit sense organs) in their legs. The most sensitive compound slit sense organ is the metatarsal organ situated on the metatarsus close to the tarsometatarsal joint (Barth 1982). An ambushing *D. triton* puts the tarsi of its front legs on the water surface, indicating that water waves are important for these spiders in detecting and localizing prey. Consequently, blinded spiders attack a vibrating, wave-generating rod, even if the distance between them is up to 40 cm (Bleckmann and Barth 1984).

### Vibration Threshold

*D. triton* is generally on the order of one to two magnitudes less sensitive to water surface waves than is the surface-feeding fish *A. lineatus*. As in fish, however, the behavioral threshold of *Dolomedes* decreases with f, but now has a minimum of about 70 Hz (Fig. 24.3). It is likely that the drop in responsiveness for frequencies above 70 Hz is not due to a true loss in sensitivity, but instead reflects the rarity of high-frequency sf-stimuli at the water surface under natural conditions (Lang 1980; see Wilcox 1979 for one exception). If so, the spiders, which in contrast to surface-feeding fish, were never fed in response to an artificial wave stimulus, may not have been motivated to approach the source of a high-frequency "unnatural" sf-stimulus spontaneously. This point of view is supported by electrophysiologically obtained threshold values (Fig. 24.3) and by the fact that a polyfrequency wave stimulus becomes much more attractive if it has some low-amplitude, high-frequency components (see below). Thus, like surface-feeding fish, *Dolomedes* is well adapted to perceiving the high-frequency, low-amplitude components in which prey (insect) waves differ from most other wave types

### Responsiveness, Frequency Bandwidth, and Stimulus Amplitude

As already mentioned, the displacement amplitude of prey and nonprey wave stimuli is highly variably (Lang 1980; Bleckmann 1985b). This fits well with the observation that the stimulus amplitude, provided it is above threshold, does not affect the responsiveness of *D. triton* when noisy, irregular wave stimuli of equal spectral composition (3–65 Hz) are offered (for method, see Bleckmann and Barth 1984). Irrespective of the mean peak-to-peak amplitude, which varied between 2 and 37 μm, the responsiveness to these stimuli is 75%–76% (Bleckmann 1985b).

### Responsiveness, Frequency Bandwidth, and Stimulus Duration

Is the responsiveness of *D. triton* affected by both frequency content and duration of the stimulus? Two stimulus types, varying in duration, were produced: (1) a noninsectlike stimulus of narrow frequency bandwidth (upper frequency limit in test range of 5–10 cm, 33–39 Hz; mean peak-to-peak amplitude 12–15 μm); and (2) an insectlike stimulus with the enlarged upper frequency limit of 64–78 Hz (peak-to-peak amplitude 10–20 μm). As expected from the frequency content of prey (insect) stimuli (Fig. 24.1.), irregular wave stimuli with high frequency components are more attractive than low-frequency ones, provided the stimulus durations are equal (compare responsiveness at 1,600 ms in Fig. 24.7). Thus *D. triton* uses the amplitude spectrum for a distinction between wave stimuli. The responsiveness of *Dolomedes* to both types of stimuli increases significantly with stimulus length (Fig. 24.7). This fits well with the observation that insect-generated wave stimuli last longer than all other wave types (Bleckmann 1985b).

### Responsiveness, Frequency Bandwidth, and Stimulus Time Structure

If we present to *Dolomedes* a stimulus with an orderly time structure, such as a click that, at 5–10 cm, has an upper frequency limit of 65–79 Hz, the responsiveness is only 1.8%, even though the stimulus duration and its mean peak-to-peak amplitude in the test range is 440–600 ms and 7–21 μm, respectively (Bleckmann 1985b). Vertebrate-generated wave stimuli are often clicklike; i.e., they are short lasting with an orderly time structure (Fig. 24.1). This probably is the reason that *Dolomedes* responds to fish-generated (*Poecilia reticulata*) wave stimuli in less than 1% of the trials (Bleckmann and Lotz 1987).

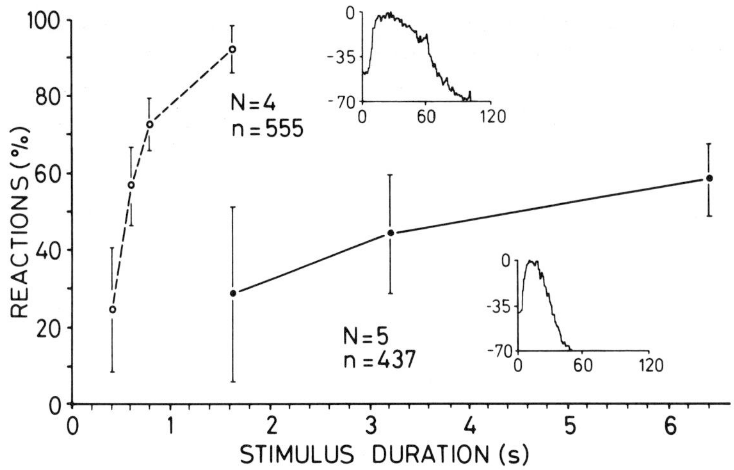


Figure 24.7. The responsiveness of *D. triton* as a function of stimulus duration and frequency bandwidth (*inset*). Power spectra refer to source distance of 5 cm. *Solid line:* responses to small bandwidth signal; *dashed line:* responses to large bandwidth signal. (From Bleckmann 1985b.)

## Responsiveness, Frequency Bandwidth, and Source Distance

Owing to the filter properties of the water surface (Eq. [24.2]), a wave stimulus will mainly lose its high-frequency components during propagation. Thus, with increasing D, a prey (insect) stimulus becomes more and more nonpreylike if only analyzed with respect to its frequency content. Because fishing spiders know the wave source distance (Bleckmann and Barth 1984), this raises the question whether they weigh the amplitude-frequency content of a wave stimulus with respect to the distance the stimulus has traveled. If so, a noisy stimulus of narrow frequency bandwidth, provided all other stimulus parameters are equal, should be more attractive if the radius (R) of the stimulus is enlarged or if the wave curvature ($C_w$) is flat (provided $C_w$ is used for the determination of D, as has been proposed for *D. triton,* Bleckmann and Barth 1984). To test this hypothesis, two stimulus types were generated: a small bandwidth stimulus (upper frequency limit of $-70$ dB bandwidth about 40 Hz) and a stimulus of enlarged frequency bandwidth (70 Hz). Stimulus duration and mean peak-to-peak amplitude for both stimulus types were 1,600 ms and 5.6 $\mu$m $\pm$ 0.8, respectively. From each stimulus type, three stimuli were produced that differed either in $C_w$ and R (e.g., 4 cm vs. 15 cm distance; R was defined as the distance between the wave source and the leg tip closest to the wave source) or in $C_w$ alone (i.e., 4 cm curved wave front vs. 4 cm flat wave front) (cf. Fig. 24.8, *top*). Irrespective of $C_w$ and R, the responsiveness of *Dolomedes* to the stimuli of enlarged frequency bandwidth is 60%–70%. As expected, the responsiveness to the small bandwidth stimulus is reduced. Responsiveness is significantly ($\chi^2$ − test, $p \leq 0.001$) increased, however, if R is 15 cm or if the wave stimulus at $R = 4$ cm has a flat wave

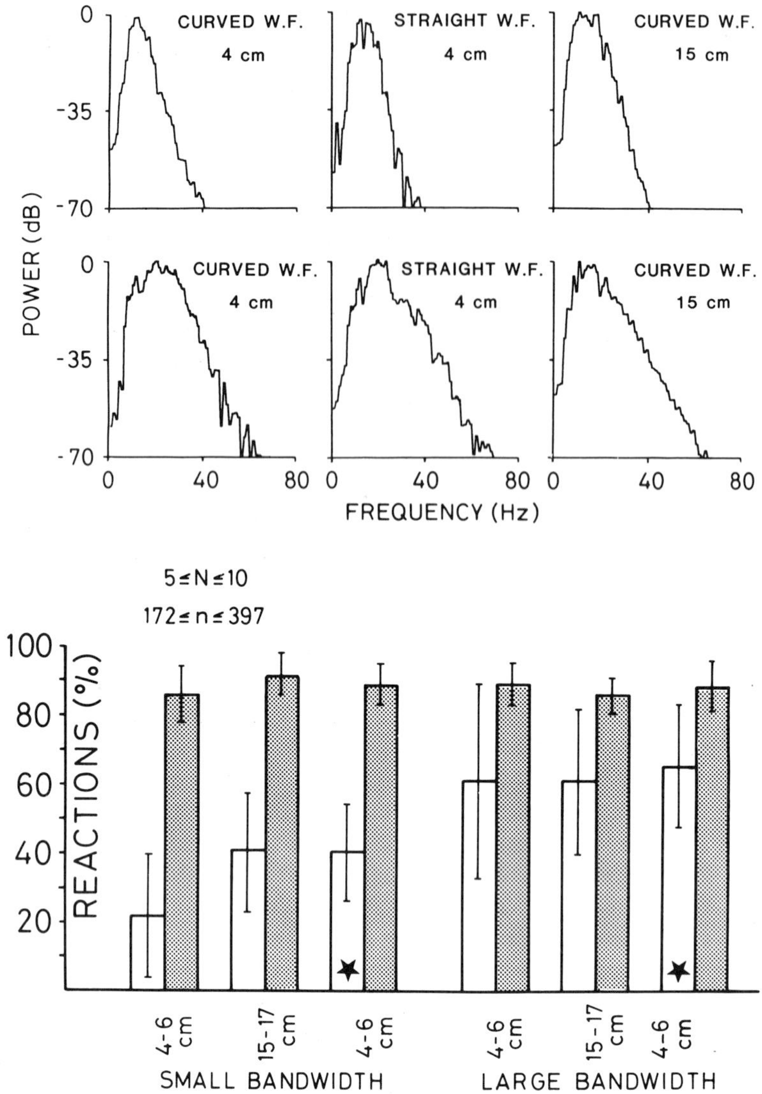


Figure 24.8. *Top:* Frequency content of wave stimuli used for discrimination test. Stimuli, which were measured with the aid of He-Ne laser beam reflected from water surface onto position-sensitive photodiode, differ in either R, $C_W$, or both. In each case, highest power value was set equal to 0 dB. Wave stimuli with flat wave front were generated with a vibrator connected to a plate resting on water surface. *Bottom:* Responsiveness (*open bars*) of *D. triton* to stimuli shown *above*, (★) = flat wave front. *Shaded bars* give the responsiveness to manual rod stimuli (controls).

front. (Fig. 24.8, *bottom*). This clearly demonstrates that *Dolomedes* considers the distance that a wave stimulus has traveled when calculating the likelihood of prey or nonprey on the basis of the wave frequency.

## Diagrammatical Model of Results

In Figure 24.9, the possible mechanisms of distance determination and stimulus discrimination in water surface wave perception are summarized. At time $t = 0$ the water surface is disturbed. The resulting pressure and surface waves may stimulate sound-sensitive (perhaps in surface-feeding fish) and water surface wave–sensitive (in surface-feeding fish and fishing spiders) mechanoreceptors, respectively (in the first step, only five receptors are drawn for clarity). In a second step, the afferent information may be processed (1) with respect to the stimulus onset (e.g., $t_0^1$ = *stimulus onset time at receptor 1*), stimulus duration ($\Delta t$) or interstimulus time ($\Delta t^*$); (2) with respect to spectral amplitude-frequency content $a_i(R_i, f)$ (where $a_i$ = spectrum recorded by *ith* receptor; and (3) with respect to the frequency modulation ($df/dt, f$) of the wave stimulus. It is further assumed that the CNS "knows" the dispersion (eq. [24.1]) and attenuation (Eq. [24.2]) functions. The *connecting lines* show in detail which minimal information must be available for a specific calculation. From the behavioral results (Figs. 24.5, 24.6, 24.7 and 24.8), it is assumed that the final calculation of D and the likelihood of prey ($L_p$) is done by building the sums $\Sigma k_i D_i$ and $\Sigma l_i L_{pi}$, respectively, where $k_i$ and $L_{pi}$ and where $\Sigma k_i$ and $\Sigma l_i$ both = 1. $D$ and $L_p$ together with the general readiness finally determine the motor output and thus the behavior of the animal.

## Summary

The struggling movements of terrestrial insects trapped at the water surface generate wave stimuli that elicit prey capture attempts from surface-dwelling predators. Fishing spiders, in addition, feed on aquatic and semiaquatic vertebrates, such as small fish, amphibians, and tadpoles. Surface-feeding fish and fishing spiders can not only discriminate between different wave stimuli, but also are able to determine the direction and distance to the wave source. This chapter gives information about the shape, amplitude, and frequency of prey and nonprey wave stimuli, the physical properties of the water surface, and the mechanisms used by surface-feeding fish and fishing spiders for stimulus discrimination and wave source (distance) localization.

*Acknowledgments.* I want to thank Drs. I. Waldner and U. Müller for permission to quote unpublished material. Further thanks are due to M. McClune and Drs. R. Käse, J. Schweitzer, J. T. Enright and J. Anders for many discussions and to Drs. W. Heiligenberg, J. Schweitzer, W. L. R. Cruce, S. Wilcox, and J. L. Leonard for critically reading early drafts of the manuscript. Dr. Leonard, in addition, deserves

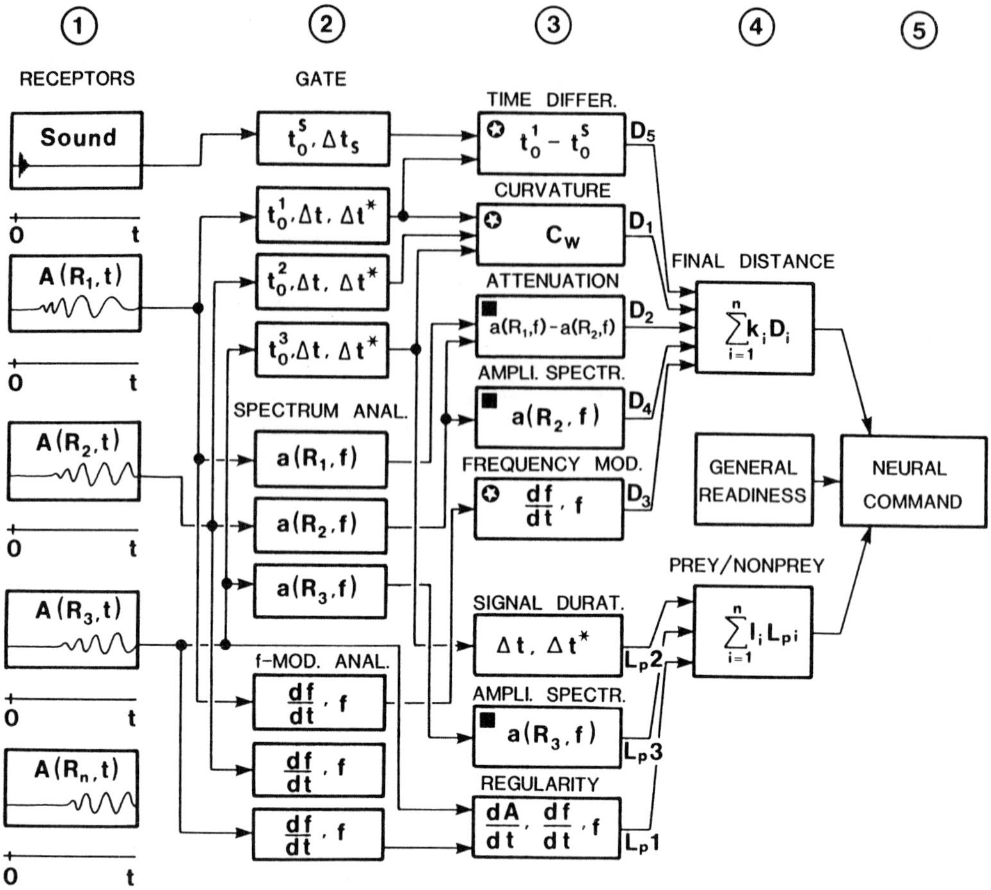


Figure 24.9. A model summarizing possible mechanisms of distance determination and stimulus discrimination with the aid of water surface waves. Calculations within a square marked with a ✪ presupposes that dispersion function (Eq. [24.1]) is known. Those marked with a ■ assume that damping function (Eq. [24.2]) is known; *Ampli. Spectr.*, amplitude spectrum; *f-Mod. Anal.*, frequency modulation analyzer; *Frequency Mod.*, frequency modulation; *Regularity*, stimulus regularity analyzer; *Signal Durat.*, stimulus duration; *Spectrum Anal.*, spectrum analyzer; $A(R_1,t)$, time-space function of wave stimulus; $a(R_1,f)$, spectral amplitude of wave stimulus as a function of R (radius) and f (frequency); $C_W$, curvature of wave front; (dA/dt, $f$), amplitude modulation of wave stimulus; (df/dt, $f$), frequency modulation of wave stimulus; $D_i$, *ith* mechanism of distance determination; $L_p i$, *ith* mechanism to calculate the likelihood of prey; $k_i$ and $l_i$, factors that depend on $D_i$ and $L_p i$, respectively; $t_0$, arrival time of wave stimulus; $t_0^s$, arrival time of sound wave; $\Delta t$, stimulus duration; $\Delta t^*$, interstimulus interval. Only for mechanisms $D_1$, $D_3$, $D_4$, and $L_p1$–$L_p3$ have we behavioral evidence. For more information see text.

thanks for improving my English. Supported by grants of the Deutsche Forschungsgemeinschaft to E. Schwartz (21,5), F. G. Barth (SFB 45), H. B. (Bl 242/1–1), and NSF and NIH grants to T. H. Bullock.

## References

Barth, F.G. (1982) Spiders and vibratory signals: sensory reception and behavioral significance, in Spider Communication. Mechanisms and Ecological Significance, Witt, P.N., and Rovner, J.S. (eds.), Princeton University Press, Princeton, pp. 67–122.

Bleckmann, H. (1980) Reaction time and stimulus frequency in prey localization in the surface-feeding fish *Aplocheilus lineatus,* J. Comp. Physiol., 140:163–172.

Bleckmann, H. (1982) Reaction time, threshold values and localization of prey in stationary and swimming surface-feeding fish *Aplocheilus lineatus,* Zool. Jahrb. Abt. Allg. Zool. Physiol. Tiere, 86:71–81.

Bleckmann, H. (1985a) Perception of water surface waves: how surface waves are used for prey identification, prey localization, and intraspecific communication, in Progress in Sensory Physiology V, Ottoson, D. (ed.), Springer-Verlag, New York, pp. 147–166.

Bleckmann, H. (1985b) Discrimination between prey and non-prey wave signals in the fishing spider *Dolomedes triton* (Pisauridae), in Acoustic and Vibrational Communication in Insects, Kalmring, K. and Elsner, N. (eds.), Paul Parey, Berlin, pp. 215–222.

Bleckmann, H. (1986) The role of the lateral line in fish behavior, In The Behaviour of Teleost Fishes, Pitcher, T.J. (ed.), Croom Helm, pp. 178–202.

Bleckmann, H., and Barth, F.G. (1984) Sensory ecology of a semiaquatic spider (*Dolomedes triton*). II. The release of predatory behavior by water surface waves, Behav. Ecol. Sociobiol., 14:302–311.

Bleckmann, H. and Lotz, T. (1987) The vertebrate-catching behavior of the fishing spider *Dolomedes triton* (Araneae, Pisauridae). Animal Behav., 35:641–651.

Bleckmann, H. and Rovner, J. (1984) Sensory ecology of a semiaquatic spider (*Dolomedes triton*). I. Roles of vegetation and wind-generated waves in site selection, Behav. Ecol. Sociobiol., 14:297–301.

Bleckmann, H. and Schwartz E. (1981) Reaction time of the topminnow *Aplocheilus lineatus* to surface waves determined by video and electromyogram recordings, Experientia, 37:362–363.

Bleckmann, H. and Schwartz E. (1982) The functional significance of frequency modulation within a wave train for prey localization in the surface feeding fish *Aplocheilus lineatus* (Cyprinodontidae), J. Comp. Physiol., 145:331–339.

Bleckmann, H. and Topp, G. (1981) Surface wave sensitivity of the lateral line system of the topminnow *Aplocheilus lineatus,* Naturwissenschaften, 68:624–625.

Bleckmann, H., Waldner, I., and Schwartz, E. (1981) Frequency discrimination of the surface-feeding fish *Aplocheilus lineatus*—a prerequisite for prey localization? J. Comp. Physiol. 143:485–490.

Crawford, A.C. and Fettiplace, R. (1981) An electrical tuning mechanism in turtle cochlear hair cells, J. Physiol., 312: 377–412.

Elepfandt, A., Seiler, B., and Aicher, B. (1985) Water wave frequency discrimination in the clawed frog, *Xenopus laevis,* J. Comp. Physiol., 157:255–261.

Fay, R.R. and Popper, A.N. (1980) Structure and function in teleost auditory system, in Proceedings in Life Sciences. Comparative Studies of Hearing in Vertebrates, Popper, A.N., and Fay, R.R. (eds.) Springer-Verlag, New York, pp. 3–42.

Flock, A. (1965) Electron microscopic and electrophysiological studies on the lateral-line canal organ, Acta Oto-Laryngol. (Stockh.) (Supp.) 199:1–90.
Franz, G.J. (1959) Splashes as sources of sound in liquids, J. Acoust. Soc. Am., 31:1080–1096.
Görner, P. (1976) Source localization with labyrinth and lateral line in the clawed toad (*Xenopus laevis),* in Sound Reception in Fish, Schuijf, A., and Hawkins, A.D. (eds.), Elsevier, Amsterdam, pp. 171–184.
Gudger, W.E. (1922) Spiders as fishermen, J. Am. Mus. Nat. Hist., 22:565–585.
Hoin-Radkovski, I., Bleckmann, H., and Schwartz, E. (1984) Determination of source distance in the surface-feeding fish *Pantodon buchholzi* (Pantodontidae), Anim. Behav., 32:840–851.
Käse R. and Bleckmann H. (1987) Prey localization by surface wave-ray tracing—fish track bugs like oceanographers track storm. Experientia (in press)
Lang, H.H. (1980) Surface wave discrimination between prey and nonprey by the backswimmer *Notonecta glauca* L. (Hemiptera, Heteroptera), Behav. Ecol. Sociobiol., 6:233–246.
Meyer, E. and Zakon, H.H. (1982) Androgens alter the tuning of electroreceptors, Science, 217:635–637.
Müller, U. (1984) Die morphologische und physiologische Anpassung des Seitenliniensystems von *Pantodon buchholzi* an den Lebensraum Wasseroberfläche, Thesis, Universität Giessen, W. Ger.
Müller, U. and Schwartz, E. (1982) Influence of single neuromasts on prey localizing behavior of surface-feeding fish, J. Comp. Physiol., 149:399–408.
Münz, H. (1986) Single unit activity in the peripheral lateral line system of the cichlid fish *Sarotherodon niloticus* L., J. Comp. Physiol., 157:555–568.
Schwartz, E. (1965) Bau und Funktion der Seitenlinie des Streifenhechtlings *Aplocheilus lineatus,* Z. Vgl. Physiol., 50:55–87.
Schwartz, E. (1970) Ferntastsinnesorgane bei Oberflächenfischen, Z. Morphol. Tiere, 67:40–57.
Schwartz, E. (1971) Die Ortung von Wasserwellen durch Oberflächenfische, Z. Vgl. Physiol. 74:64–80.
Sommerfeld, A. (1970) Vorlesungen über theoretische Physik, vol. 2, Mechanik der deformierbaren Medien, Akademische Verlagsgesellschaft, Leipzig.
Topp, G. (1983) Primary lateral line response to water surface waves in the topminnow *A. lineatus* (Pisces, Cyprinodontidae), Pfluegers Archiv. 397:62–67.
Unbehauen, H. (1980) Morphologische und elektrophysiologische Untersuchungen zur Wirkung von Wasserwellen auf das Seitenlinienorgan des Streifenhechtlings (*Aplocheilus lineatus*), Thesis, Universität Tübingen, W. Ger.
Waldner, I. (1981) Habituation von *Aplocheilus lineatus* auf Oberflächenwellen des Wassers, Thesis, Universität Giessen, W. Ger.
Wilcox, R.S. (1979) Sex discrimination in *Gerris remigis:* role of a surface wave signal, Science, 206:1325–1327.

## Abbreviations

| | |
|---|---|
| a | spectral amplitude of wave stimulus |
| A | displacement amplitude of wave stimulus |
| $C_{ph}$ | phase velocity |
| $C_w$ | curvature of wave front |

| | |
|---|---|
| D | source distance |
| f | stimulus frequency |
| g | gravitational acceleration |
| h | water depth |
| $L_p$ | likelihood of prey |
| R | radius of wave stimulus |
| sf-stimulus | single-frequency sinusoidal wave stimulus |
| t | time |
| $t_w$ | arrival time of wave stimulus |
| $t_s$ | arrival time of sound wave |
| $\Delta t$ | stimulus duration |
| $\Delta t^*$ | interstimulus time interval |
| $\lambda$ | wavelength |
| $\eta$ | surface tension |
| $\chi$ | density of water |
| æ | kinematical viscosity of water |

# Chapter 25

# Surface Wave Reception in Invertebrates and Vertebrates

R. STIMSON WILCOX*

At the surface of water, the asymmetry of intermolecular forces, coupled with strong molecular cohesion, produces an elastic, membranelike film. The film can support objects much heavier than water, such as a needle, and in nature it supports many species of surface-dwelling arthropods that have specially adapted body surfaces for moving on the surface. In contrast, the powerful surface forces often trap terrestrial insects and other small arthropods that encounter the film, resulting in a concentration of prey items in the film. This concentration has attracted a large spectrum of predators that obtain some or all their food from the surface.

Prey striking the surface or struggling in it generate concentric wave patterns, which enable predators to locate the prey and enable some prey species to avoid wave-generating predators. Also, intentional generation of wave patterns is a means by which surface dwellers can communicate. It would be expected that predators, prey, and communicators have evolved sensory and behavioral mechanisms specialized for using surface waves. These mechanisms, in the contexts of predator-prey relationships and intraspecific communication, are the focus of this chapter.

## A Note on the Use of the Term "Signal"

When an animal scans its environment with sensory receptors, it receives a wide range of stimuli, including a limited class called signals. Signals are stimuli emitted only by other organisms. As defined in this chapter, signals are designed and maintained by natural selection for information transfer and reception and thus are evolved traits that increase or maintain the inclusive fitness of signal senders and receivers (Otte 1974). Biologists studying animal communication are generally careful to use the term *signal* more or less as defined by Otte. Sensory physiologists, on the other hand, commonly refer to essentially any

---

* Department of Biology, State University of New York, Binghampton, NY 13901, USA.

environmental stimulus to which sensory structures respond as a signal, though evolved signals are sometimes referred to as "communicative signals."

Equating general stimuli with signals is more of a problem in semantics than a confusion of terms. But to scientists studying animal communication, it seems advisable to clarify the situation and label evolved signals as *signals* and other stimuli as *stimuli* or perhaps use other, more specific, terms than *signals*. I will maintain a clear distinction between signals and other stimuli in this chapter.

## Properties of Surface Waves

Surface waves generated by struggling prey and signalers tend to lie between 5 and 140 Hz, with most of the energy concentrated between 10 and 50 Hz (Wilcox 1972, 1979; Wilcox and Spence 1986; Lang 1980a). Within the frequency range of 5–140 Hz, wave propagation velocity is influenced greatly by surface tension. Above about 46 Hz, wave patterns are called *capillary waves*. Their velocity is governed mainly by surface tension, and the velocity increases with the frequency (see Fig. 24.2A, Chap. 24).

Wave patterns below about 4 Hz are called *gravity waves*. Their velocity is governed mainly by gravitational acceleration, and the velocity increases as frequency decreases. Waves between 4 and 46 Hz are gravity-capillary waves, and their velocities are governed by a combination of gravitational and surface tension forces. Surface waves travel in groups, and at most frequencies group velocities differ somewhat from those of individual waves (Fig. 24.2A, Chap. 24). Waves above 10 Hz are also called *ripples*.

Most natural wave patterns are composed of a mixture of frequencies. As a wave pattern travels, its amplitude decreases and its frequency characteristics change. Over distance, higher frequencies attenuate more rapidly than lower ones; so a complex wave pattern differentially loses its higher frequency components as it travels (Fig. 24.2B, Chap. 24). Wave shape and speed are influenced by water depth, but at the frequencies and amplitudes with which we are concerned here, water depth is rarely a significant factor.

Surface waves also have an effect below the surface. This effect is relatively small; vertical motion decreases by approximately half the peak-to-peak wave height for every depth increase of $\lambda/9$ ($\lambda$ = wavelength) (Wetzel 1975). For example, at 13.6 Hz, the frequency at which ripples travel at their minimum velocity [23.5 cm/sec. (18°C)], a ripple of 1.0 $\mu$m amplitude at the surface would have decreased to an amplitude of 0.5 $\mu$m at a depth of only 0.192 cm (= $\lambda/9$). Because of the small wavelengths (about 10 cm or less) and wave heights (about 80 $\mu$m or less) of most prey- or signal-induced waves, such surface waves have limited influence below the surface.

As previously mentioned, the concentration of food at so identifiable a place as a water surface has resulted in a wide range of invertebrate and vertebrate predators' developing specialized mechanisms for exploiting this food bonanza. Table 25.1 summarizes these groups. We shall see below that direction, amplitude, and frequency characteristics of surface waves, and changes in amplitude and frequency over distance, constitute information that enables predators, prey, and communicators to operate effectively.

## Surface Wave Sensory Structures

The structures animals use for sensing the surface wave substrate are essentially the same structures used by ecologically or phylogenetically similar species on any other substrate (Markl 1983).

### Invertebrates

The backswimmer *Notonecta* senses surface waves with apparently two types of receptors, situated in the legs and on the abdomen. Wiese (1972, 1974) and Wiese and Schmidt (1974) present morphological, behavioral, and physiological evidence suggesting that in *N. glauca,* clusters of phasic and phasic-tonic scolopidial organs sense waves especially from in front and to the side. These sense organs are located near, and are stretched by, movements of the tarsal claws of the forelegs and middle legs. Wiese (1974) also suspected the presence of a stretch receptor proximal to the tarsal claws. Murphey and Mendenhall (1973) used ablation experiments to further this suggestion by proposing that *N. undulata* possesses unidentified organs near the tibiofemoral joints of the fore- and midlegs, as well as a system of mechanoreceptive hairs located along the margins of the three most posterior segments of the abdomen and on the genitalia.

The receptors for surface wave reception in water striders have been localized by ablation and electrophysiological experiments to the tibiotarsal joint of the legs (Murphey 1971). Murphey notes, however, that in light of the results for *Notonecta, Gerris* should be reexamined for a more important role of more proximal sensory receptors. Also, Lawry (1973) suggests that the pertinent receptors may be mechanosensory hairs on the distal tibiae and tarsal joints.

Spiders possess sensitive substrate vibration sensory organs called *slit sensilla* in various parts of all their legs. The most sensitive organ is a compound set of slit sensilla located on the metatarsus near the tarsometatarsal joint. Distortion of the joint stimulates the slit sensilla (Barth 1982). Some spiders of the families Pisauridae and Lycosidae habitually capture prey at or just under the surface (see Chap. 24). In *Dolomedes triton* (Pisauridae), the orientation response after unilateral ablation of all metatarsal organs was only slightly affected (Bleckmann and Barth 1984). This implies that other receptors are also used in surface wave reception. Possible candidates include other slit organs and leg joint receptors (Bleckmann and Barth 1984).

Whirligig beetles possess the Johnston's organ, a group of chordotonal stretch receptors at the joint between the second antennal segment and the rest of the antenna. In whirligigs, the respective antennal segments are called the pedicellus, which lies on the water surface, and the flagellum, which angles upward in the air. Electrophysiological studies have demonstrated that surface waves produce

Table 25.1. Frequency Range, Best Frequency, Sensitivity, and Discrimination Values for Species that Are Sensitive to Surface Wave Input (b = behavioral studies; e = electrophysiological studies)

| Species | Freq. Tested (Hz) | Best Freq. (Hz) | Lower Threshold (μm) | Freq. Discrimination (Hz) | Kind of Expt. | References |
|---|---|---|---|---|---|---|
| | | | | Invertebrates | | |
| Bryozoa | | | | | | |
| *Bowerbankia imbricata* | 10–200 | 40–80 | 30 | — | b | Wiese, Wollnik, and Jebram 1980 |
| Backswimmers | | | | | | |
| *Notonecta glauca* | 5–300 | 100–150 | 1 | — | b | Markl and Wiese 1969 |
| *N. glauca* | 5–300 | 60–300 | 0.2–2.5 | — | e | Lang 1980b |
| *N. glauca* | 10–1,000 | 1,000 | 0.5 | — | b, e[a] | Wiese 1972 |
| Water striders | | | | | | |
| *Gerris lacustris* | 5–400 | 20–40 | 2.5 | — | b | Wiese 1969 |
| *Rhagadotarsus kraepelini* | 5–50 | 22 | — | 1.5 | b | Wilcox 1972 |
| Whirligig beetles | | | | | | |
| *Gyrinus substriatus* | 30–200 | 150 | 0.5 | — | e | Rudolph 1967 |
| Mosquito larvae | | | | | | |
| *Aedes aegypti* | 5–100 | 10–20 | 5 | — | b | Markl and Hauff 1973 |
| Fishing spiders | | | | | | |
| *Dolomedes triton* | 10–140 | 40–70 | 1–2.5 | — | b | Bleckmann and Barth 1984 |
| *D. triton* | 1–1000 | 1,000 | 0.1 | — | e[a] | Bleckmann and Barth 1984 |
| *Pirata piraticus* | 19–38 | 21–22 | — | — | b | Berestynska-Wilczek 1962 |

Table 25.1. *Continued*

| Species | Freq. Tested (Hz) | Best Freq. (Hz) | Lower Threshold (μm) | Freq. Discrimination (Hz) | Kind of Expt. | References |
|---|---|---|---|---|---|---|
| | | | | Vertebrates | | |
| Surface-foraging fish | | | | | | |
| *Aplocheilus lineatus* | 10–250 | 7–140 | 0.02 | — | b | Bleckmann 1980 |
| *A. lineatus* | 10–250 | 70–140 | 0.04 | — | e | Bleckmann and Topp 1981 |
| *A. lineatus* | 20–120 | 20[b] | — | 2.5 | b | Bleckmann, Waldner, and Schwartz 1981 |
| *Pantodon buchholzi* | 10–200 | 150 | 0.0007 | — | b | Müller 1984 |
| Aquatic amphibians | | | | | | |
| *Xenopus laevis* | — | — | 0.2 | — | b | Görner 1976 |
| *X. laevis* | 5–30 | 14[b] | — | 0.5 | b | Elephandt, Seiler, and Aicher 1986 |

[a] Legs were directly coupled to a mechanical vibrator.
[b] Indicates frequency at which frequency discrimination is best. For all other best frequencies, numbers indicate frequency at which animal responds with the lowest wave amplitude threshold.

movements between pedicellus and flagellum that the Johnston's organ senses (Rudolph 1967). In mosquito larvae, which escape surface predators with diving reactions triggered by surface waves (Markl and Hauff 1973), it also seems likely that Johnston's organ is used in sensing wave patterns.

Other invertebrates that have been studied for their response to surface waves are an aquatic leech that orients to prey in part by sensing surface waves (Young, Dedwylder, and Friesen 1981) and a bryozoan *Bowerbankia imbricata* that responds with an all-or-none protective withdrawal reflex to underwater perturbations caused by surface waves (Wiese, Wollnik, and Jebram 1980). The sensory structures of the medicinal leech, *Hirudo medicinalis,* are clusters of mechanoreceptive filiform processes projecting from segmental sensilla about 10 μm through the cuticle (Friesen 1981). The sensory structures of *B. imbricata* are not yet described, though the evidence points to displacement receptors located on the tentacles (Wiese, Wollnik, and Jebram 1980).

## Vertebrates

The vertebrates occupied with surface wave reception include various fish, several of which are highly specialized for predation at the surface; amphibians, especially the clawed toad, *Xenopus;* and some bat species that, though not aquatic animals, are in some cases as highly specialized for foraging at the water surface as are any aquatic animals.

Specialized surface-foraging fish like the topminnow *Aplocheilus lineatus* sense waves with portions of the lateral line organs, primarily on the head. The sensory structures are freestanding neuromasts or canal organs in small canals or large cavities (Schwartz 1965, 1967, 1971, 1974; Schwartz and Hasler 1966). In some species the canals are open to the water, while in others the neuromasts are closed over (for a thorough discussion of fish lateral line system morphology, see Chap. 22). The lateral line system of *Xenopus* is structurally similar to that of fish and is arranged in groups called *stitches.*

Surface-foraging bats use echolocation to find surface prey by the ripples generated by the prey and appear not to possess modifications in their emitting or sensory apparatus that can be associated in particular with water surface vs. aerial foraging (Wenstrup 1984; Wenstrup and Suthers 1984).

## Amplitude and Frequency: Range, Sensitivity, and Discrimination

Surface-dwelling species have a range of wave amplitude threshold sensitivities that varies from 0.0007 μm peak to peak in the fish *Pantodon buchholzi* (Müller 1984) to 0.02–2.5 μm in most other species (Table 25.1). There is an obvious tendency for the frequencies at which sensitivity is greatest to be higher than the 50-Hz transition area that distinguishes most prey wave patterns from nonprey patterns (Lang 1980a).

As elaborated below, surface dwellers use frequency as well as amplitude discrimination in determining the type, direction, and distance of wave patterns.

## Sensory and Behavioral Mechanisms: Predators and Prey

The main problems with surface waves that foraging predators must solve in a prey capture sequence are, in the following order: (a) identification of prey, (b) determination of direction, and (c) determination of distance.

### Prey Identification

Efficient predators should obviously be able to discriminate among surface waves from prey and nonprey species (especially conspecifics) and abiotic sources. In spite of the biological importance of such discrimination, only a few studies have been done on the topic. Potential information for such discrimination could lie in surface wave frequency, waveform, amplitude, duration, interval, sequencing, and even remembered direction in some circumstances. To date, there is evidence for use by predators of all these characteristics except direction remembered from previous experience, which requires short-term memory by predators.

Remembrance of direction seems feasible, since, for example, Elepfandt (1985) has shown that the clawed frog, *Xenopus laevis,* is capable of both short- and long-term memory of frequency discrimination and is also capable of complex learning that is within the ability range of pigeons and rats. If a predator maintained a consistent orientation at the same position when waiting for prey, the predator could easily remember repeated appearances by particular prey species approaching from the same direction. Elepfandt's results with *Xenopus* also make one wonder whether perhaps some results obtained in various experimental manipulations on species discussed in this chapter were not influenced occasionally by learning, since the investigators rarely seemed to take learning into account in their control measures, other than to accomodate for habituation.

#### Invertebrates

Prey identification by invertebrate predators has been demonstrated only in the backswimmer *Notonecta* and on the fishing spider *Dolomedes*. Lang (1980a) compared the frequency and amplitude characteristics of waves produced by the instar and adult stages of *Notonecta glauca* with waves of a number of prey species and abiotic sources and noted which patterns adult backswimmers attacked as prey.

There were interesting broad differences between prey and nonprey waves. Prey waves included suprathreshold amplitudes up to frequencies of 60–140 Hz, and their energy maxima were above 15 Hz, whereas nonprey waves included such amplitudes up to only 40–50 Hz, and their energy maxima were below 15 Hz (Figs. 25.1 and 25.2). The wave amplitudes of most aquatic and terrestrial

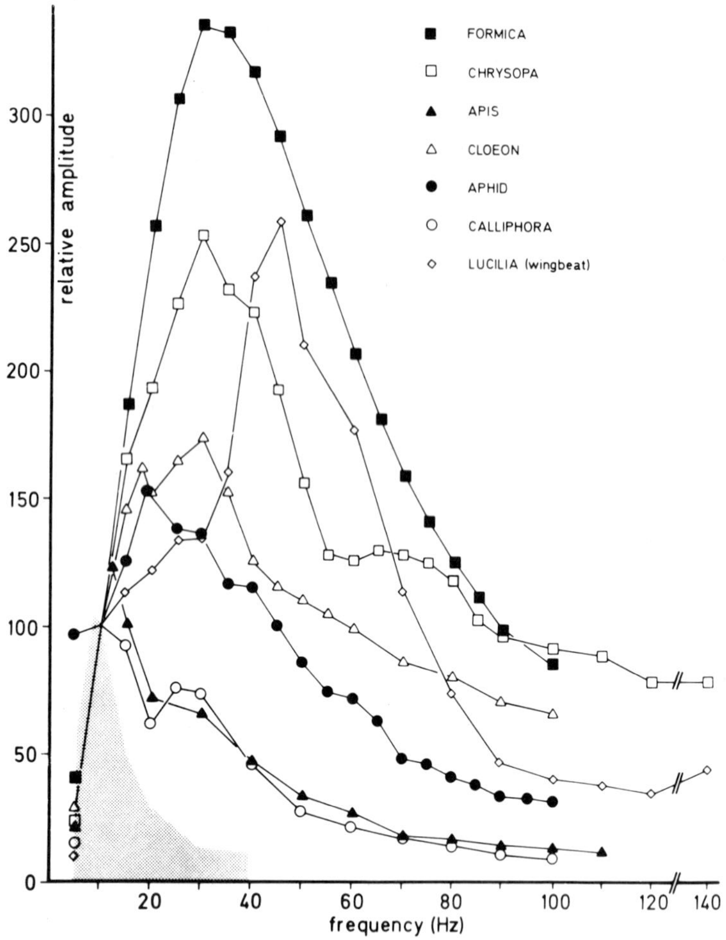


Figure 25.1. Relative spectra of waves elicited by terrestrial insects trapped on water surface. Shaded area is relative spectrum of waves from the swimming of adult backswimmers (Lang 1980a).

prey species Lang tested were considerably larger (up to 150–200 μm) than amplitudes of swimming backswimmers of any stage (up to 50 μm, except for diving and grooming, which produced amplitudes of up to ± 230 μm). *Notonecta* adults attacked all wave sources that had bandwidths above 50 Hz, including waves of first instar *Notonecta* (resulting in cannibalism), but did not attack sources with bandwidths similar to their own waves. Waves of abiotic sources, such as a fallen leaf or a wooden object, with frequencies below 50 Hz were not attacked; but waterdrops, which had frequencies above 50 Hz, were attacked. Thus *Notonecta* distinguishes rather well between prey and nonprey.

Lang (1980a) regarded higher frequency content of prey waves as the main means for discrimination by *Notonecta*. He proposed that frequencies up to about

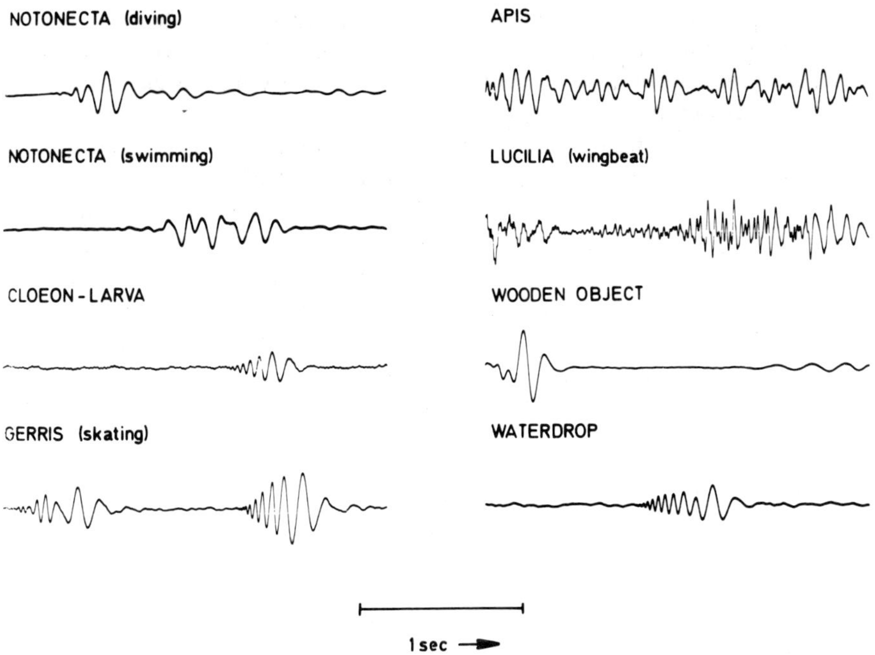


Figure 25.2. Surface waves produced by backswimmers and water striders, by some of their prey, and by some abiogenetic disturbances. Wave amplitudes are not to scale (Lang 1980a).

50 Hz are optimally perceived by sensory structures in the legs and that the abdominal sensory system responded optimally to frequencies from about 50 up to 300 Hz (Lang 1980b). Lang thought that complete prey-catching behavior in *Notonecta* occurs only when both low- and high-frequency ranges are present in a wave pattern.

Lang (1980a) did no experiments that could have demonstrated that frequency discrimination alone could be used for prey discrimination. Studies by Wilcox (1972), Bleckmann, Waldner, and Schwartz (1981), and Elepfandt, Seiler, and Aicher (1986), however, have shown frequency discrimination in water striders (1.5 Hz), surface-feeding fish (2.5 Hz), and clawed frogs (0.5 Hz), respectively (Table 1). This level of accuracy in water striders, fish, and frogs is likely to be well within the accuracy requirements for both predator-prey relationships and intraspecific communication.

Lang (1980a) noted that pattern duration is often much longer in prey-generated waves, particularly from prey's struggling consistently in the surface film, such as a fly buzzing its wings, than in nonprey generated waves, which usually consist of only a single pulse or "click." This duration could be used for prey discrimination. Some prey-generated waves are clicklike also, however, so wave pattern duration cannot always be used for discrimination. Duration between patterns, however, might be used for prey discrimination. Lang excluded absolute amplitude as a means for discrimination, since amplitude varies too much, in its original propagation and with distance, for reliability. These considerations for *Notonecta* prey discrimination hold similarly for the other species of surface predators discussed below.

Prey vs. nonprey discrimination in *Dolomedes triton* has been shown to be similar in most respects to the means of discrimination that Lang (1980a) proposed in *Notonecta* (Bleckmann and Rovner 1984; Bleckmann and Barth 1984; Bleckmann 1985a; Bleckmann and Lotz in prep.). Bleckmann (1985a) showed that the ability of *Dolomedes* to discriminate different wave patterns is based on evaluation of (1) frequency bandwidth (like *Notonecta, Dolomedes* is attracted to wave patterns with frequencies above 50 Hz); (2) pattern duration; (3) regularity with respect to amplitude and frequency modulation; and (4) frequency with which a wave train is generated (i.e., patterns are more attractive to the spiders if they have an irregular, rather than a regular, time course). Bleckmann and Lotz (in prep.) have shown that *Dolomedes* has a low (< 1%) prey capture response to surface wave patterns caused by fish, probably because these patterns lie in frequencies below 40 Hz and are thus difficult to distinguish from abiotic noise.

Prey vs. nonprey discrimination and other information about *Dolomedes* are discussed in more detail in Chapter 24.

### Vertebrates

It is likely that predatory vertebrates discriminate between prey and nonprey using surface wave patterns. There are no experimental studies yet on prey vs. nonprey discrimination in aquatic vertebrates, however.

## Prey Localization

After making a decision that a wave pattern is from potential prey, predatory surface dwellers must determine the direction of the source and then its distance.

### Locating Wave Source Direction: Accuracy and Mechanisms

Surface predators could potentially obtain directional information from the time of wave front arrival and from amplitude differences between two or more receptors. The low propagation speed and strong damping of surface waves would facilitate these sensing abilities.

### Invertebrates

When presented with surface waves of 20 or 60 Hz from directly in front, the back swimmer *Notonecta glauca* orients to within an average 2° ± 3° of the source. Orientation accuracy decreases linearly to 18° ± 10° at a presentation angle directly behind (Markl, Lang, and Wiese 1973). The fishing spiders *Dolomedes triton* and *D. fimbriata* orient to invertebrate-generated waves with 9.3°

± 13.3° and 13.6° ± 10.9° accuracy, respectively (Bleckmann and Barth 1984). Bleckmann and Lotz (in prep.) found that *D. triton* orients to within 5.2° ± 9.5° to waves generated by fish. In contrast to *Notonecta,* these spiders determine the source angle with nearly equal precision in the entire circular range. Murphey and Mendenhall (1973) found a correlation coefficient of 0.86 between source angle and response turn angle in the water strider *Gerris remigis,* and Reinig and Uhlemann (1973) found precise source location in the whirligig beetle (*Gyrinus substriatus*) also. Young, Dedwylder, and Friesen (1981) found that hungry medicinal leeches (*Hirudo medicinalis*) responded to surface waves with locomotion oriented to within 22.5° of the source in 65% of all responses.

All studies on prey location in invertebrates have proposed that surface wave sources are located by sensing wave front time of arrival at two or more sensory structures on the same animal. Time of arrival has also been expressed as a measure of difference in phase relationships within wave patterns (Wiese 1974). Murphey and Mendenhall (1973) and Murphey (1973) suggest that in *Notonecta undulata* (1) the receptors near the tibiofemoral joint determine the basic orientation response to a particular frontal sector, (2) the tarsal receptors perhaps modify this response, (3) the abdominal hairs sense waves from posterior sources, and (4) the legs and hairs interact to determine precise direction. Wiese (1974) supplemented these ideas by proposing that a wave sector (the general direction from which a wave arrives) is localized by the receptor first stimulated, with exact direction then determined by a vector gained from time of arrival at a second receptor. Additional information is supplied by other leg and abdominal receptors. Time of arrival is also proposed as a major means by which whirligig beetles determine source direction (Reinig and Uhlemann 1973).

Mechanisms of direction orientation have not been studied in other invertebrates. Presumably the mechanisms rely on time of arrival, as in *Notonecta.* Differences in stimulus amplitude do not appear to operate, at least in *Dolomedes,* since the ability of *D. triton* to determine source angle is independent of stimulus amplitude (Bleckmann and Barth 1984).

### Vertebrates

The clawed frog, *Xenopus laevis,* uses its lateral line system to locate wave sources (Kramer 1933; Dijkgraaf 1947, 1963; Görner 1973, 1976; Elepfandt 1982, 1984a, 1984b; Görner, Möller, and Weber 1984). A *Xenopus* with an intact lateral line system can locate a source to within ± 5° if it is within 90°–120° from directly in front (Elepfandt 1982). At angles greater than 120°, accuracy tends to drop off, in a fashion similar to that found in *Notonecta* (Markl, Lang, and Wiese 1973). Small *Xenopus* sometimes turned within ± 5% accuracy over 50% of the time at angles of 150°–180° (Elepfandt 1982; Görner, Möller, and Weber 1984), however, suggesting that *Xenopus* can locate all wave directions with essentially the same accuracy, as seen in the fishing spider *Dolomedes* (Bleckmann and Lotz in prep.). *Xenopus* can also orient up and down in the water to waves approaching from the surface or from below (Elepfandt 1984b). In *Xenopus* the accuracy of wave source localization is strongly dependent on individual motivational state and is associated with the type of food the animals are fed. The degree of reliance on the lateral line system and the effort the frogs were required to use in order to feed are particularly important (Elepfandt 1982). These results suggest that motivation could be influential in localization accuracy and in other components of foraging behavior in other animals as well.

Ablation experiments showed that *Xenopus* does not respond to either maximum stimulation of activated receptors or to intensity gradients among receptors, but instead responds to time-of-arrival differences among them. Information about all directions can be sensed, though with lesser accuracy, by small subsets of stitches. Located on various parts of the body, the subsets provide a multiple representation of wave directions (Elepfandt 1982; Görner, Möller, and Weber 1984). Both dorsal and ventral lateral line systems operate equally effectively (Elepfandt 1984b). Further ablation experiments have shown that if barely enough stitches are left intact for directional lateral line input, *Xenopus* will rely on this input (Elepfandt 1984a; Görner, Möller, and Weber 1984). But if all lateral line input is removed, *Xenopus* still orients to some extent, suggesting a hierarchy in input system importance and an unknown secondary localizing system (Elepfandt 1982; Görner, Möller, and Weber 1984; Elepfandt 1984b). Possibilities for this system include the inner ear; i.e., *Xenopus* might also orient to sound generated by surface waves (Görner 1976).

In contrast to *Xenopus,* in which essentially any two stitches can provide reasonable directionality, directional localization in surface-foraging fish appears to be somewhat different, in that localization is based on a topographical principle (i.e., the specific anatomical location of particular neuromasts on the head of the fish relative to other neuromasts). Müller and Schwartz (1982) showed that in *Aplocheilus lineatus* with only one neuromast left intact, the turning responses depend solely on the topographical position of that particular neuromast: a rostral position results in a small turn and a caudal one in a large turn. The fish remained sensitive to surface waves from all directions, but they were unable to determine the actual stimulus direction. Also the directional alignment of the neuromasts did not affect directionality, as has been previously suggested (Schwartz 1970), since the response angles correlated with the topographical location of neuromasts, rather than with the orientation of specific neuromasts.

Müller and Schwartz (1982) concluded that accurate directionality was achieved by sensing time-of-arrival differences in different neuromasts. This conclusion is supported by ablation experiments in which two neuromasts were left intact: when the neuromasts were almost parallel to one another and should have very small timing differences, the precision of orientation was only slightly reduced compared with that of animals with neuromasts left in less parallel alignment (Tittel, Müller, and Schwartz 1984). When all neuromasts are destroyed, *A. lineatus* unexpectedly is able to orient (Müller and Schwartz 1982), thus implying a hierarchical secondary wave sensory system in these fish, much as in *Xenopus.*

## Locating Wave Source Distance: Accuracy and Mechanisms

After orienting toward potential victims, predators must move towards and capture them. Surface wave properties that could provide distance information include a decrease in curvature of an expanding wavefront and frequency modulation. Also, frequency-amplitude content could provide distance information if compared with a commonly experienced standard, particularly since higher frequencies damp out over distance (Fig. 24.2B).

Invertebrates

Of the invertebrate predators discussed in this chapter, only fishing spiders have been reported to show a reliable and accurate distance response to wave sources (Bleckmann and Barth 1984). Perhaps this is because most invertebrate predators locomote in discrete spurts or leg strokes, usually spaced by brief pauses, thereby giving the impression that they are unaware of source distance. Water striders and back swimmers demonstrate short-range distance determination by short, accurate dashes at close-by wave sources (Wilcox casual observations). Alternatively or perhaps in addition, it may be sufficient for a predator simply to turn and move towards more distant prey inescapably trapped in the surface film, as suggested by Bleckmann (1985b). This does not explain, however, why other predators competing for the same food, such as fishing spiders and fish, move so rapidly and accurately towards potential prey. If most invertebrate predators were at a disadvantage because of not being able to sense source distance, one would have expected natural selection to have promulgated this ability long ago.

Bleckmann (1985b) points out that, since some of *Dolomedes'* common prey are fast-escaping fish, this spider may have evolved quick and accurate direction and distance predatory responses. One could also speculate that these abilities evolved before *Dolomedes* and other spiders became aquatic. For example, the grass spider *Agelinopsis apterta,* which spins a nonsticky web, often orients upon and captures prey so rapidly that the spider can hardly be followed by eye (Wilcox casual observations; Riechert pers. com.). These spiders probably evolved from lycosid-like ground predators that also use rapid prey capture (Riechert pers. com.; Riechert and Luczak 1982).

*Dolomedes triton* can locate the center of a concentric fish-generated stimulus precisely up to at least 18 cm away. *D. triton* responds in less than 1% of such trials, however, because such stimuli have a frequency bandwidth of 3–40 Hz, which makes the stimuli hard to distinguish from abiotic sources (Bleckmann and Lotz in prep.). As discussed more completely in Chapter 24, surface-feeding animals may be capable of distinguishing distance to a wave source in part by determining whether a stimulus has frequencies higher than 50 Hz. If so, this ability would apparently be based on the tendency of higher frequencies to attenuate more rapidly than do lower frequencies.

Swimming whirligig beetles may be able to determine the distance to objects by echolocation, using waves from their swimming motions or from other wave sources (Tucker 1969), but this ability has not yet been documented. Swimming *Gyrinus substriatus* can at least avoid obstacles up to 1.5 cm away by sensing the meniscus near objects (Rudolph 1967).

Vertebrates

When *Xenopus* responds to a wave stimulus, it exhibits two behavioral components, a turning response and a spatial displacement of several centimeters (Görner,

Möller, and Weber 1984). There is no experimental study that indicates the accuracy of distance determination; but it is known that blinded *Xenopus* can capture swimming guppies in midwater (Elepfandt 1984b), so presumably *Xenopus* can determine short-range distances, much as can most invertebrates.

On the other hand, surface-foraging fish have been shown to be very accurate in longer wave source distance determination. This ability is based on evaluation of (1) curvature of the wave front, (2) frequency modulation over distance, and (3) the frequency-amplitude content of the wave signal (Bleckmann 1980; Bleckmann and Schwartz 1982; Hoin-Radkovsky, Bleckmann, and Schwartz 1984). For further details on surface-feeding fish see Chapter 24.

## Surface Waves and Predator Avoidance

Surface wave patterns provide not only a means for predators to locate and capture prey, but also a means by which prey can locate and avoid predators. Since almost any motion at or near the surface will produce surface waves, one method by which prey species can avoid predation is not to produce wave patterns that attract predators (i.e., to produce patterns that are indistinguishable from nonprey or abiotic patterns). Lang (1980a) has documented that the swimming wave patterns of adults and of all instar stages of the back swimmer *Notonecta,* except the first stage, have wave characteristics (below 40–50 Hz, narrow 8–14-Hz bandwidth, maximum amplitude at 7–10 Hz), that render them indistinguishable from abiotic wave sources with respect to frequency and amplitude. Wave patterns from small vertebrates, such as fish, frogs, and tadpoles, are also below 40 Hz, and fish-generated waves elicit predatory responses from the spider *Dolomedes* in less than 1% of trials (Bleckmann and Lotz in prep.).

Many prey species do produce patterns attractive to surface predators, however. The only prey species that has been studied specifically from the viewpoint of escape from surface predators is the mosquito *Aedes aegypti,* whose larvae attempt escape from predators by a rapid diving response. The larvae are most sensitive to wave patterns in the 10–20-Hz range, which corresponds closely to the emphasized frequencies of swimming waves from such common predators as backswimmers and water striders (Markl and Hauff 1973).

The bryozoan *Bowerbankia imbricata* responds to water movements generated by surface waves with a withdrawal reflex and is most sensitive in the 40–80-Hz range, but has not been studied in relation to any known predator-produced wave patterns (Wiese, Wollnik, and Jebram 1980).

## Communication by Surface Wave Signals

The water surface is an ideal substrate for transmission of signals within and between surface-dwelling species at close range. A number of species have evolved specialized intraspecific communication systems in the contexts of mating and spacing behavior.

## Invertebrates

The most elaborately studied systems are in various species of water striders. In all species, signals are produced by vertical or lateral oscillations of the forelegs, middle legs or hind legs.

In the old-world species *Rhagadotarsus kraepelini,* signals occur during precopulation, copulation, postcopulation, and territorial aggression. Both males and females generate signals with the forelegs and middle legs. Males cling to signal sites and use 22–26-Hz calling signals (Fig. 25.3A) and single-pulse courtship calling signals to attract females, exchange courtship signals with females, and produce copulatory signals while copulating at the signal site. Females then oviposit in the signal site, while males produce postcopulatory signals and guard against male intruders with two aggressive signals of 9–13 Hz and 23–30 Hz (Wilcox 1972).

After oviposition, females leave the site, and males return to continue signalling, defending the site against takeovers by other males. Signal playback experiments showed that (1) females were attracted by the calling signal; (2) they could be induced to oviposit by the stimulus of the calling signal alone, without visual cues of any sort; and (3) they were neurophysiologically tuned to the calling signal frequency (Wilcox 1972). Playbacks of calling signals through free-swimming females to site-defending males (by the coil and magnet playback system of Wilcox 1979; Wilcox and Kashinsky 1980) induced aggressive attacks

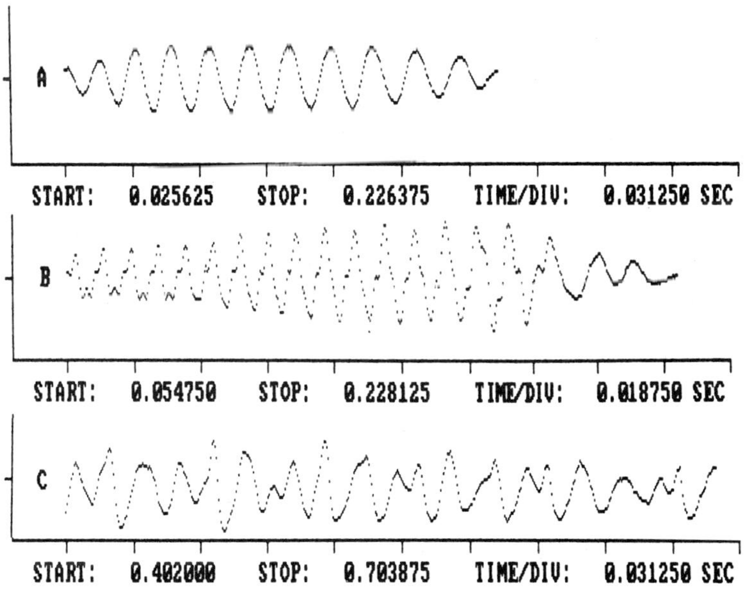


Figure 25.3. (A) Surface wave calling signal of approximately 26 Hz produced by males of water strider *Rhagadotarsus kraepelini.* (B) High-frequency (90–100 Hz) signal produced by males of water strider *Gerris remigis.* (C) Repelling signal of approximately 30 Hz produced by males of water striders *Limnoporus dissortis* and *L. notabilis.*

on the females, indicating that calling signals also function in sex identification (Wilcox unpub. data).

The accuracy of frequency discrimination is likely to be especially important in the context of intra- and interspecific communication, in which ripple signals may be important stimuli that integrate male/female reproductive and other intraspecific behavior. In the reproductive context, Wilcox (1972) found that females of the water strider *Rhagadotarsus kraepelini* can distinguish playbacks of male-produced calling signals that differ by 1.5 Hz.

Males of the North American strider species *Gerris remigis* produce high-frequency (90–100 Hz) signals with their forelegs (Fig. 25.3B), as do the males of *Limnoporus dissortis* and *L. notabilis* (Fig. 25.3C). Playback experiments through females to males in the laboratory (*G. remigis*) and field (*L. dissortis* and *L. notabilis*) show that males also use these signals for sex discrimination during mating behavior and, in *L. dissortis* and *L. notabilis,* for repelling smaller males from territorially defended areas (Wilcox 1979; Wilcox and Spence 1986). Males of *L. dissortis* and *L. notabilis* showed conditional tactics in the use of signals and territorial vs. patrolling behavior, depending on local circumstances (Spence and Wilcox 1986). Males of *G. remigis* also used signals during territoriality (Wilcox and Ruckdeschel 1982) and during copulatory guarding, in which male signals repel male intruders and enable females to forage more effectively (Wilcox 1984; Wilcox unpub. data). A number of other water strider species produce surface wave signals (e.g., Hayashi 1985), providing a rich source of material for exploring surface wave functions in this group. Surface wave communication has also recently been discovered in the fishing spider *Dolomedes triton* (Roland and Rovner 1983). In the mating season, females attract males with sex pheromones. In courtship, males produce leg waving and jerks with the first and second pairs of legs. Females often respond by drumming on the water surface with rapid vertical motions of the pedipalps. Since male signals are click-like, with an energy maximum at 8–15 Hz, they do not elicit predatory responses in the species, as one would expect (Bleckmann and Lotz in prep.).

Males of the belostomatid *Sphaerodema grassei* support themselves near the surface with vegetation and rock their bodies in bursts of 3–7 waggles, within 2–3 sec. Females are attracted to the resultant surface wave signals and, after several bouts of waggling, which may take up to over an hour, copulation may occur, after which the female lays one egg on the dorsum of the male. Usually bouts of signalling and copulation must occur in between each oviposition, increasing the paternity assurance of the male. These interactions may continue without interruption for a total of over 9 hr (Böttger 1974). Similar behavior occurs during mating behavior in the belostomatid *Abedus herberti* (Smith 1979). Males exhibit fore-and-aft "display pumping" at a rate of about 5 Hz, which Smith (1979) showed attracted females that had already been in physical contact with males.

## Vertebrates

To date, the only studies of intraspecific communication in vertebrates are on males of the fighting fish *Betta splendens,* which is an air-gulping species, and on fire-bellied toads, *Bombina bombina.* Males of *B. splendens* build bubble nests, guard fertilized eggs in the nest, and attend juveniles after hatching. Males can attract juveniles from up to 40 cm away with a surface wave signal produced by trembling motions of the pectoral fins when the male is angled upward while near the surface. When the juveniles reach the male, he sucks them into his mouth and transports them to greater safety in the nest (Kühme 1960). Juveniles show greatest sensitivity to artificial playbacks at 8–10 Hz with an amplitude of at least 13 μm. By 1–2 min. after stimulus onset, more than 70% of all juveniles gather close to the source (Kaus 1977).

In *Bombina bombina,* males defend territories during mating behavior in part by producing 50–280-μm surface waves by the emission of mating calls, by hind-legged strokes in between mating calls, or by short swimming bouts. The maximum energy of the waves is in the frequency range of 5–10 Hz at a distance of more than 10 cm from a toad. The minimal territorial distance between two calling males, around 50 cm, coincides with the threshold of detectability of surface waves of 30–50 μm (Walkowiak and Muenz 1985).

## Summary

A wide variety of invertebrate and vertebrate animals are specialized to produce and respond to water surface waves, as predators on prey, which can be located by wave patterns; as prey, which use wave patterns to escape from predators; and as individuals using ripple signals to communicate with conspecifics. Animal-induced surface waves originate from essentially a point source, propagate rather slowly in concentric circles, and have strong damping of high frequencies. Surface waves are sensed with a variety of mechanoreceptors associated with body structures that come into contact with the surface. The structures are able to analyze surface wave patterns in terms of frequency structure, amplitude, and time of arrival. These sensory abilities enable the animals to determine with reasonable accuracy the identity category, direction, and distance of a wave emitter and to send and receive information in ripple signals.

*Acknowledgments.* I thank Arthur Popper and his associate editors, and especially Horst Bleckmann, for their thorough and helpful comments on an earlier draft of this chapter. This work was supported at various stages by NSF Grants BNS-7724708 and BNS-8408483 and a Visiting Lectureship at The Australian National University.

### References

Barth, F.G. (1982) Spiders and vibratory signals: sensory reception and behavioral significance, in Spider Communication: Mechanisms and Ecological Significance, Witt, P.N. and Rovner, J.S. (eds.), Princeton University Press, Princeton, N.J., pp. 67–122.

Berestynska-Wilczek, M. (1962) Investigations of the sensitivity of the spiders *Pirata piraticus* (Clerck) to vibrations of the water surface, Acta Biol. Cracoviensia, 5:263–277.

Bleckmann, H. (1980) Reaction time and stimulus frequency in prey localization in the surface-feeding fish *Aplocheilus lineatus,* J. Comp. Physiol., 140:163–172.

Bleckmann, H. (1985a) Discrimination between prey and non-prey wave signals in the fishing spider *Dolomedes triton* (Pisauridae), in Acoustic and Vibrational Communication in Insects, Kalmring, K., and Elsner, N. (eds.), Springer-Verlag, Berlin, pp. 215–222.

Bleckmann, H. (1985b) Perception of water surface waves: how surface waves are used for prey identification, prey localization, and intraspecific communication, in Progress in Sensory Physiology 5, Autrum, H.J., and Ottoson, D. (eds.), Springer-Verlag, Berlin, pp. 147–166.

Bleckmann, H. and Barth, F.G. (1984) Sensory ecology of a semiaquatic spider (*Dolomedes triton*). II. The release of predatory behavior by water surface waves, Behav. Ecol. Sociobiol., 14:302–311.

Bleckmann, H. and Rovner, J.S. (1984) Sensory ecology of a semi-aquatic spider (*Dolomedes triton*). I. Roles of vegetation and wind-generated waves in site selection, Behav. Ecol. Sociobiol., 14:297–301.

Bleckmann, H. and Schwartz, E. (1982) The functional significance of frequency modulation within a wave train for prey localization in the surface-feeding fish *Aplocheilus lineatus* (Cyprinodontidae), J. Comp. Physiol., 145:331–339.

Bleckmann, H. and Topp, G. (1981) Surface wave sensitivity of the lateral line organs of the topminnow *Aplocheilus lineatus,* Naturwissenschaften, 62:624–625.

Bleckmann, H., Waldner, I., and Schwartz, E. (1981) Frequency discrimination of the surface-feeding fish *Aplocheilus lineatus*—a prerequisite for prey localization. J. Comp. Physiol., 143:485–490.

Böttger, K. (1974) Zur Biologie von *Sphaerodema grassei ghesguierei.* Studien an zentralafrikanishen Belostomatiden (Heteroptera, Insecta), Arch. Hydrobiol., 74:100–122.

Dijkgraaf, S. (1947) Über die Reizung des Ferntastsinns bei Fischen und Amphibien, Experientia, 3:206–208.

Dijkgraaf, A. (1963) The functioning and significance of the lateral line organs, Biol. Rev., 38:51–105.

Elepfandt, A. (1982) Accuracy of taxis response to water waves in the clawed toad (*Xenopus laevis* Daudin) with intact or with lesioned lateral line system, J. Comp. Physiol., 148:535–545.

Elepfandt, A. (1984a) Localization of water surface waves with the lateral line system in the clawed toad (*Xenopus laevis* Daudin), in Localization and Orientation Biology and Engineering, in Varju, D., and Schnitzler, H.-U. (eds.), Springer-Verlag, Berlin, pp. 63–65.

Elepfandt, A. (1984b) The role of ventral lateral line organs in water wave localization in the clawed toad (*Xenopus laevis*), J. Comp. Physiol., 154:773–780.

Elepfandt, A. (1985) Naturalistic conditioning reveals good learning in a frog (*Xenopus laevis*), Naturwissenschaften, 72:492–493.

Elepfandt, A., Seiler, B., and Aicher, B. (1986) Water wave frequency discrimination in the clawed frog, *Xenopus laevis,* J. Comp. Physiol., 157:255–267.

Friesen, W.O. (1981) Physiology of water motion detection in the medicinal leech, J. Exp. Biol., 92:255–275.

Görner, P. (1973) The importance of the lateral line system for the perception of surface waves in the clawed toad, *Xenopus laevis,* Experientia, 9:295–296.

Görner, P. (1976) Source localization with labyrinth and lateral line in the clawed toad (*Xenopus laevis*), in Sound Reception in Fish, Schuijf, A., and Hawkins, A.D. (eds.), Elsevier, Amsterdam, pp. 171–174.

Görner, P., Möller, P., and Weber, W. (1984) Lateral-line input and stimulus localization in the African clawed toad *Xenopus* sp., J. Exp. Biol., 108:315–328.

Hayashi, K. (1985) Alternative reproductive strategies in the water strider *Gerris elongatus* (Heteroptera, Gerridae), Behav. Ecol. Sociobiol., 15:301–306.

Hoin-Radkovsky, I., Bleckmann, H., and Schwartz, E. (1984) Determination of source distance in the surface-feeding fish *Pantodon bucholzi* Pantodontidae, Anim. Behav., 32:840–851.

Kaus, S. (1977) Untersuchen zum vibrotakischen Verhalten junger Kampffische (*Betta splendens* R.), Diplomarbeit Universitat Geissen, (cited in Bleckmann 1985b).

Kramer, G. (1933) Untersuchungen über die Sinnesleistungen und das Orientierungsverhalten von *Xenopus laevis,* Zool. Jahrb. Physiol., 52:629–676.

Kühme, W. (1960) Verhaltensstudien an maulbrütenden (*Betta anabatoides*) und nestbauenden Kampffischen (*Betta splendens*), Z. Tierpyschol., 18:33–55.

Lang, H.H. (1980a) Surface wave discrimination between prey and nonprey by the back swimmer *Notonecta glauca* L. (Hemiptera, Heteroptera), Behav. Ecol. Sociobiol., 6:233–246.

Lang, H.H. (1980b) Surface wave sensitivity of the back swimmer *Notonecta glauca,* Naturwissenschaften, 67:204–205.

Lawry, J.V. (1973) A scanning electron microscopic study of mechanoreceptors in the walking legs of the water strider, *Gerris remigis,* J. Anat., 116:25–30.

Markl, H. (1983) Vibrational communication, in Neuroethology and Behavioral Physiology, Huber, F., and Markl, H. (eds.), Springer-Verlag, Berlin, pp. 332–353.

Markl, H., and Hauff, J. (1973) Die Schwellenkurve des durch Vibration ausgelösten Fluchttauchens von Muckenlarven, Naturwissenschaften, 9:432–433.

Markl, H., Lang, H., and Wiese, K. (1973) Die Genauigkeit der Ortung eines Wellenzentrums durch den Rückenschwimmer *Notonecta glauca* L., J. Comp. Physiol., 86:359–364

Markl, H. and Wiese, K. (1969) die Empfindlichkeit des Rückenschwimmers *Notonecta glauca* L. für Oberflächenwellen des Wassers, Z. Vgl. Physiol., 62:413–420.

Müller, U. (1984) Die morphologische und physiologische Anpassung des Seitenliniensystems von *Pantodon buchholzi* an den Lebensraum Wasseroberfläche, Ph.D. Thesis, Universitat Giessen, W. Ger.

Müller, U. and Schwartz, E. (1982) Influence of single neuromasts on prey localizing behavior of surface feeding fish, *Aplocheilus lineatus,* J. Comp. Physiol., 149:339–408.

Murphey, R.K. (1971) Sensory aspects of the control of orientation to prey by the water-strider *Gerris remigis,* Z. Vgl. Physiol., 72:168–185.

Murphey, R.K. (1973) Mutual inhibition and the organization of a non-visual orientation in *Notonecta,* J. Comp. Physiol., 84:31–40.

Murphey, R.K., and Mendenhall, B. (1973) Localization of receptors controlling orientation to prey in *Notonecta undulata,* J. Comp. Physiol., 84:19–30.

Otte, D. (1974) Effects and functions in the evolution of signaling systems, Annu. Rev. Ecol. Syst., 5:385–417.

Reinig, H.-J. and Uhlemann, H. (1973) Uber das Ortungsvermogen des Taumelkafers *Gyrinus substriatus* Steph. (Coleoptera, Gyrinidae), J. Comp. Physiol., 84:281–298.

Riechert, S.E., and Luczak, J. (1982) Spider foraging: behavioral responses to prey, in Spider Communication, Witt, P.N., and Rovner, J.S. (eds.), Princeton University Press, Princeton, N.J., pp. 281–315.

Roland, C.H. and Rovner, J.S. (1983) Chemical and vibratory communication in the aquatic pisaurid spider *Dolomedes triton,* J. Arachnol., 11:77–85.

Rudolph, P. (1967) Zum Ortungsfahren von *Gyrinus substriatus* Steph., Z. Vgl. Physiol., 56:341–375.

Schwartz, E. (1965) Bau und Funktion der Seitenlinie des Streifenhechtlings (*Apolocheilus lineatus* Cuv. u. val.), Z. Vgl. Physiol., 50:55–87.

Schwartz, E., (1967) Analysis of surface-wave perception in some teleosts, in Lateral Line Detectors, Cahn, P. (ed.), Indiana University Press, Bloomington, pp. 123–134.

Schwartz, E. (1970) Ferntastsinne von Oberflächenfischen, Z. Morphol. Tiere, 67:40–57.

Schwartz, E. (1971) Die Ortung von Wasserwellen durch Oberflachenfische, Z. Vgl. Physiol., 50:55–87.

Schwartz, E. (1974) Lateral line mechanoreceptors in fishes and amphibians, in Handbook of Sensory Physiology, vol 3, Fessard, A. (ed.), Springer-Verlag, Berlin, pp. 257–278.

Schwartz, E. and Hasler, A.D. (1966) Perception of surface waves by the blackstripe topminnow, *Fundulus notatus*, J. Fish Res. Board Can., 23:1331–1352.

Smith, R.L. (1979) Paternity assurance and altered roles in the mating behavior of a giant water bug, *Abedus herberti* (Heteroptera: Belostomatidae), Anim. Behav., 27:716–725.

Spence, J.R. and Wilcox, R.S. (1986) The mating system in two hybridizing species of water striders (Gerridae). II. Alternative tactics of males and females, Behav. Ecol. Sociobiol., 19:87–95.

Tittel, G., Müller, U., and Schwartz, E. (1984) Determination of stimulus direction by the topminnow *Aplocheilus lineatus*, in Localization and Orientation in Biology and Engineering, Varju, D., and Schnitzler, H.U. (eds.), Springer-Verlag, Berlin.

Tucker, V.A. (1969) Wave-making by whirligig beetles (Gyrinidae), Science, 166:897–899.

Walkowiak, W. and Muenz, H. (1985) The significance of water surface-waves in the communication of fire-bellied toads, Naturwissenschaften, 72:49–51.

Wenstrup, J.J. (1984) Auditory sensitivity in the fish-catching bat, *Noctilio leporinus*, J. Comp. Physiol., 155:91–101.

Wenstrup, J.J. and Suthers, R.A. (1984) Echolocation of moving targets by the fish-catching bat, *Noctilio leporinus*, J. Comp. Physiol., 155:75–89.

Wetzel, R.G. (1975) Limnology, Saunders, Philadelphia.

Wiese, K. (1969) Wahrnehmung von Oberflachenwellen geringer Amplitude durch den Wasserlaüfer, Naturwissenschaften, 56:38.

Wiese, K. (1972) Das mechanorezeptorische Beuteortungssystem von *Notonecta*. I. Die Funktion des tarsalen Scolopidialorgans, J. Comp. Physiol., 78:83–102.

Wiese, K. (1974) The mechanoreceptive system of prey localization in *Notonecta*. II. The principle of prey localization, J. Comp. Physiol., 92:317–325.

Wiese, K. and Schmidt, K. (1974) Mechanorezeptoren im Insektentarsus. Die Konstruktion des tarsalen Scolopidialorgans bei *Notonecta*, Z. Morphol. Tiere, 79:47–63.

Wiese, K., Wollnik, F., and Jebram, D. (1980) The protective reflex on *Bowerbankia* (Bryozoa): calibration and use to indicate movements of the medium beneath a capillary surface wave, J. Comp. Physiol., 137:297–303.

Wilcox, R.S. (1972) Communication by surface waves: mating behavior of a water strider (Gerridae), J. Comp. Physiol., 80:255–266.

Wilcox, R.S. (1979) Sex discrimination in *Gerris remigis:* role of a surface wave signal, Science, 206:1325–1327.

Wilcox, R.S. (1984) Male copulatory guarding enhances female foraging in a water strider, Behav. Ecol. Sociobiol., 15:171–174.

Wilcox, R.S. and Kashinsky, W.M. (1980) A computerized method of analyzing and playing back vibratory animal signals, Behav. Res. Methods Instrum., 12:361–363.

Wilcox, R.S. and Ruckdeschel, T. (1982) Food threshold territoriality in a water strider (*Gerris remigis*), Behav. Ecol. Sociobiol., 11:85–90.

Wilcox, R.S. and Spence, J.R. (1986) The mating system of two hybridizing species of water striders (Gerridae). I. Ripple signal functions, Behav. Ecol. Sociobiol., 19: 79–85.

Young, S.R., Dedwylder, R.D., and Friesen, W.O. (1981) Responses of the medicinal leech to water waves, J. Comp. Physiol., 144:111–116.

# Chapter 26

# The Representation of Hydrodynamic Parameters in the CNS of the Crayfish *Procambarus*

KONRAD WIESE*

The sensory pathways of the nervous system form the "window" that provides organisms with essential information about events in their environment. From extensive studies of intraspecific communication systems in insects (for a comparative review see Huber 1983), it seems that these pathways are molded by natural selection to match as closely as possible the properties of the most relevant signals to be received, such as the species-specific mating call.

The adjustments most probably affect several sections of the sensory pathway, including the receptors, the response properties of interneurons, and possibly even their presence or absence. Along these lines, I should like to suggest that specific neuronal circuits, specifically those providing inhibitory contrast enhancement, have evolved at key positions of the pathway. Such circuits were first described in the visual pathways of the cat (Hartline 1940) and in the horseshoe crab *Limulus* (review by Hartline and Ratliff 1972). In the auditory pathway of crickets, one of these functions is provided by the minimum circuit of two local giant interneurons (Wohlers and Huber 1982; Wiese and Eilts-Grimm 1985).

The presence of such contrast enhancement circuits presumably indicates that the signal parameter affected by the circuit carries significant information that without the action of the circuit would be lost or would make a specific aspect of the signal ambiguous.

The mechanosensory pathway in the CNS of the crayfish *Procambarus* and its connections with the circuitry mediating the escape response have been particularly well investigated over the past twenty-five years (see reviews by Wine and Krasne 1982; Krasne and Wine 1984). This sensory system processes fluid displacement signals (Wiese, Calabrese, and Kennedy 1976) in a way similar to the lateral line system in fish (compare Münz 1985 and Chap. 23). The information extracted from water displacements by the CNS of the crayfish *Procambarus* is used to modulate escape swimming (Reichert, Plummer, and Wine 1983).

*Zoologisches Institut und Zoologisches Museum der Universität Hamburg, 2000 Hamburg 13, FRG

This sensory pathway for processing water displacement stimuli in the crayfish could provide evidence for specific properties of displacement signals. The relevant information about this pathway will be reviewed in the first section of this chapter, followed by a detailed examination of the circuits providing contrast enhancement.

## A Circuit Monitoring Water Displacements

Wiersma's discovery that a large number of ventral cord neurons in *Procambarus* responded to gentle brushing of well-defined sections of the skeleton led to his catalogue of interneurons that can be identified by their specific fields of input (Hughes and Wiersma 1960; Wiersma and Hughes 1961). Laverack (1962) and Mellon (1963) observed that the sensilla on the chelae and carapace reacted to fluid movements and were innervated by two sensory cells that responded to opposing stimulus directions. The later investigations aimed at the neuronal circuitry underlying the escape response of crayfish and focussed on the mechanosensory sensilla on the tail fan as the driving input of this circuitry. Most experiments used the ''isolated tailfan'' preparation described by Calabrese (1976), which includes the skeletal plates of the tail fan with the connected sixth abdominal ganglion and the isolated abdominal ventral cord. As a speciality of this sensory system, the fourth root of ganglion 6 in the abdominal cord contains, with the exception of one thin motor axon, exclusively sensory cell axons belonging to sensilla on the lateral parts of the telson and thus is purely sensory (Fig. 26.1).

### Receptor Level

The large number of hair-type sensilla covering the skeleton of the crayfish are not a population with a uniform morphological appearance, but an assembly of sensors with different mechanical properties (Fig. 26.1). Tautz (1979) has conducted a theoretical consideration of the role of structural modifications in the outer section of the hair sensillum for the process of mechanical coupling to movements of the medium. He concludes that short and bare hair shafts will need comparatively higher-frequency movements to achieve rigid coupling to the fluid movement as compared to hair shafts fitted with featherlike structures. Furthermore, the feathery structures potentially provide improved coupling of the sensillum to both low-velocity and high-frequency medium movements. Long shafts need minimum forces to be deflected, yet they have to make comparatively large movements at the tip.

The articulation of the crustacean mechanosensillum to the skeletal plates resembles a tendon-like hinge, the mechanics of which is known from the design of high-sensitivity relays and galvanometers (Fig. 26.2) (Wiese 1976). The articulation is a delicate hinge, fitted with guiding structures that align shaft deflections with the head-to-tail axis of the animal and with mechanical stops that allow only a few degrees of deflection of the hair shaft in the two preferred directions.

Recordings from the axons of the sensory cells made while vibrating an individ-

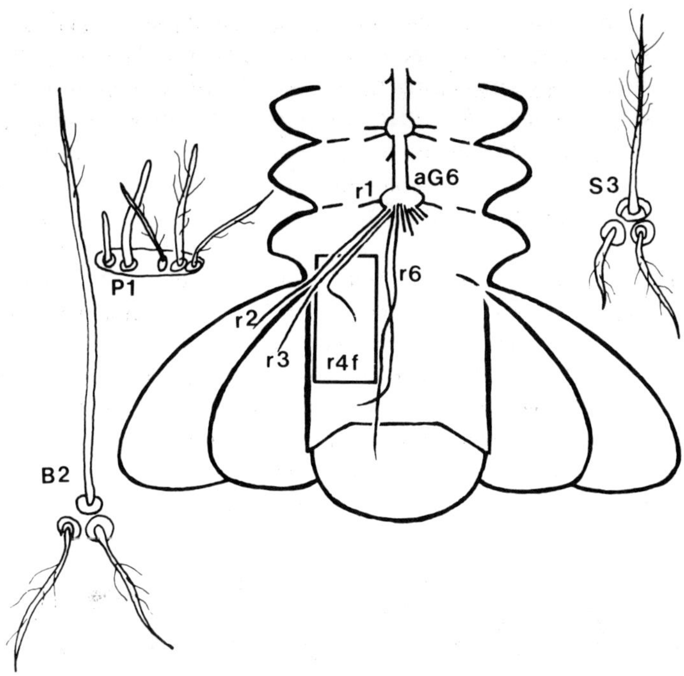


Figure 26.1. The innervation of different sections of crayfish tail fan by individual roots (*r1–r6*) of ganglion 6 (*aG6*) of abdominal ventral cord. *Rectangular frame* marked *r4f* outlines the extension of receptive field innervated by purely sensory root 4. To illustrate the diversity of receptor shapes found within this field, three groups of sensilla have been redrawn from SEM photographs, *B2, S3,* and *P1;* the nomenclature is according to the mapping done by Kennedy (1971, 1974); to give a scale: long *B2* hair measures about 700 μm in nature.

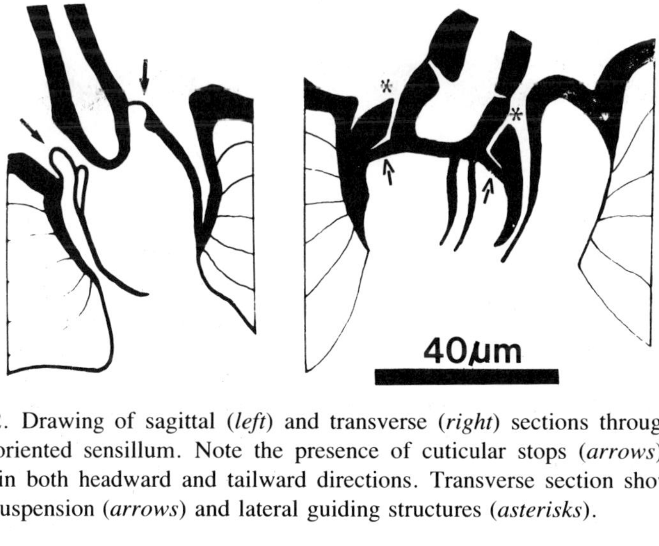


Figure 26.2. Drawing of sagittal (*left*) and transverse (*right*) sections through the base of upright-oriented sensillum. Note the presence of cuticular stops (*arrows*) that limit excursions in both headward and tailward directions. Transverse section shows tendon-type lever suspension (*arrows*) and lateral guiding structures (*asterisks*).

ual sensillum in the preferred direction show that there are two sensory cells responding to opposing directions of movement in this hair (Fig. 26.3, *upper record*). A simultaneous recording from ventral cord neurons in the connective between ganglia 5 and 6 (Fig. 26.3, *lower record*) demonstrates that there are also two interneurons following the activity of the two sensory cells. Each interneuron responds to hair deflection in one direction. The preferred directions of most sensilla in the innervation field of root 4 in ganglion 6 (*framed part of the telson* in Fig. 26.1) are aligned in parallel and sensory cells of one polarity (headward or tailward) converge onto directionally selective populations of interneurons (Wiese, Calabrese, and Kennedy 1976). The developmental mechanism for establishing this selective connectivity is unknown.

The dual innervation of the receptors in this pathway, as well as the functional dual innervation of the cupula receptors in vertebrate lateral line systems, are most likely not just other measures to improve the directionality of the receptors. This feature represents an either-or mechanism for detecting the direction of a fluid movement and presumably signals the approach or retreat of organisms in

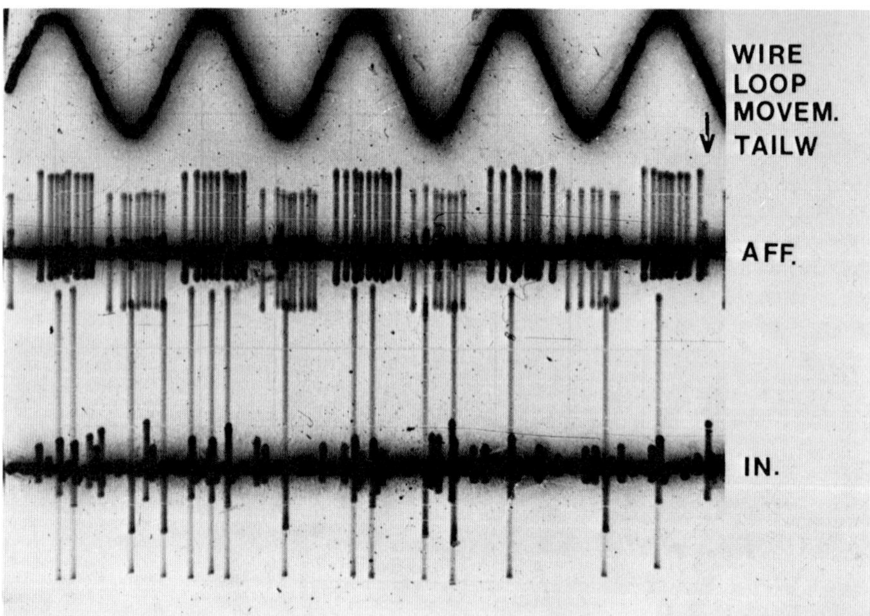


Figure 26.3. Sinusoidal movement of hair shaft in both headward and tailward directions produced by coupling the sensillum to vibrating wire loop. Extracellular recording of sensory (*afferent*) cell activity from root 4 of ganglion 6 shows the response of *two* sensory cells (look for differing positive excursions of recorded potentials) to monitor headward and tailward phases of movement, respectively. The record from interneurons (*bottom trace*) of the connective between ganglia 5 and 6 reveals that again *two* different interneurons (check top ends of recorded potentials) respond in this situation, each connected to only one of sensory cells. Two directionally selective neuronal channels in ventral cord are thus activated. Wire loop excursion: about 5 μm from rest in both directions, attachment to the hair (*S3* in Fig. 26.1) about 200 μm above the hinge.

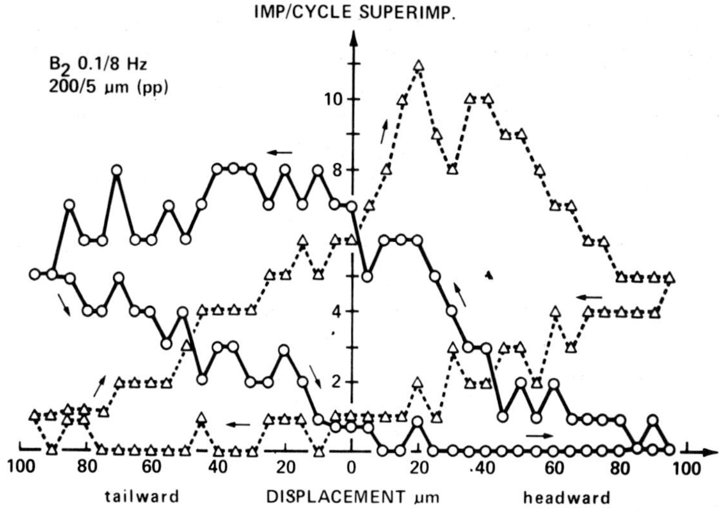


Figure 26.4. The change of response amplitude of two sensory cells of *B2* sensillum to 8-Hz superimposed vibration, depending on the state of momentaneous lasting deflection of the shaft, which is varied with a 0.1-Hz frequency. *Arrows* at response curves indicate momentaneous direction of change of slow movement. As the hair is slowly bent tailwards, tailward-sensitive cell sharply increases the response to superimposed 8-Hz oscillation of 5 μm peak-to-peak amplitude; wire loop attaches about 500 μm above hinge.

close proximity to the listening animal and will be explained later. Movements of water-displacing objects are accompanied by gross water movements that indicate the direction of progress of the moving object. On top of this event, higher-frequency vibrations of the medium occur, and these are probably significant for the nature of the moving object (e.g., swimming shrimp, walking crab, fish with paddling fins).

An experiment that attempts to simulate both the DC and AC components of a medium displacement is shown in Fig. 26.4. An oscillatory movement of the hair of 8 Hz displaces the tip of the $B_2$ hair (nomenclature from Kennedy 1971) 2.5 μm headward and 2.5 μm tailward from resting position. Superimposed onto this movement is a second, slower motion of the hair that displaces the tip by 100 μm in each direction and with a cycle duration of 10 sec. As the larger, slower movement proceeds in the headward or tailward direction, the level of response to the small 8-Hz oscillatory movement in the respective sensory cell increases sharply. During the movement in the opposite direction, the respective sensory cell has a minimum response to the 8-Hz oscillation. An underlying slow movement is thus supporting, perhaps even prerequisite of, a strong response to vibration of the medium by the sensory cells and is a hint that mixed DC and AC signals occur regularly in this environment.

## The Arrangement of Thresholds and Frequency Response

The hypothesis that the diverse morphological appearances of the receptor hairs must correlate with large variations of threshold sensitivity was tested by Ebina and Wiese (1984). By using a source of calibrated particle velocities, they recorded thresholds of sensory cells and associated interneurons from an isolated tail fan preparation (Fig. 26.5). The exact pattern of stimulus movements is shown at the *top* of the figure. The obtained threshold values for receptors and interneurons were scattered over three log units. Because sensory cells in lateral line receptors of fish are known for rather uniform morphology and uniform thresholds (Flock 1967), the canals, on the other hand, being very diverse (Chapter 22), the diversity of hair morphology in the crayfish sensilla appears to be an analogous development. Furthermore, theoretical considerations (Tautz 1979) support the notion that the structure of the coupling device in a mechanoreceptor significantly alters response properties and thresholds. It is thus obvious to assume that morphology and mechanics of the outer section of the crayfish mechanosensory sensilla accounts for most of the scattering of thresholds. The pattern of velocity stimuli used in the described experiments even suggests that receptors of high acceleration

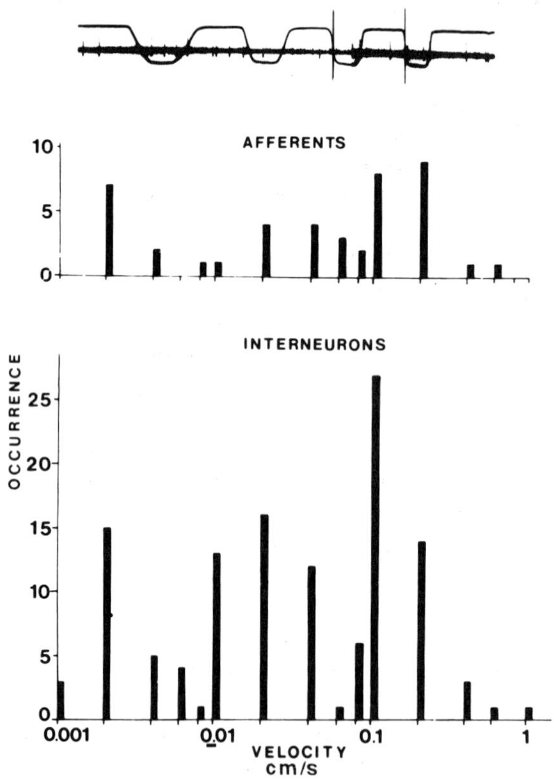


Figure 26.5. A comparison of thresholds for particle velocity in sensory cells and connected interneurons, showing the number of cells having particular threshold values. The stimulus was velocity-controlled ramp-type movement (*top*) of a sphere 10 mm in diameter (dipole source) positioned at measured distance from the center of fourth root field; it exerted maximal displacements of 1.0 mm; actual displacement at recorded sensillum was calculated.

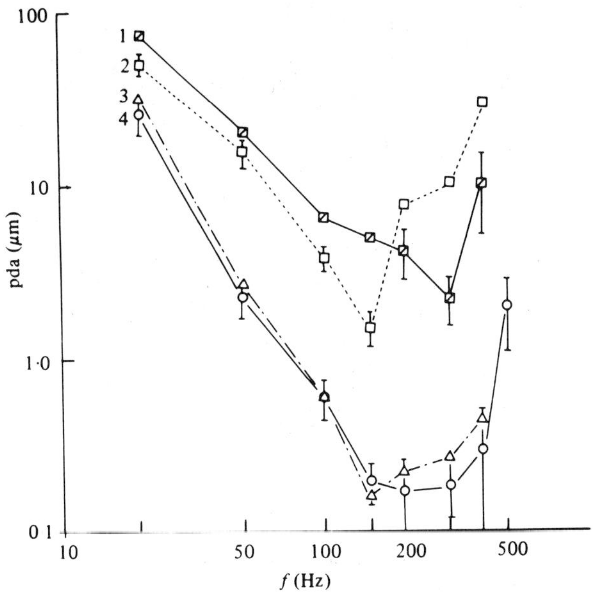


Figure 26.6. Tuning of hair-type displacement receptors on chelae of *Orconectes/Potamobius* (Tautz and Sandeman 1980); *pda*, particle displacement amplitude; *upper two curves;* two individual receptors; *lower curves;* summed activity from several receptors. Preparation was positioned inside a tube filled with saline, one end closed with a diaphragm and coupled to a vibrator and other end open and tilted upwards.

stimuli probably were not activated and thus are not included in the histograms of Fig. 26.5.

Reliable frequency-tuning curves for crayfish mechanosensory sensilla have been obtained by Tautz and Sandeman (1980) for sensors on the chelae of *Austropotamobius* (*Orconectes*) (Fig. 26.6). They found thresholds for spike response as low as 0.2 μm particle displacement amplitude from 150–300 Hz. Plummer (1986) measured the tuning curves of identified receptors and interneurons in this pathway, using carefully defined fluid oscillations. He found interneurons that responded to all frequencies from 4–400 Hz, which he termed *wide-band cells,* and others that responded only to oscillations over 40 Hz (*high-pass cells*). A third set of interneurons were active only with fluid oscillations below 100 Hz. These cells directly receive inhibitory input, which prevents their response to higher-frequency stimuli. Known as *contrast enhancement* in the frequency domain, this may be interpreted as showing that there is small-scale analysis of the frequency content of the fluid displacement stimuli in this pathway.

## The Directionality of Response in Sensory Cells and Ventral Cord Interneurons

Figure 26.7 shows the directionality of two receptive fields with sensory axons in roots 4 and 3 of ganglion 6. These were measured while recording from the

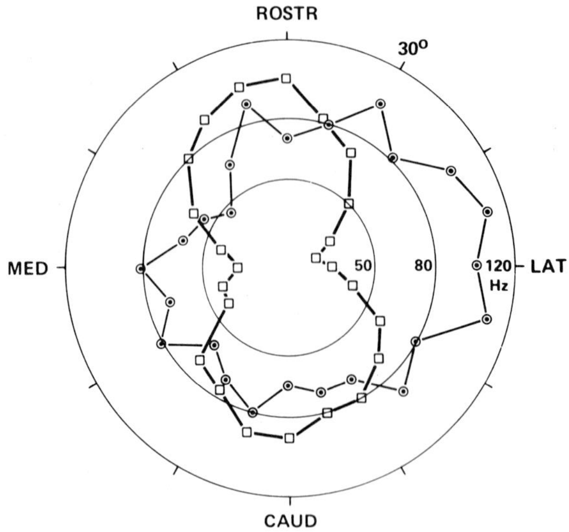


Figure 26.7. The directionality of response in whole receptive fields as measured by recording summed activity from entire sensory root. Directionality curve for root 4 of ganglion 6 (*open squares*) is figure eight-shaped in response to surface waves of 4 Hz approaching from all directions. The response of third root (*filled circles*) is maximal when waves approach laterally. This shows that receptors on inner uropod, innervated by third root, have different preferred direction from receptors in the field of root 4. Ongoing motor neuron activity in third root impedes the measurement.

whole roots, while a source of capillary surface waves was moved round the isolated tail fan. The total number of impulses from all sensory axons recorded in response to a train of surface waves is plotted against stimulus direction. The response of the sensory cells in root 4 is vigorously modulated by differences in the angle of approach of the signal. Stimuli parallel to the body are most effective in stimulating the sensilla, because the effective movements of the hair shafts are aligned in this direction. The maximum response for the sensory axons recorded in root 3 occurs when the stimulus approaches from the side. Background activity in the recording that partly occludes directionality of the receptors comes from the constantly active motorneurons in this root.

With the same stimulation technique, the directionality of the response in interneurons was measured by recording from the circumesophageal connectives (Wiese and Wollnik 1983). Fig. 26.8 shows a selection of the directional selectivities encountered, together with the respective receptive fields. Directionality of the response depends on the size of the receptive field, varying from highly specific, as in unit *L23* in Fig. 26.8, to nondirectional, as in unit *L18*, the receptive field of which consists of practically half the body surface.

The capillary surface waves used as a stimulus have a low attenuation with distance ($1/r^2$) as compared with the attenuation affecting displacements generated

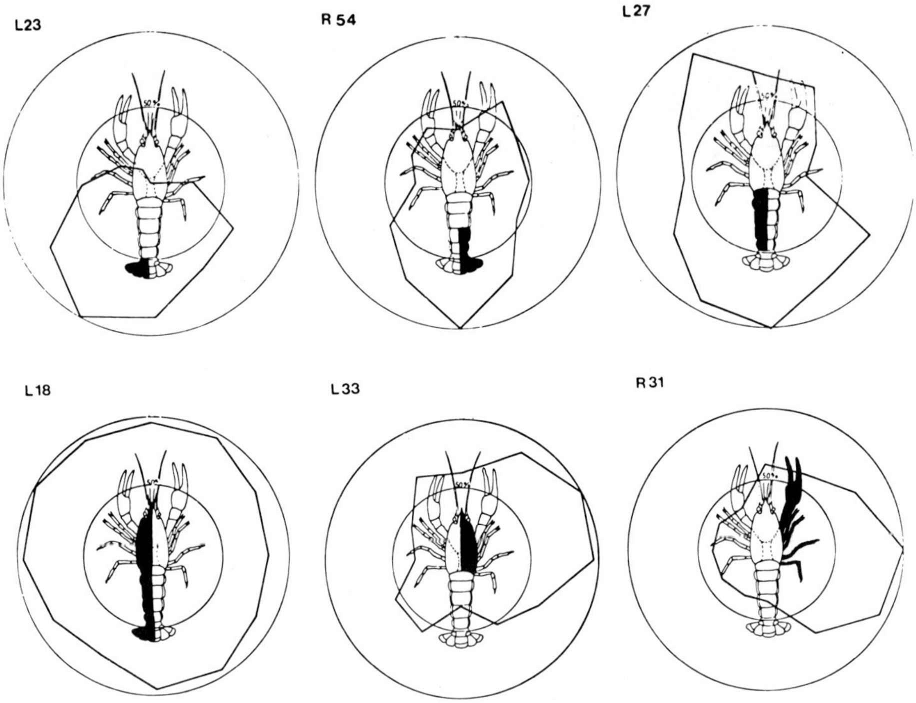


Figure 26.8. The directionality of individual tonic ventral cord interneurons of mechanosensory pathway in crayfish *Procambarus*, identified by their receptive fields (*drawn in black*). The preparation was stimulated by 8-Hz surface waves produced by a dipper that travelled full circle around the preparation. Outer circle indicates maximum response (100%) and inner circle 50%. (From Wiese and Wollnik 1983.)

inside the body of water ($1/r^3$) (see also Chapter 25). The surface wave thus assures that the whole receptive field is stimulated even when the field is very large. Surface waves also activate all possible sources of the lateral inhibition that eventually sharpens the directionality of cord neurons. In contrast, displacements inside the water body are subject to shadowing by the animal's body, while strong attenuation ($1/r^3$) may mean that only part of a given receptive field is stimulated. Neither effect necessarily increases the directional acuity of the interneuronal response.

Figure 26.9 demonstrates the reduction of the neural response when the displacement signals are not generated in the plane of the "listening" crayfish, but are propagated from sources above the animal (Heinisch and Wiese 1983). The measurements made so far indicate that when the source is between 45 and 90 degrees of elevation above the animal, the responses and directionality of the sensilla on the tail fan and of the connected interneurons are less than when the signal is generated in the same layer of water as the sense organs.

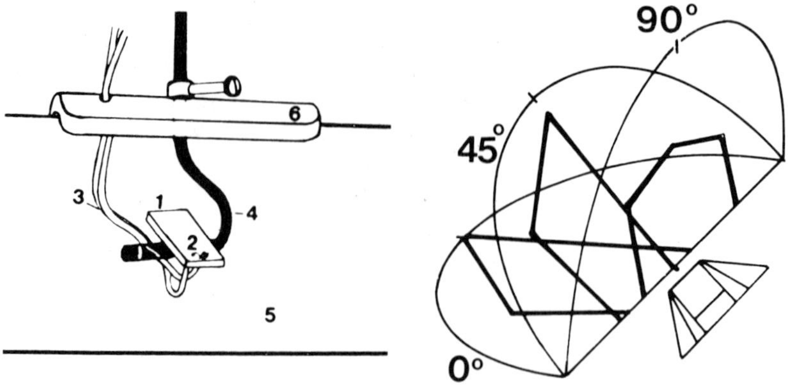


Figure 26.9. The response of displacement-sensitive interneurons to displacements generated in water layers above the plane of the crayfish. Tail fan with sensory roots and nerve cord is mounted inside Plexiglas tube, on a platform that can be tilted and rotated. Vibrator-activated diaphragm at one end of the tube produces displacements. Interneuronal response shown in the *graph* indicates 50% reduction in sensitivity if displacements occur within 45°–90° angles of incidence to the surface of tail fan; in this situation of reduced sensitivity, the directionality of response has vanished. (From Heinisch and Wiese in prep.)

## Inhibitory Contrast Enhancements

Arrays of receptors organized in distinct receptive fields are a well-known arrangement and have been previously described in the vertebrate retina (Hubel and Wiesel 1959) and in the mechanoreceptors of mammalian skin (e.g., von Békèsy 1967). Lateral inhibitory interactions have been recorded in certain sets of interneurons in the crayfish mechanosensory pathway (Fig. 26.10A, top) (Wiese and Schultz 1982; Wiese and Wollnik 1983). In the large group of multisegmental

Figure 26.10 (Top). The contribution of lateral inhibitory circuits to the directionality of response in some, but not all, mechanosensory interneurons. Reversible blocking of sensilla in receptive field adjacent to the field driving recorded interneuron removes the directionality of response. Blocking was achieved by covering the section of the skeleton with heavy-duty plastic foil; receptive field driving recorded neuron comprises two abdominal segments (marked by *x*'s; see *insets*). —●— = before block; —○— = after block. (Bottom) A test of bilateral inhibitory circuits showing that synchronous wave action in both compartments of experimental chamber (see *left-hand drawing*) reduces (*A*) sensitivity of some interneurons as compared to asynchronous (*C*) stimulation of both halves of tail fan; *tw*, interneuron response to tailward water motion. This property of the pathway demonstrates that maximum sensitivity is obtained only in response to small local sources of displacement, whereas large-scale movements produce attenuated neural responses, owing to common mode rejection, a property of presynaptic lateral inhibitory circuits (Reichert, Plummer, and Wine 1983).

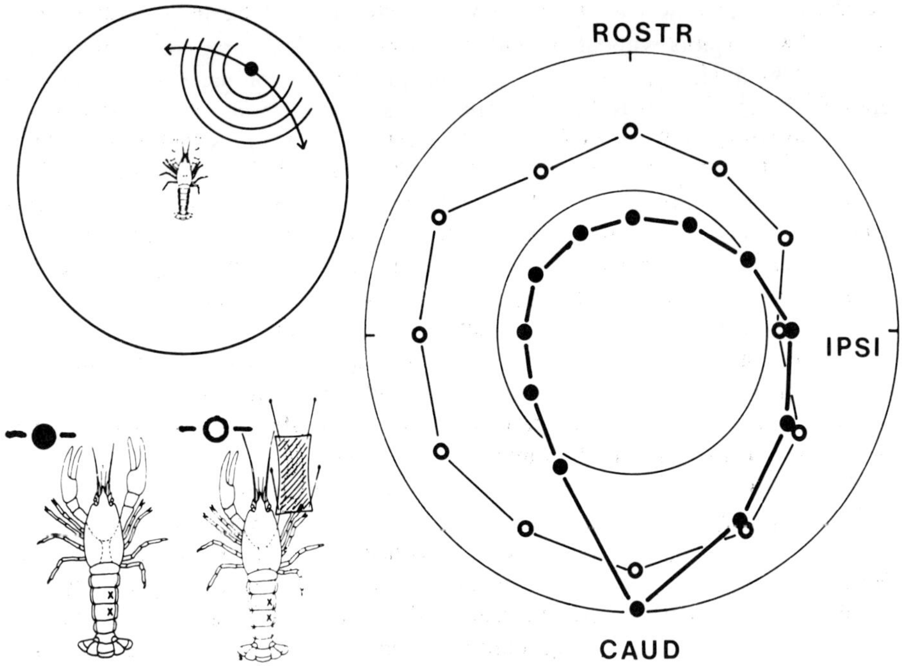
ROSTR
IPSI
CAUD

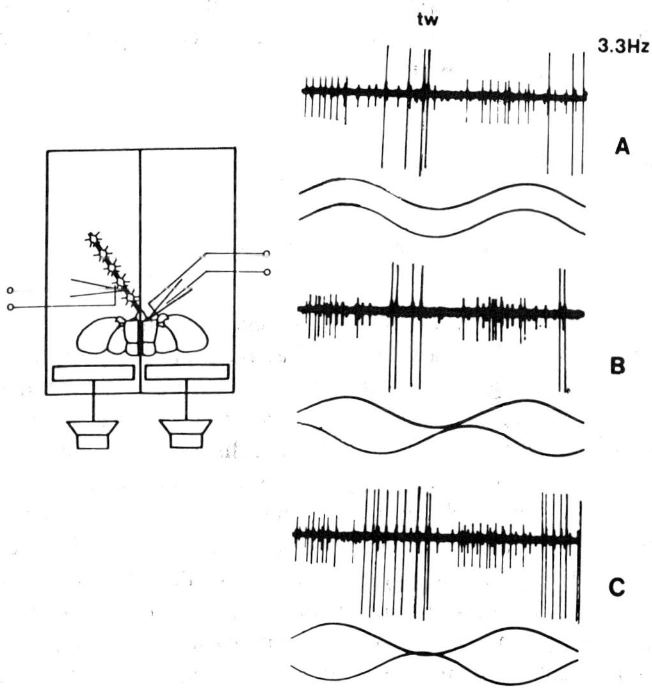
tw
3.3Hz
A
B
C

ventral cord neurons the pattern and amplitude of activity in general follows that of the receptors without modulation (Sigvardt, Hagiwara, and Wine 1982).

Common mode rejection is specifically a property of precurrent inhibitory circuits (Fig. 26.10, bottom). Evidence for such reductions of response in the mechanosensory pathway of *Procambarus* have been obtained using the experimental configuration shown in Fig. 26.10, bottom. (Wiese and Schultz 1982). A tail fan preparation is split in two, and the two halves are mounted in separate compartments of a dissection dish. When the wave motion in both compartments is synchronous, the response is reduced. It is enhanced with antiphase motion, though the motion in a single compartment is identical in both cases. This characteristic of the lateral inhibitory circuits shows not only that ''headward'' and ''tailward'' subsystems in the pathway must have their own lateral inhibitors, but also that the pathway is focussed on small, local sources of displacement, producing asymmetrical stimulation of parallel channels in this network, as has been hinted at by Reichert, Plummer, and Wine (1983).

The headward and tailward sensitive subsystems in this pathway are well separated in their responses to the two opposed directions of stimulus movement, as observed at the level of two selectively connected interneurons of the ventral cord (Fig. 26.11A). Even though the arthropod nervous system is generally economical in establishing neuronal circuitry, this clear-cut separation of subsystems is fitted with a circuit providing additional contrast between both (Fig. 26.11B).

Localization of a source of medium displacement would require only one of the subsystems. The presence of the second set of receptors and interneurons and of a circuit providing contrast enhancement between the two subsystems requires explanation. The adaptive significance of this particular arrangement has to be identified. Is it perhaps an adaptation to special properties of the near field displacement signals in water?

## Discussion

### Displacement Receptors in the Water and Air

The mechanosensitive sensilla on the crayfish telson and in the lateral line system of fish and amphibians have a common feature. In both systems the detector consists of two, or two groups of, sensory cells. These are both mechanically coupled by their common attachment to the base of the hair shaft or the cupula. This arrangement appears to be specific for the displacement-sensitive pathways that operate in water; corresponding pathways that operate in air (e.g., cercal receptors in cockroaches: Nicklaus 1965; wind-sensitive hairs on the head of the locust: Smola 1970; air movement–detecting sensilla on the body of caterpillars: Markl and Tautz 1975) have singly innervated sensilla. In systems that operate in air, the preferred direction of hair bending activating the sensory cell often is not identical from one sensillum to the next, but instead varies, so that the population of sensilla is able to provide sensitivity to displacements in all directions,

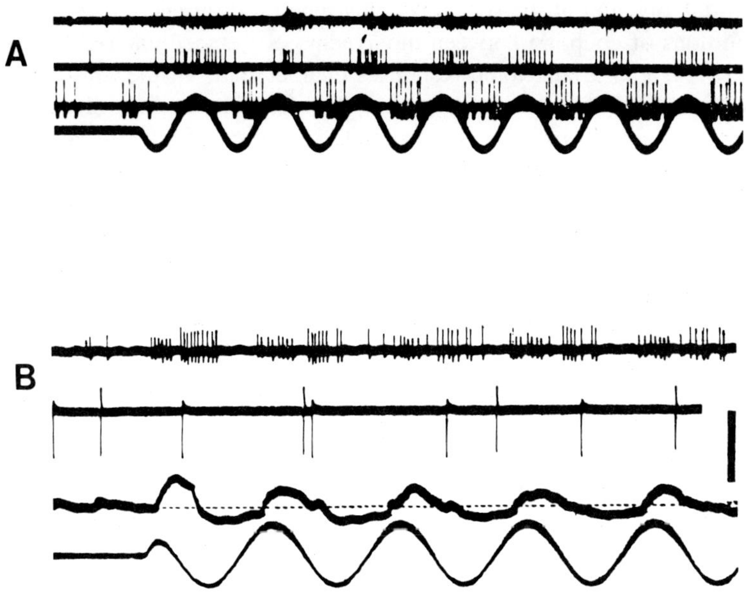


Figure 26.11(A). Directionally selective response of ventral cord interneurons in simultaneous recording from two ascending interneurons answering to opposite phases of fluid motion. *Top trace* monitors activity of sensory cells in fourth root of ganglion 6; larger spikes are in response to tailward movements of the medium. Wave motion in 5 Hz. (B) The activation of single sensillum with wire hook that vibrates at 5 Hz and a record of the activity of two connected sensory cells. Intracellular record from dendritical region of an interneuron shows depolarization in response to the activity of one sensory cell, but also shows that there is hyperpolarization during opposite phase of fluid motion. This inhibitory contrast enhancement suggests that there is special significance for discrimination between two opposed directions of medium movement; *calibration bar* at intracellular record, 10 mV.

even though the individual sensillum is distinctly deaf to specific movement directions. Only one exception to the rule has yet been reported: trichobothria in spiders have three or more sensory cells to each sensillum (Goerner and Andrews 1969).

Here I shall present a hypothesis for the functional significance of the dual innervation of sensilla designed to monitor water displacements, which has not been fully recognized.

## Stimulus Intensity Coding

To code the intensity of water movement, the sensitivity of receptors to the velocity component of the stimulus shows a wide scatter (Fig. 26.5). The response variation seems to correlate with the structure of the outer segment of the sensilla, which range from long and feathered to short and naked structures barely protruding from the opening of the socket. Such structural differences dramatically alter the conditions of coupling between movements of the medium and the hair shaft (Tautz 1979). Phasic and tonic responses characteristic of the sensory cells and variation of the efficacy of their synaptic coupling to interneurons (Kennedy 1974) are additional aids to providing appropriate increases of afferent activity as the stimulus increases from slow movements in the medium to direct vibrations of the crayfish's body.

## Signal Frequency Range in This Pathway

Tautz and Sandeman (1980) report an effective frequency bandwidth of 20–500 Hz for sensilla on the chelae of *Austropotamobius* (*Orconectes*), and a similar range of 40–1,000 Hz has been found in the wind-sensitive hairs of the caterpillars Barathra brassicae (Markl and Tautz 1975). This similarity occurs because, close to the animal, the medium, though attenuated, will follow the movements of the displacing object irrespective of its density. Direct observation of vibrating particles at defined distances from a vibrating sphere show that, up to 10 mm from the surface of the sphere, the attenuation of displacement is independent of frequency (Wiese, Wollnik, and Jebram 1980), whereas in surface waves, frequency dependence is very expressed (see Chap. 25).

## The Directionality of Response in This Pathway

The receptor cells of the sensilla of the fourth root of ganglion 6, positioned on the lateral edge of the telson, are strongly directional (Fig. 26.7), but the interneurons in general are much less so, probably because of the convergence of many receptors from many individual locations. In fact, many of the interneurons receive input from several segments, and so they show a loss of directionality of response. Parallel to this effect, they become unable to preserve the temporal pattern of the stimulus (Kennedy and Mellon 1964; Sigvardt, Hagiwara, and Wine 1982). To compensate for this lack of precision in the directionality of individual interneurons, a monitoring of more than one of these interneurons at a more central level is required to gain precise information of where in the surroundings the vibration is generated.

According to the information at hand, the crayfish is most sensitive to displacement within the same plane of the water and less sensitive to disturbances coming from above or below. Curvature of skeletal plates and differences in the inclination of the sensilla provide for variation of directionality of receptors in space. This makes it highly unlikely that a potential attacker (e.g., a kingfisher diving from above) would be able to approach unnoticed.

The same line of thought, however, reveals that the conspicuous presence of dual innervation of receptors and functional bipolarity constitutes a refinement in this sensory pathway that, though it is clearly related to directionality, is not self-evident as an a priori essential component to localize signal source. More precise reasons for its presence have to be formulated.

## Neuronal Contrast Enhancements in the Pathway

The frequency content of signals and the direction from which they approach are the focal points of sensory processing in many ascending neural pathways. As to the frequency domain, there is at least some evidence that neuronal circuitry, among other mechanisms, provides for frequency band–specific responses of auditory interneurons at some lower levels of the mammalian brain (e.g., see Pickles 1982). Konishi (1983) confirmed the essential role of binaural neural interaction in the process of sound source localization. The presence of functionally homologous circuits of contrast enhancement in the CNS of even small invertebrates (auditory pathway of insects: Suga 1963; Kalmring, Rheinlaender, and Römer 1972; Wiese and Eilts-Grimm 1985; mechanosensory pathway of crustaceans: Reichert, Plummer, and Wine 1983) supports the notion that such circuits are an essential part of the sensory pathway. With this conclusion in mind, it appears obvious that the circuit of inhibitory contrast enhancement that differentiates between headward- and tailward-directed water movements must provide some significant contribution to the sensory information processing done in this pathway.

## Dual Innervation of Receptors and Directional Subchannels in the Pathway

As to rate from the directionality of mechanosensory interneurons in this sensory system (Fig. 26.8) of the crayfish, the dual innervation of sensilla and the corresponding two subchannels of the pathway are not essential to the processing of signal source localization. The exclusive occurrence of dually innervated displacement receptors in connection with the water medium suggests that the physical properties of water are responsible for the introduction of this particular type of mechanoreceptor construction. The following consideration, though it contains many hypothetical elements, is meant to draw attention to one potential efficiency of this type of sensor. The suggestion is based on (1) the strong coupling of the sensillum to movements of high-density media, (2) the assumption that biologically relevant displacement stimuli consist of two distinct components, and finally (3) the necessity for this sensory system to prevent threatening sensory blackout.

### Strong Coupling of Movements to the Sensillum in High-Density Media

The hinges of hair-type sensilla have a certain spring tension that keeps the sensillum upright during rest. In a low-density medium, such as air, the deflecting forces provided by the frequency-dependent coupling to movements of the medium are often of a magnitude comparable to the resetting spring forces. This leads to a deflection of the sensillum during the onset of a movement of the medium. A softened coupling during ongoing movement (containing only lower frequencies) allows the hair to return soon to the resting position (see Tautz 1979). On the other hand, the greater mass of the water medium tightens the coupling between medium and sensillum. A return of the latter to the resting position during ongoing movement of the medium will be generally impossible except for very low movement velocities. Enforced, lasting deflection, however, has effects on the power of a sensillum, especially of a singly innervated one, to encode displacements (Fig. 26.4).

## Biologically Relevant Displacement Stimuli Probably Consist of Two Components

Movements of medium-displacing bodies cause several effects. Medium oscillations (AC component) will occur, generated by the motor apparatus of locomotion or by any other motion of body appendages. In a recent review of vibrational communication (Markl 1983), it has been stressed that such sources of displacements can be compared with the theoretically well-defined dipole source of medium vibration. The technical structure of such a source is the oscillating sphere moving along one axis only.

In addition to medium oscillations, mass movements of the medium (DC component) will occur. A first source of such mass movements of the medium is that each moving object drags about three times its volume of the surrounding medium to follow its motion. Motion such as the velocity of the walking crayfish or swimming fish is only detectable close to the moving object. A second source of medium movement that goes in one defined direction is the vibrating dipole source itself. This effect is known as "secondary flow" or "acoustic streaming" (Schlichting 1932, 1979) and consists of a flow of the medium away from the oscillating sphere in both directions of the oscillatory movement. This secondary flow has recently been discussed in connection with the stimulation of cochlear hair cells in the vertebrate ear (Hassan 1985). Lastly, a mass movement of the medium could occur in connection with body movements of the receiver organism during stimulus reception, and mass movements could be provided by the turbulent flow of water, as in a creek.

From all the events during the generation of medium displacements and within the habitat of the animals, each one contributing to the specific composition of displacement signals in water, I shall draw a still tentative conclusion: The dual innervation feature of some receptors in vertebrates and invertebrates monitoring water displacements reflects a need of this mechanosensory pathway to process simultaneously a lasting DC component of medium movement and a superimposed oscillatory AC component. A singly innervated receptor could encode the DC component of the stimulus movement because, even when the sensory cell hyperpolarizes (Flock 1967), the follower neurons will experience reduction of spontaneous activity as a result of the stimulus. Such reduction of spontaneous activity, however, renders impossible the encoding of the superimposed AC component. Thus a second sensory cell having opposite directional preference is required to prevent a threatening sensory blackout and to properly transform oscillatory displacements under two directions of medium flow into a corresponding neuronal activity pattern.

The prospective significance of the two components is, in the case of the DC component, the information on direction of movement of the displacing object, and in the case of the AC component, it is the information on the specific nature of the signal source.

## Summary

The highly abstracting neuronal circuitry of a sensory pathway monitoring water displacements in the crayfish *Procambarus* is described with special emphasis on the neuronal inhibitory contrast enhancements established. With respect to the dual innervation of receptors in this pathway, obviously in relation to the medium of water the hypothesis is established that naturally occurring displacement signals must be composed of a directed flow (DC) component and a superimposed vibratory (AC) component and that simultaneous recording of both components requires, for reasons of potential sensory blackout, dual innervation of receptors.

*Acknowledgments.* Financial support of the work on crayfish sensory pathways in this lab has been supplied by DFG Grants Wi-363/4,5,7-5, 7-6, and 8-1. I further want to thank Dr. R. Schultz for his energetic and productive cooperation.

### References

Calabrese, R.L. (1976) Crayfish mechanoreceptive interneurons I, J. Comp. Physiol., 105:83–102.

Ebina, Y., and Wiese, K. (1984) A comparison of neuronal and behavioural thresholds in the displacement-sensitive pathway of the crayfish *Procambarus*, J. Exp. Biol., 108:45–55.

Eilts-Grimm, K., and Wiese, K. (1984) An electrical analogue model for frequency dependent lateral inhibition, Biol. Cybern, 51:45–52.

Flock, A. (1967) Ultrastructure and function in the lateral line organs, in Lateral Line Detectors, Cahn, P. (ed.), Indiana University Press, Bloomington, pp. 163–198.

Goerner, P., and Andrews, P. (1969) Trichobothrien ein Ferntastsinnesorgan bei Webespinnen, Z. Vgl. Physiol., 64:301–317.

Hartline, H.K. (1940) The nerve messages in the fibers of the visual pathway, J. Opt. Soc. Am., 30:239–247.

Hartline, H.K., and Ratliff, F. (1972) Inhibitory interaction in the retina of *Limulus*, in Handbook of Sensory Biology VII/2, Fuortes, M.G.F. (ed.), Springer-Verlag, Berlin, 381–448.

Hassan, El-S. (1985) A suggested role for secondary flow in the stimulation of the cochlear hair cell, Biol. Cybern., 53:109–119.

Heinisch, P., and Wiese, K. (1983) Perception of space by a sensory system for water vibration in the crayfish *Procambarus*, Verh. Dtsch. Zool. Ges., 1983:191.

Hubel, D.H., and Wiesel, T.N. (1959) Receptive fields of single neurons in the cat's striate cortex, J. Physiol., 148:574–591.

Huber, F. (1983) Neural correlates of orthopteran and cicada phonotaxis, in Neuroethology and Behavioral Physiology, Huber, F., and Markl, H. (eds.), Springer-Verlag, Berlin, pp. 108–135.

Hughes, G.M., and Wiersma, C.A.G. (1960) Neuronal pathways and synaptic connections in the abdominal cord of the crayfish, J. Exp. Biol., 37:291–307.

Kalmring, K., Rheinlaender, J., and Römer, H. (1972) Akustische Neuronen im Bauchmark von *Locusta migratoria*, J. Comp. Physiol., 80:325–352.

Kennedy, D. (1971) Crayfish interneurons, Physiologist, 14:5–30.

Kennedy, D. (1974) Connections among neurons of different types in crustacean nervous systems, in The Neurosciences, Third Study Program, Schmitt, F.O., and Worden, F.G. (eds.), Massachusetts Institute of Technology Press, Cambridge, Mass., pp. 379–388.

Kennedy, D., and Mellon, DeF. (1964) Synaptic activation and receptive fields in crayfish interneurons, Comp. Biochem. Physiol., 13:275–300.

Konishi, M. (1983) Neuroethology of acoustic prey localization in the barn owl, in Neuroethology and Behavioral Physiology, Huber, F., and Markl, H. (eds.), Springer-Verlag, Berlin, pp. 303–317.

Krasne, F.B., and Wine, J.J. (1984) The production of crayfish tailflip escape responses, in Neural Mechanisms of Startle Behavior, Eaton, R.C. (ed.), Plenum, New York, pp. 179–211.

Laverack, M. (1962) Responses of cuticular sense organs in the lobster *Homarus*. I. Hair peg organs as water current receptors, Comp. Biochem. Physiol., 5:319–325.

Markl, H. (1973) Leistungen des Vibrationssinnes bei wirbellosen Tieren, Fortschr. Zool., 21:100–120.

Markl, H. (1983) Vibrational communication, in Neuroethology and Behavioral Physiology, Huber, F., and Markl, H. (eds.), Springer-Verlag, Berlin, pp. 332–352.

Markl, H., and Tautz, J. (1975) The sensitivity of hair receptors in caterpillars of *Barathra brassicae* to particle movement in a sound field, J. Comp. Physiol., 99:79–87.

Mellon, deF. (1963) Electrical responses from dually innervated tactile receptors in the thorax of the crayfish, J. Exp. Biol., 40:137–148.

Münz, H. (1985) Single unit activity in the peripheral lateral line system of the cichlid fish *Sarotherodon niloticus,* J. Comp. Physiol., 105:83–102.

Nicklaus, R. (1965) Die Erregung einzelner Fadenhaare von *Periplaneta* in Abhängigkeit von der GröBe und Richtung der Auslenkung, Z. Vgl. Physiol., 50:331–362.

Pickles, J.O. (1982) An introduction to the physiology of hearing, Academic Press, London, pp. 180–188.

Plummer, M.R. (1986) Frequency coding of water borne vibration by abdominal mechanosensory interneurons in the crayfish *Procambarus clarkii,* J. Comp. Physiol., 107 (in press).

Reichert, H., Plummer, M.R., and Wine, J.J. (1983) Lateral inhibition mediated by a non-spiking interneuron: circuit properties and consequences for behavior, J. Physiol. (Paris), 78:786–792.

Schlichting, H. (1932) Berechnung ebener periodischer Grenzschichtströmungen, Physiol. Z. 33:327–336.

Schlichting, H. (1979) Boundary layer theory, 7th ed., Mc-Graw Hill, New York.

Sigvardt, K.A., Hagiwara, G., and Wine, J.J. (1982) Mechanosensory integration in the crayfish abdominal nervous system: structural and physiological differences between interneurons with single and multiple spike initiating sites, J. Comp. Physiol., 148:143–157.

Smola, U. (1970) Untersuchungen zur Topographie, Mechanik und Strömungsmechanik der Sinneshaare auf dem Kopf der Wanderheuschrecke *Locusta,* Z. Vgl. Physiol., 67:382–402.

Suga, N. (1963) Central mechanisms of hearing and sound localization in insects, J. Insect Physiol., 9:867–873.

Tautz, J. (1979) Reception of particle oscillation in a medium: an unorthodox sensory capacity, Naturwissenschaften, 66:452–461.

Tautz, J., and Sandeman, D. (1980) The detection of water borne vibration by sensory hairs on the chelae of the crayfish, J. Exp. Biol., 88:351–356.

von Békèsy, G. (1967) Sensory Inhibition, Princeton University Press, Princeton, N.J.

Wiersma, C.A.G., and Hughes, G.M. (1961) On the functional anatomy of neuronal units in the abdominal cord of the crayfish *Procambarus clarkii* (Girard), J. Comp. Neurol., 116:209–228.

Wiese, K. (1976) Mechanoreceptors for near-field water displacements in crayfish, J. Neurophysiol., 39:816–833.

Wiese, K., Calabrese, R.L., and Kennedy, D. (1976) Integration of directional mechanosensory input by crayfish interneurons, J. Neurophysiol., 39:834–843.

Wiese, K., and Eilts-Grimm, K. (1985) Evidence for matched frequency dependence of bilateral inhibition in the auditory pathway of *Gryllus bimaculatus*, Zool. Jahrb. Physiol., 89:181–201.

Wiese, K., and Schultz, R. (1982) Intrasegmental inhibition of the displacement sensitive pathway in the crayfish (*Procambarus*), J. Comp. Physiol., 147:447–454.

Wiese, K., and Wollnik, F. (1983) Directionality of displacement sensitive interneurons in the ventral cord of *Procambarus*, Zool. Jahrb. Physiol., 87:461–475.

Wiese, K., Wollnik, F., and Jebram, D. (1980) The protective reflex of *Bowerbankia* (Bryozoa): calibration and use to indicate movements of the medium beneath a capillary surface wave, J. Comp. Physiol., 137:297–303.

Wine, J.J., and Krasne, F.B. (1982) The cellular organization of crayfish escape behavior, in The Biology of the Crustacea, vol. 4, Sandeman, D.C., and Atwood, H.L. (eds.), Academic Press, New York, pp. 241–292.

Wohlers, D., and Huber, F. (1982) Processing of sound signals by six types of neurons in the prothoracic ganglion of the cricket *Gryllus campestris*, J. Comp. Physiol., 146:161–173.

Part IIIC-2

# Hearing

It is difficult to define "hearing." It may be defined by the stimulus or by the receptor organ, and the answer will not always be the same from different investigators. Terrestrial vertebrates have specialized receptor structures for the detection of propagated pressure waves that are different from the structures found in fishes. This section deals with otolith organs of fish and the ability of the otolith organs to extract acoustic information from the environment. The key issues center around if and how sound pressure, displacement, and direction are measured. The nature of sound in water is described by Kalmijn (Chapter 3) and Rogers and Cox (Chapter 4). The behavioral responses to, and use of, sound is discussed by Blaxter (Chapter 7) and by Hopkins (Chapter 8).

Popper, Rogers, Saidel, and Cox (Chapter 27) give an overview of the morphology and innervation of the fish ear, and provide a new hypothesis concerning how the fish ear and CNS may analyze sound and extract acoustic information. A more detailed analysis of the physiological responses of the ear and eighth nerve to sound, particularly with regard to directional information, is provided by Fay (Chapter 28). McCormick and Braford (Chapter 29) give an overview and evolutionary perspective of the central connections of the ear, lateral line and electrosensory systems in fishes.

# Chapter 27

# Role of the Fish Ear in Sound Processing

ARTHUR N. POPPER,* PETER H. ROGERS,† WILLIAM M. SAIDEL,* and MARDI COX †

Despite the considerable interest in fish hearing in recent years (see Schuijf and Hawkins 1976; Tavolga, Popper, and Fay 1981), the contributions of the fish ear or brain to hearing and sound processing remain relatively unknown. Our knowledge about the fish ear is more limited as compared to our knowledge of the ears of amphibians, reptiles, birds, or mammals. Extensive studies of representatives of these classes have demonstrated that their inner ears are involved in the analysis of sounds, rather than simply as a system for transduction of mechanical to electrical energy.

Current hypotheses on ear function in fishes are limited in scope and do not encompass much of our knowledge about fish hearing, auditory physiology, or the morphology of the ear (see Schuijf and Hawkins 1976; Tavolga, Popper, and Fay 1981). It is the intent of this chapter to present hypotheses regarding, and a model of, the fish ear and auditory system that takes into consideration the behavioral, physiological, and morphological data as we know it today. The model is an approach that could serve as the basis for future experimental studies of the fish auditory system. Our approach is to provide a basic background of the fish ear and hearing and then present our hypotheses, first dealing with sound localization and then with frequency analysis. (Rogers, Popper, Cox, and Saidel in preparation for the detailed derivation of the model.)

## *The* Fish Ear!

We have stated (e.g., Platt and Popper 1981) that we are extremely cautious in discussing *the* fish ear. There are over 20,000 extant species of bony fishes (class Osteichthyes), and we are aware of considerable interspecific variation in

* Department of Anatomy and Cell Biology, Georgetown University Schools of Medicine and Dentistry, Washington, D.C.
† School of Mechanical Engineering, Georgia Institute of Technology, Atlanta, GA 30332, USA.

the morphology of the ear (Retzius 1881; Popper and Coombs 1982). Although we do not know the functional significance of the variation, it is usually associated with the auditory, rather than the vestibular, portion of the ear. Until we understand the significance of this variation, we are reluctant to generalize among species about the role of the ear in sound processing.

Despite this caveat, the ideas expressed in this chapter will refer to a ''generalized'' fish ear. In fact, as we point out toward the end of the chapter, we have reason to suggest that our hypotheses are suitable for all bony fishes, and that the morphological variations encountered among different species only represents variations on the themes presented in this chapter.

## Otolith Organs—Structure

The fish ear (Fig. 27.1) has three semicircular canals with associated cristae and three otolith organs—the saccule, lagena, and utricle (see Platt and Popper 1981 and Popper 1983 for a detailed description). Each otolith chamber contains a single otolith lying close to the sensory epithelium (macula), which is part of the medial wall of the chamber. The macula and otolith are separated by a thin gelatinous structure, the ''otolith membrane''.

The sensory epithelium is composed of sensory and supporting cells (Fig. 27.2). The apical surface of the sensory hair cells has a ciliary bundle that projects into the otolith membrane and toward the otolith. The ciliary bundle contains a single, eccentrically positioned, true cilium (the kinocilium) and a large number of stereocilia. The adequate stimulus for the sensory hair cell is the bending of the ciliary bundle (Hudspeth and Corey 1977). Bending of the stereocilia along the axis from the stereocilia towards the kinocilium results in maximum depolarization of the sensory cell and a concomitant increase in spike activity in the afferent fiber. Bending in the opposite direction causes hyperpolarization and a decrease in spike activity. Bending in any direction off this axis results in a response that is a cosine function of the direction of bending (Hudspeth and Corey 1977).

## Pathways for Sound to the Ear

Before considering the processing of sound energy by the ear, we shall briefly describe how sound reaches the ear. Sound is a pressure wave with an accompanying particle motion (see Chaps. 4 and 5). It is likely that sound reaches the ear by at least two ''pathways''. Von Frisch (1938) suggested that sound pressure sets the wall of the swim bladder, a gas-filled chamber in the abdominal cavity of many fish species, into motion, owing to the relatively high compressibility of the gas. The vibration of the swim bladder wall reradiates sound energy in the form of particle displacement, which causes motion of the otolith relative to the sensory hair cells. Otolith motion produces a shearing of the ciliary bundles and a physiological response from the sensory hair cells. This ''indirect'' path for sound to the ear is probably found in most species with a swim bladder.

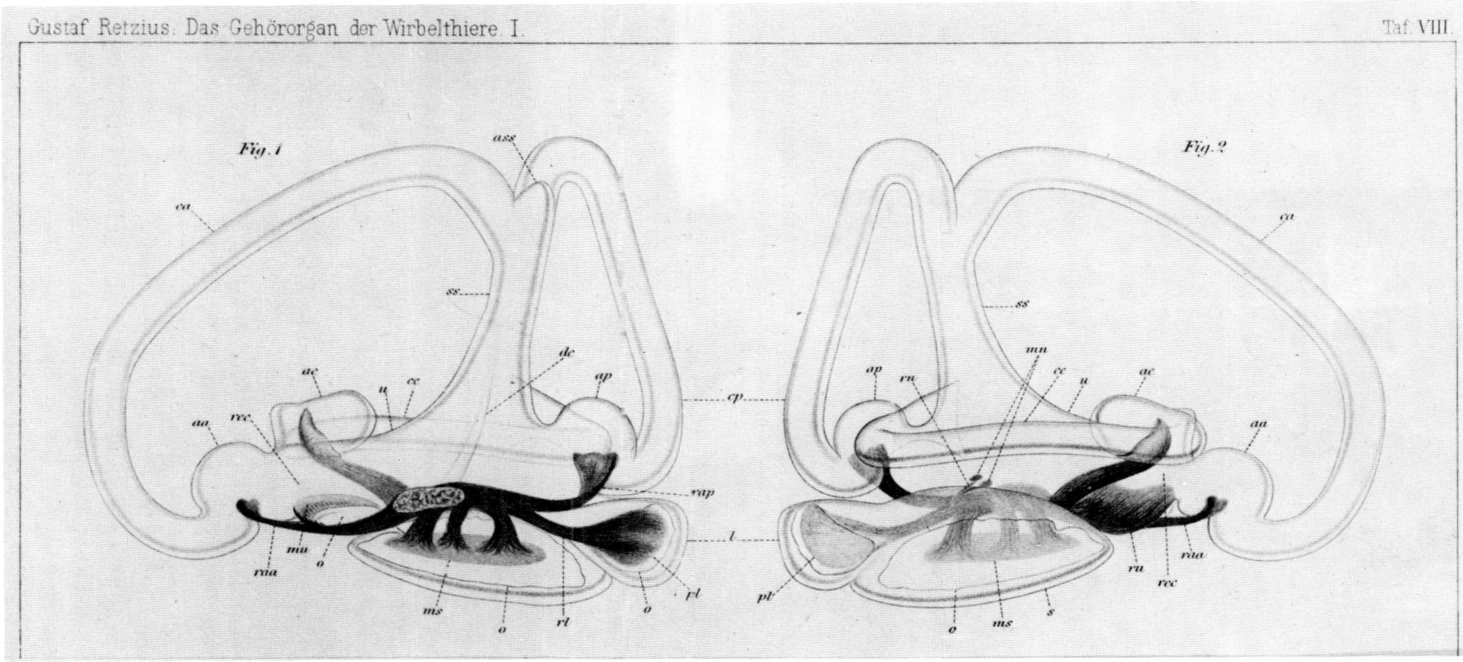


Figure 27.1. Drawing, from Retzius (1881), of medial side (*left*) and lateral side (*right*) of the ear of tuna *Scomber scomber*. Fish ears have three semicircular canals (*ca, ce, cp*) and associated ampullae (*aa, ae, ap*) and three otolith organs, the saccule (*s*), lagena (*l*), and utricle (*u*). Many species also have a small macula neglecta (*mn*), which is a sensory receptor of unknown function. Each of the otolith organs contains a single dense, calcareous otolith (*o*). A sensory epithelium (or macula) is found on the wall of each otolith chamber. Sensory maculae and cristae of the semicircular canals are innervated by branches of the eighth cranial nerve. In particular, note the large branch of the nerve innervating the saccular macula (*ms*); *aa, ae, ap*, ampullae of semicircular canals; *ass*, apex sinus superioris; *ca, ce, cp*, semicircular canals; *de*, endolymphatic duct; *l*, lagena; *mn*, macula neglecta; *ms*, saccular macula; *mu*, utricular macula; *o*, otoliths; *pl*, lagenar macula; *raa, rae, rap*, branches of eighth nerve to ampullae; *rec*, utricular recess; *rl*, lagenar branch of eighth nerve; *ru*, utricular branch of eighth nerve; *s*, saccule; *ss*, common canal; *u*, utricle.

Figure 27.2. Scanning electron micrograph of the apical surface of the lagenar macula from the goldfish, *Carassius auratus*. A ciliary bundle, consisting of a single true cilium, the kinocilium (*K*), and many stereocilia (*S*), is located at the apical end of each sensory hair cell. The ciliary bundles are organized so that all of the bundles in one region of a sensory epithelium have their kinocilia on the same side of the ciliary bundle. This figure shows the uneven dividing region between hair cells oriented in two opposing directions. The *open arrows* show the orientations of several of the ciliary bundles.

The indirect signal may be further augmented for those species which possess a connection between the swim bladder and the inner ear, such as the Weberian ossicles in members of the suborder Otophysi (goldfish, catfish, carp, and relatives), or for those species in which an air cavity is located adjacent to the ear (e.g., some squirrelfish, bubble nest builders, clupeids).

A ''direct'' method for stimulation of the ear is detection of the particle motion of an acoustic disturbance. Since the fish's density is nearly the same as that of water, the fish moves in the sound field as if it were part of the water mass. In contrast, the otolith is three times denser than water and the fish's body. Consequently, the otolith moves with a smaller amplitude and a different phase from the rest of the body. Since the otolith and sensory epithelium are not rigidly tied to one another, but are connected only by the otolith membrane, each is displaced in a sound field relative to the other. As a consequence of the coupling of the cilia to both the sensory epithelium and the otolith (either directly by contacting the otolith or via the otolith membrane), the cilia bend during any particle displacement stimulation.

The direct pathway for acoustic stimulation should operate in all species, including those with an indirect pathway. The two pathways are also complementary, since the indirect pathway is more effective above perhaps 200 Hz while the direct pathway is most effective below about 200 Hz (Fay and Popper 1974, 1975; Coombs and Fay 1984; see also Chap. 28).

## What Do We Know About Signal Analysis?

Behavioral experiments (reviewed in Popper and Fay 1973, 1984 and Fay and Popper 1983) have shown that many fish species can perform frequency and intensity discrimination for pure tone sounds, as well as detect the presence of a signal even when masked by background noise. The goldfish (*Carassius auratus*), the most widely studied species, can discriminate small differences in pulse repetition rate (Fay and Passow 1982) and detect amplitude modulation with extremely small modulation indices (Fay 1980). For these particular discriminations, Fay's data suggest that the goldfish processes only the envelope of an acoustic signal (peak picking) and determines pulse rate by counting the number of pulses in a given time period (Rogers, Popper, Cox, and Saidel in preparation). This "time domain" processing cannot explain the ability of the goldfish to discriminate sinusoids of different frequencies presented in alternating tone pulses, since the envelope of each burst is flat. Moreover, assuming that the goldfish does not employ peak picking for sinusoidal signals, the counting algorithm (which explains the pulse rate discrimination experiments) would result in a minimum detectable change in frequency that is independent of frequency. This is inconsistent with the observed increase in minimum detectable change in frequency of 6 dB/octave (Jacobs and Tavolga 1968; Fay 1970; Rogers et al. in preparation for details). The mechanism by which a goldfish discriminates frequency thus remains uncertain.

It is important to recognize the distinction between "frequency discrimination" and "spectral analysis." *Frequency discrimination* (i.e., the ability to tell the difference between two different frequencies) can be accomplished using spectral analysis (i.e., the determination of the spectral components [Fourier transform] of a complex time signal) but it can also be accomplished by means of algorithms based on counting, interval timing, autocorrelation, or, as we shall suggest further on, a comparison of a signal with its time derivative. *Spectral analysis* requires either a system of band-pass filters (such as that present in the organ of Corti) or computationally intensive numerical algorithms such as the Fast Fourier Transform (FFT). Frequency discrimination in mammals is attributed to spectral analysis by the spatially distributed band-pass filters along the organ of Corti. No species of fish is likely to have an ear that is able to support an analogous place-like mechanism. As we shall point out later, however, it is potentially possible for a fish's otolith organs to support some sort of rudimentary frequency-dependent, spatially distributed responsive areas for spectral analysis.

## Mechanisms of Inner Ear Stimulation

The remainder of this chapter will consider several ideas about signal detection and processing by the fish ear. Our hypothesis is speculative, but based on acoustical biophysics. As we shall show, much of the diverse fish hearing data are consistent with our hypothesis. It is also important to keep in mind that our hypothesis is not all-inclusive with regard to functions of the fish auditory system. Instead, it is complementary to other aspects of fish ear function, as described in the section on frequency analysis.

Put in the simplest terms, our hypothesis states that much of the analysis of sound by the fish auditory system involves detection of direct and indirect signals by the ear and the way in which the ear is ''wired'' to the CNS. We argue that the processing involved with the analysis is implicit in the ''wiring'' of the hair cells by the eighth nerve fibers. This wiring is in itself able to separate the direct and indirect sound input mathematically and provides the CNS with signals related to the acoustic pressure and each of the three vector components of the acoustic particle velocity. These signals can then be used to extract information about sound source direction and sound source distance, as well as information that can be used for frequency and intensity discrimination, and to determine the presence of signals masked by other sounds. The model we shall propose below is based on this hypothesis, and the model leads to additional suggestions regarding the way(s) in which the fish auditory system can deal with sound.

We shall concentrate on three possible contributions of fish ears to hearing–sound localization, enhancement of the signal-to-noise (S/N) ratio, and frequency discrimination. It will become apparent that our hypothesis, which was derived to encompass these three aspects of sound processing, also describes a mechanism that provides the CNS with information for more complex temporal analyses, such as discrimination of biologically significant sounds that differ only in the intervals between pulses (Spanier 1979; Myrberg and Spires 1980), or for determination of sound source distance.

## Sound Localization and Enhancement of Signal-to-Noise (S/N) Ratio

In the following sections we shall consider two aspects of sound analysis, enhancement of the S/N ratio and localization of sound. We shall also consider other hypotheses for sound localization, including the important hypotheses that have led to our current ideas.

### Enhancement of S/N Ratio

The sensitivity of biological (or engineered) underwater acoustic detectors is rarely the limiting factor in detectability of a signal. The underwater environment is characterized by a substantial level of more or less omnidirectional ambient noise

(Wenz 1964); it is the *ratio* of the signal of biological interest to this background noise (Chap. 5) that limits signal detection. Large arrays of omnidirectional hydrophones are used by engineers to detect a signal in the presence of noise and determine sound source bearing. The "directivity index" (or enhancement of S/N) of the array provides the measure of the discrimination. Since most fish are much smaller than the acoustic wavelength, conventional "beam forming" of this kind is impossible. Small receivers can be engineered to produce cosine ($\theta$) (e.g., dipole) beam patterns that, if aimed in the direction of the signal, will produce a directivity index of about 5 dB. Thus, vectorial decomposition, which is proposed as the mechanism for sound source localization, also provides a means for enhancing signal detectability.

## Sound Source Localization: Vectorial Analysis

We have only recently started to understand the mechanism(s) employed by fishes for sound source localization (see Chap. 28). It is now theoretically (van Bergeijk 1964) and experimentally (Schuijf and Buwalda 1980) clear that the mechanism by which fish localize sound must differ substantially from the mechanism(s) employed by terrestrial vertebrates. In this section, we shall discuss a number of points with regard to fish hearing and sound source localization, as well as the important models proposed by Schuijf and Buwalda (1975, 1980; also see Buwalda 1981; Schuijf 1981), which involve simultaneous detection and analysis of acoustic pressure and particle displacement for directional analysis. We shall present a major modification of this model to incorporate data from the morphology of the fish ear and from behavioral and physiological experiments on fish hearing.

### Vectorial Decomposition of Particle Motion

The forces due to the acoustic field cause motion of the otoliths. The relative motion between the otolith and the fish's body differs for each sound source direction. When the direction of the relative motion shifts, the pattern of stimulation of the hair cells does so as well.

Sensory hair cells are oriented in four directions in the majority of fish species (Fig. 27.3) (Popper and Coombs 1982). Consider only the direct signal. When the sound source is positioned on the horizontal plane of the animal, the motion of the otolith will also be along this plane. When the axis of motion coincides with the axis of the horizontally oriented cells, these cells are maximally stimulated ($\cos \theta = 1$). The vertically oriented cells would give little response. If the stimulus were along the vertical axis, the vertically oriented hair cells would give the maximum response, and the response of the horizontal cells would be minimal. Stimuli in other directions would produce responses from the hair cells that are proportional to the appropriate cosine. *The result of this vectorial analysis is that the acoustic particle acceleration is decomposed into components along the directionally polarized axes of the otolith organs.* Consequently, there is a unique

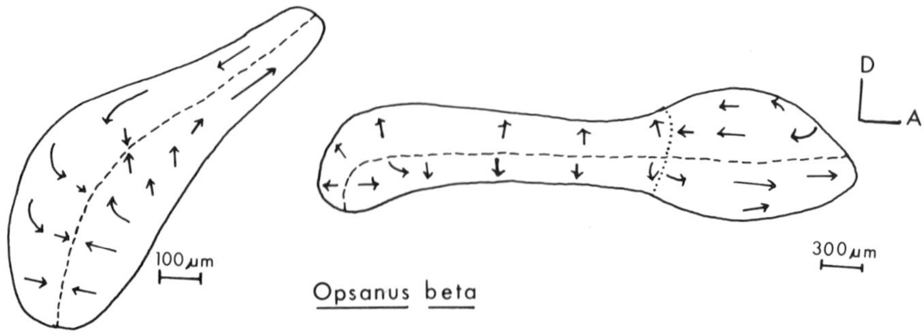


Figure 27.3. Diagramatical illustration of hair cell orientation patterns on the saccule (*right*) and lagena from the toadfish *Opsanus beta. Dashed lines* indicate approximate divisions between hair cell groups oriented in different directions. As in Figure 27.2, *arrows* indicate the position of the kinocilium relative to the stereocilia on hair cells in each region of the maculae. The saccular macula in *Opsanus,* as in many other teleost fish, has four discrete hair cell groups, two oriented rostrocaudally and two dorsoventrally. The lagena has two orientation groups.

ratio of responses from the perpendicularly oriented hair cell groups in each saccule for any sound source direction.

## The 180° Ambiguity

This simple response pattern is obviously complicated by the presence of the indirect signal and various methods for sorting out the direct and indirect components have been suggested (Buwalda 1981). In addition, though many fishes might localize a sound using vectorial analysis in this way, Schuijf and Buwalda pointed out that the ear would produce an identical response ratio for sounds from directly opposite directions. Schuijf and Buwalda demonstrated that fishes resolved this 180° ambiguity problem by comparing the phases of pressure via the indirect route and particle displacement via the direct route (e.g., Schuij f 1975; Schiujf and Buwalda 1975; Buwalda, Schuijf, and Hawkins 1983). According to Schuijf and Buwalda, the ambiguity could be resolved by comparing the phases of the pressure and particle displacement and determining which leads and which lags. Since the pressure component is detected by the indirect route and particle displacement by the direct route, the two components can always be distinguished from one another. Furthermore, since the phase difference is always greater than 90°, which leads and which lags is always apparent.

Although the basic idea for resolution of the 180° ambiguity presented by Schuijf and Buwalda is probably correct (and the behavioral data demonstrate that fishes do resolve the ambiguity [see Schuijf 1975; Schuijf and Buwalda 1975]), it should be pointed out that otolith organs do not respond to pressure, but to the relative displacement of the otolith induced by the vibration of the swim bladder in response to pressure. It follows that, for farfield signals (see Chap. 5), the direct and indirect signals are in quadrature (90° out of phase), which makes the discrimination between them, as proposed in the hypothesis of Schuijf and Buwalda, mathematically difficult when only comparing amplitude differences. Moreover, the very concept of "phase" implies sinusoidal signals, and most biologically relevant signals, including most fish communicative sounds (e.g., Spanier 1979), are nonsinusoidal. Any proposed mechanisms for resolving the 180° ambiguity must be applicable to nonsinusoidal signals (Rogers et al. in preparation for a detailed derivation of this model).

### The Role of Two Ears

Schuijf and Buwalda have shown that fish can eliminate the 180° ambiguity and presumably do so by resolving pressure and particle displacement (review in Schuijf and Buwalda 1980). They also showed that removal of one ear eliminates the ability to resolve the ambiguity and that the lateral line is not involved in this task. Although it is necessary to resolve pressure and particle displacement for a phase analysis, the specific mechanism(s) used by fishes for resolving pressure and particle velocity was postulated to be a function in the CNS (Buwalda 1981).

Schuijf and Buwalda proposed that interaural comparisons might separate the pressure and displacement components. Since the two ears in fishes are close to one another (and the interaural time is less than 10 μs in most cases), a signal reaching the fish from a sound source will be nearly identical in amplitude and time of arrival at both ears (van Bergeijk 1964). Each saccule is angled somewhat from the midline (up to 30° in many species) so that the rostral ends of the two saccules are directed away from one another (Fig. 27.4). Thus, even though the stimulus magnitude is essentially identical at each ear, the transduced responses by the two saccular sensory epithelia will differ. If the stimulus is on the horizontal plane, but not on the animal's axis of symmetry, particle displacement will cause motions that differ in the two ears (Fig. 27.4). The indirect component of the signal (reflecting the source pressure) from the swim bladder comes from behind and between the two ears. Not only would this signal be identical at the two ears, but the resulting output from the indirect signal from both ears would also be identical, since the indirect stimulus originates from the axis of symmetry of the saccule and the lagena (assuming linearity in the system so that it can decompose the direct and indirect signals [see p. 698]).

### The Role of the Central Nervous System

For vectorial analysis to function, the responses of the different hair cell groups must be maintained to a point in the CNS where the signals could be compared. Although physiological and neuroanatomical data are not extensive, it has been demonstrated that horizontally and vertically oriented cells on the saccular macula in species other than members of the order Otophysi are innervated by separate portions of the saccular branch of the eighth nerve (Saidel and Popper 1983a, 1983b; Jensen 1984; Mathiesen 1985). The data, though inferential, lead to the suggestion that this separation is maintained to the first synaptic site in the CNS (the octavolateralis region: see McCormick 1983 and Chap. 29). A single physiological study (Horner, Sand, and Enger 1980) has shown that cells in the midbrain

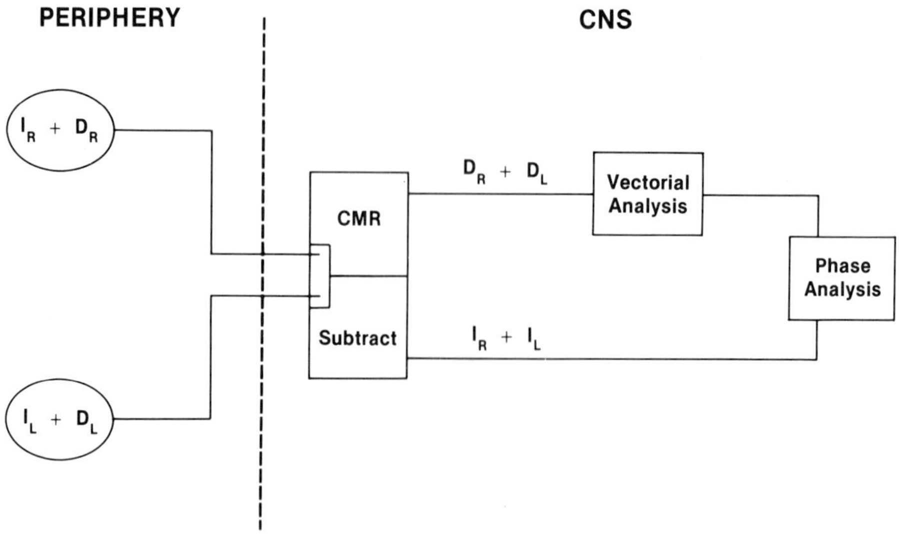


Figure 27.4. Schematic representation of the model for sound localization as proposed by Buwalda (1981). In this model, indirect (*I*) and direct (*D*) signals from each ear are combined in the *CNS* by two systems, one that allows for common mode rejection (*CMR*), producing an output that is proportional to the direct signal to the two ears, and another that does subtraction and gives an output proportional to the indirect signal from the two ears. The direct signal is used for vectorial analysis and the combined *D* and *I* signals for the phase analysis. The significant point of this model is that most of the processing occurs in the *CNS* and that the *CNS* is called on to do considerable mathematics to analyze the signal.

torus semicircularis respond to binaural signals, giving limited support to the hypothesis of comparator cells in the fish auditory system.

The output of each ear is the (assumed linear) sum of the signals from direct and indirect stimulation. Buwalda and his colleagues (Buwalda 1981; Buwalda, Schuijf, and Hawkins 1983) pointed out that, by combining the summed signal from the two ears, a fish may extract information on the direction of the sound source, as well as information sufficient to eliminate the 180° ambiguity. Only two comparisons need to be performed within the CNS, according to Buwalda (1981) (Fig. 27.5). One comparison is a common mode rejection whereby the similar aspects of the two inputs from the two ears would be eliminated. The output of this analysis (comparison) would be the difference in the direct signal between the two ears. This information could be used to determine the particle motion vector of the stimulus. A second arithmetical step resulting from the different orientations of the two ears to the midline could resolve the azimuth angle.

Another comparator could do a differential subtraction in which the difference signals from direct stimulation could be rejected. The output of this step would be proportional to the indirect signal. The phase of this signal could then be

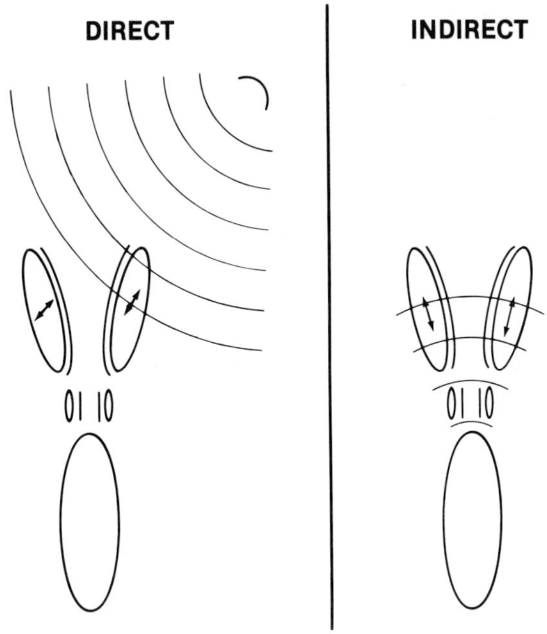


Figure 27.5. Diagrammatical representation showing the direction of sound reaching the ear via direct and indirect pathways. See the text for detailed explanation.

compared to the phase of the output of the common mode rejection system and provide the animal with the phase comparison necessary for the resolution of the 180° ambiguity.

The problem is more complicated than this, however, since the component of the particle displacement parallel to the midline is also identical for both ears. This difficulty can be overcome by a subtraction to determine the time between the saccular signals and the lagenar signals. This is possible, since the signal reaching the saccule and lagena via the indirect route is quite different from the signal detected directly. The indirect signal comes from the swim bladder, a structure that is generally closer to the lagena. The amplitude of the vertical component of the indirect signal from the swim bladder to the saccule is quite different from that to the lagena, even though these two organs are close to one another. Put into more quantitative terms, if the saccule and lagena are 1 mm apart and the lagena is 1 cm from the swim bladder, the difference in amplitudes of the signals at the two organs is 33% (by a cubic relationship). This is a consequence of the distance between the saccule and lagena being a significant fraction (one-tenth) of the distance between the lagena and the swim bladder.

In contrast, the direct signal comes from a considerably greater distance, even for ''nearfield'' sources. Consequently, the distance between the saccule and lagena is insignificant when compared to the distance to the sound source, and thus the vertical components from the direct signal will be the same (see the next section for a more extensive discussion of this concept). The appropriately weighted difference between the lagena and saccule provides an independent measure of the indirect signal and, hence, allows differentiation between the component of the direct signal along the midline and the indirect signal.

Although Buwalda and his colleagues proposed a version of the comparator mechanism in the CNS (Fig. 27.5) similar to the one described above, they did not consider the nature of the analysis that must be performed by the CNS to make the appropriate calculations. The CNS in their scheme is required to use an algorithm involving weighted sums and differences of the outputs from different regions of the two ears. These calculations must be precise to avoid subtraction error. In the next section we shall propose an alternative hypothesis: that the required high-precision arithmetic is performed, not in the CNS, but by the inherent wiring at the sensory epithelia.

Both Schuijf and Buwalda's model and ours are based on a similar premise. The major significant difference between the two models is that ours takes advantage of the "wiring" of the saccular and lagenar epithelia for the first major steps in signal processing, rather than in the CNS, as in Schuijf and Buwalda's model. In addition, our model includes various functions of the fish auditory system not addressed by Schuijf and Buwalda.

## A Model for Peripheral Processing of Directional Information

Although it is possible to place the burden of computation on the CNS, it can be demonstrated (Rogers et al. in preparation) that, with suitable wiring, the ear can perform all the calculations needed to separate the required component parts of the signal. Such a model should take into consideration the known features of the fish ear: (1) the presence of oppositely oriented sensory hair cells, (2) the large numbers of sensory hair cells, (3) the convergence of output from hair cells onto afferent fibers, (4) the increasing number of sensory hair cells with the animal's age, and (5) the generally complex morphology of the sensory epithelia and otoliths.

Let the $x$-, $y$-, and $z$-axes be the rostral-caudal (front–back), lateral (left–right) and dorsal (up–down) axes, respectively (Fig. 27.6). Any signal can be decomposed into three independent and orthogonal vectorial components, one each on the rostral, lateral, and dorsal axes of the animal.

In order to determine direction of a sound source, we want to determine $V_x$, $V_y$, $V_z$, and $p' = dp/dt$ (the pressure component) for the incident sound wave. We determine $p'$, rather than $p$, because the velocity of the otolith due to the indirect signal is proportional to the derivative of pressure, rather than pressure itself, and the otolith organs respond to this velocity (Rogers et al. in preparation for details).

We will assume a simplified ear with a saccule containing cilia oriented in the vertical and horizontal directions and a lagena with cilia oriented in the vertical direction only. We will also assume throughout that the hair cells (or groups of hair cells) respond linearly to the stimulus. (Hair cells that respond to the stimulus nonlinearly cannot be involved in the processes under consideration. Some functions of the ear, e.g., temporal or spectral analysis, however, do not require separation of the individual components of the sound field and can thus utilize information from nonlinear sensors with appropriate signal processing.)

In our model, let $S_V^{RS}$, $S_V^{LS}$, $S_H^{RS}$, and $S_H^{LS}$ be the vertical and horizontal

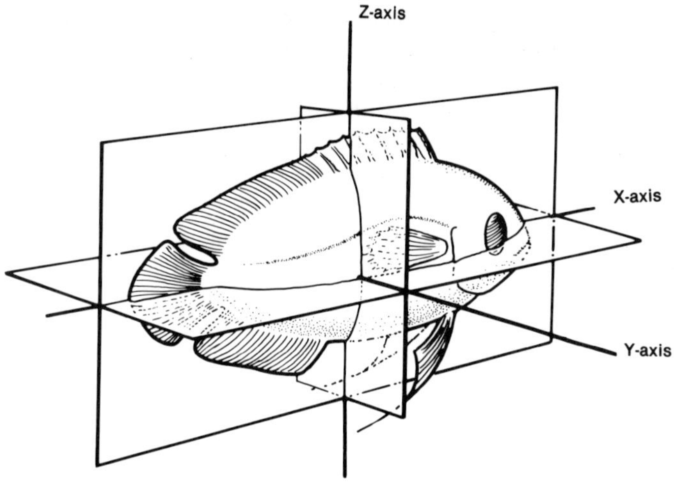


Figure 27.6. Diagrammatical representation to show three independent and orthogonal axes into which any signal can be decomposed.

signal *outputs* from the right saccule (RS) and left saccule (LS), and let $S_V^{LL}$ and $S_V^{RL}$ be the vertical signal *output* from the left lagena (LL) and right lagena (RL). From symmetry considerations, it is evident that $S_V^{RS} = S_V^{LS}$ and $S_V^{RL} = S_V^{LL}$.

It is easy to show (Rogers et al. in preparation) that the four independent signal outputs from the ear are related to the four field quantities ($V_x$, $V_y$, $V_z$, and $p'$) by a matrix equation of the form

$$\begin{pmatrix} S_H^{RS} \\ S_H^{LS} \\ S_V^{RS} \\ S_V^{RL} \end{pmatrix} = \begin{pmatrix} A & B & 0 & C \\ A & -B & 0 & C \\ 0 & 0 & D & E \\ 0 & 0 & F & G \end{pmatrix} \begin{pmatrix} V_x \\ V_y \\ V_z \\ p' \end{pmatrix} \qquad (27.1)$$

The seven independent matrix coefficients *A–G* represent responses to the *X*, *Y*, and *Z* vectors, as well as to the pressure component of the signal, and they depend only on the geometry and physical characteristics of the auditory organs (i.e., they are ''known'' quantities originating from the physiology of the peripheral ear structures).

From the form of the matrix we see that the response of the horizontally oriented sensory hair cells of both ears ($S_H^{RS}$ and $S_H^{LS}$) will be the same (*A* and *A* in the matrix), since the component of the signal on the animal's *x*-axis (along the midsagittal plane) will be the same for both ears. It also follows that the response to the *y*-axis component will differ for the horizontal cells (thus outputs *B* and $-B$). Finally, the responses of the horizontal cells of both ears will be 0 to the signal on the vertical *z*-axis, since, by definition, such cells are oriented in the horizontal plane and do not respond to vertical stimulation. It also follows that the response of the horizontal cells in both saccules will be the same ($C$) to the pressure component of the signal that comes from the swim bladder.

The vertical detectors of the ear ($S_V^{RS}$ and $S_V^{RL}$ and their equivalents in the left ear) would not detect any signals on the horizontal $x$- and $y$-axes, whereas they do respond to the $z$ component of the signal. Athough the $z$ component is essentially the same to both the saccule and lagena (see discussion on p. 695), the output from the saccule and lagena differs ($D$ and $F$), since the two organs are morphologically different from one another. Finally, since the amplitude of the indirect signal ($p'$) is different to the saccule and lagena (see discussion on p. 695), the output of the vertical cells in the two organs will differ ($E$ and $G$).

Since the matrix elements are "known constants", it follows that an inverse matrix of Equation (27.1) can be calculated. If $M_{ij}$ are the coefficients of the inverse matrix then

$$\begin{pmatrix} V_x \\ V_y \\ V_z \\ p' \end{pmatrix} = \begin{pmatrix} M_{11} & M_{12} & M_{13} & M_{14} \\ M_{21} & M_{22} & M_{23} & M_{24} \\ M_{31} & M_{32} & M_{33} & M_{34} \\ M_{41} & M_{42} & M_{43} & M_{44} \end{pmatrix} \begin{pmatrix} S_H^{RS} \\ S_H^{LS} \\ S_V^{RS} \\ S_V^{RL} \end{pmatrix}. \tag{27.2}$$

The specific form of Equation (27.1) results in some simplification of Equation (27.2), namely $M_{23} = M_{24} = M_{31} = M_{32} = M_{41} = M_{42} = 0$, $M_{11} = M_{12}$, and $M_{21} = M_{22}$.

Equation (27.2) is simply a mathematical statement of the arguments stated earlier (i.e., the required quantities are weighted sums and differences of the outputs of various hair cell groups).

To show how the fish can perform these operations we begin by noting that the sensory epithelium is composed of groups of hair cells oriented in 180° opposition. Although orthogonally oriented hair cell groups are needed for this model, a requirement for opposing hair cell groups (in each of the orthogonal directions) is not quite obvious. The presence of opposing hair cell groups provides signals that are proportional to $S_H(t)$ and $-S_H(t)$. Thus, the simple operation of signal addition (or superposition) by signals composed of $\pm S_H(t)$ permits the mathematical operations of both addition and subtraction. When one considers that a single nerve fiber innervates 50 or more hair cells in a typical fish saccular macula (Popper and Hoxter 1984), the possibility of weighted addition and subtraction of signals by the wiring of the ear emerges. For example, if nerve fibers A and B are connected to 50 hair cells oriented in the positive and negative horizontal directions, respectively, and nerve fibers C and D are connected to 30 hair cells oriented in the positive and negative vertical directions, respectively, superposing the outputs of A and C yields, to a first approximation, an output proportional to $5S_H(t) + 3S_V(t)$, whereas superposing A and D yields an output proportional to $5S_H(t) - 3S_V(t)$. When one considers the very large number of hair cells on the saccular macula of many species (20,000 or more sensory hair cells), the mathematical operations of Equation (27.2) can be carried out to great precision. If only 10 cells were affected, the greatest number one could obtain would be accurate to one decimal place. For each increase in the number of cells by one order of magnitude, however, the precision of the calculation is increased by an additional decimal place. The significance of this is apparent when we consider that the saccules of many species have very large numbers of sensory hair cells.

The inversion of the matrix (Eq. 27.1) is built into the synaptic wiring of the hair cells. The otolith organs of each ear transmit by separate channels the weighted sums and differences from each of the four hair cell regions of each ear as their contributions to the analysis of $V_x(t)$, $V_y(t)$, $V_z(t)$, and $p'(t)$ to the CNS (Fig. 27.7). Cells in the octavolateralis system then perform the needed interaural addition to produce $V_x$, $V_y$, $V_z$, and $p'$. Localization of a sound source can be accomplished by comparing, across elements in the CNS, the signs and amplitudes of $V_x$, $V_y$, and $V_z$. The CNS can also possibly recombine the $V_x$, $V_y$, and $V_z$ signals with appropriate relative weightings to construct a large number of preformed cosine ($\xi$) beams, each pointing in a different direction. This would enhance signal detectability in the presence of omnidirectional noise (as described on p. 692). The neural mechanism(s) and sites for specific determination of these functions in the CNS are not known.

A possible alternative solution to the phase discrimination problem also emerges from the preceding discussion. If we assume that the coupling between the cilia and the otolith is elastic for some groups of hair cells and viscous for other groups, then a relative phase of 45° is introduced between the signals. This phase shift would take the pressure and velocity channels out of quadrature for all source distances and enable the lead-lag discrimination to be accomplished by a simple comparison of the magnitude of the sum and difference of the two signals (see Rogers et al. in preparation for a detailed analysis that also demonstrates that the proposed algorithm is applicable to nonsinusoidal signals).

The assumption of elastic and viscous coupling between the hair cells and the otolith may be related to the different lengths of the ciliary bundles encountered on the saccular macula (i.e., as if some ciliary bundles directly contacted the otolith and others did not). Relatively longer ciliary bundles are located around the saccular margin of many species and in the caudal end of otophysan saccules (reviewed in Platt and Popper 1981). With this structural differentiation, one group of ciliary bundles may be elastically stimulated, whereas the other is stimulated in a more viscous fashion. Indirect support for this suggestion is the limited observations that some differences in the neural innervation of sensory cells with long and short ciliary bundles found in a number of species (Saidel and Popper 1983a, 1983b).

## Frequency Analysis

A number of fish species are able to discriminate frequencies (Jacobs and Tavolga 1968; Fay 1970). A critical band-like mechanism has been found in the goldfish and a few other species (Fay 1974; Tavolga 1974). In masking and critical band experiments, one frequency (the masker) interferes with the detection of another (the stimulus). In this experimental paradigm, the stimulus threshold is raised

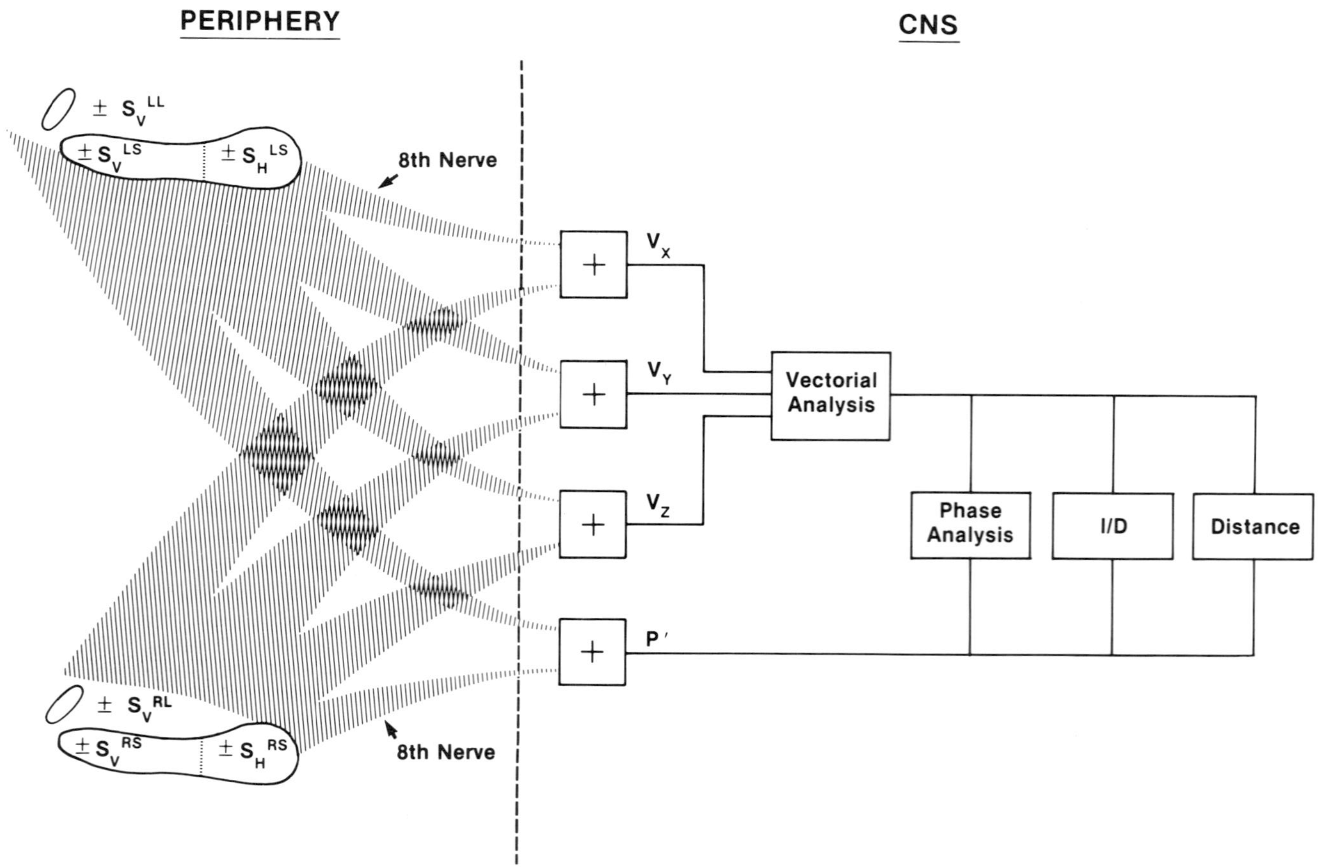


Figure 27.7. The model for signal analysis by the fish auditory system proposed in this chapter (also see Rogers et al. in preparation). The model is described in detail in the text. The significant difference between this model and the one proposed by Buwalda (1981, Figure 27.4) is that much of the processing in our proposed model occurs peripherally, while the major role of the *CNS* is to add the signals from the two ears. In addition, this model allows not only for vectorial and phase analysis, but also for frequency discrimination (*I/D*) and distance determination.

(sensitivity gets poorer) as either the frequency difference between the masker and stimulus decreases or the amplitude of the masker increases.

The next sections will examine possible mechanisms for frequency discrimination. These are (1) a temporal analysis of stimulus-generated spike rates, (2) the regionalization of sensitivity to different frequencies in different macular areas, (3) hair cell tuning to different frequencies, and (4) a comparison of direct and indirect signals. Although, as we will describe below, our model can account for much of the frequency discrimination data, we will not imply complete dependence on the proposed mechanism. In fact, we will argue that one or more of the mechanisms described below probably complements our proposed model and, consequently, provides fishes with a more comprehensive analysis of the acoustic stimulus.

## Temporal Analysis

Mechanisms of temporal analysis have been discussed by Fay (1980) (Fay and Passow 1982) and will not be examined in detail here. These mechanisms involve little more than converting sound into a spike rate that is related to the temporal dependence of the stimulus. We have discussed earlier (p. 691) some of the problems associated with this hypothesis.

## Cilia Length

Saunders and Dear (1983) demonstrated a relationship between ciliary bundle length and frequency response characteristics in different epithelial regions of amphibians, reptiles, birds, and mammals. In general, long ciliary bundles are associated with lower thresholds or higher-sensitivity detection of relatively low-frequency signals, whereas shorter bundles correlate with better sensitivity to higher frequencies.

Analysis of ciliary bundle lengths in fish (Platt and Popper 1984) showed that the ciliary bundles on the rostral end of the saccular macula of otophysans are substantially shorter than bundles at the caudal end of the epithelium and on the lagena. Limited physiological data from two otophysan species suggest that spatially distinct nerve fibers innervating the rostral end of the saccule respond to higher frequencies than do nerve fibers innervating the more caudal end of the lagena (Fay 1978; Moeng and Popper 1984). Thus, the general response pattern tentatively associated with ciliary bundle lengths in otophysans is consistent with the conclusions drawn by Saunders and Dear. In contrast, ciliary bundles in the saccule of the kissing gourami (*Helostoma temincki*) do not show a graded length (Platt and Popper 1984). There is no evidence (possibly related to the general paucity of data) for different frequency response properties along the epithelium in species other than otophysans.

## Regionalization

Considerable speculation has questioned whether fish possess a mechanism for signal analysis functionally analogous to the place mechanism found in the mammalian organ of Corti (von Békèsy 1960). In the organ of Corti, a traveling wave is combined with changes in the stiffness of the basilar membrane so that the region of maximum stimulation for different frequencies varies along the basilar membrane. The basilar membrane, in effect, performs a spectral analysis of the sound, leading to a sequential tonotopic organization along the organ of Corti.

Because the fish saccule does not physically resemble an organ of Corti, it has been difficult to analogize a mechanism for spectral analysis other than one that might result from differences in ciliary bundle lengths along the sensory epithelium (see the last section). The variation in stiffness of the basilar membrane depends on its being taut. An analogous structure is lacking in fish or frog ears, yet Lewis and his collaborators recently demonstrated the presence of a tonotopic organization in the frog amphibian papilla (Lewis, Baird, Leverenz, and Koyama 1982; Lewis, Leverenz, and Koyama 1982). It is theoretically possible that "regionalization" of responses on the saccular epithelium without a rigorous tonotopic organization might provide a rudimentary morphology for spectral analysis. Such a response regionalization may not account for the excellent frequency-resolving powers of the otophysans. However, it might provide some initial frequency analysis for them and/or a basis for limited frequency resolution in nonotophysan species.

The only direct study arguing for a sort of regionalization along the fish saccular epithelium is that of Enger (1981), who used intense tonal stimulation at different frequencies to damage different saccular regions in the cod (*Gadus morhua*). In this study, different broad regions were damaged by varying frequencies. A study by Sand and Michelsen (1978) showed limited variation in the motion of the saccular otolith of the perch (*Perca*) at different frequencies, and a study by Sand (1974) showed that the pattern of microphonic potentials changes with different frequencies.

Several possible mechanisms might account for regional response differences on the saccular epithelium. One would involve differences in otolith motion (different "motionals") for different frequencies (Popper, Platt, and Saidel 1982). In this way, a given frequency would maximally stimulate a specific region of sensory hair cells, while other regions of the epithelium would receive less stimulation by the same signal. Since innervation studies indicate that individual eighth nerve neurons innervate small areas of the saccular macula (Saidel and Popper 1983a, 1983b; Jensen 1984; Mathiesen 1985), each frequency would maximally stimulate different groups of neurons. Different patterns of neuron stimulation would indicate different frequencies!

An alternate possibility is a traveling or standing wave associated with the ear and resulting in a relatively simple tonotopic organization in which different hair cell regions (of undefinable size) would be stimulated by different frequencies.

The source of such wave motion would be the otolith, whereas the surface wave would travel in the interface between the otolith membrane and sensory epithelium. The elastic properties of the sensory macula, if present, could support the wave, while the otolith membrane would provide the interface needed for a surface wave.

A number of features of the ear would affect the nature of the traveling wave. The width and depth of the otolith groove in which the sensory epithelium lies would affect the resonance frequency at different places on the macula. The shape of the saccular macula and any gradation in the properties of the otolith membrane or its attachments to the otolith and/or epithelial surface could also affect the resonance properties of the system.

## Direct vs. Indirect Comparison

The data on fish hearing and the morphological structure of the ear provides a problem for each of the hypotheses discussed in the above sections. In addition, the hypothesis of regionalization and hair cell tuning by ciliary bundle length do not fit easily with the current ideas concerning sound localization. For example, a regionalization mechanism for frequency discrimination is inconsistent with unique otolith motionals for different frequencies when the motional of the otolith changes with sound source direction.

Our model suggests that fish can use the same sets of processed information for both detection and discrimination of frequency and for sound source localization. Our hypothesis for sound localization can potentially account for (1) frequency detection, (2) frequency discrimination, (3) masking and critical bands, and (4) determination of sound source distance without recourse to individual mechanisms for each task. Masterton and colleagues (Masterton, Heffner, and Ravizza 1969) have argued that sound source localization has been the major selective factor in the evolution of the mammalian auditory system, and the same suggestion has been applied to fishes (Popper and Coombs 1982; Popper and Northcutt 1983). That a mechanism in fish that has evolved to locate sound can also account for other auditory roles should not be surprising considering the extraordinary importance of sound source localization. After all, an animal's behavioral repertoire for finding prey and mates, defending a territory, and escaping from predators is its basis for survival.

In our discussion of sound localization (p. 698) we noted that the ear can separate the direct and indirect stimuli while maintaining their phase relationships and relative amplitudes. A fish can thus use both direct and indirect signals despite relative amplitude differences. It is significant to note that the indirect signal reaches the CNS as $dp/dt$, rather than $p$. Thus, *the amplitude observed in the CNS from the swim bladder–generated stimulus would change with frequency,* as compared to the amplitude of the direct stimulation.

Since one stimulus pathway is frequency dependent and the other is not, the amplitude ratio of direct and indirect stimuli in the CNS will be frequency dependent. Thus, fish might discriminate between two frequencies by using amplitude ratios, since the ratios would differ for any pair of frequencies. The ratio of the indirect (*I*) to the direct (*D*) signal is given by

$$I/D = f/[1 + (c/2\pi fR)^2]^{1/2} \qquad (27.3)$$

where $f$ is the frequency and $R$ is the distance to the sound source (Rogers et al. in preparation). The ratio is proportional to $f$ in the far field and $f^2R$ in the near field. Discrimination between frequencies is possible at any range, and absolute frequency identification could be made in the far field.

The discrimination of *relative* ranges (e.g., which is nearer and which farther) in the results of Schuijf and Hawkins (1983) can also be explained by this hypothesis, since $I/D$ will differ for the same frequency at different ranges. If an independent measure of the frequency were available, such as from a regionalization mechanism, the *absolute* range could also be determined from Equation (27.3), since the independent frequency could be substituted for $f$ in the equation and the equation then solved for distance ($R$). Even if the independent measure of frequency were crude, Equation (27.3) would allow for a crude measure of distance.

In addition to an $I/D$ ratio for frequency and relative or absolute distance, determination of the amplitude ratios of $I/D$ would also allow decoding of frequency-modulated (FM) signals.

Masking and critical band results can be analyzed solely by using $I/D$ measurements. Instead of a single reradiated signal whose amplitude reflects swim bladder reradiation, $I$ consists of two reradiated signals, one from the signal and the other from the masker. If the CNS has some set discrimination limen, the relative amplitudes of these reradiated signals would have to exceed the set point for the signal to be discriminated. If the signals were close in frequency, their amplitudes would have to differ greatly from the reradiated signal amplitudes to exceed the discrimination limen. If the two signals differed sufficiently in frequency, they could be similar in amplitude and still be detected as separate signals. Likewise, as in a masking paradigm, if the two signals were much different in frequency and the amplitude of one signal were raised considerably, interference, or masking, would result. For example, for a farfield source, consider the problem of detecting a signal of frequency $f_s$ and amplitude $p_s$ in the presence of a masking signal with frequency $f_m$ and amplitude $p_m$. From Equation (27.3) when the signal is present

$$(I/D)\ \textit{signal present} = (f_s p_s + f_m p_m)/(p_s + p_m), \qquad (27.4)$$

whereas in the absence of the signal

$$(I/D)\ \textit{signal absent} = f_m. \qquad (27.5)$$

The difference between these ratios, which the fish is trying to detect, is given by

$$(I/D)\ \textit{signal present} - (I/D)\ \textit{signal absent} = \frac{(f_s - f_m)p_s}{p_s + p_m} \qquad (27.6)$$

From Equation (27.6), the ability to discriminate will depend on both the amplitude ratio and the frequency separation between the signal and masker.

If our hypothesis is correct, fishes without swim bladders or fishes with minimal (or no) effective reradiation from the swim bladder should, at best, be poor in the ability to discriminate frequencies. The limited data suggest that species with swim bladders far from the ear discriminate frequency less well than do fishes with swim bladders closer to the ear (Tavolga 1974). Experimental verification of this hypothesis, as for the localization hypothesis, would come from experiments in which the swim bladder is removed to eliminate the common indirect signal or in which the swim bladder is displaced to one side so that the indirect signal differs to the two ears. Elimination of one ear should also greatly degrade frequency discrimination and localization.

Our hypothesis for frequency analysis in the fish auditory system follows elegantly from the proposed hypothesis for determining sound source direction. Alternate ideas about frequency discrimination involve mechanical stimulation of different saccular regions. This stipulation for precise frequency-dependent motion could not be compatible with direction-dependent otolith motion. Any motions of the otolith, combined with the information from the indirect pathway, is sufficient for both sound source localization and for frequency discrimination in our model.

## Additional Considerations

Our hypothesis accounts for many aspects of fish hearing, as determined by behavioral and physiological experiments. It is also consistent with the morphology of the ear, especially the directional orientation of the sensory hair cells, the differences in ciliary bundle lengths on the sensory epithelium, and the high hair cell–nerve fiber convergence ratio.

The data that have not been integrated into the model are the enormous growth in number of sensory cells that is found during the life of fish (Popper and Hoxter 1984) and elasmobranchs (Corwin 1983). However, note that any changes in the general geometry of a fish, such as that related to growth, will entail changes in the arithmetic portion of the auditory system—the saccule. The most appropriate changes would be to increase the number of sensory hair cells (and thus the hair cell–afferent fiber ratio) to do the arithmetic needed for the processing of sounds. The ability of fishes to add hair cells thereby meets the needs of our hypothesis.

The diversity in saccular morphology among the 20,000 or more species is a well-documented observation (see Retzius 1881; Platt and Popper 1981; Popper and Coombs 1982; Popper 1983). Interspecific variation may reflect the evolution of several different mechanisms to perform the same auditory tasks. Since only a few basic morphological and ultrastructural "patterns" are found among fish, however, and these patterns recur in taxonomically diverse fish groups (Popper and Coombs 1982), we suggest that the variance in structure of the auditory portions of the ear may all be related to some aspect(s) of sound processing as analyzed by the model proposed in this chapter.

Our hypotheses deal directly with localization mechanisms. It is possible that the variation in ear structure represents variants on how information is processed for the CNS, rather than different mechanisms for inner ear stimulation. Of course, the question still arises as to why different species have different processing mechanisms and why similar mechanisms keep showing up in species that are taxonomically widely diverse.

*Acknowledgments*

Portions of the work discussed in this chapter were supported by Office of Naval Research Contract N–00014–82–K–0345 and NIH Grant NS–15090 from the National Institutes of Neurological and Communicative Diseases and Stroke to A.N.P. and W.M.S. and by ONR Contract N00014–84–K–0164 to P.H.R. We are grateful to Drs. Richard R. Fay, Christopher Platt, and Joelle Presson for reading and making valuable comments on the manuscript.

## References

Buwalda, R.J.A. (1981) Segregation of directional and non-directional acoustic information in the cod, in Hearing and Sound Communication in Fishes, Tavolga, W.N., Popper, A.N., and Fay, R.R. (eds.), Springer-Verlag, New York, pp. 139–171.

Buwalda, R.J.A., Schuijf, A., and Hawkins, A.D. (1983) Discrimination by the cod of sounds from opposing directions, J. Comp. Physiol., 150:175–184.

Coombs, S., and Fay, R.R. (1984) Acoustic coding by the ear of the sunfish (*Lepomis gibbosis*): sensitivity, tuning, and directionality of the 8th nerve units, J. Acoust. Soc. Am., S75:s89.

Corwin, J.T. (1983) Postembryonic growth of the macula neglecta auditory detector in the ray, *Raja clavata*: continual increases in hair cell number, neural convergence, and physiological sensitivity, J. Comp. Neurol., 218:345–356.

Enger, P.S. (1981) Frequency discrimination in teleosts—central or peripheral? in Hearing and Sound Communication in Fishes, Tavolga, W.N., Popper, A.N., and Fay, R.R. (eds.), Springer-Verlag, New York, pp. 243–255.

Fay, R.R. (1970) Auditory frequency discrimination in the goldfish (*Carassius auratus*), J. Comp. Physiol. Psychol., 73:175–180.

Fay, R.R. (1974) Masking of tones by noise for the goldfish (*Carassius auratus*), J. Comp. Physiol. Psychol., 87:708–716.

Fay, R.R. (1978) Coding of information in single auditory nerve fibers of the goldfish, J. Acoust. Soc. Am., 63:136–146.

Fay, R.R. (1980) Psychophysics and neurophysiology of temporal factors in hearing by the goldfish: amplitude modulation detection, J. Neurophysiol., 44:312–332.

Fay, R.R., and Passow, B. (1982) Temporal discrimination in the goldfish, J. Acoust. Soc. Am., 72:753–760.

Fay, R.R., and Popper, A.N. (1974) Acoustic stimulation of the ear of the goldfish (*Carassius auratus*), J. Exp. Biol., 61:243–260.

Fay, R.R., and Popper, A.N. (1975) Modes of stimulation of the teleost ear, J. Exp. Biol., 62:379–387.

Fay, R.R., and Popper, A.N. (1983) Hearing in fishes: comparative anatomy of the ear and the neural coding of sensory information, in Hearing and Other Senses: Presentations in Honor of E.G. Wever, Fay, R.R., and Gourevitch, G. (eds.), Amorpha Press, Groton, Conn., pp. 123–148.

Horner, K., Sand, O., and Enger, P. (1980) Binaural interaction in the cod, J. Exp. Biol., 85:323–331.

Hudspeth, A.J., and Corey, D.P. (1977) Sensitivity, polarity, and conductance change in the response of vertebrate hair cells to controlled mechanical stimuli, Proc. Nat. Acad. Sci. USA, 74:2407–2411.

Jacobs, D.W., and Tavolga, W.N. (1968) Acoustic frequency discrimination in the goldfish, Anim. Behav., 16:67–71.

Jensen, J.C. (1984) On the polarization and innervation of the pars inferior sensory epithelia of the herring labyrinth, Acta Zool. (Stockh.), 65:61–74.

Lewis, E.R., Baird, R.A., Leverenz, E.L., and Koyama, H. (1982) Inner ear dye injection reveals peripheral origins of specific sensitivities, Science, 215:1641–1643.

Lewis, E.R., Leverenz, E.L., and Koyama, H. (1982) The tonotopic organization of the bullfrog auditory papilla, an auditory organ lacking a basilar membrane, J. Comp. Physiol., 145:437–445.

Masterton, B., Heffner, H., and Ravizza, R. (1969) The evolution of human hearing, J. Acoust. Soc. Am., 45:966–985.

Mathiesen, C. (1985) Structure and innervation of inner ear sensory epithelia in the European eel (*Anguilla anguilla*), Acta Zool. (Stockh.), 65:189–208.

McCormick, C. (1983) Organization and evolution of the octavolateralis area in fishes, in Fish Neurobiology, Northcutt, R.G., and Davis, R.E. (eds.), University of Michigan Press, Ann Arbor, pp. 179–214.

Moeng, R., and Popper, A.N. (1984) Auditory responses of saccular neurons of the catfish, *Ictalurus punctatus,* J. Comp. Physiol., 155:615–624.

Myrberg, A.A., Jr., and Spires, J.Y. (1980) Hearing in damselfishes: an analysis of signal detection among closely related species, J. Comp. Physiol., 140:135–144.

Platt, C., and Popper, A.N. (1981) Fine structure and function of the ear, in Hearing and Sound Communication in Fishes, Tavolga, W.N., Popper, A.N., and Fay, R.R. (eds.), Springer-Verlag, New York, pp. 3–38.

Platt, C., and Popper, A.N. (1984) Variations in lengths of ciliary bundles on hair cells along the macula of the sacculus in two species of teleost fishes, Scanning Electron Microsc. 1984, 4:1915–1924.

Popper, A.N. (1983) Organization of the inner ear and auditory processing, in Fish Neurobiology, Northcutt, R.G., and Davis, R.E. (eds.), University of Michigan Press, Ann Arbor, pp. 125–178.

Popper, A.N., and Coombs, S. (1982) The morphology and evolution of the ear in Actinopterygian fishes, Am. Zool., 22:311–328.

Popper, A.N., and Fay, R.R. (1973) Sound detection and processing by teleost fishes: a critical review, J. Acoust. Soc. Am., 53:1515–1529.

Popper, A.N., and Fay, R.R. (1984) Sound detection and processing by teleost fish: a selective review, in Bolis, L., Keynes, R.D., and Maddrell, S.H.P. (eds.), Comparative Physiology of Sensory Systems, Cambridge University Press, Cambridge, U.K., pp. 67–101.

Popper, A.N., and Hoxter, B. (1984) Growth of a fish ear. I. Quantitative analysis of sensory hair cell and ganglion cell proliferation, Hear. Res., 15:133–142.

Popper, A.N., and Northcutt, R.G. (1983) Structure and innervation of the inner ear of the bowfin, *Amia calva,* J. Comp. Neurol., 213:279–286.

Popper, A.N., Platt, C., and Saidel, W.M. (1982) Acoustic functions in the fish ear, Trends Neurosci., 5:276–280.

Retzius, G. (1881) Das Gehörorgan der Wirbelthiere, vol. 1, Samson and Wallin, Stockholm.

Sand, O. (1974) Directional sensitivity of microphonic potentials from the perch ear, J. Exp. Biol., 60:881–899.

Sand, O., and Michelsen, A. (1978) Vibration measurements of the perch saccular otolith, J. Comp. Physiol., 123A:85–89.

Saidel, W.M., and Popper, A.N. (1983a) Spatial organization in the saccule and lagena of a teleost: hair cell pattern and innervation, J. Morphol., 177:301–317.

Saidel, W.M., and Popper, A.N. (1983b) The saccule may be the transducer for directional hearing for non-ostariophysine teleosts, Exp. Brain Res., 50:149–152.

Saunders, J.C., and Dear, S.P. (1983) Comparative morphology of stereocilia, in Hearing and Other Senses: Presentations in Honor of E.G. Wever, Fay, R.R., and Gourevitch, G. (eds.), Amorpha Press, Groton, Conn., pp. 175–198.

Schuijf, A. (1975) Directional hearing of cod (*Gadus morhua*) under approximate free field conditions, J. Comp. Physiol., 98:307–332.

Schuijf, A. (1981) Models of acoustic localization, in Hearing and Sound Communication in Fishes, Tavolga, W.N., Popper, A.N., and Fay, R.R. (eds.), Springer-Verlag, New York, pp. 267–310.

Schuijf, A., and Buwalda, R.J.A. (1975) On the mechanism of directional hearing in cod (*Gadus morhua* L.), J. Comp. Physiol., 98:333–343.

Schuijf, A., and Buwalda, R.J.A. (1980) Underwater localization—a major problem in fish acoustics, in Comparative Studies of Hearing in Vertebrates, Popper, A.N., and Fay, R.R. (eds.), Springer-Verlag, New York, pp. 43–77.

Schuijf, A., and Hawkins, A.D. (eds.), (1976) Sound Reception in Fish, Elsevier, Amsterdam.

Schuijf, A., and Hawkins, A.D. (1983) Acoustic distance discrimination by the cod, Nature, 302:143–144.

Spanier, E. (1979) Aspects of species recognition by sound in four species of damselfish, genus *Eupomacentrus* (Pisces: Pomacentridae), Z. Tierpsychol., 51:301–316.

Tavolga, W.N. (1974) Signal/noise ratio and the critical band in fishes, J. Acoust. Soc. Am., 55:1323–1333.

Tavolga, W.N. (1976) Acoustic obstacle detection in the sea catfish (*Arius felis*), in Sound Reception in Fish, Schuijf, A., and Hawkins, A.D. (eds.), Elsevier, Amsterdam, pp. 185–204.

Tavolga, W.N., Popper, A.N., and Fay, R.R. (eds.), (1981) Hearing and Sound Communication in Fishes, Springer-Verlag, New York.

van Bergeijk, W.A. (1964) Directional and nondirectional hearing in fish, in Marine Bio-Acoustics, Tavolga, W.N. (ed.), Pergamon Press, Oxford, pp. 281–299.

von Békèsy, G. (1960) Experiments in Hearing, McGraw-Hill, New York.

von Frisch, K. (1938) Über die Bedeutung des Sacculus und der Lagena für den Gehörsinn der Fische, Z. Vgl. Physiol., 25:703–747.

Wenz, G.M. (1964) Curious noises and the sonic environment in the ocean, in Marine Bio-Acoustics, Tavolga, W.N. (ed.), Pergamon Press, Oxford, pp. 101–119.

# Chapter 28

# Peripheral Adaptations for Spatial Hearing in Fish

RICHARD R. FAY

We know from our own experience that hearing is a spatial sense. Sounds are perceived as sources—objects having a number of characteristics, including size, direction, and distance. In addition, ambient sound that reaches our ears contains rich information about the overall structure of the acoustical environment, including the sources of sounds and the location and characteristics of reflectors. We are often unaware of the richness of auditory space and texture, perhaps because so much of human "listening" focuses on speech and speech messages; but it is likely that auditory space perception is of greater general biological significance than the decoding of vocalized messages, since all vertebrates have clear peripheral and central adaptations for spatial hearing, whereas only some have found value in vocalization.

I suggest that the primary function of the auditory system in most animals is to help form a three-dimensional image of the local environment. The localization of specific sources is a part of this function, just as the resolution of detailed form is an aspect of vision; but just as spatial vision is more than acuity, spatial hearing is more than localization. Perceiving objects entails separating figures from "ground," and a global understanding of ground is required for the perceptual formation of objects. This chapter considers the mechanisms by which the ears of fish code some of the acoustic information required to form such a spatial image. The particular acoustic properties of the aquatic medium have helped shape the structures and functions of auditory systems among fish. These have given rise to characteristic mechanisms of acoustic analysis that are different from those of the terrestrial forms, but are partially congruent, at least in a functional sense.

## Cues Used for Directional Hearing in Air

All vertebrates have paired ears, and an outline of a theory of spatial hearing based on a two-point sampling of the acoustic waveform is now at hand for the mammals. The sensory information used in spatial hearing is in the form of monaural and binaural cues. Monaural cues arise primarily from the acoustical characteristics of the outer ear and binaural cues from differences in the acoustic waveforms impinging on the two ears. In humans and other mammals, monaural cues are provided by the ear's asymmetry in the vertical plane and by sound scattering by the torso (Kuhn 1982). These function primarily to help resolve vertical auditory space (elevation), and solve back-front confusions through direction-dependent spectral filtering (Butler and Belendiuk 1977). Binaural cues arise from the spatial separation of the two ears and the reflective and refractive characteristics of the outer ears, head, and body (Kuhn 1979). Horizontal auditory space is resolved essentially by correlating neural representations of the acoustic waveforms at the two ears to extract differences in time-of-arrival, overall intensity, ongoing phase and envelope, and frequency-dependent intensity (spectrum shape). These various cues are processed by the brain to provide both an analysis of auditory space (e.g., in isolating a given source and in directional "filtering" for enhanced signal detection) and a spatial synthesis (e.g., in imaging the global environment into which the listener fits) (Blauert 1982).

Although we know less about the auditory world of nonhuman animals, it is perhaps a reasonable assumption that all auditory systems perform similar analytical and synthetical functions in spatial hearing. We know, for example, that in the barn owl (*Tyto alba*), the brain uses binaural cues to synthesize a spatial neural map of external auditory space (Knudsen and Konishi 1978). If we extend this assumption to the fish, there exists a special challenge to understand how the information required to hear spatially in an underwater environment is encoded by fish ears.

## The Problem of Directional Hearing Underwater

The special problems fish face in their aquatic environment can be simply stated as follows: (1) The velocity of sound in water is almost five times greater than in air. This means that time, phase, and intensity differences over space are correspondingly reduced. (2) The ears of fish are quite close together, on the order of millimeters to one or two centimeters. This further reduces the potential magnitude of interaural differences. (3) The ears of many fish are not functionally independent; they may share common perilymphatic and endolymphatic fluid spaces. (4) Finally, the major acoustic input to the ears of many fish is sound reradiated from the swim bladder (or other gas bubble) responding to fluctuations in sound pressure. This means that sound arrives at the two ears equally from a source fixed in its spatial relation to the ears (van Bergeijk 1964).

## Early History

The question of spatial hearing by fish has traditionally been defined in terms of the animal's ability to discriminate between sounds originating from different locations. This is a rather reduced definition of spatial hearing, as pointed out above, but one that is readily translated into experimental manipulation and measurement. This aspect of spatial hearing has a long history, extending from the early years of this century. In summary, the results of the behavioral tests for localization were mixed, with some investigators observing a discrimination and others not. The consensus seemed to be that localization, when it was observed, only occurred for nearby, intense, low-frequency signals (von Frisch and Dijkgraaf 1935; Kleerekoper and Chagnon 1954).

These and other considerations led van Bergeijk (1964, 1967) to the hypothesis that the ears could not subserve directional hearing in fish and that other organs (of the lateral line system) must be called upon to give the information required for spatial hearing. Other considerations in favor of this lateral line hypothesis were that (1) the lateral line system was thought to be a subdivision of the auditory system, with similar embryological origin and presumed central projections (van Bergeijk 1967); (2) directional hearing had only been demonstrated behaviorally for nearby sound sources (likely within the near field) (see Chaps. 4 and 5 for a further explanation of the near field); (3) the lateral line system was known to respond within the near field (Harris and van Bergeijk 1962); and (4) the spatial extent of the lateral line system made it ideal as a processor of spatial intensity gradients characteristic of acoustic near fields (see Chap. 4).

## Contemporary Views

More recent thought and research on directional hearing in fish has changed this view to some extent. Pioneering experiments by Moulton and Dixon (1967) showed that two intact saccules (otolith organs of the ears) were necessary for the goldfish (*Carassius auratus*) to respond directionally to sound with a tail flip (presumably mediated by the Mauthner cells). Since that time, Schuijf and his colleagues have demonstrated that the cod (*Gadus morhua*) requires two intact labyrinths to make discriminations of sound source direction (Schuijf 1975) and that the lateral line system is not required (Schuijf and Siemelink 1974).

It has been further demonstrated that the cod and other species (1) are capable of learning to discriminate differences in sound source azimuth and elevation as small as a 10°–20° angle (Chapman and Johnstone 1974; Hawkins and Sand 1977), (2) can solve back-front and left-right (180°) confusions (Schuijf and Buwalda 1975, Blaxter, Chap. 8), and (3) are capable of sound source distance discrimination (Schuijf and Hawkins 1983), all at distances of several meters from the sound source. In addition, recent physiological work (Sand 1981; Denton and Gray 1983) has shown that lateral line organs respond only to a relative motion between the fish and the water surrounding it (as would occur only in the very near field) (see Chaps. 4 and 23), whereas the organs of the ear respond with greater sensitivity to any motion of the fish's head (as would occur within both the near and far fields) (Fay 1984). Finally, it is now clear that the origins and central projections of the lateral line system are not so closely related to those of the ears as was once thought (see Chap. 29).

At least for some species, however, the lateral line canals of the head are mechanically linked to the fluid systems of the ear (Denton, Gray, and Blaxter 1979; see also Chap. 22), and it seems likely that there are interactions between these two systems under some conditions. At a functional level, it is presently quite unclear whether and how the ear and lateral line may function together in a "hearing" system. It is clear, however, that the lateral line organs are optimally designed and situated for response to a relative motion between the animal's skin and the water medium, such as occurs in the near field, and are not well designed to respond to body motions with only an inertial reference (e.g., in which relative motions between the water and skin are minimal, as in the far field). As Harris and van Bergeijk (1962) observed, it is possible for the lateral line organs to respond to the near field of the swim bladder or other gas bubble as it oscillates in a pressure field. Experiments on the roach (*Rutilus rutilus*) have shown that this effect is small enough (on the trunk lateral line) to be dismissed as a factor contributing to sound detection under normal conditions (Sand 1981), but has not been ruled out in the case of suprathreshold acoustic processing.

Although the behavioral evidence indicates that fish can discriminate differences in sound source location in three-dimensional space from acoustic information transduced by the ears, the mechanisms by which this information is encoded by the ears and processed by the brain are not well understood (see Chap. 27) and have been the subject of much speculation.

## Directional and Nondirectional Response of Ears

The ears of fish include the three otolithic organs; the utricle, the saccule, and the lagena, and it is generally believed that the saccule is the primary hearing organ in most species. In the clupeids, however, the utricle is the major hearing organ (Blaxter, Denton, and Gray 1981). The otolith organs contain a patch of sensory epithelium, made up of hair cell receptors and supporting cells, which is overlain by a solid calcium carbonate otolith and an intervening otolith membrane (see Chap. 27). It is thought that stimulation occurs when a relative shearing motion is set up across the cilia, between the hair cell bodies and the otolith.

### Principles of Stimulation and Directional Coding Hypotheses

Motion of the hair cell cilia can be brought about in at least two ways. The swim bladder (or other gas bubble) expands and contracts in a pressure field and produces a displacement that can cause relative otolith motion (von Frisch 1938), given an efficient pathway for the motion to be transmitted to the ear (van Bergeijk 1967). As noted earlier, however, pressure sensitivity does not appear useful for directional hearing, since the input to the ears arrives from the swim bladder and likely stimulates both ears identically. The otolith organs respond not only to the pressure waveform via the swim bladder, however, but also to particle motion, through an inertial stimulation principle. In this case, the otolith organs act as mass-loaded accelerometers. As the animal's body accelerates along the axis of particle motion (in a near or far field), an otolith's inertia causes its motion to lag that of the surrounding tissue, and shearing forces are set up across the hair cell cilia (de Vries 1950). In contrast to the pressure waveform, which is a scalar function, the particle acceleration waveform is a vector function, with direction, as well as magnitude. As first outlined by Dijkgraaf (1960), it is presently believed that the axis of particle motion is coded within and between otolithic organs by the relative response across arrays of hair cell receptors oriented in various directions (Schuijf and Buwalda 1980).

This hypothesis has been supported by a number of recent studies on the patterns of hair cell orientation within the maculae of the various otolith organs in fish (Popper 1977). Although there is rich variation among species and organs in these orientation patterns, the general view is of a system with at least six organs (paired saccules, lagenae, and utricules) with nonparallel gross orientations, each containing arrays of hair cells with axes of greatest sensitivity pointing in various directions (Fig. 28.1). The saccules of most species contain cells generally oriented rostrocaudally and dorsoventrally, and the utricles and lagenae contain cells with a wide distribution of orientations in the horizontal and vertical planes, respectively. Many organs also show a curvature of the macula out of a single plane. Conceptual models of how the particle motion axis can be resolved using information from arrays of directionally sensitive receptors have been presented (Schuijf and Buwalda 1980; Schuijf 1981; see also Chap. 27).

## Problems to Be Solved

In its simplest form, this hypothesis raises three important questions. (1) What is the absolute sensitivity of the auditory system operating in this manner? (2) Is the response to particle motion sufficiently great relative to pressure-mediated response (e.g., via the swim bladder) to be properly coded by the ears? (3) How are confusions or ambiguities in determining sound source location from information on the axis of particle motion solved?

Several estimates of absolute behavioral sensitivity to particle movement at optimal frequencies show that displacements in the range between 1 and 100 Å may be detected (Chapman and Hawkins 1973; Chapman and Sand 1974; Fay and Patricoski 1980). At 200 Hz and well into the far field, this displacement accompanies a sound pressure level of about 1 dyne/cm$^2$ (see Chap. 4 for a definition of *far field*). This sound pressure level is above absolute detection threshold for most species. Therefore, for some species at least, there are sounds that can be detected via pressure-sensitive channels, but are below threshold for direct particle motion "vector detection" and thus may not be localized.

The second question of relative sensitivity is particularly important for those species which have efficient pathways between the swim bladder and the ear, such as the otophysans. For a far field sound well above detection threshold, the pressure (swim bladder–mediated) input to the ear would far exceed that

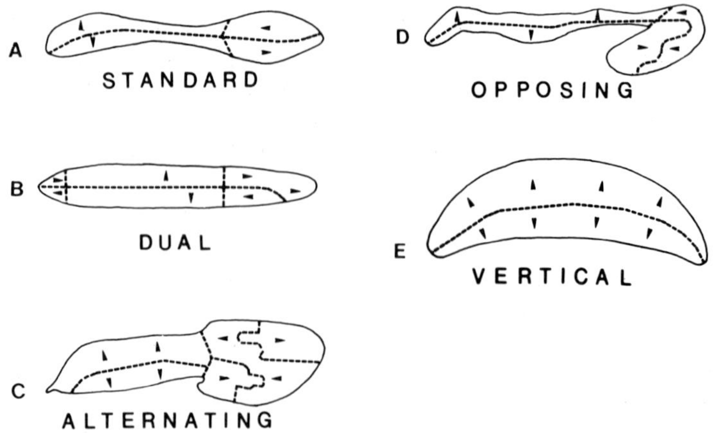


Figure 28.1. Maps of orientations of hair cells on saccular maculae of five teleost species chosen to illustrate characteristic patterns. *Arrowheads* indicate the direction of ciliary shear that would be most excitatory (movement toward the kinocilium) for hair cell populations bounded by *dashed lines*. Shown are *standard* pattern (A) from a salmoniform; *dual* pattern (B) from a gadiform, showing two separate macular regions of horizontally oriented cells; *alternating* pattern (C) from moray eel, showing complexly organized groups of cells oriented horizontally; *opposing* pattern (D) from an anabantid showing a region where horizontally oriented cells oppose each other along a vertical boundary; and, *vertical* pattern (E) from a mormyrid lacking horizontally oriented cells. This last pattern is also characteristic of otophysans, including catfish and goldfish. Anterior is to the right, and dorsal is up. (From Popper and Coombs 1982, with permission.)

due to direct particle motion, and the nondirectional input would tend to mask or otherwise interfere with the coding of the directional input (Buwalda 1981). One solution to this problem is to focus the pressure input on one otolith organ (e.g., the saccule) and to shield the other organs (the lagena and the utricle) from the swim bladder. This occurs in the Otophysi.

Another way the relative sensitivity to particle motion may be enhanced is based upon fundamental differences between the interaural effects of the swim bladder input vs. the direct motional input. The input to the two ears from the swim bladder produces inphase responses from corresponding elements of the saccules. However, acceleration of the animal's body from the side, for example, would tend to cause an asymmetrical response in the two ears (since the vector points medially for one ear and laterally for the other). Any asymmetry in input from the ears to the brain would signal a particle motion component, and the asymmetry's magnitude and polarity could contain information on the amplitude, phase, and orientation of the particle motion axis (Fay, Hillery, and Bolan 1982). In this type of model, the pressure-dependent (inphase) component is a "carrier," which could be reduced by a central subtractor mechanism (Durlach 1972). In any case, more empirical work must be done before we can progress beyond this kind of speculation.

The third question recognizes that a determination of the axis of particle motion is not enough for localization. There are two problems here. In the far field, or for monopole sources, the problem is that knowing the axis along which particles are moving does not specify which end of the axis points to the source. What is the direction of wave propagation? For more complex sources (e.g., dipoles) in the near field, the problem is even greater, because the axis of particle motion may not point to the source at all (van Bergeijk 1964; see also Chaps. 4 and 5). In empirical work on the cod (Schuijf and Buwalda 1975), it was shown that 180° confusions do not occur in a free field progressive sound wave. Further, an animal trained in a free field to orient towards the sound source could be caused to orient towards or away from a synthesized phantom "source," depending on the phase relations between the particle motion and the pressure waveforms created in a standing wave. These results led to the conclusion that 180° ambiguities are "solved" through a phase comparison of the particle motion waveform with the sound pressure waveform. For example, a compression peak in a sinusoid arriving from the right would lead a displacement peak by 90°, whereas the reverse would be true for a signal arriving from the left. Current theories of directional hearing in fish (Schuijf and Buwalda 1980; Schuijf 1981) thus require not only that the axis of motion be resolved by arrays of directionally sensitive receptor cells, but also that the motion and pressure waveforms be resolved and analyzed in the time domain. It appears that the far field pressure waveform that van Bergeijk (1967) believed contained no directional information may be useful to the fish when combined with displacement information.

Possible answers to the second aspect of this question are not yet at hand. The potential information about sound source characteristics existing within complex near fields is great. The three-dimensional pathway of particle motion and the amplitude and phase relations between pressure and particle motion are defined within the near field by such things as the mode of motion (e.g., monopole, dipole), the size and shape of the moving object, and its distance, azimuth, and elevation. It is not yet clear, however, whether this information is encoded and analyzed by the auditory system unambiguously.

## Physiological Studies of Directional Hearing

Few physiological analyses of directional hearing have been carried out in fish, since a well-controlled acoustic environment is difficult to achieve under conditions that allow recordings from auditory nerve fibers. One approach to this problem is to simulate far field particle motion by producing whole body accelerations of the animal (Fay and Olsho 1979; Hawkins and Horner 1981).

In recent studies on the saccular, lagenar, and utricular nerves of the goldfish, Fay (1984) recorded unit responses to linear sinusoidal motions produced by accelerating a small, rigid water tank (containing the fish) using a system of orthogonal mechanical drivers. Manipulations of the relative amplitudes and phases of three-axis motion input to the tank-produced linear acceleration along any three-dimensional axis. The thresholds for single units were measured for

36 axes of linear motion, with 12 axes spaced at 15° intervals in each of the horizontal, sagittal, and frontal planes of the animal. In these experiments, the assumption was made that this type of stimulus (acceleration of the water mass, surrounding and including the fish) produced minimal sound pressure levels and that the responses evoked were caused primarily by the otolith organs operating in their accelerometer modes. It is possible that sound pressures generated in the tank influenced the responses to some extent, however, particularly from the saccule, which is so well coupled to the swim bladder in the goldfish. On the other hand, the strong directionality of the responses from all three otolith organs (see below) suggest that the nondirectional pressure input did not have a major effect in this case.

The essential finding is that, in each of these planes of acceleration, thresholds plotted on polar coordinates tend to fall on a straight line having a particular orientation. (See Fig. 28.2, *left side*). This is essentially equivalent to describing a cell's directional sensitivity (the reciprocal of threshold) as a cosine function of the input angle relative to the cell's axis of best sensitivity. In polar coordinates, this function is a circle (Fig. 28.2, *right side*). These projections of directional sensitivity on a plane can be combined in three dimensions to reveal a "sensitivity solid" that is a sphere. Figures 28.3, 28.4, and 28.5 show the three planes of directional sensitivity for representative fibers from the saccular, utricular, and lagenar nerves combined in three dimensions. View the points on these "sensitivity circles" as falling on the surface of a best-fitting sphere. The vector originating at the coordinate origin and ending at the sphere's center defines a cell's directionality and sensitivity (magnitude of the vector) in three dimensions. These results suggest that peripheral nerve fibers respond as if their inputs (probably from a group of hair cells) (Furukawa 1978) were from a single hair cell with a particular three-dimensional axis of best sensitivity, at least under the conditions of this experiment.

Figure 28.6 shows distributions of these "best directions" for nerve fibers of the goldfish's three otolith organs. The view here is looking down at the north pole of a globe with the fish's ears at its center. The points plot the location on the globe at which the "best direction" vectors pass through the surface. Clearly, fibers within each organ "look" in a range of different directions, and each organ has a different distribution of fibers with regard to best direction. These distributions of best directions generally correspond with those expected on the basis of anatomical maps of hair cell orientation within the three otolith organs of the goldfish (Fig. 28.7) (Platt 1977). From the right ear, saccular fibers respond best to motions along a single axis with an azimuth of about 15° and an elevation of about 50° (see also Fig. 28.3). The utricular units respond best in the horizontal plane with a wide range of azimuths (see also Fig. 28.4). The lagenar units are generally more dispersed, but tend to cluster in a vertical plane with a wide range of elevations (see also Fig. 28.5). Saidel and Popper (1983) have demonstrated that regions of the saccular and lagenar maculae having different hair cell orientations are innervated by separate branchlets of the eighth nerve in several teleosts. If these afferents are "labelled" in the brain according to their orientations, then fish have the ingredients for a neural "place principle" for analyzing the axis of any motional input to the ears.

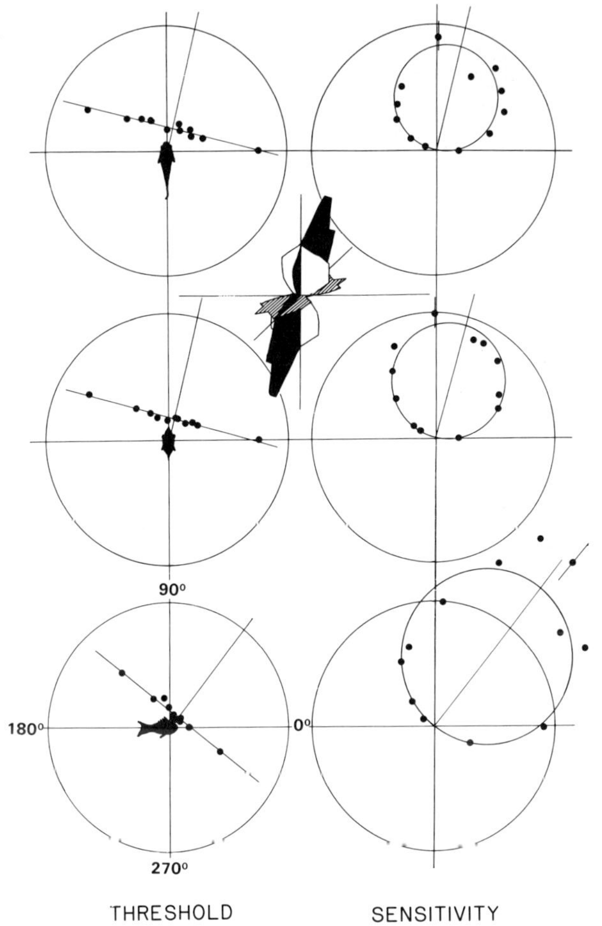


Figure 28.2. Directional sensitivity of single saccular nerve fiber from the goldfish. *Left*, polar coordinate plots of acceleration threshold as a function of stimulus axis angle in three planes: horizontal (*top*), frontal (*middle*), and sagittal (*bottom*). Each point is an independent threshold estimate based on a criterion of phase-locked spike rate. Lines through points are linear least squares best fit functions. Perpendicular thin lines show the axis of best sensitivity within each plane (polar angle at which lowest threshold occurs). *Right*, these threshold data are displayed as sensitivity measures (the reciprocal of threshold). In this case, they are fit with circles. The thin line running from coordinate origin through the center of the circle also defines best excitatory axis in each plane. The inset shows the sensitivity plotted in three dimensions, as if the fish were viewed from the back, from the right, and from above. White plane is frontal, black plane is sagittal, and striped plane is horizontal. (From Fay 1984.)

For the goldfish, it seems that information about the axis of particle motion is better represented in the profile of most active fibers within the lagena than in the saccule or utricule. The distribution of best directions for the saccule is rather tight and that for the utricle confined primarily to the horizontal plane. The homogeneity of best directions for saccular fibers is correlated with the single-

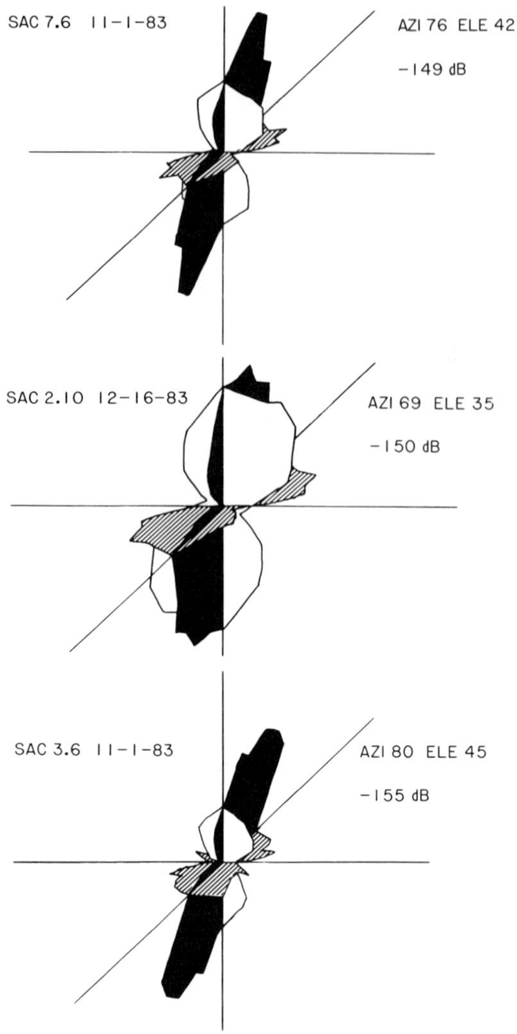


Figure 28.3. Directional sensitivity in three dimensions for three representative saccular fibers, plotted as in the inset in Figure 28.2. Azimuth and elevation values are spherical coordinate values of the vector running from coordinate origin through the center of best-fitting sphere, which is tangent at the origin. Zero degrees azimuth is to the *right,* and zero degrees elevation is *straight up*. The sensitivity for each fiber is in decibels with reference to 1 m displacement (rms).

axis pattern of hair cell orientation on the saccular macula (Fig. 28.7). This seems to be characteristic of otophysans and may be a specialization for coding the sound pressure waveform as it is transmitted along a single axis from the swim bladder. The more common pattern of hair cell orientations in the teleost saccule consists of two orthogonal orientation axes, "dorsal-ventral" and "rostral-caudal" (see Fig. 28.1). Preliminary studies on the directional characteristics of the auditory nerve fibers of the sunfish *Lepomis gibbosus* (Coombs and Fay 1984) show more variation in best direction among saccular fibers than has been seen in the goldfish, as would be expected from the hair cell orientation patterns. This more dispersed pattern of saccular fiber directionality is likely to be more characteristic of nonotophysan species.

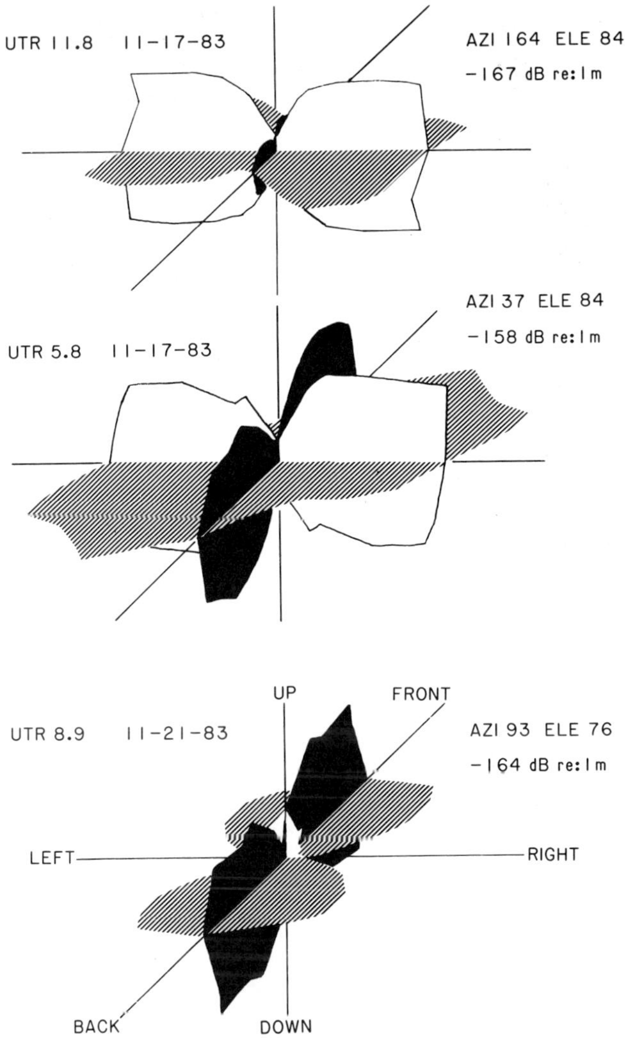


Figure 28.4. Directional sensitivity for three representative utricular fibers plotted as in Figure 28.3.

The major differences among species in auditory system organization are found in the conducting pathways between the swim bladder and the ears and in the patterns of hair cell orientation within the otolith organ receiving major swim bladder input (Fay and Popper 1980). Since both of these features are considered important for sound localization, it is possible that the mechanisms for directional hearing differ in detail across species. For example, nonotophysans may use saccular input to help resolve the axis of particle motion to a greater extent than the otophysans, since all hair cells of the latter appear to be most sensitive along a single axis of particle motion. On the other hand, since the lagena and utricle

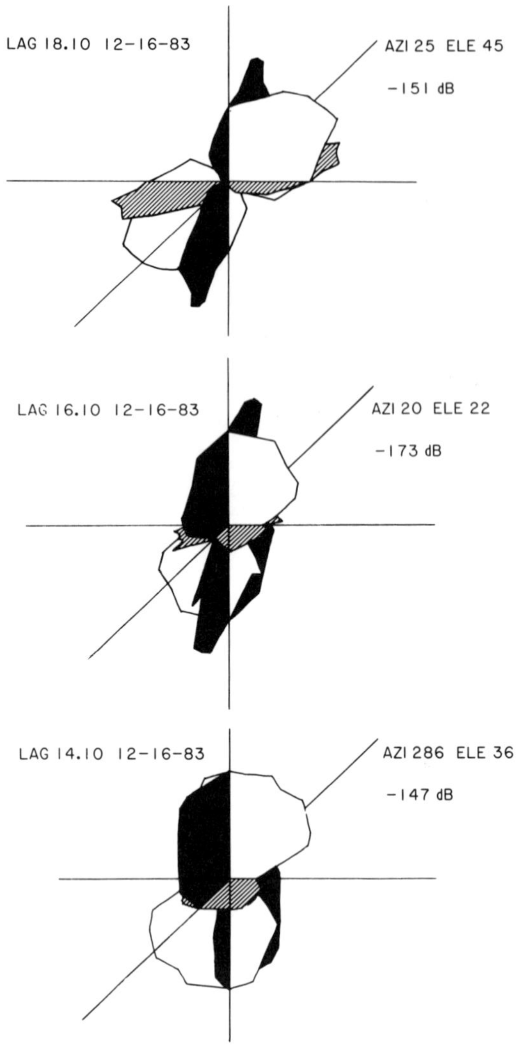


Figure 28.5. Directional sensitivity for three representative lagenar fibers plotted as in Figure 28.3.

vary little across species in their relation to the swim bladder and in their hair cell orientation patterns and since the distributions of best directions among lagenar and utricular fibers are widely dispersed, it is tempting to suggest that these organs play a major role in resolving particle motion axes in most fish species.

In addition to measures of directionality, the above studies provide data on absolute sensitivity, frequency response, and phase locking of fibers innervating the otolith organs. For 140-Hz stimuli, the lowest thresholds (for stimulation in the best excitatory direction) vary from about 1 $\mu$m to about 0.1 nm (rms) across fibers. The lowest sensitivity corresponds to a far field sound pressure level of about 2 dB re: 1 dyne $cm^{-2}$. Since the best sound pressure sensitivity for saccular fibers stimulated in a pressure field (via the swim bladder) is about $-25$ to $-35$

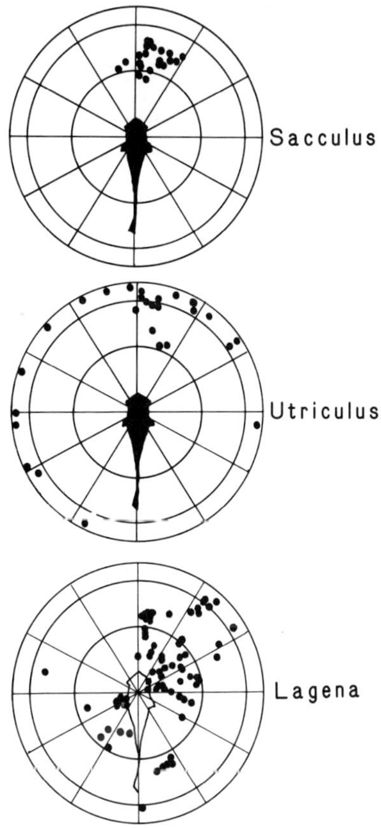


Figure 28.6. The distribution of best direction vectors for twenty-two saccular, eighty-five lagenar, and twenty-nine utricular fibers from goldfish eighth nerve. The view is down onto northern hemisphere with fish's ears at globe's center. *Dots* locate points on globe's surface at which best direction vectors point. The sensitivity is not represented. (From Fay 1984.)

dB (Fay 1981), it seems that a far field signal would have to be 25–35 dB above detection threshold before the particle motion per se would have an independent effect. Fay, Hillery, and Bolan (1982) have pointed out, however, that a potentially detectable interaction between the pressure and particle motion components is theoretically possible in the region of the sound pressure threshold.

In response to whole body acceleration, the tuning (sensitivity to particle acceleration as a function of frequency) of fibers innervating the three otolith organs tend to have the form of low-pass functions (Fay 1981; Coombs and Fay 1984). The utricular and lagenar fibers begin to roll off in sensitivity in the 100–300-Hz range, whereas some of the saccular fibers remain sensitive up to 800 Hz or so. This indicates that the wider bandwidth of the saccule is not simply determined by its association with the swim bladder, but that the saccule itself is adapted for response to relatively high frequencies.

One potentially troubling aspect of the results on fiber response to head movement is that fibers from the same otolithic organ differ considerably with regard to the stimulus phase angle at which spikes tend to occur. As noted above, present theories of directional hearing hold that the phase relations between the pressure and particle motion components are important in solving 180° confusions.

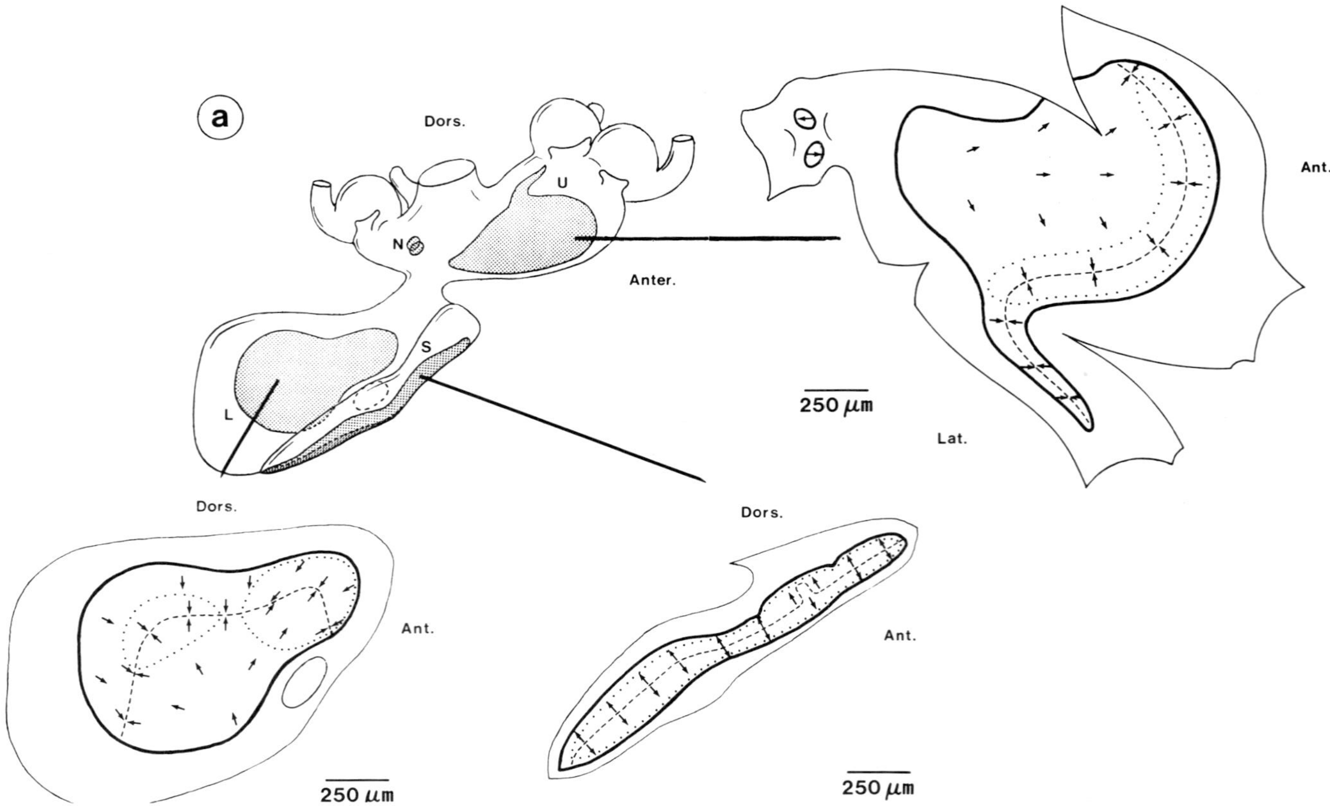


Figure 28.7. A diagram of left ear of the goldfish (looking laterally from medial location). Maculae of the saccule (*S*), lagena (*L*), utricle (*U*), and macula neglecta (*N*) are shown stippled. Surrounding are diagrams of each macula, indicating orientations of hair cells in different regions. Note that maculae of the saccule and lagena fall into approximately vertical plane, whereas that of the utricle lies near horizontal plane. (From Platt 1977, with permission.)

The physiological results show that the stimulus phase angle is not coded unambiguously in auditory nerve fibers, and it is not presently obvious how the phase information required by the theory is analyzed in the face of this ambiguity.

Finally, it must be pointed out that the measured directionality of nerve fibers corresponds qualitatively, but not quantitatively, to the maps of hair cell orientation patterns observed anatomically. For example, the best directions for saccular fibers of the goldfish are homogeneously distributed, as one would expect from the hair cell orientation map. The mean best direction defined physiologically, however, is nearly orthogonal to that predicted from the orientation of the hair cells in the animal's head. This suggests that the relative motion of the otolith and the sensory macula is not along the axis of acoustical particle motion. The motional pathway of the otolith may thus be complexly determined by a number of factors yet unknown. This would further suggest that the anatomical maps of hair cell orientation are insufficient to predict how the particle motion axis is encoded in peripheral fibers.

### Central Nervous System

Little is known about central processing of sound in the fish auditory system and less about the representation of directional information. Horner, Sand, and Enger (1980) have reported an interesting experiment on the cod in which units of the medulla and midbrain (torus semicircularis) were recorded during sound stimulation and anodal electrical block of the saccular nerve. Binaural interaction was demonstrated in some of the cells contacted at both sites, but evidence for binaural inhibition in these cells was seen only in the midbrain. Fay, Hillery, and Bolan (1982) recorded evoked potentials in the goldfish midbrain to both pressure and particle motion stimuli. Swim bladder deflation caused a large loss in pressure sensitivity across the entire frequency range of hearing, but resulted in an increase in potentials evoked by head vibration. This suggests an interaction between directional (particle motion) and nondirectional (pressure) information at the midbrain level. This interaction effect was larger at medial midbrain sites than at more lateral ones.

## Other Aspects of Directional Hearing

### Cocktail Party Effect

Another aspect of directional hearing that is well known in humans is termed the *cocktail party effect,* or the ability to use directional information to differentiate signals from noise when the source of the signal is spatially separate from the source of the noise. Terrestrial animals accomplish this through binaural processing. The interaural time and intensity differences characteristic of a signal from a given source location are used as a basis for its detection in a noisy environment.

This is an example of directional or space domain filtering, and there is good behavioral evidence that it operates for fish (Chapman 1973; Chapman and Johnstone 1974). In these experiments, a tone "signal" source is displaced in the horizontal plane relative to the source of masking noise, and the signal-to-noise ratio at threshold is measured as a function of the angle between them. A separation of 45° and more increased signal detectability by 6–8 dB relative to the condition in which signal and noise sources were coincident. An interpretation of these results is that the inputs from different directions are at least partially independent, since signals with different axes of particle motion stimulate somewhat different populations of directional receptor cells. Buwalda (1981) has also found evidence for this segregation of information in psychophysical experiments using synthesized standing wave sound fields in the lab. Thus signal detectability in fish is not only enhanced by filtering mechanisms in the frequency domain (e.g., Fay and Coombs 1983; Fay and Ream 1986), but also by filtering in the space domain.

## Reflex Orientation

Moulton and Dixon (1967) showed that directional tail flip reflexes of the goldfish to sound required an intact labyrinth. This type of escape response is common in fish and some amphibians and is believed to be mediated by Mauthner cells (M-cells), large paired neurons of the medulla that project to contralateral body musculature. There is input to the M-cell from the otolith organs, the lateral line system, and other sensory modalities. The M-cells mediate directional startle and escape responses to mechanical input such as vibration (Eaton and Kimmel 1980). Since saccular fibers provide some of the input to M-cells in goldfish, we infer that sound pressure information is input to the system. Blaxter, Denton, and Gray (1981) and Blaxter and Hoss (1981) studied sound-evoked startle responses in herring shoals (*Clupea harengus* L.) which may be mediated by the M-cells. Although both sound pressure and particle motion are probably necessary to provide the directional information, these authors concluded that the pressure waveform (mediated by the utricle in this species) triggers the response. It is not yet clear how directional information is coded in the M-cell's afferent input and how this information is processed to help the system decide which way to move.

## Cues for the Structure of Acoustic Environments

Underwater acoustic environments can be reverberant, since the bottom and surface are excellent sound reflectors. For humans, information about reflecting surfaces and about the spatial relations among sound sources and reflectors relative to the listener are contained in the waveforms reaching the ear that combine direct sound with delayed reflections (Supa, Cotzin, and Dallenbach 1944). A simple stimulus for studying the detection and processing of this type of information is "repetition noise," or the addition of a noise signal to its delayed repetition

(Bilsen and Ritsma 1970). When the "echo" is comparable in amplitude to the original signal (within 15 dB) and when delay times (t) are in the 1–50 ms range, humans judge repetition noise to have a pitch that is equal in frequency to 1/t. These coloration attributes of sound convey some information on the proximity of reflectors and on the spatial relations among sources, reflectors, and receivers.

Fay, Yost, and Coombs (1983) studied the psychophysics and neurophysiology of repetition noise processing by the goldfish. Goldfish can discriminate a change in echo delay time of about 5% for delays in the range from 1.25 to 10 ms, even when the echo is attenuated relative to the direct sound. Neurophysiological studies showed that the information used in making these discriminations is likely encoded in the time domain in saccular fibers that phase lock to the fine structure of the waveform. Although acute sensitivity to echo delay (rivaling the human) (Yost, Hill, and Perez-Falcon 1978) certainly exists for the goldfish, we do not yet know whether it provides useful information for acoustic orientation in natural environments. Tavolga (1976) has speculated that this type of processing may underlie the ability of the sea catfish *Arius felis* to detect and avoid obstacles using a simple form of echolocation.

## Summary and Conclusions

A primary function of the auditory system in fish is to help form a three-dimensional image of the local environment. Unlike most terrestrial animals, fish are unable to use interaural differences in time (or phase) and intensity at the two ears as a basis for spatial hearing. This is due to the high velocity (and long wavelengths) of sound under water and to the fact that the otolithic ears of fish do not encode spatially independent samples of the acoustic field. Information about the spatial attributes of underwater sound sources is contained in the directions of acoustic particle motion and in the amplitude and phase relations between particle motion and sound pressure. Sound pressure is indirectly detected through the ears' response to the motion of the walls of a swim bladder (or other gas sac) as it expands and contracts in a pressure field. Acoustic particle motion is detected directly by the otolithic ears, acting as accelerometers.

New physiological experiments on the coding of particle motion within single eighth nerve fibers have been carried out on the goldfish. The results show that each fiber seems to receive input from receptors having a three-dimensional directional characteristic functionally equivalent to a dipole. Each fiber can thus be characterized with a "best axis" (in three-dimensional space) for detecting particle motion. Since the best axes of fibers are distributed widely both within and between otolith organs, there is the possibility for a place principle for coding particle motion axis possibly including all three otolith organs. Although information about the axis of particle motion exists in the peripheral neural code, it is still not clear how this information is used to determine the spatial location of sound sources unambiguously. This is particularly true within the near field, where particle motion axes may not point towards the sound source.

In addition to locating particular sources of sound, auditory systems may be able to use the directional characteristics of sound fields to enhance signal-to-noise ratio for sound detection and to help achieve a more global perception of the structure of their local environment. The former depends on the essential directivity of hair cell receptors to achieve directional filtering; the latter, on the coding of complex spectral coloration characteristic of sound in reverberant environments.

## Acknowledgments

The writing of this chapter was supported by a Research Career Development Award from the NINCDS to R.F. and by a Program Project Grant from NINCDS to the Parmly Hearing Institute.

## References

Bilsen, F., and Ritsma, R. (1970) Repetition pitch and its implications for hearing theory, Acustica, 22:63–68.

Blauert, J. (1982) Binaural localization: multiple images and applications in room- and electro-acoustics, in Localization of Sound: Theory and Applications, Gatehouse, R.W. (ed.), Amphora Press, Groton, Conn., pp. 65–83.

Blaxter, J.H.S., Denton, E.J., and Gray, J.A.B. (1981) Acousticolateralis system in clupeid fishes, in Hearing and Sound Communication in Fishes, Tavolga, W.N., Popper, A.N., and Fay, R.R. (eds.), Springer-Verlag, New York, pp. 39–59.

Blaxter, J.H.S., and Hoss, D.E. (1981) Startle response in herring: the effect of sound stimulus frequency, size of fish and selective interference with the acoustico-lateralis system, J. Mar. Biol. Assoc. U.K., 61:871–879.

Butler, R.A., and Belendiuk, K. (1977) Spectral cues utilized in the localization of sound in the median sagittal plane, J. Acoust. Soc. Am., 61:1264–1269.

Buwalda, R.J.A. (1981) Segregation of directional and non-directional acoustic information in the cod, in Hearing and Sound Communication in Fishes, Tavolga, W.N., Popper, A.N., and Fay, R.R. (eds.), Springer-Verlag, New York, pp. 139–172.

Chapman, C.J. (1973) Field studies of hearing in teleost fish, Helgol. Wiss Meeresunters., 24:371–390.

Chapman, C.J., and Hawkins, A.D. (1973) A field study of hearing in the cod (*Gadus morhua*), J. Comp. Physiol., 85:147–167.

Chapman, C.J., and Johnstone, A.D. (1974) Some auditory discrimination experiments on marine fish, J. Exp. Biol., 61:521–528.

Chapman, C.J., and Sand, O. (1974) A field study of hearing in two species of flatfish—*Pleuronectes platessa* (L.) and *Limanda limanda* (L.), Comp. Biochem. Physiol., 47:371–385.

Coombs, S.L., and Fay, R.R. (1984) Acoustic coding by the ear of the sunfish (*Lepomis gibbosus*): sensitivity, tuning, and directionality of 8th nerve units, J. Acoust. Soc. Am., 75:S9.

Denton, E.J., Gray, J.A.B., and Blaxter, J. (1979) The mechanics of the clupeid acoustico-lateralis system: frequency responses, J. Mar. Biol. Assoc. UK, 59:27–47.

Denton, E.J., and Gray, J.A.B. (1983) Mechanical factors in the excitation of clupeid lateral lines, Proc. R. Soc. Lond. B Biol. Sci., 218:1–26.

de Vries, H. (1950) The mechanics of the labyrinth otoliths, Acta Oto-Laryngol, 38:262–273.

Dijkgraaf, S. (1960) Hearing in bony fishes, Proc. R. Soc. Lond. B Biol. Sci., 152:51–54.

Durlach, N.I. (1972) Binaural signal detection: equalization and cancellation theory, in Foundations of Modern Auditory Theory, Tobias, J. (ed.), Academic Press, New York, pp. 369–458.

Eaton, R.A., and Kimmel, C.B. (1980) Directional sensitivity of the Mauthner cell system to vibrational stimulation in zebrafish larvae, J. Comp. Physiol., 140:337–342.

Fay, R.R. (1981) Coding of acoustic information in the 8th nerve, in Hearing and Sound Communication in Fishes, Tavolga, W.N., Popper, A.N., and Fay, R.R. (eds.), Springer-Verlag, New York, pp. 189–222.

Fay, R.R. (1984) The goldfish ear codes the axis of acoustic particle motion in three dimensions, Science, 225:951–953.

Fay, R.R., and Coombs, S.L. (1983) Neural mechanisms of sound detection and temporal summation, Hear. Res., 10:69–92.

Fay, R.R., Hillery, C., and Bolan, K. (1982) Representation of sound pressure and particle motion information in the midbrain of the goldfish, Comp. Biochem. Physiol., 71A:181–191.

Fay, R.R., and Olsho, L.W. (1979) Discharge patterns of lagenar and saccular neurons of the goldfish eighth nerve: displacement sensitivity and directional characteristics, Comp. Biochem. Physiol., 62:377–386.

Fay, R.R., and Patricoski, M. (1980) Sensory mechanisms for low frequency vibration detection in fishes, in Abnormal Animal Behavior Preceding Earthquakes, Conference II, U.S. Geological Survey Open File Report 80–453, Buskirk, R. (ed.), pp. 63–91.

Fay, R.R., and Popper, A.N. (1980) Structure and function of teleost auditory systems, in Comparative Studies of Hearing in Vertebrates, Popper, A.N., and Fay, R.R., (eds.), Springer-Verlag, New York, pp. 3–42.

Fay, R.R., and Ream, T.J. (1986) Acoustic response and tuning in saccular nerve fibers of the goldfish (*Carassius auratus*), J. Acoust. Soc. Am. Vol 79, 1883–1895.

Fay, R.R., Hillery, C., and Bolan, K. (1982) Representation of sound pressure and particle motion information in the midbrain of the goldfish, Comp. Biochem. Physiol., 71A:181–191.

Fay, R.R., Yost, W.A., and Coombs, S.L. (1983) Psychophysics and neurophysiology of repetition noise processing in a vertebrate auditory system, Hear. Res., 12:31–55.

Furukawa, T. (1978) Sites of termination on the saccular macula of auditory nerve fibers in the goldfish as determined by intracellular injection of procion yellow, J. Comp. Neurol., 180:807–814.

Harris, G.G., and van Bergeijk, W.A. (1962) Evidence that the lateral-line organ responds to near-field displacements of sound sources in water, J. Acoust. Soc. Am., 34:1831–1841.

Hawkins, A.D., and Horner, K. (1981) Directional characteristics of primary auditory neurons from the codfish ear, in Hearing and Sound Communication in Fishes, Tavolga, W.N., Popper, A.N., and Fay, R.R. (eds.), Springer-Verlag, New York, pp. 311–328.

Hawkins, A.D., and Sand, O. (1977) Directional hearing in the median vertical plane by the cod, J. Comp. Physiol., 122:1–8.

Horner, K., Sand, O., and Enger, P. (1980) Binaural interaction in the cod, J. Exp. Biol., 85:323–332.

Kleerekoper, H., and Chagnon, E.C. (1954) Hearing in fish with special reference to *Semotilus atromaculatus* (Mitchell), J. Fish Res. Board Can., 11:130–152.

Knudsen, E.I., and Konishi, M. (1978) A neural map of auditory space in the owl, Science, 200:795–797.

Kuhn, G. (1979) The pressure transformation from a diffuse sound field to the external ear and to the body and head surface, J. Acoust. Soc. Am., 65:991–1000.

Kuhn, G. (1982) Towards a model for sound localization, in Localization of Sound: Theory and Applications, Gatehouse, G.W. (ed.), Amphora Press, Groton, Conn., pp. 51–64.

Moulton, J.M., and Dixon, R. (1967) Directional hearing in fishes, in Marine Bio-Acoustics, Tavolga, W.N. (ed.), Pergamon Press, Oxford, pp. 187–232.

Platt, C. (1977) Hair cell distribution and orientation in goldfish otolithic organs, J. Comp. Neurol., 172:283–297.

Popper, A.N. (1977) A scanning electron microscopic study of the sacculus and lagena in the ears of fifteen species of teleost fishes, J. Morphol., 153:397–418.

Popper, A.N., and Coombs, S.L. (1982) The morphology and evolution of the ear in actinopterygian fishes, Am. Zool., 22:311–328.

Saidel, W.M., and Popper, A.N. (1983) Spatial organization in the saccule and lagena of a teleost: hair cell pattern and innervation, J. Morphol., 177:301–317.

Sand, O. (1981) The lateral line and sound reception, in Hearing and Sound Communication in Fishes, Tavolga, W.N., Popper, A.N., and Fay, R.R. (eds.), Springer-Verlag, New York, pp. 459–480.

Schuijf, A. (1975) Directional hearing of cod (*Gadus morhua*) under approximate free field conditions, J. Comp. Physiol., 98:307–332.

Schuijf, A. (1981) Models of acoustic localization, in Hearing and Sound Communication in Fishes, Tavolga, W.N., Popper, A.N., and Fay, R.R. (eds.), Springer-Verlag, New York, p. 267.

Schuijf, A., and Buwalda, R.A. (1975) On the mechanism of directional hearing in cod (*Gadus morhua*), J. Comp. Physiol., 98:333–343.

Schuijf, A., and Buwalda, R.A. (1980) Sound localization: a major problem in fish acoustics, in Comparative Studies of Hearing in Vertebrates, Popper, A.N., and Fay, R.R. (eds.), Springer-Verlag, New York, pp. 43–78.

Schuijf, A., and Hawkins, A.D. (1983) Acoustic distance discrimination by the cod, Nature, 302:143–144.

Schuijf, A., and Siemelink, M.E.. (1974) The ability of cod (*Gadus morhua*) to orient towards a sound source, Experientia, 30:773–774.

Supa, M., Cotzin, M., and Dallenbach, K. (1944) "Facial vision": the perception of obstacles by the blind, Am. J. Psychol., 57:133.

Tavolga, W.N. (1976) Acoustic obstacle detection in the sea catfish (*Arius felis*), in Sound Reception in Fish, Schuijf, A., and Hawkins, A.D. (eds.), Elsevier, Amsterdam, pp. 185–204.

van Bergeijk, W.A. (1964) Directional and nondirectional hearing in fish, in Marine Bio-Acoustics, Tavolga, W.N. (ed.), Pergamon Press, Oxford, pp. 281–300.

van Bergeijk, W.A. (1967) The evolution of vertebrate hearing, in Contributions to Sensory Physiology, Neff, W.D. (ed.), Springer-Verlag, New York, pp. 1–49.

von Frisch, K. (1938) The sense of hearing in fish, Nature, 141:8–11.
von Frisch, K., and Dijkgraaf, S. (1935) Können Fische die Schallrichtung wahrnehmen? Z. Vergl. Physiol., 22:64–655.
Yost, W.A., Hill, R., and Perez-Falcon, T. (1978) Pitch and pitch discrimination of broad band signals with rippled power spectra, J. Acoust. Soc. Am., 63:1166–1173.

## Chapter 29

# Central Connections of the Octavolateralis System: Evolutionary Considerations

CATHERINE A. MCCORMICK* and MARK R. BRAFORD, JR.**

The octavolateralis system collectively comprises four classes of sensory end organs and their neuronal connections. These end organs—the auditory and vestibular receptors of the inner ear and the electroreceptors and mechanoreceptors of the lateral line system—have as their sensory elements hair cells or cells believed to be modified hair cells. Various aspects of the octavolateralis system have been repeatedly reviewed in recent years. Information regarding receptor morphology and central anatomy and physiology can be found in this volume (Chaps. 9, 22, 23, 28 and 31–33), as well as in others (e.g., Fessard 1974; Tavolga, Popper, and Fay 1981; Northcutt and Davis 1983; Bullock and Heiligenberg 1986). In this chapter we shall consider the octavolateralis system with respect to the following question: What is the relationship between major evolutionary changes in the sensory periphery and the organization of the CNS? Comparative anatomical and physiological studies have revealed that the octavolateralis system has a complicated and apparently unusual evolutionary history including the loss, invention, and "reinvention" of various receptor classes. It is thus an ideal system to use in such an analysis.

## Patterns of Organization

### Primitive Pattern of Organization

At least some components of the octavolateralis system were present in the earliest vertebrates—the jawless fishes or agnathans. Early fossil agnathans had an inner ear containing semicircular canals (e.g., Jarvik 1980) and thus had at least a vestibular sense; whether an auditory sense was present is uncertain. In fact, it is still unclear whether extant agnathans, the lampreys and hagfishes, have auditory

* Department of Biology, Oberlin College, Oberlin, OH, USA.
** Department of Anatomy, Georgetown University Schools of Medicine and Dentistry, Washington, DC 20007, USA

receptors, although such receptors are present among the jawed fishes. In terms of the lateral line, Thomson (1977) has presented evidence that both mechanoreceptors and electroreceptors were probably present in a group of fossil agnathans, the Osteostraci, and we now know that almost all living representatives of the jawless and primitive jawed fishes as well as certain amphibians, have both electro- and mechanosensory components of the lateral line system (e.g., Hetherington and Wake 1979; Fritzsch 1981; McCormick 1982; Bullock, Bodznick, and Northcutt 1983; Münz, Claas, and Fritzsch 1984). These species share a suite of brain characters associated with the inner ear and lateral line receptors that can be considered to be the primitive condition. This pattern of brain organization will be described for the elasmobranchs, the group for which the most information is available (Boord and Campbell 1977; Bodznick and Northcutt 1980; Boord and Roberts 1980; Boord and Northcutt 1982; Koester 1983; Schweitzer 1983; Bodznick and Northcutt 1984).

All of the octavolateralis receptors are innervated by bipolar neurons that form the anterior and posterior lateral line nerves and the eighth, or octaval, cranial nerve (Fig. 29.1). Depending on the modality they carry, these nerves terminate mainly in one of three longitudinal columns of cells in the medulla (Fig. 29.1). Projections of these nerves to the cerebellum and reticular formation will not be considered here.

In elasmobranchs, electroreceptors are innervated by the dorsal ramus of the anterior lateral line nerve. This ramus terminates in the dorsalmost cell column of the medulla, the nucleus dorsalis (Fig. 29.1). Lateral line mechanoreceptors

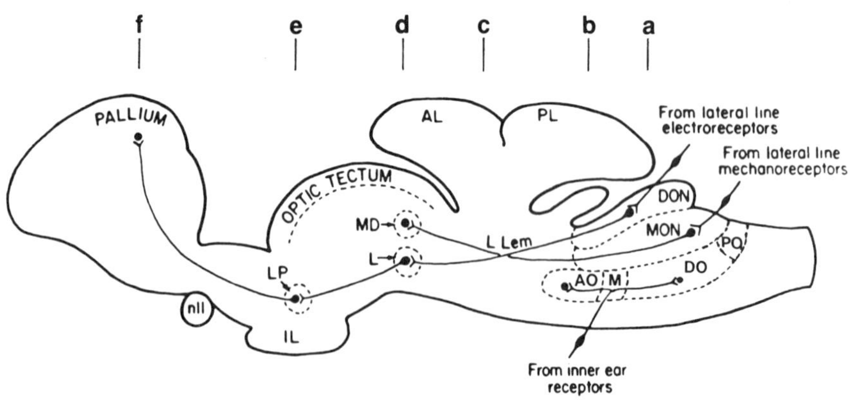


Figure 29.1. Diagrammatical lateral view of the brain of a cartilaginous fish, showing central octavolateralis pathways as presently known. Vertical lines above the figure indicate caudorostral levels of sections *a–f* shown in Figure 29.2; *AO,* anterior octaval nucleus; *AL,* anterior lobe of the cerebellum; *DO,* descending octaval nucleus; *DON,* dorsal (electrosensory) octavolateralis nucleus; *IL,* inferior lobe of the hypothalamus; *L,* lateral nucleus of lateral mesencephalic nuclear complex; *L Lem,* lateral line lemniscus; *LP,* lateral posterior thalamic nucleus; *M,* magnocellular octaval nucleus; *MD,* dorsomedial nucleus of lateral mesencephalic nuclear complex; *MON,* medial (mechanosensory) octavolateralis nucleus; *nII,* optic nerve; *PL,* posterior lobe of the cerebellum; *PO,* posterior octaval nucleus. (Modified from Boord and McCormick 1984.)

on the head are innervated by the ventral ramus of the anterior lateral line nerve; those on the trunk, by the posterior lateral line nerve. In both cases, these mechanosensory fibers terminate mainly in the intermediate longitudinal cell column, positioned ventral or ventrolateral to the electrosensory column. The major constituent of the mechanosensory column is the nucleus medialis, with a smaller nucleus caudalis forming the posterior part of the column. Finally, the eighth cranial nerve in elasmobranchs, carrying both auditory and vestibular input, terminates mainly in the ventralmost longitudinal column—the octaval column. This column contains four rostrocaudally aligned nuclei—the anterior, magnocellular, descending, and posterior octaval nuclei.

There are a small number of lateral line and octaval fibers in elasmobranchs that do not terminate in their respective medullary columns (Koester 1983). For example, lateral line mechanosensory fibers have been traced to the magnocellular octaval nucleus, and a few eighth nerve fibers course into nucleus medialis. The significance of these inputs, which have also been reported in other species, is unknown. The basic pattern in elasmobranchs, however, is one in which the *vast majority* of lateral line inputs are dorsal to the *vast majority* of inner ear inputs and in which electrosensory inputs are dorsal or dorsomedial to lateral line mechanosensory inputs. The same central organization of the two classes of lateral line input is also seen in those jawless and jawed fish and amphibians that retain the primitive state of the lateral line system (see p. 738). The projection of the eighth nerve into a ventral cell column also appears to be characteristic of many species, including lampreys and hagfish, even though in agnathans the nuclear constituents of the octaval column are not easily compared to those of jawed fish (Northcutt 1979; Amemiya et al. 1985).

Some of the higher-order centers of the octavolateralis system in elasmobranchs are schematically diagrammed in Figures 29.1, 29.2, and 29.3, and three points will be emphasized. First, the electrosensory, the mechanosensory, and the presumed auditory system form ascending lemniscal pathways that are known or believed to reach the telencephalon (e.g., Platt et al. 1974; Bullock and Corwin 1979). Second, so far as is currently known, these three modalities are processed in separate nuclei or areas at each level of the neuraxis. Although there are likely to be areas of integration of these inputs, the basic ascending pathways are largely segregated from one another. In contrast, the vestibular system in vertebrates in general and presumably in elasmobranchs (Smeets and Timerick 1981) does not display this kind of ascending lemniscal pattern (Fig. 29.4). In vertebrates, vestibular neurons of the octaval column project instead to the cerebellum, spinal cord, and motor nuclei of the extrinsic eye muscles. There thus appear to be largely separate central channels for the four end organ classes of the octavolateralis system.

## Patterns of Organization in Bony Fish

A consideration of the general organization of the central pathways of the octavolateralis system in the bony fish reveals an overall similarity to the pattern seen in elasmobranchs. Although the data are extremely limited, both in terms of the

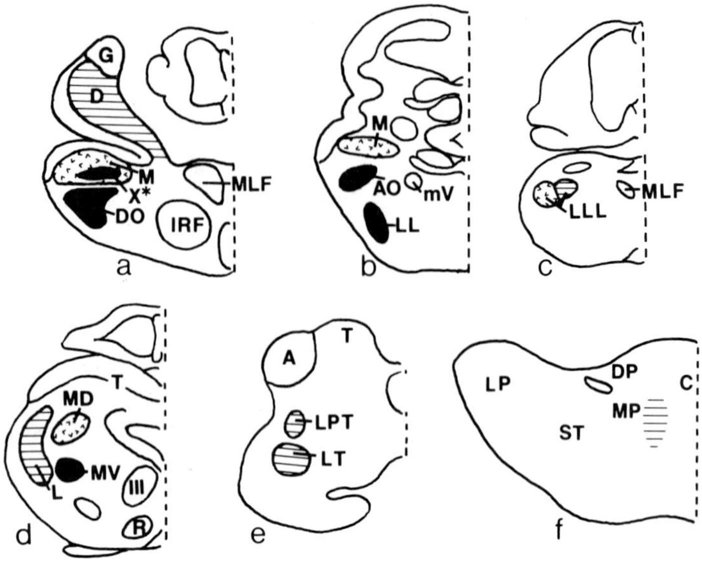


Figure 29.2. Line drawings of caudorostral series of hemisections through the brain of the cartilaginous fish *Raja eglanteria*. Mechanosensory lateral line centers are shown with floating *v's*, electrosensory lateral line centers are shown with horizontal hatching, and octaval centers are shown in black. Those octaval centers believed to process auditory information are indicated with an *asterisk; A,* anterior nucleus of lateral mesencephalic nuclear complex; *AO,* anterior octaval nucleus; *C,* central nucleus of the pallium; *D,* dorsal octavolateralis nucleus; *DO,* descending octaval nucleus; *DP,* dorsal pallium; *G,* granule cells of eminentia granularis; *IRF,* inferior reticular formation; *L,* lateral nucleus of lateral mesencephalic nuclear complex; *LL,* nucleus of lateral lemniscus; *LLL,* nucleus of lateral line lemniscus; *LP,* lateral pallium; *LPT,* lateral posterior nucleus of the thalamus; *LT,* lateral tuberal nucleus; *M,* medial octavolateralis nucleus; *MD,* dorsomedial nucleus of lateral mesencephalic nuclear complex; *MLF,* medial longitudinal fasciculus; *MP,* medial pallium; *MV,* ventromedial nucleus of lateral mesencephalic nuclear complex; *mV,* motor nucleus of trigeminal nerve; *R,* nucleus ruber; *ST,* striatum; *T,* optic tectum; *X,* cell plate x; *III,* oculomotor nucleus. (Compiled from Bodznick and Northcutt 1984; Boord and Northcutt 1982; Corwin and Northcutt 1982; Schweitzer 1983; and Northcutt 1978.)

number of species studied and in the extent to which the complete pathways are known, it appears that second-order vestibular neurons conform to the pattern shown in Figure 29.4 (Bell 1981a; McCormick unpublished observations), and that the electrosensory, lateral line mechanosensory, and presumed auditory pathways can be described by Figure 29.3 (e.g., Finger 1980; Echteler 1985; Bullock and Heiligenberg 1986). For the ascending lemniscal pathways then, second-order neurons in the medulla project primarily contralaterally to the midbrain (directly or indirectly via a relay nucleus), the midbrain projects primarily ipsilaterally to the thalamus, and the thalamus projects primarily ipsilaterally (or, in the case of the elasmobranch electroreceptive pathway, contralaterally [Bodznick and Northcutt 1984]) to the telencephalon. For the vestibular, presumed auditory,

Figure 29.3. Basic pattern of ascending lemniscal pathways shared by lateral line and auditory systems.

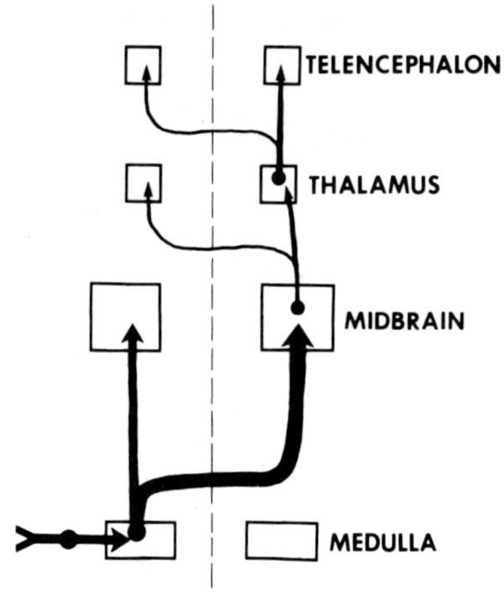


and lateral line mechanosensory systems, this similarity across species is not surprising, because all available evidence suggests that the lateral line mechanoreceptors and the inner ear receptors of fish each arose just once in vertebrate history. The basic central pathways for these senses were established early and, like the peripheral receptors, were retained in later radiations. In other words, the similarity of the mechanosensory pathways across species is evidence of their homology, or inheritance from a common ancestor, and likewise for the other two modalities. In the case of electroreception, however, the similarity between teleost and nonteleost central electrosensory pathways, as well as the similarity in these pathways among the distantly related electroreceptive teleost groups, is remarkable, given our current understanding of the evolution of electroreception. As described below, it has been hypothesized that electroreception in teleosts was independently evolved at least twice (Bullock 1974; Bullock, et al. 1983), and thus neither the receptors nor the central pathways of a given group are homologous to those in other teleost groups or in nonteleost electroreceptive groups.

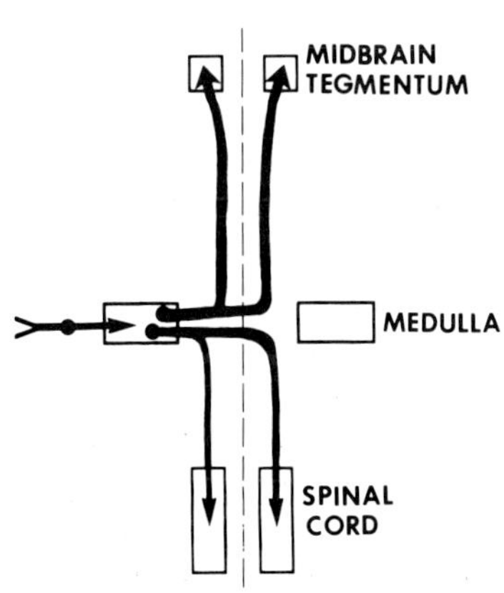


Figure 29.4. Basic pattern of vestibular system pathways.

# Electrosensory Systems

## Evolutionary History of Electroreception

Four patterns of organization are currently known for the lateral line region of the medulla in fish and amphibians (Figs. 29.5 and 29.6 [Braford, McCormick, and Saidel in prep.]). The organization of electrosensory (positioned dorsal and medial) and mechanosensory (positioned ventral and lateral) inputs described in elasmobranchs is likely the primitive pattern for the jawed and possibly the jawless fishes (Fig. 29.5). As indicated in Figure 29.6, this pattern is found in petromyzontid agnathans, chondrichthians, some early actinopterygian radiations (polypteriforms and chondrosteans), all extant sarcopterygians, and some amphibians (Fritzsch 1981; McCormick 1982; Bullock, et al. 1983). This pattern of medullary organization is indicative of the presence of the primitive electrosensory system, composed of cathodally activated ampullary receptors with input to nucleus dor-

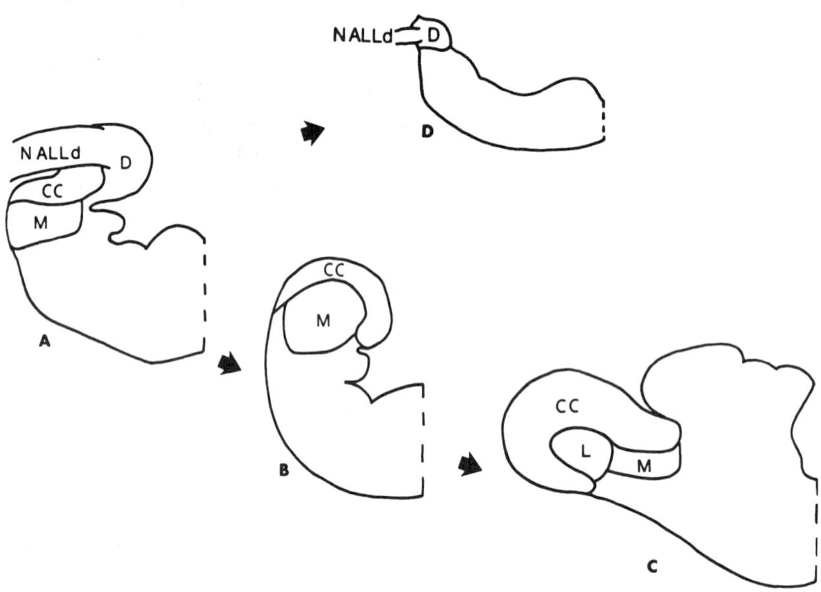


Figure 29.5. The four known patterns of organization of lateral line centers of the medulla. (A) Primitive pattern in which dorsal ramus of anterior lateral line nerve innervates electroreceptors and terminates in dorsal nucleus and ventral ramus of anterior lateral line nerve and posterior lateral line nerve innervate mechanoreceptors and terminate in medial nucleus. (B) A pattern derived from (A) by loss of the electrosensory system. (C) A pattern derived from (B) by the evolution of a new electrosensory system. (D) A pattern derived from (A) by loss of the mechanosensory system; *CC*, cerebellar crest; *D*, dorsal (electrosensory) octavolateralis nucleus; *L*, electrosensory lateral line lobe; *M*, medial (mechanosensory) octavolateralis nucleus; *NALLd*, dorsal (mechanosensory) ramus of anterior lateral line nerve.

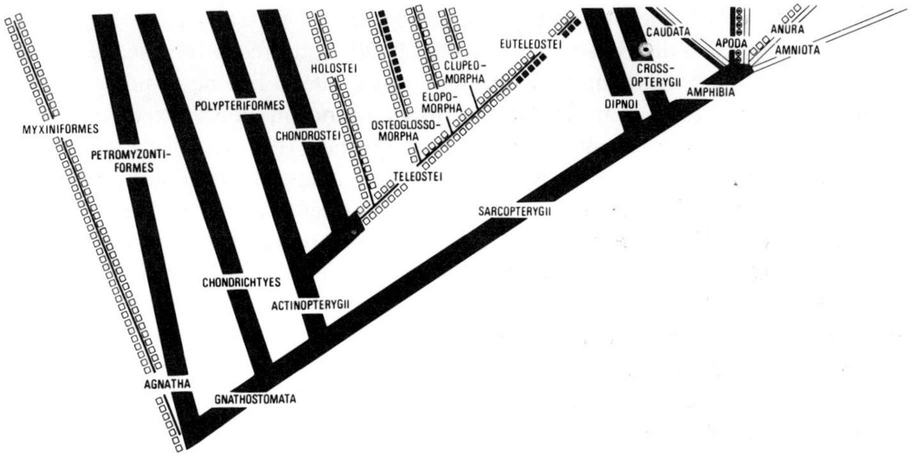


Figure 29.6. Phylogenetic distribution of the four patterns of organization of lateral line system. *Solid black* represents the primitive condition in which ampullary electroreceptors and mechanoreceptors, along with their respective medullary termination sites—nucleus dorsalis and nucleus medialis—are present. Note that electroreceptors of Petromyzontiformes are not ampullary in morphology (Ronan and Bodznick 1986). *Open squares* represent a derived condition in which only mechanoreceptors and nucleus medialis are present. Note that it is not known whether the reported absence of electroreceptors in Myxiniformes is primitive or derived. *Solid black squares* represent the derived condition of electrosensory teleosts in which mechanoreceptors and nucleus medialis are present, in addition to newly evolved electroreceptors and a medullary electrosensory lateral line lobe. *Circled dots* represent a derived condition in which only ampullary electroreceptors and nucleus dorsalis are present. *Parallel lines* represent the absence of lateral line system and central lateral line processing areas.

salis via the dorsal ramus of the anterior lateral line nerve (McCormick 1982). Likewise the presence of nucleus medialis reflects the presence of peripheral lateral line mechanoreceptors.

One derived state is seen among some groups of caecilian (the Gymnophiona) amphibians (Fig. 29.5D), in which the primitive electroreceptors and their central processing zone are retained, but the lateral line mechanoreceptors and at least nucleus medialis are absent. A second condition derived from the primitive state is present in the holosteans and most teleosts (Fig. 29.5B) and occurs independently in some amphibians. It involves the loss of the primitive electroreceptors, the exclusively electroreceptive dorsal ramus of the anterior lateral line nerve, and at least the nucleus dorsalis, but the retention of the lateral line mechanoreceptors and their central processing zones. In most of the anuran amphibians, this derived pattern is further modified by the complete loss during metamorphosis of the peripheral mechanoreceptive lateral line system, the lateral line nerves, and at least the nucleus medialis. Among holosteans and teleosts, the loss of peripheral electroreceptors and their central pathways appears to be the prevalent condition in the vast majority of species. Known electroreceptive teleosts comprise less than 1% of all teleost species and belong to two distantly related groups, the Ostariophysi and the Osteoglossomorpha. In all electroreceptive teleosts, electrosensory afferents, coursing in one or both of the lateral line nerves (depending on the species), terminate in the electrosensory lateral line lobe, a region lateral or dorsolateral to the mechanosensory nucleus medialis (Fig. 29.5C). Electrosensory teleosts thus possess ampullary receptors that are anodally activated and differ in innervation and central connections from those of earlier-derived electrosensory groups. The tuberous-type electroreceptors of electroreceptive teleosts that possess electric organs are presumably later specializations, probably derived from teleost ampullary receptors. The tuberous electroreceptors are likewise innervated by fibers in one or both lateral line nerves that terminate centrally in the electrosensory lateral line lobe.

The origin of the teleostean electrosensory system is unknown. We consider the phylogenetic distribution of the sense and the major differences in physiology and morphology of the peripheral and central receptor systems to indicate that independent evolution of electrosensitivity in teleosts from nonelectroreceptive ancestors is the most parsimonious explanation. With our current understanding of teleost phylogeny, the alternative hypothesis—that teleost electroreceptors were instead inherited from an electroreceptive nonteleost with the primitive pattern—would require (1) the multiple loss of this sense in many teleost lines and its retention in only a handful or (2) the suppression of the genome for electroreception with its reexpression in modified form in only a few species. We believe that these possibilities are considerably weakened by the morphological and physiological dissimilarities of the electroreceptors and their central pathways both between teleosts and nonteleosts and between the two electroreceptive radiations of teleosts themselves (Bullock and Heiligenberg 1986). We consider the central nuclei for the electric sense to be nonhomologous between the two teleost groups, as well as between each of these groups and electroreceptive nonteleosts.

## Possible Origins of Teleost Electrosensory Pathways

If the hypothesis of an independent evolution of electroreception in the teleosts is accepted, then what interpretation can be placed on the overall similarity in the ascending pathways for electroreception throughout the neuraxis in the different electroreceptive groups? Considering first only the teleosts, Figure 29.7 illustrates the main electro- and mechanosensory nuclei of the medulla and midbrain in several species belonging to two distantly related groups, the Ostariophysi and the Osteoglossomorpha. Each of the two groups has nonelectroreceptive species (e.g., *Carassius* and *Notopterus*), electroreceptive species having only ampullary receptors (e.g., *Ictalurus* and *Xenomystus*), and electroreceptive species having both ampullary and "tuberous" receptors (e.g., *Eigenmannia* and *Gnathonemus*). The lateral line lobes of *Ictalurus* and *Xenomystus* are relatively small and have similar, though not identical, connections with the torus semicircularis, as well as with an isthmal cell group, nucleus praeeminentialis. The species that are both electrosensitive and electric, *Eigenmannia* and *Gnathonemus,* have greatly expanded electrosensory lateral line lobes with separate projection zones for ampullary and tuberous receptors. Both species have two classes of tuberous receptors, but the two are not easily compared across species. The tuberous receptors are derived independently in each case, presumably from ampullary receptors. Interestingly, the fibers innervating the tuberous receptors form three body maps in the medulla in each species, though the details of the maps are different. In *Eigenmannia,* the central processes of neurons innervating both P and T tuberous receptors (Chap. 32) map onto each of three cortical areas in the electrosensory lateral line lobe (Carr, Maler, and Sas 1982). In fact, a given tuberous fiber trifurcates and projects to each of the maps (Heiligenberg and Dye 1982). In *Gnathonemus,* the two tuberous-type receptors are functionally segregated. The Knollenorgans (specialized for electrocommunication) project somatotopically to a nuclear zone, the nucleus of the lateral line lobe, while the mormyromasts (specialized for electrolocation), project somatotopically onto each of two cortical zones of the electrosensory lateral line lobe (Bell and Russell 1978).

In both species, the four body maps (one ampullary and three tuberous) are merged within the torus semicircularis, though again the details are different. In *Eigenmannia,* the torus semicircularis is a laminated structure containing a dorsal electrosensory zone in which the four body maps come into register (Maler 1979; Maler, Sas, and Rogers 1981). In *Gnathonemus,* the torus semicircularis has a nuclear structure, and the maps merge in nucleus lateralis, though Knollenorgan input is also separately processed in nucleus extrolateralis (Bell, Finger, and Russell 1981; Haugède-Carré 1979, 1980).

Figure 29.7 shows that the organization of the lateral line areas is more similar within each of the two distantly related groups (the ostariophysans and the osteoglossomorphs) than between them. Figure 29.7 also shows that, barring the details, there is substantial similarity between these two presumably independently derived electrosensory systems. We suggest that this similarity has its foundations, *not* in a shared electroreceptive ancestor of ostariophysans and osteoglossomorphs, but in the independent, but parallel, origins of the two electrosensory systems from the mechanosensory system. Note that the overall patterns of the two electrosensory pathways are as similar to that of the lateral line mechanoreceptive pathway as they are to each other. Since the mechanosensory pathways are homologous and virtually identical in the two groups, that similarity may account for the more general similarity of the electrosensory pathways derived from them, with the differences in their details being a result of their independent origins and/or their subsequent independent modifications.

The presence of multiple maps in the electrosensory lateral line lobe in the two groups may be a consequence of the processes by which they originated. Kaas (1982) has proposed that duplication may be responsible for the multiple sensory areas for a given modality in the mammalian neocortex. As shown in Figure 29.8, the duplicated brain area, which has appeared ''suddenly,'' is organized as a rough mirror image of the original area and might initially be innervated by collaterals of the original afferents. The redundancy would allow for subsequent modification of the duplicated area and its innervation without compromising the function of the original area.

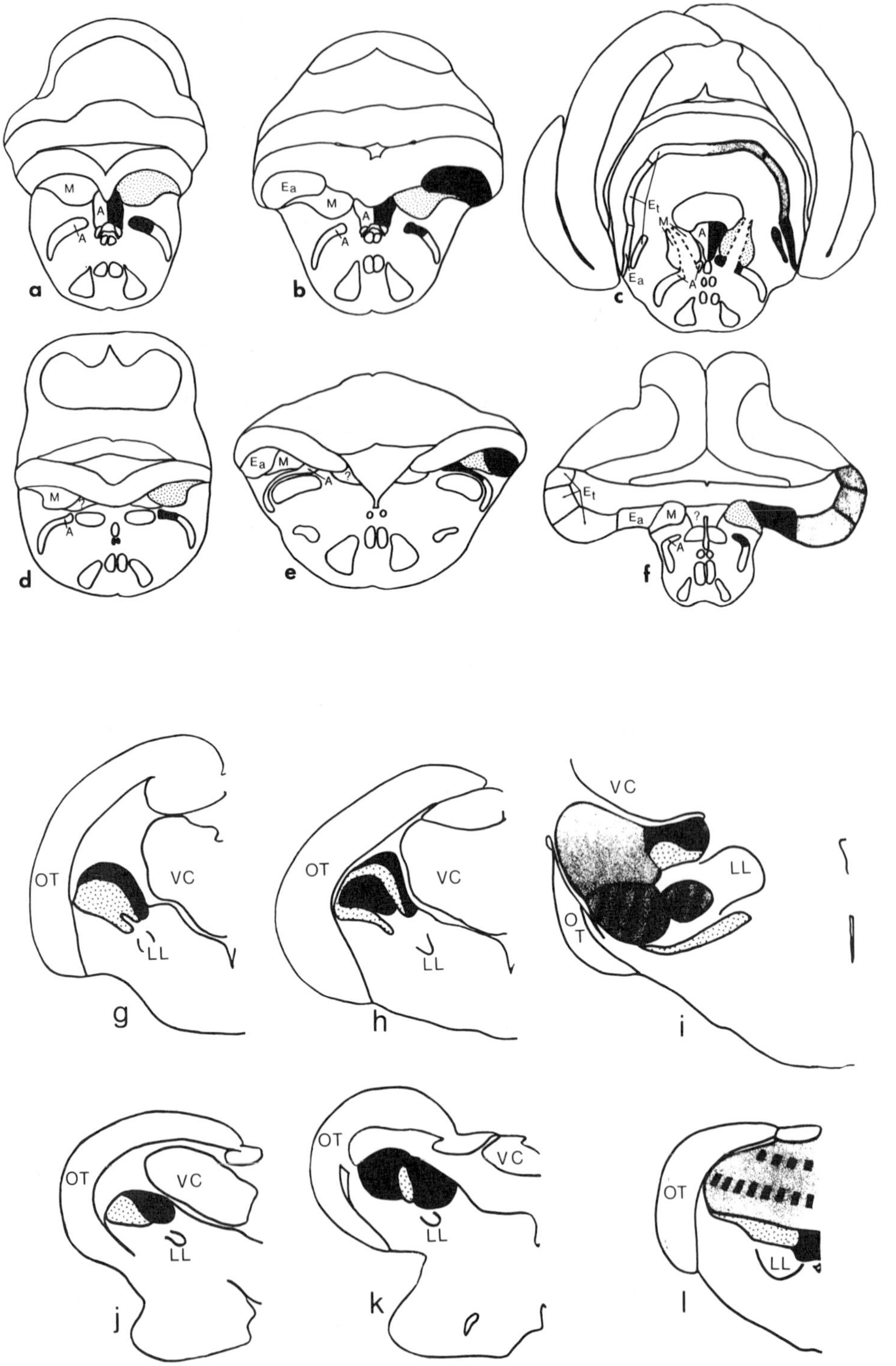
M
A
A
a
Ea
M
A
A
b
Et
M
A
Ea
A
c
M
?
A
d
Ea
M
A
?
e
Et
Ea
M
?
A
f
OT
VC
LL
g
OT
VC
LL
h
VC
LL
OT
i
OT
VC
LL
j
OT
VC
LL
k
OT
LL
l

Figure 29.8. Diagrammatical representation of duplication as an evolutionary mechanism for producing multiple sensory areas for a given sensory modality. (From Kaas 1982.)

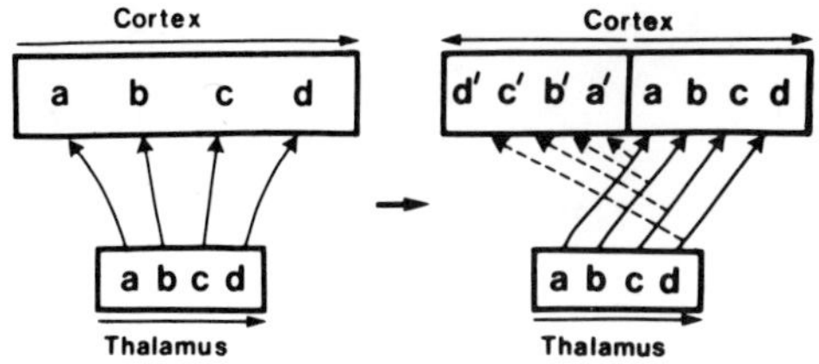


The organization of the multiple sensory maps in the medulla of *Eigenmannia* and other gymnotids is remarkably similar to the situation that would be expected under the duplication hypothesis. Gymnotids are believed to have evolved from an electroreceptive ancestor possessing only ampullary receptors (a condition like that found in *Ictalurus*). Figure 29.9 illustrates both the mirror image relationships of the three tuberous zones to each other and that of the ampullary zone to the neighboring tuberous zone. Also shown is the trifurcation of a single tuberous fiber and its projections to each of the tuberous zones, thereby producing three presumably redundant maps. Possibly, ampullary afferent collaterals originally innervated one or more duplicated ampullary areas, and these connections were subsequently replaced by newly evolved tuberous fibers. Perhaps the present-day innervation pattern of the gymnotid tuberous zones potentially sets the stage for the future functional differentiation of these areas from one another (i.e., their evolution). In *Gnathonemus*, the two mormyromast areas are organized as mirror image maps of one another, but the complicated anatomy of the mormyrid electrosensory lateral line lobe does not present the neat picture for the other receptor areas that is seen in gymnotids (Bell and Russell 1978). Whether or not the various receptor areas develop as mirror images of one another, however, a mirror image pattern of organization may not be a necessary part of the duplication process.

If teleostean ampullary electroreceptors are a derivative of lateral line mechanoreceptors, one can imagine many possible scenarios concerning the sequence of

Figure 29.7. Line drawings of transverse sections through (A) medullary and (B) midbrain levels of six species of teleosts, showing some of the important octavolateralis centers. Three osteoglossomorph species are shown: *Notopterus notopterus* (*a* and *g*), *Xenomystus nigri* (*b* and *h*), and *Gnathonemus petersii* (*c* and *i*); and three ostariophysan species are shown: *Carassius auratus* (*d* and *j*), *Ictalurus punctatus* (*e* and *k*), and *Eigenmannia* sp. (*f* and *l*). Key: *solid black,* auditory/presumed auditory centers; *floating v's,* mechanosensory lateral line centers; *dark gray,* ampullary electrosensory lateral line centers; *light gray,* tuberous electrosensory lateral line centers. Abbreviations: *A,* auditory/presumed auditory center; *Ea,* ampullary electrosensory lateral line center; *Et,* tuberous electrosensory lateral line center; *LL,* lateral lemniscus; *M,* mechanosensory lateral line center, *OT,* optic tectum; *VC,* valvula cerebelli. This figure represents our summary and interpretation of both published and unpublished data from a number of laboratories. Areas indicated for *Notopterus* are based on normal material and extrapolations from *Xenomystus.*

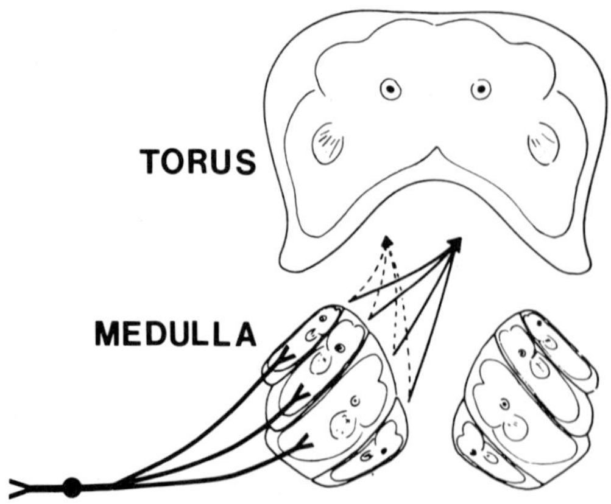


Figure 29.9. Primary afferent fiber from a tuberous electroreceptor trifurcating to terminate in three mirror image zones of the electrosensory lateral line lobe in a gymnotid teleost. These zones, along with the more medial ampullary zone, project to the torus semicircularis bilaterally with contralateral predominance. (Modified from Heiligenberg and Dye 1982 and Rose and Heiligenberg 1985.)

origin of the electrosensors and their central processing areas. Following Kaas's scheme, a duplication of a CNS structure (i.e., nucleus medialis), might have occurred, with the subsequent replacement of mechanoreceptor afferents to the duplicated, laterally positioned nucleus by electroreceptor afferents. In neither gymnotids nor mormyrids is nucleus medialis somatotopically mapped as precisely as the ampullary zone of the electrosensory lateral line lobe, but this does not necessary detract from the scenario, since the somatotopic ampullary maps may have been subsequently refined.

The peripheral anlage that forms ampullary electroreceptors would, in the above case, originate subsequent to the appearance of the duplicated central zone. Alternatively, its origin could have been concomitant with the origin of the new central zone or perhaps have been part of a mechanism forming the new central zone. In the latter case, the two sets of afferents would both initially supply nucleus medialis, and the ampullary zone would gradually segregate out of this structure (Kaas 1982). As yet, no electrosensory species have been shown to have common medullary terminal zones for first-order electro- and mechanosensitive fibers, though this arrangement might have occurred in a now extinct ancestral population.

## The Origin of the Octavolateralis System

In the previous section, the similarities in the electrosensory pathways in distantly related teleosts are attributed to the independent origins of these pathways from the lateral line mechanosensory system, which is believed to be homologous in all fish. That the ascending connections of the *primitive* electrosensory system (i.e., as in elasmobranchs), the mechanosensory system, and the auditory pathways in fish also have a common central pattern of organization (Fig. 29.3) is suggestive of evolutionary interrelationships.

The lateral line mechanoreceptors are traditionally thought of as the most primitive component of the octavolateralis system, giving rise to the ampullary organs and also to the receptors of the inner ear (traditionally considered as vestibular in fish) (e.g., van Bergeijk 1967). The central vestibular projections obviously do not follow the general pattern of organization seen in lateralis and auditory pathways (compare Figs. 29.3 and 29.4). Furthermore, the fossil evidence does not allow us to decide which, if either, of the two lateral line systems appeared first in vertebrate history, whether the inner ear preceded or followed either of the lateralis systems, or whether vestibular receptors preceded auditory receptors, as is often stated. Thus, no basis exists at present for arranging these receptors in a sequential series. As Northcutt and Gans (1983) point out, the origin of these receptors, as well as that of other vertebrate special senses, likely preceded the development of integumentary armor and thus is not preserved in the fossil record. Northcutt and Gans suggest that vertebrate special sensory organs may have their evolutionary origin in epidermal ciliary fields, such as those in larval deuterostomes. Again, one can imagine a number of possible scenarios, including one in which different populations of these mechanoreceptors gave rise to each of the four octavolateralis components.

## A Summary

At this point, two types of evolutionary change in the CNS have been discussed. First, the loss of peripheral receptors, such as electroreceptors or mechanoreceptors, can be correlated with the loss of central structures processing input from those receptors, at least at the first-order level. In amniotes, both lateral line systems are absent, and there is no recognizable nucleus dorsalis or nucleus medialis. Second, the evolution of new receptor classes, such as teleost electroreceptors, is correlated with the appearance of new nuclei and pathways that extend throughout the neuraxis. The evolution of the auditory system in fish and during the fish-amphibian transition also presents us with some interesting considerations concerning variations in end organ function and addition of new auditory receptors.

## Auditory Systems

### The Evolution of Central Auditory Pathways

It is unknown whether any of the extant agnathans possess inner ear auditory receptors, and, as already mentioned, the agnathan octaval nuclei are not easily compared to those of jawed fishes. Although the agnathan inner ear is very different from that of gnathostomes (Lowenstein, Osborne, and Thornhill 1968), the similarity between the inner ears of cartilaginous and bony fishes suggests that this type of inner ear first appeared in an ancestor common to both of these classes. This inner ear primitively contains seven receptors. The three ampullae of the semicircular canals act as angular accelerometers, but definitive information on the functional role of the otolithic end organs—the saccule, lagena, and utricle—and the macula neglecta is limited to a very few species. In elasmobranchs, auditory function is thought to reside in the macula neglecta, the saccule, and perhaps a part of the utricle (the lacinia) (Lowenstein and Roberts 1951; Corwin 1981, 1983). In bony fishes, the saccule and lagena are considered to be auditory receptors, though there are clear examples of the utricle also playing this role (i.e., in clupeids [Denton, Gray, and Blaxter 1979]). In addition, Platt (1973) has demonstrated that the saccule acts as a gravistatic end organ after metamorphosis in several species of flatfish. Platt and Popper (1981) have recently discussed the possibility that a given otolithic end organ may have dual functions (i.e., both auditory and vestibular). No physiological or behavioral studies have addressed the functional role of the macula neglecta in bony fish, though in *Amia* it shares many structural similarities with the ampullae of the semicircular canals (Saidel and McCormick 1985).

## Inner Ear Receptor Projections

The first-order projections of individual inner ear end organs are known in two species of nonteleosts, the elasmobranch *Raja eglanteria* (Barry 1983) and the holostean *Amia calva* (McCormick 1983). As seen in Figure 29.10, end organ projections in these two species are virtually identical, suggesting that this pattern is a primitive gnathostome character. Since each of the octaval nuclei receives projections from all seven end organs, there are no *nuclei* that are exclusively auditory, as in land vertebrates, though there may be auditory subdivisions. If some or all of the otolithic end organs have dual functions, however (Platt and Popper 1981), some of the octaval nuclei may be exclusively vestibular. Experimental studies *Amia* (McCormick unpublished) and of some teleosts (Bell 1981a; Echteler 1985; Finger and Tong 1984) indicate that lemniscal projections to the torus semicircularis arise from the anterior and descending nuclei.

It is impossible at the present to interpret the functional implications of this pattern of end organ terminations further, since there is no experimental information on the function of the otolithic end organs and macula neglecta in *Amia.* It is curious that, in *Raja,* the macula neglecta projects in close association with the posterior semicircular canal, since the neglecta has been implicated in hearing in other elasmobranchs (Corwin 1981, 1983). Apart from the possibility that the macula neglecta in *Raja eglanteria* is not an auditory receptor, it is also possible that the terminal zones of the receptor branches do not, at this level of analysis (horseradish peroxidase tract tracing), reflect their function. This could apply, for example, to a situation in which macula neglecta fibers terminate in a region in which second-order auditory and vestibular neurons are intermingled.

Otic receptor projections in teleosts are known in detail in the cichlid *Astronotus* (Meredith and Butler 1983) and the mormyrids *Gnathonemus* and *Brienomyrus* (Bell 1981b), although macula neglecta projections were not determined. Some information is also available on the connections of some subsets of the otolithic end organs in the osteoglossomorph *Xenomystus* (Braford and McCormick unpublished observations), the ostariophysans *Carassius* and *Ictalurus* (Braford and

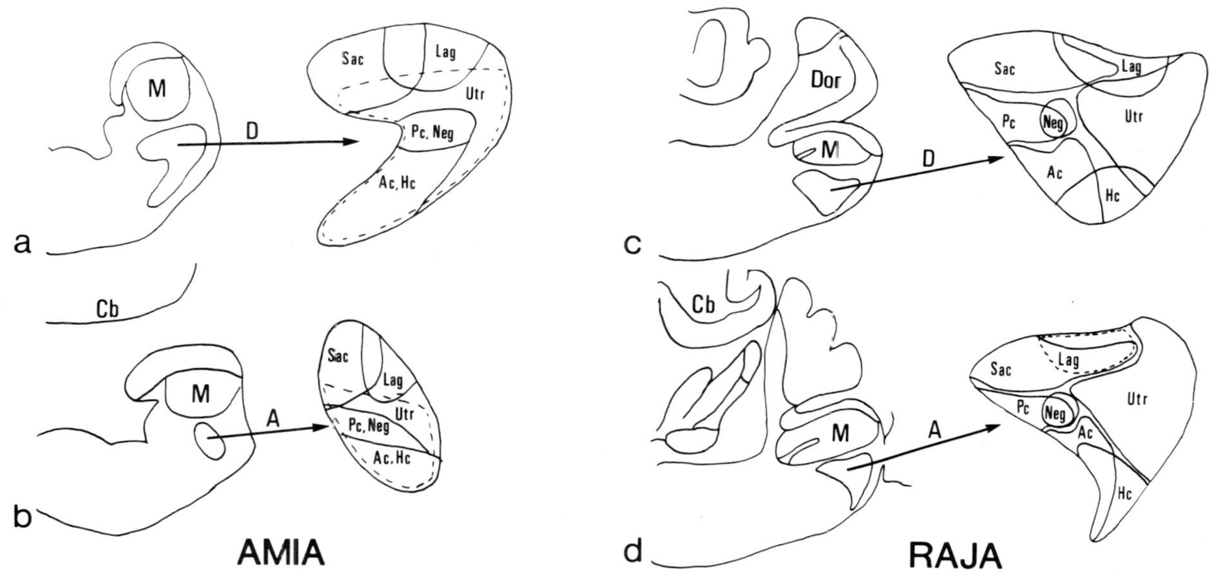


Figure 29.10. A comparison between central projections of individual end organs of the inner ear to the anterior and descending octaval nuclei of the medulla in *Raja eglanteria* and *Amia calva*. Projection fields of individual end organs are indicated as follows: *Ac*, anterior canal; *Hc*, horizontal canal; *Lag*, lagena; *Neg*, macula neglecta; *Pc*, posterior canal; *Sac*, saccule; *Utr*, utricle. Other abbreviations: *A*, anterior octaval nucleus; *Cb*, cerebellum; *D*, dorsal nucleus; *Do*, descending octaval nucleus; *M*, medial nucleus. (*Raja:* Barry 1983, 1985; *Amia:* McCormick 1983.)

McCormick unpublished observations; Finger and Tong 1984), and the clupeid *Clupea harengus* (Meredith 1985). The general morphology of the inner ear of *Astronotus* is similar to that in *Amia,* and the end organ projections are generally similar, with the exception that the saccular and lagenar projections are lateral and medial, respectively, rather than medial and lateral, as in *Amia.* On the other hand, the other species listed above (the mormyrids, *Xenomystus, Carassius, Ictalurus,* and *Clupea*) all have specializations associated with the labyrinth that link it to an air-filled chamber and/or to the head canals of the mechanosensory lateral line system. All of these species deviate to some degree from the hypothesized primitive gnathostome pattern of receptor projections.

In all the above species except the clupeid, the saccule is believed to be the otolithic receptor most likely to contribute significantly to hearing. Figure 29.7A illustrates saccular projections at the level of the descending nucleus and indicates that, not only do they supply the dorsal part of this nucleus, but also course in a dorsomedial direction to supply either an expanded portion of the descending nucleus or perhaps an additional octaval cell group. It is known in some of these species that these dorsomedial cells are a major (but not necessarily the only) source of lemniscal fibers to the known or proposed auditory portion of the torus semicircularis. In *Xenomystus* and *Gnathonemus,* the dorsomedially located cells are bilaterally innervated by the saccule, with the majority of input from the ipsilateral side. It is interesting that in *Clupea,* which is known to use the utricle for sound reception, some utricular fibers take a similar dorsomedial course (Meredith 1985). The projections of other otolithic end organs in this species have not been reported.

The functional significance of this additional dorsomedial projection zone is unknown. Perhaps its presence is related to the specialized auditory peripheries in these species. Since the ostariophysans, osteoglossomorphs, and clupeids are not closely related groups, these peripheral and central specializations are probably independent evolutionary events that have occurred several times.

In mormyrids, saccular fibers also have an exclusive bilateral projection to the dorsomedial portion of the anterior nucleus. In general, inner ear projections in mormyrids are quite different from the hypothesized primitive gnathostome pattern, and the reasons for the differences are unknown.

### Higher-Order Auditory Projections

Although the full pattern of otic receptor projections in ostariophysans is unknown, Echteler's (1985) physiological and anatomical study on the carp, *Cyprinus carpio,* provides the most complete information on the organization of higher-order central auditory pathways in teleosts. Figure 29.11 illustrates contributions to the auditory portion of the lateral lemniscus from dorsomedially positioned cells similar to those discussed above and from the descending and anterior nuclei. Following the general scheme shown in Figure 29.3, the lemniscal projections are primarily to the contralateral nucleus centralis of the torus semicircularis, which relays the information primarily to the ipsilateral central posterior nucleus. The ascending

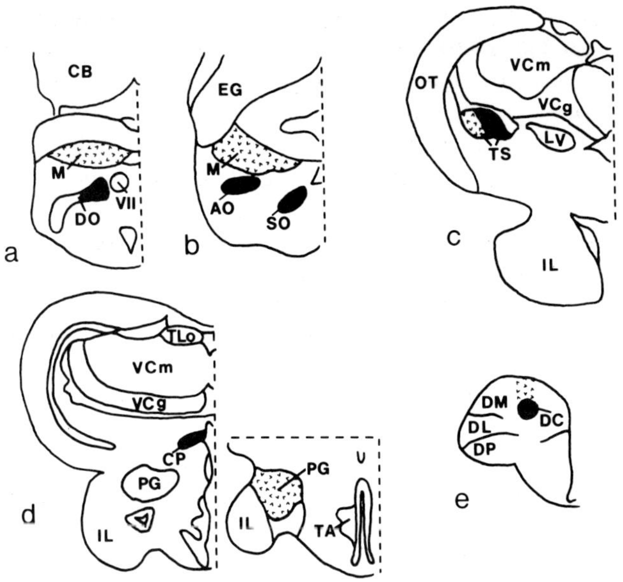


Figure 29.11. Diagrammatical representation of auditory and mechanosensory lateral line areas in the brain of the carp, *Cyprinus carpio*. Key: *solid black*, auditory centers; *floating v's*, mechanosensory lateral line centers. Abbreviations: *AO*, anterior octaval nucleus; *CB*, corpus cerebelli; *CP*, central posterior thalamic nucleus; *DC*, dorsocentral telencephalic area; *DL*, dorsolateral telencephalic area; *DM*, dorsomedial telencephalic area; *DO*, descending octaval nucleus; *DP*, dorsoposterior telencephalic area; *EG*, eminentia granularis of cerebellum; *IL*, inferior lobe of hypothalamus; *LV*, nucleus lateralis valvulae; *M*, medial nucleus; *OT*, optic tectum; *PG*, preglomerular nuclear complex of diencephalon; *SO*, superior olivary nucleus; *TA*, anterior tuberal nucleus of hypothalamus; *TLo*, torus longitudinalis; *VCg*, granule layer of valvula cerebelli; *VCm*, molecular layer of valvula cerebelli; *VII*, seventh cranial nerve fibers. (Based on Echteler 1985.)

projections of this thalamic group are primarily to the ipsilateral central zone of the caudal dorsal telencephalon.

## Audition and the Fish–Amphibian Transition

The evolution of hearing in terrestrial vertebrates is highlighted by the appearance of two ‘‘new’’ auditory receptors, the amphibian papilla and the basilar papilla. The amphibian papilla has been identified in all amphibians examined thus far, and in spite of its location in the saccular, rather than the utricular, chamber, it has been considered by Retzius (1881) and others to be the homologue of the macula neglecta. Caecilians (Gymnophiona), however, possess a macula neglecta proper (and are the only extant amphibians to do so) in the posterior utricular chamber, as well as an amphibian papilla. Although the amphibian papilla is likely a primitive amphibian feature, it is unknown whether it is derived from any of the inner ear end organs in fish or whether it is an independent specialization of the inner ear epithelium. Some have considered it to be derived from part (caecilians) or all (other amphibians) of the macula neglecta (Baird 1974; White and Baird 1982).

The origin of the basilar papilla is likewise obscure, and this receptor has no obvious counterpart in fishes. Unlike the amphibian papilla, the basilar papilla is not ubiquitous in amphibians, being present only in primitive species of urodeles and apodans and in all anurans. Recent controversies concerning this end organ focus on whether it is (White 1986) or is not (Wever 1985) homologous among amphibians and its probable nonhomology to the basilar papilla (cochlea in mammals) of amniotes (Wever 1976, 1985). Developmental studies would likely contribute to our understanding of the origin of the phylogenetically new amphibian auditory receptors, and reinvestigation of the sarcopterygian inner ear might also prove fruitful.

The evolution of new auditory receptors in amphibians might have been paralleled by the evolution of new central structures receiving first-order input exclusively from these receptors. Our current understanding of the amphibian eighth nerve medullary termination zone, though incomplete, neither strongly supports nor refutes this interpretation. Amphibians exhibit two different patterns of organization within the eighth nerve medullary termination zone. This zone in nonanuran amphibians is a single column of cells similar to the single column present in fishes, though definitive comparisons of the specific nuclear groups have not yet been made. Thus, in nonanurans, fibers from the phylogenetically new auditory receptors terminate in close association with those from phylogenetically old auditory receptors (the saccule and lagena).

In contrast, anurans possess two columns of cells that receive eighth nerve input, a dorsal and a ventral one. The ventral column, similar to the single column of nonanuran amphibians and fishes, has been considered to be a vestibular area. The dorsal column, comprising the dorsal acoustic nucleus, was reported to receive first-order input only from the new auditory receptors (Gregory 1972). Although the dorsal acoustic nucleus is sometimes compared to the cochlear nuclei of amniotes, there is no evidence to suggest homology between these structures; the anuran dorsal acoustic nucleus may be an independent specialization related to the anurans' capacity for airborne sound reception and/or their specialized communicative behaviors. Moreover, the distinction between a ventral vestibular column and a dorsal acoustic column in anurans is apparently inappropriate, since it has recently been shown in three species of anurans that the phylogenetically new auditory receptors terminate in close association with the phylogenetically old auditory receptors (saccule and lagena) in *both* the dorsal and ventral columns (Matesz 1979; Fuzessery and Feng 1981; Will, Luhede, and Gorner 1985).

If modern amphibians are considered to be monophyletic, the single column pattern of organization of medullary eighth nerve input can be hypothesized to have been present in the rhipidistian ancestor (Fig. 29.12). This ancestor would be expected to retain the central auditory pathways established earlier in fishes and would also be expected to have both the primitive electroreceptive sense

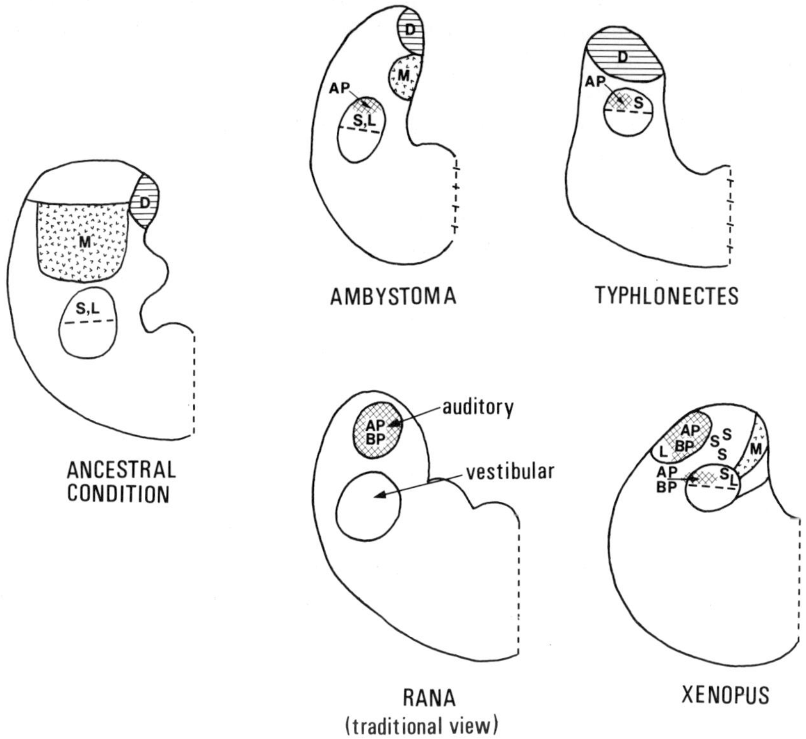


Figure 29.12. Possible evolutionary transformations of octavolateralis area of the medulla in fish-amphibian transition. Ancestral condition hypothesized to be present in rhipidistian fishes that gave rise to amphibians is shown at the *left*. The condition in nonanuran amphibians is shown *above* in the urodele *Ambystoma* and the gymnophionan *Typhlonectes*. The condition in anurans is shown *below* with traditional notion of the situation in *Rana* and more recent data obtained for *Xenopus* (Will, Luhede, and Görner 1985). Key: *horizontal hatching*, electrosensory lateral line center; *floating v's*, mechanosensory lateral line center; *cross-hatching*, areas of termination of new auditory receptors—amphibian and basilar papillae. Projection fields of end organs of inner ear are indicated as follows: *AP*, amphibian papilla; *BP*, basilar papilla; *L*, lagena; *S*, saccule. Other abbreviations: *D*, dorsal (electrosensory lateral line) nucleus; *M*, medial (mechanosensory lateral line) nucleus.

and lateral line mechanoreceptors. In nonanuran amphibians, in which all inner ear receptors terminate in a single cell column, it would be of interest to know the relationship between the terminal zones of the phylogenetically old and new auditory receptors, as well as whether the general pattern of otic inputs is similar to that of nonteleost fish.

Figure 29.12 illustrates the two medullary octaval cell columns present in anurans and shows schematically the otic end organ projections of *Xenopus* (Will, et al, 1985). The evolutionary relationship between the single column in nonanuran amphibians and the two columns of anurans is unknown. Larsell (1934) proposed, on the basis of developmental studies in *Rana,* that a lateral line nucleus (the dorsal electrosensory nucleus) loses its lateral line connections during development and is invaded by acoustic fibers from the new auditory receptors. Jacoby and Rubinson (1983), however, have not confirmed Larsell's observations. Fritzsch, Nikundiwe, and Will (1984) proposed that the anuran dorsal acoustic nucleus is derived both from an embryonic cell population that in other amphibians forms the dorsal electrosensory nucleus and from cells that migrate from the underlying ventral column, but that the proportion of cells from each of the two sources may vary in different species. Other possibilities include (1) the derivation of the dorsal acoustic nucleus from a different, perhaps newly evolved embryonic field and (2) the derivation of the dorsal acoustic nucleus entirely from the ventral cell column, either as a migrated subset of cells or as a duplicated part of the ventral column. In any case it is of interest that in anurans, lagenar, and to some extent saccular termination zones are closely apposed to those of the amphibian and basilar papillae. This apposition occurs both in the dorsal acoustic nucleus and surrounding regions and within portions of the ventral cell column. These observations reinforce the notion that the phylogenetically old receptors may retain at least some auditory function in anurans (Moffat and Capranica 1976) and also leads to the interesting speculation that the newly evolved auditory receptors ''tapped into'' the central auditory pathways that were established in the fishes.

## Summary

If we assume that a central auditory pathway following the scheme of Figure 29.3 was established with the origin of gnathostomes, then we can pose some interesting questions concerning the evolution of hearing in anamniotic vertebrates. First, are there modifications in this pathway that reflect differing contributions of particular end organs to hearing in different species? Although no definitive information is yet available concerning this question, such modifications could include changes in the projection patterns of the otic end organs or changes in the populations of cells that contribute to the lateral lemniscus. Along these lines, Graf and Baker (1983) demonstrated unique second-order semicircular canal projections in an adult flatfish (*Pseudopleuronectes americanus*) that relate to postural aspects of its metamorphosis. Second, are there modifications in the ascending auditory pathway that reflect specialized auditory capabilities? Such capabilities appear to be present in species in which there is a physical coupling between the inner ear and a gas-filled space or the lateral line system. At present, it is known only that such species have a dorsomedial eighth nerve zone apparently not present in other fishes, and information concerning the specific function of this zone is not available. Finally, is the primitive central auditory pathway retained, retained with modifications, or lost in amphibians? Because the saccule and possibly the lagena retain auditory function in at least some amphibians, it is likely that the primitive auditory pathway is not lost, and future studies on the anatomy and development of these pathways in amphibians may help us decide between the other two possibilities.

## Conclusion

We have considered here the central connections of the octavolateralis system largely as they relate to the evolutionary modifications that have occurred in the system as a whole. Recent investigators who have considered the four functional components of the octavolateralis system from a comparative viewpoint have gone a long way towards establishing a coherent understanding of their evolutionary history. We have touched here on some of the issues central to this understanding that remain unresolved.

*Acknowledgments.* Our research was supported by National Science Foundation Grants BNS–19378 to C.A.M. and BNS–10961 to M.R.B. and National Institutes of Health Grant NS21040 to C.A.M. We acknowledge the expert technical assistance of Ms. Amy Kinkler and Mr. John O'Reilly in some of our studies.

### References

Amemiya, F., Kishida, R., Goris, R.C., Onishi, H., and Kusunoki, T. (1985) Primary vestibular projections in the hagfish *Eptatretus burgei,* Brain Res., 337:73–79.

Baird, I.L. (1974) Anatomical features of the inner ear in submammalian vertebrates, in Handbook of Sensory Physiology, vol. V/1, Keidel, W.D., and Neff, W.D. (eds.), Springer-Verlag, New York, pp. 159–212.

Barry, M.A. (1985) Afferent and efferent connections of the primary octaval nuclei in the clearnose skate, *Raja eglanteria,* Thesis, University of Delaware.

Bell, C.C. (1981a) Some central connections of medullary octavolateral centers in a mormyrid fish, in Hearing and Sound Communication in Fishes, Tavolga, W.N., Popper, A.N., and Fay, R.R. (eds.), Springer-Verlag, New York, pp. 383–392.

Bell, C.C. (1981b) Central distribution of octavolateral afferents and efferents in a teleost (Mormyridae), J. Comp. Neurol., 195:391–414.

Bell, C.C., Finger, T.E., and Russell, C.J. (1981) Central connections of the posterior lateral line lobe in mormyrid fish, Exp. Brain Res., 42:9–22.

Bell, C.C., and Russell, C.J. (1978) Termination of electroreceptor and mechanical lateral line afferents in the mormyrid acousticolateral area, J. Comp. Neurol., 182:367–382.

Bodznick, D., and Northcutt, R.G. (1980) Segregation of electro- and mechanoreceptive inputs to the elasmobranch medulla, Brain Res., 195:313–321.

Bodznick, D., and Northcutt, R.G. (1984) An electrosensory area in the telencephalon of the little skate, *Raja erinacea,* Brain Res., 298:117–124.

Boord, R.L., and Campbell, C.B.G. (1977) Structural and functional organization of the lateral line system of sharks, Am. Zool., 17:431–441.

Boord, R.L., and McCormick, C.A. (1984) Central lateral line and auditory pathways: a phylogenetic perspective, Am. Zool., 24:765–774.

Boord, R.L., and Northcutt, R.G. (1982) Ascending lateral line pathways to the midbrain of the clearnose skate, *Raja eglanteria,* J. Comp. Neurol., 207:274–282.

Boord, R.L., and Roberts, B.L. (1980) Medullary and cerebellar projections of the statoacoustic nerve of the dogfish *Scyliorhinus canicula,* J. Comp. Neurol., 193:57–68.

Bullock, T.H. (1974) An essay on the discovery of sensory receptors and the assignment of their functions together with an introduction to electroreceptors, in Handbook of Sensory Physiology, vol. III/3, Fessard, A. (ed.), Springer-Verlag, New York, pp. 1–12.

Bullock, T.H., Bodznick, D.A., and Northcutt, R.G. (1983) The phylogenetic distribution of electroreception: evidence for convergent evolution of a primitive vertebrate sense modality, Brain Res. Rev., 6:25–46.

Bullock, T.H., and Corwin, J.T. (1979) Acoustic evoked activity in the brain of sharks, J. Comp. Physiol., 129:223–234.

Bullock, T.H., and Heiligenberg, W. (1986) Electroreception, John Wiley and Sons, New York.

Carr, C.E., Maler, L., and Sas, E. (1982) Peripheral organization and central projections of the electrosensory nerves in gymnotiform fish, J. Comp. Neurol., 211:139–153.

Corwin, J.T. (1981) Peripheral auditory physiology in the lemon shark: evidence of parallel otolithic and non-otolithic sound detection, J. Comp. Physiol., 142:379–390.

Corwin, J.T. (1983) Postembryonic growth of the macula neglecta auditory detector in the ray, *Raja clavata:* continual increases in hair cell number, neural convergence, and physiological sensitivity, J. Comp. Neurol., 217:345–356.

Corwin, J.T., and Northcutt, R.G. (1982) Auditory centers in the elasmobranch brain stem: deoxyglucose autoradiography and evoked potential recording, Brain Res., 236:261–273.

Denton, E.J., Gray, J.A.B., and Blaxter, J.H.S. (1979) The mechanics of the clupeid acousticolateralis system: frequency responses, J. Mar. Biol. Assoc. UK, 59:27–47.

Echteler, S. (1985) Organization of central auditory pathways in a teleost fish, *Cyprinus carpio,* J. Comp. Physiol. A Sens. Neural Behav. Physiol., 156:267–280.

Fessard, A. (1974) Handbook of Sensory Physiology, vol. III/3, Electroreceptors and Other Specialized Receptors in Lower Vertebrates, Springer-Verlag, New York.

Finger, T.E. (1980) Non-olfactory sensory pathway to the telencephalon in a teleost, Science, 210:671–673.

Finger, T.E., and Tong, S.-L. (1984) Central organization of eighth nerve and mechanosensory lateral line systems in the brainstem of ictalurid catfish, J. Comp. Neurol., 229:129–151.

Fritzsch, B. (1981) The pattern of lateral-line afferents in urodeles, Cell Tissue Res., 218:581–594.

Fritzsch, B., Nikundiwe, A.M., and Will, U. (1984) Projection patterns of lateral-line afferents in anurans: a comparative HRP study, J. Comp. Neurol., 229:451–469.

Fuzessery, Z.M., and Feng, A.S. (1981) Frequency representation in the dorsal medullary nucleus of the leopard frog *Rana p. pipiens,* J. Comp. Physiol., 143:339–347.

Graf, W., and Baker, R. (1983) Adaptive changes of the vestibulo-ocular reflex in flatfish are achieved by reorganization of central nervous pathways, Science, 221:777–779.

Gregory, K.M. (1972) Central projections of the eighth nerve in frogs, Brain Behav. Evol., 5:70–88.

Haugedé-Carré, F. (1979) The mesencephalic exterolateral posterior nucleus of the mormyrid fish *Bryenomyrus niger:* efferent connections studied by the HRP method, Brain Res., 178:179–184.

Haugedé-Carré, F. (1980) Contribution à l'étude des connexions du torus semicircularis et du cervelet chez certains mormyrides, Thèse, L'Université Pierre et Marie Curie, Paris.

Heiligenberg, W., and Dye, J. (1982) Labelling of electroreceptive afferents in a gymnotid fish by intracellular injection of HRP: the mystery of multiple maps, J. Comp. Physiol., 148:287–296.

Hetherington, T.E., and Wake, M.H. (1979) The lateral line system in larval *Ichthyophis* (Amphibia: Gymnophiona), Zoomorphologie, 93:209–225.

Jacoby, J., and Rubinson, K. (1983) The acoustic and lateral line nuclei are distinct in the premetamorphic frog, *Rana catesbeiana,* J. Comp. Neurol., 216:152–161.

Jarvik, E. (1980) Basic structure and evolution of vertebrates, Academic Press, New York.

Kaas, J.H. (1982) The segregation of function in the nervous system: Why do sensory systems have so many subdivisions? Contrib. Sens. Physiol., 7:201–240.

Koester, D.M. (1983) Central projections of the octavolateralis nerves of the clearnose skate, *Raja eglanteria,* J. Comp. Neurol., 221:199–215.

Larsell, O. (1934) The differentiation of the peripheral and central acoustic apparatus in the frog, J. Comp. Neurol., 60:473–527.

Lowenstein, O., Osborne, M.P., and Thornhill, R.A. (1968) The anatomy and ultrastructure of the labyrinth of the lamprey (*Lampetra fluviatilis* L.), Proc. R. Soc. Lond. Series B Biol. Sci., 170:113–134.

Lowenstein, O., and Roberts, T.D.M. (1951) The localization and analysis of the responses to vibration from the isolated elasmobranch labyrinth: a contribution to the problem of the evolution of hearing in vertebrates, J. Physiol. (Lond.), 114:471–489.

Maler, L. (1979) The posterior lateral line lobe of certain gymnotoid fish: quantitative light microscopy, J. Comp. Neurol., 183:323–363.

Maler, L., Sas, E., and Rogers, J. (1981) The cytology of the posterior lateral line lobe of high frequency weakly electric fish (Gymnotidae): dendritic differentiation and synaptic specificity in a simple cortex, J. Comp. Neurol., 195:87–141.

Matesz, C. (1979) Central projection of the VIIIth cranial nerve in the frog, Neuroscience, 4:2061–2071.

McCormick, C.A. (1982) The organization of the octavolateralis area in actinopterygian fishes: a new interpretation, J. Morphol., 171:159–181.

McCormick, C.A. (1983) Central projections of inner ear endorgans in the bowfin, *Amia calva,* Am. Zool., 23:895.

Meredith, G.E. (1985) The distinctive central utricular projections in the herring, Neurosci. Lett., 55:191–196.

Meredith, G.E., and Butler, A.B. (1983) Organization of eighth nerve afferent projections from individual endorgans of the inner ear in the teleost *Astronotus ocellatus,* J. Comp. Neurol., 220:44–62.

Moffat, A.J.M., and Capranica, R.R. (1976) Auditory sensitivity of the saccule of the American toad (*Bufo americanus*), J. Comp. Physiol., 105:1–8.

Münz, H., Claas, B., and Fritzsch, B. (1984) Electroreceptive and mechanoreceptive units in the lateral line of the axolotl *Ambystoma mexicanum,* J. Comp. Physiol., 154:33–44.

Northcutt, R.G. (1978) Brain organization in the cartilaginous fishes, in Sensory Biology of Sharks, Skates, and Rays, Hodgson, E.S., and Mathewson, R.F. (eds.), Office of Naval Research, Arlington, Va., pp. 117–194.

Northcutt, R.G. (1979) Central projections of the eighth cranial nerve in lampreys, Brain Res., 167:163–167.

Northcutt, R.G., and Davis, R.E. (1983) Fish Neurobiology, vol. 1, Brain Stem and Sense Organs, University of Michigan Press, Ann Arbor.

Northcutt, R.G., and Gans, C. (1983) The genesis of neural crest and epidermal placodes: a reinterpretation of vertebrate origins, Q. Rev. Biol., 58:1–28.

Platt, C. (1973) Central control of postural orientation in flatfish. I. Postural change dependence on central neural changes, J. Exp. Biol., 59:491–521.

Platt, C., Bullock, T.H., Czéh, C., Kovačević, N., Konjević, D.J., and Gojković, M. (1974) Comparison of electroreceptor, mechanoreceptor, and optic evoked potentials in the brains of some rays and sharks, J. Comp. Physiol., 95:323–355.

Platt, C., and Popper, A.N. (1981) Fine structure and function of the ear, in Hearing and Sound Communication in Fishes, Tavolga, W.N., Popper, A.N., and Fay, R.R. (eds.), Springer-Verlag, New York, pp. 3–38.

Retzius, G. (1881) Das Gehörorgan der Wirbelthiere, vol. 1, Samson and Wallin, Stockholm.

Ronan, M.C., and Bodznick, D. (1986) End buds: non-ampullary electroreceptors in adult lampreys, J. Comp. Physiol., 158:9–15.

Rose, G., and Heiligenberg, W. (1985) Structure and function of electrosensory neurons in the torus semicircularis of *Eigenmannia:* morphological correlates of phase and amplitude sensitivity, J. Neurosci., 5:2269–2280.

Saidel, W., and McCormick, C.A. (1985) Morphology of the macula neglecta in the bowfin, *Amia calva,* Soc. Neurosci. Abst., 11:1312.

Schweitzer, J. (1983) The physiological and anatomical localization of two electroreceptive diencephalic nuclei in the thornback ray, *Platyrhinoidis triseriata,* J. Comp. Physiol., 153:331–341.

Smeets, W.J.A.J., and Timerick, S.J.B. (1981) Cells of origin of pathways descending to the spinal cord in two chondrichthyans, the shark *Scyliorhinus canicula* and the ray *Raja clavata,* J. Comp. Neurol., 202:473–491.

Tavolga, W.N., Popper, A.N., and Fay, R.R. (1981) Hearing and Sound Communication in Fishes, Springer-Verlag, New York.

Thomson, K.S. (1977) On the individual history of cosmine and a possible electroreceptive function of the pore-canal system in fossil fishes, in Problems in Vertebrate Evolution, Andrews, S.M., Miles, R.S., and Walker, A.D. (eds.), Academic Press, New York, pp. 247–271.

van Bergeijk, W.A. (1967) The evolution of vertebrate hearing, in Contributions to Sensory Physiology, vol. 2, Neff, W.D. (ed.), Academic Press, New York, pp. 1–49.

Wever, E.G. (1976) Origin and evolution of the ear of vertebrates, in Evolution of Brain and Behavior in Vertebrates, Masterton, R.B., Bitterman, M.E., Campbell, C.B.G., and Hotton, N. (eds.), Lawrence Erlbaum Assoc., Princeton, N.J., pp. 89–106.

Wever, E.G. (1985) The Amphibian Ear, Princeton University Press, New York.

White, J.S. (1986) Comparative features of the surface morphology of the basilar papilla in five families of salamanders (Amphibia: Caudata), J. Morphol., 187:201–217.

White, J.S., and Baird, I.L. (1982) Comparative morphological features of the caecilian inner ear with comments on the evolution of amphibian auditory structures, Scanning Electron Microsc., 3:1301–1312.

Will, U., Luhede, G., and Görner, P. (1985) The area octavo-lateralis in *Xenopus laevis,* Cell Tissue Res., 239:147–161.

Part IIIC-3

# Equilibrium

Fish otolith organs are also involved in equilibrium responses. The otolith organs extract information about the direction of the gravity force and provide the CNS with information needed for organizing coordinated body movements in three spatial dimensions. Semicircular canals also serve as part of the equilibrium system for detection of rotational motions of the body.

While little is known regarding the role of the statocyst in hearing by aquatic invertebrates, a good deal is known about invertebrate equilibrium detectors. What is most striking is how the structure of the invertebrate equilibrium receptors parallels that of equilibrium receptors in vertebrates. The morphological similarities are explained for invertebrates by Budelmann (Chapter 30) and for vertebrates by Platt (Chapter 31). Both chapters give a comparative morphological and ultrastructural overview of equilibrium receptor systems that allows not only for comparisons within vertebrates and invertebrates, but also between the equilibrium receptor systems of both groups.

# Chapter 30

# Morphological Diversity of Equilibrium Receptor Systems in Aquatic Invertebrates

BERND-ULRICH BUDELMANN *

For proper behavior and orientation in space, any organism with the capacity to move needs information about its attitude and movement. This information is provided by equilibrium receptor systems. These enable the organism to control its position and motor activities in the three dimensions of space, using the gravitational field as a reference system. Compared with all other reference systems, such as light, sound, odor, and taste, the gravitational field is characterized by a unique feature: during the life span of an organism it is nearly constant in both magnitude and direction. Thus, the gravitational field is an ideal reference system, and during the course of evolution all organisms that use locomotion have developed specialized receptor systems for equilibrium orientation that make use of this reference system.

The structure and function of the equilibrium receptor systems are different in the various animal groups, depending on their life-style and the environment in which they live, rather than on their level of evolution. Swift moving animals require a more sophisticated system to control their position and movement than slow-moving animals do. It is a general rule that the more complex the animal's behavior in space, the more complex and specialized are the sensory structures it possesses. On the other hand, though the equilibrium receptor systems vary in complexity, as well as in number, size, and location, they all are composed of only two basic structural elements: (1) a mass, whose instantaneous position depends on the forces applied, and (2) sensory elements (cell organelles, hair cells, or cuticular sensory hairs), that are mechanically affected by the position of the mass. During the course of evolution, these two structural elements have found many different kinds of expression. This is most obvious among the invertebrates, since this animal group is not an evolutionary homogeneous monophyletic group. Consequently, one cannot expect a uniform invertebrate solution to the problem of equilibrium orientation (see Markl 1974).

For any equilibrium receptor systems, the adequate stimulus is mechanical energy in the form of acceleration. This can either be linear acceleration, including

* Marine Biomedical Institute, University of Texas Medical Branch, Galveston, Galveston, TX 77550, USA

gravity, or angular acceleration. Based on morphological differences, the respective receptor systems for the detection of these accelerations are gravity (or linear) and angular acceleration receptor systems. This separation, however, is not a clear and distinct one. Recent experiments with cephalopods (Budelmann and Wolff 1973), crustacea (Fraser and Sandeman 1975), and vertebrates (e.g., Estes, Blanks, and Markham 1975; Goldberg and Fernandez 1975; Macadar et al. 1975; Lowenstein and Compton 1978) have shown that, at the level of the receptor cells, the angular acceleration receptor systems are additionally sensitive to linear acceleration. Similarly, linear acceleration receptor systems may be sensitive to angular acceleration (see Goldberg and Fernandez 1975; Precht 1979). These additional properties do not invalidate the classic separation into linear and angular acceleration receptor systems, however.

In aquatic animals, orientation with respect to gravity has been reported in nearly all invertebrate phyla, but responses to angular acceleration are known only in decapod crustaceans and in the higher cephalopods (see Plate 1924; Vinnikov et al. 1971; Markl 1974; Vinnikov 1974; Budelmann 1976; Sandeman 1976). This correlates well with the fact that species of both these groups are the most swift moving of all the (aquatic) invertebrates.

The following chapter does not claim to be a complete review. It briefly outlines the morphological diversity of the equilibrium receptor systems of aquatic invertebrates. Using schematic diagrams, it tries to draw particular attention to the key components of these receptor systems. Furthermore, the chapter does not follow the usual line of taxonomy, but tries to arrange the various types of receptor systems according to the morphological differences of its receptor elements. With the exception of the aquatic arthropods, the arrangement employed is almost in line with an increasing complexity of the receptor elements and the whole receptor system, as well as with an increasing complexity of the animal's orientation and motor behavior in space.

## Gravity Receptor Systems

Gravity receptor systems are characterized by a mass of (generally) higher specific weight than that of the surrounding fluid. This mass stimulates the sensory elements of the receptor by its mere weight. In invertebrates, gravity receptor systems are known as *statocysts* (for vertebrate examples, see Chap. 31). The most common type of statocyst is a fluid-filled cavity that contains a single statolith or numerous statoconia, which stimulate the underlying sensory hair cells. Formerly these structures were called respectively otocysts, otoliths, and otoconia (and still are in vertebrates). Gravity receptor systems, in general, orient an animal's body axes with respect to gravity when at rest or during movement and initiate and control compensatory reactions, such as eye and head movements. Phylogenetically, the gravity receptor systems of coelenterates have been considered to be the first true sense organs (Bullock and Horridge 1965).

## Gravity Orientation Without Specialized Gravity Receptor Systems

Gravity-oriented behavior (geotaxis) without specialized gravity receptor systems is known for motile protozoa, and it has been suggested that the principal cause of their geotactic behavior could be a body shape-dependent, gravity-induced hydrodynamic torque (Roberts 1970; Winet and Jahn 1974). This may also explain the characteristic shape of many of the protozoa (Bean 1977).

Gravity-oriented behavior without specialized receptor systems has also been described in coelenterate polyps (see Markl 1974), in bryozoan larvae (Pires and Woollacott 1983), and in many echinoderms (see Markl 1974). The underlying mechanism of this behavior is still unknown, but one has to consider that basically any internal structure or external appendage that has a different specific weight from its surrounding medium may serve as an additional gravity receptor system if there is an associated sensory structure.

## Gravity Receptor Systems Using Cell Organelles

The simplest gravity receptor system most probably is a single-celled organism with cell inclusions of higher density. These inclusions then could act as statoliths and thus make the cell sensitive to gravity. This situation has been suggested for the ciliate *Paramecium* (see Markl 1974) (Fig. 30.1B). Although the mechanism is not yet understood, this arrangement resembles that of the root cells of higher plants in which starch grains (''amyloplasts'') serve as statoliths. In the plant cell (Fig. 30.1A), the differential mechanical contact between the starch grains and the membrane complex of the endoplasmic reticulum is generally accepted to be the main cause for the plant's gravity perception (Schröter, Läuchli, and Sievers 1975; Sievers and Volkmann 1977; Iversen and Rommelhoff 1978; Perbal and Rivière 1980).

Another cellular mechanism for gravity perception could be a mechanical contact of cell organelles with ciliary rootlets. Some ciliate endocommensals have vacuoles that are lined by ciliary rootlets and contain heavy calcium concrements (Fig. 30.1C) (Anderson and Dumont 1966). Also, an asymmetry in the position of cell organelles, caused by a shift of a vacuole, is discussed as a primary mechanism of gravity perception (Dennison and Shropshire 1984).

## Gravity Receptor Systems Lacking Hair Cells or Cuticular Sensory Hairs

A few gravity receptor systems have been described in multicellular organisms that are composed of, or contain, specialized receptor cells that are neither hair cells nor cuticular sensory hairs. The process of stimulus perception of these cells is still unknown.

In the coelenterate hydroid polyp *Corymorpha palma* (Fig. 30.2A), the two terminal entoderm cells of the tip of each stolonlike, filamentous holdfast that

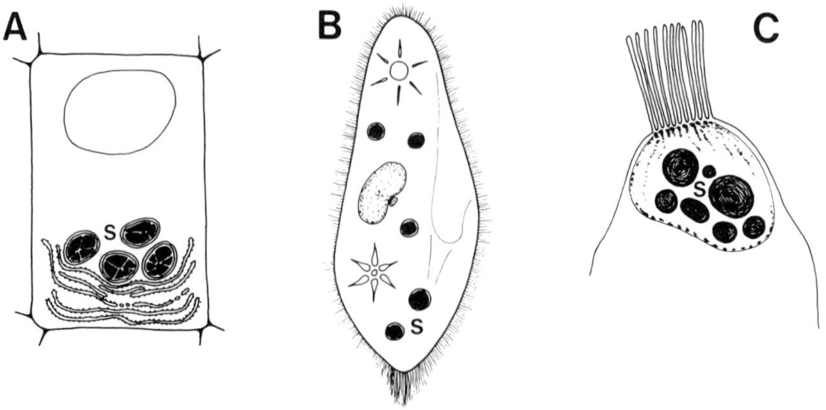


Figure 30.1. Gravity receptor systems using cell organelles. (A) Root cell of higher plants, in which starch grains (*S*) are in contact with cell's endoplasmic reticulum (based on Sievers and Volkmann 1977). (B) Possible gravity receptor system in ciliate *Paramecium,* with cell organelles of higher density (*S*). (C) Vacuole in the anterior portion of the ciliate endocommensal *Paraisotrichia;* calcium concrements (*S*) are in contact with ciliary rootlets (based on Anderson and Dumont 1966).

anchor the polyp to the substrate form the statocyst cavity. Inside the cavity, a freely moveable statolith rests directly on the membrane of the cup-shaped terminal entoderm cell. The statolith is composed of numerous crystalline particles and mitochondria that are bound together by a unit membrane. The question of transfer of the gravity information to the contractile effector system is still unclear, since no nerve has been seen to be associated with this organ (Campbell 1972).

In the two statocysts of the nemertine worm *Ototyphlonemertes pallida* (Fig. 30.2B), a single statolith cell carries 2–6 statoliths, each enclosed in a separate cell cavity. The statolith cell is in electrical synaptic contact with nerve fibers (Brüggemann and Ehlers 1981).

In the turbellarian flatworm *Convoluta psammophila,* two cells form the statocyst cavity and enclose a freely moveable specialized cell ("lithocyte") that carries a statolith of a calcium compound (Fig. 30.2C). There are no clearly defined sensory elements inside the cyst cavity. There is a specialized cell and nerve cushion located ventrally underneath the cyst, however, in the most favorable position to be mechanically affected by the statolith-carrying cell. The function of this receptor system is complicated, since the position of the statocyst is not fixed to the cerebral ganglion, but can be moved by muscles (Ferrero 1973).

In the ascidian tunicate *Ciona intestinalis,* a gravity receptor organ is present only in the free-swimming tadpole larvae (the adults become sessile after metamorphosis). The organ is a single cell with a large pendulum-like projection into the cerebral cavity (Fig. 30.2D). The projection is club shaped and carries the statolith. The small neck of this cell is surrounded by a collar that is filled with distended vesicles of the cell's endoplasmic reticulum. The neck and the collar are thought to make up the transducer region (Dilly 1961, 1962).

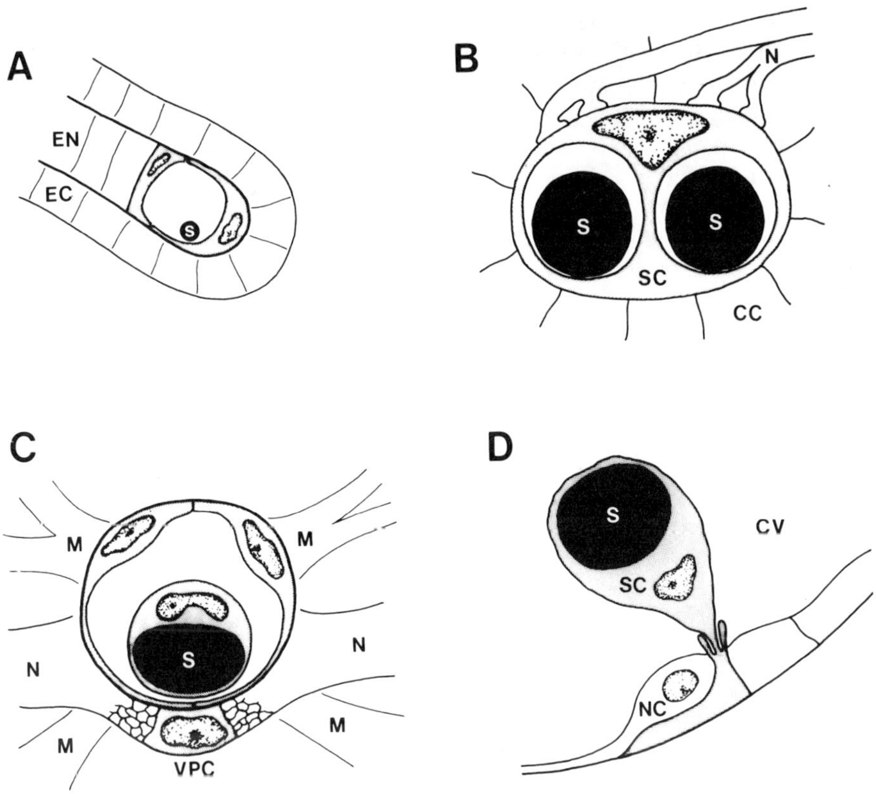


Figure 30.2. Gravity receptor systems lacking hair cells or cuticular sensory hairs. (A) Statocyst in filamentous holdfast of coelenterate polyp *Corymorpha palma*, formed by the two terminal entoderm cells (*stippled*); *EC*, ectoderm; *EN*, entoderm; *S*, statolith (longitudinal section) (based on Campbell 1972). (B) Statocyst of the nemertine worm *Ototyphlonemertes pallida*. In each of two statocysts, single statolith cell (*SC*, *stippled*) carries 2–6 statoliths (*S*), each enclosed in a separate cell cavity; *CC*, covering cell; *N*, nerve fiber (transverse section) (based on Brüggemann and Ehlers 1981). (C) Statocyst of turbellarian flatworm *Convoluta psammophila; M*, muscle; *N*, nerve; *S*, statolith; *VPC*, ventral polar cell (transverse section) (based on Ferrero 1973). (D) Gravity receptor system of tadpole larvae of ascidian tunicate *Ciona intestinalis; CV*, cerebral vesicle; *NC*, nerve cell; *S*, statolith; *SC*, statolith cell (based on Dilly 1961).

## Gravity Receptor Systems with Hair Cells

The gravity receptor systems of most invertebrates and all vertebrates are characterized by the presence of sensory hair cells. In invertebrates, these cells usually are primary sensory cells (with an axon extending from the base) but secondary sensory cells (without an axon) have been described in ctenophores (Aronova 1974) and in cephalopods (Budelmann 1977; Budelmann and Thies 1977; Colmers 1977, 1981; Budelman, Sachse, and Staudigl 1987). Each hair cell carries, species specifically, between 1 and 700 kinocilia at its distal end (see Budelmann 1976; Dorsett 1976); these are arranged close together, or up to 2 μm apart. Each kinocilium is morphologically polarized in a particular direction, as determined by the position of its basal foot structure and the orientation of its internal 9 × 2 + 2 tubuli content (see Budelmann 1979). This morphological polarization, in turn, correlates with a physiological polarization (Fig. 30.3). Electrophysiological experiments (Wolff 1973; Janse 1980; Williamson and Budelmann 1985) and data on the vertebrate vestibular and lateral line receptors (see Barber 1968; Budelmann 1979; Platt 1984) support the hypothesis that deflection of a kinocilium of an invertebrate hair cell in the direction of its morphological polarization causes maximal excitation of the cell and maximal inhibition when deflection occurs in the opposite direction.

As to the physiology of a sensory hair cell, it is interesting to note that most invertebrate hair cells lack stereocilia. The only exception is some coelenterates in which the stereocilia are symmetrically arranged around a single kinocilium (Horridge 1969). Stereocilia have recently been shown to play a key role in the transduction process of the vertebrate vestibular hair cell (Hudspeth and Jacobs 1979; Hudspeth 1983).

When a hair cell carries more than one kinocilium, it is particularly important, from the functional point of view, how the directions of polarization of all the kinocilia of that single cell are arranged. Based on this arrangement, hair cells can be characterized as either of the two morphological types: nonpolarized or polarized. In a nonpolarized hair cell (Fig. 30.4), the kinocilia are arranged in

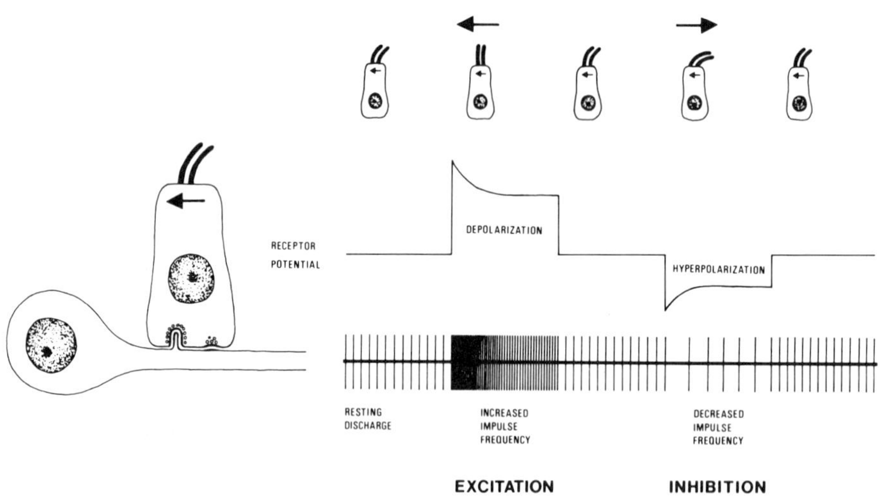


Figure 30.3. The theory of function of polarized invertebrate (cephalopod) hair cell (by analogy with the results of the vertebrate vestibular and lateral line receptors). The diagram shows the relationship between the direction of kinocilia deflection (*arrows above cells*) in, or opposite to, the direction of the cell's morphological polarization (*arrows in cells*) and receptor potential and nerve impulse frequency, respectively. (Physiological part of this figure is based on Flock 1965.)

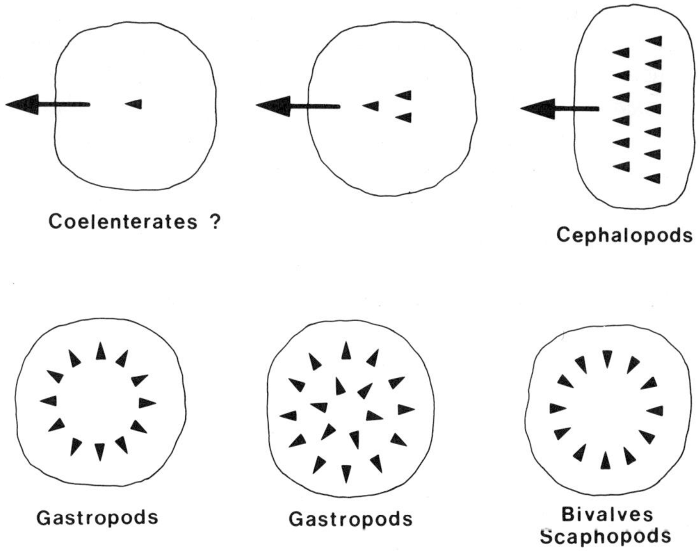


Figure 30.4. Morphological and physiological polarization of hair cells. Diagrams of three polarized (*upper row*) and three nonpolarized (*lower row*) hair cells, illustrating different arrangements of directions of polarization of cell's kinocilia (*arrowheads*) and the animal groups in which they dominantly occur. When one or all the kinocilia of single cell are polarized in same direction (*upper row*), then hair cell is polarized in that direction (*large arrows*). When the kinocilia of single cell are polarized in either of the radial patterns (*lower row*), then hair cell is nonpolarized.

concentric rings or a spiral and are polarized radially either towards the periphery, as in most gastropods (see Budelmann 1976; Kuzirian, Alkon, and Harris 1981) or towards the center of the ring as in most bivalves (Tsirulis 1981, personal communication) and in scaphopods (Budelmann and Andrade unpublished). Such a nonpolarized cell is not differentially sensitive to the direction from which the stimulus occurs (Wolff 1973). Its response can, however, change, depending on the amount of surface area that is stimulated by a freely moveable statolith or statoconia, but it is independent of the direction from which the stimulus occurs (Wolff 1970; Janse 1983). In contrast, in a polarized hair cell (Fig. 30.4), all kinocilia are polarized in the same direction, thus making the cell differentially sensitive to the direction of stimulation. In invertebrates, such polarized hair cells are common in the cephalopods (Barber 1968; Budelmann 1976, 1979; Budelmann, Sachse, and Staudigl 1987), but are rare in other invertebrates.

## Gravity Receptor Systems with Hair Cells with Single Kinocilium

In invertebrates, gravity receptor systems having hair cells with a single kinocilium are known only in the free-swimming cnidarian coelenterate medusae. Single kinocilia may also occur in the archannelid worm *Protodrilus* (Merker and Vaupel von Harnack 1967). In the medusae, the organs are located at the margin of the bell-shaped body and vary in number from 4 to 300 (see Markl 1974). They are characterized by either a single- or many-celled pendulum-like projection (Fig. 30.5). This pendulum contains one or a few intracellular statoliths. The position in which this pendulum hangs down is monitored by a single, or several, receptor hair cells that surround it (Horridge 1969; Singla 1975; see also Markl 1974).

It is still unknown whether the cnidarian coelenterate hair cell with its single kinocilium is physiologically polarized and thus differentially sensitive to stimulus direction. In contrast to the vertebrate hair cell (see Hudspeth and Jacobs 1979; Hudspeth 1983), stereocilia are absent or surround the single kinocilium symmetrically (Horridge 1969; Singla 1975). Information about stimulus direction, however, can certainly be obtained by the animal on the basis of the location of the sense organs along the margin of the bell-shaped body, and this may also explain the unusually high number of these organs in this animal group.

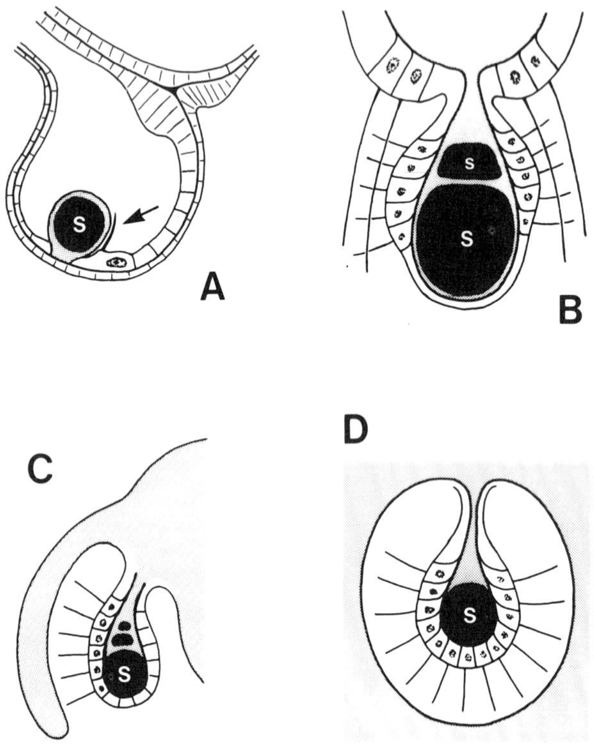


Figure 30.5. Gravity receptor systems with hair cells with single kinocilium. Radial sections of statocysts at the margin of the bell-shaped body of cnidarian coelenterate medusae; *S*, statolith. (A) Statocyst of *Tiaropsis*, with single hair cell (the *arrow* points to its kinocilium) and a statolith carrying cell ("lithocyte") (based on Singla 1975). (B) Statocyst of *Aegina citrea*, with two types of hair cells and statoliths within pendulum-like ("lithostyle") projection (based on Singla 1975). (C) Statocyst ("rhopalium") of *Nausithoe albida* (based on Hertwig and Hertwig 1878). (D) Statocyst of *Geryonia proboscidalis*, with the lithostyle enclosed in a cavity of the mesogloea (based on Horridge 1969).

## Gravity Receptor Systems with Hair Cells with Several Kinocilia

Hair cells with several (5–700) kinocilia are known in the statocysts of ctenophores and annelids, as well as in scaphopod, gastropod, bivalve, and cephalopod molluscs. The statocysts are formed by ectodermal invaginations and sometimes remain open via a canal that represents the embryonic invagination. The statocysts contain either a single statolith or numerous statoconia, which act as a single mass or move independently (see Markl 1974; Vinnikov 1974; Budelmann 1976).

*Gravity Receptor Systems with Nonpolarized Hair Cells.* Nonpolarized hair cells have been described in the statocysts of scaphopod, gastropod, and bivalve molluscs (Fig. 30.6). Each animal has two statocysts, each of which is a spherelike cavity lined by hair cells. The kinocilia are motile in several species (see Budel-

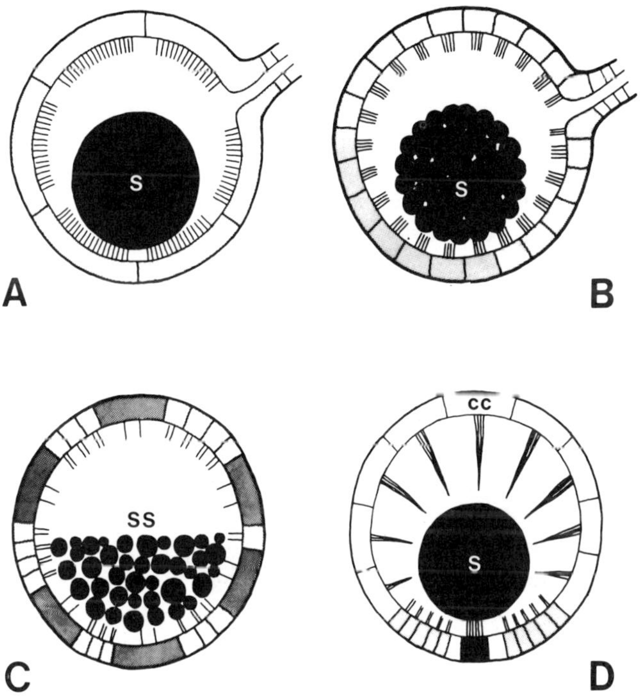


Figure 30.6. Gravity receptor systems with nonpolarized hair cells (transverse sections); *S,* statolith; *SS,* statoconia. (A) Statocyst of pulmonate and opisthobranch gastropods, with thirteen large hair cells (data from Budelmann 1976). (B) Right (lower) statocyst of the bivalve scallop *Pecten* and statocyst of other bivalves, with numerous small hair cells (data from Barber and Dilly 1969; Tsirulis 1981, personal communication). (C) Statocyst of scaphopod tooth shell *Dentalium vulgare,* with large (*dark-stippled*) and small (*light-stippled*) nonpolarized hair cells (Budelmann and Andrade unpublished). (D) Statocyst of prosobranch gastropod *Pterotrachea,* with receptor cells (*stippled*) arranged in a macula at the lower part of the statocyst; *CC,* ciliated cells (based on Tschachotin 1908).

mann 1976). The statoliths are free to move within the cyst cavity under the influence of gravity, rather than being fixed to the kinocilia. Thus they are normally in contact with the kinocilia of only those hair cells which, in a given position of the statocyst, are the lowest.

The number of hair cells in the statocysts varies. In pulmonate and opisthobranch gastropods, the cyst cavity is lined by only 13 large hair cells, each having 500–700 kinocilia (Fig. 30.6A) (see Budelmann 1976). In the right (lower) statocyst of the scallop *Pecten* and in both statocysts of other bivalves, the hair cells are smaller and more numerous, but have only 5–10 kinocilia (Fig. 30.6B) (Barber and Dilly 1969; Tsirulis 1981, personal communication). In the scaphopod tooth shell *Dentalium,* the cyst cavity is lined by about 80 hair cells. These are of two types: (1) small hair cells, with 10–20 kinocilia arranged in a small ring and (2) large hair cells, with 25–50 kinocilia arranged in a large ring (Fig. 30.6C) (Budelmann and Andrade unpublished).

Since all hair cells in these statocysts are primary sensory cells of the nonpolarized type, the information about position is encoded in the spatial pattern of those hair cells which are in mechanical contact with the free-moving statolith or statoconia (i.e., which of the total number of hair cells are active at any one time). The information about position becomes more accurate with an increasing number and density of hair cells. The information about the direction of movement may be obtained by the sequence in which the cells become active (Wolff 1970; Janse 1983). There is some evidence that, in gastropods, the hair cells of only the pulmonates receive efferent innervation (see Budelmann 1976; Budelmann and Bonn 1982).

In *Pterotrachea,* which is a free-swimming prosobranch gastropod, the statocysts are of a more complicated type (Fig. 30.6D). The nonpolarized sensory hair cells are concentrated in a limited region, and the statolith contacts the hair cells only at intervals. Normally it floats in the middle of the cyst cavity because of a fluid current that occurs during the beat of the dorsal ciliated cells (Tschachotin 1908; Barber and Dilly 1969).

*Gravity Receptor Systems with Both Nonpolarized and Polarized Hair Cells.* Statocysts with both nonpolarized and polarized hair cells in one and the same organ have been described in two molluscan species. In the bivalve scallop *Pecten,* which is capable of some jet-propelled swimming, the left (upper) statocyst is completely lined by both nonpolarized hair cells (each with about 10 kinocilia: type A) and polarized hair cells (each with about 30 kinocilia: type B) (Fig. 30.7A) (Barber and Dilly 1969). A comparable type of statocyst may be found in *Nautilus,* a primitive cephalopod (Young 1965; Budelmann unpublished). The functional significance of this double equipment is still unclear. One interpretation could be that the nonpolarized hair cells sense position and the polarized ones monitor the direction of movement. Interestingly, in the pediveliger larva of the scallop *Pecten maximus,* only nonpolarized hair cells occur in both statocysts (Cragg and Nott 1977).

In the pelagic larvae of the opisthobranch gastropod *Rostanga pulchra,* 10 hair cells have been described, 4 of which, at the bottom of the cyst, are of the

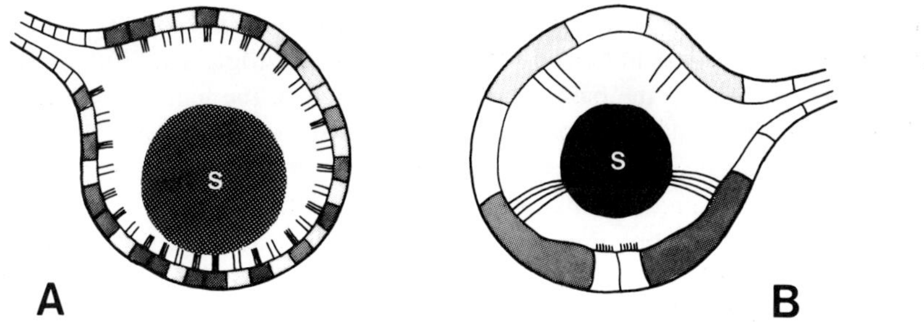


Figure 30.7. Gravity receptor systems with both nonpolarized (*light-stippled*) and polarized (*dark-stippled*) hair cells (transverse sections); *S,* statolith. (A) Left (upper) statocyst of the bivalve scallop *Pecten* (data from Barber and Dilly 1969). (B) Statocyst of larvae of opisthobranch gastropod *Rostanga pulchra* (based on Chia, Koss, and Bickell 1981).

polarized type and monitor the animal's tilt relative to gravity. The other 6, at the upper part of the cyst, are nonpolarized and are thought to monitor sinking, rather than gravity orientation (Fig. 30.7B) (Chia, Koss, and Bickell 1981).

*Gravity Receptor Systems with Polarized Hair Cells.* Statocysts with only polarized hair cells have been described in one gastropod, in ctenophores, and in great detail in the cephalopods. They vary in complexity, having either primary or secondary sensory hair cells and a well-documented efferent innervation.

In the prosobranch gastropod *Pomacea paludosa,* the statocyst has a rather simple structure (Fig. 30.8A). Its spherelike cavity is lined by 2,000–3,000 primary sensory hair cells of about 10 μm in surface diameter. The cavity contains

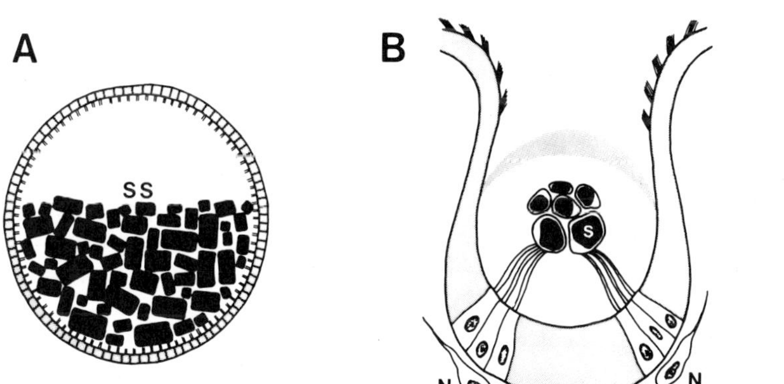


Figure 30.8. Gravity receptor systems with polarized hair cells. (A) Statocyst of the prosobranch gastropod *Pomacea paludosa,* with up to 3,000 hair cells and rectangular statoconia (*SS*) (data from Schmidt 1912; McClary 1968; Stahlschmidt and Wolff 1972). (B) Statocyst (''apical organ'') of the ctenophore comb jelly *Pleurobrachia.* Ciliary bundles of four groups of secondary sensory hair cells carry the intracellular statoliths (*S*); *N,* neuron (transverse section) (based on Krisch 1973; Aronova 1974).

numerous, roughly rectangular, statoconia that vary greatly in size (20–100μm) and move more or less independently (Schmidt 1912; McClary 1968; Stahlschmidt and Wolff 1972). On the basis of the size relation of the hair cells and statoconia and the latter's rectangular form and independent movement, however, the functional significance of the presence of a large number of polarized hair cells is difficult to understand and may require reinvestigation.

In the pelagic ctenophores, the single statocyst ("apical organ") has a unique structure (Fig. 30.8B). It has four groups of hair cells, each carrying a bundle of kinocilia that balances the statolith in the middle of the organ. The statolith is composed of cells with calcareous inclusions and is firmly attached to the kinocilia. The whole organ is roofed by a transparent dome that consists of a network of cells and specialized cilia. The receptor hair cells are secondary sensory cells and are supplied by efferent innervation (Krisch 1973; Aronova 1974). In each of the four groups, the hair cells are polarized in only one direction (it is still unclear whether this is inwards or outwards towards the rows of ciliary comb plates). The apical organ is known to control the ctenophore's equilibrium by regulating the ciliary beat of the comb plate rows during swimming and compensatory reactions (Horridge 1971; see also Markl 1974).

The cephalopod statocysts are the most sophisticated equilibrium receptor organs of all those found in the invertebrates. They show a wide range of complexity and levels of differentiation in the various genera, and such differences have been correlated with the different forms of locomotion employed by these very mobile animals (Budelmann 1976; Maddock and Young 1984). Cephalopod statocysts (with the exception of *Nautilus*) show a remarkable degree of convergence with the vertebrate vestibular apparatus. The two organs in cephalopods are situated in cavities of the cranial cartilage. They have receptor systems for both linear and angular accelerations (for details of the latter see p. 773). Although they differ in details in the various genera, the majority can be categorized in either of the two basic morphological types: the "octopod" type (in octopods, e.g., *Octopus*) and the "decapod" type (in the sepioidea, e.g., the cuttlefish *Sepia*, and in the teuthoidea, e.g., the squid *Loligo*) (Young 1960; Budelmann 1976; Stephens and Young 1982). An intermediate organ has been described in the bathypelagic *Vampyroteuthis*, which phylogenetically is a cephalopod relic, somewhere near the ancestor of octopods and decapods (Stephens and Young 1976).

The octopod type of statocyst (Fig. 30.9A) is a spherelike sac that is firmly kept in position in its cranial cavity by strands and vessels. Its gravity receptor system is an oval, vertically oriented macular plate to which a compact calcareous statolith is firmly attached. In *Octopus*, the macula consists of about 5,000 hair cells, each with up to 200 kinocilia arranged in an elongated ciliary group. All hair cells are polarized radially outwards towards the macular periphery (Fig. 30.9B) (Young 1960; Barber 1968; Budelmann 1976). The macula's cellular organization is highly complex. The hair cells are secondary sensory cells that are in afferent synaptic contact with two types of first-order afferent neurons. Hair cells and neurons receive many efferent endings (Fig. 30.9C) (Budelmann and Thies 1977; Colmers 1977, 1981).

The decapod type of statocyst (Fig. 30.10) is present in the more swift-moving

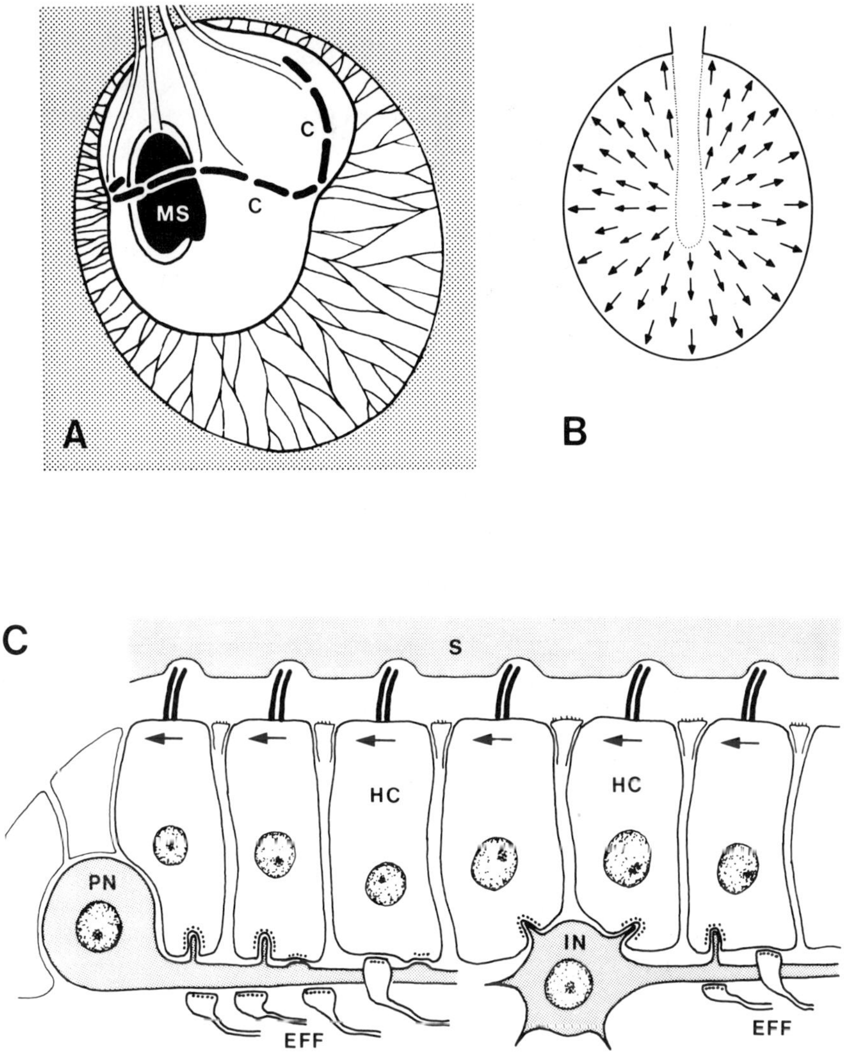


Figure 30.9. Gravity receptor system with polarized hair cells. ''Octopod'' type of statocyst of the cephalopod *Octopus vulgaris*. (A) Lateral view of the left statocyst, with the gravity receptor system (macula-statolith, *MS*) and the course of the nine sections of angular acceleration receptor system (crista-cupula, *C*) (based on Young 1960; see also Fig. 30.14B). (B) Polarization pattern of macula hair cells (based on Budelmann 1976). (C) Cellular organization of the macula. Secondary sensory hair cells (*HC*) are in afferent synaptic contact with two types of first-order afferent neurons (*PN*, perimacular neuron; *IN*, intramacular neuron). *Arrows* in the distal end of hair cells indicate the direction of the cell's polarization; *EFF*, efferent synaptic endings; *S*, mucous layer of statolith (based on Colmers 1981).

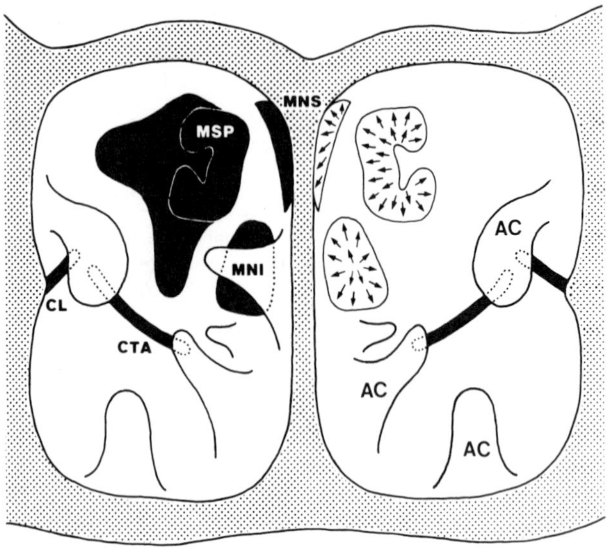


Figure 30.10. Gravity receptor system with polarized hair cells. ''Decapod'' type of statocyst of cephalopod cuttlefish *Sepia officinalis*. Anterior part (forward view) of left and right statocyst, showing (*left*) the three gravity receptor systems (*MSP, MNI, MNS*), and (*right*) polarization patterns of the maculae hair cells. The course of angular acceleration receptor system (crista-cupula: *CTA, CL*) is only partly to be seen and continues in posterior part of the statocyst (compare Fig. 30.14C); *AC*, anticrista lobe (based on Budelmann 1978).

cephalopods (e.g., cuttlefish and squids). Its cavity is more irregular in form because of cartilaginous protrusions (''anticristae''). In each statocyst, the gravity receptor system is subdivided into three separate organs. Their maculae are arranged almost at right angles to each other and thus resemble the utricle, saccule, and lagena of fish (see Chap. 31). To one macula (*MSP*), a compact statolith is attached and to the other two (*MNI, MNS*) a statoconial layer (Fig. 30.10) (Budelmann 1976; Stephens and Young 1982). In *Sepia* and *Loligo,* the three maculae contain about 9,000 and 5,000 secondary sensory hair cells respectively, each with 50–150 kinocilia arranged in an elongated ciliary group. In two maculae (*MSP, MNI*), the hair cells are polarized in an almost radial, and in one (*MNS*), in a fanlike, outward direction towards the macular periphery (Fig. 30.10) (Budelmann 1979). The cellular organization of all three maculae is basically similar to that of the macula of the octopod type of statocyst (Budelmann unpublished).

The principle of function of the cephalopod gravity receptor systems has been studied in great detail. In *Octopus,* as compared to the vertebrates and crustaceans, the information about the position in space is encoded in the orientation of the excitation pattern of all hair cells relative to macula-fixed coordinates. It is not encoded in the strength of stimulation of the hair cells (Budelmann 1970, 1975, 1976).

## Gravity Receptor Systems with Cuticular Sensory Hairs

Statocysts are well known in crustaceans and in a few aquatic insects. The organs are formed by an invagination of the exoskeleton and often remain open to the exterior by a canal. Since the cavity is lined by a cuticular layer, the receptor elements are basically different from the hair cells. In crustaceans, they are stiff cuticular hairs with an asymmetric basal joint (Fig. 30.11). On one side of this joint, a cuticular chorda thread is attached. As shown in the crayfish *Astacus* statolith hairs, the chorda is connected to three sense cells. Each hair is polarized in one direction, which is towards its inclination. Shearing of the hair in that direction causes mechanical stress on the chorda thread, which in turn stimulates the sense cells (Schöne and Steinbrecht 1968; Stein 1975; Takahata and Hisada 1979; Janse 1980).

The gravity receptor systems in the various crustaceans are basically similar in gross morphology, but vary in location. In the isopods, they are situated in the telson of the tail (Fig. 30.12A) and, in the mysids, in the base of the uropod endopodites (Fig. 30.12B). In all decapod crustaceans, they are found in the basal segment of the antennule (Fig. 30.12C). The statolith is either secreted by the animal itself or is composed of external material (e.g., sand grains). In all crustaceans, it is firmly attached to the statolith hairs. The number of these hairs varies from 3 (in the isopod *Cyathura*) to about 400 (in the decapod *Homarus*), and they are normally arranged in 2–4 rows in a crescent pattern. The directions of polarization of the statolith hairs are always towards the center of the crescent (Fig. 30.12) (Cohen 1955, 1960; Sandeman and Okajima 1972; Neil 1975; Rose

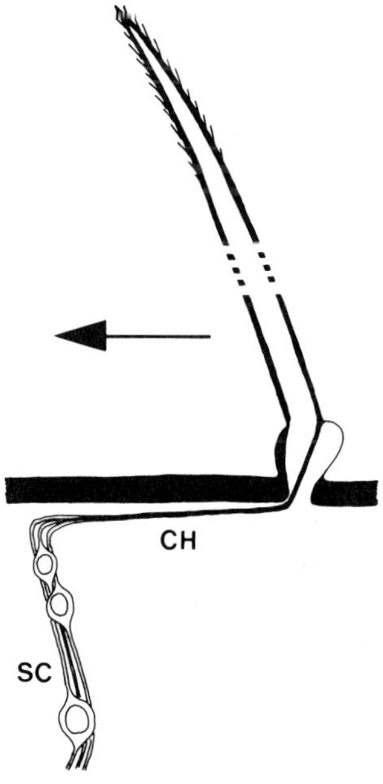


Figure 30.11. Cuticular sensory hair of crustacean statocyst. Section through statolith hair of the crayfish *Astacus fluviatilis*. Three sense cells (*SC*) are connected to the cuticular hair by chorda thread (*CH*). *Large arrow* indicates the direction of polarization of the hair (based on Schöne and Steinbrecht 1968).

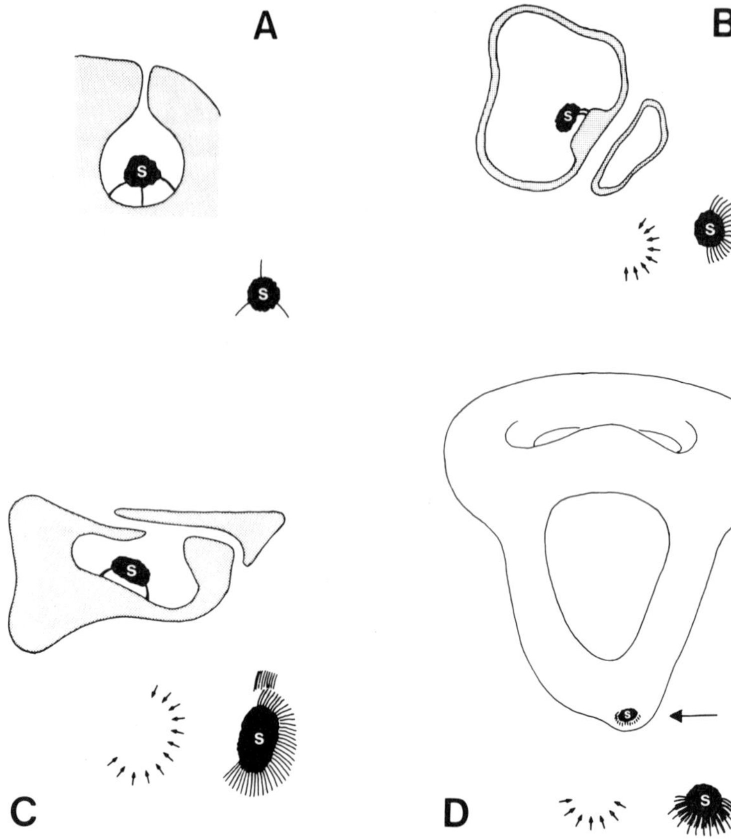


Figure 30.12. Gravity receptor systems with cuticular sensory hairs. Transverse sections of crustacean statocysts, forward views (A–C), mediofrontal view (D). *Small insets* show dorsal views of the statoliths (*S*), with the arrangements of statolith hairs and their directions of polarization. (A) Statocyst in the telson of isopod *Cyathura polita* (based on Rose and Stokes 1981). (B) Statocyst in right uropod of mysid shrimp *Praunus flexuosus* (based on Neil 1975). (C) Statocyst in the basal segment of the right antennule of the crayfish and lobster (based on Schöne 1954; Takahata and Hisada 1979). (D) Statolith organ (*large arrow*) in vertical canal of crab statocyst (based on Sandeman and Okajima 1972). In (C) and (D), sensory hairs of angular acceleration receptor systems are not shown (see Fig. 30.15).

and Stokes 1981; for the function of these organs see Schöne 1954, 1971; Sandeman 1976, 1983; Janse 1980).

In the aquatic hemipteran insects, the water scorpion *Nepa* and the water stick insect *Ranatra,* the gravity receptor organs are quite unusual. They are uniquely adapted to the animal's aquatic environment, since their function is based on the buoyancy of air. The organs in the larvae are different from those in the adults. The larvae have two ventral abdominal grooves in which bristles trap an air supply. In four abdominal segments, spots with no bristles allow

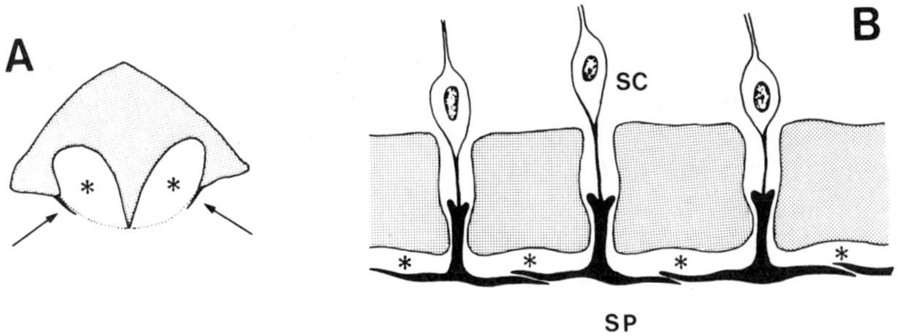


Figure 30.13. Gravity receptor systems with cuticular sensory hairs in the aquatic hemipteran insects, water scorpion *Nepa* and the water stick insect *Ranatra*. (A) Transverse section through the air-filled abdominal grooves (*) of the larvae. *Arrows* point to the sensory hairs that monitor the position of the air-water boundary (*dotted line*) (based on Baunacke 1912). (B) Section through spiracle epithelium of the imagos. Air (*) is trapped under umbrella-like distal ends of cuticular sensory hairs, the movement of which is monitored by sense cells (*SC*) (based on Plate 1924).

contact of the air with the water. At these spots, sensory hairs monitor the position of the air-water boundary during positional changes of the larvae (Fig. 30.13A) (Baunacke 1912; see also Markl 1974). In the adults, the organs are found in the spiracles of three segments. There, air is trapped under the overlapping umbrella-like distal ends of the cuticular sensory hairs. Because of the animal's position in space, the air-water content in the spiracle changes, causing the umbrellas to move and stimulate the sense cells (Fig. 30.13B) (Thorpe and Crisp 1947; Markl 1974).

## Angular Acceleration Receptor Systems

Angular acceleration receptor systems are present only in the most swift-moving of all the aquatic invertebrates (i.e., the decapod crustaceans and the higher cephalopods). They occur together with the gravity receptor systems in the same statocyst. Angular acceleration receptor systems are basically similar in structure to the gravity receptor systems. However, the mass that stimulates the sensory elements has a specific weight that is (almost) identical with that of its surrounding fluid. In aquatic invertebrates, the stimulating mass is composed of cupulae or cuticular hairs that are moved by a relative fluid movement inside a cyst cavity or inside canals. The general purpose of the system is to stabilize an animal's motor behavior during rotational accelerations and to initiate and control compensatory reactions, such as rotational nystagmic eye movements.

### Angular Acceleration Receptor Systems with Hair Cells

In invertebrates, angular acceleration receptor systems with hair cells occur only in the statocysts of higher cephalopods. There the hair cells form a crista ridge that winds inside a cyst cavity such that it covers three almost orthogonal planes (Figs. 30.9A and 30.14). The crista is subdivided into sections along its length: in the decapod type of statocyst, into four sections; in the octopod type of statocyst, into nine sections. They all have a complex pattern of oppositely polarized hair cells (Figs. 30.9A and 30.14B and C).

The neuronal and synaptic organization of the crista is highly complex. In the octopods *Octopus* and *Eledone,* it is composed of primary sensory hair cells, as well as two types of secondary sensory hair cells that are in afferent synaptic contact with two types of first-order afferent neurons. All these elements receive many efferent endings (Budelmann 1977; Budelmann, Sachse, and Staudigl 1987). A cupula is attached to each crista section; it protrudes freely into the cyst cavity and acts like a swinging door. In the octopod type of statocyst, in contrast to the decapod type, the crista-cupula system is further complicated in that the cupulae vary in form and size. Large and small cupulae alternate regularly along the length of the nine crista sections (Fig. 30.14A) (Budelmann 1977; Budelmann, Sachse, and Staudigl 1987). It has recently been demonstrated that both crista-cupula systems differ in their sensitivity ranges by a factor of about 10 (Williamson and Budelmann 1985). This dual sensitivity can be understood as an adaptation to the octopod's two forms of locomotion: slowly crawling around and fast swimming by jet propulsion. This accords well with slow-moving cephalopods' having statocysts with angular acceleration receptor systems of large dimensions (because sensitivity to slow movements requires a large inertial mass of fluid), whereas fast-moving cephalopods have systems of small dimensions. Such correlation is also found in the vertebrates (Maddock and Young 1984).

In the decapod type of cephalopod statocyst, the function of the angular acceleration receptor system is further complicated by the protrusion of several cartilaginous lobes ''anticristae'' (Fig. 30.10) into the cyst cavity. In the different decapod genera, the lobes vary in number and size (see Budelmann 1976; Young 1984) and certainly influences the fluid (''endolymph'') flow inside the cyst cavity. In slow-moving deep-sea decapod cephalopods, the lobes are reduced to provide a large inertial mass of fluid, whereas in the fast-swimming squid *Loligo,* they are well developed and almost restrict the cyst cavity to canals (Stephens and Young 1978).

## Angular Acceleration Receptor Systems with Cuticular Sensory Hairs

Decapod crustaceans are the only aquatic invertebrates that possess angular acceleration receptor systems with cuticular sensory hairs. These ''thread'' hairs are long and thin and can easily be moved by a relative movement of the fluid in the cyst cavity or in the canals. In crabs, the thread hairs are mechanically connected to two sense cells (Sandeman 1983).

The angular acceleration receptor systems of crustaceans can be categorized into two basic types: the ''lobster'' type, present in lobsters, crayfish, and some shrimp, and the ''crab'' type. In the lobster type of system (Fig. 30.15A), the thread hairs are found only at the posteromedian wall of the bowl-shaped cyst

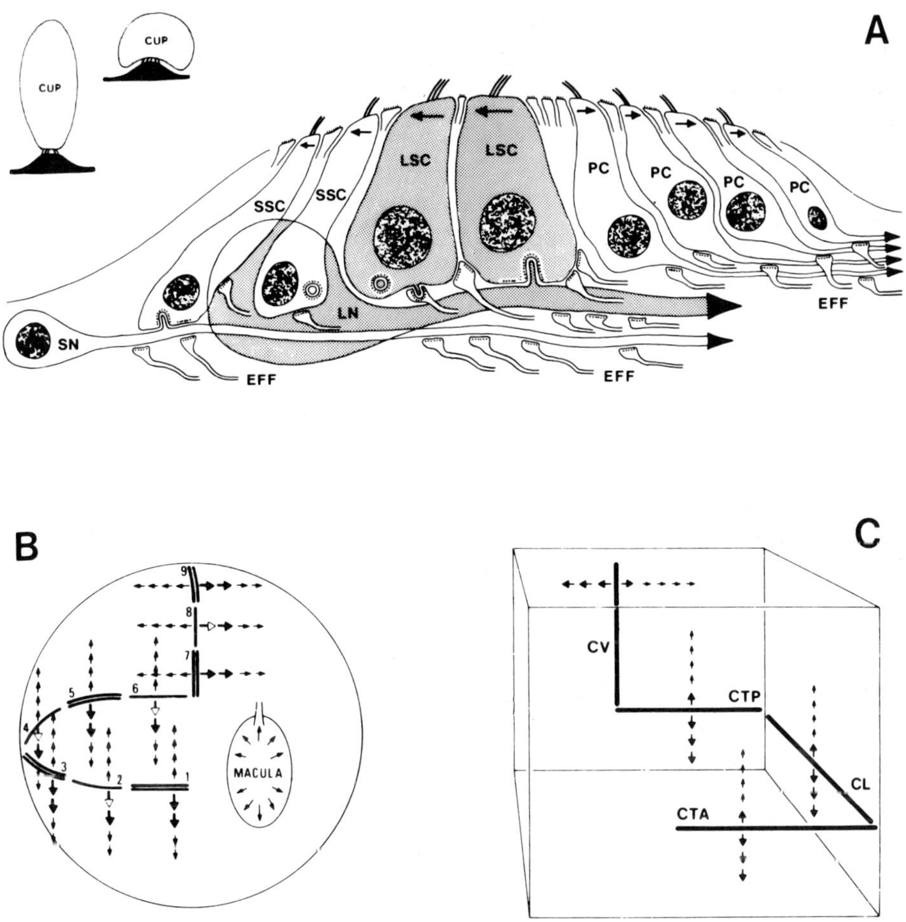


Figure 30.14. Angular acceleration receptor systems of cephalopods with polarized hair cells. (A) Cross section of the crista of *Octopus vulgaris,* showing primary sensory hair cells (*PC*), large (*LSC*) and small (*SSC*) secondary sensory hair cells, which are in afferent synaptic contact with large (*LN*) and small (*SN*) first-order afferent neurons, and many efferent synaptic endings (*EFF*). *Arrowheads* in distal ends of the hair cells indicate the direction of the cell's polarization. Insets in the *upper left* show cross sections of the two types of angular acceleration receptor systems: the crista with large and with the small cupula (*CUP*) attached to it (based on Budelmann, Sachse, and Staudigl 1987). (B) The course of the nine crista sections and the pattern of morphological polarization of crista hair cells in octopod type of statocyst of *Octopus vulgaris* (frontal view) (from Budelmann 1977. Reprinted with permission from the Zoological Society of London). (C) The course of the four crista sections and the pattern of morphological polarization of crista hair cells in decapod type of statocysts of the cuttlefish *Sepia officinalis* and the squid *Loligo vulgaris* (frontal view). (From Budelmann 1977. Reprinted with permission from the Zoological Society of London.).

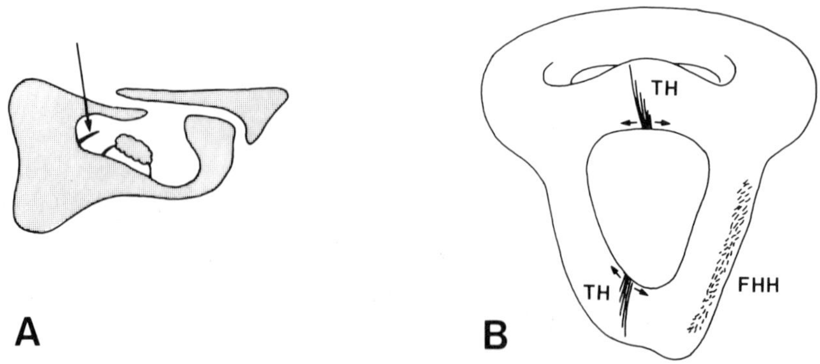


Figure 30.15. Angular acceleration receptor systems with cuticular sensory hairs. Statocysts of decapod crustaceans in basal segment of the right antennule. (A) Transverse section (forward view) of the statocyst of the crayfish and lobster. The *arrow* points to the group of thread hairs (based on Cohen and Dijkgraaf 1961). (B) Mediofrontal view of the canal system of crab statocyst, with bundles of oppositely polarized (*small arrows*) thread hairs (*TH*) in horizontal and vertical canal and the group of "free hook" hairs (*FHH*). The statolith organ in the vertical canal is not shown in this diagram (based on Sandeman and Okajima 1972).

cavity. In the lobster *Homarus,* about 70 hairs, each 600 μm long and 5 μm wide, form an array and project almost horizontally into the cavity. Each single hair has no apparent directionality, since its receptor cells respond equally well to displacements about all three major body axes (Cohen 1955, 1960). The crab type of system (Fig. 30.15B) of the more swift-moving crabs is more complex. Depending on the species, the lumen of the statocyst is divided into two more or less complete canals: a horizontal and a vertical canal. In each one, thread hairs (300 μm long and 2μm wide) form a curtain of oppositely polarized hairs. Since the plane of the vertical canal is oriented 45° to the transverse axis of the animal, the lower group of thread hairs responds to both pitching and rolling movements (Sandeman and Okajima 1972; Fraser 1977). In the vertical canal is a field of about 100 additional sensory hairs. These "free hook" hairs are much shorter (40–60 μm long and 1–2 μm wide) than the thread hairs and are auxiliary, but less sensitive, rotation receptors (Sandeman and Okajima 1972; Janse 1980).

## Conclusions

The equilibrium receptor systems of (aquatic) invertebrates show a variety of morphological expressions and levels of complexity, and there is no doubt that, in this animal group, they have evolved independently several times during the course of evolution (Markl 1974; Vinnikov 1974). Their morphological diversity can thus be understood as an evolutionary solution of a heterogeneous polyphyletic animal group to the different demands of their life-styles and the environments in which they live. The receptor systems of the aquatic invertebrates, however, do not show special adaptations to their aquatic environment (with the exception of the aquatic insects *Nepa* and *Ranatra*), since equilibrium receptor organs, in general, are "internal" organs (i.e., their basic function is independent of a direct contact of the organ with the aquatic medium). This is in contrast to most of the other sense organs of aquatic animals.

In equilibrium receptor systems, essential differences occur at the level of the sensory structures employed. These structures can be classified into (1) cell organelles; (2) cell membranes, with presumably differentially sensitive membrane areas; and (3) flexible cell appendages (cilia, cuticular hairs) with a different ultrastructure and a different arrangement on the surface of the cell and the receptor epithelium. In the present survey, details of these structures have been used as criteria for classification. Any further investigation of equilibrium receptor organs should focus on these criteria, since their detailed knowledge will certainly help in understanding the basic principle of function of these receptor organs, including details of the transduction process of the receptor cells (see Hudspeth 1983; Neugebauer and Thurm 1985; Osborne, Comis, and Pickles 1984).

The level of complexity of the equilibrium receptor organ clearly does not correlate with the animal's taxonomic position or level of evolution. Instead, it mirrors the animal's level of motor activity: the more differentiated the organ, the more complex the animal's motor behavior in space. This is true even in closely related species, such as the free-swimming gastropod *Pterotrachea,* as opposed to other gastropods, and the bivalve scallop *Pecten,* which is capable of some jet-propelled swimming, as opposed to other bivalves.

Among the aquatic invertebrates, the equilibrium receptor systems reach the highest level of complexity within the decapod crustaceans and the higher cephalopods. These are the most swift-moving of all the (aquatic) invertebrates and the only ones with receptor systems for both linear and angular accelerations. No doubt the most sophisticated systems occur in octopod and decapod cephalopods (Budelmann 1976, 1977; Colmers 1981; Stephens and Young 1982; Budelmann, Sachse, and Staudigl 1987), and it has recently been shown that, in many morphological and functional details, their equilibrium receptor organs, as well as the associated oculomotor system, display a high degree of evolutionary convergence with the vertebrate vestibulo-oculomotor system (Budelmann and Young 1984; Maddock and Young 1984; Williamson and Budelmann 1985; see also Chapter 31).

## Summary

The morphological diversity of equilibrium receptor systems of aquatic invertebrates is briefly outlined and presented in schematic diagrams. The various receptor systems are classified, not taxionomically, but on the basis of morphological differences at the level of the sensory structures employed. According to the animal's level of motor activity, the receptor systems range from simple structures in the single-cell organisms to highly sophisticated organs in the octopod and decapod cephalopods; there they reach strong parallels with the vertebrate vestibular system.

*Acknowledgments*

Work on this chapter and the author's work mentioned in the text were supported in part by grants of the Deutsche Forschungsgemeinschaft (Bu 404/1–3 and SFB 4/A2).

## References

Anderson, E., and Dumont, J.N. (1966) A comparative study of the concrement vacuole of certain endocommensal ciliates—a so-called mechanoreceptor, J. Ultrastruct. Res., 15:414–450.

Aronova, M. (1974) Electron microscopic observations on the aboral organ of ctenophora, Z. Mikrosk. Anat. Forsch. 88:401–412.

Barber, V.C. (1968) The structure of mollusc statocysts, with particular reference to cephalopods, Symp. Zool. Soc. Lond., 23:84–96.

Barber, V.C., and Dilly, P.N. (1969) Some aspects of the fine structure of the statocysts of the molluscs *Pecten* and *Pterotrachea*, Z. Zellforsch. Mikrosk. Anat., 94:462–478.

Baunacke, W. (1912) Statische Sinnesorgane bei Nepiden, Zool. Jahrb. Anat., 34:197–342.

Bean, B. (1977) Geotactic behavior of *Chlamydomonas*, J. Protozool., 24:394–401.

Brüggemann, J., and Ehlers, U. (1981) Ultrastruktur der Statocyste von *Ototyphlonemertes pallida* (Keferstein 1862) (Nemertini), Zoomorphology, 97:75–87.

Budelmann, B.U. (1970) Die Arbeitsweise der Statolithenorgane von *Octopus vulgaris*, Z. Vgl. Physiol., 70:278–312.

Budelmann, B.U. (1975) Gravity receptor function in cephalopods with particular reference to *Sepia officinalis*, Fortschr. Zool., 23:84–96.

Budelmann, B.U. (1976) Equilibrium receptor systems in molluscs, in Structure and Function of Proprioceptors in the Invertebrates, Mill, P.J. (ed.), Chapman and Hall, London, pp. 529–566.

Budelmann, B.U. (1977) Structure and function of the angular acceleration receptor systems in the statocysts of cephalopods, Symp. Zool. Soc. Lond., 38:309–324.

Budelmann, B.U. (1978) The function of the equilibrium receptor systems of cephalopods, Proc. Neurootol. Equilibriomet. Soc., 6:15–63.

Budelmann, B.U. (1979) Hair cell polarization in the gravity receptor systems of the statocysts of the cephalopods *Sepia officinalis* and *Loligo vulgaris*, Brain Res., 160:261–270.

Budelmann, B.U., and Bonn, U. (1982) Histochemical evidence for catecholamines as neurotransmitters in the statocysts of *Octopus vulgaris*, Cell Tissue Res., 227:475–483.

Budelmann, B.U., Sachse, M., and Staudigl, M. (1987) The angular acceleration receptor system of the statocyst of *Octopus vulgaris:* morphometry, ultrastucture, and neuronal and synaptic organization, Philos. Trans. R. Soc. Lond. B Biol. Sci. (in press)

Budelmann, B.U., and Thies, G. (1977) Secondary sensory cells in the gravity receptor system of the statocyst of *Octopus vulgaris*, Cell Tissue Res., 182:93–98.

Budelmann, B.U., and Wolff, H.G. (1973) Gravity response from angular acceleration receptors in *Octopus vulgaris,* J. Comp. Physiol., 85:283–290.

Budelmann, B.U., and Young, J.Z. (1984) The statocyst-oculomotor system of *Octopus vulgaris:* extraocular eye muscles, eye muscle nerves, statocysts nerves and the oculomotor centre in the central nervous system, Philos. Trans. R. Soc. Lond. B Biol. Sci., 306:159–189.

Bullock, T.H., and Horridge, G.A. (1965) Structure and Function of the Nervous Systems of Invertebrates, Freeman, San Francisco.

Campbell, R.D. (1972) Statocyst lacking cilia in the coelenterate *Corymorpha palma,* Nature, 238:49–51.

Chia, F.S., Koss, R., and Bickell, L.R. (1981) Fine structural study of the statocysts in the veliger larva of the nudibranch, *Rostanga pulchra,* Cell Tissue Res., 214:67–80.

Cohen, M.J. (1955) The function of receptors in the statocysts of the lobster *Homarus americanus,* J. Physiol., 130:9–34.

Cohen, M.J. (1960) The response patterns of single receptors in the crustacean statocyst, Proc. R. Soc. Lond. B Biol. Sci., 152:30–49.

Cohen, M.J., and Dijkgraaf, S. (1961) Mechanoreception, in The Physiology of Crustacea, vol 2, Waterman, T. (ed.), Academic Press, New York, pp. 65 108.

Colmers, W.F. (1977) Neuronal and synaptic organization in the gravity receptor system of the statocyst of *Octopus vulgaris,* Cell Tissue Res., 185:491–503.

Colmers, W.F. (1981) Afferent synaptic connections between hair cells and the somata of intramacular neurons in the gravity receptor system of the statocyst of *Octopus vulgaris,* J. Comp. Neurol., 197:385–394.

Cragg, S.M., and Nott, J.A. (1977) The ultrastructure of the statocysts in the pediveliger larvae of *Pecten maximus* (L) (Bivalvia), J. Exp. Mar. Biol. Ecol., 27:23–36.

Dennison, D.S., and Shropshire, W. (1984) The gravireceptor of *Phycomyces:* its development following gravity exposure, J. Gen. Physiol., 84:845–859.

Dilly, N. (1961) Electron microscope observations of the receptors in the sensory vesicle of the ascidian tadpole, Nature, 191:786–787.

Dilly, N. (1962) Studies on the receptors in the cerebral vesicle of the ascidian tadpole. I. The otholith, Q. J. Microsc. Sci., 103:393–398.

Dorsett, D.A. (1976) The structure and function of proprioceptors in soft-bodied invertebrates, in Structure and Function of Proprioceptors in the Invertebrates, Mill, P.J. (ed.), Chapman and Hall, London, pp. 443–483.

Estes, M.S., Blanks, R.H.I., and Markham, C.H. (1975) Physiological characteristics of vestibular first-order canal neurons in the cat. I. Response plane determination and resting discharge characteristics, J. Neurophysiol., 38:1232–1249.

Ferrero, E. (1973) A fine structural analysis of the statocyst in turbellaria acoela, Zool. Scr., 2:5–16.

Flock, A. (1965) Transducing mechanisms in the lateral line canal receptors, Cold Spring Harbor Symp. Quant. Biol., 30:133–145.

Fraser, P.J. (1977) How morphology of semicircular canals affects transduction, as shown by response characteristics of statocyst interneurons in the crab *Carcinus maenas* (L), J. Comp. Physiol., 115:135–145.

Fraser, P.J., and Sandeman, D.C. (1975) Effects of angular and linear accelerations on semicircular canal interneurons of the crab *Scylla serrata,* J. Comp. Physiol., 96:205–221.

Goldberg, J.M., and Fernandez, C. (1975) Vestibular mechanisms, Annu. Rev. Physiol., 37:129–162.

Hertwig, O., and Hertwig, R. (1878) Das Nervensystem und die Sinnesorgane der Medusen, Vogel, Leipzig.

Horridge, G.A. (1969) Statocysts of medusae and evolution of stereocilia, Tissue & Cell, 1:341–353.

Horridge, G.A. (1971) Primitive examples of gravity receptors and their evolution, in Gravity and the Organism, Gordon, S.A., and Cohen, M.J. (eds.), University of Chicago Press, Chicago, pp. 203–221.

Hudspeth, A.J. (1983) Mechanoelectrical transduction by hair cells in the acousticolateralis sensory system, Annu. Rev. Neurosci., 6:187–215.

Hudspeth, A.J., and Jacobs, R. (1979) Stereocilia mediate transduction in vertebrate hair cells, Proc. Nat. Acad. Sci. U.S.A., 76:1506–1509.

Iversen, T.H., and Rommelhoff, A. (1978) The starch statolith hypothesis and the interaction of amyloplasts and endoplasmatic reticulum in root geotropism, J. Exp. Bot., 29:1319–1328.

Janse, C. (1980) The function of the statolith-hair and free-hook-hair receptors in the statocyst of the crab, *Scylla serrata,* J. Comp. Physiol., 137:51–62.

Janse, C. (1983) The function of the statocyst sensory cells in *Aplysia limacina,* J. Comp. Physiol., 150:359–370.

Krisch, B. (1973) Über das Apikalorgan (Statocyste) der Ctenophore *Pleurobrachia pileus,* Z. Zellforsch. Mikrosk. Anat., 142:241–262.

Kuzirian, A.M., Alkon, D.L., and Harris, L.G. (1981) An infraciliary network in statocyst hair cells, J. Neurocytol., 10:497–514.

Lowenstein, O., and Compton, G.J. (1978) A comparative study of the responses of isolated first-order semicircular canal afferents to angular and linear acceleration, analysed in the time and frequency domains, Proc. R. Soc. Lond. B Biol. Sci., 202:313–338.

Macadar, O., Wolfe, G.E., O'Leary, D.P., and Segundo, J.P. (1975) Response of the elasmobranch utricle to maintained spatial orientation, transition, and jitter, Exp. Brain Res., 22:1–12.

Maddock, L., and Young, J.Z. (1984) Some dimensions of the angular acceleration receptor systems of cephalopods, J. Mar. Biol. Assoc. U.K., 64:55–79.

Markl, H. (1974) The perception of gravity and of angular acceleration in invertebrates, in Handbook of Sensory Physiology, vol. 6, Vestibular System, Part I, Basic Mechanisms, Kornhuber, H.H. (ed.), Springer-Verlag, Berlin, pp. 17–74.

McClary, A. (1968) Statoliths of the gastropod *Pomacea paludosa,* Trans. Am. Microsc. Soc., 87:322–328.

Merker, G., and Vaupel von Harnack, M. (1967) Zur Feinstruktur des Gehirns und der Sinnesorgane von *Protodrilus rubropharyngeus* Jaegersten (Archiannelida), Z. Zellforsch. Mikrosk. Anat., 81:221–239.

Neil, D.M. (1975) The mechanism of statocyst operation in the mysid shrimp *Praunus flexuosus,* J. Exp. Biol., 62:685–700.

Neugebauer, D.C., and Thurm, U. (1985) Interconnections between the stereovilli of the fish inner ear, Cell Tissue Res., 240:449–453.

Osborne, M.P., Comis, S.D., and Pickles, J.O. (1984) Morphology and cross-linkage of stereocilia in the quinea-pig labyrinth examined without the use of osmium as a fixative, Cell Tissue Res., 237:43–48.

Perbal, G., and Rivière, S. (1980) Ultrastructure des cellules perceptrices de la gravité dans l'épicotyle d' *Asparagus officinalis* L., Biol. Cell., 39:91–98.

Pires, A., and Woollacott, R.M. (1983) A direct and active influence of gravity on the behavior of a marine invertebrate larva, Science, 220:731–733.

Plate, L. (1924) Allgemeine Zoologie und Abstammungslehre. 2.Teil. Die Sinnesorgane der Tiere, Fischer, Jena.

Platt, C. (1984) Cellular bases for gravistatic reception by invertebrates and vertebrates, in Comparative Physiology of Sensory Systems, Bolis, L., Keynes, R.D., and Maddrell, S.H.P. (eds.), Cambridge University Press, pp. 563–586.

Precht, W. (1979) Vestibular mechanisms, Annu. Rev. Neurosci., 2:265–289.

Roberts, A.M. (1970) Geotaxis in motile micro-organisms, J. Exp. Biol., 53:687–699.

Rose, R.D., and Stokes, D.R. (1981) A crustacean statocyst with only three hairs: light and scanning electron microscopy, J. Morphol., 169:21–28.

Sandeman, D. (1976) Spatial equilibrium in the arthropods, in Structure and Function of Proprioceptors in the Invertebrates, Mill, P.J. (ed.), Chapman and Hall, London, pp. 485–527.

Sandeman, D. (1983) The balance and visual systems of the swimming crab: their morphology and interaction, Fortschr. Zool., 28:213–229.

Sandeman, D.C., and Okajima, A. (1972) Statocyst-induced eye movements in the crab *Scylla serrata*. I. The sensory input from the statocyst, J. Exp. Biol., 57:187–204.

Schmidt, W. (1912) Untersuchungen über die Statocysten unserer einheimischen Schnecken, Z. Med. Naturwiss., 48:515–562.

Schöne, H. (1954) Statocystenfunktion und statische Lageorientierung bei decapoden Krebsen, Z. Vgl. Physiol., 36:241–260.

Schöne, H. (1971) Gravity receptors and gravity orientation in Crustacea, in Gravity and the Organism, Gordon, S.A., and Cohen, M.J. (eds.), University of Chicago Press, Chicago, pp. 223–235.

Schöne, H., and Steinbrecht, R.A. (1968) Fine structure of statocyst receptor of *Astacus fluviatilis*, Nature, 220:184–186.

Schröter, K., Läuchli, A., and Sievers, A. (1975) Mikroanalytische Identifikation von Bariumsulfat-Kristallen in den Statolithen der Rhizoide von *Chara fragilis*, Desv. Planta, 122:213–225.

Sievers, A., and Volkmann, D. (1977) Ultrastructure of gravity-perceiving cells in plant roots, Proc. R. Soc. Lond. B Biol. Sci., 199:525–536.

Singla, C.L. (1975) Statocysts of hydromedusae, Cell Tissue Res., 158:391–407.

Stahlschmidt, V., and Wolff, H.G. (1972) The fine structure of the statocyst of the prosobranch mollusc *Pomacea paludosa*, Z. Zellforsch. Mikrosk. Anat., 133:529–537.

Stein, A. (1975) Attainment of positional information in the crayfish statocyst, Fortschr. Zool., 23:109–119.

Stephens, P.R., and Young, J.Z. (1976) The statocyst of *Vampyroteuthis infernalis* (Mollusca: Cephalopoda), J. Zool. Lond., 180:565–588.

Stephens, P.R., and Young, J.Z. (1978) Semicircular canals in squids, Nature, 271:444–445.

Stephens, P.R., and Young, J.Z. (1982) The statocyst of the squid *Loligo*, J. Zool. Lond., 197:241–266.

Takahata, M., and Hisada, M. (1979) Functional polarization of statocyst receptors in the crayfish *Procambarus clarkii* Girard, J. Comp. Physiol., 130:201–207.

Thorpe, W.H., and Crisp, D.J. (1947) The orientation responses of *Aphelocheirus* (Hemiptera, *Aphelocheiridae* [*Naucoridae*]) in relation to plastron respiration; together with an account of specialized pressure receptors in aquatic insects, J. Exp. Biol., 24:310–328.

Tschachotin, S. (1908) Die Statocysten der Heteropoden, Z. Wiss. Zool., 90:343–422.

Tsirulis, T.P. (1981) The ultrastructural organization of statocysts of some bivalvia molluscs (*Ostrea edulis, Mytilus edulis, Anodonta cygnea*), Tsitologiya, 23:631–637.

Vinnikov, Y.A. (1974) Evolution of the gravity receptor, Minerva Otorinolaringol., 24:1–48.

Vinnikov, Y.A., Gazenko, O.G., Titova, L.K., Bronshtein, A.A., Pevzner, R.A., Govardovskii, V.I., Gribakin, F.G., Ivanov, V.P., Aronova, M.Z., and Chekhondaskii, N.A. (1971) Retseptor gravitatsii: evolyutsiya strukturnoi, tsitokhimicheskoi i funktsionalnoi organizatsii, V serii "Problemy Kosmicheskoi Biologii" tom 12, Izd. Nauka, Leningrad.

Williamson, R., and Budelmann, B.U. (1985) The response of the *Octopus* angular acceleration receptor system to sinusoidal stimulation, J. Comp. Physiol. A Sens. Neural Behav. Physiol., 156:403–412.

Winet, H., and Jahn, T.L. (1974) Geotaxis in protozoa. I. A propulsion-gravity model for *Tetrahymena* (Ciliata), J. Theor. Biol., 46:449–465.

Wolff, H.G. (1970) Statocystenfunktion bei einigen Landpulmonaten (Gastropoda), Z. Vgl. Physiol., 69:326–366.

Wolff, H.G. (1973) Multi-directional sensitivity of statocyst receptor cells of the opisthobranch gastropod *Aplysia limacina*, Mar. Behav. Physiol., 1:361–373.

Young, J.Z. (1960) The statocysts of *Octopus vulgaris*, Proc. R. Soc. Lond. B Biol. Sci., 152:3–29.

Young, J.Z. (1965) The central nervous system of *Nautilus*, Philos. Trans. R. Soc. Lond. B Biol. Sci., 249:1–25.

Young, J.Z. (1984) The statocysts of cranchiid squids (Cephalopoda), J. Zool. Lond., 203:
1–21.

# Chapter 31

# Equilibrium in the Vertebrates: Signals, Senses, and Steering Underwater

CHRISTOPHER PLATT *

## Posture and Equilibrium

A freely swimming aquatic vertebrate must orient itself, move, and navigate in the presence of gravitational, buoyant, and hydrodynamic forces. The posture assumed by the animal depends on its responses to these forces and on other sensory inputs, such as the direction of light or the presence of nearby substrate. Together, these various sensory inputs are integrated by the brain to control what is known as ''equilibrium'' orientation. The sensory inputs have different central importance, or ''weighting,'' for influencing motor centers and may induce turning movements in different directions. By analogy to a simple physical model of torque forces around a pulley, the equilibrium position results from central balancing of these weighted inputs (von Holst 1950). Such equilibrium is crucial to survival, since rapid striking or escape movements require reproducibility to be successful, and reproducibility depends on starting such a complex movement from a known spatial orientation. Locomotor function must be controlled for rotational components of roll, pitch, and yaw about the longitudinal, transverse, and vertical body axes, respectively, and linear components along the same axes.

Several reviews of the vestibular system are available (Lowenstein 1971, 1974; Kornhuber 1974; Goldberg and Fernandez 1975; Precht 1979; Cohen 1981; Lewis, Leverenz, and Bialek 1985); some address visual-vestibular interactions (Henn, Cohen, and Young 1980) and some review the literature on equilibrium in fish in particular (Pfeiffer 1964; Lowenstein 1971; Platt 1983a).

This chapter describes structures and functions of receptors used by vertebrates to maintain equilibrium underwater. In all vertebrates, specialized structures in the vestibular system of the inner ear detect gravitational and other accelerations. Because nearly all aquatic vertebrate species are fishes, the main discussion here concerns how teleost fishes utilize gravity as a major cue for postural control. For many fishes postural orientation also is affected by directional light, so the role of vision in posture will be considered, along with some postural modulation by other senses.

---

* Department of Anatomy, Georgetown University Schools of Medicine and Dentistry, Washington, DC 20007, USA.

## Signals and Senses

There are several potential cues useful as references for orientation for an underwater animal. These include the direction of gravity, or light, or buoyancy, the presence of a nearby substrate (whether a level bottom or a vertical wall), or the presence of the water surface when it is close above. A moving animal also has cues from hydrodynamic lift and drag and from linear and angular accelerations produced by self-movement and by underwater surge or currents. From the point of view of gravistatic signal reception, these other linear accelerations can be considered a form of ''noise'' that interferes with the linear acceleration of gravity.

Several senses are useful for detecting such signals. The vestibular system of the inner ear detects both linear and angular accelerations. The eyes detect light directionality and, possibly, its polarization vector. Lateral line organs can detect nearby displacements and water velocities relative to the skin, probably also giving information about wave ''reflections'' off of nearby objects. Those fish with electroreceptors can use them also to detect certain nearby objects as reference points for hovering behavior. When direct contact with a substrate occurs, tactile sense in the skin or fins can be used to adjust posture. In free-swimming animals that use a gas bladder for buoyancy regulation, internal stretch receptors might be used for detecting the direction of ''up'' by the buoyancy of the bladder within the fish. Even chemoreceptors may have some role in postural control because, by alerting the animal to the presence of food, there may be central modulation of specialized locomotor and equilibrium activity.

## Vestibular Apparatus

The inner ear of all vertebrates contains mechanosensory receptors—the hair cells (see Lowenstein 1974; Hudspeth 1983). The hair cells each have a tuft of sensory cilia extending several micrometers upwards from the cell surface (Fig. 31.1A). This bundle contains a single kinocilium, which is a true cilium with internal microtubules, and an array of stereocilia, which are stiff rods containing mostly actin and considered to be related to microvilli. (Similar ciliated hair cells occur in many invertebrate equilibrium systems, but the cilia there are all kinocilia [see Chap. 30].) All vertebrate ears contain hair cells in the form known as type II, with an elongated cylindrical form and synaptic contacts with both afferent and efferent boutons; birds and mammals have an additional form known as type I, shaped like a chalice, enclosed in a single deeply cupped afferent nerve terminal (Lowenstein 1974).

When the ciliary tuft is bent by a displacement of the overlying fluid or mass, the hair cell produces receptor potentials. Depolarizing receptor potentials are coupled to release of a neurotransmitter at the synapses at the base of the cell. Although some studies have suggested that the afferent transmitter is GABA, recent pharmacological studies on the isolated frog labyrinth instead implicate glutamate or a related acidic amino acid (Annoni, Cochran, and Precht 1984), and controversy continues. The spiking activity in afferent fibers of cranial nerve

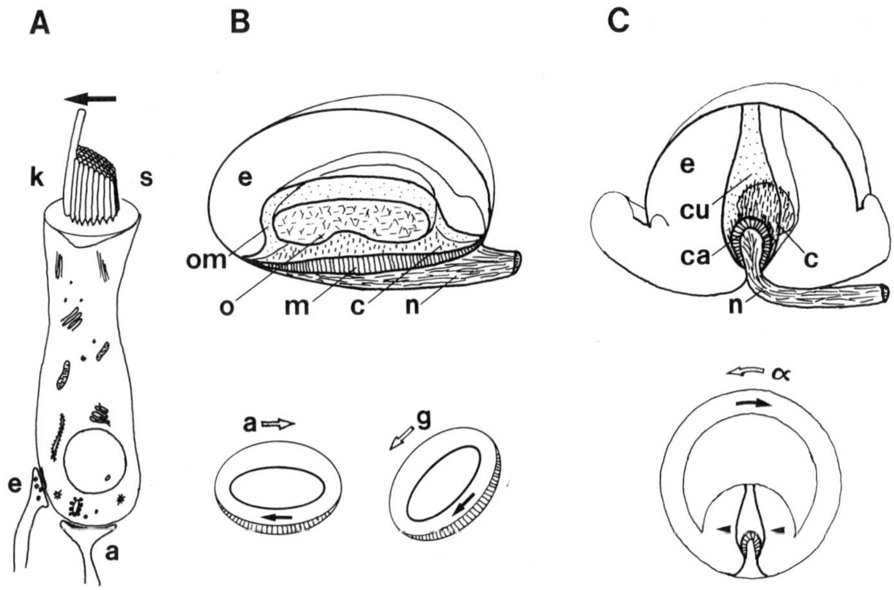


Figure 31.1. Hair cell, otolith organ, and semicircular canal crista in schematic views. (A) Hair cell of the cylindrical type II of vertebrates. Ciliary bundle containing the kinocilium (*k*) and several stereocilia (*s*) protrudes from cuticular plate at the apical end of the cell, on the surface of sensory epithelium facing into the lumen of the organ. Synaptic terminals of afferent (*a*) and efferent (*e*) neurons innervate the basal end of the cell. The direction of displacement for maximally depolarizing the hair cell is indicated by *arrow*. (B) Otolith organ in cutaway view. The pouch contains fluid endolymph (*e*) with massive otolith (*o*) imbedded in gelatinous otolithic membrane (*om*); sensory macula (*m*) lies beneath the otolith, with the ciliary bundles (*c*) extending from the apical ends of the hair cells to detect displacement and the nerve branch (*n*) innervating the basal ends of the sensory cells. The lower *small sketches* show displacement direction (*solid arrows*) for the sensory cilia of the macula under two conditions: when external linear acceleration (*a, open arrow*) moves the head, the otolith lags; or when gravitational shearing component (*g, open arrow*) is produced by fixed tilt of the head, the otolith is pulled. (C) Semicircular canal ampulla in cutaway view. Duct and ampulla contain fluid endolymph (*e*), with gelatinous cupula (*cu*) and sensory crista ampullaris (*ca*) forming a diaphragm across the canal, again with cilia (*c*) detecting displacement and nerve (*n*) innervating the hair cells. *Small sketch below* shows inertial lag of fluid endolymph (*solid arrow*) producing displacement of cupular diaphragm in response to rotation (*open arrow*) of canal.

VIII can be increased or decreased by synaptic activity from the hair cells. Often efferent fibers synapse on the base of the hair cells and release acetylcholine as neurotransmitter, which can further modulate the responses of the hair cell.

The parts of the inner ear responsible for postural and locomotor control form the vestibular labyrinth, consisting of the otolithic organs and the semicircular canal organs (see Lowenstein 1974). These structures contain a fluid, the endolymph, and additional acellular material that acts as an inertial mass when the head is subject to accelerations. An otolithic organ (Fig. 31.1B) is a pouch containing fluid and a matrix of dense calcium crystals, the otoconia, or otolith, imbedded in a gelatinous "otolith membrane." On one wall of the pouch is the sensory macula that contains the hair cells. Shearing force across the sensory macula, produced by movement of the overlying mass, deflects the sensory ciliary bundles. A semicircular canal organ is a membraneous duct containing fluid (Fig. 31.1C); this duct is enclosed within the semicircular canal of the bony labyrinth. (Proper nomenclature (Nomina Anatomica) distinguishes the duct from the canal, but in this chapter I shall follow general usage in calling the ducts the semicircular canals.) The canal at one point has a bulge, the ampulla, that contains a diaphragm-like partition across it. The lower part of this partition is formed by a ridge across the canal; this ridge, or crest, of sensory tissue is the crista ampullaris, containing hair cells. The upper part of this partition is formed by a gelatinous curtain, the cupula, extending above the crista to the roof of the ampulla. When the fluid in the canal is displaced by a rotational movement, the cupula is deflected and bends the hair cell tufts, causing a response in the cells and in the nerve.

All vertebrates except agnathan fishes and placental mammals have three otolith organs in each ear (Fig. 31.2) (Retzius 1881, 1884; Lowenstein 1974). The utriculus ("little pouch") is the most dorsal, lying in the tubular base of the semicircular duct junctions. The utricular sensory macula itself is a round or slightly crescentic patch on the floor of the utricle, curving slightly up the rostrolateral wall. With the head in the normal level position, the otolith of the utricle lies above the sensory macula. The utriculus and canals together constitute the pars superior of the inner ear. Beneath them lie two other otolith organs, the sacculus ("little sack") and lagena ("flask"), both with their sensory maculae lying roughly in the vertical plane on the medial wall. These otolithic organs form the pars inferior. In many fish species, there is also a macula neglecta (Retzius 1881), which may contain only a few hundred hair cells and lie near the utriculus, as in the goldfish (*Carassius auratus*) (Platt 1977), or may contain over one million hair cells and lie near the sacculus, as in many elasmobranchs (Corwin 1983).

The common base of the three semicircular canals is formed by the utriculus. The ampullae of both the anterior vertical canal and the horizontal (or lateral) canal are attached to the rostrolateral part of the utriculus, and the ampulla of the posterior vertical canal is at the caudal end. The two vertical canals join to form a common tube, the crus, which arises nearly vertically from the dorsomedial wall of the utriculus. The tubular structures of the utriculus and the crus enlarge the small internal radius of the canals for nearly half the circumference of each canal itself.

The pars superior usually is considered to be the peripheral vestibular system, sensing postural change, whereas the pars inferior is considered to be the peripheral auditory system. The afferent neurons in branches to the two different parts have differences in their central projections to nuclei in the brainstem (Chap. 29). The sacculus, lagena, and sometimes other accessory organs, such as the mammalian cochlea, may respond especially well to sound or vibration (see Lewis, Leverenz, and Bialek 1985 and Chap. 22 and 27). The large macula neglecta found in some elasmobranchs is believed to have an auditory function (Corwin

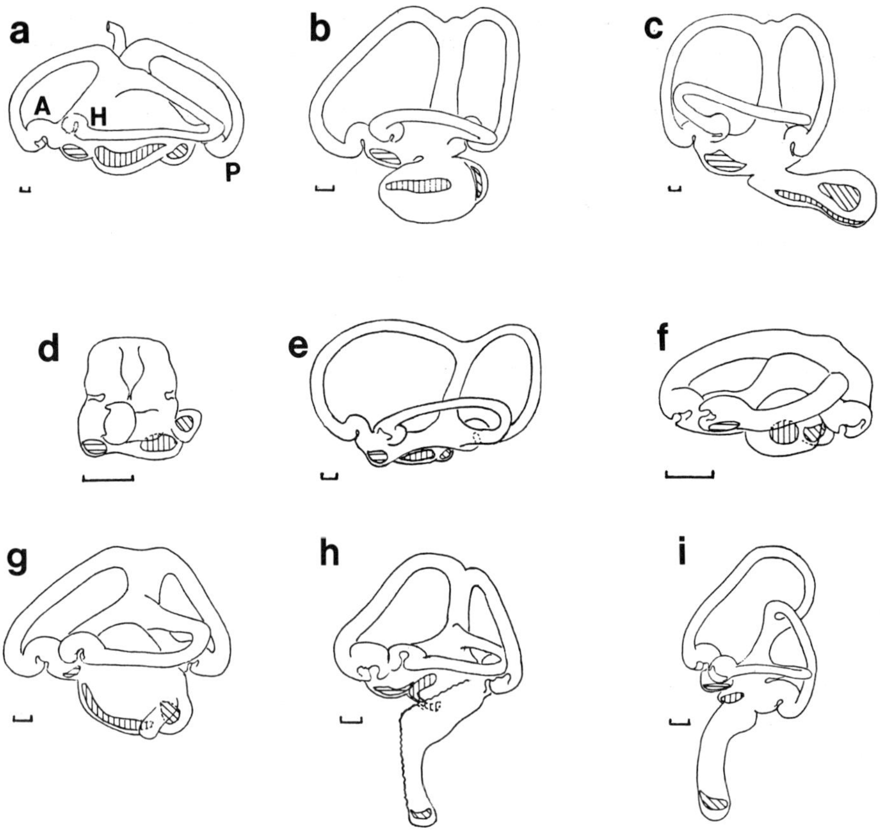


Figure 31.2. Inner ears of nine aquatic vertebrates, for comparisons of gross structure (after Retzius 1881, with his nomenclature). (a) Dogfish shark *Acanthias vulgaris*. (b) Salmon *Salmo salar*. (c) Carp *Cyprinus idus*. (d) Sea horse *Hippocampus hippocampus*. (e) Flying fish *Exocetus volitans*. (f) Salamander *Salamandra maculosa*. (g) Snapping turtle *Chelydra serpentina*. (h) Alligator *Alligator mississippiensis*. (i) Merganser *Mergus merganser*. Ampullae for horizontal (*H*), anterior vertical (*A*), and posterior vertical (*P*) semicircular canal organs are labelled for the dogfish and are in similar positions in the other species. The maculae of the three otolith organs are shaded differentially, the utriculus shown by horizontal lines, the sacculus by vertical lines, and the lagena by diagonal lines. (Scale bars = 1 mm.)

1983), but its function in other vertebrates remains unclear. The evolutionary development of the cochlea as a distinct structure for hearing, with the loss of the lagenar otolith organ in placental mammals, leaves the sacculus of mammals as a pars inferior otolith organ that may have only vestibular, rather than auditory, function (Goldberg and Fernandez 1975).

The vestibular nuclei in the brainstem occupy a strategic location for modulating many motor functions. The output of the vestibular nuclear neurons ascends to the extraocular muscles, the cerebellar cortex, and other centers in the brain and descends to affect spinal motoneurons used for both postural and locomotor control of trunk and fins or limbs.

## Variation in Vestibular Structure

Within this general plan of three otolith organs and three canals, there is some notable variation among the vertebrates. Comparisons between species show differences in the relative dimensions and shapes of the otolith organs and of the canals (Retzius 1881, 1884; Lowenstein 1971, 1974; Platt and Popper 1981; Platt 1983a; Lewis, Leverenz, and Bialek 1985). Attempts have been made to correlate the structural variation with particular postural or auditory capabilities of various animals (Jones and Spells 1963; Melvill Jones 1974; Mayne 1974; Ramprashad et al. 1984).

Comparative studies over the last century have shown that the gross structure of the pars superior retains a basic plan from cartilaginous fish through mammals. Variation has been considered relatively minor compared to the extensive variation in the pars inferior, which includes the development of the cochlea. This relative constancy in the structure of the canals and utriculus has been suggested to correlate with the relatively similar range of head movements used by most vertebrates compared to the extensive variation in auditory structures and acoustic functions of the ear throughout the vertebrates (Chap. 27).

For example, Figure 31.2 shows the inner ear in nine different species of aquatic vertebrates. (These sketches are simplified from the wonderful lithographs of Retzius (1881, 1884), and I shall use the species names given by him, though some have since been changed.) This figure suggests that the sensory macula of the utriculus varies less in size and location than the maculae of the sacculus and lagena. All the species shown have a relatively small utriculus, despite the obvious differences in size and shape of pouches and maculae of the sacculus and lagena.

There are only two cases yet found in which the utriculus does not have a macula shaped basically like a dish under the otolith. A marine catfish, *Arius felis,* has a macula that is instead a narrow band around the horizontal equator of a huge, nearly spherical otolith; this utriculus may detect very low-frequency sound (Popper and Tavolga 1981). The exceptional utriculus in the herring and allies (Clupeidae) has three separate macular patches and is adjacent to a gas-filled bulla, and so this utriculus also may detect sounds (Denton and Gray 1980; see also Chaps. 8, 22, and 23).

The semicircular canals also show a basic plan for all these species. The horizontal semicircular canal usually lies near the earth-horizontal plane of the head in the normal posture. The anterior vertical canal lies vertically in a plane angled roughly at 45° away from the rostrocaudal axis, so that the anterior canal planes of the two sides form a V opening forward. The posterior vertical canal lies vertically in a plane roughly orthogonal to the anterior canal, angled roughly at 135° from the rostrocaudal axis, so that the posterior canal planes of the two sides form a V opening rearwards. The two vertical canal planes of one side thus form a V opening laterally. Despite a wide range in size, the internal diameter of the ducts appears to be less than 1 mm (Fig. 31.2) (Jones and Spells 1963).

However, there are some noticeable differences among the shapes of the semicircular canals in different vertebrates (Fig. 31.2). The range of elongation vs. circularity in different canals is great. The dogfish shark (*Acanthias vulgaris*), salmon (*Salmo salar*), and carp (*Cyprinus idus*) have roughly circular canals, but the sea horse (*Hippocampus hippocampus*) has canals that hardly loop away from the base at all, forming a compressed rectangular shape, and the flying fish (*Exocetus volitans*) has canals that arc out very widely, almost outlining rectangles. The snapping turtle (*Chelydra serpentina*) and alligator (*Alligator mississippiensis*) have vertical canals that form nearly triangular arcs. The merganser (*Mergus merganser*), a diving duck, has canals formed in roughly circular arcs, but the anterior vertical canal forms a far larger loop than the others. These anatomical features seem to indicate a range of variation greater than that usually discussed. For canals in a wide range of vertebrate species and body sizes, the reported changes in duct internal radius and radius of curvature show a tiny, but significant, increase, believed to provide a functional match of sensitivity to the change in head movement frequencies with increasing size (Jones and Spells 1963; see also below).

Data from both gross and ultrastructural studies on the utriculus and canals and their maculae and cristae are needed to compare with postural and locomotor abilities. A constancy of structure in the utriculus and canals might be expected if postural functions are basically the same throughout the vertebrates. Extensive comparative studies on the sacculus have shown that, along with the greatest variation in size among all the otolith organs, the saccular otolith has a remarkable range of shapes (Retzius 1881, 1884; Platt and Popper 1981; Lewis, Leverenz, and Bialek 1985). The sacculi of different fish also differ at the level of cellular organization, even within a taxonomic family (Chap. 27). These structural differences in teleost sacculi have provided a basis for analyzing the extensive functional differences among teleosts in hearing (see Popper, Platt, and Saidel 1982 and Chap. 27).

## Vestibular Ultrastructural Organization

Each hair cell has a direction of maximal sensitivity to shear, determined by the location of the single kinocilium at one edge of the bundle. Within a single sensory macula, there may be hundreds of neighboring hair cells that are morphologically aligned so that a whole group shares a similar orientation. Also within a single macula, it is common for two adjacent groups to have quite different, often even directly opposing, general orientations, so that the direction of maximal depolarization for one population is 180° relative to the other (Fig. 31.3). In these cases, there often is a boundary with little interdigitation among differently oriented cells. Agnathans (Lowenstein 1971) and elasmobranchs (Lowenstein 1971; Barber and Emerson 1980) do not show clearly divided orientation patterns, although a mixed orientation pattern can be mapped; but bony fishes and all

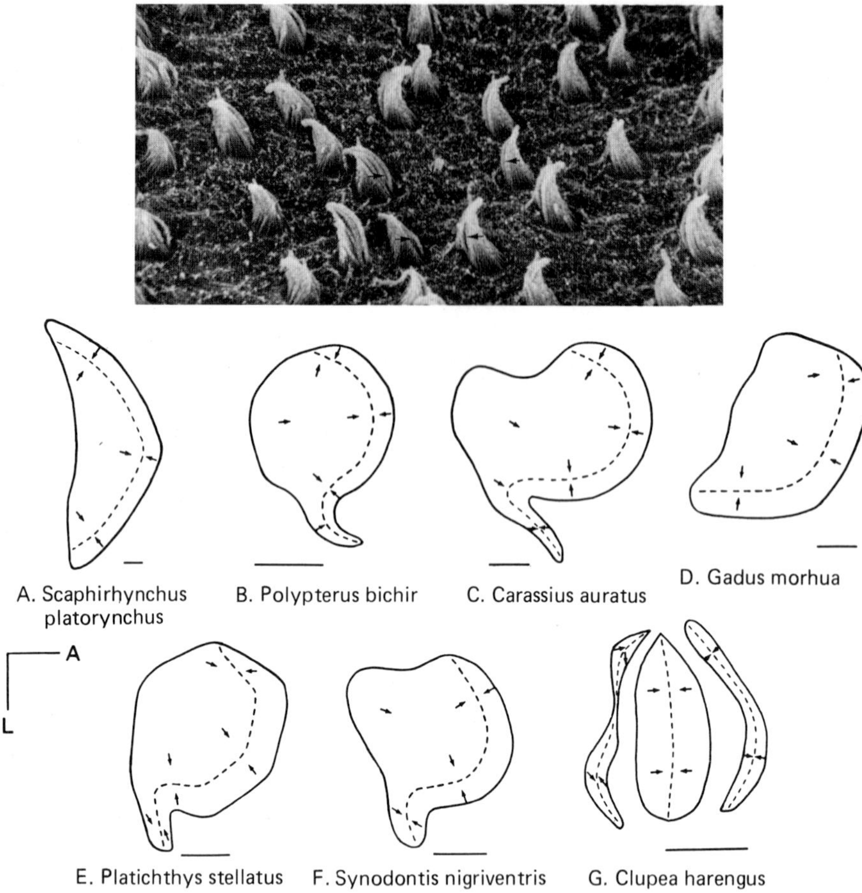


Figure 31.3. Orientation patterns of hair cells in otolith organs. *Arrows* indicate the direction of maximal depolarization sensitivity for ciliary bundles. *Top,* hair cell bundles on macular surface showing oppositely oriented groups in the utriculus of a goldfish. (Scale bar = 2 μm.) *Below,* map of utricular pattern of cell orientations in several fishes: *top row* shows (*a*) a sturgeon, (*b*) a reedfish, (*c*) a goldfish, and (*d*) a cod; and *lower row* shows (*e*) a flatfish, (*f*) upside-down catfish, and (*g*) a herring. (Scale bar = 250 μm.) (Lower figure from Platt and Popper 1981.)

tetrapods have maculae with hair cells organized into clearly separable groups of opposing orientation (see Lowenstein 1974; Platt and Popper 1981; Platt 1983a; Lewis, Leverenz, and Bialek 1985).

In the utriculus, there are usually two oriented groups, one more medial with each cell having its kinocilium on the side of the bundle facing the external margin of the macula, and one more anterolateral with each cell having its kinocilium on the side facing inward from the margin (Fig. 31.3). As shown in Figure 31.3 for a variety of fishes, a curved dividing line can be drawn to separate the inward-and outward-oriented populations. In the utriculus the two groups are said to be "facing toward" each other, since the kinocilia of the cells on either side of the dividing line are directly opposite each other. This dividing line extends roughly along the middle of a region of the macula known as the striola (stripe), a curving band in which the hair cells have larger cell bodies and larger ciliary bundles than elsewhere in the macula.

In contrast, the sacculus of ostariophysine fishes (Gonorynchiformes, Cypriniformes, Characiformes, Siluriformes) has a pattern in which two groups of cells with opposite orientations are "facing away" from each other. Cells in the dorsal part of the macula have their kinocilium on the side of the bundle facing the dorsal margin of the macula, and those in the ventral part have their kinocilium on the side facing the ventral margin (Platt and Popper 1981; Popper and Platt 1983). Here the dividing line extends rostrocaudally, thereby dividing the macula roughly in half longitudinally. The saccular macula has no striola in fishes.

The macula of the sacculus shows this bidirectional pattern in all tetrapods, as well as in these ostariophysines, and for a long time it was considered the standard vertebrate pattern (Lowenstein 1974); but comparative studies have shown that the sacculus of many fishes has a more complex macular pattern of at least four distinctly oriented cell groups. The caudal portion of the macula has the usual division, with cells in the dorsal half oriented so the kinocilium faces the dorsal margin and cells in the ventral half oriented with the kinocilium facing the ventral margin; but in the rostral portion of the macula, the cells in the dorsal half are oriented facing caudally, whereas cells in the ventral half are oriented facing rostrally (see Popper, Platt, and Saidel 1982). Although many teleost fish that are not ostariophysines have this "quadrant" pattern of orientation, some species have even more involved patterns. As discussed elsewhere in this volume, it is now believed that the multiple saccular populations may be specialized for directional sensitivity to acoustic input (Chaps. 4, 27, and 28).

The lagena is an embryonic and phylogenetic outgrowth from the sacculus; but it has been shown in fishes to have some role with the utriculus in the control of posture (Schoen and von Holst 1950), and, as in the utriculus, it usually contains a bidirectional opposition of only two populations that face "toward" each other (see Lewis, Leverenz, and Bialek 1985). In fishes, the macula of the lagena often lies nearly vertically, and the bidirectional opposition line is nearly vertical, though often curving; the cell orientations generally are parallel to the rostrocaudal axis of the fish (Platt and Popper 1981).

These patterns imply that each otolithic organ is maximally sensitive along certain specified axes of rotation. The utriculus has cells oriented mainly in the mediolateral and rostrocaudal directions, so it is expected to have maximum sensitivity to tilting about the roll and pitch axes. The sacculus and the lagena have cells oriented mainly in the dorsoventral and rostrocaudal directions, with maximum sensitivity to tilting about the pitch axis. It is important to note that these maculae are not usually planar, but have varying amounts of curvature. A deeply cupped utricular macula might, for example, have cells oriented in a pattern fanning up the walls in all three planes and thus be responsive to all possible pitch, roll, and yaw movements (see Schöne 1975). The exceptional utriculus in the herring family (Clupeidae) has three separate macular patches, each with two oppositely oriented populations, with opposing sensitivity shown by recorded microphonics (Denton and Gray 1980). These fish have a gas-filled bulla adjacent to these maculae, giving the utriculus acoustic sensitivity (Chap. 8).

A second aspect of ultrastructure concerns the variety, even within a single macula, of ciliary bundle forms (Fig. 31.4) (see Hudspeth 1983; Lewis, Leverenz, and Bialek 1985). The bundles at the margin of most maculae and sometimes covering additional wide areas are tiny; those near, but not at, the margin often have an extremely tall kinocilium and short stereocilia, whereas those more centrally located, particularly in the striolar regions when present, have tall stereocilia, often reaching nearly to the tip of the kinocilium. The tallest bundles of all are found in the semicircular canal cristae, where kinocilia sometimes are over 50 μm long. The different forms of bundles often are associated with similar neighbors, rather than being intermixed, and this organized pattern or "mosaic" suggests a functional grouping of innervated hair cells (Platt 1983a). Length of the ciliary bundles has been correlated with sensitivity to different frequencies by recording activity in single neurons, which then are filled with dye to trace to the hair cells they innervate (Lewis et al. 1982; see also Lewis, Leverenz, and Bialek 1985), or with intracellular recordings from individual hair cells (Hudspeth 1983).

As in the many other anatomical studies, it is not yet clear which features of the ultrastructure, if any, relate specifically to aquatic life.

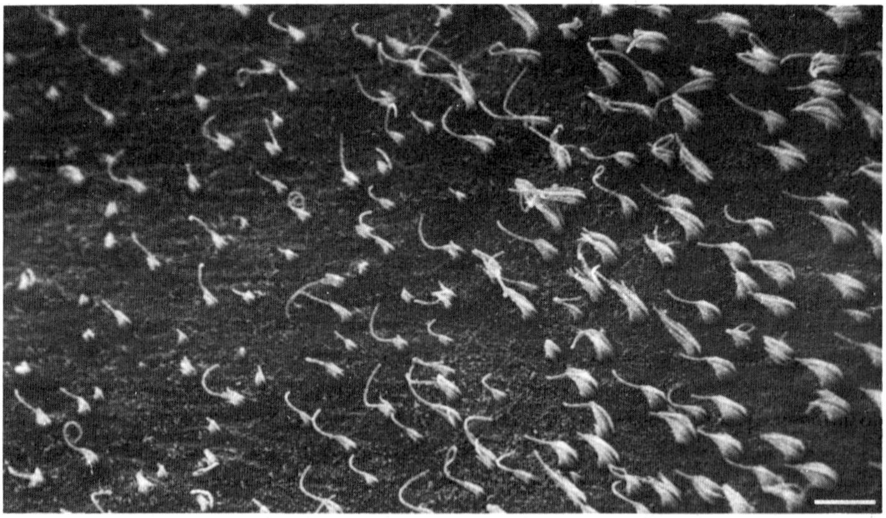

Figure 31.4. The variety in structure of sensory cell bundles. Epithelial surface near the striola of the utriculus in goldfish shows a transition from small ciliary bundles in one region to much larger bundles in the striola. The bundles between the small and large ones have a form that is different from either, with very long kinocilum and relatively short stereocilia. (Scale bar = 5 μm.)

## Vestibular Physiology

The vestibular organs are extremely sensitive to displacement of the cilia on the hair cells, but these cells are associated with accessory structures in a variety of ways. Whether a cell responds better to a maintained position of the head or to an oscillation, to higher or lower frequencies of oscillation, and to peak acceleration or peak velocity of a movement depends partly on the viscosity and elasticity of overlying structures mechanically coupled to the hairs. The hair cells themselves may also have membrane properties that provide "tuning" to preferred frequencies (Hudspeth 1983) (Chap. 32). The afferent neurons may have their own adaptation characteristics for firing rates.

### Semicircular Canal Sensitivity

The semicircular canal organs provide a physical system using fluid inertia to allow detection of angular head motions. Since the first mathematical modelling of the system by Steinhausen (1931), working on the pike (*Esox lucius*), there have been several analyses of increasing sophistication (van Egmond, Groen, and Jongkees 1949; Mayne 1974; Oman 1981; Ramprashad et al. 1984). All take into account the fundamental properties of inertia, viscosity, and elasticity found in the cupula-endolymph system. The fluid endolymph has inertia, and when the head is rotated, force must be applied to accelerate the fluid. The rotation of the canal wall accelerates the lagging fluid by friction (a viscosity term times a flow velocity term). The cupular partition across the canal acts like a flexible paddle, also producing a force accelerating the fluid (a spring constant term times a displacement term).

Usually the system is considered to be analagous to a torsion pendulum that has sufficient damping to make it act as a velocity detector, rather than a simple acceleration detector, for a wide range of head movements (Steinhausen 1931; van Egmond, Groen, and Jongkees 1949). Where $\alpha(t)$ = angular acceleration of the head in the plane appropriate for the canal, $\xi(t)$ = angular displacement of endolymph, $\theta$ = endolymph moment of inertia, $\pi$ = viscous damping term, and $\Delta$ = elastic restoring term, the behavior of the system is approximated by a second-order differential equation with two coefficients (Goldberg and Fernandez 1975):

$$\theta(d^2\xi/dt^2) + \pi(d\xi/dt) + \Delta\xi = \theta\alpha(t).$$

The viscosity effects dominate in this heavily damped system, and the response of the cupula-endolymph couple can be characterized by two time constants. The "long" constant, $t_1 = \pi/\Delta$, determines the time course of endolymph displacement; the "short" constant, $t_2 = \theta/\pi$, determines the sensitivity to rotational velocity (Goldberg and Fernandez 1975).

With this general model, two important parameters for estimating $t_2$ are the internal radius of the narrow canal and the radius of curvature of the canal. The internal radius determines the cross section through which the viscous endolymph must flow and so partly determines the frictional drag term. The radius of curvature of the whole canal partly determines the effective moment of inertia. Different models and derived equations have incorporated effects of such factors as elliptical, instead of circular, cross section for the canal and the fraction of the fluid ring that is occupied by the relatively large utricular volume (Oman 1981; Ramprashad et al. 1984). The effective radius of curvature is difficult to address in canals with very noncircular form, but is usually estimated as half the maximum center-to-center measurement between the utricle and the most distant part of the canal ring (Jones and Spells 1963).

Calculation of $t_1$ requires measurement of the elasticity of the cupula, which has not yet been made. Physiological recordings of nerve activity and motor responses have been used to try to confirm estimates of the long time constant, but various adapting processes are hard to eliminate (Goldberg and Fernandez, 1975).

The canal is considered to deliver primarily a signal that is following rotational velocity of the head, rather than acceleration itself, over a wide range of frequencies of head oscillation. Calculations indicate the canal response has a gain of roughly 1.0 for angular oscillations between $1/t_1$ and $1/t_2$ (Mayne 1974; Goldberg and Fernandez 1975). Although there appears to be general agreement that the short time constant is on the order of less than 10 ms, there is little agreement on the long time constant, except that it is on the order of several seconds (Goldberg and Fernandez 1975). Thus the system acts as a faithful velocity transducer for head oscillations ranging from around 0.025 to 25 Hz (Goldberg and Fernandez 1975).

The cupula has been found to be fixed to the roof of the ampulla under normal conditions (Hillman 1974), so the cupular displacement is not in the form of a ''swinging door'' model, which had been accepted for decades. Instead the cupula now is considered to act more like a diaphragm or drumhead (Flock and Goldstein 1978; Hillman and McLaren 1979; McLaren and Hillman 1979; Oman, Frishkopf, and Goldstein 1979). The main displacement appears to be not at the apex, but near the middle of the ampulla, right above the lowest point in the crest of the crista (Flock and Goldstein 1978; Hillman and McLaren 1979; McLaren and Hillman 1979; Oman, Frishkopf, and Goldstein 1979). Injection of fine droplets of dye into the cupula showed that the cupula might show regional changes in shape between small and larger amplitude displacements (Hillman and McLaren 1979). It is interesting that, although this changed concept is dramatic from the anatomical viewpoint, it has no major effect on the mathematical model: the displacement, viscous drag, and restoring force from elasticity are still the major terms to consider and can be handled with the same equations.

The displacement of the cupula is extremely small during normal head movements. The threshold angular acceleration for detection has been estimated to be less than $0.5°/s^2$ and the cupular deflection at threshold to be in the range of 0.01 μm (Hillman and McLaren 1979). Even at extreme rotational velocities up to 800°/s the calculated deflection is less than 1° (Melvill Jones 1974), deflecting the center of the cupula less than 10 μm (Hillman and McLaren 1979). All the hair cells in a crista have their directional polarization in the same sense relative to the fluid movement in the canal, so that a rotation in one direction is excitatory and in the other direction inhibitory for all receptors, unlike the case in the complex otolith organ maculae (Lowenstein 1974). This large population would permit an averaging mechanism that would substantially enchance signal-to-noise ratio during small deflections.

## Otolith Organ Sensitivity

The otolith organs have their cilia often coupled to the otolith membrane, which encloses the otolith (Lewis, Leverenz, and Bialek 1985). By analogy to the model for the semicircular canal, the stimulation of an otolith organ macula by linear accelerations can be physically described by equations considering the applied force on the otolith and otolith membrane mass, the displacement, the viscous drag, and the elasticity of the coupling (Mayne 1974). As in the canals, though the stimulus to the head may be acceleration (including gravity), the properties of the coupling could allow some receptors to detect primarily velocity for some ranges of head movements and other receptors to respond primarily to maintained displacement (tilt) (Mayne 1974).

The high sensitivity of the hair cell depolarization during deflection of the ciliary bundle is shown by the saturation with deflections of the ciliary tips of less than 1 μm in the frog sacculus (Hudspeth 1983). In frogs, dye injections into single afferent neurons that have their activity recorded allow tracing the neurons to the hair cells on which they terminate. The form of ciliary bundle has been correlated from these studies with the frequency sensitivity and regularity of firing of the afferent neurons (Lewis et al. 1982).

## Neural Responses

Aquatic animals have provided some extremely useful vestibular preparations. Lowenstein and his co-workers pioneered the use of the isolated elasmobranch vestibular preparation for physiological recording (see Lowenstein 1971, 1974). Several frogs (*Rana* sp.), of course included as aquatic, are major sources of data on vestibular function, posture, and vestibular compensation for lesions (Dieringer and Precht 1979a, 1979b; Galiana, Flohr, and Melvill Jones 1984). But most recordings of eighth nerve activity from teleosts have been for auditory studies. There are only a few cases of recording responses to tilt or rotation from teleost peripheral neurons innervating canals (Schmidt, Wist, and Dichgans 1972; Highstein and Politoff 1978), otolith organs (Enger 1963; Platt 1973), or central neurons in the vestibular nuclei (Schoen 1957; Dichgans, Schmidt, and Graf 1973; Allum et al. 1976).

The vestibular neural responses in aquatic animals show few apparent differences from those in other vertebrates (see Lowenstein 1974; Goldberg and Fernandez 1975; Platt 1983a). Many neurons have a "resting discharge" firing rate ranging from a few spikes (action potentials) per second up to more than 100/s. The generally lower rates found in nonmammalian species may depend partly on lower body temperature (Goldberg and Fernandez 1975; Precht 1979). This activity is present when the head is in the normal position relative to gravity and stationary. At first such neurons were believed to have a pacemaker-like endogenous property, but it now is generally thought that continuing transmitter release from hair cells drives such activity (Highstein and Politoff 1978). The responses to movement often show directionally sensitive changes, usually increasing firing rate when the head is tilted or rotated in one direction and decreasing when the head is moved in the opposite direction. Both the resting activity and the activity during response may be extremely regular or irregular; an often-used criterion to divide the two categories is the interspike coefficient of variation (CV), with units having a CV of less than 0.5 (or other stated level) termed "regularly" firing (Goldberg and Fernandez 1975). Neurons with more branched terminals are believed to be those with more regular firing rates, acting as rhythmical oscillators driven constantly towards depolarization by the nearly continuous barrage of synaptic activity over many terminals (Highstein and Politoff 1978). The response pattern to a maintained tilt or to constant velocity rotation may be tonic or phasic or some combination. Tonically responding neurons are more commonly found in frogs and fish than in mammals (Precht 1979).

Correlations between some of these properties have been suggested in some cases (Lowenstein and Roberts 1949; Enger 1963; Platt 1973; Goldberg and Fernandez 1975; Macadar et al. 1975; Budelli and Macadar 1979; Plassman 1983). Tonic units often are regular and phasic units often irregular. Phasic units may be more sensitive to vibration than tonic units and often are silent in the "resting" position of the head. A caution in interpreting these responses was found in an elasmobranch, the guitarfish *Rhinobatos productus,* with which introduction of a tiny amount of "jitter" in the position of the testing platform produced a change in some of the units from phasic to tonic and a dramatic increase in the firing rate of some fibers (Macadar et al. 1975). Correlation of response properties with anatomical features has been suggested for canal fibers in elasmobranchs (O'Leary, Dunn, and Honrubia 1976) and teleosts (Highstein and Politoff 1978), in which large fibers are believed to be those with fewest terminals, with irregular activity, and lowest thresholds. Since these fibers have the highest conductive velocities for afferent signals to the brain, they may have a crucial role in fast reflexes.

One of the most useful concepts in categorizing such units remains that of Lowenstein (Lowenstein and Roberts 1949). Some units were found in the thornback ray (*Raja clavata*) that responded only when the head was within a relatively small angle ($< 30°$) of its normal position; these units were characteristic of the lagena in the thornback ray and termed "into-level" units. Several units of another sort were found that had a large phasic component and responded when the head was tilted away from the normal position with an increase in firing rate whatever the directional sense; these were termed "out-of-position" units. The presence of both such units suggests that at least some peripheral mechanisms are related to the "set point" establishment of the normal upright position, in addition to the central mechanisms available for comparing the symmetry of inputs from the two sides (Graf and Meyer 1983).

## Steering: Posture and Locomotor Control

All animals have the problem of keeping the gravity vector as a reference while subjecting the vestibular system to locomotor and passive other accelerations. The canals may provide a useful additional signal that allows an animal to distinguish whether a change in the acceleration vector on an otolith organ is caused by a linear acceleration or by a tilt, and the speed with which such changes occur may be important cues (Mayne 1974).

As pointed out by Mayne (1974), it is probably more important for aquatic animals to monitor the correct direction of horizontal, rather than vertical. Underwater, the horizontal plane is that of equal hydrostatic pressure, which is crucial to any animal using a compressible gas for buoyancy. Since the vertical is not associated with direction of falling, as in terrestrial animals, it may be a less important cue. Of course the animal can use the vestibular system in either frame of reference. When blinded goldfish were placed in a tank and the tank was accelerated and decelerated, the goldfish showed nose-up and nose-down pitching behavior that exactly matched the subjective change in ''horizontal'' that would be produced by addition of the changing acceleration vector to the constant gravity vector (von Baumgarten, Baldrighi, Atema, and Shillinger, 1971).

Two features of aquatic life allow postural attitudes with the head oriented in directions that would be difficult for land animals larger than a few centimeters long to maintain. First, buoyancy is possible, so that posture is not determined by the requirement for a stance to prevent falling. Second, when part of the body is below the heart in a large terrestrial animal (such as a bat hanging upside down or a giraffe drinking), there is a hydrostatic pressure difference that requires circulatory compensation to return blood to the heart. But underwater the external pressure increases with depth, so any body parts below the heart are also deeper in the water and under greater external pressure, which thus assists antigravity return blood flow. Indeed, there are several fish known to show quite unusual posture, different from the usual horizontal head position. Some of these fish have been examined to see whether there are peripheral structural features that might correlate with the behavior, suggesting evolutionary adaptations.

## Unusual Posture and Otolith Organ Structure

Figure 31.5 shows some examples of fish that have been studied for comparison with the known vestibular structure of ''standard'' upright teleosts. Whereas standard fish swim with their heads in the usual posture for vertebrates, these unusual fish adopt a posture with a tilt in either the pitch or roll axis that may be extreme. First, the whole order of flatfish (Pleuronectiformes or Heterosomata) live as adults that lie on one side, having rolled 90° during metamorphosis from a standard upright larval stage. Second, the upside-down catfish (*Synodontis* sp.) feed on the underside of vegetation and continue to swim upside down in open water; these fish thus are rolled 180° relative to the standard. Third, some freshwater fish among the tetras and allies (Characinoidei) are called head standers because they swim and rest with the long axis of the body pitched nose downward, sometimes to more than 30°. Fourth, there are a few other characinoid species called tail standers because they rest and swim with the body's long axis pitched nose upwards by about 25°.

In most of these fish, the utricular macula itself is oriented in the head just

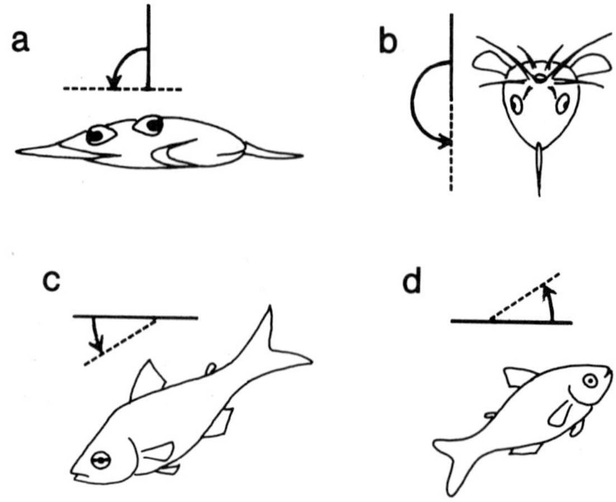


Figure 31.5. Fish with unusual posture. *Above,* head-on views of two fish that have rolled (*arrow*) so dorsal fin lies in new relation (*dashed line*) to vertical (*solid line*). (a) Flatfish (order Pleuronectiformes) lies on one side, with both eyes uppermost after one eye migrates across head; this roll orients vestibular system 90° from that in standard fish. (b) Upside-down catfish (*Synodontis* sp.) swims inverted in open water; this roll orients vestibular system 180° from standard fish. *Below,* side views of two fish that have pitched (*arrow*) so the longitudinal body axis (*dashed line*) lies in new relation to the horizontal (*solid line*). (c) Head stander *Chilodus* hovers nose downward; this pitch orients vestibular system 30° from that in standard fish. (d) Tail stander *Thayeria* swims nose upward; here vestibular system has changed so the utriculus is roughly level when the fish is pitched 25° upward.

as it is in standard fish, lying across the ventral part of the utriculus. The inner ear of several flatfish is not much different from that in perches (Perciformes), even in its ultrastructural pattern of cell orientations (Fig. 31.3) (Platt 1983b). Since a flatfish lies on its side, the utricular maculae lie nearly vertically when the fish is in its normal resting posture; here the postural adjustments are central, not peripheral. Several experiments on flatfish have shown that they depend more on their now-horizontal sacculi, rather than the utriculi, for static postural behavior (Schöne 1964; Platt 1973). Central pathways of neurons from the canal cristae to the oculomotor nuclei in flatfish appear to be unique compared to the bilaterally symmetrical projections in other vertebrates, adapting to the new relations between the vestibular system and the eyes (Graf and Baker 1985a, 1985b).

The upside-down catfish *Synodontis nigriventris* has an inner ear much like that of the other ostariophysines (Fig. 31.3) (Meyer, Platt, and Distel 1976; Popper and Platt 1983); but in this catfish, the utriculus usually is functioning with the otolith hanging below the upside-down macula. There is some behavioral evidence that the upside-down catfish may use not just its utriculus, but may rely largely on its canals for long-term postural control, again suggesting a central change in interpretation of peripheral input (Meyer, Platt, and Distel 1976).

One unique peripheral adaptation is known. The freshwater penguin fish, a tail-standing tetra (*Thayeria boehlkei* (*obliqua*)), has its utriculus oriented differently from other fish, so that the macula is roughly level when the fish is pitched nose up about 25° (Braemer and Braemer 1958). When tested in a large centrifuge, this fish did not change its pitch under increased "gravity." The utricular macula was indeed functioning as though it were not tilted when the fish was in its normal tilted posture. Other related species, however, including the tail stander pencil fish *Poecilobrycon* (*Nannobrycon*) *eques* and the head stander *Chilodus punctatus* have utriculi oriented normally in the head, thus operating at a tilt when the fish is in normal resting position and causing a change in nose-up pitch angle when centrifuged to increased g. The tail stander *Thayeria* remains the unique exception in modifying utricular orientation for postural control. It would be interesting to examine other fish with steeply pitched posture, such as the shrimpfish *Aeoliscus strigatus* and a clingfish *Diademichthys deversor* of tropical seas. These fish hover among the long spines of sea urchins and hang there vertically nose down in the water; the shrimpfish also form small schools in open water, where instead they hang vertically nose upwards (Norman and Greenwood 1963; Marshall 1966). Another potentially interesting posture is shown by fish that spend many hours at a time asleep on their sides, such as many species of wrasse (Labridae).

Despite the unusual postures, most of these fish have inner ears that are not notably different from those in standard fishes. Thus certain extreme changes in posture can occur without substantial change in the peripheral vestibular system. Instead, the central nervous system is likely to require changes in vestibular input processing, perhaps beginning with changes in projections to and from the vestibular nuclei, as shown in the flatfish (Graf and Baker 1985b).

### Canal Sensitivity and Head Movements

Some authors have tried to make simple correlations of canal form with habits of life. For example, the flying fish with its large canal system was thought by some authors to need such canals for the demands of aerial escape, whereas the more conventional salmon could work with small canals (see discussion in Lowenstein 1957). But a fish with very different habits is the goosefish, or monkfish (*Lophius piscatorius*), an anglerfish that spends most of its time buried in the sand, and occasionally makes a quick head movement to capture food. Yet this fish has canals as elaborate as those of the flying fish (Retzius 1881). Besides, the flying fish's aerial escape is really done with a stiff posture during a glide propelled by fast, but small in amplitude, caudal fin beats; in fact, it is not nearly so acrobatic or tumbling as the salmon, which swims up turbulent waterfalls. The correlations of canal form and function do not appear to be simple.

The seimcircular canals of several aquatic, as well as terrestrial, vertebrates have been examined to provide data for comparative modelling studies (Jones and Spells 1963; ten Kate, van Barnefeld, and Kuiper 1970; ten Kate 1973; Melvill Jones 1974; Goldberg and Fernandez 1975). Over a tremendous range of sizes of animal, the size of the canals does not change much. The internal radius of the canal duct ranged from only 0.07 to 0.36 mm (5-fold) and the radius of curvature from 0.9 to 11 mm (12-fold) in eighty-seven species ranging in mass from 0.015 to 450 kg (30,000-fold) (Jones and Spells 1963). The observed changes in these two radii were considered to be consistent with the canal time constants theoretically needed for the changes in head movements scaled for changes in body size and shape (Jones and Spells 1963).

There are few interesting features of canals in fish that appear significantly different from those of other vertebrates. Comparing seventeen fish species with forty-six mammals, seventeen birds, and seven reptiles, the fish had consistently larger values for both internal radius (most > 0.20, none < 0.15 mm) and radius of curvature (most > 4, none < 3.2 mm) (Jones and Spells 1963). The authors argued that the fish data predicting velocity responses at lower frequencies are consistent with behavior; most fish have no necks to produce rapid head turning, and they live in a dense medium, which slows down turning motions. Although the two seals, sea lion, and dugong they measured have relatively high values for both radii, supporting their argument, it seems paradoxical that the common porpoise has a small internal radius of the canal and the smallest radius of curvature of any of the mammals, including mice and a mole.

Work on the guitarfish has suggested a correlation of anatomy, physiology, and behavior in the use of the horizontal canal (O'Leary, Dunn, and Honrubia 1976). First, center of the crest of the crista is innervated mostly by large fibers, while smaller fibers represent a greater fraction of innervation out towards the walls. Second, responding to a range of different acceleration directions and magnitudes as a stimulus sequence, the fibers from the center of the crest typically show a lower threshold to acceleration, a shorter latency, and a faster conduction velocity than the fibers nearer the walls, suggesting both greater sensitivity and faster response (Fig. 31.6). Third, there is evidence from video films of rapid escape behavior that the predominant frequencies of head oscillations fall well within the range of canal organ sensitivities.

An interesting correlate to this physiology concerns behavioral deficits. If the nerve from the horizontal canal is cut, normal swimming and posture are not noticeably affected; but if a resting operated animal is surprised and executes the rapid escape maneuver, an immediate problem appears. Although the rapid flip of the tail drives the animal upwards and turning away from the stimulus, the animal does not brake the turn with the next tail beat. Instead of swimming rapidly away in a straight line, it continues to circle, as though missing a signal about the fast rotational action (Dunn and Koester 1984).

## Extra-Vestibular Modulation of Posture

There are a variety of senses that interact with the vestibular input to modulate postural control. The most prominent of these is vision, but there is evidence that some free-swimming animals use tactile cues as well. The presence of food in the area also has been shown to modify the way the central nervous system "weights" the different sensory inputs to determine posture.

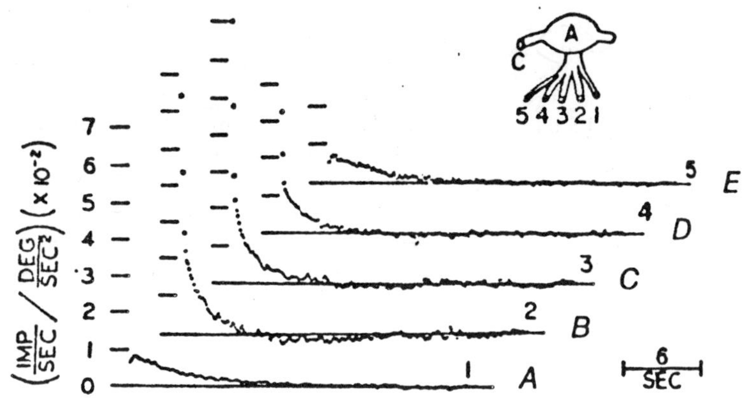


Figure 31.6. The tuning of acceleration sensitivity in semicircular canal afferents. *Upper sketch* shows that the ampulla (*A*) of the horizontal canal (*C*) in guitarfish *Rhinobatos productus* is innervated by a nerve that fans into 5 peripheral branches. Recordings from these individual branches showed activity analyzed by plotting latency to peak (*horizontal axis*) and the magnitude of peak sensitivity (*vertical axis*) to a range of rotational accelerations. The most central bundles (*2, 3,* and *4*) have shorter average latency and much higher average peak sensitivity to rotational acceleration than those toward the margins (*1* and *5*). The central bundle (*3*) predominantly innervates the crest of the ridge of sensory crista. (From O'Leary, Dunn, and Honrubia 1976.)

## Visual-Vestibular Interaction in Posture

Visual, as well as vestibular, input is important for postural control in many teleost fishes. The clever experiments of von Holst (1935, 1950) established the presence of the "dorsal light reflex" in the wrasse *Crenilabrus rostratus* and in characins such as the tetra *Gymnocorymbus ternezii* and angelfish *Pterophyllum eimeckei*. In this behavior, the fish keeps its body oriented so that its back is toward a directional light source. Normally light comes from above, and both the vestibular and visual input keep the fish in the normal posture in the water, with the dorsal fin uppermost. But if the light is presented from one side of the fish, it rolls toward that side, tilting its back toward the light. The posture that results is considered an equilibrium posture, because at this angle the influence of the light in tilting the fish on its side is exactly balanced by the influence of gravity at that tilted position to return the fish toward the upright (Fig. 31.7) (von Holst 1950).

The amount of tilt at this equilibrium is a quantitative central weighting of the visual, compared to the vestibular, input (von Holst 1950). Normally the force of gravity is one g, and it stimulates the utricular macula maximally when the fish is tilted to 90°. The effect of the directional light is at a maximum when the light enters the eye at 90° from its vertical and depends on light intensity. If the fish tilts to an equilibrium position of 45° when light is from 90°, the central weighting of the two influences is approximately equal. The gravistatic

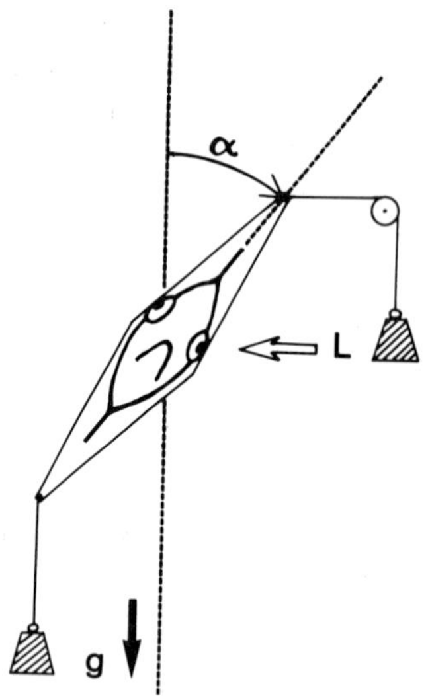


Figure 31.7. The equilibrium between gravity and light direction in dorsal light reflex. Light from the side (*L, open arrow*) causes a tilt of the fish to a position in which righting response to gravity (*g, solid arrow*) becomes strong enough to prevent further tilting response to the light. The model of the response indicates stimuli as weights affecting the posture as balance arm; gravity (*g*) pulls one way, light (*L*) the other, to roll the beam to equilibrium position α. (After von Holst 1950.)

component of this response was tested quantitatively over a range from 0 to 2 g. To increase g, the fish chamber was centrifuged, and at 2 g the vestibular stimulus was relatively greater than the visual stimulus, and the tilt angle was decreased. To decrease the effect of gravity, the fish was forced to swim nose upward into a water current; in this position, the optic component alone induced body roll around the longitudinal axis, without changing the shear of gravity on the otoliths.

The quantitative interaction of light and gravity depends on a variety of factors. By physically manipulating the eyes of the fish tested, the angle at which light entered the eye, rather than simply the angle of light relative to the vertical, was found to be the important parameter of the optic stimulus (von Holst and Schoen 1954). Increasing the intensity of the optic stimulus increased the optic component of the response. In some species the visual input has a greater influence on posture than the vestibular input (see Pfeiffer 1964). Histological analysis of the eyes of several behaviorally tested species showed that variation in the effectiveness of light correlates with the density and arrangement of rods and cones in the retina (Bogenschütz 1961).

A further principle discovered by von Holst was that the weighting relation between gravistatic and optic components depended on the state of the fish's alertness (von Holst 1949). If a fish of a species that was a visual feeder was hungry and food extract was put in the water, the dorsal light response showed an increased tilt, indicating a relative increase in the optic component weighting. This alerting function may be one way in which chemosensory cues can be considered to affect postural control.

Similar studies have been done on the response of carp and goldfish. Carp will pitch nose downward when feeding on the bottom, and the eyes counterroll in response to this tilt (Traill and Mark 1970). These rotational eye responses to tilt were measured to evaluate visual-vestibular interactions. Directional light from below decreased the angle of ocular counterrolling as compared to that for light from above. Again, hungry fish with food in the water showed an increased response to the visual input as compared to the vestibular input. Similarly, light alone can induce ocular asymmetry in a restrained goldfish, and this visually induced asymmetry is modified when the fish is rolled to various angles of tilt or if the labyrinths are removed, indicating that central integrative mechanisms use the vestibular input in part to modulate the visual response (Graf and Meyer 1983). It is likely that in fish a major structure in the midbrain, the torus semicircularis, is one of the sites of visual-vestibular integration (Schellart 1983).

### Other Sensory Influences on Posture

Aside from visual input, a new class of response has been found in some species and called the "ventral substrate response" (Meyer, Heiligenberg, and Bullock 1976). Some free-swimming fish live in areas where there are a lot of underwater objects presenting a variety of surface orientations relative to the horizontal. Some electric fishes and the upside-down catfish already discussed will tilt around the roll axis to keep the ventral body surface close to a nearby substrate. Thus, near a vertical wall, such a fish rolls on its side to keep the belly turned toward the wall, in a manner analogous to the dorsal light reflex's keeping the back toward the light. The effect depends on distance from the substrate and may be mediated in part by lateral line organs (Chaps. 4, 8, 22, and 23).

Such extra-vestibular influences on posture suggest some of the variety of ways in which postural equilibrium can be maintained. It may be that more mechanisms remain to be discovered. For example, the remora, or shark sucker, *Echeneis naucrates* sometimes, when kept in aquaria, will spend a lot of time swimming with its back tilted toward the vertical aquarium wall. When it stops swimming, it simply attaches to the wall with the specialized suction plate on the top of its head, just as it would attach to a large fish in open water. This anecdotal evidence suggests there might be a "dorsal substrate response" in some fish.

Another candidate sense that might be suspected of affecting posture is the electromagnetic sense in those aquatic animals which have it (Chap. 6); but I know of no reports on whether electric fish may maintain postures other than vertical in response to the local electromagnetic field. However, magnetic particles have been found incorporated into the otolithic mass in some elasmobranchs (Vilchestroya, Dunn, and O'Leary 1984), leading to speculation about a further role for the vestibular system in navigation, as well as simple body orientation (Chap. 6).

## Future Questions

Central integration of vestibular and extravestibular input is an important topic because it is the linkage in the system that uses the variety of sensory inputs to produce appropriate postural and locomotory responses. Modern neuroanatomical techniques will enable the tracing of details of the pathways from individual end organs into the brain. Physiological recording will clarify the cellular excitatory and inhibitory interactions. Two major areas of interest are the integration of vestibular inputs from the two sides of the head and the integration of visual with vestibular input to produce appropriate oculomotor and locomotor responses. Both deal with the question of how a central "set point" is established and maintained as a reference for behavior (Graf and Meyer 1983) (Chap. 10).

A related issue is central control over vestibular input via efferent pathways to the peripheral vestibular system by which the brain modulates its own input. The distribution of efferent endings in relation to the cell orientations or to the afferent innervation is not yet well understood. This anatomical organization is presumed to have functional significance in terms of input sensitivity for the afferent fibers of the eighth nerve (Goldberg and Fernandez 1975; Precht 1979; Lewis, Leverenz, and Bialek 1985). Behaviorally, corollary discharge in vestibular efferents has been suggested as a mechanism to prevent unwanted compensatory reflexes during voluntary movements (Chap. 10).

Development remains a large issue for vestibular function. The inner ear of several fishes and amphibians are unlike those in mammals because hair cells and neurons are added throughout life (Corwin 1983, 1985; Popper and Hoxter 1984). As a sensory macula of an otolithic organ grows, the growth changes the "angle of repose" for a hair cell sited at a particular point on the curved macula. Somehow the central nervous system continually must update the set point for the appropriate postural equilibrium (Platt 1983a). Similarly, the sensitivity of the canals has been calculated to change with growth in fish because of physical considerations (ten Kate 1973; Howland and Masci 1973).

A few aquatic vertebrates should be mentioned as worth further comparative study, representing key examples because they are the most adapted to aquatic life in their respective classes. First, there are two aquatic marine reptiles, the sea turtles and the sea snakes, and many of the latter never emerge on land. Second, there is one group of birds that have adapted more completely than any others to aquatic, not aerial, life, the penguins. Third, the fully aquatic mammals include the whales and dolphins (Cetacea), the sea cows (Sirenia) and the seals, sea lions, and walruses (Pinnepedia). Because of the great interest in their hearing, there are several figures on the cochleas of these animals, but details of their vestibular apparatus are limited; there is a paper on the ultrastructure of the inner ear of the harp seal (*Pagophilus groenlandicus*) indicating its similarity to other mammalian labyrinths (Ramprashad, Money, and Ronald 1972). Among the cetaceans, the "spinner" dolphin (*Stenella longirostris*) is known for its acrobatic leaps out of the water, spinning around the long axis and also often somersaulting. Whether the structure of the canals or otolith organs of these animals will suggest correlates with behavior remains to be seen. If such correlations are found, hypotheses about their adaptive significance must be tested (Chap. 34).

## Conclusions

A major conclusion is that the peripheral vestibular system shows few gross changes that are unique to aquatic life. The canals and otolithic organs vary tremendously in details of proportions, but the basic structure shows no more variation than that found in terrestrial vertebrates. It may be that some fish have exploited the physical properties of the aquatic medium to adopt particularly unusual postures. The one tail stander with the modified utriculus represents one case of a peripheral anatomical change, but generally there seem to be no major peripheral modifications necessary for marked postural changes.

At the ultrastructural level, the hair cells of the receptor organs have similarities in all vertebrates, though the fishes so far studied lack the chalice-shaped type I hair cells found in birds and mammals. The ciliary bundles on the hair cells are morphologically polarized, giving each bundle an axis of maximum sensitivity to a directional displacement of the bundle. Cells with similar orientations are grouped into organized populations on each sensory surface. The individual ciliary bundles have a variety of forms, and these are also found in organized populations. Thus each sensory surface consists of a ''receptor mosaic'' of cellular orientations and ciliary bundles. It should again be noted that cellular orientations suggesting inhibitory activity may be just as important as those suggesting excitatory activity, as noted for some cephalopod molluscs (Budelmann 1979).

Finally, the striking similarity should be noted between the vertebrate vestibular system with the statocyst-canal system in some decapod crustaceans and cephalopod molluscs. This convergent evolution suggests that statoliths and canals are useful adaptations for rapidly mobile animals (see Schöne 1975; Platt 1984; see also Chap. 30). As in the terrestrial vertebrates, the ubiquitous presence of gravitational force on the otolith organs appears to provide a useful reference sense for the ''postural platform'' and the fluid mechanism of the canals to provide an inertial guidance system for locomotory reflexes and visual stabilization.

## References

Allum, G., Graf, W., Dichgans, J., and Schmidt, C. (1976) Visual-vestibular interactions in the vestibular nuclei of the goldfish, Brain Res., 26:463–485.

Annoni, J.M., Cochran, S.L., and Precht, W. (1984) Pharmacology of the vestibular hair cell—afferent fiber synapse in the frog, J. Neurosci., 4:2106–2116.

Barber, V.C., and Emerson, C.J. (1980) Scanning electron microscopic observations on the inner ear of the skate, *Raja ocellata*, Cell Tissue Res., 205:199–215.

Bogenschütz, H. (1961) Vergleichende Untersuchungen über die optische Gleichgewichtshaltung bei Fischen, Z. Vgl. Physiol., 44:626–655.

Braemer, W., and Braemer, H. (1958) Orientation of fish to gravity, Limnol. Oceanogr., 3:362–372.

Budelli, R., and Macadar, O. (1979) Statoacoustic properties of utricular afferents, J. Neurophysiol., 42:1479–1493.

Budelmann, B.U. (1979) Hair cell polarization in the gravity receptor systems of the statocysts of the cephalopods *Sepia officinalis* and *Loligo vulgaris*, Brain Res., 160:261–270.

Cohen, B. (ed.) (1981) Vestibular and oculomotor physiology, Ann. N.Y. Acad. Sci., 374:1–388.

Corwin, J.T. (1983) Postembryonic growth of the macula neglecta auditory detector in the ray, *Raja clavata:* continual increases in hair cell number, neural convergence, and physiological sensitivity, J. Comp. Neurol., 217:345–356.

Corwin, J.T. (1985) Auditory neurons expand their terminal arbors throughout life and orient toward the site of postembryonic hair cell production in the macula neglecta in elasmobranchs, J. Comp. Neurol., 239:445–452.

Denton, E.J., and Gray, J.A.B. (1980) Receptor activity in the utriculus of the sprat, J. Mar. Biol. Assoc. U.K., 60:717–740.

Dichgans, J., Schmidt, C., and Graf, W. (1973) Visual input improves the speedometer function of the vestibular nuclei in the goldfish, Exp. Brain Res., 18:319–322.

Dieringer, N., and Precht, W. (1979a) Mechanisms of compensation for vestibular deficits in the frog. I. Modification of the excitatory commissural system, Exp. Brain Res., 36:311–328.

Dieringer, N., and Precht, W. (1979b) Mechanisms of compensation for vestibular deficits in the frog. II. Modification of the inhibitory pathways, Exp. Brain Res., 36:311–328.

Dunn, R.F., and Koester, D.M. (1984) Primary afferent projections of the octavus nerve to the inferior reticular formation and adjacent nuclei in the elasmobranch, *Rhinobatos* sp., Brain Res., 323:354–359.

Enger, P.S. (1963) Single unit activity in the peripheral auditory system of a teleost fish, Acta Physiol. Scand., 59 (Suppl.) 210:1–48.

Flock, A., and Goldstein, M.H. (1978) Cupular movement and nerve impulse response in the isolated semicircular canal, Brain Res., 157:11–19.

Galiana, H.L., Flohr, H., and Melvill Jones, G. (1984) A reevaluation of intervestibular nuclear coupling: its role in vestibular compensation, J. Neurophysiol., 51:242–259.

Goldberg, J.M., and Fernandez, C. (1975) Vestibular mechanisms, Annu. Rev. Physiol., 37:129–162.

Graf, W., and Baker, R. (1985a) The vestibuloocular reflex of the adult flatfish. I. Oculomotor organization, J. Neurophysiol., 54:887–899.

Graf, W., Baker, R. (1985b) The vestibuloocular reflex of the adult flatfish. II. Vestibulo-oculomotor connectivity, J. Neurophysiol., 54:900–916.

Graf, W., and Meyer, D.L. (1983) Central mechanisms counteract visually induced tonus asymmetries: a study of ocular responses to unilateral illumination in goldfish, J. Comp. Physiol. A, 150:473–482.

Henn, V.B., Cohen, B., and Young, L.R. (1980) Visual-vestibular interaction in motion perception and the generation of nystagmus, Neurosci. Res. Prog. Bull., 18:459–651.

Highstein, S.M., and Politoff, A.L. (1978) Relation of interspike baseline activity to the spontaneous discharges of primary afferents from the labyrinth of the toadfish, *Opsanus tau,* Brain Res., 150:182–187.

Hillman, D.E. (1974) Cupular structure and its receptor relationship, Brain Behav. Evol., 10:53–68.

Hillman, D.E., and McLaren, J.W. (1979) Displacement configuration of semicircular canal cupulae, Neuroscience, 4:1989–2000.

Howland, H.C., and Masci, J. (1973) The functional allometry of semicircular canals, fins and body dimensions in the juvenile centrarchid fish, *Lepomis gibbosus* (L.), J. Embryol. Exp. Morphol., 29:721–743.

Hudspeth, A.J. (1983) Mechanoelectrical transduction by hair cells in the acousticolateralis sensory system, Annu. Rev. Neurosci., 6:187–215.

Jones, G.M., and Spells, K.E. (1963) A theoretical and comparative study of the functional dependence of the semicircular canal upon its physical dimensions, Proc. R. Soc. Lond. B Biol. Sci., 157:403–419.

Kornhuber, H.H. (ed.) (1974) Vestibular System, Parts 1 and 2, Handbook of Sensory Physiology, vol. 6, Springer-Verlag, Berlin.

Lewis, E.R., Baird, R.A., Leverenz, E.L., and Koyama, H. (1982) Inner ear: dye injection reveals peripheral origins of specific sensitivities, Science, 215:1641–1643.

Lewis, E.R., Leverenz, E.L., and Bialek, W. (1985) The vertebrate inner ear, CRC Press, Boca Raton, La.

Lowenstein, O. (1957) The acoustico-lateralis system, in The Physiology of Fishes, vol. 2, Brown, M.E. (ed.), Academic Press, New York, pp. 168–182.

Lowenstein, O. (1971) The labyrinth, in Fish Physiology, vol. 5, Sensory Organs, Hoar, W.S., and Randall, D.J. (eds.), Academic Press, New York, pp. 207–240.

Lowenstein, O. (1974) Comparative morphology and physiology, in Vestibular System: Basic Mechanisms, Handbook of Sensory Physiology, vol. 6, Part I, Kornhuber, H.H. (ed.), Springer-Verlag, Berlin, pp. 75–120.

Lowenstein, O., and Roberts, T.D.M. (1949) The equilibrium function of the otolith organs of the thornback ray (*Raja clavata*), J. Physiol. (Lond.), 110:392–415.

Macadar, O., Wolfe, G.E., O'Leary, D.P., and Segundo, J.P. (1975) Response of the elasmobranch utricle to maintained spatial orientation, transition, and jitter, Exp. Brain Res., 22:1–12.

Marshall, N.B. (1966) The life of fishes, Universe Books, New York.

Mayne, R. (1974) A systems concept of the vestibular organs, in Vestibular System: Psychophysics, Applied Aspects and General Interpretations, Handbook of Sensory Physiology, vol. 6, Part II, Kornhuber, H.H. (ed.), Springer-Verlag, Berlin, pp. 493–580.

McLaren, J.W., and Hillman, D.E. (1979) Displacement of the semicircular canal cupula during sinusoidal rotation, Neuroscience, 4:2001–2008.

Melvill Jones, G. (1974) The functional significance of semicircular canal size, in Vestibular System: Basic Mechanisms, Handbook of Sensory Physiology, vol. 6, Part II, Kornhuber, H.H. (ed.), Springer-Verlag, Berlin, pp. 171–184.

Meyer, D.L., Heiligenberg, W., and Bullock, T.H. (1976) The ventral substrate response—a new postural control mechanism in fishes, J. Comp. Physiol. A, 109:59–68.

Meyer, D.L., Platt, C.J., and Distel, H.J. (1976) Postural control mechanisms in the upside-down catfish (*Synodontis nigriventris*), J. Comp. Physiol. A, 110:323–331.

Norman, J.R., and Greenwood, P.H. (1963) A history of fishes, 2d ed., Ernest Benn, London.

O'Leary, D.P., Dunn, R.F., and Honrubia, V. (1976) Analysis of afferent responses from isolated semicircular canal of the guitarfish using white-noise inputs. I. Correlation of response dynamics with receptor innervation, J. Neurophysiol., 39:631–644.

Oman, C.M. (1981) The influence of duct and utricular morphology on semicircular canal response, in The Vestibular System: Function and Morphology, Gualtierotti, T. (ed.), Springer-Verlag, New York, pp. 251–274.

Oman, C.M., Frishkopf, L.S., and Goldstein, M.H. (1979) Cupula motion in the semicircular canal of the skate *Raja erinacea,* Acta Oto-laryngol., 87:528–538.

Pfeiffer, W. (1964) Equilibrium orientation in fish, Int. Rev. Gen. Exp. Zool., 1:77–111.

Plassmann, W. (1983) Sensory modality interdependence in the octaval system of an elasmobranch, Exp. Brain Res., 50:283–292.

Platt, C. (1973) Central control of postural orientation in flatfish. I. Dependence on central rather than peripheral changes, J. Exp. Biol., 59:491–521.

Platt, C. (1977) Hair cell distribution and orientation in goldfish otolith organs, J. Comp. Neurol., 172:283–298.

Platt, C. (1983a) The peripheral vestibular system of fishes, in Fish Neurobiology, vol. 1, Brainstem and Sensory Organs, Northcutt, R.G., and Davis, R.E. (eds.), University of Michigan Press, Ann Arbor, pp. 89–123.

Platt, C. (1983b) Retention of generalized hair cell patterns in the inner ear of the primitive flatfish *Psettodes*, Anat. Rec., 207:503–508.

Platt, C. (1984) Cellular bases for gravistatic reception by invertebrates and vertebrates, in Comparative Physiology: Sensory Systems, Bolis, L., Keynes, R.D., and Maddrell, S.P. (eds.), Cambridge University Press, Cambridge, U.K., pp. 563–586.

Platt, C., and Popper, A.N. (1981) Fine structure and function of the ear, in Hearing and Sound Communication in Fishes, Tavolga, W.N., Popper, A.N., and Fay, R.R. (eds.), Springer-Verlag, Berlin, pp. 3–38.

Popper, A.N., and Hoxter, B. (1984) Growth of a fish ear. I. Quantitative analysis of hair cell and ganglion cell proliferation, Hear. Res., 15:133–142.

Popper, A.N., and Platt, C. (1983) Sensory surface of the saccule and lagena in the ears of ostariophysan fishes, J. Morphol., 176:121–129.

Popper, A.N., Platt, C., and Saidel, W.M. (1982) Acoustic functions in the fish ear, Trends Neurosci., 5:276–280.

Popper, A.N., and Tavolga, W.N. (1981) Sound detection and inner ear structure in the marine catfish, *Arius felis*, J. Comp. Physiol., 144:27–34.

Precht, W. (1979) Vestibular mechanisms, Annu. Rev. Neurosci., 2:265–289.

Ramprashad, F., Landolt, J.P., Money, K.E., and Laufer, J. (1984) Dimensional analysis and dynamic response characterization of mammalian peripheral vestibular structures, Am. J. Anat., 169:295–313.

Ramprashad, F., Money, K.E., and Ronald, K. (1972) The harp seal, *Pagophilus groenlandicus* (Erxleben 1777), XXI. The structure of the vestibular apparatus, Can. J. Zool., 50:1357–1361.

Retzius, G. (1881) Das Gehörorgan der Wirbelthiere. I. Das Gehörorgan der Fische und Amphibien, Samson and Wallin, Stockholm.

Retzius, G. (1884) Das Gehörorgan der Wirbelthiere. II. Das Gehörorgan der Reptilien, Vögeln und Säugethieren, Samson and Wallin, Stockholm.

Schellart, N.A.M. (1983) Acousticolateral and visual processing and their interaction in the torus semicircularis of the trout, *Salmo gairdneri*, Neurosci. Lett., 42:39–44.

Schmidt, C.L., Wist, E.R., and Dichgans, J. (1972) Efferent frequency modulation in the vestibular nerve of goldfish correlated with saccadic eye movements, Exp. Brain Res., 15:1–14.

Schoen, L. (1957) Mikroableitungen einzelner zentraler Vestibularisneurone von Knochenfischen bei Statolithenreizen, Z. Vgl. Physiol., 39:399–417.

Schoen, L., and von Holst, E. (1950) Das Zusammenspiel von Lagena und Utriculus bei der Lageorientierung der Knochenfische, Z. Vgl. Physiol., 32:552–571.

Schöne, H. (1964) Über die Arbeitsweise der Statolithenapparate bei Plattfischen, Biol. Jahresh., 4:135–156.

Schöne, H. (1975) On the transformation of the gravity input into reactions by statolith organs of the "fan" type, Fortschr. Zool., 23:120–127.

Steinhausen, W. (1931) Über den Nachweis der Bewegung der Cupula in der intakten Bogengangsampulle des Labyrinthes bei der natürlichen rotatorischen und calorischen Reizung, Pfluegers. Arch. Gesamte. Physiol. Menschen. Tiere, 228:322–328.

ten Kate, J.H. (1973) The mechanics of the growing semicircular canal, J. Exp. Biol., 58:351–366.

ten Kate, J.H., van Barneveld, H.H., and Kuiper, J.W. (1970) The dimensions and sensitivities of semicircular canals, J. Exp. Biol., 53:501–514.

Traill, A.B., and Mark, R.F. (1970) Optic and static contributions to ocular counter-rotation in carp, J. Exp. Biol., 52:109–124.

van Egmond, A.A.J., Groen, J.J., and Jongkees, L.B.W. (1949) The mechanics of the semicircular canal, J. Physiol. (Lond.), 110:1–17.

Vilchestroya, J., Dunn, R.F., and O'Leary, D.P. (1984) Relationship of the vestibular hair-cells to magnetic particles in the otolith of the guitarfish sacculus, J. Comp. Neurol., 226:489–494.

von Baumgarten, R.J., Baldrighi, G., Atema, J., and Shillinger, G.L. (1971) Behavioral responses to linear accelerations in blind goldfish, Space Life Sci., 3:25–33.

von Holst, E. (1935) Über den Lichtrückenreflex bei Fischen, Pubbl. Stn. Zool. Napoli, 15:143–158.

von Holst, E. (1949) Quantitative Untersuchungen über Umstimmungsvorgänge im Zentral-nervensystem. I. Der Einfluss des "Appetits" auf das Gleichgewichtsverhalten bei Pterophyllum, Z. Vgl. Physiol., 31:134–148.

von Holst, E. (1950) Die Arbeitsweise des Statolithenapparates bei Fischen, Z. Vgl. Physiol., 32:60–120.

von Holst, E., and Schoen, L. (1954) Der Einfluss mechanisch veränderter Augenstellungen auf die Richtungslokalization bei Fischen, Z. Vgl. Physiol., 36:433–442.

## Part IIID

# Electroreception

The electrosensory systems of fishes have come to serve as a model for the analysis of behavioral, physiological, and neuroanatomical aspects of other sensory systems. Because electroreception serves such a model, several chapters that would otherwise appear in this part are found in other parts of this volume. Kalmijn (Chapter 5) presents an historic overview of the study of electrosensory systems in fishes, and provides insight into electrosensory behavior and the nature of the electrical stimulus pertinent to all other chapters on electroreception. Behavioral aspects of electroreception and the acquisition of electric information by fishes are discussed by Hopkins (Chapter 9) and by Bullock (Chapter 10). Comparative morphology and physiology of electroreceptors, as well as electrosensory behavior, are described by Zakon (Chapter 32). Heiligenberg (Chapter 33) gives an overview of the physiology of the electrosensory system, and presents an analysis of the central processing of electrosensory information. The neuroanatomy of the electrosensory system is described by McCormick and Braford in the previous section (Chapter 29).

# Chapter 32

# The Electroreceptors: Diversity in Structure and Function

Harold H. Zakon[*]

## Introduction

Sensory systems that operate in an aquatic habitat face different environmental constraints than their terrestrial counterparts in the detection of stimuli. The different transmission properties of air and water for sound, light, and chemical stimuli have resulted in habitat-specific differences in the structure and function of many sensory receptor organs. Electroreception, on the other hand, is unique to the aquatic habitat and, since air behaves as an insulator, rather than a conductor, is physically restricted to it.

Electroreceptors are known only in vertebrates. Their absense in invertebrates is puzzling, as there is no obvious intrinsic physical limitation rendering them useless. The reason may be more sociological than biological, in that no one has explicitly and systematically examined the invertebrates for the presence of electroreception. Nevertheless, the speculation has recently been advanced that electroreceptors might have been derived from the ciliated epidermal mechanoreceptors of protochordates (Northcutt and Gans 1983; Jørgensen 1982). Certainly the occurrence of ampullary-like canals in the dermal plates of Devonian fishes ancestral to both agnathan and gnathostome lines (Thomson 1977), the apparent presence of electroreceptors in the coelacanth *Latimeria* (Northcutt 1980), and their widespread occurrence among the more primitive extant fishes indicate an ancient lineage. Inasmuch as the first fishes appear to have had both a mechanosensory lateral line, as well as an electrosensory system, both systems may be considered to be of equal age and probably as old as the vertebrate radiation (Bodznick and Northcutt 1981).

Electroreceptors may be naturally classified as either ampullary or tuberous on morphological and physiological grounds (see below). However, numerous subtypes are recognized and species differences abound; their taxonomic distribution and morphological peculiarities make assigning homologies difficult. This chapter will discuss both ampullary and tuberous systems from the standpoint of

---

[8] Department of Zoology, University of Texas, Austin, TX 78712, USA.

their distribution, morphology, relevant stimuli, sources of noise and "strategies" of noise minimization, biophysics, and stimulus encoding. Although this review will focus on the receptor organs and their afferents, aspects of CNS processing will be discussed where relevant.

### Taxonomic Distribution

Ampullary receptors are found in the lampreys (Petromyzoniformes), but not in the hagfishes (Myxiniformes) (Ronan and Bodznick 1984; Bullock, Bodznick, and Northcutt 1983). Either these two lines do not form a single natural monophyletic group (Agnatha) and electroreceptors evolved after the hagfish line appeared or, instead, hagfishes are highly derived and "simplified" for a parasitic lifestyle and may have secondarily lost their electrosensory systems. The characteristics of the lamprey receptor give no aid in settling the question, for they are morphologically unlike all other ampullary organs (Ronan and Bodznick 1984).

Sharks, skates, rays, and chimeras (Chondrichthyes) all possess ampullary receptors, as do almost all other nonteleost fishes. They have been described in the lungfishes (Dipnoi), the sturgeon and paddlefish (Chondrostei), the polypteriform fishes (*Polypterus* and *Calamoichthys*), and as mentioned above, in the coelacanth (Crossopterygii) (Bullock, Bodznick, and Northcutt 1983; Northcutt in press; Jorgensen 1982).

The Holostei, thought to have arisen during the late Paleozoic, attained prominence during the Mesozoic (Nelson 1983). Whatever adaptations allowed them to dominate at this time, an electrosensory system was probably not among them and appears to have been lost, as the extant Holostei, the bowfin, *Amia,* and the gar *Lepisosteus* lack electroreceptors (Chap. 29).

Teleost fishes are believed to have split off from holostean stock during the mid to late Mesozoic. Of the more than 30 orders of teleosts, only four are known to be electroreceptive. Two closely related orders, the Siluriformes and Gymnotiformes, both possess ampullary receptors. Specialized tuberous receptors and weakly electric organs are found only in the gymnotiforms. Ampullary electroreceptors are also found in two other closely related groups: the Mormyriformes and the family *Xenomystinae* (Osteoglossiformes, Notopteridae) (Chap. 29). However, only mormyriforms have a specialized tuberous receptor–electric organ system. Presumably, ampullary receptors have reappeared at least twice, and tuberous receptors have evolved independently in the two orders of weakly electric fish.

Ampullary receptors are not restricted to fish; they are additionally found in the larvae of salamanders (Urodela) and apodans, but not frogs and toads (anura) (Fritzsch and Münz 1986).

## Anatomy of Ampullary Electroreceptors

The distinguishing morphological characteristics of ampullary receptor organs are a transepidermal canal that terminates in a single or multilobed pouch (the ampulla proper) at the base of which is an epithelium containing the receptor cells (Figs. 32.1, 32.2, and 32.4A). The receptor cells, whose number within an ampulla may vary from a few or tens in aquatic species (one in *Gymnarchus*) to thousands in marine species, are contacted by one or several nerve fibers of bipolar ganglion-type neurons that run in a branch of the anterior lateral line nerve (Szabo 1965, 1974).

Ampullary organs are structurally diverse. In fact, the ampullary organs of lampreys, more properly called end buds, are not even "ampullary"; their construction is so dissimilar to "conventional" ampullary receptor organs that, based on their morphology, they were thought to be taste receptors earlier in the century (Whitear and Lane 1983). Yet their frequency response and physiological sensitivity is in keeping with other freshwater animals (Bodznick and Preston 1983; Ronan and Bodznick 1984).

## Elasmobranchs

Ampullary receptor organs of elasmobranchs are characterized by long canals that open via a pore on the skin surface, penetrate the epidermis, run some distance in the dermis and subdermal tissues, and terminate in a cluster of pouches (Fig. 32.1). This bulblike termination is the ampulla proper, and each of its pouches, called alveoli (Fig. 32.1B), contains a complement of hundreds of receptor cells giving a total of hundreds to thousands of receptor cells in each ampulla. In the rays the canals may be up to 20 cm long and up to 1 mm in diameter, although many are only 1 or a few centimeters in length (Waltman 1966; Murray 1974). Given their dimensions, it is not surprising that they were first described in the seventeenth century (Lorenzini 1678), though their function remained obscure until the mid–twentieth century (Dijkgraaf and Kalmijn 1966, Murray 1974).

The ampullae of all the organs are clumped into a few bunches (Fig. 32.1A) surrounded by a loose connective tissue sheath. These few clusters of receptor organs are located on or near the head, as they are in most nonteleost fishes (Northcutt 1986). The canals of sharks are restricted to the head, whereas the canals of skates and rays are distributed over the body as well (Fig. 32.1A).

The canals of rays radiate in all directions from the ampullary clusters, so that each one is "pointed" toward a different location around the animal (Fig. 32.1A). It has been proposed that this arrangement allows an animal to sample directionally the electric field surrounding it (Bennett 1971; Kalmijn 1974; Murray 1974). Indeed, studies of central nuclei have demonstrated that the receptors project somatotopically within the CNS and, because canal orientation imparts directionality, central receptive fields are also organized by electric field orientation (Schweitzer 1985).

A mucopolysaccharide jelly fills the canal (Murray and Potts 1961), and its walls are composed of a double layer of flattened cells, those of the inner layer being joined by a network of tight junctions (Waltman 1966). At the beginning of the ampulla, the canal walls give way to the single-layered epithelium that forms the walls and base of the ampulla (Fig. 32.1C). The network of tight junctions found in the canal wall continues into the ampulla, so that each cell is

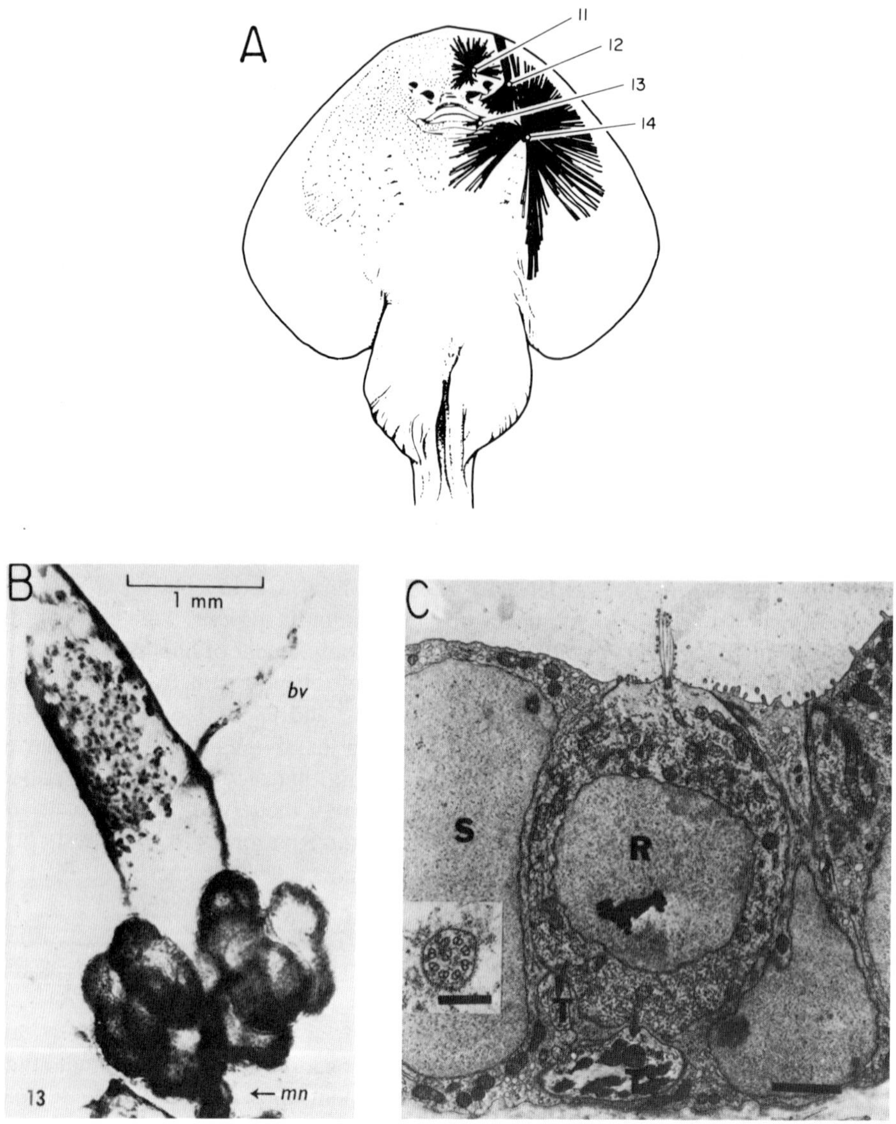


Figure 32.1. Ampullary receptor organs of elasmobranchs. (A) Ventral surface of the ray *Platyrhinoides*, illustrating the pattern of canal openings (*left side*) and canals proper (*right side*) (Montgomery 1984b). The numbers (*11–14*) refer to clusters of ampullae. (B) Terminal portion of a canal with an ampulla from *Raja* with a bundle of myelinated nerve fibers (*mn*) entering from below *bv* (blood vessel) (Waltmann 1966). (C) Cross section of ampullary epithelium of freshwater ray *Potamotrygon*. The *inset bottom left* is cross section through single kinocilium showing typical ciliary organization of fibrils (*inset calibration bar* = 0.5 μm); *R*, receptor cell; *S*, support cell. *Calibration bar* = 1 μm (Szamier and Bennett 1980).

ringed by tight junctions producing a barrier to current flow in the extracellular space and, as is typical of epithelia, dividing the cell surface membrane into apical and basal compartments (Waltman 1966; Sejnowski and Yodlowski 1982). Only about 1% of the cell membrane is apical, and this small patch supports a single kinocilium (Szabo 1974; Sejnowski and Yodlowski 1982) (Fig. 32.1C).

Receptor cells are spherical, about 30 μm in diameter, and, like other receptor cells of the octavolateralis system, make a chemical synapse of the ribbon type (Sejnowski and Yodlowski 1982). The hundreds to thousands of receptor cells of the epithelium of a single elasmobranch ampullary organ are contacted by a few to a dozen afferent fibers (Szabo 1974). In contrast to most other hair cell organs, there is no evidence of efferent synapses on these or any other electrosensory receptors.

The ampullary receptors of other nonteleost groups (see the section above on taxonomic distribution) are similar, though they typically possess many fewer receptor cells per organ and shorter canals and their apical surfaces may carry microvilli, rather than a kinocilium. For more extensive comparisons, see Zakon (1985).

### Teleosts

The ampullary receptor organs of teleosts, though similar in general construction to those of elasmobranchs, are diminutive in comparison, being not more than 100 μm long and barely penetrating the epidermis (Fig. 32.2). They have many fewer receptor cells (one in gymnarchids, twenty in the catfish *Kryptopterus*), more cells composing the capsule wall (up to half a dozen), and typically only one afferent fiber to each organ or cluster of organs (Bennett 1971; Szabo 1965, 1974; Teeter, Szamier, and Bennett 1981; Wachtel and Szamier 1969; Szamier and Wachtel 1970; Zakon 1984a). The canals are also filled with a jelly that, in the marine catfish *Plotosus,* contains high levels of $K^+$ in the ampullary lumen (Okitsu, Umekita, and Obara 1978).

The receptor cells are likewise embedded within a support cell epithelium with only a small (15%) area of apical membrane. One trait common to teleost receptor cells is that their apical surfaces are covered with microvilli, rather than a single kinocilium (Fig. 32.2C). When only ampullary receptors of elasmobranchs and a few teleosts had been studied, this difference in cilia type was suspected to predict the polarity of the receptors' response (kinocilium-cathodically excited; microvilli-anodally excited); but with the subsequent discovery of excited receptors with microvilli (lampreys, salamander larvae) and receptor cells that have both microvilli and a kinocilium (*Polypterus,* apodan amphibia), this idea is no longer viable. At the moment, the distribution and function of these organelles remains unexplained.

Ampullary morphology does not strictly follow taxonomic lines but, rather, correlates with habitat. The freshwater ray, *Potamotrygon,* has small "microampullae," like freshwater teleosts (Szabo et al. 1972; Szamier and Bennett 1980). Similarly the ampullae of the marine catfish *Plotosus* are more like those of the

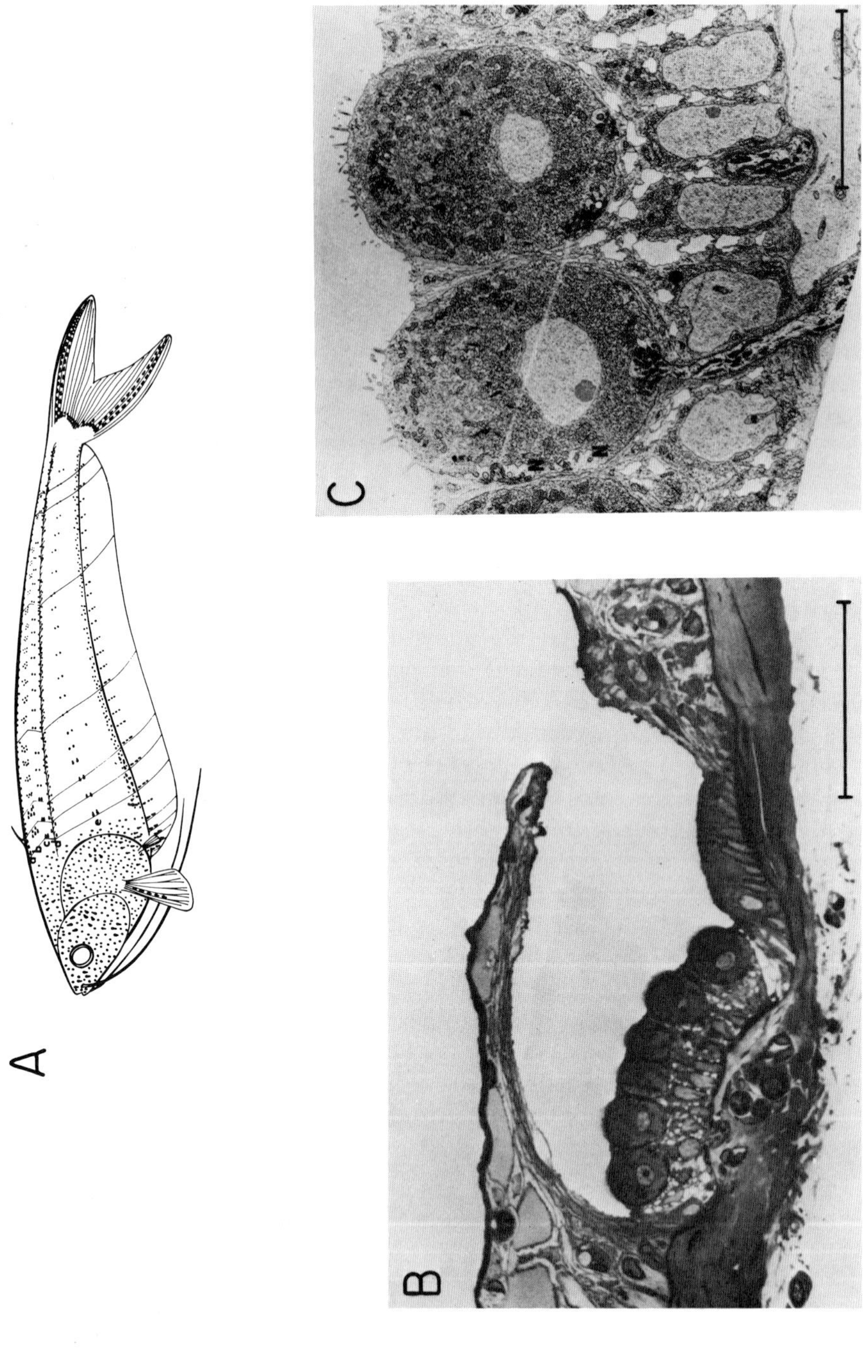
A
B
C

marine elasmobranchs, having long canals, clustering of ampullae, large numbers of receptor cells, and multiple innervation (Obara 1976). The functional consequences of these morphological differences have been elucidated (see pp. 820–821).

## Stimulus and Receptor Frequency Sensitivity

Ampullary receptors are specialized to detect low-frequency (< 50 Hz) to DC electric fields primarily of biological origin (i.e., biopotentials produced by prey species). Another demonstrated function of these receptors, at least in elasmobranchs, is the detection of the earth's magnetic field for orientation purposes (Chap. 6). Behavioral studies have demonstrated orientation to, or prey capture guided by, electrosensory cues in catfish, sharks, and salamanders (Kalmijn 1974; Himstedt, Kopp, and Schmidt 1982). Direct measurement of the fields emanating from known or putative vertebrate and invertebrate prey reveals a substantial DC component with superimposed low frequencies (< 10 Hz) and field strengths of a few to hundreds of microvolts per centimeter (Kalmijn 1974). The origin of these fields, though not well understood, appears to be standing DC potentials produced by epithelial tissues, such as gills and pharyngeal linings, which separate fluid compartments of differing ionic composition. The modulations of these fields during respiration or movement generates the AC components (Kalmijn 1974).

Since the behaviorally relevant stimuli for an ampullary electroreceptor are weak DC TO low-frequency electric fields, the frequency content of which is broad and unpredictable at any instant, one might expect ampullary receptors to be broadly sensitive to these frequencies. In fact, a defining characteristic of these receptors is their low-frequency sensitivity. Elasmobranch receptors are most sensitive to frequencies of about 1–8 Hz (Montgomery 1984b) and those of silurids to a slightly higher range (6–12 Hz) (Peters and Buwalda 1972), whereas those of the weakly electric fish are maximally sensitive to higher frequencies of 30–50 Hz (Dunning 1973). The receptors of lampreys, with no canal or capsule, are most sensitive to frequencies of 1 Hz or less (Bodznick and Preston 1983).

These receptors are not sharply tuned, but fall off about 3 dB/octave and 6 dB/octave on the low- and high-frequency slopes, respectively. Bretschneider, Kroesbergen, and Beijnink (1979) have modelled the ampullary receptor cell by a simple RC filter network that adequately describes its region of maximum sensitivity, its mild roll-off characteristics, and its phase response. Certainly the high resistance and capacitance of the basal membrane (Fig. 32.4A) would give

Figure 32.2. Ampullary receptor organs of the catfish *Kryptopterus*. (A) The distribution of receptors on the head and trunk. Each *dot* represents receptor organ. (B) Cross section through ampullary epithelium. Note the lack of real canal, which is typical of this species. *Calibration bar* = 25 μm. (C) Receptor cells in ampullary epithelium, each with its single innervating nerve fiber cupping its base. *Calibration bar* = 10 μm. (Wachtel and Szamier 1969.)

the receptor cell a large time constant that would provide a substantial low-pass effect (see below).

## Functional Anatomical Considerations and Biophysics

The sensitivity of the ampullary system to extremely weak low-frequency electric fields (5 nV/cm, behaviorally tested, Chap. 6) is the result of a number of adaptations working in series. A uniform gradient of this magnitude results in a potential of about 0.1 μV across an ampulla with a canal 20 cm long, which is longer than the average (a single 1-cm mandibular canal would have a voltage drop of about 0.005 μV across it). A number of processes seem to work in concert to endow the system with this marvelous sensitivity.

As in most sensory receptors, the non-neural accessory tissues play an important role in determining sensitivity. Waltmann (1966) has demonstrated that the canal walls of the long ampullae of elasmobranchs behave passively and that their properties confer upon the organ a great sensitivity to extrinsic electric fields of low frequency. The high resistance provided by the tight junctions of the canal walls and the low resistance of the jelly in the core of the canal result in a large space constant (as much as 80 cm). Thus, a voltage at the pore opening of even the longest canal is expressed nearly unattenuated in amplitude down to the ampullary lumen.

The flattened cells that compose the canal walls result in appreciable capacitance in the longest canals that provides a large time constant as well. The time constant of a given canal will vary inversely with length, however, since net capacitance increases with the additional membrane of a longer and wider canal. Although no study has explicitly focused on the relationship between receptor tuning and canal morphology in elasmobranchs, the disparity between the large range of canal lengths and the rather small and uniform range of maximum frequency sensitivity (1–8 Hz) (Montgomery 1984b) suggests that it is more likely the filter properties of the receptor cell that predominate in determining frequency sensitivity (Montgomery 1984b; Schweitzer 1985).

Canal length, however, varies significantly between fresh and marine forms and can be explained as an adaptation for increasing sensitivity (Bennett 1971; Szabo et al. 1972; Kalmijn 1974). Freshwater fishes have a high skin resistance to aid in osmoregulation and, by comparison, a nearly negligible internal resistance. The resistance of the water is much less than that of the skin, but greater than the inside of the fish. If placed in a uniform electric field, a fish draws little current, and the skin has the largest resistance in the current path. Consequently the voltage drop will fall mostly across the skin, and virtually none will fall across the internal tissues. An ampulla that spans the epidermis will "see" a large voltage drop, and any extension of the organ deeper into the body tissues gives it no greater sensitivity.

Circumstances are quite different for a marine fish with a lower skin resistance living in a highly conductive environment. Kalmijn (1974) has shown that the skin resistance of the ray *Platyrhinoides triseriata* is about the same as that offered by its body tissues. Thus, given the low resistance of the seawater and the skin, the internal tissues now pose a relatively larger resistance, and an appreciable fraction of the voltage drop appears across them. An ampulla with a long canal that stretches across an expanse of internal tissues enhances its sensitivity by increasing the voltage gradient between its pore opening and the base of its ampulla. It is also important to bear in mind, as Kalmijn (1974) emphasizes, that for a local dipole stimulus, like a prey item, most of the voltage drop will fall across the skin.

Since the internal resistance of the freshwater forms is so low, it can be considered as isopotential, and all the ampullae can be referenced to the same potential merely by protruding through the epidermis and contacting the low-resistance interior (Kalmijn 1974). In a marine fish, on the other hand, specifically because a voltage gradient can be built up in its internal body tissues, ampullary organs would not be referenced to the same potential unless it was adjacent. This is critical for common mode noise rejection (p. 825).

## Polarity and Sensitivity

Ampullary receptors of teleosts are excited by outside anodal stimuli, whereas those of all other groups so far investigated are excited by cathodal stimuli. Response polarity is a function of whether it is solely the basal membrane or both apical and basal membranes that are excitable (Bennett, 1967, 1971a). In marine rays, both membranes are excitable, but the conductance change of the apical membrane is much larger than that of the basal membrane. The conductance change of the basal membrane is small enough that it behaves virtually linearly (Clusin and Bennett 1979). Further, the resistance of the apical membrane is greater than that of the basal membrane, so that most of the voltage drop across the receptor cell falls across the apical membrane. When a cathodal stimulus is presented in the lumen, the voltage drop across the apical membrane depolarizes it, while the basal membrane is slightly hyperpolarized by the smaller drop. The large depolarizing conductance change of the apical membrane overpowers the small conductance change in the basal membrane and depolarizes it in turn, thereby causing transmitter release (Clusin and Bennett 1979).

The apical membrane of teleosts, the resistance of which is lower than the basal membrane, appears to be inexcitable so that it is solely the basal membrane that determines the conductance changes across the receptor cell. Here the greatest fraction of any imposed voltage appears across the basal membrane and lumen-cathodal stimuli hyperpolarize the basal membrane whereas lumen-anodal stimuli depolarize it (Obara 1976; Bennett and Obara 1986). For most teleosts, the conductance change of the basal face is still not large, so that the receptor still behaves linearly (Bennett 1971).

Sensitivity varies by at least one order of magnitude between marine and freshwater species. Behavioral tests indicate that marine elasmobranchs are sensitive to voltage gradients of at least 5 nV/cm (Chap. 6), whereas the behavioral threshold of the freshwater ray, *Potamotrygon,* is about 50–100 μV/cm (Szabo et al. 1972). Similarly, the marine catfish *Plotosus* is at least one order or magnitude more sensitive (0.5 mV/cm) than freshwater teleosts (5 mV/cm) (Kalmijn 1974).

A number of factors determine sensitivity; the noise inherent in the medium (salt vs. freshwater); the sensitivity of the receptor cell (excitable vs. inexcitable membrane); the gain between receptor and afferent impulses (synaptic gain); convergence of afferents on central neurons; and modes of noise cancellation (see the next section). Through the work of Bennett and Obara and their colleagues (Obara 1976; Clusin and Bennett 1977, 1979; Bennett and Obara 1986), many of these factors have been studied in the marine elasmobranchs and *Plotosus,* though quantitative data on freshwater forms is lacking. A quantitative accounting of the differences in sensitivity must await a broad comparative study of peripheral biophysics and central processing.

The high sensitivity of elasmobranch receptors is attributed to an amplification step engendered by the excitable apical membrane. Depolarization of the membrane causes a large $Ca^{2+}$ influx, which depolarizes the basal membrane, resulting in an inward $Ca^{2+}$ current across that membrane and an outward $K^+$ current that repolarizes the membrane potential. These conductances are spontaneously, regeneratively oscillating in the resting, unstimulated state. A weak excitatory stimulus may strongly increase the inward $Ca^{2+}$ current and additionally synchronize the oscillations of the population of receptors, much increasing their activity.

As mentioned above, this is in contrast to the teleost receptors in which stimulus voltages directly and linearly modulate the much smaller conductances of the basal secretory membrane. Thus, *Plotosus* must rely on a mechanism other than a regenerative apical membrane for its enhanced sensitivity (see below).

As with other hair cells, a large resting conductance keeps the receptors partially depolarized (Hudspeth and Corey 1977). In the case of elasmobranchs, this is due to a resting $Ca^{2+}$ conductance across the apical membrane, whereas in teleosts, it is due to an inward $K^+$ current across the apical membrane due to the high levels of $K^+$ in the ampullary lumen (Obara 1976). In both cases, this results in an increased $Ca^{2+}$ conductance in the basal membrane and transmitter release. This ongoing release of transmitter, which allows the postsynaptic membrane to be depolarized past its spiking threshold, accounts for the tonic afferent activity.

Figure 32.3 indicates the spike rate curve of an afferent from *Plotosus*. Note that, at rest, it is discharging at 25% of maximum spike rate and that its spike rate function is sigmoidal with a linear portion of steep slope. Because of this tonic activity, the threshold of the receptor cells is poised in this linear region of steep slope in which slight modulations of receptor cell membrane potential lead to large changes in afferent discharge frequency. Furthermore, because of this tonic activity, both depolarizing and hyperpolarizing directions of the stimulus will result in modulations of neural activity doubling the sensitivity over that allowed with no spontaneous activity, when only receptor cell depolarizations resulted in transmission. This strategy is used by other octavolateralis receptor systems, such as lateral line, vestibular, and auditory (Goldberg and Fernandez 1971; Hudspeth and Corey 1977).

Synaptic gain and noise reduction can be enhanced by spatial summation of

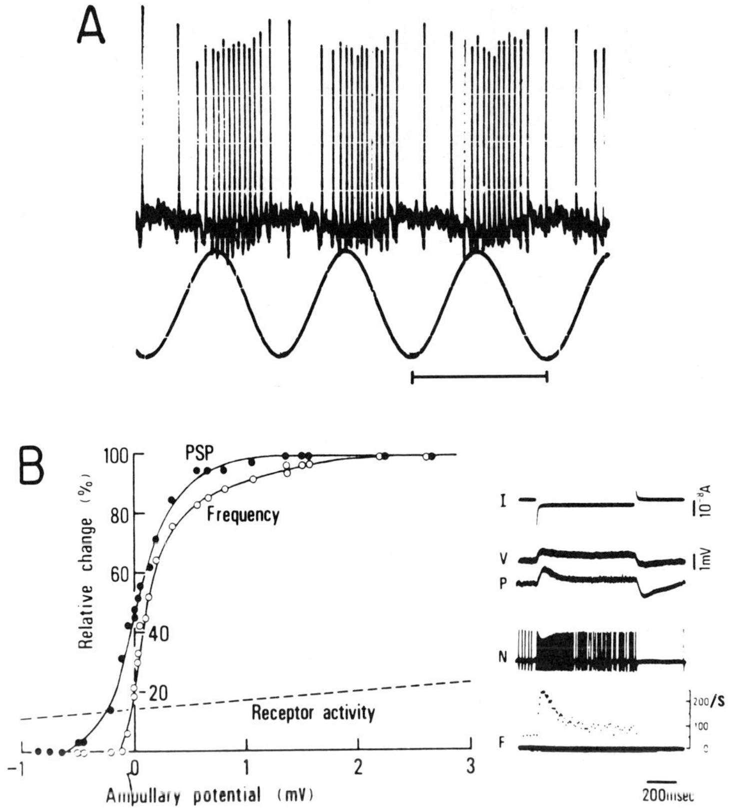


Figure 32.3. Responses of ampullary afferents. (A) Single afferent from weakly electric teleost *Sternopygus* demonstrating the moduation of ongoing spike activity by a stimulus of 7 Hz and 10 mV/cm. *Calibration bar* = 15 ms. (B) Spike rate curve of an afferent from marine catfish *Plotosus*. Raw data on the *right* indicates that constant current pulse imposed across the ampulla (*I*) produces voltage drop (*V*) across the epithelium, which leads to postsynaptic potential (*P*) and a barrage of spikes (*N*). *Left side* illustrates relative amplitude of the PSP and the frequency of spikes as a function of ampullary potential. (Obara 1976.)

many receptors. Small stimulus-induced voltage modulations of the membrane of one cell, which might otherwise be swamped by noise, would be brought out by averaging the responses of many receptors, resulting in a much enhanced postsynaptic response with a steeper spike rate curve. This mechanism is undoubtedly at work in marine fishes. Compared to *Kryptopterus,* a freshwater silurid with about 20 receptor cells per afferent, *Plotosus,* with about 250–750 receptors per afferent, has about 10–40 times more receptors per nerve fiber. It is important to bear in mind, however, that since receptors in one ampulla all ''see'' the same potential within the lumen, this mechanism cannot average out noise that is common to and coherent across all the receptors, but only noise due to fluctuations in membrane voltage (Kalmijn 1984).

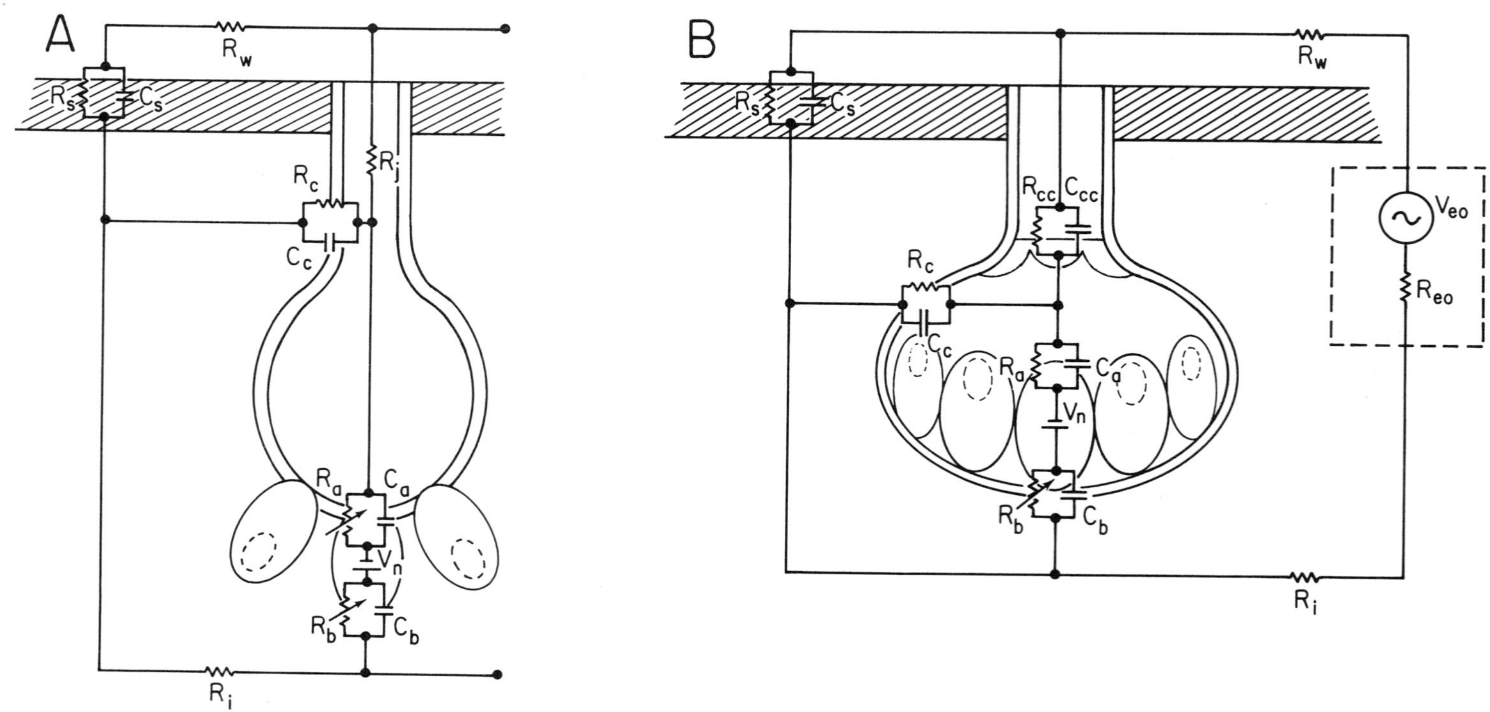


Figure 32.4. Electrical equivalent circuits of ampullary (A) and tuberous (B) receptor organs; $C_a$, capacitance of apical membrane; $C_b$, capacitance of basal membrane; $C_c$, capacitance of the canal and capsular wall; $C_{cc}$, capacitance of covering cells; $C_s$, capacitance of the skin; $R_a$, resistance of apical membrane (including parallel linear and variable elements where variable resistance is indicated here or elsewhere); $R_b$, resistance of basal membrane; $R_c$, resistance of the canal and capsular walls; $R_{cc}$, resistance of covering cells; $R_{eo}$, internal resistance of electric organ; $R_i$, internal resistance of the fish; $R_j$, resistance of ampullary jelly; $R_s$, resistance of the skin; $R_w$, resistance of the water; $V_{eo}$, internal voltage of the electric organ; $V_n$, Nernst potentials of receptor cells. Resistances and capacitances of support cells are not included, but they are believed to be entirely passive. (Modified from Bennett 1967.)

## Noise Cancellation

Any sensory system is subject to environmental noise that degrades signal quality. The ampullary system faces a particularly difficult problem, in that the respiratory or swimming movements of a fish possessing ampullary receptors will stimulate its own receptors, interfering with its detection of extrinsic signals (Peters and Buwalda 1972; Kalmijn 1974, 1984; Montgomery 1984a). This has been solved in a number of different ways in different groups of fish.

Montgomery (1984a) has recently demonstrated in the thornback ray *Platyrhinoides triseriata* that noise reduction and signal enhancement is effected by a "common mode rejection" mechanism, as predicted by Kalmijn (1974). Afferent fibers from canals whose openings are oppositely oriented, but with ampullae in the same cluster, converge on a common postsynaptic neuron in the dorsal nucleus. As mentioned earlier, since these two ampullae are housed in the same cluster, they will both see the same internal potential. Thus, internal potentials, as respiratory potentials, will be expressed identically across both ampullae, driving afferent fibers from both. One afferent makes an excitatory contact with the postsynaptic cell, however, whereas the other, possibly through an interneuron, is inhibitory. Thus the effects of any large internally generated potentials are effectively cancelled, whereas smaller local external fields that uniquely or preferentially stimulate one ampulla affects central neurons. This synaptic arrangement has the additional advantage of reducing the signal-to-noise ratio further by adding gain to the system, much like a "push-pull" amplifier.

Freshwater electroreceptive fishes are subject to the same self-produced interference (Peters and Buwalda 1972). Whether they share a common mode rejection mechanism for cancellation of self-produced fields, is not known. A point of interest is that most hair cell receptor organs have a well developed efferent system that generally reduces sensitivity at the receptor cell by hyperpolarizing it. A reduction in sensitivity of up to 15 dB has been demonstrated for lateral line hair cells (Flock and Russell 1973). Although the function of the efferent system is a much-debated topic, the efferents to the lateral line system in the dogfish have been shown to reduce the sensitivity of the hair cells of its lateral line during swimming movements, thus suppressing any self-stimulation of the hair cells during swimming (Russell and Roberts 1974). It is a continuing mystery why neither ampullary nor tuberous receptors have efferents that inhibit or modify receptor activity during an active body movement.

The ampullary receptors of many weakly electric pulse-type fish face an additional source of noise. The brief, pulsatile electric organ discharge (EOD) of many species contains broadband energy, including frequencies in the range of ampullary receptor organs (below 50 Hz, often down to DC) that may cause a spike train in ampullary afferents (Bell and Russell 1978). Mormyrids have developed a corollary discharge mechanism by which afferent input is inhibited at the first central synapse (lateral line lobe) during and shortly after the EOD pulse (Zipser and Bennett 1976). Since numerous factors, as conductivity of the water, impedance, and position of nearby objects, will change the waveshape and intensity of the EOD around a stationary fish, however, the ampullary system must be able to cancel out the incoming afferent volley with its "negative copy." Bell (1982) has demonstrated that the corollary discharge signal is dynamic, changing with time under varying external circumstances to match and cancel the incoming ampullary afferent discharge (Chap. 10).

## Tuberous Receptors

Tuberous receptors, which have independently evolved in gymnotiforms and mormyriforms, share certain functional similarities, but otherwise are morphologically quite different in the two orders (Figs. 32.4B and 32.5) (Szabo 1965, 1974; Zakon, 1986). Like the ampullary receptors of freshwater teleosts, they are composed of a capsule situated just below the epidermis, the interior of which is in contact with the aqueous environment via a small canal, and containing sensory receptor cells. However, ampullary and tuberous organs differ in numerous significant details that provide clues to their respective modes of operation. Rather than being filled with a jelly, the canal of tuberous organs is filled with a plug of loosely coupled epithelial cells (Fig. 32.5). An extensive extracellular space provides channels between these cells to the base of the canal, where a second type of cell, the covering cell, is imposed over the opening to the capsule. Whereas ampullary receptors typically have capsule walls only 2–4 cell layers thick, the capsule of tuberous receptors is composed of many layers of flattened cells.

### Tuberous Receptor Morphology

A diversity of types and subtypes of tuberous receptors exists (Table 32.1 and Fig. 32.5). For example, whereas mormyrids have evolved two distinct and parallel brain pathways for electrolocation and communication (Zipser and Bennett 1976), beginning with different receptor types, the single anatomical type of tuberous receptor of gymnotiforms is used for both purposes. Within gymnotiforms, however, this type of tuberous organ can be further classified into subtypes based on morphology and function.

The receptor organ believed to underlie communication in mormyrids is the Knollenorgan (or K unit). It has a typical tuberous morphology (capsule of many layers, epithelial-filled canal with covering cells at its base, free-standing receptor cells), except that each receptor cell is encased in its own capsule (Fig. 32.5A and 32.11A). The number of encapsulated receptor cells may vary from a single one per organ to a dozen or so, the number depending on the size of the fish (Quinet 1971). In a few species, large receptor organs containing up to 60 sensory cells are known (*Petrocephalus:* Quinet 1971). Bass and Hopkins (1980, 1984) distinguish between two types of Knollenorgan: the KI organ, which is by far the most common, being distributed on the head and the dorsal and ventral portion of the flanks, and the larger KII, exemplified by those described above from *Petrocephalus,* few and always located only around the operculum or eyes.

Knollenorgan receptor cells are large (30 μm in diameter), spherical and, in cross section, possess a thin band of mitochondria just below the apical membrane.

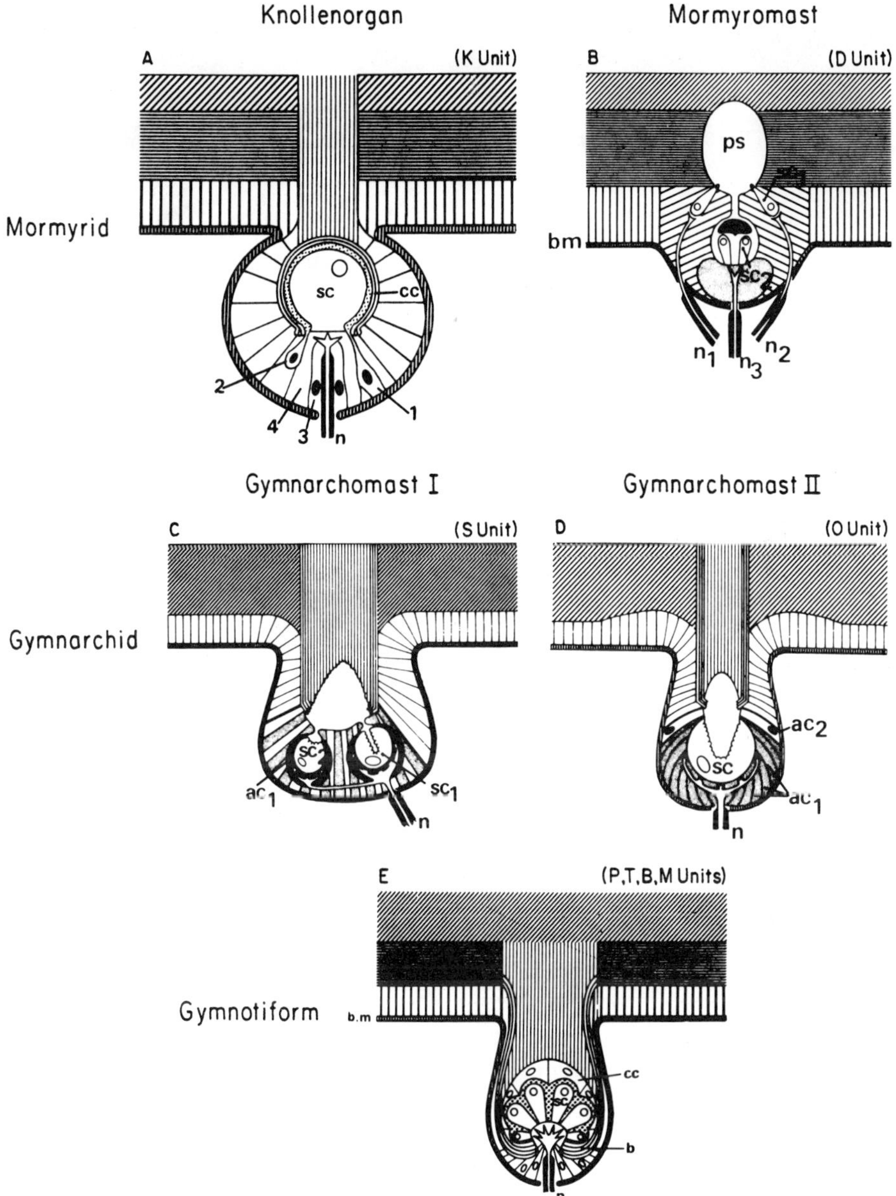


Figure 32.5. Types of tuberous organs. (Szabo 1974.)

The apical surface is greatly invested with microvilli (Fig. 32.11). As distinct from most electroreceptors, the receptor cell is capable of spiking, and this spike is thought to be conducted to the afferent via electrotonic synapses (Bennett 1971), though gap junctions have not yet been described, and only classic ribbon synapses have been observed (Derbin and Szabo 1968).

Table 32.1. EOD and Receptor Type Found in Each Family of the Two Orders of Weakly Electric Fish.

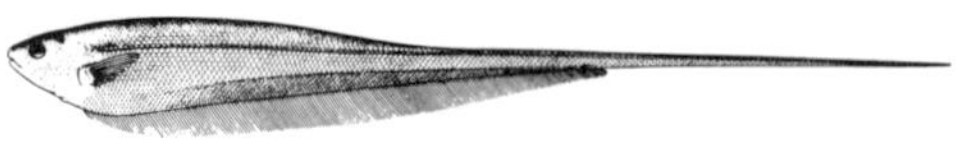

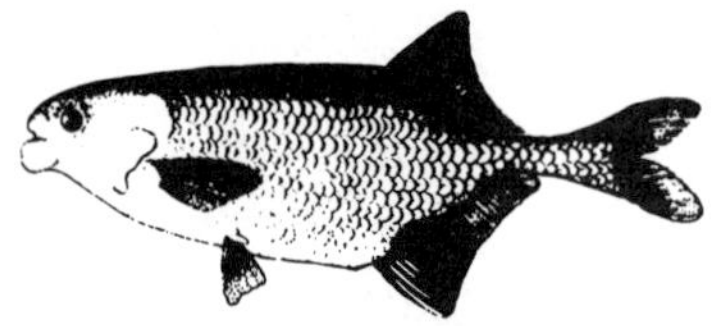

| Gymnotiforms | | | Mormyriforms | | |
|---|---|---|---|---|---|
| Family | EOD | Recpt. Types | Family | EOD | Recpt. Types |
| Sternopygidae | W | P,T | Gymnarchidae | W | O,S |
| Apteronotidae | W | P,T | Mormyridae | P | K,D |
| Rhamphichthyidae | P | B,M | | | |
| Hypopomidae | P | B,M | | | |
| Electrophoridae | P | ? | | | |
| Gymnotidae | P | B,M | | | |

Note: For EOD type, W = wave type; P = pulse type. Descriptions of the receptor types are given in the text.

Mormyromasts (or D units), on the other hand, are a more complex organ, bearing two distinct types of sensory cells, each within its own encapsulated chamber and each with its own source of innervation (Szabo and Wersäll 1970) (Fig. 32.5B). The deeper chamber contains 3–5 sensory receptors standing upright within the lumen with their extensive apical surfaces covered with microvilli. One nerve fiber innervates these cells. This chamber narrows at its top and connects to an upper chamber via a small canal. Within this superficial chamber are a number (2–10, depending on fish size) of receptor cells embedded in the support cell layer, more like ampullary receptor cells, in a ring around the lumen. These cells have no microvilli, but rather, a small, dense conical structure capping their apices. They are innervated by two additional nerve fibers, each of which contacts a semicircle of adjacent receptor cells (Szabo and Wersäll 1970; Zakon and Bell unpublished).

Gymnarchid receptors are very different morphologically from mormyrid receptors. They will be discussed briefly here, as little is known about their functioning, with the exception of one study (Bullock, Behrend, and Heiligenberg 1975). Two types are recognized, gymnarchomast types I (S unit) and II (O unit), both of which are embedded within their epithelia, more like ampullary receptors, and have a patch of long microvilli on their surfaces (Fig. 32.5C and D). Type I organs possess two distinct cell types of differing size, one of which has a deep groove and the other a shallow indentation of its apical membrane. Type

II organs are composed of a group of large and similar sensory cells. All the receptors of both organs are contacted by a single nerve fiber (Szabo 1965, 1974).

Most gymnotiform electroreceptor organs are organized similarly to each other (Fig. 32.5E), with subtle species differences. Most contain about 30 receptor cells per organ, though certain subtypes may have up to 100 (Szabo 1965, 1974; Zakon 1986, 1987). A single afferent fiber penetrates the base of the capsule, repeatedly branches, and supplies each receptor cell with a single afferent bouton with which it makes a series of chemical synapses via synaptic ribbons. The major differences between subtypes seem to be in either the morphology of the afferent terminals or the number of capsules contacted by a single afferent.

The tuberous organs of pulse-type fish, like *Hypopomus, Gymnotus,* and *Rhamphichthyes,* have two distinct axon terminal morphologies that correspond to two physiologically distinct receptors: the so-called pulse marker units (M units) and burst duration coders (B units) (see Fig. 32.9 and the section on gymnotiform encoding). The morphological differences between them are that the afferent terminal of the M unit is thick and myelinated up to the site of synaptic contact with receptors, where it swells into a spherical knob, which contacts each receptor, whereas the terminals of the B unit are thin, often bifurcating before reaching the capsule, and are unmyelinated within it (Szabo 1970; Szamier and Wachtel 1970; Zakon, 1987).

The tuberous organs of the wave gymnotiforms of the family Sternopygidae (*Sternopygus, Eigenmannia*) vary in the number of receptor organs that constitute a receptor unit (Fig. 32.10A and B). A single nerve fiber may have its own receptor organ, or an afferent fiber may contact a cluster, or "rosette," of organs, and this changes with age (Zakon 1984a, 1987). Newly hatched larvae and young juveniles have only a single receptor organ per afferent. As fish age, many existing receptor organs divide and form two daughter organs. Tuberous receptor units of older fish may have ten or more receptor organs, all presumably derived from one precursor organ. Not all receptors undergo division, however, so that older fish have two populations of receptor units: those with only one, or at most two, receptor organs and those with many. In addition, there is a tendency for the terminals of afferents contacting a single receptor organ to be bulbous and thick, while those which innervate multiple organs are often thinner and rarely have thick branches within a single capsule (Zakon 1984a).

## Electric Organ Discharge

Tuberous receptors are used in detecting electrical emissions generated from electric organs, and their function cannot be appreciated without first considering these discharges. It is important to take into account the influence of the EOD on the receptor, for by analogy with the ampullary system, the EOD provides a steady, ongoing bias generating a backdrop of stable and regular activity in the receptors, which is modulated in turn by modulations of the EOD.

With the exception of one group (Apteronotidae), the EOD is generated by modified muscle tissues forming an electric organ usually on the tail or posterior portion of the body. The timing of firing of the electric organ is determined by a medullary nucleus, the pacemaker nucleus, whereas the waveshape of each discharge is determined by the membrane properties of the electrically excitable effector cells of the electric organ (the electrocytes), the polarity of innervation of the electrocytes (anterior or posterior face), and the synchrony of their firing (Dye and Meyer 1986; Bass 1986).

Weakly electric fish can be classified according to the type of discharge that they produce. Wave-type fish produce ongoing, nearly sinusoidal, EODs with a power spectrum characterized by a distinct fundamental and higher harmonics (Fig. 32.10C and D). Pulse-type fish, on the other hand, produce brief EOD pulses spaced out with long, usually irregular silent intervals (Fig. 32.7A and B). The power spectrum of these pulses is typically broad with a peak region, though some complex waveforms may have multiple peaks, and many, especially if monophasic, may extend down to DC (Bastian 1976, 1977; Hopkins 1981). About half of the studied gymnotiforms are of each type, whereas, with the exception of the monogeneric family Gymnarchidae (Bullock, Behrend, and Heiligenberg 1975), all mormyrids are pulse-type fish.

These fields are used for electrolocation and social communication. An object within the fish's electric field will distort it in precise ways, depending on the object's size and complex impedance; specifically, EOD amplitude varies with resistive components, whereas phase (with respect to other regions of its body surface) varies with capacitative reactance (Heiligenberg 1977; Meyer 1982). Relative movement between object and fish will impart amplitude modulations of the EOD, which move across the fish's body (Bastian 1981). Detection and recognition of objects requires that a fish be ''informed'' of these local changes in amplitude and phase of its own EOD (Bastian 1981).

Each species possesses an EOD waveform that is species typical and distinct from the EODs of sympatric species; the EOD may additionally show age- or sex-dependent differences that subserve intraspecific communication (Hopkins 1972, 1981; Westby and Kirschbaum 1978; Meyer 1983). Transient stereotyped modulations of the EOD frequency (in wave-type fish) or pulse pattern (in pulse-type fish) may also serve as social signals during courtship or aggression (Hopkins 1974, 1981, Chap. 9; Hagedorn and Heiligenberg 1985).

## Enhancing Sensitivity: Impedance Matching and Noise Reduction

An important environmental factor that will influence the sensitivity and range of the electrosensory system is the resistivity of the water. Because of the large resistance posed by the skin in series with the electric organ, a fish acts as a constant current source as long as the resistance of the water does not approach that of its skin (Bell, Bradbury, and Russell 1976). Since the voltage produced by a constant current electric organ is proportional to the resistance of the water, the electric field penetrates farther away from the body if the resistance of the water is high; it is ''short-circuited'' in highly conductive water. Thus, within limits, the range of the electrosensory system is increased in relatively less conductive water (Heiligenberg 1977; Knudsen 1975).

If the resistance of the water were to exceed that of the fish, a greater fraction of the total EOD voltage would appear across the resistance of the water and less across the skin. On the other hand, if the resistance of the skin were significantly greater than that of the water, the changes in electric field voltage caused by objects around the fish would be small, resulting in small changes in the transepithelial potential. In the ideal case, the resistance of fish and water should be about the same, so that the fish is impedance matched to the environment for maximum power transfer (Heiligenberg 1977).

Since the electrosensory system functions primarily in the near field, transmission losses are frequency independent (Hopkins and Heilgenberg 1978; see also Chap. 6). Sources of environmental noise would be the volume resistance of water (Bennett 1971; Kalmijn 1974) and lightning that causes transients with peak spectral energy around 2 kHz (Hopkins 1973). The greatest and most serious source of noise for the electrosensory system is the EODs of neighboring fish, however, both con- and heterospecific.

Numerous strategies are used to minimize interference from signalling neighbors. Perhaps the simplest, used by most wave fish, is for each species to ''broadcast'' over a private species-specific frequency band. This becomes increasingly imperative in areas of great diversity of sympatric species (Hopkins 1976; Hopkins and Heiligenberg 1978). Even with each species broadcasting over its own band, conspecific fish with similar or identical EOD waveforms might produce highly destructive interference, or ''jamming,'' of each other, especially in highly social species (Hopkins and Heiligenberg 1978; Hopkins 1981). This is minimized, since each individual possesses a characteristic EOD frequency within its species-specific range. In addition, many species perform a highly stereotyped reflexive behavior, the so-called jamming avoidance response, in which the EOD frequency is rapidly shifted to a new value when jammed by conspecific fish (Chap. 33).

Pulse fish, in contrast, do not suffer the same type of ongoing interference and therefore use different mechanisms to avoid jamming each other. Since each discharge is relatively brief, jamming occurs only when a neighboring fish produces a pulse that overlaps in time with its own. An individual fish may decrease either its pulse rate or pulse duration to minimize the probability of overlap. Since lowering the discharge rate lowers sampling rate, it is most efficient to decrease the duration of each pulse. In fact, those species which school interspecifically have briefer pulses than solitary species (Hopkins 1981).

As pulse duration decreases and the EOD becomes more impulselike, its frequency composition is widened, and the difference in frequency spectrum among EODs of sympatric pulse species lessens. Coding of species-specific EODs must then rely on temporal, rather than purely spectral, cues (Hopkins and Bass 1980, 1981). On the other hand, a brief broad-spectrum pulse may have additional advantages for better appreciation of frequency-dependent phase shifts of EOD during electrolocation of objects with complex impedances (Meyer 1982).

Like wave fish, pulse fish also have specific behaviors associated with avoidance of jamming. Fish with a regular discharge rate may speed up or slow down as the discharges of another individual come closer in time to its own (Heiligenberg 1974).

## Frequency Tuning of Tuberous Receptors

Considering the high ambient noise level from the discharges of neighboring fish and the predictability of the pertinent signal, the EOD, the optimally designed tuberous electroreceptor should be most sensitive to the frequency components of the fish's own EOD. Generally speaking, that is what is observed when tuning curves are derived from afferent electroreceptive nerve fibers. Figure 32.6A depicts a family of tuning curves taken from two of the wave-type species *Sternopygus*. Note that the best frequencies of these tuning curves cluster around the EOD frequency. The receptors of wave species are well tuned to match the species-specific EOD frequency range, and this is true for the gymnotiform fishes, as well as the only wave mormyriform, *Gymnarchus* (Hopkins 1976; Scheich, Bullock, and Hamstra 1973; Bullock, Behrend, and Heiligenberg 1975; Zakon and Meyer 1983) (Fig. 32.6B). Moreover, the receptors of each individual are best tuned to its own discharge frequency, as each has a unique "signature" EOD frequency within its species's range (Hopkins 1976; Viancour 1979; Meyer and Zakon 1982; Zakon and Meyer 1983).

What constitutes "optimal design" of the electrosensors is different in pulse fish, since the power spectrum of a pulsatile EOD is continuous and broad, and contamination from the EODs of neighbors occurs only during occasional overlap of pulses in time, rather than from overlap in a frequency channel. The receptors of an individual pulse gymnotiform, vary in the extent of tuning and best frequency (BF) (Bastian 1976, 1977). In most species, some receptors may be sharply tuned, usually to the peak or a major component of the EOD power spectrum, whereas other receptors are broadly sensitive to a greater range of the EOD spectrum (Fig. 32.7). Receptors in one species of *Hypopomus*, in which the local EOD power spectrum varies along the body, also systematically vary in BF along the length of the body (Bastian 1977). In *Gymnotus*, on the other hand, BFs of receptors in a small patch of skin may vary by over an octave (Watson and Bastian 1979).

The receptors of mormyrids, on the other hand, are typically very broadly tuned to the EOD power spectrum. Their tuning encompasses a portion, often, but not always, the peak of the EOD power spectrum. When two classes of Knollenorgans (KI and KII) are found, each is tuned to separate components of

Figure 32.6. Tuning characteristics of tuberous receptors of wave-type gymnotiforms. (A) Tuning curves from two individuals (one illustrated by *open*, the other by *solid symbols*). EOD frequency of each is given by a triangle on the abscissa; 0 dB = 20 mV/cm (Zakon and Meyer 1983). (B) Best frequencies of tuning curves from three species of gymnotiforms. BFs from each fish are graphed relative to its EOD frequency. If receptors of each individual were best tuned to its own EOD frequency, they would fall along the 45° line (Hopkins 1976).

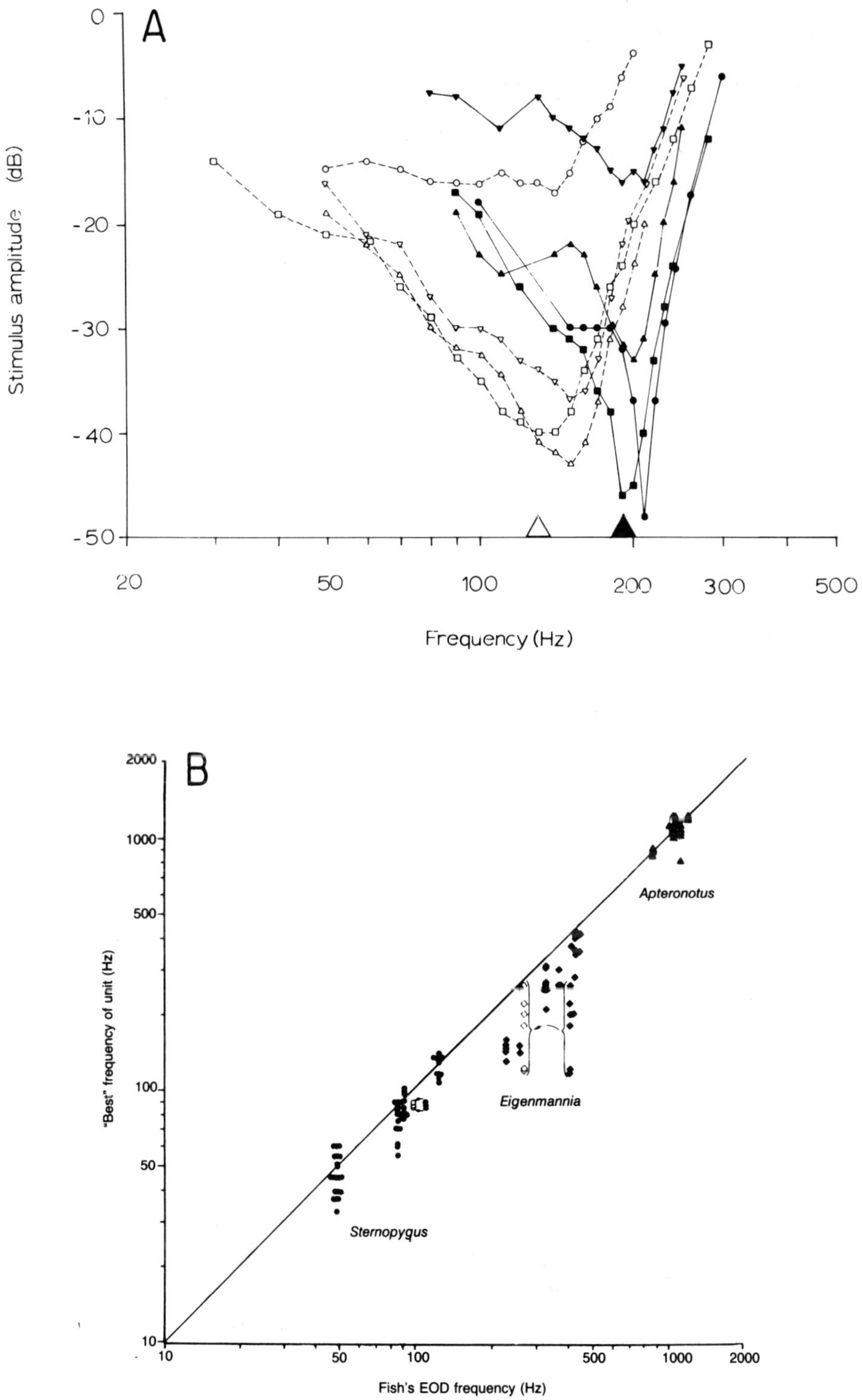
A
0
-10
-20
-30
-40
-50
Stimulus amplitude (dB)
20
50
100
200
300
500
Frequency (Hz)
B
2000
1000
500
100
50
10
"Best" frequency of unit (Hz)
10
50
100
500
1000
2000
Fish's EOD frequency (Hz)
Apteronotus
Eigenmannia
Sternopygus

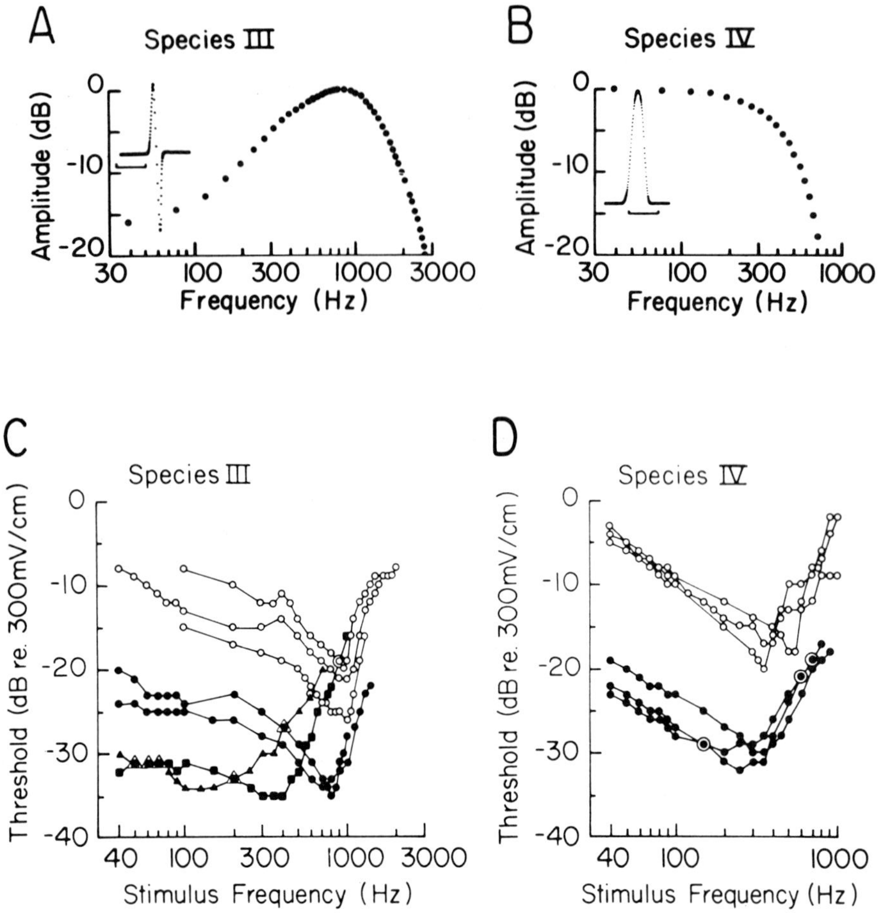


Figure 32.7. EOD waveform and spectra (A and B) and receptor tuning curves (C and D) from two species of gymnotiform genus *Hypopomus*. Tuning curves with *open circles* are from pulse markers (M units); those with *closed circles*, from burst duration coders (B units). (Bastian 1977.)

the EOD (Bass and Hopkins 1980). Tuning to the EOD power spectrum appears to be less critical in mormyrids as the response of their Knollenorgan receptors is much more dependent on temporal features of the pulse, like the rise and fall time, than it is on spectral features (Hopkins and Bass 1981) (Fig. 32.11C). The receptors of mormyrids are often sensitive to higher frequencies than those of gymnotiforms, as their EOD power spectra may peak in the range of a few kilohertz. The receptors of *Pollimyrus isidori*, for example, may be tuned to frequencies as high as 18 kHz (Hopkins 1981).

In fish of both orders, the best frequencies of tuberous receptors vary dramatically over a range of eight octaves (50 Hz in *Sternopygus* to 18 kHz in *Pollimyrus*). Although the details of the biophysics of the tuning have yet to be worked out, this extraordinary range of tuning promises to provide insights into the mechanisms and development of membrane tuning by a comparative approach.

Sexual dimorphisms in EOD frequency (wave fish) and power spectrum (pulse fish) have been identified in many species and may be the rule, rather than the exception (Hopkins 1972, 1974; Hagedorn and Heiligenberg 1985; Bass and Hopkins 1983). Where examined, the tuberous receptors also show sexual dimorphisms in tuning. These naturally occurring EOD sexual dimorphisms can be mimicked in the laboratory by the application of sex steroid hormones (Fig. 32.8). Typically, when females or juveniles are injected with male sex steroids, the EOD shifts until it comes to resemble that of the male, which is usually of lower-frequency composition (Meyer 1983; Meyer and Zakon 1982; Bass and Hopkins 1984). Concurrently, receptor tuning shifts to stay in alignment with the changed EOD during steroid treatment.

A series of experiments with *Sternopygus* has shown that, when the EOD is surgically eliminated by a high spinal transection, receptors retain their proper tuning for weeks and indicate that an ongoing EOD is unnecessary to shift receptor tuning after hormone treatment (Meyer, Zakon, and Heiligenberg 1984; Keller, Zakon, and Sanchez 1986). Thus, though the mechanism of hormonal action is now still far from understood, increasing evidence indicates that its effect on tuning is independent of the EOD. Similarly, newly generated receptors in regenerated skin become tuned to the EOD frequency of each fish in the absence of the electric field (Zakon 1986).

Studies on the sexually dimorphic pulse mormyrid *Brienomyrus brachyiustius* (triphasic) have shown that Knollenorgans, as well, shift their tuning to become more closely tuned to the EOD after dihydrotestosterone (DHT) administration. These receptors also remain tuned in the absence of an ongoing EOD, and when electrically silenced fish are treated with DHT, though receptor BFs decrease slightly, they are unable to become completely retuned during hormone treatment in the absence of an electric field (Bass and Hopkins 1984).

These examples underscore the importance of a matched receptor and EOD signal. One might ask why there is such a diversity in EOD and tuning in the first place, especially if there is no obvious selective frequency attenuation by habitat. Hopkins (1981) has emphasized that the diversity of EOD waveforms is most likely driven by such biological factors as the need for reproductive isolation and noise immunity.

## Biophysics and the Origin of Tuning

Intracellular recordings have yet to be made from tuberous receptors, and most of the knowledge of their biophysics comes from extracellular recordings of their receptor potentials or inferences from their morphology (Bennett 1971; Watson and Bastian 1979; Meyer and Zakon 1982; Zakon 1984b).

The canal walls of ampullary organs are composed of only a few layers of flattened cells, whereas those of the tuberous receptors are composed of many layers. The added capacitances of numerous unit membranes in series probably

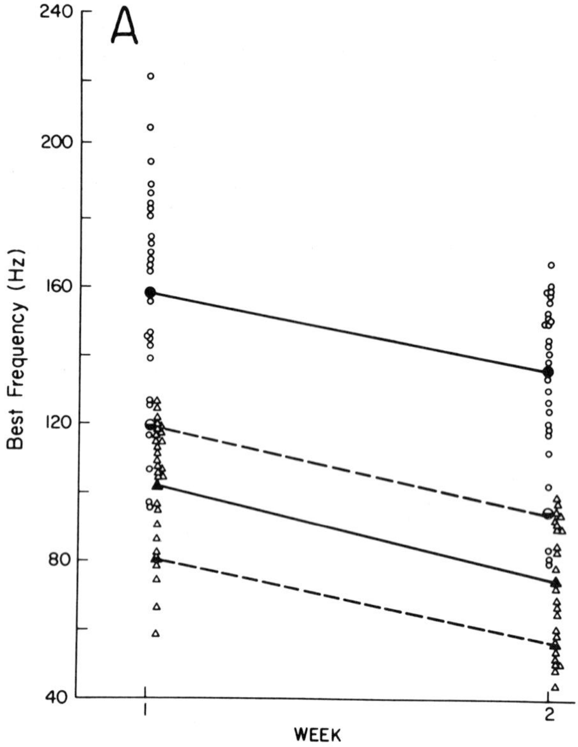
A
Best Frequency (Hz)
240
200
160
120
80
40
1
2
WEEK

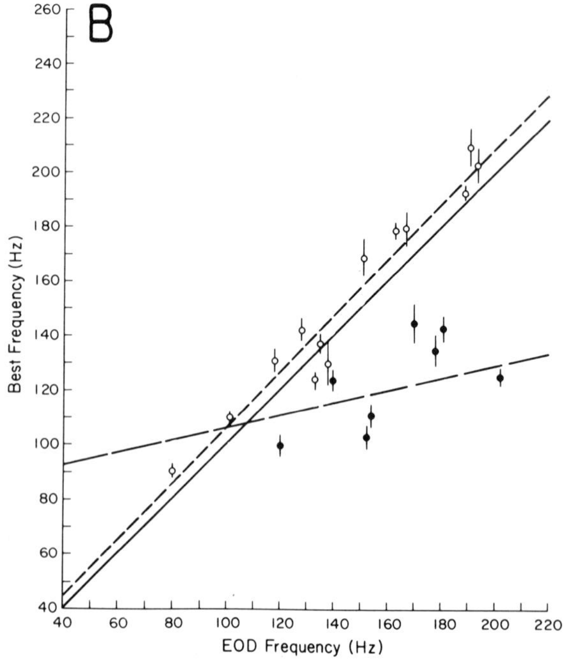
B
Best Frequency (Hz)
260
240
220
200
180
160
140
120
100
80
60
40
40
60
80
100
120
140
160
180
200
220
EOD Frequency (Hz)

reduces the total capacitance, thereby decreasing the shunting of high frequencies across the capsular walls (Bennett 1967, 1971) (Fig. 32.4B). A relationship exists between the number of capsular layers, as determined by electron microscopy (Lissmann and Mullinger 1968), and the EOD frequency composition in a number of gymnotiforms.

The covering cells at the base of the canal are connected by tight junctions (Szabo 1974), which pose a barrier to movement of ions into the capsule (Zakon 1984a) and capacitatively couple the interior of the receptor organ to the outside medium, thus acting as a high-pass filter. Although the constituents of the intracapsular fluid are unknown for any tuberous organ, they are likely to be quite different from those of the stream water surrounding the fish. Such a barrier may serve to lessen any metabolic costs associated with maintaining a stable intracapsular milieu. Whether the fluid facing the apical surface of the receptor cells is similar to extracellular fluid or whether, like endolymph, it is $K^+$ rich and provides a substantial driving force for an inward $K^+$ current (Sellick and Johnstone 1975) remains to be determined. Ampullary receptors, on the other hand, cannot afford the "luxury" of covering cells, for they must remain DC coupled to the environment (Fig. 32.4A).

The extensive apical membrane of these receptors, with its dense mat of microvilli, probably increases input capacitance and decreases its input resistance. Bennett (1967, 1971) has suggested that the large capacitance most likely acts as a blocking capacitor, that the apical membrane resistance is low compared to the basal resistance, and that the apical face is inexcitable. As above, it is important to consider the time constant of the apical membrane, which will limit the receptor's frequency response. In order that the apical membrane pass frequencies of hundreds of hertz, especially with such a large capacitance, it would be necessary for the resistance to be quite low. As Bennett (1967, 1971) has also pointed out, for greatest sensitivity the basal membrane resistance must be higher than that of the apical membrane, so that most of the voltage drop across the receptor will be expressed across the excitable basal face. Thus, given its relatively greater resistance, the small basal membrane surface area with its low capacitance will reduce its time constant.

The particular morphology of the tuberous receptor explains its high-pass character, but not its sharp filtering. Receptor tuning is most likely accomplished by the basal membrane as a result of active mechanisms. It is known that the electroreceptor rings on stimulation with a brief impulsive stimulus and it is the basal membrane that is the site of generation of this oscillation (Bennett 1971; Zakon 1984b). Pharmacological agents that are known to block $Ca^{2+}$ and $K^+$, but not $Na^+$, channels in other membranes will reduce the frequency of oscillation or block them entirely (Zipser and Bennett 1973; Zakon 1984b). Thus one may

---

Figure 32.8. The influence of androgens on receptor tuning. (A) BFs and EOD frequencies of two fish one week after daily injections of DHT. (B) The distribution of BFs in two groups of *Sternopygus*. *Open symbols* are from intact group; *solid symbols*, from those treated with DHT plotted against baseline EOD frequency (Zakon and Meyer 1983).

suppose that the kinetics (time constants) of an inward $Ca^{2+}$ current and an outward $K^+$ current may interact so as to produce conductance changes of the basal membrane that will be optimal at a certain frequency. By this line of reasoning, the kinetics of one or both current must vary to account for the observed range in tuning. It will be interesting to see what mechanisms are operative in mormyrids, in which Knollenorgan membranes may be tuned up to 18 kHz.

## Coding Mechanisms of Tuberous Receptors

Ampullary receptors convert amplitude modulations of the DC field into frequency modulations of an ongong spike train. Because the period of low-frequency stimuli is long, an ongoing spike train of tens of spikes per second can be frequency modulated. Tuberous receptors are unable to use the same encoding schema, as the frequencies that they must follow are too high. Accordingly, different encoding mechanisms are called for: receptors independently encode either amplitude or timing of the EOD.

Both pulse and wave fish have evolved physiologically and morphologically distinct receptor types for monitoring EOD amplitude and timing. One type of receptor responds to differences in EOD amplitude either with differential probabilities of spiking (in wave fish) or the number of spikes (pulse fish), whereas the other type marks the timing of the EOD by the strict phase locking of a single spike. Initially distinguished by Szabo (1970), these two physiological types have recently been labelled as "amplitude-modulated" and "rapid-timing" units, respectively (Hopkins 1984).

### Encoding in Gymnotiforms

As an example, two types of units are recognized in a number of pulse gymnotiforms, such as *Hypopomus*. M units, found in the greatest abundance on the head, fire a single spike at a short latency for each EOD pulse with minimal latency shift (~20%) as a function of EOD amplitude (Bastian 1976, 1977; Watson and Bastian 1979; Hopkins and Heiligenberg 1978). It is thought that this receptor type signals the occurrence of an EOD pulse, but gives no information about its amplitude (Fig. 32.9C). B units, on the other hand, fire a variable number of spikes (1–20) with varying latencies, depending on EOD amplitude, and can therefore encode EOD amplitude even though they give poor information on EOD timing (Fig. 32.9D). Szabo (1970) has shown that the M unit corresponds to the receptor organ with the large myelinated afferent ending, whereas the B unit has a thinly branching unmyelinated terminal (Fig. 32.10A and B) (see p. 829).

Unlike mormyrids with distinct receptor systems for electrolocation and communication, gymnotiform fishes must use the same receptors for both functions. Heiligenberg and Bastian (1984) have explained how the differentiation of a fish's pulses from the pulses of its neighbor is accomplished. When a fish fires its own electric organ, outward current will flow everywhere across the skin, exciting M units simultaneously over the body surface. When a neighbor discharges, on the other hand, current from its EOD will flow in one side of the body and out of the other, thus exciting M units on opposite sides of the body asynchronously. Pulse fish can accurately detect conspecific EODs, and they are not only sensitive to the frequency composition of the EOD, but to phase information as well (Heiligenberg and Altes 1978).

During electrolocation, B units signal changes in local EOD amplitude caused by nearby objects. Objects that are more resistive than the water push current lines away, whereas those which are more conductive draw them in (Fig. 32.9D). Many biologically relevant stimuli will also pose reactance (vegetation, other fish), as well as resistance. The effect of capacitative reactance is to phase shift the local EOD, and this would be detected by M units. By simultaneously comparing the number of spikes produced by B units over the body surface and the shift in timing of the spike of M units, location and identification of objects could be based on their complex impedance.

Two physiological classes are known in the gymnotiform wave fish as well. One type of unit, the T unit (timing coder) is well phase locked to the EOD, firing only one spike per cycle with a jitter of less than 100 μs whereas the P unit (probability coder) fires less than 1:1, its probability of firing a function of EOD amplitude, and with a jitter of more than 500 μs (Scheich, Bullock, and Hamstra 1973; Feng and Bullock 1977) (Fig. 32.10 and Table 32.1). The high jitter and probabilistic nature of P unit spiking makes them unsuitable to code for phase. Most P units are in the linear portion of their spike rate curve at the normal EOD intensities, however, "sputtering" to the resting EOD and increasing or decreasing their probability of firing more or less linearly with modulations of EOD amplitude, providing informaton on EOD amplitude.

Because T units are so well phase locked to the EOD, they provide information on EOD phase. A unit that reports on timing events unambiguously must have an intensity-independent response; some T units decrease latency of firing (phase advance) to sine wave stimuli of increasing intensity (Scheich, Bullock, and Hamstra 1973). Owing to central mechanisms starting in the first synapse in the electric lateral line lobe (Scheich 1977), however, both the jitter, which is less than 100 μs in the afferent fiber, and the slight amplitude dependence are improved.

As in pulse gymnotiforms, resistive and reactive components of objects may be coded differentially by these two classes of units. T units code for the timing of zero crossings of each EOD cycle, including phase shifts of the EOD due to rectances, whereas P units code for the amplitude changes induced in the EOD by the resistances of objects or by the summation of the EOD with that from a conspecific (see Chap. 33).

It is not certain whether these two physiological types of receptors correspond with morphologically different types. As mentioned above, it is known that the low- and mid-frequency wave fish *Sternopygus* and *Eigenmannia,* both of which have P and T cells, also have two morphologically distinct classes of receptors based on the number of receptor organs contacted by a single afferent (Zakon 1984a, 1987). It is thought that units correspond to the afferent type with thick

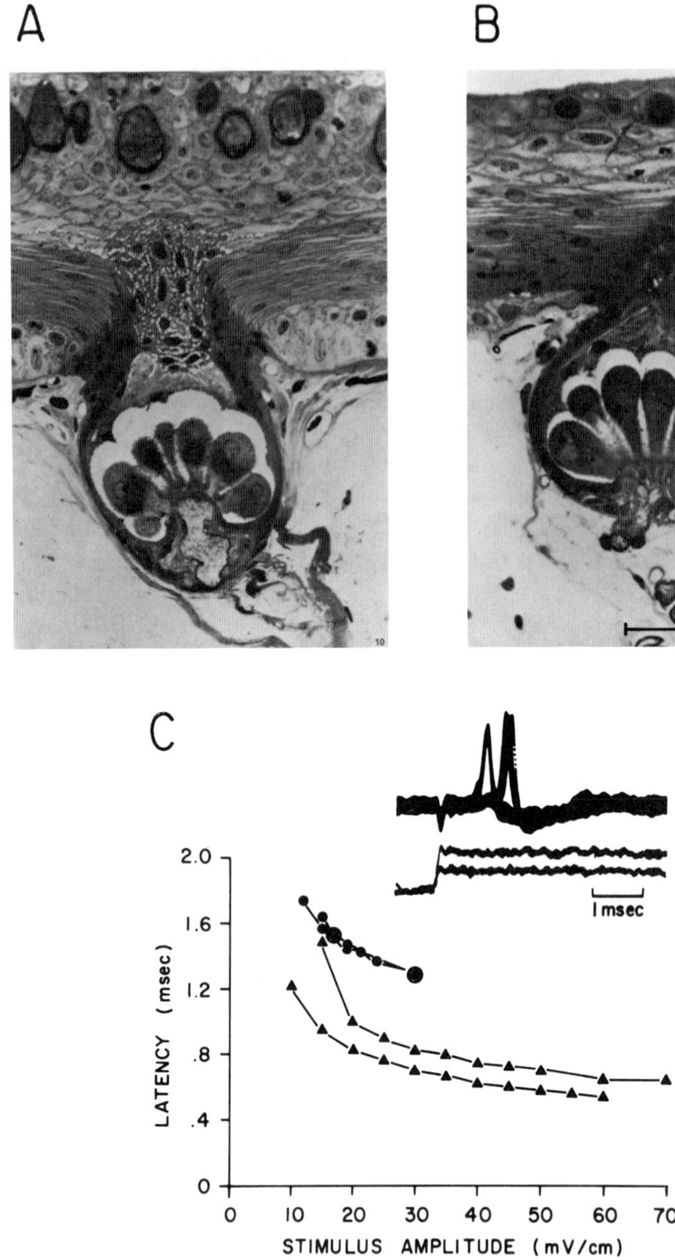


Figure 32.9. Pulse marker (M unit) and burst duration coder (B unit) receptor types from the gymnotiform *Hypopomus*. Pulse marker (A) has large axon terminal (Szamier and Wachtel 1970) and fires one spike per EOD pulse (C) (Bastian 1976). Burst duration coder, which has small axon terminal (B), fires multiple spikes to each EOD pulse in the absence of objects (D).

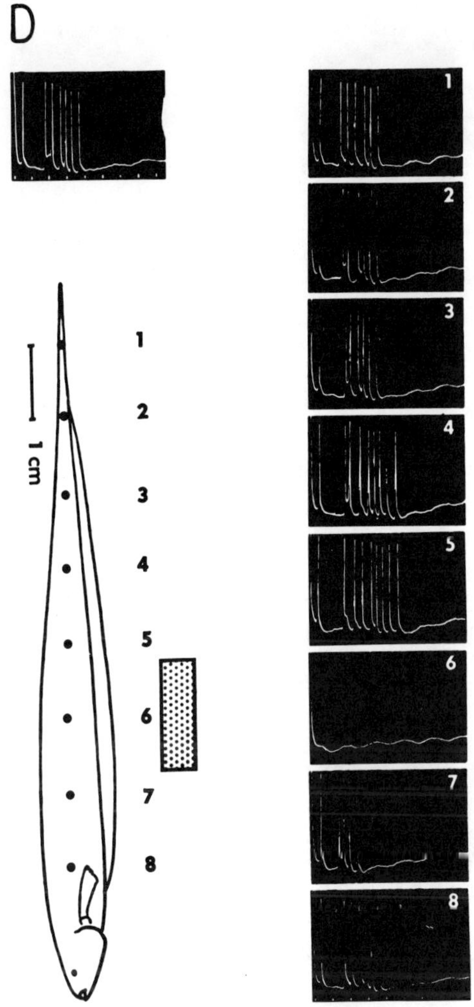


Figure 32.9 (*continued*). The number of spikes (D) varies as a function of the position of conductive object near its receptive field. (Silver platc positioned at locations *1–8* along body, corresponding to numbered neural responses given on the *right*) (Hagiwara, Kusano, and Negishi 1962).

terminals that innervates a single receptor organ. The T type, on the other hand, corresponds to those units which innervate multiple capsules through thinner branches (Sanchez and Zakon 1986).

## Encoding in Mormyrids

Mormyrids also possess rapid-timing and amplitude-modulated types of receptors. The Knollenorgan, used exclusively for social communication, fires a single

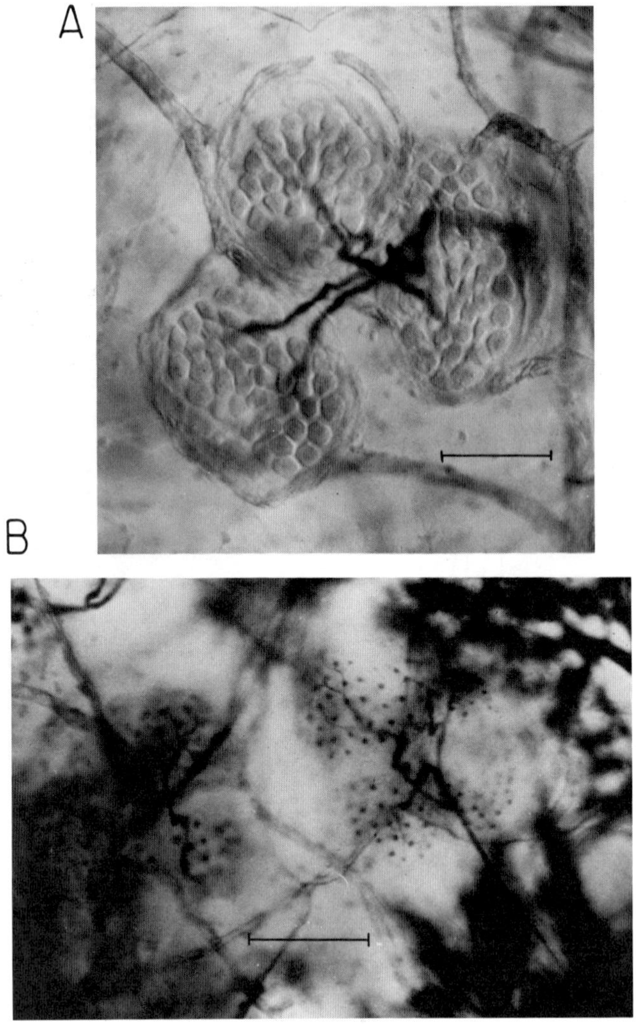


Figure 32.10. Tuberous receptors of wave-type gymnotiforms. (A) Three receptor organs innervated by single axon; silver stain visualized by Nomarski optics *(Eigenmannia)*. Calibration = 20 μm. (B) Silver-stained axon terminals of a receptor from *Sternopygus*. Calibration bar = 40 μm (Zakon 1984a).

sharply time-locked spike to each EOD pulse (Bennett 1967, 1971) (Fig. 32.11C). A recent study of the KI Knollenorgan of *Brienomyrus brachyistius biphasic* (Hopkins and Bass 1981) highlights the importance of timing of the rise-fall characteristics of the EOD for exciting these receptors; only synthesized EODs with positive-negative transitions like their own species-specific EOD are capable of reliably eliciting a time-locked spike. EODs that contain a phase-shifted version of the same power spectrum are poor at evoking a time-locked spike from the receptor or a behavioral response from a conspecific.

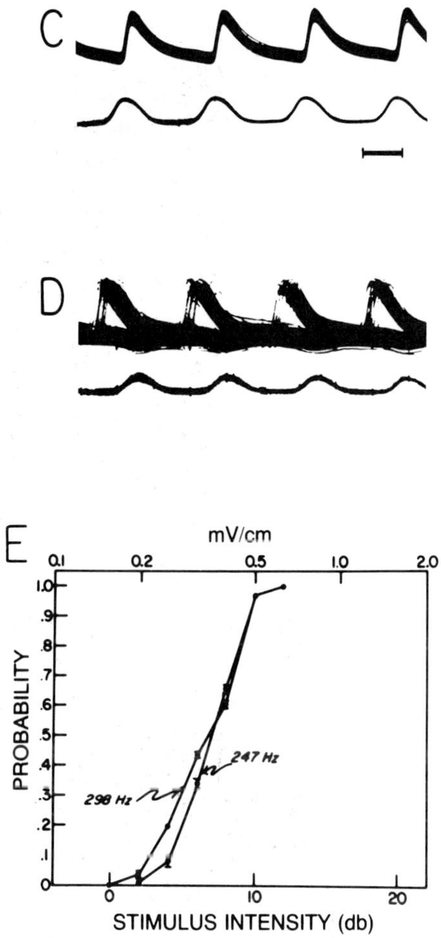


Figure 32.10. (*continued*)
(C and D) Multiple superimposed oscilloscope traces of T-type (C) and P-type (D) units responding to natural EOD (*Eigenmannia*). T units fire on every cycle with little jitter; P units often miss and fire with less precision. Calibration bar = 1 ms (Hagiwara and Morita 1963). (E) The probability of T unit's firing as a function of stimulus intensity. At stimuli greater than 0.5 mV /cm, this receptor fires 1 : 1. P receptor would have similar spike rate curve, though it would be shifted to the right so that it would still be unsaturated in the working range of the EOD (0.5–2.0 mV/cm in this species) (Hopkins 1976).

Because Knollenorgans are broadly tuned and different classes (KI, KII) may be sensitive to different portions of the EOD spectrum, Hopkins and Bass (1981) have suggested that conspecific EODs will produce a unique temporal interval between the firing times of these two classes on one side of the body. Additionally, KI receptors on opposite sides of the body will fire in a specified interval during the particular pattern of inward and outward current flow generated by an EOD. Species recognition begins at the receptors, where only the conspecific EOD will trigger the correct pattern of spikes across two classes of receptors on opposite sides of the body. The CNS may then compare the timing of Knollenorgan firings to differentiate conspecific from heterospecific EODs (Hopkins and Bass 1981).

Mormyromasts, which are facilitated centrally during the fish's own EOD (Zipser and Bennett 1976), are used in electrolocation. They have a steep spike rate curve, with a small dynamic range of about 6 dB, so that small changes in the amplitude of the EOD, induced by nearby objects of high or low conductivity, markedly enhance or decrease the number of spikes (4–7 spikes maximum) pro-

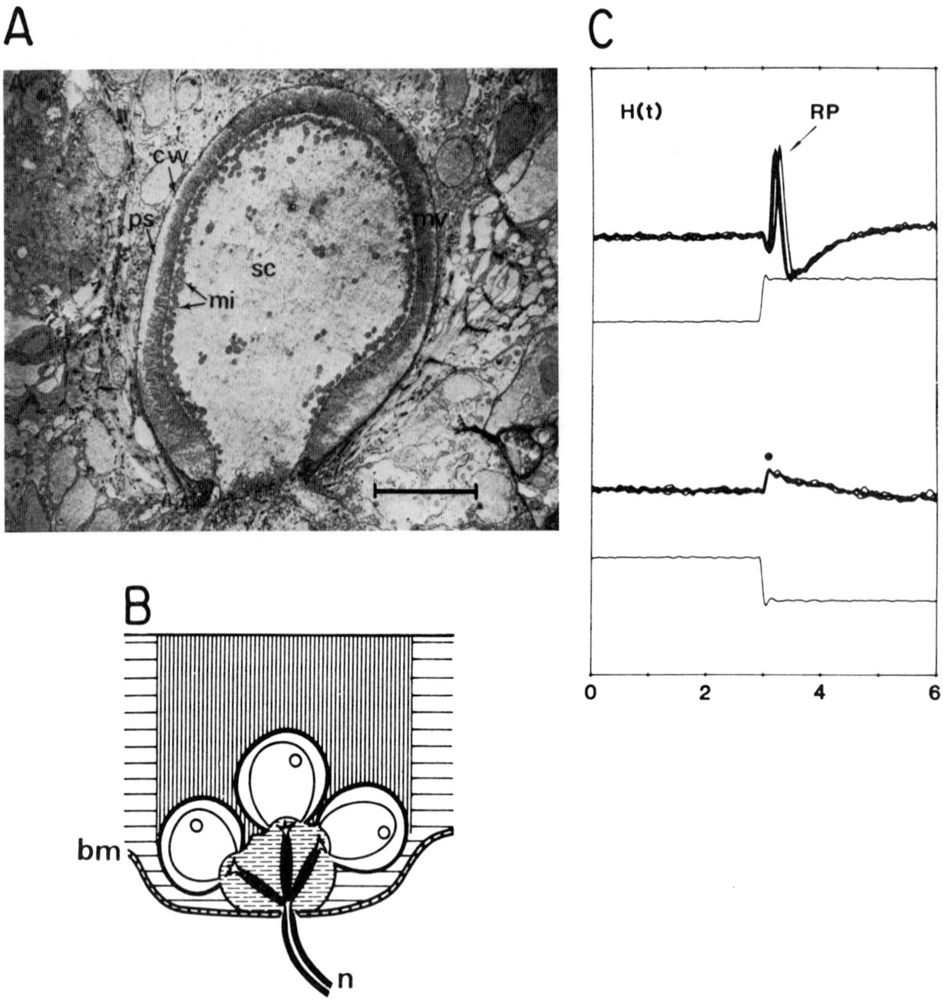


Figure 32.11. Knollenorgans from mormyrids. (A) Cross section of single sensory receptor cell (*sc*) illustrating the band of mitochondria (*mi*) directly below microvillous apical surface, capsule wall (*cw*), and perisensory space (*ps*) around receptor cell. Calibration = 20 μm. Each receptor cell is housed in its own capsule (B) at the base of common epidermal cell plug (vertically hatched lines); *bm*, basement membrane; *n*, nerve (Szabo 1974). (C) Receptor spike (*RP*) elicited from a Knollenorgan by positive-going step pulse, *H*(*t*) (*top trace*). Negative-going pulse produces no spike (*bottom trace*) (Hopkins and Bass 1981).

duced by each pulse (Szabo and Hagiwara 1967). In addition to an increasing number of spikes with increasing stimulus amplitude, the latency of the first spike in the volley from a mormyromast reliably and dramatically lessens. Both latency and spike number may also be influenced by changes in EOD phase induced by capacitative shunts (Meyer 1982).

It is known that mormyrids detect complex impedances (Meyer 1982). This information must be encoded solely with mormyromasts, as their Knollenorgans are "tied up" with other functions. Meyer (1982) suggests that, since both latency and spike number may be changed with shifts in EOD phase and amplitude, differential changes across an array of differently tuned receptors might serve as a basis for discrimination of frequency-independent resistance from frequency-dependent capacitative reactance. One possibility is that the two receptor types within a mormyromast participate differentially in the encoding of EOD amplitude and phase. Thus, by central comparisons of the outputs of mormyromasts distributed over the body surface, including perhaps, comparisons between those within the same mormyromast, reactive and resistive parameters of objects may be extracted.

## Summary

All fish that possess electroreceptors have ampullary-type receptors. Ampullary receptors, which lie at the base of a long (especially in marine forms) jelly-filled canal detect DC and low-frequency electric fields. The afferents innervating them are spontaneously and regularly discharging, and their ongoing frequency of firing is modulated by low-frequency electrical stimuli. These receptors, which are used for locating prey and for orientation, may be exquisitely sensitive in marine fishes. In sharks, for example, behavioral thresholds may be as low as 5 nV/cm. The high sensitivity of these receptors is all the more remarkable in that they must work against a high background noise level of the electric potentials associated with self-generated breathing and movement. In elasmobranchs, at least, the signal-noise ratio is greatly enhanced by a common mode rejection system in the medulla.

Tuberous receptors, in contrast, are found only in those fish which also possess a weakly discharging electric organ (gymnotiforms and mormyriforms) and are used for electrolocation and communication. In general, tuberous receptors may be divided into a type that responds with a single spike precisely time locked to each EOD pulse or cycle or a type that is poorly time locked, but responds over, and codes for, a range of stimulus intensities. Thus, impedances of surrounding objects, as well as the EODs of nearby conspecific fish, may be differentially encoded by the phase and amplitude distortions that they impose on the fish's own EOD and are encoded by each receptor type.

Tuberous receptors are typically tuned to the frequency components of the species-specific discharge. In species with sexual dimorphisms of the EOD power spectrum, receptors are tuned differently in the two sexes such that the receptors of each sex are best matched to its own EOD power spectrum.

### References

Bass, A.H. (1986) Electric organs revisited: evolution of a vertebrate communication and orientation organ, in Electroreception, Bullock, T.H., and Heiligenberg, W. (eds.), John Wiley and Sons, New York, pp. 13–70.

Bass, A.H., and Hopkins, C.D. (1980) Coding of species-specific signals in mormyrid electric fish. I. Frequency characteristics. Soc. Neurosci. Abst., 6:604.

Bass, A.H., and Hopkins, C.D. (1983) Hormonal control of sexual differentiation: changes in electric organ discharge waveform, Science, 220:971–974.

Bass, A.H., and Hopkins, C.D. (1984) Shifts in frequency tuning of electroreceptors in androgen-treated mormyrid fish, J. Comp. Physiol., 155:713–724.

Bastian, J. (1976) Frequency response characteristics of electroreceptors in weakly electric fish (Gymnotoidei) with a pulse discharge, J. Comp. Physiol., 112:165–180.

Bastian, J. (1977) Variations in the frequency response of electroreceptors dependent on receptor location in weakly electric fish (Gymnotoidei) with a pulse discharge, J. Comp. Physiol., 121:53–64.

Bastian, J. (1981) Electrolocation. I. How the electroreceptors of *Apteronotus albifrons* code for moving objects and other electrical stimuli, J. Comp. Physiol., 144:465–479.

Bell, C.C. (1982) Properties of a modifiable efference copy in an electric fish, J. Neurophysiol., 47:1043–1056.

Bell, C.C., Bradbury, J., and Russell, C.J. (1976) The electric organ of a mormyrid as a current and voltage source, J. Comp. Physiol., 110:65–88.

Bell, C.C., and Russell, C.J. (1978) Effect of electric organ discharge on ampullary receptors in a mormyrid, Brain Res., 145:85–96.

Bennett, M.V.L. (1967) Mechanisms of electroreception, in Lateral Line Detectors, Cahn, P. (ed.), Indiana University Press, Bloomington, pp. 313–393.

Bennett, M.V.L. (1971) Elecroreception in Fish Physiology, vol. 5, Hoar, W.S., and Randall, D.S. (eds.), Academic Press, New York, pp. 493–574.

Bennett, M.V.L., and Obara, S. (1986) Ionic and pharmacological mechanisms of electroreception, in Electroreception, Bullock, T.H., and Heiligenberg, W. (eds.), John Wiley and Sons, New York, pp. 157–181.

Bodznick, D.A., and Northcutt, R.G. (1981) Electroreception in lampreys: evidence that the earliest vertebrates were electroreceptive, Science, 212:465–467.

Bodznick, D.A., and Preston, D.G. (1983) Physiological characterization of electroreceptors in the lampreys *Icthyomyzon unicuspis* and *Petromyzon marinus,* J. Comp. Physiol., 152:209–217.

Bretschneider, F., Kroesbergen, G., and Beijnink, F.B. (1979) Functioning of catfish electroreceptors: relation between skin potential and receptor activity, J. Physiol. (Paris), 75:343–347.

Bullock, T.H., Behrend, K., and Heiligenberg, W. (1975) Comparison of the jamming avoidance response in gymnotoid and gymnarchid electric fish: a case of convergent evolution of behavior and its sensory basis, J. Comp. Physiol., 103:97–121.

Bullock, T.H., Bodznick, D.A., and Northcutt, R.G. (1983) The phylogenetic distribution of electroreception: evidence for convergent evolution of a primitive vertebrate sense modality, Brain Res. Rev., 6:25–46.

Clusin, W.T., and Bennett, M.V.L. (1977) Calcium activated conductance in skate electroreceptors: voltage clamp experiments, J. Gen. Physiol., 69:145–182.

Clusin, W.T., and Bennett, M.V.L. (1979) The oscillatory responses of skate electroreceptors to small voltage stimuli, J. Gen. Physiol., 73:685–702.

Derbin, C., Szabo, T. (1968) Ultrastructure of an electroreceptor (knollenorgan) in the mormyrid fish, *Gnathonemus petersii,* Int. J. Ultrastruct. Res., 22:469–484.

Dijkgraaf, S., and Kalmijn, A.J. (1966) Versuche zur biologischen Bedeutung der Lorenzinischen Ampullen bei den Elasmobranchiern, Z. Vgl. Physiol., 53:187–194.

Dunning, B.B. (1973) A quantitative and comparative analysis of the tonic electroreceptors of Gnathonemus, Gymnotus and Kryptopterus. Thesis, U. Minnesota, Minneapolis.

Dye, J., and Meyer, J.H. (1986) The central nervous control of the electric organ discharge in weakly electric fish, in Electroreception, Bullock T.H., Heilingenberg W. (eds.), John Wiley, New York.

Feng, A.S., and Bullock, T.H. (1977) Neuronal mechanisms for object discrimination in the weakly electric fish, Eigenmannia virescens, J. Exp. Biol., 66:141–158.

Flock, Å., and Russell, I.J. (1973) The postsynaptic action of efferent fibers in the lateral line organ of the burbot *Lota lota,* J. Physiol. 235:591–605.

Fritsch, B., and Münz, H. (1985) Amphibian electroreceptors, in Electroreception. Bullock T.H., Heiligenberg, W. (eds.), John Wiley, New York.

Hagedorn, M., and Heiligenberg, W. (1985) Court and spark: electric signals in the courtship and mating of gymnotoid fish, Anim, Behav. 33:254–265.

Hagiwara, S., and Morita, H. (1963) Coding mechanisms of electroreceptors fibers in some electric fish, J. Neurophysiol. 26:551–567.

Hagiwara, S., Kusano, K., and Negishi, K. (1962) Physiological properties of electroreceptors of some gymnotids, J. Neurophysiol., 25:430–449.

Heiligenberg, W. (1974) Electrolocation and jamming avoidance in a Hypopygus (Rhamphthychthyidae, Gymnotoidei), an electric fish with pulse-type discharges, J. Comp. Physiol. 91:233–240.

Heiligenberg, W. (1977) Principles of Electrolocation and Jamming Avoidance in Electric Fish, Studies of Brain Function I, Springer-Verlag, Berlin.

Heiligenberg, W., and Altes, R.A. (1978) Phase sensitivity in electroreception, Science, 199:1001–1004.

Heiligenberg, W., and Bastian, J. (1984) The electric sense of weakly electric fish, Annu. Rev. Physiol., 46:561–583.

Himstedt, W., Kopp, J., and Schmidt, W. (1982) Electroreception guides feeding behavior in amphibians, Naturwissenschaften, 69:552.

Hopkins, C.D. (1972) Sex differences in signalling in an electric fish, Science 176:1035–1037.

Hopkins, C.D. (1973) Lightening as background noise for communication among electric fish, Nature, 242:268–270.

Hopkins, C.D. (1974) Electric communication: functions in the social behavior of *Eigenmannia virescens,* Behaviour, 50:270–305.

Hopkins, C.D. (1976) Stimulus filtering and electroreception: tuberous electroreceptors in three species of gymnotoid fish, J. Comp. Physiol., 111:–171–207.

Hopkins, C.D. (1981) On the diversity of electric signals in a community of mormyrid electric fish in West Africa, Am. Zool., 21:211–222.

Hopkins, C.D. (1984) Functions and mechanisms in elecroreception, in Fish Neurobiology vol. 1, Brain Stem and Sense Organs, Northcutt, R.G., and Davis, R.E. (eds.), University of Michigan Press, Ann Arbor, pp. 215–259.

Hopkins, C.D., and Bass, A.H. (1980) Coding of species-specific signals in mormyrid electric fish. II. Temporal characteristics, Soc. Neurosci. Abst., 6:604.

Hopkins, C.D., and Bass, A.H. (1981) Temporal coding of species recognition signals in an electric fish, Science, 212:85–87.

Hopkins, C.D., and Heiligenberg, W.H. (1978) Evolutionary designs for electric signals and electroreceptors in gymnotoid fishes of Surinam, Behav. Ecol. Sociobiol., 3:113–134.

Hudspeth, A.J., and Corey, D.P. (1977) Sensitivity, polarity, and conductance change in the response of vertebrate hair cells to controlled mechanical stimuli, Proc. Nat. Acad. Sci. U.S.A., 74:2407–2411.

Jørgensen, J.M. (1982) Fine structure of the ampullary organs of the bichir *Polypterus senegalus* Cuvier, 1829 (Pisces: Brachiopterygii) with some notes on the phylogenetic development of electroreceptors, Acta Zool., 63:211–217.

Kalmijn, A. (1974) The detection of electric fields from inanimate and animate sources other than electric organs, in Handbook of Sensory Physiology, vol. III/3, Fessard, A. (ed.), Springer-Verlag, New York, pp. 147–200.

Kalmijn, A.J. (1984) Theory of electromagnetic orientation: a further analysis, in Comparative Physiology of Sensory Systems, Bolis, L. Keyenes, R.D., and Maddrell, S.H.P. (eds.), Cambridge University Press, Cambridge, U.K., pp. 525–560.

Keller, C., Zakon, H.H., and Sanchez, D.Y. (1986) Evidence for a direct effect of androgens on electroreceptor tuning, J. Comp. Physiol. 158:301–310.

Knudsen, E.I. (1975) Spatial aspects of the electric fields generated by weakly electric fish, J. Comp. Physiol., 99:103–118.

Lissmann, H.W., and Mullinger, A.M. (1968) Organization of ampullary electric receptors in Gymnotidae (Pisces), Proc. R. Soc., 169:345–378.

Lorenzini, S. (1678) Osservazioni intorno alle torpedini, Florence, Italy.

Meyer, J.H. (1982) Behavioral responses of weakly electric fish to complex impedances, J. Comp. Physiol., 145:459–470.

Meyer, J.H. (1983) Steroid influences upon the discharge frequency of a weakly electric fish, J. Comp. Physiol., 153:29–38.

Meyer, J.H., and Zakon, H.H. (1982) Androgens alter the tuning of electroreceptors, Science, 217: 635–637.

Meyer, J.H., Zakon, H.H., and Heiligenberg, W. (1984) Steroid influences upon the electrosensory system of weakly electric fish: direct effects upon discharge frequency with indirect effects upon electroreceptor tuning, J. Comp. Physiol., 154:625–631.

Montgomery, J. (1984a) Noise cancellation in the electrosensory system of the thornback ray: common mode rejection of input produced by the animal's own ventilatory movement, J. Comp. Physiol., 155:102–112.

Montgomery, J. (1984b) Frequency response characteristics of primary and secondary neurons in the electrosensory system of the thornback ray, Comp. Biochem. Physiol., 79a:189–195.

Murray, R.W. (1974) The ampullae of Lorenzini, in Handbook of Sensory Physiology vol. III/3, Fessard, A. (ed.), Springer-Verlag, New York, pp. 125–146.

Murray, R.W., and Potts, T.W. (1961) The composition of the endolymph, and other body fluids of elasmobranchs, Comp. Biochem. Physiol., 2:65–75.

Nelson, J.S. (1983) Fishes of the World, John Wiley and Sons, New York.

Northcutt, R.G. (1980) Anatomical evidence of electroreception in the coelacanth (*Latimeria chalumnae*), Zentralbl. Veterinaer Med. Reihe C, 9:289–295.

Northcutt, R.G. (1986) Electroreception in non-teleost bony fishes, in Electroreception, Bullock, T.H., and Heiligenberg, W. (eds.), John Wiley and Sons, New York. pp. 257–285.

Northcutt, R.G., and Gans, C. (1983) The genesis of neural crest and epidermal placodes: a reinterpretation of vertebrate origins, Q. Rev. Biol., 58:1–28.

Obara, S. (1976) Mechanisms of electroreception in ampullae of Lorenzini of the marine catfish *Plotosus,* in Electrobiology of Nerve, Synapse and Muscle, Reuben, J.P., Purpura, D.P., Bennett, M.V.L., and Kandel, E.R. (ed.), Raven Press, New York, pp.128–147.

Obara, S. (1981) High sensitivity processes in the sensory transduction of the *Plotosus* electroreceptors, in Advances in Physiological Sciences, vol. 31, Sensory Physiology of Aquatic Lower Vertebrates, Szabo, T., and Czeh, G. (eds.), Akadamiai Kaido, Budapest, pp. 41–56.

Okitsu, S., Umekita, S., and Obara, S. (1978) Ionic compositions of the media across the sensory epithelium in the ampullae of Lorenzini of the marine catfish, *Plotosus*, J. Comp. Physiol., 126:115–121.

Peters, R.C., and Buwalda, R.J.A. (1972) Frequency response of the electroreceptors of the catfish, *Ictalurus nebulosus* LeS., J. Comp. Physiol., 79:29–38.

Quinet, P. (1971) Étude systématique des organes sensoriels de la peau des Mormyriformes, Ann. Mus. R. Afr. Cent. Ser. Quarto Zool., 190:1–97.

Ronan, M., and Bodznick, D. (1984) Identification of electroreceptors in lamprey, Soc. Neurosci. Abst., 10:853.

Russell, I.J., and Roberts, B.L. (1974) Active reduction of lateral-line sensitivity in swimming dogfish, J. Comp. Physiol., 94:7–15.

Sanchez, D.Y., and Zakon, H.H. (1986) Neural correlates of cell addition in the electrosensory system. Soc. Neurosci. Abstr. 12:313.

Scheich, H. (1977) Neural basis of communication in the high frequency electric fish, *Eigenmannia virescens* (jamming avoidance response). III. Central integration in the sensory pathway and control of the pacemaker, J. Comp. Physiol., 113:229–255.

Scheich, H., Bullock, T.H., and Hamstra, R.H. (1973) Coding properties of two classes of afferent nerve fibers: high frequency electroreceptors in the electric fish, *Eigenmannia*, J. Neurophysiol., 36:39–60.

Schweitzer, J. (1985) Functional organization of the electroreceptive midbrain in an elasmobranch (*Platyrhinoidis triseriata*): a single unit study, J. Comp. Physiol. (in press).

Sejnowski, T.J., and Yodlowski, M.L. (1982) A freeze-fracture study of the skate electroreceptor, J. Neurocytol., 11:897–912.

Sellick, P., and Johnstone, B.M. (1975) Production and role of inner ear fluid, Prog. Neurobiol., 5:337–362.

Szabo, T. (1965) Sense organs of the lateral line system in some electric fish of the gymnotidae, mormyridae and gymnarchidae, J. Morphol., 117:229–250.

Szabo, T. (1970) Morphologische and Funktionelle Aspekte bei Elektrorezeptoren, Verh. Dtsch. Zool. Ges., 64:141–148.

Szabo, T. (1974) Anatomy of the specialized lateral line organs of electroreception, in Handbook of Sensory Physiology., vol. III/3, Fessard. A. (ed.), Springer-Verlag, New York, pp. 13–58.

Szabo, T., and Hagiwara, S. (1967) A latency-change mechanism involved in sensory coding of electric fish (Mormyrids), Physiol. & Behav., 2:331–335.

Szabo, T., Kalmijn, A.J., Enger, P.S., and Bullock, T.H. (1972) Microampullary organs and a submandibular sense organ in the fresh water ray, *Potamotrygon*, J. Comp. Physiol., 79:15–27.

Szabo, T., and Wersäll, J. (1970) Ultrastructure of an electroreceptor (Mormyromast) in a mormyrid fish, *Gnathonemus petersi* II, J. Ultrastruct. Res., 30:473–490.

Szamier, R.B., and Bennett, M.V.L. (1980) Ampullary electroreceptors in the fresh water ray, *Potamotrygon*, J. Comp. Physiol., 138:225–230.

Szamier, R., and Wachtel, A.W. (1970) Special cutaneous receptor organs of fish. VI. Ampullary and tuberous organs of *Hypopomus*, J. Ultrastruct. Res., 30:450–471.

Teeter, J.H., and Bennett, M.V.L. (1981) Synaptic transmission in the ampullary electroreceptor of the transparent catfish, *Kryptopterus*, J. Comp. Physiol., 142:371–377.

Teeter, J.H., Szamier, R.B., and Bennett, M.V.L. (1981) Ampullary electroreceptors in the sturgeon *Scaphyrynchus platorhynchus* (Rafinesque), J. Comp. Physiol., 138:213–223.

Thomson, K.S. (1977) On the individual history of cosmine and a possible electroreceptive function of the pore-canal system in fossil fishes, in Problems in Vertebrate Evolution, Andrews, S.M., Miles, R.S., and Walker, A.D. (ed.), Academic Press, New York, pp. 247–272.

Viancour, T.A. (1979) Electroreceptors of a weakly electric fish. II. Individually tuned receptor oscillations, J. Comp. Physiol., 133:327–338.
Wachtel, A.W., and Szamier, R.B. (1966) Special cutaneous receptor organs of fish: the tuberous organs of *Eigenmannia,* J. Morphol., 119:51–80.
Wachtel, A.W., and Szamier, R.B. (1969) Special cutaneous receptor organs of fish. IV. Ampullary organs of the nonelectric catfish, *Kryptopterus,* J. Morphol., 128:291–202.
Waltman, B. (1966) Electrical properties and fine structure of the ampullary canals of Lorenzini, Acta. Physiol. Scand., 66 (supp. 264):1–60.
Watson, D., and Bastian, J. (1979) Frequency response characteristics of electroreceptors in the weakly electric fish *Gymnotus carapo,* J. Comp. Physiol., 134:191–202.
Westby, G.W.M., and Kirschbaum, F. (1978) Emergence and development of the electric organ discharge in the mormyrid fish, *Pollimyrus isidori.* II. Replacement of the larval by the adult discharge, J. Comp. Physiol., 127:45–59.
Whitear, M., and Lane, E.B. (1983) Multivillous cells: epidermal sense cells of unknown function in lamprey skin, J. Zool. (Lond.), 201:259–272.
Zakon, H.H. (1984a) Postembryonic changes in the peripheral electrosensory system of a weakly electric fish: addition of receptor organs with age, J. Comp. Neurol., 228:557–570.
Zakon, H.H. (1984b) The ionic basis of the oscillatory receptor potential of tuberous electroreceptors in *Sternopygus,* Soc. Neurosci. Abst., 10:193.
Zakon, H.H. (1986) Emergence of tuning in newly-generated electroreceptors, J. Neurosci. (submitted).
Zakon, H.H. (1986) The electroreceptive periphery, in Electroreception, Bullock, T.H., and Heiligenberg, W. (eds.), John Wiley and Sons, New York. pp. 103–156.
Zakon H.H. (in press) Variation in the mode of receptor cell addition in the electrosensory system of gymnotiform fishes. J. Comp. Neurol.
Zakon, H.H., and Meyer, J.H. (1983) Plasticity of electroreceptor tuning in the weakly electric fish *Sternopygus dariensis,* J. Comp. Physiol., 153:477–487.
Zipser, B., and Bennett, M.V.L. (1973) Tetrodotoxin resistant electrically exitable responses of receptor cells, Brain Res., 62:253–259.
Zipser, B., and Bennett, M.V.L. (1976) Interaction of electrosensory and electromotor signals in the lateral line lobe of a mormyrid fish, J. Neurophysiol., 39:713–721.

# Chapter 33

# The Neuronal Basis of Electrosensory Perception and Its Control of a Behavioral Response in a Weakly Electric Fish

WALTER HEILIGENBERG*

Although electroreception is a sensory modality alien to us and, therefore, not easily accessible to our intuition, it has become a favorite system for the exploration of a wide spectrum of issues in sensory physiology and perception, involving disciplines as diverse as receptor biophysics and neuroethology. A number of reasons may be responsible for this development. First, electric stimuli can easily be measured and controlled experimentally. Second, electroreceptors are distributed over the body surface and are thus individually accessible. Third, the central nervous anatomy of some electrosensory systems has been explored in great detail and appears to be more transparent than the neuronal substrate of more commonly studied sensory systems, such as audition and vision in mammals. Fourth, a number of rather simple behavioral responses controlled by electroreception are sufficiently robust to function in neurophysiological preparations. This allows for the exploration of individual neurons under simultaneous monitoring of an intact behavioral response. It thus appears that the electric sense offers a suitable model system for the study of principles in the neuronal organization of perception and its control of adaptive behavioral responses, both central issues of neuroethology.

## Neuroethological Research Strategies

The study of any processor of complex information requires that one know the nature of the patterns of information that the system is designed to handle. Without this knowledge, one is likely to be sidetracked by the exploration of epiphenomena and may fail to see the relevant functional properties of the system altogether. The most complex information processors known are perceptual mechanisms in

* Neurobiology Unit, Scripps Institution of Oceanography, UCSD, La Jolla, CA, 92093, USA.

animals, and any study of the neuronal basis of perception is doomed unless one first explores the natural set of stimulus patterns for which this neuronal substrate has evolved. The presentation of pure tones or white-noise signals, for example, may be a suitable tool for the study of the biophysics of auditory receptors. But the use of such standard stimuli will reveal very little about the central processing of auditory information in, for example, the brain of a bat, which is designed to evaluate spectral and temporal patterns so exceedingly complex and specific (Suga 1982) as to be unimaginable to a physiologist unfamiliar with echolocation.

The study of an animal's ethology and its ecological adaptations is therefore a necessary prerequisite for neurophysiological explorations of its sensory physiology and perception. Of particular value is the comparison of related species that differ in behavioral and ecological specializations, since it aids in the identification and interpretation of neuronal specializations that have evolved in parallel and thus are adapted to specific functions. Following the example of early ethologists, one should first identify biologically relevant stimulus patterns and then proceed to simplify and to modify such patterns to determine those physical attributes which the animal evaluates in a particular behavioral context. One may learn, in this manner, that the animal disregards certain stimulus attributes, though it has the sensory equipment to perceive them, and yet it may well respond to these attributes in a different behavioral context.

While identifying relevant features in a given stimulus pattern, one may also begin to speculate about logical rules that govern the central nervous evaluation of this pattern. Insofar as a theory predicts behavioral responses to stimulus patterns that differ in their composition from their natural models in specific ways, one can test this theory by behavioral experimentation. To the extent that one is then able to "fool" animals in a predictable manner by unusual forms of stimulus patterns, confidence is gained in such theories of perception.

After the relevant stimulus attributes and abstract computational rules of their central nervous evaluation have been identified, specific neuronal implementations of stimulus coding and computation can be postulated. Starting out at the receptor level, one advances to higher-order neurons, testing their responses to stimulus patterns for which the relevance has been demonstrated by prior behavioral experimentation. The combination of intra- and extracellular recording and labelling techniques will then lead deeper into central nervous mechanisms of sensory perception. This experimental strategy has guided our research on neuronal mechanisms of perception and its control of adaptive behavioral responses in electric fish. The focus of these studies has been the jamming avoidance response (JAR) in the weakly electric fish *Eigenmannia*. Because of its relative simplicity and robustness, this behavior is particularly suitable for neurophysiological approaches.

Since the experimental data discussed in this chapter have recently been reviewed in great detail (Heiligenberg 1986), only a summary of the results will be presented here and greater attention given to more general issues of these studies.

## Neural Mechanisms of Jamming Avoidance Response (JAR)

### The Study of Central Nervous Rules of Computation by Behavioral Experiments

The weakly electric fish *Eigenmannia* produces a continual, nearly sinusoidal electric organ discharge (EOD) at a rather stable individual frequency in the range of 250 to 700 Hz. In the presence of a signal of similar frequency from another source, the fish shifts its discharge frequency to increase the magnitude of the difference frequency (Df) (Watanabe and Takeda 1963; Bullock, Hamstra, and Scheich 1972). Since the fish promptly lowers its frequency for a positive Df and raises it for a negative Df, it must be able to discriminate the sign of Df. By silenceing the animal's own EOD and replacing it by an artificial sine wave, $S_1$, the experimenter gains total control over electroreceptive afferent information and can thus test the mechanism of Df sign discrimination (Scheich and Bullock 1974). A series of behavioral experiments has revealed that the animal can solve this problem without knowledge of its own pacemaker frequency and therefore purely on the basis of afferent information (Heiligenberg, Baker, and Matsubara 1978).

Theoretically the animal could discriminate the sign of Df by simultaneous evaluation of modulations in the instantaneous amplitude and frequency that characterize the signal resulting from the interference of its own EOD with that of its neighbor. Behavioral tests, however, show that the animal ignores frequency modulations in this context (Heiligenberg, Baker, and Matsubara 1978). This is surprising, since the animal has central neurons that are extremely sensitive to frequency modulations (Bastian and Heiligenberg 1980b), and it employs frequency modulations in the context of social communication during reproduction (Hopkins 1974; Hagedorn and Heiligenberg 1985; see also Chap. 9).

The only theory of Df-sign discrimination that has survived all behavioral tests and experimental challenges postulates that the animal obtains this information by simultaneous evaluation of the modulations in amplitude and timing of zero crossings, or phase, of the nearly sinusoidal signal that results from the addition of the interfering EOD, $S_2$, with the animal's own EOD, $S_1$ (Fig. 33.1) (see reviews by Heiligenberg 1980, 1983, 1986). Phase information is obtained by a comparison of the phase in areas of the body surface that experience different mixing ratios between the interfering EODs. If one records a local amplitude modulation on the ordinate and the simultaneous modulation of the difference in phase with reference to some other area of the body surface less strongly affected by $S_2$ on the abscissa, one obtains a circular graph that rotates at a rate equal to the magnitude of Df, and the sense of rotation is clockwise or counterclockwise for negative or positive Dfs, respectively (Fig. 33.1). If *Eigenmannia* is able to read the sense of rotation in this amplitude-phase plane it immediately knows in which direction it must shift its own frequency to increase the magnitude of Df. The discovery of a neuronal machinery designed and tuned for this form of

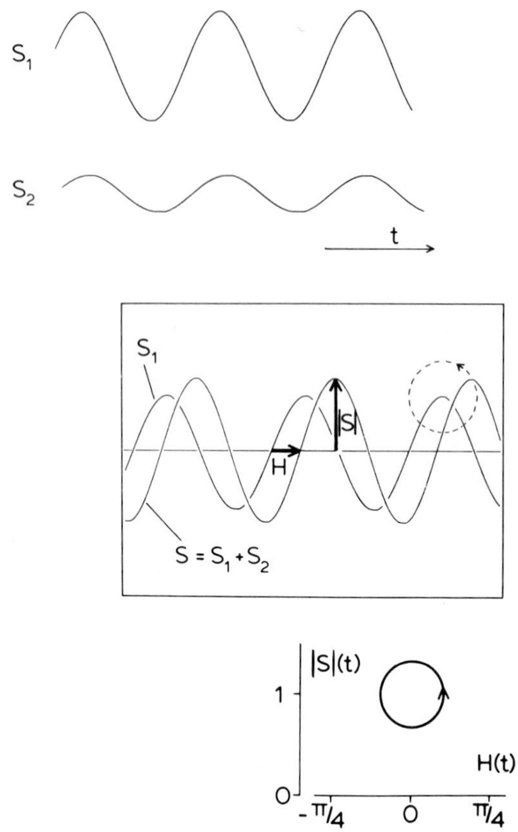


**a**

Figure 33.1a. The addition of two sine waves simulating EODs of two fish with slightly different frequencies. *Top:* Two sine waves ($S_1$ and $S_2$) plotted as functions of time ($t$). Stronger signal ($S_1$) represents animal's own EOD. $S_2$ is the EOD of a neighbor. *Center:* Oscilloscope display, triggered by $S_1$ so that $S_1$ appears stationary. Simultaneous display of added signals ($S_1 + S_2$) yields nearly sinusoidal signal whose momentary amplitude ($|S|$) and phase angle ($H$) are modulated at difference frequency ($Df = f_2 - f_1$). As a consequence of joint modulation of $|S|$ and $H$, the peak of $S_1 + S_2$ is seen to rotate around stationary peak of $S_1$, as indicated by *dashed circle. Bottom:* Lissajous figure display of $|S|$ vs. $H$ in two-dimensional state plane. Nearly circular graph rotates in counterclockwise sense for positive $Df$'s. Mean amplitude of $|S|$ is the amplitude of dominant signal ($S_1$) which is considered unity. Mean value of $H$ is 0, which corresponds to the timing of the positive zero crossing of $S_1$. The radius of the graph is equal to amplitude ratio $|S_2|/|S_1|$. Positive $H$ implies that the zero crossing of $S_1 + S_2$ lags with respect to that of $S_1$. The period of $S_1$ cycle, which is 2 ms if $f_1$ is 500 Hz, is 2 $\pi$. The width of the abscissa in *lower right* would then be 0.5 ms (from Heiligenberg 1980).

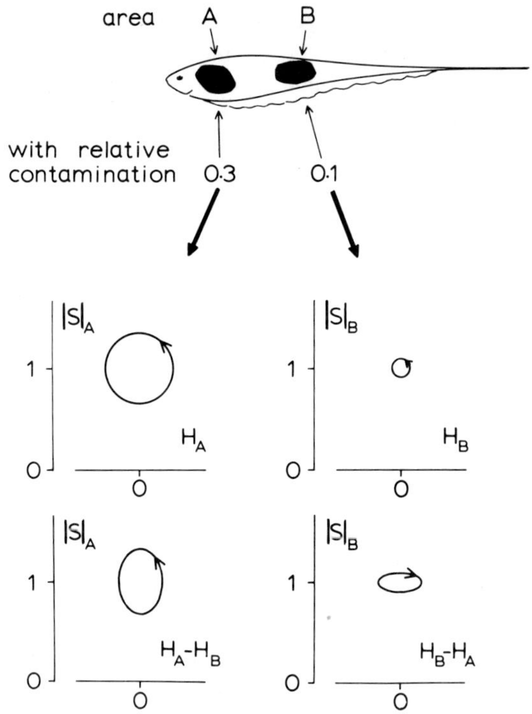


Figure 33.1b. Information about phase can be obtained by comparing areas of body surface with different amplitude ratios between interfering signals ($S_1$ and $S_2$). Let animal's EOD or its substitute ($S_1$) be contaminated by foreign signal ($S_2$) by 30% and 10% of the amplitude of $S_1$ in area $A$ and $B$, respectively. Stimulus modulation graphs $|S|_A$ vs. $H_A$ and $|S|_B$ vs. $H_B$ have radii equal to local amplitude ratios between $S_2$ and $S_1$ and rotate synchronously and in counterclockwise direction for positive Df's. By replacing local phase modulations ($H_A$ and $H_B$) by differential phase modulations $H_A - H_B$ and $H_B - H_A$, elliptical graphs with opposite rotations are obtained. Whereas counterclockwise rotation of the graph in area $A$ lowers pacemaker frequency, clockwise rotation in area $B$ raises pacemaker frequency. As can be shown experimentally, area $A$ wins in this competition because of its larger amplitude modulation. Note that the animal has no direct access to either $H_A$ or $H_B$, but that it can evaluate their difference by comparing timings of spike arrivals from T-type receptors from these two parts of body surface (from Heiligenberg 1980).

stimulus evaluation convinces us that the animal determines the sign of Df exactly in this manner.

## The Exploration of Neuronal Implementations

Electrosensory information relevant for the JAR is coded by two types of receptors distributed over the animal's body surface (Scheich, Bullock, and Hamstra 1973).

T units fire one spike per EOD cycle, phase locked to the timing of the zero crossing of the stimulus, and thus code phase by the timing of their spikes. P units fire irregularly, increasing their discharge rate with stimulus amplitude, and thus code amplitude by the rate of their firing. P units and T units show some overlap in their response characteristic, and their distinction has even been doubted (Viancour 1979). When driven at natural stimulus intensities, however, these two types can normally be distinguished safely on the basis of their firing patterns (Bastian and Heiligenberg 1980a; Heiligenberg and Partridge 1981). These two receptor types project via primary afferents to different targets of the electrosensory lateral line lobe (ELL) (Maler, Sas, and Rogers 1981; Mathieson, Heiligenberg, and Maler 1987) which contains three somatotopically ordered laminated maps with identical primary afferent input (Heiligenberg and Dye 1982; Carr, Maler, and Sas 1982; see also Chap. 29). A fourth map of the ELL receives input exclusively from ampullary electroreceptors, which code low-frequency signals not relevant for the JAR.

Within the ELL, amplitude information is further processed by basilar and nonbasilar pyramidal cells that receive direct, as well as indirect, input from P-type receptors (Maler, Sas, and Rogers 1981; Saunders and Bastian 1984). Basilar pyramidal cells, or E neurons (Bastian and Heiligenberg 1980a), are excited by a rise in stimulus amplitude, whereas nonbasilar pyramidal cells, or I neurons, are excited by a fall in stimulus amplitude. These two types of pyramidal cells are, therefore, designed to detect the direction of vertical movements in the amplitude-phase plane, critical information for the control of the JAR. Both types of pyramidal cells project to various laminae of the somatotopically organized torus semicircularis, pars dorsalis, of the midbrain (Carr et al. 1981). Phase information is relayed via the spherical cells of the ELL to lamina 6 of the torus, and here the difference in phase between areas of the body surface is computed (Fig. 33.2). Small cells in lamina 6 are driven by the differential phase between specific parts of the body surface, being excited either by a phase lead or a phase lag (Heiligenberg and Rose 1985a). The underlying circuitry has been explored ultrastructurally by Carr, Taylor, and Maler (1986) (see Fig. 33.3).

The information conveyed by the pyramidal cells of the ELL and by the small cells of lamina 6 of the torus suffices for the control of the JAR. Logical "and" operations between these forms of information would allow the animal to determine whether a local rise or fall in stimulus amplitude coincides with a lead or lag in differential phase, and the sense of rotation of graphs in the amplitude-phase plane could thus be determined. To our surprise, however, we have labelled a total of only three single cells in the torus that appear to compute amplitude and phase information in this manner. Instead we have found a multitude of cell types that process information in forms not obviously suitable for the control of the JAR but potentially useful for the analysis of electric images of moving objects (Heiligenberg and Rose 1985a).

Further computations of electrosensory information occur in the tectum opticum, which receives a massive topographic projection from the torus (Carr et al. 1981). The tectum receives input from cells that code either differential phase or local amplitude modulations or code both stimulus qualities in a nearly additive

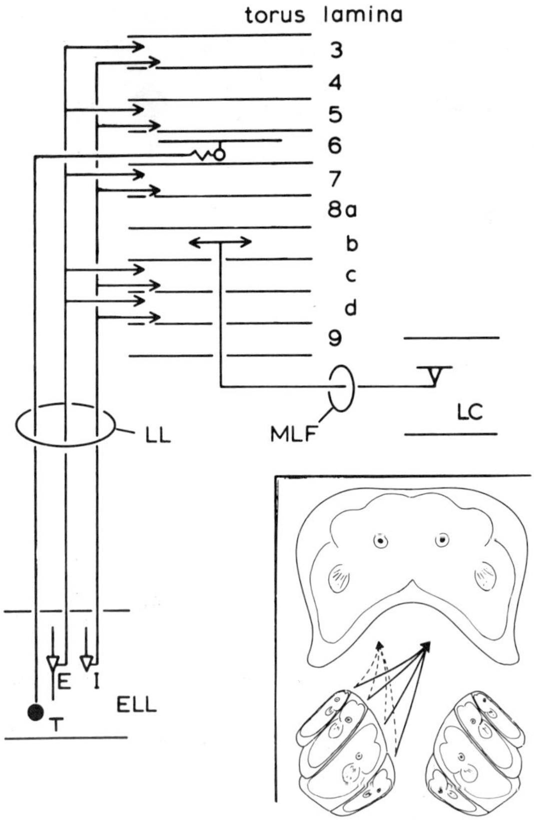


Figure 33.2. Electrosensory afferents to torus semicircularis. Two types of pyramidal cells of electrosensory lateral line lobe (*ELL*), *E* and *I* units, relay information about amplitude modulations to laminae *3, 5, 7, 8c,* and *8d* via lateral lemniscus (*LL*). Spherical cells of the *ELL* (*T, solid circle*) relay phase information electrotonically to giant cells in lamina *6* (*open circle*). The significance of projections of eurydendroid cells of lobus caudalis (*LC*) to lamina *8b* via medial longitudinal fasciculus (*MLF*) is still not known. The *insert* in *lower right* shows horizontal section through the torus (*top*) and the two ELLs. Four maps of the body represented in the ELL, with left side of the body in left ELL, project predominantly to contralateral side of the torus to a single map (note *symbols for eyes and fins*). Tuberous electroreceptor input is limited to three lateral maps of the ELL, with identical neurons projecting to all three maps. Medial map of the ELL receives ampullary receptor input.

manner (Heiligenberg and Rose 1985a). Most cell types in the tectum respond with directional specificity to moving objects and are often multimodal, i.e., integrate different sensory representations of space (Bastian 1982). We also find a relatively small number of neurons that apparently combine amplitude and phase information by ''and'' gating and thus respond to jamming signals in a Df sign—selective manner (Rose and Heiligenberg 1984, 1986). By responding

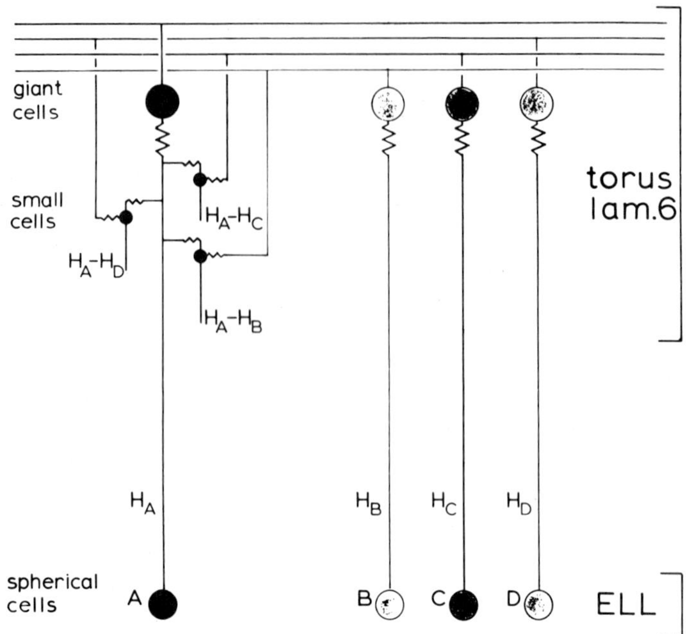


Figure 33.3. Neuronal substrate for the computation of differential phase. *Spherical cells* in the *ELL* receive somatotopically ordered information about the timing of zero crossings of stimuli on body surface. This local phase information ($H_A$, $H_B$, $H_C$) is relayed to somatotopically ordered somata of *giant cells* and nearby dendrites of *small cells* of *lamina 6 of the torus*. Somata of *small cells*, in turn, receive input from giant cell axon whose timing of spikes reflects phase in that part of body surface which projects to the soma of this particular giant cell. Firing rate of small cell is thus controlled by the difference in timing between spike arrivals on their dendrites and spike arrivals on their somata (i.e., by the difference between the phase ($H_A$) in receptive field corresponding to their location in somatotopical map and reference phase ($H_B$, $H_C$, or $H_D$) from some other part of body surface. *Small cells* within given column of *lamina 6* all receive dendritical phase input from same region of body surface, whereas the source of phase input to their somata appears to be chosen randomly from passing axon collaterals of *giant cells* (after Carr, Taylor, and Maler 1986).

preferentially to one particular sign of Df, these cells are suitable candidates for the control of the JAR. They project to the region of the nucleus electrosensorius complex of the diencephalon, where cells with similar properties have been found (Bastian and Yuthas 1984). Projections from the nucleus electrosensorius region to the prepacemaker nucleus of the midbrain, the only known source of neuronal input to the pacemaker in the medulla (Heiligenberg et al. 1981), apparently close the neuronal loop that controls the JAR (see Figs. 33.4 and 33.5).

## General Lessons from the Study of the JAR

### Technical Difficulties in the Study of the CNS of Vertebrates

When one argues that a particular type of neuron participates in a certain behavior, one normally has to rely on indirect evidence. Only in exceptional instances, such as the Mauthner system (Eaton and Hakett 1984) or the medullary pacemaker of the electric organ (Bennett 1971), can the functional significance of a neuron be demonstrated by intracellular stimulation and direct observation of predictable behavioral consequences. Normally, however, if a neuron responds to stimulus features that are critical for a given behavior, one can only assume that this neuron plays a role in the control of this behavior. This assumption will then find further support if neuroanatomical studies demonstrate that this neuron is part of a pathway whose functional importance has been suggested by lesioning or electrical stimulation experiments.

Great difficulties plague efforts at constructing functional neuronal networks from anatomical and physiological data obtained from studies of individual neurons. Ultrastructural reconstructions of synaptic connections of intracellularly labelled neurons are time consuming, and it is virtually impossible to demonstrate the functional nature of such connections by simultaneous recording from pairs of cells. This forces one to propose patterns of neuronal interactions on the basis of physiological response characteristics and light-microscopic evidence. If, for example, one type of cell responds to stimulus feature A and another to stimulus feature B, and if both types project to an area where a third cell type is found with response characteristics that can only be explained by a certain logical combination of features A and B, then it may be assumed that the two former cell types make direct or indirect contact with this third type.

Further support for an assumption about the functional significance of a certain cell type can be derived from the observation that its dynamic properties, such as time constants or stimulus-dependent response characteristics, closely match such properties at the behavioral level. Although all such evidence is indirect, its gradual accumulation eventually builds into a compelling argument.

### The Nature of Distributed Parallel Processor

The neuronal organization of the JAR shows no small group of "pontifical," or decision, neurons, the loss of which would jeopardize the performance of the behavior. Instead, one finds a distributed, anonymous network of neurons that controls the behavior in a cumulative and parliamentary manner. Local evaluations of P unit activity yield information about amplitude modulations, and pairwise comparisons of T unit inputs from small areas of the body surface yield information about differential phase. The joint evaluation of these two forms of information is sufficient for the control of the JAR. Although a particular partner in these

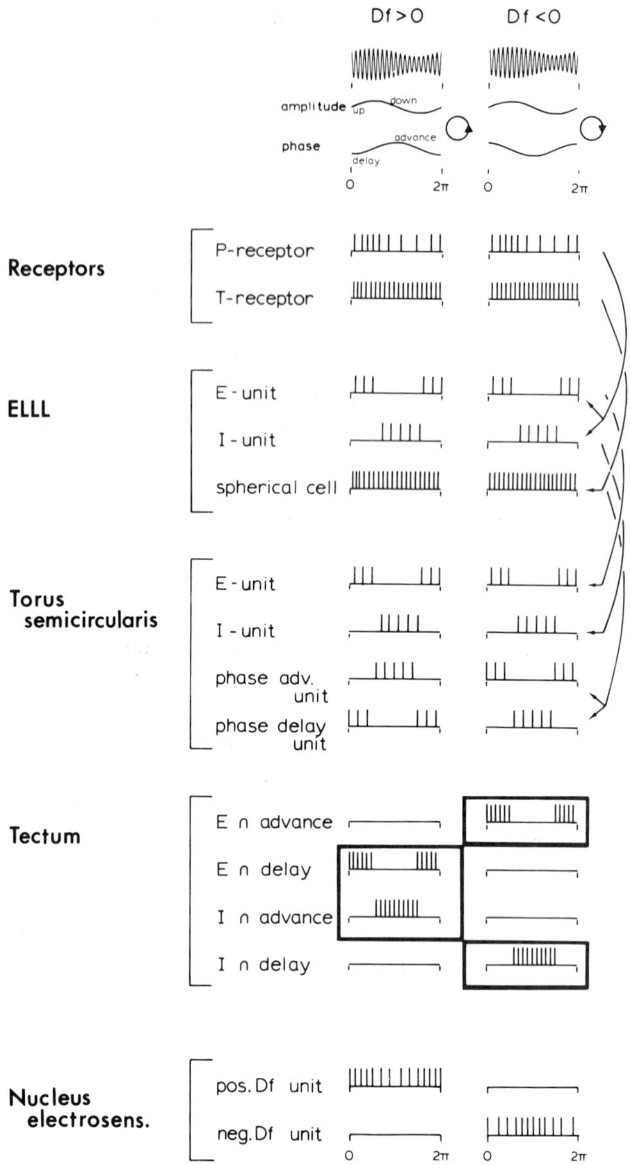


Figure 33.4. The computation of the sign of *Df* along electrosensory pathway, from the *receptor* to *nucleus electrosensorius* complex. *Top:* Two beat patterns with opposite signs of *Df* and their modulations of *amplitude* and *phase*. Modulation graphs with opposite senses of rotation are obtained if amplitude is plotted against phase in each case. Spike patterns of *E-* and *I-units* of *ELL* mark rise and fall, respectively, in stimulus amplitude. Their input originates from *P-receptors*. Higher-order *E-* and *I-units* of the *torus* relay amplitude information to the *tectum*. *Phase advance* and *phase delay units* of the *torus* code differential phase. This information is obtained by comparison of arrival times of *T-receptor* spikes relayed from different parts of body surface via *spherical cells* of the

pairwise interactions, such as area *B* in Figure 33.1b, might cast the wrong vote and drive the pacemaker frequency in the wrong direction, it will always be overruled by a stronger contribution resulting from the reciprocal interaction. This pooling of spatially distributed information conveys two advantages. First, the system is invulnerable to even large losses of afferent information due to physical injury. Second, by pooling information from large populations of neurons, the fish can improve the signal-noise ratio. *Eigenmannia* can discriminate the sense of rotation of graphs with a peak-to-peak phase modulation as small as 400 ns (Carr, Heiligenberg, and Rose, 1986), provided a sufficiently large area of body surface is exposed to the interfering stimuli (Rose and Heiligenberg 1985b). Phase discriminations of 400 ns require convergence of phase information from many T-type receptors, since the phase code of individual receptors is too jittery. Even the most regular primary afferents from T receptors show a standard deviation in the phase locking of their spike in the range of 10–30 μs, and, with a response latency of 300 ms, the animal can evaluate only approximately 100 spikes from any single afferent before it executes the appropriate behavior.

The extraordinary temporal resolution displayed in the JAR is all the more surprising in view of the rather limited precision in the design of the underlying neuronal "hardware." A correct evaluation of differential phase requires that simultaneous zero crossings of signals in different parts of the body surface also be interpreted as simultaneous in the animal's brain. Owing to differences in conduction time, however, spikes from distant T receptors should arrive later in the brain than spikes from proximal T receptors. Although the animal has largely compensated for such differences by higher conduction speeds in T afferents from more distant receptors (Heiligenberg and Dye 1982), small differences in arrival time still remain. T receptor spikes from the tail end of the body arrive 0.5–1 ms later in the brain than do T receptor spikes from the head. As a consequence of this difference, a phase value of 0 between head and tail is reported as a phase lag of the tail in reference to the head, and the animal produces incorrect JARs if its evaluation of differential phase is experimentally limited to the comparison of widely separated areas of body surface (Heiligenberg and Bastian 1980). Since, however, comparisons between more closely spaced areas of body surface contribute more strongly to the JAR, the animal always produces a correct JAR as long as inputs from the whole body surface are available. The animal can, therefore, tolerate a certain amount of imprecision in the performance of its T system. Since the resulting JAR functions with such perfection, there is no evolutionary pressure for still finer tuning of the T system.

---

*ELL*. Logical combinations ("and" gating) of amplitude and differential phase information at the level of the tectum yield neurons that fire preferentially for particular sign of *Df* (*heavily framed*). Higher-order neurons of this type can be recorded from in the *nucleus electrosensorius* complex, which appears to form the link to prepacemaker nucleus in the midbrain. This prepacemaker nucleus provides only known input to medullary pacemaker, which ultimately controls EOD frequency by triggering each EOD cycle. The classification of neurons of the torus has been simplified. Neurons that combine phase and amplitude inputs in nearly additive manner have been left out.

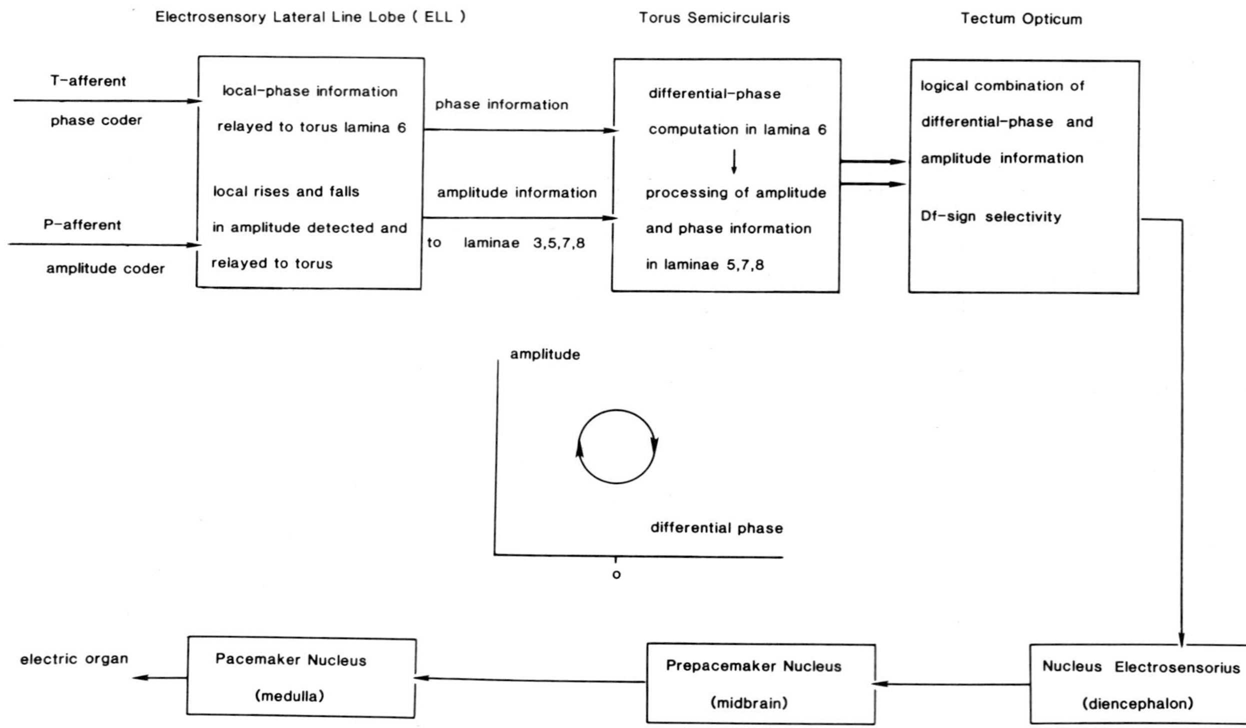


Figure 33.5. A summary of neuronal substrate of the JAR, from receptor input to control of electric organ. Modulations of stimulus *amplitude* and *differential phase* that drive the JAR are shown in *central insert*. *Circular graph* has clockwise sense of rotation (see Fig. 33.1) and thus reflects negative Df value. A rise in amplitude is paired with phase lead, and a fall in amplitude is paired with phase lag. Reverse combination and therefore reverse direction of rotation reflects positive Df value. *Df-sign*–selective cells are differentially recruited by one sign of *Df* and are thus suitable candidates for the control of pacemaker frequency in the JAR. Such cells are found in the *tectum* and in the region of *nucleus electrosensorius*. Cells of *prepacemaker nucleus* have not yet been explored physiologically. Electrotonically coupled cells of the pacemaker trigger each cycle of *electric organ* by joint action potential. Physiological mechanism of frequency shifts of the pacemaker is still unknown.

A particular neuron type may respond to stimulus features that control different behavioral responses. Certain neurons of the torus, for example, are driven by modulations of stimulus amplitude and phase that characterize moving objects, as well as the interference of the animal's EOD with that of a neighbor. Since moving objects elicit physical responses of the body, whereas EOD interference elicits a shift in pacemaker frequency (i.e., a JAR), it is conceivable that such neurons contribute to the control of both forms of behavior. The responses of a single neuron are not uniquely determined by a particular stimulus regimen and are not diagnostic for the particular form of behavior that is elicited. Only by monitoring a larger set of neurons may one be able to learn whether the animal faces a moving object or interference with a neighbor's EOD. It thus appears that the composition of the whole set of neurons being recruited determines the nature of the behavioral response and that a single neuron is of limited significance.

## The Emergence of Feature Detection and Behavioral Stimulus Specificity

As even high-order neurons may respond to a variety of stimulus features and thus potentially participate in the control of very different behavioral responses, it is difficult to identify such neurons with the detection of complex, biologically relevant stimulus patterns. A higher-order electrosensory neuron, for example, may respond most strongly to a local rise in stimulus amplitude paired with a lead in stimulus phase, and this particular pattern can be generated through interference with a neighbor's EOD, as well as by a moving object with certain ohmic and capacitive properties. As nuclei of even higher-order neurons, still closer to the generation of specific behavioral response, are examined, however, one may find a stricter specialization and stimulus pattern specificity of neurons. It appears that these cells, by sampling information supplied by larger sets of lower-order neurons, are able to respond more selectively to specific temporal and spatial configurations of stimulus features. Neurons closer to the control of the JAR may less likely be "fooled" by moving objects.

Even dynamic response properties of a behavior may emerge gradually along a chain of neuronal processing stations. The JAR, for example, is driven most strongly by Df magnitudes between 2 and 6 Hz (Bullock, Hamstra, and Scheich 1972). Examination of neurons in the ELL and torus does not reveal an absolute preference for modulation rates in this range. Torus units that are driven most strongly in this range, however, are more common than others that are driven more strongly by rates outside this range (Partridge, Heiligenberg, and Matsubara 1981). At the level of the nucleus electrosensorius, however, all cells that discriminate the sign of Df appear to be most responsive to modulation rates between 2 and 6 Hz (Bastian and Yuthas 1984).

## The Relation Between Neuronal Structure and Function

In order to explore functional properties of higher-order neurons it is necessary that one employ a broad spectrum of biologically relevant stimulus patterns and consider a sufficient variety of behavioral contexts. Without a sufficient diversity in stimulus forms, it is difficult to recognize functional specializations among neurons. As an example, consider the various types of neurons in laminae 5 and 7 of the torus. These laminae sandwich lamina 6, in which differential phase information is computed. Laminae 5 and 7 receive input from the amplitude-coding pyramidal cells of the ELL, and all neurons of these laminae also respond vigorously to modulations in stimulus amplitude. Some of these neurons also respond to modulations in differential phase, and these same neurons extend their dendrites into lamina 6 (Rose and Heiligenberg 1985a). One thus finds a strong link between the laminar distribution of dendrites and the functional properties of these cells. This particular functional specialization could easily have escaped detection had the behavioral significance of differential phase modulations and the role of lamina 6 in its computation not been known. Carr and Maler (1985) identified approximately 50 cell types in the electrosensory torus by Golgi studies. Not more than 30 of these types have been identified physiologially by intracellular labelling (Heiligenberg and Rose 1985a; Rose and Heiligenberg 1985a). Extensive studies at the behavioral, neurophysiological, and neuroanatomical level will be required to understand the structural and functional organization of the torus in its entirety.

## Similarities to Neuronal Organization of Directional Hearing

The separate and parallel representation of phase and amplitude information in the CNS of *Eigenmannia* resembles that in the auditory system of the owl (Sullivan and Konishi 1984; Takahashi, Moiseff, and Konishi 1984; Konishi 1986). Separate pathways originate in the hindbrain and then converge at the level of the midbrain. Different aspects of phase and amplitude information, however, are evaluated by these two animals. In the case of the owl, the azimuth of a sound source is coded by interaural phase differences, whereas the elevation is coded by interaural amplitude differences. A point in the phase-amplitude plane thus signals the spatial location of prey and thereby controls a directional behavioral response. A stationary point in the phase-amplitude plane, however, has no significance for *Eigenmannia,* which instead requires a periodical motion of a point along curved contours within this plane to arrive at a behavioral decision. *Eigenmannia,* therefore, has to solve a task of higher complexity, comparable, maybe, to the tracking of a moving sound source by the owl.

## Speculations About the Evolution and Ontogeny of Neuronal System

The more one learns about details of the neuronal organization of the JAR, the more one wonders about its seemingly unnecessary complexity. The logical computations of the JAR are extremely simple, and a few "and" gates at the level of the torus would allow the fish to determine the sign of Df immediately. Instead, the ultimate computation of the sign of Df is delayed until the level of the tectum is reached, and the JAR appears to emerge gradually from a network of neurons that also participates in the control of other behavioral responses. Modulations in phase and amplitude are also caused by moving objects, and physical responses to such objects are guided by this information. The phase and amplitude modulations that characterize a jamming situation are, however, of a particular spatial and temporal structure, and it appears that a subset of neurons originally, and perhaps still, used in the detection of moving objects has been specialized to process the particular forms of phase and amplitude modulations that are relevant for the JAR. This impression is confirmed by the anatomical similarity of motion-sensitive tectal neurons that project to premotor areas of the reticular formation and those which decode the sign of Df (Heiligenberg and Rose 1985b; Rose and Heiligenberg 1986). The neuronal substrate of the JAR thus must have evolved by gradual modifications of, and additions to, a neuronal hardware that was adapted to the more general task of detecting movements of objects. Owing to evolutionary constraints, a neuronal system may, in the eyes of a human engineer, not always be designed in a straightforward manner.

The study of neuronal interconnections invites speculations about the underlying rules of ontogenetic self-organization. Since genetic information is rather limited, one expects neuronal networks to develop according to simple sets of instructions. Therefore, among the possible patterns of neuronal interconnections that could achieve a particular task of processing information, those which can be built with a minimal set of simple developmental rules are the most appealing. The organization of lamina 6 of the torus (Carr, Taylor, and Maler 1986) presents a fascinating example. This lamina receives somatotopically structured input from the spherical cells of the ELL. Small cells within this lamina respond to differences in phase between the area of body surface corresponding to their somatotopical location within lamina 6 and some reference area that appears to be chosen randomly. The rules for the organization of this lamina could be few and simple (see Fig. 33.3):

1. Spherical cell axons invade lamina 6 by maintaining their somatotopical neighborhood relations as they originate in the ELL.
2. Upon arrival in lamina 6, these axons synapse with local somata of giant cells and dendrites of small cells.
3. Giant cells send their axonal arborizations across wide areas of lamina 6 and synapse on somata of small cells. Only one such synapse will form on any one small cell soma, and the probability of such contacts falls off with the distance from the soma of the giant cell. As a consequence of this rule, small cells will pick random phase references, and a phase comparison between two areas of body surface will be made more likely the closer these areas are to each other.

This pattern of organization yields the functional properties of phase evaluation governing the JAR. This system is pleasingly parsimonious in its design and elegant in its performance.

## Summary

Neural mechanisms of perception and the control of a simple behavior, the jamming avoidance response (JAR), have been studied in the weakly electric fish *Eigenmannia*. Essential stimulus variables and central rules for their evaluation were identified by behavioral tests of systematically modified stimulus configurations. The neuronal implementations of these perceptual mechanisms were analyzed by joint neurophysiological and anatomical approaches, particularly by intracellular labelling of physiologically identified cells. General conclusions are as follows:

Stimulus patterns are processed in distributed systems of neurons with rather general response properties, and behavioral decisions originate from the nature of the set of neurons recruited by a given stimulus situation. There is no evidence for the existence of so-called pontifical, or decision, neurons that operate like command elements. The stimulus specificity of the behavioral response emerges gradually and statistically on the pathway from lower-order to higher-order neurons. Despite certain deficiencies and the improvisational character of its design, the neuronal system of the JAR excels in its acuity of performance at the behavioral level.

The functional analysis of neuronal systems encourages speculations about their ontogeny and evolution.

*Acknowledgments.* I thank John Dye and Gary Rose for their kind criticism of this chapter in manuscript. This research is supported by grants from NIMH and NSF.

### References

Bastian, J. (1982) Vision and electroreception: integration of sensory information in the tectum of the weakly electric fish *Apteronotus allbifrons,* J. Comp. Physiol., 147:287–297.

Bastian, J., and Heiligenberg, W. (1980a) Neural correlates of the jamming avoidance response in *Eigenmannia,* J. Comp. Physiol., 136:135–152.

Bastian, J., and Heiligenberg, W. (1980b) Phase sensitive midbrain neurons in *Eigenmannia:* Neural correlates of the jamming avoidance response, Science, 209:828–831.

Bastian, J., and Yuthas, J. (1984) The Jamming Avoidance Response of *Eigenmannia:* properties of a diencephalic link between sensory processing and motor output, J Comp. Physiol., 154:895–908.

Bennett, M.V.L. (1971) Electric organs, in Fish Physiology, vol. 5, Hoar, W.S. and Randall, D.J. (eds.), Academic Press, New York, pp. 347–491.

Bullock, T.H., Hamstra, R.H., and Scheich, H. (1972) The Jamming Avoidance Response of high-frequency electric fish, J. Comp. Physiol., 77:1–48.

Carr, C.E., Heiligenberg, W., and Rose, G.J. (1986) A phase-comparison circuit in the electric fish midbrain. I. Behavior and physiology, J. Neurosci., 6:107–119.

Carr, C.E., and Maler, L. (1985) Cell types of the dorsal torus semicircularis of the electric fish *Eigenmannia:* a Golgi study, J. Comp. Neurol., 235:207–240.

Carr, C.E., Maler, L., Heiligenberg, W., and Sas, E. (1981) Laminar organization of the afferent and efferent systems of the torus semicircularis of gymnotiform fish: morphological substrates for parallel processing in the electrosensory system, J. Comp. Neurol., 203:649–670.

Carr, C.E., Maler, L., and Sas, E. (1982) Peripheral organization and central projections of the electrosensory nerves in gymnotiform fish, J. Comp. Neurol., 211:139–53.

Carr, C.E., Taylor, B., and Maler, L. (1986) A phase-comparison circuit in the electric fish midbrain. II. Functional morphology, J. Neurosci., 6:1372–1383.

Eaton, R.C., and Hackett, J.T. (1984) The role of the Mauthner cell in fast starts involving escape in teleost fishes, in: Neural Mechanisms of Startle Behavior, Eaton, R.C. (ed.), Plenum Press, New York, pp. 213–266.

Hagedorn, M., and Heiligenberg, W. (1985) Court and spark: electric signals in the courtship and mating of gymnotoid fish, Anim. Behav. 33:254–265.

Heiligenberg, W. (1980) The jamming avoidance response in the weakly electric fish *Eigenmannia,* Naturwissenschaften, 67:499–507.

Heiligenberg, W. (1983) the Jamming Avoidance Response in an electric fish: algorithms in sensory information processing and their neuronal organization, in: Advances in Vertebrate Neuroethology, Ewert, J.P., Ingel, D., and Capranica, R. (eds.), Plenum Press, New York, pp. 669–699.

Heiligenberg, W. (1986) Jamming Avoidance Responses: model systems for neuroethology, in: Electroreception, Bullock, T.H., and Heiligenberg, W. (eds.), John Wiley and Sons, New York, pp. 613–649.

Heiligenberg, W., Baker, C.L., and Matsubara, J.A. (1978) The Jamming Avoidance Response in *Eigenmannia* revisited: the structure of a neuronal democracy, J. Comp. Physiol., 127:267–286.

Heiligenberg, W., and Bastian, J. (1980) The control of *Eigenmannia's* pacemaker by distributed evaluation of electroreceptive afferences, J. Comp. Physiol., 136:113–133.

Heiligenberg, W., and Dye, J.C. (1982) Labelling of electroreceptive afferents in a gymnotoid fish by intracellular injection of HRP: the mystery of multiple maps, J. Comp. Physiol., 148:287–296.

Heiligenberg, W., Finger, T., Matsubara, J.A., and Carr C.E. (1981) Input to the medullary pacemaker nucleus in the weakly electric fish, *Eigenmannia* (Sternopygidae, Gymnotiformes), Brain Res., 211:418–423.

Heiligenberg, W., and Partridge, B.L. (1981) How electroreceptors encode JAR-eliciting stimulus regimes: reading trajectories in a phase amplitude plane, J. Comp. Physiol., 142:295–308.

Heiligenberg, W., and Rose, G.J. (1985a) Phase and amplitude computations in the midbrain of an electric fish: intracellular studies of neurons participating in the Jamming Avoidance Response (JAR) of *Eigenmannia,* J. Neurosci. 5:515–531.

Heiligenberg, W., and Rose, G.J. (1985b) Neural correlates of the Jamming Avoidance Response (JAR) in the weakly electric fish *Eigenmannia* Trends Neurosci., 8:442–449.

Hopkins, C.D. (1974) Electric communication: functions in the social behavior of *Eigenmannia virescens,* Behaviour, 50:270–305.

Konishi, M. (1986) Centrally synthesized maps of sensory space, Trends Neurosci., 9:163–168.

Maler, L., Sas, E., and Rogers, J. (1981) The cytology of the posterior lateral line lobe of high-frequency weakly electric fish (Gymnotoidei): dendritic differentiation and synaptic specificity in a simple cortex, J. Comp. Neurol. 195:87–140.

Mathieson, W.B., Heiligenberg, W., and Maler L. (1986) Ultrastructural studies of physiologically identified electrosensory afferent synapses in the gymnotiform fish, *Eigenmannia,* J. Comp. Neurol. (submitted).

Partridge, B.L., Heiligenberg, W., and Matsubara, J.A. (1981) The neural basis of a sensory filter in the Jamming Avoidance Response: no Grandmother Cells in sight, J. Comp. Physiol., 145:153–168.

Rose, G.J., and Heiligenberg, W. (1984) Neural correlates of the Jamming Avoidance Response: Df-sign-specific neurons in the optic tectum of the electric fish *Eigenmannia*. Neurosci. Abst. 10:403.

Rose, G.J., and Heiligenberg, W. (1985a) Structure and function of electrosensory neurons in the torus semicircularis of *Eigenmannia:* morphological correlates of phase and amplitude sensitivity, J. Neurosci. 5:2269–2280.

Rose, R.J., and Heiligenberg, W. (1985b) Temporal hyperacuity in the electric sense of fish, Nature, 318:178–180.

Rose, R.J., and Heiligenberg, W. (1986) Neural coding of difference frequencies in the midbrain of the electric fish *Eigenmannia:* reading the sense of rotation in an amplitude-phase plane, J. Comp. Physiol. (submitted).

Saunders, J., and Bastian, J. (1984) The physiology and morphology of two types of eletrosensory neurons in the weakly electric fish *Apteronotus leptorhynchus,* J. Comp. Physiol., 154:199–209.

Scheich, H., and Bullock, T.H. (1974) The role of electroreceptors in the animal's life. II. The detection of electric fields from electric organs, in: Handbook of Sensory Physiology, vol. III/3, Fessard, A. (ed.), Springer-Verlag, Berlin, pp. 201–256.

Scheich, H., Bullock, T.H., and Hamstra, R.H. (1973) Coding properties of two classes of afferent nerve fibers: high frequency electroreceptors in the electric fish, *Eigenmannia,* J. Neurophysiol., 36:39–60.

Suga, N. (1982) Functional organization of the auditory cortex: representation beyond tonotopy in the bat, in Cortical Sensory Organization, vol. 3, Woollsey, C.N. (ed.), Humana Press, Clifton, N.J., pp. 157–218.

Sullivan, W.E., and Konishi, M. (1984) Segregation of stimulus phase and intensity coding in the cochlear nucleus of the barn owl, J. Neurosci. 4:1787–1799.

Takahashi, T., Moiseff, M., and Konishi, M. (1984) Time and intensity cues are processed independently in the auditory system of the owl, J. Neurosci., 4:1781–1786.

Viancour, T. (1979) Electroreceptors of a weakly electric fish: characterization of tuberous receptor organ tuning, J. Comp. Physiol., 133:317–325.

Watanabe, A., and Takeda, K. (1963) The change in discharge frequency by AC stimulus in a weakly electric fish, J. Exp. Biol., 40:57–66.

Part IV

# Adaptation and Sensory Systems

A major theme running through this volume is evolution and the adaptation of sensory systems to the environment. In this part, Northcutt (Chapter 34) deals with adaptation of animal traits in response to environmental factors, and with ways for investigating adaptation. He also argues for the suitability of the comparative approach for studies of biological systems, another theme that is reflected in many chapters of this volume. However, while many of the earlier chapters deal with one sensory system and one or just a few groups of animals, the principles discussed by Northcutt are pertinent to all sensory systems and all taxa, and thus to all of the chapters of this volume. Consequently, this final chapter fittingly ties together the major themes of adaptation and of the comparative principles that so influence the thinking not only of investigators represented in this volume; but of all investigators interested in the function of sensory systems in aquatic animals; and in particular, the outstanding scientists to whom this book is dedicated.

# Chapter 34

# Sensory and Other Neural Traits and the Adaptationist Program: Mackerels of San Marco?

R. GLENN NORTHCUTT*

Comparative biologists normally use two principles, phylogeny and adaptation, to explain the origin and maintenance of patterns of organic diversity. That is to say, we believe that organisms and their traits change through time and that these changes are the result of complex interactions between traits and various environmental pressures (natural selection). Not surprisingly, many biologists view Darwin's scientific explanation of trait design as the greatest triumph of his theory of natural selection. The emergence and maintenance of the form and function of a given trait by natural selection may, however, be one of the most complex biological phenomena to demonstrate.

Most comparative studies are descriptive in the sense that one or more traits are surveyed across a number of species in order to recognize patterns of diversity. Although this method is an effective way to reveal patterns (Gans 1974; Frazzetta 1975; Ulinski 1980; Huey and Bennett 1986), it is limited in demonstrating causal relationships. The results of most comparative studies reveal differences in a particular pattern of traits between species, and these differences are correlated with a particular set of environmental factors. In fact, correlations may or may not be causal, but too frequently such correlations between a trait and some environmental factor are *assumed* to be causal, and we are asked to believe that a given trait exhibits a particular form and function because of natural selection. In a seminal paper, Gould and Lewontin (1979) described this type of reasoning as the ''adaptationist program'' and faulted this approach on the following grounds: (1) natural selection is viewed as an optimizing agent, so that traits are optimally designed by natural selection for their functions; (2) such explanations rely on plausibility, rather than tests of alternate hypotheses of design; (3) such explanations do not distinguish current utility from the reasons for origin.

Many conclusions of comparative neurobiology, like those of other biological disciplines, are riddled with such fallacious reasoning. Recently, it has become fashionable to view many sensory receptors and their associated central circuits as being optimally designed for the detection and analysis of certain stimuli.

*Neurobiology Unit, Scripps Institute of Oceanography, University of California, San Diego, La Jolla, CA 92093, USA.

Sensitivity and other physical properties are cited as being at, or close to, theoretical limits. For example, we are impressed by the extreme sensitivity of the ampullae of Lorenzini as receptors of electrical fields, but we forget that they are also extremely temperature sensitive and mechanosensitive (Hensel 1955; Murray 1960; see Bullock 1974 for a review of the problem of recognizing the biological role of receptors). How can multiple sensitivities be reconciled with selection for optimizing ampullary organs as electroreceptors?

Similarly, comparative studies frequently examine the organization of various receptors and circuits in different species and correlate traits with particular environmental factors, the assumption being that these factors are responsible for the design traits of the receptors and circuits (Walls 1942; Okada et al. 1969; Northcutt and Butler 1974; Dartnall 1975; Ali, Croll, and Jaeger 1978; Calford et al. 1985). Rarely do such studies determine the actual biological role of the trait under field conditions or demonstrate that the trait is inherited and that possessing it confers greater fitness. Such studies do exist (Endler 1978, 1986; Hopkins 1983; Konishi 1985, as examples), but they are too infrequent.

Finally, most neurobiological studies of the origin of receptors and circuits fail to distinguish those traits which may have arisen from chance or selection, but are not now under selection, from those that are currently maintained by selection. For example, the anatomy of the electroreceptors in all anamniotic vertebrates is generally interpreted as evidence that these electroreceptors evolved from mechanoreceptors, such as neuromasts, owing to increased selection for their labile sensitivity to electrical fields (Lissmann and Mullinger 1968; Bennett 1971; Szabo 1974). However, other possible hypotheses of the origin of electroreceptors have not been posed. It is possible that these structures arose by chance (genetic drift or founder effect) and were only subsequently selected for as electroreceptors. It is also possible that electroreceptors arose as the result of new pleiotropic connections or as changes in preexisting pleiotropic connections linking the lateral line with other traits. In any of these three cases, selection would not have operated to produce electroreceptors, and we would not be justified in stating that these receptors arose as an adaptation for electroreception.

These criticisms of the application of the comparative method and its failure to demonstrate adaptive traits do not necessarily imply that the method is incapable of demonstrating adaptation nor that adaptive traits are rare. They do suggest that comparative biologists have generally failed (1) to recognize the limitations of a predominantly descriptive discipline, (2) to consider alternate hypotheses of design, and (3) to generate a formal methodology for analyzing such hypotheses. With these problems in mind, current definitions of adaptation and alternative explanations of design are examined below. Methodologies for evaluating hypotheses of design are then discussed, with particular emphasis on the limitations of the comparative method in discriminating adaptive traits.

## Definitions of Adaptation

The term *adaptation* is applied to several different biological phenomena and has been used to describe processes, as well as a state of being (Bock and von

Wahlert 1965; Williams 1966; Bock 1980; Gould and Vrba 1982). Generally, two different types of responses of organisms to their environment have been termed adaptive. The ability of organisms to adjust phenotypically to short-term changes in the environment is called *physiological adaptation* (somatic or phenotypic plasticity). These responses are generally shorter than the life span of the organism and are not heritable. The capacity to develop such responses may be heritable, however, and for this reason, phenotypic plasticity is frequently viewed as a special case of evolutionary adaptation. Long-term, hereditary changes that occur in species in response to a particular set of environmental factors are termed *evolutionary adaptation.* It is these long-term responses that are primarily of interest to evolutionary biologists, and it is those with which I shall deal.

An examination of several definitions of evolutionary adaptation reveals a number of common themes.

> An adaptation is a characteristic of an organism advantageous to it or to the conspecific group in which it lives. (Simpson 1953, 160)
>
> Adaptation is ''the condition of showing fitness for a particular environment, as applied to characteristics of a structure, function, or entire organism; also the process by which such fitness is acquired. . . .'' (Mayr 1969, 397)
>
> Adaptation represents the general or specific way (or aspect) in which an organism survives in an environment. The term adaptation also defines the process by which this state arose. (Gans 1974, 6)
>
> An adaptation, the state of being, can be defined as a feature having properties of form and function which permit the organism to maintain successfully the synerg between a biological role of that feature and a stated selection force. (Bock 1980, 221)

There is a general consensus that an adaptation, as a state of being, entails a correlation between a trait and an environmental factor and that the relationship is maintained by natural selection. There is little agreement, however, regarding a number of questions: whether individual organisms and populations, as well as traits of organisms, are adaptive; how well adapted a given trait or organism is compared to others; whether the term should be restricted solely to traits that arose and have been maintained by selection or should include any trait that enhances current fitness, regardless of its phyletic origin. Bock (1980) argued that evolutionary adaptation applies only to phenotypic traits of individuals, not to individuals, populations, or species. On the other hand, Simpson (1953) and Mayr (1969, 1983) imply that individuals and species can be judged to be more or less adapted. Excluding the possible implications of kin and group selection, it seems preferable to restrict adaptive considerations to traits of individuals, recognizing that individuals and not traits reproduce.

Until recently, little consideration has been given to the details of form–function relationships, their matching to environmental factors, or how to evaluate the degree of fit between a trait and a given environmental factor (Bock and von Wahlert 1965; Gans 1974; Bock 1977; Huey and Bennett 1986). Morphologists have come to realize that most traits or features of an organism possess both form and function and both aspects of a trait may change during the life history of an organism. Changes in form are obvious; changes in function, less easy to identify. We frequently fail to realize that any given trait usually has multiple functions, or at least the potential for multiple functions, only some of which are actually used by an individual animal during its life. These unrealized or underused functions represent a major source of variability for further evolution. Because most traits do have multiple functions, it is necessary to determine how a trait is actually used by an organism in the real world—its biological role—when trying to evaluate its adaptiveness. For example, the laboratory observation that ampullae of Lorenzini are extremely temperature sensitive and mechanosensitive does not tell us that the ampullae are used behaviorally to detect electrical fields. It is the interaction of the biological role and the selective force of an environmental factor that determines an adaptation. This interaction was termed a *synerg* by Bock and von Wahlert (1965), and they suggested that the relative amount of energy an organism exerts to maintain a synerg is a reasonable measure of how good an adaptation it represents. A perfect adaptation would require the minimum of energy, whereas a poor one would be far less energy efficient. Although energy efficiency appears to be an appropriate measure, practical considerations aside, for traits with obvious mechanical properties, such as those of the skeletomuscular system, it is not clear that such a measure can be similarly applied to all types of traits. Neural tissues are some of the most metabolically expensive tissues to maintain; are the adaptations of relatively large-brained organisms less fit than those of smaller-brained organisms? The answers to this and similar questions are not obvious, but these questions must be examined if we are to develop a truly comparative approach to adaptations. For the present, most studies will continue to compare biological traits with "design features" of other natural and artificially occurring chemical and physical systems for goodness of fit.

There is considerable disagreement whether adaptive traits are only those that arose and have been maintained by selection or are also traits that enhance current fitness regardless of their history. Williams (1966) restricted the term to those traits whose origin and perfection can be attributed to a long period of selection for effectiveness in a particular biological role. He argued that only adaptive traits exhibit functions, whereas useful traits that have not arisen because of selection for their current roles exhibit "effects," (i.e., accidental consequences). Gould and Vrba (1982) support Williams's distinction by recognizing two classes of "adaptive" traits. They suggest that traits exhibiting the static phenomenon of being fit should be termed *aptations* (*aptus,* fit), and they recognize two categories of aptive traits: *adaptive* and *exaptive*. Adaptive (*ad* + *aptus,* towards a fit) traits are those that arose and have been maintained by the same selective pressures, whereas exaptive (*aptus,* fit, *ex,* by reason of their form) traits are those that arose by chance or were shaped by selective pressures different from those currently operating on them.

Bock (1980) argued that adaptive traits need not have evolved under the control of the same pressures; it is unclear whether he would describe as adaptive those traits that arose by chance and were subsequently selected for. An important issue in this disagreement concerns the prevalence of nonadaptive (*nonaptive,* in Gould and Vrba's terminology) traits and their subsequent fate. Gould and Vrba (1982) suggest that nonaptive traits are common and that they represent an important pool of variability for evolution, as such traits may be subsequently coopted by selection (co-optations). The more prevalent opinion is that nonadaptive traits do exist, but they are rare (Bock 1980; Gans 1985). This view suggests that continuity of adaptation and natural selection is the predominant factor underlying the phyletic divergence of traits.

It is presently impossible to determine how widespread nonadaptive traits may be. We must analyze traits within the context of such questions, however, formulate hypotheses based on such a framework, and generate a methodology that will separate adaptive from nonadaptive traits. Our failure to do so will preclude our determining the distribution of nonadaptive traits; equally important, it may lead to the construction of phyletic scenarios that bear little or no relationship to reality. For these reasons, it seems preferable to consider adaptive traits within Gould and Vrba's framework and to suggest that adaptive traits are those that are correlated with an aspect of the environment and postulated to have arisen and been subsequently maintained by the same selective pressure(s).

## Alternate Explanations of Design

Historically, at least five different factors have been considered to cause evolutionary changes in traits: inheritance of acquired characters, intrinsic directional forces, saltational evolution, selection, and chance (Mayr 1983). An extensive literature has discredited the first three factors, and most biologists consider selection and chance as the only mechanisms of evolutionary change. Although the categories and operations of selective forces have been extensively examined (Wright 1949; Mayr 1963; Williams 1966; Frazzetta 1975; Calow 1983), there are few comparable discussions of the role of chance in evolutionary change (Gould and Lewontin 1979; Conrad 1983; Ridley 1983). Gould and Lewontin (1979) provide an extensive list of situations in which an explanation of design does not reflect an immediate adaptation due to selection, but four of their five categories do include selection at some level.

Their first category includes traits that are not adaptive and not under selection. Such traits may arise in small populations by changes in the frequencies of alleles due to genetic drift or founder effect. In such small populations, many loci for polygenetic traits may exhibit random fixation of unfavorable alleles in the absence of, or in spite of, natural selection. Thus, it is possible for traits that are not adaptative to arise and be maintained in a population. Equally important, it is possible for adaptive traits to be lost as a consequence of genetic drift or founder effect. For example, it is possible that the loss of electroreceptors in primitive neopterygian fishes and anuran amphibians occurred by chance, rather than selection.

Gould and Lewontin's second and third categories involve trait design that is not directly due to selection or that is "adaptive" in the absence of selection. A trait may be highly correlated with a particular environmental factor, but that correlation might be due to pleiotrophy (i.e., produced by the same alleles as a second trait that is under selection). For example, there is considerable evidence that brain–body allometric patterns involve pleiotrophy (Lande 1979), and though brain size and body size are subject to selection, many increases in relative brain volume in closely related taxa may be the result of selection's increasing body size, rather than behavioral complexity. In this case, behavioral complexity may increase, but it does so as a consequence of direct selection of other traits and therefore cannot be said to be adaptive. Similarly, adaptive traits can be lost if they are pleiotropically linked to traits that are being strongly selected against. Again, it is possible that electroreceptors in primitive neopterygian fishes involved alleles that also affected dermal bones and scales (Pehrson 1944). Selection for increased efficiency in feeding and locomotion drastically altered the pattern of the dermal skeleton (Romer 1966) and may have also resulted indirectly in the loss of electroreceptors. In this case, electroreceptors would not have been selected against but have been lost as a consequence of selection against other traits. Similarly, changes in heterochrony may result in pleiotropic effects that are not due to selection. For example, many paedomorphic neural traits in salamanders and lepidosirenid lungfishes (Northcutt 1986) may be a by-product of selection's acting for rapid turnover of generations. Therefore a thin-walled and highly undifferentiated nervous system in these animals is probably a result of truncated maturation, rather than an adaptation for shallow, often anaerobic, freshwater habitats, as suggested by Herrick (1948).

Short-term, nonheritable responses of organisms to their environment (phenotypic plasticity) constitute an additional class of trait design to which the term *adaptation* is applied. In cases such as respiratory responses to high altitudes or neural responses to low ambient temperatures in many anamniotes, the adaptations per se are not heritable, but the capacity to develop such responses is believed to be. In one sense the traits are adaptive, but they are not selected for; at most, the capacity to develop these traits may be selected for. The widespread occurrence of phenotypic plasticity is one more reminder that a correlation between a trait and some environmental factor is insufficient reason to infer a causal relationship due to natural selection.

The fourth and fifth categories of Gould and Lewontin concern trait design that entails adaptation and selection, but in which there is no selective basis for the differences among adaptations or in which the adaptation is secondary to historical and biomechanical constraints. The differences between traits may be nonadaptive owing to the existence of multiple solutions to the same or similar selective pressures. The interaction between genes and their influence on fitness (Wright 1931) suggests that not all allelic combinations can be equally successful. A plot of average fitness against the frequency of two or more alleles can be visualized as a pattern of valleys and peaks (adaptive landscape), in which the fittest populations lie at the tops of the peaks. Not all adaptive peaks will be occupied, as there are more potential combinations of alleles in a species than there are species. Selection tends to force a population up a peak, however, whereas genetic drift will force the population down the peak. Which peak is climbed will depend on the proximity of a population to given peaks in the adaptive landscape. Therefore it is possible that different populations in the same environment, starting with different alleles, will climb different peaks. Each solution may be adaptive, but the differences between the solutions may be due to chance. For example, the percentage of nondecussating axons of retinal ganglion neurons in the optic tracts of vertebrates is generally believed to be positively correlated with the development of stereopsis. However, many nonmammalian vertebrates possess excellent stereopsis and have completely crossed optic tracts. In these species, optic tract fibers or higher-order pathways redecussate in commissures other than the optic chiasm, so that ultimately visual centers do receive information from both eyes. Thus, bilateral optic inputs to the visual centers appear to be required for stereopsis and are an adaptation for it, but bilaterality appears to have been provided for in different ways in different groups of vertebrates. There is no evidence that any one solution is superior, and the way in which each group has solved the problem may be due as much to chance or historical constraints or both of them as to selection. A similar interpretation may explain differences in tectal lamination and cerebellar foliation in different vertebrate radiations.

Nonadaptive differences may also occur due to sexual selection or as a result of *evolutionarily stable strategies*, a term coined by Maynard Smith (1972). If a given trait in one sex becomes linked with fitness, the other sex is selected to mate with individuals possessing that trait, which consequently becomes more prominent. If different characters in different populations become associated with fitness, different exaggerated characters will result in the two populations. The cause of these different characters could be a random historical association of the characters and fitness, with environmental differences having no effect.

Similarly, certain combinations of traits may possess functions so fit (evolutionarily stable strategies) that they cannot be replaced by an invading mutant. If there are several strategies (similar to Wright's adaptive landscapes) the stable strategies exhibited by different populations may be due to chance.

Traits may have current utility and be selected for; thus they may be said to be adaptive, but the adaptation might be secondary (i.e., the current function may be a secondary use of form for reasons of architecture or history). Gould and Lewontin (1979) analogize such secondary adaptations to the spandrels of St. Mark's Cathedral in Venice. The central dome is supported by rounded arches, and surfaces (spandrels) are formed by the intersections of two arches. The surface of each spandrel contains a painted mural that admirably fits the space. It is possible to be so impressed by the form and "function" of the spandrels as to believe that the architects designed the arches and dome to create spandrels; in fact, spandrels are a necessary by-product of mounting a dome on rounded arches. Similar explanations of biological trait design, based solely on current function, frequently assume that this function was the cause of origin, though, in fact, current function may have nothing to do with the primary reason for the existence of the trait. For example, most textbooks describe the fore-, mid- and hindbrain as constituting the major divisions of all vertebrate brains and usually state that each of these divisions arose from the influence of olfactory, optic, and otic sense organs, respectively (Ariëns Kappers, Huber, and Crosby 1936). In fact, the earliest divisions of the brain that occur embryonically do not match the three divisions that occur later in development (von Kupffer 1906), and all three divisions are used in processing most types of sensory information in all vertebrate species that have been examined. Although there is little doubt that neural centers in vertebrate brains are adaptive and selected for, there is little reason to believe that each brain division arose as an adaptation to a specific sensory modality. Hypertrophy of the cranial neural tube in ancestral vertebrates was probably strongly selected for, but the actual number of divisions may have occurred by chance or have been due to historical or ontogenetic constraints. Similarly, the electroreceptors of anamniotic vertebrates possess either kinocilia (nonteleost gnathostomes) or microvilli (lampreys and teleosts) on their apical surfaces. The adaptive significance of these organelles, if any, has eluded physiologists (Szabo 1972; Murray 1974; Teeter, Szamier, and Bennett 1980). Both teleostean and nonteleostean electroreceptors are believed to have evolved independently from mechanoreceptive neuromasts (Lissmann and Mullinger 1968) whose receptors possess both kinocilia and microvilli that are used in mechanical transduction. It is possible that the existence of these organelles in electroreceptors may be due to the historical origin of electroreceptors from neuromasts, and the kinocilia and microvilli may be nonadaptive in electroreceptors. It is also possible that these organelles are important functional elements in the detection of electric fields by electroreceptors, but understanding their present function would not explain their origin, form, or present phyletic distribution.

## Evaluation of Hypotheses of Design

No evolutionary biologist seriously believes that adaptive traits do not exist; the question is how to test hypotheses of causal relationships between traits and environmental factors. Generally, there appear to be three possible hypotheses that can be applied to any trait: (1) the trait is adaptive in that it arose and has been maintained by the same selective pressure; (2) the trait is exaptive in that it arose by chance or owing to a selective pressure not currently operating and has been co-opted by a new, or different selective pressure for its present use; (3) the trait is nonaptive in that it arose by chance or because of a selective pressure not currently operating and is not under current selective pressure. Initially, these hypotheses can be examined in the context that traits are either aptive or nonaptive. It is probably impossible to demonstrate that a trait arose by chance or that it does not possess a function at some point in the life of an individual or in some individuals during the history of the species. For this reason, it has been argued that the hypothesis proposing a trait to be aptive should be treated as the null hypothesis and be rejected only after all attempts to demonstrate origin and maintenance due to natural selection have failed (Mayr 1983; Gans 1985). At least four different approaches have been used to demonstrate adaptive traits: (1) the correlational approach, (2) the comparative approach, (3) a predictive approach, and (4) a synthetic approach in the sense of Bock (1980).

The correlational approach, as generally used, assumes that a close statistical correlation between a trait and some environmental factor indicates a functional (causal) association. Obviously, two factors can be associated statistically without one's being the cause of the other. However, adaptation is a cause-and-effect relationship: a selective pressure is the cause, and the modification of a trait is the effect. A correlational approach is a first step in demonstrating an adaptive trait, but it cannot demonstrate that selection is occurring or has occurred.

The comparative approach appears to be a specific application of the more general correlational approach in that it comprises two types of correlations. The comparative approach usually begins by noting a possible correlation between a trait in a given taxon and a certain factor in the environment. It is then hypothesized that this correlation is due to natural selection. To test this hypothesis, other taxa exposed to the same environmental factor are examined to see if they also exhibit a similar trait. This search for correlations among additional taxa is based on the fact that convergence should provide evidence of adaptation, as differences among species that are not adaptive will not correlate with environmental variables. Clearly, the comparative approach is not applicable to all cases. It can only be applied to cases in which a suspected aptive trait in a given taxon can be compared to a similar trait in a second taxon. Furthermore, the traits being compared must be homoplasous, or there is no test that the same selective pressure has independently caused changes in the traits. The comparative approach has been criticized on two grounds: (1) it fails because it is not lawlike in evaluating aptive traits because of multiple adaptive solutions (Bock 1980); (2) it can test, in some cases, whether different traits in different species are adaptations to the same environmental factor, but it assumes that the relationship of the traits and the factor are due to natural selection (Ridley 1983).

Bock (1980) argued that the comparative approach to recognizing aptive traits is not valid because there are exceptions to the general rule that a given selective pressure will produce the same changes in traits (i.e., the same selective pressure may produce morphologically different traits because of multiple adaptive solutions (see the last section), or morphological convergence of traits might occur owing to different selective pressures. Although these considerations do complicate the recognition of cases of adaptation, it is possible to test alternate hypotheses that similarities in two or more traits are due to different selective pressures or that the same selective pressure can cause different traits by examining the corollaries associated with each hypothesis.

Ridley's (1983) objection regarding the comparative approach is more serious. He argued that the comparative method is a special type of correlational approach in which the association of independently occurring similar traits with the same environmental factor allows one to test a hypothesis regarding *how* species might be adapted, but not *whether* they are adapted. Mayr's (1983) answer to this objection is that one can deduce the probability of causation by selection if one shows that a particular trait would be favored by selection. Although it is true that the likelihood of a correlation's existing between a particular trait and an environmental factor by chance decreases with the number of independent cases (convergence) that occur, probability statements cannot demonstrate natural selection; they only support a hypothesis that such a correlation is probably due to selection. The problem is particularly vexing when the number of possible solutions to a particular selective pressure is unknown, as is the possible number of independent "trials" that may have occurred historically. It would seem that the strength of the comparative approach lies in recognizing patterns of variation and in testing hypotheses of how species might be adapted; it cannot, however, demonstrate causality.

A predictive (a priori) approach is not concerned with explaining correlations between traits and environmental factors; rather, it would predict correlations that should exist, based on the assumption that natural selection is an optimizing process (Rosen 1967; Milsum and Roberge 1973) and on applications of game theory (Maynard Smith 1972). This approach uses mathematical methods that allow one to generate the best solutions to particular mechanical and economical problems; it can be applied only to problems in which all possible solutions are known and each possible solution can be assigned a numerical value and function. Although it appears to be a successful approach in predicting the mechanical properties of some tissues, as well as some behavioral and ecological properties of species, it clearly assumes optimal solutions without regard to historical or other constraints. This approach may be useful in predicting certain broad categories of adaptive traits, but it is unlikely to be useful in more specific applications, as solutions are not always optimal and restraints often do exist.

Bock (1980) argues that the only valid approach to demonstrating an adaptive trait is a synthetic approach, using direct analysis of all aspects of a trait, including a determination of its biological role and measurement of its fitness in field studies. Although such an approach represents an extreme amount of time and may require researchers in more than one discipline, it appears to be the only way to actually demonstrate that natural selection is operating on a trait and, therefore, that the trait is adaptive. Certainly the most well-documented cases of adaptation have used such an approach, one example being industrial melanism in the peppered moth *Biston betularia*. Breeding experiments have demonstrated the genotypic control of melanism; predation rates and fitness have been measured in field studies. In this specific case, the recent emergence of melanism and its restriction to heavily industrialized areas suggests that this trait is not only aptive, but also adaptive.

The major limitation of a synthetic approach to adaptation is that it is not directly applicable to cases in which experimental studies cannot be performed, as is likely with rare or inaccessible living species and is certainly true of fossils. The comparative approach is the only possible one in these circumstances, and it can provide evidence that a particular trait is adaptive, though it cannot actually demonstrate adaptation. In cases that cannot be experimentally manipulated, one can only compare these traits to traits demonstrated to be adaptive to see if similar environmental correlations exist. As the number of cases increases, it becomes increasingly probable that the correlations are due to selection and not chance.

Except in rare cases in which an aptive trait has arisen recently, statements regarding the probability of a trait's being adaptive or exaptive must also be based on the comparative approach. Since the distinction between adaptive and exaptive traits is based on the history of these traits and their environments, which are not accessible, it is impossible to use a synthetic approach to demonstrate selection; statements regarding whether a trait is adaptive or exaptive will necessarily be probabilistic.

## Summary

An evolutionary adaptation, as a state of being, entails a correlation between a trait and an environmental factor; if we are to call the trait adaptive, this association must be due to natural selection. Historically the term *adaptive* has been used to refer to any trait that enhances current fitness, regardless of its history, as well as those traits whose emergence and maintenance are due to natural selection. Failure to distinguish between these two categories of traits can result in the construction of erroneous phyletic scenarios and the failure to falsify hypotheses of evolutionary mechanisms that would be corollaries of such scenarios. For these reasons, it seems preferable to describe as adaptive only those traits that arose and are maintained by the same selective pressures and to describe as exaptive those traits that arose by chance or were shaped by selective pressures different from those currently operating to maintain them (Gould and Vrba 1982).

A correlation between a given trait and an environmental factor may be due to causality (natural selection), or the correlation may be the result of chance (the incidental by-product of stochastic processes). There are at least three ways by which traits not being directly selected for can exist in a population and be correlated with environmental factors: fixation, pleiotropy, and phenotypic plasticity. There are at least two ways in which the form of a trait selected for can differ without the differences' being due to selective pressures: multiple adaptive solutions and historical or biochemical constraints.

Traits can arise in small populations owing to genetic drift or founder effect, rather than as a result of natural selection. Similarly, such traits can be maintained subsequently in the absence of selection for them, or even in spite of selection against them, because of fixation. A trait's form might be highly correlated with a particular environmental factor, but that correlation might be due to pleiotropy and not directly to selection. Short-term, nonheritable responses of organisms to their environment (phenotypic plasticity) are certainly correlated, but these correlations are not due to selection. A trait can be selected for, in terms of its biological role, but the form of this trait can differ in two different populations. In each case, the differences in the traits may be correlated with an environmental factor, but the differences are not necessarily due to differences in selective pressures. In such multiple adaptive solutions to selective pressures, the differences in the traits are a by-product of the initial gene pools in the two populations before the traits were selected for. Finally, a trait can be selected for, but the form of the trait is due to historical or biochemical constraints and not directly due to current selective pressures.

Because correlations between a trait and an environmental factor may result from natural selection or be a by-product of chance and because traits have a history during which both types of correlations may exist, it is necessary to test multiple hypotheses regarding the nature of the correlations. Four approaches have been utilized to test hypotheses that traits are adaptive: (1) the correlational approach, (2) the comparative approach, (3) the predictive approach, and (4) the synthetic approach.

The correlational approach fails because of the assumption that a close statistical correlation between a trait and an environmental factor indicates a causal association.

The comparative approach is a specific application of the more general correlational approach, but involves correlations among convergent traits. These correlations allow one to make statements regarding the probability of such correlations' occurring among independently converging traits resulting from natural selection or from chance. Thus, though the comparative approach does not allow one to test a hypothesis regarding *whether* a trait is adaptive, it does allow tests regarding *how* a trait might be adaptive, as well as statements regarding the probability that a trait is adaptive.

The predictive (a priori) approach is primarily concerned with predicting what types of correlations should exist between traits and environmental factors, rather than explaining the bases of already discovered correlations. Although this approach appears to be useful in generating theories regarding the types of adaptations that should exist, it does not appear to be generally applicable to specific cases.

The synthetic approach that entails direct analysis of all aspects of a trait, including a determination of its biological role and measurement of its fitness in field studies, appears to be the only valid approach to demonstrating an adaptation (Bock 1980). This approach, because it requires experimental manipulations, is not applicable to rare or inaccessible living species or to fossils. Similarly, it is not generally applicable to tests of hypotheses regarding adaptive vs. exaptive traits, as this distinction involves the history of a trait. In those cases in which the synthetic approach cannot be applied, the comparative approach is generally the most appropriate method, and statements regarding such traits will necessarily be probabilistic.

*Acknowledgments.* I am grateful to William P. Anderson, Andrew H. Bass, William Hodos, Walter Wilczynski, and Mario F. Wullimann for discussions bearing on the topics discussed in this chapter. Mary Sue Northcutt also discussed many of the issues with me and has assisted with my research and with preparation of the manuscript. Specific research projects relating to the broader issues covered herein have been supported in part by Grants NS-11006 and EY-02485, National Institutes of Health.

## References

Ali, M.A., Croll, R. P., and Jaeger, R. (1978) Phylogenetic survey of sensory functions, in Sensory Ecology, NATO Advanced Study Institute Series A: Life Sciences, vol. 18, Ali, M.A. (ed.), Plenum Press, New York, pp. 11–29.

Ariëns Kappers, C.U., Huber, G.C., and Crosby, E.C. (1936) The Comparative Anatomy of the Nervous System of Vertebrates, Including Man, reprinted in 1960, Hafner, New York.

Bennett, M.V.L. (1971) Electroreception, in Fish Physiology, Hoar, W.S., and Randall, D.J. (eds.), Academic Press, London, pp. 493–574.

Bock, W.J. (1977) Adaptation and the comparative method, in Major Patterns in Vertebrate Evolution, NATO Advanced Study Institute Series A: Life Sciences, vol. 14, Hecht, M., Goody, P.C., and Hecht, B.M. (eds.), Plenum Press, New York, pp. 57–82.

Bock, W.J. (1980) The definition and recognition of biological adaptation, Am. Zool., 20:217–227.

Bock, W.J., and von Wahlert, G. (1965) Adaptation and the form-function complex, Evolution, 19:269–299.

Bullock, T.H. (1974) An essay on the discovery of sensory receptors and the assignment of their functions together with an introduction to electroreceptors, in Electroreceptors and Other Specialized Receptors in Lower Vertebrates, Fessard, A. (ed.), Handbook of Sensory Physiology, vol. III/3, Springer-Verlag, Berlin, pp. 1–12.

Calford, M.B., Graydon, M.L., Huerta, M.F., Kaas, J.H., and Pettigrew, J.D. (1985) A variant of the mammalian somatotopic map in a bat, Nature, 313:477–479.

Calow, P. (1983) Evolutionary Principles, Blackie, Glasgow.

Conrad, M. (1983) Adaptability: the Significance and Variability from Molecule to Ecosystem, Plenum Press, New York.

Dartnall, H.J.A. (1975) Assessing the fitness of visual pigments for their photic environments, in Vision in Fishes, NATO Advanced Study Institute Series A: Life Sciences, vol. 1, Ali, M.A. (ed.), Plenum Press, New York, pp. 543–578.

Endler, J.A. (1978) A predator's view of animal color patterns, Evol. Biol., 11:319–364.

Endler, J.A. (1986) Defense against predators, in Predator-Prey Relationships, Feder, M.E., and Lauder, G.V. (eds.), University of Chicago Press, Chicago, pp. 109–134.

Frazzetta, T.H. (1975) Complex adaptations in evolving populations, Sinauer, Sunderland, Mass.

Gans, C. (1974) Biomechanics, Lippincott, Philadelphia.

Gans, C. (1985) Scenarios: why? in Evolutionary biology of primitive fishes, NATO Advanced Science Institute Series A: Life Sciences, vol. 103, Foreman, R.E., Gorbman, A., Dodd, J.M., and Olsson, R. (eds.), Plenum Press, New York, London, pp. 1–9.

Gould, S.J., and Lewontin, R.C. (1979) The spandrels of San Marco and the Panglossian paradigm: a critique of the adaptationist programme, Proc. R. Soc. Lond. B, 205:581–598.

Gould, S.J., and Vrba, E.S. (1982) Exaptation—a missing term in the science of form, Paleobiology, 8:4–15.

Hensel, H. (1955) Quantitative Beziehungen zwischen Temperaturreiz und Aktionspotentialen der Lorenzinischen Ampullen von Selachiern, Z. Vgl. Physiol., 37:509–526.

Herrick, C.J. (1948) The brain of the tiger salamander, University of Chicago Press, Chicago.

Hopkins, C.D. (1983) Functions and mechanisms in electroreception, in Fish Neurobiology, vol. 1, Brain Stem and Sense Organs, Northcutt, R.G., and Davis, R.E. (eds.), University of Michigan Press, Ann Arbor, pp. 215–259.

Huey, R.B., and Bennett, A.F. (1986) A comparative approach to field and laboratory studies in evolutionary biology, in Predator-Prey Relationships, Feder, M.E., and Lauder, G.V. (eds.), University of Chicago Press, Chicago, pp. 82–98.

Konishi, M. (1985) Birdsong: from behavior to neuron, Annu. Rev. Neurosci. 8:125–170.

Lande, R. (1979) Quantitative genetic analysis of multivariate evolution, applied to brain: body size allometry, Evolution, 33:402–416.

Lissmann, H.W., and Mullinger, A.M. (1968) Organization of ampullary electric receptors in Gymnotidae (Pisces), Proc. R. Soc. B, 169:345–378.

Maynard Smith, J. (1972) On evolution, Edinburgh University Press, Edinburgh.

Mayr, E. (1963) Animal Species and Evolution, Harvard University Press, Cambridge.

Mayr, E. (1969) Principles of Systematic Zoology, McGraw-Hill, New York.

Mayr, E. (1983) How to carry out the adaptationist program? Am. Nat., 121:324–334.

Milsum, J.H., and Roberge, F.A. (1973) Physiological regulation and control, in Foundations of Mathematical Biology III, Rosen, R. (ed.), Academic Press, London, pp. 1–95.

Murray, R.W. (1960) The response of the ampullae of Lorenzini of elasmobranchs to electrical stimulation, J. Exp. Biol., 37:417–424.

Murray, R.W. (1974) The ampullae of Lorenzini, in Electroreceptors and Other Specialized Receptors in Lower Vertebrates, Handbook of Sensory Physiology, vol. III/3, Fessard, A. (ed.), Springer-Verlag, Berlin, pp. 125–146.

Northcutt, R.G. (1986) Lungfish neural characters and their bearing on sarcopterygian phylogeny, J. Morphol. Suppl. 1:277–297.

Northcutt, R.G., and Butler, A.B. (1974) Evolution of reptilian visual systems: retinal projections in a nocturnal lizard, *Gekko gecko* (Linnaeus), J. Comp. Neurol., 157:453–465.

Okada, Y., Aoki, M., Sato, Y., and Masai, H. (1969) The brain patterns of sharks in relation to habitat, J. Hirnforsch., 11:347–365.

Pehrson, T. (1944) The development of latero-sensory canal bones in the skull of *Esox lucius,* Acta Zool., 25:1–23.

Ridley, M. (1983) The explanation of organic diversity, Clarendon Press, Oxford.

Romer, A.S. (1966) Vertebrate paleontology, University of Chicago Press, Chicago.

Rosen, R. (1967) Optimality principles in biology, Butterworth, London.

Simpson, G.G. (1953) The major features of evolution, Columbia University Press, New York.

Szabo, T. (1972) Ultrastructural evidence for a mechanoreceptor function of the ampullae of Lorenzini, J. Microsc. 14:343–350.

Szabo, T. (1974) Anatomy of the specialized lateral line organs of electroreception, in Electroreceptors and Other Specialized Receptors in Lower Vertebrates, Handbook of Sensory Physiology, vol. III/3, Fessard, A. (ed.), Springer-Verlag, Berlin, pp. 13–58.

Teeter, J.H., Szamier, R.B., and Bennett, M.V.L. (1980) Ampullary electroreceptors in the sturgeon *Scaphirhynchus platorynchus* (Rafinesque), J. Comp. Physiol., 138:213–223.

Ulinski, P.S. (1980) Functional morphology of the vertebrate visual system: an essay on the evolution of complex systems, Am. Zool., 20:229–246.

von Kupffer, K.V. (1906) Die Morphogenie des Zentralnervensystems, in Handbuch der vergleichenden und experimentellen Entwicklungslehre der Wirbeltiere 2:1–272, Hertwig, O. (ed.), Gustav Fischer, Jena.

Walls, G.L. (1942) The vertebrate eye and its adaptive radiations, Cranbrook Institute of Science, Bloomfield Hills, Mich.

Williams, G.C. (1966) Adaptation and natural selection: a critique of some current evolutionary thought, Princeton University Press, Princeton.

Wright, S. (1931) Evolution in Mendelian populations, Genetics, 16:97–159.
Wright, S. (1949) Adaptation and selection, in Genetics, Paleontology, and Evolution, Jepsen, G., Mayr, E., and Simpson, G.G. (eds.), Princeton University Press, Princeton, pp. 365–389.

# Author Index

Graf, W., 752, *754*, 795, 796, 798, 799, 803, 804, *805*, *806*

# Animal Index

# Subject Index